UNDERSTANDING

BIOLOGY

THIRD EDITION

PETER H. RAVEN

Director, Missouri Botanical Garden;
Engelmann Professor of Botany
Washington University

GEORGE B. JOHNSON

Professor of Biology
Washington University

Volume 1 General Principles

Volume 2 The Biological World

WCB **Wm. C. Brown Publishers**

Dubuque, IA   Bogota   Boston   Buenos Aires   Caracas   Chicago
Guilford, CT   London   Madrid   Mexico City   Sydney   Toronto

Book Team

Editor *Carol J. Mills*
Developmental Editor *Diane E. Beausoleil*
Production Editor *Sue Dillon*
Designer *Lu Ann Schrandt*
Art Editor *Jodi K. Banowetz*
Photo Editor *Janice Hancock*
Permissions Coordinator *LouAnn K. Wilson*
Visuals/Design Developmental Coordinator *Donna Slade*

Wm. C. Brown Publishers
A Division of Wm. C. Brown Communications, Inc.

Vice President and General Manager *Beverly Kolz*
Vice President, Publisher *Kevin Kane*
Vice President, Director of Sales and Marketing *Virginia S. Moffat*
Vice President, Director of Production *Colleen A. Yonda*
National Sales Manager *Douglas J. DiNardo*
Marketing Manager *Craig Johnson*
Advertising Manager *Janelle Keeffer*
Production Editorial Manager *Renée Menne*
Publishing Services Manager *Karen J. Slaght*
Royalty/Permissions Manager *Connie Allendorf*

Wm. C. Brown Communications, Inc.

President and Chief Executive Officer *G. Franklin Lewis*
Senior Vice President, Operations *James H. Higby*
Corporate Senior Vice President, President of WCB Manufacturing *Roger Meyer*
Corporate Senior Vice President and Chief Financial Officer *Robert Chesterman*

Copyedited by Mary Monner

Cover photo © Frans Lanting/Minden Pictures

Photo research by Marty Levick

The credits section for this book begins on page C-1 and
is considered an extension of the copyright page.

Publisher's Note to the Instructor

Binding Options and Recycled Paper

Understanding Biology—available in the binding options listed here—is printed on **recycled paper stock.** All of its ancillaries, as well as all advertising pieces, are also printed on recycled paper. Whenever possible, soy inks are used for printing.

Our goal in offering these products on recycled paper is to take an important step toward minimizing the environmental impact of our products. If you have any questions about recycled paper or inks, *Understanding Biology,* its package, any of its binding options, or any of our other biology texts, feel free to call us at 1–800–553–4920.

Carol Mills
Biology Editor

Understanding Biology, Third Edition

Casebound

The full-length text with hardcover binding (chapters 1–45).

Volume 1

Volume 1 features the first six parts (chapters 1–24). The Introduction and units on cell biology, energy, genetics, evolution, and ecology are covered. The material in this volume fits the way many instructors teach their first semester of the course. Volume 1 is available to students at a significant savings over the cost of the entire text.

Volume 2

Volume 2 covers the last three parts of the text (chapters 25–45). It includes biological diversity, and plant and animal biology. The material covered in this volume often fits the way instructors teach the second semester of the course. This volume also costs less than the entire text.

Chapter-by-Chapter Customization

Customize *Understanding Biology* to fit your ideal course. Select only the chapters you use and WCB will bind them in full color. Depending on the number of chapters selected, receive significant savings over the cost of the full-length casebound version.

A personal library is a lifelong source of enrichment and distinction. Consider this book an investment in your future and add it to your personal library.

UNDERSTANDING

BIOLOGY

Learning System

Chapter opening paragraphs preview the chapter, engaging student interest.

A videotape icon placed beside figures throughout the text indicates that those figures appear animated in the new Wm. C. Brown *Life Science Animations* series videotapes/videodisk.

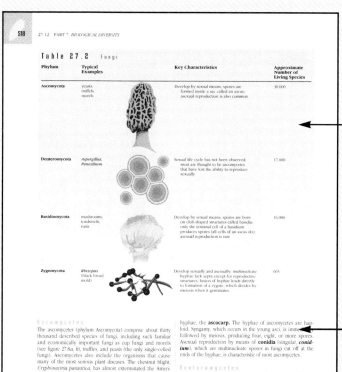

Table 27.2 Fungi

Phylum	Typical Examples	Key Characteristics	Approximate Number of Living Species
Ascomycota	yeasts, truffles, morels	Develop by sexual means; spores are formed inside a sac called an ascus; asexual reproduction is also common	30,000
Deuteromycota	*Aspergillus, Penicillium*	Sexual life cycle has not been observed; most are thought to be ascomycetes that have lost the ability to reproduce sexually	17,000
Basidiomycota	mushrooms, toadstools, rusts	Develop by sexual means; spores are borne on club-shaped structures called basidia; only the terminal cell of a basidium produces spores (all cells of an ascus do); asexual reproduction is rare	16,000
Zygomycota	*Rhizopus* (black bread mold)	Develop sexually and asexually; multinucleate hyphae lack septa except for reproductive structures; fusion of hyphae leads directly to formation of a zygote, which divides by meiosis when it germinates	665

Numerous tables summarize important information, making it readily accessible for efficient study.

Boldfaced key terms are explained in the narrative and defined again in the glossary at the end of the book.

Sidelight boxes cover relevant human biological issues, for example, cancer and AIDS.

In-text summaries highlight important concepts as they appear throughout each chapter, helping students focus on important material.

Key Experiment boxes feature many of the classic experiments upon which the study of biology is founded.

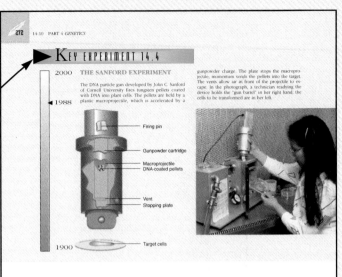

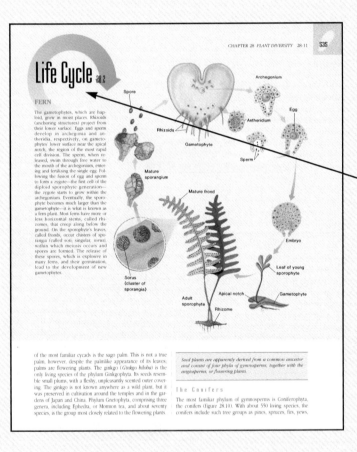

Life Cycle boxes illustrate and describe major biological life cycles.

Evolutionary Viewpoints appear at the end of each chapter and reinforce the evolutionary theme of the text.

Summaries at the end of each chapter recap the chapter's material and facilitate student learning.

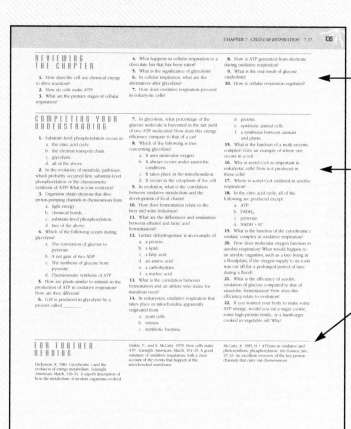

New *Reviewing the Chapter* questions are objective-type questions that test students' ability to recall what they have read.

New *Completing Your Understanding* questions require students to think critically about the material in the chapter. Answers to these questions appear in the *Instructor's Manual* and in the *Student Study Guide*.

For Further Reading ends each chapter and provides annotated references to articles and texts that supplement the chapter material.

Modules from the new WCB *Explorations* CD-ROM, indicated by a CD-ROM icon, are noted at the ends of chapters that contain relevant, correlating material. A brief description of the module and several questions are included.

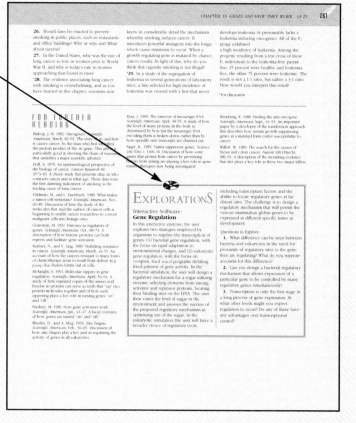

Brief
CONTENTS

CONTENTS

Sidelights

Life Science Animations Figures

Animated versions of the following illustrations are available on the *Life Science Animations* videotapes/videodisks, produced by Wm. C. Brown Publishers.

Figure	Process/concept
2.7	Ionic Bond Formation
4.8	Journey into a Cell
4.16	Cellular Secretion
5.11	Endocytosis
5.13	Sodium-Potassium Pump
6.7	Catalytic Cycle of an Enzyme
6.13	ATP/ADP + P_i Cycle
7.5	Glycolytic Pathway
7.9	Citric Acid Cycle
7.10	Electron Transport Chain
8.10	Chemiosmotic Synthesis of ATP
8.12	Calvin Cycle
8.13	Carbon Fixation in C_4 Plants
9.8	Mitosis
9.15	Meiosis
9.16	Crossing-Over
9.17	Chiasmata
12.8	Replication Fork
13.5	Transcription
13.6	Protein Synthesis
13.16	*lac* Region of *E. Coli*
17.5	Continental Drift
17.6	Plate Tectonics
30.13	Journey into a Leaf
34.10	Formation of Myelin Sheath
34.11	Saltatory Nerve Conduction
34.15	Signal Integration
35.15	Organs of Equilibrium
	Organ of Corti Senses
35.28	Reflex Arcs
36.10	Levels of Muscle Structure
36.12	Actin and Myosin Microfilament Interaction
36.14	Sliding Filament Model of Muscle Contraction
37.5	Peptide Hormone Action
38.1	Digestion of Carbohydrates
38.8	Peristalsis
38.12	Digestion of Proteins
40.15	Blood Circulation
40.17	Production of an Electrocardiogram
40.19	Common Congenital Defects of the Heart
41.12	Helper T-Cells and Killer T-Cells
41.13	Types of T-Cells
41.14	B-Cell Immune Response
41.16	Stucture and Function of Antibodies

Explorations in Human Biology and Explorations in Cell Biology, Genetics, and Molecular Biology CD-ROMs

Interactive software modules from the *Explorations in Human Biology, Cell Biology, Genetics, and Molecular Biology* CD-ROMs are referenced in the following chapters.

Chapter	Topic
4	Cell Size
5	Active Transport
	Cell-Cell Interaction
	Cystic Fibrosis
6	What Enzymes Do
	Enzyme Kinetics
7	Oxidative Respiration
8	Photosynthesis
9	Genetic Reassortment during Meiosis
10	Three-Point Genetic Cross
11	Heredity in Families
	Find that Gene
	Gene Segregation within Families
12	Smoking and Cancer
	Reading DNA
13	Gene Regulation
14	Genetic Engineering
	Restriction Site Mapping
24	Pollution of a Freshwater Lake
26	AIDS
33	Life Span and Lifestyle
34	Nerve Conduction
	Synaptic Transmission
35	Drug Addiction
36	Muscle Contraction
37	Hormone Action
38	Diet and Weight Loss
40	Evolution of the Heart
41	Immune Reponse

PREFACE

Biology in a Rapidly Changing World

A lot has happened since we first put pen to paper in 1986. The biological sciences, especially, have had a major impact on our world in myriad important ways. For example, there were fewer than twenty thousand reported cases of AIDS in the United States in 1986. Currently, the total number of Americans who have contracted AIDS has passed the half-million mark! Attempts to find effective treatment for AIDS have been largely unsuccessful, and the vaccines developed to date appear effective against only the particular HIV strains from which they were developed. Because of the increasing impact of this disease on our society, it is imperative that students gain a clear understanding of it. Hence, in this edition, we have expanded an already extensive treatment of HIV and AIDS.

In the last eight years, genetic engineering has increasingly influenced American life. Many drugs are now produced using this approach, and reports of exciting medical advances that rely on genetic engineering appear almost daily in the newspapers. Gene replacement therapy is being used with encouraging results in attempts to cure cystic fibrosis and other hereditary diseases. Efforts to fully sequence the human genome are moving rapidly ahead. Techniques such as PCR and gene sequencing have filtered into everyday parlance, due in part to novels and movies such as *Jurassic Park*. However, while the public is becoming increasingly aware of genetic engineering, many people are far from comfortable with the impact it is beginning to have on their lives. Hence, there was considerable public discussion when the first foods from genetically engineered organisms hit the nation's supermarkets in 1994—milk from cows injected with hormones from genetically engineered bacteria, and more tasty tomatoes by virtue of a fish gene that delays ripening so that the tomatoes can be left on the vine longer. Because students will have to evaluate issues such as these to form effective policies in the years ahead, we have updated and added material in this edition to help students understand modern genetic engineering technology.

The environmental crisis that we emphasized in past editions continues to be one of the most serious problems facing our world, and plays a major role in shaping our future. In the past few years, there have been encouraging signs that these problems are being addressed.

Representatives from the countries of our world gathered in South America in 1993 to discuss the impact of development on the environment, and in Egypt in 1994 to reflect on the consequences of rapid increases in the world's human population (which has increased by more than half a BILLION people since the first edition of this book was published). While curbing population growth cannot in itself solve our world's problems (over-consumption is another culprit that will have to be dealt with), we cannot hope to succeed without slowing our unbridled growth. We try to face these issues squarely in this edition, for every one of our students will have to deal with these major concerns in the coming decades.

The science of biology forms the conceptual basis for addressing the various problems just mentioned. In the last eight years, a lot has happened to advance our understanding of biology. Biological scientists have dissected the molecular mechanisms underlying cancer and drug addiction; learned a great deal about how development regulates pattern formation; fought ferocious battles over the proposition that our species, *Homo sapiens*, emerged from Africa to replace *H. erectus*; and located the spot in Central America where the impact of a comet led to the extinction of the dinosaurs sixty-five million years ago. In this edition we have devoted significant space to describing these advances, both for their own importance, and because in many cases the advances actually make the teaching of a particular point easier to understand!

How This Edition Is Organized to Teach Biology

Like the first two editions, this third edition of *Understanding Biology* is organized into nine parts that are divided into two broad areas: **basic biological principles**—cell biology, biochemistry, genetics, evolution, and ecology (the first 24 chapters); and **structure** and **function of organisms**—diversity, plant biology, and vertebrate anatomy and physiology (the final 21 chapters). **In this third edition it is possible for the first time to obtain these two halves of the text separately,** as volumes 1 and 2, if your course is limited in focus to one or the other.

The Organization of Understanding Biology, Third Edition

Volume 1

Part 1 Introduction
The text begins with an introduction to the nature of science.
General Principles
Part 2 Cell Biology
Part 3 Energy
Part 4 Genetics
Part 5 Evolution
Part 6 Ecology
The first half of the text is devoted to principles shared by all organisms.

This organization reflects what we believe to be the clearest way to teach biology. Giving the students an overview of the principles of biological processes and the workings of the biosphere prepares them to understand the form and function of organisms.

The text presents a progression of basic principles, which gives students the information needed to understand the basic properties of all living things. A student moving through the book in this order comes away with the understanding that ecology is the end result of evolution acting on genes.

In the second half of the text, using the framework provided by the principles of biology, students can explore the chapters on form and function using principles that apply to every living creature.

Those familiar with the first two editions of *Understanding Biology* know that evolution has been the grounding feature of this text. Evolution remains the guiding perspective of this third edition, and our organization of the form and function chapters, along with new *Evolutionary Viewpoints* at the end of each chapter, reflect this perspective.

A second theme found throughout the third edition of *Understanding Biology* is its experimental emphasis. In addition to a thorough discussion of the scientific method in the introduction, students can read boxed *Key Experiments* that feature many of the classic experiments upon which the study of biology is founded.

What's New in This Edition

Evolution, the major theme of *Understanding Biology*, has even greater prominence in this third edition, with new *Evolutionary Viewpoints* at the end of every chapter. The experimental theme is emphasized with newly created *Key Experiment* boxes placed throughout the text.

In addition to the increased emphasis on these two fundamental themes in biology, several other major changes have been made to the textbook, including new material, updating to reflect new information, and careful editing to correct errors.

Some of the major content changes in *Understanding Biology,* third edition, include:

Chapter 14: Gene Technology It is not surprising that this chapter has evolved significantly in four years. The third edition contains many new biotechnological experiments, presented in detail within *Key Experiment* features.

Chapters 18 and 19: The Story of Vertebrate Evolution and **How Humans Evolved** The previous edition of this text presented the entire discussion of evolution in one chapter. In this edition, human evolution is discussed separately from the evolution of our vertebrate ancestors.

Chapter 24: Our Changing Environment As the title of the chapter suggests, the problems and promises of our environment have not remained static since 1991. Key environmental topics covered in this edition include rain forest destruction, nuclear power, pollution, recycling, CFCs (chlorofluorocarbons) and the ozone, and the critical importance of being "biologically aware."

Chapters 25 and 26: The Six Kingdoms of Life and **The Invisible World: Bacteria and Viruses** *Six* kingdoms? Has a whole new kingdom been discovered since 1991?! No, however, enough evolutionary information has recently been collected that the consensus among taxonomists is to split the former Kingdom **Monera** (the bacteria) into the **Archaebacteria** (very primitive forms that can live in astoundingly harsh environments), and the **Eubacteria** (the more familiar kind that decompose things and give you strep throat). These new chapters have been completely reorganized and updated to reflect this major taxonomic change.

Chapter 27: Protists and Fungi The protists are such a diverse kingdom that many students dread studying them. In this edition we grouped them according to shared characteristics, making a muddy picture a little clearer.

Chapter 31: Flowering Plant Reproduction This new edition contains an expanded discussion of the formation of sex cells (pollen and eggs). Updated discussions of seeds, fruits, and plant growth complement the chapter.

Chapter 34: How Animals Transmit Information Helping students understand how smelling a rose or throwing a baseball is a consequence of ions bouncing back and forth across a long, skinny cell can be a challenge. This edition concisely, logically, and *painlessly* walks students through the process of nervous transmission. And, in case some of the more adventurous students are considering experimenting with their nervous systems, this edition contains a major new discussion of the effects of drugs on the nervous system and the mechanics of drug addiction.

Chapter 35: The Nervous System This totally reworked section contains the obligate essential organizational information on the nervous system, and also discusses some of the more ill-defined cortical functions. Interpretation of sensation, speech, pleasure, rage (emotion), memory, and sleep are all newly-addressed topics. And what does the *reticular system* do, anyway? Do we really need a part of our brain to remind us if we're awake or asleep? This just might be something to which a sleepy student can relate.

Chapter 40: Circulation The critical plumbing system that keeps us alive is discussed in a systematic, well-organized manner. The third edition contains expanded discussions, particularly of the blood vessels themselves (arteries, veins, and capillaries), and the consequences of damage to them—a whole new section on **cardiovascular disease.** Angina, heart attack, arteriosclerosis, stroke, and hypertension will become personally significant topics later in many students' lives. Rather than blithely announcing that the readers' years are numbered, however, the authors present all the available evidence on how to extend those years.

New Technology and Understanding Biology

Several new state-of-the-art technology products are available that are correlated to this textbook. These useful and enticing supplements can assist you in teaching and can improve student learning.

Explorations in Human Biology and Explorations in Cell Biology, Genetics, and Molecular Biology CD-ROMs

Each of these two interactive CDs by Dr. George Johnson comprises sixteen modules, each a fascinating topic in biology. These interactive investigations are correlated to appropriate topical material in *Understanding Biology.* Many are referenced at the ends of chapters along with a brief explanation of the module and several relevant study questions.

Chapter		Exploration Topic
4	Cells	Cell Size
5	How Cells Interact with the Environment	Active Transport/ Cystic Fibrosis Cell-Cell Interactions
6	Energy and Metabolism	What Enzymes Do/ Enzyme Kinetics
7	Cellular Respiration	Oxidative Respiration
8	Photosynthesis	Photosynthesis
9	How Cells Reproduce	Genetic Reassortment during Meiosis
10	Mendelian Genetics	Three-Point Genetic Cross
11	Human Genetics	Heredity in Families/ Find that Gene/ Gene Segregation within Families
12	DNA: The Genetic Material	Smoking and Cancer
13	Genes and How They Work	Gene Regulation
14	Gene Technology	Genetic Engineering/ Restriction Site Mapping
24	Our Changing Environment	Pollution of a Freshwater Lake
26	The Invisible World: Bacteria and Viruses	AIDS
33	The Vertebrate Body	Life Span and Lifestyle
34	How Animals Transmit Information	Nerve Conduction/ Synaptic Transmission
35	The Nervous System	Drug Addiction
36	How Animals Move	Muscle Contraction
37	Hormones	Hormone Action
38	How Animals Digest Food	Diet and Weight Loss
40	Circulation	Evolution of the Heart
41	How the Body Defends Itself	Immune Response

Life Science Animations Videotapes/Videodisk

Approximately fifty animations of key physiological processes are available on videotape as well as a videodisk. The animations, correlated to this text by a videotape icon, bring visual movement to biological processes like active transport across a cell membrane and cellular secretion, processes that are difficult to understand on the text page.

Other Technology Products that are Excellent Complements to Understanding Biology

BioSource videodisc, by Wm. C. Brown and Sandpiper Multimedia, Inc., features twenty minutes of moving animations and nearly ten thousand full-color illustrations and photos, many from leading WCB biology textbooks.

Biology StartUp, a five-disk set of Macintosh tutorials by Myles C. Robinson and Kathleen Hakola Pace, Grays Harbor College, is designed to help nonmajor students master challenging biological processes like chemistry and cell biology. This set can be a valuable addition to a resource center and is especially helpful to students enrolled in developmental education courses or those who need additional assistance to succeed in an introductory biology course.

Supplemental Materials Available with Understanding Biology

In addition to the "all new" technology products, a comprehensive, revised ancillary package is presented with the third edition of *Understanding Biology.*

One of the newest ancillaries is the *Student Study Art Notebook,* which contains all the images in the main transparency set (in full color with space for student notes), and is packaged free with every new textbook from WCB.

The *Instructor's Manual with Test Item File,* prepared by Dr. Bernard Frye of the University of Texas-Arlington, features a chapter outline; overview; list of key terms; suggested lecture outline, which is also available on a disk; a section called "Presentation of Material" that offers suggestions for presenting the chapter's material in class; suggested group activities; suggestions for research projects; supplemental readings; a list of relevant audio-visual materials; and the answers to the "Completing Your Understanding" questions for each chapter. MicroTest III, a computerized test bank of an test items, is also available.

The *Student Study Guide,* prepared by Professor Ann Vernon of St. Charles Community College, contains all of the answers to the "Completing Your Understanding" questions for each text chapter; key terms exercises; chapter outline; chapter overview; a section called "Concept Check" that includes critical thinking-based questions relevant to the text; self-quizzes; perforated, removable flashcards with terms and their definitions at the back of the study guide; and a section called "Test Your Identification Skills," which asks students to correctly label parts of figures. *New!* The Study Guide is available in a Spanish translation.

The *General Biology Laboratory Manual,* written by Dr. Darrell Vodopich of Baylor University and Dr. Randall Moore of the University of Akron, is now in color and customizable. It has a new exercise on ecology and three new human exercises. The lab manual is accompanied by an *Instructor's Manual,* which includes instructions for setting up each lab and a list of laboratory suppliers.

A set of 250 full-color transparencies, which includes figures and key tables from the text, is free to all adopters. The figures in this set are in the Student Study Art Notebook.

New electronic acetates include nearly all of the illustrations from the entire text on disk, in full color. This new supplement is free to qualified adopters.

MediSim computerized study guide, prepared by Professor Ann Vernon, is a question bank available for student purchase.

Other Titles of Related Interest from Wm. C. Brown Publishers

You Can Make a Difference
by Judith Getis
This short, inexpensive supplement offers students practical guidelines for recycling, conserving energy, disposing of hazardous wastes, and other pollution controls. It can be shrink-wrapped with the text at minimal additional cost. (ISBN 0–697–13923–9)

How to Study Science
by Fred Drewes, Suffolk County Community College
This excellent new workbook offers students helpful suggestions for meeting the considerable challenges of a college science course. If offers tips on how to take notes, how to get the most out of laboratories, and how to overcome science anxiety. The book's unique design helps students develop critical thinking skills while facilitating careful note taking. (ISBN 0–697–14474–7)

The Life Science Lexicon
by William N. Marchuk, Red Deer College
This portable, inexpensive reference helps introductory-level students quickly master the vocabulary of the life sciences. Not a dictionary, it carefully explains the rules of word construction and derivation, in addition to giving complete definitions of all important terms. (ISBN 0–697–12133–X)

Biology Study Cards
by Kent Van De Graaff, R. Ward Rhees, and Christoper H. Creek, Brigham Young University
This boxed set of 300 two-sided study cards provides a quick yet thorough visual synopsis of all key biological terms and concepts in the general biology curriculum. Each card features a masterful illustration, pronunciation guide, definition, and description in context. (ISBN 0–697–03069–5)

The Gundy-Weber Knowledge Map of the Human Body
by G. Craig Gundy, Weber State University
The 13-disk Mac-Hypercard program is for use by instructors and students alike. It features carefully prepared computer graphics, animations, labeling exercises, self-tests, and practice questions to help students examine the systems of the human body. Contact your local Wm. C. Brown representative or call 1–800–351–7671.

The Knowledge Map Diagrams
1. Introduction, Tissues, Integument System (ISBN 0–697–13255–2)
2. Viruses, Bacteria, Eukaryotic Cells (ISBN 0–697–13257–9)
3. Skeletal System (ISBN 0–697–13258–7)
4. Muscle System (ISBN 0–697–13259–5)
5. Nervous System (ISBN 0–697–13260–9)
6. Special Senses (ISBN 0–697–13261–7)
7. Endocrine System (ISBN 0–697–13262–5)
8. Blood and the Lymphatic System (ISBN 0–697–13263–3)
9. Cardiovascular System (ISBN 0–697–13264–1)
10. Respiratory System (ISBN 0–697–13265–X)
11. Digestive System (ISBN 0–697–13266–8)
12. Urinary System (ISBN 0–697–13267–6)
13. Reproductive System (ISBN 0–697–13268–4)

Demo—(ISBN 0–697–13256–0)
Complete Package—(ISBN 0–697–13269–2)

Acknowledgments
Although authors wrestle with the actual writing of a textbook, there are many who contribute to its creation. First and foremost, we authors need and use the comments and criticisms of you—the instructors and students who use the text. Information about new advances in research, interesting current stories, and careful scrutiny of content to improve accuracy are a major assist to us as we tell the fascinating story of biology. While we can't identify all of the people who have assisted us here, we hope they will recognize their contributions to this edition.

In the publishing house, the developmental editor starts the process, working closely with the authors to keep them on schedule and providing feedback from instructors who use the text and who have carefully reviewed it. The photo editor and art editor do yeoman's work in responding to the enormous challenge of finding the right image to help a student visualize a difficult process or detailed body system. The designer works to create a design that is accessible and will assist student learning. The permissions editor, is, of course, essential to the process. And finally, the production editor interfaces with all these folks and makes sure the end product consists of all the pieces in all the right places.

The first two successful editions of this text were carefully shepherded through the developmental, production, and manufacturing processes at C. V. Mosby Publishing. The third edition of *Understanding Biology* is in the hands of Wm. C. Brown Publishers, a company composed of people who helped the authors fine-tune each and every page. For their diligence and hard work, we thank Kevin Kane, Vice President and Publisher; Michael D. Lange, Executive Editor; Carol Mills, Acquisitions Editor; Diane Beausoleil, Developmental Editor; Kennie Harris, Production/Developmental Coordinator; Sue Dillon, Production Editor; Janice Hancock, Photo Editor; Marty Levick and Kathy Husemann, photo researchers; Jodi Banowetz, Art Editor; Lou Ann Wilson, Permissions Editor; and Lu Ann Schrandt, Designer.

Contributors

Important contributions to the content of *Understanding Biology* were made by instructors who reviewed sections of the book pertinent to their areas of expertise. These contributors played a significant role in helping us update content, improve accuracy, and add new material:

George Cox
San Diego State University

Ronald Hoham
Colgate University

Spencer Lucas
New Mexico Museum of Natural History

Jan Pechenik
Tufts University

Lansing Prescott
Augustana College

Roy Scott III
Ohio State University

David Sonneborn
University of Wisconsin-Madison

Robert Tamarin
Boston University

Henry Tedeschi
The University of Albany

And finally, to Dr. Paul Tabor, *Clarke College,* who scrutinized every word on every page along with the illustrations and their labels, our special thanks.

Reviewers

A sincere thank you to the many instructors who have reviewed *Understanding Biology* or participated in focus groups. Their comments have been invaluable to us and are greatly appreciated.

Reviewers of the First Edition

Ann Antelfinger
University of Nebraska at Omaha

Mary Berenbaum
University of Illinois at Urbana

Brenda C. Blackwelder
Central Piedmont Community College

Richard K. Boohar
University of Nebraska at Lincoln

John S. Boyle
Cerritos College

Donald Collins
Orange Coast College

Joyce Corban
Wright State University

Michael Corn
College of Lake County

John Crane
Washington State University

Ronald S. Daniel
California State Polytechnic University

Rose Davis
St. Louis Community College

Katherine Denniston
Towson State University

William Dickison
University of North Carolina

Fred Drewes
Suffolk County Community College

Paul Elliott
Florida State University

Larry Friedman
University of Missouri at St. Louis

Judy Goodenhough
University of Massachusetts

Gene Goselin
Diablo Valley College

Lane Graham
University of Manitoba

Thomas Gray
University of Kentucky

John P. Harley
Eastern Kentucky University

Holt Harner
Broward Community College

Terry Harrison
Central State University

Fred Hinson
Western Carolina University

Wilfred Iltis
San Jose State University

Alan Journet
Southeast Missouri State University

Peter Kareiva
University of Washington

Ann Lumsden
Florida State University

Constance Murray
Tulsa Junior College-Metro Campus

Steve Murray
California State University at Fullerton

Jim Peck
University of Arkansas at Little Rock

Gary Peterson
South Dakota State University

Ronald Rak
Morraine Valley Community College

Jonathan Reiskind
University of Florida

Martin Rochford
Fullerton College

Samuel Rushforth
Brigham Young University

Larry St. Clair
Brigham Young University

Ed Samuels
Los Angeles Valley College

Donald Scoby
North Dakota State University

Erik Scully
Towson State University

Russell Skavaril
Ohio State University

Gerald Summers
University of Missouri at Columbia

Jay Templin
Widener University

Richard R. Tolman
Brigham Young University

Richard Van Norman
University of Utah

David Calvin Whitenberg
Southwest Texas State University

Dana L. Wrensch
Ohio State University

Focus Group Participants

Ann Antelfinger
University of Nebraska at Omaha

Brenda Blackwelder
Central Piedmont Community College

Fred Drewes
Suffolk County Community College

Holt Harner
Broward Community College

Alan Journet
Southeast Missouri State University

Steve Murray
California State University at Fullerton

Richard Tolman
Brigham Young University

Richard Van Norman
University of Utah

Reviewers of the Second Edition

John Adler
Michigan Technological University

L. Rao Ayagari
Lindenwood College

Robert Beckman
North Carolina State University

Marlin Bolar
California State University-Sacramento

James Botsford
New Mexico State University

Clyde Bottrell
Tarrant County Junior College

David Bruck
San Jose State University

Warren Burrgren
University of Massachusetts-Amherst

John Clamp
North Carolina Central University

Roy Clarkson
University of West Virginia

Donald Collins
Orange Coast College

Roger Denome
University of North Dakota

Ronald Downey
Ohio University

Tom Emmel
University of Florida

Elizabeth Gardner
Pine Manor College

Gregory Grove
Pennsylvania State University

Elizabeth Gulotta
Nassau Community College

Joyce Hardin
Central State University

Holt Harner
Broward Community College

Stephen Hedman
University of Minnesota-Duluth

Robert Hersch
University of Kansas

George Hudock
Indiana University

Sylvia Hurd
Southwest Texas State University

Leonard Kass
University of Maine

Robert Kaul
University of Nebraska

Jay Kunkle
Ann Arundel Community College

Army Lester
Kennesaw State College

Ben Liles
University of Maine

Charles Lytle
North Carolina State University

Steven McCullagh
Kennesaw State College

Clifton Nauman
University of Tennessee-Knoxville

W. Brian O'Connor
University of Massachusetts-Amherst

Kevin Patton
St. Charles Community College

Jim Peck
University of Arkansas

Douglas Reynolds
Eastern Kentucky University

Michael Rourke
Bakersfield College

James Smith
California State University-Fullerton

Major James Swaby
United States Air Force Academy

Thomas Terry
University of Connecticut

Kathy Thompson
Louisiana State University

F. R. Trainor
University of Connecticut

James Traniello
Boston University

Rob Tyser
University of Wisconsin-La Crosse

Nancy Webster
Prince George's Community College

Dana Wrensch
Ohio State University

John Zimmerman
Kansas State University

Steve Ziser
Austin Community College

Focus Group Participants

Focus Group I (Chapters 1 through 31)

James Botsford
New Mexico State University

John Clamp
North Carolina Central University

Tom Emmel
University of Florida

Bill Glider
University of Nebraska-Lincoln

Stephen Hedman
University of Minnesota-Duluth

Focus Group II (Chapters 32 through 44)

Clyde Bottrell
Tarrant County Junior College

Judy Goodenough
University of Massachusetts-Amherst

Ed Joern
University of Missouri-St. Louis

James Smith
California State University-Fullerton

Rob Tyser
University of Wisconsin-La Crosse

Art Reviewers

Steve Dina
St. Louis University

Ed Joern
University of Missouri-St. Louis

Reviewers of the Third Edition

Jane Aloi
Saddleback College

F. N. Bebe
Kentucky State University

Charles H. Bennett
Kentucky State University

Carol A. Brewer
University of Montana

John R. Crooks
Arkansas State University-Beebe

Terrence Davin
Penn Valley Community College

Donald Dorfman
Monmouth College

Susan K. Dutcher
University of Colorado-Boulder

John E. Frey
Mankato State University

Philip F. Ganter
Tennessee State University

Richard Gross
Motlow State Community College

Elizabeth Gulotta
Nassau Community College

John J. Heise
Georgia Institute of Technology

Robert Hiskey
Wright State University-Lake Campus

D. M. Ivey
University of Arkansas-Fayetteville

Craig T. Jordan
University of Texas-San Antonio

Daniel J. Klionsky
University of California-Davis

Andrew H. Lapinski
Reading Area Community College

Roger M. Lloyd
Florida Community College-South Campus

Jeanne Marie Lust
College of St. Benedict

Stephen Manning
Arkansas State University-Beebe

Anthony G. Moss
Auburn University

Murray W. Nabors
Colorado State University

Judy H. Niehaus
Radford University

Charles L. Ralph
Colorado State University

Irwin Rehm
Adirondack Community College

Julia Riggs
Victoria College

Stephen G. Saupe
College of St. Benedict

Delbert M. Shankel
University of Kansas

Philip C. Shelton
Clinch Valley College

Paul Keith Small
Eureka College

James Smith
California State University

Sarah Anne Staples
Andrew College

Mark V. Sutherland
Hendrix College

Sheila Summers Thompson
University of Denver

Michael E. Toliver
Eureka College

Dr. Peter H. Raven is the Director of the Missouri Botanical Garden and Engelmann Professor of Botany at Washington University. Dr. Raven oversees the Garden's internationally recognized research program in tropical botany, one of the most active in the world in the study and conservation of imperiled habitats. A distinguished scientist, Dr. Raven is a member of the National Academy of Science, the National Research Council, and is a MacArthur and a Guggenheim fellow. He has been the recipient of numerous honors and awards for his botanical research and work in tropical conservation.

In addition to co-authoring this text and *Biology,* third edition, with Dr. George Johnson, Dr. Raven has authored fifteen other books and more than 450 articles.

Dr. George B. Johnson is Professor of Biology at Washington University in St. Louis. Also Professor of Genetics at Washington University's School of Medicine, he is a recognized authority on population genetics and evolution and is renowned for his pioneering studies on genetic variability. He has authored more than fifty scientific publications and has teamed with his long-time friend, Dr. Peter H. Raven, to publish two best-selling college texts, *Biology* and *Understanding Biology.* They have recently completed *Environment,* a college text on environmental biology. Dr. Johnson has also recently authored a college-level human biology text and the concept-oriented high school biology texts *Biology: Visualizing Life* and *Biology: Principles and Explorations.*

Part

1

INTRODUCTION

Biology is the study of life. This green frog, peering out from the plant-covered surface of a pond, is so perfectly suited to its environment that one has to look carefully to see it. The ways that organisms achieve such seemingly well-designed harmony lie at the heart of biology.

THE SCIENCE OF BIOLOGY

This bristlecone pine has lived for hundreds of years in a harsh climate—the successful result of eons of evolution. Evolution is the core of the science of biology.

The green turtle in figure 1.1 has laid eggs during the night on the upper beach. As dawn breaks, she is returning to the sea, laboriously dragging her 400-pound body across the sand. She is on Ascension Island, in the middle of the Atlantic Ocean, far from her feeding grounds on the coast of Brazil, 1,400 miles away. Every year, great numbers of green turtles cross this great distance to Ascension Island. Both males and females swim to the island, but the males remain offshore while the females scramble up onto the beaches to lay their eggs. The journey lasts about two months: two months of churning through strong equatorial currents with no food. The turtles seek the seven small, sandy beaches on this tiny, rocky island as if drawn by a beacon. By tagging the turtles, scientists have learned that individual females usually return faithfully to the same beach, or even the same stretch of beach, for nesting in successive years. Some evidence suggests that they swim back to the beaches where they were born. Once the eggs have been laid, both male and female turtles begin the long swim home. After the eggs hatch, the newborn turtles find their way back over open sea to Brazil.

In Brazil, where the green turtles feed and remain most of the year, there are countless miles of warm, sandy beaches. Why are the turtles so drawn to the isolated, cold strips of sand on Ascension Island? How can they plow head down through the waves for long days and weeks and still find this island, which is a mere speck in the ocean, hard for an airplane or ship to find even with modern navigational equipment? How can the turtles know that the island is out there, over the horizon, more than a thousand miles away? And how do the newborn turtles know that they must swim to a Brazil that they have never seen?

At first glance, the turtle in figure 1.1 appears to be just a turtle on a beach. Only observation and study reveal the mystery that is so much a part of this turtle's life. Similar mysteries are common in human life, and many of them have a biological basis. Thus, the more we know about the biology of other animals, the better we can understand ourselves. The science of biology is devoted to this larger vision. It provides knowledge about the living world of which we are a part, while ceaselessly bombarding us with questions. Slowly, a little at a time, it sometimes provides answers. This text weaves a tapestry of questions and answers, a picture of life on our planet.

The Characteristics of Life

Biology is a science that attempts to understand the teeming diversity of life on earth, a diversity of which we all are a part. Humans must learn how to live in harmony with earth's other residents, and the science of biology has much to contribute to this effort.

Focus for a moment on biology's subject—life. What is life? What do we mean when we use this term? Life is not a simple

Figure 1.1 End of a long migration.
A female green turtle leaves Ascension Island in the South Atlantic Ocean after laying her eggs in the sand.

concept. A definition of the term must cite the properties that characterize *all* living things. This task sounds easier than it is because some nonliving entities exhibit several characteristics of life and thus appear at first glance to be alive. For instance, viruses, which are pieces of genetic material packaged in a capsule, exhibit some criteria for life, such as a complex chemical organization. However, they lack the ability to reproduce themselves outside of a living host organism. To reproduce, viruses must take over the metabolic machinery of host cells and insert their own genetic program so that the cells stop their usual function and begin producing more viruses. Because viruses cannot reproduce on their own, scientists characterize them as nonliving. As this example demonstrates, what constitutes life is not always cut-and-dried.

The best way to determine whether something is alive or not is to determine the ways in which it resembles living organisms. All known organisms share certain general properties that were probably derived from the first organisms that evolved on earth. These properties allow us to recognize other living things, and to a large degree, define what we mean by the process of life. The fundamental properties shared by all organisms on earth are: (1) complex molecular organization, (2) hierarchical organization, (3) metabolism, (4) reproduction, (5) development, and (6) heredity.

Complex Molecular Organization

All living things are composed of **molecules,** the smallest unit of a chemical compound that still displays the chemical properties of that compound. Four different types of molecules are found in living things: carbohydrates, proteins, lipids, and nucleic acids. The sugars within the stems of plants are assemblages of carbohydrates; the enzymes that control the rate of chemical reactions in our bodies are composed of proteins; the cell membranes of living organisms are made up of lipids; and the hereditary molecules in which living things store genetic information are long chains of nucleic acids.

All living things contain these four types of molecules, but in different forms and configurations. For instance, the only difference between an enzyme in a plant and an enzyme in the human body is in the organization of the enzyme's molecules—the number of chemical bonds, for instance, or the molecule's complexity. *All* enzymes, whether found in a plant or animal, are proteins, and *all* proteins, no matter what kind, are chains of building blocks called amino acids. A plant enzyme might have five amino acids of a certain type, while an animal enzyme might have seven amino acids of different types. But both enzymes are proteins.

All living organisms are composed of carbohydrate, protein, lipid, and nucleic acid molecules. Yet, despite this display of biochemical unity, living organisms have the potential for great chemical diversity.

Hierarchical Organization

The simplest entity of both living and nonliving matter is the atom. Living and nonliving things can also be organized into molecules. At this point, however, the organization of nonliving things stops—there are no higher levels of complexity. Living organisms, on the other hand, have a hierarchical organization of increasing levels of complexity. The four types of molecules—carbohydrates, proteins, lipids, and nucleic acids—discussed in the previous section, are organized in living things into **organelles,** which can be defined as the different "organs" of a cell, such as the chloroplast and the nucleus (figure 1.2). The organelles are themselves organized into **cells,** which are the smallest units of a living organism that can live independently. Cells, in turn, are organized into **tissues,** tissues are organized into **organs,** and organs are organized into **organ systems.** The **organism** is made up of the organ systems. Organisms live in **populations,** which form **communities,** which are part of **ecosystems,** which are part of the **biosphere,** the area on earth where life is found (figure 1.3).

An interesting characteristic of the hierarchy of living things is that of **emergent properties.** An emergent property is a characteristic that "emerges" (hence the name) at a particular level of the hierarchy but that cannot be inferred even by knowing and understanding all the properties of previous

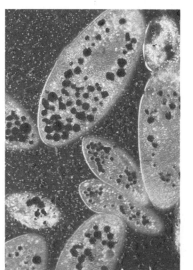

Figure 1.2 Organelles.
These Paramecia *have just ingested several yeast cells, which are stained red in this photograph. The yeasts are enclosed within the* Paramecia *in organelles called digestive vacuoles.*

levels. For instance, the cell, as defined in the previous paragraph, is the smallest unit of living organisms that can exist independently. The characteristic of living independently cannot be inferred from the characteristics inherent in the three previous levels: organelles, molecules, and atoms. This is not to say that the property of cells being able to live independently is not *influenced* by the previous levels. In fact, biologists are trying to understand how the different hierarchical levels influence and prompt the emergent properties of subsequent levels.

This goal also has implications for you, the beginning biology student, since the next chapter in this text is chemistry. Why do you need to study chemistry in a biology course? The answer involves emergent properties. Chemistry, the study of atoms and molecules, examines the first two levels of the hierarchy of life. An understanding of chemistry will help you begin to see how chemistry influences subsequent levels of the hierarchy.

Metabolism

All living things use chemical reactions to obtain the energy they need to grow, synthesize and repair tissues, and reproduce. This process is called **metabolism.** Different kinds of organisms use different types of reactions to produce energy. Plants, algae, and some bacteria make their own food by using the energy of sunlight to convert relatively simple molecules of water and carbon dioxide into more complex carbohydrates, a process called **photosynthesis.** These carbohydrates are then used as energy sources to fuel activities. Animals, on the other hand, do not make their own food and thus obtain their energy sources by eating plants or other animals that eat plants. Animals then liberate the energy from the food sources in a process called **cellular respiration** (figure 1.4). Thus, all living organisms drive the processes of life within themselves by using chemical energy first captured by photosynthesis.

Another common element found in the metabolism of living things is that the energy, once broken down, exists in the same form in all living organisms. In all living things, the "currency" in which energy is stored, transferred, and converted is an **ATP** (adenosine triphosphate) molecule.

Reproduction

All living things reproduce themselves (figure 1.5). Beginning with cells, the lowest rung of the biological hierarchy that is considered "alive," almost each level of the hierarchy is capable of reproduction. Cells divide to form new "daughter" cells. Organisms reproduce to produce offspring. Populations can become fragmented in the process known as speciation, in

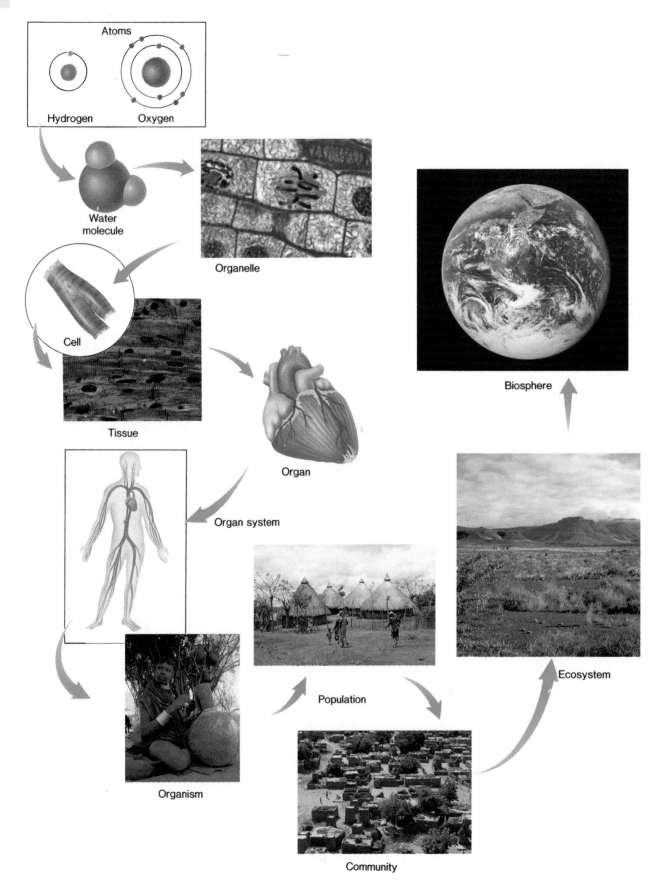

Figure 1.3 The hierarchical organization of life for human beings.
From the simplest level—the atom—the levels become increasingly more complex until the ultimate level—the biosphere, the area on earth where life can be found.

Figure 1.4 Metabolism.
These cedar waxwing chicks obtain the energy they need to grow and develop from the food they eat. They metabolize the food using the process of cellular respiration.

Figure 1.5 Reproduction.
These snakes hatching from their eggs in a Costa Rican rain forest represent successful reproduction. All organisms reproduce, although not all hatch from eggs. Some organisms reproduce several generations each hour; others only reproduce once in a thousand years.

which new species of organisms are formed. Reproduction is an important hallmark of life and drives such important biological themes as evolution and diversity.

Development

Imagine what life would be like if there was no **development,** which can be rather loosely defined as the metamorphosis through a life cycle, the passage through different developmental stages until an organism reaches adulthood. All living things, no matter how simple, exhibit development. Even a bacterium, which appears to be immutable, changes throughout its life span: It grows in size and also duplicates its internal parts in preparation for replication.

Other organisms show a more pronounced development, in which the life-cycle stages are noticeably different in appearance and function. For example, frogs hatch as tadpoles from eggs. These tadpoles, which are solely aquatic, slowly develop into mature frogs, which can live both on land and in water.

The mature frog looks nothing like the immature tadpole, and both are so different that they live in separate environments. Human development is also dramatic. From a completely dependent, helpless infant, the human being slowly progresses through childhood and adolescence, finally reaching independent adulthood.

Heredity

All organisms on earth possess a "genetic" system that is based on the replication (duplication) of a complex linear molecule called **DNA (deoxyribonucleic acid).** The order of the subunits making up the DNA contains, in code, the information that determines what an individual organism will be like, just as the order of letters on this page determines the sense of what you are reading. These subunits of DNA are called **genes.** Because DNA is copied faithfully from one generation to the next, any *change* in a gene is also preserved and passed on to future generations.

The essence of being alive, then, is the ability to reproduce permanently the results of change. **Heredity,** the transmission of characteristics from parent to offspring, along with reproduction, gives rise to adaptation and evolution, two great themes of biology.

An **adaptation** is any peculiarity of structure, physiology (life processes), or behavior that promotes the likelihood of an organism's survival and reproduction in a particular environment. Organisms are remarkably well suited to their environments. Their progressive adaptation to their environment is a process known as **evolution.** Organisms that are less well suited to particular environments do not persist. The ones found on earth today are the "winners" for the moment. The features of any living organism, therefore, are a record of its history. Not only does life evolve, but evolution reflects the very essence of *life.*

All living things on earth are characterized by complex molecular organization, hierarchical organization, metabolism, reproduction, development, and heredity. These six characteristics define the term life.

The Nature of Science

As a science, biology is devoted to understanding biological diversity and its consequences. Like all other scientists, biologists achieve understanding by observing nature and by drawing deductions from these observations. In doing so, biologists also attempt to explain the unity of structure and function that underlies the diversity, and the consequences of that unity. The special relationship between unity and diversity underlies and distinguishes the science of biology.

What is *science?* The word conjures up images of people in white lab coats peering at instruments and shaking test tubes. What do scientists do, and why?

Science is a particular way of investigating the world. Not all investigations are scientific. For example, when you want to know how to get to Chicago from St. Louis, you do not conduct a scientific investigation—instead, you look at a map to determine a route. Making individual decisions by applying a "map" of general principles is called **deductive reasoning.** Deductive reasoning is the reasoning of mathematics, of philosophy, of politics, and of ethics; it is the way in which a computer thinks. All of us rely on deductive reasoning to make everyday decisions. We use general principles as the basis for examining and evaluating these decisions.

Where do the general principles come from? Religious and ethical principles often have a religious foundation; political principles reflect social systems. Some general principles, however, are not derived from religion or politics, but from observation of the physical world around us. If you drop an apple, it will fall, whether or not you wish it to and despite any laws you may pass forbidding it to do so. Science is devoted to discovering the general principles that govern the operation of the physical world.

How Scientists Work: Inductive Reasoning

How do scientists discover such general principles? They find them in stone and air and fire, in a butterfly's wing and a tiger's stare—in short, wherever they look in the world around them. Scientists are, above all, observers: They look at the world to understand how it works. From their observations, scientists determine the general principles that govern our physical world.

This way of discovering general principles by careful examination of specific cases is called **inductive reasoning.** Inductive reasoning first became popular about four hundred years ago, when Isaac Newton, Francis Bacon, and others began to conduct experiments, and from experiment results, to infer general principles about how the world operates. The experiments were sometimes quite simple. Newton's consisted simply of releasing an apple from his hand and watching the apple fall to the ground. This simple observation is the stuff of science. From a host of particular observations, each no more complicated than the falling of an apple, Newton inferred a general principle—that all objects fall toward the center of the earth. This principle was a possible explanation, or **hypothesis,** about how the world works. Like Newton, scientists today formulate hypotheses, and observations are the materials on which they build them.

Testing Hypotheses

Scientists distinguish among hypotheses that are actually true and the many hypotheses that might be true by attempting systematically to demonstrate that certain hypotheses are *not* valid—not consistent with what has been learned from experimental observation. Construction of hypotheses that account for the available facts, observations, and experiments

KEY EXPERIMENT 1.1

2000
◄ 1994

DESIGNING A CONTROLLED EXPERIMENT

Suppose that you want to test whether yeast, a single-celled organism, produces carbon dioxide as it extracts energy from sugar. To test this hypothesis, you design an experiment in which different amounts of a sugar solution are added to separate yeast suspensions, while in a control yeast suspension, no sugar is added. All other variables are the same: the amount and temperature of the water in which the yeast is dissolved, the amount of light the yeast receives, the type of test tubes used, and so on. At the end of the experiment, you look for any differences among the various yeast suspensions. These differences can be assumed to result from the addition of the sugar.

Specifically, your experiment design might look like the following:

OBSERVATION

1900   Yeast produces carbon dioxide.

concerning a particular area of science requires careful and creative thinking. Hypotheses that scientists are not yet able to disprove are retained—at least for the time being—because they fit the known facts. For example, Newton's hypothesis that all objects fall toward the center of the earth has yet to be rejected, even today. However, even longstanding hypotheses could ultimately be rejected if, in the light of new information, they are found to be inconsistent with observation.

Scientists test hypotheses by experimentation. An **experiment** evaluates alternative hypotheses. For example, suppose that you face two closed doors. Four possible hypotheses might be:

1. "There is a tiger behind the door on the left."
2. "The door on the right has a tiger behind it."
3. "There is no tiger behind either door."
4. "There is a tiger behind both doors."

A successful experiment eliminates one or more of the hypotheses. For example, to test these alternative hypotheses, you might open the door on the right and find that a tiger leaps out at you. Thus, your experiment disproves the third hypothesis: It was clearly incorrect to say that there was no tiger behind either door. Note that this experiment does not prove that only one alternative is true; rather, it demonstrates that one of the alternatives is *not* true. In this instance, the fact that a tiger is behind the door on the right does not rule out

QUESTION

What causes yeast to produce carbon dioxide?

POTENTIAL HYPOTHESES

1. Yeast produces carbon dioxide as a by-product of its extraction of energy from sugar.

2. Yeast contains large amounts of carbon dioxide and thus excretes it.

3. Yeast produces carbon dioxide by using the components of water molecules and combining them with other components to yield carbon dioxide.

EXPERIMENT

The following experiment will test hypothesis 1:

1. Assemble four test tubes and label them 1 through 4. To test tube 1, add 2.5 milliliters of glucose (sugar) solution; to test tube 2, add 3 milliliters of glucose solution; to test tube 3, add 3.5 milliliters of glucose solution; to test tube 4, add no glucose solution.

2. Add 5 milliliters of water and 5 millimeters of yeast suspension to all the test tubes.

3. Test tubes 1, 2, and 3 begin to bubble. You test the contents of the test tube bubbles and find that they are carbon dioxide bubbles. You also notice that the test tubes that have more sugar have more bubbles. Test tube 4 has no bubbles.

CONCLUSIONS

1. You can eliminate hypothesis 3.

2. You can state that, when glucose is added to a yeast suspension, yeast produces carbon dioxide.

3. You must design an experiment that tests hypothesis 2. You could conduct a similar experiment, in which a yeast suspension is put into a test tube without water, and then watch for carbon dioxide

bubbles. (What would be the control for this experiment?) You predict, however, that the presence of sugar in a solution is the agent that causes yeast to produce carbon dioxide.

the possibility that a tiger also lurks behind the door on the left. The next step is to test each of the remaining hypotheses—in this case, by opening the door on the left. The existence or nonexistence of a tiger behind the left door successfully pinpoints the correct hypothesis.

An experiment is successful when one or more of the alternative hypotheses are demonstrated to be inconsistent with experimental observation and thus are rejected. Scientific progress is something like sculpturing a marble statue, by chipping away unwanted bits.

The process of science consists of demonstrating that one or more hypotheses are not consistent with experimental observation.

As you proceed through this text, you will encounter a great deal of information, often coupled with explanations. These explanations are hypotheses that have stood the test of experiment. Many will continue to do so; others will be revised. Biology, like all healthy sciences, is in a constant state of ferment, with new ideas bubbling up and replacing old ones.

Controls

Processes often are influenced by many factors. Each factor that influences a process is called a **variable.** Evaluation of alternative hypotheses about one variable requires holding all the other vari-

ables constant so that these other influences do not confuse the findings. This is done by conducting two parallel experiments: In the first experiment, one variable is altered in a known way to test a particular hypothesis; in the second **controlled experiment,** that variable is not altered. In all other respects, the two experiments are the same. Then, any difference between the outcomes of the two experiments must result from the influence of the altered variable, since all other variables remained constant (the same in both experiments). Much of the challenge of experimental science lies in designing controlled experiments—in successfully isolating a particular variable from all other effects that might influence a process. Key Experiment 1.1 provides a detailed analysis of a controlled experiment.

The Importance of Prediction

A successful scientific hypothesis must be valid and also useful—it needs to tell us something that we want to know. A hypothesis is most useful when it makes predictions. Such predictions provide an important way to further test hypothesis validity.

A hypothesis that your experiment does not reject, but that makes a prediction the experiment *does* reject, must itself be rejected. On the other hand, a hypothesis that makes verifiable predictions becomes more demonstrably valid. For example, Einstein's hypothesis of relativity was at first provisionally accepted because no one could think of an experiment that invalidated it. Acceptance soon became far stronger because the theory made a clear prediction: that the sun would bend the path of light passing by it. When this prediction was tested in a total eclipse, the light of background stars was indeed bent. Because this result was not known ahead of time (when Einstein's proposal was being formulated), it strongly supported his hypothesis.

Theories

Hypotheses that stand the test of time—their predictions often tested and never rejected—are called **theories.** Thus, Newton's general principle—that all objects fall toward the center of the earth—is now known as the theory of gravity. While theories are the solid ground of science, there is no absolute truth in science, only varying degrees of uncertainty. Future evidence could always cause a theory to be revised. Therefore, a scientist's acceptance of a theory must be provisional.

Scientists' use of the word *theory* is very different from that of the general public. To scientists, a theory represents that of which they are most certain; to the general public, the word *theory* implies a *lack* of knowledge, or a guess. Confusion often results from these different interpretations. In this text, the word *theory* is always used in its scientific sense and refers to a generally accepted scientific principle.

> *A theory is a hypothesis that is supported by a great deal of evidence.*

Some theories are so strongly supported that future rejection is highly unlikely. For example, most of us are willing to bet that the sun will rise in the east tomorrow or that an apple, when dropped, will fall. In physics, the theory of the atom is universally accepted, although until recently, no one had ever seen one. In biology, the theory of evolution by natural selection is so broadly supported by different lines of inquiry that biologists accept it with as much certainty as they do the theory of gravity. The theory of evolution is particularly important to biologists because it provides the conceptual framework that unifies biology as a science. Later in this chapter, the development of evolutionary theory is examined to demonstrate how an idea develops into a hypothesis, is tested, and eventually is accepted as a theory.

The Scientific Method

The way in which scientists investigate hypotheses, called the **scientific method,** is outlined in figure 1.6. As you can see in the figure and as already outlined in previous chapter discussion, a number of potential explanations (hypotheses) are suggested in answer to a question. Experiments are conducted in an attempt to eliminate one or more of these hypotheses. Predictions are made based on the remaining hypotheses, and further experiments test these predictions. The result of this process is the selection of the most likely hypothesis. If it is validated by numerous experiments and stands the test of time, the hypothesis may eventually lead to formation of a more general statement—a theory.

Following these steps rigorously and exactly in every scientific investigation is impossible, however. An orderly sequence of logical "either/or" steps, with each step rejecting one of two mutually incompatible alternatives, simply does not exist for most scientific investigations. Instead, as British

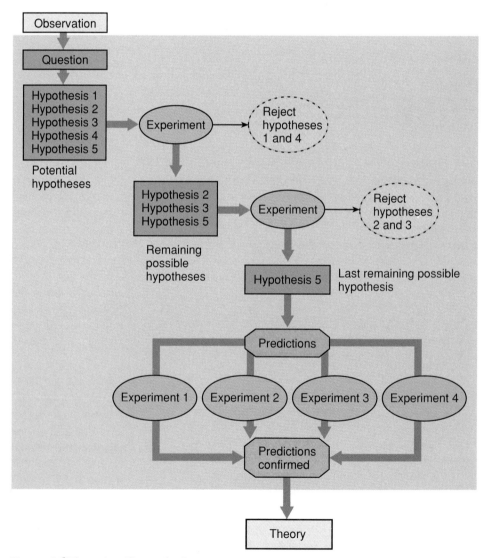

Figure 1.6 The scientific method.
This diagram illustrates how scientific investigations proceed. Not all scientific investigations proceed in this orderly manner, however. Creativity plays a large role in the process of science.

philosopher Karl Popper pointed out, successful scientists design their experiments with a pretty fair idea of how the experiments are going to come out—they use what Popper called an "imaginative preconception" of what the truth might be. Successful scientists test hypotheses that are "hunches" or educated guesses into which these scientists have integrated all that they know and imagine in an attempt to get a sense of what *might* be true. It is because insight and imagination play such a large role in scientific progress that some scientists are so much better at science than others. Beethoven and Mozart stand out among most other composers for precisely the same reason.

> *The scientific method is the experimental testing of a hypothesis formulated after the systematic, objective collection of data. Hypotheses are not usually formulated simply by rejecting a series of alternative possibilities; instead, they often involve creative insight.*

History of a Biological Theory: Darwin's Theory of Evolution

The theory of **evolution**—the idea that living things on earth change gradually from one form into another over the course of time—provides a good example of how an educated guess is developed into a hypothesis, tested, and eventually accepted as a theory.

Darwin and the Voyage of the HMS *Beagle*

Charles Robert Darwin (1809–1882; figure 1.7) was an English naturalist who, at the age of fifty, after thirty years of study and observation, completed one of the most famous and influential books of all time. The full title of this book, *On the Origin of Species by Means of Natural Selection, or the Preservation of Favoured Races in the Struggle for Life,* expressed both the nature of its subject and the way in which Darwin treated it. The book created a sensation when first published in 1859, and the ideas expressed in it have played a central role in the development of human thought ever since.

In Darwin's time, most philosophers believed that the various kinds of organisms and their individual structures resulted from the direct actions of the Creator. Species were held to be specially created and unchangeable over the course of time. In contrast, a number of other scholars, both before and

Figure 1.7 Charles Darwin.
This portrait was painted when Darwin was twenty-nine years old, two years after his return from the voyage on HMS Beagle. Darwin had just married his cousin Emma Wedgwood and was hard at work studying the materials he had gathered on the voyage.

during Darwin's time, believed that living things had changed during the course of the history of life on earth and that some form of evolution from simple to complex forms had occurred.

Darwin, though, was the first to present a coherent, logical explanation for this process—natural selection—and the first to bring the notion of evolution to wide attention. His book, as evidenced by its title, presented a conclusion that differed sharply from conventional wisdom. Darwin argued that the evolution of organisms—the origin of the vast and diverse array of life on earth—followed a series of more or less orderly steps that could be studied and understood and that produced continual change and improvement. Although his theory did not challenge the existence of a divine Creator, Darwin's views put him at odds with most people of his time, who believed in a literal interpretation of the Bible and accepted the idea of a fixed and constant world. Darwin's revolutionary theory troubled many of his contemporaries, and Darwin himself, deeply.

The son of a wealthy father, Darwin had a sketchy education, spending more time outdoors than in school. As a medical student in Edinburgh, Scotland, for example, he cut lectures, spending his time collecting beetles! In desperation, his father sent him to Cambridge to train for the ministry. There, in 1831, when Darwin was twenty-two years old, one of his professors recommended him for a post as naturalist on a five-year voyage (from 1831–1836) around the coasts of South America on the HMS *Beagle* (figure 1.8a). Darwin was offered the position. His father at first refused to allow him to go, and Darwin regretfully declined, but his girlfriend's father interceded at the last moment, and Darwin was off on a voyage that would change forever how we think of ourselves.

During his long journey, Darwin had the chance to study plants and animals on continents and islands and in far-flung seas (figure 1.8b). He was able to experience firsthand the biological richness of the tropical forests, the extraordinary fossils of huge extinct mammals in Patagonia at the southern tip of South America, and the remarkable series of related but distinct life-forms on the Galápagos Islands off the west coast of South America. Such opportunities clearly played an important role in the development of his thoughts about the nature of life on earth.

When Darwin returned from the voyage, at the age of twenty-seven, he began a long life of study and contemplation. During the next ten years, he published important books on several different subjects, including the formation of oceanic islands from coral reefs and the geology of South America. He also devoted eight years of study to barnacles, a group of marine animals, writing a four-volume work on their classification and natural history. In 1842, Darwin and his

Figure 1.8 The voyage of the HMS Beagle.
(a) *A replica of the HMS* Beagle *off the southern coast of South America. Darwin set forth on the HMS* Beagle *in 1831 at the age of twenty-two.* (b) *The five-year voyage of the HMS* Beagle *took Darwin around the world. Most of the time, however, was spent exploring the coasts and coastal islands of South America, such as the Galápagos Islands. Darwin's studies of Galápagos animals played a key role in his eventual development of the theory of evolution by means of natural selection.*

(a)

(b)

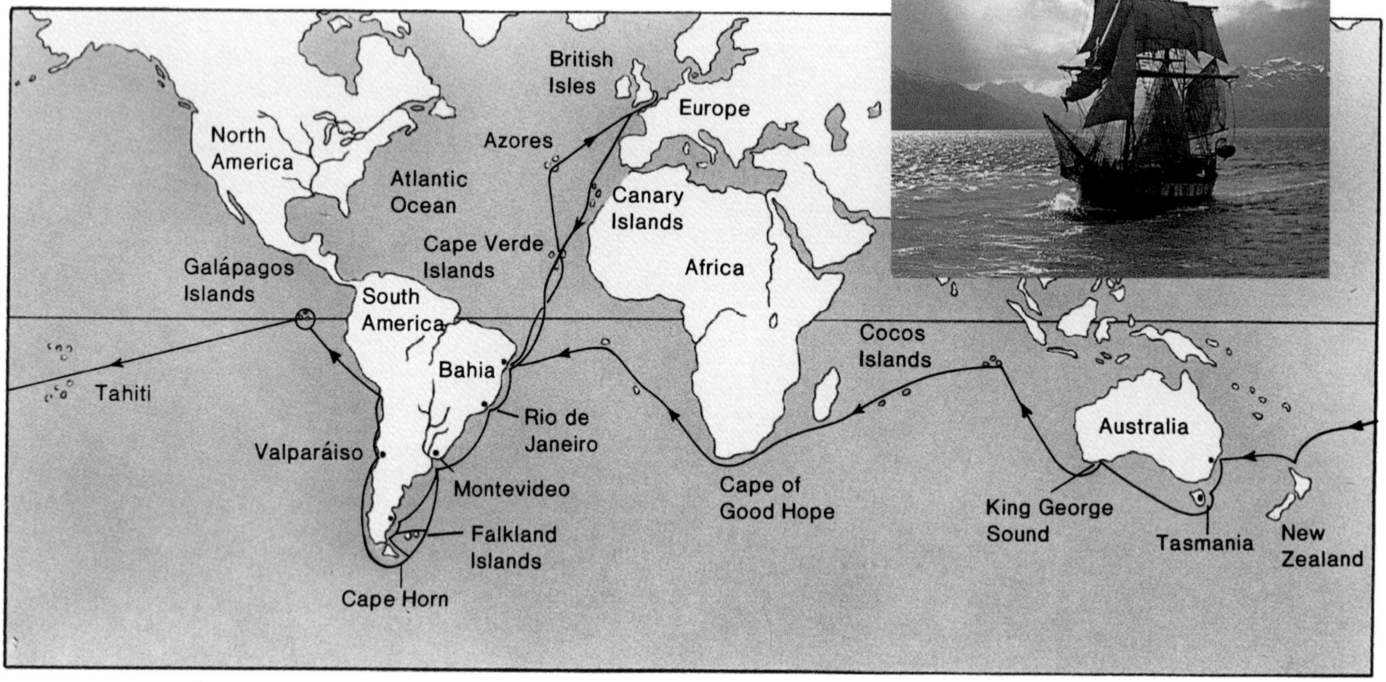

family moved a short distance out of London to a country home at Downe, in Kent. In these pleasant surroundings, he lived, studied, and wrote for the next forty years.

Darwin presented his preliminary ideas on evolution and natural selection in a manuscript shown only to trusted colleagues in 1842. This manuscript proposed, for the first time, a theory of evolution that explained *why* evolution occurred. Presented with convincing detail, Darwin's ideas also logically explained the diversity of life on earth, the intricate adaptations of living things, and the ways in which living things are related to one another. Then he put the manuscript aside and worked on other subjects for the next seventeen years.

Darwin's Evidence

So much information had accumulated by 1859 that acceptance of the theory of evolution now seems, in retrospect, to have been inevitable. Darwin was able to arrive at a successful theory, where many others had failed, because he rejected supernatural explanations for the phenomena he was studying. One of the obstacles that blocked the acceptance of any natural theory of evolution was the incorrect notion, still widely

believed at that time, that the earth was only a few thousand years old. However, discoveries of thick layers of rocks, evidence of extensive and prolonged erosion, and the increasing numbers of diverse and unfamiliar fossils found during Darwin's time were making this assertion seem less and less likely. In addition, the great geologist Charles Lyell (1797–1875), whose works Darwin read eagerly while sailing on the HMS *Beagle,* outlined for the first time the story of an ancient world of plants and animals in flux. In this world, some species were constantly becoming extinct while others were emerging. It was this world that Darwin sought to explain.

Darwin's Observations

When the HMS *Beagle* had set sail, Darwin had been fully convinced that species were unchanging and immutable. Indeed, he later wrote that it was not until two or three years after his return that he began to consider seriously the possibility that they could change. Nevertheless, during his five years on the ship, Darwin observed a number of phenomena that were of central importance to his ultimate conclusion (table 1.1). For example, in rich beds of fossils in southern South America, he observed fossils of extinct armadillos (called glyptodonts) that

Table 1.1 Darwin's Evidence That Evolution Occurs

Fossils

1. Extinct species, such as the fossil armadillos of figure 1.9, most closely resemble living ones in the same area, suggesting that one gave rise to the other.

2. In rock strata (layers), progressive changes in characteristics can be seen in fossils from progressively older layers.

Geographical Distribution

3. Lands that have similar climates, such as Australia, South Africa, California, and Chile, have unrelated plants and animals, indicating that differences in environment are not creating the diversity directly.

4. The plants and animals of each continent are distinctive, although there is no reason why special creation (that is, creation by a supernatural entity) should create this association. For example, all South American rodents belong to a single group, structurally similar to the guinea pigs, whereas most of the rodents found elsewhere belong to other groups.

Oceanic Islands

5. Although oceanic islands have few species, those they have are often unique ("endemic") and show relatedness to one another, such as the tortoises of the Galápagos (see figure 1.10). This suggests that the tortoises and other groups of endemic species formed after their ancestors reached the islands and are therefore directly related to one another.

6. Species on oceanic islands show strong affinities to those on the nearest mainland. Thus, the finches of the Galápagos closely resemble a finch seen on the western coast of South America. The Galápagos finches do *not* resemble birds of the Cape Verde Islands, islands in the Atlantic Ocean off Africa that are very similar to the Galápagos. Darwin visited the Cape Verde Islands and many other island groups personally and was able to make such comparisons on the basis of his own observations.

were directly related to the armadillos that still lived in the same area (figure 1.9). Why would there be living and fossil organisms, directly related to one another, in the same area, unless one had given rise to the other?

Repeatedly, Darwin saw that the characteristics of closely related species varied from place to place. These patterns suggested to him that organisms change gradually as they migrate from one area to another. On the Galápagos Islands off the coast of Ecuador, Darwin encountered giant land tortoises. Surprisingly, these tortoises were not all identical. Indeed, local residents and the sailors who captured the tortoises for food could tell which island a particular animal had come from just by looking at it (figure 1.10). This pattern of variation suggested that all of the tortoises were related but had changed slightly in appearance after becoming isolated on the different islands.

In a more general sense, Darwin was struck by how the relatively young volcanic islands supported a profusion of living things that resembled those of the nearby coast of South America. If each one of these plants and animals had been created independently

and simply placed on the Galápagos Islands, why did they not resemble the plants and animals of faraway Africa, for example? Why did they resemble those of the adjacent South American coast instead?

> *The patterns of distribution and relationship among organisms that Darwin observed on the voyage of the HMS* Beagle *ultimately made him certain that a process of evolution had been responsible for these patterns.*

The Connection between Darwin and Malthus

It is one thing to observe evolution, another to understand how it happens. Darwin's great achievement was his perception that evolution occurs because of natural selection. Of key importance to the development of Darwin's insight was his study of Thomas Malthus's *Essay on the Principles of Population*. In this book, Malthus pointed out that populations of plants and animals, including humans, tend to increase geometrically, whereas food supply increases only arithmetically. In a geometric progression, the elements progress by a constant factor, as in 2, 6, 18, 54, and so forth; in this example, each number is three times the preceding one. In an arithmetic progression, in contrast, elements increase by a constant difference, as in 2, 6, 10, 14, and so forth; in this case, each number is four more than the preceding one.

According to Malthus, virtually any animal or plant, if left to reproduce unchecked, would cover the entire surface of the world within a surprisingly short period. In fact, however, this does not occur. Malthus found instead that species populations remain more or less constant year after year because death intervenes and limits population numbers.

Malthus's findings provided the key ingredient that Darwin needed to develop his hypothesis that evolution occurs by natural selection.

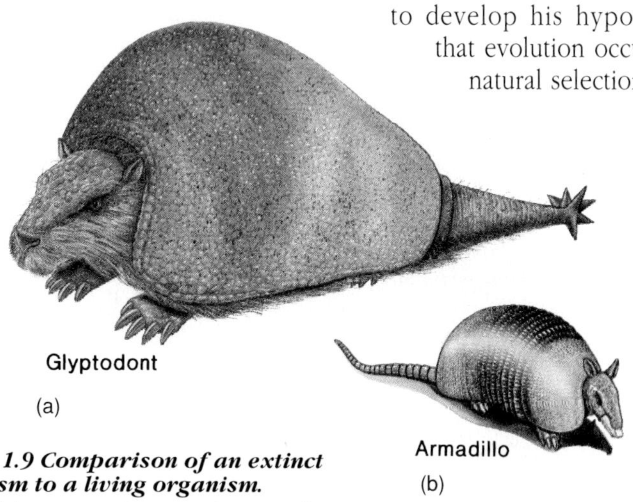

Glyptodont

(a)

Armadillo

(b)

Figure 1.9 Comparison of an extinct organism to a living organism.
(a) *On his journey, Darwin discovered fossils of the glyptodont, an ancient, two-ton South American armadillo.*
(b) *Darwin noted the similarity of the glyptodont fossils to the living armadillo found in the same region. This similarity between ancient extinct species and related living species led Darwin to his theory of evolution.*

(a)

(b)

Figure 1.10 Galápagos tortoises.
(a) *Tortoises with large, domed shells are found in relatively moist habitats. (b) The lower saddleback-type shells, in which the front of the shell is bent up, exposing the head and part of the neck, are found on tortoises that live in dry habitats. Differences of these kinds make it possible to identify the species of tortoises that inhabit the different islands of the Galápagos.*

A key contribution to Darwin's thinking was Malthus's concept of geometric population growth. The fact that real populations do not expand at this rate implies that nature acts to limit population numbers.

The Key Association: Natural Selection

Sparked by Malthus's ideas, Darwin saw that, although every organism has the potential to produce more offspring than are able to survive, only a limited number of offspring actually survive and produce their own offspring. Combining this observation with what he had seen on the voyage of the HMS *Beagle*, as well as with his own experiences in breeding

domestic animals, Darwin made the key association: *Those individuals that possess superior physical, behavioral, or other attributes are more likely to survive than those that are not so well endowed.* In surviving, these individuals have the opportunity to pass on their favorable characteristics to their offspring. Because these characteristics then increase in the population, the nature of the population as a whole gradually changes. In his book *On the Origin of Species*, Darwin called this process **natural selection** and referred to the driving force he had identified as **"survival of the fittest"**:

> *Can we doubt . . . that individuals having any advantage, however slight, over others, would have the best chance of surviving and of procreating their kind? On the other hand, we may feel sure that any variation in the least degree injurious would be rigidly destroyed. This preservation of favorable variations, I call Natural Selection.*

Natural selection is the increase in succeeding generations of the traits of those organisms that leave more offspring. Its operation depends on the traits being inherited. The nature of the population gradually changes as more and more individuals with those traits appear.

Darwin was thoroughly familiar with variation in domesticated animals and began *On the Origin of Species* with a detailed discussion of pigeon breeding. He knew that varieties of pigeons and other animals, such as dogs, could be selected to exhibit certain characteristics. Once this had been done, the animals would breed true for the characteristics that had been concentrated in them. Darwin had also observed that the differences that could be developed between domesticated races or breeds in this way were often greater than those that separated wild species. The breeds of domestic pigeons are much more different from one another in various ways than are all of the hundreds of wild species of pigeons found throughout the world. Such relationships suggested to Darwin that evolutionary change could occur very rapidly under the right circumstances.

Publication of Darwin's Theory

As mentioned earlier, Darwin drafted the overall argument for evolution by natural selection in 1842 and continued to enlarge and refine it for many years. The stimulus that finally brought it into print was an essay that he received in 1858 from a young English naturalist named Alfred Russel Wallace (1823–1913) (figure 1.11). Wallace's essay concisely set forth the theory of evolution by means of natural selection! Like Darwin, Wallace had been greatly influenced in his development of this theory by reading Malthus's 1798 essay. After receiving Wallace's work, Darwin arranged for a joint presentation of their ideas at a seminar in London and proceeded to complete his own book, on which he had been working for so long, for publication in what he considered an abbreviated version.

Darwin's book appeared in November 1859 and caused an immediate sensation. For example, many people were deeply disturbed by the idea that human beings were closely related to apes. Darwin did not discuss this idea in his book, but it followed directly from the principles he outlined. It had long been accepted that humans closely resembled apes, but the possibility of a direct evolutionary relationship between apes and humans was unacceptable to many people (figure 1.12). Darwin's arguments for the theory of evolution by natural selection were so compelling, however, that his views were almost completely accepted within the intellectual community of Britain after the 1860s.

Figure 1.11 Alfred Russel Wallace in 1902.
Wallace and Darwin simultaneously formulated the theory of evolution.

Figure 1.12 Darwin greeting his "monkey ancestor."
In his time, Darwin was often portrayed unsympathetically, as in this drawing from an 1874 publication.

fossil record is known to a degree that would have been unthinkable in the nineteenth century. Recent discoveries of microscopic fossils have extended the known history of life on earth back to more than 3.5 billion years ago. Other fossil discoveries have shed light on the ways in which organisms have evolved from simple to complex over this enormous time span. For vertebrate animals—those with backbones—especially, the fossil record is rich and exhibits a graded series of changes in form, with the evolutionary parade visible for all to see.

The Age of the Earth

In Darwin's day, some physicists argued that the earth was only a few thousand years old. This bothered Darwin because the evolution of all living things from some single original ancestor would have required a great deal more time. Evidence obtained by studying rates of radioactive decay now shows that the earth formed some 4.5 billion years ago.

Evolution after Darwin: Testing the Theory

Darwin did more than propose a mechanism that explains how evolution has generated the diversity of life on earth. He also assembled masses of facts, otherwise seemingly without logic, that began to make sense when viewed in the light of his theory. After publication of Darwin's book, other biologists continued this process, and it soon became evident that the theory of evolution was supported by a wide variety of biological information gathered by many investigators. Three general observations resulted: (1) Members of different biological groups often share common features, (2) embryos (organisms developing from fertilized eggs) are more similar at earlier embryo stages and diverge through later stages, and (3) the fossil record is increasingly complex over time.

In the century since Darwin, the evidence supporting his theory has grown progressively stronger, and scientists' understanding of how evolution works has advanced significantly. Evolution has become the main unifying theme of the biological sciences, providing one of the most important insights that human beings have achieved into their own nature and that of the earth on which they have evolved. Specifically, evidence supporting Darwin's theory has come from the fossil record, studies of the age of the earth, the science of genetics, comparative studies of organisms, and the field of molecular biology.

The Fossil Record

Darwin predicted that the fossil record would show intermediate links between the great groups of organisms—for example, between fishes and the amphibians thought to have arisen from them and between reptiles and birds (figure 1.13). Today, the

Figure 1.13 Fossil of an early bird, Archaeopteryx.
This well-preserved fossil of the approximately 150-million-year-old bird was discovered within two years of Darwin's publication of On the Origin of Species. *The fossil provides an indication of the evolutionary relationship between birds and reptiles.*

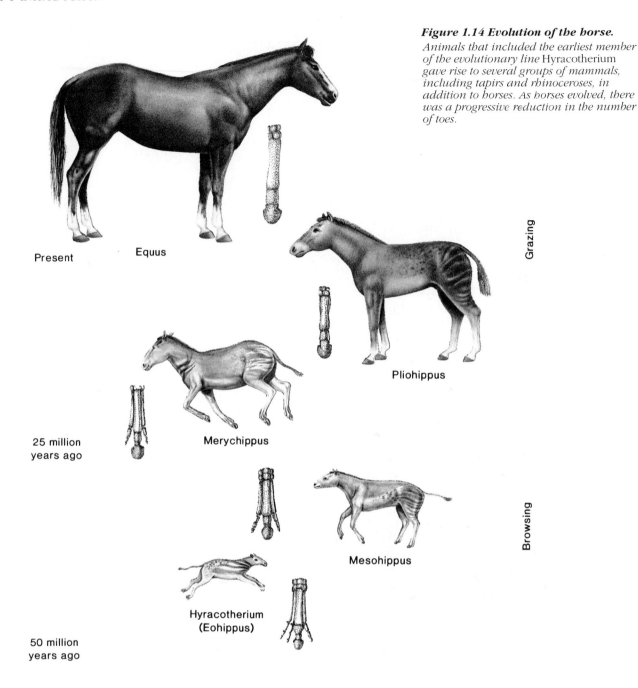

Figure 1.14 Evolution of the horse.
Animals that included the earliest member of the evolutionary line Hyracotherium *gave rise to several groups of mammals, including tapirs and rhinoceroses, in addition to horses. As horses evolved, there was a progressive reduction in the number of toes.*

Present   Equus

Grazing

Pliohippus

25 million years ago   Merychippus

Browsing

Mesohippus

Hyracotherium (Eohippus)

50 million years ago

The Mechanism of Heredity

Darwin received some of his sharpest criticism in the area of heredity. Since at that time no one had any concept of how heredity works, Darwin could not explain completely how evolution occurs. Theories of heredity current in Darwin's day seemed to rule out the possibility of genetic variation in nature, a critical requirement of Darwin's theory. Genetics was established as a science only at the start of the twentieth century, forty years after the publication of Darwin's *On the Origin of Species*. When the laws of inheritance became understood, problems with Darwin's theory vanished because heredity (discussed in chapter 11) clearly explains the variations in nature required by Darwin's theory.

Comparing Different Kinds of Organisms

Comparative studies of organisms also have provided strong evidence for Darwin's theory. For example, as vertebrates have evolved, the same bones have sometimes been put to different uses—and yet they can still be seen, betraying their evolutionary past (figure 1.14). Thus, the forelimbs in figure 1.15 are all constructed from the same basic array of bones, modified in one way in the wings of bats, in another way in the fins of porpoises, and in yet other ways in the legs of frogs, horses, and humans. The bones have the same evolutionary origin but now differ in structure and function.

Some more good stuff.
Better grades. Easier class.

A formula for success

 1 OK. Here's what you should study.

[Student Study Guide]
ISBN 0-697-22216-0

Could you use a little help with your studying? Most of us could. Inside the *Student Study Guide* are removable

 flash cards with key terms and definitions. The guide contains fill-in-the-blank and multiple-choice questions, so you can quiz yourself. And with the "Test Your Identification Skills" feature, you can practice labeling diagrams such as the human heart and nerve cell. This guide also contains all the answers to the text's "Completing Your Understanding" questions.

2 Conserve your energy.

[Biology Study Cards]
ISBN 0-697-03069-5

Why work harder than you have to? There's a set of study cards already prepared for you. These two-sided, 3-by-5-inch cards contain the important information you'll need to study. On them you'll find complete definitions of key terms, clearly labeled illustrations, pronunciation guides, and much more.

3 May we help you?

[Biology Startup software]
ISBN 0-697-24864-X

 Learn at your own pace, and have fun while you're at it. This set of *interactive Macintosh tutorials* portray concepts in a new way. Click the mouse, and up come definitions, color illustrations and animations, and pronunciations spoken by a human voice. Quiz yourself at the end of major sections. Students like you say this software is fun and easy to use —and that they actually *understand* concepts they never did before.

4 See human life processes—up close and dynamic.

[Explorations in Human Biology]
ISBN 0-697-22964-5 • Macintosh
ISBN 0-697-22963-7 • IBM/Windows

CD-ROM

Action, reaction, interaction, and exploration—that's what human biology is about. Investigate vital processes as they should be explored—with movement, color, sound, and interaction. *Explorations in Human Biology* CD-ROM is a set of 16 interactive animations that allow you to set— and reset—variables and then evaluate results. On it you'll find clear information; colorful graphics and animated illustrations; and a glossary with written and oral definitions in both English and Spanish. Topics explored include diet and weight loss, drug addiction, and AIDS.

RAVEN 22213

Such a deal—
cut your study time
and improve
your grades.

[
Computerized Study Guide

IBM: ISBN 0-697-28315-1

Macintosh: ISBN 0-697-28314-3
]

Focus your study time on only those areas where you need help. This easy-to-use software helps you identify and target those areas. Simply test yourself on each text chapter. The program not only gives you the correct answers, but creates *your own personalized study plan* based on your incorrect answers—with page references for each chapter.

This great program is available for use with Macintosh and IBM or compatible computers.

To order the *Computerized Study Guide,* call 800-338-5578.

---

 m. C. Brown Publishers **has all** *the* **tools you need to**

nail down **general biology and other science courses—**

materials designed to *make your life easier* **and not cost you a**

small  *fortune.* **To order, contact your** *local bookstore.*

Or call our Customer Service Department **at** 800-338-5578.

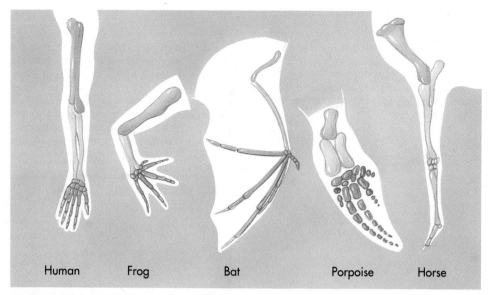

Figure 1.15 Comparing vertebrate limbs.
The forelimbs of these five vertebrate animals have the same evolutionary origin, but over time, the limbs have been put to different uses.

Human   Frog   Bat   Porpoise   Horse

Molecular Biology

Biochemical tools have become very important in efforts to better understand evolution. Within the last few years, for example, evolutionists have begun to "read" genes, much as you read this page, by recognizing in order the "letters" of the long DNA molecules that store genetic information. When the DNA sequences of different groups of animals or plants are compared, the degree of relationship between the groups can be specified more precisely than by any other means. In many cases, detailed "family trees" can be constructed. When DNA studies of two different molecules lead to the same family tree, this provides strong evidence that the family tree accurately represents the group's evolutionary history. Indeed, the rates at which evolution is occurring in different groups can often be measured.

> *In the century since Darwin proposed his theory, a large body of evidence, contributed by many branches of science, has supported his view of evolution driven by natural selection.*

Why Is Biology Important to You?

Biologists do more than simply write books about evolution. They live with gorillas, collect fossils, and listen to whales. They isolate viruses, grow mushrooms, and grind up fruit flies. They read the message encoded in the long molecules of heredity and count how many times a hummingbird's wings beat each second. In its broadest sense, biology is the study of living organisms, of the diverse array of living things that blanket our earth. Biologists try to understand the sources of this rich diversity of life and, in many cases, to harness particular life-forms to perform useful tasks. Even the narrowest study of a seemingly unimportant life-form represents one more brush stroke in the painting of biological diversity that biologists have labored over for centuries.

Biology is fun and interesting because of its great variety, but it is also a body of information that will affect your future in many ways. The knowledge that biologists are gaining is fundamental to our ability to manage the world's resources in a sustainable manner, to prevent or cure diseases, and to improve the quality of our lives and those of our children and grandchildren (see Sidelight 1.1). An understanding of biology provides the key to the perpetuation of life on earth.

How This Text is Organized to Teach You Biology

In the century since Darwin's publication of *On the Origin of Species*, biology has exploded as a science, presenting today's student with a wealth of information and theory. A beginner can be introduced to biology and can see what is available to learn in many different ways. An introductory biology course offers a variety of experiments and observations that you *could* learn, and from this you must select a small body of information that you *will* learn. Your target is the basic body of principles that unite biology as a science.

A good way to begin your examination of basic biological principles is to focus on complexity. Arranging the wealth of biological information in terms of levels of complexity leads to "levels-of-organization" approach to the field. This approach mirrors the hierarchy of organization that is one of the hallmarks of life (see figure 1.3).

The first half of this text is devoted to a description of basic biological principles. It uses a levels-of-organization framework to introduce different principles at the level where each principle is most easily understood. At the molecular, subcellular, and cellular levels of organization, you are introduced to the principles of cell biology and learn how cells are constructed and how they grow, divide, and communicate. At the organismal level, you learn the principles of genetics, which deals with the way in which an individual's traits are transmitted from one generation to the next. At the population level, you study evolution, a field concerned with the nature of population changes from one

Sidelight 1.1
URGENT PROBLEMS AND CHALLENGES FOR BIOLOGISTS

Biologists are working on many problems that critically affect our lives—from dealing with the demands of the world's rapidly expanding population to trying to find ways to prevent cancer and acquired immune deficiency syndrome (AIDS). Because the activities of biologists alter our lives in so many ways, an understanding of biology is becoming increasingly important for any educated person.

HUMAN POPULATION GROWTH

About ten million people were alive in the world when humans first spread across North America. Today there are almost six billion, and the earth's ability to support them is being strained.

INFECTIOUS DISEASE

AIDS is a serious and growing health problem worldwide. Other diseases are even more serious: Mosquito-spread malaria will kill over three million people this year, most in the underdeveloped countries of the world. Scientists are struggling to produce vaccines against these scourges.

USE OF ADDICTIVE DRUGS

The use of dangerous addictive drugs, particularly powerful heroin and cocaine derivatives, is creating a nightmare in many cities. Other dangerous but legal drugs are also in widespread use, such as cigarettes and other tobacco products that kill thousands by causing lung cancer.

INDUSTRIAL ALTERATION OF THE ENVIRONMENT

The destruction of the atmosphere's ozone (as shown in the accompanying photograph), the creation of acid rain, the greenhouse effect caused by increasing concentrations of carbon dioxide in the atmosphere, the pollution of rivers and underground water supplies, the dilemma of how to dispose of radioactive wastes—all of these problems require urgent attention.

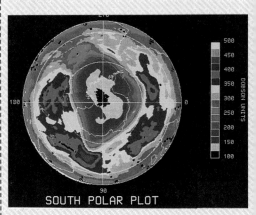

SOUTH POLAR PLOT

---

generation to the next as a result of selection, and the way in which this has led to the biological diversity we see around us. Finally, at the community and ecosystem levels, you study ecology, which deals with how organisms interact with their environments and with one another to produce the complex communities characteristic of life on earth.

The second half of this book is devoted to an examination of organisms, the products of evolution. Organisms are classified into six kingdoms (figure 1.16). Members of the Eubacteria and Archaebacteria have a simple cellular structure and no true nucleus; organisms in the other four kingdoms have nuclei and a much more complex structure. Among organisms with nuclei, Protists include a very diverse series of several dozen distinct evolutionary lines, most of them predominantly single-celled. From ancestors that would have been classified as Protists,

three primarily multicellular kingdoms—the plants (kingdom Plantae), the animals (kingdom Animalia), and the molds and mushrooms (kingdom Fungi)—have been derived independently. The diversity of living organisms is *incredible*. At least ten million different kinds of plants, animals, and microorganisms are estimated to exist. The last section of the book examines the vertebrates, the group of animals of which humans are members, and focuses on the vertebrate body and how it functions (especially the human body).

As you proceed through this text, what you learn at one level of the hierarchical organization of life will give you the basic tools to tackle the next. Chapter 2 examines some simple chemistry. To understand lions and tigers and bears, you first need to know the basic chemistry that makes them tick, for they are chemical machines, as are you.

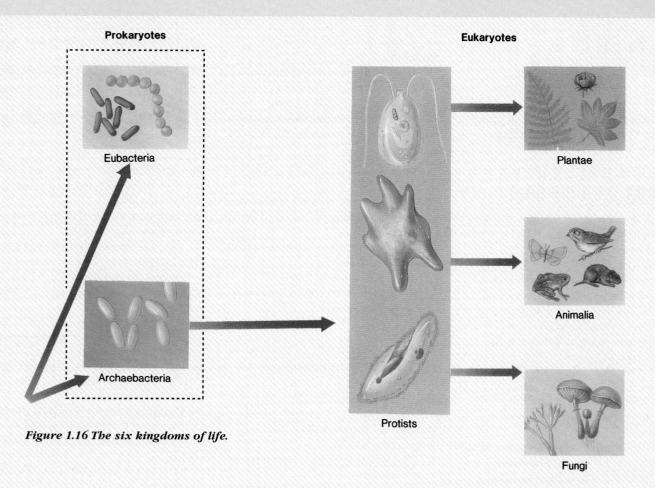

Prokaryotes

Eubacteria

Archaebacteria

Eukaryotes

Plantae

Animalia

Protists

Fungi

Figure 1.16 The six kingdoms of life.

EVOLUTIONARY VIEWPOINT

The core explanatory principle of the science of biology is evolution by natural selection, a theory first advanced by Charles Darwin over a hundred years ago. Diverse lines of evidence convince biologists of the general validity of this theory, although lively discussions continue about the details. There is essentially no support among biologists for so-called "scientific creationism," which holds that the biblical account of the origin of the earth is literally true, that the earth is much younger than most scientists believe, and that all species of organisms were individually created just as they are today.

SUMMARY

1. Biology is a science that attempts to describe and understand both the unity and the diversity of life.

2. Complex molecular organization, hierarchical organization, metabolism, reproduction, development, and heredity characterize all living things, and together they define life. Of these, heredity is perhaps the key characteristic.

3. Science is the determination of general principles from observation and experiment.

4. Scientists test the validity of alternative hypotheses about a given phenomenon by attempting to reject some of the alternatives on the basis of experimentation.

5. Hypotheses that are supported by a large body of evidence are called theories. Unlike the everyday use of the word, the term *theory* in science refers to what scientists are most sure about.

6. Because even a theory is accepted only provisionally, there are no sure truths in science, no propositions that are not subject to change.

7. A basic principle of biology is Darwin's theory that evolution occurs by natural selection. Proposed over a hundred years ago, this theory has withstood a century of testing and questioning.

8. On a five-year voyage aboard the HMS *Beagle*, during which he studied the animals and plants of oceanic islands, Darwin accumulated a wealth of evidence that evolution has occurred.

9. Sparked by Malthus's ideas on population, Darwin proposed that evolution occurs as a result of natural selection. Some individuals have superior heritable traits that let them produce more offspring in a given kind of environment than do other individuals who lack these traits. As a result, the frequency of these traits gradually increases in a population.

10. A wealth of evidence since Darwin's time has supported his twin proposals: that evolution occurs and that its agent is natural selection. Together, these two hypotheses are now usually referred to as Darwin's theory of evolution.

11. Biology may be considered at many levels of organization. This text examines basic principles first and then considers biological diversity.

REVIEWING THE CHAPTER

1. What are the characteristics of life shared by all organisms?

2. What is the hierarchical organization of life?

3. What is the concept of emergent properties?

4. What are the differences between deductive and inductive reasoning?

5. What is a controlled experiment?

6. What is the role of prediction, and how does it relate to the formation of hypotheses?

7. How does the word *theory* apply to science?

8. How is the scientific method employed?

9. What is Darwin's theory of evolution, and how did it progress?

COMPLETING YOUR UNDERSTANDING

1. Why is a virus not a living organism?

2. Why should a student in biology study chemistry?

3. What is the smallest unit of a living organism that can live independently?

4. What is the process of metabolism, and what are some examples in living organisms?

5. Give an example of an organism where the life-cycle stages are different in appearance and function. What is an example where these stages are essentially not different?

6. What is the basis of heredity?

7. How does a hypothesis relate to science?

8. Why do scientists conduct experiments?

9. What is a controlled experiment?

10. How does a theory and a hypothesis differ from one another?

11. What is the creative nature of the scientific method?

12. How did Darwin's book *On the Origin of Species* revolutionize the biological world in 1859?

13. Which of the following best describes Darwin as a student?

   a. He was very responsible.

   b. He rarely went outside because he preferred studying.

   c. He usually followed his father's advice.

   d. He frequently did not go to classes.

14. How does the HMS *Beagle* relate to biology?

15. What was Darwin's evidence that evolution occurs?

16. How did Thomas Malthus influence Darwin's thinking?

17. Who was Alfred Wallace, and how did he influence Darwin?

18. How do embryos, the sharing of common features, and the fossil record support evolution?

19. What is the time span between the formation of the earth and the first cells from the fossil record?

20. How does molecular biology assist scientists who study evolution?

21. What are some of the major concerns on earth today because of the increasing pressures of the human population?

22. Why do biologists not support "scientific creationism"?

23. On the Galápagos Islands, Darwin saw a variety of different kinds of finches, but few other small birds. Imagine that you are visiting another group of islands about as far away from the South American mainland as are the Galápagos, but upwind, so that no birds travel between the two island groups. Do you expect that on your visit to this second island group you will find a variety of finches? (Comment on how your knowledge of the birds of the Galápagos, as discussed in this text, aids you in predicting what you will find on the second island group, if indeed it does.)

FOR FURTHER READING

Bowen, B. W., A. B. Meylan, and J. C. Avise. 1989. An odyssey of the green sea turtle: Ascension Island revisited. *Proceedings of the National Academy of Sciences USA* 86:573–76. An advanced but clear presentation of the latest research results on the fascinating annual migration of the green sea turtle.

Darwin, C. R. 1962 [reprint]. *The voyage of the Beagle.* Garden City, N.Y.: Natural History Press. Darwin's own account of his observations and adventures during his famous five-year voyage.

Darwin, C. R. 1975 [reprint]. *On the origin of species by means of natural selection, or the preservation of favoured races in the struggle for life.* New York: Cambridge University Press. One of the most important scientific books of all time, but still comprehensible and interesting to modern readers.

Futuyma, D. 1983. *Science on trial: The case for evolution.* New York: Pantheon Books. An excellent exposition of the basic reasons why the creationist argument is flawed by serious errors.

Gould, S. 1987. Darwinism defined: The difference between fact and theory. *Discover.* Jan., 64–70. A clear account of what biologists do and do not mean when they refer to the theory of evolution.

Gould, S. 1989. *Wonderful life. The Burgess shale and the nature of history.* New York: W. W. Norton. A marvelous book about the early evolution of animals and about evolution in general.

Irvine, W. 1954. *Apes, angels, and Victorians.* New York: McGraw-Hill. The story of Darwin and the early years of the theory of evolution. Beautifully written.

Moore, J. A. 1993. *Science as a way of knowing— The foundations of modern biology.* Cambridge, Mass.: Harvard University Press. An outstanding exposition of evolution and of the whole field of biology.

This E. coli bacterium has burst open, so that the DNA within spilled out. It is hard to conceive that so much DNA could be packed into that one cell. How macromolecules are packaged is but one of the fascinating topics studied by cell biologists.

THE CHEMISTRY OF LIFE

All of life is conditioned by the chemistry of water.

You and lions and butterflies and daisies are all alike in that you are alive. What differentiates living things from nonliving things, like a car or computer, is that living things are made up of organic chemicals, not metal and plastic. All organisms are chemical machines. To understand them requires an understanding of chemistry.

Atoms: The Stuff of Life

All matter is composed of small particles called **atoms** (figure 2.1). Atoms are very small and hard to study. Early in the twentieth century, however, Danish physicist Niels Bohr proposed a simple view of atomic structure that provided a good starting point. According to Bohr, every atom possesses an orbiting cloud of tiny subatomic particles called **electrons** that whiz around the core like planets of a miniature solar system. Each electron carries a negative (–) charge. At the center of each atom is a small, dense **nucleus** formed of two other kinds of subatomic particles, **protons** and **neutrons** (figure 2.2).

Within the nucleus, the cluster of protons and neutrons is held together by subatomic forces that work only over very short distances. Each proton carries a positive (+) charge. The number of charged protons (called the **atomic number**) determines the atom's chemical character. Neutrons are similar to protons in mass, but as their name implies, they are neutral and possess no charge. The **atomic mass** of an atom consists of the combined weight of all of its protons and neutrons. Atoms that

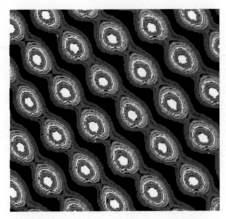

Figure 2.1 All matter is composed of atoms.
This remarkable photo of individual atoms on the surface of a silicon crystal was made with a newly developed technique called tunneling microscopy.

occur naturally on earth contain from 1 to 92 protons and up to 146 neutrons.

All atoms now in the universe are thought to have been formed long ago, as the universe itself evolved. Thus, every atom in your body probably was created in a star.

Different kinds of atoms are called **elements.** Technically, an element is any substance that cannot be broken down to any other substance by ordinary chemical means. Each element is made up of one kind of atom but may contain several versions of it.

Isotopes

Atoms that have the same number of protons but different numbers of neutrons are called **isotopes.** Isotopes of an atom differ in atomic mass but have similar chemical properties. Most elements in nature exist as mixtures of different isotopes. For example, there are three isotopes of the element carbon, all of which possess six protons (figure 2.3). The most common isotope of carbon has six neutrons.

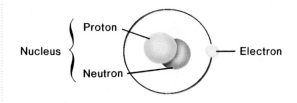

Figure 2.2 The basic structure of an atom.
The nucleus of an atom is made up of protons and neutrons and is orbited by electrons. The element shown here is deuterium, which possesses one neutron and one proton in the nucleus, and a single orbiting electron.

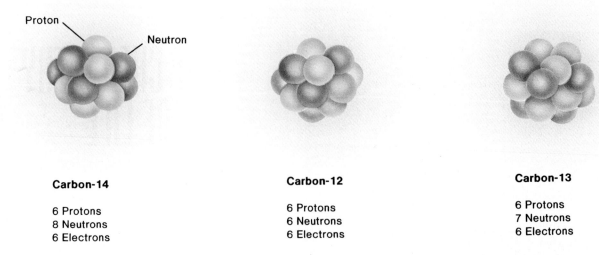

Carbon-14

6 Protons
8 Neutrons
6 Electrons

Carbon-12

6 Protons
6 Neutrons
6 Electrons

Carbon-13

6 Protons
7 Neutrons
6 Electrons

Figure 2.3 The three most abundant isotopes of carbon: carbon-14, carbon-12, and carbon-13.
The yellow "clouds" in the diagrams represent the orbiting electrons, whose numbers are the same for all three isotopes.

Because its total mass is twelve (six protons plus six neutrons), it is referred to as carbon-12. Over 99 percent of the carbon in nature is carbon-12. Most of the rest is carbon-13, with seven neutrons. The third isotope, carbon-14, is rare in nature. Unlike the other two isotopes of carbon, carbon-14 is unstable. Its nucleus tends to break up into particles with lower atomic numbers, a process called radioactive decay. Isotopes such as carbon-14 that decay in this fashion are called **radioactive.** By determining the ratios of the different isotopes of carbon and other elements in samples of biological origin and in rocks, scientists are able to determine with certainty when these materials formed.

Scientists use radioactive isotopes to track physiological processes in the human body. Because their atomic numbers are the same, isotopes can "stand in" for elements in chemical reactions. For example, the human thyroid gland needs the element iodine to make thyroid hormone. A patient with thyroid disease might be injected with an isotope of iodine to see if the thyroid processes iodine effectively. Because the iodine isotope is radioactive, doctors can trace how the thyroid uses the iodine with special detectors that measure the emission of the protons and neutrons from the radioactive iodine. Radioactive isotopes are also used in the treatment of some cancers and in some imaging procedures, such as positron emission tomography (PET) scanning.

Electrons and Ions

The positive charges in the nucleus of an atom are counterbalanced by negatively (–) charged electrons that orbit the atomic nucleus at various distances. The negative charge of one electron exactly balances the positive charge of one proton. Thus, atoms with the same number of protons and electrons have no net charge and are known as neutral atoms. Atoms in which the number of electrons does not equal the number of protons are known as **ions,** and ions *do* carry an electrical charge.

Electrons have very little mass (only 1/1,840 of the mass of a proton). Of all the mass contributing to your weight, the portion contributed by electrons is less than the mass of your eyelashes.

Electrons stay in their orbits because they are attracted to the positive charge of the nucleus. This attraction is sometimes overcome by other forces, and one or more electrons fly off, lost to the atom. For example, an atom of sodium (Na) that has lost an electron becomes a positively charged sodium ion (Na^+) because the positive charge of one of the protons is not balanced by the negative charge of an electron. Conversely, atoms sometimes gain additional electrons, transforming a neutrally charged atom into a negatively charged ion. Chloride, for example, becomes negatively charged (Cl^-) when it gains an electron from another atom.

An atom is a core (nucleus) of protons and neutrons surrounded by a cloud of electrons. Protons carry a positive charge, electrons a negative charge, and neutrons no charge. Atoms that have the same number of protons but different numbers of neutrons are called isotopes. The number of electrons and protons determines the charge of an atom. Atoms in which the number of electrons does not equal the number of protons are called ions.

Energy Within the Atom

Because electrons carry negative charges, they are attracted to the positively charged nucleus, and it takes work to keep them in orbit, just as it takes work to hold an apple in your hand when gravity is pulling the apple down toward the ground. The apple in your hand is said to possess energy (the ability to do work) because of its position—if you were to release it, the apple would fall. Similarly, electrons have energy in relation to their proximity to the nucleus. Energy of this sort is called potential energy. It takes work to oppose the attraction of the nucleus and to move the electron farther out. Thus, moving an electron away from the nucleus requires an input of energy and results in an electron with greater potential energy. For the same reason, a bowling ball released from the top of a building hits with greater force than one dropped from only a meter. Moving an electron in toward the nucleus has the opposite effect: Energy is released, and the electron ends up with less potential energy.

The different levels of potential energy surrounding the nucleus of an atom are called **energy levels,** or **electron shells.** In diagrams, energy levels are represented by concentric circles that ring the atomic nucleus. Electrons in the energy level closest to the nucleus contain the least amount of potential energy. Electrons in the outermost energy level contain the most amount of potential energy (figure 2.4). The lowest energy

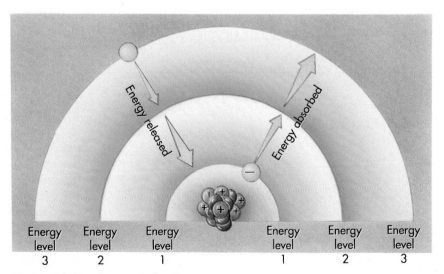

Energy level	Energy level	Energy level		Energy level	Energy level	Energy level
3	2	1		1	2	3

Figure 2.4 Atomic energy levels.
When an electron absorbs energy, it moves to higher energy levels farther away from the nucleus. When an electron releases energy, it falls inward to lower energy levels closer to the nucleus.

level or electron shell, nearest the nucleus, is called level *K.* The next highest level is called *L,* and so on.

Sometimes, an electron is transferred from one atom to another during a chemical reaction. The loss of an electron is called **oxidation;** the gain of an electron is called **reduction** (figure 2.5). An easy way to remember these terms is to keep in mind that *adding* negatively charged electrons *reduces* the atom's charge. Chemical reactions that involve the transfer of electrons from one atom to another are called **oxidation/reduction reactions,** or **redox reactions** for short. Redox reactions are important in such cellular processes as photosynthesis and cellular respiration.

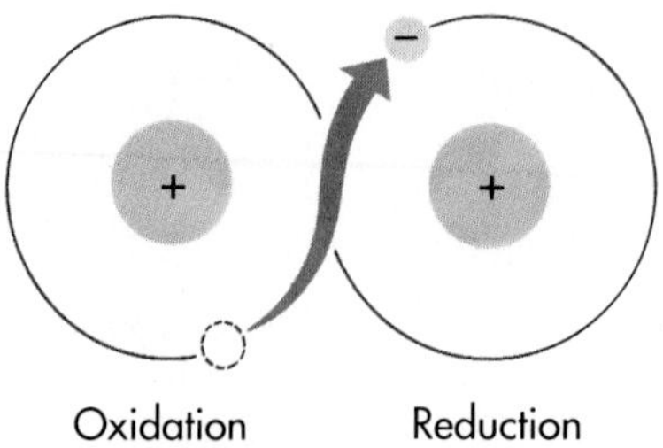

Electron Orbitals

While the energy levels of an atom are often visualized as well-defined circular orbits around a central nucleus, the way Earth, Mars, and Venus circle the sun, such a simple picture is not realistic. Specifying the position of any individual electron at any given time is impossible. In fact, theories predict that, at any given instant, a particular electron can be located anywhere from close to the nucleus to infinitely far away from it.

However, a particular electron is not equally likely to be located at all positions. Some locations are much more probable than others. For this reason, it is possible to say where an electron is *most likely* to be located. The volume of space around a nucleus where an electron is most likely to be found is called the **orbital** of that electron. Atoms can have many electron orbitals. Some are simple spheres that enclose the nucleus like a wrapper. Others resemble dumbbells and other complex shapes.

Orbitals are located in each energy level, and each energy level has a specific number of orbitals. Each orbital can hold up to two electrons. The first energy level in any atom (level *K*) contains one orbital. Helium, for example, has one energy level with one orbital that contains two electrons (figure 2.6). In

Oxidation   Reduction

Figure 2.5 Oxidation/reduction.
Oxidation is the loss of an electron. Reduction is the gain of an electron.

atoms with more than one energy level, the second energy level (level *L*) contains four orbitals. Therefore, energy level *L* can contain up to eight electrons. Notice that in figure 2.6 nitrogen has two energy levels: The single orbital in energy level *K* is completely filled with two electrons; the four orbitals in energy level *L,* however, are not completely filled, since nitrogen's energy level *L* contains only five electrons. Of the four orbitals in energy level *L,* one orbital contains its full complement of two electrons, and the other three orbitals contain only one electron. In atoms with more than two energy levels, subsequent energy levels also contain up to four orbitals and a maximum of eight electrons.

Electrons, then, can be visualized as whizzing around the nucleus in complex shapes, each electron containing a certain amount of potential energy, depending on how close it is to the nucleus. The term *close,* however, is relative. Electrons are actually located quite far away from the nucleus. For example, if the nucleus of an atom was the size of an apple, the orbit of the nearest electron would be more than a mile out. The nuclei of two atoms, therefore, never come close enough to each other in nature to interact. The electrons of two atoms, however, do come in contact with each other, and it is for this reason that the electrons of an atom determine the atom's chemical behavior. This is also why isotopes of an element, all of which have the same arrangement of electrons, behave the same way chemically.

> *Energy levels contain electrons of different potential energy and are located in concentric circles around the nucleus of an atom. Oxidation is the loss of an electron; reduction is the gain of an electron. Each energy level of an atom contains orbitals, in which electrons spend most of their time. The first energy level of an atom contains one orbital; subsequent energy levels contain four orbitals. An orbital can contain two electrons.*

Elements and Molecules and The Nature of Chemical Bonds

Much of the earth's core is thought to consist of atoms of iron, nickel, and other heavy elements. As mentioned earlier in the chapter, elements cannot be separated into different substances by ordinary chemical methods. However, they can be combined with **chemical bonds** into stable associations of atoms called **molecules.**

Atoms enter into chemical bonds to complete their outer energy levels with the full complement of electrons. In every type of chemical bond, atoms either gain or lose electrons according to how many electrons are needed to complete its outer shell. Atoms with many electrons in the outer shell tend to donate electrons; atoms with few electrons tend to gain electrons.

The atoms in a molecule are held together by energy. The force holding two individual atoms together is the chemical

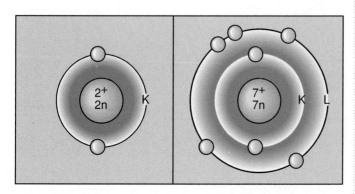

Helium   Nitrogen

Figure 2.6 Energy levels and orbitals for helium and nitrogen.
Helium has one energy level (L), and this energy level has one orbital. Nitrogen has two energy levels (K and L): The first energy level has one orbital, and the second has four orbitals. As you can see, the orbitals in the second energy level of nitrogen are not completely filled with electrons.

Emily Duroin

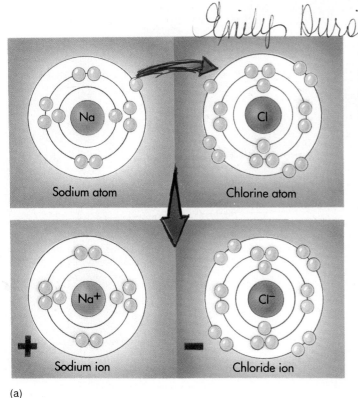

(a)

(b)

▶ *Figure 2.7 The formation of the ionic bond in sodium chloride.*
(a) *When a sodium atom donates an electron to a chlorine atom, the sodium atom, lacking that electron, becomes a positively charged sodium ion. The chlorine atom, having gained an extra electron, becomes a negatively charged chloride ion.* (b) *Sodium chloride forms a highly regular lattice of alternating sodium ions and chloride ions. You are familiar with these crystals as table salt.*

bond. The force can result from the attraction of opposite charges, called an **ionic bond,** or from the sharing of one or more pairs of electrons, a **covalent bond.** Other, weaker kinds of bonds also occur.

Ionic Bonds: Attractions of Opposite Charges

Ionic bonds form when atoms are attracted to one another by opposite electrical charges. Common table salt—sodium chloride (NaCl)—is a lattice of ions in which atoms are held together by ionic bonds. Sodium (Na) atoms have eleven electrons. Two of these are in the inner energy level, eight are at the next energy level, and one is at the outer energy level. This outer electron is unpaired ("free") and has a strong tendency to form a pair. Loss of the outer electron results in a stable configuration as well as in the formation of a positively charged sodium ion (Na^+) (figure 2.7a).

The chlorine (Cl) atom faces a similar dilemma. It has seventeen electrons: two at the inner energy level, eight at the next energy level, and seven at the outer energy level. The outer energy level of the chlorine atom has an unpaired electron. The addition of an electron to the outer level causes the formation of a negatively charged chloride ion (Cl^-) (figure 2.7a).

When placed together, metallic sodium and gaseous chlorine react swiftly and explosively, with the sodium atoms donating electrons to the chlorine atoms. The result is production of Na^+ and Cl^- ions. Because opposite charges attract, the association of these ions with each other results in a neutral balance of all charges. The ions aggregate, or come together, and form a crystal matrix with a precise geometry known as a salt crystal (figure 2.7b). If a salt such as Na^+Cl^- is placed in water, the electrical attraction of the water molecules

(for reasons discussed later in this chapter) disrupts the forces holding the salt ions in their crystal matrix, causing the salt to dissolve into a roughly equal mixture of free Na^+ and free Cl^- ions. Approximately 0.06 percent of the atoms in your body—about 40 grams, the weight of your fingernails—are free Na^+ or Cl^- ions.

> *An ionic bond is an attraction between ions of opposite charge.*

Covalent Bonds: The Sharing of Electrons

Covalent bonds form when two atoms share electrons. For example, each hydrogen (H) atom has an unpaired electron. When two hydrogen atoms are close enough to one another, however, each unpaired electron can orbit both nuclei. In effect, nuclei in close proximity share electrons. The result is a diatomic molecule (one with two atoms) of hydrogen gas (H_2) (figure 2.8).

The diatomic hydrogen gas molecule that forms as a result of this sharing of electrons is not charged. Both of the hydrogen atoms still contain two protons and two electrons and are considered to have two orbiting electrons in the outer electron shell. Each shared outer-shell electron orbits both nuclei and therefore is included in the outer shell of *both* atoms. Because the relationship results in the pairing of the two free electrons, the two hydrogen atoms thus form a stable molecule. This stability, however, is conferred by the electrons that orbit *both* nuclei and occurs only when the nuclei are very close. For this reason, the strong chemical forces tending to pair electrons and to fill the outer energy level with the maximum number of

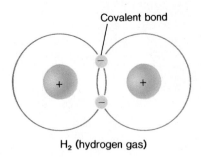

Covalent bond

H₂ (hydrogen gas)

(a)

(b)

Figure 2.8 Hydrogen gas can be a highly volatile molecule.
(a) *Hydrogen gas is a molecule composed of two hydrogen atoms linked by a covalent bond.* (b) *The flash of fire that consumed the* Hindenburg *occurred when the hydrogen gas used to inflate the airship combined explosively with oxygen gas in the air to form water.*

electrons act to keep the two hydrogen nuclei near one another. The bond between the two hydrogen atoms of diatomic hydrogen gas is an example of a covalent bond.

The bonds that hold the molecules of your body together are covalent bonds. Covalent bonds are far better suited to this task than are ionic ones because covalent bonds are *directional*—that is, a covalent bond is formed between two specific atoms, whereas an ionic bond is formed between a charged atom and the electrical field contributed by all nearby atoms of opposite charge. Ionic bonds can form regular crystals, but such crystals dissolve in water. Complex stable shapes require the more specific associations made possible by covalent bonds.

A covalent bond is a chemical bond formed by the sharing of one or more pairs of electrons.

Covalent bonds can be very strong—that is, difficult to break. Covalent bonds that share *two* pairs of electrons, called **double bonds,** are stronger than covalent bonds that share only one electron pair, called **single bonds.** Energy is required to form larger aggregations of atoms from smaller ones because of the need to establish new orbitals for the electrons. This energy is released when the bonds are broken. Covalent bonds are represented in chemical formulations as lines connecting atomic symbols. Each line between two bonded atoms represents the sharing of one pair of electrons. Hydrogen gas is thus symbolized H—H and oxygen gas, O=O.

Molecules are often made up of more than two atoms. One reason larger molecules may form is that a given atom is able to

share electrons with more than one other atom. An atom that requires two, three, or four additional electrons to fill its outer energy level completely may acquire them by sharing its electrons with two or more other atoms. For example, carbon (C) atoms (atomic number 6) contain six electrons, two of them at the inner level and the other four in the outer shell. A carbon atom can form the equivalent of four covalent bonds. Because four covalent bonds may form in many ways, carbon atoms are able to participate in many different kinds of molecules, making carbon ideal for constructing the many molecules of living things.

Covalent bonds can form between a wide variety of different atoms. However, of the ninety-two kinds of atoms (elements) that form the earth's crust, only eleven are common in living organisms. Table 2.1 lists the frequency with which various elements occur in the earth's crust and in the human body. Unlike the elements that occur most abundantly in the earth's crust, all of the elements common in living organisms are light. Each has an atomic number of less than 21 and thus a low mass. The great majority of the atoms in living things (for example, 96 percent of the atoms in the human body) are either nitrogen, oxygen, carbon, or hydrogen (figure 2.9). You can remember these elements by their first letters, NOCH. Other elements such as sodium, phosphorus, calcium, potassium, and sulfur, although present in lower amounts, also play important roles.

The Cradle of Life: Water

The most common atoms in living things are oxygen and hydrogen atoms; the great majority of these are combined together in water molecules. The chemical formula (a list of the atoms in a molecule with a subscript to indicate how many of each) for

Table 2.1   The Most Common Elements on Earth and Their Distribution in the Human Body

Element	Symbol	Atomic Number	Percent of Human Body by Weight	Importance or Function
Oxygen	O	8	65.0	Required for cellular respiration; component of water
Silicon	Si	14	Trace	—
Aluminum	Al	13	Trace	—
Iron	Fe	26	Trace	Critical component of hemoglobin in the blood
Calcium	Ca	20	1.5	Component of bones and teeth; triggers muscle contraction
Sodium	Na	11	0.2	Principal positive ion bathing cells; important in nerve function
Potassium	K	19	0.4	Principal positive ion in cells; important in nerve function
Magnesium	Mg	12	0.1	Critical component of many energy-transferring enzymes
Hydrogen	H	1	9.5	Electron carrier; component of water and most organic molecules
Manganese	Mn	25	Trace	—
Fluorine	F	9	Trace	—
Phosphorus	P	15	1.0	Backbone of nucleic acids; important in energy transfer
Carbon	C	6	18.5	Backbone of organic molecules
Sulfur	S	16	0.3	Component of most proteins
Chlorine	Cl	17	0.2	Principal negative ion bathing cells
Vanadium	V	23	Trace	—
Chromium	Cr	24	Trace	—
Copper	Cu	29	Trace	Key component of many enzymes
Nitrogen	N	7	3.3	Component of all proteins and nucleic acids
Boron	B	5	Trace	—
Cobalt	Co	27	Trace	—
Zinc	Zn	30	Trace	Key component of some enzymes
Selenium	Se	34	Trace	—
Molybdenum	Mo	42	Trace	Key component of many enzymes
Tin	Sn	50	Trace	—
Iodine	I	53	Trace	Component of thyroid hormone

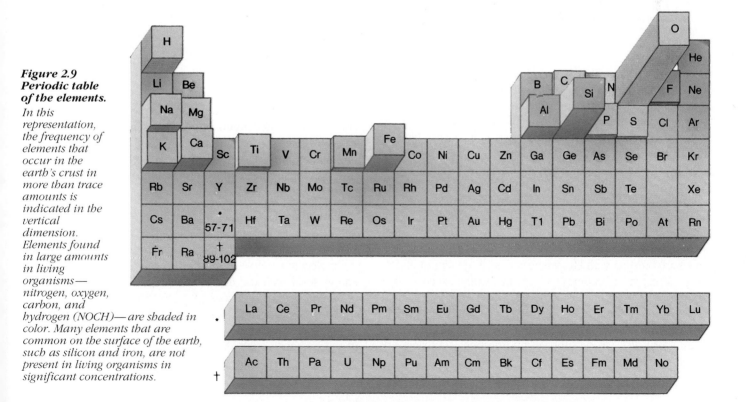

Figure 2.9 Periodic table of the elements.

In this representation, the frequency of elements that occur in the earth's crust in more than trace amounts is indicated in the vertical dimension. Elements found in large amounts in living organisms—nitrogen, oxygen, carbon, and hydrogen (NOCH)—are shaded in color. Many elements that are common on the surface of the earth, such as silicon and iron, are not present in living organisms in significant concentrations.

Figure 2.10 Many forms of water.

(a) *As a liquid, water fills our rivers and runs down over the land to the sea, sometimes falling in great cascades.* (b) *The icebergs on which the penguins are holding their meeting were formed in Antarctica from huge blocks of ice breaking away into the ocean.*

(c) *When water cools below 0 degrees Celsius, it forms beautiful crystals, familiar to us as snow and ice.* (d) *Water, however, is not always plentiful. In a dry creek bed, there is no hint of water, save for the broken patterns of dry mud.*

water is H_2O. This seemingly simple molecule has many surprising properties. For example, of all the common molecules on earth, only water exists as a liquid at the relatively cool temperatures prevailing at the earth's surface (figure 2.10). When life on earth was beginning, water, because it is a liquid at such temperatures, provided a medium in which other molecules could move around and interact without being bound by strong covalent or ionic bonds. Life evolved as a result of these interactions.

Life on earth is inextricably tied to water. Three-fourths of the earth's surface is covered by water. You yourself are about two-thirds water, and you cannot exist long without it. All other organisms also require water. It is no accident that tropical rain forests are bursting with life, whereas deserts are almost lifeless except when water becomes temporarily plentiful, such as after a rainstorm. Desert organisms have evolved complex adaptations that allow them to conserve water in times of drought. Farming is possible only in areas of the earth where rain is plentiful or water can be supplied by irrigation. The chemistry of life, then, is water chemistry.

The Polarity of Water and Hydrogen Bonds

The single most outstanding chemical property of water is its ability to form **hydrogen bonds,** weak chemical associations with only 5 to 10 percent of the strength of covalent bonds. This one property of water, which derives directly from its structure, is responsible for much of the organization of living chemistry.

The Polar Structure of the Water Molecule

Water has a simple atomic structure: One oxygen atom bound by covalent single bonds to two hydrogen atoms. The resulting molecule is stable: It satisfies the tendency to fill the outer energy level with the maximum number of electrons, has no unpaired electrons, and does not carry a charge.

But the electron-attracting power of the oxygen atom (referred to by chemists as its electronegativity) is much greater than that of the hydrogen atom. As a result, the electron pair shared in each of the two single oxygen-hydrogen covalent bonds of a water molecule is more strongly attracted to the oxygen nucleus than to either of the hydrogen nuclei. Although electron levels encompass both the oxygen and hydrogen nuclei, the negatively charged electrons are far more likely, at any given moment, to be found near the oxygen nucleus than near one of the hydrogen nuclei. This relationship has a profoundly important result: The oxygen atom acquires a partial negative charge. It is as if the electron cloud is more dense in the neighborhood of the oxygen atom and less dense around the hydrogen atoms. This charge separation within the water molecule creates negative and positive electrical charges on the ends of the molecule. These partial charges are much less than the unit charges of ions.

A water molecule thus has distinct "ends," each with a partial charge, like the two poles of a magnet. Molecules such as water that exhibit charge separation are called **polar molecules** because of these magnetlike poles (figure 2.11). Water is one of the most polar molecules known. *The polarity of water underlies its chemistry and thus the chemistry of life.*

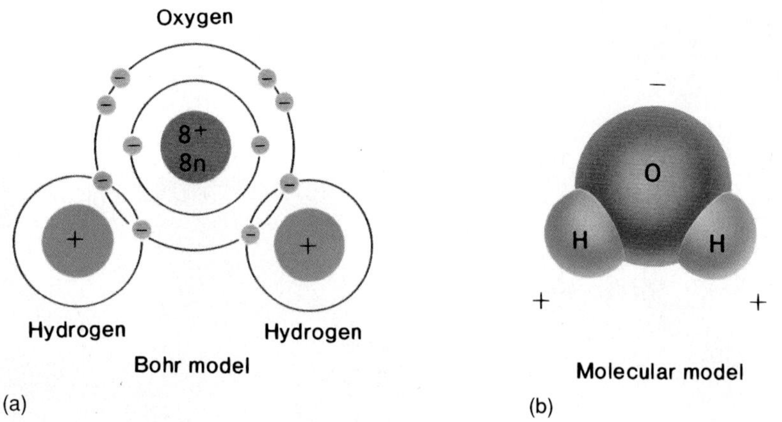

Figure 2.11 The polar structure of water.
(a) *The Bohr model (a type of diagram named for Niels Bohr) shows atoms and molecules with their electrons and bonds in place. It demonstrates that the oxygen atom of water shares a pair of electrons with each hydrogen atom.* (b) *The molecular model shows that this sharing of electrons results in a polar molecule, in which the hydrogen ends of the oxygen molecule carry a slightly positive charge, and the oxygen end of the molecule a slightly negative charge.*

> *Much of the biologically important behavior of water results because the oxygen atom attracts electrons more strongly than do the hydrogen atoms, with the result that the water molecule has electron-rich (−) and electron-poor (+) regions, giving it positive and negative poles.*

Hydrogen Bonds

Polar molecules interact with one another. The partial negative charge at one end of a polar molecule is attracted to the partial positive charge of another polar molecule. As already mentioned, this weak attraction is called a hydrogen bond. Water forms a lattice of such hydrogen bonds. Each individual hydrogen bond is weak and transient. A given bond lasts only $1/100,000,000,000$ of a second. Although each bond is transient, a large number of such hydrogen bonds can form, with enormous cumulative effects. Such cumulative effects are responsible for many of water's important physical properties (figure 2.12).

Although hydrogen bonds are an important property of water, they are not exclusive to water. Ammonia, with a chemical formula of NH_3, also forms hydrogen bonds with water. The nitrogen (N) part of the ammonia molecule has a small negative charge. When an ammonia molecule comes in close contact with a water molecule, the negatively charged nitrogen is attracted to the positively charged hydrogen of the water molecule, forming a hydrogen bond.

Water: A Powerful Solvent

When you pour salt into a glass of water, the salt dissolves in the water, and the mixture of salt and water is uniform throughout the glass. Any agent used to dissolve a substance is called a **solvent.**

When water is used as a solvent, the water molecules gather closely around any molecule that exhibits an electrical charge, whether the molecule carries a full charge (ion) or a charge separation (polar molecule). For example, salt is composed of a negatively charged chloride ion and a positively charged sodium ion (see figure 2.7). A salt crystal dissolves rapidly when placed in water because the polar ends of the water molecules are attracted to the oppositely charged salt ions. The positively charged hydrogen portions of each water molecule are attracted to the chloride ions and pull these ions away from the crystal. Similarly, the negatively charged oxygen portion of each water molecule attracts and pulls away the sodium ions. As water molecules cluster around the ions, they form a "water shell," or hydration shell, around the individual sodium and chloride ions, which prevents the ions from associating with each other. When all the ions are separated by hydration shells, the salt crystal is said to be dissolved (figure 2.13).

Molecules that dissolve in water in this way are said to be **soluble** in water; the only criterion for solubility is that the molecules are polar—that is, exhibit an electrical charge. Oil is an example of a nonpolar molecule without an electrical charge; thus, it is not soluble in water. Life originated in water not only because water is a liquid, but also because so many molecules are polar or ionized and thus are water-soluble.

How Water Organizes Nonpolar Molecules

Water molecules always tend to form the maximum number of hydrogen bonds possible. When nonpolar molecules, which do not form hydrogen bonds, are placed in liquid water, water molecules act to exclude them by preferentially forming hydrogen bonds with other water molecules. The nonpolar molecules are forced to associate with one another, minimizing their disruption of the hydrogen bonding of water; they are thus

(a)

(b)

(c)

Figure 2.12 Physical properties of water that depend on hydrogen bonding.

(a) *Ice formation. When water cools below 0 degrees Celsius, it forms a regular crystal structure in which the partial charges of each atom in the water molecule interact with opposite charges of atoms in other water molecules to form hydrogen bonds.* (b) *Cohesion. The ability of water to "stick together" is a result of the collective force of its hydrogen bonds. Water can form droplets that stick together, a property called cohesion.* (c) *Surface tension.*

Some insects, such as this water strider, literally walk on water. In this photograph, you can see the dimpling the strider's feet make on the water as its weight bears down on the surface. Surface tension is a property derived from cohesion—water has a "strong" surface due to the force of its hydrogen bonds. Because the surface tension of the water is greater than the force of one "foot" of the water strider, the water strider does not sink, but rather slides along.

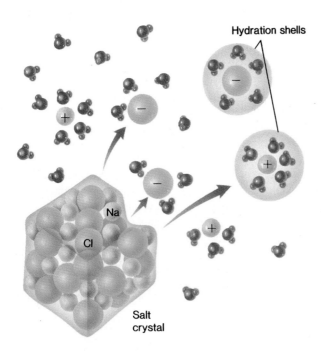

Figure 2.13 How salt dissolves in water.

The partial charges on water molecules are attracted to the charged sodium and chloride ions. The water molecules surround the ions, forming hydration shells. When all of the ions have been separated from the crystal, the salt is said to be dissolved.

crowded together. This is why oil and water do not mix but always separate. Nonpolar compounds seem to shrink from contact with water, and for this reason, they are called **hydrophobic** (from the Greek *hydros,* meaning "water," and *phobos,* meaning "hating"; "water-hated" might be a more apt description). The tendency for nonpolar molecules to band together in water solution is called **hydrophobic bonding.**

Polar molecules, on the other hand, form hydrogen bonds and are "welcomed" by water molecules. For example, as mentioned earlier, ammonia spontaneously forms hydrogen bonds with water molecules. Polar molecules such as ammonia are called **hydrophilic** (from the Greek *hydros,* meaning "water," and *philic,* meaning "loving") or "water-loving" molecules.

The interaction between hydrophobic and hydrophilic molecules is extremely important in biology. Cell membranes protect the cells they surround because the outside portion of the cell membrane is composed of hydrophobic molecules. This hydrophobic shell protects the interior of the cell from the outside environment. The inside of the cell membrane is composed of hydrophilic molecules so that *some* selected polar substances can be transferred across the cell membrane into the cell.

Many of the important interactions between molecules in biological systems involve hydrophobic bonding, the tendency for nonpolar molecules to bond together in a polar environment.

Water Ionization: Acids and Bases

The covalent bonds of water sometimes break spontaneously. When this happens, one of the protons in the hydrogen atom dissociates from the molecule. Because the dissociated proton lacks the negatively charged electron that it shared in the covalent bond with oxygen, its own positive charge is not counterbalanced; it is a positively charged hydrogen ion (H^+). The remaining bit of the water molecule retains the shared electron from the covalent bond and has one less proton to counterbalance it; it is a negatively charged hydroxyl ion (OH^-). This process of spontaneous ion formation is called **ionization** and can be diagrammed by the following chemical reaction:

$$H_2O \rightarrow OH^- + H^+$$

When water dissociates, the result is an equal number of H^+ ions and OH^- ions. Some substances, however, called **acids,** donate H^+ ions to the solution when dissolved in water. Acids thus increase the concentration of H^+ ions in a solution. **Bases,** on the other hand, are substances that, when dissolved in water, decrease the H^+ concentration in a solution. Bases do this in a number of ways. For example, bases commonly dissociate to form OH^-, which then combines with H^+ ions to form water. Since the H^+ ions in the solution are used to form water, they are "used up," and their concentration is lowered.

The **pH scale** is a way to determine the acidic or basic nature of a solution compared to pure water. As we have seen, pure water dissociates into equal numbers of H^+ and OH^- ions. The pH of water is determined by taking the negative value of the exponent of the H^+ ion concentration in the solution. In other words, since pure water has an H^+ concentration of 10^{-7}, the pH of pure water is the negative value of the exponent, or $-(-7)$, or, more simply, 7 (figure 2.14). Remember that a negative exponent such as 10^{-7} can be written out as 0.0000001. When a negative exponent decreases, there are fewer zeros, and the numerical value of the exponent actually increases. An acid, for example, produces more H^+ ions than pure water, and its pH is thus lower than pure water. Hydrochloric acid, which is abundant in your stomach, has a concentration of 10^{-1}, or 0.1 H^+ ions. Its pH, then, is 1. Some acids, such as nitric acid, are even stronger, although such strong acids are rarely found in living things. The pH of champagne, which bubbles because of the carbonic acid dissolved in it, is about 2.

Bases, because they lower H^+ concentration, have higher pH values than acids. Sodium hydroxide, a very strong base, has a pH value of 14.

pH is extremely important in the maintenance of living organisms. Many important chemical reactions in your body, such as the functioning of your heart and the exchange of carbon dioxide for oxygen, are affected by pH. Even slight variations can cause these chemical reactions to cease, resulting in grave harm.

Outside the body, pH and its effects can be destructive. Over the past ten years, **acid rain** has become a major environmental concern. Acid rain is formed when coal that contains a high percentage of sulfur is burned for fuel, a practice common in many industries. The smoke from this burning is released from

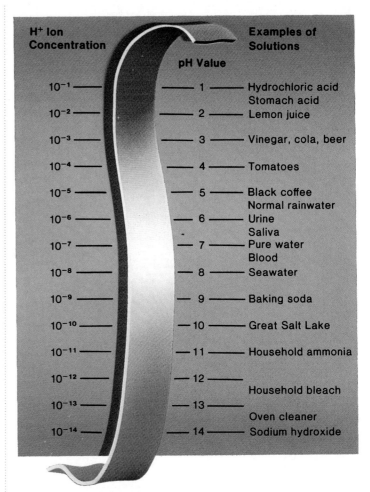

Figure 2.14 The pH scale.
A solution is assigned a pH value according to the number of hydrogen (H^+) ions present in the solution. The scale is logarithmic, so that a change of only one means a tenfold change in the concentration of H^+ ions. Thus, lemon juice is one hundred times more acidic than tomato juice, and seawater is ten times more basic than pure water.

tall smokestacks high in the atmosphere so that the sulfur stench and pollution do not disturb local inhabitants but rather are blown great distances away. In the air, the sulfur dioxide (SO_2) released from the coal combines with water (H_2O) to form sulfuric acid (H_2SO_4)! Falling back to earth, this acid rain disrupts biological systems on a regional scale, a problem discussed further in chapter 24 (figure 2.15). Sidelight 2.1 focuses on scientists' current search for "green" chemicals that will not harm the environment.

pH refers to the relative concentration of hydrogen (H^+) ions in a solution. The numerical value of the pH is the negative value of the exponent of the H^+ ion concentration of a substance in pure water. Low pH values indicate solutions with high concentrations of H^+ ions (acids), and high pH values indicate solutions with low concentrations of H^+ ions (bases).

The Chemical Building Blocks of Life

The basic chemical building blocks of organisms—like the mortar and bricks used to build a house—are made of molecules. The molecules formed by living organisms contain carbon and are called **organic molecules.** An organic molecule can be visualized as a carbon-based core with special bits (groups of atoms with definite chemical properties) attached. These groups of atoms are called **functional groups.** For example, a hydrogen atom bonded to an oxygen atom (—OH) is a hydroxyl functional group. Most chemical reactions that occur within organisms involve transferring a functional group from one molecule to another or breaking a carbon-carbon bond. Proteins called kinases, for example, transfer phosphate groups from one kind of molecule to another.

Figure 2.15 The effects of acid rain.
These trees at Camel's Hump Mountain in Vermont are diseased and dying. Acid rain and air pollution weaken trees, making them more susceptible to pests and predators.

Sidelight 2.1

ENVIRONMENTAL CHEMISTRY—THE NEW SEARCH FOR "GREEN" CHEMICALS

Back in the 1920s, scientists introduced what they thought was a group of revolutionary chemicals. These chemicals had a variety of uses: They could be used as coolants in air conditioners, as propellants for aerosol sprays, and as structural components in lightweight packaging materials. They were inexpensive to manufacture, and, promised scientists, chemically inert, and so would not pollute the environment. Almost twenty years later, these chemicals, called chlorofluorocarbons, are blamed for most of the destruction of the world's ozone layer, and many countries have banned their manufacture and use.

What went wrong? Scientists around the world are asking themselves this question and are now trying to combat chemical pollution not by the traditional method of merely cleaning up the mess, but by preventing it in the first place (Sidelight figure 2.1). Environmental chemistry, once an obscure subfield of the discipline, has suddenly taken the spotlight, and more scientists than ever before are committed to designing chemicals that will not harm the environment.

Sidelight figure 2.1
Often, air pollution is the product of industrial activity or the burning of coal for energy production. The Clean Air Act of 1990 encourages industries to install "scrubbers" on their stacks, trapping the pollutants before they are released.

To encourage the research of environmental chemists, the National Science Foundation (NSF) issued a report that lists projects that it feels deserve funding. Among the high-priority challenges issued by the NSF are: developing a more energy-efficient, less polluting fuel for automobiles; finding replacements for chlorofluorocarbons; and designing enzymes that can destroy current pollutants, such as PCBs (chemical compounds with industrial applications that are poisonous—environmental pollutants that tend to accumulate in animal tissues). Some scientists, under their own initiative, are already researching "green" chemical solutions to old problems by replacing traditional, polluting elements in chemicals with environmentally friendly elements. For example, some enzymes used to drive chemical reactions in industry contain mercury and silver, which cannot be broken down and thus become hazardous waste. Scientists are researching the possibility of using pigment molecules and sunlight to drive these chemical reactions, a solution that is borrowed from photosynthesis.

Although environmental chemistry is a young field, experts are encouraged by the growing interest among chemists to protect the environment. Industries, as well, are beginning to see the sense of using more environmentally benign chemicals. After all, these chemicals do not require expensive disposal and storage measures. The greening of chemistry has finally arrived, and scientists will continue to seek green solutions to old environmental dilemmas.

Emily

Some molecules that occur in organisms are simple organic molecules, often with a single reactive functional group protruding from a carbon chain. Other molecules, called **macromolecules,** are far larger and often play a structural role in organisms or store information. Most of these macromolecules are themselves composed of simpler components, just as a wall is composed of individual bricks. The similar components that are linked together to form a macromolecule are called subunits. Macromolecules fall into the following four classes: carbohydrates, lipids, proteins, and nucleic acids (table 2.2).

The molecules formed by living organisms all contain carbon and are called organic molecules. Large organic molecules, or macromolecules, play a structural role in organisms or store information.

Table 2.2 Macromolecules

Macromolecule	Subunit	Function
Carbohydrates		
Glucose (monosaccharide)	—	Energy storage
Starch, glycogen (polysaccharides)	Glucose	Energy storage
Cellulose (polysaccharide)	Glucose	Component of plant cell walls
Chitin (polysaccharide)	Modified glucose	Cell walls of fungi; outer skeleton of insects and related groups
Lipids		
Fats	Glycerol + three fatty acids	Energy storage
Phospholipids	Glycerol + two fatty acids + phosphate	Component of cell membranes
Steroids	Four carbon rings	Message transmission (hormones)
Terpenes	Long carbon chains	Pigments in photosynthesis
Proteins		
Globular	Amino acids	Catalysis
Structural	Amino acids	Support and structure
Nucleic Acids		
DNA	Nucleotides	Encoding of hereditary information
RNA	Nucleotides	Blueprint of hereditary information
ATP	Nucleotides	Energy transmission and conversion

Polymers

Many macromolecules are polymers. A **polymer** is a molecule built of a long chain of similar molecules called **monomers,** like railway cars coupled together to form a train. Complex carbohydrates, for example, are polymers of simple monomers called sugars. Enzymes, membrane proteins, and other proteins are polymers of monomers called amino acids. DNA and RNA are two versions of a long-chain molecule called a nucleic acid, which is a polymer composed of a long series of monomers called nucleotides.

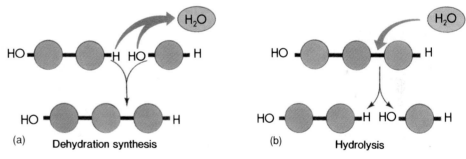

Figure 2.16 Dehydration synthesis and hydrolysis.
(a) *Biological molecules are formed by linking subunits. The covalent bond between subunits is formed in dehydration synthesis, a process during which a water molecule is eliminated.* (b) *Breaking such a bond requires the addition of a water molecule, a reaction called hydrolysis.*

Making (And Breaking) Macromolecules

Although as shown in table 2.2, the four different kinds of macromolecules are assembled from different kinds of subunits, they all put their subunits together in the same way: A covalent bond is formed between two subunits in which a hydroxyl group (OH) is removed from one subunit and a hydrogen (H) is removed from the other. This process is called **dehydration** (water-losing) **synthesis** because, in effect, the removal of the OH and H groups constitutes removal of a molecule of water (figure 2.16*a*). In the synthesis of a polymer, one water molecule is removed for every link in the chain of monomers. Energy is required to break the chemical bonds when water is extracted from the monomers, and so cells must supply energy to assemble polymers. The process also requires that the two monomers be held close together and that the correct chemical bonds be stressed and broken. This process of positioning and stressing is facilitated by helper molecules that do not themselves change, in a process called **catalysis.** In cells, catalysis is carried out by a special class of proteins called **enzymes.**

> *Polymers are large molecules formed of long chains of simi-*
> *lar molecules called monomers that are joined by dehydra-*
> *tion synthesis, in which a hydroxyl (OH) group is removed*
> *from one monomer and a hydrogen (H) group is removed*
> *from the other. In organisms, monomers are held together*
> *and their bonds stressed by enzymes (specialized proteins) in*
> *a process called catalysis.*

No molecule lasts forever. While cells are building up some macromolecules, they are disassembling others. For example, the protein and fat that you consume when eating a steak are broken down into their subunit parts by enzymes of your digestive system. The process of tearing down a polymer is essentially the reverse of dehydration synthesis: Instead of removing a molecule of water, one is added. A hydrogen is attached to one subunit and a hydroxyl to the other, breaking the covalent bond. The breaking up of a polymer in this way is an example of a **hydrolysis reaction** (from the Greek *hydro,* meaning "water," and *lyse,* meaning "break"; literally, "to break with water") (figure 2.16*b*).

Carbohydrates

As mentioned earlier, four classes of macromolecules make up the bodies of organisms. This discussion of the first class—carbohydrates—is followed by an examination of the other three classes of macromolecules: lipids, proteins, and nucleic acids.

Carbohydrates are a loosely defined group of molecules that contain the elements carbon, hydrogen, and oxygen, usually in the ratio 1:2:1. Some carbohydrates are simple, small monomers. Others are long polymers. The chemical formula for a carbohydrate is $(CH_2O)_n$, where n is the number of carbon atoms. Because they contain many carbon-hydrogen (C—H) bonds, carbohydrates are well suited for energy storage. Such C—H bonds are the ones most often broken by organisms to obtain energy.

Sugars: Simple Carbohydrates

Among the simplest of the carbohydrates are the simple sugars or **monosaccharides** (from the Greek *monos,* meaning "single," and *saccharon,* meaning "sweet"). As their name implies, monosaccharides taste sweet. Simple sugars may have as few as three carbon atoms, but the monosaccharide molecules that play the central role in energy storage have six carbon atoms and the following chemical formula:

$$C_6H_{12}O_6 \quad \text{or} \quad (CH_2O)_6$$

Sugars can exist in a straight-chain form, but in water solution, they almost always form rings. The primary energy-storage molecule in living organisms is glucose, a six-carbon sugar with seven energy-storing C—H bonds (figure 2.17).

> *Among the most important energy-storage molecules in or-*
> *ganisms are sugars. Many simple sugars contain six carbon*
> *atoms and seven energy-storing C—H bonds.*

Many organisms transport sugars within their bodies. In human beings, glucose circulates in the blood. In many other organisms, glucose is converted to a transport form before it is moved from place to place. Glucose is less readily consumed (metabolized) while in transport form. Transport forms of sugars commonly develop when two monosaccharide molecules link to form a **disaccharide** (from the Greek *di,* meaning "two"). Sucrose (table sugar), a disaccharide formed by linking a molecule of glucose to a molecule of fructose, is the common transport form of sugar in plants. Much of common table sugar is refined from the sap of sugarcane and sugar beets. If a glucose molecule is linked to galactose, the resulting disaccharide is lactose, which is the molecule many mammals feed their babies. Maltose, two glucose subunits linked together, lends a sweet taste to the barley seeds that brewers ferment into alcohol in the process of making beer.

Starches: Chains of Sugars

Organisms store the metabolic energy contained in glucose by converting glucose to an insoluble form and depositing it in specific storage areas. Sugars are made insoluble by joining

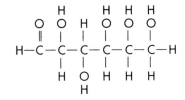

 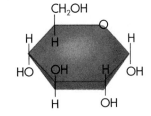

(a)

(b)

Figure 2.17 Simple sugars, or monosaccharides.
(a) *The structure of glucose. Glucose is a linear six-carbon molecule that forms a ring when added to water.* (b) *Many animals consume sugar. This hungry butterfly has its mouthparts extended down into nectar, a solution rich in glucose, produced in the flower shown here.*

them together into long polymers called **polysaccharides,** which are composed of monosaccharide sugar subunits. If the polymers are branched—that is, if they have side chains coming off of a main chain—the molecules are even less soluble. **Starches** are polysaccharides formed from glucose.

The starch with the simplest structure is amylose. Amylose is made up of many hundreds of glucose molecules linked together in long, unbranched chains. Potato starch is about 20 percent amylose. When you eat a potato, digestive proteins called enzymes first break the potato starch into fragments of random length. These shorter fragments are soluble and thus easier to digest. Baking or boiling potatoes has the same effect.

Most plant starch, including 80 percent of potato starch, is a more complicated variant of amylose called amylopectin (figure 2.18a, b). Pectins are branched polysaccharides. Amylopectin is a form of amylose with short, linear amylose branches consisting of twenty to thirty glucose subunits. Most humans consume a great deal of plant starch. The seeds of rice, wheat, and corn supply about two-thirds of all the calories used by humankind.

Animals also store glucose in branched amylose chains. However, the average chain length is much longer in animals, and there are more branches. This results in a highly branched animal starch called glycogen (figure 2.18c, d).

> Starches are storage polysaccharides formed from glucose. Because they form long chains, starches are relatively insoluble and thus function well for storage.

Cellulose: Difficult Starch to Digest

Imagine that you could draw a line down the central axis of a starch molecule, like threading a rope through a pipe. Because all the glucose subunits (CH_2OH groups) of the starch chain are joined in the same orientation, they all would fall on the same side of the line. Another way to build a chain of glucose molecules, however, involves the glucose subunit orientations switching back and forth (the CH_2OH groups alternating on opposite sides of the line). The resulting polysaccharide is **cellulose,** the chief component of plant cell walls (figure 2.19).

Cellulose is chemically similar to amylose, but with one important difference: The starch-degrading enzymes that occur in most organisms cannot break the bond between two sugars in opposite orientation. It is not that the bond is stronger, but rather that the bond's cleavage requires the aid of a different protein, one not usually present. Because cellulose cannot

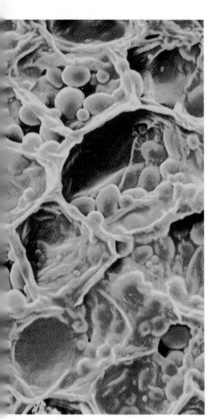

(a)

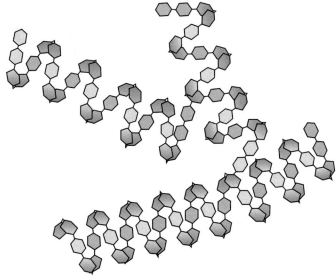

(c)

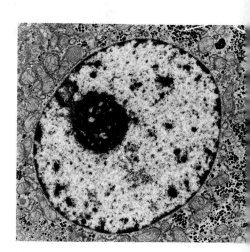

(d)

(b)

Figure 2.18 Starches.
(a) *Amylopectin is the form of starch that plants use to store energy.* (b) *The photomicrograph of these plant cells shows the packets of amylopectin contained in the chloroplasts, the organelles in which photosynthesis takes place.* (c) *Glycogen is a highly branched starch that animals use to store glucose.* (d) *Glycogen granules can be seen in this micrograph of a liver cell.*

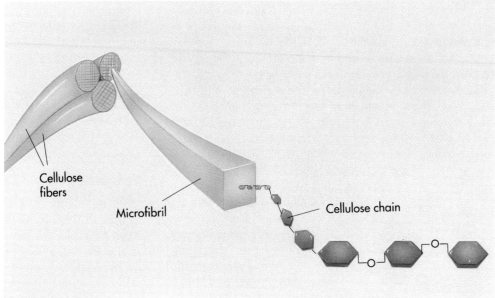

(a)   (b)

Figure 2.19 A journey into wood: The structure of cellulose.
(a) *Cellulose fibers from a ponderosa pine. Cellulose fibers are found in the cell walls of plants and are very strong and resistant to metabolic breakdown, which is one reason why wood is such a good building material.* (b) *This diagram shows the increasingly microscopic structure of cellulose fibers. As you can see, each fiber is composed of cellulose chains, which, in turn, are composed of individual glucose subunits.*

readily be broken down, it works well as a biological structural material and occurs widely in this role in plants. For those few animals able to break down cellulose, it provides a rich source of energy. Certain vertebrates, such as cows, do not themselves produce the enzymes necessary to digest cellulose, but are able to break it down by means of the bacteria and protists they harbor in their intestines. In humans, cellulose is a major component of dietary fiber that is necessary for the proper functioning of the digestive system.

The structural material in insects and their relatives (figure 2.20), many fungi, and certain other organisms is called **chitin.** Chitin resembles cellulose, but a nitrogen group has been added to the glucose units. Chitin is a tough, resistant surface material, and few organisms are able to digest it.

Lipids

Lipids are a loosely defined group of macromolecules that are insoluble in water but soluble in oil. Fats are one kind of lipid. Oils, such as olive oil, corn oil, and coconut oil, are also lipids, as are waxes, such as beeswax and earwax.

Fats: Efficient Energy Storage

When organisms store glucose molecules for long periods, they usually convert the glucose into insoluble molecules that contain more C—H bonds than do carbohydrates. These storage molecules are called **fats.** The ratio of hydrogen to oxygen in carbohydrates is 2:1, but in fat molecules, the ratio is much higher. Like starches, fats are insoluble and can therefore be deposited at specific storage locations within the organism. Starches are insoluble because they are long polymers. In

Figure 2.20 Chitin, a protective carbohydrate.
Chitin, which is a modified form of cellulose with nitrogen groups added to the sugar subunits, is the principal structural element in the external skeletons of many animals, including this lobster.

contrast, fats are insoluble because they are nonpolar. Unlike the H—O bonds of water, the C—H bonds of carbohydrates and fats are nonpolar and cannot form hydrogen bonds. Because fat molecules contain a large number of C—H bonds, they are hydrophobically excluded by water because water molecules tend to form hydrogen bonds with other water molecules. The result is that the fat molecules cluster together and are insoluble in water.

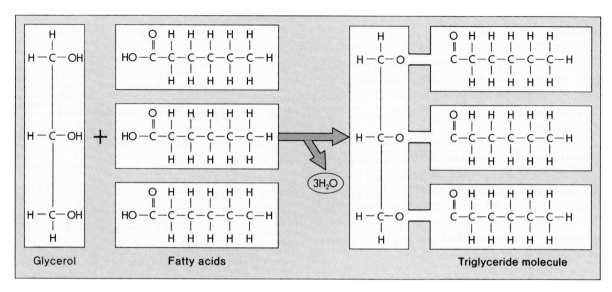

Figure 2.21 Structure of a fat molecule.
This fat molecule, a triglyceride, is formed by dehydration synthesis, in which the glycerol is attached to three fatty acids.

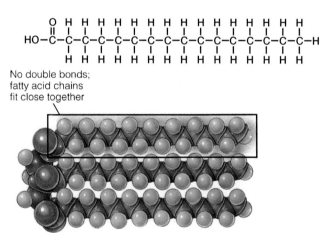

No double bonds; fatty acid chains fit close together

(a) **Saturated fat**

Double bonds present; fatty acid chains do not fit close together

Unsaturated fat

(b)

Figure 2.22 Saturated and unsaturated fats.
(a) *Many animal fats are saturated—that is, each carbon is "saturated" with the maximum number of hydrogens. Because their fatty acid chains can fit closely together, these fats form immobile arrays called hard fats.* (b) *Unsaturated fats, however, are characterized by double bonds that replace some of the hydrogen atoms. These bonds produce "kinks" in the fatty acid chains that prevent close association of the chains. Unsaturated fats are thus more fluid than the hard saturated fats.*

Fats are composite molecules. Each molecule is built from two different kinds of subunits (figure 2.21):

1. *Glycerol* A three-carbon alcohol with each carbon bearing a hydroxyl (—OH) group. The three carbons form the backbone of the fat molecule, to which three fatty acids are attached.

2. *Fatty acids* Long **hydrocarbon** chains (chains consisting only of carbon and hydrogen atoms) ending in a carboxyl (—COOH) group. Three fatty acids are attached to each glycerol backbone.

The structure of an individual fat molecule, such as the one diagrammed in figure 2.21, consists simply of a glycerol molecule with a fatty acid joined to each of its three carbon atoms. Because there are three fatty acids, the resulting fat molecule is called a triglyceride.

Fatty acids vary in length. The most common are even-numbered chains of fourteen to twenty carbons. Fatty acids with all internal carbon atoms having two hydrogen side groups contain the maximum number of hydrogen atoms possible. Fats composed of these fatty acids are said to be **saturated** (figure 2.22*a*). Some fatty acids have double bonds between one or more pairs of successive carbon atoms, and these double bonds replace some of the hydrogen atoms. Fatty acids with double bonds therefore contain fewer than the maximum number of hydrogen atoms. Fats composed of these fatty acids are said to be **unsaturated** (figure 2.22*b*). Many plant fatty acids, such as oleic acid (a vegetable oil) and linolenic acid (a linseed oil), are unsaturated. Animal fats, in contrast, are often saturated and occur as hard fats.

If a given fat has more than one double bond, it is said to be **polyunsaturated.** Polyunsaturated fats have low melting points because their chains bend at the double bonds and the fat molecules cannot be closely aligned. Consequently, the fat may be fluid. A liquid fat is called an **oil.** An oil can be converted into a hard fat by adding hydrogen. For example, peanut butter is usually hydrogenated to prevent the peanut fatty acids from separating out as oils while the jar sits on the store shelf.

Fats are very efficient energy-storage molecules because of their high concentration of C—H bonds. Most fats contain more than forty carbon atoms. The ratio of energy-storing C—H bonds to carbon atoms is more than twice that of carbohydrates, making fats much more efficient vehicles for storing chemical energy. Fats usually yield about twice the amount of chemical energy per gram that carbohydrates yield. As you might expect, the more highly saturated fats are richer in energy than are the less saturated ones.

In the past twenty years, researchers have discovered a link between the ingestion of saturated fats and the incidence of cardiovascular disease. Cholesterol is a type of lipid called a **steroid.** Excess saturated fat intake can cause plugs of cholesterol, called plaques, to form in the blood vessels, which may lead to blockage, high blood pressure, stroke, or heart attack. Experts advise limiting intake of saturated fats to prevent these plaques from forming. Saturated fats are most commonly found in animal fats (red meat), but some plant fats, such as palm oil, are also highly saturated. Fats derived from vegetables, such as corn oil and canola oil, are better choices for those who want to protect their arteries from cholesterol.

Other Kinds of Lipids

Fats are just one example of the oily or waxy class of molecules called lipids. Your body contains many different kinds of lipids. As discussed in the preceding section, cholesterol is a kind of lipid called a steroid. Some hormones, or chemical messengers, such as the male and female sex hormones, also are steroids. The membranes of your cells are composed of another kind of lipid called a phospholipid. Lipids called terpenes form many of the biologically important pigments, such as the photosynthetic pigment carotene found in plants and the light-absorbing pigment retinol found in your eyes.

Lipids are macromolecules that provide long-term storage for energy (fats), provide the structural basis for cell membranes (phospholipids), and act as chemical messengers (steroids).

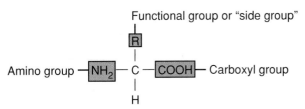

Figure 2.23 The structure of an amino acid.
The amino and carboxyl groups are found in every amino acid. The only variable is the functional group or "side group."

Proteins

Proteins are the third major group of macromolecules that make up the bodies of organisms. Perhaps the most important proteins are enzymes, which, as mentioned earlier in the chapter, are proteins capable of speeding up specific chemical reactions. Enzymes lower the energy required to activate or start a chemical reaction but are unaltered themselves in the process. Enzymes are biological catalysts, a more general term for substances that affect chemical reactions in this way. Other kinds of proteins also have important functions. Cartilage, bones, and tendons all contain a protein called collagen. Keratin, another protein, forms both the horns of a rhinoceros and the feathers of a bird. The fluid within your eyeballs contains still other proteins. Short proteins called peptides are chemical messengers within your brain and throughout your body. Despite their diverse functions, all proteins have the same basic structure: a long polymer chain of amino acid subunits linked end to end.

Amino Acids: Building Blocks of Proteins

Amino acids are small molecules with a simple basic structure: They contain an amino group (—NH₂), a carboxyl group (—COOH), a hydrogen atom (H), and a functional group, or "side group," designated *R*, all bonded to a central carbon atom (C) (figure 2.23).

The identity and unique chemical properties of each amino acid are determined by the nature of the *R* group linked to the central carbon atom. An amino acid can potentially have any of a variety of different *R* groups. Although many different amino acids occur in nature, only twenty are used in proteins. These twenty "common" amino acids and their side groups are illustrated in figure 2.24. The different side groups of the twenty amino acids give each amino acid distinctive chemical properties. For example, when the side group is —H, the

---

Figure 2.24 The twenty common amino acids. ▶
Each amino acid has the same chemical backbone but can be differentiated from other amino acids by its side group. Six of the amino acid R groups are nonpolar. Some of these are more bulky than others, particularly the ones containing ring structures, which are called the aromatic amino acids. Another six are polar but uncharged, and these differ from one another in their polarity. Five more are polar and are capable of ionizing to a charged form. The remaining three have special chemical (structural) properties that are important in forming links between protein chains or in forming kinks in their shapes.

Special structural property

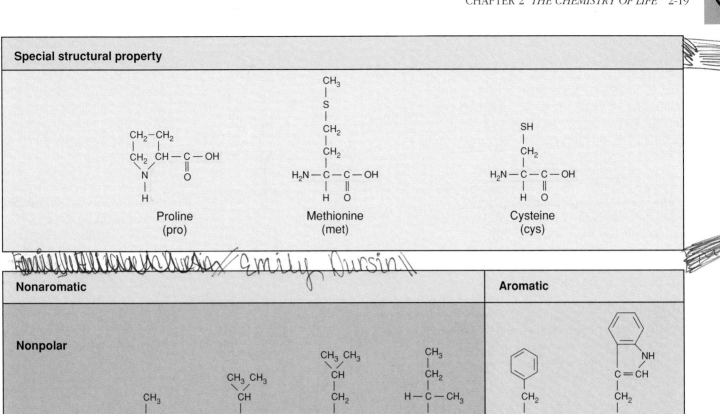

Proline
(pro)

Methionine
(met)

Cysteine
(cys)

Nonaromatic

Aromatic

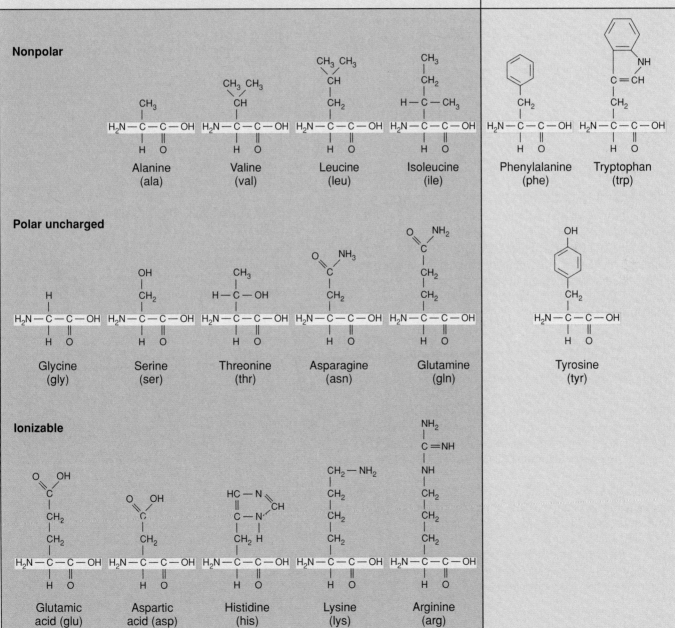

Nonpolar

Alanine
(ala)

Valine
(val)

Leucine
(leu)

Isoleucine
(ile)

Phenylalanine
(phe)

Tryptophan
(trp)

Polar uncharged

Glycine
(gly)

Serine
(ser)

Threonine
(thr)

Asparagine
(asn)

Glutamine
(gln)

Tyrosine
(tyr)

Ionizable

Glutamic
acid (glu)

Aspartic
acid (asp)

Histidine
(his)

Lysine
(lys)

Arginine
(arg)

amino acid (glycine) is polar, whereas when the side group is —CH₃, the amino acid (alanine) is nonpolar. The twenty amino acids that occur in proteins are commonly grouped into five chemical classes based on the chemical nature of their side groups, as shown in figure 2.24.

The way that each amino acid affects the shape of a protein depends on the chemical nature of the amino acid's side group. For example, portions of a protein chain with many nonpolar amino acids tend to be shoved into the interior of the protein by hydrophobic interactions because polar water molecules tend to exclude nonpolar amino acid side groups.

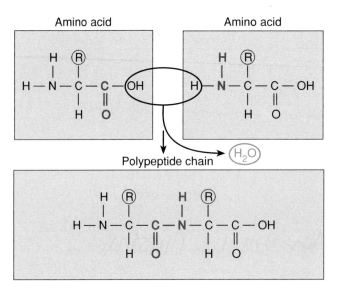

Figure 2.25 The peptide bond.
Amino acids are linked together by peptide bonds to form polypeptides, or proteins.

Proteins can contain up to twenty different kinds of amino acids. These amino acids fall into five chemical classes, each with different properties. These differences determine what the proteins are like.

In addition to its *R* group, each amino acid, when ionized, has a positive (amino, or NH₃⁺) group at one end and a negative (carboxyl, or COO⁻) group at the other end. These two groups can undergo a chemical (dehydration) reaction, losing a molecule of water and forming a covalent bond between two amino acids. A covalent bond linking two amino acids is called a **peptide bond** (figure 2.25).

Polypeptides: Chains of Amino Acids

As mentioned previously, a protein is composed of a long chain of amino acids linked end to end by peptide bonds. The general term for chains of this kind is **polypeptide.** Thus, proteins are long, complex polypeptides. The sequence of amino acids that make up a particular polypeptide chain is termed the polypeptide's **primary structure** (figure 2.26a). Because the *R* groups that distinguish the various amino acids play no role in the peptide backbone of proteins, a protein can be composed of any sequence of amino acids. A protein made up of one hundred amino acids linked together in a chain might have any of 20¹⁰⁰ different amino acid sequences. The great variability possible in the sequence of amino acids is perhaps the most important property of proteins, permitting great diversity in the kinds and, therefore, functions of specific proteins.

Each amino acid of a polypeptide interacts with its neighbors, forming hydrogen bonds. Because of these near-neighbor interactions, polypeptide chains tend to fold spontaneously into sheets or wrap into coils, or helices. The form that a region of a polypeptide assumes is called the polypeptide's **secondary structure** (figure 2.26b).

The three-dimensional shape, or **tertiary structure,** of a protein depends heavily on its secondary structure (figure 2.26c). Proteins made up largely of sheets often form fibers that have a structural function, whereas proteins that have regions forming coils frequently fold into globular shapes. The shape of a globular protein is very sensitive to the order and nature of amino acids in the sequence. A change in the identity of a single amino acid can have either subtle or profound effects. Globular proteins are extremely effective biological catalysts because they can assume so many different shapes (see chapter 6).

When two polypeptides associate to form a functional unit, the chains are termed subunits. The subunits need not all be the same, although they can be. For example, the protein hemoglobin is composed of four subunits: two identical subunits of one kind and two identical subunits of a second kind. How a protein's subunits are assembled into a whole is called the protein's **quaternary structure** (figure 2.26d). Proteins are discussed in detail in chapter 6.

A protein's shape is determined by the sequence of amino acids in the polypeptide. Because different amino acid R groups have different chemical properties, the shape of a protein may be altered by a single amino acid change.

Proteins perform many functions in your body. The thousands of different enzymes that carry out your body's chemical reactions are **globular proteins.** So are the antibodies that protect you from infection and cancer. **Structural proteins** play structural roles. The keratin in your hair is a structural protein, as are the actin and myosin that make up your muscles and the fibrin that aids in blood clotting. Indeed, the most abundant protein in your body—collagen—is a fibrous structural protein. Collagen forms the matrix of your skin, ligaments, tendons, and bones (figure 2.27).

Figure 2.26 How the primary structure determines a protein's shape.

The amino acid sequence, or primary structure, of the enzyme protein lysozyme encourages the formation of hydrogen bonds between nearby amino acids, producing coils and fold-backs called the secondary structure. The lysozyme protein assumes a three-dimensional shape like a spoon; this is called its tertiary structure. Many proteins (not lysozyme) aggregate in globular clusters called the quaternary structure of the protein.

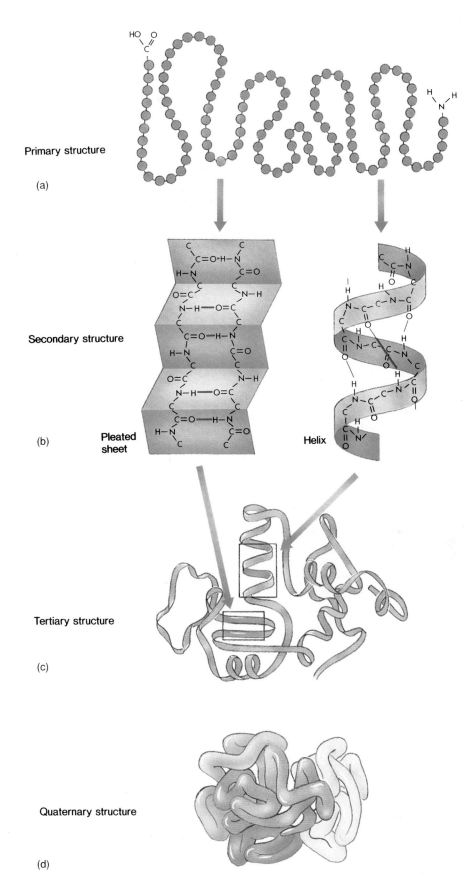

Primary structure

(a)

Secondary structure

(b) Pleated sheet   Helix

Tertiary structure

(c)

Quaternary structure

(d)

Figure 2.27 Some of the more common structural proteins.
(a) ***Fibrin.*** *This electron micrograph shows a red blood cell caught in threads of fibrin. Fibrin is important in the formation of blood clots.* (b) ***Collagen.*** *The so called "cat-gut" strings of a tennis racket are made of collagen.* (c) ***Keratin.*** *This type of protein makes up bird feathers, such as this peacock feather.* (d) ***Spider silk.*** *The web spun by this agile spider is made of protein.* (e) ***Hair.*** *Hair is also a protein.*

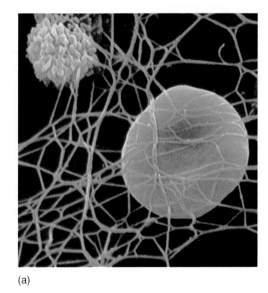

(a)

(b)

(c)

(d)

(e)

Nucleic Acids

All organisms store the information specifying the structures of their proteins in nucleic acids. **Nucleic acids** are long polymers of repeating subunits called **nucleotides**. Each nucleotide, the basic repeating unit, is a composite molecule made up of three smaller building blocks (figure 2.28):

1. A five-carbon sugar
2. A phosphate group (PO_4)
3. An organic nitrogen-containing (nitrogenous) base

In the formation of a nucleic acid chain, the individual sugars are linked together in a line by the phosphate groups. The phosphate group of one sugar binds to the hydroxyl group of another, forming an —O—P—O bond, called a **phosphodiester bond.** A nucleic acid is simply a chain of five-carbon sugars

Figure 2.28 The structure of a nucleotide.

Nucleotides are composed of three parts: a five-carbon sugar, a phosphate group, and an organic nitrogenous base.

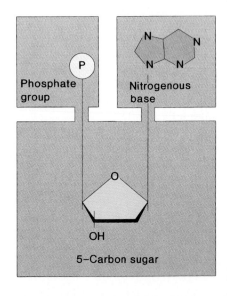

Phosphate group

Nitrogenous base

5–Carbon sugar

(called ribose sugars) linked by phosphodiester bonds, with a nitrogenous base protruding from each sugar. Each of the repeating phosphate-sugar-base links in the chain is a nucleotide.

Nucleotides play many critical roles in the life of the cell. For example, the nitrogenous base adenine is a key component of the nucleotide ATP, which as mentioned in chapter 1, is the energy currency of the cell, and of the molecules NAD and FAD, which are called cofactors, molecules whose function is to carry energy-rich electrons from one place to another during metabolism.

In addition, the nucleic acid **DNA (deoxyribonucleic acid)** stores important information in its sequence of nucleotides. DNA is a double chain of nucleotides. The chains wind around each other like the outside and inside rails of a circular staircase. Such a winding shape is called a helix, and one composed of two chains winding about one another, as in DNA, is called a double helix.

Each chain is made up of millions of nucleotides, but these millions are of four types that differ only in their nitrogenous bases. The four bases found in the nucleotides that compose DNA are adenine, guanine, cytosine, and thymine. The steps of the DNA helical staircase are hydrogen bonds between the nitrogenous bases in one polymer chain and those opposite them in the other chain. This hydrogen bond formation is called base pairing and holds the two chains together as a duplex (figure 2.29).

The order of the four different nucleotides of DNA comprises a coded set of instructions that directs all of an organism's activities. DNA is able to do this because the DNA sequence of nucleotides codes for the sequence of amino acids in proteins. As we have seen, proteins are the enzymes that catalyze

Figure 2.29 The structure of DNA.

(a) *Hydrogen bonds form between the nitrogenous bases (C = cytosine; G = guanine; A = adenine; T = thymine) and cause the two chains of the DNA molecule to bind to each other. This hydrogen bond formation is called base pairing. (b) The DNA molecule is not a straight chain, but rather a graceful double helix, rising like a circular staircase. (c) This beautiful computer-enhanced photograph of DNA was made by a new microscopic photographing technique called scanning-tunneling microscopy. If you look carefully, you will be able to discern the individual strands of DNA, and no wonder—this new technique can magnify an object a million times!*

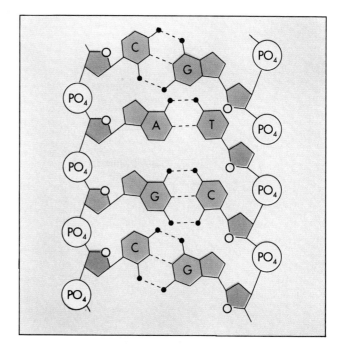

(a)

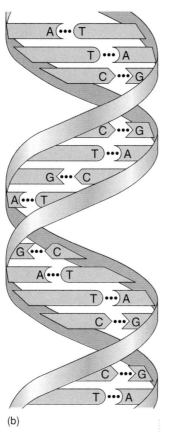

(b)

(c)

organisms' chemical reactions. Every physiological process that an organism performs, then, is directed by a protein. And the structure of the thousands of proteins in an organism's body is dictated by the sequence of nucleotides in the organism's DNA.

Even more importantly, the structure of DNA permits transmission of hereditary information to offspring. Organisms store hereditary information in two forms of nucleic acid: DNA is the basic storage vehicle, or master plan, and is found in the cell nucleus. **Ribonucleic acid (RNA)** is similar in structure and is made as a template copy of portions of the DNA. This

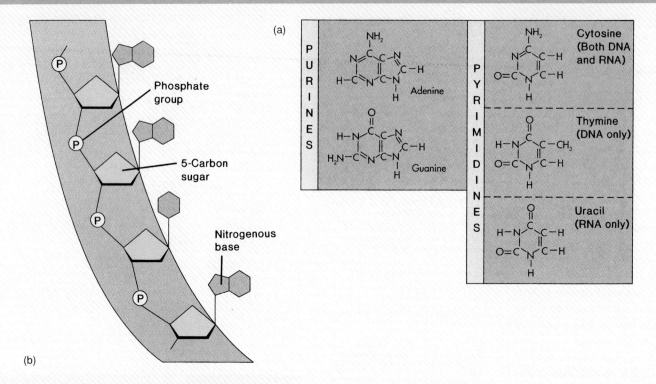

(a)

(b)

Figure 2.30 The five nitrogenous bases of nucleic acids.
(a) *Adenine and guanine are large, double-ring compounds called purines. Cytosine, thymine, and uracil are smaller, single-ring compounds called pyrimidines. In DNA, thymine replaces the* uracil found in RNA. (b) *One half of a DNA helix. In a nucleic acid such as DNA, the nucleotides that contain the nitrogenous bases are linked to one another by phosphodiester bonds.*

copy passes from the nucleus out into the rest of the cell, where it provides a blueprint specifying the amino acid sequence of proteins.

Two of the four nitrogenous bases that make up DNA and RNA—adenine and guanine—are large, double-ring compounds called **purines.** The other nitrogenous bases that occur in these molecules—cytosine (in both DNA and RNA), thymine (in DNA only), and uracil (in RNA only)—are smaller, single-ring compounds called **pyrimidines** (figure 2.30). The details of DNA structure and of how DNA interacts with RNA in the production of proteins are presented in chapter 12.

Organisms store and use hereditary information by encoding the sequence of the amino acids of each of their proteins as a sequence of nucleotides in nucleic acids.

SUMMARY

1. The smallest stable particles are protons, neutrons, and electrons. These particles associate, forming atoms. The core (nucleus) of an atom is composed of protons and neutrons. Electrons orbit the core.

2. The chemical behavior of an atom is largely determined by the distribution of its electrons, particulary the number of electrons in its outermost energy level.

3. A molecule is a stable collection of atoms. The forces holding atoms together in a molecule are called chemical bonds.

4. The force of a chemical bond can result from the attraction of opposite charges, as in an ionic bond, or from the sharing of one or more pairs of electrons, as in a covalent bond.

5. The chemistry of life is the chemistry of water. In water molecules, the oxygen atom more strongly attracts the electrons shared between oxygen and the hydrogen atoms. As a result, the oxygen atom carries a partial negative charge, and the hydrogen atoms carry a partial positive charge. Because of this charge separation, water is termed a "polar" molecule.

6. A hydrogen bond is formed by the attraction of the partial positive charge of one hydrogen atom of a water molecule with the partial negative charge of the oxygen atom of another. Water molecules tend to form the maximum number of hydrogen bonds and to exclude nonpolar molecules.

7. The molecules formed by living organisms all contain carbon and are called organic molecules. Macromolecules, which are large organic molecules, often play a structural role in organisms or store information. Four classes of macromolecules are: carbohydrates, lipids, proteins, and nucleic acids.

8. Organisms store energy in carbon-hydrogen (C—H) bonds. The most important of the energy-storing carbohydrates is glucose, a six-carbon sugar.

9. Excess energy resources may be stored in complex sugar polymers called starches, especially in plants. Glycogen, a comparable storage polymer occurring frequently in animals, is characterized by complex branching.

10. Fats are molecules containing many more C—H bonds than carbohydrates do, thereby providing more efficient energy storage.

11. Proteins are linear polymers of amino acids. Because the twenty amino acids that occur in proteins have side groups with very different chemical properties, the function and shape of a protein are critically affected by the protein's particular sequence of amino acids.

12. Hereditary information is stored as a sequence of nucleotides in a linear nucleotide polymer called deoxyribonucleic acid, or DNA. DNA is a double helix. A second form of nucleic acid—ribonucleic acid, or RNA—is similar in structure and is made as a template copy of portions of the DNA. In the cell, RNA provides a blueprint specifying the amino acid sequence of proteins.

REVIEWING THE CHAPTER

1. What is the structure of an atom?

2. What is the definition of an isotope?

3. What are the concepts of electron energy levels and electron orbitals?

4. What are molecules?

5. How do ionic bonds and covalent bonds differ?

6. What are the most abundant elements in living things?

7. How is the water molecule significant to the evolution of life on earth?

8. What is significant about the polarity of water?

9. What are hydrogen bonds?

10. Why is water a powerful solvent?

11. How does water organize nonpolar molecules?

12. What is the association between water, acids, and bases?

13. What does the pH scale measure?

14. What are the chemical building blocks of life?

15. How does one distinguish a polymer from a monomer?

16. What are carbohydrates, and how are they significant to living organisms?

17. What are lipids, and why are they important in living organisms?

18. What are proteins, and why are they necessary for living things?

19. What are nucleic acids, and what is their role in living organisms?

COMPLETING YOUR UNDERSTANDING

1. Carbon-14 is an example of
   a. a radioactive isotope.
   b. an ion.
   c. an oxidation reaction.
   d. a neutron.
   e. an ionic bond.
   f. a reduction reaction.

2. What is an ion?

3. What is the difference between a molecule and an element?

4. When two atoms share electrons, they form a _____ _____ .

5. The diatomic hydrogen gas molecule is an example of
   a. an ionic bond.
   b. a covalent bond.
   c. a radioactive isotope.
   d. a double bond.
   e. an element.
   f. all of the above.

6. If molecule A is placed in water, what is the correlation between its polarity and solubility in water?

7. How do the terms *hydrophilic* and *hydrophobic* relate to biology?

8. What is the difference between an acid and a base?

9. How many times more acidic is a pH of 5 than a pH of 7?

10. What is the correlation between pH and the maintenance of living things?

11. Why is acid rain a major environmental concern? What steps should be taken to curtail acid rain? What are the costs to your suggested solutions?

12. Which of the following is not a macromolecule?
   a. Protein
   b. Lipid
   c. Carbohydrate
   d. Nucleic acid
   e. Water
   f. All the above are macromolecules.

13. What is a hydrolysis reaction?

14. What is the source of about two-thirds of all calories consumed by humans?

15. What are the differences and similarities between animal and plant starches?

16. What benefits do cellulose provide humans, even though it is not readily broken down during digestion?

17. What are the functions of glucose, starch, cellulose, and chitin in living organisms?

18. What are saturated and unsaturated fatty acids?

19. Why do some companies in the peanut butter industry hydrogenate the peanut oil prior to selling their product at the grocery store? Healthwise, which would be better to purchase: the nonhydrogenated or hydrogenated jar of peanut butter?

20. If you went to a bakery to purchase fresh bread, which would be better for you: bread made with vegetable oil or lard? Why?

21. Why is it necessary for the human body to maintain a certain level of steroids?

22. Which of the following is not a lipid?
   a. Fat
   b. Enzyme
   c. Steroid
   d. Phospholipid
   e. Terpene
   f. All of the above are lipids.

23. What is significant about the *R* group in amino acids?

24. How do the primary, secondary, tertiary, and quaternary structures of proteins function?

25. What are the functions of globular and structural proteins?

26. What are nucleotides, and where are they found in living organisms?

27. RNA differs from DNA because RNA
   a. is single-stranded.
   b. contains the sugar ribose.
   c. has uracil.
   d. All of the above are correct.

28. The basic difference between purines and pyrimidines is that the former is a _____ -ringed compound.

FOR FURTHER READING

Fitzgerald, J., and G. Taylor. 1989. Carbon fibres stretch the limits. *New Scientist,* May, 48–53. A fascinating discussion of the ways in which the principles discussed in this chapter are used to produce modern, synthetic materials.

Jandacek, R. J. 1991. The development of olestra, a noncaloric substitute for dietary fat. *Journal of Chemical Education,* June, 476–79. Describes how scientists went about designing a "nonfattening" fat for use as a fat substitute.

Karplus, M., and A. McCammon. 1986. The dynamics of proteins. *Scientific American,* April, 42–51. Explains why flexibility is critical to protein function.

Olson, A., and D. Goodsell. 1992. Visualizing biological molecules. *Scientific American,* Nov., 76–81. Discussion of computer-generated images that are aiding research in molecular structure, helping scientists to understand how the form of a protein influences its function.

Sharon, N., and H. Lis. 1993. Carbohydrates in cell recognition. *Scientific American,* Jan., 82–89. Examination of telltale surface sugars that act like dog tags and enable cells to identify and interact with one another.

Sutton, C. 1989. Subatomic forces. *New Scientist,* Feb., supplement, 1–4. Contemporary review of the interactions between particles within the atom.

THE ORIGIN OF LIFE

Ionizing radiation in the form of lightning hits the earth in a spectacular display over the desert in Tucson, Arizona. Was lightning an electrical "switch" that triggered the dawn of life over 3.5 billion years ago?

Spinning around the sun are a host of dead planets and moons, their surfaces bare and bleak. Alone among them, the earth teems with life. However, when the earth first formed, no life existed here—no grass grew, and no fish swam in the sea.

By studying the distribution of radioactive isotopes in ancient rocks, scientists have determined that the earth was formed about 4.5 billion years ago. At first the earth was molten, but soon a thin crust of rock, the shell on which we live, solidified over the hot core. Early earth was a land of molten rock and violent volcanic activity. The oldest rocks that have survived on earth are about 3.9 billion years old. These ancient rocks contain no definite traces of life, or at least none that can be recognized with current technology.

Today the earth is very different, and life exists in profusion in every crack and crevice. Where did all of this life come from? Studies of ancient rocks provide some clues, but the record of events is incomplete and often silent (figure 3.1). Perhaps the most fundamental question concerns the nature of the agency or force that led to the origin of life. In principle, there are at least three possibilities:

1. *Extraterrestrial origin* Life may not have originated on earth at all but instead may have been carried to it, perhaps as an extraterrestrial infection of spores that originated on a planet of a distant star. How life came to exist on *that* planet is a question that then must be answered.

2. *Special creation* Life-forms may have been put on earth by supernatural or divine forces. This viewpoint, common to most Western religions, is the oldest hypothesis and is widely accepted. It forms the basis of the "scientific creationism" viewpoint discussed in chapter 15.

3. *Evolution* Life may have evolved from inanimate matter, with associations among molecules becoming more and more complex. In this view, the force leading to life was natural selection—that is, changes in molecules that increased their stability caused the molecules to persist longer.

This book deals only with the third possibility because it is the only hypothesis that lends itself to scientific investigation. The first possibility—extraterrestrial origin—cannot be scientifically investigated because scientists currently lack the necessary tools. The second possibility—special creation—is a matter of religious faith, and faith is a separate entity altogether from the scientific endeavor. There is no way to scientifically test whether a divine force created life on earth, and therefore, special creation is not a valid scientific hypothesis. However, with the third possibility—evolution—scientists can attempt to understand whether the forces of evolution could have led to the origin of life, and if so, how the process might have occurred.

This search for an understanding of how life evolved on earth requires looking back to the time before life appeared, to when the earth was just starting to cool. The investigation must go back at least that far because fossils of bacteria exist in rocks that are about 3.5 billion years old, demonstrating that life existed on earth within no more than one billion years of the planet's origin. A determination of how the first organisms originated means first considering the mode of origin of organic molecules, which are the building blocks of organisms. The process by which organic molecules originated is called **chemical evolution.** The second phenomenon requiring investigation is how organic molecules might have become organized into living cells.

Figure 3.1 A fossil fish.
Scientists use fossils, such as this fossil fish from about 590 million years ago, to reconstruct the earth's past. Until recently, much older fossils of microscopic organisms were not helpful because they were simply too small to be seen with the naked eye. Now, with microscopes, scientists can trace the fossil record back 3.5 billion years.

Life on earth may have evolved from inorganic substances, and the process by which this may have occurred can be studied directly. If life came to earth from another planet, it probably originated in some similar way, but scientists currently lack the tools to investigate this possibility. The suggestion that life was created by a supreme being lies beyond the scope of scientific investigation.

The Origin of Organic Molecules

Scientists who study the conditions of the primitive earth are called geochemists. Theorists surmise that the current universe was at one time condensed into an incredibly small space, which then exploded in the so-called "big bang" about fourteen billion years ago. Researchers theorize that the dust and gas from this explosion continued to erupt to form stars and other planets, including the earth about 4.5 billion years ago. Recent data from a special satellite that measures heat in terms of radio waves supports the "big bang" hypothesis, since it shows that the universe still is awash with heat waves left over from the "big bang" (figure 3.2).

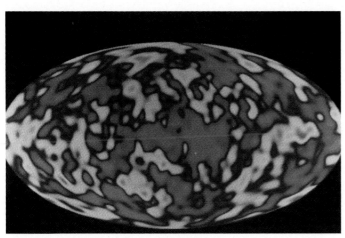

Figure 3.2 Evidence of the "big bang."
Careful measurement of patterns of very-low-energy radiation in the sky by the COBE satellite in 1993 indicates a very uniform distribution of residual heat waves from the "big bang."

The Atmosphere of the Early Earth

Geochemists believe that, as the primitive earth cooled and its rocky crust formed, many gases were released from the molten core and blasted skyward by volcanoes. These gases formed a cloud around the earth and were held as an atmosphere by the earth's gravity. Earth's current atmosphere is very different from what it used to be; as we shall see later, it has been changed by the activities of organisms. Yet, geochemists have been able to learn what the early atmosphere must have been like by studying the gases released by volcanoes and by deep sea vents in the earth's crust. While a consensus on the exact composition of this original atmosphere has not been reached, geochemists do agree that it was principally composed of nitrogen gas and also contained significant amounts of carbon dioxide and water. It is probable, although not certain, that compounds in which hydrogen atoms were bonded to other light elements, such as sulfur, nitrogen, and carbon, were also present in the earth's early atmosphere. These compounds would have been hydrogen sulfide (H_2S), ammonia (NH_3), and methane (CH_4).

The atmosphere of early earth was probably rich in hydrogen, although debate continues on this point. Such an atmosphere is called a "reducing" one because of the ample availability of hydrogen atoms and associated electrons. (As discussed in chapter 2, in chemistry the donation of electrons to a molecule is called reduction, and the removal of electrons is oxidation.) Little if any oxygen gas was present. In a reducing atmosphere, it takes little energy to form the carbon-rich molecules from which life evolved.

Later, the earth's atmosphere changed as living organisms began to carry out photosynthesis, which involves harnessing the energy in sunlight to split water molecules and form complex carbon-containing molecules, giving off gaseous oxygen molecules in the process. The earth's atmosphere is now approximately 21 percent oxygen. In today's oxidizing atmosphere, complex carbon-containing molecules cannot form spontaneously.

Therefore, the first step in the evolution of life seems to have occurred in a reducing atmosphere that was devoid of gaseous oxygen. The early earth was awash with energy: solar radiation, lightning from intense electrical storms, violent volcanic eruptions, and heat from radioactive decay (figure 3.3). Living on earth today, shielded from the effects of solar ultraviolet radiation by a layer of ozone gas (O_3) in the upper atmosphere, most humans cannot imagine the enormous flux of ultraviolet energy to which the early earth's surface was exposed. Subjected to ultraviolet energy and to other energy sources as well, the gases of the early earth's atmosphere underwent chemical reactions with each other and formed a complex assemblage of molecules (figure 3.4). In the covalent bonds of these molecules, some of the abundant energy present in the atmosphere was captured as chemical energy.

Figure 3.3 Lightning.
Before life evolved, the simple molecules in the earth's atmosphere combined to form more complex molecules. The energy that drove some of these chemical reactions came from lightning and other forms of geothermal energy.

The Miller-Urey Experiment: Re-creation of the Primeval Atmosphere

What kinds of molecules might have been produced in the atmosphere of the primitive earth? One way to answer this question is to re-create early earth conditions: (1) Assemble an atmosphere similar to the one thought to exist on early earth; (2) place this atmosphere over water, which was present on the surface of the cooling earth; (3) exclude gaseous oxygen from the atmosphere because none was present in the earth's early atmosphere; (4) maintain this mixture at a temperature somewhat below 100 degrees Celsius; and (5) bombard it with energy in the form of electrical sparks. When Harold C. Urey and his student Stanley L. Miller performed this experiment in 1953, they found that, within one week, 15 percent of the carbon that was originally present as methane gas had been converted into more complex carbon-based molecules (see Key Experiment 3.1).

Among the first substances produced in the Miller-Urey experiment were molecules derived from the breakdown of methane, including formaldehyde and hydrogen cyanide. These

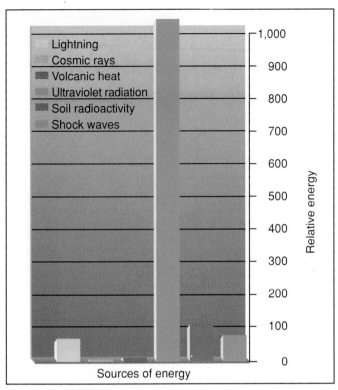

Legend:
- Lightning
- Cosmic rays
- Volcanic heat
- Ultraviolet radiation
- Soil radioactivity
- Shock waves

(y-axis: Relative energy — 0 to 1,000)

Sources of energy

Figure 3.4 Sources of energy for the synthesis of complex molecules in the atmosphere of the primitive earth.
Of the ultraviolet (UV) radiation, only the very short wavelengths (less than 100 nanometers) would have been effective in promoting chemical reactions. These UV rays came from the sun, just as they do today, but in the primitive atmosphere, there was no protective layer of ozone to shield the earth from these intense, shorter wavelengths.

molecules then combined to form more complex molecules containing carbon-carbon bonds, including the amino acids glycine and alanine. Amino acids are the basic building blocks of proteins, which are one of the major kinds of molecules of which organisms are composed. About 50 percent of the dry weight of each cell in your body consists of amino acids—alone or linked together in protein chains.

Later experiments identified more than thirty complex carbon-containing molecules, including the amino acids glycine, alanine, glutamic acid, valine, proline, and aspartic acid. The production of amino acids indicates that proteins could have formed under conditions similar to those that probably existed on the early earth. Other biologically important molecules, including the purine base adenine, which is a constituent of both DNA and RNA, were also formed in these later experiments. Thus, at least some of the key molecules from which life evolved were created in the atmosphere of early earth as a by-product of its birth.

Molecules that form spontaneously under conditions thought to be similar to those of primitive earth include some of those molecules that form the building blocks of organisms.

KEY EXPERIMENT 3.1

2000

1953

1900

THE MILLER-UREY EXPERIMENT

The apparatus used by Miller and Urey to re-create the primitive atmosphere consisted of a closed tube connecting two chambers. The upper chamber contained a mixture of gases thought to resemble the earth's primitive atmosphere. Any complex molecules formed in the atmosphere chamber dissolved in droplets and were carried to the lower "ocean" chamber, from which samples were withdrawn for analysis. The photo shows Dr. Urey posing with his famous apparatus in 1991.

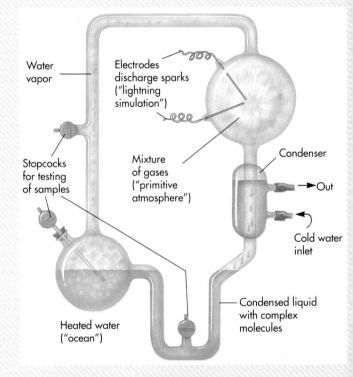

Labels:
- Water vapor
- Electrodes discharge sparks ("lightning simulation")
- Stopcocks for testing of samples
- Mixture of gases ("primitive atmosphere")
- Condenser
- Out
- Cold water inlet
- Condensed liquid with complex molecules
- Heated water ("ocean")

As the earth cooled, much of the water vapor present in the atmosphere condensed into liquid water and accumulated in the oceans. The Miller-Urey experiments indicate that the water droplets carried nucleotides, amino acids, and other compounds produced by chemical reactions in the atmosphere.

The earth's primitive oceans were dilute, hot, smelly soups of ammonia, formaldehyde, formic acid, cyanide, methane, hydrogen sulfide, and organic hydrocarbons. Yet, within such oceans arose the organisms from which all later life-forms, including ourselves, were derived. How did organisms evolve from complex molecules? What is the origin of life? Sidelight 3.1 suggests alternative theories to the one presented here.

Origin of the First Cells

Many different kinds of molecules gather together or aggregate in water, much as foreigners tend to aggregate with others from their country when living in a large city. Sometimes, the aggregation of molecules of one kind forms a cluster big enough to see. If you shake up a bottle of oil-and-vinegar salad dressing, you can see this happen: Small, spherical bubbles of oil appear, grow in size, and fuse with one another. Small **coacervates** of this sort—spherical aggregations only 1 to 2 micrometers in diameter—form spontaneously from lipid molecules suspended in water (figure 3.5). Similar coacervates that formed in the primeval ocean "soups" may have been the first step in the evolution of cellular organization. Coacervates have several remarkably cell-like properties:

1. Coacervates form an outer boundary that has two layers (a bilayer membrane) and thus resembles a biological membrane, as discussed in chapter 4.

2. Coacervates grow by accumulating more subunit molecules from the surrounding medium.

3. Coacervates form budlike projections and divide by pinching in two, as do bacteria.

4. Coacervates may contain amino acids and use them to facilitate several kinds of chemical reactions that are mainly found in living cells.

A process of chemical evolution involving coacervate microdrops of this sort may have taken place before the origin of life. The early oceans must have contained untold numbers of these microdrops—billions in a spoonful—each one forming spontaneously, persisting for a while, and then dispersing. Some of the droplets would by chance have contained amino acids with side groups that were better able than others to catalyze growth-promoting reactions. These droplets would have survived longer than the others because the persistence of both protein and lipid coacervates is greatly increased when they carry out metabolic reactions, such as glucose degradation (breakdown), and when they are actively growing.

Over millions of years, those complex microdrops that were better able to incorporate molecules and energy from the lifeless oceans of early earth would have tended to persist more than the others. Also favored would have been those

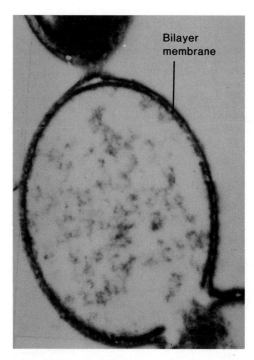

Bilayer membrane

Figure 3.5 Coacervate droplet.
This hollow microsphere possesses many of the characteristics of living cells. If you look carefully, you can see that it is bounded by a bilayer membrane.

microdrops that could incorporate molecules to expand in size, growing large enough to divide into daughter microdrops with features similar to those of the parent microdrop. The daughter microdrops would have been able to use the same favorable combination of characteristics as their parent, and to grow and divide as well. When a means occurred to facilitate this transfer of new ability from parent to offspring, heredity—and life—began.

There is considerable discussion among biologists as to how the first cells may have evolved. The discovery made in the 1980s that RNA can act like an enzyme to assemble new RNA molecules on an RNA template raised the interesting possibility that coacervates may not have been the first step in the evolution of life. Nucleotides were also produced in the Miller-Urey experiments. Perhaps the first macromolecules were RNA molecules, and the initial steps on the evolutionary journey were ones leading to more complex and stable RNA molecules. Later, stability might have been improved by surrounding the RNA within a coacervate.

Still other scientists reject the notion of an "RNA world" entirely, pointing out that some RNA components are too complex to have been present on the primitive earth. For instance, ribose, the sugar that makes up the backbone of RNA, cannot be synthesized without creating other sugars that would stop RNA synthesis. "Anti-RNA" scientists are trying to come up with alternative replicating compounds, but so far, none of their experiments have yielded convincing results.

Coacervates may have been the basis for the first cells.

Sidelight 3.1
ALTERNATIVE THEORIES ABOUT THE ORIGIN OF LIFE

The scenario presented in this chapter about how life originated on earth is not the only one accepted by scientists. Scientists are currently investigating literally dozens of other theories about how life began, from the plausible to the outlandish. Many of these theories are based on the hypothesis that Drs. Miller and Urey were wrong in one way or another about the components of the primeval atmosphere. Some scientists proposed that instead of a reducing atmosphere rich in hydrogen, the primitive atmosphere was awash with carbon dioxide and nitrogen spewed by volcanoes. These scientists suggest that the intense ultraviolet radiation striking the earth destroyed any hydrogen and that the primeval earth was constantly bombarded by meteors, which generated intense amounts of heat. Such an atmosphere, these scientists claim, would hardly be conducive to the delicate synthesis of amino acids and other organic molecules. If not in the "primordial soup" of the primitive oceans, where then do these scientists propose that life began? Here are some alternative theories:

"GIVING VENT" TO ORGANIC MOLECULES

In the late 1970s, scientists studying the ocean depths off the Galápagos Islands made a remarkable discovery. Deep in the ocean floor were *hydrothermal vents,* cracks in the ocean floor that spewed heated water into the cold depths of the ocean (Sidelight figure 3.1). Because the ocean depths are cold and dark, few organisms are usually found there. But the hydrothermal vents hosted a remarkable variety of organisms, from bacteria to clams to exotic-looking tube worms. Some scientists have proposed that these hydrothermal vents could have also hosted the origin of life. They point out that archaebacteria favor hot environments, and hydrothermal vents can certainly be hot enough (some vents spew water up to 300 degrees Celsius!). These vents, nestled as they are in the ocean floor, would also have been protected from meteors. Research for this scenario is ongoing, and quite a number of scientists ascribe to the vent theory as a plausible explanation for life's beginnings.

PYRITE SITES AND CLAY CREATION

An alternative theory points to pyrite, or fool's gold, as the site of the origin of life. This theory is based on the facts that certain solids are made of crystals and that new crystals are formed by a process of replication that mimics cell replication. The first cell could have been a grain of pyrite enclosed by a membrane of organic molecules. Theoretically, when the pyrite grain replicated, the cell also replicated and become encapsulated in its own membrane. This cell could have eventually broken free from the pyrite and "learned" to replicate on its own. While this "crystal replication" theory sounds far-fetched, it forms the basis of a host of similar theories. For instance, some scientists think that, instead of pyrite, clay provided the surface for the origin of life (Sidelight figure 3.2). These scientists suggest that clay naturally attracts organic molecules and that these organic molecules would "go along for the ride" when the clay crystals replicated, eventually starting to replicate on their own.

SPACE INVADERS

Among the more imaginative theories explaining the origin of life is the proposition that life did not begin on earth but was carried to it from outer space. For example, recent meteorites that have crashed into earth have been shown to contain amino acids and other organic molecules. Some researchers theorize that meteors passing very near the primeval earth could have left a trail of organic compounds in their wake, which then fell to earth. Still other scientists think that intact living organisms, such as bacteria, fell to earth from outer space. While many scientists scoff at this theory, others point out that, although outer space has been shown to be devoid of life, microbes can exist in this environment if they become encapsulated in a protective layer of ice.

Whatever the correct scenario, the only thing that scientists know for sure is that the origin of life on earth is a thing of wonder that is currently beyond certain determination.

Sidelight figure 3.1
A hydrothermal vent off the Galápagos Islands. The high temperatures of these vents and their seclusion on the ocean floor make them, some scientists believe, ideal sites for the origin of life.

Sidelight figure 3.2
Clay magnified nearly ten thousand times. Is this the site for the origin of life?

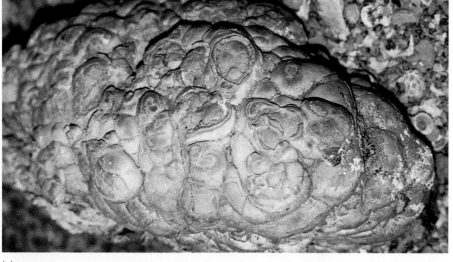

(a)

(b)

(c)

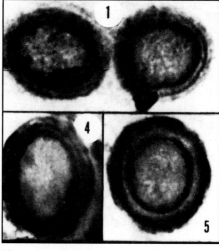

(d)

Figure 3.6 Ancient rocks yield fossil treasures.
These ancient rocks in South Africa (a) contain a fossil bacterium (b) dated at about 3.4 billion years old. Stromatolites, like these in Australia (c), have yielded fossil bacteria about 850 million years old (d), shown in cross section.

The Earliest Cells

Many fossils have been found in ancient rocks and in **stromatolites,** formations of fossilized bacteria (figure 3.6). These fossils show an obvious progression from simple to complex organisms during the vast period of time that began no more than one billion years after the earth's origin. Living things may have been present earlier, but rocks of such great antiquity are rare, and fossils have not yet been found in them.

What is known about these early life-forms? Study of early microfossils indicates that, for most of the history of life, all living organisms resembled living bacteria in their physical characteristics, although some ancient forms cannot be matched exactly. The ancient bacteria were small (1 to 2 micrometers in diameter) and single-celled, lacked external appendages, and had little evidence of internal structure.

Simple organisms with a body plan of this sort are called **prokaryotes** (from the Greek words for "before" and "kernel" [or "nucleus"]). The name reflects their lack of a nucleus, which is a spherical organelle (structure) characteristic of the more complex **eukaryotes** (from the Greek words for "true" and "nucleus") that evolved much later. Collectively, the prokaryotes are known as **bacteria.** The fossil record indicates that eukaryotes did not appear until about 1.5 billion years ago. Therefore, for at least two billion years—nearly half the age of the earth—bacteria were the only organisms that existed.

Living Fossils

Most organisms living today resemble one another fundamentally, having the same kinds of membranes and hereditary systems and many similar aspects of metabolism. However, organisms occasionally found in uncommon environments, such as the oxygenless depths of the Black Sea or the boiling waters of hot springs, often differ in form and metabolism from most other living things. Sheltered from evolutionary alteration in unchanging habitats that resemble those of earlier times, these "living fossils" are the surviving representatives of the first ages of life on the earth.

In those ancient times, biochemical diversity was the rule, and living things did not resemble each other in their metabolic features as closely as they do today. Thus, in some of these preserved ancient environments, scientists can still find bacteria living without oxygen and displaying a bewildering array of metabolic strategies. For example, some of these bacteria use sulfur compounds to fuel their metabolic activities. Some also have shapes similar to those of the fossils of bacteria that lived two or three billion years ago.

Because of their fundamental differences from any other form of bacteria, these ancient living relics have been named **archaebacteria** and have been recently separated from the **eubacteria,** or "true bacteria," to form their own prokaryotic kingdom. (The six kingdoms of life are discussed in more detail in chapter 25.) Methane-producing bacteria and photosynthetic bacteria are two examples of archaebacteria.

Methane-Producing Bacteria

Among archaebacteria that still exist today are methane-producing bacteria. These organisms are typically simple in form and are able to grow only in an oxygen-free environment. For this reason, they are said to grow "without air," or **anaerobically** (from the Greek *an,* meaning "without," *aer,* meaning "air," and *bios,* meaning "life") and are poisoned by oxygen. Methane-producing bacteria convert carbon dioxide (CO_2) and hydrogen gas (H_2) into methane gas (CH_4). They resemble all other bacteria in that they possess hereditary machinery based on DNA, a cell membrane composed of lipid molecules, an exterior cell wall, and a metabolism based on an energy-carrying molecule called ATP. However, the resemblance ends at that point.

The details of membrane and cell wall structure of methane-producing bacteria are different from those of all other bacteria, as are their fundamental biochemical processes of metabolism. Methane-producing bacteria are survivors from an earlier time, when there was considerable variation in the mechanisms of cell wall and membrane synthesis, in the reading of hereditary information, and in energy metabolism.

Photosynthetic Bacteria

Another group of bacteria—photosynthetic eubacteria—can, like plants and algae, capture the energy of light and transform it into the energy of chemical bonds within cells. The pigments used to capture light energy vary in different groups

Figure 3.7 A look toward the past.
The earth probably looked something like this area in Yellowstone National Park as life began. The brownish streaks are masses of bacteria, indicating that the scene closely resembles conditions that occurred billions of years ago, soon after the origin of life.

of photosynthetic bacteria. When these bacteria are massed, they often color the earth, water, or other areas where they grow with characteristic hues (figure 3.7).

One group of photosynthetic bacteria that is very important in the history of life on earth is the **cyanobacteria,** formerly called "blue-green algae." Cyanobacteria have the same kind of chlorophyll pigment that is most abundant in plants and algae, plus other pigments that are blue or red. They produce oxygen as a result of their photosynthetic activities. When cyanobacteria appeared at least three billion years ago, they played the decisive role in increasing the concentration of free oxygen in the earth's atmosphere from below 1 percent to the current level of 21 percent (figure 3.8). As oxygen concentration increased, so also did the amount of ozone in the atmosphere's upper layers, thus affording protection from most of the sun's ultraviolet radiation—radiation that is highly destructive to proteins and nucleic acids.

The Evolution of Bacteria

The early stages of the history of life on earth appear to have been rife with evolutionary metabolic experimentation. Novelty abounded, and many biochemical possibilities were apparently represented among the organisms alive at that time. From the array of different early living forms, a very few became the ancestors of the great majority of organisms alive today. Several of the other "evolutionary experiments," such as the methane-producing bacteria, have survived locally or in unusual habitats, but others probably became extinct millions or even billions of years ago.

Most organisms now living are descendants of a few lines of early bacteria. Many other diverse forms of bacteria did not survive.

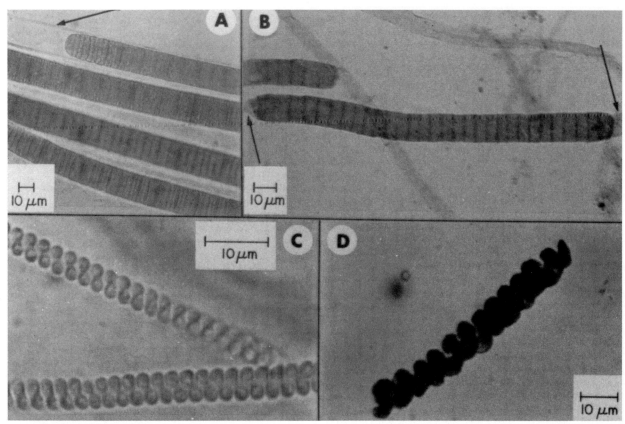

Figure 3.8 Cyanobacteria and the advent of oxygen.
Prior to the appearance of cyanobacteria, the earth's atmosphere did not contain oxygen. With their origin, however, cyanobacteria increased the oxygen concentration of the earth's atmosphere by *employing photosynthesis, which utilizes carbon dioxide and the sun's energy to produce oxygen and carbohydrates. These photos show living cyanobacteria (a, c) and fossil cyanobacteria (b, d).*

Bacteria, for the most part, seem to have stemmed from a tough, simple little cell whose hallmark was adaptability. For at least two billion years, bacteria were the only form of life on earth (table 3.1). All of the eukaryotes, including animals, plants, fungi, and protists, are their descendants.

The Appearance of Eukaryotic Cells

All fossils that are more than 1.5 billion years old are generally similar to one another structurally. They are small, simple cells: Most measure 0.5 to 2 micrometers in diameter, and none is more than about 6 micrometers thick (figure 3.9).

Microfossils that are noticeably different in appearance from the earlier, simpler forms are first found in rocks about 1.5 billion years old. These cells are much larger than bacteria and have internal membranes and thicker walls (figure 3.10). Cells more than 10 micrometers in diameter rapidly increased in abundance. Some fossil cells that are 1.4 billion years old are as much as 60 micrometers in diameter. Others, 1.5 billion years old, contain what appear to be small, membrane-bound structures. Many of these fossils have elaborate shapes, and some exhibit highly branched filaments, icosahedral (nine-sided) configurations, or spines.

Table 3.1   The Geological Time Scale[*]

Fossil Evidence	Millions of Years Ago	Ancient Life-Forms
Oldest multicellular fossils	570–600	Origin of multicellular organisms
Oldest compartmentalized fossil cells	1,500	Appearance of first eukaryotes
		Appearance of aerobic respiration
Disappearance of iron from oceans and formation of iron oxides	2,500	Appearance of oxygen-forming photosynthesis (cyanobacteria)
		Appearance of chemoautotrophs (sulfate respiration)
Oldest definite fossils	3,500	Appearance of life—anaerobic (methane-producing) bacteria and anaerobic (hydrogen sulfide) photosynthesis
Oldest dated rocks	4,500	Formation of the earth

*This time scale is based on the dating of rocks and the fossils within them.

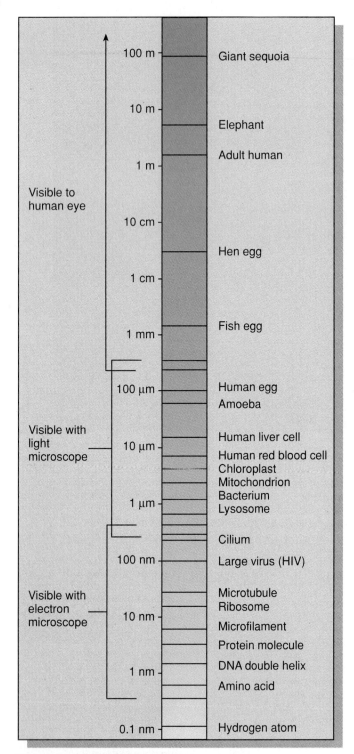

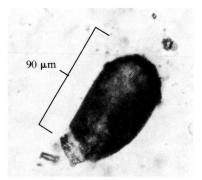

Figure 3.10 Fossil unicellular eukaryote from about 800 million years ago.
All life was unicellular until about the past 700 million years.

Figure 3.9 Scale of object sizes.
To help you visualize the size of ancient organisms, this scale compares the sizes of various objects. Bacteria are generally 1 to 2 micrometers (μm) thick, while human cells are typically much larger. The scale goes from nanometers (nm) to micrometers (μm) to millimeters (mm) to centimeters (cm), and finally, to meters (m).

These early fossil traces mark a major event in the evolution of life: A new kind of organism—eukaryotes—had appeared. As mentioned earlier, eukaryotes possess an internal chamber called the cell nucleus and comprise all organisms other than the bacteria. Eukaryotes evolved to produce all of the diverse multicellular organisms that inhabit the earth today, including humans (figure 3.11). The next chapter explores in detail the structure of eukaryotes in relation to the factors involved in their origin.

> *For at least the first two billion years of life on earth, all organisms were bacteria. The first eukaryotes appeared about 1.5 billion years ago.*

Is There Life on Other Planets?

The life-forms that evolved on the earth closely reflect the nature of this planet and its history. If the earth were farther from the sun, it would be colder, and chemical processes would be greatly slowed down. For example, water would be a solid, and many carbon compounds would be brittle. If the earth were closer to the sun, it would be warmer, chemical bonds would be less stable, and few carbon compounds would persist. Apparently, the evolution of a carbon-based life-form is possible only within the narrow range of temperatures that exists on the earth, and this range of temperatures is directly related to the distance from the sun.

The size of the earth also played an important role in the evolution of life-forms on this planet because the earth's size permitted a gaseous atmosphere. If the earth were smaller, it would not have sufficient gravitational pull to hold an atmosphere. If it were larger, it might hold such a dense atmosphere that all solar radiation would be absorbed before it reached the earth's surface.

Has life evolved on other planets? In the universe, many planets undoubtedly have physical characteristics like those of the earth (figure 3.12). The universe contains some 10^{20} stars with physical characteristics that resemble those of our sun; at least 10 percent of these stars are thought to have planetary systems. If only one in ten thousand planets is the right size and at the right distance from its star to duplicate the conditions in which life originated on the earth, the "life experiment" will have been repeated 10^{15} times (that is, a million billion times). It seems likely that we are not alone.

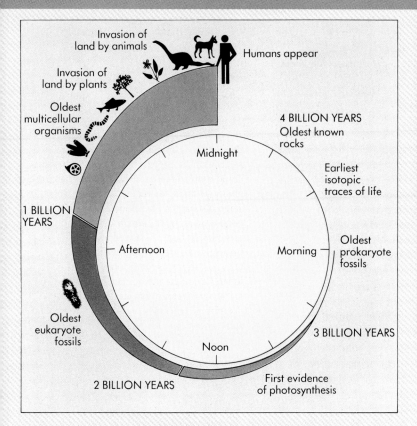

Figure 3.11 A clock of biological time.
A billion seconds ago, it was 1957, and most students using this text had not yet been born. A billion minutes ago, Jesus was alive and walking in Galilee. A billion hours ago, the first human had not yet been born. A billion days ago, no biped walked on earth. A billion months ago, the first dinosaurs had not yet been born. A billion years ago, no creature had ever walked on the face of the earth.

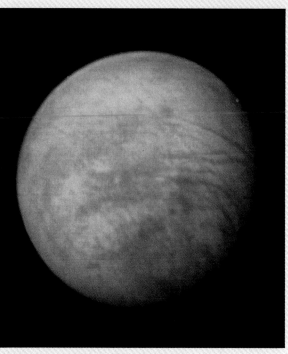

Figure 3.12 Is there life elsewhere?
Currently, the most likely candidate for life within the solar system is Europa, one of the many moons of the large planet Jupiter. Although most of Jupiter's moons resemble the earth's moon—they are pockmarked by meteors and devoid of life—Europa is covered with ice. Perhaps below its frozen surface, the pressures of gravity create enough heat to maintain water in a liquid form. Life may have evolved in such an environment. The conditions on Europa now are far less hostile to life than the conditions that existed in the oceans of the primitive earth.

EVOLUTIONARY VIEWPOINT

While many scenarios, from clay to ocean bubbles, are possible, scientists have no direct evidence of how life first evolved on earth. What *is* clear, however, is that it happened fast: The earth, which formed 4.5 billion years ago, did not have a solid surface until less than four billion years ago—and the oldest fossil bacteria are 3.5 billion years old!

SUMMARY

1. Evolution is a scientifically testable hypothesis concerning the origin of life on earth. The possibility of extraterrestrial origin of life cannot be investigated experimentally because scientists lack the necessary tools. The religious belief that life on earth was specially created rests on faith and also cannot be scientifically tested.

2. The experimental re-creation of atmospheres, energy sources, and temperatures similar to those thought to have existed on primitive earth leads to the spontaneous formation of amino acids and other biologically significant molecules.

3. The first cells are thought to have arisen through a process by which aggregations of molecules (coacervates) that were more stable persisted longer.

4. Microscopic fossils of bacteria (prokaryotes) are found continuously in the fossil record as far back as 3.5 billion years in the oldest rocks suitable for the preservation of organisms.

5. Bacteria are metabolically, but not structurally, diverse. Some distinctive ones, probably of ancient origin, survive in unusual habitats that may resemble those of the early earth.

6. Cyanobacteria, formerly called "blue-green algae," greatly increased the concentration of oxygen in the earth's atmosphere, paving the way for the aerobic organisms on the earth today.

7. Bacteria were the only life-form on earth for two billion years or more. The first eukaryotes appear in the fossil record about 1.5 billion years ago. At first, all eukaryotic organisms were unicellular. All organisms other than bacteria are descendants of these first eukaryotes.

8. Life evolved on the earth because the earth is the right size and distance from the sun. It is likely that the earth is not the only "living" planet in the universe.

REVIEWING THE CHAPTER

1. What are the different possibilities for the origin of life?
2. How did organic molecules originate?
3. What was the makeup of the atmosphere of the early earth?
4. What is the significance of the Miller-Urey experiments?
5. What are some alternative theories for the origin of life?
6. How did the first cells originate?
7. What were the first organisms on the earth?
8. What is a "living fossil"?
9. When did eukaryotic cells appear?
10. Is there life on other planets?

COMPLETING YOUR UNDERSTANDING

1. Of the three possibilities suggested for the origin of life, which one lends itself to scientific investigation?
2. The process by which organic molecules originated is called _____ _____ .
3. What is the "big bang" theory, and how does it relate to the earth?
4. How did the atmosphere of the early earth differ from that of today?
5. What are the suggested hypotheses for the makeup of the early earth's atmosphere?
6. What is a reducing atmosphere, and what type of evolution would occur under such conditions?
7. Where did the oxygen that was added later to the earth's atmosphere originate?
8. What were some of the molecules produced from the Miller-Urey (and subsequent) experiments? From these observations, what is the correlation with life?
9. How does a coacervate relate to the origin of life?
10. How would clay act as a template for the origin of life?
11. If life-forms such as bacteria originated from outside the earth, how would these organisms have survived such a long journey through space to arrive here?
12. Why do some scientists suggest that the formation of RNA molecules may have been an important step in forming the first cells? What are some of the counterarguments to this?

13. What are stromatolites? How old are they? What is their significance?
14. The first eukaryotes appeared on the earth approximately _____ years after the first prokaryotes.
15. In what kind of environment do methane-producing bacteria live?
16. How did cyanobacteria change the history of life on the earth?
17. Which organisms are the oldest life-forms?

    a. Fungi   d. Animals
    b. Protists   e. Plants
    c. Bacteria   f. Viruses

18. According to the fossil record, which event has occurred most recently?

    a. The origin of multicellular organisms
    b. The appearance of methane-producing bacteria
    c. Photosynthesis
    d. Formation of the earth

19. In Fred Hoyle's science fiction novel *The Black Cloud,* the earth is approached by a large interstellar cloud of gas. As the cloud orients around the sun, scientists discover that the cloud is feeding, absorbing the sun's energy through the excitation of electrons in the outer energy levels of cloud molecules, a process similar to the photosynthesis that occurs on the earth. Different portions of the cloud are isolated from each other by associations of ions created by this excitation. Electron currents pass between these sectors, much as they do on the surface of the human brain, and endow the cloud with self-awareness, memory, and the ability to think. Using electricity produced by static discharges, the cloud is able to communicate with human beings and to describe its history, as well as to maintain a protective barrier around itself. The cloud tells our scientists that it once was smaller, having originated as a small extrusion from an ancestral cloud, but has grown by absorbing molecules and energy from stars such as our sun, on which it has been grazing. Eventually, the cloud moves off in search of other stars. Is the cloud alive? Which of its features would you consider important in deciding whether or not the cloud is alive?

20. The nearest galaxy to ours is the spiral galaxy Andromeda. It contains millions of stars, many of which resemble our sun. The universe contains more than a billion galaxies. Each galaxy, like Andromeda, contains countless thousands of stars. It is interesting to speculate: On planets orbiting these stars, are there students speculating on *our* existence? If one in ten of the stars that are like our sun has planets, if one in ten thousand of these planets is capable of supporting life, and if one in each million life-supporting planets evolves an intelligent life-form, how many planets in the universe support intelligent life? Can you think of any objections to this estimate?

FOR FURTHER READING

De Duve, C. 1992. *Blue print for a cell—The nature and origin of life.* Portland, Oreg.: Patterson Press. Presentation of many of the arguments being contested by those interested in how cells evolved. The author is best known for discovering the true nature of lysosomes within eukaryotic cells.

Gray, J., and W. Shear. 1992. Early life on land. *American Scientist,* Sept., 444–56. Discusses how minute fossils suggest that life may have invaded the land millions of years earlier than previously thought.

Horgan, J. 1991. In the beginning . . . .*Scientific American,* Feb., 116–25. A readable, engaging article that summarizes the myriad theories of a variety of scientists about the origin of life. Some theories are plausible (the "RNA world"), while others are simply harebrained (life, in the form of fully formed organisms, was dropped on earth from outer space).

Margulis, L., and D. Sagan. 1986. *Microcosmos: Four billion years of evolution from our microbial ancestors.* New York: Summit Books, Simon and Schuster. A beautifully written, highly recommended essay in which this mother-son team outlines the evolution of life on earth, showing how all the features seen today are derived from the early evolution of bacteria.

Tennesen, M. 1989. Mars. Remembrance of life past. *Discover,* July, 88. A lively discussion of the way life may have originated, and then disappeared, on another planet of our solar system.

York, D. 1993. The earliest history of the earth. *Scientific American,* Jan., 90–96. A discussion of how increasingly sophisticated radioactive dating techniques are providing a clearer look at the earth's first billion and a half years.

CELLS

There are trillions of cells in the human body, representing an astonishing range of structure and function. These red blood cells in the circulatory system deliver oxygen to tissues.

FOR REVIEW

Here are some important terms and concepts that have been discussed in previous chapters and that you will encounter again in this chapter. Review them before proceeding if necessary.

Proteins (*chapter 2*)
Lipids (*chapter 2*)
Distinction between prokaryotic and eukaryotic cells (*chapter 3*)
Evolution of eukaryotes (*chapter 3*)

Hold up your hand and look at it closely. What do you see? An irregular surface of skin is all our eyes can discern. But if you could look more closely, with eyes that worked like microscopes, you would see that the skin of your hand is actually a mosaic of tiny, flat "paving stones," fitted tightly together like tiles on a bathroom floor. Biologists call these tiny elements **cells.** All organisms are composed of cells. Some are composed of a single cell, and some, such as us, are composed of many cells. The gossamer wing of a butterfly is a thin sheet of cells, and so is the glistening layer covering your eyes. The hamburger you eat is composed of cells, whose contents will soon become part of your cells. Orange juice, the wood in your pencil, and your eyelashes and fingernails—all were produced by or consist of cells. Cells are so much a part of life as we know it that we cannot imagine an organism that is not cellular in nature. This chapter looks more closely at cells and their internal structure. The chapters that follow focus on cells in action—how they communicate with their environment, grow, and reproduce.

Overview of Cell Structure

An overview of what a typical cell is like is useful before examining cell structure in detail. There are two major kinds of cells: prokaryotes and eukaryotes. Prokaryotes, or bacterial cells, are enclosed by an outer cell wall that surrounds a **plasma membrane,** a lipid bilayer (double layer) with embedded proteins that controls the permeability of the cell to water and dissolved substances. Within, prokaryotes are relatively uniform in appearance. They do not have **organelles** (specialized parts found in eukaryotic cells) or a defined nucleus. Their **flagella,** if present, are uniform, threadlike, protein structures that move prokaryotes by rotating. Eukaryotic cells are far more complex. Eukaryotes contain various organelles that perform different functions. Within every eukaryotic cell is a **nucleus,** which is its control center. The power that drives a eukaryotic cell comes from internal bacteria-like inclusions called **mitochondria.** A eukaryotic cell is further subdivided into separate compartments by a winding membrane system called the **endoplasmic reticulum.** The flagella that propel motile eukaryotic cells are much more complex than those of prokaryotes and are not related to prokaryotic flagella in an evolutionary sense. They propel the cell through its medium by undulating rapidly.

Despite their structural differences, prokaryotic and eukaryotic cells share four basic characteristics:

1. *Plasma membrane* A plasma membrane surrounds every cell, isolating it from the outside. Chapter 5 describes the many different kinds of passageways and communication channels that span these membranes and provide the only connection between the cell and the outside world.

2. *Nuclear region* The nuclear region directs cell activities. In prokaryotes (bacteria), most of the genetic material is included in a closed, circular molecule of DNA that resides in the **nucleoid region,** a portion of the cell not bounded by membranes. In eukaryotes, by contrast, a double membrane— the **nuclear envelope**—surrounds the nucleus, which contains the DNA.

3. *Cytoplasm* A semifluid matrix called the **cytoplasm** fills prokaryotic cells; in eukaryotic cells, the cytoplasm occupies the space between the nuclear region and the cell membrane. The cytoplasm contains the sugars, amino acids, and proteins with which the cell carries out its everyday activities of growth and reproduction. The cytoplasm of a eukaryotic cell also contains numerous organized structures called organelles, or "little organs." Many of these organelles are created by the membranes of the endoplasmic reticulum, which close off compartments within which different activities occur. The cytoplasm of eukaryotic cells also contains organelles called mitochondria, which provide power.

4. *Ribosomes* Both eukaryotic and prokaryotic cells contain **ribosomes,** which function in the manufacture of proteins. In eukaryotes, ribosomes are attached to internal membranes and are also found in the cytoplasm. In prokaryotes, ribosomes are not attached to any structure and are found dispersed throughout the cytoplasm.

All cells share this architecture (figure 4.1). However, the general plan is modified in various ways in different classes of cells. For example, the cells of some kinds of organisms— plants, bacteria, fungi, protists—possess a rigid outer cell wall that provides structural strength; animal cells, and those of a few other kinds of organisms, do not. Eukaryotic cells generally possess a single nucleus, whereas the cells of fungi, some other groups of organisms, and particular kinds of tissues have several nuclei. Most cells derive all their power from the kind of organelles called mitochondria. But plant and algal (**algae** are photosynthetic protists) cells contain a second kind of powerhouse—**chloroplasts** (organelles that carry out photosynthesis)—in addition to their mitochondria. Despite their differences, all cells are fundamentally similar (figure 4.2).

A cell is a membrane-bound unit containing hereditary machinery and other components. Because of these components, cells are able to metabolize, grow, and reproduce.

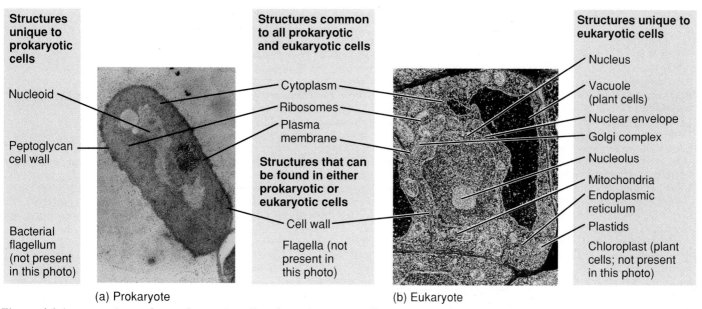

Figure 4.1 A comparison of a prokaryotic cell and a eukaryotic cell.
Both (a) *prokaryotic* (Bacillus megaterium) *and* (b) *eukaryotic cells share four characteristics: a plasma membrane, a nuclear region, cytoplasm, and ribosomes. Prokaryotic cells, however, lack the membrane-bound organelles found in eukaryotic cells.*

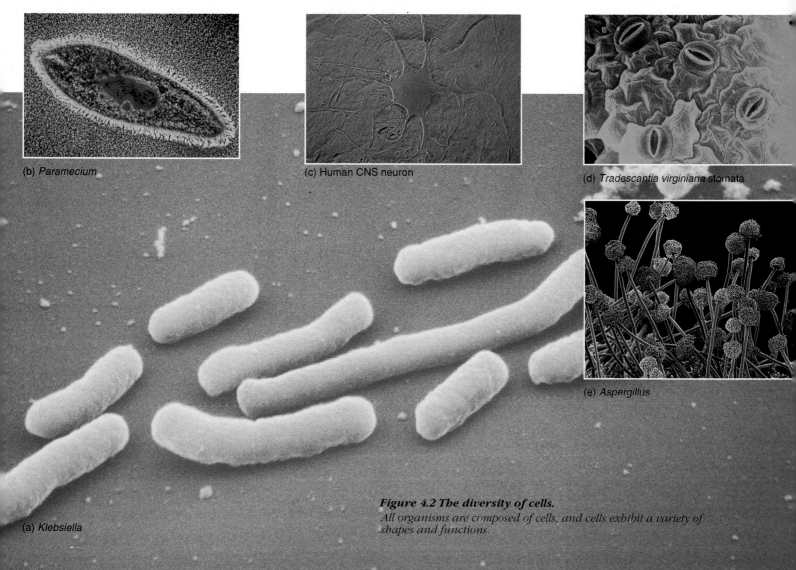

Figure 4.2 The diversity of cells.
All organisms are composed of cells, and cells exhibit a variety of shapes and functions.

Sidelight 4.1
HOW DOES A MICROSCOPE MAGNIFY?

Most cells are so small that you cannot see them with the naked eye. Most eukaryotic cells are between 10 and 30 micrometers in diameter. You cannot see such small objects because, when two objects are closer together than about 100 micrometers, the two light beams reflected from these objects fall on the same "detector" cell at the rear of the eye. Only when two dots are farther apart than 100 micrometers do the beams fall on different cells, and only then can your eye tell that there are two objects and not one.

Robert Hooke and Antonie van Leeuwenhoek used simple microscopes to magnify cells so that the cells appeared larger than the 100-micrometer limit imposed by the structure of the human eye. The microscopes magnified images of cells by bending light through a glass lens. Sidelight figure 4.1 shows how a single-lens microscope is able to magnify an image. The size of the image that falls on the screen of detector cells lining the back of your eye depends on how close the object is to your eye—the closer the object, the bigger the picture. However, because it is limited by the size and thickness of its lens, your eye is unable to comfortably focus on an object closer than about 25 centimeters. Hooke and van Leeuwenhoek helped the eye out

by interposing a glass lens between the object and the eye. The glass lens added additional focusing power, producing an image of the close-up object on the back of the eye. But because the object is closer, the image on the back of the eye is bigger than it would have been had the object been 25 centimeters away from the eye. It is as big as a *much larger* object placed 25 centimeters away would have appeared without the lens. You perceive the object as magnified, or bigger.

Van Leeuwenhoek's microscope consisted of (1) a plate with a single lens, (2) a mounting pin that held the specimen to be observed, (3) a focusing screw that moved the specimen nearer or farther from the eye, and (4) a

specimen-centering screw. Although simple in construction, this microscope is very powerful (Sidelight figure 4.2a). One of van Leeuwenhoek's original images—of a thin slice of cork—was recently discovered among his papers. In this image obtained with van Leeuwenhoek's own microscope, the magnification is 266 times, as good as many modern microscopes. The finest structures visible are less than 1 micrometer (1,000 nanometers) in diameter (Sidelight figure 4.2b).

Modern light microscopes use two magnifying lenses (and various correcting lenses) that act like back-to-back eyes. The first lens focuses the image of the object on the second lens. The second lens magnifies the image again and

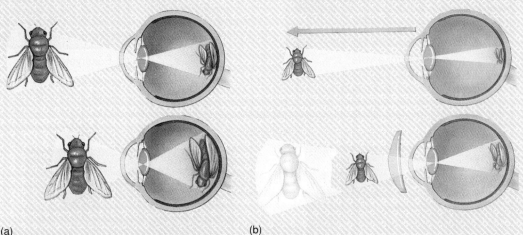

(a)   (b)

Sidelight figure 4.1 How does a microscope magnify?
(a) *The closer an object is to the eye, the larger the image that falls on the back of that eye.* (b) *The eye does not comfortably focus on an object closer than about 25 centimeters because the lens of the eye must change shape to focus and cannot exceed this limit. The glass lens aids the eye in focusing the close object. Because the object is closer, it produces a much larger image on the back of the eye, and so the image appears "larger."*

Cell Size

One of cells' most striking traits is their very small size. Cells are not like shoe boxes, big and easy to study. Instead, they are so small that you cannot see a single one of your body's cells with the naked eye. Your body contains about one hundred trillion cells. If each cell was the size of a shoe box and they were lined up end to end, the line would extend to Mars and back, over 500 million kilometers.

The Cell Theory

Because cells are so small, they were not observed until microscopes were invented in the mid-seventeenth century (see Sidelight 4.1). In 1665, Robert Hooke was the first to describe cells. Using a microscope he had built to examine a thin slice of cork, Hooke observed a honeycomb of tiny, empty compartments, which he called *cellulae* (from the Latin word for "a small room"). The first living cells were observed by Dutch naturalist Antonie van Leeuwenhoek a few years later. Van Leeuwenhoek called the tiny organisms that he observed *animalcules* ("little

animals"). For another century and a half, however, biologists did not appreciate cells' general importance. In 1838, the German Matthias Schleiden, after a careful study of plant tissues, made the first statement of what is now known as the cell theory. Schleiden stated that all plants "are aggregates of fully individualized, independent, separate beings, namely the cells themselves." The following year, Theodor Schwann reported that all animal tissues are also composed of individual cells.

In its modern form, the cell theory includes three principles:

1. All organisms are composed of one or more cells.

2. Cells are the smallest living things, the basic unit of organization of all organisms, within which the life processes of metabolism and heredity occur.

3. All cells arise from previously existing cells.

All the organisms on earth are cells or aggregates of cells, and all of us are descendants from the first cells.

focuses it on the back of the eye. Microscopes that magnify in stages by using several lenses are called *compound light microscopes*. The finest structures visible with modern compound light microscopes are about 200 nanometers in thickness. A contemporary light micrograph is shown in Sidelight figure 4.3*a*.

Compound light microscopes are not powerful enough to resolve many structures within cells. A membrane, for example, is only 5 nanometers thick. Why not just add another magnifying lens to the microscope and so increase the microscope's *resolving power*, the ability to distinguish two lines as separate? This approach does not work because, when two objects are closer than a few hundred nanometers, the light beams reflected from the two objects start to overlap. A light beam vibrates like a plucked string, and the only way two beams can get closer together and still be resolved is if the "wavelength" is shorter.

The wavelength of light ranges from about 0.4 micrometer for violet light to about 0.7 micrometer for red light. This limits the best light microscopes to a resolving power of 0.2 micrometer; they are thus able to improve on the naked eye about five hundred times. They are able to distinguish the structures of eukaryotic cells and individual prokaryotic cells, but they are not able to visualize the internal structure of prokaryotic cells or of bacteria.

One way to achieve greater magnifications is by using a beam of electrons rather than a light beam. Elec-trons have a much shorter wavelength, and a microscope employing electron beams has four hundred times the resolving power of a light microscope. *Transmission electron microscopes* today are capable of resolving objects only 0.2 nanometer apart—just five times the diameter of a hydrogen atom (Sidelight figure 4.3*b*).

Scanning electron microscopes beam electrons onto the surface of the specimen as a fine probe, which passes back and forth rapidly. The electrons reflected back from the specimen surface, together with other electrons that the specimen itself emits as a result of the bombardment, are amplified and transmitted to a television screen, where the image can be viewed and photographed. Scanning electron microscopy yields striking three-dimensional images and has proven to be very useful in understanding many biological and physical phenomena (Sidelight figure 4.3*c*).

(a)   (b)

Sidelight figure 4.2 An early microscope.

(a) Antonie van Leeuwenhoek's microscope. (b) An image of Leeuwenhoek's sample of cork, obtained with his microscope, which is preserved at Utrecht, in the Netherlands. In this image, the cork is magnified 266 times; it compares well with modern images of thin sections of cork. The finest structures that are visible are less than 1 micrometer across.

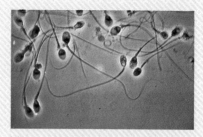

(a)

(b)

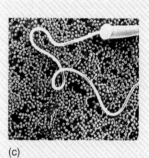

(c)

***Sidelight figure 4.3
Modern microscopy.***

*(a) Image of sperm cells taken with a light microscope (×400).
(b) Transmission electron micrograph of a sperm cell (×15,000).
(c) Scanning electron micrograph of a sperm cell (×8,500).*

Advantages of Small Cell Size

Cells are not all the same size. Individual cells of the marine alga *Acetabularia*, for example, are up to 5 centimeters long (figure 4.3). In contrast, the cells of your body are typically from 5 to 20 micrometers in diameter. If a typical cell in your body were the size of a shoe box, an *Acetabularia* cell to the same scale would be about 2 kilometers high! The cells of bacteria are much smaller, usually about 1 to 10 micrometers thick.

Each cell must maintain centralized control to function efficiently. The nucleus must send commands to all parts of the cell via molecules that direct the synthesis of certain enzymes, the entry of ions from the exterior, and the assembly of organelles. These molecules must pass by diffusion, a relatively slow process of random molecular movement, from the nucleus to all parts of the cell, and reaching the periphery of a large cell takes a long time. For this reason, an organism made up of relatively small cells has an advantage over one composed of larger cells.

Figure 4.3 Some cells are very large.

Acetabularia, a marine protist, is a relatively large single-celled organism. Each individual cell can be up to 5 centimeters in length!

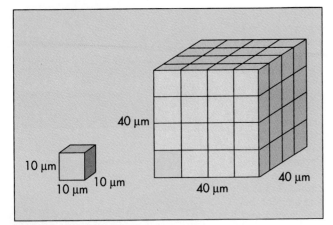

Figure 4.4 Surface-to-volume ratio.
Since cells depend on their surface area to interact with their environment, large organisms are divided into smaller cells to provide the maximum surface area. The smaller cube in the illustration symbolizes a single cell, the larger cube a multi-celled organism. When the larger cube is divided into smaller cubes, surface area is maximized.

Another advantage of small cell size involves the cell's **surface-to-volume ratio.** As cell size increases, volume grows much more rapidly than does the cell's surface area (figure 4.4). For a round cell, surface area increases as the square of the diameter, whereas volume increases as the cube. Thus, a cell with 10 times greater diameter would have 10^2 (squared) or 100 times the surface area but 10^3 (cubed) or 1,000 times the volume. A cell's surface provides the cell's only opportunity to interact with the environment, and large cells have far less surface per unit of volume than do small ones. All substances must enter and exit a cell via the plasma membrane. The plasma membrane plays a key role in controlling cell function, a role that it performs more effectively when cells are relatively small. The plasma membrane is discussed in more detail in chapter 5.

> *An organism made up of many small cells, rather than few large ones, has an advantage because small cells can be commanded more efficiently and have a greater opportunity to communicate with their environment.*

The Structure of Prokaryotes

Prokaryotes (bacteria) are the most simply organized cellular organisms. Over twenty-five hundred species are recognized, while many times that number undoubtedly exist but have not yet been described properly (figure 4.5). Although these species are diverse in form, their organization is fundamentally similar: small cells about 1 to 10 micrometers thick; enclosed, like all living cells, by a plasma membrane; and encased within a rigid cell wall, with no distinct interior compartments (figure 4.6). Sometimes, prokaryotic cells adhere in

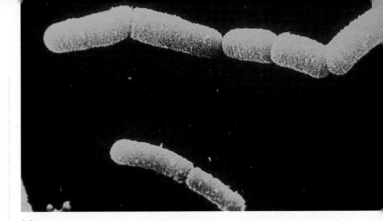

(a)

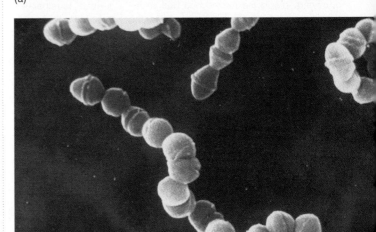

(b)

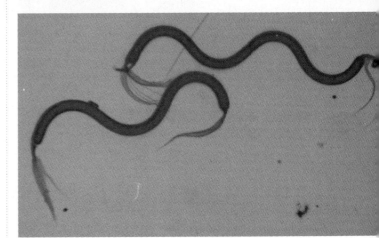

(c)

Figure 4.5 An assortment of prokaryotes.
The cells of Pseudomonas *(a) are cylindrical, with daughter cells often adhering in short clusters.* Streptococcus *(b) has spherical cells that link together in long chains.* Spirilla *(c) has long, twisted shapes with terminal flagella.*

chains or masses, but fundamentally, the individual cells are separate from one another. Some prokaryotes also possess flagella, which they use to propel themselves.

> *Compared with eukaryotic cells, which evolved from them, prokaryotic (bacterial) cells are smaller and have a simpler interior organization.*

Figure 4.6 The structure of a prokaryotic cell.

Not all prokaryotic (bacterial) cells have flagella, but all do have a nucleoid, ribosomes, a plasma membrane, cytoplasm, and a cell wall.

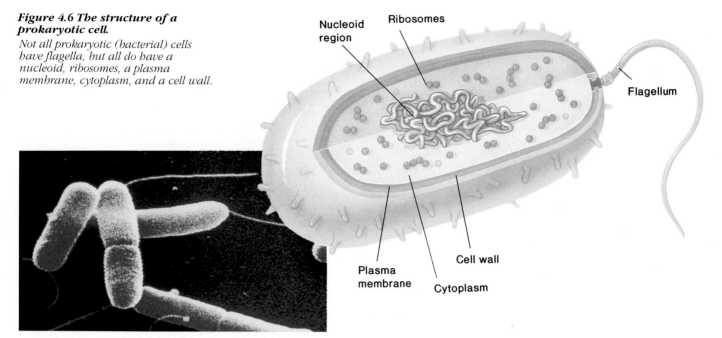

Strong Cell Walls

Prokaryotes (bacteria) are encased by a strong **cell wall,** in which a framework made up of carbohydrates provides the basic structure. These carbohydrates are cross-linked into a rigid structure by short peptide units—chains of amino acids, like segments of proteins. No eukaryote possesses cell walls with this kind of structure.

Simple Interior Organization

Prokaryotic (bacterial) cells have a simple interior organization. There are no internal compartments bounded by membranes and no membrane-bounded organelles—the kinds of distinct structures characteristic of eukaryotic cells. The entire cytoplasm of a prokaryotic cell is one unit, with no internal support structure; thus, the cell's strength comes primarily from its rigid wall.

The plasma membrane of prokaryotic cells often intrudes into the cell's interior, where it may play an important role. In some photosynthetic bacteria, the cell membrane is extensively folded, with the folds extending into the cell's interior (figure 4.7). The pigments connected with photosynthesis are located on these folded membranes.

Scattered throughout the cytoplasm of prokaryotic cells are small structures

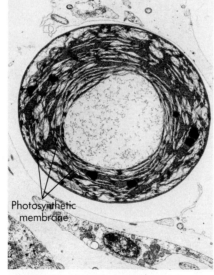

Figure 4.7 Electron micrograph of a photosynthetic bacterial cell, Prochloron.

Extensive folded photosynthetic membranes are visible. The cellular DNA is located in the clear area in the cell's central region.

called **ribosomes,** the sites where proteins are made. Ribosomes are composed of two subunits, and each subunit, in turn, is composed of a special type of RNA called **ribosomal RNA** (abbreviated **rRNA**) and proteins. Ribosomes are not considered organelles because they lack a membrane boundary. The hereditary material of prokaryotic cells—the cells' DNA—is located in a specific region called the **nucleoid region,** which, again, is not a true organelle because of its lack of a membrane boundary.

Because there are no membrane-bounded compartments within a prokaryotic cell, both the DNA and the enzymes within the prokaryote have access to all parts of the cell. Reactions are not compartmentalized, as they are in eukaryotic cells.

Prokaryotes (bacteria) are encased by an exterior cell wall composed of carbohydrates cross-linked by short peptides. They lack a true nucleus and other interior membrane-bound compartments.

The Structure of Eukaryotes

Although eukaryotic cells are diverse in form and function, they share a basic architecture (figures 4.8 and 4.9). All are bounded by a plasma membrane, all contain a supporting matrix of protein called a **cytoskeleton,** and all possess numerous organelles. Before taking a detailed tour of the eukaryotic cell, let us first examine the origin of eukaryotes to see if any structures were derived from prokaryotes.

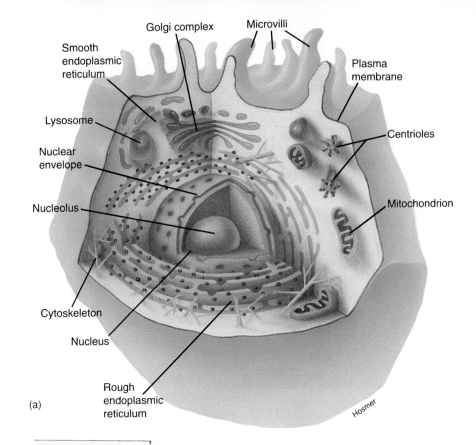

Figure 4.8 The structure of a eukaryotic cell: animal example.
(a) *An idealized animal cell.*
(b) *Micrograph of animal cell with drawings detailing organelles.*

Golgi complex

Microvilli

Smooth endoplasmic reticulum

Plasma membrane

Lysosome

Centrioles

Nuclear envelope

Mitochondrion

Nucleolus

Cytoskeleton

Nucleus

Rough endoplasmic reticulum

Hosmer

(a)

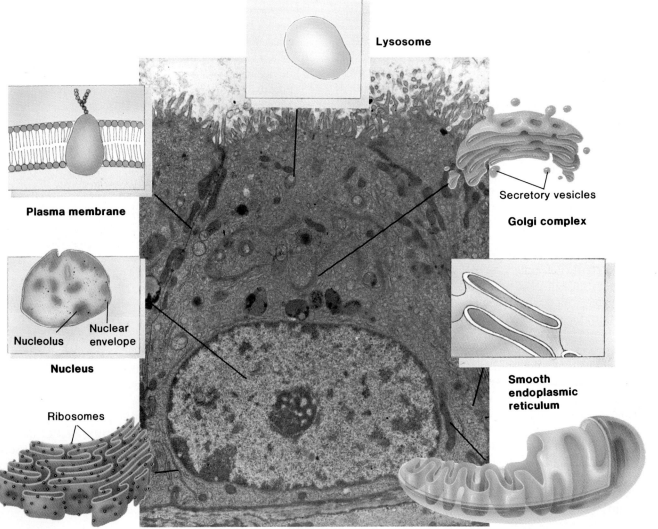

Lysosome

Plasma membrane

Secretory vesicles

Golgi complex

Nucleolus   Nuclear envelope

Nucleus

Smooth endoplasmic reticulum

Ribosomes

Rough endoplasmic reticulum

(b)

Mitochondrion

(a)

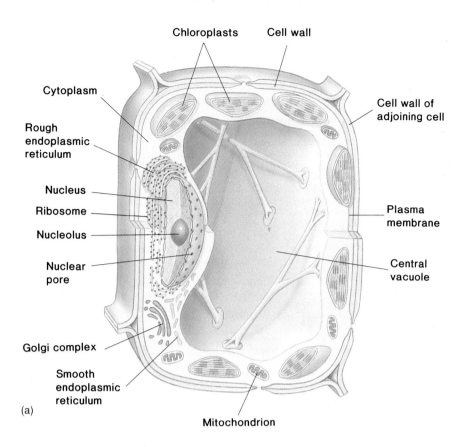

Chloroplasts

Cell wall

Cytoplasm

Rough endoplasmic reticulum

Nucleus

Ribosome

Nucleolus

Nuclear pore

Golgi complex

Smooth endoplasmic reticulum

Mitochondrion

Cell wall of adjoining cell

Plasma membrane

Central vacuole

Figure 4.9 The structure of a eukaryotic cell: plant example.
(a) *A plant cell. The large central vacuole of a plant cell occupies most of the cell's space, providing an expanded surface area.*
(b) *Micrograph of plant cell with drawings detailing organelles.*

(b)

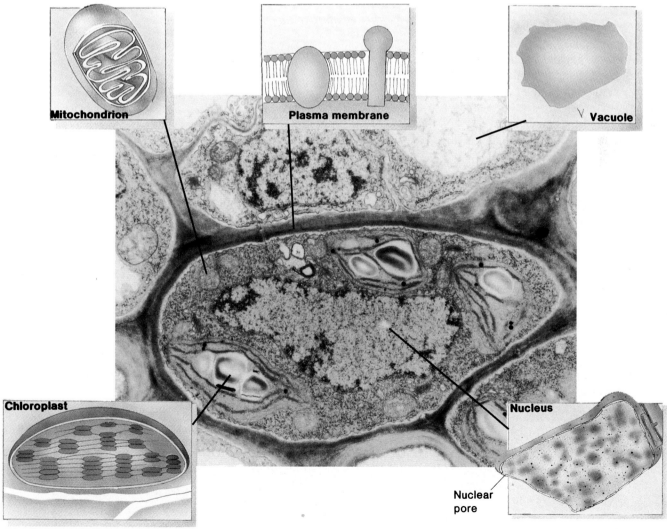

Mitochondrion

Plasma membrane

Vacuole

Chloroplast

Nucleus

Nuclear pore

Endosymbiosis and the Origin of Eukaryotes

The first eukaryotes had a cell structure that was radically different from that of all earlier organisms. Eukaryotic cells are far more complex internally than their prokaryotic (bacterial) ancestors. Little is known about how this increase in internal organization first evolved 1.5 billion years ago—with one exception: Within the cells of virtually all eukaryotes are organelles that resemble intact bacterial cells in both size and appearance. Most biologists interpret this resemblance as evidence that these organelles were once bacteria that eventually became incorporated into eukaryotic cells, providing their host the advantages associated with their special metabolic abilities. Mitochondria (the energy factories of eukaryotic cells), chloroplasts (the organelles in which photosynthesis takes place), and perhaps centrioles (organelles that participate in cell division) were, at one point in time, separate bacterial cells that became incorporated into eukaryotic cells.

Two organisms living together in close association is called **symbiosis.** The incorporation of ancient bacteria into eukaryotes is called **endosymbiosis** (*endo* means "within") to convey the idea of one organism incorporating another through symbiosis (figure 4.10). The ancient bacteria that assumed the role of eukaryotic organelles are called **endosymbionts.** (An endosymbiont is usually the smaller of the two organisms associated in a symbiotic relationship.) The tour of the eukaryotic cell that follows points out the resemblance of some organelles to their bacterial (prokaryotic) endosymbionts.

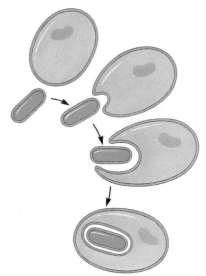

Figure 4.10 A simplified diagram of endosymbiosis.
The smaller green object represents a bacterium, the larger purple object a eukaryotic cell. This particular figure shows how a double membrane, like those seen in mitochondria, could have been created during the endosymbiotic origin of mitochondria.

> The endosymbiotic theory states that ancient bacteria (prokaryotes) became incorporated into eukaryotic cells as organelles. Mitochondria, chloroplasts, and centrioles are all believed to have been, at one time, ancient bacteria.

The Plasma Membrane: A Cell's Boundary

All **plasma membranes** are bilayers about 7 nanometers thick with embedded proteins. Viewed with the electron microscope in cross section, such membranes appear as two dark lines separated by a lighter area (figure 4.11). Each bilayer is composed of a special kind of lipid called a **phospholipid.** Tail-to-tail packing of these phospholipid molecules gives plasma membranes their distinctive appearance.

The major proteins of a membrane are hydrophobic; therefore, they associate with and become embedded in the phospholipid matrix. These embedded proteins selectively control the passage of molecules across the plasma membrane. In other words, they control the cell's interactions with its environment. The major kinds of proteins embedded in the plasma membrane are:

1. *Channels* Channel proteins act as doors that admit specific molecules to the cell. For example, membranes possess specific channels for sodium ions and others for glucose.

2. *Receptors* Receptor proteins transmit information rather than molecules. Receptors induce changes within the cell when they come in contact with particular molecules on the cell surface. For example, some hormones induce changes in cells by binding to such surface receptors.

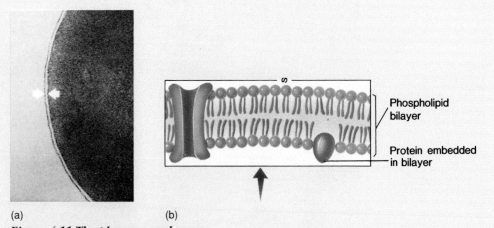

(a)   (b)

Figure 4.11 The plasma membrane.
The plasma membrane of a cell functions as the cell's boundary and controls the cell's interaction with its environment. (a) This electron micrograph of a red blood cell (×200,000) clearly shows the double nature of the plasma membrane (indicated by arrows). (b) Plasma membranes are composed of a phospholipid bilayer, in which proteins are embedded. The proteins act as "doors" or "gates" for the passage of materials into and out of the cell.

3. *Markers* Marker proteins identify a cell as being of a particular type. This is very important in a multicellular individual because cells must be able to recognize one another for tissues to form and function correctly.

Chapter 5 presents a more detailed look at the structure and functions of cell membranes.

The Nucleus: Control Center of the Cell

The largest and most easily seen of the organelles within eukaryotic cells is the **nucleus,** first described by English botanist Robert Brown in 1831 (figure 4.12*a*). The word *nucleus* is derived from the Greek word meaning "nut," an object that nuclei somewhat resemble. Nuclei are usually spherical. In animal cells, they generally occur near the center of the cell, sometimes appearing to be cradled in this position by a network of fine filaments. The nucleus is the repository of the genetic information that directs all the activities of a living cell. It is also where the subunits of ribosomes are made. Some kinds of cells, such as mature red blood cells, lose their nuclei at the final stage of their development. After this loss, the cells can no longer grow, change, and divide, and they become merely vessels for the substances they contain.

The Nuclear Envelope: Getting In and Out

The surface of the nucleus is bounded by *two* membranes, the outer and inner membranes of the **nuclear envelope.** Scattered over the surface of this envelope, like the craters of the moon, are shallow depressions called **nuclear pores** (figure 4.12*b*). These pores, 50 to 80 nanometers apart, form at locations where the two membrane layers of the nuclear envelope pinch together. A nuclear pore is not an empty opening like the hole in a doughnut. It contains many embedded proteins that act as molecular channels, permitting certain molecules to pass into and out of the nucleus. Passage is restricted primarily to two kinds of molecules: (1) proteins moving into the nucleus, where they are incorporated into nuclear structures or catalyze nuclear activities, and (2) ribosomal subunits that are manufactured in the nucleus and then exported to the cytoplasm.

> *The nucleus of a eukaryotic cell contains the cell's hereditary apparatus, isolated from the rest of the cell.*

Complex Chromosomes of Eukaryotes

Both in prokaryotes (bacteria) and eukaryotes, all the hereditary information specifying cell structure and function is encoded in DNA. However, unlike prokaryotic DNA, the DNA of eukaryotes is divided into several segments and associated with protein and RNA, forming **chromosomes.** Association with protein enables eukaryotic DNA to wind up into a highly condensed form during cell division (figure 4.13). Under a light microscope, these condensed chromosomes are readily seen in dividing cells as densely staining rods. After cell division, eukaryotic chromosomes uncoil and can no longer be distinguished individually with a light microscope. Uncoiling the chromosomes into a more extended form permits the enzymes that make RNA copies of DNA to gain access to the DNA molecule. Only by means of these RNA copies can the hereditary information be used to direct protein synthesis.

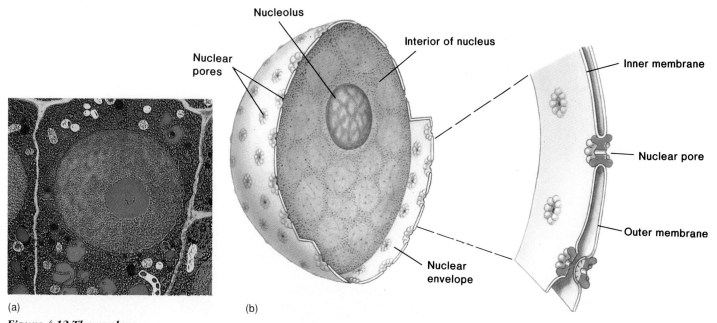

(a)   (b)

Figure 4.12 The nucleus.
The nucleus is the cell's control center. It contains the cell's hereditary information and manufactures the subunits of ribosomes, important structures in the synthesis of proteins. (a) The nucleus is enclosed by two membranes, which together are called the nuclear envelope. (b) The envelope is studded with pores through which materials leave and enter the nucleus. The nucleolus is the site of ribosome subunit synthesis and appears darker than the rest of the nucleus.

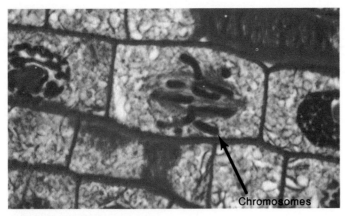

Figure 4.13 Chromosomes.
These chromosomes in an onion root-tip cell were caught in their condensed state directly before cell division.

> *A distinctive feature of eukaryotes is the organization of their DNA into chromosomes. Chromosomes can be condensed into compact structures during eukaryotic cell division. Later, the chromosomes can be unraveled so that their hereditary information can be used.*

Ribosome Subunits: Made in the Nucleolus

One region of the nucleus appears darker than the rest; this darker region is called the **nucleolus** (see figure 4.12). The nucleolus is where the ribosome subunits—the rRNA and proteins associated with ribosomes—are manufactured. These subunits leave the nucleus through the nuclear pores and enter the cytoplasm, where final assembly of the ribosomes takes place.

The Endoplasmic Reticulum: The Internal Membrane System Within the Cell

When viewed with a light microscope, the interiors of eukaryotic cells exhibit a relatively featureless matrix within which various organelles are embedded. But with the advent of electron microscopes, it became strikingly evident that the interior of a eukaryotic cell is packed with membranes, membranes so thin that they are not visible with the relatively low resolving power of light microscopes. These membranes fill the eukaryotic cell, dividing it into compartments, channeling the transport of molecules through the cell's interior, and providing the surfaces on which enzymes act.

The extensive system of internal membranes within the cells of eukaryotic organisms is called the **endoplasmic reticulum (ER)**. The term *endoplasmic* means "within the cytoplasm," and the term *reticulum* comes from a Latin word meaning "a little net." The ER constitutes the most fundamental distinction between eukaryotes and prokaryotes. Like the plasma membrane, the ER is composed of a double layer of lipid, with various enzymes attached to its surface. Weaving in sheets through the cell's interior, the ER creates a series of channels and interconnections between its membranes that isolates some spaces as membrane-enclosed sacs called **vesicles** (figure 4.14).

The surface of the ER is where cell ribosomes manufacture proteins intended for export from the cell, such as enzymes. These new proteins are then passed out across the ER membrane into the vesicle-forming system called the Golgi complex (discussed in the next section). They travel within vesicles to the cell's inner surface, where they are released to the outside of the cell in which they were produced. From the time a protein is first synthesized on the ER-bound ribosome and crosses into these channels, it is, in a sense, already located outside the cell.

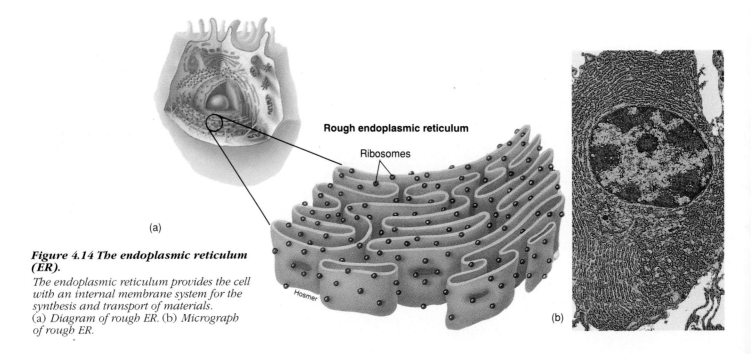

Rough endoplasmic reticulum

Ribosomes

Hosmer

(a)

(b)

Figure 4.14 The endoplasmic reticulum (ER).
The endoplasmic reticulum provides the cell with an internal membrane system for the synthesis and transport of materials.
(a) Diagram of rough ER. (b) Micrograph of rough ER.

The surface of those regions of the ER devoted to the synthesis of such transported proteins are heavily studded with ribosomes and appear pebbly, like the surface of sandpaper, when seen through an electron microscope. For this reason, these regions are called **rough ER.** Regions in which ER-bound ribosomes are relatively scarce are correspondingly called **smooth ER.**

Many of the cell's enzymes cannot function when they are free in the cytoplasm; they are active only when associated with a membrane. Smooth ER membranes have many different enzymes embedded within them, and the enzymes carry out their functions from these positions. Enzymes anchored within the smooth ER catalyze the synthesis of a variety of carbohydrates and lipids. In cells that carry out extensive lipid synthesis, such as the cells of the testicles, smooth ER is particularly abundant. Intestinal cells, which synthesize triglycerides, and brain cells are also rich in smooth ER. In the liver, enzymes embedded within the smooth ER are involved in the detoxification of such drugs as amphetamines, morphine, codeine, and phenobarbital.

> *The endoplasmic reticulum (ER) is an extensive system of membranes that divides the interior of eukaryotic cells into compartments and channels. Rough ER functions mainly in the synthesis and transport of proteins across the ER membrane, whereas smooth ER organizes the synthesis of lipids and other molecules.*

The Golgi Complex: The Delivery System of the Cell

At various locations in the cytoplasm of eukaryotic cells are flattened stacks of membranes called **Golgi bodies,** named for Camillo Golgi, the nineteenth-century Italian physician who first called attention to them. Animal cells contain ten to twenty Golgi bodies each (they are especially abundant in glandular cells, which manufacture the substances that Golgi bodies secrete), whereas plant cells may contain several hundred. Collectively, the Golgi bodies are referred to as the **Golgi complex** (figure 4.15).

Golgi bodies function in the collection, packaging, and distribution of molecules synthesized in the cell. The proteins and lipids that are manufactured on the rough and smooth ER membranes and that are destined for use outside the cell are transported through the channels of the ER, or as **transport vesicles** budded off from it, into the Golgi bodies. Within the Golgi bodies, many of these molecules are bound to polysaccharides to form compound molecules, such as **glycoproteins** (consisting of a polysaccharide bound to a protein) and **glycolipids** (consisting of a polysaccharide bound to a lipid). The newly formed glycoproteins and glycolipids collect at the ends of the membranous folds of the Golgi bodies; these folds are given the special name **cisternae** (from the Latin word meaning "collecting vessels"). **Secretory vesicles,** filled with either glycolipids or glycoproteins, pinch off from the cisternae and are carried to the cell membrane. The membrane of the secretory vesicle fuses to the cell membrane, and the contents of the vesicle are released outside the cell (figure 4.16).

> *The Golgi complex is the delivery system of the eukaryotic cell. It collects, modifies, packages, and distributes molecules that are synthesized at one location within the cell and used at another.*

Other Vesicles Within the Eukaryotic Cell

Peroxisomes are membrane-bound spherical bodies, 0.2 to 0.5 micrometers in diameter, that are present in almost all eukaryotic cells. Apparently derived from the smooth ER, peroxisomes

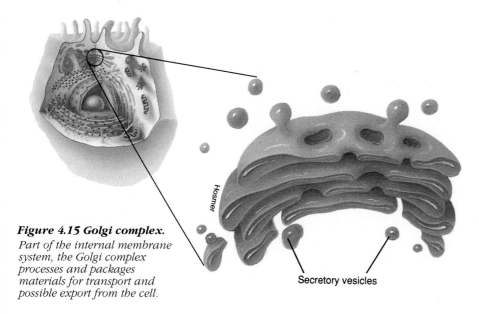

Figure 4.15 Golgi complex.
Part of the internal membrane system, the Golgi complex processes and packages materials for transport and possible export from the cell.

Secretory vesicles

Golgi complex

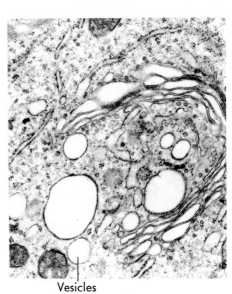

Vesicles

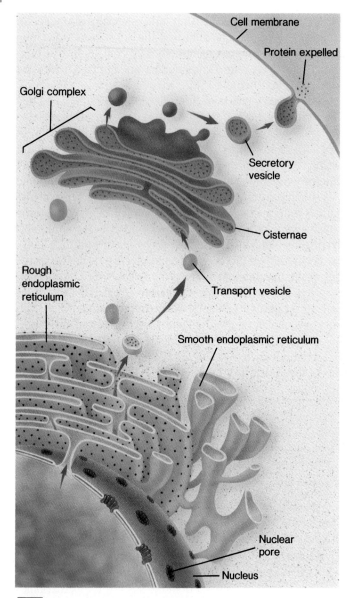

Cell membrane

Protein expelled

Golgi complex

Secretory vesicle

Cisternae

Rough endoplasmic reticulum

Transport vesicle

Smooth endoplasmic reticulum

Nuclear pore

Nucleus

Figure 4.16 How the internal membrane system of a cell packages a protein for export.

The instructions for making a protein that is destined for export from a cell, such as a digestive enzyme that is made by a pancreas cell, are first transcribed from DNA by RNA in the nucleus. The RNA then leaves the nucleus through a nuclear pore and proceeds to a ribosome located on the rough endoplasmic reticulum (ER). There, it provides instructions for the correct sequence of amino acids for synthesizing that particular digestive enzyme. When enzyme synthesis is complete, the enzyme travels through the ER and is then encapsulated in a transport vesicle. The transport vesicle fuses with a Golgi body, releasing the enzyme. In the Golgi complex, the enzyme is further modified and is then shunted to the ends of the Golgi complex, or cisternae. There, the enzyme waits for a secretory vesicle, which will carry it to the perimeter of the cell, the cell membrane. The secretory vesicle membrane then fuses with the cell membrane, and the enzyme is released outside the cell.

carry one set of enzymes active in converting fats to carbohydrates and another set that detoxifies the various potentially harmful molecules formed by this conversion.

Lysosomes, another class of membrane-bound organelles, are about the same size as peroxisomes. They provide an impressive example of the metabolic compartmentalization achieved by the activity of the Golgi complex. They contain in a concentrated mix the cell's digestive enzymes—enzymes that catalyze the rapid breakdown of macromolecules (proteins, nucleic acids, lipids, and carbohydrates).

Lysosomes digest worn-out cellular components, making way for newly formed ones while recycling the materials locked up in the old ones. Cells can persist for a long time only if their components are constantly renewed. Otherwise, the ravages of use and accident chip away at their metabolic capabilities and slowly degrade their ability to survive. Cells age for the same reason that people do—because of a failure to renew themselves. Throughout the lives of eukaryotic cells, lysosomes break down the organelles and recycle their component proteins and other molecules at a fairly constant rate. For example, mitochondria are replaced in some tissues every ten days, with lysosomes digesting the old ones as new ones are produced.

Lysosomes that are actively engaged in digestive activities keep their battery of hydrolytic enzymes—those that catalyze the hydrolysis of molecules—fully active by maintaining a low internal pH. They do this by pumping hydrogen ions into their interiors. Only at such acid pH values are the hydrolytic enzymes maximally active. Lysosomes that are not functioning actively do not maintain such an acid internal pH. A lysosome in such a "holding pattern" is called a **primary lysosome.** When a primary lysosome fuses with a food vacuole or other organelle, its pH falls and the arsenal of hydrolytic enzymes is activated. When it becomes active, it is called a **secondary lysosome.**

What prevents lysosomes from digesting *themselves* is not known, but the prevention of autodigestion clearly requires energy, which is why metabolically inactive eukaryotic cells die. Without a constant input of energy, the hydrolytic enzymes of lysosomes digest their membranes from within. When these membranes disintegrate, lysosomes' digestive enzymes pour out into the cell cytoplasm and destroy it. In contrast to eukaryotes, prokaryotes (bacteria) do not possess lysosomes and do not die when metabolically inactive. Instead, they are able to remain quiet but alive until altered conditions restore their metabolic activity, a property that allows them to survive unfavorable environmental conditions. For humans, the very process that repairs the ravages of time within our cells eventually also leads to their destruction. Our dependency on a constant supply of energy is the price that we pay for our long lives.

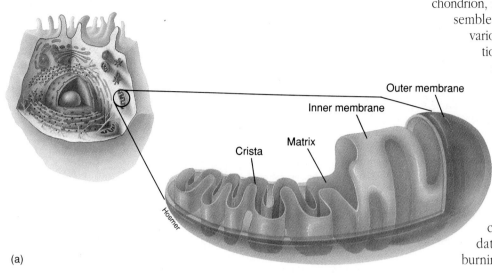

chondrion, is folded into numerous **cristae** that resemble the folded membranes that occur in various groups of bacteria. The cristae partition the mitochondrion into two compartments, an inner **matrix** and an **outer compartment.** On the surfaces of the membranes that divide these compartments, and also submerged within them, are the enzymes that carry out **oxidative metabolism,** the process by which organisms extract energy from macromolecules in the presence of oxygen (figure 4.17). Mitochondria are known as the cell's chemical "furnaces" because oxidation is also the chemical process in burning fossil fuels.

(a)

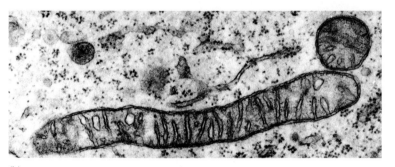

(b)

Figure 4.17 Mitochondria.

The mitochondria of a cell are the sites of oxidative metabolism, whereby energy is extracted from food using oxygen. A mitochondrion has a double membrane. The inner membrane is shaped into folds called cristae. The space within the cristae is called the matrix. The cristae greatly increase the surface area for oxidative metabolism. (a) Drawing and (b) micrograph.

> *Peroxisomes contain enzymes that convert fats to carbohydrates and that destroy the harmful by-products of this conversion. Lysosomes are membrane-bound organelles that contain digestive enzymes. The isolation of these enzymes in lysosomes protects the rest of the cell from the enzymes' digestive activity.*

Mitochondria: The Cell's Chemical "Furnaces"

Mitochondria (singular, *mitochondrion*) are tubular or sausage-shaped organelles 1 to 3 micrometers long; thus, they are about the same size as most bacteria. Mitochondria are bounded by two membranes: The outer membrane is smooth and was apparently derived from the ER of the host cell, whereas the inner one, which was apparently the original plasma membrane of the bacterium that gave rise to the mito-

> *Mitochondria are the chemical "furnaces" of the eukaryotic cell since the enzymes that carry out oxidative metabolism occur within them.*

Most biologists believe that the mitochondria that occur in all but a few eukaryotic cells originated as symbiotic, **aerobic** (oxygen-requiring) bacteria. According to this theory, the bacteria that became mitochondria were engulfed by ancestral eukaryotic cells early in eukaryotic cells' evolutionary history. Before acquiring these bacteria, the host cells had been unable to carry out the metabolic reactions necessary for living in an atmosphere that contained increasing amounts of oxygen. The symbiotic bacteria solved this problem by carrying out oxidative metabolism. Over the course of time, these bacteria became mitochondria.

During the 1.5 billion years in which mitochondria have existed as endosymbionts in eukaryotic cells, most of their genes have been transferred to the chromosomes of their host cells. For example, the genes that produce the enzymes involved with the oxidative metabolism characteristic of mitochondria are located in the cell nucleus. But mitochondria still have some of their original genes, contained in a circular, closed molecule of DNA that resembles those found in bacteria. On this mitochondrial DNA are located several genes that produce some of the proteins essential for the mitochondrion's role as the site of oxidative metabolism.

Genes located on mitochondrial DNA are copied into RNA within the mitochondria and used there to make proteins. In this process, the mitochondria use small RNA molecules and ribosomal components that are also encoded within the mitochondrial DNA. These ribosomes are generally smaller than those of eukaryotes and resemble bacterial ribosomes in size and structure.

When a eukaryotic cell divides, its mitochondria split into two by simple fission. The products of this division are

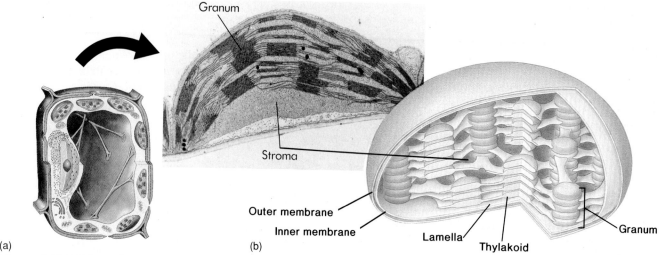

Figure 4.18 A chloroplast.
Chloroplasts (a) are the sites of photosynthesis in photosynthetic eukaryotes, and like mitochondria, have a complex system of internal membranes on which chemical reactions take place. The inner membrane of a chloroplast (b) is fused to form stacks of *closed vesicles called thylakoids. Photosynthesis occurs within these thylakoids. Thylakoids are stacked one on top of the other in columns called grana. The interior of the chloroplasts is bathed in a semiliquid substance called the stroma.*

partitioned between the new eukaryotic cells. All mitochondria are produced by the division of existing mitochondria in the same way that all bacteria are produced from existing bacteria. In both mitochondria and bacteria, the circular DNA molecule is replicated during the division process.

Chloroplasts: Where Photosynthesis Takes Place

Chloroplasts are the sites of photosynthesis and are found in photosynthetic eukaryotes (plants and algae). Symbiotic events similar to those postulated for the origin of mitochondria also seem to have been involved in the origin of chloroplasts. Chloroplasts apparently were derived from symbiotic, **anaerobic** (requiring no oxygen), photosynthetic bacteria, which continue to carry out photosynthesis within the cells in which they occur. Chloroplasts bring an obvious advantage to the organisms that possess them: These organisms can manufacture their own food.

A chloroplast is bounded, like a mitochondrion, by two membranes, the inner membrane apparently derived from the symbiotic bacterium and the outer membrane from the host cell's ER. Chloroplasts are larger than mitochondria, and their inner membranes have a more complex organization. The inner membranes are fused to form stacks of closed vesicles called **thylakoids.** Photosynthesis takes place within the thylakoids. The thylakoids are stacked on top of one another to form a column called a **granum** (plural, **grana**). The interior of a chloroplast is bathed with a semiliquid substance called the **stroma** (figure 4.18).

Like mitochondria, chloroplasts also have a circular DNA molecule, but it is larger than the DNA molecules found in mitochondria. Many of the genes that specify chloroplast components are located in the cell nucleus; they apparently were transferred from the chloroplasts to the nucleus over time, in a

process similar to that described for mitochondria. But the specific RNA and proteins necessary for photosynthesis are synthesized entirely within the chloroplast.

Photosynthetic cells typically contain from one to several hundred chloroplasts, depending on the organism involved or the particular kind of cell. Neither mitochondria nor chloroplasts can be grown in a cell-free culture: They are dependent on the cells in which they occur for their survival.

> *Chloroplasts are the sites of photosynthesis in photosynthetic eukaryotes. They are believed to have been derived from anaerobic photosynthetic bacteria.*

Special Structures and Organelles Associated with Plants

In addition to chloroplasts, plants have several other special structures and organelles: cell walls, plastids, and vacuoles.

Cell Walls

Plant cells share a characteristic with bacteria that is not shared with animal cells: That is, plants have cell walls, which protect and support the plant cell. Although bacteria also have cell walls, plant cell walls are chemically and structurally different from that of bacterial cell walls.

Plastids: Storage and Energy-Harvesting Organelles

Plastids are organelles that manufacture and store food in plant cells. Chloroplasts are just one type of plastid. Other plastids include leucoplastids, which store starch, lipids, or proteins. *Leuco* means "white," and thus these plastids appear colorless. Chromoplastids, on the other hand, store various light-absorbing

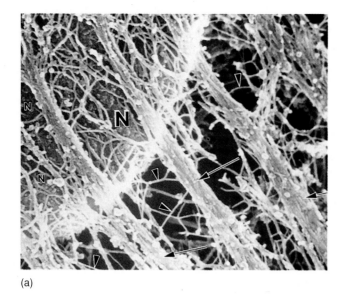

(a)

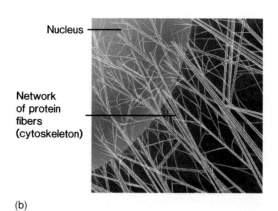

Nucleus

Network
of protein
fibers
(cytoskeleton)

(b)

Figure 4.19 The cytoskeleton.
(a) *This photo shows the nucleus (N) of a cell being held in place by a network of protein fibers called the cytoskeleton. These fibers are attached at the other end to the interior of the cell membrane.* (b) *Interpretative drawing of cytoskeleton.*

pigments of different colors, such as yellow, red, and orange. *Chromo* means "color," and the pigments in chromoplastids give many fruits and vegetables their characteristic colors.

Vacuoles: A Central Storage Compartment
The center of a plant cell usually contains a large, apparently empty space called the **central vacuole.** This vacuole is not really empty; it contains large amounts of water and other materials, such as sugars, ions, and pigments. The central vacuole functions as a storage center for these important substances and also helps to increase the surface-to-volume ratio of the plant cell by applying pressure on the cell membrane. The cell membrane expands outward under this pressure, thereby increasing its surface area.

> *Plant cells have cell walls that are chemically and structurally unique. They also contain special organelles, called plastids, that manufacture and store food, and a large central vacuole, which functions as a storage compartment and also helps the plant cell to maintain a large surface area with respect to volume.*

The Cytoskeleton: Interior Framework of the Cell

The cytoplasm of all eukaryotic cells is crisscrossed by a network of protein fibers, called the **cytoskeleton,** that support the shape of the cell and anchor organelles such as the nucleus to fixed locations (figure 4.19). The cytoskeleton cannot be seen with a light microscope because the fibers are single chains of protein, much too fine for microscopes to resolve. Cytoskeleton fibers are a dynamic system, constantly being formed and disassembled. Individual fibers form by **polymerization,** a process in which identical protein subunits are attracted to one another chemically and spontaneously assemble into long chains. Fibers are disassembled in the same way, by the removal from one end of first one subunit and then another.

Plant and animal cells contain three different kinds of cytoskeleton fibers, each formed from a different kind of subunit: (1) long protein fibers called **actin filaments,** (2) hollow tubes called **microtubules,** and (3) ropes of protein called **intermediate fibers** (figure 4.20). Both actin filaments and intermediate fibers are anchored to proteins embedded within the plasma membrane; they provide the cell with mechanical support. Intermediate fibers act as intracellular tendons, preventing excessive stretching of cells, whereas actin filaments play a major role in determining cell shape. As discussed previously, animal cells lack rigid cell walls; because actin filaments can form and dissolve readily, the shape of such cells can change rapidly. The surface of an animal cell is alive with motion: Projections shoot outward from the surface and then retract, only to shoot out elsewhere moments later (figure 4.21).

The cytoskeleton is not only responsible for cell shape; it also provides a scaffold on which the enzymes and other macromolecules are located in defined areas of the cytoplasm. Many of the enzymes involved in cell metabolism bind to actin filaments, as do the ribosomes that carry out protein synthesis. By anchoring particular enzymes near one another, the cytoskeleton participates with the ER in organizing cell activities.

Actin filaments and microtubules also play important roles in cell movement. Your own muscle cells use actin filaments to contract their cytoskeletons. Indeed, all cell motion is tied to these same processes. The fluttering of an eyelash, the flight of an eagle, and the awkward crawling of a baby all depend on the movements of actin filaments in the cytoskeletons of muscle cells. Many single-celled eukaryotes move not by contracting their cytoskeleton but rather by moving long, slender extensions of the cytoskeleton, much as oars propel a rowboat.

Flagella and Cilia: Slender, Waving Threads
Flagella (singular, *flagellum*) are fine, long, threadlike organelles protruding from cell surfaces; they are used in locomotion and feeding. The structures called flagella in prokaryotes (bacteria) are totally distinct in structure and origin from those in eukaryotic organisms, but they have a similar function.

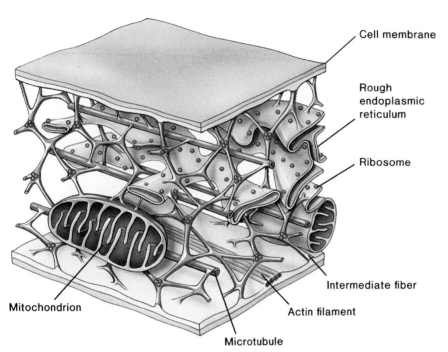

Figure 4.20 How the cytoskeleton anchors organelles.
In this diagrammatic cross section of a eukaryotic cell, the mitochondria, ribosomes, and endoplasmic reticulum are all supported by a fine network of filaments, through which pass microtubules linking various portions of the cell.

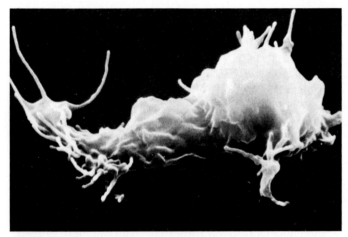

Figure 4.21 The surfaces of animal cells are in constant motion.
Projections called pseudopods enable single-celled organisms to move. The projections are the work of actin filaments rapidly forming and dissolving. This Amoeba, *a single-celled protist, is advancing with its leading edges extended and pseudopods outward.*

In the bacteria that possess them, flagella are long protein fibers. They rotate and are so efficient in locomotion that bacteria with flagella can move as much as twenty cell diameters per second (figure 4.22). Imagine trying to run twenty body lengths per second! One or more flagella trail behind each swimming bacterial cell, depending on the bacteria species. Each flagella moves like a propeller, driven by a complex rotary "motor" embedded in the cell wall and membrane. Rotary motion is mainly characteristic of bacteria; only a few eukaryotes have flagella that rotate.

Eukaryotic cells have a completely different kind of flagellum, based on a cable made up of microtubules. Such flagella are sometimes called **undulipodia** to underscore their complete distinctiveness from bacterial flagella, which they resemble only to a limited degree in external form and function. Eukaryotic flagella, which arise from structures called **basal bodies,** consist of a circle of nine microtubule pairs surrounding two central ones; they are called **9 + 2 flagella** (figure 4.23). These complex structures are a fundamental feature of, and evidently evolved early in, the history of eukaryotes. Even in eukaryotic cells that lack flagella, derived structures with the same 9 + 2 structure often occur.

The arrangement of flagella differs greatly in different eukaryotes. If the flagella are numerous and organized in dense rows, they are called **cilia** (singular, *cilium*), but cilia do not differ from flagella in their structure—that is, they have a 9 + 2 arrangement of microtubules. In many multicellular organisms, cilia perform tasks far removed from their original function of propelling cells through water. For example, in several kinds of human tissues, the beating of rows of cilia moves water over a tissue surface. The 9 + 2 arrangement of microtubules also occurs in the sensory hairs of the human ear, where the bending of these hairs by pressure constitutes the initial sensory input of hearing. Throughout the evolution of eukaryotes, 9 + 2 flagella have been elements of central importance.

Flagella are slender, threadlike structures that aid cellular motion and other functions. Bacterial flagella, which are long protein fibers, are distinct from eukaryotic flagella (undulipodia), which are formed of microtubules in a characteristic 9 + 2 arrangement. Cilia have the same structure as flagella, but their arrangement on the cell surface is different.

Centrioles: Producing the Cytoskeleton

Centrioles are organelles associated with the assembly and organization of microtubules in the cells of animals and most protists. They are the factories that assemble many of the key

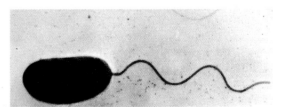

Figure 4.22 Bacterial flagella in action.
Flagella are long protein fibers used in locomotion and feeding.

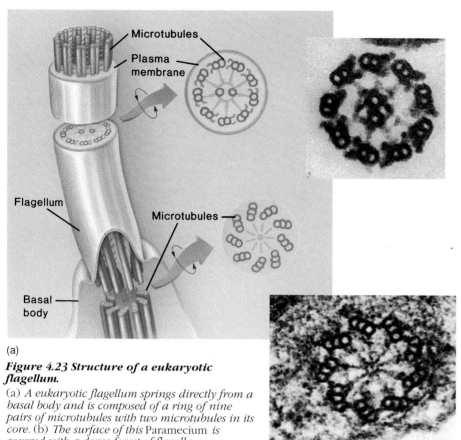

(b)

(a)

Figure 4.23 Structure of a eukaryotic flagellum.

(a) *A eukaryotic flagellum springs directly from a basal body and is composed of a ring of nine pairs of microtubules with two microtubules in its core.* (b) *The surface of this* Paramecium *is covered with a dense forest of flagella.*

structural components of the cytoskeleton. Microtubules are long, hollow cylinders about 25 nanometers in diameter and are composed of the protein **tubulin.** They influence cell shape, move the chromosomes in cell division, and provide the functional internal structure of cilia and flagella, as discussed in the previous section.

Centrioles occur in pairs within the cytoplasm of eukaryotic cells and are usually located at right angles to one another near the nuclear envelope. They resemble tubes and are among the most structurally complex organelles of the cell. As mentioned earlier, in cells that contain flagella or cilia, each flagellum is anchored by a basal body. Basal bodies are a form of centriole. In addition, a pair of centrioles is also associated with the microtubules that move duplicated chromosomes apart in cell division. These centrioles do not actually *control* the movement of the microtubules; this control is the domain of a set of proteins and other materials called the **microtubule organizing center (MTOC).** In those cells that have centrioles, however, the centrioles may play a role in selecting the plane along which the cell divides (figure 4.24).

Most animal and protist cells have both centrioles and basal bodies. The cells of plants and fungi lack centrioles and basal bodies, and their microtubules are organized by amorphous structures.

In many respects, centrioles resemble a type of bacteria called spirochaete bacteria. Spirochaetes are slender, spiral, mobile bacteria that can expand and contract. Researchers have determined that at least some centrioles within eukaryotes contain DNA, which is apparently involved in the production of their structural proteins. This DNA closely resembles spirochaete DNA. The resemblance of centrioles to spirochaetès, combined with the fact that some centrioles have DNA, lends strong support to the hypothesis that centrioles were derived from endosymbiotic spirochaete bacteria.

Structural Differences of Prokaryotic and Eukaryotic Cells

As we have seen, eukaryotic cell structure is much more complicated and diverse than prokaryotic (bacterial) cell structure (table 4.1). The most distinctive difference between these two cell types is the extensive subdivision of the interior of eukaryotic cells by membranes. The most visible of these membrane-bound compartments—the nucleus—gives eukaryotes ("true nucleus") their name. Prokaryotic cells have no equivalent membrane-bound compartments.

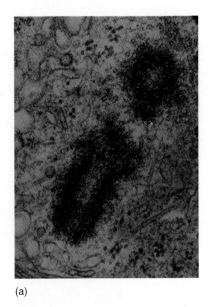

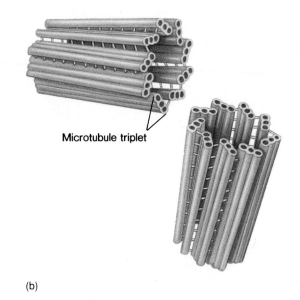

Microtubule triplet

(a)

(b)

Figure 4.24 Centrioles.

Centrioles anchor and organize microtubules. In their anchoring capacity, they can be seen as the basal bodies of eukaryotic flagella. In their organizing capacity, they function during cell division to indicate the plane along which the cell separates. (a) A pair of centrioles in a cell are about to divide.

Centrioles usually occur in pairs and are located in the cell in characteristic planes. One centriole usually lies parallel to the cell surface, while another lies perpendicular to the surface. (b) Centrioles are composed of nine triplets of microtubules.

Țable 4.1   A Comparison of Bacterial, Animal, and Plant Cells

	Bacterium	**Animal**	**Plant**
Exterior Structure			
Cell wall	Present	Absent	Present
Cell membrane	Present	Present	Present
Flagella	May be present (one strand)	Many be present	Absent except in sperm of a few species
Interior Structure			
Endoplasmic reticulum	Absent	Usually present	Usually present
Microtubules	Absent	Present	Present
Centrioles	Absent	Present	Absent
Golgi bodies	Absent	Present	Present
Ribosomes	Present	Present	Present
Organelles			
Nucleus	Absent	Present	Present
Mitochondria	Absent	Present	Present
Chloroplasts	Absent	Absent	Present
Chromosomes	A single circle of naked DNA	Multiple units, DNA associated with protein	Multiple units, DNA associated with protein
Lysosomes	Absent	Usually present	Equivalent structures called "spherosomes"
Vacuoles	Absent	Absent or small	Usually a large single vacuole in mature cell

The membranes of some photosynthetic bacteria are extensively folded inwardly, but they do not isolate any one portion of the cell from any other portion. A molecule can travel unimpeded from any location in a prokaryotic cell to any other location.

The chapters that follow consider the consequences of these structural differences and how they influence the metabolism and biochemistry of eukaryotes. The metabolic processes that occur within eukaryotic cells differ from those of prokaryotic cells, and these differences, like the structural ones discussed in this chapter, are substantial.

EVOLUTIONARY VIEWPOINT

Mitochondria and chloroplasts, two of the most complex organelles of eukaryotic cells, seem almost certainly to have evolved by a process of endosymbiosis from ancient symbiotic bacteria capable of carrying out oxidative metabolism (mitochondria) and photosynthesis (chloroplasts).

SUMMARY

1. Cells are the smallest units of life. They are composed of a nuclear region, which contains the hereditary apparatus, within a larger space called the cytoplasm. The nuclear region supervises the cell's day-to-day activities within the cytoplasm. The cytoplasm is bounded by a plasma (phospholipid) membrane in all cells.

2. According to the cell theory, all organisms are composed of one or more cells, cells are the smallest living things, and all cells arise from previously existing cells.

3. The cells of bacteria, which are prokaryotic in structure, do not have membrane-bound organelles. Their DNA is located in a nucleoid region. Ribosomes are scattered throughout the prokaryotic cytoplasm.

4. The endosymbiotic theory states that some eukaryotic organelles were originally bacteria that became incorporated into the eukaryotic cell. Organelles thought to have originated in this way are mitochondria, chloroplasts, and centrioles.

5. The nucleus is the cell's control center. Within it, the cell's DNA is separated from the rest of the cytoplasm.

6. The endoplasmic reticulum (ER) is a series of membranes that subdivide the interior of eukaryotic cells into separate compartments. Rough ER is studded with ribosomes and manufactures proteins. Smooth ER synthesizes lipids and other molecules.

7. In addition to rough and smooth ER, a third key membrane-associated organelle is the Golgi complex, which serves as a cellular "express package service," packaging molecules within special membrane vesicles and transporting them to various locations in the cell.

8. Peroxisomes contain enzymes that digest fats and neutralize the harmful by-products of this process in cells.

9. One class of vesicle created by the Golgi complex consists of lysosomes, organelles that contain high concentrations of enzymes that constantly digest the cell's macromolecules and thus enable their renewal. A cell continuously expends energy to prevent lysosomes from digesting themselves. Without a constant input of energy, lysosomes soon burst, digesting and killing the cell.

10. Mitochondria are tubular or sausage-shaped organelles, 1 to 3 micrometers long. Bounded by two membranes, they closely resemble the aerobic bacteria from which they were originally derived. As chemical "furnaces" of the cell, they carry out its oxidative metabolism.

11. Chloroplasts, which occur in plants and algae (photosynthetic protists), were apparently derived from anaerobic photosynthetic bacteria. They resemble mitochondria but are larger and have more DNA.

12. Many, but not all, of the genes that originally were present in mitochondrial and chloroplast DNA seem to have been transferred to, or had their functions taken over by, the DNA in the chromosomes of the host cell. However, both classes of organelles have retained the genes necessary to create the distinctive structures related to their particular functions.

13. Plant cells have cellulose cell walls, a characteristic that differentiates them from animal cells. Chloroplasts (which are one type of plastid) manufacture and store food in plant cells. Vacuoles are storage compartments and also help the plant cell to increase its surface-to-volume ratio.

14. Every eukaryotic cell includes a cytoskeleton of microtubules, which helps to determine the cell's shape and also fixes the location of many internal components.

15. Flagella are threadlike structures by virtue of which some cells move. In many multicellular organisms, flagella (cilia) perform secondary functions acquired during the course of evolution. Bacterial flagella are threads of protein that propel the bacteria that possess them by rotating. Eukaryotic flagella, sometimes called undulipodia, have a distinctive and complex 9 + 2 structure of microtubules and are one of the characteristic features of eukaryotes.

16. Centrioles organize microtubules in animals and most protists, but not in plants or fungi. Centrioles occur in pairs and might be derived from spirochaete bacteria.

REVIEWING THE CHAPTER

1. What are the major components of cell structure?

2. How big are cells?

3. What is the history of the cell theory?

4. How are different types of microscopes used to study cells?

5. What are some advantages of small cell size?

6. What is the structure of a prokaryotic cell?

7. What is the structure of a eukaryotic cell?

8. How did eukaryotic cells originate?

9. What is the plasma membrane, and how does it function?

10. How does the nucleus control the cell?

11. What is the function of the nuclear envelope?

12. What are chromosomes?

13. Where are ribosomes made?

14. What is the endoplasmic reticulum, and how does it function?

15. How does the Golgi complex transport materials?

16. What are the functions of peroxisomes and lysosomes?

17. What are mitochondria, and how do they function?

18. How does chloroplast structure relate to photosynthesis?

19. What are the special structures and organelles associated with plants?

20. How does the cytoskeleton function in a cell?

21. What are flagella and cilia, and how do they function?

22. What are centrioles and basal bodies?

23. What are the major structural differences between prokaryotic and eukaryotic cells?

COMPLETING YOUR UNDERSTANDING

1. What is common to hamburger, eyelashes, fingernails, orange juice, and pencils?

2. How does a flagellum in a bacterial cell differ from that in a eukaryote?

3. What are the three principles of the cell theory?

4. Why are the best light microscopes limited to a resolving power of 0.2 micrometers? How does this compare to the resolving power of the human eye?

5. Why does a transmission electron microscope have a better resolving power than a light microscope?

6. What is the advantage of a high surface-to-volume ratio in a cell?

7. How do prokaryotes differ from eukaryotes in cell size?

8. What is the function of a cell wall?

9. Why is a ribosome not an organelle?

10. When did eukaryotes first evolve?

11. How and where do channel, receptor, and marker proteins function?

12. What happens to a cell, such as a red blood cell, when it loses its nucleus?

13. Why do nuclear pores occur in the nuclear envelope?

14. In chromosomes, how does the protein function with the DNA?

15. If you were injured in a sporting event and your physician placed you on a prescription of codeine, how would your body get rid of the codeine?

16. How does the Golgi complex coordinate with the endoplasmic reticulum?

17. How do primary and secondary lysosomes differ?

18. In plant cells, how does the vacuole increase the surface-to-volume ratio of the cell?

19. How do actin filaments, microtubules, and intermediate fibers function? How do these structures differ?

20. What is an undulipodium?

21. What are the functional differences among flagella, basal bodies, and centrioles? In which organisms are these structures present or absent?

22. What is a cilium, and how does it differ from a flagellum?

23. What is the function of a microtubule organizing center?

FOR FURTHER READING

De Duve, C. 1986. *A guided tour of the living cell,* vols. 1 and 2. New York: Scientific American Books. A classic review with excellent illustrations and presenting a wide variety of the techniques used to study cells.

Dingwall, C., and R. Laskey. 1992. The nuclear membrane. *Nature,* Nov. 6, 942–47. A scholarly article that describes all aspects of the nuclear membrane.

Glover, D. et al. 1993. The centrosome. *Scientific American,* June, 62–68. Discussion of how, by directing the assembly of a cell's cytoskeleton, the centrosome controls a cell's shape, when and how it divides, and how it moves.

Goodsell, D. 1992. A look inside the living cell. *American Scientist,* Sept., 457–65. How computer enhancement and quantitative calculations allow biologists to paint the portraits of proteins in action.

McDermott, J. 1989. A biologist whose heresy redraws earth's tree of life. *Smithsonian,* Aug., 71–81. Engaging account of the scientific approach taken by Lynn Margulis, leading contemporary advocate of the endosymbiotic theory of the origin of some organelles.

Murray, M. 1991. Life on the move. *Discover,* March, 72–75. A general account of how eukaryotic cells crawl, powered by delicate protein filaments and powerful motors.

Rapoport, T. A. 1992. Transport of proteins across the endoplasmic reticulum membrane. *Science,* vol. 258, Nov. 6, 931–35. A summary of the mechanism that transports proteins made on ribosomes located in the cytoplasm of cells into the endoplasmic reticulum.

Stossel, T. P. 1993. On the crawling of animal cells. *Science,* vol. 260, May 21, 1086–94. Sophisticated article that outlines the mechanism of animal cell movement.

Warren, G. 1992. Bridging the gap. *Nature,* Nov. 6, 297–98. Encapsulates the latest findings about how vesicles operate within cells. Special attention is paid to how the Golgi complex moves molecules from one Golgi body to another.

EXPLORATIONS

Interactive Software:
Cell Size

In this interactive, the user explores a cell in cross section, monitoring metabolites entering and wastes leaving with a "molecular speedometer" that measures how long it takes a molecule to travel from surface to center of the cell. The key variable is the diameter of the cell, which the user is free to vary. The user may also vary the shape of the cell in ways that do not alter the surface-to-volume ratio.

Questions to Explore:

1. How does the rate of *cell volume increase* compare to rate of *cell surface increase* ?

2. What happens to the velocity of materials going into and out of the cell when the surface-to-volume ratio *increases* (for example, a S:V of 1:2 increasing to 10:1)? Why?

3. What happens to the velocity of materials going into and out of the cell when the surface-to-volume ratio *decreases* (for example, a S:V of 20:1 *decreasing* to 2:1)? Why?

4. What kind of cell would have a high surface-to-volume ratio?

5. How is surface-to-volume ratio related to metabolism? What kind of organism typically has a faster metabolism—a large animal or a small one? What kind of S:V ratio does this fast-metabolism animal have?

HOW CELLS INTERACT WITH THE ENVIRONMENT

Energy is an essential part of every organism's life. It powers the movement of your eyes as they scan this page—and the graceful leap of this European tree frog.

FOR REVIEW

Here are some important terms and concepts that have been discussed in previous chapters and that you will encounter again in this chapter. Review them before proceeding if necessary.

Polar nature of water (*chapter 2*)
Hydrogen bonds (*chapter 2*)
Functional groups (*chapter 2*)
Structure of fat molecules (*chapter 2*)
Types of membrane proteins (*chapter 4*)
Glycolipids (*chapter 4*)

If you look at a drop of pond water through a microscope, you see a vibrant swarm of cells darting about, every cell ceaselessly exploring the environment around it (figure 5.1). The cells are constantly feeding on food they encounter, ingesting molecules and sometimes entire cells. They dump their wastes back into the environment, together with many other kinds of molecules. Cells continuously garner information about the world around them, responding to a host of chemical clues and often passing on messages to other cells. This constant interplay with the environment is a fundamental characteristic of all cells. Without it, life could not exist.

If you were to coat a living cell in plastic, giving it a hard, impermeable shell, all of the cell's transactions with the environment would stop. No molecules could pass in or out; nor could the cell learn anything about the molecules around it. The cell might as well be a rock. Life in any meaningful sense would cease—unless, of course, there were doors and windows in the shell.

In actuality, every cell is encased within a lipid membrane, an impermeable shell through which no water-soluble molecules and little information (data about the cell's surroundings) can

pass, but the shell contains doors and windows made of protein. Molecules and information pass in and out of a cell through these passageways. A cell interacts with the world through a delicate skin of protein molecules embedded in a thin sheet of lipid. This assembly of lipid and protein is called a **plasma membrane;** alternative terms are *cell membrane* or *plasmalemma.* The structure and function of this membrane is the focus of this chapter.

The Lipid Foundation of Membranes

The plasma membranes that encase all living cells are sheets only a few molecules thick; it would take more than ten thousand of these sheets, which are about 7 nanometers thick, piled on top of one another, to equal the thickness of this sheet of paper. The sheets are made up of diverse collections of proteins enmeshed in a lipid framework. Regardless of the kind of cells or organelles that they enclose, all plasma membranes have a similar molecular structure.

Phospholipids

As is true for all biological membranes, the lipid layer that forms the foundation of a plasma membrane is composed of molecules called **phospholipids.** Like the fat molecules examined in chapter 2, a phospholipid has a backbone derived from a three-carbon molecule called glycerol, with long chains of carbon atoms called fatty acids attached to this backbone. In diagrams, these chains look like tails, and thus the fatty acid region of the phospholipid molecule is sometimes called the "tail." A fat molecule has three such chains, one attached to each carbon of the backbone. Because these chains are nonpolar (do not form hydrogen bonds with water), the fat molecule

Figure 5.1

The waters of a pond are teeming with living organisms that are too tiny to see with the naked eye, but are vibrantly alive when viewed under the microscope.

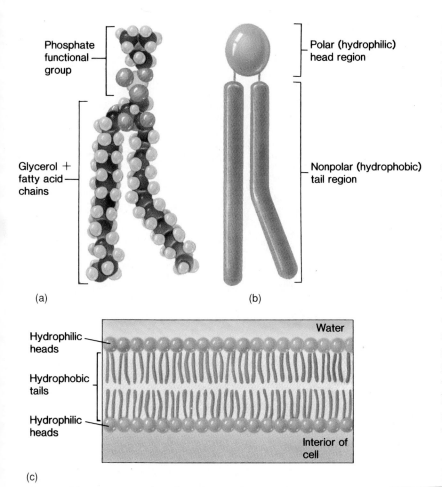

Phosphate functional group

Glycerol + fatty acid chains

Polar (hydrophilic) head region

Nonpolar (hydrophobic) tail region

(a)   (b)

Hydrophilic heads

Hydrophobic tails

Hydrophilic heads

Water

Interior of cell

(c)

Figure 5.2 Structure of a phospholipid molecule.
(a) Each phospholipid molecule consists of a phosphate functional group and two fatty acid chains attached to a glycerol molecule. (b) The fatty acid chains and glycerol form nonpolar, hydrophobic "tails," and the phosphate functional group forms the polar, hydrophilic "head" of the phospholipid molecule. (c) When placed in water, the hydrophobic tails of the molecule face inward, away from the water, and the hydrophilic head faces outward, toward the water.

Formation of Lipid Bilayer

Imagine what happens when a collection of phospholipid molecules is placed in water. The long, nonpolar tails of phospholipid molecules are pushed away by the water molecules that surround them because the water molecules seek partners that can form hydrogen bonds. Water molecules always tend to form the maximum number of hydrogen bonds. The long nonpolar tails that cannot form hydrogen bonds get in the way, like too many chaperones at a party. The best way to rescue the party is to put all the chaperones together in a separate room, and that is what water molecules do—they shove all the long, nonpolar tails of the lipid molecules together, out of the way. The polar heads of the phospholipids are "welcomed," however, because they form good hydrogen bonds with water. What happens is that every phospholipid molecule orients so that its polar head faces water and its nonpolar tails face away (see figure 5.2*c*). This results in the formation of *two* layers, with the tails facing each other and no tails being in contact with water. The resulting structure is called a **lipid bilayer** (figure 5.3). Lipid bilayers form spontaneously in water, driven by the forceful way in which water tends to form hydrogen bonds.

> The basic foundation of all plasma membranes is a lipid bilayer that forms spontaneously. In such a layer, the nonpolar tails of phospholipid molecules point inward, forming a nonpolar zone in the bilayer's interior.

Because the interior of a lipid bilayer is completely nonpolar, it repels any large, charged, or polar molecules that attempt to pass through it, just as a layer of oil stops the passage of a drop of water (which is why ducks do not get wet). This barrier to the passage of water-soluble molecules is the key biological property of the lipid bilayer. Because of their small size, however, water and other small polar molecules can traverse phospholipid bilayers through small imperfections between the individual phospholipids. In addition, the membranes of every cell also contain proteins that extend across the lipid bilayer, providing passage across the membrane.

Fluid Nature of Lipid Bilayer

A lipid bilayer is very stable because water's hunger for hydrogen bonding never stops. Although water continually urges phospholipid molecules into this orientation, it is indifferent to the location of individual phospholipid molecules. Water forms just as many hydrogen bonds, regardless of whether a particular phospholipid molecule is located here or there. As a result, individual phospholipid molecules are free to move about

is insoluble in water. A phospholipid, by contrast, has only two such chains attached to its backbone (figure 5.2). The third position, sometimes called the "head" of the phospholipid, is occupied instead by a highly polar phosphate functional group that readily forms hydrogen bonds with water.

The plasma membranes of mammalian cells contain four types of phospholipids that differ only in their associated phosphate functional groups. The chemical nature and structure of these functional groups are not important at this point. Remember, however, that the "head" of the phospholipid molecule is highly polar and contains phosphate.

Thus, the tail end of a phospholipid molecule is strongly nonpolar (water insoluble), while the head is extremely polar (water soluble). The two nonpolar fatty acids extend in one direction, roughly parallel to each other, and the polar phosphate functional group points in the other direction. Because of this structure, phospholipids are often diagrammed as a (polar) head with two dangling (nonpolar) tails (see figure 5.2*b*).

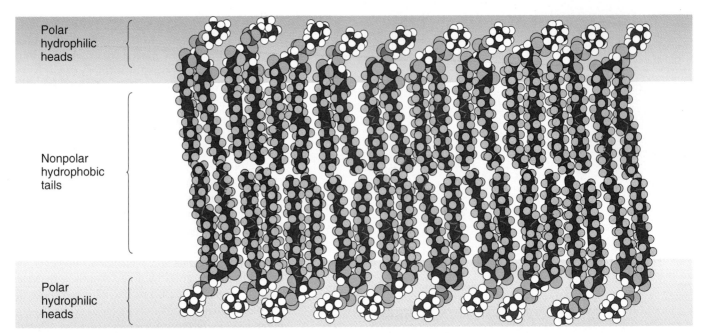

Polar hydrophilic heads

Nonpolar hydrophobic tails

Polar hydrophilic heads

Figure 5.3 A phospholipid bilayer.
The long, nonpolar tails of phospholipids orient toward one another, away from water. Because some of the tails contain double bonds, which introduce kinks in their shape, the tails do not align perfectly, and the membrane is "fluid"—that is, individual phospholipid molecules can move laterally from one place to another within the membrane.

within the membrane. For that reason, the lipid bilayer is not a solid, like a rubber balloon, but rather, a liquid, like the "shell" of a soap bubble. The bilayer itself is a fluid, with the viscosity of olive oil. Just as surface tension holds a soap bubble together, even though the bubble is made of a liquid, so the hydrogen bonding of water holds a membrane together.

Some membranes are more fluid than others. The tails of individual phospholipid molecules attract one another when lined up close together. This stiffens the membrane because aligned molecules must pull apart from one another before they can move about in the membrane. The more alignment, the less fluid the membrane is. Some phospholipids have tails that do not align well because they contain one or more double bonds between successive carbon atoms (C=C), which introduce kinks in the tail. Membranes containing phospholipids of these sorts are more fluid than those that lack them. Sometimes, membranes contain other lipids, such as cholesterol, that prevent phospholipid tails from coming into contact with one another. This also results in the membrane being more fluid.

Architecture of the Plasma Membrane

A eukaryotic cell contains several kinds of plasma membranes that are similar in their lipid bilayers but that differ in the nature of the molecules embedded in them. All such membranes are assembled from four components (figure 5.4):

1. *A phospholipid bilayer foundation* Every plasma membrane has as its basic foundation a phospholipid bilayer. The other components of the membrane are enmeshed within the bilayer, which provides a flexible matrix (interior), while also establishing a barrier to permeability.

2. *Membrane proteins* A major component of every plasma membrane is a collection of proteins that float within the lipid bilayer. These proteins provide channels through which molecules and information pass. Membrane proteins are not fixed in position; instead, they move about freely (see Key Experiment 5.1). Chains of carbohydrates are attached to some membrane proteins; these proteins are called **glycoproteins.** Other membrane proteins have separate proteins attached to them that act as anchors for the cytoskeleton. Not all membranes have the same number and distribution of proteins: Some membranes are crowded with proteins, while others contain only a few. Figure 5.5 diagrams some of the functions of plasma membrane proteins.

3. *Network of supporting fibers* A plasma membrane is structurally supported by a scaffold of proteins on its inner surface that reinforces the membrane's shape (figure 5.6). This is why, for example, a red blood cell is shaped like a disk rather than being irregular. Membranes use networks of other proteins to control the lateral movements of some key membrane proteins, anchoring them to specific sites so that they do not simply drift away. Unanchored proteins have been observed to move as much as 10 micrometers in one minute.

4. *Exterior glycolipids* In some kinds of membranes, many carbohydrate chains extend outward from the cell membrane, like a thicket of brush. These carbohydrate chains are actually attached to lipid molecules, and the entire molecule is called a **glycolipid.** The carbohydrate portion of a glycolipid extends

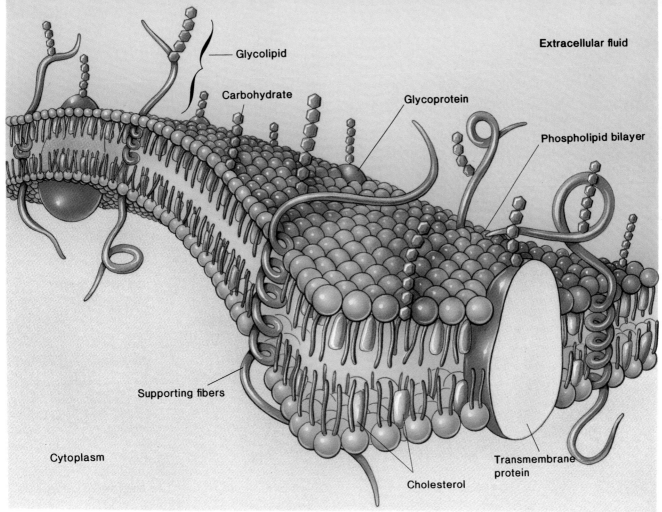

Extracellular fluid

Glycolipid

Carbohydrate

Glycoprotein

Phospholipid bilayer

Supporting fibers

Cytoplasm

Cholesterol

Transmembrane protein

Figure 5.4 The plasma membrane.
The plasma membrane consists of four elements: a phospholipid bilayer, proteins embedded in the phospholipid bilayer, supporting fibers that anchor the plasma membrane in place, and glycolipids that project outward from the membrane surface.

KEY EXPERIMENT 5.1

2000

1994

1900

DEMONSTRATION THAT PROTEINS MOVE WITHIN MEMBRANES

Protein movement within membranes can be easily demonstrated by labeling the proteins of a mouse cell with fluorescent molecules and then fusing that cell with a human cell. Within one hour, the labeled and unlabeled proteins are intermixed throughout the fused cell's membranes.

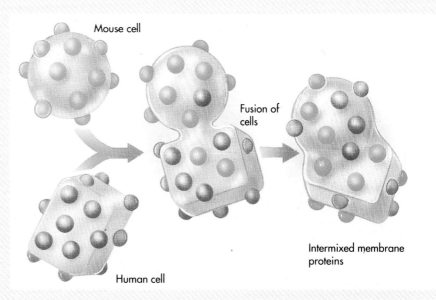

Mouse cell

Fusion of cells

Intermixed membrane proteins

Human cell

Figure 5.5
Functions of plasma membrane proteins.

The plasma membrane proteins illustrated here give you an idea of the variety of functions performed by the different types of plasma membranes.

Outside

Plasma membrane

Inside

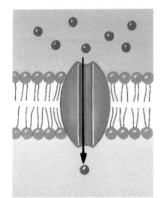

Transport channel

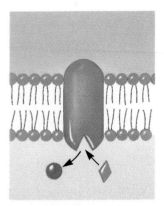

Enzyme

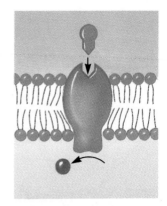

Cell surface receptor

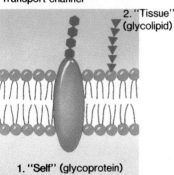

2. "Tissue" (glycolipid)

1. "Self" (glycoprotein)

Cell surface markers

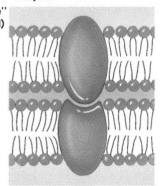

Cell adhesion

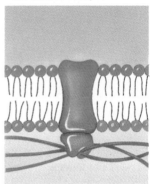

Attachment of cytoskeleton

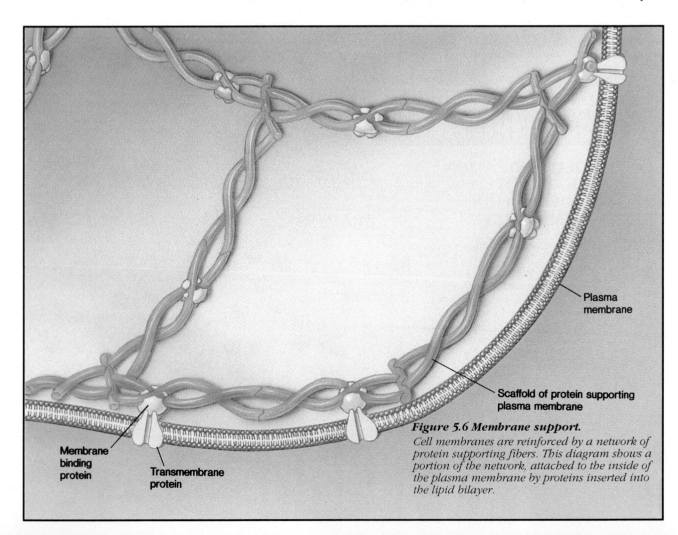

Plasma membrane

Scaffold of protein supporting plasma membrane

Membrane binding protein

Transmembrane protein

Figure 5.6 Membrane support.
Cell membranes are reinforced by a network of protein supporting fibers. This diagram shows a portion of the network, attached to the inside of the plasma membrane by proteins inserted into the lipid bilayer.

outward, while the lipid portion is embedded within the lipid bilayer. Glycolipids act as cell identity markers, since different cell types exhibit different kinds of carbohydrate chains on their plasma membrane surfaces.

> *Plasma membranes are assembled from four components: a phospholipid bilayer foundation, membrane proteins, a network of supporting fibers, and exterior glycolipids.*

How the Plasma Membrane Regulates Interactions With the Cell's Environment

The plasma membrane consists of a complex assembly of many floating proteins that are loosely anchored to the membrane. A list of all the different kinds of proteins in one cell's plasma membrane would run to many pages. The enormous flexibility of plasma membrane design permits a broad range of interactions with the environment. A plasma membrane is like a rack that can hold many different tools. With these tools (proteins), the cell can interact with its environment in many ways, admitting a particular molecule here, sensing the presence of a hormonal signal there. Like the tools of a busy factory, a plasma membrane's proteins are highly active.

Six of the many ways in which cell membranes regulate cell interactions with the environment are:

1. *Passage of water* Plasma membranes are freely permeable to water, but the spontaneous movement of water into and out of cells sometimes presents problems.

2. *Bulk passage into the cell* Cells sometimes engulf other cells or large pieces of such cells, or gulp liquids.

3. *Selective transport of molecules* Plasma membranes are picky about which molecules they allow to enter or leave the cell.

4. *Reception of information* Plasma membranes can identify chemical messages with exquisite sensitivity.

5. *Expression of cell identity* Plasma membranes carry molecular name tags that tell other cells who they are.

6. *Physical connections with other cells* In forming tissues, plasma membranes make special connections with each other.

The remainder of the chapter explores these interactions in more detail.

Passage of Materials Into and Out of Cells: Passive Processes

Molecules dissolved in liquid are in constant motion, moving about randomly. This random motion causes a net movement of molecules toward zones where the concentration of molecules is lower. Driven by random motion, molecules always "explore" the space around them, diffusing until they fill the space uniformly. The difference in molecule concentration between zones of few molecules and zones with many molecules is called a **concentration gradient.** Molecules always try to move into zones where there are fewer molecules, a process that is sometimes called "moving down a concentration gradient." This movement does not require an input of energy and is therefore called **passive transport.** In cells, there are three types of passive transport: diffusion, osmosis, and facilitated diffusion.

Diffusion

A simple experiment demonstrates the process of diffusion: Fill a small jar with ink, screw a lid on the jar, place the jar at the bottom of a bucket full of water, and remove the jar's lid. The ink molecules will slowly diffuse out of the jar until there is a uniform concentration of ink molecules in the bucket of water and in the jar. **Diffusion** is the random movement of molecules from an area of high concentration to an area of low concentration. Diffusion ends when the molecules are uniformly distributed (figure 5.7).

> *Diffusion is the random movement of molecules from an area of high concentration to an area of low concentration. Diffusion tends to distribute molecules uniformly.*

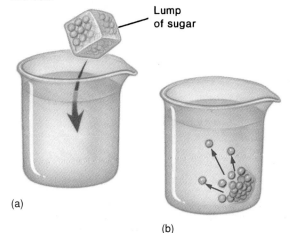

Lump of sugar

(a)

(b)

(c)

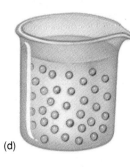

(d)

Figure 5.7 Diffusion.
If a lump of sugar is dropped into a beaker of water (a), *the sugar molecules dissolve* (b) *and diffuse* (c). *Eventually, diffusion results in an even distribution of sugar molecules throughout the water* (d).

Osmosis

The cytoplasm of a cell consists of molecules such as sugars, amino acids, and ions dissolved in water. The mixture of these molecules and water is called a **solution.** Water, the most common molecule in the mixture, is the **solvent,** and the other kinds of molecules dissolved in the water are **solutes.**

Because of diffusion, both solvent and solute molecules in a cell move from regions of higher concentration to regions of lower concentration. When two regions of different concentrations are separated by a membrane, molecules may not be able to pass freely through that membrane. Plasma membranes allow only some types of substances to pass across them, and for this reason, they are described as **selectively permeable membranes.** Because sugars, amino acids, and other solutes are water soluble and not lipid soluble, they are imprisoned within the cell, unable to cross the membrane's lipid bilayer. Water molecules, in contrast, can pass through slight imperfections in the sheet of lipid molecules and so diffuse across the membrane into the cell. Water molecules stream into the cell across the membrane, thereby diluting the high concentration of solutes within the cell so that this concentration matches more and more closely the lower concentration in the outside solution. This form of water diffusion into or out of a cell through a selectively permeable membrane is called **osmosis** (figure 5.8).

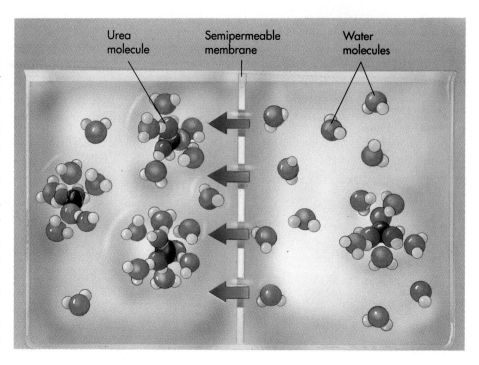

Figure 5.8 Osmosis.

Charged or polar molecules, such as urea, are soluble in water because they form hydrogen bonds with water molecules clustered around them. When such a polar solute is added to one side of a membrane, the water molecules that gather around each urea molecule are no longer free to diffuse across the membrane—in effect, the polar solute has reduced the number of free water molecules on that side of the membrane. Because the other side of the membrane (on right, with less solute) has more unbound water molecules than the side with more solute, water moves by diffusion from the right to the left.

Osmosis is the diffusion of water across a selectively permeable membrane that permits the free passage of water but not that of one or more solutes.

The fluid content of a cell immersed in pure water is said to be **hypertonic** (from the Greek *hyper,* meaning "more than") with respect to its surrounding solution because it has a higher concentration of solutes than does the water. The surrounding solution, which has a lower concentration of solutes than does the cell, is said to be **hypotonic** (from the Greek *hypo,* meaning "less than") with respect to the cell. A cell with the same concentration of solutes as its environment is said to be **isotonic** (from the Greek *iso,* meaning "the same") (figure 5.9).

As water molecules continue to diffuse inward toward an area of lower *water* concentration (the concentration of the water is lower inside than outside the cell because of the dissolved solutes in the cell), water pressure within the cell increases. This kind of pressure is called **osmotic pressure** (see Key Experiment 5.2).

Because osmotic pressure opposes the inward diffusion of water, such diffusion will not continue indefinitely. The cell will eventually reach an equilibrium—a point at which the osmotic pressure driving water inward is counterbalanced exactly by the pressure driving water out. In practice, the osmotic pressure at equilibrium is typically so high that an unsupported cell membrane cannot withstand it and will burst like an overinflated balloon. Cells whose membranes are surrounded by cell walls, in contrast, can withstand high internal osmotic pressures.

Within the closed volume of a cell that is hypertonic to its surroundings, the movement of water inward, which tends to lower the relative concentration difference of water, will at the same time increase the internal water pressure. The net movement of water stops when an equilibrium condition is reached or the cell bursts.

Facilitated Diffusion

In addition to the transport processes of diffusion and osmosis, cells must have a ready means of preventing the buildup of unwanted molecules within the cell and the ability to glean from the external environment the molecules necessary for survival. Cells perform these two tasks with a process called

KEY EXPERIMENT 5.2

2000

◀ 1994

A DEMONSTRATION OF OSMOSIS

The end of a tube containing a 3 percent salt solution is closed by stretching a selectively permeable membrane across its face that will allow water molecules to pass through but not salt molecules. When this tube is immersed in a beaker of distilled water, the salt cannot cross the membrane; however, water can. The added water causes the salt solution to rise in the tube. Water continues to enter the tube from the beaker until the weight of the column of water in the tube exerts a downward force equal to the force drawing water molecules upward into the tube. This force is referred to as osmotic pressure.

1900

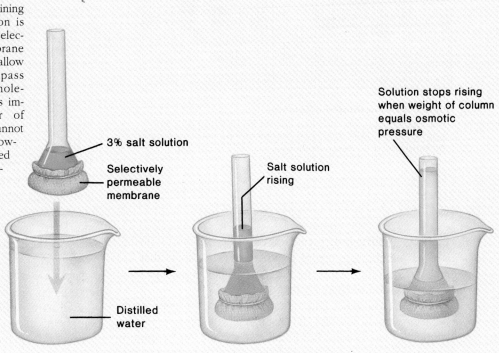

3% salt solution

Selectively permeable membrane

Distilled water

Salt solution rising

Solution stops rising when weight of column equals osmotic pressure

---

Figure 5.9 Osmosis in animal and plant cells.
(a) *In an animal cell, when the surrounding solution is hypotonic with respect to the cell, water will move in, swelling the cell. When the surrounding solution is hypertonic, water will move out of the cell. (b) In plant cells, the large central vacuole contains a high concentration of solutes. Therefore, when a plant cell is placed in a hypotonic solution, water tends to move inward, causing the cell to swell outward against its rigid cell wall. However, if a plant cell is immersed in a hypertonic solution, water will leave the cell, causing the cytoplasm to shrink and pull in from the cell wall.*

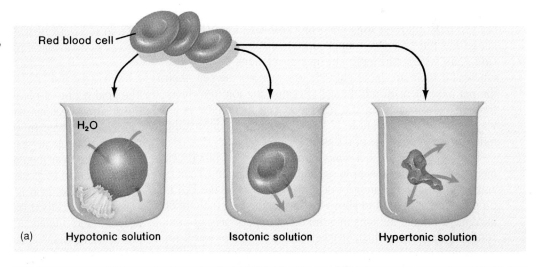

Red blood cell

H_2O

(a)   Hypotonic solution   Isotonic solution   Hypertonic solution

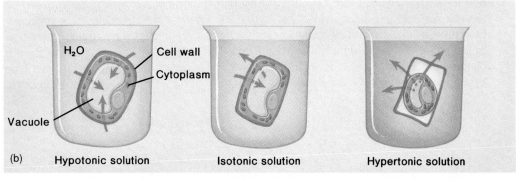

H_2O Cell wall
Cytoplasm

Vacuole

(b)   Hypotonic solution   Isotonic solution   Hypertonic solution

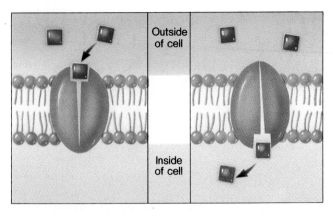

Figure 5.10 Facilitated diffusion.
This type of transport utilizes a carrier protein and does not require energy. Researchers are still unsure about what happens to the transported molecule while it is "inside" the carrier.

facilitated diffusion. Facilitated diffusion capitalizes on the plasma membrane's use of highly specific proteins that are embedded in the membrane to transport molecules across the plasma membrane (figure 5.10).

Some of the most important channels in a plasma membrane are highly selective, facilitating *only* the passage of specific molecules or ions, but in either direction. An example is the channel of vertebrate red blood cell membranes that transports negatively charged ions, or **anions.** This channel plays a key role in the oxygen-transporting function of these cells. The anion channels of the red blood cell membrane readily pass chloride ions (Cl⁻) or carbonate ions (HCO₃⁻) across the red blood cell membrane. If there are more chloride ions within the cell, the net movement is outward, whereas if there are more chloride ions outside the cell, then the net movement is into the cell. Because the ion movement through the channels in the red blood cell membrane is always toward the direction of lower ion concentration, the transport process is one of diffusion.

These channels are not simply holes in the red blood cell membrane, that are somehow specific for these anions. For example, if the concentration of chloride ions outside the cell is progressively increased to above that inside the cell, the rate of movement of chloride ions into the cell increases only up to a certain point, after which it levels off and will proceed no faster, despite increases in the concentration of exterior chloride ions. The diffusion rate can increase no further because the chloride ions are being transported across the membrane by a protein "carrier," and all available carriers are in use. The capacity of the carrier system has been saturated. The transport of these anions is a diffusion process facilitated by a carrier; hence, the apt term *facilitated diffusion.*

Facilitated diffusion has two essential characteristics: (1) it is *specific,* with only certain molecules being able to traverse a given channel, and (2) it is *passive,* the direction of net movement being determined by the relative concentrations of the transported molecule inside and outside the membrane.

Facilitated diffusion is the transport of molecules across a membrane by a carrier protein in the direction of lowest concentration.

Passage of Materials Into and Out of Cells: Processes That Require Energy

The transport processes examined so far in this chapter have all been passive processes, involving the movement of molecules from an area of high concentration to an area of low concentration, or "down a concentration gradient." This next section examines active transport processes, which *do* require an input of energy. These processes involve the movement of molecules from areas of low concentration to areas of high concentration, or "against a concentration gradient." Or, in the case of endocytosis and exocytosis (discussed in the next paragraph), energy is required to get polar food molecules across the nonpolar, hydrophobic interior of the plasma membrane. The energy input that cells use to fuel their active transport processes is in the form of adenosine triphosphate or ATP (see chapter 1). ATP is made when a phosphate group (P_i) is added to an adenosine diphosphate molecule (ADP) during oxidative respiration. ATP formation is discussed in chapters 6, 7, and 8; for now, it is necessary only to recognize that ATP is the "energy currency" that drives the cell's physiological processes, including transport processes.

Endocytosis and Exocytosis

The lipid nature of plasma membranes raises a second problem, in addition to getting materials in and out, for growing cells. The molecules required by cells as food are mostly polar molecules; they will not pass across the hydrophobic barrier interposed by a lipid bilayer. How then are organisms able to get food molecules into their cells? Particularly among single-celled eukaryotes, the dynamic cytoskeleton is employed to extend the cell membrane outward toward food particles, such as bacteria. The membrane encircles and engulfs a food particle. Its edges eventually meet on the other side of the food particle, where, because of the fluid nature of the lipid bilayer, the membranes fuse together, forming an enclosed chamber called a *vesicle* around the food particle. This process is called **endocytosis.** Endocytosis involves the incorporation of a portion of the exterior medium into the cytoplasm of the cell by capturing it within a vesicle.

There are two forms of endocytosis: phagocytosis and pinocytosis. **Phagocytosis** (from the Greek *phagein,* meaning

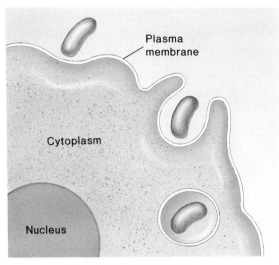

PHAGOCYTOSIS

(a)

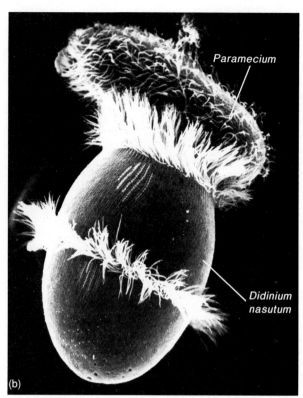

(b)

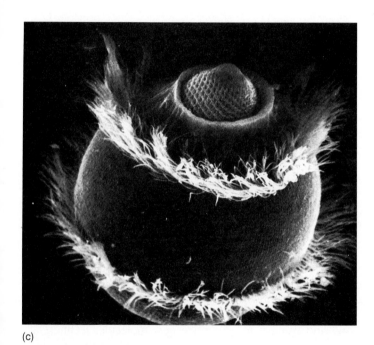

(c)

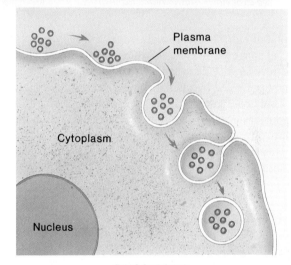

(d)   PINOCYTOSIS

📼 **Figure 5.11 Endocytosis.**
(a) *Phagocytosis is a particular kind of endocytosis and involves incorporating an organism or some other relatively large fragment of organic matter into the cytoplasm of the cell.*
(b) *Here, the egg-shaped protist* Didinium nasutum *illustrates phagocytosis by ingesting the smaller protist* Paramecium.
(c) *The* Didinium's *meal is almost over.* (d) *Pinocytosis, the second form of endocytosis, involves the engulfing of relatively small particles.*

"to eat," and *cytos,* meaning "cell") is when the material brought into the cell is an organism or some other fragment of organic matter (figure 5.11*a–c*). **Pinocytosis** (from the Greek *pinein,* meaning "to drink") is when the material brought into the cell is relatively small, such as a liquid that contains dissolved molecules (figure 5.11*d*). Pinocytosis is common among the cells of multicellular animals. Human egg cells, for example, are "nursed" by surrounding cells that secrete nutrients that the maturing egg cell takes up by pinocytosis.

> *Endocytosis is a process in which cells engulf organisms or fragments of organisms, enfolding them within vesicles. There are two forms of endocytosis: Phagocytosis is the engulfing of large particles, and pinocytosis is the engulfing of small particles.*

Virtually all eukaryotic cells are constantly carrying out endocytosis, trapping extracellular fluid in vesicles and ingesting it. Rates of endocytosis vary from one cell type to

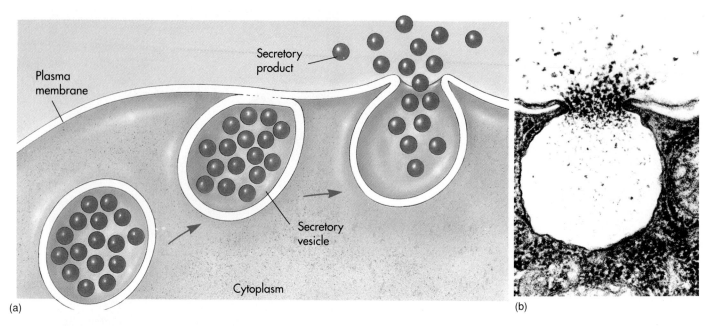

Plasma membrane

Secretory product

Secretory vesicle

Cytoplasm

(a)

(b)

Figure 5.12 Exocytosis.
(a) *Proteins and other molecules are secreted from cells in small pockets called secretory vesicles, whose membranes fuse with the cell membrane, thereby allowing the secretory vesicles to release their contents to the cell surface. (b) In this photomicrograph, you can see exocytosis taking place explosively.*

another but can be surprisingly large. Some types of white blood cells ingest 25 percent of their cell volume each hour.

The reverse of endocytosis is **exocytosis,** the extrusion of material from a cell by discharging it from secretory vesicles at the cell surface (figure 5.12). In plants, vesicle discharge constitutes a major means of exporting the materials used in the construction of the cell wall through the plasma membrane. In animals, many cells are specialized for secretion using the mechanism of exocytosis.

Active Transport

Many molecules that a cell admits across its membrane are maintained within the cell at a concentration different from that of the surrounding medium. In all such cases, the cell must expend energy to maintain the concentration difference. Transport that requires the expenditure of energy is called **active transport.** Active transport may maintain molecules at a higher concentration inside the cell than outside it by expending energy to pump more molecules in than would enter by diffusion, or it may maintain molecules at a lower concentration inside the cell by expending energy to pump them out actively.

Active transport is the transport of a solute across a membrane to a region of higher concentration by the expenditure of chemical energy.

Active transport is one of the most important functions of any cell because it allows a cell to concentrate molecules. Without active transport, the cells of your body would be unable to

harvest glucose—a major source of energy—from the blood because the concentration of glucose molecules is often already higher in the cells than it is in the blood from which it is extracted. Imagine how difficult it would be to survive as a beggar if you could only obtain money from those who had less than you did! Active transport permits a cell, by expending energy, to take up additional molecules of a substance that is already present in its cytoplasm in concentrations higher than those found in the cell's environment.

A cell takes up or eliminates many molecules against a concentration gradient. Some molecules, such as sugars and amino acids, are simple metabolites that the cell extracts from its surroundings and adds to its internal stockpile. (Metabolites are molecules involved in metabolism—in extracting energy from food and using it to synthesize new molecules.) Others are ions, such as sodium and potassium, that play a critical role in such functions as the conduction of nerve impulses. Still others are the nucleotides that the cell uses to synthesize DNA.

These many kinds of molecules enter and leave cells by way of a wide variety of different kinds of selectively permeable transport channels. Some of the channels are permeable to one or a few sugars, others to a certain size of amino acid, and still others to a specific ion or nucleotide. You might suspect that active transport occurs at each of these channels, but you would be wrong. In animal cells, *one* major active transport channel in plasma membranes transports sodium and potassium ions; all the others tie their activity to this all-important **sodium-potassium pump.** Some molecules enter the cell by tying their transport to a channel that admits hydrogen ions (protons) into the cell. The many channels in the membrane that the cell uses to concentrate metabolites and ions are called **coupled channels.** Three types of active transport mechanisms—the sodium-potassium pump, the proton pump, and coupled channels—are discussed next.

The Sodium-Potassium Pump

A cell that is not actively dividing expends more than a third of all its energy to actively transport sodium (Na^+) and potassium (K^+) ions. The remarkable channel by which these two ions are transported across the plasma membrane is referred to as the **sodium-potassium pump.** Most animal cells have a low internal concentration of Na^+ ions and a high internal concentration of K^+ ions relative to their surroundings. They are able to maintain these concentration differences because the sodium-potassium pump actively transports Na^+ ions out of the cell and K^+ ions in. For every three Na^+ ions that are pumped out of the cell, two K^+ ions are pumped into the cell.

The sodium-potassium pump is a highly specific transmembrane protein channel. Some membranes contain large numbers of sodium-potassium channels, whereas other have few. Passage through this channel entails rapid changes in the shapes of the proteins within it. Each channel is capable of transporting as many as three hundred Na^+ ions per second.

The sodium-potassium pump is an active transport process, transporting Na^+ and K^+ ions from areas of low concentration to areas of high concentration. This transport into a zone of higher concentration is the opposite of what occurs spontaneously in diffusion; it is achieved only by the constant expenditure of metabolic energy (figure 5.13). The energy used in the process is obtained from adenosine triphosphate (ATP), the functioning of which is explained in chapter 6.

The Proton Pump

A second channel whose importance in the life of the cell equals that of the sodium-potassium pump is the **proton pump.** The proton pump is not necessarily located on the plasma membrane. More often, it is found on the internal membranes of the two energy-harvesting organelles, the mitochondria and chloroplasts.

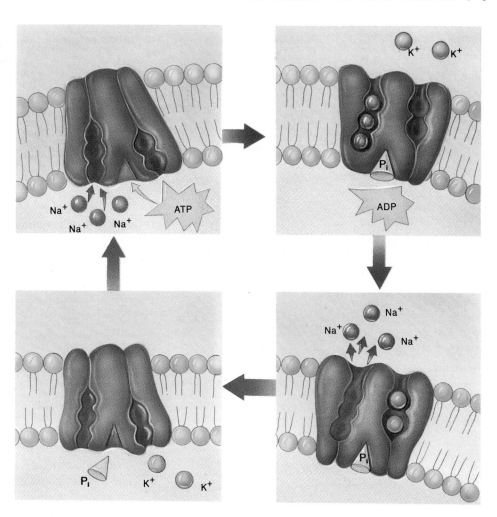

Figure 5.13 The sodium-potassium pump.
The protein channel known as the sodium-potassium pump transports sodium (Na^+) and potassium (K^+) ions across the cell membrane. For every three Na^+ ions that are transported out of the cell, two K^+ ions are transported into the cell. The sodium-potassium pump is fueled by ATP (see chapter 6).

The proton pump involves two special transmembrane protein channels: The first pumps protons (H^+ ions) outside of the membrane, using energy derived from energy-rich molecules or from photosynthesis to power the active transport. This creates a proton gradient in which the concentration of protons outside the membrane is higher than inside. As a result, diffusion drives protons back across the membrane toward a zone of lower proton concentration. But biological membranes are impermeable to protons, so the only way protons can diffuse back in is through a second channel, which couples the transport of protons to the production of ATP (figure 5.14). The net result is the expenditure of energy derived from metabolism or photosynthesis and the production of ATP. This mechanism, called **chemiosmosis,** is responsible for the production of almost all the ATP that you harvest from food that you eat and for all the ATP produced by photosynthesis (see chapter 7). ATP provides the cell with a usable energy source for its many activities.

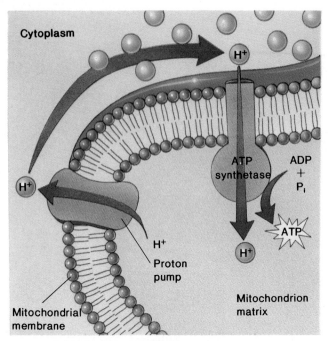

Figure 5.14 The proton pump.
Proton pumps are usually located in the membranes of cell organelles, especially in mitochondria and chloroplasts. The movement of protons (H⁺ ions) is tied to the formation of ATP, making the proton pump an extremely important part of a cell's energy metabolism. In the diagram, H⁺ is pumped out of the cell through a protein channel. It then enters the cell through another channel, triggering the formation of ATP (see chapter 6).

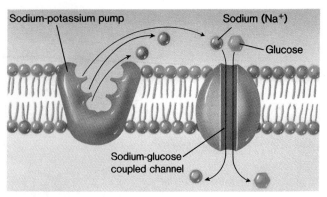

Figure 5.15 A coupled channel.
In this coupled channel, the sodium-potassium pump maintains the high concentration of sodium (Na^+) ions outside the cell. Glucose can enter the cell through a second channel that is especially designed for the transport of Na^+ and glucose. Because the concentration of Na^+ is high outside the cell, Na^+ tends to diffuse into the cell through the second channel, bringing glucose along with it.

Coupled Channels

To grow and carry out their metabolic reactions, cells need to import materials from the external environment. Cell cytoplasm is usually rich in amino acids, sugars, and other materials. Therefore, additional molecules of these that the cell needs must be harvested from a surrounding environment in which the concentration of these materials is much lower than it is inside the cell. The active transport of these materials is accomplished by a **coupled channel.** "Coupled" implies "two," but only one of the channels actually transports the necessary molecule into the cell. The other channel is usually either a sodium-potassium pump or a proton pump that functions only indirectly in the coupled channel—it provides the driving force of a concentration gradient that the necessary molecule can take advantage of by linking itself to the transport of sodium (Na^+) or H^+ ions across the cell membrane. For example, some plant cells couple the transport of molecules to H^+ ions that are diffusing back into the cell through the proton pump. Other cells, such as kidney and intestinal cells, couple the transport of "hitchhiker" molecules to special carrier proteins that admit Na^+, which was pumped outside the cell by the sodium-potassium pump.

A detailed look at a kidney cell demonstrates how the coupled channel works. To function, a kidney cell must import glucose, but the surrounding environment of the kidney cell is low in glucose. Embedded in the kidney cell's plasma membrane is a special channel that will admit glucose as long as it is accompanied by Na^+. The concentration of Na^+ outside the kidney cell is high, due to the activity of the sodium-potassium pump. Therefore, Na^+ tends to diffuse back into the cell through this special channel, and as it does, it brings glucose along with it (figure 5.15). The two channels involved in this coupled channel are the sodium-potassium pump and the special sodium-glucose channel. The sodium-potassium pump functions in an indirect way be creating a concentration gradient in which Na^+ concentrations are higher outside the cell than inside. As a result, Na^+ will passively diffuse back into the cell, and glucose comes along for the ride (see Sidelight 5.1).

Table 5.1 summarizes the mechanisms for transport across plasma membranes considered in this chapter.

Reception of Information

So far, this chapter has focused on membrane proteins that act as channels—doors across a membrane through which only particular molecules may pass. But cells also interact with their environments in a way that does not involve the passage of molecules across membranes, but instead, the transmission of information into cells. **Cell-surface receptors,** a second general class of membrane proteins, are proteins embedded within the plasma membrane that bind specifically to particles but do *not* themselves provide a transport channel for particles; what receptors transmit into the cell is information (see figure 5.5).

In general, a cell-surface receptor is an information-transmitting protein that extends across a plasma membrane. The end of the receptor protein exposed on the cell surface has a shape that fits to specific hormones or other "signal" molecules. When such molecules encounter the receptor on the cell surface, they bind to it. This binding produces a change in the shape of the other end of the receptor protein—the end protruding into the cell's interior—and this shape change, in turn, causes a change in cell activity in one of several ways.

Sidelight 5.1

CHLORIDE CHANNELS AND CYSTIC FIBROSIS

Cystic fibrosis is a fatal disease of human beings in which affected individuals secrete a thick mucus that clogs the airways of the lungs. These same secretions block the ducts of the pancreas and liver so that the few patients who do not die of lung disease die of liver failure. Cystic fibrosis is usually thought of as a children's disease because few affected individuals live long enough to become adults. There is no known cure.

A genetic disease resulting from a defect in a single gene that is passed down from parent to child, cystic fibrosis is the most common fatal genetic disease of Caucasians. One in twenty individuals possesses at least one copy of the defective gene. Most carriers are not afflicted with the disease. Only those children who inherit two copies of the defective gene, one from each parent, succumb to cystic fibrosis—about one in eighteen hundred Caucasian children.

Cystic fibrosis has proven to be difficult to study. Many organs are affected, and until recently, identifying the nature of the defective gene responsible for the disease was impossible. In 1985, however, the first clear clue was obtained. An investigator, Paul Quinton (Sidelight figure 5.1), seized on a commonly ob-served characteristic of cystic fibrosis patients—that their sweat is abnormally salty—and performed the following experiment: He isolated a sweat duct from a small piece of skin and placed it in a solution of salt (NaCl) that was three times as concentrated as the NaCl inside the duct. He then monitored the movement of ions. Diffusion tends to drive both the sodium and chloride ions into the duct because of the higher outer ion concentrations. In skin isolated from normal individuals, sodium ions indeed entered the duct, transported by the sodium-potassium pump; chloride ions followed, passing through a passive channel. Both ions crossed the membrane easily. In skin isolated from individuals with cystic fibrosis, the sodium-potassium pump transported sodium ions into the ducts, but no chloride ions entered. The passive chloride channels were not functioning in these individuals.

Cystic fibrosis appears to result from a defective channel within plasma membranes, one that trans-ports chloride ions across the membranes of normal individuals but not across those of affected persons. It was learned in 1986 that the genetic defect is the result of an alteration in a protein regulating the activity of the channel, rather than in the transmembrane protein itself. The defective gene was isolated in 1987, and its position on a particular human chromosome was pinpointed in 1989. Now that scientists have finally identified the primary cause of the disease, they are much more likely to be able to find a cure for it.

Sidelight figure 5.1 Dr. Paul Quinton in his laboratory.

Cell-surface receptors play a very important role in the lives of multicellular animals. Among the substances that these receptors receive are: the chemicals called neurotransmitters that pass from one nerve to another; the protein hormones, such as adrenaline and insulin, which your body uses to regulate its metabolic level; and the growth factors, such as epidermal growth factor, which regulate development. All of these substances act by binding to specific cell-surface receptors. The antibodies that your body uses to defend itself against infection are themselves free forms of receptor proteins. They are free-floating but structurally identical to those caught within membranes and called cell-surface receptors. Without receptor proteins, the cells of your body would be "blind," unable to detect the wealth of chemical signals that body tissues use to communicate with one another.

Cell-surface receptors in membranes bind specifically to particles and transmit information about the particles into the cells. Some of the information that cell-surface receptors transmit are the presence of hormones and the signals that pass from one nerve to another. Antibodies are free forms of receptor proteins.

Expression of Cell Identity

In addition to passing molecules across their membranes and transmitting information about their surroundings, cells also convey information *to* the environment. This communication of information helps to explain a variety of phenomena. For example:

1. Multicellular animals like ourselves must develop and maintain highly specialized groups of cells called tissues. Your blood is a tissue, and so is your muscle. Tissues are remarkable in that each cell within a tissue performs the functions of a member of that tissue and not some other tissue, even though all body cells have the same genetic complement of DNA and are derived from a single cell at conception. How does a cell "know" to which tissue it belongs?

2. In the course of development in human beings and other vertebrates, some cells move over others, as if they are seeking particular collections of cells with which to develop. How do they sense where they are?

3. Every day of your adult life, your immune system inspects the cells of your body, looking for cells infected by viruses. How does it recognize them?

T a b l e   5 . 1   *Mechanisms For Transport Across Plasma Membranes*

Process	Passage through Membrane	How It Works	Example
Processes That Do Not Require Energy			
Diffusion	Imperfections in lipid bilayer	Random movement of molecules from areas of high concentration to areas of low concentration	Movement of oxygen into cells
Osmosis	Imperfections in lipid bilayer	Diffusion of water across a selectively permeable membrane	Movement of water into plant leaf cells
Facilitated diffusion	Carrier protein	Molecule binds to carrier protein and is transported across; movement is in direction of lowest concentration	Movement of calcium into cells
Processes That Require Energy			
Endocytosis	Membrane vesicle	Large particle (phagocytosis) or small particle (pinocytosis) is engulfed by membrane, which forms vesicle around it	Ingestion of bacteria by white blood cells (phagocytosis); "nursing" of human egg cells (pinocytosis)
Exocytosis	Membrane vesicle	Vesicle fuses with plasma membrane and ejects its contents	Secretion of mucus
Active Transport Processes			
Sodium-potassium pump	Protein channel	Export of three Na^+ ions for every import of two K^+ ions	Found in all cells
Proton pump	Protein channel	Export of protons (H^+ ions) against a concentration gradient	Chemiosmotic generation of ATP; found in chloroplasts and mitochondria
Coupled channels	Protein channels	Import of molecule with Na^+ or H^+ using the concentration gradient established by the pumps of these ions	Import of glucose into cell

4. If foreign tissue is transplanted into your body, your immune system rejects it. How does the immune system know that the transplanted tissue is foreign?

The answer to all of these questions is the same: During development, every cell type in your body makes a "banner" proclaiming its identity, a set of proteins called **cell-surface markers** that are unique to it alone. Cell-surface markers are the tools a cell uses to signal to the environment what kind of cell it is (see figure 5.5).

Some cell-surface markers are glycoproteins anchored in plasma membranes (see figure 5.4). The immune system uses such marker proteins to identify "self." All the cells of a given individual have the same "self" marker, called a **major histocompatibility complex (MHC)** protein. Practically every individual makes a different version of the MHC marker protein; thus, each MHC protein is a distinctive marker for a particular individual.

Other cell-surface markers are glycolipids, lipids with carbohydrate tails (see figure 5.4). These cell-surface markers differentiate the various organs and tissues of the vertebrate body. For example, glycolipids on the surfaces of red blood cells distinguish different blood types, such as A, B, and O. Over the course of development, the cell population of glycolipids changes dramatically as the cells divide and differentiate.

The cells that make up specific kinds of tissues are marked by proteins called cell-surface markers. These are either glycoproteins anchored in plasma membranes or glycolipids.

Physical Connections Between Cells

In addition to passing molecules to and from the environment, acquiring information from the environment, and conveying information about their identity to the environment, cells also have physical interactions with other cells. Most of the cells of multicellular organisms are in contact with other cells, usually as members of organized tissues in such organs as the lungs, heart, or gut. The immediate environment of any one cell is the mass of other cells clustered around it.

Figure 5.16 The three principal types of intercellular connections.

These connections help cells to communicate with other cells.

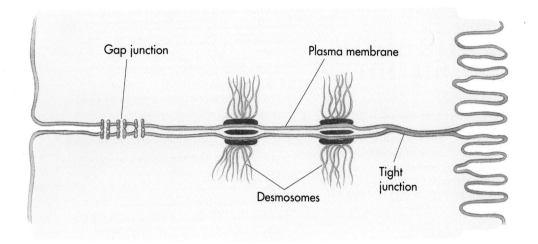

Table 5.2 Types of Intercellular Connections

Type	Name	Characteristic Specializations in Cell Membrane	Width of Intercellular Space	Function
Adhering junctions	Desmosomes	Buttonlike welds joining opposing cell membranes	Normal size 24 nanometers	Hold cells tightly together
Organizing junctions	Tight junctions	Belts of protein that partition the plasma membrane	Intercellular space disappears because the two membranes are adjacent	Form a barrier separating surfaces of cells
Communicating junctions	Gap junctions (animal cells), plasmodesmata (plant cells)	Channels or pores through the two cell membranes and across the intercellular space	Intercellular space greatly narrowed to 2 nanometers	Provide for electrical communication between cells and for flow of ions and small molecules

The nature of the physical connections between a cell and the other cells of a tissue largely determines that cell's contribution to what the tissue will be like.

The locations on the cell surface where cells of a tissue adhere to one another are called **cell junctions** (figure 5.16). There are three general classes of cell junctions in animals (table 5.2):

1. *Adhering junctions* Adhering junctions called **desmosomes** hold cells together as if they were welds constructed of protein.

2. *Organizing junctions* Organizing junctions partition the plasma membrane into separate compartments. **Tight junctions** are belts of protein that girdle each cell and act like fences, preventing any membrane proteins afloat in the lipid bilayer from drifting across the boundary from one side of the cell to the other.

3. *Communicating junctions* Communicating junctions called **gap junctions** are large enough to permit the passage of small molecules, such as sugar molecules and amino acids, from one animal cell to another, but small enough to prevent the passage of larger molecules, such as proteins. In plants, the plasma membranes of adjacent cells come together through pairs of holes in the walls. The cytoplasmic connections that extend through such holes are called **plasmodesmata.**

How a Cell Communicates With the Outside World

Every cell is a prisoner of its lipid envelope, unable to communicate with the outside world except by means of the proteins that traverse its lipid shell. Anchored within the lipid layer by nonpolar segments, these proteins are the "senses" of the cell. They detect the presence of other molecules, often initiating responses within the cell. They provide doors into the cell through which food molecules, ions, and other molecules may pass, but like protective doormen, they are picky about whom they admit. They form the shape of a cell and bind one cell to another. They provide a cell with its identity by means of surface name tags that other cells can read. This diverse collection of proteins, together with the lipid shell within which they are embedded, constitute the cell's membrane system, one of a cell's most fundamental features.

EVOLUTIONARY VIEWPOINT

The "motor" that drives the spinning of the tail-like flagella that bacteria use to swim is powered by pumping protons and may be the evolutionary ancestor of the eukaryotic proton pumps that use the energetic electrons harvested by oxidative metabolism to make ATP.

SUMMARY

1. Every cell is encased within a plasma membrane, a bilayer sheet of phospholipids in which proteins and glycolipids are embedded. The proteins act as "doors" for the transport of molecules into and out of the cell.

2. Three types of transport across the plasma membrane do not require energy: diffusion, osmosis, and facilitated diffusion. These types of passive transport involve the movement of molecules with a concentration gradient, from areas of high concentration to areas of low concentration.

3. Diffusion is the random movement of molecules from an area of high concentration to an area of low concentration.

4. Osmosis is the diffusion of water across a selectively permeable membrane.

5. Because cells contain significant concentrations of sugars, amino acids, and other solutes and thus constitute hypertonic environment, water tends to diffuse into them. In the process, osmotic pressure builds until equilibrium is reached or cells lacking a cell wall or other means of support rupture.

6. Facilitated diffusion is the transport of materials across the plasma membrane in the direction of lowest concentration, using a carrier protein specific for that material.

7. Some types of transport across the plasma membrane require energy. These include endocytosis and exocytosis, as well as the active transport mechanisms: the sodium-potassium pump, the proton pump, and coupled channels.

8. Endocytosis is the transport of materials into a cell by enfolding the materials within vesicles. Phagocytosis is the transport of a large particle into the cell; pinocytosis is the transport of smaller particles into the cell.

9. Exocytosis is the expulsion of materials from secretory vesicles at the cell surface.

10. The sodium-potassium pump is a plasma membrane protein channel that constantly pumps three sodium (Na^+) ions out of the cell for every two potassium (K^+) ions that it pumps into the cell. The sodium-potassium pump operates against a concentration gradient and maintains very low concentrations of Na^+ ions within the cell.

11. Proton pumps, which are located on the inner membranes of mitochondria and chloroplasts, transport protons (H^+ ions) out of the plasma membrane. The entrance of H^+ ions into the cell through special channels is tied to the formation of ATP.

12. Proton and sodium-potassium pumps also function indirectly in coupled channels. They provide the concentration gradients necessary so that a molecule required by the cell can link itself to the transport of sodium (Na^+) or hydrogen (H^+) ions across the cell membrane.

13. Instead of transporting molecules, many proteins embedded within the plasma membrane transmit information into the cell. These proteins, called cell-surface receptors, initiate chemical activity inside the cell in response to the binding of specific molecules on the cell surface.

14. Cells use both proteins and glycolipids as cell-surface markers. These permit cells of a given tissue to identify one another and also provide your body with a means of identifying foreign cells.

15. There are three general classes of cell junctions (connections between cells) in animal cells: (1) adhering junctions, or desmosomes, which hold cells together; (2) organizing junctions, or tight junctions, which partition the plasma membrane into separate compartments; and (3) communicating junctions, called gap junctions, which pass small molecules from one cell to another. In plants, cell walls have openings that allow cytoplasmic connections, called plasmodesmata, to connect adjacent cells.

REVIEWING THE CHAPTER

1. What is the molecular structure of membranes?

2. What is the role of phospholipids in cell membranes?

3. What is a lipid bilayer?

4. What is the fluid nature of the lipid bilayer?

5. How does the architecture of the plasma membrane vary?

6. How does the cell membrane regulate interactions with its environment?

7. What are the passive transport processes for materials passing into and out of cells?

8. What are the energy-requiring processes for materials passing into and out of cells?

9. How do endocytosis and exocytosis differ?

10. How is active transport an energy-expending process?

11. How does the sodium-potassium pump work?

12. What does the proton pump do in a cell?

13. What is the role of coupled channels in a cell?

14. How do cells receive information?

15. How do cells express their identity?

16. How are cells physically connected?

17. How do cells communicate with the outside world?

COMPLETING YOUR UNDERSTANDING

1. How is the cell membrane correlated with the existence of life?

2. How fluid is the lipid bilayer?

3. What is the relationship between the lipid bilayer and water-soluble molecules?

4. How do glycoproteins and glycolipids differ?

5. The difference in molecule concentration between zones of few molecules and zones with many molecules is called

  a. active transport.

  b. passive transport.

  c. diffusion.

  d. osmosis.

  e. a concentration gradient.

  f. None of the above is correct.

6. Molecules that are dissolved in water are called _____.

7. How does a selectively permeable membrane relate to diffusion and osmosis?

8. Give an example of where active and passive transport occur in a cell.

9. If a freshwater protist floated down the Mississippi River into the saline Gulf of Mexico, which of the following would apply to the cell once it entered the Gulf?

  a. The cell would be hypertonic to its surrounding solution.

  b. The cell would be hypotonic to its surrounding solution.

  c. Water would diffuse out of the cell.

  d. Water would diffuse into the cell.

  e. The cell would be isotonic to its surrounding solution.

10. How would you design an experiment to show the movement of proteins within cell membranes?

11. Facilitated diffusion is a process that utilizes _____ to transport molecules across the plasma membrane.

12. If a single-celled amoeba ingests an entire yeast cell, enclosing it within a vesicle, this process is called

  a. osmosis.

  b. phagocytosis.

  c. pinocytosis.

  d. exocytosis.

  e. facilitated diffusion.

  f. active transport.

13. What are the different location sites of the sodium-potassium and proton pumps in cells?

14. The energy needed to run the sodium-potassium pump in cells comes from _____.

15. Which of the following means of moving things into or out of cells requires an expenditure of energy? [*two answers required*]

  a. Diffusion

  b. Osmosis

  c. Facilitated diffusion

  d. Active transport

  e. Sodium-potassium pump

16. Which of the following can move molecules *against* a concentration gradient? [*two answers required*]

  a. Diffusion

  b. Osmosis

  c. Facilitated diffusion

  d. Active transport

  e. Sodium-potassium pump

17. The mechanism by which ATP is produced in cells is called

  a. chemiosmosis.

  b. diffusion.

  c. passive transport.

  d. exocytosis.

  e. coupled channels.

  f. active transport.

18. What mechanism is responsible for the production of almost all the energy that you obtain from food?

19. How do chloride channels relate to the human disease cystic fibrosis?

20. How do cell-surface receptors function in cells?

21. What is an MHC, and how does it relate to the human immune system?

22. If your blood type is AB, how can it be distinguished from other blood types in the human population?

23. In plants, the cytoplasmic connections that extend through holes between cells are called

  a. desmosomes.

  b. gap junctions.

  c. tight junctions.

  d. adhering junctions.

  e. organizing junctions.

  f. plasmodesmata.

24. Cells maintain many internal metabolite molecules at high concentrations by coupling their transport into the cell to the transport of sodium and potassium ions by the sodium-potassium pump. What happens to all the potassium ions that are constantly being pumped into the cell?

FOR FURTHER READING

Bretscher, M. S. 1985. The molecules of the cell membrane. *Scientific American*, Oct., 100–108. A good description of the structure of the cell membrane and of how transmembrane proteins are anchored within the lipid bilayer.

Edelson, E. 1990. Conduits for cell/cell communication. *Mosaic* 21: 48–56. Discussion of how rings of protein embedded in cell membranes provide critical links between cells.

Hakomori, S. 1986. Glycosphingolipids. *Scientific American*, May, 44–53. Discusses the important role that carbohydrate chains attached to lipid molecules play in cell-to-cell recognition, an area of intensive present-day research.

Husten, L. 1990. Cholesterol to go. *Discover*, Dec., 30–31. A report on ongoing research on the human cell's cholesterol-removing system and how it lowers the risk of heart disease.

Luna, E. J., and A. L. Hitt. 1992. Cytoskeleton and plasma membrane interactions. *Nature*, Nov. 6, 955–62. A longish article that looks in detail at the cytoskeleton fibers that anchor the plasma membrane. This is an advanced article, but has some very clear illustrations.

McNeil, P. 1991. Cell wounding and healing. *American Scientist*, May, 222–35. A report on how the opening of cell membranes by wounds provides a route for the release of molecules that maintain and repair tissue.

Unwin, N., and R. Henderson. 1984. The structure of proteins in biological membranes. *Scientific American*, Feb., 78–94. A lucid account of how proteins are anchored within membranes by means of nonpolar segments.

EXPLORATIONS

Interactive Software: **Cystic Fibrosis**

This interactive exercise allows students to explore the way in which transport proteins influence the passage of water in and out of cells by examining the effects of a mutation that disables a particular transport protein, the one responsible for chloride ion transport. Students can explore the consequences of changing extracellular chloride ion concentrations on transport of water into lung cells, learning that high extracellular concentrations lead to inhibition of water transport into cells because of osmosis. Students can then invoke the CF mutation, which prevents the chloride channel from opening and thus leads to a buildup of chloride ion outside lung cells. The failure of water transport into lung cells that results is the direct cause of the symptoms of cystic fibrosis.

Questions to Explore:

1. What is the effect of increased extracellular chloride ion concentration on water movement into the cell?

2. What is the effect of disabling the chloride channel upon water movement?

3. Can artificially decreasing the chloride ion concentration in extracellular fluids counteract cystic fibrosis?

Interactive Software: **Active Transport**

This interactive exercise allows students to explore how substances are transported across membranes against a concentration gradient (that is, toward a region of higher concentration). The exercise presents a diagram of a coupled channel within a membrane through which amino acids are pumped into the cell. By altering ATP concentrations, and relative inside/outside concentrations of the amino acids, the user can explore the consequences of cellular ATP expenditure on amino acid accumulation by the cell. Because the amino acid transport channel is coupled to the ATP-driven sodium-potassium pump, users will discover that both ATP and amino acid levels have important influences.

Questions to Explore:

1. How does the level of ATP influence the operation of the sodium-potassium pump?

2. Does a fall in cellular levels of ATP inhibit cellular uptake of amino acids?

3. Does an increase in extracellular levels of amino acid always lead to an increased expenditure of ATP?

4. As long as ATP is readily available, is there any condition under which an increase in extracellular levels of amino acid does not result in increased transport of amino acid into the cell?

5. How is the ATP-driven sodium pump affected by increased levels of sugar?

6. What would happen if the amount of ATP supplied to the cell was suddenly decreased?

Interactive Software: **Cell-Cell Interactions**

The key role of cell surface receptors in communication between cells is explored in this interactive exercise. The user alters the design of a hypothetical receptor and its intracellular signaling system, and assesses the consequences on intercellular communication. The user explores the critical roles of receptor design, G-proteins, and phosphorylation cascades in amplifying a signal, evaluating how changes in these elements alter the communication between two cells.

Questions to Explore:

1. Could this cell communicate with adjoining cells without G-proteins? How?

2. What effect does altering the structure of the receptor have on cell-to-cell communication?

3. What is the relationship between phosphorylation cascades and signal amplification?

4. What is the optimum arrangement of variables for the most efficient cell-to-cell communication?

5. Is any one variable more critical than others for cell-to-cell communication to occur? Which one? Why?

Part

3

ENERGY

When looking at a spectacular sunset like this, we rarely think that this light energy is ultimately responsible for most life on earth.

CHAPTER 6

ENERGY AND METABOLISM

Which came first—the need for energy to find food, or the need for food to find energy? This cheetah exerts a considerable amount of energy in pursuit of its prey, energy that will be replenished when the cheetah sits down to its dinner.

FOR REVIEW

Here are some important terms and concepts that have been discussed in previous chapters and that you will encounter again in this chapter. Review them before proceeding if necessary.

Oxidation/reduction reactions (*chapter 2*)
Nature of chemical bonds (*chapter 2*)
Nucleotides (*chapter 2*)
Protein structure (*chapter 2*)
pH scale (*chapter 2*)

"You are what you eat" is a common saying, and an accurate one. Your body is built entirely from materials you eat or drink. What you eat is also the source of the energy that drives everything you do.

If you stopped eating, you would soon begin to lose weight as your body used up its stored energy. Without some outside source of energy, you would eventually die. The same is true of all living things. Deprived of a source of energy, life stops. This happens because most of the significant properties by which life is defined—metabolism, reproduction, development, and heredity—use energy (figure 6.1). Once energy is used, it is dissipated as heat and cannot be used again. To keep life going, more energy must be supplied, like putting more logs on a fire.

Life can be viewed as a constant flow of energy, which is channeled by organisms to do the work of living. This chapter focuses on energy—what it is, and how organisms capture, store, and use it. Chapters 7 and 8 explore in more detail the energy-capturing and energy-using engines of cells, a network of chemical reactions that is the highway system for the energy of your body. This living chemistry—the total of all the chemical reactions that an organism performs—is called **metabolism.**

What Is Energy?

Energy is defined as the ability to bring about change or, more generally, as the capacity to do work. Instinctively, we all know something about energy. It is "work," such as the force of a falling boulder, the pull of a locomotive, or the swift dash of a horse; it is also "heat," such as the blast from an explosion or the warmth of a fire. Energy can exist in many forms: as mechanical force, heat, sound, an electrical current, light, radioactivity, and the pull of a magnet. All are able to create change, to do work.

Energy exists in two states. Some energy is actively engaged in doing work, such as driving a speeding bullet or lifting a brick. This form of energy is called **kinetic energy,** or energy of motion. Other energy is not actively doing work but has the capacity to do so, just as a boulder perched on a hilltop has the capacity to roll downhill. This form of energy is called **potential energy,** or stored energy. Much of the work performed by living organisms involves the transformation of potential energy to kinetic energy (figure 6.2).

Figure 6.1 All life processes use energy.
These Japanese monkeys are huddled together sharing their body warmth, which they generate by converting body fat to ATP and then expending ATP to generate heat. Although their fur provides them with excellent insulation, they are constantly losing heat to the cold around them and would not survive long if they could not generate heat to replace that which is lost.

Figure 6.2 The energy of motion.
Transforming potential energy into kinetic energy, this cheetah has just sprung into action.

> *Energy is the capacity to do work, and it can exist in many forms. Energy that is actively engaged in doing work is called kinetic energy. Energy that is stored for later use is called potential energy.*

Because energy exists in many forms, there are many ways to measure it. The most convenient way is in terms of heat because all other forms of energy can be converted into heat. Indeed, the study of energy is called **thermodynamics**—that is, "heat changes." The unit of heat most commonly employed in biology is the **kilocalorie (kcal).** One kilocalorie is equal to 1,000 calories, and one calorie is the heat required to raise the temperature of 1 gram of water 1 degree (from 14.5 to 15.5 degrees Celsius). Another term for kilocalorie is Calorie (with a capital C), which is the "calorie" often referred to in discussions of diet and nutrition.

The Laws of Thermodynamics: How Energy Changes

All of the changes in energy that take place in the universe, from nuclear explosions to the buzzing of a bee, are governed by two laws called the laws of thermodynamics. The **first law of thermodynamics** concerns amounts of energy. It states that energy can change from one form to another and can transform from potential energy to kinetic energy, but that it can neither be created nor destroyed. The total amount of energy in the universe is constant.

> *The first law of thermodynamics states that energy cannot be created or destroyed; it can only undergo conversion from one form to another.*

A ground squirrel gnawing on a nut is busy acquiring energy. He is not creating new energy, but rather is transferring the potential energy stored in the nut's tissues to his own body, where it will fuel running, digging, and all his other daily activities.

The potential energy in the nut's tissues is stored in chemical bonds. A covalent chemical bond, as discussed in chapter 2, is created when two atomic nuclei share electrons, and breaking such a bond requires energy to pull the nuclei apart. Indeed, the strength of a covalent bond is measured by the amount of energy required to break it. For example, it takes 98.8 kilocalories of energy to break a mole (the atomic weight of a substance expressed in grams) of carbon-hydrogen (C—H) bonds.

What happens to the nut's potential energy after the squirrel eats the nut? Some of the potential energy is transferred to other forms of potential energy—for example, stored as fat. Another portion accomplishes mechanical work, such as bending blades of grass and running. Almost half is dissipated to the environment as heat, where it speeds up the random motions of molecules. This energy is not lost, but rather is converted to a nonuseful form: random molecular motion.

The **second law of thermodynamics** concerns this transformation of potential energy to the kinetic energy of random molecular motion—that is, to heat. It states that all objects in the universe tend to become more disordered and that the disorder in the universe is continuously increasing. The idea behind this law is easy to understand and part of everyone's experience. For example, it is much more likely that a stack of six soda cans will tumble over than six cans will spontaneously leap one onto another and form a stack. Stated simply, disorder is more likely than order. This is true of a child's room, of the desk where you study, of a waiting crowd of people—and of molecules.

> *The second law of thermodynamics states that disorder in the universe is constantly increasing. Energy spontaneously converts to less organized forms.*

At normal temperatures, all molecules dance about randomly. As energy is transferred from one molecule to another, some always leaks away as kinetic energy of motion, thus increasing the random motion of molecules. This form of kinetic energy is called heat energy.

> *Heat is the energy of random molecular motion.*

With every transfer of energy, more potential energy is dissipated as heat. Although heat can be harnessed to do work when there is a gradient (that is how a steam engine works), it is generally not a useful source of energy for biological systems. Thus, from a biological point of view, although the total amount of energy does not change, the amount of *useful* energy available to do work decreases as progressively more energy is degraded to heat.

When energy becomes so randomized and uniform in a system that it is no longer available to do work, the energy lost to disorder is referred to as **entropy.** Entropy is a measure of the disorder of a system. Sometimes, the second law of thermodynamics is stated simply as "Entropy increases." When the universe was formed about fourteen billion years ago, it had all the potential energy it will ever have. It has become progressively more disordered ever since, with every energy exchange frittering away useful energy and increasing entropy. Someday, all of the energy will be random and uniform in distribution; the universe will have wound down like an abandoned clock. No stars will shine, no waves will break upon a beach. But scientists speculate that this final state will not be reached for perhaps another 100 billion years.

The first law of thermodynamics states that the universe as a whole is a closed system: Energy does not come in or go out. The earth, however, is not a closed system; it is constantly receiving energy from the sun. Estimates indicate that every year the earth receives in excess of 13×10^{23} calories of energy from the sun, which is equal to two million trillion calories per second. Much of this energy heats up the oceans and continents, but some is captured by photosynthetic organisms; plants, algae (photosynthetic protists), and photosynthetic bacteria (figure 6.3).

In photosynthesis, energy acquired from sunlight is converted to chemical energy, transforming small molecules (water and carbon dioxide) into ones that are more complex (sugars). The energy is stored as potential energy in the bonds of the sugar molecules. This energy then can be shifted to other molecules by forming different chemical bonds or can be converted into motion, light, electricity, and heat. During each shift or conversion, more energy is dissipated as heat. Energy continuously flows through the biological world, with new energy from the sun constantly flowing in to replace the energy dissipated as heat.

Life converts energy from the sun to other forms of energy that drive life processes. The energy is never lost, but as it is used, more and more is converted to heat energy, a form of energy that is not useful in performing biological work.

Figure 6.3 All the energy that powers life is captured from sunlight.
These trees in a Michigan forest are the first of many plants to snare sunlight as it falls toward the forest floor.

Energy stored in chemical bonds can be transferred to new chemical bonds, with the electrons shifting from one energy level to another. In some (but not all) of these chemical reactions, electrons actually pass from one atom or molecule to another. This class of chemical reaction is called an oxidation/reduction reaction (see chapter 2). **Oxidation/reduction (or redox) reactions** are critical to the flow of energy through living systems.

Upon losing an electron, an atom or molecule is said to be **oxidized,** and the process by which this occurs is called **oxidation.** The name reflects the fact that, in biological systems, oxygen, which strongly attracts electrons, is the most frequent electron acceptor.

Upon gaining an electron, an atom or molecule is said to be **reduced,** and the process by which this occurs is called **reduction.** Oxidation and reduction always take place together because every electron that is lost by one atom (oxidation) is gained by some other atom (reduction) (see figure 2.5).

Oxidation is the loss of an electron; reduction is the gain of one.

Oxidation/reduction reactions play a key role in energy flow through biological systems because the electrons that pass from one atom to another carry their own potential energy—that is, they maintain their distance from the nucleus. For example, in photosynthesis, energy that originally enters the system when light boosts the electrons in an energy-trapping pigment to higher-energy levels is passed from one molecule to another by these electrons. Until the electrons return to their original lower-energy level, they continue to store potential energy.

Atoms can store potential energy by means of electrons that orbit at higher-than-usual energy levels. When such an "energetic" electron is removed from one atom (oxidation) and donated to another (reduction), it carries the energy with it and orbits the second atom's nucleus at the higher-energy level.

Oxidation/Reduction: The Flow of Energy in Living Things

The flow of energy into the biological world comes from the sun, which shines a constant beam of light on the earth and its moon. Life exists on the earth because some of that continual flow of energy can be captured and transformed into chemical energy, which can be transferred from one organism to another and used to create cattle, and fleas, and you.

Where is the energy in sunlight, and how is it captured? Answers to these questions require looking more closely at the atoms on which sunlight shines. As described in chapter 2, an atom is composed of a central nucleus surrounded by one or more orbiting electrons, and different electrons possess different amounts of energy, depending on how far from the nucleus they are and how strongly they are attracted to it. Light (and other forms of energy) can boost an electron to a higher energy level. The effect of this is that the added energy is stored as potential energy—that is, as chemical energy that the atom can later release by dropping the electron back to its original energy level.

In biological systems, electrons often do not travel alone from one atom to another, but rather in the company of a proton. Recall that a proton and an electron together make up a hydrogen atom (H). Thus, oxidation/reduction in a chemical reaction usually involves the removal of hydrogen atoms from one molecule (oxidation) and the addition of hydrogen atoms

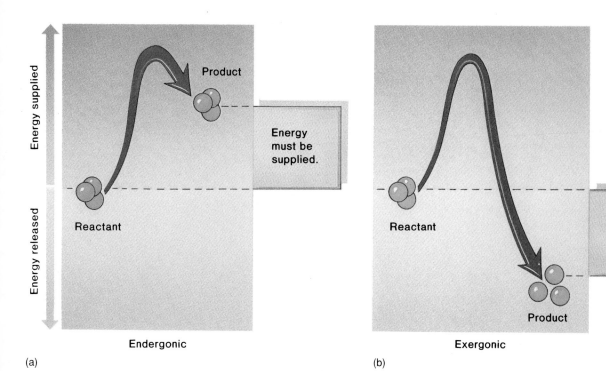

Endergonic (a)

Exergonic (b)

Figure 6.4 Energy in chemical reactions.
(a) *In an endergonic reaction, the products of the reaction contain more energy than do the reactants, so the extra energy must be supplied for the reaction to proceed.* (b) *In an exergonic reaction, the products contain less energy than do the reactants, and the excess energy is released.*

to another molecule (reduction). For example, in photosynthesis, hydrogen atoms are transferred from water to carbon dioxide, reducing the carbon dioxide to form glucose:

$$6CO_2 + 6H_2O + Energy \rightarrow C_6H_{12}O_6 + 6O_2$$
Carbon Water   Glucose Oxygen
dioxide

In this reaction, electrons move to higher energy levels. The conversion of carbon dioxide to glucose stores 686 kilocalories of energy in the chemical bonds of the glucose.

The energy stored in a glucose molecule is released in a process called **cellular respiration,** in which the glucose is oxidized. Hydrogen atoms are lost by glucose and gained by oxygen:

$$C_6H_{12}O_6 + 6O_2 \rightarrow 6CO_2 + 6H_2O + Energy$$
Glucose Oxygen   Carbon   Water
                      dioxide

The oxidation of glucose releases 686 kilocalories of energy, the same amount that was stored in making it.

Activation Energy: Preparing Molecules For Action

As the laws of thermodynamics predict, all chemical reactions tend to proceed spontaneously toward a state of maximum disorder and minimum energy. Reactions in which the products contain more energy than do the reactants require an input of usable energy from an outside source before they can proceed. These reactions are not spontaneous and are called **endergonic** (figure 6.4*a*). Reactions in which the products contain less energy than do the reactants release the excess usable energy (called "free energy"). These reactions are spontaneous and are called **exergonic** (figure 6.4*b*).

> *Any reaction producing products that contain less free energy than that possessed by the original reactants tends to proceed spontaneously.*

If all chemical reactions that release free energy tend to occur spontaneously, why have not all such reactions already occurred? When gasoline is ignited, the resulting chemical reaction releases free energy. So why does not the world's gasoline supply burn up right now? The explanation is that most reactions require an input of energy to get started, like the heat from the flame of a match. Before new chemical bonds with less energy can form, the existing bonds must be broken. Extra energy, called **activation energy,** is required to destabilize existing chemical bonds and to initiate a chemical reaction (figure 6.5*a*).

The speed, or reaction rate, of an exergonic reaction does not depend on how much energy the reaction releases, but rather on the amount of activation energy required for the reaction to begin. Reactions that involve larger activation energies

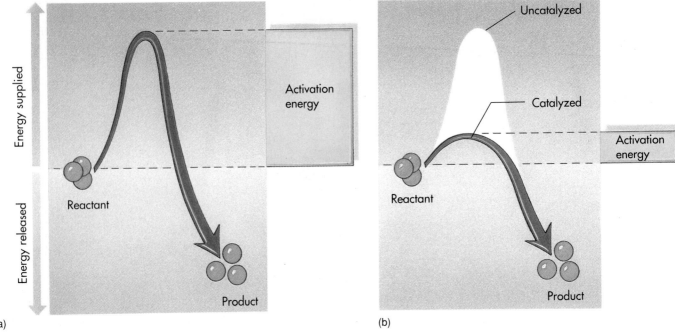

(a)   (b)

Figure 6.5 Exergonic reactions do not proceed spontaneously because energy is required to get them going.
(a) *Activation energy must be supplied to destabilize existing chemical bonds.*
(b) *Enzymes catalyze particular reactions by lowering the amount of activation energy required to initiate the reactions.*

tend to proceed more slowly because fewer molecules succeed in overcoming the initial energy hurdle. However, activation energies are not fixed constants. Putting stress on particular chemical bonds can make them easier to break. The process of influencing chemical bonds in a way that lowers activation energies is called **catalysis** (see chapter 2), and substances that perform catalysis are **catalysts.** Catalysts cannot violate the basic laws of thermodynamics; for example, they cannot make an endergonic reaction proceed spontaneously. Only exergonic reactions proceed spontaneously, and catalysis cannot change that fact. What catalysts *can* do is make a reaction rate much faster.

The speed of a reaction depends on the activation energy necessary to initiate it. Catalysts reduce the amount of activation energy required and thus speed up reactions.

Enzymes: The Catalysts of the Cell

The chemistry of living things—metabolism—is organized by controlling the points at which catalysis takes place. Therefore, life is a process regulated by **enzymes,** which are agents that perform catalysis in living organisms. Enzymes are globular proteins whose shapes are specialized to form temporary associations with the molecules that are reacting. By putting stress on

particular chemical bonds, enzymes lower the amount of activation energy required for new bonds to form (figure 6.5b). The reaction thus proceeds much faster than it would otherwise. Because the enzymes themselves are not changed, they can be used over and over.

To understand how an enzyme works, consider the joining of carbon dioxide (CO_2) and water (H_2O) to form carbonic acid (H_2CO_3):

$$CO_2 \quad + \quad H_2O \quad \rightleftarrows \quad H_2CO_3$$
Carbon   Water   Carbonic
dioxide   acid

This reaction can proceed in either direction, but in the absence of an enzyme, it proceeds very slowly because there is a significant need for activation energy. Only about two hundred molecules of carbonic acid form in one hour. Given the speed at which events usually occur within cells, this reaction rate is like a snail racing in the Indianapolis 500. Cells overcome this problem by employing an enzyme called carbonic anhydrase (enzymes are usually given names that end in *-ase*), which accelerates the reaction dramatically. In the presence of carbonic anhydrase, an estimated six hundred thousand molecules of carbonic acid form every second! The enzyme speeds up the reaction rate about ten million times.

Cells employ proteins called enzymes as catalysts.

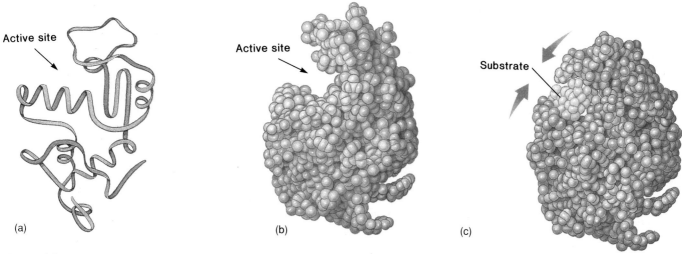

Active site

Active site

Substrate

(a)   (b)   (c)

Figure 6.6 The induced-fit model of enzyme action.
The tertiary structure of an enzyme (in this case, lysozyme) is shown as a ribbon in (a) *and as a three-dimensional structure in* (b) *and* (c). *The active site of the enzyme is a groove through the middle of the enzyme* (a *and* b). *This active site fits the shape of the substrate. The entry of the substrate into the active site induces*

the enzyme to alter its shape slightly to embrace the substrate more intimately (c). *This induced fit positions an amino acid in the enzyme right next to the bond between two components of the substrate. The amino acid "steals" an electron from the bond, causing it to break.*

Thousands of different kinds of enzymes have been described, each catalyzing a different chemical reaction. By facilitating particular chemical reactions, the enzymes in a cell determine the course of metabolism in that cell, much as traffic lights determine the flow of traffic in a city. Not all cells contain the same enzymes, which is why there is more than one type of cell. The chemical reactions within a red blood cell are very different from those within a nerve cell because the cytoplasm and membranes of a red blood cell contain a different array of enzymes.

How Enzymes Work

Enzymes are globular proteins with one or more pockets, or clefts, on their surface that resemble deep creases in a prune. These surface depressions, called **active sites,** are where catalysis occurs. The molecule on which the enzyme acts, called a **substrate,** must fit precisely into the active site so that many of its atoms nudge up against atoms of the enzyme, much like a foot in a tight-fitting shoe. Enzymes are not rigid, however, and in some cases, the binding of the substrate may induce the enzyme to adjust its shape slightly, allowing a better fit. This adjustment is called the **induced-fit model** of enzyme action (figure 6.6).

When a substrate molecule binds to the active site of an enzyme, amino acid side groups of the enzyme are placed against certain bonds of the substrate, just as when you sit in a chair, certain parts of you press against the seat (figure 6.7). These amino acid side groups chemically interact with the substrate, usually by stressing or distorting a particular bond, lowering the activation energy needed to break the bond.

Enzymes typically catalyze only one or a few different chemical reactions because they are very "picky" in their choice of substrate. Each enzyme's active site is shaped so that only a certain substrate molecule will fit into it.

Factors Affecting Enzyme Activity

Enzyme activity is affected by any change in condition that alters the enzyme's three-dimensional shape. These changes include alterations in temperature and pH, as well as the presence of specific chemicals that bind to the protein.

Changes in Temperature

The shape of an enzyme is determined by hydrogen bonds that hold its peptide chains in particular positions and also by the tendency of noncharged (nonpolar) segments of the enzyme to avoid water. Chemists call interactions of this second kind hydrophobic, or water-hating, interactions (see chapter 2). Both hydrogen bonds and hydrophobic interactions are easily disrupted by slight changes in temperature. Most human enzymes function best within a relatively narrow temperature range: between 35 and 40 degrees Celsius (close to body temperature). Below this temperature range, the bonds that determine enzyme shape are not flexible enough to permit the induced-fit change sometimes necessary for catalysis; above this temperature range, the bonds are too weak to hold the enzyme's peptide chains in the proper position (figure 6.8a). In contrast, bacteria that live in hot springs have enzymes with stronger bonding between their peptide chains and therefore can function at temperatures of 70 degrees Celsius or higher.

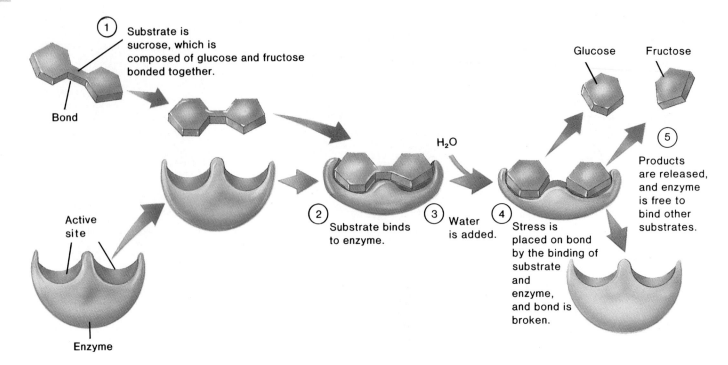

Figure 6.7 The catalytic cycle of an enzyme.
Enzymes increase the speed with which chemical reactions occur but are not themselves altered in the process. In the reaction illustrated here, the enzyme splits the sugar sucrose (steps 1, 2, 3, and 4) into its two parts, the simpler sugars glucose and fructose. After the enzyme releases the resulting glucose and fructose, it is then ready to bind another molecule of sucrose (5) and begin the catalytic cycle again.

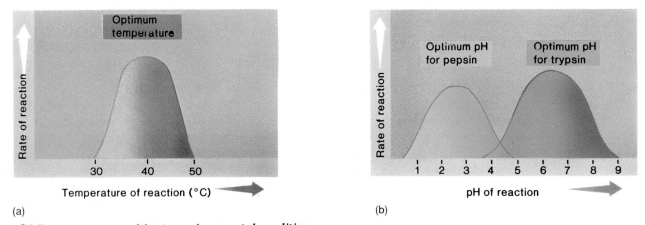

Figure 6.8 Enzymes are sensitive to environmental conditions.
The activity of an enzyme is influenced by both temperature and pH. (a) The optimum temperatures at which most human enzymes function. (b) The range of pH values in which two human digestive enzymes—pepsin and trypsin—work best. As you can see, pepsin operates at a much lower (acidic) pH than trypsin.

Changes in pH

A third kind of bond that acts to hold the peptide chains of enzymes in position is the bond that forms between oppositely charged amino acids, such as glutamic acid (–) and lysine (+). These bonds are sensitive to hydrogen ion (H^+) concentration. The more H^+ ions available in the solution, the fewer negative charges and the more positive charges are present in the enzymes. This is because the H^+ ions combine with the negative charges and neutralize them. For this reason, most enzymes have a pH optimum, usually in the range of pH 6 to 8, just as they have a temperature optimum. Enzymes that are able to function in very acidic environments have amino acid sequences that maintain their ionic and hydrogen bonds even in the presence of high levels of hydrogen ions. For example, the enzyme pepsin digests proteins in your stomach at pH = 2, a very acidic level (figure 6.8*b*).

> *Temperature and pH have an important influence on chemical reactions within cells because the enzymes that catalyze these reactions require a specific shape to be effective, and that shape is easily changed by alterations in pH and temperature.*

How Enzyme Activity is Regulated

Enzyme activity is sensitive not only to temperature and pH but also to the presence of specific chemicals that bind to the enzyme and cause changes in its shape. By means of these specific chemicals, a cell is able to regulate which enzymes are active and which are inactive at a particular time. When the binding of the chemical alters the shape of the enzyme and thus shuts off enzyme activity, the chemical is called an **inhibitor;** when the change in the enzyme's shape is necessary for catalysis to occur, the chemical is called an **activator.**

Biochemical Pathways

Your body contains over a thousand kinds of enzymes, which catalyze a bewildering variety of reactions. Many of the reactions occur in sequences, called **biochemical pathways,** in which the product of one reaction becomes the substrate for another (figure 6.9). Biochemical pathways are the organizational units of metabolism, just as the many metal parts of an automobile are organized into distinct subassemblies such as the carburetor, transmission, and brakes.

How Enzymes Regulate Biochemical Pathways

The change in shape that occurs when an activator or inhibitor binds to an enzyme is called an **allosteric change** (from the Greek *allos,* meaning "other," and *steros,* meaning "shape"). Enzymes usually have special binding sites for the activator and inhibitor molecules that affect them, and these binding sites are different from their active sites (figure 6.10). The enzyme catalyzing the first step in a biochemical

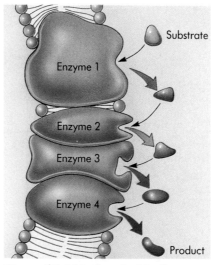

Figure 6.9 A biochemical pathway.
The original substrate is acted on by enzyme 1, changing the substrate to a new form, which is then acted on by enzyme 2. Each enzyme in a pathway acts on the product of the previous stage. Not all pathways are embedded in membranes, although many are.

pathway often has an inhibitor-binding site to which the molecule produced by the last step in the series binds. As the amount of this molecule builds up in the cell, it begins to bind to the initial enzyme in the biochemical pathway, thus inhibiting the activity of that enzyme. By this process, the biochemical pathway is shut down when it is no longer needed. Such **end-product inhibition** (or, alternatively, **feedback inhibition**) is a good example of the way many enzyme-catalyzed processes within cells are self-regulating (figure 6.11).

> *Enzyme activity is regulated by allosteric changes in enzyme shape. These changes result when specific, small molecules, molecules that are not substrates of that enzyme, bind to the enzyme.*

Coenzymes: Tools Enzymes Use to Aid Catalysis

Enzymes often use additional chemical components called **cofactors** as tools to aid catalysis. For example, many enzymes have metal ions locked into their active sites, and these ions help draw electrons from substrate molecules. The enzyme carboxypeptidase chops up proteins by using a zinc ion to draw electrons away from the bonds being broken. The zinc in

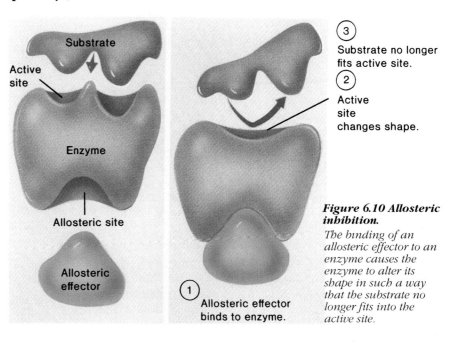

Figure 6.10 Allosteric inhibition.
The binding of an allosteric effector to an enzyme causes the enzyme to alter its shape in such a way that the substrate no longer fits into the active site.

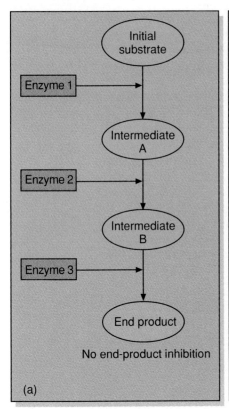

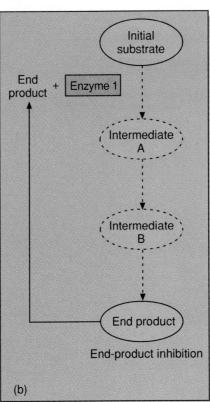

Figure 6.11 End-product inhibition.
(a) *A biochemical pathway with no end-product inhibition.* (b) *A biochemical pathway in which the final end product becomes the allosteric effector for the first enzyme in the pathway. In other words, the formation of the pathway's final end product stops the pathway.*

carboxypeptidase is an example of a cofactor. Many of the trace elements, such as molybdenum and manganese, which are necessary for your health, also use metal ions in this way. A cofactor that is a nonprotein organic molecule is called a **coenzyme.** Many of the vitamins that your body requires are parts of coenzymes.

Many enzymes employ metal ions or organic molecules, called cofactors, to facilitate their activity. Cofactors that are nonprotein organic molecules are called coenzymes.

Coenzymes As Electron Acceptors

In many enzyme-catalyzed oxidation/reduction reactions, energy-bearing electrons are passed from the active site of the enzyme to a coenzyme that serves as the electron acceptor. The coenzyme then carries the electrons to a different enzyme that is catalyzing another reaction, releases the electrons (and the energy they bear) to that reaction, and then returns to the original enzyme for another load of electrons. In most cases, the electrons are paired with protons as hydrogen atoms. Just as armored cars transport cash around a city, so coenzymes shuttle energy, in the form of hydrogen atoms, from one place to another in a cell.

Examples of Coenzymes in Biological Systems

One of the most important coenzymes is the electron acceptor **nicotinamide adenine dinucleotide,** usually referred to by the abbreviation **NAD+.** When NAD+ acquires a hydrogen atom (along with the hydrogen atom's electron) from an enzyme's active site, it becomes reduced to NADH. The energetic electron of the hydrogen atom is then carried by the NADH molecule, like money in your wallet. The oxidation of foodstuffs in your body, from which you get the energy to drive your life, takes place by the cell's stripping of electrons from food molecules and donating them to NAD+, thus forming a wealth of NADH. This wealth is the principal energy income of your cells. However, much of the energy in NADH is eventually converted to another currency when the NADH, in turn, donates its acquired electrons to another carrier, becoming oxidized in this process back to NAD+.

Another important coenzyme is **FAD (flavin adenine dinucleotide).** FAD, an electron carrier used in cellular respiration, can carry two hydrogen atoms and the two electrons in the hydrogen atoms. When FAD accepts these two hydrogen atoms, it becomes reduced to FADH$_2$. The energy contained in the electrons that FADH$_2$ carries is then passed along to another carrier in the electron transport chain. Once FADH$_2$ donates its "package" of electrons, it then becomes oxidized back to FAD and is once again ready to pick up more hydrogen atoms.

Coenzymes and their roles in biochemical pathways are discussed further in chapters 7 and 8. For now, it is only necessary to recognize NAD+ and FAD as important coenzymes that accept hydrogen atoms and their electrons and carry the energy contained in the electrons to other carriers in the pathway.

NAD+ and FAD are coenzymes that accept hydrogen atoms and their electrons and carry the energy contained in the electrons to other carriers. In accepting the electrons, these coenzymes become reduced. In donating the electrons to other carriers, these coenzymes become oxidized.

ATP: The Energy Currency of Life

The chief energy currency of all cells is a molecule called **adenosine triphosphate (ATP)** (see chapter 1). Just as much of the energy that plants harvest during photosynthesis is channeled into ATP production, so is most of the NADH that soaks up the energy resulting from the oxidation of food. The energy

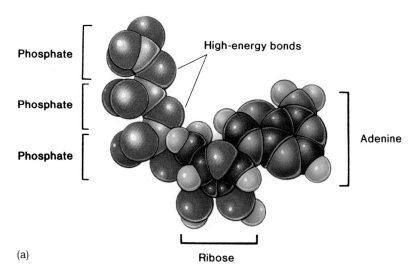

Phosphate

Phosphate

High-energy bonds

Phosphate

Adenine

(a)

Ribose

(b)

Triphosphate group

O⁻

O = P — O⁻

O

High-energy bonds

O = P — O⁻

O

Adenine

NH₂

N

AMP core

O = P — O — CH₂

O⁻

H   H

H   H

OH   OH

Ribose

(b)

Figure 6.12 The structure of ATP, the energy currency of the cell.

ATP consists of a ribose sugar, adenine, and a triphosphate group. The three phosphates are linked by high-energy bonds. Breaking these high-energy bonds releases energy from the ATP molecule. (a) Diagram of a molecule of ATP. (b) Chemical structure of ATP. (AMP = adenosine monophosphate.)

Because ATP plays this central role in *all* organisms, its role as the major energy currency of cells clearly evolved early in the history of life.

The Structure of ATP

Each ATP molecule is composed of three subunits (figure 6.12). The first subunit is a five-carbon sugar called ribose, which serves as the backbone to which the other two subunits are attached. The second subunit is adenine, an organic molecule composed of two carbon-nitrogen rings. Each of the nitrogen atoms in the ring has an unshared pair of electrons that weakly attract hydrogen ions. Adenine therefore acts as a chemical base and is usually referred to as a nitrogenous base. As described in chapter 2, adenine plays another major role in the cell: It is one of the four nitrogenous bases that are the principal components of the genetic material DNA. The two subunits together make up the AMP (adenosine monophosphate) core of the ATP molecule. The same AMP core is found in the NAD⁺ molecule.

The third subunit of ATP is a **triphosphate group** (three phosphate groups linked in a chain). The covalent bonds linking these three phosphates are usually indicated by a squiggle (~) and are sometimes called **high-energy bonds.** When one is broken, slightly more than 7 kilocalories of energy are released per mole of ATP. These phosphate bonds possess what a chemist would call "high transfer potential"; that is, they are bonds that have a low activation energy and are broken easily, which releases their energy. In a typical energy transaction, only the outermost of the two high-energy bonds is broken, breaking off the phosphate group on the end. When this happens, ATP becomes **adenosine diphosphate (ADP),** and 7 kilocalories of energy are expended per mole of ATP (figure 6.13).

Cells use ATP to drive endergonic reactions, reactions whose products possess more energy than their substrates. Such reactions will not proceed unless the reactants are supplied with the necessary energy, any more than a boulder will roll uphill. But as long as the cleavage of ATP's terminal high-energy bond is more exergonic than the other reaction is endergonic, the overall energy change of the two "coupled" reactions is exergonic, and the reaction will proceed. Because almost all endergonic reactions in the cell require fewer than 7 kilocalories of energy per mole, ATP is able to power all of the cell's activities.

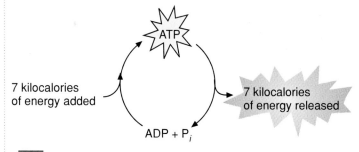

7 kilocalories of energy added

ATP

7 kilocalories of energy released

ADP + P_i

Figure 6.13 The ATP/ADP + P_i cycle.

ATP is formed with an input of 7 kilocalories of energy, which bonds an inorganic phosphate (P_i) onto ADP, forming ATP. When the phosphate is cleaved from ATP, 7 kilocalories of energy are released.

stored in the chemical bonds of fat and starch is converted to ATP, as is the energy carried by the sugars circulating in your blood. Cells then use their supply of ATP to drive active transport across membranes, to power movement, to provide activation energy for chemical reactions, and to grow. Almost every energy-requiring process that cells perform is powered by ATP.

ATP is the universal energy currency of all cells.

How Cells Form ATP

Cells contain a pool of ATP, ADP, and phosphate. ATP is constantly being cleaved into ADP plus phosphate to drive the cell's endergonic, energy-requiring processes. An individual on a typical diet of 2,000 Calories/day goes through about 125 pounds of ATP a day. But cells do not maintain large stockpiles of ATP, just as most people do not carry large amounts of cash with them. Instead, cells are constantly recycling their ADP, withdrawing from their energy reserves to rebuild more ATP. Using the energy derived from foodstuffs and from stored fats or starches (or, in the case of plants, from photosynthesis), ADP and inorganic phosphate (P_i) recombine to form ATP, with 7 kilocalories of energy per mole contributed to each newly formed high-energy bond. If every ATP molecule in your body at one instant in time could be viewed, they would be gone in a flash. Most cells maintain a particular molecule of ATP for only a few seconds before using it.

The Evolution of Metabolism

The way in which contemporary organisms harvest energy and employ it to drive chemical reactions is different from the way in which the earliest organisms probably operated. For most of the period in which life has existed on earth, metabolic processes took place in what was essentially an oxygen-free, or anaerobic, environment. In fact, as discussed in chapter 7, anaerobic metabolism still exists.

The first step in cellular respiration, a process called *glycolysis,* does not require oxygen to proceed. *All* organisms, whether a bacterium, fern, or human being, use glycolysis as the first step in the oxidation of organic compounds. With the advent of an atmosphere rich in oxygen, some organisms evolved a process of *oxidative respiration,* in which organic compounds are oxidized in the presence of oxygen. These organisms did not do away with glycolysis; they merely tacked this new aerobic process onto the old anaerobic process. Some organisms, such as yeasts, can still extract energy from organic compounds in the absence of oxygen, a process called *fermentation.*

Chapters 7 and 8 first examine cellular respiration, in which organisms extract energy from food and convert this energy into a usable form that fuels their physiological activities. Chapter 8 explores photosynthesis, the process that plants use to make their own food from inorganic compounds. The oxygen given off as a result of photosynthesis has had profound implications for the biology of all aerobic organisms: Without photosynthesis, oxygen would not be a primary part of our atmosphere, and we might all be anaerobic organisms.

▼

EVOLUTIONARY VIEWPOINT

All living creatures share the same basic metabolism, using ATP to power the making of molecules by driving coupled reactions. All organisms have also retained the ancient ability to produce the necessary ATP by glycolysis, although the more recently evolved process of oxidative respiration now accounts for the bulk of ATP production in most organisms. Metabolism has evolved by adding new capabilities onto existing ones, like putting on extra clothes.

SUMMARY

1. Energy is the capacity to bring about change, to do work. Kinetic energy is actively engaged in doing work, whereas potential energy has the capacity to do work. Most energy transformations in living things involve conversion of potential energy to kinetic energy.

2. The first law of thermodynamics states that the amount of energy in the universe is constant and that energy cannot be destroyed or created. However, energy can be converted from one form to another.

3. The second law of thermodynamics says that disorder in the universe is continuously increasing. As a result, energy spontaneously converts to less organized forms.

4. The least organized form of energy is heat, which is random molecular motion. Heat energy can be used to do work when heat gradients exist, as in a steam engine, but cannot accomplish work in cells.

5. An oxidation/reduction (redox) reaction occurs when an electron is taken from one atom (oxidation) and donated to another (reduction). The electrons in redox reactions often travel in association with protons, as hydrogen atoms.

6. If the electron transferred in a redox reaction is an energetic electron, then the energy is transferred with the electron, and the electron moves to a higher-energy orbital in the reduced atom. Many energy transfers in cells take place by means of redox reactions.

7. Any chemical reaction whose products contain less free energy than the original reactants tends to proceed spontaneously.

8. The speed of a reaction depends on the amount of activation energy required to break existing bonds. Catalysis is the process of lowering the amount of activation energies needed by stressing chemical bonds. Enzymes are the catalysts of cells.

9. Cells contain many different enzymes, each of which catalyzes a different reaction. A given enzyme is specific because its active site fits only one or a few potential substrate molecules.

10. Coenzymes are nonprotein cofactors that aid enzymes in their catalytic activities. The coenzymes NAD^+ and FAD function as electron carriers in metabolic reactions.

11. Cells focus all of their energy resources on the manufacture of ATP from ADP and phosphate. Cells must supply 7 kilocalories of energy obtained from photosynthesis or from electrons stripped from foodstuffs to form 1 mole of ATP. Cells then use this ATP to drive endergonic reactions.

REVIEWING THE CHAPTER

1. What is energy?

2. How are the laws of thermodynamics concerned with energy?

3. How does energy flow in living things?

4. How does activation energy prepare molecules for action?

5. How do enzymes regulate life processes in a cell?

6. How do enzymes work?

7. What factors affect enzyme activity?

8. How is enzyme activity regulated?

9. How do coenzymes aid in catalysis?

10. Why is ATP the energy currency of life?

11. Historically, how did metabolism evolve?

COMPLETING YOUR UNDERSTANDING

1. The total of all the chemical reactions that an organism performs is called

   a. thermodynamics.

   b. potential energy.

   c. metabolism.

   d. catalysis.

   e. allosteric change.

   f. feedback inhibition.

2. How does kinetic energy differ from potential energy?

3. What are the differences between the first and second laws of thermodynamics?

4. An increasing entropy applies to which law of thermodynamics?

5. The passing of electrons from one atom or molecule to another is called an _____ reaction.

6. What does it mean when an atom or molecule is reduced or oxidized?

7. A reaction in which the reactants contain more energy than the products is called

   a. exocytosis.

   b. exergonic.

   c. entropic.

   d. endergonic.

   e. a catalyst.

   f. an activator.

8. Enzymes are examples of or do all the following except

   a. catalysis.

   b. catalysts.

   c. globular proteins.

   d. complex carbohydrates.

9. How do active sites, substrates, and induced-fit models relate to enzymes?

10. Most human enzymes function best between [*two answers required*]

   a. 35 to 40 degrees Celsius.

   b. 25 to 35 degrees Celsius.

   c. 20 to 25 degrees Celsius.

   d. pH 5 and 7.

   e. pH 6 and 8.

   f. pH 7 and 9.

11. How do activators and inhibitors influence enzymatic activity?

12. If, in a metabolic pathway, a particular sugar molecule builds up in a cell, how does the cell "know" when to "shut down" the production of that sugar?

13. If you bought some vitamins at a health-food store, what do these vitamins do in your body after being taken?

14. How do NAD^+ and FAD function in cells?

15. Which of the following is not part of the ATP molecule?

   a. An amino acid

   b. A sugar

   c. A nitrogenous base

   d. A phosphate group

16. Why is ATP able to power most of the cell's activities?

17. If a human on a strict diet ate only a 1,500-calorie meal consisting of a hamburger and some french fries, would this meal provide enough ATP energy to get through the entire day? If not, where would the additional needed energy come from? What could happen after extending a strict diet such as this for a long time period?

18. In the evolution of metabolism, what is the evidence that fermentation preceded oxidative respiration?

FOR FURTHER READING

Alberts, B. et al. 1994. *Molecular biology of the cell.* 3d ed. New York: Garland Press. A clear and comprehensive discussion of energy and metabolism in chapter 2.

Atkins, P. 1984. *The second law.* San Francisco: Freeman. A basic, quite understandable explanation of thermodynamics and its biological implications.

Hinkle, P. C., and R. E. MacCarty. 1978. How cells make ATP. *Scientific American*, March, 104–23. Written more than fifteen years ago, but still provides one of the clearest expositions of how cells use electrons stripped from foodstuffs to make ATP.

Koshland, D. 1973. Protein shape and biological control. *Scientific American,* Oct. A discussion of how changes in protein shape regulate enzyme activity.

Stryer, L. 1988. *Biochemistry.* 3d ed. San Francisco: Freeman. An excellent introduction to enzymes and metabolism in chapters 10 and 12.

EXPLORATIONS

Interactive Software:
What Enzymes Do

This interactive exercise presents the user with a diagram of the catalyase enzyme-catalyzed reaction, showing the free-energy-of-activation "hill" to be overcome. The user keeps track of what is going on with a "molecular speedometer" that measures how fast the reaction is going. The key variables manipulated by the user are identity of the substrate, the enzyme concentration, the substrate concentration, and temperature. The user is able to plot 1/V (velocity) vs. 1/S (substrate concentration), and assess the strength with which the enzyme binds its substrate (the slope of the line).

Questions to Explore:

1. How does temperature increase affect the rate of the reaction? What happens if the temperature is increased to the maximum amount? Why? What is the temperature doing to these molecules to affect the rate of the reaction?

2. What effect does increasing enzyme concentration have on the rate of the reaction? Can the concentration of the enzyme be increased indefinitely with indefinite results?

3. What are the optimum conditions for lowering activation energy? How are the variables most favorably aligned to achieve this?

4. As the concentration of substrate increases and all other variables remain the same, what happens to the rate of the reaction? Why?

5. How would denaturation interfere with the catalytic (enzymatic) process?

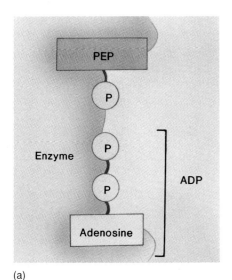

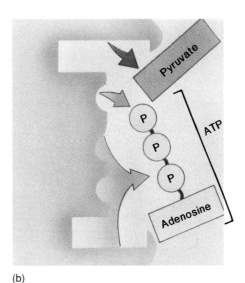

(a)   (b)

Figure 7.2 Substrate-level phosphorylation.
(a) *Some molecules, such as phosphoenol pyruvic acid (PEP), possess a high-energy phosphate bond (P) similar to those of ATP.* (b) *When the phosphate group of PEP is transferred enzymatically to ADP, the energy in the bond is conserved, and ATP is created. The reaction takes place because the phosphate bond of PEP has higher energy than does the terminal phosphate bond of ATP.*

2. *Chemiosmotic generation of ATP* The chemical formation of ATP powered by a diffusion force similar to osmosis is called **chemiosmosis.** All organisms possess transmembrane protein channels that pump protons across the membranes of cell organelles (see chapter 5). These proton-pumping channels use a flow of excited electrons extracted from food molecules to induce a shape change in the channel protein, which in turn causes protons to pass outward, not unlike going through a turnstile. As the proton concentration outside the membrane increases relative to that inside, the outside protons are driven inside by diffusion. However, they can only come in through special proton channels that use their passage to induce the formation of ATP from ADP plus P_i (figure 7.3).

> *Chemical energy is harvested in one or both of two ways: (1) Substrate-level phosphorylation, which involves a reshuffling of chemical bonds to couple ATP formation to a highly exergonic reaction, and (2) The transport of electrons to a membrane, where they drive a proton pump and thus power the chemiosmotic synthesis of ATP.*

Substrate-level phosphorylation was probably the first of the two ATP-forming mechanisms to evolve. **Glycolysis,** which is the most basic of all ATP-generating processes and is present in every living cell, employs this mechanism. (Glycolysis is examined in detail later in the chapter.)

However, most of the ATP that organisms make is produced by chemiosmosis. Organisms obtain the electrons that drive proton-pumping channels involved in chemiosmosis from two sources:

1. *Light* In photosynthetic organisms, light energy boosts electrons to higher-energy levels, and these electrons are ferried to proton pumps.

2. *Chemical bonds* In all organisms—photosynthetic and nonphotosynthetic alike—high-energy electrons are extracted from chemical bonds and carried by coenzymes to proton pumps.

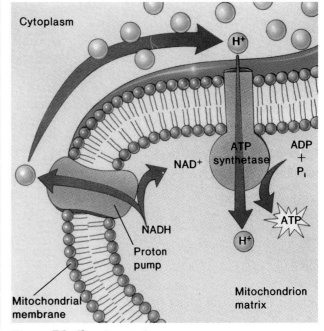

Figure 7.3 Chemiosmosis.
High-energy electrons harvested from food molecules during metabolism are transported by NADH carrier molecules (see chapter 6) to proton pumps, which use the energy to pump protons out across the membrane. As a result, the concentration of protons on the outside of the membrane rises, encouraging protons to diffuse back. The diffusing protons pass through the only channel open to them: a special ATP-forming protein complex. As each proton passes, an ATP molecule is formed.

The second of these two chemiosmotic processes—the extraction of electrons from chemical bonds—is an oxidation reduction process. **Oxidation** is defined as the removal of electrons (see chapter 6). When electrons are taken away from the chemical bonds of food molecules to drive proton pumps, the food molecules are being oxidized chemically.

Electron-harvesting processes are often referred to as **oxidative respiration** because they use oxygen. Oxidative respiration is the oxidation of food molecules to obtain energy. Do not confuse this with your body's breathing of oxygen gas, which is also called respiration.

Overview of Cellular Respiration

As mentioned earlier, **cellular respiration** is the conversion of chemical energy to ATP. In most organisms, cellular respiration is carried out in three stages. The first stage is a biochemical pathway called **glycolysis.** The second and third stages involve two pathways of oxidative respiration: the oxidation of pyruvate and the citric acid cycle.

Glycolysis

In glycolysis, ATP is made by substrate-level phosphorylation within the cytoplasm of the cell. However, during this process, a few energetic electrons are stripped away from food molecules. The food molecules are said to have been oxidized because electrons are removed from them. These electrons carry a great deal of energy and can be used to make ATP via chemiosmosis, but only if oxygen is present. Oxygen is required because, while glycolysis will produce ATP by substrate-level phosphorylation whether or not oxygen is present, the energetic electrons harvested during glycolysis cannot drive the chemiosmotic synthesis of ATP unless oxygen is available to serve as the final acceptor of the harvested electrons:

$$C_6H_{12}O_6 + 6O_2 \rightarrow 6CO_2 + 6H_2O + Energy$$

In the absence of oxygen, cells are restricted to the substrate-level phosphorylation reactions of glycolysis to obtain ATP.

Oxidative Respiration

All cells carry out glycolysis, but glycolysis in the absence of oxygen is not efficient at extracting energy from food molecules: Only a small percentage of the energy in the chemical bonds of food molecules is converted by substrate-level phosphorylation to ATP. In the presence of oxygen, however, two other stages of cellular respiration—both of them oxidative pathways that extract far more energy—also occur. The first oxidative respiration stage is the oxidation of **pyruvate,** while the second is called the **citric acid cycle,** after the six-carbon citric acid molecule formed in its first step. Alternatively, the citric acid cycle is also known as the Krebs cycle, after British biochemist Sir Hans Krebs, who discovered it. (Less commonly, it is called the tricarboxylic acid cycle because citric acid has three carboxyl groups.)

Pyruvate oxidation and the reactions of the citric acid cycle take place only within the cell organelles called mitochondria. Thus, all organisms that possess mitochondria, including plants, as well as many forms of bacteria (chapter 4 recounts how mitochondria are thought to have evolved from bacteria), carry out oxidative respiration. While plants produce ATP in their leaves by photosynthesis, plant roots and stems also produce ATP in this other oxidative fashion, just as you do.

> *All cells with mitochondria make ATP by substrate-level phosphorylation and by oxidative respiration. Some organisms also make ATP by photosynthesis.*

The Fate of a Chocolate Bar: Cellular Respiration in Action

When you metabolize food and thus obtain the ATP that powers your life, you employ both substrate-level phosphorylation and oxidative respiration. To better understand what goes on, let's see what happens to a chocolate bar after you eat it.

A chocolate bar, like many of the foodstuffs that you consume, is a complex mixture of sugars, lipids, proteins, and other molecules. The first thing that happens in its journey toward ATP production is that the complex molecules are broken down to simple ones. Disaccharides such as sucrose are split into simple sugars—either glucose or sugars that are converted to glucose. Proteins are split into amino acids, and complex lipids are broken into smaller bits. These initial steps usually yield no usable energy, but they assemble the energy wealth of a diverse array of complex molecules into a smaller number of simple molecules, such as glucose.

For simplicity, let's assume that the chocolate bar is entirely broken down to molecules of the six-carbon sugar glucose. Glucose is important in metabolism because many different foodstuffs are converted to glucose and because it is the starting point for ATP production.

The first stage of extracting energy from glucose is the ten-reaction biochemical pathway called glycolysis. In glycolysis, ATP is generated in two ways. For each glucose molecule, two ATP molecules are used up in preparing the glucose molecule, and four ATP molecules are formed by substrate-level phosphorylation, for a net yield of two ATP molecules. In addition, four electrons are harvested, which in the presence of oxygen, can be used to form ATP molecules by oxidative respiration. But the total yield of ATP molecules is small. When glycolysis is complete, the two molecules of pyruvate that are left still contain most of the energy that was present in the original glucose molecule.

The second stage of extracting energy from glucose, after glycolysis, is the conversion of pyruvate to a two-carbon molecule called acetyl-CoA. Because this process involves stripping electrons from pyruvate, the pyruvate is said to be oxidized. If oxygen is present, these high-energy electrons can be used to make ATP.

The third stage of extracting energy from glucose consists of a cycle of nine reactions—the citric acid cycle. The acetyl-CoA left over from the oxidation of pyruvate feeds into this cycle. In the cycle, two more ATP molecules are extracted by substrate-level phosphorylation, and a large number of electrons are removed and donated to electron carriers. The harvesting of these electrons leads, in the presence of oxygen, to the greatest amount of ATP formation.

When all three stages of cellular respiration have been completed, the six-carbon glucose molecule has been divided into six molecules of CO_2, and thirty-eight ATP molecules have been generated, six by substrate-level phosphorylation and thirty-two by oxidative respiration. Even though two of the ATP molecules must be expended to transport electrons out into the cytoplasm (carried by the molecule NADH—see chapter 6), the net yield of thirty-six ATP molecules is still very good (figure 7.4). The overall process for extracting energy from glucose can be summarized as:

Substrate-level phosphorylation	+	Oxidative respiration	−	Cost of NADH transport	=	Net ATP production
6		32		2		36

This brief overview gives some sense of how cells organize their production of ATP from a food source such as a chocolate bar. In-depth examinations of glycolysis and the citric acid cycle as processes that direct the flow of energy follow.

Glycolysis: A More Detailed Look

Although many simple molecules are available as a consequence of degradation, the metabolism of early bacteria clearly focused on the simple six-carbon sugar glucose, possibly because glucose was a major constituent of the cell's carbohydrates. Glucose molecules can be dismantled in many ways, but early bacteria evolved the ability to do it in a way that includes reactions that release enough free energy to drive the synthesis of ATP in coupled reactions. The process involves a sequence of ten reactions that convert glucose into two three-carbon molecules of pyruvate. For each molecule of glucose that passes through this transformation, the cell acquires two ATP molecules. The overall process is called glycolysis.

The Glycolytic Pathway

Glycolysis consists of two very different processes, one wedded to the other: First, glucose is converted to two molecules of the three-carbon compound **glyceraldehyde 3-phosphate (G3P),** with the expenditure of ATP. Second, ATP is generated from G3P.

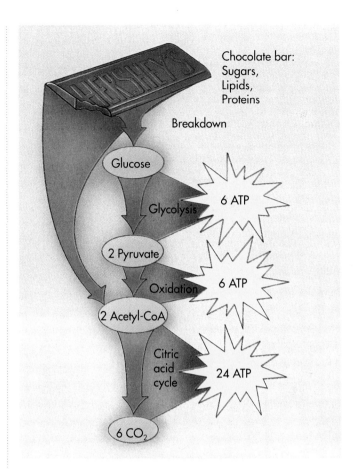

Figure 7.4 The fate of a chocolate bar.
When you eat a chocolate bar, its complex molecules are first broken down to simple molecules: the sugars to glucose molecules, and the proteins, fats, and other lipids to two-carbon molecules called acetyl-CoA. These breakdowns produce little or no energy, but they prepare the way for three major energy-producing processes: (1) glycolysis, a process that converts glucose to two molecules of pyruvate; (2) the oxidation of this pyruvate to two molecules of acetyl-CoA; and (3) the oxidation of acetyl-CoA molecules in the citric acid cycle.

These two processes require ten sequential chemical reactions, which are most easily understood as consisting of four sequential stages (figure 7.5):

Stage A: Stage A is composed of three reactions that change glucose into a compound that can readily be split into three-carbon phosphorylated units. Two of these reactions require the cleavage of an ATP molecule, so this stage requires that the cell invest two ATP molecules.

Stage B: Stage B is cleavage, in which the six-carbon product of Stage A is split into two three-carbon molecules. One is G3P, and the other is converted to G3P by another reaction.

Stage C: Stage C is oxidation, in which a hydrogen atom carrying a pair of electrons is removed from G3P and donated to nicotinamide adenine dinucleotide (NAD^+)

124

(□□) **Figure 7.5 The glycolytic pathway.**

The first five reactions convert a molecule of glucose into two molecules of glyceraldehyde 3-phosphate (G3P). This process is endergonic and requires the expenditure of two ATP molecules to drive it. The next five reactions convert G3P molecules into pyruvate molecules and generate four molecules of ATP for each two molecules of G3P. These reactions also generate two molecules of NADH, which, as you will see, are further oxidized in your body to produce four more molecules of ATP. After subtracting the two ATP molecules expended in driving the initial endergonic reactions, the net yield in your body is thus six ATP molecules for each molecule of glucose. When cells are forced to operate without oxygen, the two molecules of NADH cannot be used to produce ATP, and the net yield is then only two molecules of ATP per molecule of glucose.

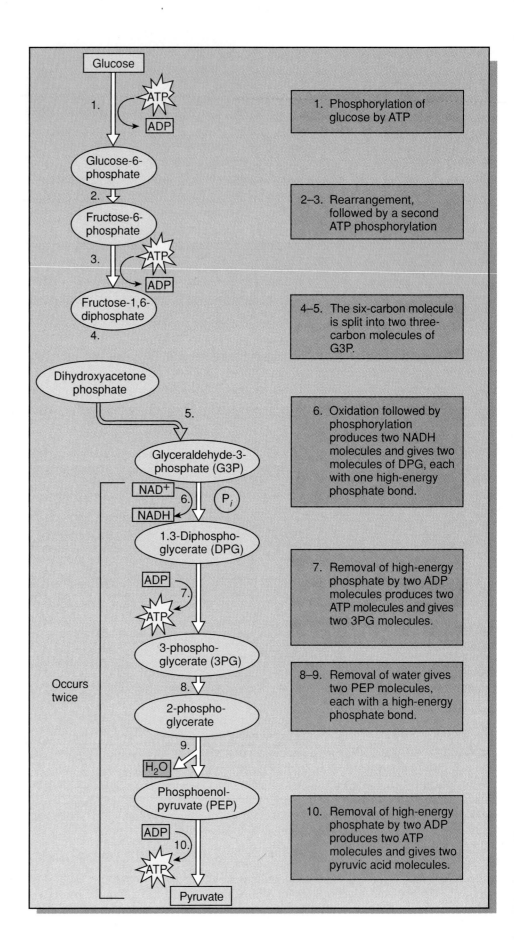

(see chapter 6). NAD$^+$ acts as an electron carrier in the cell, in this case accepting the proton and two electrons from G3P to form NADH. Note that NAD$^+$ is an ion and that *both* electrons in the new covalent bond come from G3P.

Stage D: Stage D, called ATP generation, is a series of four reactions that convert G3P into another three-carbon molecule—pyruvate—and in the process generate two ATP molecules for each G3P.

Because each glucose molecule is split into *two* G3P molecules, the overall net reaction sequence yields two ATP molecules and two molecules of pyruvate:

$$\begin{array}{l} -2 \ \text{ATP Stage A} \\ +2 \ \ (2 \ \text{ATP) Stage D} \\ \hline \ \ 2 \ \text{ATP} \end{array}$$

Each ATP molecule represents the capture of 7.3 kilocalories of energy per mole of glucose. Therefore, the net yield of two ATP molecules represents a total capture of 14.6 kilocalories per mole. The total energy content of the chemical bonds of glucose is 686 kilocalories per mole. Thus, in the absence of oxygen, only 2 percent of the available energy is harvested (see Sidelight 7.1). By contrast, a car converts about 25 percent of the energy in gasoline into useful energy.

> *The glycolytic reaction sequence generates a small amount of ATP by reshuffling the bonds of glucose molecules. Glycolysis is a very inefficient process, capturing only about 2 percent of the available chemical energy of glucose.*

Evolution of Glycolysis

The glycolytic reaction sequence is thought to have been among the earliest of all biochemical processes to evolve. It uses no molecular oxygen and therefore occurs readily in an anaerobic environment. All of its reactions occur free in the cytoplasm; none is associated with any organelle or membrane structure. Every living creature is capable of carrying out the glycolytic sequence. However, most present-day organisms are able to extract considerably more energy from glucose molecules than does glycolysis. Of the thirty-six ATP molecules you obtain from each glucose molecule that you metabolize, only two are obtained by substrate-level phosphorylation during glycolysis. Why is glycolysis still maintained when its energy yield is comparatively meager?

This simple question has an important answer: Evolution is an incremental process. Adaptation occurs during evolution by improving on past successes. In catabolic metabolism (the breakdown of complex molecules into simpler ones), glycolysis satisfied the one essential evolutionary requirement: It was an improvement over trying to glean energy from minerals. Cells that could not carry out glycolysis were at a competitive disadvantage. Studies of the metabolism of contemporary organisms show that only those cells that were capable of glycolysis survived the early competition of life. Later improvements in catabolic metabolism built on this success. Glycolysis was not discarded during the course of evolution but rather was used as the starting point for the further extraction of chemical energy. Nature did not, so to speak, go back to the drawing board and design a different and better metabolism from scratch. Rather, catabolic metabolism evolved as one layer of reactions added to another, just as successive layers of paint can be found in an old apartment. We all carry glycolysis with us—a metabolic memory of our evolutionary past.

Closing The Metabolic Circle: Regenerating NAD$^+$

Inspect for a moment the net reaction of the glycolytic sequence:

$$\text{Glucose} + 2\text{ADP} + 2\text{P}_i + 2\text{NAD}^+ \rightarrow$$
$$2 \ \text{Pyruvate} + 2\text{ATP} + 2\text{NADH} + 2\text{H}^+ + 2\text{H}_2\text{O}$$

You can see that three changes occur in glycolysis: (1) glucose is converted to pyruvate, (2) ADP is converted to ATP, and (3) NAD$^+$ is converted to NADH.

As long as foodstuffs that can be converted to glucose are available, a cell can continually churn out ATP to drive its activities, except for one problem: the NAD$^+$. As a result of glycolysis, the cell accumulates NADH molecules at the expense of the pool of NAD$^+$ molecules. A cell does not contain a large amount of NAD$^+$. For glycolysis to continue, the cell must recycle the NADH that it produces back to NAD$^+$. Some other home must be found for the hydrogen atom taken from G3P, some other molecule that will accept the hydrogen and be reduced.

What happens after glycolysis depends on the fate of this hydrogen atom. One of two things generally happens:

1. *Oxidative respiration* Oxygen is an excellent electron acceptor, and in the presence of oxygen gas, the hydrogen atom taken from G3P can be donated to oxygen, forming water. Because air is rich in oxygen, this process is referred to as **aerobic metabolism.** Oxidative respiration is examined in detail later in the chapter.

2. *Fermentation* When oxygen is not available, another organic molecule must accept the hydrogen atom instead. Such a process is called **fermentation.** This process, which is what happens when bacteria grow without oxygen, is referred to as **anaerobic metabolism,** or metabolism without oxygen.

Heterotrophs carry out many sorts of fermentations, each employing some form of carbohydrate molecule to accept the hydrogen atom from NADH and thus reform NAD$^+$:

$$\text{Carbohydrate} + \text{NADH} \rightarrow \text{Reduced carbohydrate} + \text{NAD}^+$$

Ethanol Fermentation

More than a dozen fermentation processes have evolved among bacteria, each using a different carbohydrate as the hydrogen atom acceptor. By contrast, only a few kinds of fermentations occur among eukaryotes. In one fermentation

Sidelight 7.1
METABOLIC EFFICIENCY AND THE LENGTH OF FOOD CHAINS

The simplest heterotrophs are those bacteria that first evolved glycolysis early in the history of life, before the earth's atmosphere contained any oxygen gas. These bacteria derive chemical energy by breaking carbon-hydrogen bonds, use the energy to promote ATP production through coupled reactions, and dispose of the hydrogen atoms by fermentations. The inefficiency of the process, which extracts only a small percentage of the chemical energy available in organic compounds, probably placed an important constraint on early organisms: Most heterotrophs must have lived by consuming photosynthetic organisms rather than other heterotrophs.

For example, a heterotroph preserves only 2 percent of the energy of a photosynthetic organism that it consumes, then any other population of heterotrophs that consumes this kind of heterotroph has only 2 percent of the original energy available to it and can harvest from this by glycolysis only 2 percent of *that* energy, or 0.04 percent of the original amount available. Therefore, a very large base of autotrophs would be needed to support a small number of heterotrophs.

When organisms became able to extract energy from organic molecules by oxidative metabolism, this constraint became far less severe because oxidative processes are much more efficient. The efficiency of oxidative respiration is 38 percent, meaning that about two-thirds of the available energy is lost at each level of consumption. Thus, animals that eat plants can obtain no more than approximately one-third of the energy in the plants, and other animals that eat the first animals obtain no more than a third of the energy in the original plant-eaters, or *herbivores*. But losses of this magnitude still result in the transmission of far more energy from one *trophic level* (defined as a step in the movement of energy through an ecosystem) to another than does anaerobic glycolysis, in which more than 98 percent of the energy is lost at each step.

The improved efficiency of oxidative metabolism made possible the evolution of *food chains*, in which some heterotrophs consume plants, and others consume the plant-eaters. You will read more about food chains in chapter 22.

The length of a food chain is limited by the efficiency of oxidative metabolism; most food chains involve only three and rarely four levels (Sidelight figure 7.1). Too much energy is lost at each transfer point to allow the chain to become much longer. For example, you could not support a large human population with the meat of lions captured from the Serengeti Plain of Africa—the amount of grass there will not support enough zebras to maintain a large population of lions. Thus, the earth's ecological complexity is fixed in a fundamental way by the chemistry of oxidative respiration.

Sidelight figure 7.1 The four levels of food chains.
At each of these four levels, only about a third or less of the energy present is utilized by the recipient.

Level 1: Photosynthesizer. The grass growing under these palms grows actively during the hot, rainy season, and capturing the energy of the sun abundantly, converts it into molecules of glucose that are stored in the grass plants as starch.

Level 2: Herbivore. These large antelopes, known as wildebeests, consume the grass and convert some of its stored energy into their own bodies.

Level 3a: Carnivore. The lion feeds on wildebeests and other animals, converting part of their stored energy to its body.

Level 3b: Scavenger. This hyena and the vulture occupy the same stage in the food chain as the lion. Like the lion, they are consuming the body of the dead wildebeest, which the lion has abandoned.

Level 4: Refuse utilizer. These butterflies, mostly Precis octavia, are feeding on the material left in the hyena's dung after the material the hyena consumed passed through its digestive tract.

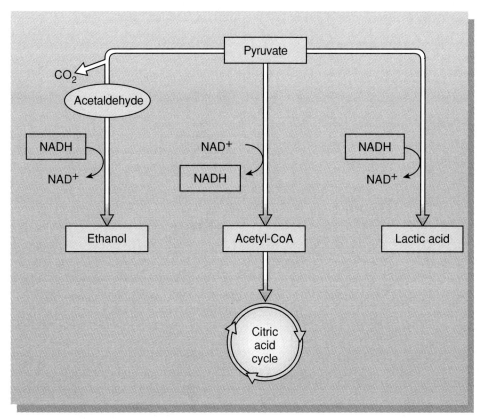

Figure 7.6 What happens to pyruvate, the product of glycolysis?

In the presence of oxygen, pyruvate is oxidized to acetyl-CoA and enters the citric acid cycle. In the absence of oxygen, pyruvate instead is reduced, accepting the electrons extracted during glycolysis and carried by NADH. When pyruvate is reduced directly, as it is in your muscles, the product is lactic acid. When CO_2 is first removed from pyruvate and the remainder reduced, as it is in yeasts, the product is ethanol.

process, which occurs in the single-celled fungi called yeasts, the carbohydrate that accepts the hydrogen from NADH is pyruvate, the end product of the glycolytic process. Yeasts remove a terminal CO_2 group from pyruvate, producing a toxic two-carbon molecule called acetaldehyde and CO_2. Because of this CO_2 production, bread with yeast rises, and "unleavened" bread—that is, bread without yeast—does not. Acetaldehyde then accepts the hydrogen from NADH, producing NAD^+ and ethyl alcohol, also called ethanol (figure 7.6).

Beer is manufactured by fermenting barley into ethyl alcohol in this fashion. Wine is made when the same fermentation occurs in grapes (figure 7.7).

Lactic Acid Fermentation

Most multicellular animals regenerate NAD^+ without removing a carbon atom. The processes that they use involve the production of by products that are less toxic than alcohol. For example, your muscle cells use an enzyme called lactate dehydrogenase to add the hydrogen of the NADH produced by glycolysis back to the pyruvate that is the end product of glycolysis, converting pyruvate plus NADH into lactic acid plus NAD^+.

Lactic acid fermentation closes the metabolic circle, allowing glycolysis to continue for as long as the glucose holds out. Blood circulation removes excess lactic acid from muscles. When the lactic acid cannot be removed as fast as it is produced, your muscles cease to work well. Subjectively, they feel tired or leaden. Try raising and lowering your arm rapidly a hundred times, and you will soon experience this sensation. A more efficient circulatory system developed through training will let you run longer before the accumulation of lactic acid becomes a problem—this is why people train before they run marathon races—but lactic acid production always exceeds lactic acid removal at some point. This limit is the primary reason why the world record for running a mile is just under 4 minutes and not significantly less.

In fermentations, which are anaerobic processes, the electron generated in the glycolytic breakdown of glucose is donated to an oxidized organic molecule. In contrast, in aerobic metabolism such electrons are transferred to oxygen, generating ATP in the process.

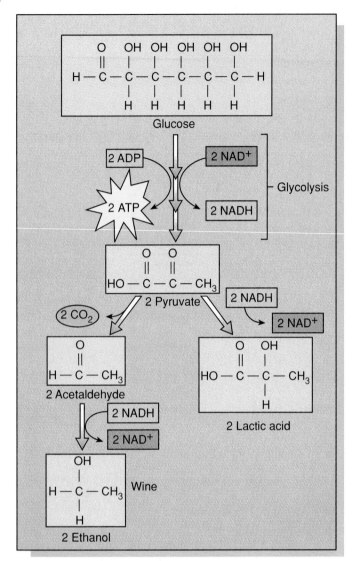

Figure 7.7 How wine is made.
Yeasts convert pyruvate to ethanol. The conversion takes place naturally in grapes left to ferment on vines, as well as in fermentation vats of crushed grapes. When ethanol concentration reaches about 12 percent, the alcohol's toxic effects kill the yeasts, leaving wine. In your muscles, pyruvate is instead converted to lactic acid, which is less toxic than ethanol.

Oxidative Respiration: A More Detailed Look

Even in aerobic organisms, not all cellular respiration is aerobic; glycolysis and fermentation play important roles in the metabolism of most organisms. But in all aerobic organisms, the oxidation of glucose, which began in Stage C of glycolysis, is continued where glycolysis leaves off—with pyruvate. The evolution of this new biochemical process was conservative, as is almost always the case in evolution; the new process was simply tacked onto the old one. In eukaryotic organisms, oxidative respiration takes place exclusively in the mitochondria, which apparently originated as symbiotic bacteria. Oxidative

respiration occurs in two stages: (1) the oxidation of pyruvate to form an intermediate product—acetyl-CoA—and (2) the later oxidation of the acetyl-CoA.

The Oxidation of Pyruvate

The first stage of oxidative respiration is a single oxidative reaction in which one of the three carbons of pyruvate is split off, departing as CO_2 (chemists call a reaction of this kind **decarboxylation**) and leaving behind two remnants: (1) a pair of electrons and their associated hydrogen, which reduce NAD^+ to NADH; and (2) a two-carbon fragment called an acetyl group, which is added to a carrier molecule called coenzyme A (CoA), forming a compound called acetyl coenzyme A, or **acetyl-CoA** for short:

$$Pyruvate + NAD^+ + CoA \rightarrow Acetyl\text{-}CoA + NADH + CO_2$$

This reaction is very complex, involving three intermediate stages, and is catalyzed by an assembly of enzymes called a **multi-enzyme complex.** A multi-enzyme complex organizes a series of reactions so that the chemical intermediates do not diffuse away or undergo other reactions. Within such a complex, component polypeptides pass the reacting substrate molecule from one enzyme to the next in line without ever letting go of it. The multi-enzyme complex called **pyruvate dehydrogenase** removes the CO_2 from pyruvate and is one of the largest enzymes known—it contains forty-eight different polypeptide chains.

This reaction produces a molecule of NADH, which is later used to produce ATP. Of far greater significance than the reduction of NAD^+ to NADH, however, is the residual fragment acetyl-CoA (figure 7.8). Acetyl-CoA is important because it is produced, not only by the oxidation of the product of glycolysis, as just described, but also by the metabolic breakdown of proteins, fats, and other lipids. Acetyl-CoA thus provides a single focus for the many catabolic processes of the eukaryotic cell, all the resources of which are channeled into this single molecule.

Although acetyl-CoA is formed by many catabolic processes in the cell, only a limited number of processes use acetyl-CoA. Most acetyl-CoA either is directed toward energy storage (it is used in lipid synthesis) or is oxidized to produce ATP. Which of these two processes occurs depends on the amount of ATP in the cell. When ATP levels are high, acetyl-CoA is channeled into making fatty acids, which is why people get fat when they eat too much. When ATP levels are low, the oxidative pathway is stimulated, and acetyl-CoA flows into energy-producing oxidative metabolism.

The Oxidation of Acetyl-CoA: The Citric Acid Cycle

For every glucose molecule, two molecules of pyruvate are produced, and thus two molecules of acetyl-CoA. The oxidation of acetyl-CoA begins with the binding of the acetyl group to a four-carbon carbohydrate. The resulting six-carbon molecule is then passed through a series of electron-yielding

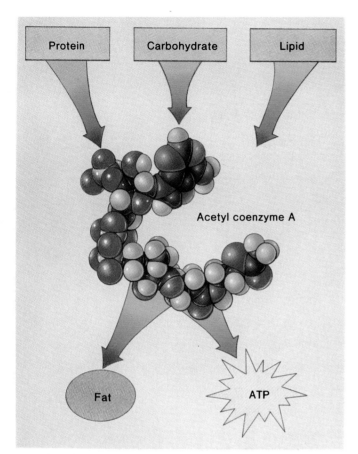

Figure 7.8 Acetyl coenzyme A (acetyl-CoA), the central molecule of energy metabolism.
Almost all of the molecules that you consume as foodstuffs are converted to acetyl-CoA when you metabolize them. The acetyl-CoA is then channeled into fat synthesis or into ATP production, depending on your body's energy requirements.

oxidation reactions, during which two CO_2 molecules are split off, regenerating the four-carbon carbohydrate, which is then free to bind another acetyl group. The process is a continuous, cyclical flow of carbon. In each turn of this citric acid cycle, a new acetyl group replaces the two CO_2 molecules that are lost, and more electrons are extracted.

The Reactions of the Citric Acid Cycle

The citric acid cycle, which oxidizes acetyl-CoA, consists of nine reactions, diagrammed in figure 7.9. The cycle has two stages:

Stage A: Three preparation reactions set the scene. In the first reaction, acetyl-CoA joins the cycle, and in the other two reactions, chemical groups are rearranged (Steps 1 and 2 in figure 7.9).

Stage B: Energy extraction occurs during this stage. Four of the six reactions are oxidations in which electrons are removed, and one generates an ATP equivalent directly by substrate-level phosphorylation (Steps 3 through 7 in figure 7.9).

Together, the nine reactions constitute a cycle that begins and ends with oxaloacetate. At every turn of the cycle, acetyl-CoA enters and is oxidized to CO_2 and H_2O, and the electrons are channeled off to drive proton pumps that generate ATP.

The Products of the Citric Acid Cycle

In the process of oxidative respiration, the glucose molecule is consumed entirely. Its six carbons are first split into three-carbon units during glycolysis. One of the carbons of each three-carbon unit is then lost as CO_2 in the conversion of pyruvate to acetyl-CoA, and the other two are lost during the oxidations of the citric acid cycle. All that is left to mark the passing of the glucose molecule is its energy, which is preserved in four ATP molecules and the reduced state of twelve electron carriers.

The oxidative consumption of one molecule of glucose proceeds in three stages: glycolysis (which can be anaerobic), the oxidation (decarboxylation) of pyruvate, and the citric acid cycle. Both glycolysis and the citric acid cycle produce two ATP molecules by substrate-level phosphorylation. All three processes harvest electrons. The extracted electrons are temporarily housed within NADH molecules. In one reaction, the extracted electrons are not energetic enough to reduce NAD^+, and a different coenzyme, **flavin adenine dinucleotide, (FAD),** is used to carry these less energetic electrons, as reduced $FADH_2$. Table 7.1 accounts for the number of molecules of ATP and of electron carriers that have been generated from one molecule of glucose.

> *The systematic oxidation of the pyruvate remaining after glycolysis generates two ATP molecules by substrate-level phosphorylation, which is as many as glycolysis produced. More important, however, is that this process harvests many energized electrons, which can then be directed to the chemiosmotic synthesis of ATP.*

Using The Electrons Generated By Oxidative Respiration To Make ATP

The NADH and $FADH_2$ molecules formed during glycolysis and the subsequent oxidation of pyruvate each contain a pair of electrons gained when NADH was formed from NAD^+ and when $FADH_2$ was formed from FAD. The NADH molecules carry their electrons to the inner membrane of the mitochondria (the $FADH_2$ is already attached to it), and there they transfer the electrons to a complex, membrane-embedded protein called **NADH dehydrogenase,** or sometimes, ATPase. The electrons are then passed on to a series of respiratory proteins called **cytochromes** and other carrier molecules, one after the other, losing much of their energy in the process by driving several transmembrane proton pumps. This series of membrane-associated electron carriers is collectively called the **electron transport chain** (figure 7.10). At the terminal step of the electron transport chain, the electrons are passed to the cytochrome c oxidase complex, which uses four of the electrons to reduce a molecule of oxygen gas and form water:

$$O_2 + 4H^+ + 4e^- \rightarrow 2H_2O$$

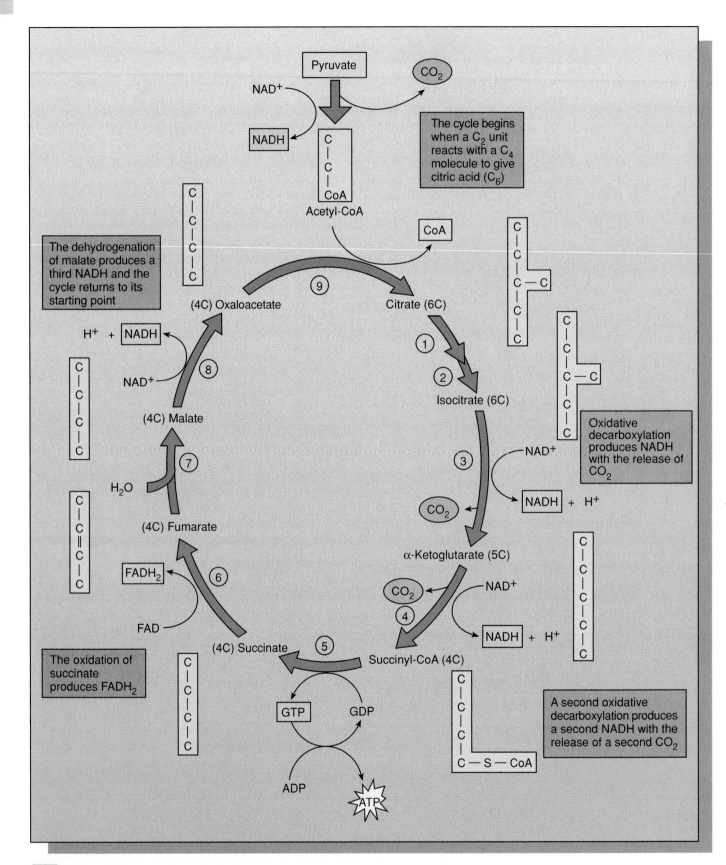

Figure 7.9 The citric acid cycle. *The cycle, which consists of nine (numbered) reactions, oxidizes acetyl-CoA.*

> *The electron transport chain puts the electrons harvested from the oxidation of glucose to work driving proton-pumping channels. The ultimate acceptor of the electrons is oxygen gas, which is reduced to form water.*

The availability of a plentiful electron acceptor (that is, an oxidized molecule) is what makes oxidative respiration work. Without such a molecule, oxidative respiration is not possible.

The electron transport chain used in aerobic respiration is similar to, and probably evolved from, the one employed in aerobic photosynthesis.

Thus, the final product of oxidative metabolism is water, which is not an impressive result in itself. But recall what happened in the process of forming that water: The electrons contributed by NADH molecules passed down the electron transport chain, activating three proton-pumping channels. Similarly, the passage of electrons contributed by $FADH_2$, which entered the chain later, activated two of these channels. The electrons harvested from the citric acid cycle have thus been used to pump a large number of protons out across the membrane, and *that* result is the payoff of oxidative respiration.

In eukaryotes, oxidative metabolism takes place within the mitochondria, which are present in virtually all

Table 7.1   The Output of Aerobic Metabolism

Metabolic Process	Substrate-Level Phosphorylation	Oxidative Respiration	
Glycolysis	2 ATP	2 NADH	
Oxidation (decarboxylation) of pyruvate (×2)		2 NADH	
Citric acid cycle (×2)	2 ATP	6 NADH	2 FADH$_2$
Total	4 ATP	10 NADH	2 FADH$_2$

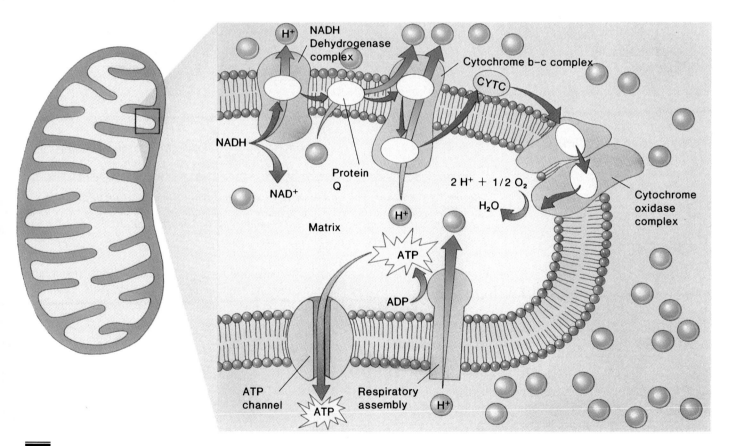

📼 Figure 7.10 The electron transport chain: How mitochondria use electrons to make ATP.

High-energy electrons harvested from food molecules (red arrows) are transported along a chain of membrane proteins, three of which use portions of the electron's energy to pump protons (blue arrows) out of the matrix into the other compartment. The electrons are finally donated to oxygen to form water. ATP is formed by chemiosmosis as protons reenter the matrix, driven by the high outside concentration.

eukaryotic cells. The internal compartment, or matrix, of a mitochondrion contains the enzymes that perform the reactions of the citric acid cycle. Electrons harvested there by oxidative respiration are used to pump protons out of the matrix into the outer compartment, the space between the two mitochondrial membranes. As proton concentration increases, protons cross the inner mitochondrial membrane back into the matrix, driven by diffusion. The only way they can get in is through special channels called respiratory assemblies that traverse the membrane. At the inner boundary of the membrane, the channel is linked by a stalk to a large protein complex that synthesizes ATP within the matrix when protons travel inward through the channel. The ATP then passes out of the mitochondrion by facilitated diffusion and into the cell's cytoplasm.

Electrons harvested from glucose and transported to the mitochondrial membrane by NADH drive protons out across the inner membrane. Return of the protons by diffusion generates ATP.

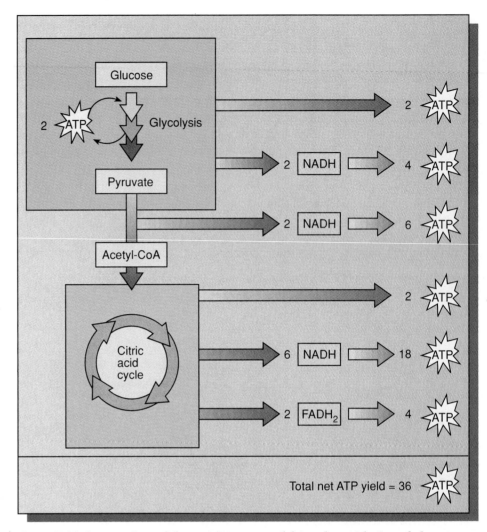

Figure 7.11 An overview of the energy extracted from the oxidation of glucose.
Most of the energy is extracted as energetic electrons (oxidation is the removal of electrons), carried by NADH and FADH$_2$ and used to produce ATP via chemiosmosis.

Overview of Glucose Catabolism: The Balance Sheet

How much metabolic energy does the chemiosmotic synthesis of ATP produce? One ATP molecule is generated chemiosmotically for each activation of a proton pump by the electron transport chain. Thus, each NADH molecule that the citric acid cycle produces ultimately causes the production of three ATP molecules because its electrons activate three proton pumps. Each FADH$_2$, which activates two proton pumps, leads to the production of two ATP molecules.

However, as mentioned earlier, eukaryotes carry out glycolysis in their cytoplasm and the citric acid cycle within their mitochondria. This separation of the two processes within the cell requires that the electrons of the NADH created during

glycolysis be transported across the mitochondrial membrane, consuming one ATP molecule per NADH. Thus, each glycolytic NADH produces only two ATP molecules in the final tally, instead of three.

The overall reaction for glucose catabolism is:

$$C_6H_{12}O_6 + 10NAD^+ + 2FAD^+ + 36ADP + 36P_i + 14H^+ + 6O_2$$
$$\rightarrow 6CO_2 + 36ATP + 6H_2O + 10NADH + 2FADH_2$$

As seen in table 7.1, four of the thirty-six ATP molecules result from substrate-level phosphorylation. The remaining thirty-two ATP molecules result from chemiosmosis.

The overall efficiency of glucose catabolism is very high: The aerobic oxidation of glucose yields thirty-six ATP molecules (a total of $-7.3 \times 36 = -263$ kilocalories per mole) (figure 7.11). The aerobic oxidation of glucose thus has an efficiency of 263/686,

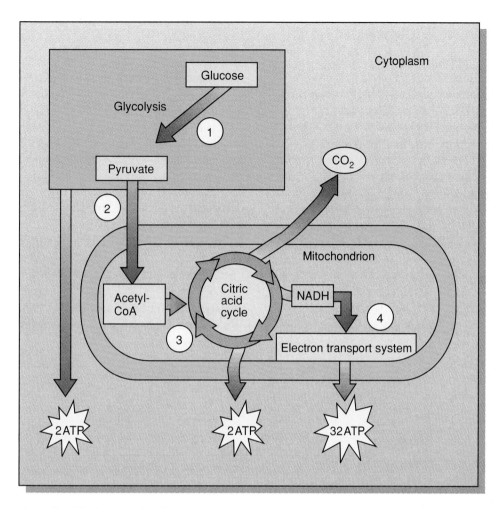

Figure 7.12 *An overview of oxidative respiration.*
There are four basic processes:
(1) The initial stage of glucose metabolism, glycolysis, does not require oxygen and occurs in the cell cytoplasm. During it, each glucose molecule is converted to two molecules of pyruvate.
(2) In eukaryotic cells, the pyruvate molecules enter the mitochondria, where they are converted to acetyl-CoA.

(3) The acetyl-CoA enters the citric acid cycle and is broken down to carbon dioxide (CO_2), harvesting chemical bond energy to drive the production of NADH.
(4) The NADH, in turn, powers the fourth and final stage of respiration, the electron transport system that chemiosmotically produces ATP. Most of the ATP that results from oxidative respiration is at the final chemiosmotic stage.

or 38 percent. (The total energy content of the chemical bonds of glucose is 686 kilocalories per mole.) Compared with the two ATP molecules generated by glycolysis, oxidative respiration is eighteen times more efficient.

Oxidative respiration is eighteen times more efficient than is glycolysis at converting the chemical energy of glucose into ATP. It produces thirty-six molecules of ATP from each glucose molecule consumed, compared with the two ATP molecules produced by glycolysis.

Figure 7.12 presents an overview of the oxidative respiration process.

Regulation of Cellular Respiration

When cells possess plentiful amounts of ATP, the ATP molecules within the cell act to shut down key enzymes of glycolysis and the citric acid cycle and thus slow down ATP production. Conversely, when ATP levels in the cell are low, ADP molecules (left over after ATP is used by the cell) activate enzymes in the biochemical pathway to stimulate more ATP production. Regulation of these biochemical pathways by levels of ATP is an example of feedback inhibition, discussed in chapter 6.

Almost all of the biochemical pathways in cells are regulated by the sensitivity of key enzymes to particular metabolites. The cell's metabolic machinery operates like a well-designed automobile engine, every part working in concert.

VOCABULARY OF CELLULAR RESPIRATION

aerobic respiration   That portion of cellular respiration that requires oxygen as an electron acceptor. Includes pyruvate oxidation, the citric acid cycle, and electron transport.

anaerobic respiration   That portion of cellular respiration that can occur in the absence of oxygen. Includes glycolysis and fermentation.

cellular respiration   The oxidation of organic molecules in cells. ATP is produced both by substrate-level phosphorylation and by chemiosmosis. Includes glycolysis, pyruvate oxidation, and the citric acid cycle.

chemiosmosis   ATP production by the transport of high-energy electrons to a membrane, where the electrons drive proton pumps and thus power ATP synthesis via the osmotic passage of protons back across the membrane.

fermentation   Respiration in which the final electron acceptor is an organic molecule; this process is typically an anaerobic metabolic process.

oxidation   The loss of an electron. In cellular respiration, high-energy electrons are stripped from food molecules, thereby oxidizing the food molecules.

oxidative respiration   Respiration in which the final electron acceptor is molecular oxygen.

photosynthesis   The light-fueled synthesis of organic molecules from carbon dioxide. ATP provides the energy to drive the synthesis. In photosynthesis, ATP is generated chemiosmotically by high-energy electrons, just as in cellular respiration. In photosynthesis, however, the energy is derived from light, whereas in cellular respiration, it is provided by electrons stripped from chemical bonds.

reduction   The gain of an electron. In cellular respiration, electrons from food molecules are added to oxygen molecules, reducing the oxygen molecules to form water.

substrate-level phosphorylation   The generation of ATP by coupling its synthesis to a strongly exergonic (energy-yielding) reaction.

EVOLUTIONARY VIEWPOINT

A key event in the evolution of metabolism was the invention of chemiosmosis, which provided a way to use energetic electrons to power ATP production. The spent electrons (and their associated protons) are then combined with oxygen to form water.

SUMMARY

1. Organisms acquire ATP from photosynthesis or by harvesting the chemical energy in the bonds of organic molecules.

2. Metabolism is driven by the application of energy to endergonic reactions.

3. Chemical energy can be harvested in two ways: (a) substrate-level phosphorylation, in which some reactions involving a large decrease in free energy are coupled to ATP formation; or (b) oxidative respiration, in which electrons are used to drive proton pumps, resulting in the chemiosmotic synthesis of ATP.

4. Glycolysis harvests chemical energy using substrate-level phosphorylation. The chemical bonds of glucose are rearranged to form two molecules of pyruvate and two molecules of ATP.

5. Organisms living in anaerobic environments require a mechanism to dispose of the electron and the associated hydrogen produced in the oxidation step of glycolysis. They are donated to one of a number of carbohydrates in a process called fermentation.

6. Aerobic organisms direct the electrons to the inner mitochondrial membrane, where they drive a proton pump. Pyruvate is further oxidized, yielding many additional electrons, all of which are also channeled to the mitochondrial membrane to drive proton pumps.

7. As the protons transferred to the outer mitochondrial compartment by the proton pumps increase in concentration, they begin to diffuse back across the inner mitochondrial membrane through special channels, and the passage of these protons drives ATP production.

8. Within all eukaryotic cells, oxidative respiration of pyruvate takes place within the matrix of mitochondria. The mitochondria act as closed osmotic compartments from which protons are pumped and into which protons pass by diffusion to create ATP. The ATP then leaves the mitochondria by facilitated diffusion.

9. In total, the oxidation of glucose results in the net production of thirty-six ATP molecules, all but four of them produced chemiosmotically.

10. Almost all of the stages of cellular respiration in cells are regulated by the sensitivity of key enzymes to particular metabolites.

REVIEWING THE CHAPTER

1. How does the cell use chemical energy to drive reactions?

2. How do cells make ATP?

3. What are the primary stages of cellular respiration?

4. What happens in cellular respiration to a chocolate bar that has been eaten?

5. What is the significance of glycolysis?

6. In cellular respiration, what are the alternatives after glycolysis?

7. How does oxidative respiration proceed in eukaryotic cells?

8. How is ATP generated from electrons during oxidative respiration?

9. What is the end result of glucose catabolism?

10. How is cellular respiration regulated?

COMPLETING YOUR UNDERSTANDING

1. Substrate-level phosphorylation occurs in
   a. the citric acid cycle.
   b. the electron transport chain.
   c. glycolysis.
   d. all of the above.

2. In the evolution of metabolic pathways, which probably occurred first: substrate-level phosphorylation or the chemiosmotic synthesis of ATP? What is your evidence?

3. Organisms obtain electrons that drive proton-pumping channels in chemiosmosis from
   a. light energy.
   b. chemical bonds.
   c. substrate-level phosphorylation.
   d. two of the above.

4. Which of the following occurs during glycolysis?
   a. The conversion of glucose to pyruvate
   b. A net gain of two ADP
   c. The synthesis of glucose from pyruvate
   d. Chemiosmotic synthesis of ATP

5. How are plants similar to animals in the production of ATP in oxidative respiration? How are they different?

6. G3P is produced in glycolysis by a process called _____.

7. In glycolysis, what percentage of the glucose molecule is harvested in the net yield of two ATP molecules? How does this energy efficiency compare to that of a car?

8. Which of the following is true concerning glycolysis?
   a. It uses molecular oxygen.
   b. It always occurs under anaerobic conditions.
   c. It takes place in the mitochondrion.
   d. It occurs in the cytoplasm of the cell.

9. In evolution, what is the correlation between oxidative metabolism and the development of food chains?

10. How does fermentation relate to the beer and wine industries?

11. What are the differences and similarities between ethanol and lactic acid fermentations?

12. Lactate dehydrogenase is an example of
   a. a protein.
   b. a lipid.
   c. a fatty acid.
   d. an amino acid.
   e. a carbohydrate.
   f. a nucleic acid.

13. What is the correlation between fermentation and an athlete who trains for marathon races?

14. In eukaryotes, oxidative respiration that takes place in mitochondria apparently originated from
   a. yeast cells.
   b. viruses.
   c. symbiotic bacteria.
   d. protists.
   e. symbiotic animal cells.
   f. a symbiosis between animals and plants.

15. What is the function of a multi-enzyme complex? Give an example of where one occurs in a cell.

16. Why is acetyl-CoA so important in eukaryotic cells? How is it produced in these cells?

17. Where is acetyl-CoA oxidized in aerobic respiration?

18. In the citric acid cycle, all of the following are produced except
   a. ATP.
   b. $FADH_2$.
   c. pyruvate.
   d. $NADH + H^+$.

19. What is the function of the cytochrome *c* oxidase complex in oxidative respiration?

20. How does molecular oxygen function in aerobic respiration? What would happen to an aerobic organism, such as a tree living in a floodplain, if the oxygen supply to its roots was cut off for a prolonged period of time during a flood?

21. What is the efficiency of aerobic oxidation of glucose compared to that of anaerobic fermentation? How does this efficiency relate to evolution?

22. If you wanted your body to make some ATP energy, would you eat a sugar cookie, some high-protein lentils, or a hamburger cooked in vegetable oil? Why?

FOR FURTHER READING

Dickerson, R. 1980. Cytochrome *c* and the evolution of energy metabolism. *Scientific American*, March, 136–54. A superb description of how the metabolism of modern organisms evolved.

Hinkle, P., and R. McCarty. 1978. How cells make ATP. *Scientific American*, March, 104–25. A good summary of oxidative respiration, with a clear account of the events that happen at the mitochondrial membrane.

McCarty, R. 1985. H + ATPases in oxidative and photosynthetic phosphorylation. *Bio Science*, Jan., 27–33. An excellent overview of the key protein channels that carry out chemiosmosis.

EXPLORATIONS

Interactive Software:
Oxidative Respiration

This interactive exercise presents a diagram of a mitochondrial membrane in cross section. The user can explore how electrons garnered from food molecules are used to drive proton pumps and how, by chemiosmosis, this produces ATP. The key variables manipulated by the user are oxygen levels, the amounts of food supplied, and existing levels of ATP (high levels shut down electron extraction from food, favoring storage instead).

Questions to Explore:

1. What is the amount of fuel (food) supplied to the mitochondrion for optimum performance? Does the mitochondrion make as much ATP as the amount of food it is supplied?

2. How does the preexisting supply of ATP affect energy extraction from food?

3. How would chemiosmosis be affected by a drop in oxygen supply to the mitochondrion?

4. Where does the energy to operate the proton pump come from?

5. How does ATP concentration influence the operation of the proton pump?

CHAPTER
8

PHOTOSYNTHESIS

Gesneria, *a South American perennial, is one of roughly 266,000 species of plants that cover the surface of the earth, performing the photosynthesis upon which almost all life on earth ultimately depends.*

FOR REVIEW

Here are some important terms and concepts that have been discussed in previous chapters and that you will encounter again in this chapter. Review them before proceeding, if necessary.

Electron energy levels (*chapter 2*)
Chloroplast structure (*chapter 4*)
Chemiosmosis (*chapters 5 and 7*)
Oxidation/reduction (*chapter 6*)
Glycolysis (*chapter 7*)

We all depend on the process of photosynthesis, which is the means by which the energy that ultimately builds our bodies is captured from sunlight. Through photosynthesis, plants, algae, and photosynthetic bacteria harvest solar energy and thereby generate oxygen for the air we breathe and food to fuel our lives. All life on earth depends on solar energy captured by these organisms.

Photosynthesis is only one aspect of plant biology—although an important one—and chapters 30 through 32 examine plants in detail. Photosynthesis is discussed here because it evolved long before plants did and because all organisms depend directly or indirectly on photosynthesis for the energy that powers their lives.

The total amount of radiant energy that reaches the earth from the sun each day is equivalent to the energy of about one million Hiroshima-sized atomic bombs. Approximately one-third of this energy is immediately radiated back into space, and most of the remainder is absorbed by the earth and converted to heat. Less than 1 percent is captured in the process of photosynthesis to provide the energy that drives all life activities on earth (figure 8.1).

Photosynthesis is a set of reactions in which light energy from the sun is converted into the chemical bond energy of glucose and ATP. During photosynthesis, light energy from the sun rearranges the carbon, hydrogen, and oxygen atoms present in water and atmospheric carbon dioxide into molecules of glucose, water, and oxygen. In other words, photosynthesis converts inorganic molecules (carbon dioxide and water), which have low potential energy because of their lack of carbon-hydrogen (C—H) bonds, into an organic molecule (glucose), which has high potential energy because of its many C—H bonds. This extraordinary process drives all life on earth. Any fuel that you ingest—be it plant or animal—was originally manufactured by the photosynthetic process.

Figure 8.1 The energy that drives photosynthesis comes from the sun.

Less than 1 percent of all the energy that reaches the earth from the sun is captured in the process of photosynthesis. Yet, this 1 percent drives all the activities of life on earth.

Photosynthesis also has played a critical role in the evolution of life on earth. Before photosynthesis, the earth's atmosphere contained little oxygen. Photosynthesis yields oxygen as a by-product, and thus the advent of photosynthesis millions of years ago greatly increased the atmosphere's oxygen concentration. Without photosynthesis, the world as we know it today would be populated with very different kinds of organisms, organisms that would not depend on oxygen for life.

Before turning to the intricacies of photosynthesis, the chapter first examines the form of energy that allows photosynthesis to take place—the energy of light from the sun. Without this energy, photosynthesis would come to a halt—and all life would quickly follow.

The Biophysics of Light

Where is the energy in light? What is there about sunlight that a plant can use to create chemical bonds? An answer to these questions begins with a consideration of the physical nature of light itself.

The Photoelectric Effect

In a laboratory in Germany in 1887, a young physicist named Heinrich Hertz was attempting to verify a mathematical theory that predicted the existence of electromagnetic waves. To see whether such waves existed, Hertz constructed a spark generator in his laboratory—a machine composed of two shiny metal spheres standing near each other on slender rods. When a very high static electrical charge built up on one sphere, sparks would jump across to the other sphere. To investigate whether the sparking would create invisible electromagnetic waves, as predicted by the mathematical theory, Hertz placed a thin metal hoop that was not quite a closed circle on top of an insulating stand on the other side of the room. When he turned on the spark generator, tiny sparks could be seen crossing the gap in the hoop! This was the first demonstration of radio waves. But Hertz noted a curious side effect as well. When light was shone on the ends of the hoop, the sparks crossed the gap more readily. This unexpected effect, called **photoelectric effect,** puzzled investigators for many years.

Especially perplexing was the fact that the strength of the photoelectric effect depended not only on the brightness of the light shining on the gap in the hoop but also on the light's wavelength. Short wavelengths were much more effective than long in producing the photoelectric effect. Albert Einstein finally explained this effect as a natural

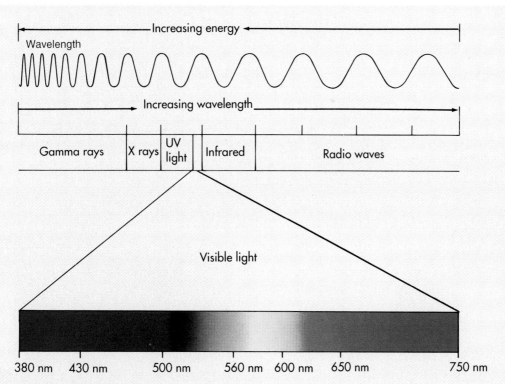

Figure 8.2 The electromagnetic spectrum.
Light is a form of electromagnetic energy and is conveniently thought of as a wave. The shorter the wavelength of light, the greater the energy. Visible light represents only a small part of the electromagnetic spectrum, that between 380 and 750 nanometers (nm). (UV stands for ultraviolet light.)

consequence of the physical nature of light: The light was literally blasting electrons from the metal surface at the ends of the hoop, creating positive ions and thus facilitating the passage of the electronic spark induced by the radio waves. Light consists of units of energy called **photons,** and some of these photons were being absorbed by the hoop's metal atoms. In this process, some of the electrons of the metal atoms were being boosted into higher energy levels and so ejected from the metal atoms into the gap.

Photons Not All the Same

All photons do not possess the same amount of energy. Some contain a great deal of energy, others far less. Photons of short-wavelength light contain higher energy than photons of long-wavelength light. For this reason, the photoelectric effect is more pronounced with short-wavelength light.

Sunlight contains photons of many energy levels, only some of which our eyes perceive as visible light. The highest energy photons, which occur at the short wavelength end of the **electromagnetic spectrum,** are gamma rays with wavelengths of less than 1 nanometer; the lowest energy photons, with wavelengths of thousands of meters, are radio waves (figure 8.2).

Ultraviolet Light

Sunlight contains a significant amount of ultraviolet (UV) light. Because of its shorter wavelength, UV light possesses considerably more energy than does visible light. Scientists believe that UV light was an important energy source on primitive earth when life originated, before the oxygen-rich atmosphere developed.

Today's atmosphere contains ozone (derived from oxygen gas), which absorbs most of the UV-energy photons in sunlight, but anyone who has been sunburned knows that UV-energy photons manage to penetrate this protective ozone layer. Thin places, or "holes," in the ozone layer have been caused by human activities and are associated with a general decrease in ozone in the atmosphere's upper layers (see chapter 24). The general decline in UV-absorbing ozone significantly increases the risk of skin cancers in humans who do not take protective measures.

Absorption Spectra

As described in chapter 2, electrons occupy distinct energy levels in their orbits around the atomic nucleus. Boosting an electron to a different energy level requires just the right amount of energy—no more, no less. Similarly, when climbing a ladder, you must raise your foot just so far to climb a rung. Therefore,

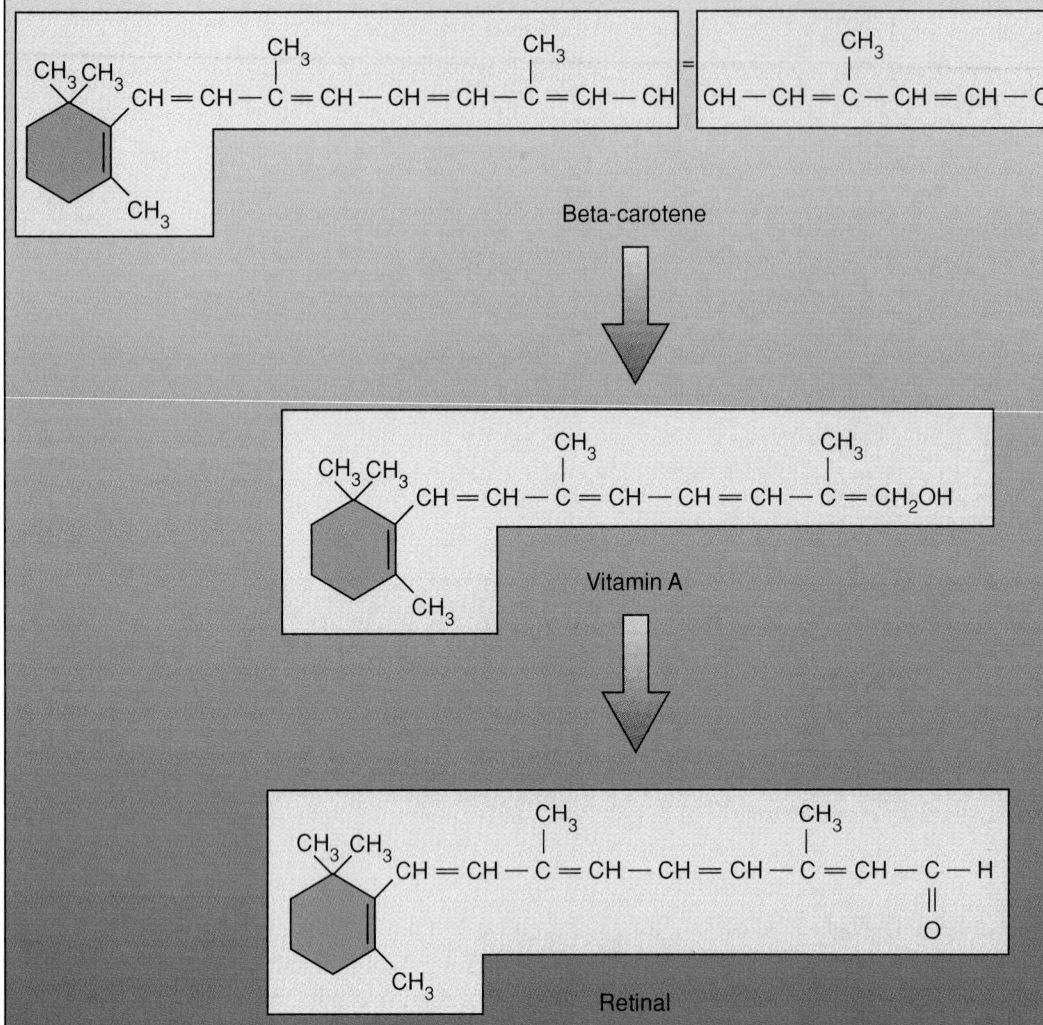

Figure 8.3 What do carrots have to do with vision?

The pigment beta-carotene is what makes carrots look orange. When the double bond of the two halves of beta-carotene is broken, two molecules of vitamin A are produced. Retinal, which is the key visual pigment in your eyes, is produced from vitamin A.

specific atoms can absorb only certain photons of light—those that correspond to available energy levels. A given atom or molecule has a characteristic range, or **absorption spectrum,** of photons that it is capable of absorbing, depending on the electron energy levels it has available.

Photons are packets of energy. The shorter the wavelength of light, the more energetic the photons.

Capturing Light Energy in Chemical Bonds: Pigment Molecules

When light energy is "captured" or absorbed by a molecule, the photon of energy boosts an electron of the molecule to a higher energy level. Molecules that absorb light are called **pigments.** Organisms have evolved a variety of pigments, two general categories of which are:

1. *Carotenoids* Carotenoids consist of carbon rings linked to chains of alternating single and double bonds. Carotenoids can absorb photons of a wide range of energies, although not always with high efficiency. They are responsible for most of the yellow and orange colors seen in plants, where they play an important role in capturing light energy and transferring it to chlorophyll in the process of photosynthesis.

A typical carotenoid is **beta-carotene,** the orange pigment found in carrots. In beta-carotene, two carbon rings are linked by a chain of eighteen carbon atoms connected alternately by single and double bonds. Splitting a molecule of beta-carotene into equal halves results in the production of two molecules of **vitamin A.** When vitamin A is subsequently oxidized, the pigment **retinal** is produced.

Retinal is the same pigment used by the human eye to trap photons of light. Thus, the claim that eating carrots improves vision is based on fact (figure 8.3). When retinal absorbs a photon of light, the resulting electron excitation causes a change in the shape of a pigment located in the membrane of certain cells in the eye and thus triggers a nerve impulse. This nerve impulse travels to the brain, causing us to "see" the image. Retinal absorbs photons that produce light ranging from violet (380 nanometers) to red (750 nanometers) and so determines the range of colors—the **visible light**—that we can see.

Some other organisms use different light-absorbing pigments for vision and thus "see" a different portion of the electromagnetic spectrum. For example, most insects have eye pigments that absorb at lower wavelengths than does retinal. As a result, bees can perceive ultraviolet light, which is produced by photons with a shorter wavelength than violet has, but cannot see red, which is produced by photons with a relatively long wavelength.

2. *Chlorophylls* Other biological pigments called **chlorophylls** absorb photons by means of an excitation process analogous to the photoelectric effect. These pigments use a metal atom (magnesium), which lies at the center of a complex ring structure, called a **porphyrin ring,** that consists of alternating single- and double-carbon bonds. Photons absorbed by the pigment molecule excite electrons of the magnesium atom, which are then channeled away from the magnesium through the carbon-bond system. Several small side groups attached outside the porphyrin ring alter the absorption properties of the pigment in different kinds of chlorophylls.

The two kinds of chlorophylls that occur in plants are called chlorophyll *a* and *b*. Figure 8.4*a* shows the chemical structure of chlorophyll *a*.

Unlike retinal, the different kinds of chlorophylls absorb only photons of a narrow energy range. As you can see from their absorption spectra, chlorophylls *a* and *b* absorb primarily violet-blue and red light (figure 8.4*b*). The light between wavelengths of 500 and 600 nanometers is not absorbed by chlorophyll pigments and therefore is *reflected* by plants. The light reflected from a chlorophyll-containing plant has had all of its photons except those in the 500- to 600-nanometer range absorbed by the chlorophyll. When these photons are subsequently absorbed by the retinal in our eyes, we perceive them as green (figure 8.5).

A pigment is a molecule that absorbs light. The wavelengths absorbed by a particular pigment depend on the energy levels available in that molecule to which light-excited electrons can be boosted. Two general categories of pigments are carotenoids and chlorophylls.

Figure 8.4 Chemical structure and absorption spectra for chlorophylls.

(a) *The chemical structure of chlorophyll* a. *The complex ring structure that forms the center of the molecule is called a porphyrin ring. Magnesium, a metal ion, lies at the center of the porphyrin ring. For the sake of simplicity, this diagram does not show the twenty-carbon alcohol that is attached to the chlorophyll* a *molecule; however, the position of this alcohol is marked by R.* (b) *Absorption spectra for chlorophylls* a *and* b. *The peaks represent the wavelengths that the respective chlorophylls absorb best. As you can see, the chlorophylls absorb predominantly violet-blue and red light in two narrow bands of the spectrum. The chlorophylls reflect the greenish-yellow light in the middle of the spectrum.*

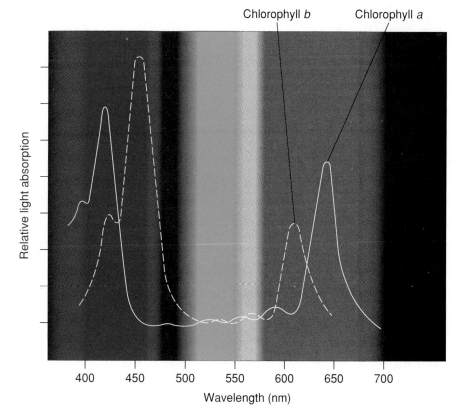

(a)

(b)

All green plants and algae and all but one primitive group of photosynthetic bacteria use chlorophylls as their primary light-gatherers. Why don't these photosynthetic organisms use a pigment like retinal, which has a very broad absorption spectrum and can harvest light in the 500- to 600-nanometer wavelength range as well as at other wavelengths? The most likely hypothesis involves photoefficiency. Retinal absorbs a broad spectrum of light wavelengths but does so with relatively low efficiency; in contrast, chlorophylls absorb in only two narrow bands, violet-blue and red, but do so with very high efficiency. Therefore,

by using chlorophylls, plants and most other photosynthetic organisms achieve far higher overall photon capture rates than would be possible with a pigment that allows a broader but less efficient spectrum of absorption.

Overview of Photosynthesis

Photosynthesis consists of a complex series of events involving three kinds of chemical processes. The first process is the chemiosmotic generation of ATP by electrons, using energy captured from sunlight. The reactions involved in this process are called the **light-dependent reactions** of photosynthesis because the resultant ATP synthesis takes place only in the presence of light. Second, the light-dependent reactions are followed by a series of enzyme-catalyzed reactions that use this newly generated ATP to drive the formation of organic molecules from atmospheric carbon dioxide. These reactions are called the **light-independent reactions** of photosynthesis because, as long as ATP is available, they occur as readily in the absence of light as in its presence. Third, the pigment that absorbed the light in the first place is rejuvenated and made ready to initiate another light-dependent reaction (figure 8.6).

Absorbing Light Energy

Light-dependent reactions occur on photosynthetic membranes. In photosynthetic bacteria, these membranes are the cell membrane itself. In plants and algae, photosynthetic membranes are called **thylakoids,** which are located within a special organelle called a **chloroplast** (see chapter 4). Thylakoids are stacked one on top of another to form columns called **grana** (singular, *granum*). Grana are held in place by buttresses, or **lamellae** (singular, *lamella*). A semi-liquid substance called the **stroma** bathes the interior of the chloroplast and contains the enzymes that catalyze light-independent reactions (figure 8.7).

Light-dependent reactions occur in three stages:

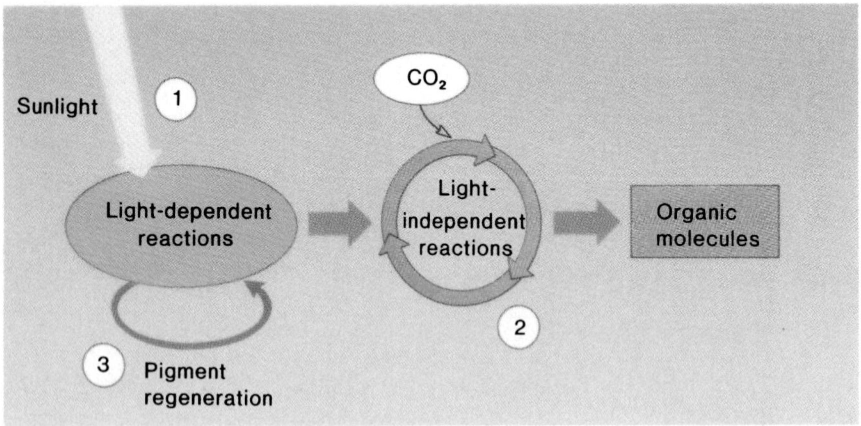

1. A photon of light is captured by a pigment molecule that is bound to proteins embedded in the thylakoid. This photon capture results in the excitation of an electron within the pigment.

2. The excited electron is shuttled along a series of electron-carrier molecules embedded within the thylakoid to a transmembrane proton pump, where the electron's arrival induces the transport of a proton inward across the thylakoid membrane. The thylakoid membrane is impermeable to protons and most molecules; therefore, transport across it occurs almost exclusively through these proton pumps. The excited electron is then passed on to an electron acceptor.

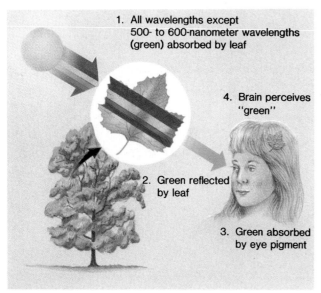

Figure 8.5 Why are plants green?
As shown in figure 8.4b, a green leaf containing chlorophylls absorbs all the colors in the spectrum except for those in the 500- to 600-nanometer range, which are the greenish-yellow colors. The leaf reflects these colors. These reflected wavelengths are absorbed by the visual pigments in our eyes, and our brains perceive the reflected wavelengths as "green."

Within figure 8.5:
1. All wavelengths except 500- to 600-nanometer wavelengths (green) absorbed by leaf
2. Green reflected by leaf
3. Green absorbed by eye pigment
4. Brain perceives "green"

Figure 8.6 The three processes of photosynthesis.
(1) Light-dependent reactions require light from the sun to excite electrons. (2) Light-independent reactions can take place in the presence or absence of light. (3) The pigment that first absorbed the light is rejuvenated and prepared to initiate another light-dependent reaction.

Within figure 8.6:
Sunlight — 1 — Light-dependent reactions — Light-independent reactions — CO_2 — 2 — Organic molecules — 3 Pigment regeneration

Figure 8.7 Journey into a leaf: The structure of a chloroplast.

Chloroplasts are bounded by a double membrane and contain photosynthetic membranes called thylakoids. Stacked one on top of the other, a column of thylakoids is called a granum. Grana are held in place by lamellae, and the interior of the entire chloroplast is bathed by a semi-liquid called the stroma.

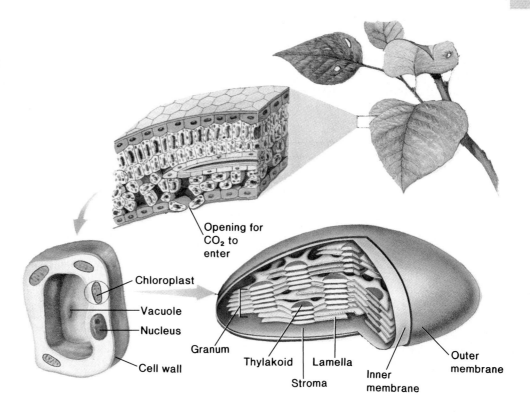

Opening for CO_2 to enter

Chloroplast

Vacuole

Nucleus

Cell wall

Granum   Thylakoid   Lamella   Inner membrane   Outer membrane

Stroma

Figure 8.8 A simplified version of chemiosmosis in a plant cell.

During the light-dependent reactions, the arrival of an excited electron at a proton pump causes a proton to be transported through the pump into the interior of the thylakoid. Later, the proton is ejected into the stroma through a transmembrane enzyme called ATPase. The proton ejection drives the synthesis of an ATP molecule.

3. Later, the proton passes back through the thylakoid membrane into the stroma through a special enzyme channel called an ATPase that is embedded in the thylakoid membrane. This exiting of the proton drives the chemiosmotic synthesis of ATP (figure 8.8).

Fixing Carbon

At this stage in the photosynthetic process, ATP has been generated, and thus energy, in a "raw" form, is now available. But how should this energy be stored so that photosynthetic organisms can readily get to and use it when necessary? The answer is to store the energy in the C—H bonds of glucose. The light-independent reactions of photosynthesis use the carbon from carbon dioxide and the hydrogen atoms from the boosting of electrons to build glucose, an organic molecule. Glucose is a relatively large molecule, compared to carbon dioxide and water, and has many C—H bonds. The first step in

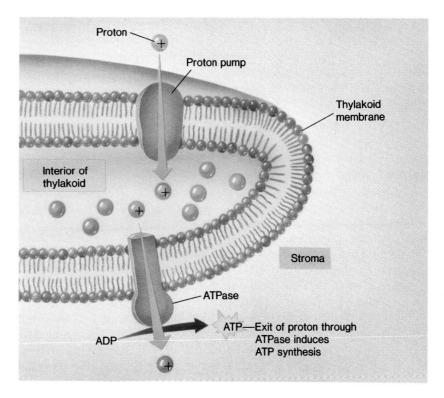

Proton

Proton pump

Thylakoid membrane

Interior of thylakoid

Stroma

ATPase

ADP

ATP—Exit of proton through ATPase induces ATP synthesis

constructing a glucose molecule is **carbon fixation:** taking the carbons in atmospheric carbon dioxide and adding them to a smaller carbon-containing molecule, thus building a larger carbon-containing organic molecule. Carbon-carbon (C—C) bonds, however, have little potential energy; C—H bonds contain higher amounts of energy. Therefore, hydrogen atoms that can form the high-energy bonds in glucose need to be generated.

As discussed in chapter 6, when hydrogen is added to any molecule, that molecule is said to be reduced. Also, you might recall from chapter 2 that hydrogen atoms consist of one proton and one electron, and that electrons travel with protons in the form of hydrogen atoms. Thus, the generation of hydrogen atoms from the boosting of electrons to higher energy levels is sometimes called the generation of **reducing power.** Organisms have devised many ways—some more efficient than others—to generate this reducing power during light-dependent reactions.

Replenishing the Pigment

The third kind of chemical event that occurs during photosynthesis involves the electron that was stripped from the chlorophyll at the beginning of the light-dependent reactions. This electron must be returned to the pigment, or another source of electrons must be used to replenish the pigment's electron supply. Continual electron removal would cause the pigment to become deficient in electrons (bleached), and no longer able to trap photon energy by electron excitation. Various organisms have evolved different approaches to solving this problem during the course of their evolutionary history.

> *Photosynthesis involves three processes: (1) the use of light-ejected electrons to drive the chemiosmotic synthesis of ATP; (2) the use of the ATP to fix carbon; and (3) the replenishment of the photosynthetic pigment.*

The overall process of photosynthesis may be summarized by a simple oxidation/reduction equation:

$$\text{Light}$$
$$6CO_2 + 12H_2{}^*O \rightarrow C_6H_{12}O_6 + 6H_2O + 6{}^*O_2$$

| Atmospheric carbon dioxide | Water vapor | Glucose | Water | Oxygen gas from the original water molecules |

The H_2O that appears on both sides of the equation does *not* represent the same water molecule. This can be demonstrated by carrying out photosynthesis with water vapor in which the oxygen atom is a heavy isotope. Most of the heavy oxygen atoms (indicated in the equation by the symbol $^*O$) end up in oxygen gas and not in water.

The remainder of the chapter presents a more detailed discussion of photosynthesis.

Light-Dependent Reactions: How Light Drives Chemistry

Photosynthesis in plants, algae, and those bacteria in which it occurs is the result of a long evolutionary process. As this evolution progressed, new reactions were added to older ones, thus making the overall series of reactions more complex. Much of the evolution has centered on the light-dependent reactions, which apparently have changed considerably since they first evolved.

Light-Dependent Reactions in Bacteria

The simplest form of photosynthesis is carried out by bacteria. In evolutionary terms, bacteria were the first to use photosynthesis to make organic molecules. As evolution progressed, plants evolved a more complex form of photosynthesis that incorporated elements of the simpler bacterial process. The discussion that follows examines not only the details of the photosynthetic process, but also how the process changed over time to accommodate the different needs of plants and other photosynthetic organisms.

Evolution of the Photosystem

In chloroplasts and all but one group of photosynthetic bacteria, light is captured by a network of chlorophyll pigments working together. Each chlorophyll molecule within the network, which is called a **photosystem,** is capable of capturing photons efficiently. Chlorophyll molecules are held on a lattice of protein within the photosystem. This arrangement permits the channeling of excitation energy from anywhere in the array to a central point. The assembly of chlorophyll molecules thus acts as a sensitive "antenna" to capture and focus photon energy. Its mode of operation is similar to how a magnifying glass that is focusing light can generate enough heat energy at the point of focus to burn paper. Similarly, the photosystem channels the excitation energy gathered by any one of the chlorophyll molecules to one which in green plants is called P_{700}. P_{700} is associated with a membrane-bound protein called **ferredoxin,** which acts as an electron acceptor.* P_{700} channels the energy out of the photosystem to the ferredoxin, providing the ferredoxin with many more electrons than would otherwise be possible (figure 8.9).

When light of the proper wavelength strikes any chlorophyll pigment molecule of the photosystem, the resulting excitation passes from one chlorophyll molecule to another. The excited electron does not transfer physically from one chlorophyll molecule to the next. Instead, the chlorophyll molecule passes the *energy* along to an adjacent molecule of the photosystem, after which the chlorophyll's electron returns to the low-energy level it had before the photon was absorbed. A crude analogy to this form of energy transfer exists in the initial "break" in a game of pool. If the cue ball squarely hits the point of the triangular array of fifteen pool balls, the two balls at the far corners of the triangle fly off, and none of the central balls move at all. The energy is transferred through the central balls to the most distant ones. The photosystem's protein lattice, in which the molecules of chlorophyll are embedded, serves as a sort of scaffold, holding individual chlorophyll molecules in orientations that are optimal for energy transfer. In this way, the process eventually channels excitation energy in the form of electrons to the membrane-bound ferredoxin.

---

*Recent research suggests that the primary electron acceptor may not be ferredoxin, as has been commonly thought, but rather, molecules of chlorophyll (photosystem I) and pheophytin (photosystem II). The matter is a subject of intensive research.

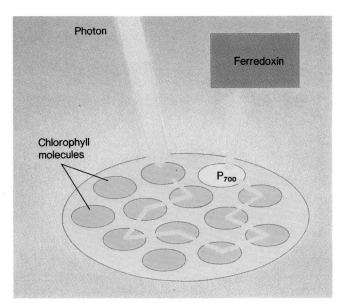

Figure 8.9 How photosystems work.
When light of the proper wavelength strikes any chlorophyll molecule within a photosystem, the molecule absorbs the photon's energy. The energy passes from one chlorophyll molecule to another until it encounters P_{700}, which channels the energy out of the photosystem to ferredoxin, an electron acceptor.

A photosystem, which is an array of chlorophyll pigment molecules, acts as a light antenna, directing photon energy captured by any of its members toward a single chlorophyll molecule and thus amplifying the light-gathering powers of individual chlorophyll molecules.

Where Does the Electron Go?

Photosystems probably evolved more than three billion years ago in bacteria similar to the current green sulfur bacteria. In green sulfur bacteria, the photosystem's absorption of a photon of light results in the transmission of an electron from the P_{700} molecule to ferredoxin. As in many oxidation/reduction processes, the electron is accompanied by a proton, traveling as a hydrogen atom. The proton is extracted by green sulfur bacteria from hydrogen sulfide (H_2S) through a process that produces elemental sulfur as a residual by-product.

The ejection of an electron from P_{700} and its donation to ferredoxin leaves P_{700} short one electron. Before the photosystem of the green sulfur bacteria can function again, the electron must be returned. These bacteria channel the electron back to the pigment through an electron transport system (see chapter 7), in which the electron's passage drives a proton pump and thus promotes the chemiosmotic synthesis of ATP. Therefore, the path of the electron originally extracted from P_{700} is circular. Chemists call the process **cyclic photophosphorylation.** *Cyclic* means that the process is circular, and *photophosphorylation* means merely that the energy from sunlight is used to add a phosphate onto ADP to form ATP.

Cyclic photophosphorylation, however, is not truly circular. The electron that left P_{700} was a high-energy level electron, boosted to its high-energy level by the absorption of a photon of energy. The electron that returns has only as much energy left as it had before the photon absorption. The difference in energy levels is the photosynthetic payoff—that is, the energy used to drive the proton pump.

In bacteria, the electron ejected from the pigment by light travels a circular path in which it powers a proton pump and then returns to the photosystem where it originated. This process is called cyclic photophosphorylation.

Light-dependent reactions of all other photosynthetic systems, including those in other groups of bacteria, evolved by adding to the simple cyclic photophosphorylation process of the green sulfur bacteria. Just as glycolysis is retained as a fundamental component of the respiratory metabolism of all organisms, so cyclic photophosphorylation remains a fundamental component of photosynthesis in the chloroplasts of all plants, algae, and most photosynthetic bacteria.

Light-Dependent Reactions in Plants

For more than one billion years, cyclic photophosphorylation was the only form of photosynthetic light reaction that organisms used. But it has a fundamental limitation: It is geared only toward ATP production and not toward the synthesis of organic molecules, such as glucose. The ultimate objective of photosynthesis is *not* to generate ATP but rather to fix carbon—to incorporate atmospheric carbon dioxide into new carbon compounds. Because the carbon-containing molecules (glucose) produced during carbon fixation are more reduced (have more hydrogen atoms) than their precursor (carbon dioxide—CO_2), a source of hydrogens must be provided. Cyclic photophosphorylation does not provide hydrogen atoms. Thus, bacteria that use this process must scavenge hydrogens from other sources—a very inefficient undertaking.

The Advent of Photosystem II

At some point after the appearance of the green sulfur bacteria, other kinds of bacteria evolved an improved version of the photosystem that solved the reducing power problem simply and neatly. These bacteria grafted a second, more powerful photosystem onto the original one, using a new form of chlorophyll called chlorophyll *a*. This great evolutionary advance originated with the cyanobacteria, no less than 2.8 billion years ago.

In this second photosystem, called **photosystem II,** molecules of chlorophyll *a* are arranged so that more of the shorter-wavelength photons of higher energy are absorbed than in the more ancient bacterial photosystem (called **photosystem I** in algae and plants). As in photosystem I, energy is transmitted from one chlorophyll pigment molecule to another until it encounters a particular pigment molecule that is positioned near a strong membrane-bound electron acceptor. In photosystem II, the absorption peak of this pigment molecule is 680 nanometers; therefore, the molecule is called P_{680}.

How the Two Photosystems Work Together: Noncyclic Photophosphorylation

Plants, algae, and cyanobacteria, in contrast to other photosynthetic bacteria, use both photosystems. Photosystem II acts first. When a photon jolts an electron from photosystem II, the excited electron is donated to a series of electron acceptors, which pass it along to photosystem I. In its journey to photosystem I, the electron drives a proton pump and thus generates an ATP molecule chemiosmotically (figure 8.10).

When the electron reaches photosystem I, it has already expended its excitation energy in driving the proton pump and thus contains only the same amount of energy as the other electrons of this photosystem. However, its arrival does give the photosystem an electron that it can afford to lose. Photosystem I now absorbs a photon, boosting one of its pigment electrons to a high-energy level. This electron is then channeled to ferredoxin and is used to generate reducing power. In plants, algae, and cyanobacteria, ferredoxin contributes two electrons to reduce **NADP⁺ (nicotinamide adenine dinucleotide phosphate)**, generating **NADPH.** The hydrogen atom in NADPH is used later, in the formation of glucose molecules in the light-independent reactions of photosynthesis. Because this molecule is used instead of the NAD⁺ used in oxidative respiration, the flow of electrons in the two processes is kept separate (figure 8.10).

> *Plants, algae, and cyanobacteria use two photosystems. First, a photon is absorbed by photosystem II, which passes an electron to photosystem I. During this process, the electron uses its photon-contributed energy to drive a proton pump and thus generate a molecule of ATP. Then another photon is absorbed, this time by photosystem I, which also passes on a photon-energized electron. This second electron is channeled away to provide reducing power.*

Thus, the energy produced in photosystem II is spent in ATP synthesis; the energy produced by photosystem I creates reducing power. These two processes together comprise the light-dependent reactions of eukaryotic photosystems. Because two photosystems are used, this process is sometimes called **noncyclic photophosphorylation** to distinguish it from the quasi-cyclic process used by bacteria.

The Splitting of Water and the Formation of Oxygen Gas

How does P_{680}, the pigment of photosystem II that starts the photosynthetic process by donating an electron, make up for this loss if the electron is not returned but is instead expended to synthesize NADPH? As you might expect, an electron is obtained from another source. The loss of the excited electron from photosystem II converts P_{680} into a powerful oxidant (electron-seeker) that obtains the required electron from a protein called Z. The removal of this electron from Z renders Z a strong electron-acceptor in turn, and it obtains electrons from water to funnel to P_{680}. Z catalyzes a complex series of reactions in which water is split into electrons (passed to P_{680}), hydrogen ions (H⁺), and free OH⁻ groups. The free OH⁻ groups (chemists call them "free radicals") are collected and reassembled as water and oxygen gas. The H⁺ ions (protons) are exported across the membrane, thus augmenting the proton concentration gradient that was established during the passage of electrons to photosystem I.

> *The electrons and associated protons that oxygen-forming photosynthesis employs to form reduced organic molecules are obtained from water. The leftover oxygen atoms of the water molecules combine to form oxygen gas.*

The use of two photosystems solves in a simple way the evolutionary problem of how to obtain reducing power for biosynthetic reactions. Even though the cyclic photophosphorylation of green sulfur bacteria provides ATP, it does not provide a ready means of generating NADPH. Therefore, organisms that use this form of photosynthesis must make NADPH in a roundabout way and, in doing so, expend a lot of ATP.

Comparison of Plant and Bacterial Light-Dependent Reactions

A comparison of the dual P_{680}/P_{700} photosystems with the P_{700} photosystem from which it evolved is useful. The removal of an electron from P_{700} yields enough energy to extract hydrogen from hydrogen sulfide (H_2S), which requires 78 kilocalories, but not from water (H_2O), which requires 118 kilocalories. By contrast, the removal of an electron from P_{680} yields considerably more energy, and that energy is adequate to split water molecules, producing gaseous oxygen as a by-product. In cyanobacteria, algae, and plants, all of which use two photosystems, there is no cyclic flow of electrons. Instead, electrons and associated hydrogen atoms are extracted continually from water and are eventually used to reduce NADP⁺ to NADPH. Because hydrogens are stripped from water, oxygen gas is continuously generated as a product of the reaction. This photosynthetic process generates all of the oxygen in the air that we breathe.

> *Every oxygen molecule in the air you breathe was once split from a water molecule by an organism carrying out oxygen-forming photosynthesis.*

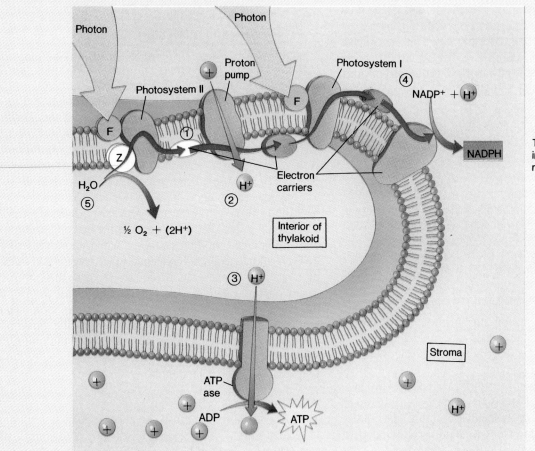

Figure 8.10 A detailed look at the chemiosmotic synthesis of ATP in plants.

This diagram shows an electron along the photosynthetic path (follow the red *arrow) and the mechanism of chemiosmosis. (1) When a photon of light strikes a pigment molecule (P_{680}) in photosystem II, it excites an electron. This electron is coupled to a proton stripped from water by a Z protein and passes along a chain of membrane-bound cytochrome electron carriers (*red *arrow) to a proton pump. (2) There, the energy supplied by the photon is used to transport a proton across the membrane into the thylakoid. (3) The resulting proton gradient drives the chemiosmotic synthesis of ATP. (4) The spent electron then passes to photosystem I. Photosystem I absorbs a new photon of light, boosting one of its pigment electrons to a high-energy level. This electron eventually causes the reduction of NADP to NADPH. (5) The splitting of water also contributes to the proton gradient inside the thylakoid space. More hydrogens (protons) build up inside the space, prompting the protons to leave through ATPase.*

Light-Independent Reactions

The preceding section focused on the light-dependent reactions of photosynthesis. These reactions use light energy to produce metabolic energy in the form of ATP and reducing power in the form of NADPH. But this is only half the story. Photosynthetic organisms employ the ATP and NADPH produced by the light-dependent reactions to build organic molecules from atmospheric carbon dioxide (CO_2). This later phase of photosynthesis, which comprises the so-called light-independent reactions, is carried out by a series of enzymes found in the stroma of the chloroplast.

How the Products of Light-Dependent Reactions are Used to Build Organic Molecules from Carbon Dioxide

The light-independent reactions consist of a set of reactions known as the **Calvin cycle.** In the first stage of the Calvin cycle, carbon is fixed, or captured, from atmospheric carbon dioxide and attached to a carbon-containing molecule. As discussed earlier, to build a large, complex, carbon-containing molecule like glucose, carbon atoms need to be added to a smaller carbon-containing molecule. The carbon fixation stage of the Calvin cycle accomplishes this by attaching the carbon in atmospheric carbon dioxide to a five-carbon sugar called **ribulose 1,5 bisphosphate (RuBP).** This reaction is catalyzed by an enzyme called RuBP carboxylase. The six-carbon intermediate molecule that is formed is unstable and immediately splits into two three-carbon molecules of phosphoglycerate (PGA) (figure 8.11).

In the second stage of the Calvin cycle, a phosphate group is added to the PGA molecules, forming glyceraldehyde phosphate (PGAL). Most of the PGAL is rearranged to reform the RuBP that was used during carbon fixation. But one molecule of PGAL is used to form glucose. With the hydrogens from NADPH, the PGAL is converted into simple glucose molecules that will become the building blocks of such complex molecules as starch and cellulose.

The reactions of the Calvin cycle that are diagrammed in figure 8.12 show an input of three carbon dioxide molecules and the formation of six PGAL molecules, five of which are used to reform RuBP and one of which is used to construct glucose. These processes are shown in this manner for simplicity's sake. In reality, it takes six molecules of PGAL to yield enough carbons to build a six-carbon sugar like glucose. Therefore, it takes six turns of the Calvin cycle to produce one six-carbon sugar.

The energy (ATP) and reducing power (NADPH) produced by the light-dependent reactions are used to fix carbon and construct simple sugars in a series of light-independent reactions called the Calvin cycle.

Photosynthesis is not Perfect

One of the ironies of evolution is that processes evolve to be only as good as they need to be, rather than as good as they potentially might be. Evolution favors not optimum solutions, but rather, workable ones that can be derived from others that already exist. New reactions are often grafted onto old ones, as the citric acid cycle was grafted onto glycolysis. Photosynthesis is no exception.

Photorespiration: The "Undoing" of Photosynthesis

Several stages of the glycolytic pathway are used in the Calvin cycle, and one of the carryover enzymes, RuBP carboxylase (the enzyme that catalyzes the key carbon-fixing reaction of photosynthesis) provides a decidedly nonoptimum solution. This enzyme has a second activity that interferes with the successful performance of the Calvin cycle: RuBP carboxylase also initiates the *oxidation* of RuBP. In this process, called **photorespiration,** carbon dioxide is released without the production of ATP or NADPH. Because it produces neither ATP nor NADPH, photorespiration acts to undo the work of photosynthesis.

RuBP carboxylase initiates its undesirable oxidative reaction at the same active site that carries out the carbon-fixation reaction so important to photosynthesis. When photosynthesis first evolved, the atmosphere contained little oxygen, and because the decarboxylation reaction requires oxygen, there was little or no photorespiration. Under these conditions, the fact that the active site of RuBP carboxylase was capable of carrying out both reactions presented no problem. Only after millions of years of oxygen buildup in the atmosphere did the competition of carbon dioxide and oxygen gas for the same site lead to the problem that photorespiration now poses.

C_3 Photosynthesis

The loss of fixed carbon caused by photorespiration is not trivial. Plants that use the Calvin cycle to fix carbon engage in **C_3 photosynthesis,** so named because the Calvin cycle in these plants yields a three-carbon molecule, PGAL. C_3 plants lose between a quarter and a half of their photosynthetically fixed carbon in the Calvin cycle. The extent of carbon loss depends largely on temperature because the oxidative activity of the RuBP carboxylase enzyme increases far more rapidly with increases in temperature than does its carbon-fixing activity. In tropical climates (those in which the temperature is often above 28 degrees Celsius), the problem is severe and has a major limiting effect on tropical agriculture.

C_4 Photosynthesis

Some plants in tropical environments have evolved two principal ways to deal with this problem. One approach, taken by a number of grasses, including corn, sugarcane, and sorghum, as well as members of about two dozen other plant groups, is called **C_4 photosynthesis.** In this process, the first product of carbon dioxide fixation to be detected is not a three-carbon molecule, as in the Calvin cycle, but rather a four-carbon molecule, oxaloacetate. C_4 plants concentrate carbon dioxide by adding a carbon to a three-carbon molecule called **phosphoenolpyruvate (PEP).** The resulting four-carbon molecule, oxaloacetate, is, in turn, converted to the intermediate

Figure 8.11 How carbon is fixed.
Ribulose 1,5 bisphosphate (RuBP) and carbon dioxide are combined to form an unstable six-carbon molecule (not shown). This molecule immediately splits into two three-carbon molecules of phosphoglycerate (PGA).

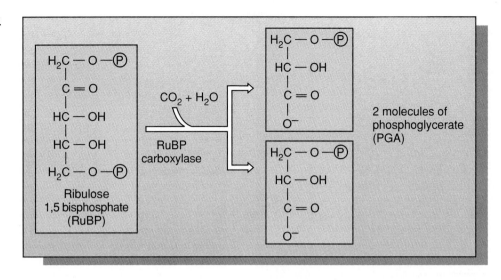

Figure 8.12 The Calvin cycle.
The NADPH and the ATP that were generated by the light-dependent reactions are used in the Calvin cycle, or light-independent reactions, to build carbon molecules. The Calvin cycle produces one molecule of glyceraldehyde phosphate (PGAL) for every three molecules of carbon dioxide that enter the cycle. Try counting the number of carbon atoms at each stage (indicated in parentheses) as you go around the cycle. It takes six turns of the cycle to make one molecule of glucose.

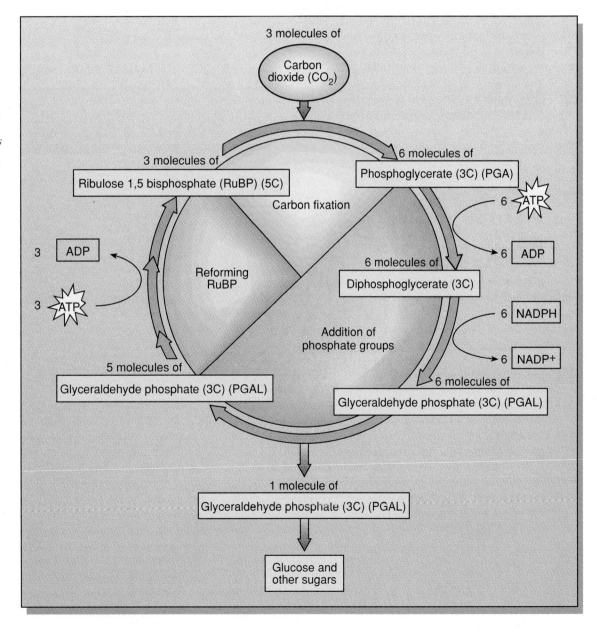

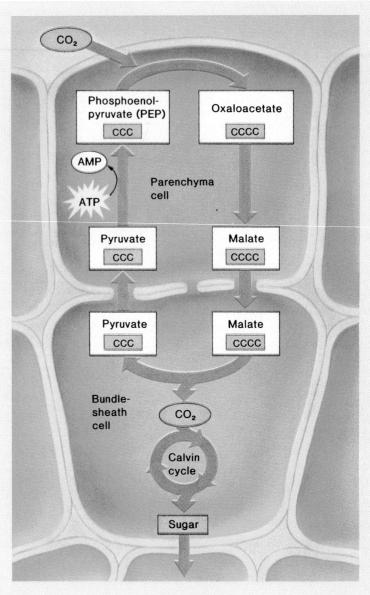

[■□■] **Figure 8.13 The path of carbon fixation in C_4 plants.**
C_4 plants shuttle a four-carbon molecule (malate) to bundle-sheath cells, where carbon dioxide can be concentrated. In this way, C_4 cells conserve carbon dioxide.

molecule malate. Malate is then transported from the **parenchyma cell,** where carbon fixation takes place, to an adjacent **bundle-sheath cell,** located in the leaves of the plant near the veins. Such cells are impermeable to carbon dioxide and therefore hold carbon dioxide within them. Carbon and oxygen atoms are removed from the malate, which forms pyruvate. This process also releases a carbon dioxide molecule into the interior of the bundle-sheath cell. Pyruvate returns to the parenchyma cell, where two high-energy bonds are split from ATP to form adenosine monophosphate (AMP) in converting the pyruvate back to PEP, thus completing the cycle.

C_4 plants expend a considerable amount of ATP to concentrate carbon dioxide within the cells that carry out the Calvin cycle. Because carbon dioxide binds to the same place within the RuBP carboxylase active site that oxygen gas does, high concentrations of carbon dioxide act to commandeer the available enzyme for carbon fixation.

The path of carbon fixation in C_4 plants is diagrammed in figure 8.13. The enzymes that carry out the Calvin cycle are located within the bundle-sheath cells, where the increased carbon dioxide concentration inhibits photorespiration. Because each carbon dioxide molecule is transported into the bundle-sheath cells at a cost of two high-energy ATP bonds, and because six carbons must be fixed to form a molecule of glucose, twelve additional molecules of ATP are required to form a molecule of glucose. The unique leaf structure in which these processes occur is illustrated in figure 8.14.

In C_4 photosynthesis, the energetic cost of forming glucose is almost doubled—from eighteen to thirty molecules of ATP. However, in a hot climate in which photorespiration would otherwise remove more than half of the carbon fixed, it is the best compromise available. For this reason, C_4 plants are more abundant in warm regions than in cooler ones (figure 8.15).

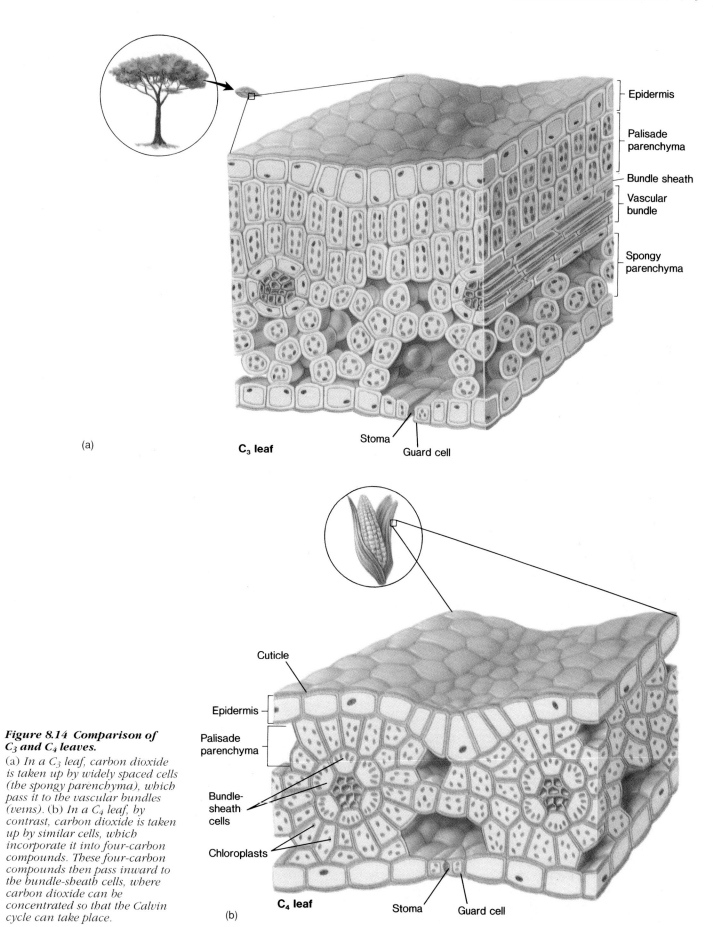

Figure 8.14 Comparison of C₃ and C₄ leaves.

(a) *In a C₃ leaf, carbon dioxide is taken up by widely spaced cells (the spongy parenchyma), which pass it to the vascular bundles (veins). (b) In a C₄ leaf, by contrast, carbon dioxide is taken up by similar cells, which incorporate it into four-carbon compounds. These four-carbon compounds then pass inward to the bundle-sheath cells, where carbon dioxide can be concentrated so that the Calvin cycle can take place.*

Epidermis

Palisade parenchyma

Bundle sheath

Vascular bundle

Spongy parenchyma

Stoma

Guard cell

(a)

C₃ leaf

Cuticle

Epidermis

Palisade parenchyma

Bundle-sheath cells

Chloroplasts

Stoma

Guard cell

C₄ leaf

(b)

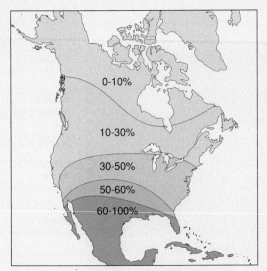

Figure 8.15 The distribution of C₄ grasses in North America.

Many more C₄ grasses occur in the South, where the average temperatures during the growing season are higher. At higher temperatures, photorespiration wastes more of the products of photosynthesis, and the ability of C₄ plants to counteract photorespiration is more of an advantage than it is in cooler regions, where C₃ grasses predominate. This map shows the percentage of species of C₄ grasses among all grass species.

(a)   (b)

Figure 8.16 Examples of CAM plants.
(a) *Pineapple and* (b) *a hedgehog cactus.*

CAM Photosynthesis

A second strategy to facilitate photosynthesis in hot regions has been adopted by many succulent (water-storing) plants, such as cacti, pineapples, and some members of about two dozen other plant groups (figure 8.16). This mode of carbon fixation is called **crassulacean acid metabolism (CAM) photosynthesis,** after the plant family Crassulaceae (the stonecrops, or hen-and-chickens) in which it was first discovered. In these plants, **stomata** (singular, *stoma*)—specialized openings in the leaves of all plants, through which carbon dioxide enters and water vapor is lost—open during the night and close during the day, which is the reverse of how plants normally behave. When stomata are closed during the day, photorespiration is reduced because carbon dioxide is prevented from entering the leaves. The carbon dioxide necessary for producing sugars is instead provided from organic molecules made the night before. Like C₄ plants, CAM plants use both C₄ and C₃ pathways. They differ from C₄ plants in that the C₄ pathway operates at night and the time of operation of the two pathways differs, with the C₃ pathway operating *within the same cells* during the day. In C₄ plants, on the other hand, the two cycles take place in different, specialized cells.

Photorespiration releases carbon dioxide without the production of ATP and NADPH and so short-circuits photosynthesis and carbon fixation pathways. C₄ plants and CAM plants circumvent this waste through modifications in leaf architecture.

SUMMARY

1. Photosynthesis is a set of reactions in which light energy from the sun is converted into the chemical bond energy of glucose and ATP.

2. Light consists of units of energy called photons. The shorter the wavelength of light, the more energy in the light's photons.

3. When light strikes a pigment, photons are absorbed by boosting an electron to a higher energy level. Most biological pigments are carotenoids, such as the retinal of human eyes, or chlorophylls, such as those that make grass green.

4. Photosynthesis seems to have evolved in organisms similar to the green sulfur bacteria, which use a network of chlorophyll molecules (a photosystem) to channel energy to one pigment molecule, referred to as P_{700}. P_{700} then donates an electron to a series of electron acceptors, which drives a proton pump and returns the electron to P_{700} in a process called cyclic photophosphorylation.

5. Cyanobacteria, algae, and plants are descendants of these ancient bacteria. They developed a two-stage process in which a new photosystem, called photosystem II, was grafted onto the old one. Photosystem II employs a new pigment, chlorophyll *a;* it is able to generate enough energy to use water (H_2O) rather than hydrogen sulfide (H_2S) as a hydrogen source.

6. In organisms with both photosystems, light is first absorbed by photosystem II, which jolts an electron out of one chlorophyll molecule, P_{680}. This has two effects: (1) the absence of the high-energy electron causes photosystem II to seek another electron actively, which results in the eventual splitting of water to obtain it, with oxygen gas as the by-product; and (2) the high-energy electron is passed to photosystem I, driving a proton pump in the process and thus bringing about the chemiosmotic synthesis of ATP.

A cell's metabolism betrays its
evolutionary past perhaps
more than any other aspect of
its life. This is particularly true
of cells that perform
photosynthesis. The two-stage
photosystem of plants, algae,
and cyanobacteria has as its
second stage a photosystem
that evolved hundreds of
millions of years earlier in
anaerobic bacteria and that
uses hydrogen sulfide (H_2S)
rather than water as a source
of reducing power (figure
8.17). The Calvin cycle uses
part of the ancient glycolytic
pathway, run in reverse, to
produce glucose. The
principal chlorophyll
pigments of plants and algae
evolved in cyanobacteria at
least 2.8 billion years ago;
they, in turn, are simple
modifications of other
bacterial chlorophylls that
existed still earlier in
anaerobic photosynthetic
bacteria. A modern plant
exhibits many metabolic
aspects that evolved billions
of years before any life
existed on land.

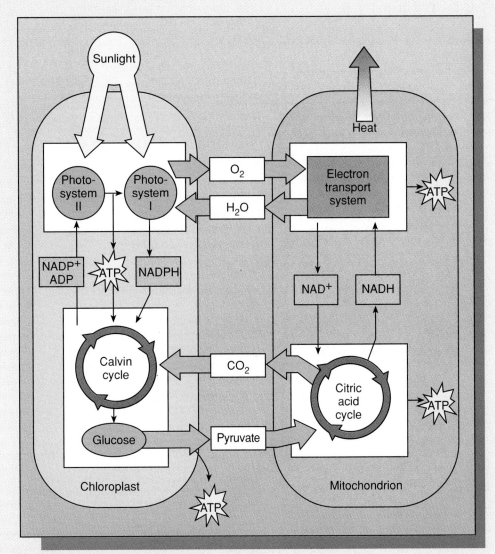

Figure 8.17 The metabolic machine.
*Within the chloroplast, sunlight drives the production of glucose, using up carbon dioxide
and water while generating oxygen. This oxygen is converted back to water in the
mitochondrion by accepting the electrons harvested from glucose molecules, after these
electrons have been used by the electron transport system to drive ATP synthesis. Water and
oxygen thus cycle between chloroplasts and mitochondria within a plant cell, as do glucose
and CO_2. Cells without chloroplasts, such as your own, require an outside source of glucose
and oxygen and generate carbon dioxide and water.*

7. When the spent electron arrives at P_{700}
from P_{680}, the P_{700} pigment absorbs a photon
of light, boosting one of its electrons to a
higher energy level. This electron is directed
to ferredoxin, where it is used to drive the
synthesis of NADPH from $NADP^+$, thus
providing reducing power.

8. The energy (ATP) and reducing power
(NADPH) produced by the light-dependent
reactions are used to fix carbon and
construct simple sugars in a series of light-
independent reactions called the Calvin
cycle. In this process, a carbon is added to
ribulose 1,5-bisphosphate (RuBP). The
resulting three-carbon molecules run through
a series of reactions to form glyceraldehyde
phosphate (PGAL) molecules, some of
which are used to reconstitute RuBP. The
remainder enters the cell's metabolism as
newly fixed carbon in glucose.

9. In hot climates, photosynthetic carbon
fixation tends to be short-circuited by
photorespiration, which releases carbon
dioxide instead of fixing it. Some plants deal
with this problem by concentrating carbon
dioxide within the cells where carbon
fixation takes place (C_4 photosynthesis) or by
closing their stomata during the day to keep
carbon dioxide in (CAM photosynthesis). It is
important to remember, however, that the
vast majority of plant species in the tropics
have C_3 photosynthesis.

REVIEWING THE CHAPTER

1. What are the reactants and products in photosynthesis?

2. In terms of different wavelengths of light, how does the electromagnetic spectrum relate to photosynthesis?

3. What are the differences in the absorption spectra of carotenoids and chlorophylls?

4. What are the stages of photosynthesis, and how do they relate to one another?

5. Specifically, where are the sites of the light-dependent and light-independent reactions in the internal structure of a chloroplast?

6. What is *carbon fixation*, and where does it occur in photosynthesis?

7. What are the differences and similarities in the light-dependent reactions in bacteria and eukaryotic organisms?

8. What is meant by *reducing power*, and how is reducing power significant in photosynthesis?

9. What are the differences between cyclic and noncyclic photophosphorylation, and which is the oldest, in terms of evolution?

10. How do the two stages of the light-independent reactions interrelate in photosynthesis?

11. What is a *Calvin cycle*, and how does it contribute to the growth of a plant?

12. What are two strategies that plants have evolved to overcome problems associated with photorespiration?

COMPLETING YOUR UNDERSTANDING

1. In the evolution of plants and animals, which group would have appeared first on land and why? How does this question relate to photosynthesis?

2. In photosynthesis, pigment molecules called carotenoids are used in addition to chlorophylls. What evolutionary advantage is there to this?

3. In a summer flower bed of mixed colors, why would a bumblebee frequently visit blue and yellow flowers and probably ignore red flowers?

4. What are thylakoids, and what role do they play in the effectiveness of photosystems I and II?

5. In a photosystem, what would happen to the efficiency of photosynthesis if all chlorophyll molecules were either P_{700} or P_{680}?

6. How is water split into electrons, hydrogen (H^+) ions, and free OH^- groups during photosynthesis? Why does this *not* happen when light strikes an open ocean or lake?

7. Why is photorespiration a wasteful process in plants?

8. What is the significance of a bundle-sheath cell, and how does a bundle-sheath cell relate to C_3 and C_4 photosynthesis?

9. What are CAM plants, where do they grow, and how do they differ from other C_4 and C_3 plants?

10. Why are there more species of C_4 grasses in southern North America than in northern North America?

FOR FURTHER READING

Barber, J. 1989. A quantum step forward in understanding photosynthesis. *Plants Today,* Sept., 165–69. An engaging account of the discovery of photosystem II, for which the Nobel prize was awarded in 1988.

Govindjee, J., and W. Coleman. 1990. How plants make oxygen. *Scientific American,* Feb., 50–58. How plants and some bacteria exploit solar energy to split water molecules into oxygen gas.

Holzenburg, A., et al. 1992. Three-dimensional structure of photosystem II. *Nature* 363 (6428): 470–72. A technical article that presents the chemical structure of photosystem II. Although the language is difficult, the article is accompanied by computer-enhanced photographs.

Kuhlbrandt, W., and D. Wang. 1991. 3-D structure of plant light-harvesting complex. *Nature,* March, 130–34. Describes for the first time the plant photocenter's molecular architecture.

Ting, I. P. 1989. Photosynthesis of arid and subtropical succulent plants. *Aliso* 12: 387–406. An excellent review of the occurrence and significance of CAM metabolism.

Youvan, D., and B. Marrs. 1987. Molecular mechanisms of photosynthesis. *Scientific American,* June, 79–91. Explores molecular aspects of photosynthesis, a hot research area.

EXPLORATIONS

Interactive Software:
Photosynthesis

This interactive exercise explores a chloroplast membrane in cross section. The user examines how photons bash electrons from chlorophyll, the path of the electrons through the two photosystems, driving proton pumps, and the chemiosmotic production of ATP and NADPH. The key variables are the wavelength of incident light and its intensity.

Questions to Explore:

1. How does light intensity affect the photosynthetic process?

2. How does the wavelength of light affect the photosynthetic process?

3. What is the optimum arrangement of intensity and wavelength of light to get the most efficient photosynthetic process?

4. Where do the electrons that travel through the photo system come from, and how?

5. How does photosynthesis actually convert *radiant energy* into *chemical energy?*

Parents shape and mold us in our early years, and much of what we become reflects this period of learning and development. Are our personalities a result of this nurturing process or the product of the particular genes we receive from our parents?

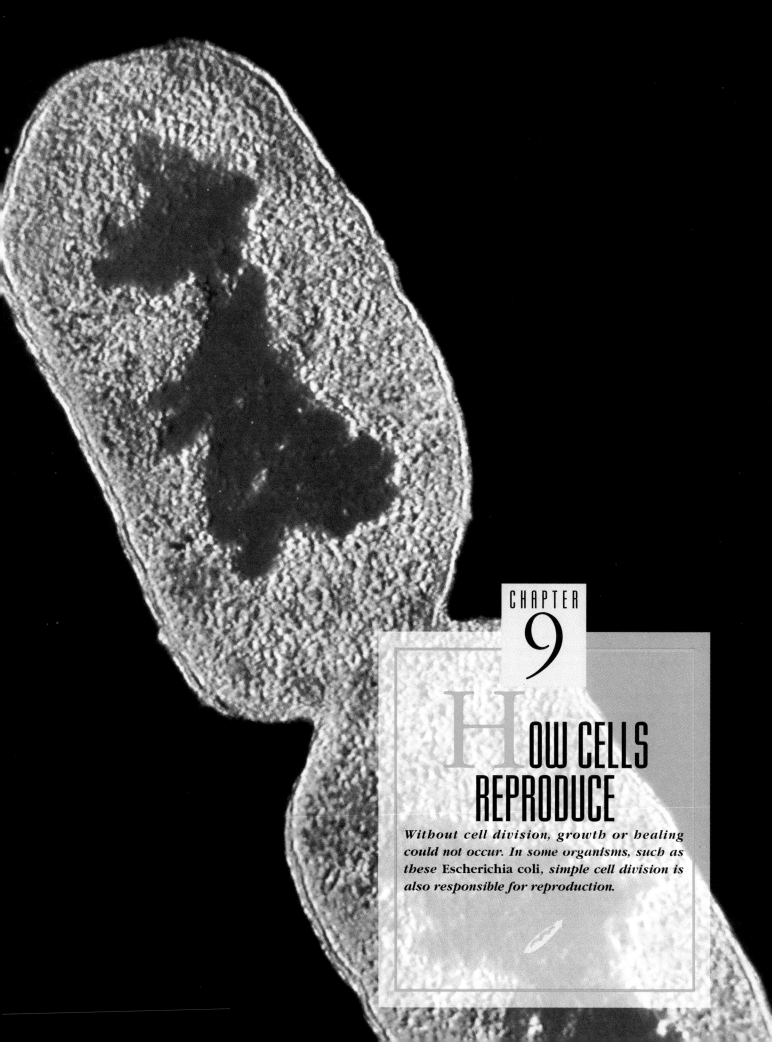

Ηow cells
reproduce

*Without cell division, growth or healing
could not occur. In some organisms, such as
these* Escherichia coli, *simple cell division is
also responsible for reproduction.*

FOR REVIEW

Here are some important terms and concepts that have been discussed in previous chapters and that you will encounter again in this chapter. Review them before proceeding if necessary.

Natural selection (*chapter 1*)
Evolution (*chapter 1*)
Chromosomes (*chapter 4*)
Centriole (*chapter 4*)
Microtubule (*chapter 4*)
Nuclear envelope (*chapter 4*)
Microtubule organizing center (MTOC) (*chapter 4*)
Cytoplasm (*chapter 4*)

Figure 9.1 All reproduction depends on cell division.
Like you, this turtle started life as a fertilized egg.

All living organisms grow and reproduce. Bacteria too small to see, alligators, the weeds growing in your lawn—all organisms produce offspring like themselves and pass on to them the hereditary information that makes them what they are (figure 9.1). All reproduction of organisms depends on the reproduction of cells. This chapter, then, begins our consideration of heredity with an examination of how cells reproduce themselves. The ways in which cells reproduce, and the biological consequences, have changed significantly during the evolution of life on earth.

Cell Division in Bacteria: Binary Fission

Among bacteria, cell division is simple. The genetic information, or **genome,** of a bacterium is a single, circular, deoxyribonucleic acid (DNA) molecule, attached at one point to the interior surface of the cell membrane. At a special site, called the **replication origin,** on the chromosome, a battery of more than twenty different enzymes begins making a complete copy of the DNA molecule early in the cell's life (figure 9.2). When replication is complete, the cell possesses two copies of the DNA molecule, attached side by side to the interior cell membrane.

The growth of a bacterial cell to an appropriate size induces cell division. First, new plasma membrane and cell wall materials are laid down between the membrane attachment sites of

the two DNA molecules. This begins the process of binary fission. **Binary fission** is the division of a cell into two equal or nearly equal halves. As new material is added in the zone between the DNA molecules, the growing plasma membrane pushes inward (invaginates), and the cell is progressively constricted. Initiation of the constriction at a point between the membrane attachment sites of the two DNA molecules ensures that each of the two new cells will contain one of the two identical genomes. Eventually, the invaginating circle of membrane reaches all the way into the cell center, pinching the cell in two. A new cell wall forms around the new membrane, and what was originally one cell is now two.

Bacteria divide by binary fission, a process in which a cell is pinched in two. Constriction begins at a point between where the two DNA molecules are bound to the cell membrane, ensuring that one DNA copy will end up in each cell.

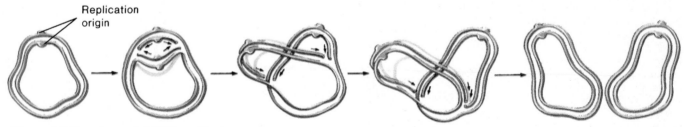

Replication origin

Figure 9.2 How bacterial DNA replicates.
The circular DNA molecule that contains the genetic information of a bacterium initiates replication at a single site, called the replication origin, moving out in both directions. When the two moving replication points meet on the far side of the DNA molecule, replication is complete.

Eukaryotic Chromosomes

The evolution of the eukaryotes introduced several additional factors into the process of cell division. Eukaryotic cells are much larger than bacteria, and they contain genomes with much larger quantities of DNA (figure 9.3). This DNA is located in several individual chromosomes, rather than in a single, circular molecule, and is associated with proteins and wound into tightly condensed coils. The eukaryotic chromosome is much more complex than the single, circular DNA molecule that plays the role of a chromosome in bacteria.

The Structure of Eukaryotic Chromosomes

In the century since their discovery, much has been learned about chromosomes, their structure, and how they function. Eukaryotic chromosomes are composed of a complex—called **chromatin**—about 40 percent DNA and 60 percent protein. The DNA exists as one very long, double-stranded fiber, called a **duplex,** that extends unbroken through the chromosome's entire length. A typical human chromosome contains more than 300 million (3×10^8) nucleotides in its DNA fiber. If nucleotides were words, the amount of information on one chromosome would fill about six hundred books of one thousand pages each, assuming that each page had about five hundred words on it. If the strand of DNA from a single eukaryotic chromosome was laid out in a straight line, it would be about 5 centimeters (2 inches) long. For it to fit into a cell, the DNA must be coiled.

How is this long DNA fiber coiled? Under an electron microscope, DNA resembles a string of beads. Every two hundred nucleotides, the DNA duplex is coiled around a complex of **histones,** which are small, very basic polypeptides, rich in the amino acids arginine and lysine. Eight of these histones form the core of an assembly called a **nucleosome.** Because so many of their amino acids are basic, histones are very positively charged. The DNA duplex, which is negatively charged, is strongly attracted to the histones and wraps tightly around the histone core of each nucleosome. The core thus acts as a "form" that promotes and guides DNA coiling. The DNA coils further when the string of nucleosomes wraps up into higher-order coils called **supercoils** (figure 9.4).

Condensed (that is, tightly coiled) portions of the chromatin are called **heterochromatin.** Some parts remain condensed permanently, so their genes can never be used. The remainder of the chromosome, called **euchromatin,** is not condensed except during cell division, when chromosome movement is made easier by such compact packaging. At all other times, the euchromatin is present in an open configuration, and its genes are active.

Number of Chromosomes in a Cell

With the exception of **gametes** (the sex cells—eggs and sperm—of an organism), all cells contain forty-six chromosomes—two nearly identical copies of each of the basic set of twenty-three chromosomes. (Gametes contain only the basic set of twenty-three chromosomes.) The two nearly identical copies of each of the twenty-three different chromosomes are called **homologous chromosomes,** or **homologues** (from the Greek word *homologia,* meaning "agreement"). Before cell division, each of the two homologues replicates, resulting in two identical copies, called **sister chromatids,** that remain joined together at a special linkage site called the **centromere.** Thus, at the beginning of cell division, a body cell contains a total of forty-six replicated chromosomes, each composed of two sister chromatids joined by one centromere. By convention, each pair of sister chromatids is counted as a single chromosome as long as the chromatids remain joined.

How Chromosomes are Examined: Karyotyping

Chromosomes may differ widely from one another in appearance. They vary in such features as the location of the centromere, the relative length of the two arms (regions on either side of the centromere), size, staining properties, and the position of constricted regions along the arms. An individual's particular array of chromosomes is called a **karyotype.** Karyotypes may differ greatly among species, or sometimes, even among individuals in a species.

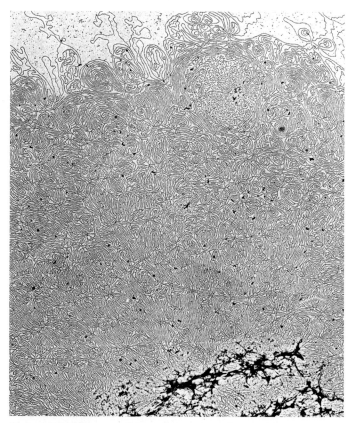

Figure 9.3 A human chromosome contains an enormous amount of DNA.

The dark element at the bottom of this photograph is part of the protein component of a single chromosome. All of the surrounding material is the DNA of that chromosome.

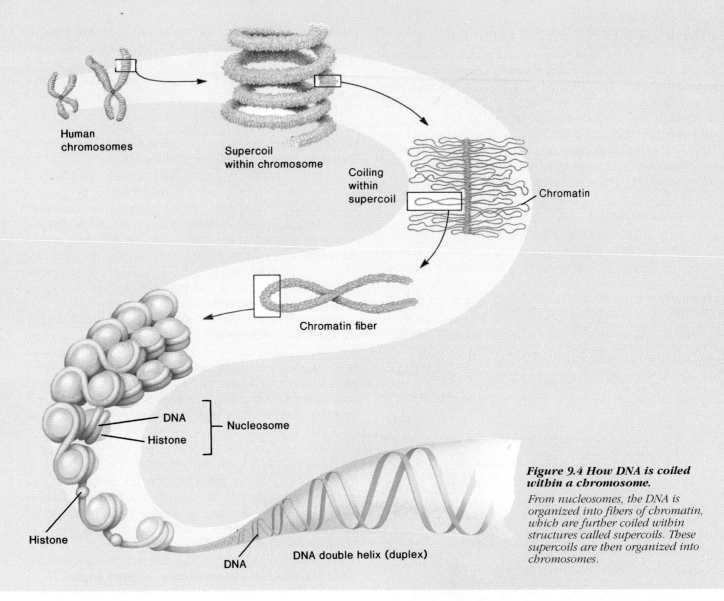

Human chromosomes

Supercoil within chromosome

Coiling within supercoil

Chromatin

Chromatin fiber

DNA

Histone

Nucleosome

Histone

DNA

DNA double helix (duplex)

Figure 9.4 How DNA is coiled within a chromosome.

From nucleosomes, the DNA is organized into fibers of chromatin, which are further coiled within structures called supercoils. These supercoils are then organized into chromosomes.

To examine human chromosomes, investigators collect a blood sample, add chemicals that induce the cells in the blood sample to divide, and then add other chemicals that stop ("fix") cell division at the stage when the chromosomes are most condensed. Investigators then break the cells to spread out their contents, including the chromosomes, and stain and examine the chromosomes. To make the karyotype easier to examine, the chromosomes may be photographed, cut out of the photograph like paper dolls, paired, and arranged in order of their size (figure 9.5).

The karyotypes of individuals are often examined to detect genetic abnormalities, a number of which arise from extra or lost chromosomes. For example, the human birth defect known as Down syndrome is associated with the presence of an extra copy of chromosome 21, which can be recognized easily in photographs of the set of chromosomes. Thus, karyotypes of fetal cells taken before birth can reveal genetic abnormalities of this sort.

The particular array of chromosomes that an individual possesses is called the karyotype. The human karyotype usually contains twenty-three pairs of chromosomes.

The Cell Cycle

The profound change in genome organization that occurred during the evolutionary transition from bacteria (a single circle of naked DNA) to eukaryotes (several segments of DNA packaged with protein) required radical changes in the way cells divide. Eukaryotic cell division is more complex than bacterial cell division and requires several predivision growth and preparation phases. The events that prepare the cell for division and the division process itself constitute a **cell cycle.** The cell cycle is, in effect, the entire life cycle of a cell. The phases of the cell cycle are:

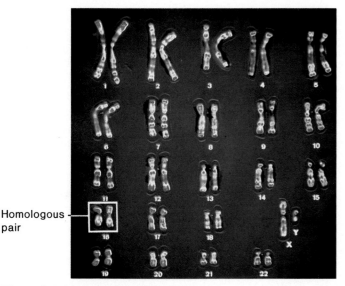

Figure 9.5 A human karyotype.

These human chromosomes have been paired with their homologues. The chromosomes were "fixed" immediately after they duplicated in preparation for cell division. Therefore, this photograph shows homologous pairs in their duplicated state.

Homologous pair

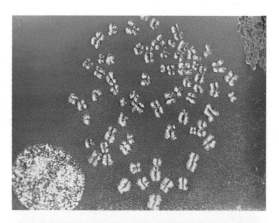

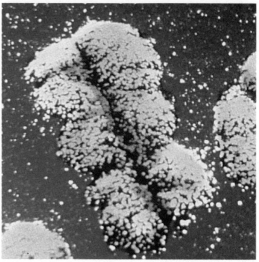

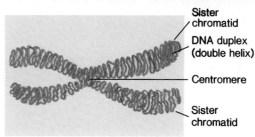

Sister chromatid

DNA duplex (double helix)

Centromere

Sister chromatid

Figure 9.6 The structure of a chromosome.

A duplicated chromosome looks somewhat like an "X" and is composed of two sister chromatids held together by a centromere. Each sister chromatid contains one DNA duplex (double helix). An unduplicated chromosome (not shown in this figure) consists of only one chromatid.

G_1 ("first gap") phase. This is the cell's growth phase. For most organisms, this phase occupies the major portion of the cell's life span.

S ("synthesis") phase. During this phase, each chromosome replicates to produce two sister chromatids with identical DNA (figure 9.6).

G_2 ("second gap") phase. In this phase, cell division preparations continue with the replication of mitochondria and other organelles, chromosome condensation, and the synthesis of microtubules.

M ("mitosis") phase. This is the phase in which a microtubular apparatus is assembled that binds to the chromosomes and moves them apart. This **mitosis** phase is essential in the separation of the two daughter genomes.

C ("cytokinesis") phase. During this phase, called **cytokinesis,** the cytoplasm divides, creating two daughter cells.

In the past, the G_1, S and G_2 phases of the cell cycle were characterized as "resting stages," because during these phases, the cell does not appear as active as it does during mitosis and cytokinesis. Scientists now know, however, that these phases comprise the major portion of the cell's life cycle and involve important metabolic activities. Collectively, the G_1, S, and G_2 phases are referred to as **interphase,** or the period between cell divisions (figure 9.7)

Cells go through a continuing cycle of division and growth.

Mitosis and cytokinesis often represent only a short portion of the cell cycle. Both phases, however, are of critical importance.

Mitosis

Mitosis (the M phase of the cell cycle) has long fascinated biologists because of its central role in bringing about the proper division of the genetic material. After mitosis, each daughter cell has the same number of chromosomes as the parent cell. In addition to maintaining the genetic integrity of eukaryotic organisms, mitosis provides for organism growth and allows for structural repairs. Mitosis is also the primary means by which an organism develops: If it were not for mitosis, you would remain a single-celled organism, without fingers and toes and a heart.

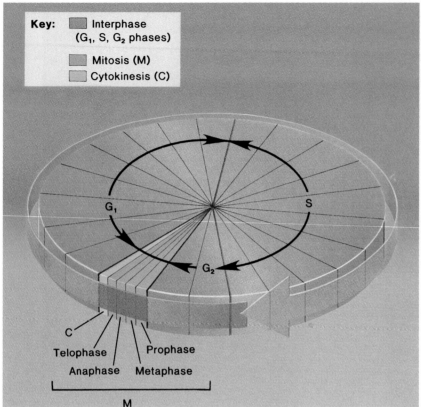

Figure 9.7 The cell cycle.
Interphase consists of the G_1 ("first gap"), S ("synthesis"), and G_2 ("second gap") phases, which take up most of the cell cycle. The M ("mitosis") and C ("cytokinesis") phases comprise a relatively small part of the cell cycle.

kinetochore is the site to which microtubules will attach to pull the sister chromatids apart in a later stage of mitosis. The centromere occurs at a specific site on any given chromosome, which is why it is useful in identifying the individual homologous chromosomes in a karyotype.

3. At the completion of the S phase, the replicated chromosomes remain fully extended and uncoiled; they are not visible under a light microscope. In the G_2 phase, the chromosomes begin the long process of **condensation,** coiling into more and more tightly compacted bodies.

4. During the G_2 phase, the cells begin to assemble the materials that they will later use to move the chromosomes to opposite poles of the cell. In the cells of animals and most protists, the centrioles replicate. Centrioles determine the plane along which the cell will divide (see chapter 4). The cells also undertake extensive synthesis of tubulin, the protein of which microtubules are formed.

> *Interphase is that portion of the cell cycle in which the chromosomes usually are not visible under a light microscope. It includes the G_1, S, and G_2 phases. During the G_2 phase, the cell mobilizes its resources for cell division.*

This description of mitosis focuses on the process as it occurs in animals and plants; its details vary in other groups of organisms. Mitosis is subdivided into four stages: prophase, metaphase, anaphase, and telophase. Such subdivision is convenient, but the process is actually continuous, with the stages flowing smoothly one into another (figure 9.8).

Interphase: Preparing for Mitosis

For the successful initiation and completion of mitosis, certain events must occur in the preceding interphase (that is, the G_1, S, and G_2 phases) (figure 9.8):

1. During G_1, the cell grows in size. The decision to commit to cell division is made during G_1—a point in the cell cycle called the restriction point. Once a cell passes the restriction point, the rest of the cell cycle will be completed. The volume of cytoplasm is an important factor in committing to cell division.

2. During the S phase, each chromosome replicates to produce two sister chromatids with identical DNA that remain attached to each other at the centromere. The centromere includes a specific DNA sequence of about 220 nucleotides, to which is bound a disk of protein called the **kinetochore.** The

Prophase: Formation of the Mitotic Spindle

When the chromosome condensation begun in the G_2 phase reaches the point at which individual condensed chromosomes first become visible with a light microscope, the first stage of mitosis, **prophase,** has begun (figure 9.8). Condensation continues throughout prophase, with individual chromosomes becoming progressively thicker.

> *Prophase is the stage of mitosis at which condensing chromosomes first become visible.*

As the chromosomes condense, the nuclear envelope begins to break down, and the microtubular apparatus that will be used to separate the sister chromatids is assembled. This apparatus, called the **mitotic spindle,** consists of two kinds of fibers: **polar fibers** and **kinetochore fibers.** As their names imply, polar fibers determine the plane of cell division, and kinetochore fibers attach to the kinetochores of the chromosomes and direct the sister chromatids to opposite poles of the cell.

 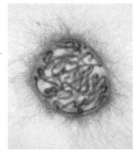   

Interphase	Mitosis			
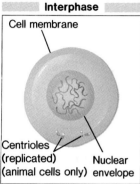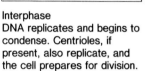	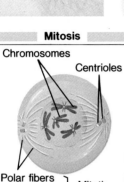	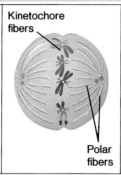		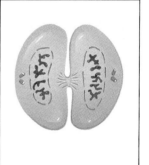

Interphase
DNA replicates and begins to condense. Centrioles, if present, also replicate, and the cell prepares for division.

Prophase
The nuclear envelope begins to break down. DNA further condenses into chromosomes. The mitotic spindle begins to form; it is complete at the end of prophase.

Metaphase
The chromosomes align on a plane in the center of the cell. The kinetochore fibers attach to the polar fibers at opposite sides. The centromeres replicate.

Anaphase
The sister chromatids separate and move to opposite poles.

Telophase
The nuclear envelope reappears. The chromosomes decondense. As telophase progresses, cytokinesis also occurs.

Figure 9.8 The stages of mitosis.
Photos show chromosomes of the African blood lily, Haemanthus katharinae. *Microtubules are stained* blue *and chromosomes are stained* red. *In the drawing, centrioles are also shown, as they are present in animal cells, although typically absent in plants.*

In animal cells and those of most protists, the polar fibers are associated with the centrioles. After the centrioles are replicated, the pairs separate and move to opposite poles of the cell. When they reach the poles, they radiate an array of microtubules called the **aster.** The aster's position determines the plane in which the cell will divide, a plane that passes through the center of the cell at right angles to the aster.

Plant cells do not have centrioles. Yet, they also form polar fibers that determine the plane of cell division. This observation has led scientists to believe that centrioles are not essential in the formation of the mitotic spindle, but that another structure, called the **microtubule organizing center (MTOC)** is. The structure and function of the MTOC are still somewhat of a mystery and are being actively investigated by cell biologists.

During prophase, the mitotic spindle is assembled. Polar fibers form between opposite poles of the cell. In animal and protist cells, these fibers form at the aster. The microtubule organizing center (MTOC) may control the formation of these microtubules.

As prophase continues, a second set of microtubules—the kinetochore fibers—attach to the kinetochores of the individual chromosomes. The kinetochores are located on each sister chromatid in the region of the centromere and face outward in opposite directions. Two kinetochore fibers extend in opposite directions from each chromosome and connect the kinetochore of each of the sister chromatids to one side of the polar fibers. The result is that one sister chromatid is attached to one side of the polar fibers, while the other sister chromatid is attached to the other side.

At the end of prophase, kinetochore fibers attach to the kinetochores of the sister chromatids. The kinetochore fibers are attached to the polar fibers.

Metaphase: Alignment of the Chromosomes

The second phase of mitosis, **metaphase,** begins when the chromosomes, each consisting of a pair of chromatids, align in the center of the cell, equidistant from the two poles (see figure 9.8). Viewed with a light microscope, the chromosomes appear to be lined up along the inner circumference of the cell in a circle perpendicular to the axis of the polar fibers (figure 9.9). An imaginary plane passing through this circle is called the **metaphase plate.**

Metaphase is the stage of mitosis characterized by the alignment of the chromosomes along the inner perimeter of a plane in the center of the cell.

Each sister chromatid has one kinetochore, to which two kinetochore fibers are attached. At the end of metaphase, each centromere replicates, freeing the two sister chromatids to be drawn to the opposite poles of the cell in the next phase by the kinetochore fibers attached to their kinetochore.

Anaphase: Separation of the Chromatids

Anaphase, during which the sister chromatids move rapidly to opposite poles, is the shortest stage of mitosis (see figure 9.8). Two forms of movement take place simultaneously, each driven by microtubules:

1. *The poles move apart.* The polar fibers slide past one another. Because the polar fibers are physically anchored to opposite poles, their sliding past one another pushes the poles apart. Because the sister chromatids are attached by kinetochore fibers to the polar fibers, and thus, the poles, they also move apart. In this process, animal cells and others that lack a rigid cell wall become visibly elongated.

2. *The centromeres move toward the poles.* The kinetochore fibers attached to the kinetochores shorten. This shortening process is not a contraction because the microtubules do not get any thicker. Instead, tubulin subunits are continuously removed from the ends of the microtubules by the microtubule organizing center (MTOC). The progressive disassembly of the kinetochore fiber renders it shorter and shorter, directing the sister chromatid ever closer to the cell's poles.

Anaphase is the stage of mitosis characterized by the physical separation of sister chromatids. The poles of the cells are pushed apart by sliding of polar fibers; the sister chromatids are drawn to opposite poles by the shortening of the kinetochore fibers attached to them.

Telophase: Re-Formation of the Nuclei

With the movement of the sister chromatids to the opposite poles during anaphase, the only tasks that remain in **telophase** are the dismantling of the stage and the removal of the props (see figure 9.8). The mitotic spindle is disassembled, with the microtubules broken back down into tubulin monomers ready for use in constructing the cytoskeleton of the new cell. A nuclear envelope forms around each set of chromosomes while they begin to uncoil into the more extended form that permits genes to be used.

Telophase is the stage of mitosis during which the mitotic spindle assembled during prophase is disassembled, the nuclear envelope is reestablished, and the chromosomes uncoil in preparation for use.

(a)

Figure 9.9 The metaphase plate.

(a) *Chromosomes line up on the metaphase plate, or plane of cell division, during metaphase.* (b) *The metaphase plate is usually located at the midpoint of the polar fibers.*

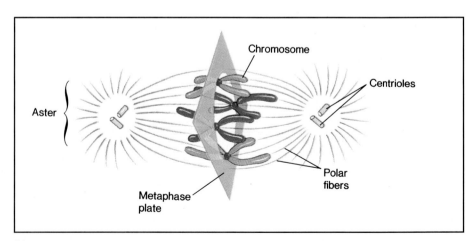

(b)

Cytokinesis

At the end of telophase, mitosis is complete. The cell has divided its replicated genome into two nuclei, which are positioned at opposite ends of the cell. While this process has been going on, the cytoplasmic organelles, such as the mitochondria and, if present, the chloroplasts, have been reassorted to the areas that will separate and become the daughter cells. Replication of the cytoplasmic organelles occurs before cytokinesis, often in the S or G_2 stage of interphase.

But the process of cell division is still not complete. Cytokinesis, the division of the cytoplasm, has not yet even begun. Cytokinesis generally involves the cleavage of the cell into roughly equal halves.

In animal cells, which lack cell walls, cytokinesis is achieved by pinching the cell in two with a contracting belt of microtubules. As contraction proceeds, a **cleavage furrow** becomes evident around the cell's circumference, where the cytoplasm is being progressively pinched inward by the decreasing diameter of the microtubule belt (figure 9.10). The cleavage furrow deepens until the cell is literally pinched in two.

Plant cells have rigid cell walls that are far too strong to be deformed by microtubule contraction. A different approach to cytokinesis has therefore evolved in plants. Plant cells assemble membrane components in their interior, at right angles to the mitotic spindle. This expanding partition, called a **cell plate,** grows outward until it reaches the interior surface of the cell membrane and fuses with it, at which point it has effectively divided the cell in two (figure 9.11). Cellulose (see chapter 2) is then laid down on the new membranes, creating two new cells. The space between the two new cells becomes impregnated with pectins (see chapter 2) and is called a **middle lamella.**

> *Cytokinesis is the physical division of the cell's cytoplasm and usually occurs after nuclear division is complete.*

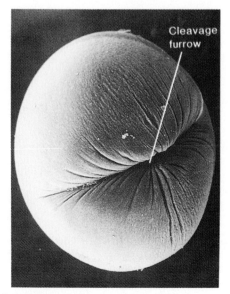

Figure 9.10 Cytokinesis in an animal cell.

A cleavage furrow is forming around this dividing sea urchin egg.

Meiosis and Sexual Reproduction

As discussed earlier in the chapter, the set of chromosomes present in the body or vegetative cells of an animal or plant consists of two homologous pairs of each individual chromosome. A

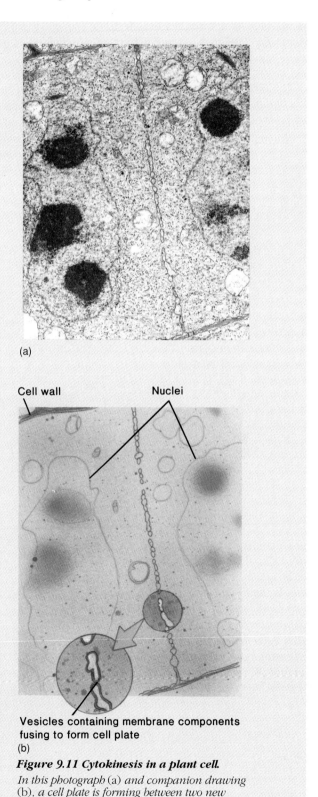

(a)

(b)

Figure 9.11 Cytokinesis in a plant cell.

In this photograph (a) and companion drawing (b), a cell plate is forming between two new plant cells.

cell, tissue, or individual with such a double set of chromosomes is said to be **diploid.** The **gametes,** or sex cells (eggs and sperm) of an organism, have one set of chromosomes each—half the number present in the body or vegetative cells—and are said to be **haploid.** During **fertilization,** gametes fuse, thereby restoring the diploid chromosome number. **Meiosis,** which alternates with fertilization in the cycle of reproduction, is the process by which the haploid gametes are produced. Meiosis is a special form of cell division in which the number of chromosomes is reduced to form haploid gametes. Without such a mechanism, the number of chromosomes in body cells would become impossibly large. Meiosis ensures that an organism's chromosome number remains consistent generation after generation.

When the gametes are differentiated into smaller, motile ones and larger, nonmobile ones, as they are in humans and other animals, the former are called sperm and the latter are called eggs. Both of these types of gametes, and all other gametes, are haploid. In humans, the diploid body cells have forty-six chromosomes—two sets of twenty-three—and the haploid gametes have twenty-three chromosomes. When these gametes fuse, the diploid number of chromosomes—forty-six—is restored.

Reproduction that involves the alternation of fertilization and meiosis is called **sexual reproduction.** Its outstanding characteristic is that an individual offspring inherits genes from two parent individuals (figure 9.12). You, for example, inherited genes from both your mother and your father, your mother's genes being contributed by the egg fertilized at your conception, and your father's by the sperm that fertilized that egg (figure 9.13).

Other Forms of Reproduction

Sexual reproduction is not the only way that reproduction can occur. For example, sponges or hydras can reproduce simply by fragmenting their bodies. In this **asexual reproduction,** a small portion of the organism divides from it and gives rise to a new individual. There is no alternation of haploid and diploid cells, no meiosis, and no gametes. Because these cell divisions are mitotic, every cell of the offspring individual has the same genetic makeup as those of the parent from which the offspring arose. Asexual reproduction is widespread among different groups of organisms. Its outstanding characteristic is that an individual offspring is *genetically identical* to its parent.

Even when meiosis and gamete production occur, organisms may reproduce without sex. The development of an adult from an unfertilized egg is called **parthenogenesis,** which occurs in some

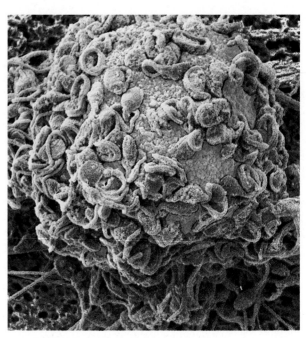

Figure 9.12 Role of meiosis in reproduction.
These haploid sperm and the haploid egg they are attempting to fertilize were produced by meiosis, a special form of cell division that reduces the number of chromosomes in the daughter cells. When egg and sperm meet, they combine their chromosomes to form the first diploid body cell of a new individual.

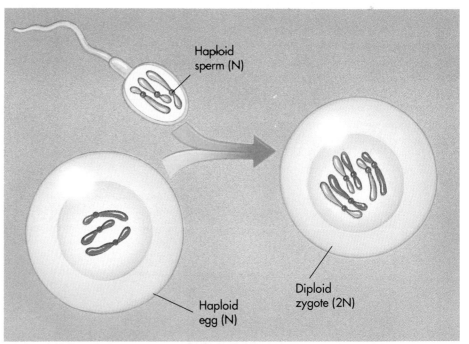

Haploid sperm (N)

Haploid egg (N)

Diploid zygote (2N)

Figure 9.13 Diploid cells are formed by combining two haploid cells.
During fertilization, two haploid cells meet and combine their chromosomes. The diploid cell that results contains two versions of each chromosome, one contributed by the haploid egg of the mother, the other by the haploid sperm of the father. In this drawing, (N) stands for haploid, and (2N) stands for diploid, since the diploid number is exactly double the haploid number.

groups of insects. For example, among bees, the development of eggs into adults does not require fertilization. Fertilized eggs develop into diploid females, unfertilized eggs into haploid males. Parthenogenesis occurs even among the vertebrates. Some fishes, amphibians, and lizards are capable of reproducing in this way, their unfertilized eggs undergoing a mitotic division, followed by the fusion of two haploid nuclei to produce a diploid cell, which then develops into a diploid adult.

The Sexual Life Cycle

The life cycles of all sexually reproducing organisms follow the same basic pattern of alternating between the diploid and the haploid chromosome number (figure 9.14). The first cell of a diploid individual is called the **zygote.** The zygote is formed by the fusion of two haploid gametes. Dividing by mitosis, a human zygote eventually gives rise to an adult body with some one hundred trillion cells, all of them genetically identical to the zygote.

In animals, the diploid cells that will eventually undergo meiosis to produce the gametes are located in the sex organs and are known as **germ line cells** (see chapter 43). These cells become obviously different from the **somatic cells—** all the other diploid body cells—early in the course of development.

In plants, the haploid cells that result from meiosis divide by mitosis, forming multicellular haploid individuals. Certain cells of this haploid phase eventually differentiate into eggs or sperm, which, when they unite, form a zygote, the first cell of the diploid phase of the life cycle.

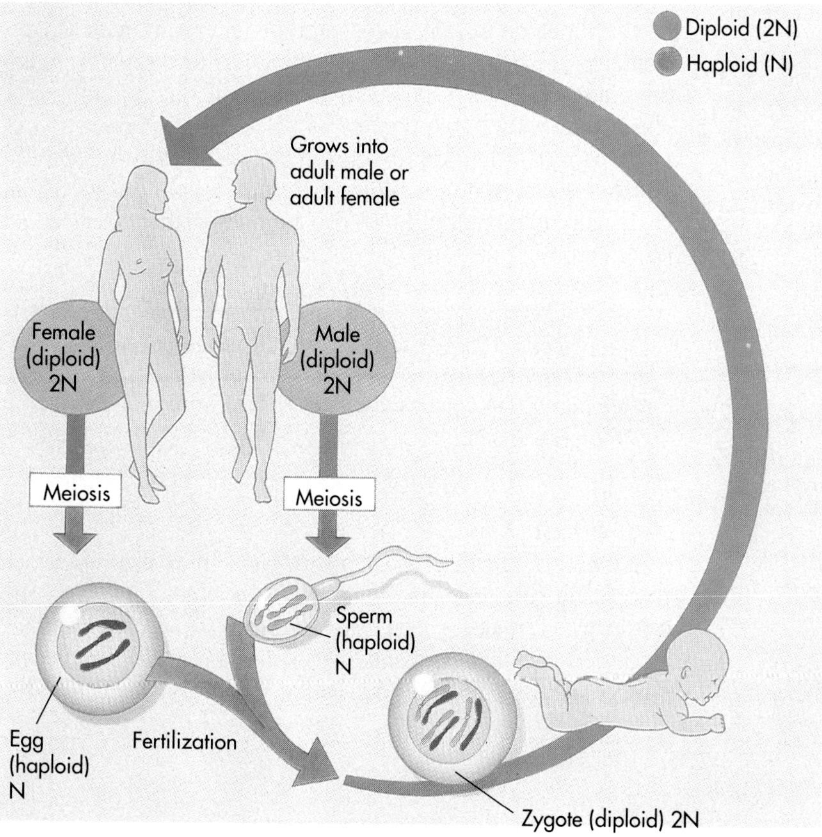

Stages of Meiosis

Although meiosis, like mitosis, is a continuous process, it is best described in terms of arbitrary stages. In a sense, meiosis consists of two rounds of nuclear division, during the course of which two unique events occur:

1. In an early stage of the first of the two nuclear divisions, the two nearly identical versions of each chromosome, the homologues, pair with each other all along their length. Each chromosome has replicated during the S phase that precedes cell division and now consists of two sister chromatids bound together by a single centromere. When chromosomes pair during meiosis, therefore, they unite four chromatids. While these four chromatids are held together, nonsister chromatids exchange portions of DNA strands in a process called **crossing-over.** Because the exchange takes place between nonsister chromatids of the different homologues, the two homologues are held together, thus forming a pair of chromosomes. When the two homologues separate later in meiosis, the nonsister chromatids have exchanged genetic material, as discussed in more detail later.

2. The second meiotic division is identical to a mitotic division, except that *the chromosomes do not replicate between the two divisions.* Therefore, during the second meiotic division, sister chromatids separate, becoming individual chromosomes.

> *Two important features distinguish meiosis from mitosis:*
> *(1) In meiosis, homologous chromosomes pair lengthwise, and nonsister chromatids exchange genetic material, and (2) the sister chromatids of each homologue do not separate from each other in the first nuclear division, and the chromosomes do not replicate between the two nuclear divisions.*

The two stages of meiosis are called **meiosis I** and **meiosis II** (figure 9.15). Each stage is further subdivided into prophase, metaphase, anaphase, and telophase, just as in mitosis. In meiosis I, however, prophase is more complex than it is in mitosis.

The First Meiotic Division

Prophase I

In **prophase I,** individual chromosomes first become visible, as viewed with a light microscope, as their DNA coils more and more tightly. Because the chromosomes (DNA) have replicated before the onset of meiosis, each of these threadlike

Diploid (2N)
Haploid (N)

Grows into adult male or adult female

Female (diploid) 2N

Male (diploid) 2N

Meiosis

Meiosis

Sperm (haploid) N

Egg (haploid) N

Fertilization

Zygote (diploid) 2N

Figure 9.14 The sexual life cycle in animals.

Animals are diploid (2N) organisms but produce special haploid (N) cells that function in reproduction.

chromosomes actually consists of two sister chromatids joined at their centromeres. The two homologous chromosomes then line up side by side, a process called **synapsis.** A lattice of protein and RNA is laid down between the chromatids of each pair of homologous chromosomes. This lattice holds the chromatids in precise relation to one another, with each gene located directly across from its corresponding sister on the homologue. The effect is similar to zipping up a zipper. Within the lattice, the DNA duplexes of each sister chromatid unwind, and each strand of DNA pairs with a complementary strand from the nonsister chromatid adjacent to it on the other homologous chromosome. Eight single strands of DNA are present in each lattice: two per DNA duplex molecule × two sister chromatids per homologue × two homologues.

Synapsis is the close pairing of homologous chromosomes that occurs early in prophase I of meiosis. During synapsis, a molecular lattice aligns the genes of the two homologous chromosomes side by side. As a result, the nonsister chromatid of one homologue can pair with the corresponding nonsister chromatid of the other homologue.

Synapsis initiates a complex series of events, called crossing-over, in which DNA is exchanged between the two nonsister chromatids of homologous chromosomes (figure 9.16). Once crossing-over is complete, the lattice breaks down. At that point, the nuclear envelope dissolves, and the sister chromatids begin to move apart. Each chromosome consists of four chromatids at this point: two homologous chromosomes, each replicated and so present twice as sister chromatids.

The points of crossing-over, where portions of chromosomes have been exchanged, can sometimes be seen under a light microscope as X-shaped structures known as **chiasmata** (singular, *chiasma*) from the Greek word meaning cross (figure 9.17). A chiasma indicates that two of the four chromatids of paired homologous chromosomes have exchanged parts, one participant from each homologue.

Late in prophase, the nuclear envelope disperses.

In prophase I, the nonsister chromatids of the two homologues pair with one another. Crossing-over occurs between these nonsister chromatids, creating the chromosomal configurations known as chiasmata.

Metaphase I

In **metaphase I,** the second phase of meiosis I, spindle apparatus forms, just as in mitosis (see figure 9.15). But because the chromosomes are paired in meiosis I, one of the kinetochores of each centromere is inaccessible to kinetochore fibers because the homologous chromosomes are held tightly together by chiasmata. In mitosis, on the other hand, *both* kinetochores are attached to kinetochore fibers, and the sister chromatids therefore separate from one another during anaphase. The chromosomes line up double file in meiosis and single file in mitosis (figure 9.18).

In meiosis, because kinetochore fibers bind to only one kinetochore of each centromere, the kinetochore of one homologue becomes attached to the polar fibers extending to one pole, whereas the kinetochore of the other homologue becomes attached to the polar fibers extending to the other pole. Each joined pair of homologues then lines up along the spindle equator. For each pair of homologues, the orientation on the spindle axis is random: Which homologue is oriented toward which pole is a matter of chance.

Anaphase I

In **anaphase I,** after spindle attachment is complete, the kinetochore fibers attached to the homologues begin to slide past one another and shorten. The movement of the kinetochore fibers breaks the chiasmata apart and directs the centromeres toward the two poles, dragging the chromosomes along with them. Because the kinetochore fibers are attached to only one of the kinetochores of each centromere, the individual centromeres are not pulled apart as they are in mitosis. Instead, the entire centromere proceeds to one pole, taking both sister chromatids with it (see figure 9.15). Because the orientation of each pair of homologous chromosomes along the spindle equator is random, the chromosome that a pole receives from each pair of homologues is also random with respect to all other chromosome pairs.

In metaphase I, the two similar chromosomes of homologous pairs orient randomly along the spindle equator. In anaphase I, the homologues then separate and move toward opposite poles.

Telophase I

At the end of anaphase I, each pole has half as many chromosomes and half as many centromeres as were present in the cell in which meiosis began. Remember that the chromosomes replicated and thus contained two sister chromatids before the start of meiosis. The purpose of the meiotic stages up to this

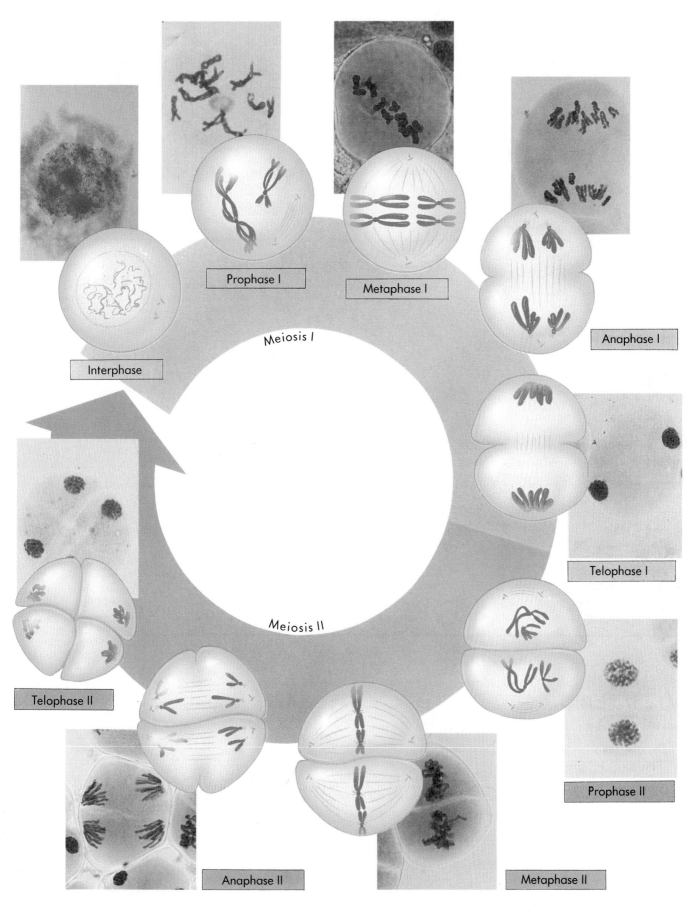

Figure 9.15 The stages of meiosis.

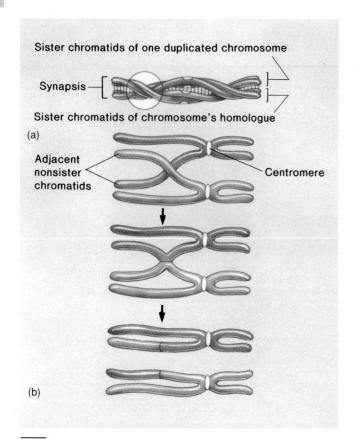

Sister chromatids of one duplicated chromosome

Synapsis

Sister chromatids of chromosome's homologue

(a)

Adjacent nonsister chromatids

Centromere

(b)

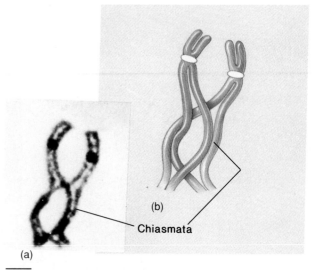

(b)

Chiasmata

(a)

📼 *Figure 9.17 Chiasmata.*

Chiasmata are the X-shaped structures that form during crossing-over and are visible in this (a) *micrograph and* (b) *companion drawing of a homologous pair in synapsis.*

📼 *Figure 9.16 Crossing-over, the cornerstone of meiosis.*

(a) *The open circle highlights the complex series of events called crossing-over.* (b) *During the crossing-over process, nonsister chromatids that are next to each other exchange genetic information.*

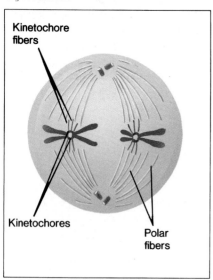

Kinetochore fibers

Kinetochores

Kinetochores

Polar fibers

(a) Metaphase during mitosis

Kinetochores

Sister chromatids

Polar fibers

Kinetochore fibers

(b) Metaphase I during meiosis

Figure 9.18 Comparison of chromosome movement in mitosis and meiosis I.

(a) *In mitosis, the homologous chromosomes line up on the metaphase plate in single file, and kinetochore fibers are attached to the kinetochores on each sister chromatid. When the polar fibers shorten, the sister chromatids are pulled apart.* (b) *In meiosis I, on the other hand, the homologous pairs line up on the metaphase plate in double file. Therefore, the kinetochores of the "inner" sister chromatids are not facing outward and so cannot have kinetochore fibers attached to them. Therefore, when the polar fibers shorten, the sister chromatids are* not *pulled apart, but the* homologous pairs *are.*

point has not been to reduce the number of chromosomes, but to allow for the exchange of genetic material in crossing-over. The number of chromosomes is reduced in meiosis II.

In **telophase I,** the chromosomes gather at their respective poles to form two chromosome clusters (see figure 9.15). After an interval of variable length, meiosis II occurs.

Meiosis I is traditionally divided into four stages:
Prophase I   Homologous chromosomes pair, and nonsister chromatids exchange segments.
Metaphase I   Homologous chromosomes align randomly along spindle equator.
Anaphase I   Homologous chromosomes move toward opposite poles.
Telophase I   Individual chromosomes gather at the two poles.

The Second Meiotic Division

Meiosis II is simply a mitotic division involving the products of meiosis I, except that the sister chromatids are not genetically identical, as they are in mitosis, because of crossing-over. At the end of anaphase I, each pole has a haploid complement of chromosomes, each of which is still composed of two sister chromatids attached at the centromere. The main purpose of the four stages of meiosis II—**prophase II, metaphase II,**

anaphase II, and **telophase II**—is to separate these sister chromatids. At both poles of the original cell, the chromosomes divide mitotically. Kinetochore fibers bind to each of the two kinetochores at the centromere region. The centromeres separate and move to opposite ends of each polar region. The result of this division is four haploid complements of chromosomes. At this point, the nuclei are reorganized, and nuclear envelopes form around each of the four haploid complements of chromosomes. The cells that contain these haploid nuclei may function directly as gametes, as they do in animals, or they may divide again by mitosis, as they do in plants, fungi, and many protists, eventually producing gametes after further mitotic divisions.

Each of the four haploid products of meiosis contains a haploid set of chromosomes. These haploid cells may function directly as gametes, as they do in animals, or they may continue to divide by mitosis, as they do in plants, fungi, and many protists.

Figure 9.19 compares and contrasts the various stages of mitosis and meiosis.

The Evolutionary Consequences of Sexual Reproduction

The reassortment of genetic material that occurs during meiosis is the principal explanation for the incredible diversity of eukaryotic organisms that has developed over the past 1.5 billion years. Sexual reproduction represents an enormous advance in organisms' ability to generate genetic variability. To understand why this is true, recall that most organisms have more than one chromosome. Human beings have twenty-three different pairs of homologous chromosomes. Each human gamete receives one of the two copies of each of the twenty-three different chromosomes, but which copy of a particular chromosome it receives is random. For example, the copy of chromosome number 14 that a particular human gamete receives has no influence on which copy of chromosome number 5 it will receive. Each of the twenty-three pairs of chromosomes goes through meiosis independently of all the others, so there are 2^{23} (more than eight million) different possibilities for the kinds of gametes that can be produced.

Because the zygote that forms a new individual is created by the fusion of *two* gametes, each produced independently, fertilization squares the number of possible outcomes (8 million × 8 million = 64 trillion). Fertilization therefore creates a unique individual, a new combination of the twenty-three chromosomes that probably has never occurred before and probably will never occur again.

The exchange that occurs as a result of crossing-over between the nonsister chromatids of homologous chromosomes adds even more possibilities to the random assortment of chromosomes. Thus, the number of genetic combinations possible among gametes is virtually unlimited.

Sexual reproduction increases genetic variability through fertilization, crossing-over, and random assortment of chromosomes.

Few subjects have been so profoundly interesting to biologists as sexual reproduction. For one thing, how sexual reproduction evolved in the first place is difficult to understand. Evolutionary biologists think that it may have arisen secondarily from natural selection for better repair of genetic damage.

The evolutionary consequences of sexual reproduction have been profound. No genetic process generates diversity more quickly, and genetic diversity is the raw material of evolution, the fuel that determines how far and how rapidly natural selection can change a population's characteristics. Among bacteria and other groups of organisms in which genetic recombination is limited, a short generation time appears to play a similar role in enhancing the amount of genetic diversity available for selection during a given period. In many cases, the pace of evolution appears to be geared to the level of genetic diversity: The greater the genetic diversity, the faster evolution proceeds. Programs for selecting larger size in domesticated cattle and sheep proceed rapidly at first, but then slow as all of the existing genetic combinations are exhausted. Further progress must then await the generation of new gene combinations.

The genetic recombination associated with sexual reproduction has had an enormous evolutionary impact because of the extensive variability that it can generate rapidly within the genome.

Know for test

Mitosis

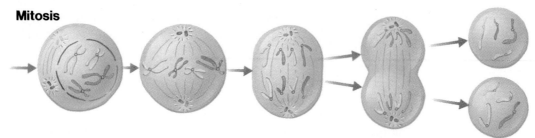

Prophase
Chromosomes condense. Spindle fibers form between centrioles, which move toward opposite poles. *Centrials act as anchors*

Nuclear membrane breaks down

Metaphase
Microtubule spindle apparatus attaches to chromosomes. Chromosomes align along spindle equator.

Kinetochore fibers form, Centromeres will replicate

Anaphase
Sister chromatids separate and move to opposite poles.

Telophase
Chromatids arrive at each pole, and new nuclear membranes form. Cell division begins.

Nuclear membranes reappear, chromosomes decondense

Daughter cells
Cell division is complete. Each cell receives chromosomes that are identical to those in original nucleus.

Meiosis I

Zygote is first body cell

Prophase I
Homologous chromosomes further condense and pair. Crossing-over occurs. Spindle fibers form between centrioles, which move toward opposite poles.

Metaphase I
Microtubule spindle apparatus attaches to chromosomes. Homologous pairs align along spindle equator.

Anaphase I
Homologous pairs of chromosomes separate and move to opposite poles.

Telophase I
One set of paired chromosomes arrives at each pole, and nuclear division begins.

Daughter cells
Each cell receives exchanged chromosomal material from homologous chromosomes.

Meiosis II

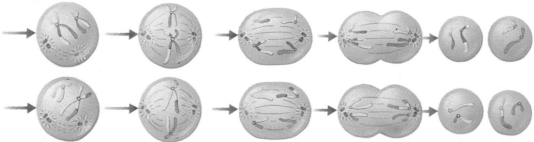

Prophase II
Chromosomes recondense. Spindle fibers form between centrioles, which move toward opposite poles.

Metaphase II
Microtubule spindle apparatus attaches to chromosomes. Chromosomes align along spindle.

Become haploid

Anaphase II
Sister chromatids separate and move to opposite poles.

Telophase II
Chromatids arrive at each pole, and cell division begins.

Daughter cells
Cell division is complete. Each cell ends up with half the original number of chromosomes.

Figure 9.19 A comparison of mitosis and meiosis.

The four primary differences between these two processes are: (1) Mitosis results in two diploid daughter cells, while meiosis results in four haploid daughter cells. (2) Because the chromosomes line up differently during mitosis than during meiosis I, sister chromatids are pulled apart in mitosis, while homologous pairs are pulled apart in meiosis I. (3) During mitosis, chromosomes do not exchange genetic material; during meiosis, they do. (4) Mitosis occurs in body cells, while meiosis occurs in reproductive cells, such as eggs and sperm.

EVOLUTIONARY VIEWPOINT

Synapsis, the close pairing of homologous chromosomes that occurs in meiotic prophase I, probably evolved as a mechanism to repair genetic damage: Damaged material was excised, and the gap filled in by comparison to the homologue. Crossing-over, which uses this close alignment to effect an exchange of chromosome arms between homologues, has been a key factor in the evolution of eukaryotes because it is a powerful mechanism for reshuffling genes.

SUMMARY

1. Bacterial cells divide by simple binary fission. The two replicated circular DNA molecules attach to the plasma membrane at different points, and cellular fission is initiated between these points.

2. Eukaryotic cells contain much more DNA than do those of bacteria. Eukaryotic DNA is coiled around a framework of histone proteins and distributed among several chromosomes. Some of the DNA is permanently condensed into heterochromatin; the rest, called euchromatin, becomes condensed only during cell division.

3. The cell cycle consists of five phases: three that prepare for cell division and two that involve the actual division process. The G_1 ("first gap"), S ("synthesis"), and G_2 ("second gap") phases are collectively referred to as interphase. The subsequent

M ("mitosis") phase is followed by a C ("cytokinesis") phase, in which the cell itself divides.

4. DNA replication is complete before mitosis begins. Immediately before the onset of mitosis, chromosome condensation and tubulin synthesis (for microtubules) begin.

5. The first stage of mitosis is prophase, during which the chromosomes become more condensed and the mitotic spindle forms. At the end of prophase, the nuclear envelope disassembles, and kinetochore fibers connect each pair of sister chromatids to the two poles of the cell.

6. The second stage of mitosis is metaphase, during which the chromosomes align around the periphery of a plane cutting through the center of the cell at right angles to the axis of the mitotic spindle. At the end of metaphase, the centromeres joining each pair of sister chromatids replicate, freeing each chromatid to be pulled to one of the poles of the cell by the kinetochore fibers attached to it.

7. The third stage of mitosis is anaphase, during which the sister chromatids physically separate, moving to opposite poles of the cell.

8. The fourth and final stage of mitosis is telophase, during which the mitotic spindle is disassembled, the nuclear envelope re-forms, and the chromosomes uncoil.

9. Following mitosis, most cells undergo cytoplasmic division, or cytokinesis. In animal and many protist cells, which lack a cell wall, the cell is pinched in two by a belt of microtubules drawing inward around the cell's midsection to form a cleavage furrow. In plant, fungal, and many protist cells, which are surrounded by a rigid cell wall, an expanding cell plate forms along the midline of the mitotic spindle.

10. Meiosis is a special form of nuclear division that precedes gamete formation in sexually reproducing eukaryotes. In meiosis, there is a single replication of the chromosomes and two chromosome separations. Meiosis results in the formation of four nuclei, each with half the original

number of chromosomes. The cells that include these nuclei may serve as gametes directly, or they may undergo further mitotic divisions before the gametes differentiate.

11. Meiosis involves a pair of serial nuclear divisions. The two unique characteristics of meiosis are synapsis, which is the intimate pairing of homologous chromosomes, and the lack of chromosome replication before the second nuclear division.

12. Crossing-over is an essential element of meiosis. It occurs during prophase I, when the chromosomes are so closely associated that the kinetochore fibers are able to bind to only one kinetochore of each homologous chromosome's centromere because the other kinetochore is covered by the opposite homologue. During crossing-over, genetic material is exchanged between the nonsister chromatids of homologous pairs.

13. A spindle apparatus forms during metaphase I, and homologous pairs line up randomly along the spindle equator. In anaphase I, the homologues separate and begin to move toward opposite poles. In telophase I, individual chromosomes gather at their respective poles.

14. At the end of meiosis I, one member of each pair of homologues is present at each of the two poles of the dividing nucleus. These chromosomes already consist of two sister chromatids, differing genetically from each other as a result of the crossing-over that occurred when the chromosomes were paired with their homologues. No further replication occurs between the next nuclear division, which is a normal mitosis that occurs at each of the two poles. The sister chromatids simply separate from each other. This results in the formation of four clusters of chromosomes, each with half the number of chromosomes that was present initially.

15. Meiotic recombination is the principal factor that has made the evolution of eukaryotic organisms possible. The number of possible genetic combinations that can occur among gametes is virtually unlimited because of meiosis and crossing-over.

REVIEWING THE CHAPTER

1. How do cells in bacteria divide?
2. What is the structure of a eukaryotic chromosome?

3. How many chromosomes are in a cell?
4. How are chromosomes karyotyped?
5. What is the cell cycle?
6. What are mitosis and cytokinesis?
7. What are the stages of mitosis?

8. What is meiosis, and how does it relate to sexual reproduction?
9. What are the stages of meiosis?
10. What are the evolutionary consequences of sexual reproduction?

COMPLETING YOUR UNDERSTANDING

1. Most eukaryotic chromosomes are a combination of about 60 percent _____ and 40 percent _____. The genome in a bacterial cell, however, is 100 percent _____.

2. Why is the DNA duplex strongly attracted to histones in eukaryotic cells?

3. What is the interrelationship between homologous chromosomes, sister chromatids, and centromeres?

4. Why don't today's biologists still characterize the G_1, S, and G_2 phases of the cell cycle as "resting stages"?

5. What determines the plane in which a cell divides?

6. In mitosis, what is the function of polar fibers and kinetochore fibers?

7. Why do scientists believe that the MTOC, and not centrioles, is essential in the formation of the mitotic spindle?

8. In mitosis, during which phase does the alignment of the chromosomes on a plane in the center of the cell occur?

9. What are the distinguishing features of anaphase during mitosis?

10. In cytokinesis, what are the distinguishing features between animals and plants?

11. Which probably evolved first—meiosis or mitosis? How would you support your answer?

12. Sexual reproduction involves an alternation of _____ and _____.

13. In meiosis I, how does prophase I differ from prophase in mitosis?

14. What is the significance of synapsis to genetic recombination?

15. What are the similarities and differences between meiosis II and mitosis?

16. In meiosis, is the chromosome number halved during meiosis I or meiosis II? Which phase would you use to support your answer?

17. What is the evolutionary advantage of sexual reproduction over asexual reproduction? If there is an evolutionary advantage to sexual reproduction, why have a number of organisms, such as some protists (amoebae, euglenas), many ferns, and some flowering plants, survived with asexual reproduction as the sole or major means of reproduction?

FOR FURTHER READING

Baserga, R. 1985. *The biology of cell reproduction*. Cambridge, Mass.: Harvard University Press. A very stimulating and insightful look at mitosis in a chatty style that is fun to read.

The cell cycle. 1989. *Science*. 246 (4930): 545–640. An entire issue devoted to a comprehensive account of all aspects of modern research into the cell cycle.

John, B. 1990. *Meiosis*. New York: Cambridge University Press. A broad survey of meiosis, with emphasis on the surprising diversity within the plant and animal kingdoms in how it is carried out.

Manuelidis, L. 1990. A view of interphase chromosomes. *Science*, Dec., 1533–40. A review of how chromosomes are organized to express their genes.

Marx, J. 1989. The cell cycle coming under control. *Science*, July, 252–55. A highly readable account of how researchers uncovered the biochemical machinery that controls cell division in all eukaryotes.

McIntosh, R., and K. McDonald. 1989. The mitotic spindle. *Scientific American*, Oct., 48–57. New information on how microtubules part the DNA of dividing cells into two equal clusters.

Moyzis, R. 1991. The human telomere. *Scientific American*, Aug., 48–55. A discussion of how a unique nucleotide sequence repeated thousands of times forms a protective cap on the ends of chromosomes.

Murray, A., and M. Kirschner. 1991. What controls the cell cycle. *Scientific American*, March, 56–63. A modern look at how cell division is controlled. One protein plays a key role in virtually all organisms.

EXPLORATIONS

Interactive Software:
Genetic Reassortment During Meiosis

This interactive exercise presents a diagram of meiosis, showing how chromosomes pass through its stages, with a variety of visible characters located on each chromosome. A "recombination frequency" meter allows the user to assess the probability that a particular combination of alleles will appear together in a gamete. The user will discover that the key variable is the number of chromosomes on which the alleles appear.

Questions to Explore:

1. How does the number of chromosomes on which the alleles appear affect recombination probabilities?

2. What recombination frequency would be required to make a 50-50 representation in the *phenotype* of any given trait?

3. How is recombination affected by the actual location (specific position) of alleles on the chromosome?

4. Why doesn't recombination occur during mitosis?

MENDELIAN GENETICS

"Some of us look more alike than others."
These identical twins have all of their genes
in common, and differ only in how they have
developed. In general, people differ from
one another in a significant fraction of all
their genes, and these so-called "heritable
differences" are responsible for much of the
differences among us in appearance and
behavior.

FOR REVIEW

Here are some important terms and concepts that have been discussed in previous chapters and that you will encounter again in this chapter. Review them before proceeding if necessary.

Chromosomes (*chapters 4, 9*)
Meiosis (*chapter 9*)
Homologous chromosomes (*chapter 9*)
Crossing-over (*chapter 9*)

Did you ever see anybody who looks exactly like you? Look at your classmates. Rarely will any of you resemble one another closely. Even your brothers and sisters are not exactly like you, unless you have an identical twin. Human beings are extremely diverse in appearance (figure 10.1).

Variation in itself is not surprising. Differences in diet during development can have great effects on adult appearance, as can variations in environments. One remarkable property in some patterns of variation, however, has been especially fascinating and puzzling: Some traits are inherited, passed down from parent to child.

For all of recorded history, patterns of resemblance among the members of particular families have been noted and discussed. Some features shared by family members are unusual, such as the protruding lower lip of the Austrian royal family, the Hapsburgs, which is evident in pictures and descriptions of that family since the thirteenth century. Other characteristics are more familiar, such as the common occurrence of red-haired children within families of red-haired parents (figure 10.2). Such inherited features are the focus of this chapter.

Figure 10.1 No two individuals look alike.

As a result of both heredity and environmental influences, these babies will grow to adulthood as separate individuals.

Early Ideas About Heredity: The Road to Mendel

Two centuries ago, English gentlemen farmers who were trying to improve varieties of agricultural plants carried out matings, called **crosses,** between different strains, selecting the most desirable of the offspring of each cross. Whereas the strains they used were **true-breeding,** producing offspring that resembled their parents in all respects, the **hybrids**—the plants that resulted from crossing different strains—often were variable. The offspring of hybrid plants had different combinations of their parents' features; not all offspring of a cross received the same selection of inherited traits. In these hybrid situations, the alternatives of a trait appeared to be "segregating" among the progeny—that is, some progeny exhibited one alternative form of a trait, while others exhibited another.

Figure 10.2 Inherited traits.

Red-haired parents often produce children with red hair.

In their crosses, the farmer found that some alternative forms of a trait appeared more often among the progeny than others. They also discovered that the less frequent forms disappeared in one generation, only to reappear unchanged in the next. In a series of experiments in the 1790s, T. A. Knight crossed two true-breeding varieties of the garden pea *Pisum sativum* (figure 10.3). One of these varieties had purple flowers; the other, white flowers. All the progeny of the cross had purple flowers. Among the offspring of these hybrids, however, some plants had purple flowers, while others, occurring less frequently, had white ones. As had been observed earlier, a trait from one of the parents (white flowers) was hidden in one generation, only to reappear in the next.

> *Two centuries ago, plant breeders learned that the hybrid plants derived from crosses between different strains exhibited offspring with variable combinations of their parents' traits. One of the alternative forms of a particular trait sometimes disappeared in the first hybrid generation, reappearing in its progeny generation.*

In these deceptively simple results were the makings of a scientific revolution. But it took another century before the process of **segregation of alternative traits** was understood. Why did it take so long? One reason was that some characteristics appeared to be *blended* in the offspring, so it was not clear that individual, distinct factors were involved. Another problem was that early workers did not quantify, or count, their results, and a numerical record of results proved to be crucial to understanding this process. For example, both Knight and later experimenters who carried out other crosses with pea plants noted that some traits had a "stronger tendency" to appear than others, but they did not record the actual numbers of the different classes of progeny. Science was young then, and it was not obvious that the numbers were important.

Mendel and the Garden Pea

The first quantitative studies of inheritance were conducted by Austrian monk Gregor Mendel (figure 10.4*a*). Born in 1822 to peasant parents, Mendel was educated in a monastery and went on to study science and mathematics at the University of Vienna, where he failed his examinations for a teaching certificate. Returning to the monastery, where he spent the rest of his life and eventually became abbot, Mendel initiated a series of experiments on plant heredity (figure 10.4*b*). The results of these experiments ultimately and irrevocably changed views of heredity.

For his experiments, Mendel chose the garden pea, the same plant that Knight and many others had studied earlier. The choice was a good one for several reasons:

1. Many earlier investigators already had produced hybrid peas by crossing different varieties. From their results, Mendel knew that he could expect to observe the segregation of alternative traits among the progeny of crosses. That is, Mendel knew that some of the progeny would exhibit the alternative characteristic of one variety, while other individuals would exhibit the alternative characteristic of the other variety.

2. Many varieties of peas were available. Mendel initially examined thirty-two varieties. Then, for further study, he selected lines that differed with respect to seven easily distinguishable traits, such as smooth versus wrinkled seeds and purple versus white flowers (a characteristic that Knight had studied sixty years earlier) (figure 10.5).

3. Pea plants are small and easy to grow, produce large numbers of offspring, and mature quickly. Thus, experiments can involve many plants and produce results relatively soon.

Figure 10.3
A pod of the garden pea, **Pisum sativum.**

Easy to cultivate and with many distinctive varieties, the garden pea was a popular experimental subject in investigations of heredity as long as a century before Gregor Mendel's experiments.

(a)   (b)

Figure 10.4 The father of modern genetics.
(a) *Gregor Johann Mendel.* (b) *Mendel's garden plot, recently restored at the abbey in Brnö, Czechoslovakia, where Mendel was abbot. This garden plot yielded the "secrets" of genetic inheritance.*

4. The pea's sexual organs are enclosed within the flower. The flowers of peas, like those of most flowering plants, are bisexual, containing the structures in which male gametes are produced (the anthers) as well as those in which female gametes are produced (the carpels and the stigma [a specialized area of the carpel that receives the pollen]). **Self-fertilization** takes place automatically within an individual pea

Figure 10.5 A Mendelian trait.

One of the differences among varieties of pea plants that Mendel studied was seed shape. In some varieties, the seeds were round, whereas in others, they were wrinkled. As you can see, wrinkled seeds look like dried-out, shrunken versions of the round ones.

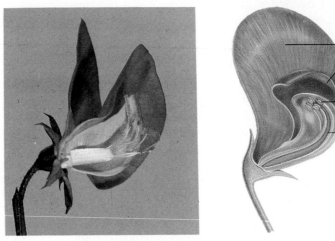

Figure 10.6 Anatomy of the pea flower.

In a pea flower, the petals enclose the male (anther) and female (stigma) parts, ensuring that self-fertilization will take place unless the flower is disturbed.

flower if the flower is not disturbed (figure 10.6). As a result, the offspring of garden peas are the progeny of a single individual. A controlled cross-fertilization involves removing the anthers before fertilization and introducing pollen (which is produced in the anthers and gives rise to the gametes) from a strain with alternative characteristics (figure 10.7).

> *Gregor Mendel chose peas for his classical genetics experiments because: the results of crosses in peas had been studied earlier; different varieties are variable in their features; the plants are relatively small, so many can be grown in a limited area, and the generation time is short; and true-breeding strains are produced by self-fertilization.*

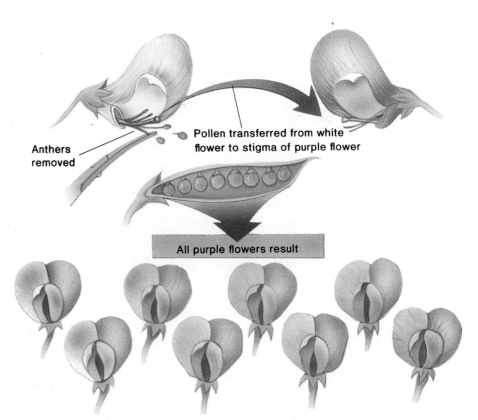

Figure 10.7 How Mendel conducted his experiments.

Mendel pushed aside the petals of a white flower, cut off the anthers, and then transferred the pollen from the white flower to the stigma of a similarly castrated purple flower. The seeds that resulted from this cross-fertilization all grew into plants with purple flowers.

Mendel's Experimental Design

Mendel usually conducted his experiments in three stages:

1. He first allowed pea plants of a given variety to produce progeny by self-fertilization for several generations, thereby ensuring that the pea plants were true-breeding—that is, were transmitted regularly from generation to generation. Pea plants with white flowers, for example, produced only plants with white flowers, regardless of the number of generations studied.

2. Mendel then conducted crosses between varieties exhibiting alternative traits. For example, he removed the anthers from a white-flowered plant and then fertilized the plant with pollen from a purple-flowered plant. He also reversed the procedure, using pollen from a white-flowered plant to fertilize a purple-flowered plant.

3. Finally, Mendel permitted the hybrid offspring produced by these crosses to self-fertilize for several generations, thereby allowing the alternative traits to segregate among the progeny.

This was the same experimental design that Knight and others had used much earlier. But Mendel added a new element: He counted the numbers of offspring in each class and in each succeeding generation. These quantitative results were essential to an understanding of the process of heredity.

What Mendel Found

When Mendel crossed two plants that differed in one characteristic, such as purple-flowered plants with white-flowered plants, the hybrid offspring did not have an intermediate flower color, instead always resembling one of the parents. These hybrid offspring are called the **first filial, or F_1, generation.** In a cross of white-flowered with purple-flowered plants, the F_1 offspring all had purple flowers, just as Knight and others had reported earlier. Mendel referred to the trait expressed in the F_1 plants (purple flowers) as **dominant,** and to the alternative trait, which was not expressed in the F_1 plants (white flowers), as **recessive.** For each pair of contrasting traits that Mendel examined, one proved to be dominant and the other recessive.

After allowing individual F_1 plants to mature and self-fertilize, Mendel collected and planted the fertilized seed from each plant to see what the offspring in this **second filial,** or **F_2, generation** would look like. He found, just as Knight had earlier, that some F_2 plants exhibited the recessive trait. Latent in the F_1 generation, the recessive alternative reappeared among some F_2 individuals.

Mendel termed those features of the parental strains that occurred in all members of the F_1 generation as dominant traits and to their alternative features, which reappeared in some individuals of the F_2 and subsequent generations, as recessive traits.

It was at this stage that Mendel instituted his radical change in experimental design: He counted the numbers of each type among the F_2 progeny to determine whether the proportions of the F_2 types would provide some clue about the mechanism of heredity. He examined a total of 929 F_2 individuals in the cross between purple-flowered F_1 plants just described. Of these F_2 plants, 705 had purple flowers and 224 had white flowers. About one-fourth of the F_2 individuals (24.1 percent) exhibited white flowers, the recessive trait.

Mendel examined each of seven pairs of contrasting traits in this way (figure 10.8). The ratio was always the same: three-fourths of the F_2 individuals exhibited the dominant trait, and one-fourth displayed the recessive trait. The ratio of dominant to recessive among the F_2 plants was always 3:1.

Mendel went on to examine how the F_2 plants behaved in later generations. He found that the one-fourth that displayed the recessive trait were always true-breeding (that is, continued to exhibit the trait in all individuals of succeeding generations). For example, in Mendel's cross of white-flowered with purple-flowered plants, the white-flowered F_2 individuals reliably produced white-flowered offspring when allowed to self-fertilize. By contrast, only one-third of the dominant F_2 individuals (one-fourth of the entire offspring) were true-breeding, whereas two-thirds were not. This latter group produced dominant and recessive F_3 individuals in the ratio 3:1. This result suggested that, for the entire sample, the 3:1 ratio that Mendel observed in the F_2 generation was really a disguised 1:2:1 ratio, with one-fourth true-breeding dominant individuals to one-half not-true-breeding dominant individuals to one-fourth true-breeding recessive individuals (figure 10.9).

How Mendel Interpreted His Results

From these experiments, Mendel learned four things about the nature of heredity:

1. The traits that Mendel studied did not produce intermediate types when plants with these traits were crossed. Instead, the alternative features were inherited intact; they either appeared or did not appear in a particular generation.

2. For each pair of traits that Mendel examined, one alternative was not expressed in the F_1 hybrids, although it reappeared in some F_2 individuals. *The "invisible" trait must therefore have been latent (present but not expressed) in the F_1 individuals.*

3. The pairs of alternative traits that Mendel examined segregated among the progeny of a particular cross, with some individuals exhibiting one trait, some the other.

4. Pairs of alternative traits were expressed in the F_2 generation in the ratio of three-fourths dominant to one-fourth recessive. This characteristic 3:1 segregation ratio is often referred to as the **Mendelian ratio.**

Mendel observed that alternative genetic traits were inherited intact and did not blend; that one of the alternatives was not expressed in the F_1 generation, but reappeared in the F_2 and subsequent generations; that the alternative traits segregated among the progeny of a particular cross; and that pairs of alternative traits were expressed in the F_2 generation in the ratio of three-fourths dominant, one-fourth recessive.

Trait	Dominant versus Recessive	F₂ Generation Results		Ratio
		Dominant Form	**Recessive Form**	
Flower color	 Purple × White	705	224	3.15:1
Seed color	 Yellow × Green	6,022	2,001	3.01:1
Seed shape	 Round × Wrinkled	5,474	1,850	2.96:1
Pod color	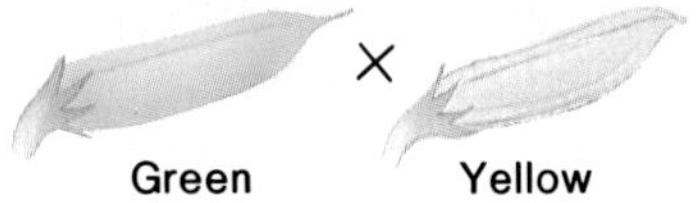 Green × Yellow	428	152	2.82:1
Pod shape	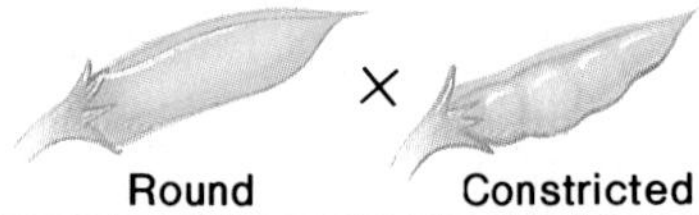 Round × Constricted	882	299	2.95:1
Flower position	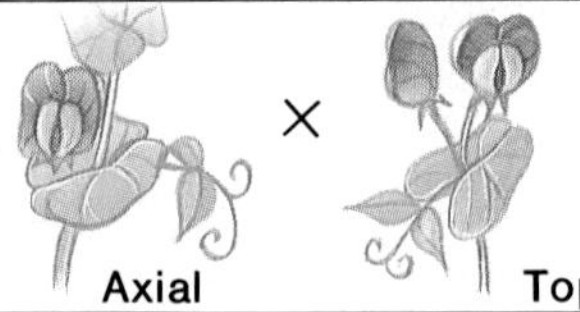 Axial × Top	651	207	3.14:1
Plant height	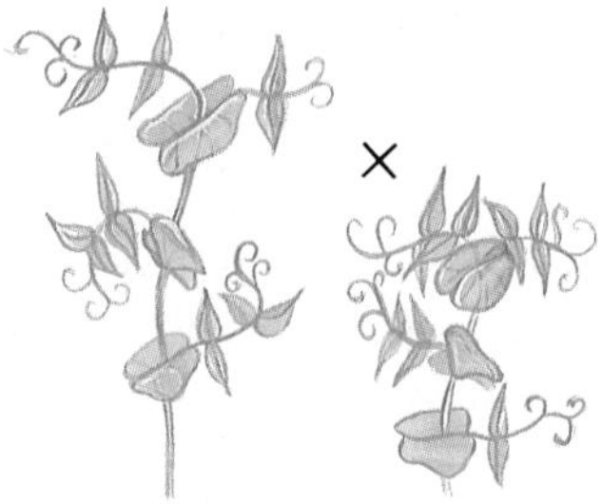 Tall × Dwarf	787	277	2.84:1

Figure 10.8 Mendel's experimental results.
Mendel studied seven pairs of contrasting traits in the garden pea and recorded his results in crosses of these traits. For each *of the seven traits, the ratio of dominant to recessive was very close to 3:1.*

Mendel synthesized his conclusions into a set of hypotheses with the five elements that follow (rephrased in modern terms where appropriate). Many scientists have since examined, tested, and reconfirmed Mendel's hypotheses, which have become one of the most important theories in biology.

1. Parents do not transmit their features directly to their offspring; instead, they transmit encoded information about the features. Mendel called these bits of information, which act in the offspring to produce the trait, **factors.** Today, they are called **genes.**

2. For every feature of a diploid organism, there are two factors, which may or may not be the same. If the factors differ from each other, the individual is said to be **heterozygous** for that characteristic. If they are the same, the individual is **homozygous.**

3. Genes are composed of a sequence of nucleotides in a DNA molecule. Alternative forms of a particular gene, such as white flowers and purple flowers, are called **alleles.** The position on a chromosome where a gene is located is called a **locus.**

4. The two alleles, one contributed by the male parent and one by the female parent, remain distinct. Mendel said that they were "uncontaminated" in that alleles do not blend with one another or become altered in any other way. When an individual's gametes (sex cells—eggs or sperm) form, the alleles segregate at random, and only one allele of each pair is present in each gamete. Both alleles have an equal chance of being included in a gamete.

5. The presence of a particular allele does not ensure that the trait specified by that allele will actually be expressed. In heterozygous individuals, only one (the dominant) allele is

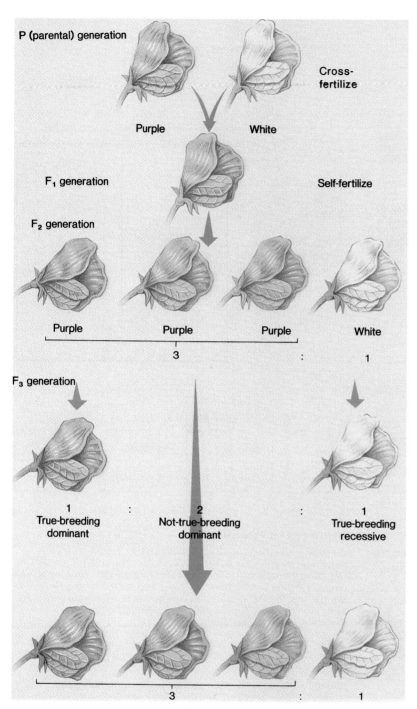

Figure 10.9 The F₃ generation: Proof that the F₂ generation was a disguised 1:2:1 ratio.

By allowing the F₃ generation to self-fertilize, Mendel found from the offspring in the F₃ generation that the ratio of F₂ plants was one true-breeding dominant, two not-true-breeding dominant, and one true-breeding recessive.

Many human traits are controlled by dominant or recessive alleles like those Mendel studied in peas (figure 10.10). Chapter 11 discusses dominant and recessive traits in more detail.

> *An individual's collection of genes is called the individual's genotype. The physical expression of an individual's traits is called the individual's phenotype. Thus, the genotype is the "blueprint"; the phenotype is the realized outcome.*

Mendel Revisited: Analysis of His Results

The five elements just described constitute the set of hypotheses by which Mendel explained heredity. Let us see if they predict the sort of result that Mendel actually obtained.

The F₁ Generation

Consider again Mendel's cross of purple-flowered with white-flowered plants. The symbol *p* will refer to the recessive allele, associated with the production of white flowers. The symbol *P* will refer to the dominant allele, associated with the production of purple flowers. As is conventional, the feature as a whole—purple versus white flowers, determined by the single gene—has been assigned a letter relating to its more common condition: thus, *P* for purple flowers. The recessive allele, associated with white flowers, is indicated by lowercase, as *p;* the alternative dominant allele, associated with purple flowers, is assigned the same symbol in uppercase, as *P.*

A genotype is indicated by two letters to represent the two alleles of a gene present at each locus. Thus, the genotype of an individual that is true-breeding (or homozygous) for the recessive white-flowered trait would be *pp.* In such an individual, both alleles specify white flowers. Similarly, the genotype of a true-breeding (or homozygous) purple-flowered individual would be *PP.* A heterozygous individual would be designated *Pp* (the dominant allele is usually written first). With these conventions and a times sign (×) to denote a cross between two breeding lines, Mendel's original cross can be symbolized as *pp × PP.*

The possible results from a cross between a true-breeding, white-flowered plant (*pp*) and a true-breeding, purple-flowered plant (*PP*) can be visualized with a **Punnett square,** named

expressed, and the features associated with that allele are the only ones seen. The recessive allele, although present, is unexpressed. Modern geneticists refer to an individual's collection of alleles as the individual's **genotype.** The physical expression of an individual's traits (that is, how the traits alter appearance, behavior, or physiological activity) is called an individual's **phenotype.**

Figure 10.10 A common recessive trait in humans.
Blue eyes are considered a recessive trait in humans, although many genes influence the exact shade of blue.

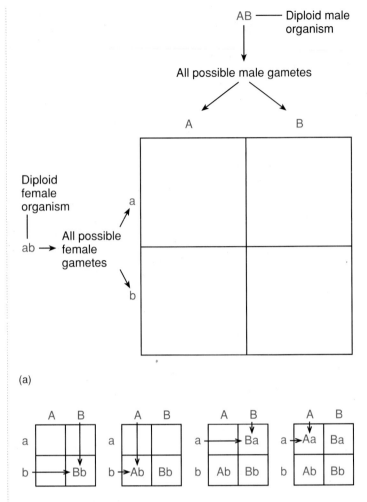

(a)

(b)

Figure 10.11 A Punnett square.
A Punnett square analysis is an easy way to determine all the possible genotypes of a particular cross. (a) The possible gametes for one parent are placed along one side of the square, and the possible gametes for the other parent are placed along the other side of the square. (b) The Punnett square can be used like a mathematical table to visualize the genotypes of all potential offspring.

after its originator, British geneticist Reginald Crundall Punnett. In a Punnett square analysis, the possible gametes of one individual are listed along the horizontal side of the square, while the possible gametes of the other individual are listed along the square's vertical side. The genotypes of potential offspring are represented by the cells within the square (figure 10.11). Punnett squares are really mathematical tables that allow all the possible genotypes to be visualized by combining all the possible alleles.

For practice, sketch the Punnett square that shows the cross between a true-breeding white flower (*pp*) and a true-breeding purple flower (*PP*). Your Punnett square should show that a homozygous-dominant (*PP*) individual can produce only *P* gametes and a homozygous-recessive (*pp*) individual can produce only *p* gametes, so all offspring *must* be heterozygous (*Pp*). This is the result that Mendel found in his F_1 generation. Because the *P* allele is dominant, all the F_1 offspring are expected to have purple flowers. The *p* allele is present in these heterozygous individuals, but it is not phenotypically expressed.

In a cross between homozygous-dominant and homozygous-recessive individuals, all of the F_1 progeny will be heterozygous and will resemble the homozygous-dominant parent in their phenotype. The outcome of any cross can be visualized with a Punnett square.

The F₂ Generation

When F_1 individuals are allowed to self-fertilize, the *P* and *p* alleles segregate at random during gamete formation. Because of the properties of meiosis (see chapter 9), any gamete has a 50 percent (that is, random) chance of obtaining either the *P* allele or the *p* allele. The subsequent union of these gametes to form F_2 individuals is also random; it is not influenced by the particular allele an individual gamete carries.

Thus, the probability of obtaining two dominant alleles is $0.5 \times 0.5 = 0.25$, or 25 percent, and the probability of obtaining two recessive alleles is the same. What will F_2 individuals look like? The possibilities can be visualized in a Punnett square that shows a cross between two heterozygous individuals (*Pp*). Such a Punnett square clearly predicts that, in the F_2 generation, three-fourths of the plants will have purple flowers (*PP* or *Pp*), and one-fourth will have white flowers (*pp*), a phenotypic ratio of 3:1.

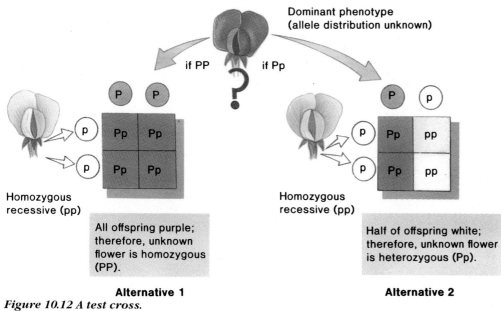

Dominant phenotype
(allele distribution unknown)

if PP   if Pp

Homozygous recessive (pp)

All offspring purple; therefore, unknown flower is homozygous (PP).

Alternative 1

Homozygous recessive (pp)

Half of offspring white; therefore, unknown flower is heterozygous (Pp).

Alternative 2

Figure 10.12 A test cross.
To determine whether an individual exhibiting a dominant phenotype, such as purple flowers, is homozygous (PP) or heterozygous (Pp) for the dominant allele, Mendel crossed the individual in question with a known homozygous recessive (pp)—in this case, a plant with white flowers.

In an F_2 generation derived from heterozygous individuals, any gamete can receive either allele. The fusion of the gametes following fertilization is random and is not influenced by the particular allele contained in a particular gamete.

. Of F_2 individuals, a fourth are true-breeding, white-flowered individuals (*pp*); half are heterozygous, purple-flowered individuals (*Pp*); and a fourth are true-breeding, purple-flowered individuals (*PP*). The 3:1 phenotypic ratio is actually a "disguised" expression of a 1:2:1 genotypic ratio because the heterozygotes are phenotypically indistinguishable from the homozygous, purple-flowered (dominant) individuals.

The Test Cross

Of the purple-flowered individuals in the F_2 generation (*PP* and *Pp*), which are homozygous, and which are heterozygous? It is not possible to tell simply by looking at them. For this reason, Mendel devised a simple and powerful procedure called the **test cross** to determine an individual's actual genotypic composition. He crossed a purple-flowered individual (with genotype *PP* or *Pp*) with a homozygous recessive (*pp*) individual. His hypothesis predicted different results for homozygous and heterozygous test plants (figure 10.12):

Alternative 1. Test individual is homozygous. *PP* × *pp*: All offspring have purple flowers (*Pp*).

Alternative 2. Test individual is heterozygous. *Pp* × *pp*: Half of the offspring have white flowers (*pp*), and half have purple flowers (*Pp*).

To ensure that his test-cross procedure was effective, Mendel crossed heterozygous F_1 individuals back to the parent homozygous for the recessive trait (for white- and purple-flowered plants, this cross is represented as *Pp* × *pp*). He found that the offspring were phenotypically half purple and half white. In other words, the dominant and recessive traits appeared in a 1:1 ratio. This observation confirmed Mendel's hypothesis that individual alleles of a gene segregate when gametes are formed.

Mendel's First Law: The Law of Segregation

Mendel's set of hypotheses thus accounted in a neat and satisfying way for the segregation ratios he had observed. His central premises—(1) that alleles do not blend in heterozygotes; (2) that alleles segregate in heterozygous individuals; and (3) that alleles have an equal probability of being included in either gamete—have since been verified in countless other organisms. These points are usually referred to as **Mendel's First Law**, or the **Law of Segregation.**. As you will see later in this chapter, the segregational behavior of alleles has a simple physical basis in meiosis, but that process was unknown to Mendel.

In modern terms, Mendel's Law of Segregation states that (1) alleles do not blend in heterozygotes; (2) in meiosis, alleles segregate from one another—when they are present in a heterozygous condition, the results of the segregation can be observed; and (3) each gamete has an equal probability of possessing either member of an allele pair.

Mendel's Experiments with Dihybrids

After Mendel had demonstrated the segregation of the different alleles of individual genes, he went on to investigate whether different *genes* segregate independently of one another or as a unit. For example, would a gamete's allele of a gene for seed shape influence which allele the gamete had for a gene affecting seed color?

Dihybrid Crosses: Mendel's Prediction

To determine whether the alleles of one gene could affect the alleles of another gene, Mendel first established a series of true-breeding lines of peas that differed from one another with respect to two of the seven pairs of characteristics he had studied. Second, he crossed contrasting pairs of the true-breeding lines. In a cross involving different seed-shape alleles (round *R*, and wrinkled, *r*) and different seed-color alleles (yellow, *Y*, and green, *y*) all the F_1 individuals were identical, each being heterozygous for both seed shape (*Rr*) and seed color (*Yy*). The F_1 individuals of such a cross are said to be **dihybrid**—that is, heterozygous for two genes.

The third step in Mendel's analysis was to allow the dihybrid individuals to self-fertilize. If the segregation of alleles affecting seed shape and seed color was independent, the probability that a particular pair of seed-shape alleles would occur together with a particular pair of seed-color alleles would be simply a product of the two individual probabilities that each pair would occur separately. For example, the probability of an individual with wrinkled, green seeds appearing in the F_2 generation would be equal to the probability of an individual with wrinkled seeds (one in four) multiplied by the probability of an individual with green seeds (one in four), or one in sixteen.

Because the genes for seed shape and seed color are each represented by a pair of alleles in the dihybrid individuals, dihybrids are expected to have four types of gametes: *RY, Ry, rY,* and *ry.* In the F_2 generation following the self-pollination of a dihybrid or a cross between two dihybrids, there are sixteen possible combinations of alleles, each of them equally probable (figure 10.13). Of the sixteen combinations, nine possess at least one dominant allele for each gene (the second allele is usually signified with a dash, *R—Y—*, which means that the second allele at the *R* locus and the second allele at the *Y* locus may be either dominant or recessive) and thus should have round, yellow seeds. Three possess at least one dominant *R* allele but are homozygous recessive for color (*R—yy*). Three other combinations possess at least one dominant *Y* allele but are homozygous recessive for shape (*rrY—*). One combination among the sixteen is homozygous recessive for both genes (*rryy*).

In summary, the hypothesis that color and shape genes assort independently thus predicts that the F_2 generation of this dihybrid cross will display the following phenotypic ratio: nine individuals with round, yellow seeds to three individuals with

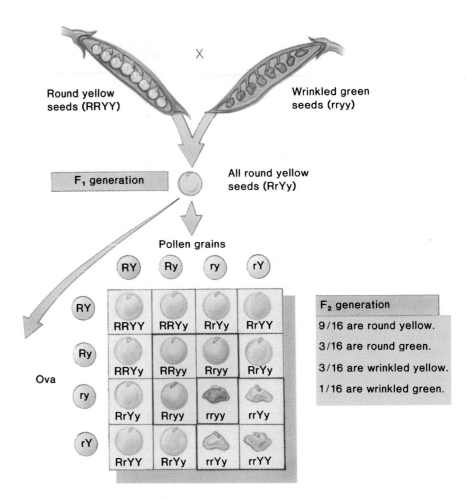

Figure 10.13 A dihybrid cross.

This dihybrid cross shows round (R) versus wrinkled (r) seeds and yellow (Y) versus green (y) seeds. The ratio of the four possible combinations of phenotypes is predicted to be 9:3:3:1, the ratio that Mendel found.

round green seeds to three individuals with wrinkled, yellow seeds to one with wrinkled, green seeds—a 9:3:3:1 ratio.

Results of Mendel's Dihybrid Crosses

Mendel examined a total of 556 seeds from dihybrid plants that had been allowed to self-fertilize and obtained the following results:

Phenotype	Genotype
315 Round, yellow	*R—Y—*
108 Round, green	*R—yy*
101 Wrinkled, yellow	*rrY—*
32 Wrinkled, green	*rryy*

This was very close to Mendel's theoretical 9:3:3:1 ratio, which for 556 seeds would have been 313:104:104:35. Thus, the two genes appeared to assort independently of one another. At the same time, round and wrinkled seeds still occurred approximately in the ratio 3:1 (423:133), as did yellow and green seeds

(416:140). Those features were segregating normally, as well as independently, of one another. Mendel strengthened his conclusions by obtaining similar results for other pairs of traits.

Mendel's Second Law: The Law of Independent Assortment

Mendel's conclusions are often referred to as **Mendel's Second Law** or the **Law of Independent Assortment.** As you will see later in the chapter, genes that assort independently of one another often do so because they are located on different chromosomes. Homologous chromosomes orient independently of one another on the metaphase I plate in meiosis, which is why the genes on those particular chromosomes also segregate independently. A modern restatement of Mendel's Law of Independent Assortment is: *Genes located on different chromosomes assort independently during meiosis.*

> *Mendel's Law of Independent Assortment states that genes located on different chromosomes assort independently of one another.*

Mendel's original paper describing his experiments, published in 1866, remains interesting reading today. His explanations are clear, and the logic of his arguments is presented in lucid detail. Unfortunately, Mendel's findings, which were published in the journal of the local natural history society, failed to arouse much interest. Only 115 copies of the journal were sent out, in addition to 40 reprints, which Mendel distributed himself. Although Mendel's results did not receive much notice during his lifetime, in 1900, sixteen years after his death, three different investigators independently rediscovered his pioneering paper. They came across it while searching the literature in preparation for publishing their own findings, which were similar to those Mendel had quietly presented more than three decades earlier.

Chromosomes: The Vehicles of Inheritance

Because chromosomes are not the only kinds of organelles that segregate regularly when eukaryotic cells divide (centrioles also divide and segregate in a regular fashion, as do the mitochondria and chloroplasts in the cytoplasm), it was not immediately obvious that chromosomes were the vehicles for the information of heredity. German geneticist Carl Correns first suggested a central role for chromosomes in 1900 in one of the papers announcing the rediscovery of Mendel's work. Soon after, observations that homologous chromosomes paired with each other in the process of meiosis led directly to the **chromosomal theory of inheritance,** first formulated by the American Walter Sutton in 1902. Sutton argued as follows:

1. Reproduction involves the initial union of only two cells: egg and sperm. If Mendel's model is correct, then these two gametes must make equal hereditary contributions. Sperm, however, contain little cytoplasm. Therefore, the hereditary material must reside within the nuclei of the gametes.

2. The two homologous chromosomes of each pair segregate during meiosis in a fashion similar to that exhibited by the alleles Mendel studied.

3. Gametes have a copy of one member of each pair of homologous chromosomes; diploid individuals have a copy of both members of each pair. Similarly, Mendel found that gametes have one allele of a gene and that diploid individuals have two.

4. During meiosis, each pair of homologous chromosomes orients on the metaphase plate independently of any other pair. This independent assortment of chromosomes is a process similar to the independent assortment of alleles postulated by Mendel.

Sutton's theory was based only on conjecture and was formulated entirely from the experimental evidence presented in Mendel's original paper. Sutton had no experimental evidence to confirm his ideas. Eight years after Sutton's theory was published, a scientist working with fruit flies provided the experimental data needed to prove that genes were indeed located on chromosomes.

Morgan and the Fruit Fly: Sex Linkage and the Chromosomal Theory of Inheritance

The proof that genes are located on chromosomes was provided by a single small fly. In 1910, American geneticist Thomas Hunt Morgan, studying the fly *Drosophila melanogaster,* detected a **mutant** fly, a male fly whose eyes were white instead of the normal red (figure 10.14).

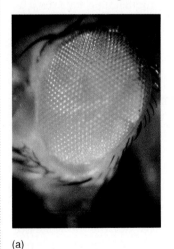 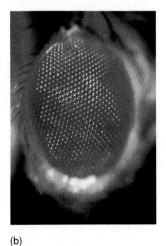

(a)   (b)

Figure 10.14 The key to the chromosomal theory of inheritance.

In Drosophila, *white eyes* (a) *are mutant, while red eyes* (b) *are normal. The white-eyed defect in eye color is hereditary, the result of a mutation in a gene located on the sex-determining X chromosome. By studying this mutation, American geneticist Thomas Hunt Morgan was able to demonstrate that genes are located on chromosomes.*

Morgan immediately set out to determine if this new trait would be inherited in a Mendelian fashion. He first crossed the mutant male with a normal female to see whether white eyes were dominant or recessive. All F_1 progeny had red eyes, and Morgan therefore concluded that red eye color was dominant over white. Following Mendel's experimental procedure, Morgan then crossed flies from the F_1 generation with each other. Eye color did indeed segregate among the F_2 progeny, as predicted by Mendel's theory. Of 4,252 F_2 progeny that Morgan examined, 782 had white eyes—an imperfect 3:1 ratio, but one that nevertheless provided clear evidence of segregation. Something was strange about Morgan's result, however, something that was totally unpredicted by Mendel's theory: All the white-eyed F_2 flies were males!

How could this strange result be explained? Perhaps it was not possible to be a white-eyed female fly; such individuals might not be viable for some unknown reason. To test this idea, Morgan test-crossed one of the red-eyed F_1 female progeny back to the original white-eyed male and obtained white-eyed and red-eyed males and females. So a female could have white eyes. Why, then, were there no white-eyed females among the progeny of the original cross?

The solution to this puzzle proved to involve chromosomes. In *Drosophila,* the sex of an individual is influenced by the number of copies of a particular chromosome, the **X chromosome,** that an individual possesses. An individual with two X chromosomes is a female. An individual with only one X chromosome, which pairs in meiosis with a larger, dissimilar partner called the **Y chromosome,** is a male. Thus, the female produces only X-containing gametes, whereas the male produces both X- and Y-containing gametes. When fertilization involves an X-containing sperm, the result is an XX zygote, which develops into a female. When fertilization involves a Y-containing sperm, the result is an XY zygote, which develops into a male.

The solution to Morgan's puzzle lies in the fact that, in *Drosophila,* the white-eye trait resides on the X chromosome but is absent from the Y chromosome. Geneticists now know that the Y chromosome carries almost no functional genes and that there is no corresponding locus for the white-eye trait on it. The eye color of males, therefore, is determined by whichever allele is present on their X chromosome. Females that are homozygous recessive for the white-eye allele would have white eyes, but a cross involving a male with red eyes could not produce a progeny that included white-eyed females. A characteristic that is determined by genes located on the sex chromosomes is said to be **sex-linked.** Because the white-eye trait is recessive to the red-eye trait, Morgan's result was a natural consequence of the Mendelian assortment of chromosomes (figure 10.15).

Morgan's experiment was one of the most important in the history of genetics because it presented the first clear evidence that Sutton was right and that the factors determining Mendelian traits do indeed reside on the chromosomes. The segregation of the white-eye trait, evident in the eye color of the flies, has a one-to-one correspondence with the segregation of the X chromosome, evident from the sexes of the flies.

The white-eye trait behaves exactly as if it were located on an X chromosome, which is indeed the case. The gene that specifies eye color in *Drosophila* is carried through meiosis as part of an X chromosome. In other words, Mendelian traits such as eye color in *Drosophila* assort independently because chromosomes do. When Mendel observed the segregation of alleles in pea plants, he was observing a reflection of the meiotic segregation of chromosomes.

> *Mendelian traits assort independently because they are determined by genes located on chromosomes, which also assort independently in meiosis.*

Linkage and the Effects of Crossing-Over

Mendel's Law of Independent Assortment predicted that alleles segregate randomly from each other—that is, that all alleles separate from each other during gamete formation and are inherited independently from one another. However, as Morgan and his associates continued to work with fruit flies, they observed that some traits seemed to be inherited as a group. For example, fruit flies that had yellow bodies almost always had white eyes. How could this observation be explained in light of Mendel's Law of Independent Assortment?

Morgan and his associates devised an explanation that explains this seeming contradiction of Mendel's second law. They proposed that genes located on the same chromosome are linked, and that when the genes are linked closely together on the chromosome, they tend to be inherited together. This **linkage,** however, does not mean that all genes located on the same chromosome are always inherited together. The farther apart the genes are on the chromosome, the more likely that crossing-over—the physical exchange of genetic information between chromosomes—will occur (see chapter 9). In other words, if the genes are located far enough apart, then they will segregate independently. This conclusion neatly explained Morgan's observation that some traits were inherited as a group, while a small but significant number of genetic combinations were new in each generation. The "grouped traits" were specified by genes that were linked closely together on a single chromosome, while the "recombined" traits were specified by genes that were located far apart on the chromosome and that were affected by the recombination of genetic information that occurs during crossing-over.

At first scientists were reluctant to believe that chromosomes were able to exchange segments during meiosis. Chromosomes seemed so solid when seen under the microscope that crossing-over seemed an unlikely hypothesis. That crossing-over does, in fact, occur was clearly demonstrated in an experiment by Curt Stern in which the alternative alleles being recombined were located on chromosomes that were visibly different—the

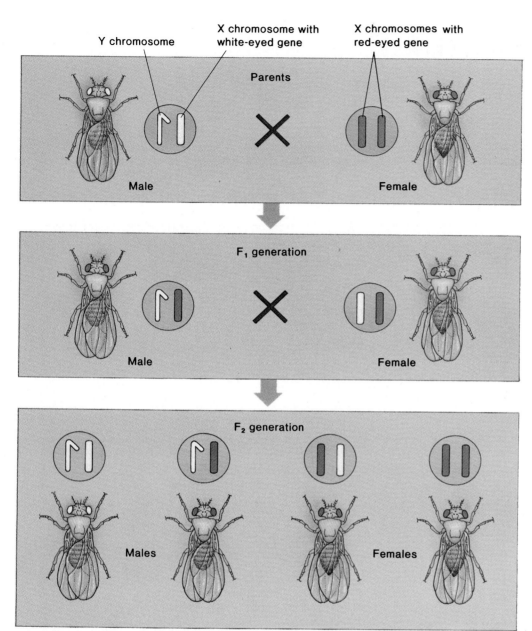

X chromosome with
white-eyed gene

Y chromosome

X chromosomes with
red-eyed gene

Parents

Male

Female

F₁ generation

Male

Female

F₂ generation

Males

Females

Figure 10.15 Morgan's experiment demonstrating sex linkage in Drosophila.

The white-eyed mutant male fly was crossed with a normal female. The F₁ generation flies all exhibited red eyes, as expected for flies heterozygous for a recessive white-eyed allele. In the F₂ generation, all white-eyed flies were male.

Genetic Maps

Because crossing-over occurs more often between two genes relatively far apart than between two genes relatively close together, the frequency of crossing-over can be used to map the relative positions of genes on chromosomes. In a cross, the proportion of progeny that exhibit recombined traits is a measure of how frequently cross-over events occur between the genes that specify these traits and thus of the distance separating them.

A **genetic map** is a diagram showing the relative positions of genes. Genetic maps can be constructed by ordering fragments of DNA, studying chromosomal alterations, or performing crosses to determine how frequently crossing-over occurs between pairs of genes. When the frequencies of crossing-over events in crosses are used to construct a genetic map, it is called a **recombination map.** Distances are measured in terms of the frequency of crossing-over. On such a map, one "map unit" is defined as the distance within which a cross-over event is expected to occur, on the average, in one of one hundred gametes. This unit, 1 percent recombination, is now called a **centimorgan,** in honor of Thomas Hunt Morgan.

In constructing a genetic map, scientists simultaneously monitor recombination among three or more genes located on the same chromosome. When such genes are close enough together, they do not segregate independently.

chromosomes could be seen to have exchanged segments in individuals with recombinant combinations of alleles (see Key Experiment 10.1).

As an interesting side note, Mendel never encountered linked genes in his experiments. Figure 10.16 shows the chromosomal locations of the seven genes studied by Mendel in the garden pea. As you can see, the only closely linked genes are those for plant height and pod shape, and by chance, Mendel did not study this particular pairing of traits in his experiments with dihybrid crosses. One wonders what Mendel would have made of the linkage he would surely have detected had he tested this pair of traits!

Using this technique, scientists have constructed a genetic map of the four chromosomes of the fruit fly and have shown the position of about one thousand of the genes. Scientists have also undertaken perhaps the most ambitious project ever in the history of genetics—the mapping of all twenty-three pairs of human chromosomes. Called the **Human Genome Project,** this effort seeks to pinpoint the exact location of every gene on every human chromosome. Scientists hope that this effort will lead to novel gene manipulation therapies that

KEY EXPERIMENT 10.1

2000

STERN'S DEMONSTRATION OF PHYSICAL EXCHANGE

Curt Stern monitored crossing-over between two genes—recessive carnation eye color (*car*) and dominant Bar-shaped eye (*B*)—on chromosomes with physical peculiarities that he could see under a microscope. Whenever genes recombined through crossing-over, chromosomes recombined as well. The recombination of genes reflects a physical exchange of chromosome arms.

1931

1900

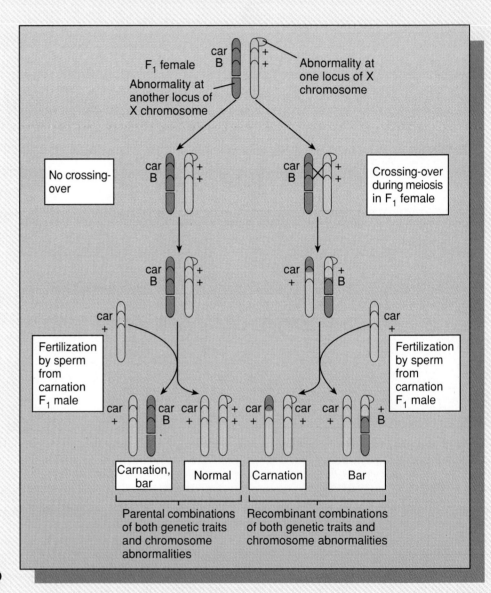

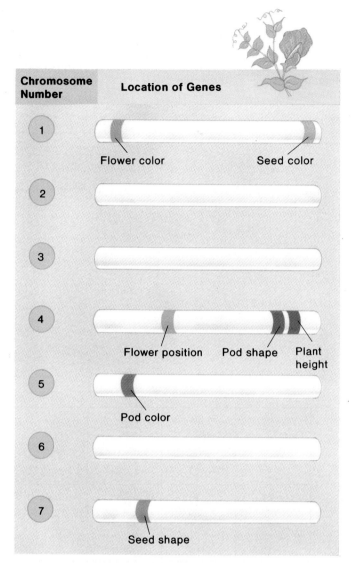

| Chromosome Number | Location of Genes |

Figure 10.16 The chromosomal locations of the seven genes studied by Mendel in the garden pea.

The genes for plant height and pod shape are very close to one another and do not recombine freely. Mendel never encountered gene linkage, however, because pod shape and plant height were not among the pairs of traits that Mendel examined in dibybrid crosses.

will cure cancer, genetic disorders such as hemophilia, and even heart disease. The Human Genome Project is discussed in more detail in chapter 11.

Genes that are located close together on chromosomes are said to be "linked" because particular combinations of their alleles tend to stay together in genetic crosses; crossing-over is less likely to occur between genes that are closer. A genetic map portraying these locations can be made by comparing how frequently the alleles of different genes recombine.

From Genotype to Phenotype: How Genes Are Expressed

Keep in mind that the "gene" of Mendelian genetics is an abstract concept that Mendel used to explain the results of crosses. Later workers determined that these elements are located on the chromosomes. In fact, as discussed in chapter 12, the Mendelian gene is no more than a segment of a DNA molecule. Genes act in ways Mendel did not understand to produce traits among progeny of crosses. How genes work is the subject of chapter 13.

The relationship between chromosomal genes and the phenotype that Mendelian traits exhibit is not always a simple one to understand, even with the extensive knowledge of DNA that scientists have today. Mendel was lucky in his choice of traits. Genes often reveal more complex patterns of inheritance than simple Mendelian 3:1 ratios. These more complex patterns may be the result of multiple alleles, epistasis, continuous variation, pleiotropy, incomplete dominance, and environmental effects.

Multiple Alleles

Although a diploid individual may possess no more than two alleles at one time, this does not mean that only two allele alternatives are possible for a given gene in the entire population. On the contrary, almost all genes exhibit several different alleles. The gene that determines the human ABO blood group, for example, has three common alleles (see chapter 11).

Epistasis

Few phenotypes are the result of the action of only one gene. Most traits reflect the action of many genes that act sequentially or jointly. When genes act sequentially, as in a biochemical pathway, an allele expressed as a defective enzyme early in the pathway blocks the flow of material through the pathway and thus makes it impossible to judge whether the later steps of the pathway are functioning properly. Such interactions between genes are the basis of the phenomenon called **epistasis.**

Epistasis is an interaction between the products of two genes in which one of them modifies the phenotypic expression produced by the other. The first example of epistasis was observed in 1918 by geneticist R. A. Emerson, who worked with particular varieties of corn. Some commercial varieties of corn contain a purple pigment called anthocyanin in the coats of their grains, which thus are purple. Other varieties do not contain the pigment, and their grains are white. Emerson crossed two true-breeding corn varieties that he thought did not contain anthocyanin in their grains and obtained a surprising result: All of the F_1 corn plants had purple grains! The two white varieties that Emerson crossed and that he believed were true-breeding, had, when crossed, produced offspring that made the purple pigment.

When Emerson crossed two of these pigment-producing F_1 plants to produce an F_2 generation, he noted that 56 percent of the F_2-generation plants were pigment-producers and 44 percent were not. What was happening? Emerson correctly deduced that *two* genes were involved in the pigment-producing process and that the second cross had thus been a dihybrid cross as described by Mendel. Recall that Mendel predicted sixteen possible genotypes in equal proportions (9 + 3 + 3 + 1 = 16), suggesting to Emerson that the total number of genotypes in his F_2 generation was also sixteen.

Emerson then set out to determine the genotypes of his F_2 corn plants. He multiplied the percentage that were pigment-producers by the total number of possible genotypes—0.56 × 16 = 9—and did the same for those plants that were not pigment-producers—0.44 × 16 = 7. Thus, Emerson had a modified ratio of 9:7 instead of the usual 9:3:3:1 ratio that Mendel predicted for the F_2 generation of a dihybrid cross. Such a modified ratio illustrates epistasis, a kind of interaction between two genes in which one of the genes changes the phenotypic expression of the other.

In this example, the pigment anthocyanin is produced from a colorless molecule by two enzymes that work one after the other. In other words, the pigment is a product of a two-step biochemical pathway (figure 10.17*a*). For pigment to be produced, a plant must possess at least one functional copy of each enzyme. If even one of the enzymes is defective, the biochemical pathway that produces the pigment comes to a halt, and no pigment is produced. The dominant alleles specify functional enzymes; the recessive alleles specify defective, nonfunctional enzymes. Of the sixteen genotypes predicted by Mendel, nine of Emerson's F_2 plants contained at least one dominant allele of both genes—these are the purple progeny. The 9:7 ratio that Emerson observed resulted from the pooling of the three phenotypic classes that lacked the dominant alleles at either or both loci (3 + 3 + 1 = 7) so that all seven looked the same: nonpigmented (figure 10.17*b*).

Continuous Variation

When multiple genes act jointly to influence a trait, such as height or weight, the contribution caused by the segregation of the alleles of one particular gene is difficult to monitor, just as following the flight of one bee within a swarm is difficult. Because all of the genes that play a role in determining phenotypes such as height or weight segregate independently of one another, there is a gradation in degree of difference when many individuals are examined (figure 10.18).

Pleiotropy

Often, an individual allele will have more than one effect on the phenotype. Such an allele is said to be **pleiotropic.** Thus, when the pioneering French geneticist Lucien Cuénot studied yellow fur in mice, a dominant trait, he was unable to obtain a true-breeding yellow strain by crossing individual yellow mice with one another—individuals that were homozygous for the yellow allele died. The yellow allele was pleiotropic: One effect was yellow color; another effect was a lethal developmental

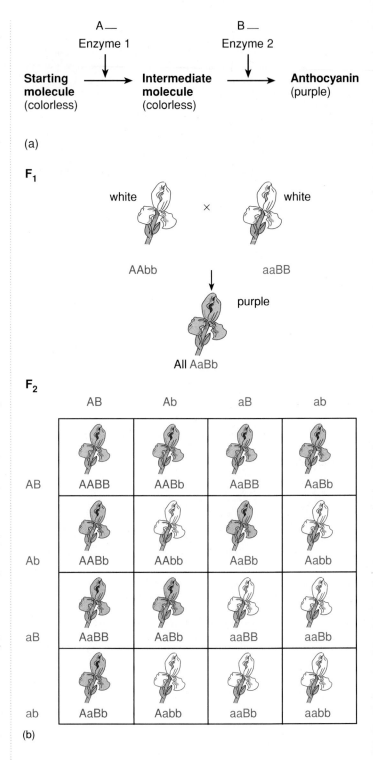

Figure 10.17 Epistasis.

(a) *The purple pigment anthocyanin found in some varieties of corn is a product of a two-step biochemical pathway. The two enzymes that govern each step are specified by two separate genes. For the enzyme to be functional, each gene must possess at least one dominant allele. The presence of the negative allele exerts an effect on the enzyme, making it dysfunctional.* (b) *Emerson's experimental design. Emerson crossed two strains of unpigmented, true-breeding corn and found that all the offspring in the F_1 generation were purple. He then allowed the F_1 generation to self-fertilize and found that the ratio of the purple varieties to the white varieties in the F_2 generation was 9:7.*

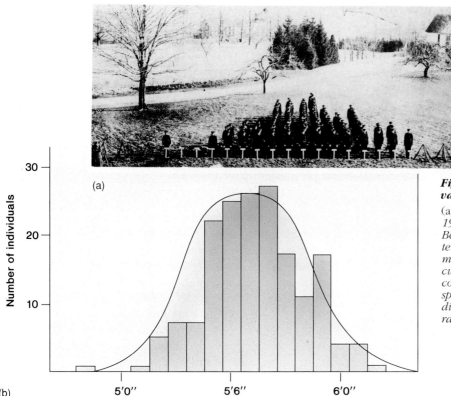

Figure 10.18 Height as a continuously varying trait.

(a) *Variation in height among students of the 1914 class of Connecticut Agricultural College. Because many genes contribute to height and tend to segregate independently of one another, many combinations are possible.* (b) *The cumulative contribution of different combinations of alleles to height forms a spectrum of possible heights—a random distribution in which the extremes are much rarer than the intermediate values.*

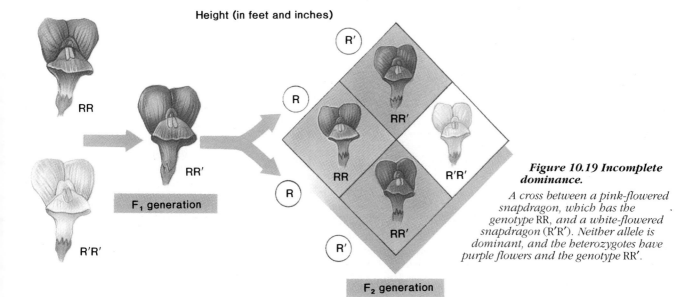

Figure 10.19 Incomplete dominance.

A cross between a pink-flowered snapdragon, which has the genotype RR, and a white-flowered snapdragon (R'R'). Neither allele is dominant, and the heterozygotes have purple flowers and the genotype RR'.

defect. A pleiotropic gene alteration may be dominant with respect to one phenotypic consequence (yellow fur) and recessive with respect to another (lethal developmental defect). Pleiotropic relationships occur because the characteristics of organisms result from the interactions of products made by genes. These products often also perform other functions about which scientists may be ignorant.

Incomplete Dominance

Not all alternative alleles are fully dominant or recessive in heterozygotes. Sometimes, heterozygous individuals do not resemble one parent precisely. Some pairs of alleles produce instead a heterozygous phenotype that (1) is intermediate between the parents (intermediate or **incomplete dominance,** figure 10.19), (2) resembles one allele closely but can be distinguished from it (partial dominance), or (3) is one in which both parental phenotypes can be distinguished in the heterozygote (co-dominance).

Environmental Effects

The degree to which many alleles are expressed depends on the environment. Some alleles encode an enzyme whose activity is more sensitive to conditions such as heat or light than are other alleles (figure 10.20).

(a)

(b)

Figure 10.20 Environmental influences on gene expression.

(a) *An arctic fox in winter has a coat that is almost white, so it is difficult to see the fox against a snowy background.* (b) *In summer, the fox's fur darkens to a red-brown, in which condition it resembles the color of the tundra over which it runs.*

While genes themselves do not interact on the chromosome, the products that they encode sometimes do. Thus, a mutation that renders the enzyme encoded by an enzyme inactive may alter the phenotype of the affected individual simply because the products that would have been produced by that enzyme are not available for other enzymes to use. The genes encoding these other enzymes have not changed, but the ability to detect their influence on the phenotype has been lost.

The Science of Genetics

After many centuries of speculation about heredity, the puzzle was finally solved within a few generations. Mendel's work and that of a generation of investigators determined to explain Mendel's results culminated in a basic outline of the mechanism of heredity. Hereditary traits are specified by genes, which are integral parts of chromosomes, and the movements of chromosomes during meiosis produce the patterns of segregation and independent assortment that Mendel reported. Two of the most important discoveries were that (1) chromosomes exchange genes during meiosis, and (2) genes located far apart on chromosomes are more likely to have an exchange occur between them. These findings allowed investigators to learn how genes are distributed on chromosomes long before they knew enough to isolate and study them.

This core of knowledge, this basic outline of heredity, has led to a long chain of investigation and questions. What is the physical nature of a gene? How is information encoded within genes? How do genes change, and why? How does a gene create a phenotype? The people who ask such questions are called geneticists, and the body of what they have learned and are learning is called genetics. Genetics is one of the most active subdisciplines of biology. The next four chapters will answer the questions just posed and many others.

EVOLUTIONARY VIEWPOINT

Eukaryotic cells contain far more genetic material than prokaryotic ones, necessitating a new way of separating copies of the genomes into daughter cells. By separating its DNA into segments and packaging the segments with proteins into discrete chromosomes, a eukaryotic cell achieves error-free segregation of daughter genomes. It is this partitioning of chromosomes that Mendel detected.

SUMMARY

1. Knight and others noted the basic facts of heredity a century before Mendel. They found that alternative traits segregate in crosses and may mask each other's appearance. Mendel, however, was the first to quantify his data, counting the numbers of each alternative type among the progeny of crosses.

2. From counting progeny types in crosses involving single traits, Mendel learned that the alternatives that were masked in hybrids appeared only 25 percent of the time when they subsequently segregated in the F_2 generation. This finding, which led directly to one of Mendel's most significant hypotheses concerning heredity, is usually referred to as the Mendelian ratio of 3:1, the ratio of dominant to recessive traits.

3. Mendel deduced from the 3:1 ratio that traits are specified by what he called discrete "factors," which do not blend. Today, Mendel's factors are called genes, and the alternative forms of his factors are referred to as alleles. Often, one number of a pair of alleles is dominant, the other recessive.

4. When two heterozygous individuals mate, an individual offspring has a 50 percent (that is, random) chance of obtaining the dominant allele from the father and a 50 percent chance of obtaining the dominant allele from the mother, so the probability of obtaining two dominant alleles is $0.5 \times 0.5 = 0.25$, or 25 percent, and the probability of obtaining two recessive alleles is the same. The remaining half of the progeny are heterozygotes; their phenotypes resemble the homozygous dominants. For this reason, the progeny thus appear as three-fourths dominant, one-fourth recessive, a ratio of 3:1 dominant to recessive.

5. The three key points in Mendel's Law of Segregation are:
(1) Alleles do not blend in heterozygotes;
(2) alleles segregate in heterozygous individuals; and (3) alleles have an equal probability of being included in either gamete.

6. Mendel's Law of Independent Assortment states that when two genes are located on nonhomologous chromosomes, the alleles included in an individual gamete are selected at random. The allele selected for one gene has no influence on which allele is selected for the other gene. Such genes are said to assort independently.

7. Thomas Hunt Morgan provided the first clear evidence that genes reside on chromosomes. Morgan demonstrated that the segregation of the white-eye trait in *Drosophila* flies was associated with the segregation of the X chromosome, the chromosome responsible for sex determination. Genetic traits located on the sex chromosomes are said to be sex-linked.

8. Genes located on the same chromosome are said to be linked. When two genes are located close together on a chromosome, the traits specified by these genes are inherited together. When genes are located far apart on a chromosome, the genes are affected by crossing-over.

9. The frequency of crossing-over between genes can be used to construct genetic maps. Such maps are representations of the physical locations of genes on chromosomes, inferred from the degree of crossing-over between particular pairs of genes.

10. Traits are often influenced by the action of several genes. Sometimes, a defect in one gene will mask the detectability of the alleles of another gene (epistasis), and the segregation of such traits often does not exhibit simple Mendelian ratios.

REVIEWING THE CHAPTER

1. What were the early ideas about heredity prior to Mendel?

2. Why did Mendel choose to work with the garden pea?

3. What was the main difference between Mendel's experimental work and the work of previous investigators?

4. From his experiments with pea plants, what did Mendel discover?

5. How did Mendel interpret his experiment results?

6. What is the post-Mendel analysis of his results?

7. What is the significance of the F_1 generation?

8. What is the significance of the F_2 generation?

9. What is the use and significance of the test cross?

10. What is the significance of Mendel's First Law?

11. What is a dihybrid?

12. What were the results of Mendel's dihybrid crosses?

13. What is the significance of Mendel's Second Law?

14. How are chromosomes the vehicles of inheritance?

15. What is the significance of Morgan's experiments with the fruit fly?

16. What is the significance of linkage and crossing-over?

17. How is a genetic map constructed, and what is its value?

18. What are different ways that genes act to produce traits?

19. What is the current status of the science of genetics?

COMPLETING YOUR UNDERSTANDING

1. What is true-breeding, and how does it apply to Mendel's experiments?

2. Why did Mendel remove the anthers (male flower parts) from some pea flowers in his experiments?

3. What is the Mendelian ratio, and how did it originate?

4. How would you apply the following terms to Mendel's experiments: *factors, heterozygous, homozygous, genes, alleles, locus?*

5. A garden pea plant that is homozygous for the recessive white-flowered trait will produce gametes of what kind?

6. What is the difference between a genotype and a phenotype?

7. The white color of a homozygous recessive plant is its

  a. phenotype.

  b. genotype.

8. How is a Punnett square used?

9. In a cross of purple- and white-flowered garden peas, Mendel observed that the F_1 offspring were all purple. A cross of two of these purple plants yielded F_2 purple- and white-flowered plants in the ratio of

_____.

10. When two heterozygous individuals are crossed, the percentage of the progeny that exhibit the recessive trait is _____.

11. When two heterozygous individuals are crossed, the proportion of the progeny that are true-breeding for the dominant trait is

_____.

12. In all of Mendel's crosses, the F_2 plants displayed a 3:1 ratio of dominant to recessive traits. Of those showing the dominant trait, what proportion were true-breeding?

  a. One-fourth

  b. One-third

  c. One-half

  d. Two-thirds

  e. All

13. Of the following, which is a test cross?

  a. $WW \times WW$ c. $Ww \times ww$

  b. $WW \times Ww$ d. $Ww \times W$

14. Which of Mendel's laws is demonstrated by the 9:3:3:1 ratio observed in the F_2 generation of a dihybrid cross?

15. What is the significance of "independent assortment" in biology, and how does it relate to meiosis?

16. Which of the following designates a male?

  a. XX   c. YX

  b. XY   d. YY

17. Morgan's experiments on white-eyed *Drosophila* clearly showed that

  a. white-eyed flies see better than normal flies.

  b. chromosomes are the carriers of genetic information.

  c. only female flies can have white eyes.

  d. diploid individuals have two copies of each trait.

  e. because eggs carry not only the genetic factors but large amounts of cytoplasm, the female has a greater influence on the genotype of the offspring.

18. The exchange of segments of chromosomes during meiosis is called

_____.

19. Under what circumstances does Mendel's Second Law (independent assortment) not hold?

20. How is a centimorgan used in genetic mapping?

21. What is the purpose of the Human Genome Project, and how might this relate to you during your lifetime?

22. How does epistasis relate to a biochemical pathway that is not functioning properly?

23. How many alleles of a given gene are possible?

    a. One   c. Three   e. Any number
    b. Two   d. Four

24. In Emerson's experiments with corn, what is the explanation for his obtaining a 9:7 ratio instead of the expected 9:3:3:1 ratio predicted for the F_2 generation in a dihybrid cross?

25. If in a biology laboratory each student is weighed and his or her height determined, what is the best explanation for the variation observed?

26. In Sweden, many more people possess blue eyes than eyes of all other colors combined. It follows that the allele determining blue eyes that occurs in Sweden is dominant to most other eye color alleles. True or false?

27. The French geneticist Cuénot studied mice. Why was he not able to develop a true-breeding population with yellow fur?

28. How can the environment influence genetic expression? Give some examples.

29. If one of your relatives was suddenly afflicted with a life-threatening disease, what approach would you use to determine if this disease was induced by heredity or the environment?

*30. Many sexually reproducing lizard species are able to generate local populations that reproduce asexually. What do you imagine the sex of the local asexual populations to be—male, female, or neuter? Explain your reasoning.

*31. The plant *Haplopappus gracilis* has only two chromosomes, whereas the adder's tongue fern has 1,262. Can you suggest a reason for this wide variation in chromosome number? There is much less variation among mammals. Can you propose an explanation?

*32. Imagine that you are constructing an artificial chromosome. What elements will you introduce into it, at a *minimum,* so that it will function normally in mitosis?

*For discussion

MENDELIAN GENETICS PROBLEMS

1. The annual plant *Haplopappus gracilis* has two pairs of chromosomes 1 and 2. In this species, the probability that two traits *a* and *b* selected at random will be on the same chromosome of *H. gracilis* is the probability that they will both be on chromosome 1 (½), multiplied by the probability that they will both be on chromosome 2 (also ½): ½ × ½ = ¼, or 25 percent. This is often symbolized:

$$\tfrac{1}{2}(\text{Number of pairs of chromosomes})$$

Human beings have twenty-three pairs of chromosomes. What is the probability that any two human traits selected at random will be on the same chromosome?

2. Among Hereford cattle, individuals with the dominant allele called *polled* lack horns. Imagine that after college, you become a cattle baron and stock your spread entirely with polled cattle. You personally make sure that each cow has no horns. Among the calves that year, however, some grow horns. Angrily, you dispose of them and make certain that no horned adult gets into your pasture. The next year, however, more horned calves are born. What is the source of your problem? What should you do to rectify it?

3. An inherited trait among Norwegians causes affected individuals to have very curly hair, not unlike that of a sheep. The trait is called *woolly.* The trait is very evident when it occurs in families; no child possesses woolly hair unless at least one parent does as well. Imagine that you are a Norwegian judge and that you have before you a woolly-haired man suing his normal-haired wife for

divorce because their first child has woolly hair but their second child has straight, long, blond hair. The husband claims that this constitutes evidence of infidelity on the part of his wife. Do you accept his claim? Justify your decision.

4. In human beings, Down syndrome, a serious developmental abnormality, results from the presence of three copies of chromosome 21, rather than the usual two copies. If a female exhibiting Down syndrome mates with a normal male, what proportion of their offspring would be expected to be affected?

5. Many animals and plants bear recessive alleles for *albinism,* a condition in which homozygous individuals completely lack any pigments. An albino plant lacks chlorophyll and is white. An albino person lacks any melanin pigment. If two normally pigmented persons heterozygous for the same albinism allele mate, what proportion of their children would be expected to be albino?

6. Your uncle dies and leaves you his race horse, Dingleberry. To obtain some money from your inheritance, you decide to put the horse out to stud. In looking over the stud book, however, you discover that Dingleberry's grandfather exhibited a rare clinical disorder that leads to brittle bones. This disorder is hereditary and results from homozygosity for a recessive allele. If Dingleberry is heterozygous for the allele, it will not be possible to use him for stud because the genetic defect may be passed on. How would you go about determining whether Dingleberry carries this allele?

7. In the fly *Drosophila,* the allele for dumpy wings (symbolized *d*) is recessive to the normal long-wing allele (symbolized *D*). The allele for white eyes (symbolized *w*) is recessive to the normal red-eyed allele (symbolized *W*). In a cross of *DDWw × Ddww,* what proportion of the offspring are expected to be normal, that is, have long wings and red eyes? What proportion are expected to have dumpy wings and white eyes?

8. In some families, children exhibit recessive traits (and therefore must be homozygous for the recessive allele specifying the trait), even though one or neither of the parents exhibits the trait. What can account for this occurrence?

9. You collect in your backyard two individuals of *Drosophila melanogaster,* one a young male and the other a young, unmated female. Both are normal in appearance, with the typical vivid red eyes of *D. melanogaster.* You keep the two flies in the same vial, where they mate. Two weeks later, hundreds of little offspring are flying around in the vial. They all have normal red eyes. From among them, you select one hundred individuals, some male and some female. You cross each individual to a fly you know to be homozygous for a recessive allele called "sepia" (*se*), which leads to black eyes when homozygous (these flies thus have the genotype *se/se*). Examining the results of your one hundred crosses, you observe that, in about half of them, only normal, red-eyed progeny flies are produced. In the other half, however, the progeny are about 50 percent red-eyed and 50 percent black-eyed. What must have been the genotypes of your original backyard flies?

FOR FURTHER READING

Blixt, S. 1975. Why didn't Gregor Mendel find linkage? *Nature* 256: 206. A classic examination of the pea strains first studied by Mendel.

Corcos, A., and F. Monaghan. 1990. Mendel's work and its rediscovery: A new perspective. *Critical Reviews in Plant Sciences* 9 (3): 197–212. An evaluation of the many myths surrounding Mendel's work.

Morgan, T. H. 1910. Sex-limited inheritance in *Drosophila. Science* 32: 120–22. Morgan's original account of his famous analysis of the inheritance of the white-eye trait.

Plumin, R. 1990. The role of inheritance in behavior. *Science* 185 (April): 183–88. Discussion of how genes affect behavior, but not in a simple Mendelian fashion.

Sutton, W. S. 1903. The chromosomes of heredity. *Biological Bulletin* 4: 213–51. The original statement of the chromosomal theory of heredity.

EXPLORATIONS

Interactive Software:
Constructing A Genetic Map

In this interactive exercise, students explore the influence of physical separation upon genetic recombination by moving the relative position of genes on a chromosome. The exercise analyzes the behavior of three alleles in a dihybrid cross. The symbols identifying the alleles are specified by the user. If recombination frequencies are entered for the eight recombinant types, the program will draw a genetic map of the three loci. If instead the user specifies the relative locations of the three genes on the chromosome, and the number of F_2 progeny to be analyzed, the program will indicate the expected number of each recombinant type among the F_2 progeny.

Questions to Explore:

1. In a 3-point cross (that is, a dihybrid cross involving three alleles), why does the sum of the two short lengths not add up to the long one (that is, for the map *a-b-c,* why do the recombination frequencies *a-b* plus *b-c* not equal *a-c*)?

2. Can you do a 3-point cross in which one parent is recessive for one gene (*a-B-C*) and the other parent recessive for the two other genes (*A-b-c*)?

3. In a 3-point cross in which two genes are close together and the third gene far away, how can you tell what side of the two genes the third gene is on [*a-b------c* or *c------a-b*]?

HUMAN GENETICS

These are the human chromosomes that determine whether a zygote will develop into a boy or girl. The larger chromosome is the X chromosome; the smaller is the Y. A zygote that is XX develops into a female, while an XY zygote develops into a male.

FOR REVIEW

Here are some important terms and concepts that have been discussed in previous chapters and that you will encounter again in this chapter. Review them before proceeding if necessary.

Cell-surface markers (*chapter 5*)
Cystic fibrosis (*chapter 5*)
Karyotypes (*chapter 9*)
Sex linkage and sex chromosomes (*chapter 10*)
Dominant and recessive traits (*chapter 10*)

The principles of genetics apply not only to pea plants and *Drosophila,* but also to you. How closely you resemble your father or mother was largely established before your birth by the chromosomes that you received from them, just as meiosis in peas determined the segregation of Mendel's traits. But many of the alleles segregating within human populations demand more serious concern than the color of a pea. Some of the most devastating human disorders result from alleles specifying defective forms of proteins that have important functions in our bodies. By studying human heredity, scientists are more able to predict which disorders parents might expect to pass on to their children, and with what probabilities (figure 11.1).

A special chapter is devoted to human heredity because, although humans pass genes to the next generation in much the same way that other organisms do, we naturally have a special

curiosity about ourselves. We know that some conditions are hereditary. If a member of our family has had a stroke, we tend to worry about our own future health because we know that the propensity to suffer strokes can be hereditary. Few parents have babies without worrying about the possibility of birth defects. Genes are also clearly involved in such conditions as adult diabetes, manic depression, and alcoholism. The ways in which genes interact with the environment to produce individuals with specific characteristics are the subject of continuing, intensive study. Because of the importance of genes in determining the course of our lives, we are all human geneticists, interested in what the laws of genetics reveal about ourselves and our families.

Human Chromosomes

Although chromosomes were discovered more than a century ago, the exact number of chromosomes that humans possess (forty-six) was not established until 1956, when new techniques for accurately determining the number and form of human and other mammalian chromosomes were developed. Biologists examine human chromosomes by collecting a blood sample, adding chemicals that induce the blood cells in the sample to divide, and then adding other chemicals that arrest cell division at metaphase. Metaphase is the stage of mitosis when the chromosomes are most condensed and thus most easily distinguished from one another. The cells are then flattened, spreading out their contents, and the individual chromosomes are separated for examination. The chromosomes are stained and photographed, and a karyotype is prepared, as discussed in chapter 9, with the photographs of the individual chromosomes arranged in order of descending size.

Of the twenty-three pairs of human chromosomes, twenty-two consist of members that are similar in size and morphology in both males and females; these chromosomes are called **autosomes.** The two members of the remaining pair—the **sex chromosomes**—are unlike each other in males and similar in females. As in *Drosophila* (see chapter 10), females are designated XX, and males are designated XY; the Y chromosome bears few functional genes. In humans, the Y chromosome is

Figure 11.1 Genetic disorders passed within families.

The last Russian czar and his family: Czar Nicholas II of Russia, his wife Alexandra, and their five children: Olga, Tatiana, Maria, Anastasia, and Alexis. Alexandra was a carrier of the genetic disorder hemophilia, a disease that results in the blood not clotting properly. She passed this disease on to her son Alexis. No one knows whether her daughters were carriers of the disease, since the entire family was killed soon after the Russian Revolution, before any of the daughters married or had children.

It sounds incredible but is true: By the twenty-first century, scientists hope to have identified and located the one hundred thousand genes contained within the twenty-three pairs of human chromosomes. Called the Human Genome Project, this effort has had its share of controversies, supporters, and critics. Knowledgeable people in the field of genetics want to know what will be done with this information and if the money necessary to fund this project is really well-spent. Others, who are perhaps not so knowledgeable, want to know why the project is taking so long.

To answer the last question first, the Human Genome Project can be likened to looking for one hundred thousand needles in forty-six haystacks. It is an evolutionary quirk that eukaryotic chromosomes contain a vast number of DNA sequences that are simply nonsense—these DNA sequences do not specify any proteins. During the transcription process, these bits of nonspecifying DNA are snipped out of the RNA transcript. Researchers working with DNA on the Human Genome Project have to wade through billions of nitrogenous base pairs to find one gene. Several solutions have been proposed to speed up this tedious process, among them an automated procedure run by robots that tests long sequences of DNA for the presence of genes. By multiplying the number of robots, one research lab in France has greatly speeded up the gene search.

Ethical questions raised about the Human Genome Project are more difficult to answer. What are researchers going to do with the information provided by the study once they have it? For example, imagine that you are applying for a job and that one of the conditions of the interview process is that you submit to an analysis of your DNA. The analysis shows that you have a gene for hypercholesterolemia, an inherited condition in which the liver does not break down cholesterol. The excess cholesterol in your blood puts you at risk for early heart attack or stroke. Your prospective employer may not want to hire someone who could have a heart attack at age thirty since it could cost the employer money in lost time and insurance benefits. Or imagine that you are applying for medical insurance and that the insurance company asks that you submit to a DNA analysis. The analysis shows that you carry the dominant gene for Huntington's disease, and the insurance company refuses to sell you insurance. What if the scientists working on the Human Genome Project discover a gene involved in determining intelligence? Will everyone be tested for this gene? Will those who do not have the correct genes be eliminated?

The Human Genome Project raises these ethical questions and many more. While the project is an exciting scientific journey and offers full knowledge of the genetic basis of human life, the ethical questions that go along with mapping the entire human genome must be anticipated and decisions made about how the information will be used.

---

much smaller than the X chromosome. The DNA of all twenty-three chromosomes is being sequenced in the Human Genome Project, an attempt to fully delineate all the genes of the human genome (see Sidelight 11.1).

Defects in Chromosome Number: Mistakes in Meiosis

Many human genetic abnormalities are a result of a condition called **aneuploidy,** or an abnormal number of chromosomes. Aneuploidy results from mistakes made during the meiotic division of the parental gametes. As discussed in chapter 9, during meiosis I, the homologous pairs of chromosomes line up on the metaphase plate and then separate. The homologues then proceed toward opposite poles of the cell. Sometimes, however, the homologous pairs remain stuck together and do not separate, resulting in gametes with an abnormal number of chromosomes. In another scenario, the sister chromatids sometimes do not separate from each other in meiosis II. The result is the same: The gametes end up with an abnormal number of chromosomes. The failure of chromosomes to separate correctly during either stage of meiosis is called **nondisjunction.**

The cause of nondisjunction is not known, but its incidence seems to increase with age, and occurs much more often in women than in men. Why is this so? The answer lies in the fact that all the eggs that a woman will ever produce are present in her ovaries at the time of her birth. Over the course of an average reproductive lifetime—say, between the ages of twelve and fifty—these eggs age, and problems of various kinds, including nondisjunction, are much more likely to accumulate over time. In contrast, men continually make new sperm cells. Each sperm cell in the male ejaculate is probably only a few days old. For this reason, the mother's age is much more critical than that of the father when a couple contemplates having children.

The offspring that result from the union of a gamete with a normal number of chromosomes and a gamete with an abnormal number of chromosomes often display severe physical and mental disabilities. Individuals who are missing one autosome, called **monosomics,** do not survive embryonic development and are usually spontaneously aborted. In all but a few cases, those who have received an extra autosome, called **trisomics,** also do not survive. Five of the smallest chromosomes—those numbered 13, 15, 18, 21, and 22—can be present in human beings as three copies and still allow the individual to survive for a time. Individuals with an extra chromosome 13, 15, or 18 have severe developmental defects and usually die within three months. In contrast, individuals who have an extra copy of chromosome 21 (a condition known as trisomy 21, or Down syndrome), and more rarely, those who have an extra copy of chromosome 22, usually survive to adulthood.

Down Syndrome

The developmental defect produced by trisomy 21 was first described in 1866 by J. Langdon Down; thus, it is called **Down syndrome** (formerly "Down's syndrome") (figure 11.2). Down syndrome occurs frequently in all human racial groups, with an approximate incidence of one in every 750 children. Similar conditions also appear in chimpanzees and other related primates. In humans, the defect is associated with a particular, small portion of chromosome 21. When this chromosomal segment is present in three copies instead of two, Down syndrome results. In 97 percent of the human cases examined, all of chromosome 21 is present in three copies. In the other 3

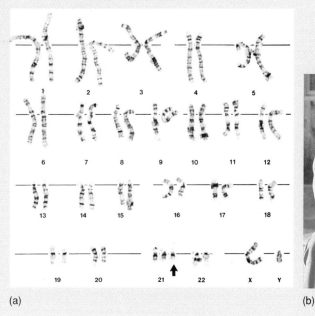

(a)

(b)

Figure 11.2 Down syndrome.

(a) *In this karyotype of the chromosomes of a male child with Down syndrome, the trisomy at position 21 can be clearly seen (arrow). (b) This child shows the physical effects of Down syndrome. Many individuals with Down syndrome can lead productive lives if they receive adequate help and support.*

percent, a small part of chromosome 21 containing the critical segment has been added to another chromosome, in addition to the normal two copies of chromosome 21.

The developmental role of the genes whose duplication produces Down syndrome is not known in detail, although clues are beginning to emerge. When human genes with alleles associated with some forms of cancer (see chapter 13) were identified and localized on the chromosomes in 1985, one of them turned out to be located on chromosome 21 at precisely the location of the segment associated with Down syndrome. Cancer is indeed more common in children with Down syndrome; for example, the incidence of leukemia is eleven times higher in these children.

Individuals with Down syndrome display a spectrum of physical and mental defects, the most obvious of which are mental impairment and a typically short stature. They also have stubby hands and feet and a characteristic heavy eye fold that gives them a somewhat Asian appearance (in fact, individuals with Down syndrome used to be called "mongoloids"). Persons with Down syndrome also have a heart defect that can be surgically corrected. The outlook for these individuals, especially with the recent advent of special programs and employment opportunities for the mentally challenged, is quite good, and many individuals with Down syndrome lead independent, productive lives.

The nondisjunction that causes trisomy 21 occurs more often in women over age thirty. In mothers less than twenty years of age, the incidence of giving birth to a child with Down syndrome is only about one in 1,700 births; in mothers twenty to thirty years old, the risk is only slightly greater, about one in 1,400. In mothers thirty to thirty-five years old, however, the risk doubles, to one in 750. In mothers older than forty-five, the risk is as high as one in 16 births (figure 11.3). Ways in which prospective parents can actually test the genetic composition of their fetus to determine whether the fetus has a nor-

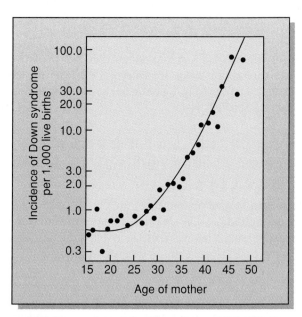

Figure 11.3 Increasing risk of bearing a child with Down syndrome.

As women age, the chance that they will bear a child with Down syndrome increases. After age thirty-five, the frequency of nondisjunction of chromosome 21 increases rapidly.

mal number of chromosomes are discussed later in the chapter. These types of tests have allowed parents to decide whether they are prepared to care for a child with this and other types of genetic defects.

Abnormal Numbers of Sex Chromosomes

As mentioned earlier, humans possess twenty-two pairs of autosomes that are the same in both sexes and one pair of sex chromosomes that differ between males and females. Females are designated XX because they have two sex chromosomes that are similar in size and appearance, although not necessarily in genetic content. Males are designated XY because their sex chromosomes differ, as shown in figure 11.2a. The Y chromosome is highly condensed, and few of its genes are used. For this reason, recessive alleles on the single X chromosome of males have no counterpart on the Y chromosome, or at least no active counterpart. Thus, the characteristics of such recessive alleles are often expressed, just as if they were present in a homozygous condition in a female.

The Y chromosome is not completely inert genetically, however; it does possess some active genes. For example, the genes that cause a zygote to develop into a male are located on the Y chromosome: Any individual with at least one Y chromosome is a male, and any individual with no Y chromosome is a female. The *number* of X chromosomes an individual possesses does not determine the individual's sex.

Even though a female has two copies of the X chromosome and a male only has one, female cells do not produce twice as much of the proteins encoded by genes on the X chromosome. In females, one of the X chromosomes is inactivated shortly after sex determination, early in embryonic development. Such inactivation is called **Lyonization.** The inactivated chromosome can be seen as a deeply staining body, the **Barr body,** which remains attached to the nuclear membrane.

Variation in the Numbers of X Chromosomes
Individuals who lose a copy of the X chromosome or gain an extra one are not subject to the severe developmental abnormalities usually associated with similar changes in autosomes. These individuals may mature, although they have somewhat abnormal features.

The occasional failure of the sister chromatids of the X chromosome to separate during meiosis II yields some gametes that are XX and others that have no sex chromosome (designated O). If the XX gamete joins an X gamete, it forms an XXX zygote that develops into a female individual with one functional X chromosome. She is sterile but usually normal in other respects. If the XX gamete instead joins a Y gamete, the resulting XXY zygote develops into a sterile male with many female body characteristics and, in some cases, diminished mental capacity. This condition, called **Klinefelter syndrome,** occurs in about one out of every thousand male births.

If an O gamete, produced when the X chromosomes fail to separate, fuses with a Y gamete, the resulting OY zygote is nonviable because it cannot survive without any of the genes on the X chromosome. If, on the other hand, the O gamete fuses with an X gamete, the resulting XO zygote develops into a sterile female with short stature, a webbed neck, sex organs that resemble those of an infant, and low to normal mental abilities. This condition, called **Turner syndrome,** occurs approximately once in every two thousand female births. Figure 11.4 diagrams the ways in which nondisjunction can result in abnormalities in the number of sex chromosomes.

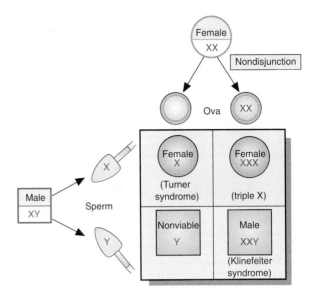

Figure 11.4 Nondisjunction and sex chromosome number abnormalities.
Nondisjunction in the female gametes can result in abnormalities in the number of sex chromosomes.

Variation in the Numbers of Y Chromosomes
The sister chromatids of the Y chromosome may also fail to separate during meiosis II. This leads to the formation of YY gametes and viable XYY zygotes, which develop into fertile males of normal appearance. The frequency of XYY among newborn males is about one per thousand. Interestingly, the frequency of XYY males in penal and mental institutions has been reported to be approximately 2 percent (that is, twenty per thousand), which is about twenty times the frequency of such individuals in the population at large. This observation has led to the controversial suggestion that XYY males may be inherently antisocial. The observation is confirmed in some studies but not in others. In any case, most XYY males do not appear to develop patterns of antisocial behavior.

When X or Y chromosomes fail to separate during meiosis, so-called "nondisjunction," gametes may be produced with two or no versions of the sex chromosomes. Such gametes produce zygotes with abnormal numbers of sex chromosomes, often with significant effects on the phenotype.

Studying Patterns of Inheritance
Imagine that you wanted to learn about an inherited trait present in your family. How would you find out if the trait is dominant or recessive, how many genes contribute to it, and how likely you might be to transmit it to your future children? If you

(a)

Figure 11.5 A pedigree of albinism.
(a) *One of three girls from a Hopi Indian family (the left-most family in generation IV of the pedigree in* b.*) is albino.*
(b) *In pedigrees, males are conventionally shown as squares, females as circles, and marriages as horizontal lines connecting them, with offspring shown below. The individuals who exhibit the trait being considered are indicated by solid symbols.*

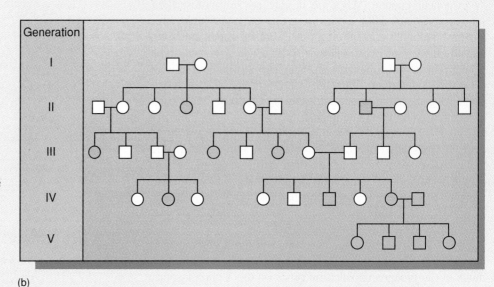

(b)

wanted to study such a trait in *Drosophila*, you could conduct crosses and examine the chromosomes of offspring. Studying your own heredity requires a less direct approach.

Analyzing Pedigrees: Albinism and Color Blindness

To study human heredity, biologists look at the results of crosses that have already been made. They study family trees, called **pedigrees,** to identify which relatives exhibit a trait. Then they can often determine whether the gene producing the trait is sex-linked or autosomal and whether the trait's phenotype is dominant or recessive. Frequently, they can infer which individuals are homozygous and which are heterozygous for the allele specifying the trait.

Albino individuals lack all pigmentation; their hair and skin are completely white (figure 11.5a). In the United States, about one in thirty-eight thousand Caucasians and one in twenty-two thousand African-Americans are albinos. In the pedigree of albinism presented in figure 11.5b, each symbol represents one individual in the family history, with the circles representing females and the squares, males. In such a pedigree, individuals that exhibit a trait being studied—in this case, albinism—are indicated by solid symbols. Marriages are represented by horizontal lines connecting a circle and a square, from which a cluster of vertical lines indicate the children, arranged from left to right in order of their birth.

The pedigree in figure 11.5b can be analyzed by asking three questions:

1. Is albinism sex-linked or autosomal? If the trait is sex-linked, it is usually seen only in males; if it is autosomal, it appears in both sexes fairly equally. In figure 11.5b, the proportion of affected males (4/12, or 33 percent) is reasonably similar to the proportion of affected females (8/19, or 42 percent). (When counting the numbers of affected individuals in a pedigree, exclude the parents in generation I, as well as any "outsiders" who marry into the family.) Thus, the trait is autosomal.

2. Is albinism dominant or recessive? If the trait is dominant, every albino child will have an albino parent; if recessive, an albino child's parents can appear normal, since both parents may be heterozygous. In figure 11.5b, parents of most of the albino children do not exhibit the trait, which indicates that albinism is recessive. Four children in one family *do* have albino parents because the allele is very common among the Hopi Indians, from which this pedigree was derived. Thus, homozygous individuals, such as these albino parents, are present among the Hopis in sufficient numbers that they sometimes marry. In this example, *both* parents are albino and *all* four children are albino, which is consistent with the finding that the trait is recessive, since both parents must be homozygous for this allele.

3. Is the albinism trait determined by a single gene or by several? If the trait is determined by a single gene, then a ratio of 3:1 (normal to albino) offspring should be born to heterozygous parents, reflecting Mendelian segregation in a cross. Thus, about 25 percent of the children should be albinos. Conversely, if the trait is determined by several genes, the proportion of albinos would be much lower, only a few percent. In this case, 8/28 (do not count the four children of the marriage between the two homozygous individuals because this is not a cross between heterozygotes), or approximately 30 percent, of the children are albinos, strongly suggesting that only one gene is segregating in these crosses.

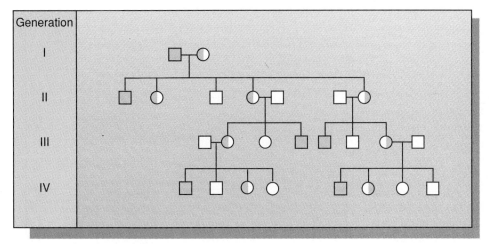

Figure 11.6 Color blindness.
In this pedigree, the half-filled symbols indicate a heterozygous individual who carries the trait governing a person's ability to perceive red and green, but does not express it.

The preceding pedigree analysis shows that albinism is an autosomal recessive trait controlled by a single gene. Other traits are studied in a similar way. For example, **red-green color blindness** is a rare, inherited trait in humans. In the pedigree shown in figure 11.6, a color-blind man has five children with a woman who is heterozygous for the allele. The solid-color symbols indicate a color-blind individual; half-filled symbols indicate a heterozygous individual who carries the trait but does not express it. The same three questions as before allow analysis of this pedigree:

1. Is red-green color blindness sex-linked or autosomal? Of the five affected individuals, all are male. The trait is clearly sex-linked.

2. Is red-green color blindness dominant or recessive? If the trait is dominant, then every color-blind child should have a color-blind parent. In this pedigree, however, that is not true in any family after that of the original male. The trait must be recessive.

3. Is the trait determined by a single gene? If it is, then children born to heterozygous parents should be color-blind in about 25 percent of the cases, reflecting a 3:1 Mendelian segregation of the trait. In this case, 4/14, or 28 percent, of the children of heterozygous parents are color-blind, indicating that a single gene is segregating. (Do not count the five children of the generation I parents because the father in this case is homozygous for the trait.)

A pedigree analysis infers the genetic nature of a trait from the pattern of trait inheritance.

Multiple Alleles

Mendel studied pairs of contrasting traits in pea plants. His plants were either tall or short, and their flowers were either purple or white. Similarly, Morgan's *Drosophila* flies had eyes that were either white or red. Many human genes also exhibit two alternative alleles. For example, as we have seen, an individual may be either albino or pigmented. But many genes possess more than two possible alleles. One such gene encodes an enzyme that adds sugar molecules to lipids on the surface of human blood cells. These sugars act as recognition markers in the human immune system and are called **cell-surface antigens.** The gene encoding the enzyme is designated I and possesses three common alleles: (1) allele I^B, which adds the sugar galactose; (2) allele I^A, which adds a modified form of the sugar, galactosamine; and (3) allele i, which does not add a sugar.

When more than one of these alleles occurs, which is dominant? Often, no one allele is dominant. Instead, each has its own effect. Thus, an individual heterozygous for the I^A and I^B alleles of the I gene produces both forms of the enzyme and adds both galactose and galactosamine to lipids on the cell surface. The cell surfaces of this individual's blood cells thus possess antigens with both kinds of sugar attached to them. Because both alleles are expressed simultaneously in heterozygotes, the I^A and I^B alleles are said to be **co-dominant.** Either is dominant over the i allele because, in heterozygotes, the I^A or I^B allele leads to sugar addition and the i allele does not.

Many genes possess multiple alleles, several of which may be common within populations.

ABO Blood Groups

Different combinations of the three possible I gene alleles occur in different individuals because each person possesses two copies of the chromosome bearing the I gene and may be homozygous for any allele or heterozygous for any two. The different combinations of the three alleles result in four different phenotypes:

1. Persons who add only galactosamine are called **type A** individuals (either $I^A I^A$ homozygotes or $I^A I^i$ heterozygotes).

2. Persons who add only galactose are called **type B** individuals (either $I^B I^B$ homozygotes or $I^B I^i$ heterozygotes).

3. Persons who add both sugars are called **type AB** individuals ($I^A I^B$ heterozygotes).

4. Persons who add neither sugar are called **type O** individuals ($I^i I^i$ homozygotes).

These four different cell-surface phenotypes are called the ABO blood groups or, less often, the Landsteiner blood groups, after the man who first described them. As Landsteiner first noted, the human immune system can tell the difference between these four phenotypes. If a type A individual receives a transfusion of type B blood, the recipient's immune system will recognize that the type B blood cells possess a "foreign" antigen (galactose) and will attack the donated blood cells. The donated type B blood cells will clump together under the attack, a reaction called **agglutination.** Agglutination can cause blood clots to form in the recipient's blood vessels and may lead to severe and lethal consequences, such as a stroke. If the donated blood is type AB, this will also happen. However, if the donated blood is type O, no attack will occur because no foreign galactose antigens are present on the surfaces of blood cells produced to the type O donor. In general, any individual's immune system will tolerate a transfusion of type O blood, so type O individuals are called **universal donors.** Because neither galactose nor galactosamine is foreign to type AB individuals (they add both to their red blood cells), they may receive any type of blood and are called **universal recipients** (figure 11.7).

In human populations, some of the ABO blood group phenotypes are more common than others (table 11.1). In general, type O individuals are the most common, and type AB individuals the least common. However, human populations differ greatly from one another. As shown in the table, among North American Indians, the frequency of type A individuals is 31 percent, whereas among South American Indians, it is only 4 percent. Figure 11.8 illustrates the frequency with which the I^B allele occurs in different parts of the world. This sort of genetic variation is an important property of genes in human and most other populations.

The Rh Blood Group

Another set of cell-surface antigens on human red blood cells are the **Rh blood group** antigens, named for the rhesus monkey in which they were first described. About 85 percent of adult humans have the Rh cell-surface marker on their red blood cells and so are called Rh-positive (Rh⁺). Rh-negative (Rh⁻) persons lack this cell-surface marker because they are homozygous recessive for the gene encoding it.

If an Rh⁻ person is exposed to Rh⁺ blood, the Rh surface antigens of that blood are treated like foreign invaders by the Rh⁻ person's immune system, which proceeds to make antibodies directed against the Rh antigens. This most commonly happens when an Rh⁻ woman gives birth to an Rh⁺ child (the father being Rh⁺). Some fetal red blood cells cross the placental barrier and enter the mother's bloodstream, where they induce the mother's production of "anti-Rh" antibodies. These antibodies persist for many years in the mother's bloodstream, and in later pregnancies, can attack a new fetus and cause its red blood cells to clump, a potentially fatal condition called **erythroblastosis fetalis.**

Prospective parents should know their Rh blood type because therapy is available for couples comprising an Rh⁺ male and an Rh⁻ female. In such cases, the mother is injected with antibodies directed against the Rh blood group antigen (called Rho gammaglobulins, or **RhoGam**) at the birth of the first child

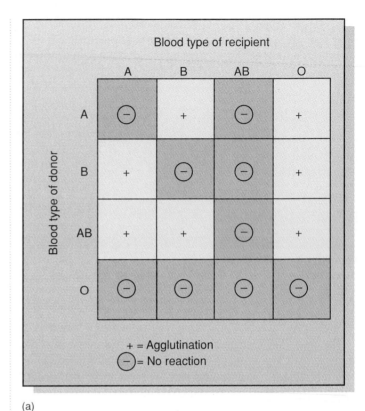

(a)

(b)

Figure 11.7
The agglutination reaction.
When a person receives blood that is not compatible with his or her blood type, clumps form that can lead to dangerous consequences. (a) *Notice the combinations in which agglutination of blood cells will occur. Notice that persons with type O blood can donate blood to anyone with any blood type and that type AB individuals can receive any blood type.* (b) *Hypothetical cross between a type AB mother and a type O father.*

T a b l e 11.1   Distribution of ABO Blood Groups in Some Human Populations

Population	Phenotype Frequency (%)			
	A	**B**	**AB**	**O**
U.S. Caucasians	39.7	10.6	3.4	46.3
U.S. African-Americans	26.5	20.1	4.3	49.1
Africans (Bantu)	25.0	19.7	3.7	51.7
North American Indians (Navaho)	30.6	0.2	0.0	69.1
South American Indians (Ecuador)	4.0	1.5	0.1	94.4
Japanese	38.4	21.9	9.7	30.1
Russians	34.6	24.2	7.2	34.0
French	45.6	8.3	3.3	42.7

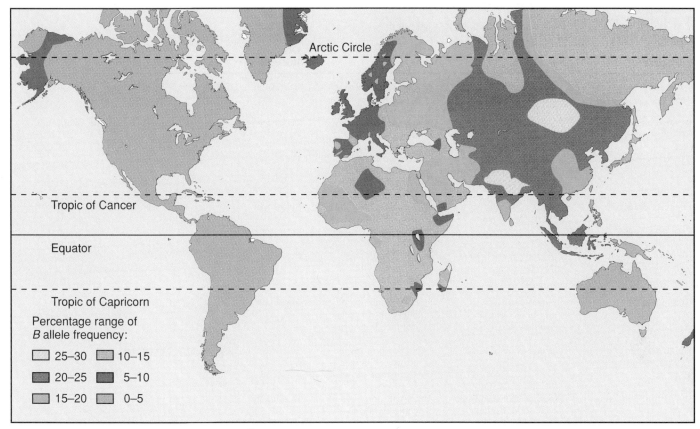

Figure 11.8 The frequency with which the I^B allele of the ABO blood group occurs throughout the world.

By studying the distribution pattern of such alleles, biologists can often trace past human migrations. The I^B allele is thought to have spread out from central Asia.

and after the birth of each successive Rh^+ child. These antibodies bind to the Rh antigens on any fetal blood cells that enter the mother's circulation, covering them so that the mother's immune system does not detect the Rh antigens and make antibodies directed toward them. A future fetus is safe because there are no anti-Rh antibodies in the mother's blood waiting to attack it.

Genetic Disorders

Humans differ greatly from one another genetically, and yet, for most loci, one allele is much more common than its alternatives. For most genes, variant alleles are rare, although there are several loci—such as those determining the blood types just discussed—for which two or more alleles are common among humans. But why should one allele of most genes be overwhelmingly the most common?

Part of the answer seems to lie in the fact that the proteins specified by most genes must function in a very precise fashion to support the many complex processes of development and to regulate bodily functions. Genes encode the sequence of amino acids in proteins, and proteins orchestrate all of the body's biochemical reactions. A defective allele in a gene may lead to the production of proteins that do not function properly, if they function at all, resulting in physical and mental abnormalities. Chapter 13 discusses the process, called mutation, that is responsible for the production of such alternative alleles.

Mutation is a random process. Random changing of a gene rarely improves the function of the encoded protein any more than randomly changing a wire in a computer is likely to improve the computer's functioning—in fact, this change will probably cause the computer to malfunction. On the other hand, without mutation, there would be no variation, no possibility for natural selection, and consequently, no evolutionary change. Thus, mutation is necessary for progressive change and adaptation, even though most individual mutations are harmful to the organisms that possess them.

Alternative alleles with detrimental effects, called genetic disorders, usually are rare in human populations. Sometimes, however, a detrimental allele can become common, as in small, isolated communities, where an individual with the allele is one of the few members. A common allele that results in unfavorable characteristics can have disastrous effects on the group of humans in which it occurs.

Most alleles that lead to the production of abnormal phenotypes are recessive. Such recessive alleles are more easily retained in a population than are dominant ones because they are not expressed in most individuals (heterozygotes) in which they occur. In populations where such alleles occur at high frequencies, the chance of two seemingly normal, heterozygous individuals mating and producing abnormal, homozygous

Table 11.2   Some Important Genetic Disorders

Disorder	Symptom	Defect	Dominant/ Recessive	Frequency among Human Births
Cystic fibrosis	Mucus clogging lungs, liver, and pancreas	Failure of chloride ion transport mechanism	Recessive	1 in 1,800 (U.S. Caucasians)
Sickle-cell anemia	Poor blood circulation	Abnormal hemoglobin molecules	Recessive	1 in 1,600 (African-Americans)
Tay-Sachs disease	Deterioration of central nervous system while person is young	Defective form of enzyme hexosaminidase A	Recessive	1 in 1,600 (Jews)
Phenylketonuria	Failure of brain to develop in infants	Defective form of enzyme phenylalanine hydroxylase	Recessive	1 in 18,000
Hemophilia (Royal)	Failure of blood to clot	Defective form of blood-clotting factor IX	Sex-linked recessive	1 in 7,000
Huntington's disease	Gradual deterioration of brain tissue in middle age	Production of an inhibitor of brain cell metabolism	Dominant	1 in 10,000
Muscular dystrophy (Duchenne)	Wasting away of muscles	Degradation of myelin coating of nerves stimulating muscles	Sex-linked recessive	1 in 10,000
Hypercholesterolemia	Excessive cholesterol levels in blood, leading to heart disease	Abnormal form of cholesterol cell-surface receptor	Dominant	1 in 500

individuals as a quarter of their progeny is correspondingly great. Learning how to avoid such tragedies is one of the principal goals of human genetics. Table 11.2 lists some of the most important human genetic disorders. Scientists know a great deal about some of them, but much less about many others.

> *The harmful effect produced by a detrimental allele is called a* **genetic disorder.**

Autosomal-Recessive Genetic Disorders

Genetic disorders that are autosomal (that is, they appear in both sexes fairly equally) and linked to a recessive allele include cystic fibrosis, sickle-cell anemia, Tay-Sachs disease, and phenylketonuria.

Cystic Fibrosis

As mentioned in chapter 5, **cystic fibrosis** is the most common fatal genetic disorder among Caucasians, among whom about one in twenty individuals is a carrier—that is, these individuals have a copy of the defective gene but show no symptoms. Homozygous recessive individuals account for about one in eighteen hundred children. In individuals with cystic fibrosis, the membrane channels that normally transport chloride ions into cells do not function, and water is prevented from passing

from their bloodstream into the passages of their lungs. As a result, the mucus that is a normal component of the lung's inner surface becomes too thick. This mucus clogs the airways of the lungs and the passages of the pancreas and liver. Individuals with cystic fibrosis inevitably die at a young age, typically before reaching adulthood (figure 11.9). Fortunately, the genetic and cellular bases of cystic fibrosis are now well understood.

Cystic fibrosis occurs when an individual is homozygous for the allele that encodes a defective version of the protein regulating the chloride-transport channel. This allele is recessive to the allele that encodes the normal-functioning version of the regulating protein. Thus, the chloride channels of heterozygous individuals function normally, and such persons do not develop cystic fibrosis.

Recently, great progress has been made in the genetic arena of cystic fibrosis research. Scientists have pinpointed the gene that specifies the protein regulating the chloride-transport channel and have identified the defective allele. They are now trying to "cure" this genetic disorder by introducing the "correct" version of the allele directly into the cells of the lungs. The correct allele is being inserted into a cold virus, using the techniques of genetic engineering. The cystic fibrosis patient inhales this virus, which carries the correct allele into the lungs. Preliminary results indicate that, once it is in the lungs, the correct allele functions normally, producing the protein needed to regulate

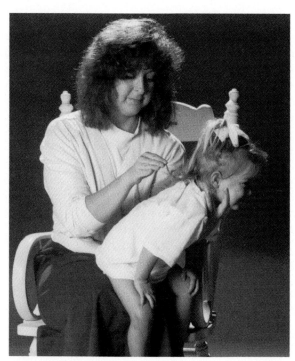

Figure 11.9 A child with cystic fibrosis.
This girl has a good chance for a normal childhood, thanks to modern therapies.

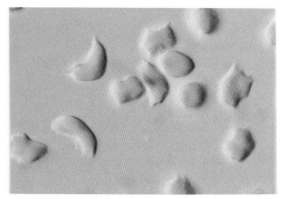

Figure 11.10 "Sickled" red blood cells.
In individuals who are homozygous for the sickle-cell trait, many of the red blood cells have such shapes.

the chloride-transport channel. Experimental trials of this novel procedure, successfully conducted with monkeys, are now underway with human subjects, and scientists are hopeful that a cure for this devastating genetic disorder will soon be available.

Sickle-Cell Anemia

Sickle-cell anemia is a genetic disorder in which the affected individuals cannot transport oxygen to their tissues properly because **hemoglobin** proteins, the molecules within red blood cells that carry oxygen, are defective. When oxygen is scarce, these defective hemoglobin molecules combine with one another, forming stiff, rodlike structures. Surprisingly, the defective hemoglobin that causes sickle-cell anemia differs from normal hemoglobin in only one out of a total of about three hundred amino acid molecules. The defect involves one molecule of valine occurring in place of the glutamic acid located in the same position in normal hemoglobin.

Red blood cells with large proportions of defective molecules become sickle-shaped and stiff (figure 11.10); normal red blood cells are disk-shaped and much more flexible. Because of their stiffness and irregular shape, sickle-shaped red blood cells have trouble moving through the smallest blood vessels and also tend to accumulate in the blood vessels, forming clots. Individuals with a large percentage of sickle-shaped red blood cells tend to have intermittent illness and far shorter life spans than those with normal red blood cells.

Individuals homozygous for the sickle-cell allele exhibit the characteristics just mentioned; those who are heterozygous for the allele are generally indistinguishable from normal persons. But in people who are heterozygous for this trait, some of the red blood cells show the sickling characteristic when exposed to low levels of oxygen. The allele responsible for the sickle-cell characteristic is particularly common among individuals of African descent. In the United States, about one in eleven African-Americans is heterozygous for this allele, and about one in five hundred is homozygous and therefore has sickle-cell anemia. In some groups of people in Africa, up to 45 percent of the individuals are heterozygous for this allele.

Individuals with sickle-cell anemia usually die before they are old enough to reproduce. Why, then, has the sickle-cell allele not been eliminated from all populations, rather than being maintained at high levels in some? The answer has proven much easier to find than has the parallel question regarding cystic fibrosis. People who are heterozygous for the sickle-cell allele are much less susceptible to falciparum malaria, which is one of the leading causes of illness and death, especially among young children, in the areas where the allele is common. Thus, the sickle-cell allele is maintained at high levels in populations where falciparum malaria is common. In addition, for reasons not understood, women who are heterozygous for this allele are more fertile than are those who lack it.

Tay-Sachs Disease

Tay-Sachs disease is an incurable genetic disorder in which the brain deteriorates. Homozygous individuals lack an enzyme necessary to break down a special class of lipids called **gangliosides,** which occur within the lysosomes of brain cells. As a result, the lysosomes fill with gangliosides, swell, and eventually burst, releasing oxidative enzymes that kill the brain cells.

Affected children appear normal at birth and usually do not develop symptoms until about the eighth month, when signs of mental deterioration become evident. Within a year after birth, affected children are blind; they rarely live beyond age five (figure 11.11). There is no known cure.

Tay-Sachs disease is rare in most human populations, occurring in one in three hundred thousand births. However, among Jews of Eastern and Central Europe (Ashkenazim) and among American Jews (90 percent of whom are descendants of Eastern and Central European ancestors), its incidence is approximately

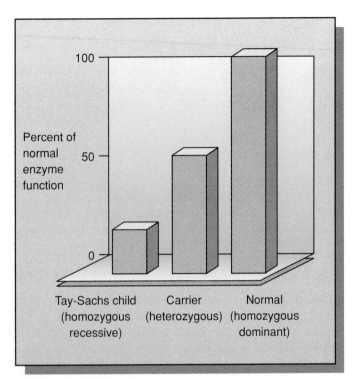

Figure 11.11 Tay-Sachs disease.
Tay-Sachs disease is a genetic disorder in which an enzyme critical to lipid metabolism does not function, leading to harmful accumulations of lipids in the lysosomes of brain cells. Homozygous individuals typically have less than 10 percent of normal levels of the enzyme, whereas heterozygous individuals have about 50 percent of normal levels, enough to prevent deterioration of the central nervous system.

one in thirty-six hundred. Because it is a recessive condition, most people who carry the defective allele do not themselves develop the characteristic symptoms. An estimated one in twenty-eight individuals in these Jewish populations is a heterozygous carrier of the allele.

Phenylketonuria

Many other hereditary disorders are not as common as cystic fibrosis in Caucasians, sickle-cell anemia in Africans, or Tay-Sachs disease in Central and Eastern European Jews. A good example of a relatively infrequent genetic disorder is **phenylketonuria (PKU),** a condition in which affected individuals cannot metabolize the amino acid phenylalanine normally.

Phenylalanine is necessary for the synthesis of many proteins and thus is an essential ingredient in human diets. Before phenylalanine can be used in protein synthesis, however, it must be converted into another form of amino acid called tyrosine. Individuals with PKU lack sufficient enzymatic activity to convert phenylalanine to tyrosine. Most humans obtain the tyrosine needed for protein synthesis through the metabolism of phenylalanine and do not require tyrosine in their diets. For PKU individuals, however, tyrosine becomes an essential dietary component.

If individuals with PKU consume more than the amount of phenylalanine required for protein synthesis, the excess

phenylalanine is converted by alternative metabolic pathways into harmful molecules called phenylketones. Phenylketones cause abnormal development of brain cells and are thus associated with the extreme mental retardation seen in untreated cases of PKU. Untreated PKU individuals rarely live beyond age thirty. Scientists are now able to design diets that are low in phenylalanine for individuals with PKU, provided that the condition is detected early enough to avoid the deleterious consequences.

Phenylalanine is not harmful to adults with PKU, presumably because their brain cells have already completed their development. Consequently, afflicted individuals who receive an appropriate diet when young mature into normal, healthy adults.

PKU is a recessive disorder caused by a defective allele of the gene encoding the enzyme that normally breaks down phenylalanine. Only individuals homozygous for the mutant allele (in the United States, about one in every fifteen thousand infants) develop the disorder.

Most genetic disorders involve autosomal recessive alleles. The result of random mutations, most such alleles are rare in human populations, although a few are common.

A Sex-Linked Recessive Genetic Disorder: Hemophilia

Hemophilia is a genetic disorder in which the blood is slow to clot or does not clot at all. When a normal person cuts his or her finger, the blood in the immediate area of the cut solidifies into a blood clot and seals the cut. A blood clot forms from the meshing together of several kinds of protein fibers that circulate in the blood. A mutation that results in the loss of activity of any of the necessary proteins leads to a form of hemophilia.

Hemophilia is a recessive disorder, expressed only when an individual does not possess at least one copy of the gene that is normal and so cannot produce one of the proteins necessary for clotting. Individuals homozygous for a defective allele do not produce any active version of the affected clotting protein and thus cannot clot blood. Most of the dozen protein-clotting genes are on autosomes. However, two of them (designated factor VIII and factor IX) are located on the X chromosome. Any male who inherits a defective allele for factor VIII or factor IX will develop hemophilia because his other sex chromosome is the Y (which is not expressed), and so he lacks a functioning allele of the protein-clotting gene.

A mutation factor IX occurred in one of the parents of Queen Victoria of England (1819–1901) (figure 11.12a). In the five generations since Queen Victoria, ten of her male descendants have had hemophilia. The British royal family escaped the disorder, often called the **Royal hemophilia,** because Queen Victoria's son King Edward VII did not inherit the defective factor IX allele. Three of Victoria's nine children, however, carried the defective allele by marriage into many of the royal families of Europe (figure 11.12b). It is still being transmitted to

(a)

Figure 11.12 The Royal hemophilia.
(a) *In this photograph taken in 1894, Queen Victoria of England is surrounded by some of her descendants. Of Victoria's four daughters who lived to bear children, two—Alice and Beatrice—were carriers of hemophilia. Two of Alice's daughters are standing behind Victoria (and both are wearing feathered boas): To the right is Princess Irene of Prussia and to the left is Alexandra, who would soon become Czarina of Russia. Both Irene and Alexandra were also carriers of hemophilia. Irene carried it into the Prussian royal house, where it affected two of her three sons. Alexandra carried it into the Russian royal house, where it affected her son Alexis.*
(b) *The Royal hemophilia pedigree. From Queen Victoria's daughter Alice, the disorder was introduced into the Russian and Prussian royal houses, and from Victoria's daughter Beatrice, it was introduced into the Spanish royal house. Victoria's son Leopold, himself a victim, transmitted the disorder in a third line of descent.*

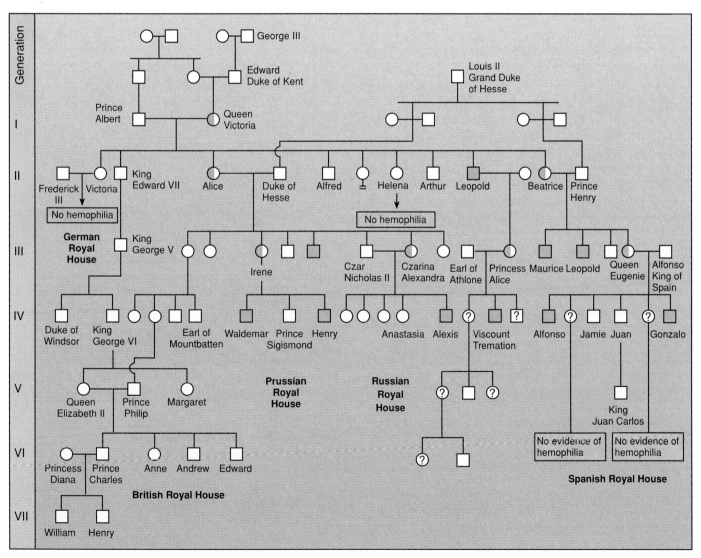

(b)

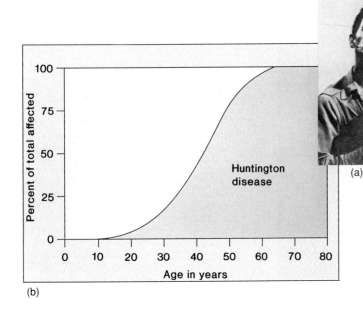

Figure 11.13 Huntington's disease, a dominant genetic disorder.

(a) *Folksinger Woody Guthrie was a victim of Huntington's disease.* (b) *The disorder persists, despite its dominance and fatal effects, because symptoms usually are not evident until individuals are over thirty, by which time many of them have had children.*

future generations along these family lines, except in Russia, where the five children of Alexandra (Victoria's granddaughter) were killed in the turbulent times soon after the Russian Revolution (see figure 11.1).

An Autosomal-Dominant Genetic Disorder: Huntington's Disease

Not all genetic disorders are caused by recessive alleles. Huntington's disease is associated with a dominant allele that causes the progressive deterioration of brain cells in the approximately one in ten thousand individuals who develop the disorder. Because Huntington's disease is a dominant condition, every individual who carries an allele expresses it. The genetic disorder does not die out because symptoms of Huntington's disease do not usually develop until individuals are over age thirty, by which time most of them have had children. Thus, the allele is transmitted before the lethal condition develops (figure 11.13).

There is good news for those who have Huntington's disease in their families and thus might be at risk to develop the disease. After ten years of searching, researchers have finally determined that the gene that causes Huntington's disease is located near the top of chromosome 4. This discovery will lead to a better test to determine whether those who have the disease in their families have actually inherited the gene. It also may lead to a cure involving replacement of the defective gene with a correct gene, in much the same way researchers are attempting to cure cystic fibrosis.

Genetic Counseling

Although most genetic disorders cannot yet be cured, researchers are learning much about them, and progress toward successful therapy is being made in many cases. In the absence of a cure, the only recourse is to try to avoid producing children with genetic disorders. The process of identifying parents at risk for producing children with genetic defects, and of assessing the genetic state of early embryos, is called **genetic counseling.**

Analyzing Family Histories

If the genetic defect is a recessive allele, how can potential parents determine if they carry the allele? Pedigree analysis is often employed as an aid in genetic counseling. For example, if one of your relatives is afflicted with a recessive genetic disorder, such as cystic fibrosis, you might be a carrier of the trait (in other words, be heterozygous for the trait). Analysis of your family history often allows estimation of that likelihood. When a couple is expecting a child and pedigree analysis indicates that both parents have a significant probability of being carriers of a recessive allele responsible for a serious genetic disorder, the pregnancy is said to be a **high-risk pregnancy.** Another class of high-risk pregnancies involves mothers who are older than age thirty-five because the frequency of Down syndrome increases dramatically after that age (see figure 11.3).

Amniocentesis and Chorionic Villi Sampling

When a pregnancy is diagnosed as being high-risk, many women elect to undergo **amniocentesis,** a procedure that permits the prenatal diagnosis of many genetic disorders (figure 11.14). In the fourth month of pregnancy, a sterile hypodermic needle is inserted into the mother's expanded uterus to withdraw a small sample of amniotic fluid. The needle's position and that of the fetus are usually observed simultaneously by means of a technique called **ultrasound** (figure 11.15). Ultrasound uses sound waves, which are not damaging to the mother or the fetus, to produce an image of the fetus. This allows the person withdrawing the amniotic fluid to avoid damaging the fetus. In addition, the fetus can be examined for the presence of major abnormalities. The amniotic fluid, which

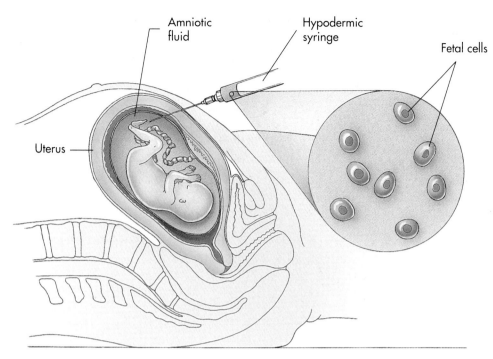

Figure 11.14 Amniocentesis.

A needle is inserted into the amniotic cavity, and a sample of amniotic fluid, containing some cells derived from the embryo, is drawn into the syringe. The fetal cells are then grown in a tissue culture so that their karyotype and many of their metabolic functions can be examined.

surrounds the fetus, contains free-floating cells derived from the fetus. Once removed, these cells can be grown as tissue cultures in the laboratory, and their karyotype and many of their metabolic functions can be examined.

In the last few years, physicians have increasingly turned to a new, less invasive procedure—**chorionic villi sampling**—for genetic screening. In this procedure, the physician removes cells from the chorion, a membrane part of the placenta that nourishes the fetus. This procedure can be used earlier in pregnancy (by the eighth week) and yields results much more rapidly than does amniocentesis.

Tissue cultures from amniocentesis or tissue from chorionic villi sampling can be tested for many of the most common genetic disorders:

1. **Enzyme activity tests.** In many cases, it is possible to test directly for the proper functioning of the enzymes involved in genetic disorders. A lack of proper activity signals the presence of the disorder. Thus, the lack of the enzyme responsible for breaking down phenylalanine signals PKU, the absence of the enzyme responsible for the breakdown of gangliosides indicates Tay-Sachs disease, and so forth.

2. **Association with genetic markers.** For sickle-cell hemoglobin, Huntington's disease, and one form of muscular dystrophy (a condition characterized by weakened muscles that do not function normally), investigators have searched and found other mutations on the same chromosome that, by chance, occur at about the same place as the disorder-causing mutation. By testing for the presence of the second mutation, an investigator can identify individuals with a high probability of possessing the disorder-causing defect. Identifying such

mutations in the first place is a little like searching for a needle in a haystack, but persistent efforts have proven successful in these cases. The associated mutations are detected because they alter the length of DNA segments produced by enzymes that cut strands of DNA at particular places. Such enzymes, called **restriction enzymes,** are discussed in chapter 14. The mutations are called **restriction fragment-length polymorphisms,** or **RFLPs.**

Genetic Therapy

When analysis of amniotic fluid indicates a severe genetic fetal disorder, the parents may consider terminating the pregnancy by means of therapeutic abortion. In some instances, other options are available. For example, if PKU is diagnosed, the defects of the disorder can be avoided by placing the mother on a low-phenylalanine diet. This provides the mother and her unborn baby with enough phenylalanine to make proteins, but not enough to lead to the buildup of damaging molecules. After birth, the child is maintained on a low-phenylalanine diet until age six. At that age, the child's brain is fully developed, and PKU is no longer a potential health problem.

Advances in gene technology are making it possible in some cases to correct undesirable genetic conditions directly by transferring genes from the cells of healthy individuals to those

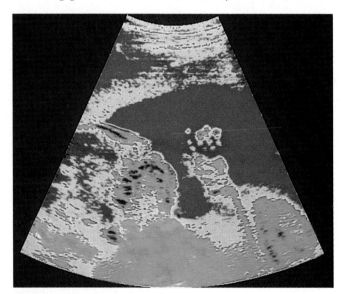

Figure 11.15 Ultrasound.

Ultrasound uses sound waves to produce an image of structures inside the human body. It is an especially useful tool in monitoring pregnancy. In this ultrasound photo, the fetus can be clearly seen.

of individuals with faulty versions. The first such use of gene therapy in humans was approved in August 1990. Other experimental trials involving gene therapy are currently underway and focus on treating such genetic disorders as cystic fibrosis, muscular dystrophy, hemophilia, and hypercholesterolemia (an inherited condition in which cholesterol is not broken down by the liver and thus accumulates in the blood). So far, gene transfers have not been carried out *in vitro* (in a test tube) with embryos recently fertilized there; in such a transfer, all of the cells of an individual would be affected. This type of gene therapy is likely in the future.

The following organizations study major genetic disorders in an effort to find new and better ways to manage the individual conditions. Support of these organizations may eventually lead to the detection of ways to alleviate the disorders completely.

Cystic Fibrosis Foundation
3379 Peachtree Road NE
Atlanta, GA 30326

National Hemophilia Foundation
25 West 39th Street
New York, NY 10018

Committee to Combat Huntington's Disease, Inc.
250 West 57th Street, Suite 2016
New York, NY 10107

Muscular Dystrophy Association
810 Seventh Avenue
New York, NY 10019

National Association of Sickle-Cell Disease, Inc.
945 South Western Avenue
Los Angeles, CA 90006

National Tay-Sachs and Allied Diseases Association
122 East 42nd Street
New York, NY 10017

When a genetic disorder occurs in a family history, analysis of the pedigree can often provide information about the likelihood of a family producing affected individuals. In some cases, therapies can mitigate the effects of the disorder.

EVOLUTIONARY VIEWPOINT

Many harmful gene defects, like sickle-cell anemia and cystic fibrosis, persist within human populations at a high frequency, despite their deleterious effects. In the case of sickle-cell anemia, the high frequency reflects the advantageous malarial resistance of heterozygous individuals. Scientists do not know why cystic fibrosis is as common as it is; different levels of incidence around the world seem to reflect the amount of salt in people's diet. Today, genetic technology is approaching a cure for many genetic defects, although gene substitution in germline cells (gametes, or cells that give rise directly to gametes) has not yet been attempted.

SUMMARY

1. Human body cells contain forty-six chromosomes: forty-four autosomes and two sex chromosomes. The autosomes form twenty-two pairs of homologous chromosomes in meiosis. The two sex chromosomes may be similar in size and appearance, as occurs in females, where the chromosomes are designated XX. The human Y chromosome is much smaller than the X chromosome. XY individuals are male because the genes that initiate the development of male characteristics are located on the Y chromosome.

2. Abnormalities in chromosome number are caused by nondisjunction, in which the homologous pairs of a chromosome do not separate during meiosis I or the sister chromatids do not separate in meiosis II. In humans, the loss of an autosome is invariably fatal. Gaining an extra autosome, which leads to a condition called trisomy, is also fatal, with a few exceptions.

3. The chromosomes that can be present in an extra copy and still allow the individual to survive to adulthood are chromosomes 21 and 22. Individuals with an extra copy of chromosome 21 (three copies in all) are mentally disabled and have Down syndrome. The nondisjunction that causes trisomy 21 occurs more often in mothers over age thirty.

4. Nondisjunction can also result in abnormalities in the number of sex chromosomes. XXX, XXY, XO, and XYY individuals usually survive to adulthood, although they may have somewhat abnormal features.

5. Patterns of inheritance observed in family histories, called pedigrees, can be used to determine the mode of inheritance of a particular trait. Such analysis often determines if the gene determining the trait is located on the X chromosome (that is, whether the trait is sex-linked or autosomal), whether the trait is associated with a dominant or recessive allele, and if the trait is specified by more than one gene.

6. Many human genes possess more than two common alleles. An example is the ABO blood group gene. This gene encodes an enzyme that adds sugars to the surfaces of blood cells. The A and B alleles add different sugars, and the O allele adds none. Human populations vary in the proportions of these three alleles that they possess.

7. Genetic disorders are often caused by alleles that encode abnormal proteins that result in physical and mental abnormalities. Some genetic disorders are relatively common in human populations, whereas others are rare. Many of the most important genetic disorders are associated with recessive alleles that lead to the production of defective versions of enzymes that normally perform critical functions. Because such traits are determined by recessive alleles and therefore are expressed only in homozygotes, the alleles are not eliminated from the human population, even though their effects in homozygotes may be lethal. Dominant alleles that lead to severe genetic disorders are less common. In those that occur more frequently, the allele often is not expressed until after the affected individuals are in their reproductive years. Thus, the allele can be transmitted to offspring before the individual is aware of the disorder.

8. Parents who suspect that their children may express a genetic disorder, such as Down syndrome, may elect to undergo amniocentesis. In this procedure, a sample of fetal cells obtained from amniotic fluid is used to establish a tissue culture, which can then be checked for the presence of various genetic disorders. Chorionic villi sampling is another procedure in which fetal cells can be sampled and checked for genetic abnormalities.

9. For a few genetic disorders, such as phenylketonuria, therapy can be initiated and the disorder's detrimental effects avoided if the disorder is diagnosed during pregnancy. Advances in gene technology have resulted in current experimental trials involving the direct transfer of genes from the cells of healthy individuals to the cells of individuals with faulty versions.

REVIEWING THE CHAPTER

1. What are the structure and number of human chromosomes?
2. How do mistakes in meiosis affect chromosome number?
3. What is Down syndrome?
4. How do the numbers of sex chromosomes become abnormal?
5. How are patterns of inheritance studied?
6. What are multiple alleles?
7. How are ABO blood groups distinguished from one another?
8. What are the causes of genetic disorders?
9. Why have genetic counseling?
10. What are some benefits of genetic therapy?

COMPLETING YOUR UNDERSTANDING

1. What is an autosome?
2. Down syndrome is an example of all the following except
    a. aneuploidy.
    b. nondisjunction.
    c. monosomics.
    d. trisomics.
3. In mothers, the highest incidence of giving birth to a child with Down syndrome is when the mother is _____ years old or older.
4. What is a Barr body, and where is it found?
5. How do Kleinfelter and Turner syndromes originate, and what are their consequences?
6. Why are sex-linked traits expressed much more frequently in males than in females?
    a. Because males have only one copy of the X chromosome.
    b. Because males have only one copy of the Y chromosome.
    c. Because males are more developmentally fragile than females.
    d. Sex-linked traits are carried on the X chromosome and therefore occur with equal frequency in males and females.
    e. These males lack an autosome.
7. What is the controversy concerning XYY males? How would you scientifically attempt to substantiate the claims made in this controversy?
8. What is a pedigree, and how does it relate to human heredity?

9. A detailed pedigree of a family with a history of a particular genetic disorder can help determine
    a. whether the disorder is sex-linked.
    b. whether the disorder is autosomal.
    c. whether the disorder is dominant.
    d. whether the disorder is determined by a single gene or by several genes.
    e. all of the above.
10. How would you prove that albinism is an autosomal trait and that red-green color blindness is a sex-linked trait?
11. An individual with type O blood can receive blood from type _____ individuals.
    a. A
    b. B
    c. AB
    d. O
    e. Two of the above
    f. All of the above
*****12.** Schizophrenia is a serious mental disorder in which mental contact is lost with the environment. To assess whether schizophrenia is hereditary, researchers studied monozygotic and dizygotic twins who had been reared apart. Monozygotic twins are genetically identical, whereas dizygotic twins are normal brothers and/or sisters who happen to be born at the same time. Investigators asked if twins reared apart develop schizophrenia more often if they are genetically identical. From data collected over forty years, they found: Of 289 sets of monozygotic twins studied, both twins developed the disorder if one did in 51 percent of the cases; of 398 sets of dizygotic twins studied, both developed the disorder if one did in 10 percent of the cases. Do you think that these results suggest that schizophrenia is a hereditary disorder?

13. What are antigens, antibodies, and agglutination, and how do they relate to blood types?
14. In a family where the mother is Rh⁻ and the father is Rh⁺, they decided to have four children. What would you predict to happen in their family planning if the mother was not injected with RhoGam?
*****15.** As you can see in table 11.1, both North American and South American Indians exhibit very low frequencies of ABO blood group allele B. Can you think of a reason why they should differ from other human populations in this regard?
16. What is the genetic basis for the human disorders of cystic fibrosis, sickle-cell anemia, Tay-Sachs disease, phenylketonuria, hemophilia, and Huntington's disease?
17. The most common fatal genetic disorder among Caucasians is
    a. cystic fibrosis.
    b. sickle-cell anemia.
    c. hemophilia.
    d. phenylketonuria.
    e. Tay-Sachs disease.
    f. Huntington's disease.
18. What is the best explanation for why 9 percent of African-Americans are heterozygous for the sickle-cell allele, and in some populations in Africa, up to 45 percent of the individuals are heterozygous for this same allele?
19. In which human populations is Tay-Sachs disease the most common, and what are the symptoms of this disorder?
20. What happens to an individual with PKU who accumulates phenylketones?

*For discussion

21. Why has hemophilia received so much media attention over the past decade?

*22. How do we know that the mutation to Royal hemophilia did not occur in one of Queen Victoria's own ova?

23. What promising news is there for those individuals who have Huntington's disease in their family?

24. Why would a pregnant woman undergo amniocentesis or chorionic villi sampling?

25. What are restriction enzymes, and how do they relate to genetic disorders?

26. What are *in vitro* procedures, and how do they apply to genetic disorders?

*For discussion

HUMAN GENETICS PROBLEMS

1. George has Royal hemophilia and marries his mother's sister's daughter Patricia. His maternal grandfather also had hemophilia. George and Patricia have five children: two daughters are normal, two sons develop hemophilia, and one daughter is a carrier. Draw the pedigree.

2. A couple with a newborn baby are troubled that the child does not appear to resemble either of them. Suspecting that a mix-up occurred at the hospital, they check the infant's blood type. It is type O. Because the father is type A and the mother is type B, they conclude that a mistake must have been made. Are they correct?

3. Mabel's sister dies as a child from cystic fibrosis. Mabel herself is healthy, as are her parents. Mabel is pregnant with her first child. If she were to consult you as a genetic counselor, wishing to know the probability of her child developing cystic fibrosis, what would you tell her?

4. How many chromosomes would one expect to find in the karyotype of a person with Turner syndrome?

5. A woman is married for the second time. Her first husband was ABO blood type A, and her child by that marriage was type O. Her new husband is type B, and their child is type AB. What is the woman's ABO genotype and blood type?

6. Two bald parents have five children, three of whom eventually become bald. Assuming that this trait is governed by a single pair of alleles, is this baldness best explained as an example of dominant or recessive inheritance?

7. In 1986, *National Geographic* magazine conducted a survey of its readers' abilities to detect odors. About 7 percent of Caucasians in the United States could not smell the odor of musk. If both parents cannot smell musk, then none of their children are able to smell it. Conversely, two parents who can smell musk generally have children who can also smell it; only a few children in each family are unable to smell it. Assuming that a single

pair of alleles governs this trait, is the ability to smell musk best explained as an example of dominant or recessive inheritance?

8. Total color blindness is a rare hereditary disorder among humans in which no color is seen, only shades of gray. It occurs in individuals homozygous for a recessive allele and is not sex-linked. A man whose father is totally color-blind intends to marry a woman whose mother is totally color-blind. What are the chances that they will produce offspring who are totally color-blind?

9. A normally pigmented man marries an albino woman. They have three children, one of whom is an albino. What is the father's genotype?

10. Four babies are born within a few minutes of each other in a large hospital, when suddenly an explosion occurs. All four babies are found alive among the rubble. None had yet been given identification bracelets. The babies prove to be of four different blood groups: A, B, AB, and O. The four pairs of parents have the following pairs of blood groups: A and B, O and O, AB and O, and B and B. Which babies belong to which parents?

11. This pedigree is of a rare trait in which children have extra fingers and toes. Which if any of the following patterns of inheritance is consistent with this pedigree?

   a. Autosomal recessive

   b. Autosomal dominant

   c. Sex-linked recessive

   d. Sex-linked dominant

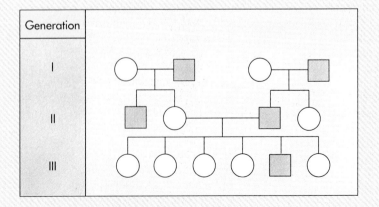

12. This pedigree is for a common form of inherited color blindness. What pattern of inheritance best accounts for this pedigree?

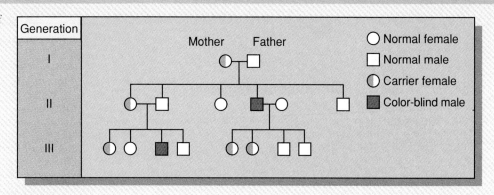

13. Of the forty-three people in the five generations of this family, over a third exhibit an inherited mental disorder. Is the trait dominant or recessive?

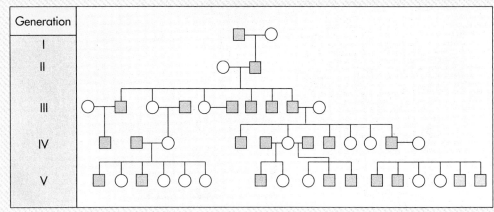

FOR FURTHER READING

Cystic fibrosis: Cloning and genetics. 1989. *Science,* Sept. An entire issue devoted to the finding of the gene responsible for this widespread genetic defect.

Diamond, J. 1989. Blood, genes, and malaria. *Natural History,* Feb., 8–18. A lucid account of the evolutionary history of sickle-cell anemia.

Diamond, J. 1991. Curse and blessing of the ghetto. *Discover,* March. A discussion of the biology and sociology of Tay-Sachs disease.

Gould, S. J. 1980. Dr. Down's syndrome. *Natural History* 89:142–48. An account of the history of Down syndrome, a relatively common chromosomal abnormality that results in severe mental retardation.

Lawn, R., and G. Vehar. 1986. The molecular genetics of hemophilia. *Scientific American,* March, 48–65. A review of the complex way in which blood clots and the many genes that affect this process.

Morrell, V. 1993. Huntington's gene finally found. *Science,* April 2, 28–30. A review of the long search for the gene, found on chromosome 4, that causes Huntington's disease.

Stone, J. 1990. The Marfan family. *Discover,* Nov. An account of a genetic disorder that may have affected Abraham Lincoln.

Verma, I. 1990. Gene therapy. *Scientific American,* Nov., 68–84. Review of the exciting results obtained from introducing healthy genes into the body of an affected person to treat genetic disorders.

EXPLORATIONS

Interactive Software: **Heredity in Families**

This interactive exercise allows students to explore how the character and location of a genetic trait influence its heritability within families. The key variables are dominance/recessiveness of the inherited allele, and X-linkage vs. autosomal location of the gene encoding the allele. From a bank of 30 actual pedigrees, one is selected at random, and the two variables analyzed. The program scores the answer selected, and then presents another pedigree from the bank. As the 30 pedigrees are examined in random order, the program keeps score of the analyses. These pedigrees are often all a human geneticist has to work with in attempting to assess the dominance and chromosomal location of a trait.

Questions to Explore:

1. In analyzing family pedigrees, why are sex-linked alleles expressed so much more often in male offspring than in females?

2. Why do you imagine some genetic disorders like sickle-cell anemia and hemophilia are common, while others are rare?

3. In a pedigree, what affect does sex linkage have on your ability to determine if a trait is being influenced by a single gene?

4. What is the minimal evidence you would accept that a trait appearing in a family is indeed hereditary (that is, caused by an allele), rather than environmentally induced (that is, not due to an allele)?

Interactive Software: **Find That Gene**

In this interactive exercise, the user explores a database of genetic maps. Included are current maps of the human genome, of the mouse, of *Drosophila,* of yeast, of *E. coli,* and of a variety of other prokaryotic and eukaryotic organisms. By specifying an organism, the user can see the appropriate overall genetic map, and by specifying a portion of the map, can explore individual gene locations in detail. When the user specifies a particular gene, its location is highlighted on the map. By exploring a particular gene's location in different organisms, the user examines how the relative positions of key genes have changed over the course of evolution. The locations of multiple copies of genes can also be explored.

Questions to Explore:

1. The genetic maps of bacteria like *E. coli* are present as circles, while the genetic maps of humans and other eukaryotes are presented as a family of straight lines. What difference in the structure of the chromosomes is responsible for this?

2. To what degree, if any, are the genetic maps of bacteria more functionally organized than the maps of eukaryotes? Are clusters of functionally related genes more common in bacteria than eukaryotes?

3. Some portions of a genetic map typically seem more "crowded" (that is, with many more gene locations specified) than other portions. Why do you imagine this is so?

4. If you focus on a particular set of genes (say, the genes involved in glucose metabolism), do you see any particular pattern among the changes that occur in gene location over the course of evolution from bacteria to simple eukaryotes like yeasts to complex eukaryotes like mice and humans?

DNA: THE GENETIC MATERIAL

A computer-generated model of DNA, winding like a double staircase.

FOR REVIEW

Here are some important terms and concepts that have been discussed in previous chapters and that you will encounter again in this chapter. Review them before proceeding if necessary.

Nature of scientific experiments (*chapter 1*)
Radioactive isotopes (*chapter 2*)
Structure of DNA (*chapter 2*)
Nucleotides (*chapter 2*)
Structure of proteins (*chapter 2*)
Structure of chromosomes (*chapter 10*)
Sickle-cell anemia (*chapter 11*)

The realization, reached at the beginning of the twentieth century, that patterns of heredity can be explained by the segregation of chromosomes in meiosis was one of the most important advances in human thought. Mendel's pea plants and Morgan's work with *Drosophila* demonstrated that traits can be transmitted from parent to offspring in many different combinations. This work led directly to the formulation of the science of genetics and thus to great progress in agriculture and medicine, and it also profoundly influenced how we think about ourselves. Because the process of heredity was no longer a mystery, the biological nature of human beings seemed much more approachable (figure 12.1). What was left to deduce, however, was the "how" of heredity—that is, how is genetic information stored in the chromosomes?

Scientists are now able to describe in considerable detail the mechanism by which the information on chromosomes is converted into the traits that pea plants, fruit flies, and human beings exhibit. Their understanding was not deduced in a single

Figure 12.1 Heredity shapes all of us.
Some of what this boy will be as an adult will be influenced by what he learns from his grandfather, but much will reflect the genes he has inherited from his grandfather.

flash of insight, but rather, was developed slowly over many years by a succession of investigators. How they did this and what they learned are the subjects of this chapter.

Where Do Cells Store Hereditary Information?

Where is hereditary information stored in the cell? Danish biologist Joachim Hammerling's approach to this question in 1943 was to cut a cell into pieces and see which pieces were able to express hereditary information. As an experimental subject, Hammerling chose the large, unicellular green alga *Acetabularia*. *Acetabularia* has distinct base, stalk, and cap regions, all of which are differentiated parts of a single cell. The cell nucleus is located in the base. As a preliminary experiment, Hammerling tried alternately amputating the caps and bases of individual cells. He found that, when the cap was amputated, a new one regenerated from the remaining portions of the cell. When the base was amputated and discarded, however, no new base was regenerated. Hammerling concluded that the hereditary information resided within *Acetabularia*'s base.

To test his hypothesis, Hammerling selected individuals from two species of *Acetabularia: Acetabularia mediterranea* (abbreviated *A. mediterranea*) has a disk-shaped cap, and *Acetabularia crenulata* (abbreviated *A. crenulata*) has a branched, flowerlike cap. Hammerling cut the stalk and cap away from an individual of *A. mediterranea;* to the remaining base, he grafted a stalk cut from a cell of *A. crenulata*. The cap that formed resembled the flower-shaped one characteristic of *A. crenulata,* although it was not exactly the same. Hammerling then cut off this regenerated cap and found that a disk-shaped one exactly like that of *A. mediterranea* formed in the second regeneration and in every regeneration thereafter. Hammerling concluded that the instructions specifying the kind of cap that is produced are stored in the base of the cell—and probably therefore in the nucleus—and that these instructions must pass from the base through the stalk to the cap (see Key Experiment 12.1). The flower-shaped cap that formed in Hammerling's initial regeneration was the result of instructions already present in the transplanted stalk when it was excised from the original *A. crenulata* cell. In contrast, all subsequent caps used new information, derived from the base of the *A. mediterranea* cell onto which the stalk had been grafted. In some unknown way, the instructions originally present in the stalk were eventually "used up" and replaced by new instructions from the base.

Hammerling's experiments identified the nucleus as the *likely* repository of the hereditary information. To prove this, American embryologists Robert Briggs and Thomas King transplanted isolated nuclei from one cell to another in 1952. Using a glass pipette with a fine tip and working under a microscope, Briggs and King removed the nucleus from a frog egg (figure 12.2). Without a nucleus, the egg would not develop. They then replaced the absent nucleus with one they isolated from a cell of a young frog embryo. The diploid nucleus derived from that embryo took over directing the development of the original

Key Experiment 12.1

2000

HAMMERLING'S *ACETABULARIA* RECIPROCAL GRAFT EXPERIMENT

Although *Acetabularia* is a large organism with clearly differentiated parts, such as the stalks and elaborate caps shown in the photo, individuals are actually single cells. To the base of each of two species of *Acetabularia*—*Acetabularia crenulata* and *Acetabularia mediterranea*—Joachim Hammerling grafted the stalk of another. In each case, the cap that eventually developed was dictated by the base, where the nucleus was located, and not the stalk.

◄ 1943

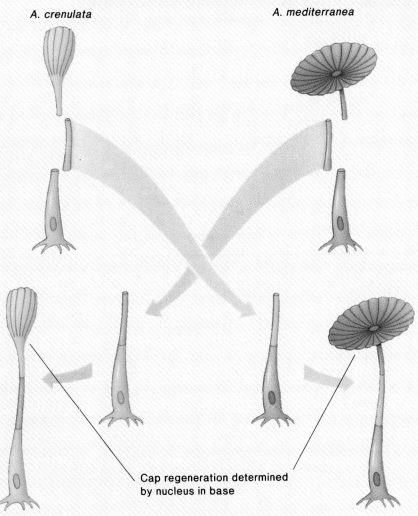

A. crenulata

A. mediterranea

Cap regeneration determined by nucleus in base

1900

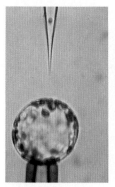

Figure 12.2 The technique of microinjection used by Briggs and King.

Working under a microscope, scientists can pierce a cell with the fine tip of a glass micropipette without rupturing the cell. These photographs show microinjection into a tobacco leaf cell.

egg cell, which ultimately grew into an adult frog. Clearly, the nucleus was the important element in bringing about this result.

Hereditary information is stored in the nuclei of eukaryotic cells.

Which Chromosome Component Contains the Hereditary Information?

The identification of the nucleus as the source of hereditary information focused attention on the chromosomes, which had already been shown to be the vehicles of Mendelian inheritance. Specifically, biologists wondered how the actual hereditary information was arranged in the chromosomes. They knew that chromosomes contain both protein and DNA, but on which of these was the hereditary information written?

Over a period of about thirty years, starting in the late 1920s, a number of groups of investigators addressed this issue, and ultimately, their collective results resolved the problem clearly. Descriptions of three very different experiments, each of which was instrumental in this chromosome research, follow.

The Griffith-Avery Experiments: DNA as the Transforming Principle

As early as 1928, British microbiologist Fred Griffith made a series of unexpected observations while experimenting with a bacterium then called pneumococcus and now called *Streptococcus pneumoniae* (abbreviated *S. pneumoniae*). This bacterium had been recognized in 1886 as a primary cause of pneumonia in humans.

Normally, *S. pneumoniae* cells are enclosed by a polysaccharide capsule (see chapter 2) and are highly pathogenic (disease-causing); if even a few are injected into a mouse, the animal will die in a day or two as a result of the infection. However, mutant strains of the bacterium that lack polysaccharide capsules also exist; even if a large quantity of such mutant cells is injected into a mouse, the animal will remain healthy. If cells of the normal, pathogenic strain are killed by exposure to high temperatures before they are injected into a mouse, they have no effect, indicating that living bacteria are necessary to produce the harmful effects associated with this strain.

Building on these results, Griffith mixed heat-killed, pathogenic bacteria, which had polysaccharide capsules, with living, mutant, capsuleless bacteria and injected the mixed bacteria into healthy mice. Unexpectedly, the injected mice developed disease symptoms, and many of them died. The blood of the dead mice was found to contain high levels of normal, capsulated, pathogenic *S. pneumoniae!* Somehow, the information specifying the polysaccharide capsule had passed from the dead bacteria to the live but capsuleless bacteria in the mixture, transforming them into pathogenic, capsulated bacteria that infected and killed the mice (see Key Experiment 12.2).

The agent responsible for transformation in *S. pneumoniae* was discovered in a classic series of experiments conducted by Oswald Avery and his colleagues at Rockefeller University in New York during the 1930s. These scientists succeeded in characterizing what they referred to as the "transforming principle." Its properties resembled those of DNA rather than of protein in that the activity of the transforming principle was not affected by protein-destroying enzymes but was lost completely in the presence of the DNA-destroying enzyme DNAse.

When the transforming principle was purified, it indeed consisted predominantly of DNA. Later experiments showed that all but trace amounts of protein (0.02 percent) could be removed without reducing the transforming activity. The conclusion was inescapable: DNA is the hereditary material in bacteria. It has since proven possible to use purified DNA to change the genetic characteristics of eukaryotic cells in tissue culture and to inject pure DNA into fertilized *Drosophila* eggs and thereby alter the genetic characteristics of the resulting adult.

The Hershey-Chase Experiments: Bacterial Viruses That Direct Their Heredity with DNA

Avery's results were not widely appreciated at first. Many biologists stubbornly preferred to believe that proteins were the source of hereditary information. This preference for proteins was probably due to the chemical interest being paid to proteins at this time and to the lack of experiments being conducted on the chemical nature of nucleic acids. However, convincing experiments with viruses soon made DNA difficult to ignore.

Viruses are simply RNA or DNA packaged within a protein coat; they are described in more detail in chapter 26. These new experiments focused on bacteriophages, which are viruses that infect bacteria. When a bacteriophage infects a bacterial cell, it first binds to the cell's outer surface and then injects its

KEY EXPERIMENT 12.2

2000

1928

1900

GRIFFITH'S DISCOVERY OF THE "TRANSFORMING PRINCIPLE"

In a series of experiments with mice, British microbiologist Fred Griffith arrived at some surprising results. Griffith suspected that the pathogenic bacterium *Streptococcus pneumoniae* would kill many of the mice into which it was injected, but only if the bacterial cells were alive and enclosed by a polysaccharide capsule. First, he injected mice with heat-killed, capsulated, pathogenic cells of *S. pneumoniae*. The mice lived. When he injected the mice with live, capsulated, pathogenic cells, the mice died.

When the injections contained live, capsuleless, nonpathogenic cells, the mice lived.

However, when Griffith injected mice with a mixture of live, capsuleless, nonpathogenic bacteria and dead, capsulated, pathogenic bacteria—both of which should have been harmless, given previous experiment results—the mice died. Griffith concluded that the live cells had been "transformed" by the dead ones—that the genetic information specifying the polysaccharide capsule had passed from the dead cells to the living ones. What he could not assume from his experiments was that the transforming principle was DNA.

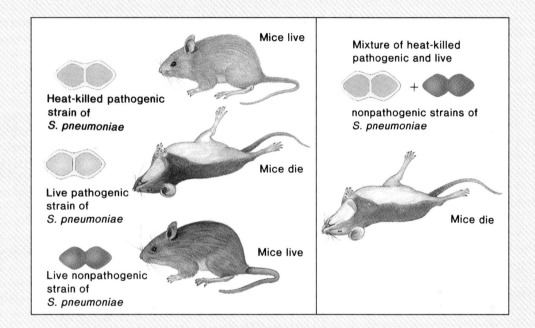

hereditary information into the cell. There, the hereditary information directs the production of thousands of new virus particles within the bacterial cell. The host bacterial cell eventually bursts, or lyses, releasing the newly made viruses.

In 1952, Alfred Hershey and Martha Chase set out to identify the material that a bacteriophage injects into a bacterial cell at the start of an infection. They used a strain of bacteriophage known as T2, which contains DNA rather than RNA, and designed an experiment to determine whether the genetic material was DNA or protein. Hershey and Chase used a technique known as labeling, which is the introduction of a radioactive isotope (see chapter 2) of a particular element into a molecule so that the source of that molecule can be identified and its pathways can be followed. They labeled the DNA of some of the bacteriophages with a radioactive isotope of phosphorus, ^{32}P, and the protein coats of other viruses with an isotope of sulfur, ^{35}S. The labeled viruses were permitted to infect bacteria. The bacterial cells were then agitated violently (in an ordi-

nary Waring blender!) to shake the protein coats of the infecting viruses loose from the bacterial surfaces to which they were attached. Then the solution was spun at high speed in a centrifuge so that the heavier bacteria were pulled to the bottom, forming a pellet. The lighter virus protein coats hovered in a solution, called the supernatant, over the pellet.

Using this technique, Hershey and Chase conducted two separate experiments. In the first experiment, they labeled the viral protein coats with ^{35}S. They found that the ^{35}S-labeled protein coats were in the supernatant, and not with the bacteria in the pellet. In the second experiment, Hershey and Chase labeled the viral DNA with ^{32}P. They found that the ^{32}P-labeled DNA was in the pellet. Investigating further, they found that the ^{32}P-labeled DNA was not just in the pellet, but within the infected bacterial cells. Hershey and Chase confidently concluded that the hereditary information that viruses use to direct the replication of new viruses in bacterial cells is composed of DNA, not protein (see Key Experiment 12.3).

KEY EXPERIMENT 12.3

2000

THE HERSHEY-CHASE EXPERIMENT WITH BACTERIAL VIRUSES

The T2 bacterial viruses that Hershey and Chase employed have a simple structure: They are composed of a protein envelope within which DNA is packaged. Hershey and Chase labeled the T2 virus particles with either of two radioactive isotopes: In one experiment, the protein coats were labeled with ^{35}S (sulfur occurs in protein in the amino acids cysteine and methionine but does not occur in DNA), and in the other experiment, the DNA molecules were labeled with ^{32}P (phosphorus occurs in the phosphate groups of DNA but does not occur in proteins). These T2 particles were then allowed to infect bacterial cells.

As shown in part 1 of the illustration, each virus bound to the outside of the cell but, instead of entering, injected its DNA into the cell. As shown in part 2, individ-

ual DNA strands entered the cell from virus particles bound to its surface. Within the cell, the injected DNA commandeered the machinery of the cell and directed the synthesis of all the parts necessary to make new viruses. As shown in part 3, new virus particles were assembled from parts within the infected cell. Eventually, these new viruses ruptured the cell and were released into the surroundings.

When Hershey and Chase used a Waring blender to knock the virus particles off the bacterial cells after the initial injection and separated the bacterial cells (with the injected viral DNA) from the liquid medium (with the dislodged viral protein coats in it), they found that when ^{35}S-labeled protein coats were used, the bulk of the radioactivity (and thus the virus protein) was in the medium. When ^{32}P-labeled DNA was used, the radioactivity (and thus the virus DNA) was present in the *interior* of the bacterial cells. Hershey and Chase concluded that the virus DNA, not the virus protein, was responsible for directing the production of new viruses.

1951

1900

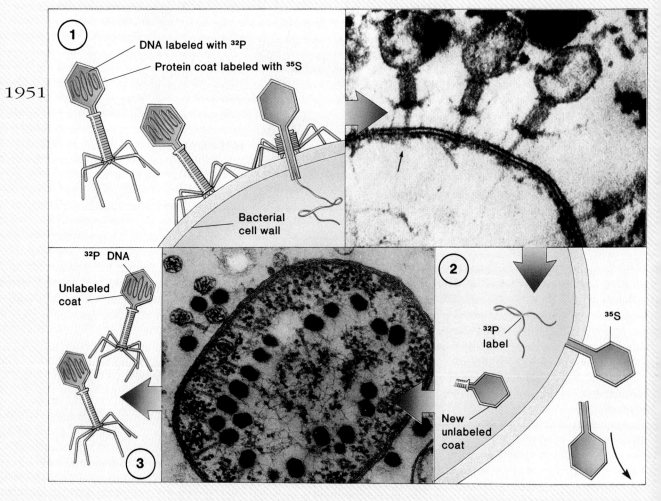

(1) DNA labeled with ^{32}P
Protein coat labeled with ^{35}S
Bacterial cell wall

(2) ^{32}P label
^{35}S
New unlabeled coat

(3) ^{32}P DNA
Unlabeled coat

The Fraenkel-Conrat Experiment: Viruses That Direct Their Heredity with RNA

Even with the evidence provided by the Hershey-Chase experiments, one objection could still be raised about the hypothesis that DNA is the genetic material: Some viruses contain *no* DNA and yet manage to reproduce themselves. What is their genetic material?

In 1957, Heinz Fraenkel-Conrat and coworkers isolated tobacco mosaic virus (TMV) from tobacco leaves. From ribgrass (*Plantago lanceolata*), a common weed, they isolated a second, rather similar kind of virus, Holmes ribgrass virus (HRV). In both TMV and HRV, the viruses consist of protein and a single strand of RNA. After isolating these viruses, the scientists broke them up, separating their protein from their RNA. By putting the protein component of one virus with the RNA of another, they were able to produce hybrid virus particles.

The payoff of the experiment was in its next step. To determine whether the genetic material of viruses is protein or RNA, Fraenkel-Conrat infected healthy tobacco plants with a hybrid virus composed of TMV protein coats and HRV RNA, being careful not to include any nonhybrid virus particles. The tobacco leaves that were infected with the hybrid virus particles developed the lesions that were characteristic of HRV and that normally formed on infected ribgrass. Clearly, the virus's hereditary properties were determined by the RNA in its core and not by the protein in its coat.

Later studies have shown that many kinds of viruses have RNA as their genetic material, rather than the DNA that is found universally in cellular organisms. When these viruses infect a cell, they may either multiply themselves directly or, in the case of the group called **retroviruses,** make DNA copies of themselves, which can then be inserted into the cellular DNA as if they were cellular genes.

> *DNA is the genetic material for all cellular organisms and most viruses. RNA is the genetic material for some viruses.*

The Chemical Nature of Nucleic Acids

Before continuing our sleuthing into the hereditary function of DNA, let us look for a moment at the chemical nature of DNA. Ultimately, DNA's chemical composition and the special chemical structures of its four nitrogenous bases provide the key clues to DNA's hereditary role.

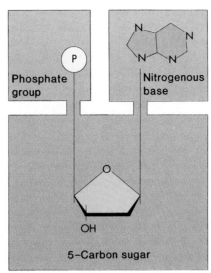

Figure 12.3 Nucleotide subunit of DNA.

The nucleotide subunits of DNA are composed of three elements: a nitrogenous base, a phosphate group, and a five-carbon sugar.

DNA was discovered only four years after the publication of Mendel's work. In 1869, German chemist Friedrich Miescher extracted a white substance from the cell nuclei of human pus and from fish sperm nuclei. The proportion of nitrogen and phosphorous in this substance was very different from that of any other known constituent of cells, which convinced Miescher that he had discovered a new biological substance. He called this substance "nuclein" because it seemed to be associated specifically with the cell nucleus.

Levene's Chemical Investigations

Because Miescher's nuclein was slightly acidic, it came to be called nucleic acid. For fifty years, little research was done on nucleic acid because nothing was known of its function in cells, and there seemed little to recommend it to investigators. In the 1920s, biochemist P. A. Levene worked out the basic chemistry of nucleic acids, discovering that there were two sorts of nucleic acid: ribonucleic acid, or RNA, and deoxyribonucleic acid, or DNA. Levene found that DNA contained three basic components (figure 12.3):

1. Phosphate (PO_4) groups

2. Five-carbon sugars, called deoxyribose sugars (RNA contains five-carbon sugars called ribose sugars)

3. Four **nitrogenous** (or nitrogen-containing) **bases: adenine** (A) and **guanine** (G) (double-ring compounds called **purines**), and **thymine** (T) and **cytosine** (C) (single-ring compounds called **pyrimidines**) (RNA contains the pyrimidine **uracil** (U) in place of thymine)

From the roughly equal proportions of the three elements, Levene correctly concluded that DNA and RNA molecules are composed of units of these three elements, strung one after another in a long chain. Each unit—a five-carbon (ribose) sugar to which is attached a phosphate group and a nitrogen-containing base—is called a **nucleotide** (see chapter 2). The identity of the nitrogenous base is the only factor that distinguishes one nucleotide in a nucleic acid from another.

The chain of nucleotides that forms DNA has other important chemical properties that are essential to its structure. One end of a DNA strand always terminates with a phosphate group that is attached to the number five carbon of the deoxyribose sugar; thus, this end of the DNA strand is called the **5′ end.** The other end of a DNA strand terminates with a hydroxyl (OH) group that is attached to the number three carbon of the deoxyribose sugar; thus, this end of the DNA strand is called the **3′ end.** With these ends as a guide, the strands can be described as having either a 5′-to-3′ orientation, or a 3′-to-5′ orientation, depending on how the sugars are arranged on the strand.

These two orientations of DNA strands are important to discussions of DNA replication because the strands are replicated in different ways.

The presence of the 5′-phosphate and 3′-hydroxyl groups is also what allows DNA and RNA to form a polymer, or long chain, of nucleotides since the 5′ and 3′ groups can react chemically with one another. The chemical reaction between the phosphate group of one unit and the hydroxyl group of another causes the elimination of a water molecule and the formation of a covalent bond linking the two groups together. The linkage is called a **phosphodiester bond** because the phosphate group is now linked to the two sugars by means of two ester (—O—) bonds. Further linking up can occur in the same way: The two-unit polymer resulting from the condensation reaction just described still has a free phosphate group at one end (the 5′ end) and a free hydroxyl (OH) group at the other (the 3′ end) (figure 12.4). In this way, many thousands of nucleotides can be linked together in long chains.

A strand of DNA or RNA is a long chain of nucleotide subunits joined together like cars in a train.

Levene's early studies indicated that all four types of DNA nucleotides (that is, the nucleotides with the four different nitrogenous bases) were present in roughly equal amounts. This result, which later proved to be an error, led to the mistaken belief that DNA was a simple repeating polymer in which the four nucleotides occurred together in a long series of identical units (for instance, GCAT . . . GCAT . . . GCAT . . . GCAT . . .). In the absence of sequence variation in such a repeating chain, it was difficult to see how DNA might contain the hereditary information necessary to specify even the simplest organism. This was why the results of Avery's experiments on DNA as the transforming principle, although crystal clear, were not readily accepted at first. It seemed more

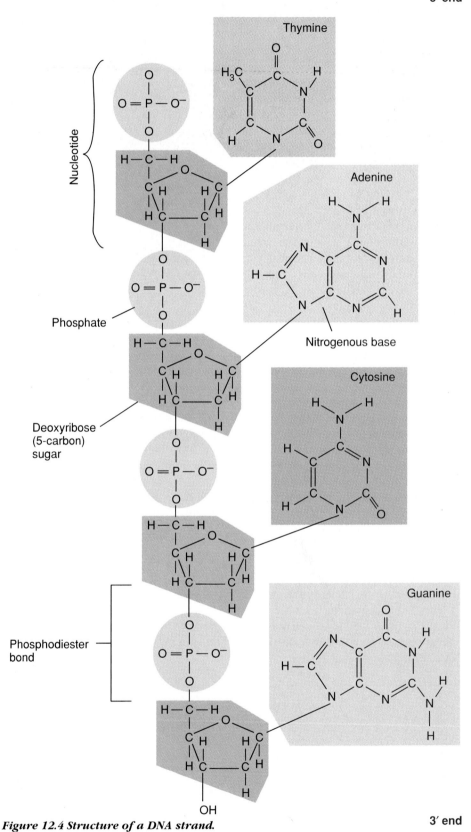

Figure 12.4 Structure of a DNA strand.

A DNA strand is composed of a chain, or polymer, of nucleotides held together by phosphodiester bonds. One end of the strand terminates in a hydroxyl (OH) group, and this end is called the 3′ end; the other end terminates in a phosphate group, and this end is called the 5′ end.

plausible to some biologists that DNA was no more than a structural element of the chromosomes, with proteins playing the central genetic role.

Chargaff's Rule

The key advance came after World War II, when Levene's chemical analysis of DNA was repeated, using more accurate techniques than had previously been available and obtaining quite a different result. A careful study by Erwin Chargaff showed that the four nucleotide bases were *not* present in equal proportions in DNA molecules after all. Chargaff found that the nucleotide composition of DNA molecules varied in complex ways, depending on the source of the DNA. This strongly suggested that DNA was not a simple repeating polymer and that it might have the information-encoding properties required of genetic material. Despite DNA's complexity, Chargaff observed an important underlying regularity: The amount of adenine present in DNA molecules is always equal to the amount of thymine, and the amount of guanine is always equal to the amount of cytosine. Chargaff's finding is commonly referred to as **Chargaff's rule.**

> *Chargaff pointed out that, in all natural DNA molecules, the amount of adenine (A) equals the amount of thymine (T), and the amount of guanine (G) equals the amount of cytosine (C).*

The Structure of DNA: A Synthesis of Ideas

As it became clear that the DNA molecule was the repository of hereditary information, investigators began to puzzle over how such a seemingly simple molecule could carry out such a complex function. The significance of the regularities pointed out by Chargaff was not immediately obvious but soon became clear.

British chemist Rosalind Franklin attempted X-ray crystallographic analysis of DNA fibers. In this process, the DNA molecule is bombarded with an X-ray beam. When individual rays encounter atoms, their path is bent, or diffracted; the pattern created by the total of all these diffractions can be captured on a piece of photographic film. Such a pattern resembles the ripples created by tossing a rock into a smooth lake. Careful analysis of the diffraction pattern allows the development of a three-dimensional image of the molecule.

Franklin's studies were severely handicapped in that it had not proven possible to obtain true crystals of "natural" DNA, so she had to work with DNA in the form of fibers. Although the DNA molecules in a fiber are all aligned with one another, they do not form the perfectly regular crystalline array required to take full advantage of X-ray diffraction. One of Franklin's colleagues, biochemist Maurice Wilkins, was able to prepare more uniformly oriented DNA fibers than had been possible previously. Using these fibers, Franklin was able to obtain crude diffraction information on natural DNA. The diffraction patterns she obtained suggested that the DNA molecule was a helical coil. Her photographs also made it possible to determine some of the molecule's basic structural parameters. The diffraction pattern indicated that the helix had a diameter of about 2 nanometers and makes a complete turn every 3.4 nanometers (figure 12.5).

Two young investigators at Cambridge University—Francis Crick and James Watson—learned about Franklin's results before they were published and quickly worked out a likely structure of the DNA molecule that is now known to be substantially correct (see Sidelight 12.1). They analyzed the problem deductively: First, they built models of the nucleotides, and then they tested how these could be assembled into a molecule that fits what they knew about DNA structure. They tried various possibilities, first assembling molecules with three strands of nucleotides wound around one another to stabilize the helical shape. None of these early efforts proved satisfactory. They finally hit on the idea that the molecule might be a simple **double helix,** or duplex, in which the bases of two strands pointed inward toward one another (figure 12.6). If a purine, which is large, is always paired with a pyrimidine, which is small, the diameter of the duplex stays the same— 2 nanometers. Because hydrogen bonds can form between the two strands, the helical form is stabilized.

The Watson-Crick DNA structure immediately showed why Chargaff's rule was viable: The purine adenine (A) will not form proper hydrogen bonds in this structure with cytosine (C), but will with thymine (T); thus, every A is paired with T. Similarly, the purine guanine (G) will not form proper hydrogen bonds with thymine, but will with cytosine; thus, every G is paired with C (figure 12.7).

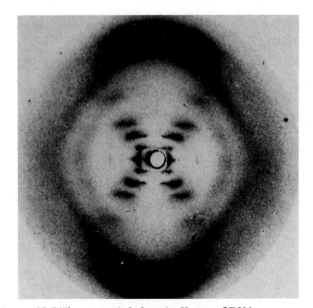

Figure 12.5 The essential clue: An X ray of DNA.

This X-ray diffraction photograph of fibers of DNA was made in 1953 by Rosalind Franklin in the laboratory of Maurice Wilkins. It suggested to Watson and Crick that the DNA molecule was a helix, like a winding staircase.

Sidelight 12.1
WATSON, CRICK, AND THE DOUBLE HELIX

The discovery of the structure of the DNA molecule is undoubtedly one of the most important breakthroughs in the history of science. Given the import of Watson and Crick's discovery, most beginning biology students assume that Watson and Crick spent long hours in the lab, designing elaborate experiments and collecting quantitative data. But this was not the case. James Watson, in his thoroughly readable and entertaining memoir *The Double Helix*, states that "science seldom proceeds in the straightforward logical manner perceived by outsiders. . . . its steps forward (and sometimes backward) are often very human events in which personalities and cultural transitions play major roles." Watson's penchant for understatement comes into play with this declaration, for he and his colleague Francis Crick both possessed formidable personalities that sometimes irritated those around them, and the "cultural traditions" of which Watson writes were, more often than not, long talks with Crick over three-hour lunches or pints of beer at the local pub.

Watson's *The Double Helix* has fascinated readers since its publication in 1968. In the book, Watson describes himself as an "uneducated Ph.D." who could not grasp chemistry. Bouncing from one postdoctoral fellowship to another, Watson tried to align himself with someone who could teach him chemistry. He was convinced that he could figure out the structure of DNA if only he could grasp some basic chemical principles. At this time, scientists were still debating whether DNA or protein was the genetic material. Although some experimental evidence suggested that DNA was the genetic molecule, many scientists were more interested in proteins. At the age of twenty-four, Watson finally finagled a postdoctoral fellowship at the Cavendish Laboratory in Cambridge, England, to work with plant viruses and hopefully find someone who would talk to him about DNA. He found that person in Francis Crick.

The loquacious and brilliant Crick, a thirty-five-year-old Ph.D. student, was involved in a number of projects, but was as fascinated with DNA as was Watson. Unfortunately, to hear Watson tell it, hardly anyone at the Cavendish Lab liked to talk to Crick, because as Watson states, "I have never seen Francis Crick in a modest mood." Reminiscing years later at a celebration of the fortieth anniversary of the discovery of the structure of DNA, Watson recalled that the head of the Cavendish Lab—Sir Lawrence Bragg—could not stand the sight of Crick and that Bragg had "just one ambition, which was to get Francis out of Cambridge." Nevertheless, the young, somewhat aimless, American postdoctoral fellow and the eccentric British Ph.D. student with an eye for the ladies got along famously. Within a half hour of their first meeting, Watson and Crick discussed the structure of DNA.

Watson and Crick did not conduct any experiments; nor did they collect any of their own quantitative data. Instead, they relied on experimental information published in the leading scientific journals. They also had a friend in Maurice Wilkins, a scientist who was working on X-ray diffraction studies of crystals of DNA. Maurice Wilkins's young colleague, Rosalind Franklin, was a leading expert in this new field, but Watson and Franklin took immediate dislike to one another. The unkind words that Watson has recorded in his memoir about Franklin's plain appearance and uncooperative attitude shocked the scientific community when the book was first published and still shock modern readers. (In the epilogue of *The Double Helix*, Watson apologizes for his vindictive comments about Rosalind Franklin and lauds her for her pioneering work with DNA X-ray studies. Watson has stated that Franklin could have discovered DNA's structure if "she'd just talked to Francis for an hour" and acknowledges that Franklin's X-ray diffraction studies were the key evidence that the DNA molecule was a helix.)

Watson attended a seminar given by Franklin at Oxford, at which time Franklin reported the results of her X-ray diffraction studies. For months, Watson and Crick tried to puzzle out what Franklin's results might mean. The results seemed to imply a helix, but details could not be seen. Finally, they constructed a tentative metal model of the DNA molecule. This model was completely wrong—it had the bases on the outside, and the sugar phosphate component was on the inside of the helix. Proudly and somewhat pompously displaying their model to their colleagues, including Wilkins and Franklin, Watson and Crick were crushed when the X-ray diffraction experts rejected the model as absolutely implausible. Watson and Crick, disappointed but not defeated, went back to the drawing board for more pints and more conversation.

By this time, other scientists had become interested in DNA. Several key pieces of experimental evidence were unequivocally pointing toward DNA as the hereditary material, and suddenly, the most famous scientists in the world were hot on the trail. One of these was Linus Pauling, who in 1952 had demonstrated the helical structure of an important protein. Now, it seemed, he had turned his attention to DNA. Linus's son Peter was also at the Cavendish Lab at this time, and he helpfully shared his father's letters with Watson and Crick, keeping them apprised of his father's progress with his experiments. In fact, Linus Pauling published a paper on DNA structure, but his structure left out a chemical component that was so important that, as Watson states, "Pauling's nucleic acid was not an acid at all." Relieved, he went off to report gleefully to Wilkins that Pauling's model was "far off base."

The key breakthrough in the search for DNA's structure came when Wilkins showed Watson a new X-ray diffraction study of a different form of

Sidelight figure 12.1 The discovery of DNA's structure.

A key breakthrough in genetics occurred in 1953, when James Watson, a young American postdoctoral student (he is the one peering up, as if afraid their homemade model of DNA will topple over) and the English scientist Francis Crick (pointing) deduced the structure of DNA, the molecule that stores the hereditary information.

DNA, the so-called B form. As soon as Watson saw the photo, he resolved to try placing the bases on the inside of the helix instead of on the outside. With the sugar-phosphate backbone in place, Watson and Crick now turned to the problem of how the bases fit inside the helix. They were familiar with Chargaff's rule, but they could not figure out how to make the bases fit together. They found that when two purines were placed together, there was a gap between the two bases. When two pyrimidines were placed together, the bases overlapped. Frustrated, Watson spent a morning cutting out cardboard models of the purines and pyrimidines (the metal models that he had ordered from the lab's metal shop had not yet arrived). As he fooled with his cardboard cutouts, he suddenly became aware that "an adenine-thymine pair held together by two hydrogen bonds was identical in shape to a guanine-cytosine pair." Buoyed by this discovery, he ran to get Crick. At first skeptical, Crick soon saw that Watson's purine-pyrimidine base pairing made sense. The two then went out to lunch, where Watson "felt slightly queasy when . . . Francis winged into the Eagle (restaurant) to tell everyone within hearing distance that we had found the secret of life."

The rest of the story is history (Sidelight figure 12.1). After perfecting their rickety model with solder and confirming their measurements, Watson and Crick wrote their famous 1953 paper that contained perhaps one of the greatest understatements in scientific history: "It has not escaped our notice that the specific pairing we have postulated immediately suggests a possible copying mechanism for the genetic material." Watson and Crick went on to receive the Nobel Prize in December 1962 for deducing the structure of DNA. And Watson has left behind a blunt, funny, and sometimes disconcerting look into the nature of scientific discovery.

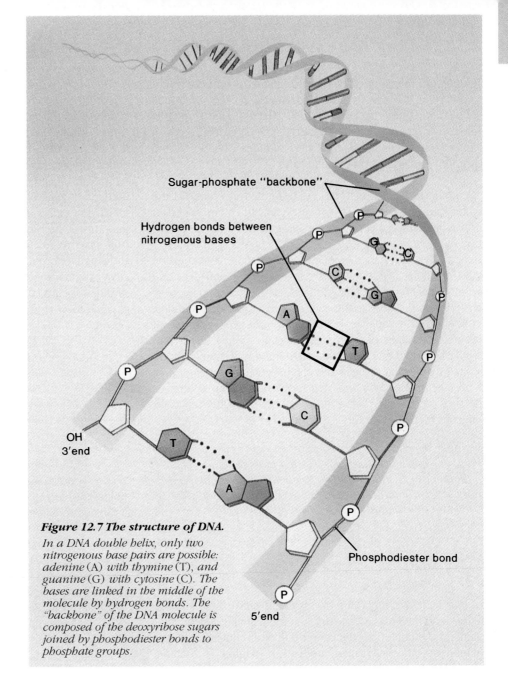

Sugar-phosphate "backbone"

Hydrogen bonds between nitrogenous bases

OH
3'end

Figure 12.7 The structure of DNA.

In a DNA double helix, only two nitrogenous base pairs are possible: adenine (A) with thymine (T), and guanine (G) with cytosine (C). The bases are linked in the middle of the molecule by hydrogen bonds. The "backbone" of the DNA molecule is composed of the deoxyribose sugars joined by phosphodiester bonds to phosphate groups.

Phosphodiester bond

5'end

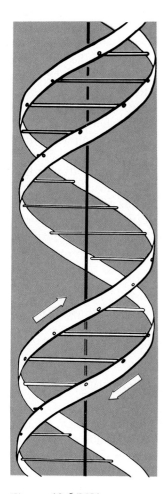

Figure 12.6 DNA as a double helix.

This drawing, executed by Francis Crick's wife Odile, appeared in Watson and Crick's original 1953 paper that presented the structure of DNA.

The Watson-Crick model of DNA is a double helix, a double staircase composed of two polynucleotide chains hydrogen-bonded to each other and wrapped around a central axis.

Complementarity and DNA Replication

The Watson-Crick model of DNA structure immediately suggested that the basis for copying the genetic information is **complementarity.** One chain of the DNA molecule may have any conceivable base sequence, but this sequence completely determines that of its partner in the double helix. For example, if the sequence of one chain is ATTGCAT, the sequence of its partner in the double helix *must* be TAACGTA. Each chain in the double

helix is a complementary image of the other. Copying the DNA molecule requires only "unzipping" it and constructing a new complementary chain along each naked single strand.

Semiconservative Replication

The form of DNA replication suggested by the Watson-Crick model is called **semiconservative replication** because, after one round of replication, the original double helix is not conserved; instead, each strand of the original double helix becomes part of another double helix.

The complementary nature of the DNA double helix provides a ready means of duplicating the molecule. If the molecule were unzipped, only the appropriate complementary nucleotides on the exposed single strands would need to be assembled to form two daughter double helixes of the same sequence. This prediction of the Watson-Crick model was tested in 1958 by Matthew Meselson and Frank Stahl of the California

Institute of Technology. These two scientists grew bacteria for several generations in a medium containing the heavy nitrogen isotope ^{15}N; thus, the DNA of their bacteria was eventually denser than normal. They then transferred the growing cells to a new medium containing the lighter nitrogen isotope ^{14}N and harvested the DNA at various intervals.

At first, the bacteria manufactured DNA that was all heavy. But as the new DNA that was being formed incorporated the lighter nitrogen isotope, DNA density fell. After one round of DNA replication, the density of the bacterial DNA had decreased to a value intermediate between all-light isotope and all-heavy isotope DNA. After another round of replication, centrifugation was used to separate two density classes—one intermediate and the other light (corresponding to DNA that included none of the heavy isotope). These results indicated that, after one round of replication, each daughter DNA double helix possessed one of the labeled strands of the parent molecule. When this hybrid double helix replicated, it contributed one heavy strand to form another hybrid double helix and one light strand to form a light double helix. Meselson and Stahl's experiment thus confirmed the prediction of the Watson-Crick model that DNA replicates semiconservatively (see Key Experiment 12.4).

The basis for the great accuracy of DNA replication is complementarity. A DNA molecule is a double helix, containing two strands that are mirror images of each other, so either one can be used as a template to reconstruct the other.

How Does the DNA Molecule Copy Itself?

A DNA molecule replicates by separating into single strands. Each strand then acts as a template for assembling a new complementary strand. The separation is typically initiated at one or more specific sites called **replication origins.** In bacteria, whose DNA is circular, replication proceeds from one origin in both directions until the two growing points meet at the far side of the circle. In eukaryotes, chromosomes typically possess multiple replication origins.

The enzyme that catalyzes this process is called **DNA polymerase.** DNA polymerase adds nucleotides to the 3′ end of DNA strands, using an existing strand as a template to guide selection of the proper nucleotide. Because the enzyme adds only to the 3′ end, the two strands of the DNA molecule are assembled in opposite directions. At every growing point, sometimes called a **replication fork,** one new strand is built by simply adding nucleotides to its 3′ end, the strand growing inward toward the Y-junction as the double helix unwinds. The other strand, which ends in a 5′ nucleotide, cannot grow in this fashion because DNA polymerase does not add to 5′ nucleotides.

How is this strand replicated? The DNA polymerase allows a gap of one to two thousand nucleotides to form and then proceeds to fill in the gap, moving away from the replication fork until the short new segment meets the rest of the growing strand. The segment is sealed to the end of the strand with a special enzyme called a **ligase,** and the polymerase returns to the growing point to build another short segment that can be attached in turn. This mode of replication, in which the polymerase copies one strand in short segments that are stitched to the growing strand, is called **discontinuous synthesis.** Careful analysis of electron micrographs showing DNA replication in progress indicates that one of the DNA strands behind the polymerase indeed appears single-stranded for about a thousand nucleotides, just as this mode of replication would predict.

A DNA molecule copies (replicates) itself by separating its two strands and using each as a template to assemble a new complementary strand, thus forming two daughter double helixes. The two strands are assembled in different directions.

Eukaryotic Replication of Chromosomes

Chapter 9 described how bacteria replicate their single, circular molecule of DNA (see figure 9.2). The DNA within a eukaryotic chromosome is not naked, as in bacteria; as discussed in chapter 9, it is clothed with a protein wrapping that makes it much easier to pack compactly. If you were to look at a eukaryotic chromosome under the electron microscope, you would see numerous replication forks spaced along the chromosome, rather than the single replication fork seen in bacterial chromosomes. Each individual zone of the chromosome replicates as a discrete unit, called a **replication unit.** Replication units vary in length, from ten thousand to one million base pairs; most are about one hundred thousand base pairs in length. They have been described for many different eukaryotes. Because each eukaryotic chromosome possesses so much DNA (see figure 9.3), the orderly replication of DNA in eukaryotes undoubtedly requires sophisticated controls, which are, as yet, largely unknown.

Genes: The Units of Hereditary Information

In 1902, British physician Archibald Garrod noted that certain diseases were prevalent in particular families. Indeed, when he examined several generations within such families, he found that some of these disorders behaved as if they were controlled by simple recessive alleles. Garrod concluded that these disorders were Mendelian traits and that they had resulted from hereditary information changes in an ancestor of the affected families.

KEY EXPERIMENT 12.4

THE MESELSON-STAHL EXPERIMENT

Matthew Meselson and Frank Stahl transferred bacterial cells containing heavy nitrogen isotopes to a new medium containing only light nitrogen isotopes. At various times thereafter, they took samples, and the DNA was spun at a high speed in a cesium chloride solution. Because the cesium ion is so massive, the cesium chloride tends to settle in the rapidly spinning tube, estab-

lishing a gradient of cesium concentration. DNA molecules sink in the gradient until they reach a place where the cesium concentration has the same density as the DNA density; the DNA then "floats" at that position. Because DNA built with heavy nitrogen isotopes is denser than normal DNA, it sinks to a lower position on the cesium gradient. By sampling DNA at regular intervals and then placing the samples in the cesium gradient, Meselson and Stahl were able to tell what the strands of newly synthesized DNA were composed of: the light isotope or the heavy isotope, or eventually, no isotope.

2000

1958

1900

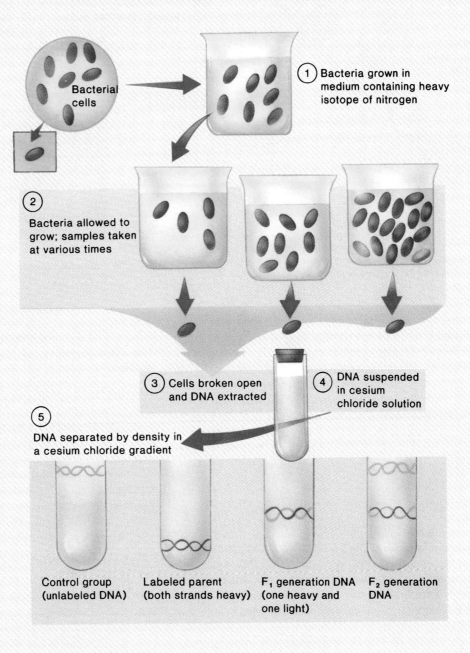

Bacterial cells

1. Bacteria grown in medium containing heavy isotope of nitrogen

2. Bacteria allowed to grow; samples taken at various times

3. Cells broken open and DNA extracted

4. DNA suspended in cesium chloride solution

5. DNA separated by density in a cesium chloride gradient

Control group (unlabeled DNA)

Labeled parent (both strands heavy)

F₁ generation DNA (one heavy and one light)

F₂ generation DNA

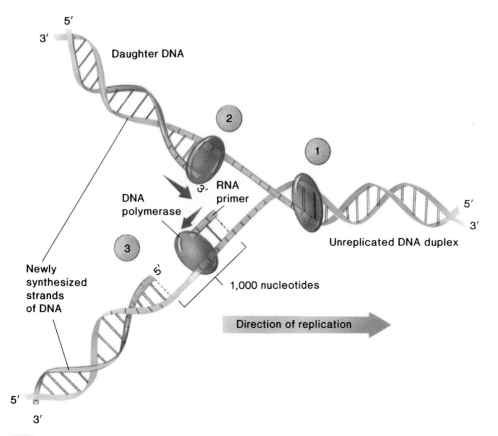

Figure 12.8 A replication fork.

DNA "unzips," and nucleotides are added to the exposed strands. Replication occurs in three stages: (1) Unwinding. Special proteins separate and stabilize the strands of the double helix. (2) Continuous synthesis. Nucleotides are added by DNA polymerase to the end of the "leading strand" (signified by 3'). (3) Discontinuous synthesis. A short RNA primer is added about a thousand nucleotides ahead of the end of the "lagging strand" (signified by 5'). DNA polymerase then adds nucleotides to the primer until the gap is filled in.

Garrod examined several of these disorders in detail. In one, **alkaptonuria,** the patients passed urine that rapidly turned black on exposure to air. Such urine contains homogentisic acid (alkapton), which air oxidizes. In normal individuals, homogentisic acid is broken down into simpler substances, but the affected patients are unable to carry out that breakdown. With considerable insight, Garrod concluded that the patients suffering from alkaptonuria lacked the enzyme necessary to catalyze this breakdown and, more generally, that many inherited disorders might reflect enzyme deficiencies.

The One-Gene/ One-Enzyme Hypothesis

From Garrod's finding, it is but a short leap of intuition to surmise that the information encoded within the DNA of chromosomes is used to specify particular enzymes. But this point was not actually established until 1941, when a series of experiments by Stanford University geneticists George Beadle and Edward Tatum finally provided definitive evidence. Beadle and Tatum deliberately set out to create mutations in the chromosomes; they then studied the effects of these mutations on the organism.

Creating Genetic Mutations

One reason why Beadle and Tatum's experiments produced clear-cut results—and one characteristic of most successful laboratory experiments in biology—is that they chose an excellent experimental organism. They selected the bread mold *Neurospora,* a fungus that can be readily grown in the laboratory on a **defined medium** (a medium that contains only known substances, such as glucose and sodium chloride, rather than some uncharacterized cell extract, such as ground-up yeasts). Beadle and Tatum then exposed the *Neurospora* spores to X rays and allowed the progeny to grow on a **complete medium** (a medium that contained all necessary nutrients and would therefore supply these nutrients to the growing fungi, whether or not individual strains could manufacture the nutrients for themselves). In this way, the investigators were able to preserve strains that, as a result of the earlier irradiation, had experienced damage to their DNA in a region encoding the ability to make one or more of the compounds that the fungus needed for normal growth. Change of this kind is called **mutation,** and the strains that have lost the ability to use one or more compounds are called **mutant strains.**

Identifying Mutant Strains

The next step was to test the progeny of the irradiated spores to see if any mutations leading to metabolic deficiency actually had been created by the X-ray treatment. Beadle and Tatum did this by attempting to grow subdivisions of individual fungal strains on a **minimal medium,** which contained only sugar, ammonia, salts, a few vitamins, and water. A cell that had lost the ability to make a necessary nutrient would not grow on such a medium. Using this approach, Beadle and Tatum succeeded in identifying and isolating many deficient mutants.

Pinpointing the Problem

To determine the nature of each deficiency, Beadle and Tatum tried adding various chemicals to the minimal medium to find one that would make it possible for a given strain to grow. In this way, they were able to pinpoint the nature of the biochemical problems that many of their mutants had developed (see Key Experiment 12.5). Many of the mutants proved unable to synthesize a particular vitamin or amino acid. The

KEY EXPERIMENT 12.5

THE BEADLE-TATUM EXPERIMENT: ISOLATING NUTRITIONAL MUTATIONS IN *NEUROSPORA*

2000

George Beadle and Edward Tatum used the procedure shown here to isolate nutritional mutations in the bread mold, *Neurospora*. *Neurospora* grows easily on an artificial medium in test tubes. In this experiment, spores were first irradiated (exposed to X rays) to increase the frequency of mutation and then were placed on a complete medium and allowed to grow. Any mutation that might have occurred in genes that were normally used by the fungus to produce its necessary amino acids or vitamins would not prevent growth, since all of these substances were present in the complete medium.

Once the colonies were established, individual spores were taken and tested to see whether they would grow on a minimal medium, which lacks the amino acids and vitamins that the fungus normally manufactures. Any strains that will not grow on a minimal medium but will grow on a complete medium contain one or more mutations in the genes that are necessary to produce one of the substances in the complete but not the minimal medium.

1941

To find out which one, the line is tested for its ability to grow on a minimal medium supplemented with particular substances. The mutation illustrated here is an arginine mutant, a cell line that has lost the ability to produce arginine. It will not grow on a minimal medium but will grow on a minimal medium to which only arginine has been added.

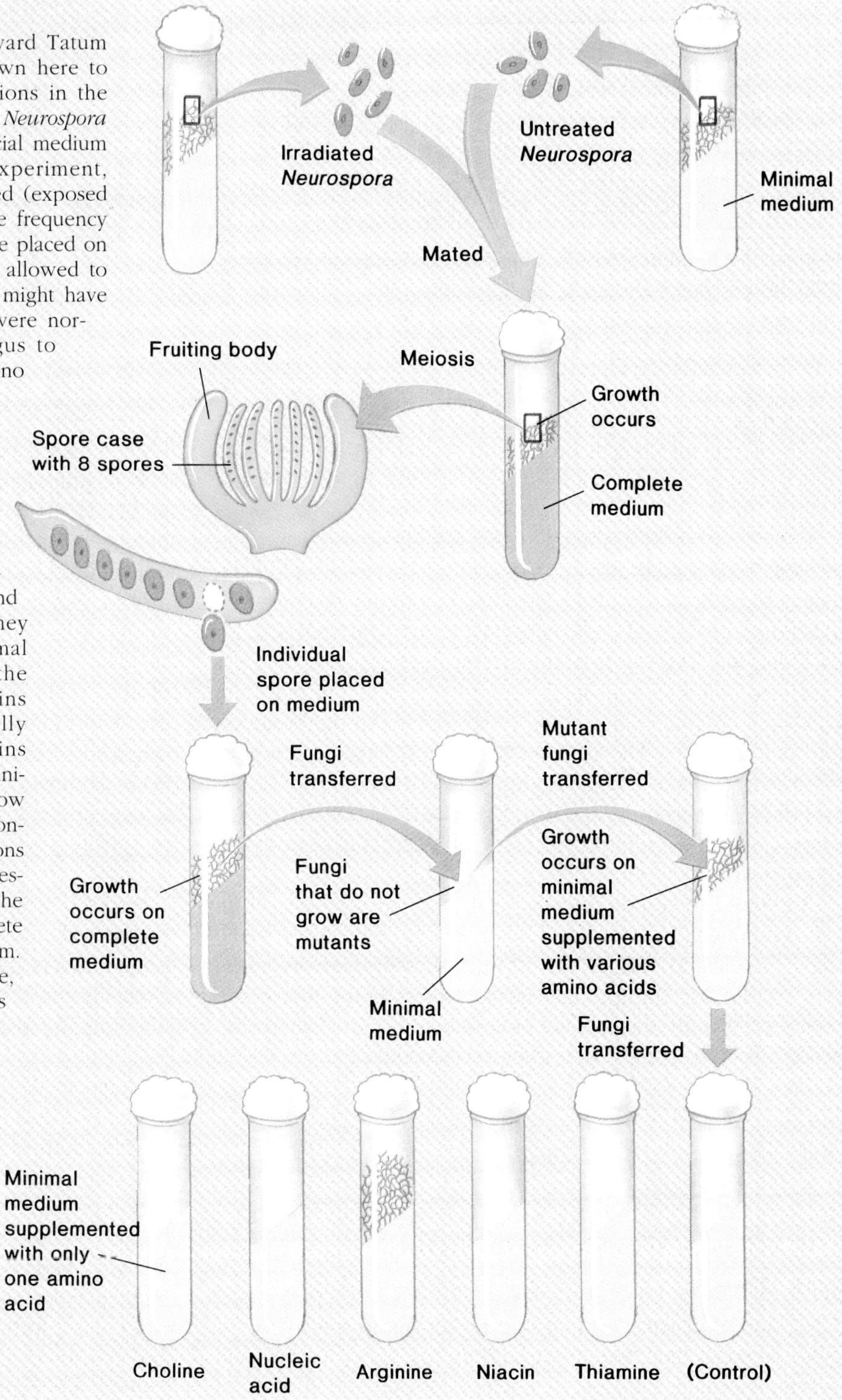

1900

Irradiated *Neurospora*

Untreated *Neurospora*

Minimal medium

Mated

Fruiting body

Meiosis

Growth occurs

Complete medium

Spore case with 8 spores

Individual spore placed on medium

Fungi transferred

Mutant fungi transferred

Growth occurs on complete medium

Fungi that do not grow are mutants

Growth occurs on minimal medium supplemented with various amino acids

Minimal medium

Fungi transferred

Minimal medium supplemented with only one amino acid

Choline   Nucleic acid   Arginine   Niacin   Thiamine   (Control)

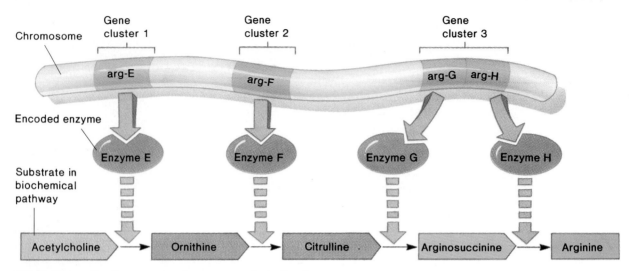

Figure 12.9 Evidence for the one-gene/one-enzyme hypothesis.
The chromosomal locations of the many arginine (arg) mutations isolated by Beadle and Tatum are found at three gene clusters.

addition of arginine, for example, permitted the growth of a group of mutant strains, dubbed *arg* mutants. The chromosomal positions of the mutant *arg* genes were found to cluster in three areas.

For each enzyme in the arginine biosynthetic pathway, Beadle and Tatum were able to isolate a mutant strain with a defective form of that enzyme. The mutation always proved to be located at *one* specific chromosomal site, a different site for each enzyme. Thus, each of the mutants that Beadle and Tatum examined could be explained in terms of a defect in one (and only one) enzyme, which could be localized at a single site on one chromosome. The geneticists concluded that genes produce their effects by specifying the structure of enzymes and that each gene encodes the structure of a single enzyme. They called this relationship the **one-gene/one-enzyme hypothesis** (figure 12.9).

Enzymes are responsible for catalyzing the synthesis of all cell components. They control the assembly of themselves and other proteins, as well as that of nucleic acids, carbohydrates, fats, and lipids. From the hair on your head to the toenails of your feet, you are the product of enzyme-directed chemical reactions. By specifying your enzymes, DNA specifies you.

> *Genetic traits are expressed largely as a result of enzyme activities. Organisms store hereditary information by encoding the structures of enzymes in the DNA of their chromosomes.*

DNA Encoding of Proteins

What kind of information must a gene contain to specify a protein? For some time, the answer was not clear because protein structure seemed impossibly complex. For example, it was not evident whether or not the molecules of a particular kind of protein had a consistent, identical sequence of amino acids. The picture changed in 1953, the same year in which Watson and Crick unraveled the structure of DNA. The great English biochemist Frederick Sanger, after many years of work, announced the complete sequence of amino acids in the protein insulin. Sanger's achievement demonstrated for the first time that proteins consist of definable sequences of amino acids, instead of random or meaningless ones. For any given form of insulin, each molecule has the same amino-acid sequence as every other, and this sequence can be learned and written down. All enzymes and other proteins are strings of amino acids arranged in a certain, definite order (see chapter 2 for a review of the structure of proteins). The information necessary to specify an enzyme, therefore, is an ordered list of amino acids.

In 1956, Sanger's pioneering work was followed by Vernon Ingram's analysis of the molecular basis of sickle-cell anemia, a protein defect inherited as a genetic disorder (see chapter 11). By analyzing the structure of normal and sickle-cell hemoglobin, Ingram, working at Cambridge University, showed that sickle-cell anemia was caused by the change of the amino acid from glutamate to valine at a *single* position in the protein. The alleles of the gene encoding hemoglobin differed only in their specifications of this one amino acid in the hemoglobin amino acid chain (figure 12.10).

These experiments, and other related ones, have finally brought scientists to a clear understanding of the nature of the unit of heredity. Like the dots and dashes of Morse code, the *sequence* of nucleotides in DNA is a code. The sequence provides the information that specifies the identity and order of amino acids in a protein. The sequence of nucleotides that encodes this information is called a **gene.**

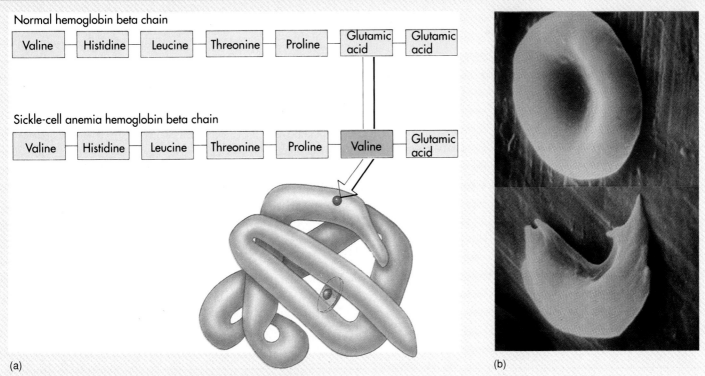

Normal hemoglobin beta chain

| Valine | Histidine | Leucine | Threonine | Proline | Glutamic acid | Glutamic acid |

Sickle-cell anemia hemoglobin beta chain

| Valine | Histidine | Leucine | Threonine | Proline | Valine | Glutamic acid |

(a)

(b)

Figure 12.10 The molecular basis of a hereditary disease.
(a) *Sickle-cell hemoglobin is produced by a recessive allele of the gene encoding the beta chain of the protein hemoglobin. It represents a single amino acid change—from glutamic acid to valine at the sixth position in the chain. In this model of a hemoglobin molecule, the position of the mutation can be seen near the end of the upper arm.* (b) *Top cell is normal; bottom cell is sickled.*

> *The amino-acid sequence of a particular protein is specified by a corresponding sequence of nucleotides in the DNA. This nucleotide sequence is called a gene.*

The next two chapters focus on the molecular nature of genes, exploring how genes work and what happens when one is changed. As you shall see, cancer is one example of what can happen when a gene is altered. Other changes—deliberately induced by genetic engineers—have proven beneficial. For example, most of the insulin used today by diabetics is the product of a human gene introduced into bacterial cells. The understanding that genes are the units of heredity represents one of the high-water marks of biology as a science. The intellectual path biologists have followed in their pursuit of this understanding has not always been a straight one, the best questions not always obvious. But however erratic and lurching the experimental journey, the picture of heredity has become progressively clearer and more sharply defined.

EVOLUTIONARY VIEWPOINT

DNA is one of the great constants of life. All cells use it to encode their hereditary instructions. Many biologists believe that the first organisms encoded their hereditary information in RNA, with DNA evolving only later as a more stable repository of the genetic instructions.

SUMMARY

1. Eukaryotic cells store hereditary information within the nucleus. When the nucleus is transplanted, so also are the hereditary specifications of an organism.

2. In viruses, bacteria, and eukaryotes, the hereditary information resides in DNA. The transfer of pure DNA can lead to the transfer of hereditary traits. In some viruses, the genetic material is RNA.

3. A DNA molecule is a long chain made up of repeating units. Each unit is called a nucleotide and is comprised of a five-carbon sugar, a phosphate group, and one of four nitrogen-containing bases (adenine, guanine, cytosine, and thymine).

4. DNA has a double helix structure, with the two chains held to each other by hydrogen bonds between nucleotides. Within the nucleotides, adenine bonds with thymine, and guanine bonds with cytosine.

5. During cell division, the hereditary message is duplicated with great accuracy because of complementarity. The DNA molecule is organized as a double helix with two strands that are complementary images of one another. By adding complementary nitrogenous bases, either of the strands can be used to re-form the helix.

6. DNA is replicated by a battery of enzymes, including a DNA polymerase and a variety of other proteins. Replication begins with the double helix unwinding. Each of the single strands is then used as a template from which to assemble a complementary new strand. The two strands are assembled in different directions: the new 5′-to-3′ oriented strand by the continuous addition of nucleotides to the growing end; the new 3′-to-5′ oriented strand in the opposite direction in thousand-nucleotide segments, which are then joined to the end of the growing strand.

7. Most hereditary traits reflect the actions of enzymes. The traits are hereditary because the information necessary to specify these enzymes is encoded within the DNA.

8. Enzymes are encoded in the DNA of genes, which are nucleotide sequences that specify the identity and order of enzymes' amino acids.

9. Changes in individual nucleotides can alter the identity of a particular amino acid in a protein. When such an alteration in DNA, called a mutation, results in the production of a protein with altered biological activity, or no activity at all, the mutation may cause a significant change in appearance or abilities.

REVIEWING THE CHAPTER

1. How did Hammerling's *Acetabularia* experiment contribute to the search for the structure of DNA?

2. Which chromosome component contains the hereditary information?

3. How did the Griffith-Avery experiments with *Streptococcus pneumoniae* contribute to the search for the structure of DNA?

4. How did the Hershey-Chase experiment with labeled viruses contribute to the search for the structure of DNA?

5. How did the Fraenkel-Conrat experiment with viruses differ from previous experiments?

6. What is the chemical nature of nucleic acids?

7. What is the significance of Chargaff's rule to the search for the structure of DNA?

8. What is the structure of DNA?

9. How does DNA replicate?

10. What is the significance of the Meselson-Stahl experiment?

11. How does the DNA molecule copy itself?

12. How does a eukaryote replicate its chromosomes?

13. What are the units of hereditary information?

14. How did the Beadle-Tatum experiment lead to the one-gene/one-enzyme hypothesis?

15. How does DNA encode proteins?

COMPLETING YOUR UNDERSTANDING

1. In Hammerling's experiments with *Acetabularia,* he grafted a nucleus-containing basal portion of an *A. mediterranea* individual (with a disk-shaped cap) to the stalk of an *A. crenulata* individual (with a flower-shaped cap). The cap that formed was flower-shaped, like that of *A. crenulata.* When he removed that cap, another one formed, but this one was disk-shaped, like that of *A. mediterranea.* What was responsible for the change in morphology of these two regenerated caps?

2. Why was *Acetabularia* a good research organism for Hammerling's experiments?

3. What is the transforming principle, and who introduced it?

4. Why are viruses such excellent models to use in nucleic acid experiments?

5. From an extract of human cells growing in tissue culture, you obtain a white fibrous substance. How would you distinguish whether it was DNA, RNA, or protein?

6. How were radioactive isotopes employed in the search for the structure of DNA?

7. How does DNA differ from RNA?

8. In analyzing DNA obtained from your own cells, you are able to determine that 15 percent of the nucleotide bases it contains are thymine. What percentage of the bases are cytosine?

9. From a hospital patient afflicted with a mysterious illness, you isolate and culture cells and then purify DNA from the culture. You find that the DNA sample obtained from the culture contains two quite different kinds of DNA: One is double-stranded human DNA, and the other is single-stranded virus DNA. You analyze the base composition of the two purified DNA preparations, with the following results:

Tube 1: 22.1% A : 27.9% C : 27.9% G : 22.1% T

Tube 2: 31.3% A : 31.3% C : 18.7% G : 18.7% T

Which of the two tubes contains single-stranded virus DNA?

10. The Watson-Crick model of DNA structure suggested that the basis for the faithful copying of the genetic material is complementarity. This means that, if you know that the base sequence of one strand is AATTCG, the sequence of the other strand must be

a. AATTCG.

b. TTGGAC.

c. TTAACG.

d. TTAAGC.

e. Do not have enough information.

11. What is a replication fork, and how does it relate to a 3′ end and a 5′ end in DNA?

12. What are the major contributions of Rosalind Franklin, Maurice Wilkins, Francis Crick, and James Watson in the search for the structure of DNA?

13. What was Linus Pauling's involvement in the search for the hereditary material?

14. What are the functions of the enzymes ligase and DNA polymerase?

15. What are the relationships among defined, complete, and minimal media and the identification of mutant strains of microbes, such as the bread mold *Neurospora?*

FOR FURTHER READING

Crick, F. H. C. 1988. The discovery of the double helix was a matter of selecting the right problem and sticking to it. *The Chronicle of Higher Education,* Oct. 5, 1–9. Francis Crick's own recollections of the hectic days when he and James Watson deduced that the structure of DNA is a double helix.

Freedman, D. 1991. Life's off switch. *Discover,* July. An account of how scientists are trying to learn what turns off DNA synthesis.

Hall, S. S. 1990. James Watson and the search for biology's "Holy Grail." *Smithsonian,* Feb., 41–49. Discussion of the effort to provide a complete sequence of base pairs for the human genome.

Judson, H. F. 1979. *The eighth day of creation.* New York: Simon & Schuster. The definitive historical account of the experimental unraveling of the mechanism of heredity, based on personal interviews with the participants. This book is full of the feel of how science is really conducted.

Olson, A., and D. Goodsell. 1992. Visualizing biological molecules. *Scientific American,* Nov., 76–81. Computer-generated pictures of how proteins that interact with DNA are powerful tools helping scientists to understand how genes are regulated.

Radman, M., and R. Wagner. 1988. The high fidelity of DNA duplication. *Scientific American,* Aug., 40–47. A well-illustrated account of how errors are avoided during DNA replication.

Rhodes, D., and A. Klug. 1993. Zinc fingers. *Scientific American,* Feb., 56–65. New findings that shed light on how "transcription faction" proteins interact with the groove of DNA to regulate when genes are active and how they are transcribed.

Watson, J. D. 1968. *The double helix.* New York: Atheneum. A lively, often irreverent account of what it was like to discover the structure of DNA, recounted by someone in a position to know.

Watson, J. D., and F. H. C. Crick. 1953. A structure for deoxyribose nucleic acid. *Nature* 171:737. The original report of the double helical structure of DNA. Only one page long, this paper marks the birth of molecular genetics.

EXPLORATIONS

Interactive Software: **Smoking and Cancer**

This interactive exercise allows the student to explore the relationship between smoking and cancer. The exercise presents a diagram of a human chromosome, showing the location of four genes that regulate cell growth. When their activities are disabled, they actively promote growth. In this exercise, all four must be disabled for cancer to be initiated. Students investigate the relationship between smoking and cancer by varying the amount an individual smokes and seeing how long it takes before all four genes have been mutated to a cancer-causing state.

Questions to Explore:

   1. What role does dose play in the probability that smoking will lead to cancer?

   2. Can you discover a "safe" amount of smoking?

   3. Is the twentieth cigarette smoked in a day more or less dangerous than the first?

   4. How much does smoking one pack of cigarettes a day increase the likelihood that you will get cancer?

Interactive Software: **Reading DNA**

In this interactive exercise, the user explores what might at first seem a puzzling contradiction: How can regulatory proteins "read" the DNA double helix without unzipping it, when the base pairs of the two strands point *inward* toward the center of the helix? As the user can see from the animation, the proteins slide along the major groove of the helix, feeling the edges of the hydrogen bonds for clues. The user investigates the nature of this protein-DNA interaction by designing proteins with different structural motifs, including leucine zippers and homeodomains. These hypothetical proteins are tested against particular DNA sequences, including promotors and homeoboxes.

Questions to Explore:

   1. Can you observe any pattern to which motifs are most effective in binding to particular kinds of DNA sequences? Can you suggest a reason that might account for such a pattern?

   2. What is the effect of repeating a particular motif twice, side by side, within the DNA-reading portion of the hypothetical protein you are designing? When, if ever, would you expect to find such motif dimers in nature?

   3. How would you go about demonstrating in a real laboratory that a particular regulatory protein is not in fact locally "unzipping" the helix to get a better look at the sequence of base pairs?

   4. Is there any instance among the DNA sequences examined here that a dimer made up of different motifs is more effective than any single motif?

GENES AND HOW THEY WORK

*Each of these human chromosomes,
already duplicated, contains hundreds
of genes that govern development and
living. The exact number of chromosomes is critical—a lost or extra copy
of any chromosome leads to abnormal
development, and in most instances,
death prior to birth.*

FOR REVIEW

Here are some important terms and concepts that have been discussed in previous chapters and that you will encounter again in this chapter. Review them before proceeding if necessary.

Structure of DNA and RNA (*chapters 2, 12*)
Ribosomes (*chapter 4*)
Enzymes and enzyme activity (*chapter 6*)
Structure of eukaryotic chromosomes (*chapter 9*)

The discoveries that genes are composed of DNA within chromosomes, and that genes act by directing the production of specific proteins, left unanswered the question of how this is accomplished. Solving this puzzle has been one of the great achievements of modern science.

Why did it prove so difficult to understand how the DNA within cells functions? For one thing, there is a great deal of DNA to understand. If the total DNA from a single cell of your body were stretched out, with all forty-six chromosomes lined up, one after the other, the DNA would extend as high as you are tall. All the DNA in your body, stretched end-to-end, would extend about 200 billion kilometers! Its instructions specify that you will have arms and not fins, hair and not feathers, two eyes and not one. The color of your eyes, the texture of your fingernails, whether or not you dream in color—all of the many traits that you receive from your parents are recorded in the DNA present in every cell of your body (figure 13.1).

The information in DNA is arrayed in little blocks, like entries in a dictionary. Each block is a gene specifying a particular polypeptide. Some of these polypeptides function as entire proteins as soon as they are formed, whereas many other proteins are formed of two or more polypeptides produced individually by separate genes. Proteins are the tools of heredity. Many of them are enzymes that carry out reactions within cells: What you are is the result of what they do.

The essence of heredity is a cell's ability to faithfully copy its meters of DNA-encoded instructions, making few errors, and to use these instructions to bring about the production of particular polypeptides and so affect what the cell will be like. This chapter examines how this happens.

RNA's Role in Polypeptide Synthesis

Where in the cell are proteins made? By placing cells in a medium containing radioactive amino acids, cell biologists were able to demonstrate that proteins are assembled not in the nucleus, where the chromosomal DNA resides, but rather in the cytoplasm, on large clusters of protein and ribonucleic

acid (RNA) called ribosomes. Ribosomes are composed of two subunits, with the smaller subunit fitting into a depression on the surface of the larger one (figure 13.2). These polypeptide-making factories are very complex, containing over fifty different proteins as well as RNA. As explained in chapter 2, RNA is very similar to DNA, and its presence in ribosomes hints at its importance in polypeptide synthesis (figure 13.3).

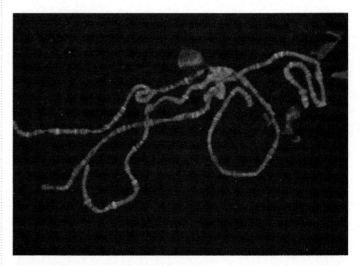

Figure 13.1 Genes on a chromosome.
In the giant chromosomes of the fly Drosophila melanogaster, *active genes appear as brighter bands. At any one time, only a fraction of a chromosome's many genes are actively being transcribed to produce RNA molecules.*

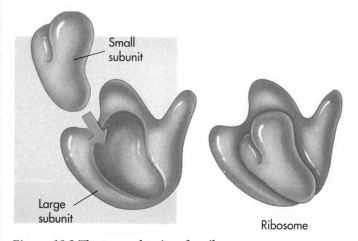

Figure 13.2 The two subunits of a ribosome.
The smaller subunit fits into a depression on the surface of the larger one.

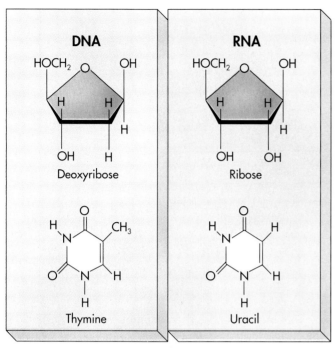

Figure 13.3 How RNA is different from DNA.
There are two important differences: First, in place of the sugar deoxyribose, RNA contains ribose, which has an additional oxygen atom. Second, RNA contains the pyrimidine uracil (U) instead of thymine (T). Another difference is that RNA does not have a regular helical structure and is usually single-stranded.

A cell contains many kinds of RNA. There are three major classes:

1. Ribosomal RNA. The class of RNA found in ribosomes, where it occurs with characteristic proteins, is called **ribosomal RNA** or **rRNA.** During polypeptide synthesis, rRNA molecules provide the site on the ribosome where the polypeptide is assembled.

2. Transfer RNA. A second class of RNA, called **transfer RNA,** or **tRNA,** occurs as much smaller molecules than rRNA. Cells contain more than sixty kinds of tRNA molecules, which float free in the cytoplasm. During polypeptide synthesis, tRNA molecules transport the amino acids to the ribosome for use in building the polypeptide and position each amino acid at the correct place on the elongating polypeptide chain.

3. Messenger RNA. A third class of RNA is **messenger RNA,** or **mRNA.** Each mRNA molecule is a long, single strand of RNA that passes from the nucleus to the cytoplasm. During polypeptide synthesis, mRNA molecules bring information from the chromosomes to the ribosomes to direct which polypeptide is assembled.

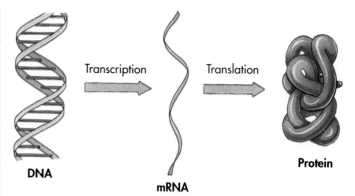

Figure 13.4 Transcription and translation.
These two phases of gene expression occur together in bacteria, with the mRNA beginning to be translated before all of it is transcribed. In eukaryotes, transcription takes place in the nucleus, while translation occurs in the cytoplasm.

These RNA molecules, together with ribosome proteins and certain enzymes, constitute an apparatus that reads the genetic message and produces the polypeptide that the particular message specifies. They are the cell's code-readers, what the cell uses to translate its hereditary information. This hereditary information is "written" in code and specified by the sequence of nucleotides in the DNA. The cell's polypeptide-producing apparatus reads this message one gene at a time, translating the genetic code of each gene into a particular polypeptide. As we shall see, biologists have also learned to read this code. In so doing, they have learned a great deal about what genes are and how they dictate what a protein will be like and when it will be made. Breaking the genetic code is one of the greatest achievements of modern biology.

The Central Dogma of Gene Expression

The hereditary apparatus of all organisms operates basically the same way: An RNA copy of each active gene is made, and the RNA copy directs the sequential assembly of a chain of amino acids at a ribosome. This whole process is known as **gene expression.** Stated simply, gene expression is the conversion of DNA sequence information into the amino-acid sequences of particular proteins.

There are many minor differences in the details of gene expression between bacteria and eukaryotes and a single major difference that is discussed later in the chapter. But the basic apparatus used in gene expression appears to be the same in all organisms and has apparently persisted virtually unchanged since early in the history of life. Gene expression occurs in two phases, called transcription and translation (figure 13.4).

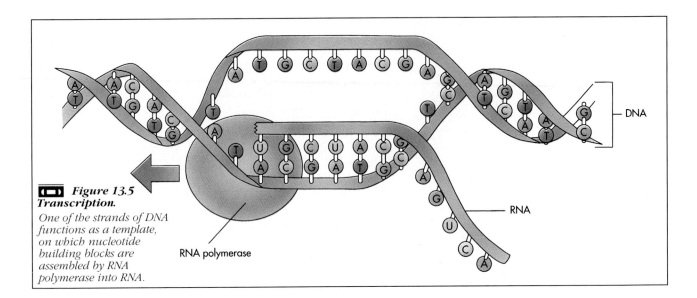

□□ *Figure 13.5*
Transcription.

One of the strands of DNA functions as a template, on which nucleotide building blocks are assembled by RNA polymerase into RNA.

RNA polymerase

DNA

RNA

Transcription

The first stage of gene expression is the production of an RNA copy of the gene, called messenger RNA or mRNA. Like all classes of RNA that occur in cells, mRNA is formed on a DNA template (figure 13.5). The production of mRNA is called **transcription;** the mRNA molecule is said to have been transcribed from the DNA.

Transcription is initiated when a special enzyme, called an RNA polymerase, binds to a particular sequence of nucleotides on one of the DNA strands, a sequence located at the starting edge of a gene. Starting at that end of the gene, the RNA polymerase assembles a single strand of mRNA with a nucleotide sequence complementary to that of the DNA strand to which it has bound. **Complementarity** refers to the way in which the two single strands of DNA that form a double helix relate to one another, with A (adenine) pairing with T (thymine) and G (guanine) pairing with C (cytosine). An RNA strand complementary to a DNA strand has the same relationships, but with U (uracil) in place of thymine.

As the RNA polymerase moves along the gene, encountering each DNA nucleotide in turn, it adds the corresponding complementary RNA nucleotide to the growing mRNA strand. Upon arriving at the far end of the gene, the RNA polymerase disengages from the DNA, releasing the newly assembled mRNA chain. This chain is complementary to the DNA strand from which the polymerase assembled it; thus, it is an RNA copy or "transcript" of the gene, called the **primary mRNA transcript.** In eukaryotes, this mRNA transcript passes out of the nucleus into the cytoplasm before it is used to direct the synthesis of a specific protein in the process of translation.

Translation

The second stage of gene expression is the synthesis of a polypeptide by a ribosome, which uses the information contained on an mRNA molecule to direct the choice of amino acids. This process is called **translation** because nucleotide-sequence information is translated into amino-acid-sequence information. Translation begins when an rRNA molecule within a ribosome binds to one end of an mRNA molecule. The ribosome then proceeds to move down the mRNA in steps of three nucleotides each. At each step, the ribosome adds an amino acid to a growing polypeptide chain. It continues until it encounters a "stop" signal on the mRNA, which indicates the end of the polypeptide. The ribosome then disengages from the mRNA and releases the newly assembled polypeptide. An overview of protein synthesis is presented in figure 13.6.

The information encoded in genes is expressed in two stages: transcription, in which a polymerase enzyme assembles an mRNA molecule whose sequence is complementary to the DNA; and translation, in which a ribosome assembles a polypeptide, using the mRNA to specify the amino acids.

The Genetic Code

How does the order of nucleotides in a DNA molecule encode the information that specifies the order of amino acids in a protein? What, in other words, is the nature of the DNA genetic code? The answer came in 1961 as the result of an experiment led by Francis Crick. By treating virus DNA with chemicals that add or delete single nucleotides, Crick was able to show that genes are read in increments of three consecutive nucleotides. Each block of three nucleotides, a **codon,** codes for one amino acid. Because there are four kinds of nucleotides in RNA (cytosine, guanine, adenine, and uracil instead of thymine), 4^3 or 64 different three-letter codons are possible.

Just which three-nucleotide code words correspond to which amino acids (the **genetic code**) was soon worked out,

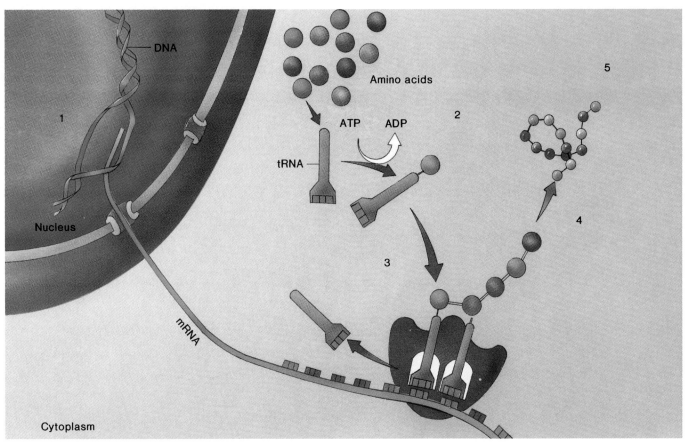

Figure 13.6 An overview of protein synthesis.
(1) *mRNA is transcribed from DNA.* (2) *tRNA binds amino acids.*
(3) *Loaded tRNA binds to ribosome.* (4) *Peptide chain elongates.*
(5) *Completed protein is released.*

the first results coming within a year of Crick's experiment. Researchers had developed mixtures of RNA and protein isolated from ruptured cells ("cell-free systems") that would synthesize proteins in a test tube. To determine which three-nucleotide sequences specify which amino acids, researchers added artificial RNA molecules to these cell-free systems and then watched to see what proteins were made. For example, when Marshall Nirenberg of the National Institutes of Health added polyU (an RNA molecule consisting of a string of uracil nucleotides) to such a system, polyphenylalanine (a protein consisting of a string of phenylalanine amino acids) was synthesized. This indicated that the three-nucleotide codon specifying phenylalanine was UUU. In this and other ways, all sixty-four possible triplets were examined, and the full genetic code was determined.

How Ribosomes Work

Working out the genetic code was a great step forward in removing the mystery from the process of gene expression. However, an important question was still unanswered: How is the information stored in a sequence of nucleotides, such as UUU, used to identify a specific amino acid, such as phenylalanine? How is the genetic code deciphered?

Answering this question requires looking more carefully at the first events of the translation process. Translation occurs on the ribosomes. First, the initial portion of the mRNA transcribed from a gene binds to an rRNA molecule interwoven in the ribosome. The mRNA lies on the ribosome in such a way that only a three-nucleotide portion of the mRNA molecule—the codon—is exposed at the polypeptide-assembly site (figure 13.7). As each codon of the mRNA message is exposed in turn, a molecule of tRNA with the complementary three-nucleotide sequence, or anticodon, binds to it (figure 13.8). Because this tRNA molecule carries a particular amino acid, that amino acid and no other is added to the polypeptide in that position. Protein synthesis occurs as a series of tRNA molecules bind one after another to the exposed portion of the mRNA molecule as it moves through the ribosome. Each tRNA molecule has an amino acid attached to it, and the amino acids are added, one after another, to the end of a growing polypeptide chain (figure 13.9).

Activating Enzymes

How does a particular tRNA molecule come to possess the amino acid that it does, and not just any amino acid? The correct amino acid is placed on each tRNA molecule by a collection of twenty enzymes called **activating enzymes.** There is one

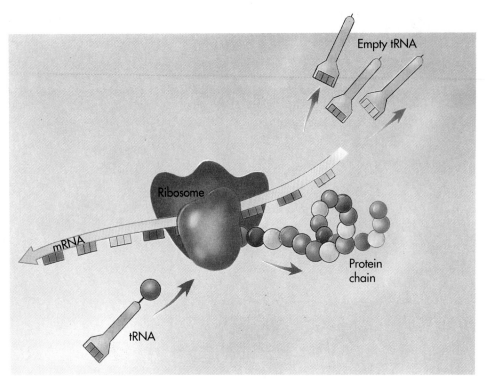

Figure 13.7 Translation.
The mRNA strand acts as a template for tRNA molecules. The appropriate tRNA is selected and positioned by the ribosome, which moves along the mRNA in three-nucleotide steps.

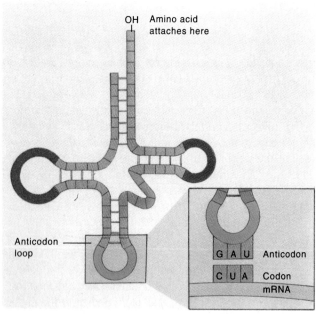

Figure 13.8 Structure of a tRNA molecule.
Loops called the T ψ C loop and the D loop function in binding to the ribosomes during polypeptide synthesis. The third loop contains the anticodon sequence. The activating enzyme adds an amino acid to the free, single-stranded —OH end.

activating enzyme for each of the twenty common amino acids. An activating enzyme binds the amino acid that it recognizes to a tRNA molecule (figure 13.10). If the nucleotide sequence of mRNA is considered a coded message, then the twenty activating enzymes are the cell's codebooks—the instructions for decoding the message. An activating enzyme recognizes both nucleotide-sequence information (a specific anticodon sequence of a tRNA molecule) and protein-sequence information (a particular amino acid).

Examining the Genetic Code

The code word recognized by an activating enzyme is three nucleotides long. As mentioned earlier, sixty-four different three-letter code words, or codons, are possible. Some of the activating enzymes recognize only one tRNA molecule, corresponding to one of these code words; others recognize two, three, four, or six different tRNA molecules, each containing a different anticodon. The base sequences of the tRNA anticodons are complementary to the associated sequences of mRNA and relate to the same amino acid as do those of their partner. Table 13.1 lists the different mRNA codons specific for each of the twenty amino acids—the genetic code.

The genetic code is the same in all organisms, with only a few exceptions. A particular codon, such as AGA, corresponds to the same amino acid (arginine) in bacteria as in humans. The genetic code's universality is strong evidence that all living things share a common evolutionary heritage.

Three of the sixty-four codons (UAA, UAG, and UGA) are not recognized by any activating enzyme. These codons serve as "stop" signals in the mRNA message, marking the end of a polypeptide. The "start" signal, which marks the beginning of a polypeptide amino-acid sequence within a mRNA message, is the codon AUG, a three-base sequence that also encodes the amino acid methionine. The ribosome uses the first AUG that it encounters in the mRNA message to signal the start of its translation.

All organisms possess a battery of twenty enzymes, called activating enzymes, one or more of which recognize a particular three-base anticodon sequence in a tRNA molecule. The mRNA codons specific for the twenty common amino acids constitute the genetic code.

Figure 13.9 Genetic shorthand.
Common representations of tRNA, mRNA, ribosomes, and proteins are shown.

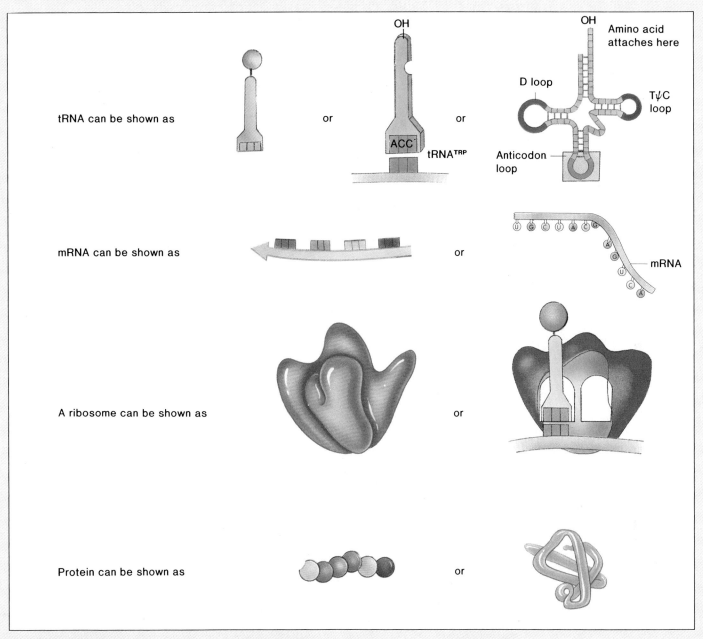

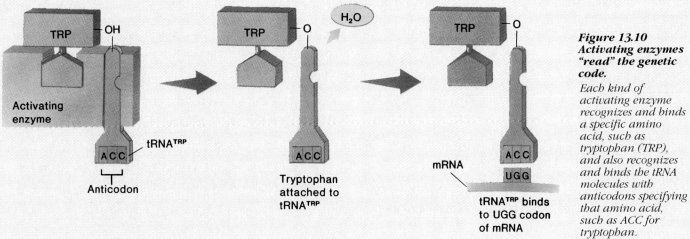

Figure 13.10 Activating enzymes "read" the genetic code.

Each kind of activating enzyme recognizes and binds a specific amino acid, such as tryptophan (TRP), and also recognizes and binds the tRNA molecules with anticodons specifying that amino acid, such as ACC for tryptophan.

Table 13.1   The Genetic Code

First Letter	Second Letter				Third Letter
	U	**C**	**A**	**G**	
U	Phenylalanine	Serine	Tyrosine	Cysteine	U
	Phenylalanine	Serine	Tyrosine	Cysteine	C
	Leucine	Serine	Stop	Stop	A
	Leucine	Serine	Stop	Tryptophan	G
C	Leucine	Proline	Histidine	Arginine	U
	Leucine	Proline	Histidine	Arginine	C
	Leucine	Proline	Glutamine	Arginine	A
	Leucine	Proline	Glutamine	Arginine	G
A	Isoleucine	Threonine	Asparagine	Serine	U
	Isoleucine	Threonine	Asparagine	Serine	C
	Isoleucine	Threonine	Lysine	Arginine	A
	(Start); Methionine	Threonine	Lysine	Arginine	G
G	Valine	Alanine	Aspartate	Glycine	U
	Valine	Alanine	Aspartate	Glycine	C
	Valine	Alanine	Glutamate	Glycine	A
	Valine	Alanine	Glutamate	Glycine	G

To read table: A codon consists of three nucleotides read in the sequence indicated by the column heads. For example, ACU codes threonine. The first letter, A, is read in the first-letter column; the second letter, C, from the second-letter column; and the third letter, U, from the third-letter column. Each codon is recognized by a corresponding anticodon sequence on a tRNA molecule. Some tRNA molecules recognize more than one codon sequence but always for the same amino acid. Most amino acids are encoded by more than one codon. For example, threonine is encoded by four codons (ACU, ACC, ACA, and ACG), which differ from one another only in the third position.

How Translation Occurs

Polypeptide synthesis begins with the formation of an initiation complex. Special proteins called **initiation factors** position the initial tRNA on the ribosomal surface. The proper positioning of this first amino acid is critical because it determines the **reading frame**—the particular groups of three bases with which the nucleotide sequence will be translated into a polypeptide.

After the tRNA has been positioned on the ribosomal surface, this initiation complex, guided by another initiation factor, then binds to mRNA. The complex must bind to the beginning of a gene, so that all of the gene will be translated (figure 13.11). In bacteria, the beginning of each gene is marked by a sequence that is complementary to one of the rRNA molecules on the ribosome. This ensures that genes are read from the beginning; each mRNA binds to the ribosomes that read it by base-pairing between the sequence at its beginning and the complementary sequence on the rRNA, which is a part of the ribosome.

An initiation complex consists of a ribosome, mRNA, and a tRNA molecule.

After the initiation complex has been formed, polypeptide synthesis proceeds as follows (figure 13.12):

1. The ribosome exposes the mRNA codon that is next to the initiating AUG codon, positioning it for interaction with another incoming tRNA molecule. When a tRNA molecule with the appropriate anticodon appears, it binds to the mRNA molecule at the exposed codon position. Special proteins called **elongation factors** (because they aid in making the mRNA molecule longer) help to position the incoming tRNA. Binding the incoming tRNA to the ribosome in this fashion places the amino acid held by the incoming tRNA molecule next to the initial amino acid, which is held by the initiating tRNA molecule still bound to the ribosome.

2. The two amino acids undergo a chemical reaction in which the initial amino acid is released from its tRNA and is attached instead by a peptide bond to the adjacent incoming amino acid. The abandoned tRNA falls from its site on the ribosome, leaving that site vacant.

3. In a process called **elongation** (figure 13.12), the ribosome now moves along the mRNA molecule a distance corresponding to three nucleotides, guided by other elongation factors. This movement repositions the growing chain, at this point containing two amino acids, and exposes the next mRNA codon. This is the same situation that existed in step 1. When a tRNA molecule that recognizes this next codon appears, the anticodon of the incoming tRNA binds this codon, placing a new amino acid adjacent to the growing chain. The growing chain transfers to the incoming amino acid, as in step 2, and the elongation process continues (figure 13.13).

4. When a stop codon is encountered, no tRNA exists to bind to it. Instead, the stop codon is recognized by special **release factors,** proteins that release the newly made polypeptide from the ribosome.

Figure 13.11 Formation of the initiation complex.

Proteins called initiation factors (for purposes of illustration, they are called initiation factors a *and* b *here) play key roles in positioning the small ribosomal subunit and the tRNA molecule carrying methionine (met-tRNA) at the beginning of the mRNA message. When the met-tRNA is positioned over the first AUG codon sequence of the mRNA, the large ribosomal subunit binds, forming the A-and P-sites, and polypeptide synthesis begins.*

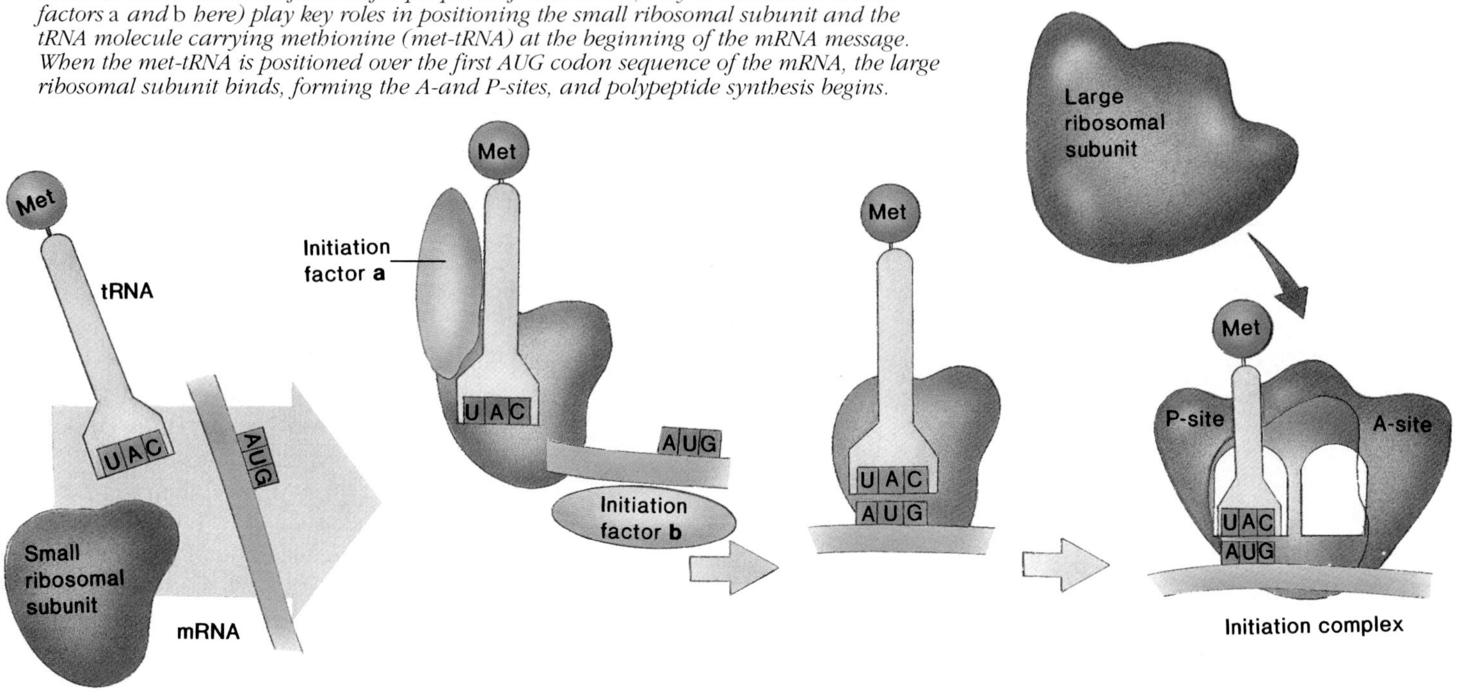

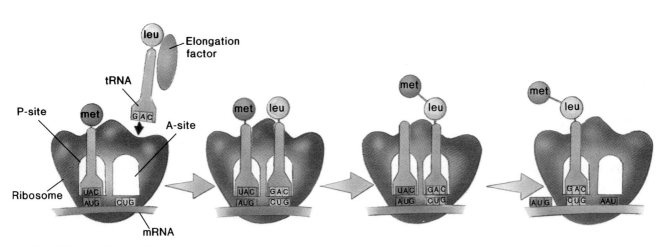

Figure 13.12 How polypeptide synthesis proceeds.

The A-site becomes occupied by a tRNA with an anticodon complementary to the mRNA codon exposed there. A special protein called an elongation factor helps to position the incoming tRNA. The growing polypeptide chain (which at this point, consists of methionine [met]) is transferred to this incoming amino acid (here, leucine [leu]), and the ribosome moves three nucleotides to the right.*

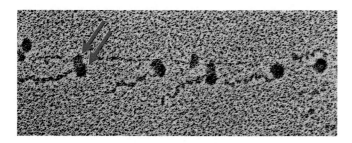

Figure 13.13 Translation in action.

These ribosomes are reading along an mRNA molecule of the fly Chironomus tentans *from top to bottom, assembling polypeptides that dangle behind them like the tail of a tadpole. Clearly visible are the two subunits (arrows) of each ribosome translating the mRNA.*

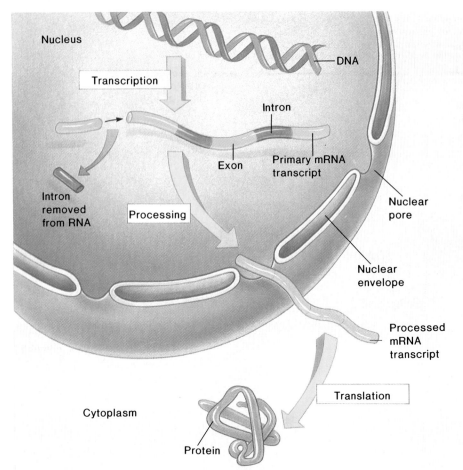

Figure 13.14 Gene processing in a eukaryotic cell.
Long stretches of nucleotides, called introns, are cut out of the primary mRNA transcript before the processed mRNA transcript is used for protein synthesis.

The ribosome sites where mRNA codons are exposed play critical roles in polypeptide synthesis. The site where incoming tRNA molecules first bind to the mRNA is called the attachment site, or A-site. The site next to it, to which the growing chain of amino acids is attached, is called the polypeptide chain site, or P-site. Recent data suggest the existence of a third release site, or R-site, that is past the P-site and functions in releasing the tRNA after it has donated its amino acid to the growing chain.

Polypeptide synthesis is carried out by ribosomes, which bind to sites at one end of the mRNA and then move down the mRNA in increments of three nucleotides. Each step of the ribosome's progress exposes a three-base sequence to binding by a tRNA molecule with the complementary nucleotide sequence. Ultimately, the amino acid carried by that tRNA molecule is added to the end of the growing polypeptide chain.

Comparing Translation in Prokaryotes and Eukaryotes

Translation is similar in both prokaryotes (bacteria) and eukaryotes, although one difference in the initial transcript is of particular importance: Unlike bacterial genes, most eukaryotic genes are much larger than they need to be, containing long stretches of nucleotides that are cut out of the mRNA transcript before it is used in polypeptides synthesis. These removed sequences are called **introns** and are not translated into polypeptides (figure 13.14) (see Key Experiment 13.1).

The remaining segments of the gene—the nucleotide sequences that encode the amino-acid sequence of the polypeptide—are called **exons.** Exons are scattered among larger, noncoding, intron sequences. In a typical human gene, the nontranslated (intron) portion of a gene is ten to thirty times larger than the coding (exon) portion. Despite these differences, eukaryotes make polypeptides in much the same way as do bacteria.

The Need to Regulate Gene Expression

An organism must be able to control which of its genes are being transcribed, and when. If a cell produces an enzyme when the enzyme's substrate—the target of the enzyme's activity—is not present in the cell, a great deal of energy is expended for nothing. Cells avoid this waste, conserving energy until the appropriate substrate is encountered and the enzyme's activity is of use to the cell.

From a broader perspective, the growth and development of many multicellular organisms, including human beings, entails a long series of biochemical reactions, each of which is delicately tuned to achieve a precise effect. Specific enzyme activities are called into play to bring about a particular developmental change. After the change, those particular enzyme activities cease, lest they disrupt other activities that follow. During development, genes are transcribed in a carefully prescribed order, each gene for a specified period. The hereditary message is played like a piece of music on a grand organ, in which particular proteins are the notes and the hereditary information that regulates their expression is the score.

KEY EXPERIMENT 13.1

2000

CHAMBON'S DISCOVERY OF INTRONS

◀ 1980

Virtually every nucleotide within the transcribed portion of a bacterial gene participates in an amino-acid—specifying codon, and the order of amino acids in the protein is the same as the order of the codons in the gene. It was assumed for many years that all organisms would naturally behave in this logical way. In the late 1970s, however, biologists were amazed to discover that this relationship, one with which they had become completely familiar, did not in fact apply to eukaryotes. Instead, French researcher Piere Chambon discovered that eukaryotic genes are encoded in segments that are excised from several locations along the transcribed mRNA and subsequently stitched together to form the mRNA that is eventually translated in the cytoplasm. With the benefit of hindsight, designing an experiment that reveals this unexpected mode of gene organization is not difficult:

1. Isolate the mRNA corresponding to a particular gene. Much of the mRNA of red blood cells, for example, is related to the production of the proteins hemoglobin and ovalbumin, making it easy to purify the mRNAs from the genes related to these proteins.

2. Using an enzyme called reverse transcriptase, make a DNA version of the mRNA that has been isolated. Such a version of a gene is called "copy" DNA (cDNA).

3. Using genetic engineering techniques (chapter 14), isolate from the nuclear DNA the portion that corresponds to one of the actual hemoglobin genes. This procedure is referred to as "cloning" of the gene in question.

4. Mix single-strand forms of this hemoglobin cDNA and nuclear DNA and permit them to pair with each other ("hybridize") and form a duplex.

When this experiment was conducted by Chambon for the ovalbumin gene and the resulting duplex DNA molecules were examined with an electron microscope, the hybridized DNA did not appear as a single duplex. Instead, unpaired loops were observed.

As shown in the illustration (1), the ovalbumin gene and its primary transcript contain seven segments not present in the mRNA version, which the ribosomes use to direct protein synthesis. These segments, called *introns,* are removed by enzymes that cut out the introns and splice together the remaining coding segments, called *exons*. The seven loops in the photo (2) and accompanying schematic drawing (3) are the seven introns of the DNA and the primary mRNA transcript.

The conclusion is inescapable: Nucleotide sequences are removed from within the gene transcript before the cytoplasmic mRNA is translated into protein. Because introns are removed from the mRNA transcript before it is translated into tRNA, they do not affect the structure of the protein that is encoded by the gene in which they occur.

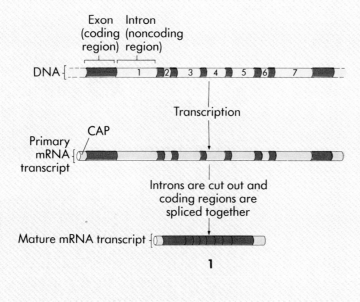

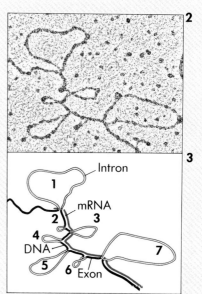

1900

Regulatory Sites and Regulatory Proteins

Organisms regulate gene expression largely by controlling when the transcription of individual genes begins. Most genes possess special nucleotide sequences called **regulatory sites** that act as points of control. These nucleotide sequences are recognized by specific **regulatory proteins** that bind to the sites.

Cells use gene regulatory sites to control which genes are transcribed by influencing the shape of the regulatory proteins that bind these sites. Regulatory proteins not only possess surfaces that fit gene regulatory sites, they also possess pockets on their surface that fit small molecules within cells. The binding of one of these small molecules into such a pocket can change the shape of a regulatory protein and thus destroy or enhance the regulatory protein's ability to bind to the gene. In some cases, the protein in its new shape may no longer recognize the gene's regulatory site. In other cases, the recontoured regulatory protein may begin to recognize a regulatory site that it previously ignored.

The cell thus uses the presence of particular "signal" molecules within the cell to incapacitate particular regulatory proteins or to mobilize them for action. These regulatory proteins, in turn, repress or activate the transcription of particular genes. The pattern of metabolites in the cell sets "on/off" protein regulatory switches, thereby achieving a proper configuration of gene expression.

> *Organisms control the expression of their hereditary information by selectively inhibiting the transcription of some genes and facilitating the transcription of others. Transcription is controlled by modifying the shape of regulatory proteins, thereby affecting their tendency to bind to sites on the gene that influence the initiation of transcription.*

Negative Control

Regulatory sites often function to shut off transcription, a process called **negative control.** In these cases, the site at which a regulatory protein binds to the DNA is located between the site at which RNA polymerase binds and the beginning edge of the gene that the polymerase is to transcribe. When a regulatory protein is bound to its regulatory site, its presence blocks the movement of the polymerase toward the gene. The effect is something like preparing to shoot a cue ball at the eight ball on a pool table, only to have someone place a brick on the table between the cue ball and the eight ball. Functionally, this brick is like the regulatory protein that binds to the DNA: Its placement blocks movement of the cue ball to the eight ball, just as placement of the regulatory protein between the polymerase and the gene blocks polymerase movement to the gene. The process of blocking transcription in this way is called repression, and the regulatory protein responsible for the blockage is called a **repressor protein** (figure 13.15).

The *lac* Operon

A specific example will help to demonstrate how repressor proteins work. The set of genes to be examined is the *lac* system of the bacterium *Escherichia coli.* The *lac* system is a cluster of genes that encode three proteins that bacteria use to obtain energy from the sugar lactose. These proteins include two enzymes and a membrane-bound transport protein (a permease). Researchers have found this cluster to be typical of how genes are organized in bacteria. Within the cluster are four different regions (figure 13.16):

1. Coding sequences. Three coding sequences specify the three lactose-utilizing enzymes. All three sequences are transcribed onto the same piece of mRNA and constitute part of an operational unit called an **operon.** An operon consists of one or more structural genes and the associated regulatory elements—the operator and the promoter—which are discussed shortly. This particular operon is called the *lac* operon because the three structural genes that it includes are all involved in lactose utilization. The clustering of coding sequences onto a single operon is common among bacteria, but rare in eukaryotes.

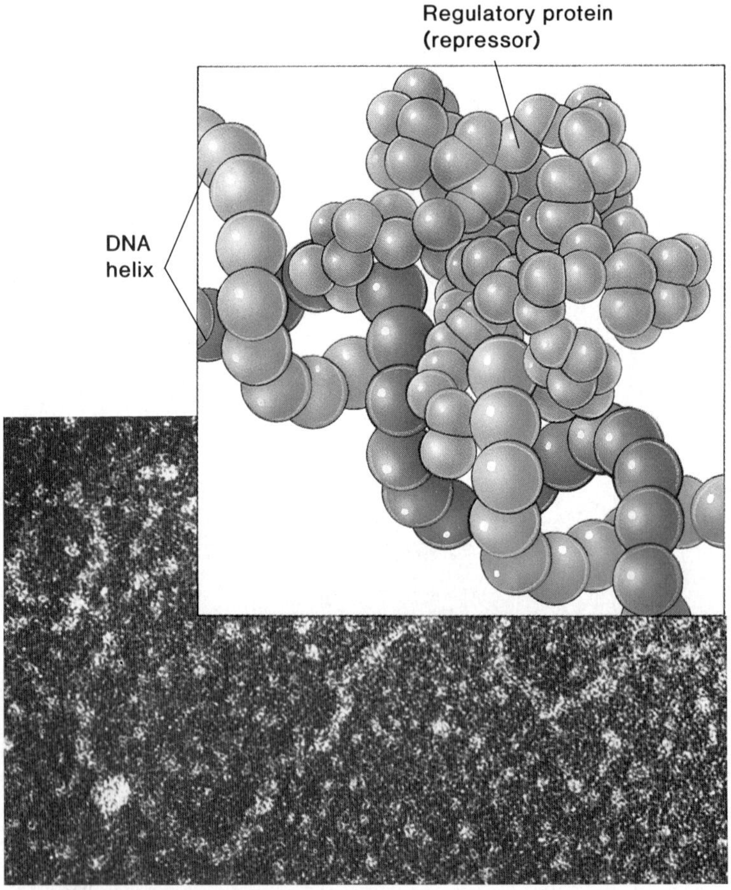

Figure 13.15 Repression: Blocking transmission.
The regulatory protein controlling transcription of the lactose genes is the large whitish sphere bound to the DNA strand at the lower left. Because this regulatory (repressor) protein fills the groove of the DNA double helix, the RNA polymerase cannot attach there, and the repressor blocks transcription.

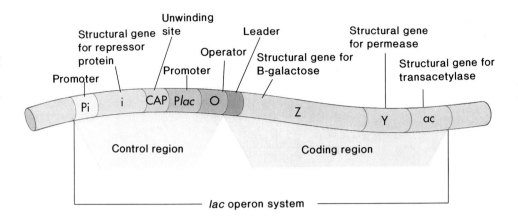

Figure 13.16 The **lac** *region of the* **Escherichia coli** *chromosome.*

All of the key elements of gene structure can be clearly identified: The coding region contains the protein-encoding genes; the leader region is the ribosome recognition site; the promoter is the RNA polymerase-binding site; the operator and CAP sites are negative and positive regulatory protein-binding sites.

2. Ribosome recognition site. In front of the three coding sequences is the binding site for the ribosome, a series of nucleotides that lies within an initial untranslated portion of the mRNA and sometimes called a leader region. Each mRNA molecule transcribed from the cluster is composed of the leader region and the three coding sequences, transcribed in order. The cluster of coding sequences and the ribosome recognition site (leader region) are often referred to as a transcription unit.

3. RNA polymerase-binding site. Still farther in front of the coding sequences specifying the three lactose-utilizing enzymes is a specific DNA nucleotide sequence that the polymerase recognizes and to which it binds. Such polymerase-recognition sites are called **promoters** because they promote transcription.

4. Regulatory protein-binding site. Between the promoter and the ribosome recognition site is a regulatory site, the **operator,** where a repressor protein binds to block transcription.

Genes encoding enzymes also possess regulatory regions. The segment that is transcribed into mRNA is called a transcription unit; it consists of the elements that are involved in the translation of the mRNA—the ribosome-binding site and the coding sequences. In front of the transcription unit on the DNA are the elements involved in regulating its transcription—binding sites for the polymerase and for regulatory proteins.

How Negative Control Works in the *lac* Operon

The *lac* operon's complex system of regulatory sites works to ensure that mRNA is copied from the three structural genes only when the cell can effectively utilize the proteins that the genes encode—that is, when there is lactose present. The sugar lactose is encountered by bacteria only occasionally, so the enzymes that metabolize lactose usually lack a substrate on which to act. In this situation, the *lac* repressor protein usually blocks

the proper functioning of RNA polymerase, and the cell is said to be repressed with respect to *lac* operon transcription (see Key Experiment 13.2).

The *lac* repressor protein is capable of changes in shape. When lactose binds to the repressor protein, the protein assumes a different shape, one that does not recognize the operator sequence! Therefore, if the cell contains much lactose, the *lac* repressor proteins become inactive. This removes the block from in front of the polymerase and so permits transcription of the *lac* genes to begin. Thus, if lactose is added to a growing bacterial culture, there is a burst of synthesis of the lactose-utilizing enzymes. The transcription of the enzymes is said to have been induced by the lactose. This element of the control system ensures that the *lac* operon is transcribed only in the presence of lactose.

Positive Control

Regulatory sites may also serve to turn on transcription, a process called **positive control.** In these situations, a regulatory protein must bind to the DNA before the transcription of a particular gene can begin. This regulatory protein is called an **activator protein,** and this way of turning on the transcription of specific genes is called activation.

Activation can be achieved by various mechanisms. In some cases, the activator protein's binding promotes the unwinding of the DNA duplex. This facilitates the production of an mRNA transcript of a gene because the polymerase, although it can bind to a double-stranded DNA duplex, cannot produce an mRNA transcript from such a duplex: mRNA is transcribed from a single strand of the duplex.

How Positive Control Works in the *lac* Operon

The *lac* operon has positive as well as negative regulatory sites. The principal positive regulatory site—where the activator protein binds—is located just in front of the promoter. In this instance, the activator protein is called the **catabolite activator protein,** or **CAP,** and the positive regulatory site is called the CAP site (see figure 13.16).

Like most other activator proteins, CAP works by making the polymerase's task easier. In this instance, binding of the

KEY EXPERIMENT 13.2

2000

1961

1900

THE JACOB-MONOD EXPERIMENT

François Jacob and Jacques Monod analyzed how the genes encoding lactose-utilizing enzymes (what they called the *lac* operon) became turned on ("induced") by obtaining mutations that blocked the process in various ways. One kind of mutation, in a region they called the *operator,* left the genes turned on at all times. It turned out that this was the site at which a negative regulatory protein (they called it the *repressor*) binds to the DNA, blocking movement by the polymerase toward the *lac* genes. Deletion of the DNA located just in front of the operator (a site called the *promotor*) turned all the genes "off." This proved to be the site where RNA polymerase binds the DNA. Jacob and Monod postulated that the *lac* operon is shut off ("repressed") when the repressor protein is bound to the operator site. Biologists now know

why: Further studies have shown that the promoter and operator sites overlap, so that the polymerase and the repressor cannot bind at the same time, any more than two people can sit in one chair. In the theory advanced by Jacob and Monod, the *lac* operon is transcribed ("induced") when lactose binding to the repressor protein changes the repressor protein's shape so that it can no longer sit on the operator site and block polymerase binding.

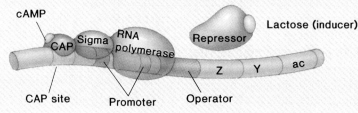

lac operon is "repressed"

lac operon is transcribed ("induced")

CAP protein to the CAP site of the *lac* operon facilitates the unwinding of the DNA duplex and so enables the polymerase to bind to the nearby promoter.

The CAP protein stimulates the transcription of the *lac* operon when the cell is low in energy. The CAP protein is able to sense when the cell is low in energy because it, like the repressor protein, is able to alter its shape when particular small molecules bind to it. In this case, the small molecule is a cyclic molecule made from ATP called cAMP (see the illustration in Key Experiment 13.2). Only when cAMP is attached to it can the CAP protein bind to DNA at the CAP site.

Cellular levels of cAMP are the key to how the positive control regulates the cell's transcription of lactose-utilizing genes. When cell glucose levels are high, cAMP levels are low; as a consequence, few CAP proteins have cAMP bound to them. Because few CAP proteins are around to activate transcription under these "well-fed" conditions, the *lac* genes are not transcribed often. The opposite situation occurs when the cell needs energy. Low glucose levels cause cAMP levels to be high. Under these circumstances, many CAP proteins have cAMP bound to them. Under these "starved" conditions, many CAP proteins activate transcription of the *lac* genes.

The advantage of having a positive control system in the *lac* operon is clear: The enzymes needed for the metabolism of lactose are produced only when the cell requires the energy that lactose would provide.

Positive and Negative Controls Working Together

The *lac* operon is thus controlled at two levels: (1) the lactose-utilizing enzymes are not produced unless the sugar lactose is available, and (2) even if lactose is available, the enzymes are not produced unless the cell has need of the energy. Similarly precise control mechanisms are known in eukaryotes, but the *lac* operon example illustrates the complex nature of cellular control of protein synthesis.

Gene Mutation

An enormous amount of DNA resides within the cells of your body. This DNA represents a long series of DNA replications, starting with the DNA of a single cell—the fertilized egg. The DNA replication required to produce the cells of your body is equivalent to producing a length of DNA nearly 97×10^9 kilometers long from an original 0.9-meter piece.

Living cells have evolved many mechanisms to avoid errors during DNA replication and to preserve the DNA from damage. These mechanisms "proofread" the strands of each daughter cell against one another for accuracy and correct any mistakes. But the proofreading is not perfect. If it were, no mistakes would occur, no variation in gene sequence would result, and evolution would come to a halt.

In fact, living cells *do* make mistakes, and changes in the genetic message do occur, although only rarely (figure 13.17). If changes were common, the genetic instructions encoded in DNA would soon degrade into meaningless noise. Typically, a particular gene is altered in only one out of a million gametes. Limited as it might seem, this small amount of change is the stuff of evolution. Every single difference between the genetic message specifying you and the one specifying your cat, or the fleas on your cat, arose as the result of genetic change.

A change in a cell's genetic message is called a **mutation.** Some mutational changes affect the message itself, altering the sequence of DNA nucleotides. These alterations in the coding sequence are called **point mutations** because they usually involve only one or a few nucleotides. Other classes of mutation involve changes in how the genetic message is organized. In both prokaryotes (bacteria) and eukaryotes, individual genes may move from one place on the chromosomes to another by a process called **transposition.** When a particular gene moves to a different location, there is often an alteration in its expression or in that of neighboring genes. In eukaryotes, large seg-

ments of chromosomes may change their relative location or undergo duplication. Such chromosomal rearrangement often drastically affects the expression of the genetic message. Sources and types of mutations are summarized in table 13.2.

Point mutations, involving only one or a few nucleotides, result from (1) either chemical or physical damage to the DNA or (2) spontaneous pairing errors that occur during DNA replication. The first of these is of particular practical importance because modern industrial societies produce and release into the environment many chemicals capable of damaging DNA, chemicals called **mutagens.**

Point mutations are changes in the hereditary message of an organism. They may result from physical or chemical damage to the DNA or from spontaneous errors during DNA replication.

DNA Damage

Although a DNA duplex can be damaged in many ways, three primary sources of damage are: (1) ionizing radiation, (2) ultraviolet radiation, and (3) chemical mutagens.

Ionizing Radiation

High-energy radiation, such as X rays and gamma rays, is highly mutagenic. When such radiation reaches a cell, it is absorbed by the atoms that it encounters, imparting energy to the electrons of their outer shells and causing these electrons to be ejected from the atoms. The ejected electrons leave behind ionized atoms with unpaired electrons, called **free radicals.**

Because the great majority of free radicals created by ionizing radiation are produced from water molecules and not DNA, most radiation damage to DNA is indirect. It occurs because free radicals are highly reactive chemically with other cell molecules, including DNA. The action of free radicals on a chromosome is like that of shrapnel from a grenade blast tearing into a human body.

DNA locations where damage occurs are random, and the damage is often severe. Cells can repair some of this damage. Chemical changes can be repaired by excising altered nucleotides, and single ruptured bonds can be re-formed. However, this **mutational repair** is not always accurate, and some of the mistakes are incorporated into the genetic message.

Ultraviolet Radiation

Ultraviolet (UV) radiation, the component of sunlight that leads to suntan (and sunburn), is much lower in energy than are X rays. In DNA, the principal absorption of UV radiation is by the pyrimidine bases thymine and cytosine. When these bases absorb UV energy, the electrons in their outer shells become reactive. If one of the nucleotides on either side of the absorbing pyrimidine is also a pyrimidine, a double covalent bond is formed between the two pyrimidines. This type of cross-link between adjacent bases of the DNA strand is called

Figure 13.17 Mutation.
Fruit flies normally have one pair of wings, extending from the thorax. This fly is a mutant, bithorax. Due to a mutation in a gene regulating a critical stage during development, it possesses two thorax segments and thus two sets of wings.

Table 13.2 Sources and Types of Mutation

Source	Primary Effect	Type of Mutation
Mutational		
Ionizing radiation	Two-strand breaks in DNA	Deletions, translocations
Ultraviolet radiation	Pyrimidine dimers	Errors in nucleotide choice during repair
Chemical mutagens	Base analogue mispairing	Single nucleotide substitution
	Modification of a base that leads to mispairing	Single nucleotide substitution
Spontaneous	Isomerization of a base	Single nucleotide substitution
	Slipped mispairing	Frameshift, short deletion
Recombinational		
Transposition	Insertion of transposon into gene	Insertional inactivation
Mispairing of repeated sequences	Unequal crossing-over	Deletions, addition, inversions
Homologue pairing	Gene conversion	Single nucleotide substitution

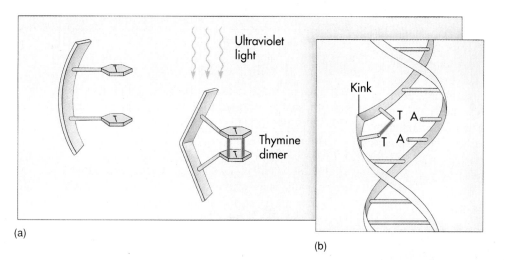

(a)

(b)

Figure 13.18 Making a thymine dimer.
(a) *When two thymines (T) are adjacent to one another in a DNA strand, the absorption of ultraviolet radiation can cause the formation of a covalent bond between them—a thymine dimer.* (b) *Such a dimer introduces a "kink" into the double helix, which prevents replication of the duplex by DNA polymerase. (A = adenine.)*

a **pyrimidine dimer** (figure 13.18). If a cross-link is left unrepaired, it can block DNA replication, in which case the damage is lethal. Removal of DNA cross-links often leads to errors in base selection.

Fortunately, your body has very efficient systems for detecting and repairing DNA cross-links. Without such systems, sunlight would wreak havoc on skin cells. Individuals with the rare hereditary disorder **xeroderma pigmentosum** are homozygous for a mutation that destroys body cells' ability to repair UV damage. These individuals develop extensive skin tumors after exposure to sunlight (figure 13.19).

Chemical Mutagens

Many mutations result from direct chemical modification of the DNA bases. The chemicals that act on DNA fall into three classes: (1) chemicals that look like DNA nucleotides but pair incorrectly when incorporated into DNA; (2) chemicals that remove the amino group from adenine or cytosine, causing them to mispair; and (3) chemicals that add hydrocarbon groups to nucleotide bases, also causing them to mispair. This last group includes many particularly potent mutagens that are commonly used in the laboratory, as well as compounds that are sometimes released into the environment, such as mustard gas.

> *The three major sources of mutational damage to DNA are: (1) high-energy radiation, such as X rays, which physically breaks DNA strands; (2) low-energy radiation, such as ultraviolet light, which creates DNA cross-links whose removal often leads to errors in base selection; and (3) chemicals that modify DNA bases and thus alter the base-pairing behavior of those bases.*

Mutation and Cancer

Cancer is perhaps the most pernicious human disease. Of the children born this year, a third will contract cancer; a fourth of the males and fully a third of the females will someday die of the disease. Each of us has had family or friends affected by cancer, and many of us will die of it. Not surprisingly, a great deal of effort has been expended to learn the cause of this disease. With the new techniques of molecular biology, much progress has been made, and the rough outlines of understanding are now emerging.

Cancer is a growth disorder of cells. It starts when an apparently normal cell begins to grow in an uncontrolled and invasive way (figure 13.20). The result is a ball of cells called a **tumor,** that constantly expands in size. When this ball remains a hard mass, it is called a **sarcoma** (if connective tissue, such as muscle, is involved) or a **carcinoma** (if epithelial tissue, such as skin, is involved). Cancer cells often develop abnormally and can no longer recognize the tissue where they originated; such cells leave the tumor mass and spread throughout the body, forming new tumors at distant sites. The spreading cells are called **metastases.**

The search for a cause of cancer has focused, in part, on suspected environmental factors, and many cancer-causing agents **(carcinogens)** have been identified, including ionizing radiation, such as X rays, and various chemicals (figure 13.21). Many carcinogens have in common the property of being potent mutagens. This observation led to

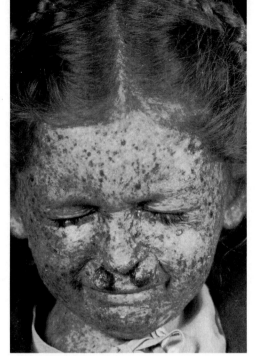

Figure 13.19 The inherited genetic disorder xeroderma pigmentosum.
Those who have the disease develop extensive malignant skin tumors after exposure to sunlight.

the suspicion that cancer might be caused, at least in part, by the creation of mutations.

Mutagens were not the only carcinogens found, however. Some tumors seemed almost certainly to be the result of viral infection. As early as 1910, an RNA virus was associated with sarcoma cancers in chickens. Isolation and study of this virus showed that the virus contained a growth-promoting gene normally found (shut off) on the chicken chromosome. The virus had picked up the chicken gene at some point in the past, but not the regulatory DNA to keep it shut off. When the virus infected healthy chickens, the growth-promoting gene was introduced without control, leading to unrestrained growth—cancer (see Sidelight 13.1).

What do mutagens and viruses have in common, that they can both induce cancer? At first, it seemed that they had little in common: Mutagens alter genes directly, whereas viruses introduce foreign genes into cells. However, as discussed in Sidelight 13.1, this seemingly fundamental difference is less significant than it might first appear, and these two causes of cancer have, in fact, proven to be one and the same.

The Molecular Basis of Cancer

Investigators have used a variety of techniques to study what happens to a human cell to make it cancerous. The most important of these techniques, called **transfection,** consists of (1) isolating the nuclear DNA from human tumor cells, (2) cleaving it into random fragments by using appropriate enzymes, and (3) testing the fragments individually for the ability of any particular fragment to induce cancer in the cells that assimilate it.

With transfection techniques, researchers found that only a single gene isolated from a cancer cell is needed to transform cells growing normally in tissue culture into cancerous ones. The cancerous cells differ from the normally growing ones only with respect to this one gene. The hypothesis that cancer results from the action of one or more specific tumor-inducing ***onc genes*** is called the oncogene (from the Greek word *oncos,* meaning "cancer") theory.

Onc genes are normal genes gone wrong. By identifying *onc* genes by transfection and then isolating them, investigators have been able to compare them with their normal counterparts to determine why normal genes are converted into cancer-causing ones. The analysis of a number of *onc* genes associated with cancers of various tissues has led to the following conclusions:

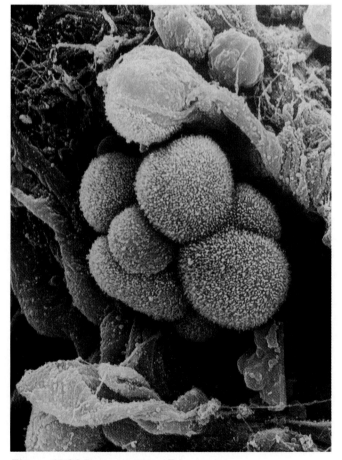

Figure 13.20 Lung cancer cells.
These cells are from a tumor located in the alveolus of a lung.

Sidelight 13.1
THE STORY OF CHICKEN SARCOMA

In 1910, American medical researcher Peyton Rous reported the presence of a virus—Rous avian sarcoma virus (RSV)—associated with chicken sarcomas; he was awarded the 1966 Nobel Prize in physiology or medicine for his discovery. RSV proved to be a particular kind of RNA virus called a retrovirus. Among the RNA viruses, retroviruses are unusual in how they replicate themselves. When they infect a cell, they make the DNA copy of their RNA—a copy that can be inserted into the animal DNA (Sidelight figure 13.1). The RSV is able to initiate cancer in chicken fibroblast (connective tissue) cells growing in culture; from these transformed cells, more virus can be isolated.

How does RSV act to initiate cancer? When RSV was compared with a closely related virus, RAV-O, which was not able to transform chicken cells into cancer cells, the two viruses proved to be identical, except for one gene that was present in the RSV virus but absent from the RAV-O virus. The cancer-causing gene was called the *src* gene (for sarcoma) (Sidelight figure 13.2).

What might be the nature of a virus gene that causes cancer? An essential clue came in 1970, when RSV mutants that were temperature-sensitive were isolated. These mutants could transform tissue culture cells into cancer cells at 35 degrees Celsius, but not at 41 degrees Celsius. Temperature sensitivity of this kind is almost always associated with proteins. Therefore, it seemed likely that the *src* gene was actively transcribed by the cell, rather than serving as a recognition site for some unknown regulatory protein. This exciting result suggested that the protein specified by the *src* gene could be isolated and its properties studied.

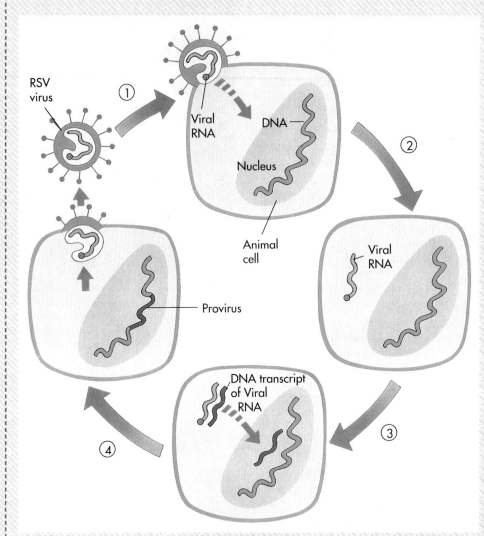

Sidelight figure 13.1 Life cycle of the RSV virus.
(1) *The genome of RSV is composed of RNA; that of animal cells is composed of DNA.* (2) *On infecting an animal cell, the RSV RNA is added to the animal cell.* (3) *There, a DNA copy of it is made.* (4) *Some or all of this DNA copy can become incorporated into the animal cell's DNA (provirus).*

1. The induction of many cancers involves changes in cellular activities that occur at the inner surface of the plasma membrane. In a normal cell, these activities are associated with the initiation of cell division: A cellular regulatory protein—epidermal growth factor (EGF)—binds to a specific receptor on the inner plasma membrane, acts as a tyrosine kinase, and triggers cell division. In this process, a protein encoded by a gene called *ras* is associated with the membrane EGF receptor site and acts to determine what cellular levels of EGF are adequate to initiate cell division (figure 13.22). In several forms of human cancer, the cancerous, or *onc*, version of the *ras*-encoded protein activates the receptor site in response to much lower levels of EGF than does the normal version of the protein.

2. The difference between a normal gene encoding the proteins that carry out cell division and a cancer-inducing *onc* version need only be a single-point mutation in the DNA. For example, in a human bladder carcinoma induced by *ras*, a single nucleotide alteration from guanine (G) to thymine (T) converts a glycine in the normal *ras* protein into a valine in the cancer-causing protein. This is the only difference between the normal and cancer-inducing forms of the *ras* gene.

3. The mutation of a gene such as *ras* to an *onc* form in different tissues can lead to different forms of cancer. There are probably no more than thirty to fifty genes whose mutation can lead to cancer.

The *src* protein was first isolated by Raymond Ericson and coworkers in 1977. It proved to be an enzyme of moderate size that acts to phosphorylate (add a phosphate group to) the tyrosine amino acids of proteins; such an enzyme is called a tyrosine kinase. Tyrosine kinases are not common in animal cells. Among the few known to occur in cells is epidermal growth factor, a protein that signals the initiation of cell division by binding to a special receptor site on the plasma membrane and phosphorylating key protein components. This raised the exciting possibility that RSV perhaps causes cancer by introducing into cells an active form of a normally quiescent enzyme. This indeed proved to be the case.

To determine whether the *src* gene actually integrates into the host chromosome with the RSV virus, investigators prepared a radioactive version of the *src* gene and allowed it to bind to complementary sequences on the chicken genome. They then examined where the chicken chromosomes became radioactive. Sites of radioactivity were sites where a sequence complementary to *src* occurred. As expected, radioactive *src* DNA bound to the site where RSV was inserted into the chicken genome—but unexpectedly, it also bound to a second site, where there was no RSV.

From similar experiments, investigators learned that the *src* gene is not a virus gene at all, but rather a growth-promoting gene that

evolved in and occurs normally in chickens. This chicken gene is the second site where *src* binds to chicken DNA. Somehow, a copy of the normal chicken gene was picked up by an ancestor of the RSV virus in some past infection. Now part of the virus, the gene is active in a pernicious new way, escaping the normal regulatory controls of the chicken genome; its transcription is governed instead by virus promoters, which are actively transcribed during infection. Thus, the study of RSV has shown that cancer results from the inappropriate activity of growth-promoting genes that are normally less active or completely inactive.

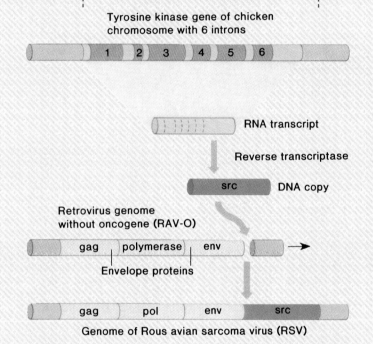

Tyrosine kinase gene of chicken chromosome with 6 introns

1 2 3 4 5 6

RNA transcript

Reverse transcriptase

src — DNA copy

Retrovirus genome without oncogene (RAV-O)

gag | polymerase | env

Envelope proteins

gag | pol | env | src

Genome of Rous avian sarcoma virus (RSV)

Sidelight figure 13.2
Structure of the Rous avian
sarcoma virus (RSV).

The virus contains only a few virus genes, encoding the virus envelope proteins (called the env *and* gag *genes), and reverse transcriptase, which produces a DNA copy of its RNA genome (called* pol *for polymerase). It also contains the gene* src *(for sarcoma). The RAV-O virus shown here lacks this* src *gene, but otherwise is identical to the RSV retrovirus. RSV causes cancer in chickens; RAV-O does not.*

4. The induction of many cancers involves the action of two or more different *onc* genes. Particular *onc* genes can be isolated by transfection because, in the strains being studied, the other *onc* genes necessary for cancer are already turned to the "cancer" mode. The initiation of cancer may require changes both at the plasma membrane and in the nucleus. This may be why most cancers occur in people over age forty (figure 13.23). It is as if human cells accumulate mutational changes, and time is required for several such mutations to occur in the same cells.

The emerging picture of cancer involves aborted regulation of the genes that normally signal the onset of cell proliferation. Cancer seems to occur when several of the controls that cells

normally impose on their own growth and division become inoperative. Among the examples that have been studied in detail, the specific means whereby these controls are evaded vary, but many of them involve one of two general causes. In the first case, cancer may result from the mutation of a gene with a growth-promoting regulatory function, as in the case of the *ras* gene, whose mutant *onc* form induces human bladder carcinoma. In the second case, cancer may result from the mutation of a so-called "tumor suppressor" gene with a growth-inhibiting regulatory function; a variety of human genes act to restrain cell growth, and when they are inactivated, cancer is more likely.

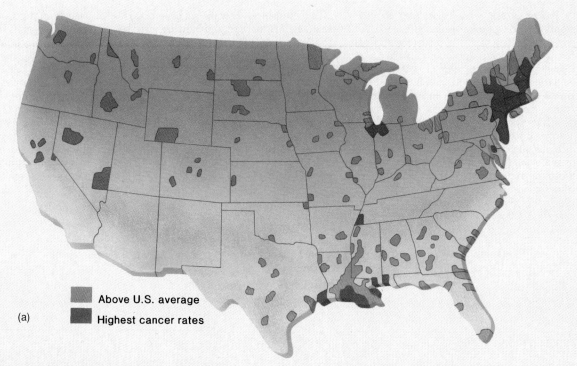

(a)

[legend] Above U.S. average
[legend] Highest cancer rates

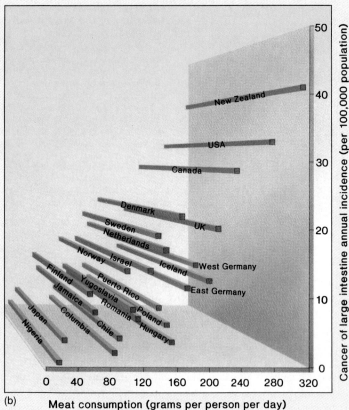

(b)   Meat consumption (grams per person per day)

Cancer of large intestine annual incidence (per 100,000 population)

New Zealand
USA
Canada
Denmark
UK
Sweden
Netherlands
Norway   Israel   Iceland
West Germany
Finland   Puerto Rico   East Germany
Yugoslavia
Jamaica   Romania   Poland
Japan   Columbia   Chile   Hungary
Nigeria

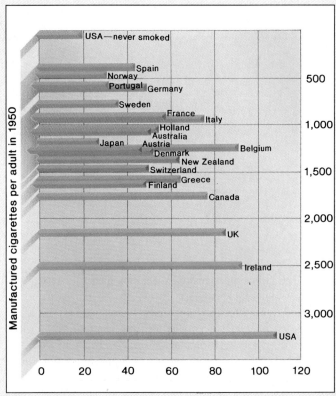

(c)   Lung cancer at ages 35 to 44 in the early 1970s
(per 100,000 population)

Manufactured cigarettes per adult in 1950

USA—never smoked
Spain
Norway
Portugal   Germany
Sweden
France
Italy
Holland
Australia
Japan   Austria   Belgium
Denmark
New Zealand
Switzerland
Greece
Finland
Canada
UK
Ireland
USA

Figure 13.21 Potential cancer-causing agents.

(a) *The incidence of cancer per one thousand people is not uniform throughout the United States. Rather, the highest cancer rates are centered in cities and in the Mississippi Delta. This suggests that pollution and pesticide runoff may contribute to cancer.* (b) *One of the most deadly cancers in the United States, that of the large intestine, is not at all common in many other countries, such as Japan. Its incidence appears to be related to the amount of meat an individual consumes—a high-meat diet slows* passage of food through the intestine, prolonging the exposure of the intestinal wall to digestive wastes. (c) *The biggest killer among cancers is lung cancer, and the most important environmental agent producing lung cancer is cigarettes. When lung cancer levels in many countries are compared, the incidence of lung cancer among adult males between forty and fifty years of age is strongly correlated with the cigarette consumption in that country twenty years earlier.*

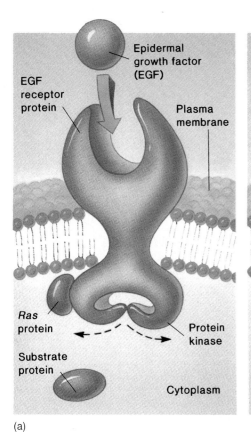

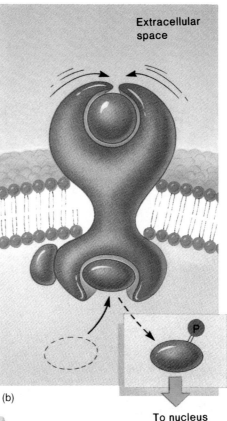

(a)   (b)

To nucleus

Figure 13.22 How a mutation can cause cancer.

Mutations that lead to cancer often involve proteins of the plasma membrane associated with cell division. (a) In a normal cell, cell division is triggered by a protein called epidermal growth factor (EGF), which binds to an EGF receptor protein on the cell's exterior surface. (b) This alters the shape of the portion of the receptor protruding into the cell, initiating a signal that passes to the cell nucleus and starts cell division. The level of EGF necessary to begin this process is affected by another protein, ras. *A mutation increasing receptor efficiency or* ras *signal facilitation may trigger more frequent cell division—cancer.*

> *Cancer is a growth disease of cells in which the controls that normally restrict cell proliferation do not operate, transforming cells to a state of cancerous growth.*

Smoking and Cancer

How can we prevent cancer? The most obvious strategy is to minimize mutational insult to our genes. Anything we do to increase our exposure to mutagens will result in an increased incidence of cancer for the unavoidable reason that such exposure increases the probability of mutating a potential *onc* gene. It is no accident that the most reliable tests for carcinogenic substances are those that measure their mutagenic ability.

Of all the environmental mutagens to which we are exposed, cigarettes are perhaps the most tragic—tragic because the cancers they cause are largely preventable. About a third of all U.S. cancer cases can be attributed directly to cigarette smoking. The association is particularly striking for lung cancer. The dose-response curve for male smokers shows a highly positive correlation, with the risk of lung cancer rising with increasing amounts of smoking (figure 13.24). For those who smoke two or more packs a day, the risk of contracting lung cancer is forty times or more greater than it is for nonsmokers.

Note that the curve extrapolates back approximately to zero. Clearly, an effective way to avoid lung cancer is to not smoke. Life insurance companies have computed that, on a statistical basis, smoking a single cigarette lowers your life expectancy 10.7 minutes, which is more than the time it takes to smoke the cigarette! Every pack of twenty cigarettes bears an unwritten label:

> "The price of smoking this pack of cigarettes is 3½ hours of your life."

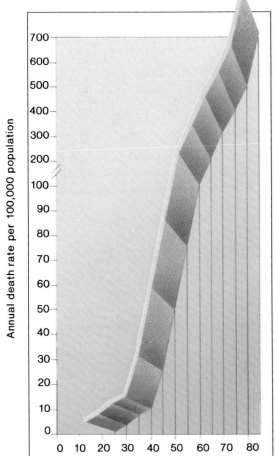

Figure 13.23 The annual death rate from cancer is a function of age.

A plot of age versus death rate is linear, suggesting that several independent events are required to give rise to cancer.

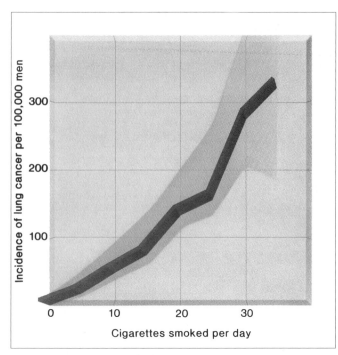

Figure 13.24 Smoking causes cancer.
The annual incidence of lung cancer per one hundred thousand men is clearly related to the number of cigarettes smoked per day. Of every one hundred American college students who smoke cigarettes regularly, one will be murdered, about two will be killed on the roads, and about twenty-five will be killed by tobacco.

Figure 13.25 Evidence that smoking causes cancer.
(a) *Photograph of lung cancer in an adult human. The bottom half of the lung is normal. The top half has been taken over completely by a cancerous growth.* (b) *Incidence of smoking since 1900 correlated with the incidence of lung cancer during the same period for both men and women.*

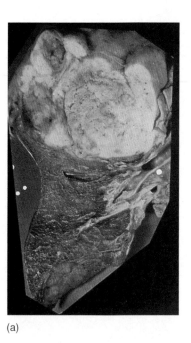

(a)

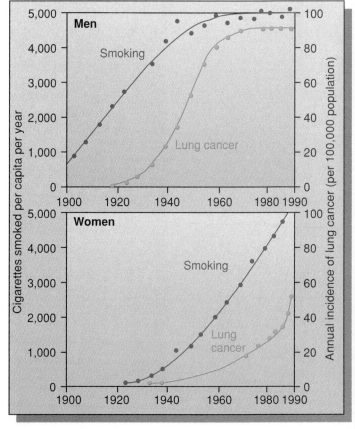

(b)

Of the 1.7 million people who died in the United States in 1993, a fourth died of cancer, and a third of these—140,000 people—died of lung cancer. About 160,000 cases of lung cancer were diagnosed each year in the 1980s, and 90 percent of these persons died or will die within three years. Of those who die, 96 percent will be cigarette smokers.

Smoking is a popular pastime in the United States. A third of the U.S. population smokes. American smokers consumed over 500 billion cigarettes in 1993. These cigarettes emit in their tobacco smoke some three thousand chemical components, among them vinyl chloride, benzo (*a*) pyrenes, and nitroso-*nor*-nicotine, all of them potent mutagens. Smoking introduces these mutagens to lung tissues, and figure 13.25*a* illustrates the result. As you might imagine, a lung in this condition does not function well, but difficulty in breathing is rarely the cause of death in lung cancer. As the cancer grows within the lung, its cells invade the surrounding tissues and eventually break through into the lymph and blood vessels. From there, the cancer cells spread rapidly through the body, lodging and growing at many locations, particularly in the brain. Death soon follows.

Cigarette manufacturers have often argued that the causal connection between smoking and cancer has not been proven, that somehow the relationship is coincidental. Look carefully at the data presented in figure 13.25*b* and see if you agree. The upper graph presents data collected for American males, including the incidence of smoking from the turn of the century until now and the incidence of lung cancer over the same period. Note that, as late as 1920, lung cancer was a rare disease. With a lag of some twenty years behind the increase in smoking, it became progressively more common.

The lower graph in figure 13.25*b* presents data on American females. Because of social mores, significant numbers of American females did not smoke until after World War II, when many social conventions changed. As late as 1963, when lung cancer among males was near current levels, this disease was still rare in females. In the United States that year, only 6,588 females died of lung cancer. But as women's smoking increased, so did their incidence of lung cancer, with the same inexorable lag of about twenty years. American females today have achieved equality with their male counterparts in the numbers of cigarettes that they smoke, and their lung cancer death rates are now approaching those for males. In 1990 more than 49,000 women died of lung cancer in the United States.

Among smokers, the current rate of deaths resulting from lung cancer is 180 per 100,000, or about 2 of each 1,000 smokers each year. Smoking is like going into a totally dark room, standing still, and then calling in someone with a gun and closing the door. The person with the gun cannot see you, does not know where you are, and so just shoots once in a random direction and leaves the room. Every time individuals smoke a cigarette, they are "shooting" mutagens at their genes. Just as in the dark room, a hit is unlikely, and most shots will miss potential *onc* genes. As they keep shooting, however, the odds of eventually scoring a hit get better and better. Nor do statistics protect any one individual; nothing says that the first shot will not hit. Older people are not the only ones to die of lung cancer.

Except for eating powerful radioisotopes, there is probably no more certain way to develop cancer than to smoke.

EVOLUTIONARY VIEWPOINT

The basic architecture of gene expression has remained largely unchanged from early times, although eukaryotes have introduced a few complications, such as introns. Gene expression is regulated in all organisms by proteins that fit into the grooves of the DNA double helix; cancer results from damage that prevents such regulation from moderating cell growth.

SUMMARY

1. The expression of hereditary information in all organisms takes place in two stages. First, an mRNA molecule with a nucleotide sequence that is complementary to a particular segment of the DNA is synthesized by the enzyme RNA polymerase. Second, a ribosome assembles an amino-acid chain, using the mRNA sequence to direct its choice of amino acids. The first process is called transcription, the second, translation.

2. A ribosome reads the coding sequence of a gene in increments of three nucleotides, called a codon, from the mRNA. The ribosome positions the three-nucleotide segments of the message so that a tRNA molecule with the complementary base sequence can bind to it. Attached to the other end of the tRNA molecule is an amino acid, which is added to the end of the growing polypeptide chain.

3. Most eukaryotic genes contain additional sequences called introns embedded within the coding sequences of their transcription units. Introns are removed from the transcript before it is translated.

4. Gene expression is controlled largely by regulating the initiation of transcription. Control is both negative, in which other cellular signals prevent transcription except under appropriate circumstances, and positive, in which cellular signals are required before transcription can start.

5. Transcription is regulated largely by controlling the ability of RNA polymerase molecules to initiate transcription. In some instances, a nucleotide sequence located between the promoter and the transcription unit acts as a binding site for a repressor protein. When bound to the site, this repressor protein prevents the read-through of the RNA polymerase, just as a tree fallen across a road blocks traffic.

6. In other instances, nucleotide sequences located near the promoter site act as binding sites for activator proteins. The binding of the activator protein to the promoter site causes the unwinding of the DNA duplex. Because RNA polymerase can translate only a single strand of DNA, this forced unwinding greatly facilitates the translation process.

7. A mutation is any change in the cell's hereditary message. Mutations that change one or a few nucleotides are called point mutations. They may result from physical damage from ionizing radiation, chemical damage, or mistakes made during correction of damage caused by absorption of ultraviolet light.

8. Cancer is a growth disease of cells in which the regulatory controls that usually restrain cell division do not work.

9. The testing of DNA fragments from tumor cells to identify those fragments capable of inducing cancer has led to the isolation of various cancer-causing genes. In every case, the cancer-causing gene is a normal gene whose product has a role in cell proliferation but becomes active when it should not.

10. In some cases, cancer induction results from a single nucleotide mutation in a normal gene. In other cases, a normal gene is acquired by a virus and is transcribed at the high rate characteristic of the virus.

11. The best way to avoid cancer is to avoid those things that cause mutation, notably smoking.

REVIEWING THE CHAPTER

1. How do cells use RNA to make proteins?
2. What are the three major classes of RNA?
3. What is gene expression?
4. How is information transcribed?
5. How is information translated?
6. What is the genetic code?
7. How do ribosomes function in the cell?
8. What is the function of activating enzymes?
9. What are the combinations of the genetic code?
10. How does translation occur?
11. How does translation compare between prokaryotes and eukaryotes?
12. How is gene expression regulated?
13. How do regulatory proteins work?
14. What is the function of negative control?
15. What is the *lac* operon?
16. How does negative control work in the *lac* operon?
17. What is the function of positive control?
18. How does positive control work in the *lac* operon?
19. How do positive and negative controls work together?
20. How frequent is gene mutation?
21. How is DNA damaged?
22. What are the effects of ionizing radiation?
23. What are the effects of ultraviolet radiation?
24. What are the effects of chemical mutagens?
25. What is the correlation between mutations and cancer?
26. What is the molecular basis of cancer?
27. What is the correlation between smoking and cancer?

COMPLETING YOUR UNDERSTANDING

1. How do rRNA, tRNA, and mRNA differ in their functions?
2. What is complementarity, and how does it relate to transcription?
3. What is the first step in RNA transcription?
   a. Nucleotides are assembled into RNA.
   b. The DNA strands open up.
   c. The hydrogen bonds of the DNA re-form.
   d. RNA polymerase attaches to the DNA.
   e. The DNA replicates.
*4. You are provided with a sample of aardvark DNA, obtained at great risk to life. As part of your investigation of this DNA, you transcribe messenger RNA from the DNA and purify it. You then separate the two strands of DNA and analyze the base composition of each strand, and then analyze that of the mRNA. You obtain the following results:

	A	G	C	T	U
DNA strand 1	19.1	26.0	31.0	23.9	0
DNA strand 2	24.2	30.8	25.7	19.3	0
mRNA	19.0	25.9	30.8	0	24.3

   Which strand of the DNA is the "sense" strand, serving as the template for mRNA synthesis?
5. What is the primary mRNA transcript, and how is it assembled?
6. What is the difference between a codon and an anticodon?
7. In translation, what is the function of activating enzymes, initiation factors, elongation factors, and release factors?
8. In eukaryotes, the nontranslated portion of a gene is called
   a. an operon.   d. an operator.
   b. an exon.   e. a repressor.
   c. a promoter.   f. an intron.
9. Two key differences between bacterial and eukaryotic mRNA are [*two answers required*]
   a. bacterial mRNA molecules do not need ribosomes for transcription.
   b. eukaryotic mRNA molecules rarely contain more than one gene, whereas bacterial mRNAs often do.
   c. ribosomes of eukaryotes are much larger than those of bacteria.
   d. the primary mRNA transcripts of eukaryotes contain introns that are cut out before translation.
   e. in bacterial mRNA, the nucleotide sequence is not arranged in codons.
10. What is the interaction between regulatory sites and regulatory proteins?
11. In the *lac* region of the bacterium *Escherichia coli*'s chromosome, where does the repressor protein bind?
   a. At the activator protein site
   b. At the promoter site
   c. At the operator site
   d. At the transcription unit
   e. At the permease gene site
12. What is the function of repressor and activator proteins?
*13. In medical research, mice are often used as model systems in which to study the immune system and other physiological systems important to human health. Many medical centers maintain large colonies of mice for these studies. In one such colony under your supervision, a hairless mouse is born. What minimum evidence would you accept that this variant represents a genetic mutation?
14. What is a point mutation and what are its consequences?
15. What is transposition, and what are its consequences?
16. In the next several decades, what are the potentials for and consequences of increases in ionizing radiation, ultraviolet radiation, and chemical mutagens to the human population?
17. What is xeroderma pigmentosum, and how is it related to a pyrimidine dimer?
18. Ultraviolet radiation damages cells by
   a. burning their plasma membranes.
   b. breaking DNA molecules into nonfunctional segments.
   c. forming pyrimidine dimers that can block DNA replication.
   d. causing xeroderma pigmentosum.
   e. causing the cells to proliferate and become cancerous.
19. What is the correlation between meat consumption and cancer of the large intestine?
20. The rapid spreading of a cancer in a human body from point of origin to distant sites is caused by
   a. carcinomas.   d. carcinogens.
   b. sarcomas.   e. tumors.
   c. metastases.   f. transfections.
21. The highest cancer rates in the United States are found in three geographical regions (see figure 13.21*a*). How would you conduct a study to pinpoint the causes of these high rates of cancer?
22. Retroviruses use_____instead of DNA as their hereditary information molecule.
23. At this point in time, why has cancer not been eradicated? What problems do you envision in the future with curing cancer?
24. What is the oncogene theory?
25. Why do most cancers occur in humans over age forty?

*For discussion

26. Should laws be enacted to prevent smoking in public places, such as restaurants and office buildings? Why or why not? What about taverns?

27. In the United States, why was the rate of lung cancer so low in women prior to World War II, and why is today's rate in women approaching that found in men?

*28. The evidence associating lung cancer with smoking is overwhelming, and as you have learned in this chapter, scientists now know in considerable detail the mechanism whereby smoking induces cancer. It introduces powerful mutagens into the lungs, which cause mutations to occur. When a growth-regulating gene is mutated by chance, cancer results. In light of this, why do you think that cigarette smoking is not illegal?

*29. In a study of the segregation of leukemia in several generations of laboratory mice, a line selected for high incidence of leukemia was crossed with a line that never develops leukemia (it presumably lacks a leukemia-inducing oncogene). All of the F_1 progy exhibited
a high incidence of leukemia. Among the progeny resulting from a test cross of these F_1 individuals to the leukemia-free parent line, 25 percent were healthy and leukemia-free; the other 75 percent were leukemic. The result is not a 1:1 ratio, but rather a 3:1 ratio. How would you interpret this result?

*For discussion

FOR FURTHER READING

Bishop, J. M. 1982. Oncogenes. *Scientific American*, March, 82–92. The story of *src* and how it causes cancer, by the man who first identified the protein product of the *src* gene. This article is particularly good at showing the chain of reasoning that underlies a major scientific advance.

Doll, R. 1978. An epidemiological perspective of the biology of cancer. *Cancer Research* 38: 3573–83. A classic study that presents data on who contracts cancer and at what age. These data were the first damning indictment of smoking as the leading cause of lung cancer.

Feldman, M., and L. Eisenbach. 1989. What makes a tumor cell metastatic? *Scientific American,* Nov., 60–85. Discussion of how the study of the molecules that stud the surface of cancer cells is beginning to enable cancer researchers to convert malignant cells into benign ones.

Grunstein, M. 1992. Histones as regulators of genes. *Scientific American*, Oct., 68–74. A description of how histone proteins can both repress and facilitate gene activation.

Kartner, N., and V. Ling. 1989. Multidrug resistance in cancer. *Scientific American,* March, 44–51. An account of how the cancers resistant to many forms of chemotherapy seem to result from defects in a pump that flushes toxins out of cells.

McKnight, S. 1991. Molecular zippers in gene regulation. *Scientific American,* April, 54–64. A study of how repeated copies of the amino acid leucine in proteins can serve as teeth that "zip" two protein molecules together and of how such zippering plays a key role in turning genes "on" and "off."

Ptashne, M. 1989. How gene activators work. *Scientific American,* Jan., 41–47. A lucid overview of how genes are turned "on" and "off."

Rhodes, D., and A. Klug. 1993. Zinc fingers. *Scientific American,* Feb., 56–65. Discussion of how zinc fingers play a key part in regulating the activity of genes in all eukaryotes.

Ross, J. 1989. The turnover of messenger RNA. *Scientific American,* April, 48–55. A study of how the level of many proteins in the body is determined by how fast the messenger RNA encoding them is broken down, rather than by how speedily new transcripts are churned out.

Sager, R., 1989. Tumor suppressor genes. *Science* 246 (Dec.): 1406–16. Discussion of how some genes that protect from cancer by preventing tumors from arising are playing a key role in gene-transfer therapies now being investigated.

Weinberg, R. 1988. Finding the anti-oncogene. *Scientific American*, Sept., 44–51. An important paper by a developer of the transfection approach that describes how certain growth-suppressing genes in a mutated form confer susceptibility to cancer.

Willett, W. 1989. The search for the causes of breast and colon cancer. *Nature* 338 (March): 389–94. A description of the mounting evidence that diet plays a key role in these two major killers.

EXPLORATIONS

Interactive Software:
Gene Regulation

In this interactive exercise, the user explores two strategies employed by organisms to regulate the transcription of genes: (1) bacterial gene regulation, with the focus on rapid adaptation to environmental changes, and (2) eukaryotic gene regulation, with the focus on complex, hard-wired programs dictating fixed patterns of gene activity. In the bacterial simulation, the user will design a regulatory mechanism for a sugar-utilizing enzyme, selecting elements from among activator and repressor proteins, locating their binding sites on the DNA. The user then varies the level of sugar in the environment and assesses the success of the proposed regulatory mechanism in optimizing use of the sugar. In the eukaryotic simulation, the user will have a broader choice of regulatory tools, including transcription factors and the ability to locate regulatory genes at far distant sites. The challenge is to design a regulatory mechanism that will permit the various mammalian globin genes to be expressed at different specific times in development.

Questions to Explore:

1. What difference can be seen between bacteria and eukaryotes in the need for proximity of regulatory sites to the gene they are regulating? What do you suppose accounts for this difference?

2. Can you design a bacterial regulatory mechanism that allows expression of a particular gene to be controlled by many regulatory genes simultaneously?

3. Transcription is only the first stage in a long process of gene expression. At what other levels might you expect regulation to occur? Do any of these have any advantages over transcriptional control?

CHAPTER

14

GENE TECHNOLOGY

A molecular model of taxol, an anticancer drug originally derived from the yew tree. Transferring the taxol gene to another organism is making the drug available in greater quantities while protecting the yew from extinction.

FOR REVIEW

Here are some important terms and concepts that have been discussed in previous chapters and that you will encounter again in this chapter. Review them before proceeding if necessary.

Enzymes (*chapter 6*)
Genetic therapy (*chapter 11*)
DNA (*chapters 2 and 12*)

Genetic engineering is revolutionizing biology and affecting all of our lives. During the last twenty years, new and powerful techniques for studying and manipulating DNA have allowed biologists for the first time to intervene directly in organisms' genetic fate (figure 14.1). This chapter discusses these techniques and considers their application to specific problems of agriculture and medicine (see Sidelight 14.1). Few areas of biology will have as great an impact on our lives, and the potential advances are just beginning to be understood.

Plasmids and the New Genetics

Interferon is a rare protein, difficult to purify in any appreciable amounts, that increases human resistance to viral infection. It may prove effective in improving human resistance to cancer. At first, this possibility was difficult to explore because purification of the substantial amounts of interferon required for large-scale testing was prohibitively expensive. In 1980, however, geneticists succeeded in introducing the human gene that encodes the protein interferon into a bacterial cell, thus making it possible to inexpensively produce large amounts of interferon.

Within the bacterial cell, the gene directs the production of copious amounts of interferon. Meanwhile, the bacterial cell with the introduced gene continues to grow and divide. Soon, the culture contains many millions of such cells, all descended from the original altered bacterial cell and all producing human interferon. This procedure, called **cloning,** succeeds in making every cell in the culture a miniature factory for interferon production.

In a similar way, the successful cloning of bacteria able to produce insulin, a polypeptide, has led to the development of new and relatively inexpensive ways to produce large amounts of the hormone. Cloning also has proven a valuable technique in molecular biology because it allows the multiplication of particular genes and gene products that are being studied. Collectively, these techniques of transferring genes from one kind of organism to another and multiplying them are termed **genetic engineering.**

Genetic engineering involves cutting up DNA into recognizable pieces and rearranging these pieces in different ways. In the first interferon experiment, the DNA segment carrying the gene was inserted into a plasmid. **Plasmids** are small, usually

Figure 14.1 Genetically engineered tomato.
The normal tomato on the right is harvested and shipped to market before it ripens. The tomato on the left possesses a gene isolated from an Antarctic fish that delays its ripening, allowing it to remain on the vine longer. The gene promotes the breakdown of ethylene, which hastens the ripening of plants.

circular fragments of DNA that replicate independently outside of the main bacterial chromosome. They make up about 5 percent of the DNA of many bacteria but are rare in eukaryotic cells. Genetic engineers usually employ plasmids or viruses to carry a particular gene into a recipient cell.

Restriction Enzymes

The initial step in a genetic engineering experiment is the key to the whole procedure. Success depends on being able to cut up the source DNA (for example, human DNA in the interferon experiment) and the plasmid DNA so that the desired fragment of source DNA can be spliced permanently into the plasmid genome. This cutting is performed by a special kind of enzyme called a **restriction endonuclease.** Restriction enzymes are able to recognize specific nucleotide sequences within a DNA strand, bind to DNA strands at sites where these sequences occur, and cut a bound strand of DNA at a specific place in the recognition sequence. They are the basic tools of genetic engineering.

Other bacterial enzymes called **methylases** recognize the same sequences in bacterial DNA as restriction enzymes, bind to them, and add methyl (CH_3) groups to the nucleotides. When the recognition sites of bacterial DNA have been modified with methyl groups in this way, they are no longer recognized by the restriction enzymes. Consequently, bacterial DNA is protected from being degraded—that is, the restriction enzymes leave bacterial DNA unharmed. Viral DNA is not so lucky. Because it has not been methylated, viral DNA is chopped up when it encounters restriction enzymes.

Restriction enzymes recognize sequences that are typically four to six nucleotides long and symmetrical. Their symmetry is of a special kind, called **twofold rotational symmetry.** The nucleotides at one end of the recognition sequence are complementary to those at the other end so that the two

Sidelight 14.1
RAPE AND DNA "FINGERPRINTS": THE MOLECULAR WITNESS

The twenty-seven-year-old woman never saw her assailant. She was asleep on 22 February 1987, when a man broke into her house in Orlando, Florida, covered her head with a sleeping bag, and raped her in bed. On 3 November 1987, a man named Tommie Lee Andrews was placed on trial for this rape. The case against him did not, at first, seem strong. Two indistinct fingerprints on a window screen resembled his, but the woman could not identify Andrews as her attacker—her head covered, she had never seen the man who raped her. Andrews' girlfriend and sister swore that he had never left home that night. Like many rape cases, this one presented the jury with conflicting testimony and evidence open to more than one interpretation.

Then the prosecuting attorney introduced a new line of evidence: There was a witness. The man who raped this woman could be clearly identified, without ambiguity and beyond the shadow of a doubt. The witness was not someone who had seen what had happened. Such a witness could always be doubted, since people sometimes make mistakes in identification. The witness was DNA.

The chromosomes of every human cell contain scattered throughout their DNA short, often repeated fifteen-nucleotide segments called "minisatellites." The locations and number of repeats of any particular minisatellite are so highly variable that no two people are alike. The

probability of two unrelated individuals having the same pattern of location and repeat-number of minisatellite is one in ten billion. With a world population of just over 5.4 billion, no two people are ever alike. If one analyzes two minisatellites to be sure, the probability is a minuscule 5×10^{-19}, which for all practical purposes is zero.

Sidelight figure 14.1 shows the evidence presented by the prosecuting attorney. It consisted of autoradiographs, parallel bars on X-ray film resembling the line patterns of the universal price code found on groceries. Each bar represents the position of a minisatellite, detected by techniques similar to those de-

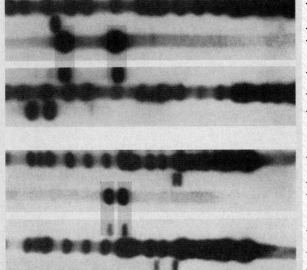

Variable Sequence A

———— Sample 1 Control

——→ Sample 2 Victim
——→ Sample 3 Rapist's semen

——→ Sample 4 Suspect's blood

——→ Sample 5 Control

——→ Sample 6 Control

Variable Sequence B

———— Sample 1 Control

——→ Sample 2 Victim
——→ Sample 3 Rapist's semen

——→ Sample 4 Suspect's blood

——→ Sample 5 Control

——→ Sample 6 Control

Sidelight figure 14.1 The two DNA comparisons used to convict.
In each case, the pink color highlights bands shared by the suspect's blood and the rapist's semen.

scribed in figure 14.4. A vaginal swab had been taken from the victim within hours of her attack. From it, semen was collected, and the semen DNA was analyzed for its minisatellite patterns. These minisatellite patterns, labeled "Rapist's semen" in Sidelight figure 14.1, are beyond question those of the rapist.

When a minisatellite pattern of the rapist is compared to that of the suspect Andrews, lining the two samples up with a standardized control lane that has many bands, the suspect's pattern is identical to that of the rapist (and not at all like that of the victim). And all the pat-

terns are similarly the same. Clearly, the semen collected from the rape victim and the blood collected from Tommie Lee Andrews came from the same person.

On November 6th, the jury returned a verdict of guilty. Andrews became the first man in the United States to be convicted of a crime based on DNA evidence.

Since the Andrews verdict, hundreds of rape and murder trials in the United States have resulted in convictions based on DNA evidence. Just as fingerprinting revolutionized forensic evidence in the early 1900s, so DNA "fingerprinting" is revolutionizing it today.

A hair, a minute speck of blood, a drop of semen—they all can serve as sources of DNA, to damn or clear a suspect. If you are innocent, DNA "fingerprinting" can be your salvation. Over a dozen inmates have been freed from prison because of evidence provided by DNA testing. In June 1993, for example, a man named Kirk Bloodsworth was freed after serving almost nine years in prison. He had been convicted of the rape and murder of a nine-year-old girl on the testimony of three witnesses, and after two trials and exhausting all appeals, faced imminent execution. During these nine long years, he steadfastly maintained his innocence. In a last-minute attempt to reopen the case, a new attorney asked that key physical evidence be examined, and a tiny semen spot was discovered on the girl's underpants. Amplified by PCR (polymerase chain reaction—discussed later in the chapter), enough material was recovered to conclusively demonstrate that Bloodsworth could not have been the rapist.

---

strands of DNA duplex have the same nucleotide sequence running in opposite directions for the length of the recognition sequence.

This arrangement has two consequences. The first is of great importance to bacteria (it provides the weapon they use to attack invading viruses); the second is of little significance to bacteria, but of paramount importance to humans (it makes genetic engineering possible):

1. Because the same recognition sequence occurs on both strands of the DNA duplex (running in opposite directions), the restriction enzyme is able to recognize and cleave both strands of the duplex, effectively cutting the DNA duplex in half.

2. The sites at which the two strands of a duplex are cut are offset from one another. This occurs because the position of the bond cleaved by a particular restriction enzyme typically is not in the center of the recognition sequence to which it binds.

The sequence runs in opposite directions on the two strands, and the cleavage sites are therefore offset. Because of this, after cleavage, the two fragments of DNA duplex each possess a short, single strand a few nucleotides long dangling from the end. As shown in figure 14.2, the two single-stranded tails are complementary to one another.

The hundreds of bacterial restriction enzymes recognize a variety of four- to six-nucleotide sequences. Six-nucleotide sequences are the most common. Every cleavage by a particular restriction enzyme occurs at the same recognition sequence. By chance, this sequence probably occurs somewhere in any given sample of DNA; thus, a restriction endonuclease can cut DNA from any source into fragments. The shorter the sequence, the more often it will arise by chance within a genome. Each of these fragments will have the dangling sets of complementary nucleotides (sometimes called "sticky ends") characteristic of that endonuclease.

Because the two single-stranded ends produced at a cleavage site are complementary, they can pair with each other. Once they do, the two strands can then be joined back together with the aid of a sealing enzyme called a **ligase,** which re-forms the phosphodiester bonds—the bonds between the sugars and phosphates of DNA. This latter property makes restriction endonucleases the invaluable tools of genetic engineers: Any two fragments produced by the same restriction enzyme can be joined together. Thus, fragments of elephant and ostrich DNA cleaved by the same bacterial restriction enzyme can be joined to one another just as readily as can two bacterial fragments because they have the same complementary sequences at their ends.

A restriction enzyme cleaves DNA at specific sites, generating in each case two fragments whose ends have one strand of the duplex longer than the other. Because the trailing strands of the two cleavage fragments are complementary in nucleotide sequence, any pair of fragments produced by the same restriction enzyme, from any DNA source, can be joined together.

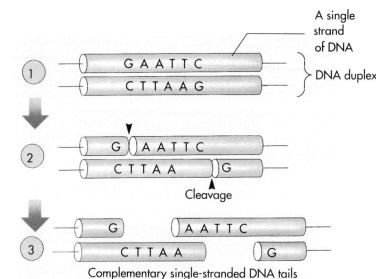

Figure 14.2 How restriction enzymes produce DNA fragments with "sticky ends."

(1) *The same recognition sequence occurs on both strands of the DNA duplex.* (2) *The restriction enzyme Eco R1 always cleaves sequence GAATTC at the same spot, after the first G. Because the same sequence occurs on both strands, both are cut. But the position of the G is not the same on the two strands; the sequence runs the opposite way on the other strand. As a result, single-stranded tails are produced.* (3) *Because of the twofold rotational symmetry of the sequence, the single-stranded tails are complementary to each other, or "sticky."*

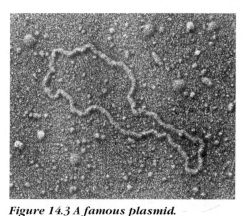

Figure 14.3 A famous plasmid.
The circular molecule in this electron micrograph was the first plasmid—pSC101—used successfully to clone a vertebrate gene. Its name refers to it being the 101st plasmid isolated by Stanley Cohen.

Constructing Chimeric Genomes

A chimera is a mythical creature with the head of a lion, the body of a goat, and the tail of a serpent. No chimera ever existed in nature. But human beings have made them—not the lion-goat-snake variety, but chimeras of much greater importance to the future of humanity.

Recombinant DNA

The first actual chimera was a bacterial plasmid that American geneticists Stanley Cohen and Herbert Boyer made in 1973. Cohen and Boyer used a restriction endonuclease to cut up a large bacterial plasmid called a resistance transfer factor. From the resulting DNA fragments, they isolated a fragment nine thousand nucleotides long that contained both the sequence necessary for replicating the plasmid—the **replication origin**—and a gene that conferred resistance to the antibiotic tetracycline.

Because both ends of this fragment were cut by the same restriction enzyme (called *Escherichia coli*-restriction endonuclease 1, or **Eco R1**), the ends could join together to form a circle, a small plasmid that Cohen dubbed pSC101 (figure 14.3). Cohen and Boyer also used Eco R1 to cut up DNA isolated from an adult amphibian, the African clawed toad, *Xenopus laevis*. They then mixed the toad DNA fragments with open-circle molecules of pSC101, allowed

KEY EXPERIMENT 14.1

THE COHEN-BOYER EXPERIMENT

◀ 1973

In 1973, Stanley Cohen and Herbert Boyer inserted an amphibian gene encoding rRNA into a bacterial plasmid. This plasmid—the 101st plasmid isolated by Stanley Cohen (plasmid Stanley Cohen #101, or pSC101)—contains a single site cleaved by the restriction enzyme Eco R1, as well as a gene for tetracycline resistance (Tc^R gene). The rRNA-encoding region was inserted into pSC101 at the cleavage site by cleaving the rRNA region with Eco R1 and allowing the complementary sequences to pair.

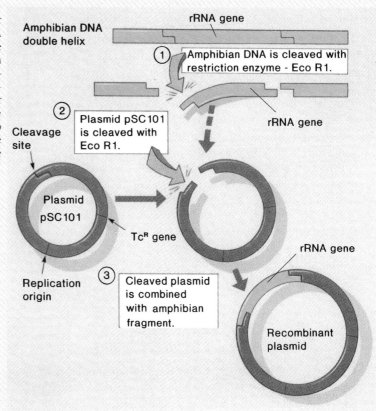

1900

---

bacterial cells to take up DNA from the mixture, and selected bacterial cells that had become resistant to tetracycline. From among these pSC101-containing cells, they were able to isolate ones containing the toad ribosomal RNA gene. These versions of pSC101 had the toad gene spliced in at the Eco R1 site. Instead of joining to one another, the two ends of the pSC101 plasmid had joined to the two ends of the toad DNA fragment that contained the ribosomal RNA gene (see Key Experiment 14.1).

The pSC101 containing the toad ribosomal RNA gene is a true chimera. It is an entirely new kind of molecule that never existed in nature and would never have evolved there. It is a form of **recombinant DNA,** a DNA molecule created in the laboratory by molecular geneticists who joined together bits of several genomes into a novel combination.

In 1973, Stanley Cohen and Herbert Boyer inserted an amphibian ribosomal RNA gene into a bacterial plasmid. By doing so, they initiated the age of genetic engineering.

Employing plasmids or viruses to carry fragments of foreign DNA into bacterial cells is now a technique that scientists use frequently. Newer-model plasmids can be induced to make many hundreds of copies of themselves and thus of the foreign genes that are included in them within bacterial cells.

Inserting the foreign DNA fragment into the genome of a bacterial virus, such as the lambda virus, instead of into a plasmid, makes for even easier entry into bacterial cells. The infective genome that harbors the foreign DNA and carries it into the target cell is called a **vector.** Not all vectors have bacterial targets. Animal viruses have been used as vectors to carry bacterial genes into monkey cells, and animal genes have even been carried into plant cells.

Safeguards Necessary for Recombinant DNA Experiments

What is the potential danger of inadvertently creating an undesirable life-form in the course of a recombinant DNA experiment? What if the DNA of a cancer cell was fragmented and then incorporated at random into viruses propagated within bacterial cells? Could one of the resulting bacteria be capable of causing an infective form of cancer?

Even though most recombinant DNA experiments are not dangerous, such concerns should be taken seriously. Both scientists and governments monitor genetic engineering experiments to detect and forestall any such hazard. In addition, researchers have established appropriate experimental safeguards. For example, the bacteria used in many recombinant DNA experiments are unable to live outside of laboratory conditions; many of them can survive only in an oxygen-free atmosphere. Experiments such as the cancer cell experiment just described, which are clearly dangerous, are prohibited.

How to Do Genetic Engineering Experiments

The movement of genes from one organism to another is often termed recombinant DNA technology or genetic engineering. Each experiment in this field presents unique problems, but all share the same general strategy.

The first stage of any genetic engineering experiment is the generation of specific DNA fragments by cleavage of a genome with restriction endonuclease enzymes. Different "libraries" of fragments may be obtained by using enzymes that recognize different sequences. The fragments are usually separated by electrophoresis, which permits their relative size to be estimated (figure 14.4).

In the second stage of genetic engineering experiments, the fragments are amplified in number in order to obtain enough material for further work. This can be done in either of two ways:

1. Cloning. DNA fragments are incorporated into plasmids or virus "vectors," which are then introduced into bacterial cells. Each cell then reproduces, forming a clone of cells that all contain the fragment-bearing plasmid or virus. Each of the cell lines is maintained separately; together, they constitute what is called a **clone library.**

2. PCR amplification. The **polymerase chain reaction (PCR)** can be employed to generate large numbers of copies of shorter DNA sequences (see Key Experiment 14.2).

In the final stage of genetic engineering experiments, the amplified fragment is transferred into a living organism, changing the organism's genetic makeup.

Genetic Engineering: Producing a Scientific Revolution

Genetic engineering has already revolutionized biology and is changing the lives of every one of us. The last twenty years have seen an explosion of interest in applying genetic engineering techniques to practical human problems.

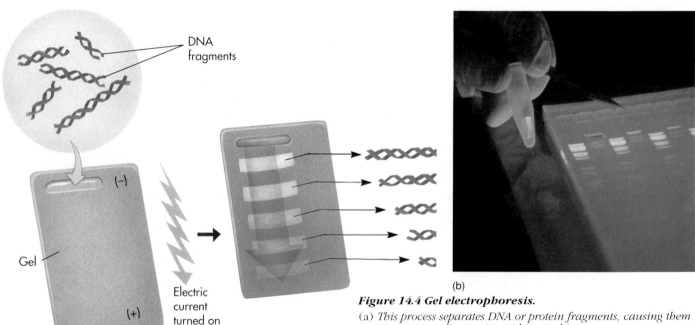

(b)

Figure 14.4 Gel electrophoresis.

(a) *This process separates DNA or protein fragments, causing them to migrate according to size within a gel in response to an electrical field.* (b) *The fragments can be visualized easily because the migrating bands glow in fluorescent light.*

KEY EXPERIMENT 14.2

2000

THE MULLIS PCR EXPERIMENT

◄ 1986

Invented by K. Mullis (see photo) less than a decade ago, **polymerase chain reaction,** or **PCR,** allows billions of identical DNA fragments to be created within a few hours. The PCR technique has five steps:

1. Tagging. First, to each end of the DNA fragment is added a different short, single-stranded, nucleotide sequence. Called *oligonucleotides,* such short twenty to thirty nucleotide segments can be manufactured in large quantities by machine. When this step is completed, the fragment is a little longer than it was, and each of its ends has a unique short tag.

2. Heating. Now a solution of the tagged fragment is heated to about 98 degrees Celsius, just below boiling and a large excess of the oligonucleotide duplexes is added to the solution. At this high temperature, the DNA duplex of the fragment and oligonucleotides dissociates into single strands.

3. Priming. Next, the solution is allowed to cool to about 60 degrees Celsius. As it does, the single strands of DNA reassociate into double strands, *but because there is a large excess of oligonucleotides, each fragment strand picks an oligonucleotide as a double-strand partner.* The oligonucleotide binds at the end, to its complementary sequence tag.

4. Copying. Now DNA polymerase is added, a very heat-stable type extracted from a bacterium that lives in hot springs. Using the oligonucleotide as a starting point, or primer, the polymerase proceeds to copy the rest of the fragment as if it were replicating DNA. When it is done, what used to be the oligonucleotide primer is now lengthened into a complementary copy of the entire single-stranded fragment—and because *both* single strands behave this way, there are now two copies of the original fragment!

5. Repeating the cycle. Steps 2 through 4 are now repeated, and the two copies

1900

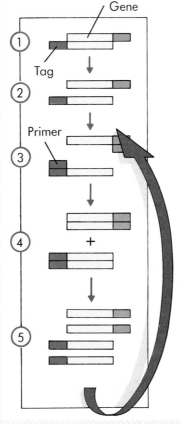

become four. It is not necessary to add any more polymerase, as the heating step does not harm this bacterial enzyme. Each heating and cooling cycle can be as short as one or two minutes, and each cycle doubles the number of DNA target molecules. After twenty cycles, a single fragment produces 2^{20} copies—over a million! In a few hours, one hundred billion copies of the fragment can be easily manufactured.

PCR has revolutionized many aspects of molecular biology because it allows the investigation of minute samples of DNA. In criminal investigations, DNA fingerprints can be prepared from the cells in a tiny speck of dried blood or at the base of a single human hair. Doctors can detect genetic defects in very early embryos by collecting a few sloughed-off cells and amplifying their DNA with PCR, or PCR can be used to examine the DNA of Abraham Lincoln. Mullis received the Nobel Prize in 1993 for his discovery.

George Poinar, Jr. of the University of California at Berkeley has suggested a novel PCR use, popularized in the book *Jurassic Park:*

1. Find a bead of amber that contains a blood-sucking insect from the age of dinosaurs.

2. Extract DNA from blood cells of a bitten dinosaur and amplify with PCR.

3. Process and inject into the embryo of an alligator.

4. Wait until it hatches!

This recipe makes many as-yet-untested assumptions but emphasizes the great impact this technique is having on all aspects of biology.

Dr. Kary Mullis

Pharmaceuticals

Perhaps the most obvious commercial application for genetic engineering, and one seized on first, was to introduce human genes that encode clinically important proteins into bacteria. Because they can be grown cheaply in bulk, bacteria that incorporate human genes can produce the human proteins that the genes specify in large amounts. This method has been used to produce several forms of human interferon, to place rat growth-hormone genes into mice, and to manufacture many commercially valuable nonhuman enzymes.

Among the many medically important proteins now being produced by these techniques is tissue plasminogen activator, a protein that causes blood clots to dissolve and that is normally produced by the body in minute amounts. Plasminogen activator is now being produced in quantities large enough to test its effectiveness in preventing heart attacks and strokes. Also being produced are atrial peptides, small proteins that regulate blood pressure and kidney function and that are produced in the hearts of mammals. Genetically engineered versions of atrial peptides are being tested as possible new ways to treat high blood pressure or kidney failure.

> *Gene technology is having a major impact on the pharmaceutical industry, making practical the production of a wide range of new drugs.*

Agriculture

A second major area of genetic engineering activity involves manipulating the genes of crop plants. In broadleaf plants, such as tomatoes, tobacco, and soybeans, the **Ti plasmid** of the bacterium *Agrobacterium tumefaciens,* which causes crown gall in plants, is the main vehicle used to introduce foreign genes. A part of this Ti plasmid integrates into the plant DNA, carrying with it whatever genes have been incorporated by earlier manipulation. The characteristics of a number of plants have been altered with this technique, which will be valuable in improving crops and forest trees. At one industrial lab alone, more than forty-five thousand independent "transgenic" plant lines have been produced in this way. Among the features that scientists are attempting to affect are: resistance to disease, frost, and other forms of stress; nutritional balance and protein content; and herbicide resistance (see Key Experiment 14.3).

Unfortunately, *Agrobacterium* generally does not infect the cereals, such as corn, rice, and wheat, and alternative methods have been sought to introduce new genes into them. Investigators first tried removing plant cell walls (producing "protoplasts") to facilitate DNA entry, but success rates with this procedure were very low. Nor did microinjection techniques work, since the ultra-fine needle tips required to penetrate single cells break and clog easily. On the average, only one in ten thousand microinjection attempts succeeds.

What *has* succeeded, beyond any expectation, is the novel approach of *shooting* DNA into plant cells! DNA is coated onto tiny metal particles, only 1 to 2 microns in diameter, and then shot with blank .22 caliber cartridges into cells (see Key Experiment 14.4). The metal particles penetrate the walls of intact cells and deliver the DNA inside, and the holes are so small that they rapidly close themselves, leaving the plant cell undamaged. Corn and wheat are now routinely "transformed" in this way (figure 14.5).

Herbicide Resistance

Some genetically engineered plants are resistant to the herbicide **glyphosate,** the active ingredient in *Roundup,* a powerful, biodegradable herbicide that kills most actively growing plants (figure 14.6). Glyphosate kills plants by inhibiting an enzyme called EPSP synthetase, which plants need to produce the so-called "aromatic" amino acids.

Glyphosate is an excellent herbicide for several reasons:

1. It is safe. Humans do not make aromatic amino acids but obtain them from food, and so are unaffected by glyphosate.

2. Unlike many other chemical herbicides, which accumulate in the environment, glyphosate poses no environmental risk. When applied, it acts quickly and then rapidly biodegrades into harmless natural compounds.

To obtain plants resistant to *Roundup,* agricultural scientists first inserted extra copies of the EPSP genes into plants, carrying them in on Ti plasmids. The engineered plants "overproduced" the enzyme to twenty times the normal levels and so were able to withstand EPSP suppression caused by glyphosate—that is, the extra amounts of EPSP enzyme permitted the plants to continue protein production and grow. In later experiments, a bacterial form of the EPSP synthetase gene that differed from the plant form by a single nucleotide was introduced into plants via Ti plasmids with the same result—the plants became resistant to *Roundup.*

This advance is of great interest to farmers. A crop resistant to *Roundup* would require little weeding, since treating the field with *Roundup* would kill most weeds, but have no effect on the crop. This would also reduce the number of herbicides needed for a single crop (most herbicides kill only a few kinds of weeds, while *Roundup* kills most growing plants). When natural selection brings about the development of glyphosate resistance in a particular species of weed, as will inevitably happen, that weed can be eliminated by the short-term use of another herbicide. EPSP genes are being successfully introduced into corn, wheat, and other cereal crops to make them *Roundup* (glyphosate)-resistant.

Virus Resistance

Ti plasmids have been used to introduce a variety of other genes into broadleaf crop plants. In one of the most interesting experiments, Roger Beachy, working at Washington University in St. Louis, developed plants immune to a virus (see Key Experiment 14.5). Unlike humans, plants have no immune system

KEY EXPERIMENT 14.3

2000

THE CHILTON EXPERIMENT

Mary Dell Chilton was one of the first to use the Ti plasmid of *Agrobacterium tumefaciens* in plant genetic engineering. Her procedure consisted of seven steps:

1. The Ti plasmid is removed and cut open with a special enzyme.

2. The gene is cut out of the chromosome of another organism, using the same special enzyme.

3. The new gene is inserted into the Ti plasmid.

4. The Ti plasmid is put back into *Agrobacterium*.

5. When mixed with plant cells, *Agrobacterium* duplicates the Ti plasmid.

6. The bacterium transfers the new gene into the chromosome of the plant cell.

7. The plant cell divides, and each daughter cell receives the new gene, giving the whole plant a new trait.

◄ 1981

1900

of their own. Beachy introduced a single virus gene—the tobacco mosaic virus (TMV) gene encoding the virus protein coat—into a tobacco cell chromosome, using the Ti-plasmid transfer system. He then grew the plasmid-infected plant cell in tissue culture, and from the culture grew a whole tobacco plant. Every cell of this new plant was derived from the Ti-infected cell an271d contained the TMV gene. When the genetically engineered tobacco plant was reinfected with TMV virus, it did not contract the disease! Transfer of the single gene had made the plant immune to TMV virus infection.

Genetic engineers are vigorously pursing this exciting development because it offers hope of engineering resistance to many viral diseases of commercial crops. Recent experiments demonstrate that expression of the coat protein of almost any

KEY EXPERIMENT 14.4

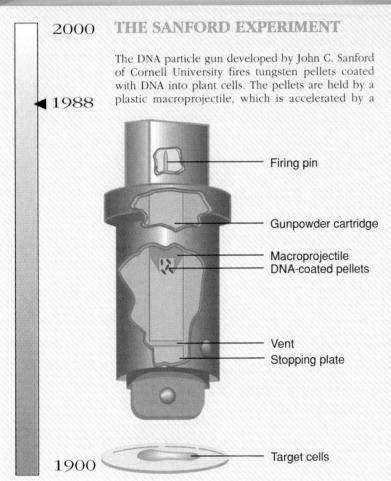

THE SANFORD EXPERIMENT

2000

◄ 1988

1900

The DNA particle gun developed by John C. Sanford of Cornell University fires tungsten pellets coated with DNA into plant cells. The pellets are held by a plastic macroprojectile, which is accelerated by a gunpowder charge. The plate stops the macroprojectile; momentum sends the pellets into the target. The vents allow air in front of the projectile to escape. In the photograph, a technician readying the device holds the "gun barrel" in her right hand; the cells to be transformed are in her left.

- Firing pin
- Gunpowder cartridge
- Macroprojectile
- DNA-coated pellets
- Vent
- Stopping plate
- Target cells

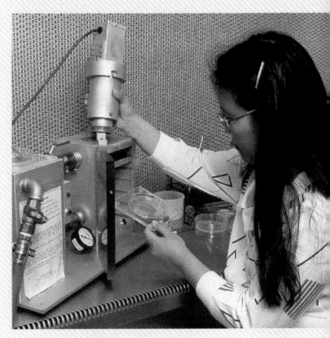

---

plant virus protects against infection by that virus. A broad range of crop species have already been engineered to be tolerant to more than a dozen different plant viruses.

Immunity to Insects

Many commercially important plants are attacked by insects. Over 40 percent of the chemical insecticides used today are employed to kill boll weevil, bollworm, and other insects that eat cotton plants (figure 14.7). Genetic engineers may be able to produce cotton plants that are naturally resistant to these insect pests. This would allow reductions in the doses of chemical insecticides that are applied to cotton fields and that negatively affect a number of organisms other than those they are intended to control.

One already successful application makes use of the bacterium *Bacillus thuringiensis,* which normally lives in soil and produces proteins **(Bt proteins)** that are toxic to the larvae (caterpillars) of moths and butterflies but not to other organisms. Bt proteins bind to specific receptors on the gut membranes of these insects, blocking ion transport so that the insects cannot feed or retain body moisture. These natural insecticides are not toxic to any other animals because no other animals possess these specific receptors. When the genes that produce Bt proteins are transferred via the Ti plasmid to such plants as tomatoes, potatoes, and cotton, the plants become toxic to serious pests, such as tomato hornworms, which die after eating them.

Plants containing Bt protein genes have been successfully tested in the field for more than two years. For example, cotton with Bt protein genes is protected against all major caterpillar pests, including the bollworm, allowing for a 40 to 60 percent reduction in the use of insecticide on cotton.

The Bt proteins are ideal insecticides for several reasons:

1. They do not wash off the plant; they are produced *inside* the crop.

2. They are active only against the target insect. Other insects (many of which are beneficial) are not harmed.

3. They are fully biodegradable.

Agrobacterium method

DNA particle gun method

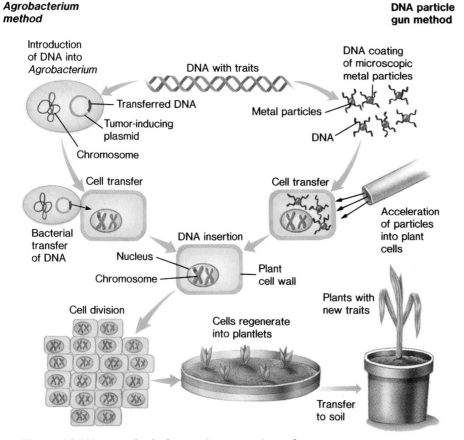

Figure 14.5 Two methods for getting genes into plants.

In the Agrobacterium *method, the genes to be transferred are placed into the Ti plasmid, which then infects the plant cell, carrying the gene in with it piggyback. In the DNA particle gun method, a solution of DNA containing the gene is painted onto particles, which are fired explosively into the plant cell. With either method, the target plant cell obtains a copy of the gene, which it incorporates into its own chromosome. The bioengineered plant cell can then be induced to form a full plant, all of whose cells will now carry the transferred gene.*

Figure 14.6 Genetically engineered herbicide resistance.

All four of these petunia plants were exposed to equal doses of the potent herbicide Roundup. *The two on top were genetically engineered to be resistant to glyphosate, the active ingredient of* Roundup, *whereas the dead ones were not.*

Bt insecticides are now being engineered to suit many other situations. Bt genes active against mosquitoes have been identified, and genetic engineers are currently trying to insert them into algae as a way of controlling malaria. Many important plant pests attack roots, and to counter this threat, biologists have introduced the Bt insecticidal protein gene into root-colonizing bacteria, especially strains of *Pseudomonas.*

Delayed Ripening of Fruit

Researchers have identified and isolated several genes that slow the biosynthesis of ethylene, the signal molecule that triggers the ripening of fruits. Such genes isolated from Antarctic fish (in which the gene has a very different function, producing a kind of antifreeze) have been successfully introduced into tomatoes, delaying spoilage for weeks. Thus, the tomatoes can be harvested at a later stage—an extra period of growth that improves both their flavor and nutritional value—and still arrive at markets in peak condition.

Environmental alarmists have objected ferociously to transgenic crops, such as these delayed-ripening tomatoes, citing fears of unknown consequences when mixing genes from different species. However, there appears to be no danger, and the U.S. government has approved a variety of transgenic approaches for improving commercial crops. Some of these crops are now appearing in local groceries (table 14.1).

Nitrogen Fixation

A long-range goal of agricultural genetic engineering is to increase yield and plant size in crop plants. However, scientists do not yet clearly know which genes are responsible for these complex characteristics.

Another long-range goal is to introduce into key crop plants the genes that enable symbiotic bacteria living in nodules on the roots of soybeans, other legumes, and certain other plants to "fix" nitrogen. **Nitrogen fixation,** discussed in detail in chapter 22, is the process by which such bacteria obtain nitrogen from the atmosphere, where it is abundant, and convert it into a form that living organisms can use in their metabolism—a form that is scarce. Most plants do not form associations with nitrogen-fixing bacteria and must obtain their nitrogen from the soil. Farmland, in which crops are grown repeatedly, soon becomes depleted of nitrogen unless treated with nitrogen fertilizers. Worldwide, farmers applied over 60 million metric tonnes of nitrogen fertilizers in 1987, an expensive undertaking. Farming would be much cheaper if major crops, such as wheat and corn, could be engineered to carry out nitrogen fixation. So far,

KEY EXPERIMENT 14.5

2000

◀ 1988

THE BEACHY EXPERIMENT

Roger Beachy used the Ti plasmid to produce plants resistant to a pathogenic virus. Working with tobacco plants, he introduced one gene from the tobacco mosaic virus (TMV) into a tobacco cell chromosome, using the Ti plasmid. The gene he selected for transfer was the one encoding the virus coat protein. He then grew the plasmid-infected tobacco cell in tissue cul-

ture, and from the culture grew a whole tobacco plant. Every cell of this bioengineered plant contained the TMV gene. When the genetically engineered tobacco plant was reexposed to the TMV virus, it did not contract the disease! The TMV virus does not infect cells that are already infected, and Beachy's experiment tricked the TMV virus into behaving as if the bioengineered plant was already infected with TMV. In effect, Beachy made the plant immune to the virus.

1900

Ti plasmid

Agrobacterium tumefaciens

(1) Plasmid is removed and cut open.

(2) Gene for TMV coat is isolated from virus.

(3) Gene is inserted in plasmid.

(4) Plasmid is reinserted in bacterium.

Tobacco mosaic virus

(5) TMV gene is transferred into plant cell chromosome, providing "immunity" to virus.

(6) Cells are grown into culture.

(7) Cultured cells are grown into plant resistant to TMV.

Figure 14.7 Genetic engineering to fight insect pests.
(a) *Crop plants have been engineered to produce a naturally occurring protein that kills certain insect pests, like this caterpillar. Advances like this could reduce farmers' dependence on chemical insecticides.* (b) *Both of these tomato plants were exposed to destructive caterpillars under laboratory conditions. The nonengineered plant on the left has been completely eaten, whereas the engineered plant on the right shows virtually no signs of damage.*

(a)

(b)

however, introducing the nitrogen-fixing genes from bacteria into plants has proven difficult because these genes do not seem to function properly in their new eukaryotic environment. Various nitrogen-fixing bacteria currently are being tested to determine whether their genes might function well in plants.

Bigger and Better Farm Animals

One of the first genes to be cloned successfully was that for the growth hormone **somatotropin.** Instead of being extracted from cow pituitaries at great expense, bovine somatotropin (BST) is produced in large amounts by a gene introduced into bacteria. The hormone, injected into dairy cows, improves the animal's milk production efficiency (see Key Experiment 14.6). This has no effect on humans because the amount of BST in milk is not increased. Other experiments that use somatotropin to increase the weight of cattle and pigs are underway. The human version of the same growth hormone is being used as a treatment for disorders like dwarfism, which are the result of inadequate levels of somatotropin.

From increased milk production to crops that require less pesticides because they are resistant to insects, gene technology is revolutionizing agriculture.

The Human Genome Project

A third area in which genetic engineering techniques are revolutionizing science involves what these techniques are revealing about the human genome. Any cloned gene can now be traced to a specific chromosomal location by determining where a radioactive version of the cloned gene sticks on the chromosome. The radioactive label accumulates only at the site where the probe binds to the chromosome DNA. Using large restriction fragments, geneticists have assembled several clone libraries of the human genome and are constructing maps that indicate the location of these fragments on the chromosomes.

Efforts to sequence the entire human genome, which contains some 3×10^9 nucleotide base pairs, are underway. This work, which is known as the **Human Genome Project,** will result in a detailed map of the human genome.

The Human Genome Project is significant because many genetic diseases are associated with specific restriction fragments. When the genes associated with particular diseases have been cloned, they can be used as probes to screen the clone library and to identify the appropriate restriction fragment. Mutations often affect the way in which the fragment moves on an electrophoresis gel, and they can always be detected as a change in nucleotide sequence of the fragment. Genetic screening of restriction fragments is a real possibility in the near future.

KEY EXPERIMENT 14.6

2000

◄ 1990

THE GROWTH HORMONE TRANSFER EXPERIMENT

The gene encoding bovine somatotropin (cattle growth hormone) was one of the first genes from an agricultural animal produced in quantity by genetically engineered bacteria. A cow DNA fragment containing the gene was inserted by Monsanto scientists into a bacterial plasmid, which was, in turn, inserted into the DNA of living bacteria. These bacteria became miniature hormone-producing factories, and because very large numbers of bacteria can be grown cheaply, the bacterial culture provided a plentiful and inexpensive source of bovine somatotropin. Injected into cows, the hormone improves milk production efficiency. In 1994, the U.S. government approved the use of bioengineered bovine somatotropin to improve commercial milk production. The hormone can also dramatically increase cattle body weight, an improvement of considerable commercial importance.

1900

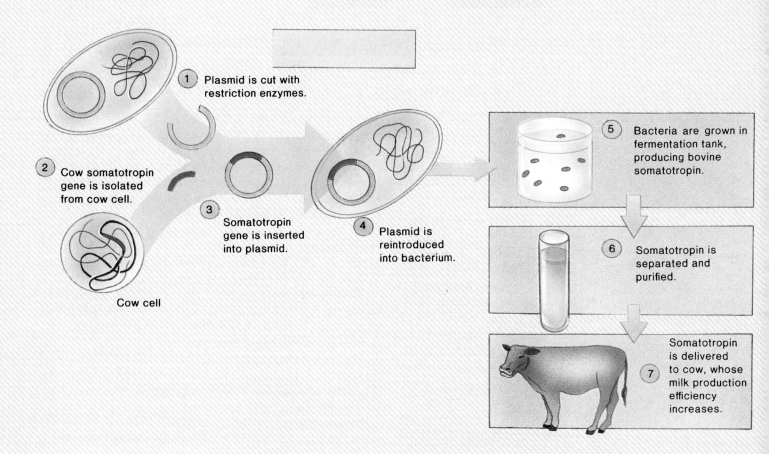

1 Plasmid is cut with restriction enzymes.

2 Cow somatotropin gene is isolated from cow cell.

3 Somatotropin gene is inserted into plasmid.

Cow cell

4 Plasmid is reintroduced into bacterium.

5 Bacteria are grown in fermentation tank, producing bovine somatotropin.

6 Somatotropin is separated and purified.

7 Somatotropin is delivered to cow, whose milk production efficiency increases.

Table 14.1 Crop Species Improved by Genetic Engineering

Alfalfa	Corn	Raspberry	Plum	Sugarcane
Apple	Cotton	Kiwi	Potato	Sunflower
Asparagus	Cranberry	Lettuce	Tobacco	Sweet
Broccoli	Cucumber	Muskmelon	Rice	potato
Cabbage	Eggplant	Oilseed rape	Rye	Tomato
Carrot	Flax	Papaya	Soybean	Walnut
Cauliflower	Grape	Pea	Strawberry	Wheat
Celery	Horseradish	Pepper	Sugarbeet	

Analysis of one restriction fragment at a time will eventually reveal the entire DNA sequence (figure 14.8). Clearly, the information gained will be of great value to medicine as well as to science. Sidelight 11.1 looks at some of the ethical issues that the Human Genome Project raises.

Subunit Vaccines

A fourth area of potential genetic engineering significance involves using genetic engineering techniques to help produce **subunit vaccines.** For example, gene-splicing techniques have resulted in the recent manufacture of vaccines against herpesvirus and hepatitis viruses. Genes specifying the protein-polysaccharide coat of the herpes simplex virus or hepatitis B virus have been spliced into a fragment of the DNA genome of the cowpox, or vaccinia, virus—the same virus that British physician Edward Jenner used almost two hundred years ago in his pioneering vaccinations against smallpox.

In these procedures, live vaccinia virus is introduced into a mammalian cell culture, along with spliced fragments of vaccinia virus DNA into which have been incorporated genes encoding the protein-polysaccharide coat of the herpes simplex virus genome (or, in other experiments, of the hepatitis B virus genome). The spliced herpes simplex gene is able to recombine into the vaccinia virus genome, producing a recombinant vaccinia virus. When injected into a mouse or rabbit, this recombinant virus dictates the production not only of the proteins that the vaccinia virus genome specifies, but also of the protein-polysaccharide coat specified by the herpes genes that it carries. The viruses that it produces are thus somewhat like sheep in wolf's clothing: They have the interior of a benign vaccinia virus and the exterior surface of herpesvirus. Infected individuals contract cowpox, which is relatively harmless, instead of herpes. At the same time, contact with the coat of the herpesvirus causes infected individuals to make antibodies against herpes and so become immune to it. Vaccines produced in this way are harmless because only a small DNA fragment of the disease-associated virus is introduced via the recombinant virus (figure 14.9).

In a subunit vaccine, a gene encoding one or a few proteins from a pathogen is added to the DNA of a harmless virus, which carries it piggyback into vaccinated people. They respond by making antibodies to the protein(s), defending themselves from the pathogen.

Gene Therapy

Similar "piggyback" approaches are being used to correct genetic defects in humans—a process called **gene therapy.** On 14 September 1990, a young girl became the first being to undergo gene therapy (figure 14.10). She suffered from a rare genetic disorder called "severe combined immune deficiency." Certain critical white blood cells, called T-cells, of her immune

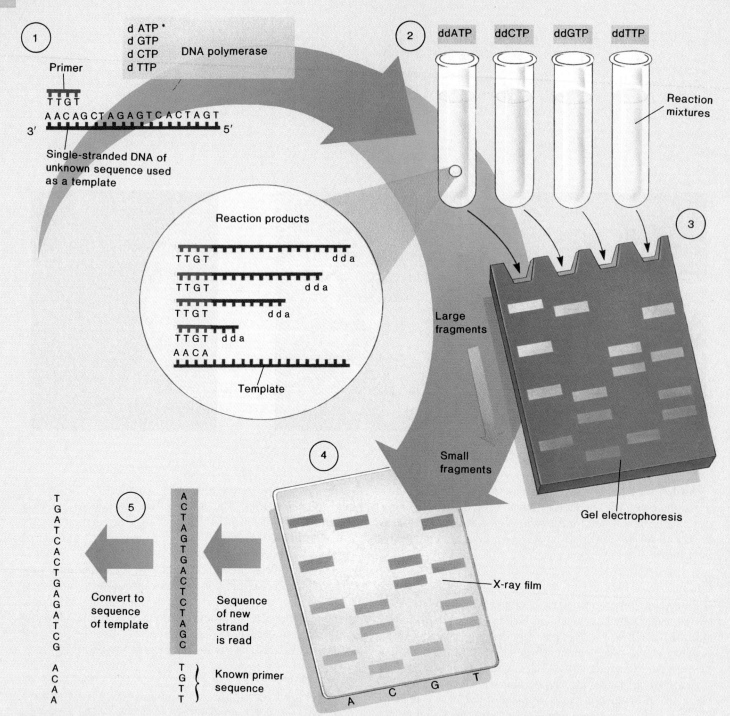

Figure 14.8 Steps in sequencing DNA.

Most DNA sequencing is currently done using the technique of "primer extension." (1) A short, two-stranded primer is added to the end of a single-stranded fragment to be sequenced. This provides a 3′ end for DNA polymerase to add to. (2) The primed fragment is placed in four in vitro synthesis tubes, each containing a different synthesis-stopping dd-nucleotide. The first tube, for example, contains ddATP, and synthesis stops whenever ddATP is incorporated instead of dATP. Thus, this tube will come to contain a series of fragments, corresponding to the different lengths the polymerase can travel from the primer before A nucleotides are encountered (ddATP is added at random and will not always hit every site). (3) Electrophoresis separates these fragments according to size. (4) A radioactive label (here ddATP) allows the fragments to be visualized on X-ray film, and (5) the newly made sequence can be read directly. Try it. The fragment you started with has the complementary sequence.*

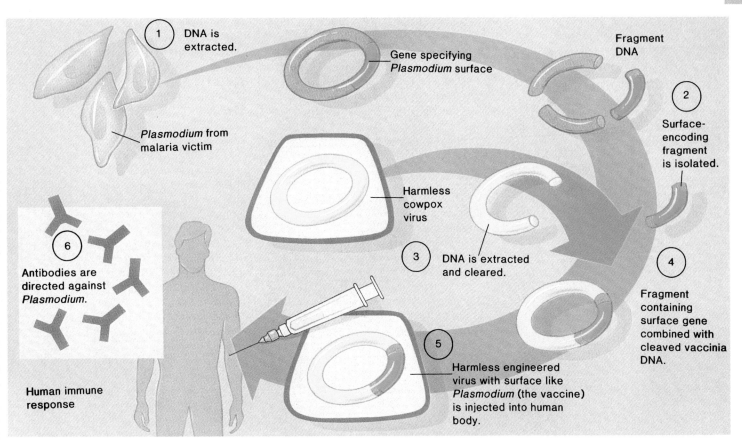

Figure 14.9 Constructing a subunit vaccine against malaria.
*One of the most exciting possibilities in the search for new
vaccines is the recent progress in developing a subunit vaccine for
malaria. Genes encoding surface proteins from various phases of*
the Plasmodium *life cycle are inserted into vaccinia virus, which
carries them "piggyback" into humans. The vaccinia-infected
individuals develop antibodies against* Plasmodium.

system were not working for lack of a necessary enzyme called
ADA (adenosine deaminase), leaving her no defense against infection. Doctors removed some of her blood and from it cultured T-cells. They then added to the cell culture viruses that
possessed a good working copy of the human ADA gene that
the girl lacked. These viruses carried the ADA gene with them
as they infected the T-cells and spliced it along with their own
genes into T-cell chromosomes. Doctors grew these altered T-
cells in the laboratory until they numbered in the billions, and
then injected them into a vein in the child's hand.

Over the next two years, the genetically repaired cells multiplied, populating the girl's immune system with gene-corrected
T-cells that produce the necessary enzyme. If the treatment
continues to be successful, she will be the first human "cured"
of a genetic disorder. Investigators are actively trying to cure
other genetic disorders—notably cystic fibrosis and muscular
dystrophy—with the same general approach. The first gene
transfers of cystic fibrosis genes were carried out on human patients in 1993.

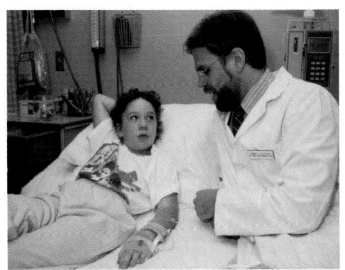

Figure 14.10 Gene therapy.
*One of two young girls who participated in a pioneering test of
gene therapy in 1990 and 1991, and her doctor.*

EVOLUTIONARY VIEWPOINT

Genetic engineering, which has revolutionized biology in the twenty years since Cohen and Boyer's experiment, is only possible because of so-called restriction enzymes that cut DNA at particular sequences. These enzymes evolved in bacteria as a defense against invading viruses. Co-evolution of the same sort is encountered in chapter 21 between insects and plants.

SUMMARY

1. Genetic engineering involves the isolation of specific genes and their transfer to new genomes.

2. The key to genetic engineering technology is a special class of enzymes called restriction endonucleases, which cleave DNA molecules into fragments. These DNA fragments always have short, one-stranded tails that are complementary in nucleotide sequence to each other.

3. Because the nucleotide sequences of the tails of the two DNA fragments resulting from the cleavage of DNA by restriction enzymes are complementary, any pair of fragments produced by the same restriction enzyme can be joined together. DNA fragments from very different genomes can be combined in this way.

4. Stanley Cohen and Herbert Boyer achieved the first combination of DNA fragments from different genomes (recombinant DNA) in 1973. They inserted a ribosomal RNA gene from a toad into a bacterial plasmid.

5. The three stages of genetic engineering experiments are: (1) generating specific DNA fragments by cleavage of a genome with restriction endonuclease enzymes, (2) using either cloning or polymerase chain reaction (PCR) techniques to make large numbers of copies of the fragments, and (3) transferring the fragments into living organisms.

6. Cloning is a technique in which fragment-containing plasmids or viruses are introduced into bacterial cells, which then reproduce, forming a clone of cells that all contain the fragment-bearing plasmid or virus. The PCR technique takes a more direct approach, using DNA polymerase to make billions of identical copies of the genes of interest.

7. Genetic engineering has fostered many advances in the production of affordable drugs, including insulin, interferon, tissue plasminogen activator, and atrial peptides.

8. Agriculture has been a major focus of genetic engineering activity. Successes have included incorporation of genes for herbicide and virus resistance, immunity to insects, delayed fruit ripening, nitrogen fixation, and increased milk production in cows.

9. The Human Genome Project involves efforts to produce a detailed map of the entire human genome.

10. Gene splicing holds great promise as a clinical tool, particularly in the prevention of disease. Several subunit vaccines are being prepared by inserting genes specifying the protein coats of disease-causing viruses or cells into the genomes of harmless viruses to create a recombinant form that is harmless and has the outside coat of the disease-causing form. The recombinant form can be used as a vaccine to induce antibody formation against the disease because its protein coat produces a response from the immune system.

REVIEWING THE CHAPTER

1. What are plasmids, and how do they relate to genetic engineering?

2. What are restriction enzymes?

3. How is a chimeric genome constructed?

4. What is PCR, and how is it used in genetic engineering?

5. What techniques are creating a scientific revolution?

6. How is genetic engineering used in pharmaceuticals?

7. What are the benefits of genetic engineering to agriculture?

8. How is genetic engineering being used in herbicide resistance?

9. How and why are plants engineered for viral resistance?

10. How does genetic engineering relate to insect immunity?

11. How does genetic engineering delay fruit ripening?

12. What is nitrogen fixation?

13. Why is genetic engineering being used on farm animals?

14. Why are scientists probing the human genome?

15. What are subunit vaccines?

16. How is gene therapy beneficial?

COMPLETING YOUR UNDERSTANDING

1. What is cloning, and what are its benefits?

2. How do restriction endonucleases and methylases differ in their functions?

3. What is the significance of a recognition sequence in a DNA duplex?

4. Why are restriction endonucleases invaluable tools to genetic engineers?

5. What is a vector, and how are vectors used in research?

6. What are the potential dangers and safeguards of recombinant DNA experiments?

7. A clone library is
   a. a collection of different lines of bacterial cells, each containing a plasmid carrying one of a series of different DNA fragments.
   b. a collection of books about clones.
   c. the particular bacterial line that contains   DNA fragment of interest.
   d. a series of restriction enzymes that can be used to produce different clones.
   e. none of the above.

8. What is electrophoresis, and how is it used in genetic engineering?

9. How can PCR aid in criminal investigations?

10. What are the potential benefits of plasminogen activator and atrial peptides to the human population?

11. Why is *Agrobacterium* used in genetic engineering? What are its limitations?

12. How has "shooting" DNA into plant cells "transformed" corn and wheat?

13. Why is glyphosate an excellent herbicide choice?

14. Why and how did agriculturists engineer plants resistant to the herbicide *Roundup?*

15. What are the benefits of using genetic engineering rather than chemical insecticides to eradicate insects?

16. What are Bt proteins, and why are they of interest to crop science?

17. What is ethylene, and why is it important biologically?

18. Why has it been so difficult to introduce nitrogen-fixing genes into cereal crops?

19. What is somatotropin, and why are genetic engineers interested in it?

*****20.** How big is the human genome, and how does this relate to rapist cases?

21. After two hundred years, why is there an interest again in the smallpox vaccine?

*****22.** You are having a genetic engineering nightmare. In your dream, a well-meaning student cleaves the DNA of reticulocyte (red blood cells) isolated from a man dying of leukemia with the restriction enzyme Eco R1. The student then mixes the resulting fragments with the Eco R1—treated plasmid pSC101 and infects a growing culture of *Escherichia coli* bacteria with the resulting mix of normal and chimeric plasmids. From this mix, the student selects for cells containing plasmids by applying the antibiotic tetracycline to the liquid culture of growing bacteria. Many cells continue to grow. The student, in an excess of joy at this intimation of success, accidently drops the test tube containing the cells in the sink. It breaks, and the liquid flows down the drain. Knowing that *E. coli* is a common inhabitant of the human intestinal tract, the student breaks out into a cold sweat—and you wake up. Was the student right to be scared? Why?

*****23.** A major focus of genetic engineering has been the attempt to produce large quantities of hormones and other molecules that are scarce in humans by placing the appropriate human gene into bacteria. Because prodigious numbers of bacteria can be produced rapidly and cheaply, large amounts of the desired molecule can be obtained readily. Human insulin is now manufactured in this way. However, this approach does not work to produce human hemoglobin (beta-globin). If an experimenter first identifies the proper clone from a clone library by using an appropriate radioactive probe, inserts the fragments containing the beta-globin gene into a plasmid, and infects bacterial cells with the chimeric plasmid, no human hemoglobin is produced by the infected cells. This negative result is obtained despite the fact that the proper fragment clone was chosen, the beta-globin gene was successfully incorporated into the plasmid, and the chimeric plasmid successfully infected the *E. coli* cells. Why isn't this experiment working?

*For discussion

FOR FURTHER READING

DiLisi, C. 1988. The Human Genome Project. *American Scientist,* 76: 438–39. Mapping and deciphering the complete sequence of human DNA that will stimulate research in fields ranging from computer technology to theoretical chemistry.

Erlich, H. et al. 1991. Recent advances in the polymerase chain reaction. *Science,* June, 1643–50. An overview of how this revolutionary new technique is being used in genetic engineering.

Gasser, C., and R. Fraley, 1992. Transgenic crops. *Scientific American,* June, 62–70. A clear account of how biotechnology is creating plants that withstand pests and fruits that resist spoilage.

McElfresh, K. et al. 1993. DNA-based identity testing in forensic science. *BioScience,* March, 149–57. A review of the challenges to court admissibility of DNA data—it has survived five years of strong challenges.

Montgomery, G. 1990. The ultimate medicine. *Discover,* March, 60–68. An account of how virus vectors are being used in human gene transfer therapy.

Mullis, K. 1990. The unusual origin of the polymerase chain reaction. *Scientific American,* April, 56–65. A charming account of how a young maverick molecular biologist thought up a revolutionary procedure for making unlimited copies of DNA fragments.

Neufeld, P. J., and N. Colman. 1990. When science takes the stand. *Scientific American,* May, 46–53. Article discusses how DNA and other evidence is increasingly being applied to the solution of criminal cases but must be used with caution.

Verma, I. M., 1990. Gene therapy. *Scientific American,* Nov., 68–84. Gene technology's newest advance: Healing genetically inherited diseases by inserting healthy genes into patients.

EXPLORATIONS

Interactive Software:
Genetic Engineering

The user attempts to clone a gene in this interactive. The key variables are: (1) the size of the desired gene and the number of times it appears on a typical chromosome, and (2) the choice of restriction enzyme. A variety of real gene sequences will be examined, and the consequences of selecting different restriction enzymes to cleave the DNS are explored in terms of the efficiency of a shotgun cloning experiment.

Questions to Explore:

1. Can you identify one size of restriction site that is optimal for the full range of genes examined? If not, what relationship do you find between size of restriction site and size of gene being sought?

2. Is there any relationship between size of restriction site and the number of copies of a gene on the chromosome?

3. Can you identify any pattern within the sequences of the restriction sites, as portrayed in this interactive? Are odd and even numbers of nucleotides at a site equally effective?

4. What range of restriction site sizes do you find to be the most effective? Why do you suppose this particular range, and not another, is the more effective?

Interactive Software:
Constructing a Restriction Site Map

In this interactive exercise, the student constructs a restriction map by entering measured band position data from a set of electrophoresis gels. The data may be supplied by the user from real lab experiments, or the user may choose to analyze one of several data sets provided by the interactive exercise. Maps such as these, based on determining the relative positions of overlapping DNA restriction fragments, are the principal way in which today's geneticists construct physical maps of genes.

Questions to Explore:

1. Do you believe that restriction maps such as these have greater solutions than maps prepared by scoring the frequency with which recombinants appear among the progeny of genetic crosses? Why?

2. What great advantage do restriction maps have over genetic maps? What great advantage do genetic maps have over restriction maps?

3. In preparing your restriction map, do you find that there is any particular advantage to employing enzymes with different size restriction sites? Why do you think this is so?

Part
5
EVOLUTION

A lone coconut begins what could become a long voyage from its native island in the Caribbean. Washed ashore on some faraway island, it may sprout and form a new tree. The survival of the species depends on such "long shots" as this.

THE EVIDENCE FOR EVOLUTION

Few locations on earth are more readily identified with the theory of evolution than the Galápagos Islands. Darwin formed his key ideas about species by studying specimens collected on these isolated islands, located 1,000 kilometers out into the Pacific Ocean from South America.

FOR REVIEW

Here are some important terms and concepts that have been discussed in previous chapters and that you will encounter again in this chapter. Review them before proceeding, if necessary.

Darwin's theory of natural selection (*chapter 1*)
Origin of life (*chapter 3*)
Alleles (*chapter 10*)
Heterozygosity and homozygosity (*chapter 10*)
Punnett square (*chapter 10*)
Mutation (*chapter 13*)

Almost everybody has heard of Darwin's theory of evolution—not because of a widespread interest in biology but because many people regard this theory as a challenge to their religious beliefs. The theory of evolution by natural selection is routinely condemned in churches and attacked at local school board meetings, an astonishing state of affairs given that this theory is uniformly accepted by biological scientists and indeed forms the heart of biology as a science!

Within recent years, controversial attempts have been made to require the teaching of the set of religious dogmas known as scientific creationism alongside evolution theory in science classes. The theory of evolution has faced similar highly publicized challenges since the time of Darwin, and others are likely in the future. For this reason, this chapter addresses the issue squarely. Just what *is* the evidence for evolution?

As pointed out in chapter 1, Darwin deduced that **evolution,** the progressive change that occurs in organisms' characteristics through time, could best be explained as being the result of natural selection (figure 15.1). **Natural selection** is the process whereby some individuals in a population have genetic attributes that give them features promoting their survival and that allow them to produce more surviving offspring. Survival is not random; rather, it is related to the genes of the individuals who do and do not survive. The result of natural selection is a change in allele frequencies from generation to generation.

This chapter first explores the factors believed to bring about evolutionary change. It then looks at individual cases in which evolutionary change in populations can be seen to be adaptive to a particular environment, a key tenet of Darwin's theory of evolution by means of natural selection. Finally, the chapter reviews the evidence that supports the general validity of Darwin's proposal.

The word *evolution* often conjures up images of dinosaurs, woolly mammoths frozen in blocks of ice, or Darwin confronting a monkey. Traces of ancient life-forms, now extinct, survive as fossils that help to piece together the history of life on earth. Thus, evolution is usually interpreted to mean changes in the kinds of animals and plants on the earth, changes that take place over long periods, with new forms replacing old ones. Actually, this kind of evolution is called **macroevolution.** Macroevolution is evolutionary change on a grand scale, encompassing the origin of novel designs, evolutionary trends, new kinds of organisms penetrating new habitats, and major extinction episodes.

Much of Darwin's theory of evolution, however, focuses not on the way in which new species are formed from old ones, but rather on how changes occur within species. As a result of natural selection, a population gradually comes to include more and more individuals with advantageous characteristics, assuming that the characteristics have a genetic basis. In this way, the population evolves. Change of this sort within populations—the progressive change in allele frequencies—is called **microevolution.** Natural selection is the process by which microevolutionary change occurs; the result of the process—the features that promote the likelihood of survival and reproduction by an organism in a particular environment—is called **adaptation.** The essence of Darwin's explanation of evolution is that progressive adaptation by natural selection is responsible for evolutionary changes *within* a species. These changes, when they accumulate, lead to the creation of new kinds of organisms, new species. In short, microevolution prompts macroevolution. Chapter 16 discusses the ways in which new species originate in more detail.

Genes within Populations: The Hardy-Weinberg Principle

According to Darwin, evolution is a progressive series of adaptive changes brought about by natural selection. The raw material available for the selection process is the genetic variation among individuals within a species, from which natural selection chooses the best-suited alleles. In the early twentieth century, the science of genetics—the study of genes—contributed relatively little to scientific understanding of the process of evolution. At first, geneticists were involved primarily with understanding the actions of individual genes. Evolutionists, in turn, could not understand how such observations would have a bearing on the evolution of a complex structure, such as an eye.

The gap between geneticists and evolutionists finally started to close with the development of the field of **population genetics**—simply defined as the study of the properties of genes in populations. From the 1920s onward, scientists began to formulate a comprehensive theory of how alleles behave in populations and the ways in which changes in gene frequencies lead to evolutionary change. The fundamental tool of population genetics is the Hardy-Weinberg principle.

Variation within natural populations puzzled Darwin and his contemporaries. The patterns by which certain features were inherited were poorly understood, and the discovery of the principles of genetics, as well as of how meiosis produces genetic segregation among the progeny of a hybrid, lay in the future. The concept of blending inheritance, by which the features of parent organisms can merge with one another in offspring, was widely accepted. However, even when the nature of dominant and recessive alleles was understood, in the early twentieth century, dominant alleles were believed to drive recessive alleles out of populations, with selection favoring an optimal form.

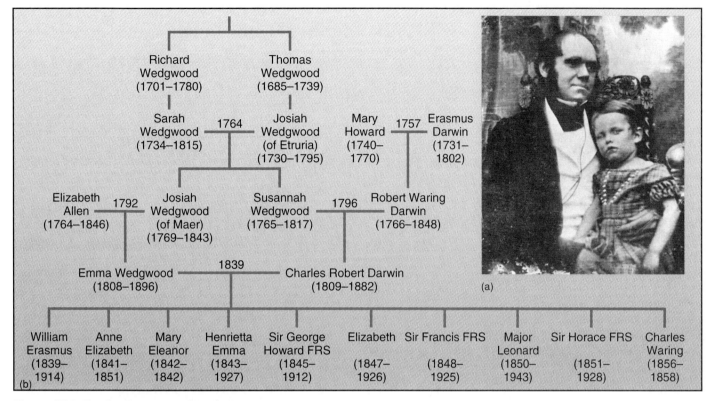

Figure 15.1 Charles Darwin and evolution.
(a) *Charles Darwin and his eldest son William in 1842, the year he wrote the preliminary version of his ideas on evolution by natural selection. (b) As shown in this family tree, Darwin had ten children before he published his ideas in* On the Origin of Species *in 1859. Three of his children went on to become scientists and Fellows of the Royal Society. Note that not only did Darwin marry his first cousin, but so did his grandfather, Josiah Wedgewood.*

The solution to the puzzle of why genetic variation persists was developed independently and published almost simultaneously in 1908 by G. H. Hardy, an English mathematician, and G. Weinberg, a German physician. Hardy and Weinberg pointed out that, in a large population in which there is random mating, and in the absence of forces that change the proportions of the alleles at a given locus (these forces are discussed later in the chapter), the original genotype proportions will remain constant from generation to generation. Dominant alleles do not, in fact, replace recessive ones. Because their proportions do not change, the genotypes are said to be in **Hardy-Weinberg equilibrium.**

In algebraic terms, the Hardy-Weinberg principle is written as an equation. Its form is what is known as a binomial expansion. For a gene with two alternative alleles, called A and a, the frequency of allele A can be expressed as p and that of the alternative allele a as q. By convention, the dominant and usually more common of the two alleles is designated p; the rarer allele, q. Because there are only two alleles, $p + q$ must always equal one. The equation looks like this:

$$(p + q)^2 \quad = \quad p^2 \quad + \quad 2pq \quad + \quad q^2$$

Individuals homozygous for allele A	Individuals heterozygous for alleles A and a	Individuals homozygous for allele a

In statistics, **frequency** is defined as the proportion of individuals falling in a certain category, relative to the total number of individuals being considered. For example, in a population of ten cats, of which seven are white and three are black, the respective phenotypic frequencies are 0.7 (or 70 percent) and 0.3 (or 30 percent). If the white cats are assumed to be homozygous recessive for a gene designated b, and the black cats are therefore either homozygous dominant BB or heterozygous Bb, the frequencies of the two alleles in the population can be calculated from the proportion of black and white cats.

Another population of one hundred cats, with the genetics of coat color as stated in the previous paragraph, might have a phenotypic ratio of sixteen white cats to eighty-four black ones. If q is the frequency of the allele b, then the Hardy-Weinberg equation states that $q^2 = 0.16$ and $q = 0.4$. The frequency of allele B—p—must therefore be $1 - 0.4 = 0.6$. In addition to the sixteen white cats, which have a bb genotype, there are $2pq$, or $2 \times 0.6 \times 0.4 \times 100$ (the number of individuals in the whole population), or forty-eight heterozygous individuals of genotype Bb. It can also be easily calculated that there are $p^2 = (0.6)^2 \times 100$, or thirty-six homozygous dominant BB individuals.

Figure 15.2 traces genetic reassortment during sexual reproduction to see how it affects the frequencies of the B and b alleles during the next generation. The Punnet square in figure 15.2b assumes that the union of sperm and egg in these cats is random, so all combinations of B and b alleles are equally likely. For this reason, the alleles are, in effect, mixed randomly and are represented in the next generation in proportion to their original representation; there is no inherent reason for them to change in frequency from one generation to the next. Each individual in each generation has a 0.6 chance of receiving a *B* allele and a 0.4 chance of receiving a *b* allele.

In the next generation, and for each succeeding generation, the chance of combining two *B* alleles is 0.36 (that is, 0.6 × 0.6), and an average of 36 percent of the population will continue to have the *BB* genotype. The frequency of *bb* individuals (0.4 × 0.4) will continue to be about 16 percent, and the frequency of *Bb* individuals (2 × 0.6 × 0.4) will continue to be 48 percent.

This simple relationship has proven extraordinarily useful because it permits biologists to predict readily what proportion of a population will be homozygous for a given allele and what proportion will be heterozygous, simply from a knowledge of that allele's frequency. For example, the recessive allele responsible for cystic fibrosis is present in North American Caucasians at a frequency of about twenty-five per one thousand individuals, or 0.025. What proportion of North American Caucasians are expected to express this trait? The frequency of homozygous-recessive individuals (q^2) is expected to be 0.025 × 0.025, or one in every sixteen hundred individuals. What proportion is expected to be heterozygous carriers? If the frequency of the recessive allele *q* is 0.025, then the frequency of the dominant allele *p* must be 1 − 0.025, or 0.975. The frequency of heterozygous individuals ($2pq$) is thus expected to be 2 × 0.975 × 0.025, or forty-seven in every one thousand individuals.

How valid are these calculated predictions? For many genes, they prove to be very accurate. Most human populations are large and randomly mating and thus are similar to the "ideal" population envisioned by Hardy and Weinberg. For some genes, however, the calculated predictions do *not* match the actual values. The reasons why explain a great deal about evolution.

The Hardy-Weinberg principle states that the proportions of different alleles will stay the same in a large population if mating occurs at random and in the absence of outside forces.

Why Do Allele Frequencies Change?

According to the Hardy-Weinberg principle, the proportions of homozygotes and heterozygotes in a large, random-mating population will remain constant, *as long as the individual allele*

Phenotypes			
Genotypes	BB	Bb	bb
Frequency of genotype in population	0.36	0.48	0.16
Frequency of gametes	0.36 + 0.24 → 0.6B		0.24 + 0.16 → 0.4b

(a)

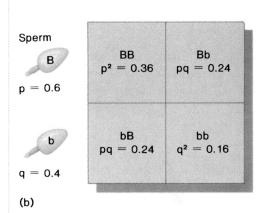

(b)

Figure 15.2 The Hardy-Weinberg equilibrium.
(a) *In the absence of factors that alter them, the frequencies of alleles, genotypes, and phenotypes remain constant generation after generation. The example shown here involves a population of one hundred cats, in which sixteen are white and eighty-four are black. White cats are* bb, *and black cats are* BB *or* Bb. (b) *A Punnett square analysis of the potential crosses in this cat population.*

frequencies do not change. This last qualifier is the key to the importance of the Hardy-Weinberg principle for biology because individual allele frequencies *are* changing all the time in natural populations, with some alleles becoming more common than others. The Hardy-Weinberg principle establishes a convenient baseline against which to measure such changes. By looking at how various factors alter the proportions of homozygotes and heterozygotes in populations, scientists can identify those forces that are affecting particular situations.

Many factors can alter allele frequencies. But only five alter the proportions of homozygotes and heterozygotes enough to produce significant deviations from the proportions predicted by the Hardy-Weinberg principle: (1) mutation, (2) migration (including both immigration into and emigration out of a given population), (3) genetic drift (random loss of alleles, which is more likely to occur in small populations), (4) non-random mating, and (5) selection (table 15.1). Of these, only selection produces adaptive evolutionary change because only in selection does the result depend on the nature of the environment. The other factors generally operate independently of the environment, although some of them—mutation, for instance—can be induced by environmental factors in special circumstances. Environmental influence on these factors, however, is unusual and should be viewed as an observation rather than a normal consequence.

> *Five factors can cause a deviation from the genotype frequencies predicted by the Hardy-Weinberg principle: mutation, migration, genetic drift (random loss of alleles), nonrandom mating, and selection.*

Mutation

Mutation from one allele to another obviously can change the proportions of particular alleles in a population. But mutation rates are generally too low to significantly alter Hardy-Weinberg proportions of common alleles. Many genes mutate one to ten times per one hundred thousand cell divisions. Because most environments are constantly changing, populations stable enough to accumulate differences in allele frequency produced this slowly are rare. Nonetheless, mutation is the ultimate source of genetic variation and thus makes evolution possible.

Migration

Migration, defined in genetic terms as the movement of individuals from one population into another, can be a powerful force in upsetting the genetic stability of natural populations. Sometimes, migration is obvious, as when an animal moves from one place to another. If the characteristics of the newly arrived animal differ from those already there, the genetic composition of the receiving population may be altered if the newly arrived individual or individuals can adapt to survive in the new area and mate successfully.

Other important kinds of migration are not as obvious. These subtler movements include the drifting of gametes or immature stages of marine animals or plants from one place to another. For example, the male gametes of flowering plants are often carried great distances by insects and other animals that visit their flowers. Their seeds may also be blown in the wind or carried by animals or other agents to new populations far from their place of origin.

However it occurs, migration can alter the genetic characteristics of populations and prevent the maintenance of the Hardy-Weinberg equilibrium. However, the evolutionary role of migration is more difficult to assess and depends heavily on the selective forces prevailing at the different places where the species occurs.

Table 15.1 Agents of Evolutionary Change

Factor	Description
Mutation	The ultimate source of all change. Individual mutations occur so rarely that mutation alone does not change allele frequency much.
Migration	A very potent agent of change. Migration acts to promote evolutionary change by enabling populations that exchange members to converge toward one another.
Genetic drift	Statistical accidents. Usually occurs only in very small populations.
Nonrandom mating	Inbreeding is the most common form. It does not alter allele frequency but decreases the proportion of heterozygotes ($2pq$).
Selection	The only form that produces *adaptive* evolutionary changes. Only rapid for allele frequency greater than 0.01.

Genetic Drift

In small populations, the frequencies of particular alleles may be changed drastically by chance alone. The individual alleles of a given gene are all represented in few individuals, and some of them may be lost from the population if those individuals fail to reproduce. Allele frequencies appear to change randomly, as if the frequencies were drifting; thus, random loss of alleles is known as **genetic drift.** A series of small populations that are isolated from one another may come to differ strongly as a result of genetic drift.

> *Genetic drift is the change in allele frequencies due to chance events. In small populations, such fluctuations may lead to the loss of particular alleles.*

When one or a few individuals are dispersed and become the founders of a new, isolated population at some distance from their place of origin, the alleles that they carry are of special significance. Even if these alleles are rare in the source population, they will be a significant fraction of the new population's genetic endowment. This effect—by which rare alleles and combinations of alleles may be enhanced in new populations—is called the **founder effect.** The founder effect is particularly important in the evolution of organisms on islands, such as the Galápagos Islands, which Darwin visited. Most of the kinds of organisms that occur in such areas were probably derived from one or a few initial "founders." In a similar way, isolated human populations are often dominated by the genetic features that were characteristic of their founders, if only a few individuals were involved initially.

The founder principle is demonstrated in the Amish community, a small religious group that resides primarily in Pennsylvania and Ohio. Founded in the eighteenth century by a small (about two hundred!) contingent of followers, the Amish community has shown a high incidence of a rare genetic disorder called Ellis-van Creveld syndrome (figure 15.3). In this syndrome, caused by a recessive allele, the affected individual exhibits dwarfism, shortened limbs, and extra fingers and toes. Victims of this disorder usually die within a few months after birth. Geneticists researching the high frequency of Ellis-van Creveld syndrome in the Amish community have traced the disorder back to a single couple who immigrated to Pennsylvania in 1774. Geneticists surmise that either the husband or wife was heterozygous for the Ellis-van Creveld allele, and the recessive allele combined with another founder, who also, by chance, carried the recessive allele. Because the Amish are an isolated community, the recessive allele has persisted and continues to afflict modern Amish followers.

Another ramification of genetic drift is known as the **bottleneck effect,** in which a small founder population becomes the sole source of alleles for a given species. The bottleneck effect can be seen in the current cheetah population. Researchers think that the cheetah population in Africa underwent some sort of crisis about ten thousand years ago, which depleted their numbers considerably. In the last century, another population crisis occurred as cheetahs were almost hunted to extinction. Their decreased numbers have limited cheetahs' genetic variability, with serious consequences. For example, cheetahs are susceptible to a number of fatal diseases. This decrease in genetic variability and its various consequences has placed the cheetah population at risk, and scientists fear that the species may become extinct because of the lack of allele variation.

Nonrandom Mating

Individuals with certain genotypes sometimes mate with one another more commonly than would be expected on a random basis, a phenomenon known as **nonrandom mating.** Inbreeding (mating with relatives), a type of nonrandom mating that is characteristic of many groups of organisms, causes the frequencies of particular genotypes to differ greatly from those predicted by the Hardy-Weinberg principle. Inbreeding does not change the frequency of the alleles, but rather the proportion of individuals that are homozygous. Inbred populations contain more homozygous individuals than is predicted by the Hardy-Weinberg principle. For this reason, populations of self-fertilizing plants consist primarily of homozygous individuals, whereas "outcrossing" plants, which interbreed with

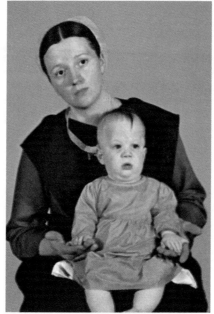

Figure 15.3 A consequence of the founder effect.

This Amish woman is holding her child, who has Ellis-van Creveld syndrome. The characteristic symptoms of this disorder are short limbs, dwarfed stature, and extra fingers. This disorder was introduced in the Amish community by one of its founders in the eighteenth century and persists to this day because of the reproductive isolation of the Amish.

individuals different from themselves, have a higher proportion of heterozygous individuals (figure 15.4).

Because inbreeding increases the proportion of homozygous individuals in a population, homozygous-recessive combinations are more likely. This is why marriages between relatives are discouraged—such marriages greatly increase the possibility of producing children homozygous for an allele associated with one or more of the genetic disorders discussed in chapter 11. Even though alleles specifying some disorders, such as Tay-Sachs disease, are rarer in Japan than elsewhere, there is a much greater likelihood that a marriage between first cousins will produce affected children there because the Japanese population is more homogeneous in general.

Selection

As Darwin pointed out, some individuals leave behind more progeny than others, and the rate at which they do so is affected by their inherited characteristics. The results of this process are called **selection,** and there is both **artificial selection** and **natural selection.** In artificial selection, the breeder selects for the desired characteristics (figure 15.5). In natural selection, the environment plays this role, with conditions in nature determining which kinds of individuals in a population are the most fit and so affecting the proportions of genes among individuals of future populations. This is the key point in Darwin's proposal that evolution occurs because of natural selection: The environment imposes the conditions that determine the results of selection and, thus, the direction of evolution.

Like mutation, migration, and genetic drift, selection causes deviations from Hardy-Weinberg proportions by directly altering allele frequencies. Darwin argued that the more successful reproduction of particular genotypes, which is how he defined selection, is the primary force that shapes the pattern of life on earth. But the selection of these genotypes is indirect: Selection acts directly on the phenotype, which is determined by the interaction of the genotype and the environment, and the linkage between particular alleles and particular characteristics of the phenotype is less direct for some features than for others.

Although selection is perhaps the most powerful of the five principal agents of genetic change, there are limits to what it can accomplish. These limits arise because alternative alleles may interact in different ways with other genes. These interactions tend to set limits on how much a phenotype can be altered. For example, selecting for large clutch size in barnyard

Figure 15.4 Nonrandom mating versus random mating.
This bee is pollinating a desert poppy in eastern Arizona. Flowers pollinated by bees and other insects or animals show more genetic variation than flowers that are self-pollinating because the pollination is random.

pressed in an organism's phenotype can affect the organism's ability to produce progeny. Selection does not operate efficiently on rare, recessive alleles simply because they do not often come together as homozygotes, and there is no way of selecting them unless they do come together. For example, when a recessive allele a is present at a frequency q equal to 0.1, 10 percent of the alleles for that particular gene will be a, but only one out of one hundred individuals (q^2) will be homozygous recessive and so display the phenotype associated with this allele. For lower allele frequencies, the effect is even more dramatic: If the frequency in the population of the recessive allele q equals 0.01, the frequency of homozygotes in that population will be only one in ten thousand.

Selection, the differential reproduction of genotypes as a result of the way that their phenotypic characteristics lead to reproductive success in the environment, is a powerful mechanism for producing deviations from Hardy-Weinberg equilibrium.

What this means is that selection against undesirable genetic traits in humans or domesticated animals is difficult unless the heterozygotes can also be detected. For example, if, for a particular undesirable recessive allele r, q equals 0.01, and none of the homozygotes for this allele are allowed to breed, it would take one thousand generations, or about twenty-five thousand human years, to lower the allele frequency by half to 0.005. At this point, the frequency of homozygotes would still be one in forty thousand, or 25 percent of what it was initially. This is the basic reason why few geneticists advocate **eugenics,** the field that deals with efforts to change the genetic characteristics of human beings by artificial selection. Aside from the moral implications, such efforts are essentially doomed to failure by the sheer difficulty of producing the desired results within any plausible human time frame.

Forms of Selection

Selection operates in natural populations of a species something like skill does in football games: It is difficult to predict the winner in any individual game because chance can play an important role in the outcome, but over a long season, the teams with the most skillful players usually win the most games. In nature, too, those individuals best suited to their environment tend to win the evolutionary game by leaving the most offspring, although chance can play a major role in the

Figure 15.5 Artificial selection.
Racehorses are bred specifically for one trait, speed.

chickens eventually leads to eggs with thinner shells that break more often than did the shells of previous eggs. Because of the limits imposed by gene interactions, strong selection is apt to result in rapid change initially, but the change soon comes to a halt as the interactions between genes increase. For this reason, farmers do not have gigantic cattle that yield twice as much meat as leading strains, chickens that lay twice as many eggs as the best layers do now, or corn with an ear at the base of every leaf, instead of at just one or a few leaves.

A second factor limits what selection can accomplish: Selection acts only on phenotypes. Only those characteristics ex-

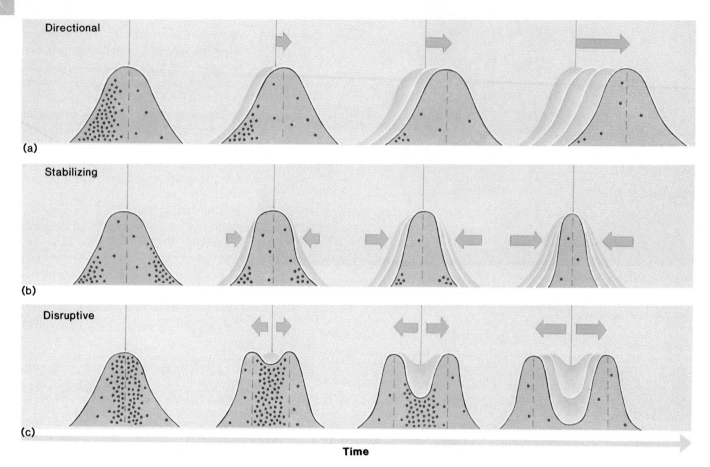

(a) Directional

(b) Stabilizing

(c) Disruptive

Time

Figure 15.6 Three kinds of natural selection.
These diagrams show how three kinds of natural selection—
(a) directional selection, (b) stabilizing selection, and (c)
disruptive selection—act on a trait, such as height, that varies in a
population. The dots represent individuals who do not contribute
to the next generation. The curves represent the trait measurements
taken from each individual in the population. All three kinds of
natural selection have the same starting point, and the three series
show how selection alters the distribution of the characteristics as
time passes, moving to the right.

life of any one individual. Thus, selection is a statistical concept, just as betting is. Although the fate of any one individual, or any one coin toss, cannot be foreseen, predicting which kind of individual will tend to become more common in populations of a species *is* possible.

In nature, many traits, perhaps most, are affected by more than one gene. Interactions between genes are typically complex, as discussed in chapter 10. For example, alleles of many different genes play a role in determining human height (see figure 10.18). In such cases, selection operates on all the genes, influencing most strongly those that make the greatest contribution to the phenotype. How selection changes the population depends on which phenotypes are favored. Three types of natural selection have been identified: (1) directional selection, (2) stabilizing selection, and (3) disruptive selection (figure 15.6).

Directional Selection

When selection acts to eliminate one extreme from an array of phenotypes, the genes determining this extreme become less frequent in the population. Thus, in the *Drosophila* population

illustrated in figure 15.7, the investigator's elimination of flies that move toward light causes the population to contain fewer individuals with alleles promoting such behavior. The result is that randomly chosen individuals from the new fly population have a lesser chance of spontaneously moving toward light than randomly chosen individuals from the old population. The population has been changed by selection in the direction of lower light attraction. This form of selection is called **directional selection** (see figure 15.6*a*).

Stabilizing Selection

When selection acts to eliminate *both* extremes from an array of phenotypes, the frequency of the intermediate type, which is already the most common, is increased. In effect, selection is operating to prevent change away from this middle range of values. In humans, infants with intermediate weight at birth have the highest survival rate (figure 15.8). In ducks and chickens, eggs of intermediate weight have the highest hatching success. This form of selection is called **stabilizing selection** (see figure 15.6*b*).

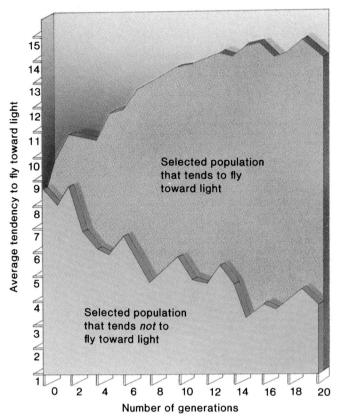

Figure 15.7 Directional selection in action.
In generation after generation, individuals of the fly Drosophila *were selected for their tendency to fly toward light more strongly than usual (the red curve). The blue curve represents flies selected against the tendency to fly toward light. The increasing red curve indicates that, when flies with a strong tendency to fly toward light were used as the parents for the next generation, their offspring had a greater tendency to fly toward light. The decreasing blue curve shows the opposite: When flies with a strong tendency to* not *fly toward light were used as the parents for the next generation, their offspring had a reduced tendency to fly toward light.*

Disruptive Selection

In some situations, selection acts to eliminate, rather than favor, the intermediate type. For example, in different parts of Africa, the color pattern of the butterfly *Papilio dardanus* is dramatically different, although in each instance it closely resembles the coloring of some other butterfly species that birds do not like to eat (*P. dardanus* is said to be a "mimic"). Birds quickly detect and eat butterflies with color patterns that do not resemble those of distasteful butterflies, so any intermediate color patterns are eliminated. In this case, selection is acting to eliminate the intermediate phenotypes. This form of selection is called **disruptive selection** (see figure 15.6c). Disruptive selection is far less common than the other two types of selection.

Directional selection acts to eliminate one extreme from an array of phenotypes, while stabilizing selection acts to eliminate both extremes. Disruptive selection acts to eliminate, rather than favor, intermediate phenotypes.

The Evidence That Natural Selection Explains Microevolution

For a century, biologists have studied genetic variation in nature to see if it is adaptive, as Darwin's theory suggests. Much of it clearly is adaptive, as evidenced by two of the best-studied examples: variation in human hemoglobin and variation in the darkness of moths.

Sickle-Cell Anemia

As discussed in chapter 11, sickle-cell anemia is a hereditary disease affecting the hemoglobin molecules in the blood. Chicago physician James B. Herrick first identified sickle-cell anemia in 1910 when he reported that a West Indies black student exhibited symptoms of severe anemia that appeared related to abnormal red blood cells: "The shape of the red cells was very irregular, but what especially attracted attention was the large number of thin, elongated, sickle-shaped and crescent-shaped forms." The disease was soon found to be common among African-Americans, affecting roughly one in every five hundred individuals. Sickle-cell anemia is associated with a particular recessive allele, whose frequency in the African-American population can be calculated with the Hardy-Weinberg equation. If one in five hundred individuals is affected, then two out of one thousand are affected. The frequency, then, is the square root of 0.002, or approximately 0.045. In contrast, the frequency of the allele among American Caucasians is only about 0.001.

Sickle-cell anemia, which is usually fatal, occurs because of a single amino acid change: A valine is substituted for the usual glutamic acid at a location on the surface of hemoglobin near the oxygen-binding site. Unlike glutamic acid, valine is nonpolar (water-hating), and its presence on the molecule's surface creates a "sticky" patch that tries to escape from the polar water environment by binding to another similar patch. As long as oxygen is bound to the hemoglobin molecule, there is no problem, because the oxygen atoms shield the critical area. But when oxygen levels fall, such as after exercise, then oxygen is not so readily bound to hemoglobin, and the exposed sticky patch binds to similar patches on other molecules, eventually producing long, fibrous clumps. The result is a sickle-shaped red blood cell.

Individuals who are heterozygous for the dominant, glutamic acid-specifying allele (designated allele *S*) are said to possess the sickle-cell trait. They produce some sickle-shaped red blood cells, but only 2 percent of the level seen in homozygous individuals.

The average incidence of the *S* allele in Central Africa is about 0.12, far higher than that found among African-Americans. Application of the Hardy-Weinberg principle results in estimates that one in five Central African individuals is heterozygous for the *S* allele, and that one in one hundred develops the fatal form of the disorder. Since people homozygous for the sickle-cell allele usually die before they reach reproductive age, why is the *S* allele not eliminated from Central Africa by selection, rather than being maintained at such high levels? Part of the reason is that people heterozygous for the sickle-cell allele are much less susceptible to malaria, which is one of the leading causes of illness and death in areas where the allele is common. Also, for reasons not understood, women who are heterozygous for this allele are more fertile than those who lack it. Consequently, even though most homozygous-recessive individuals die before they have children, the sickle-cell allele is maintained at high levels in these populations because of its role in malaria resistance in heterozygotes and its association with increased fertility in female heterozygotes.

As Darwin's theory predicts, the environment acts to maintain the sickle-cell allele at high frequency. In this case, the environmental characteristic of Central Africa that is exercising selection is the presence of malaria. For people living in malaria-prone areas, maintaining a certain level of the sickle-cell allele in the population has adaptive value (figure 15.9). Among African-Americans, many of whom have lived for some fifteen generations in a country where malaria is now essentially absent, the environment does not place a premium on malaria resistance; thus, there is no adaptive value to counterbalance the disease's ill effects. In this nonmalarial environment, selection is acting to eliminate the *S* allele.

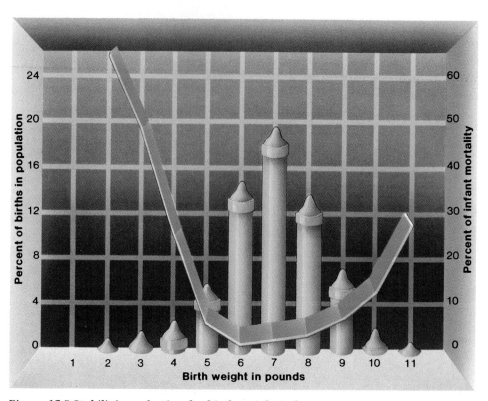

Figure 15.8 Stabilizing selection for birth weight in humans.
The death rate among babies is lowest at an intermediate birth weight between 7 and 8 pounds. Both larger and smaller babies have a greater tendency to die at or near birth. Stabilizing selection acts to eliminate both extremes from an array of phenotypes and to increase the frequency of the intermediate type.

Peppered Moths and Industrial Melanism

The European peppered moth, *Biston betularia,* rests on tree trunks during the day. Until the mid-nineteenth century, almost every captured individual of this species had light-colored wings. From that time on, however, individuals with dark-colored wings increased to almost 100 percent frequency in the moth populations near industrialized centers. The dark individuals have a dominant allele that was present in populations before 1850, although rare. Biologists noticed that, in industrialized regions, where the dark moths were common, tree trunks were darkened almost black by the soot of pollution, thus making the dark moths resting on them much less conspicuous than the light moths. In addition, the air pollution that was spreading in the industrialized regions had killed many of the light-colored lichens that had earlier occurred on tree trunks, thus making these trunks even darker than they would have been otherwise (figure 15.10).

Can Darwin's theory explain the dark allele's increased frequency? Why was it an advantage for the dark moths to be less conspicuous? Although, initially, there was no evidence, ecologist H. B. D. Kettlewell hypothesized that moths resting on tree

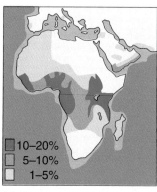
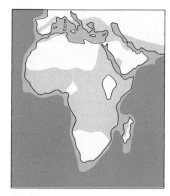

(a) Frequency of sickle-cell allele

(b) Distribution of falciparum malaria

□ 10–20%
□ 5–10%
□ 1–5%

Figure 15.9 Frequency of sickle-cell allele and distribution of falciparum malaria.
The frequency of the sickle-cell allele in Africa (a) is closely correlated with the distribution of falciparum malaria in Africa (b).

trunks during the day were eaten by birds. He tested the hypothesis by rearing populations of peppered moths in which dark and light individuals were evenly mixed. Kettlewell then released these populations into two sets of woods: one, near Birmingham, England, was heavily polluted; the other, in Dorset, England, was unpolluted. Kettlewell set up rings of traps around the woods to see how many of both kinds of moths survived. So that he would be able to evaluate his results, he marked the moths that he released with a dot of paint on their wing underside, where it could not be seen by the birds.

In the polluted area near Birmingham, Kettlewell trapped 19 percent of the light moths and 40 percent of the dark ones. This indicated that the dark moths had a far better chance of surviving in the polluted woods, where the tree trunks were dark. In the relatively unpolluted Dorset woods, Kettlewell recovered 12.5 percent of the light moths but only 6 percent of the dark ones. These results indicated that, where tree trunks were still light-colored, light moths had a better chance of survival than the dark ones. Kettlewell later solidified his argument by placing hidden blinds in the woods and filming birds eating the moths. The birds sometimes passed right over a moth that was of the "correct" color and thus well concealed (figure 15.11).

The evolutionary process in which initially light-colored organisms become dark as a result of natural selection is called **industrial melanism.** The process, which is common among moths that rest on tree trunks, takes place because the dark organisms are better concealed from their predators in habitats that have been darkened by soot and other forms of industrial pollution.

(a)

(b)

Figure 15.10 Color variation in the peppered moth, Biston betularia.

In (a), the dark moth is more easily seen against the light bark, while in (b), the dark moth is not as easily seen against the bark that has been darkened by pollution.

Dozens of other species of moths have changed in the same way as the peppered moth in industrialized areas throughout Eurasia and North America, with dark forms becoming more common from the mid-nineteenth century onward as industrialization spread. In the second half of the twentieth century, with the widespread implementation of pollution controls, the trends are reversing, not only for the peppered moth in many areas in England, but also for many other species of moths throughout the northern continents. Such examples provide some of the best documented instances of changes in allele frequencies of natural populations because of natural selection in relation to specific environmental factors.

An Overview of Microevolution

These case histories of sickle-cell anemia and industrial melanism are but two of many well-documented cases of adaptation that provide clear evidence that microevolutionary changes can be produced by natural selection. All of these cases share the same fundamental characteristic: Changes in a population's allele frequencies alter population characteristics to make the population better adapted to its environment. In every case, the *environment* dictates the direction and extent of the change. Just as Darwin's theory demands, the nature of the environment leads to natural selection and so determines the direction of evolutionary change.

Studies of how individual traits evolve within natural populations provide powerful evidence that natural selection can be a powerful agent of microevolutionary change within species.

Evidence for Macroevolution

Adaptation within natural populations such as those just described constitutes strong evidence that Darwin was right in arguing that selection could bring about genetic change within populations. These examples offer direct,

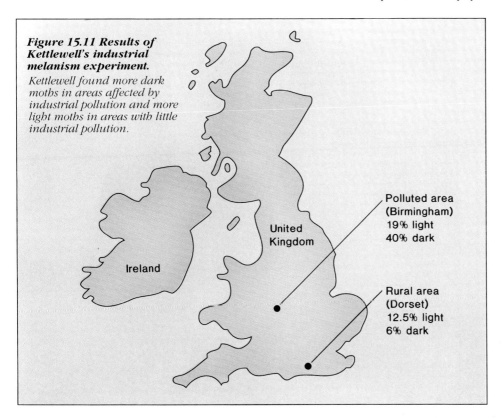

Figure 15.11 Results of Kettlewell's industrial melanism experiment.
Kettlewell found more dark moths in areas affected by industrial pollution and more light moths in areas with little industrial pollution.

Ireland

United Kingdom

Polluted area (Birmingham)
19% light
40% dark

Rural area (Dorset)
12.5% light
6% dark

compelling, and observable evidence of microevolutionary change. But what about macroevolution? What is the evidence that macroevolution has led to the diversity of life on earth?

Chapter 1 outlined the evidence that Darwin presented in favor of macroevolution, for which he was the first to provide an explanation of the mechanism. A great deal of additional evidence has accumulated since then, much of it far stronger than that available to Darwin and his contemporaries. Among the many lines of available evidence, seven are reviewed here: (1) the fossil record, (2) the molecular record, (3) homology, (4) development, (5) vestigial structures, (6) parallel adaptation, and (7) patterns of distribution (table 15.2).

The Fossil Record

The most direct evidence of macroevolution is found in the fossil record, of which a far more complete understanding currently exists than was available in Darwin's time. Fossils are created when organisms become buried in sediment, the calcium in bone and other hard tissue is mineralized, and the sediment eventually is converted to rock. Sedimentary rock layers reveal a history of life on earth in their fossils (figure 15.12).

By dating the rocks in which fossils occur, scientists can get a very accurate idea of the fossils' age. In Darwin's day, rocks were dated solely by their position with respect to one another; that is,

Table 15.2 Examples of the Evidence for Evolution

Fossil record	When fossils are arrayed according to their age, they provide evidence of a progressive series of changes.
Molecular record	The longer that organisms have been separated according to the fossil record, the more differences are seen in their DNAs.
Homology	All vertebrates contain a similar pattern of bones, muscles, nerves, blood circulation, and organs, suggesting that these organisms are derived from a common ancestor.
Development	During development, humans exhibit characteristics of other vertebrates, which suggests that humans are related to the other forms.
Vestigial structures	Many vertebrates contain structures that have no function but that resemble functional structures of other vertebrates. This suggests that the structures are inherited from a common ancestor.
Parallel adaptation	The marsupials in Australia closely resemble the placental mammals of the rest of the world, which suggests that parallel selection has occurred.
Patterns of distribution	Inhabitants of ocean islands resemble forms of the nearest mainland but show some differences, which suggests that they have evolved from mainland migrants.

Figure 15.12 Sedimentary rocks.
In this photo of a banded sedimentary cliff in Zion National Park, Utah, distinct layers of sediment are clearly visible. In general, the deeper the layers, the older the rocks and fossils found in those layers. The upper layer of rocks was deposited about 10 million years ago, the lowest layer over 250 million years ago. The prominent gray layer is from the late Triassic—about 210 million years ago—the dawn of the Age of Dinosaurs.

rocks in deeper layers of rock are generally older than those in shallow layers. Because they knew the relative positions of sedimentary rocks, and the rates of erosion of different kinds of sedimentary rocks in different environments, nineteenth-century geologists achieved a fairly accurate idea of rocks' relative ages.

Today, much more accurate ways of dating rocks provide absolute, rather than relative, dates. Currently, a rock is dated by measuring the degree to which certain isotopes that it contains have decayed since the rock formed; the older the rock, the more of its isotopes will have decayed. Because isotopes decay at a constant rate not altered by temperature or pressure, the isotopes in a rock act as an internal clock, measuring the time since the rock was formed. Decay of the isotope carbon-14 is used to date fossils less than thirty thousand years old. Carbon-14 has a **half-life**—the time it takes for half of a sample of a given size to decay—of 5,568 years. For older fossils, investigators examine the decay of radioactive potassium-40 into argon and calcium (half-life of 1.3 billion years). For very old fossils, a third isotope measure, the decay of uranium-238 into lead (half-life of 4.5 billion years) is used. An investigator has only to measure the proportion of uranium-238 to lead in the rock to estimate its age.

When fossils are arrayed according to their age, from oldest to youngest,

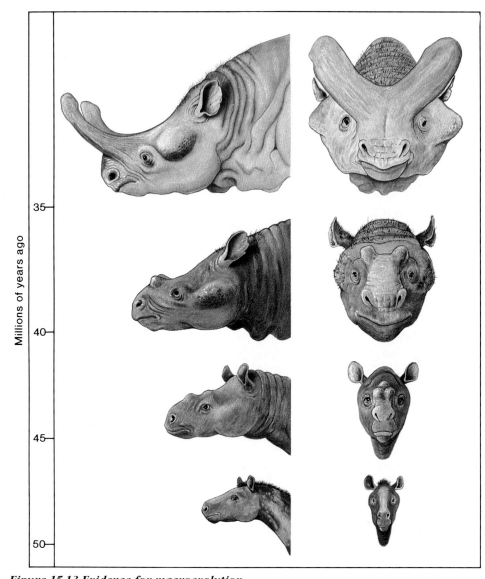

Figure 15.13 Evidence for macroevolution.
In a group of hoofed mammals known as titanotheres, small, bony, nose protuberances evolved into relatively large, blunt horns over a period of about fifteen million years.

they provide evidence of progressive evolutionary change. Among the hoofed mammals illustrated in figure 15.13, small, bony nose bumps changed progressively until they became large, blunt horns. In the evolution of horses, the number of toes on the front foot gradually reduced from four to one (see figure 1.14). About two hundred million years ago, oysters underwent a change from small, curved shells to larger, flatter ones, with progressively flatter fossils being seen in the fossil record over a period of twelve million years (figure 15.14). A host of other examples, all illustrating a record of progressive change, comprise one of the strongest lines of evidence for evolution.

The Molecular Record

That organisms have evolved progressively from relatively simple ancestors implies that a record of evolutionary change is present in the DNA of cells. According to evolutionary theory, every evolutionary change involves the substitution of new versions of genes for old ones, the new arising from the old by mutation and coming to predominance because of favorable selection. Thus, a series of evolutionary changes involves a progressive accumulation of genetic changes in the DNA, and organisms that are more distantly related have accumulated a greater number of evolutionary differences. This is

Figure 15.14 Evolution of shell shape in oysters.
During a twelve-million year portion of the early Jurassic period, the shells of a group of coiled oysters became larger, thinner, and flatter. These animals rested on the ocean floor, and the larger, flatter shells may have proven more stable against potentially disruptive water movements.

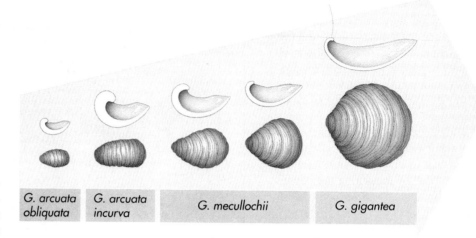

G. arcuata obliquata   G. arcuata incurva   G. mecullochii   G. gigantea

indeed what is seen when DNA sequences of various organisms are compared. For example, the longer the time since the organisms diverged, the greater the number of differences in the nucleotide sequence of the gene for cytochrome *c*, a protein that plays a key role in oxidative metabolism (figure 15.15). Again, evolutionary history is shown to involve a pattern of progressive change.

Some genes, such as the ones specifying the protein hemoglobin, have been particularly well studied, and the entire time course of their evolution can be laid out with confidence by tracing the origin of particular substitutions in their nucleotide sequences. The pattern of descent obtained, called a **phylogenetic tree,** represents the gene's evolutionary history. Note that the progressive changes seen in the hemoglobin molecule produce a tree that reflects precisely the evolutionary relationships predicted by anatomical study. Whales and dolphins cluster together, as do the primates and the hoofed animals (figure 15.16). The pattern of progressive change seen in the molecular record constitutes very strong direct evidence for macroevolution.

Homology

A third demonstration of macroevolution lies in **homology,** which is the term that describes the presence in different organisms of structures derived from a common ancestor. For example, the forelimbs of all mammals contain the same pattern of bones, although the bones now have a variety of functions (see figure 1.15). All vertebrates have the same pattern of bones, muscles, nerves, blood circulation, and organs, the pattern becoming gradually more complex as one moves from the fishes to amphibians to reptiles to mammals.

Development

In many cases, an organism's evolutionary history unfolds during its development, with the embryo exhibiting characteristics of the embryos of its ancestors. For example, early in their development, human fetuses possess gill arches like those of fish and later exhibit a tail, the vestige of which we carry to adulthood as

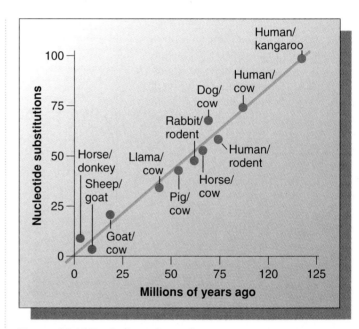

Figure 15.15 Evolution of cytochrome c.
Investigators compared various pairs of organisms and counted the number of nucleotides in the cytochrome c genes that were not the same. (Cytochrome c is a protein used as an electron acceptor in cellular respiration.) Plotting the number of such "substitutions" against the time investigators believe has elapsed since the pair of organisms diverged results in a straight line. This constant rate of nucleotide substitution suggests that the cytochrome c gene is evolving at a constant rate.

the coccyx at the end of our spine. Human embryos even possess a fine fur (called *lanugo*) during the fifth month of development. These relict developmental features argue strongly that human development has evolved, with new instructions being layered on top of old ones and the overall developmental program getting progressively longer. Several vertebrate embryos are shown in figure 15.17.

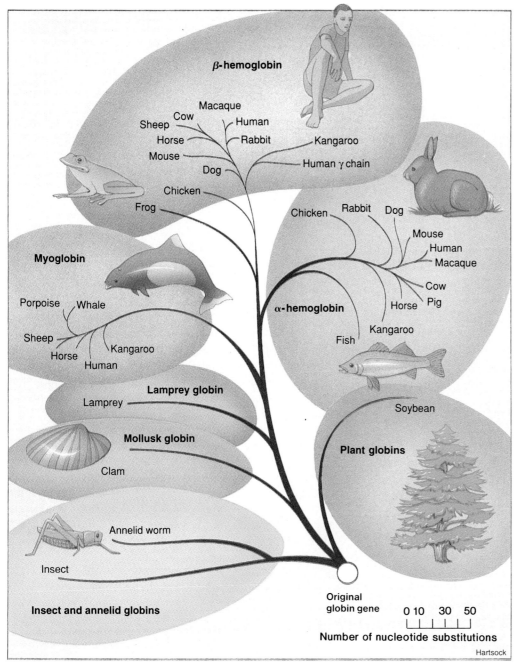

Figure 15.16 Evolution of the globin gene.
The length of the various lines is relative to the number of nucleotide substitutions in the gene—that is, the longer the line, the greater the number of substitutions.

Vestigial Structures

Many organisms possess **vestigial structures** with no apparent function that resemble the formerly functional structures of presumed ancestors. For example, you possess a complete set of muscles for wiggling your ears, just as a coyote does. Figure 15.18 shows the skeleton of a baleen whale, a representative of the group that contains the largest living mammals. The whale's skeleton contains pelvic bones, just like the skeletons of other mammals, even though such bones serve no known function in the whale. The human vermiform appendix, a hollow, wormlike appendage of the cecum, or sac, in which the large intestine begins, apparently is vestigial and represents the degenerate terminal part of the cecum. In many respects, the appendix is a dangerous organ: Quite often, it becomes infected, leading to an inflammation called *appendicitis*. If not removed surgically when inflamed, the vermiform appendix may burst, allowing the contents of the gut to come in contact with the lining of the body cavity. This condition can be fatal if unchecked. Vestigial structures such as these are difficult to understand in any way other than as evolutionary relics, holdovers from the evolutionary past. They argue strongly for the common ancestry of the organisms that share them, regardless of how different the organisms have become subsequently.

Parallel Adaptation

Plant and animal communities in widely separated areas are often similar, especially if the areas have similar climates. This may be true even if the individual kinds of plants and animals are only distantly related to one another. These similarities are difficult to explain as mere coincidence and suggest that parallel adaptation has occurred.

Perhaps the best-known example of parallel adaptation is the resemblance of Australia's marsupial mammals to the placental mammals found in the rest of the world. More than

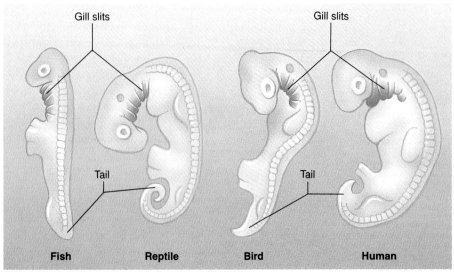

Figure 15.17 Embryos show evolutionary history.
The embryos of various groups of vertebrate animals show the primitive features that all share early in their development, such as gill slits and a tail.

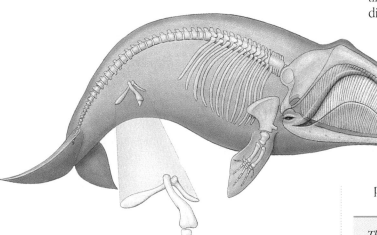

Figure 15.18 Vestigial features.
The skeleton of a baleen whale shows the pelvic bones, which resemble those of other mammals, but are only weakly developed in the whale and have no apparent function.

fifty million years ago, before placental mammals—the group that dominates throughout most of the modern world—are thought to have arrived in the area, the continent of Australia separated from Antarctica and thus from the other continents. Today, most Australian mammals are marsupials; their young are born in a very immature condition and held in a pouch until they are ready to emerge into the outside world. Marsu-

pials might have evolved earlier than placental mammals and seem certainly to have arrived in Australia before its separation from Antarctica. To an astonishing degree, some Australian marsupials resemble the placental mammals present on other continents. The similarity of some of the individual members of these two sets of mammals, in which specific kinds have similar habits and find their food in similar ways, argues strongly for evolution in different, isolated areas as a result of natural selection in relation to similar environments (figure 15.19).

Patterns of Distribution

Darwin was the first to present evidence that the animals and plants living on oceanic islands resemble most closely the animal and plant forms of the nearest continent. This relationship, which has been observed many times since Darwin's time, strongly suggests that the island forms evolved from individuals that came to the islands from the adjacent mainland at some time in the past. In many cases, the island forms are *not* identical to those that still occur on nearby continents. The Galápagos finch, for example, has a very different beak than its South American relative. In the absence of evolution, there seems to be no logical explanation for why individual kinds of island plants and animals are clearly related to, but have diverged in their features from, other kinds of plants and animals on adjacent mainlands. As Darwin argued, this relationship provides strong evidence of macroevolution.

The strongest direct evidence that macroevolution has occurred comes from the fossil record and from examination of accumulated differences in DNA and proteins. Equally strong but less direct evidence comes from comparative studies of anatomy, development, and geographical distributions.

Overall, the evidence for macroevolution is overwhelming. Almost all biologists agree that (1) macroevolutionary changes have occurred, and (2) microevolutionary changes result from natural selection. The next chapter considers Darwin's proposal that microevolutionary changes have led directly to macroevolutionary ones, the key argument in his theory that evolution occurs by natural selection.

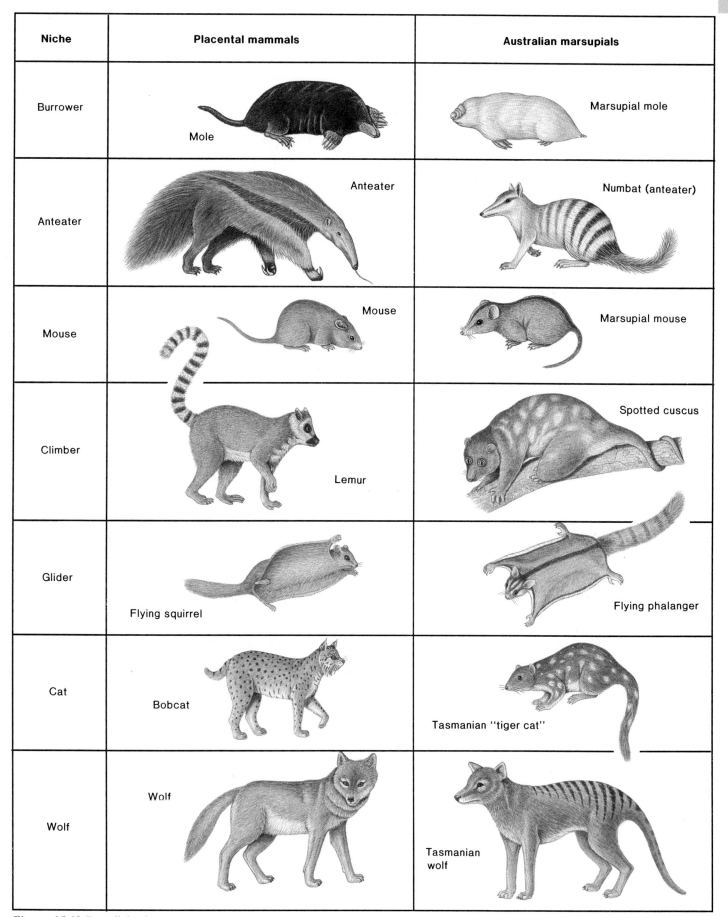

Niche	Placental mammals	Australian marsupials
Burrower	Mole	Marsupial mole
Anteater	Anteater	Numbat (anteater)
Mouse	Mouse	Marsupial mouse
Climber	Lemur	Spotted cuscus
Glider	Flying squirrel	Flying phalanger
Cat	Bobcat	Tasmanian "tiger cat"
Wolf	Wolf	Tasmanian wolf

Figure 15.19 Parallel adaptation.
The parallel adaptation of marsupials in Australia (right column) and placental mammals in the rest of the world (left column) *shows that the two groups of mammals evolved similarly in response to similar environments.*

Figure 15.20 A living relic.
The lungfish is a member of a group of vertebrates that have changed little during the past 150 million years.

The Rate and Mode of Evolution

Using the fossil record, scientists have been able to examine the rates of evolution among groups of organisms. On the basis of a relatively complete fossil record, they have estimated that most living individual species of mammals go back at least one hundred thousand years, almost none a million years. A good average value for the life of a mammal species might be about two hundred thousand years. Mammal groups, on the other hand, seem to persist several million years, on average. American paleontologist George Gaylord Simpson pointed out that certain other groups, such as the lungfishes, are apparently evolving much more slowly than are the mammals. Simpson estimated that lungfishes have experienced few evolutionary changes over the past 150 million years, and even slower rates of evolution are known in certain other groups (figure 15.20).

Not only does the rate of evolution differ greatly from group to group, but each individual group also apparently experiences rapid and relatively slow evolutionary periods. The fossil record provides evidence for such variability in evolutionary rate, and evolutionists are anxious to understand the environmental and other factors that account for it. In 1972, paleontologists Niles Eldredge of the American Museum of Natural History in New York and Stephen Jay Gould of Harvard University proposed that it was the norm of evolution to proceed in spurts. They claimed that the process of evolution includes a series of **punctuated equilibria.** According to this proposal, evolutionary innovations occur and give rise to new lines; these lines might persist unchanged for a very long time. Eventually, there is a new spurt of evolution, creating a "punctuation" in the fossil record.

Eldredge and Gould proposed that evolution is usually rapid when populations are small, possibly different from their parent populations as a result of the founder effect, and still local enough for rapid adaptation to novel ecological circumstances. In contrast, **stasis,** or lack of evolutionary change, is expected for large populations under diverse and conflicting selective pressures.

Unfortunately, the distinctions are not as clear-cut as implied by this discussion. The fossil record is seriously incomplete because of changes in the conditions under which fossils are deposited, and interpretation of many of its "gaps" is problematical. In spite of these difficulties, the punctuated equilibrium model provides a useful perspective for considering the mode of evolution. Eldredge and Gould contrasted their theory of punctuated equilibrium with that of **gradualism,** or gradual evolutionary change, which they claimed was what Darwin and most earlier students of evolution had considered normal. Whether or not Darwin and other evolutionists actually embraced gradualism is debatable.

The punctuated equilibrium model assumes that evolution occurs in spurts, between which there are long periods with little evolutionary change. The gradualism model assumes that evolution proceeds gradually, with progressive change in a given evolutionary line.

Scientific Creationism

In the century since he proposed it, Darwin's theory of evolution by natural selection has become almost universally accepted by biologists as the best available explanation of biological diversity. Its predictions have been supported by the experiments and observations of generations of scientists. That evolution has occurred and is still occurring is as well accepted a theory as is the existence of gravity, and the operation of evolution no more doubted than predictions that a dropped apple will fall or that the sun will rise tomorrow morning.

But evolution is not the only way in which the diversity of life on earth can be explained. The argument advanced in Darwin's day—that the Bible provides the correct and literal explanation of biological diversity—is still widely accepted by many people today, people who prefer a religious to a scientific explanation. Now, as in Darwin's time, the scientific perspective is only one of many worldviews. Understanding the limits of science is important in this regard. Science provides a coherent means of organizing observations, and of making predictions about how the world is going to behave. It is not a substitute for religion, which addresses a different arena of human concerns: questions of ethics and of ultimate causes. Religion and science do not preclude one another but are regarded by many as complementary ways of viewing the world.

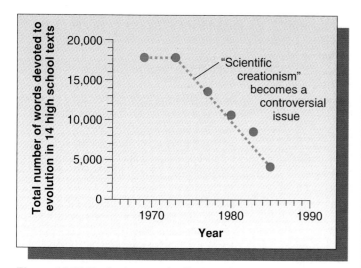

Figure 15.21 Evolution versus "scientific creationism."
Creationism loses the battle in the courts but wins in the schools.
Textbook publishers, wary of controversy, have been removing
evolution from textbooks in recent years.

The clear distinction between science and religion, however, sometimes gets muddled (figure 15.21). Thus, a number of individuals, mainly in the United States and starting largely in the 1970s, have put forward a view they call **scientific creationism,** which holds that the biblical account of the earth's origin is literally true, that the earth is much younger than most scientists believe, and that all species of organisms were individually created just as they are today. Scientific creationists are arguing in the courts that their views should be taught alongside evolution in classrooms. If both evolution and scientific creationism provide scientific explanations of biological diversity, they argue, then teachers have an obligation to present *both* views, taught side by side, so that students can choose knowledgeably between them.

This argument only makes sense if you accept the premise—that the view of the scientific creationists is indeed scientific. The confusion is not in the beliefs of the scientific creationists, which are religious beliefs that many people hold, but rather in their labeling of these beliefs as scientific. There is no scientific evidence to support the hypothesis that the earth is only a few thousand years old and none that indicates that every species of organism was created separately. These conclusions can be reached only on the basis of faith alone; they are untestable, and as such, they lie outside the realm of science.

Science, as represented by the observations of scientists, has come to different conclusions. Biologists, and indeed, nearly all scientists, agree on the following major points: (1) The earth is about 4.5 billion years old; (2) organisms have inhabited the earth for the greater part of that time; and (3) all living things, including human beings, have arisen from earlier, simpler living

things. The antiquity of the earth and the role of evolution in the production of all organisms, living and extinct, is accepted by virtually all scientists.

An even more fundamental problem associated with labeling scientific creationism as science is that scientific creationism implicitly denies the intellectual basis of science itself, the reasoning on which the operation of science depends. Science insists on acceptance of the most predictive explanation of biological diversity, which is evolution, whereas scientific creationism says, "Yes, but God just made it look that way." Perhaps so, but this is simply substituting a religious argument for a scientific one. Science consists of inferring principle from observation, and when faith is substituted for observation, the conclusion is not science.

> *Scientific creationism should not be labeled science for three reasons: (1) It is not supported by any scientific observations; (2) it does not infer its principles from observation, as does all science; and (3) its assumptions lead to no testable and falsifiable hypotheses.*

Scientific creationism implicitly denies the whole intellectual basis for the set of facts that human beings have assembled over the centuries about the nature of life on earth. It implies that a creator has made a world in which all phenomena are deceptively consistent with a great age for the earth and an evolutionary relationship among organisms. Evolutionary theory, in contrast, provides a coherent, scientific explanation for the nature of the world in which we live and enables predictions to be made about that world. Certainly, there is controversy among serious students as to the details of how evolution has occurred, just as there is controversy in every active scientific field. But there is no controversy about Darwin's basic finding that natural selection has played, and is continuing to play, the central role in the process of evolution.

The future of the human race depends largely on our collective ability to deal with the science of biology and all the phenomena that it comprises. We need the information that we have gained to deal with the problems, challenges, and uncertainties of the world in an appropriate way. We cannot afford to discard the advantages that this knowledge gives us simply because some people wish to do so as an act of what they construe as religious faith. Instead, we must use all of the knowledge that we are able to gain for our common benefit. Only then can we come to understand ourselves and our potentials better. In no way should such rational behavior be interpreted as denial of the existence of a supreme being; it should rather be considered by those who do have religious faith as a sign that they are using their God-given gifts to reason and to understand.

A VOCABULARY OF GENETIC CHANGE

adaptation A change in structure, physiology, or behavior that promotes the likelihood of an organism's survival and reproduction in a particular environment.

allele One of two or more alternative versions of a gene.

allele frequency The relative proportion of a particular allele among individuals of a population. Not equivalent to *gene frequency*, although the two terms are sometimes confused.

evolution Genetic change in a population of organisms over time (generations). Darwin proposed that natural selection was the mechanism of evolution.

fitness The genetic contribution of an individual to succeeding generations, relative to the contributions of other individuals in the population.

gene The basic unit of heredity. A sequence of DNA nucleotides on a chromosome that encodes a protein or RNA molecule or regulates the transcription of such a sequence.

gene frequency The frequency with which individuals in a population possess a particular gene. Often confused with *allele frequency.*

genetic drift Random fluctuations in allele frequencies over time.

macroevolution The creation of new species and the extinction of old ones.

microevolution Evolution within a species.

mutation A permanent change in a gene, such as an alteration of its nucleotide sequence.

natural selection The differential reproduction of genotypes in response to factors in the environment. Artificial selection, by contrast, is in response to demands imposed by human intervention.

polymorphism The presence in a population of more than one allele of a gene at a frequency greater than that of newly arising mutations.

population Any group of individuals, usually of a single species, occupying a given area at the same time.

species A kind of organism. In nature, individuals of one species usually do not interbreed with individuals of other species.

EVOLUTIONARY VIEWPOINT

The theory of evolution by natural selection forms the backbone of biology as a science. In the century since Darwin, a large body of evidence has accumulated in support of this theory. The details of how evolution operates have been and continue to be a source of lively discussion among biologists.

SUMMARY

1. Macroevolution describes the grand outlines of evolution, which is based on the evolution of species. Microevolution, also called adaptation, refers to the evolutionary process itself. Adaptation leads to species formation and thus, ultimately, to macroevolution.

2. Population genetics is the branch of genetics that deals with the properties of genes in populations. The Hardy-Weinberg principle, the baseline for all population genetic theory, illustrates that in large populations with random mating, allele and genotype frequencies—and consequently, phenotype frequencies—will remain constant indefinitely, provided that selection, net mutation or migration in one direction, nonrandom mating, and genetic drift do not occur.

3. Some organisms leave more offspring than competing members of the same population. Their genetic traits thus tend to appear in greater proportions among members of succeeding generations than do the traits of those individuals that leave fewer offspring. This process is called selection.

4. There are three principal kinds of natural selection: Directional selection acts to eliminate one extreme from an array of phenotypes. Stabilizing selection acts to eliminate both extremes. Disruptive selection acts to eliminate, rather than to favor, the intermediate type.

5. There is clear evidence of microevolutionary change in natural populations. For example, the disease called sickle-cell anemia occurs when an altered hemoglobin allele is present in homozygous form; it is almost invariably fatal. However, if the allele is present in heterozygous form, it not only does not produce anemia, it also confers resistance to malaria and increases female fertility. For these reasons, the allele has reached high frequencies in African populations that live in areas where malaria is prevalent. The industrial melanism of peppered moths is another example of changes in allele frequencies because of natural selection in relation to specific environmental factors.

6. Two direct lines of evidence argue that macroevolution has occurred: (1) the fossil record, which exhibits a record of progressive change correlated with age, and (2) the molecular record, which exhibits a record of accumulated changes, the amount of change correlated with age as determined in the fossil record.

7. Indirect lines of evidence that argue that macroevolution has occurred include progressive changes in homologous structures, relict and vestigial developmental structures, parallel patterns of evolution, and patterns of distribution.

8. The rate of macroevolution has not been constant, although biologists disagree about the reason why. The punctuated equilibrium model assumes that evolution occurs in spurts, between which there is little evolutionary change, while the gradualism model assumes that evolutionary change proceeds gradually.

9. A nonscientific theory is one whose principles are not derived from or supported by observation and whose conclusions are untestable. Scientific creationism is an example of a nonscientific theory.

REVIEWING THE CHAPTER

1. What are macroevolution and microevolution, and how do they occur?

2. How is the Hardy-Weinberg equilibrium used to predict allele frequencies in populations?

3. Which factors alter allele frequencies, and which of these factors is influenced directly by the environment?

4. What are the types of selection, and what happens to phenotypes in each type?

5. What is the evidence that natural selection explains microevolution?

6. Why has the allele for sickle-cell anemia, a dominant genetic disorder, not been eliminated from the human population?

7. What is industrial melanism, and what are its implications in natural selection?

8. What is the current evidence for macroevolution?

9. What are the rates of evolution and the proposed models for considering the mode of evolution?

10. What is scientific creationism, and why is it not a valid scientific theory?

COMPLETING YOUR UNDERSTANDING

1. For review, what were Darwin's observations that evolution is best explained by the process of natural selection?

2. How does adaptation relate to microevolution?

3. What is population genetics, and how does it relate to evolution?

4. How are the founder effect and the bottleneck effect examples of genetic drift?

5. Which of the following is *not* a factor that causes change in the proportions of homozygous and heterozygous individuals in a population?

   a. Mutation   d. Random mating

   b. Migration   e. Selection against

   c. Genetic drift   a specific allele

6. If you came across a population of plants with a surprisingly high level of heterozygosity, what would you predict about their mating system?

   a. Their pollen is dispersed by wind.

   b. Their pollen is spread by insects that visit the flowers.

   c. They probably reproduce asexually.

   d. They are predominantly outcrossing.

   e. They are predominantly self-fertilizing.

7. In a large, randomly mating population with no forces acting to change gene frequencies, the frequency of homozygous-recessive individuals for the characteristic of extra-long eyelashes is ninety per thousand, or 0.09. What percent of the population carries this very desirable trait but displays the dominant phenotype—short eyelashes?

Would the frequency of the extra-long lash allele increase, decrease, or remain the same if long-lashed individuals preferentially mated with each other and no one else?

8. What are the controlling factors for artificial selection and natural selection?

9. The North American human population is similar to the ideal Hardy-Weinberg population in that it is very large (nearly 300 million people in the United States and Canada) and generally randomly mating. Although mutation occurs, it does not lead by itself to great changes in allele frequencies. However, migration from Latin American and Asian countries occurs at relatively high levels—perhaps 1 percent per year. The following data were obtained in 1976 by geneticist A. E. Mourant about relative numbers of individuals bearing the two alleles of what is known as the *MN* blood group:

	MM	MN	NN	Total
Observed number of individuals	1,787	3,037	1,305	6,129

Do these data suggest that migration, selection, or some other factor is acting to perturb the Hardy-Weinberg proportions of the three genotypes?

10. What is eugenics, and what are some of the problems related to it?

11. Why is disruptive selection much less common than either stabilizing or directional selection?

12. The sickle-cell (*S*) allele has a relatively high frequency (0.12) in Central Africa, even though individuals homozygous for this allele usually die before they reach reproductive age. Why has this allele persisted in the

population in high frequencies when there appears to be such strong natural selection against it?

   a. Because heterozygous individuals are resistant to malaria.

   b. Because individuals homozygous for this characteristic are resistant to malaria.

   c. Because females heterozygous for this allele are more fertile than are those who lack it.

   d. Because females homozygous for this allele are more fertile than are those who lack it.

   e. Both *a* and *c*.

13. In Central Africa, there exists in low frequency a third hemoglobin allele called *C*, in addition to the normal *A* allele and the sickle-cell (*S*) allele discussed in this chapter. Individuals that are heterozygous for *C* and the normal allele *A* are susceptible to malaria, just as are *AA* homozygotes, but *CC* individuals are resistant to malaria—and do *not* develop sickle-cell anemia! Assuming that the Bantu people entered Central Africa relatively recently from land where malaria is not common (scientists think this is what actually happened), and that among the original settlers both *C* and *S* alleles were rare, can you suggest a reason why *CC* individuals have not become predominant?

14. If you found a dinosaur bone, why would you not use carbon-14 to determine its age?

15. How is molecular biology used as a tool to show similarities and differences between organisms?

16. Why do many organisms have vestigial structures, such as the appendix in humans?

17. What is the best explanation for why most mammals on the island continent of Australia are marsupials?

18. What is the best explanation for why finches on the Galápagos Islands have different beaks than those found in their South American relatives?

19. If you found that a series of small islands were each occupied by a distinctive population of land snails, what kind of evidence would you gather to attempt to determine whether the differences had resulted from natural selection or from genetic drift?

20. What do punctuated equilibrium and gradualism demonstrate?

21. What are the major points that biologists and most all scientists, but not scientific creationists, agree on?

FOR FURTHER READING

French, M. 1990. *Invention and evolution*. New York: Cambridge University Press. Evolutionary adaptation as seen through the eyes of an engineer.

Futuyma, D. 1983. *Science on trial: The case for evolution*. New York: Pantheon. An excellent exposition of the basic reasons why the creationist argument is flawed by serious errors.

Gilkey, L. 1985. *Creationism on trial: Evolution and God at Little Rock*. Minneapolis: Winston Press. Excellent book, written by a theologian, outlining the case for creationism as argued in the courts at Little Rock, Arkansas.

Gillis, A. 1991. Can organisms direct their evolution? *BioScience*, April, 202–5. The rethinking of biologists on this question in light of recent findings that challenge the randomness of bacterial mutations.

Gould, S. J. 1987. Darwinism defined: The difference between fact and theory. *Discover*, Jan., 64–70. A clear account of what biologists do and do not mean when they refer to the theory of evolution by natural selection.

Joyce, G. 1992. Directed molecular evolution. *Scientific American*, Dec., 90–97. Biochemists who harness Darwinian evolution on a molecular scale.

McDonald, J. 1990. Macroevolution and retroviral elements. *BioScience*, March, 183–91. A discussion of how insertion of DNA segments may quicken the pace of gene evolution.

Rennie, J. 1993. DNA's New Twists. *Scientific American*, March, 122–32. The influence of molecular biology on the theory of evolution in many unanticipated ways.

Sibley, C., and J. Alquist. 1986. Reconstructing bird phylogenies by comparing DNAs. *Scientific American*, Feb., 82–92. A good introduction to the way in which biologists are beginning to use the tools of molecular biology to answer questions about evolution.

Simpson, G. G. 1983. *Fossils and the history of life*. New York: Scientific American Library. A short and beautifully illustrated account by a master in the field of how fossils are used to learn about life's evolutionary past.

H OW SPECIES FORM

This South American toucan is a member of a large order of birds that also includes the woodpeckers. All of the 383 species of this order, the Piciformes, have well-developed bills.

FOR REVIEW

Here are some important terms and concepts that have been discussed in previous chapters and that you will encounter again in this chapter. Review them before proceeding if necessary.

Darwin's studies (*chapters 1 and 15*)
Meiosis (*chapter 9*)
Microevolution (*chapter 15*)
Macroevolution (*chapter 15*)
Natural selection (*chapter 15*)
Adaptation in populations (*chapter 15*)

If Darwin was right, then you and a grasshopper are related, sharing a common ancestor far back in time. The biological diversity manifested by the incredible collection of creatures with which we share the planet today is, in this view, simply history viewed over very long periods of time. Species change in response to a changing environment, and over time are replaced by other species better able to accommodate to the changing world. The case for evolution now rests solidly on three pillars of evidence, which establish (1) that macroevolution occurs, (2) that natural selection is the agent of microevolutionary change—change within populations, and (3) that microevolutionary change leads to macroevolutionary change—change in the kinds of organisms that survive on the earth.

Chapter 15 considered the evidence for macroevolution and for natural selection being the agent of microevolutionary change. Natural selection drives evolution by favoring genetic variations that better enable organisms to survive and reproduce. Different environments favor different responses from the organisms that survive in them, and organisms undergo a continuing process of change in relation to these environments. This is the key lesson of chapter 15: Evolution is not blind; instead, it is directed by the environment. Just as a football coach has the team try a variety of plays but keeps in the team's game plan only those plays that work, so a population of organisms keeps only those changes that "work." The population does not decide which changes to keep, any more than the football coach does. The pattern of success determines the outcome.

This chapter considers in detail the third pillar of evidence on which Darwin's theory rests: the evidence that microevolutionary change leads to macroevolutionary change, that adaptive changes *within* a species convert to differences *between* species. This third point is the crux of Darwin's evolutionary argument.

The Species Concept

How do adaptive changes in natural populations lead to the origin of species? Darwin was extremely interested in this question because he considered species to be the most important evolutionary units. What does the concept **species** mean, and how has this concept changed through the years?

John Ray (1627–1705), an English clergyman and scientist, was one of the first to propose a general definition of species. In about 1700, he pointed out how a species could be recognized: All the individuals that belonged to it could breed with one another and produce progeny that were still of that species (figure 16.1). Even if two different-looking individuals appeared among the progeny of a single mating, they were still considered to belong to the same species. All dogs were one species, all pigeons, and so on; carp, however, were not the same species as goldfish, nor horses the same species as donkeys, and so forth.

In an informal way, people had always recognized species. Indeed, the word *species* is simply Latin for *kind*. However, with Ray's observation, the species began to be regarded as an important biological unit that could be catalogued and understood (figure 16.2). With other scientists of his time, Ray believed that species were individually created by a supreme being and did not change, a widely held view until Darwin challenged it in 1859.

Darwin was interested in those situations in which it was not clear whether or not he was dealing with distinct species. That some individuals were intermediate in their features between two different species (the species were then said to *intergrade*) was important supporting evidence for Darwin's theory of evolution by natural selection. Darwin explained the relative constancy of species by saying that each had its own distinctive role in nature, a role that, in modern terms, would be called a **niche.** Thus, each species occurs in a particular kind of place, displays different activities at different times of the year, has different habitats, and so forth.

From the 1920s onward, with the emergence of population genetics, there was a desire to define the species category more precisely. American evolutionist Ernst Mayr defined species as "groups of actually or potentially interbreeding natural populations which are reproductively isolated from other such groups." In other words, **hybrids** (offspring that result from a cross between two dissimilar parents) between species occur rarely in nature, whereas individuals that belong to the same species are able to interbreed freely. In fact, however, there are essentially no barriers to hybridization between the species in many groups of organisms and strong barriers to hybridization between the species of other groups. In practice, scientists recognize species in different groups primarily because the species differ from one another in their visible features (figure 16.3). As a general definition, then, improving on the early view that species are kinds of organisms has not been possible.

Within the units classified as species, the populations that occur in different places may be more or less distinct from one another. But when such populations occur in the same areas, individuals with intermediate or mixed characteristics usually are frequent; in other words, the distinct-appearing populations within a species usually intergrade with one another when the populations occur together. In areas where these different-looking varieties or **subspecies** approach one another, many individuals

Figure 16.1 Variation within a single species.
All of the different breeds of dogs are members of the same species, Canis familiaris, *and all can breed successfully with each other.*

Dog breeders engaging in artificial selection are responsible for the great differences among dog breeds.

may occur in which the distinct features characteristic of each of the intergrading subspecies are combined (figure 16.4). Many of these individuals may not match either subspecies in their characteristics. In contrast, when *species* occur together, they usually do not intergrade, although they may hybridize occasionally.

In some groups of organisms, even local populations are not capable of interbreeding with one another. This pattern occurs in many annual plants. In contrast, species of trees, some groups of mammals, and fishes generally *are* able to form fertile hybrids with one another, even though they may not do so in nature. For still other kinds of plants and animals, biologists do not know whether the species can form hybrids. A species, therefore, is defined as a group of organisms that is unlike other such groups of organisms and that does not intergrade extensively with other groups in nature.

Species are groups of organisms that differ in one or more characteristics and that do not intergrade extensively if they occur together in nature.

How Species Form: The Divergence of Populations

As Darwin predicted, the driving force that creates new species is the environment. Organisms' ability to adapt and change according to environmental conditions is the cornerstone of species formation.

Local populations of a species are usually more or less separated geographically, and the conditions in which they occur are dissimilar. Populations of interbreeding individuals are often extremely small, and exchange of individuals, and thus of genetic material, between such populations may be very limited (figure 16.5). For these reasons, local populations often are able to adjust individually and effectively to the demands of their particular environment. As they do, their characteristics change. The rate at which they change depends primarily on the selective forces to which they are responding. If these forces are strong, the populations will change rapidly; they do not need to be isolated on islands to become distinct species.

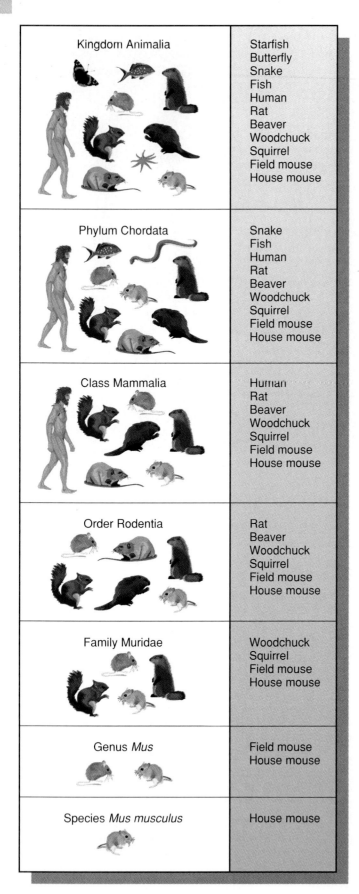

Kingdom Animalia	Starfish Butterfly Snake Fish Human Rat Beaver Woodchuck Squirrel Field mouse House mouse
Phylum Chordata	Snake Fish Human Rat Beaver Woodchuck Squirrel Field mouse House mouse
Class Mammalia	Human Rat Beaver Woodchuck Squirrel Field mouse House mouse
Order Rodentia	Rat Beaver Woodchuck Squirrel Field mouse House mouse
Family Muridae	Woodchuck Squirrel Field mouse House mouse
Genus *Mus*	Field mouse House mouse
Species *Mus musculus*	House mouse

The characteristics of populations tend to diverge more or less rapidly, depending on the features of their particular environment. Because exchange of genetic material between populations is limited, even if the populations are geographically close, all populations tend to become increasingly divergent from one another in their characteristics over time.

Many genes interact in the production of most phenotypic characteristics. Furthermore, all of the developmental processes that occur within a single organism are closely integrated with one another. In addition, every population begins with a different endowment of alleles, as noted in the discussion of the founder effect in chapter 15. Because of these three factors, the response of a given population of organisms to a combination of selective factors tends to be predictable. Even if two populations respond to similar selective forces, and even though they may be geographically close to one another, they still tend to change and to differ more and more from one another over time. If their environments are dissimilar enough or change rapidly enough, the populations may diverge rapidly and strikingly, and soon come to look very different from one another.

Ecological Races: An Intermediate Stage in the Evolution of Species

What kinds of patterns can be expected as a result of the differentiation of populations? One consequence is that the individuals of a species that occur in one part of its range often look different from those that occur elsewhere (see figure 16.4). Such groups of distinctive individuals are informally called **races,** and they may, as mentioned earlier, be classified as subspecies or varieties. Darwin considered such races to be an intermediate stage in the evolution of species. The kinds of ecologically defined races discussed in the next two sections may change over time to become clusters of species, which are considered later in the chapter. Both provide important examples of the evolution of populations in nature.

Ecotypes in Plants

Ecological races were first studied in detail in plants. As every gardener knows, the same plants may differ greatly in appearance, depending on where they are grown. This is true even for genetically identical divisions of the same plant, called *clones.* The part of an individual that is usually in the sun often produces leaves unlike those that the same plant produces in

Figure 16.2 Classification of organisms.
In the seventeenth century, scientists began to construct a hierarchical system for classifying all life on earth. Today, this system is still in use, as shown by this classification scheme for a mouse. This chapter is concerned only with the categories of genus and species.

(a)   (b)   (c)

(d)   (e)   (f)

Figure 16.3 Distinct species within one genus.
The top row shows three different species within one genus of butterflies: (a) *the so-called Western painted lady,* Vanessa carye; (b) *the American painted lady,* V. virginiensis *and* (c) *the red admiral,* V. atalanta. *The bottom row shows three different species within one genus of flowering plants:* (d) Clarkia concinna; (e) C. speciosa; *and* (f) C. rubicunda.

Figure 16.4 Subspecies.
Subspecies of the seaside sparrow, Ammodramus maritimus, *are quite local in distribution, and some of them are even in danger of extinction because of the alteration of their habitats. The widespread subspecies* Ammodramus maritimus maritimus (1, *adult;* 2, *juvenile) is the most common.* A. fisheri (3) *occurs along the Gulf coast. The Cape Sable seaside sparrow,* A. m. mirabilis (4), *occurs in a small area of southwestern Florida. The dusky seaside sparrow,* A. m. nigrescens (5), *occurred only near Titusville, Florida. The last individual, a male, died in captivity in 1987.*

(a)   (b)

Figure 16.5 Some animal populations are very small and localized.
The butterfly Euphydryas editha *(a) occurs on Jasper Ridge (b), a biological preserve in the foothills of San Francisco. The butterflies were once considered one continuous population throughout the open grassland, which occurs as an "island" surrounded by oak forest and chaparral. But extensive studies by Paul Ehrlich and colleagues over more than thirty years, involving marking butterflies with ink dots on their wings, releasing them, and recapturing them, have revealed that this species actually exists as a series of discontinuous populations between which individuals seldom move. Each of these populations, separated by dashed lines on the maps (c), is free to respond to the selective forces characteristic of its parts of the ridge. While the three populations have remained essentially constant in overall distribution since the late 1950s, the black dots show a detailed census of the population densities in three years in the 1960s. As you can see, numbers of individuals change, but the three populations remain distinct.*

the shade. For example, shade leaves are usually thinner and broader and have more internal air spaces than do sun leaves. Thus, many nineteenth-century botanists believed that environmental factors, rather than genetic factors, accounted for many of the differences between races and even between species of plants. As it turns out, however, this belief was wrong.

In the 1920s and 1930s, Swedish botanist Göte Turesson performed a series of experiments designed to test whether differences between races of plants were genetically or environmentally determined. Turesson observed that many plants had distinctive races that grew in different habitats. These races differed from one another in such characteristics as height, leaf size, leaf shape, degree of hairiness, flowering time, and branching pattern. Turesson dug up individuals representing these races and cultivated them together in his experimental garden at Lund, Sweden. In nearly every case, he found that the unique features of individual races were maintained when the plants were grown in a common environment. Therefore, most of the characteristics that he ob-

served had a genetic basis; only a few were environmental. Turesson called the ecological races that he studied and showed to have a genetic basis **ecotypes.**

Ultimately, Turesson's studies and those of others led to the conclusion that most of the differences between individuals, populations, races, and species of plants are genetically determined. Even though plants change their characteristics in relation to their environments, differences between races are usually fixed genetically in the course of their evolution.

Ecological Races in Animals

Similar patterns of variation are also found in animals. The differences may be morphological (that is, pertain to an organism's appearance) or physiological, and they have the same basis as to ecotypes in plants. The differences between subspecies may be striking. For example, the larger races of some species of birds may often consist of individuals that weigh three or four times as much as individuals of the smaller races. Races may differ from one another in their tolerance to different temperatures, in the speed of their larval development, in their behavioral characteristics—in short, in virtually any feature that can be measured and studied. Their features are almost always genetically determined.

> *In both animals and plants, natural selection causes local populations to adapt to their surroundings. Over time, this can lead to the formation of local races that differ genetically from other populations of the species in a variety of traits.*

The Aftermath of Adaptive Change: Barriers to Hybridization

As isolated populations become more different from one another in their overall characteristics, they may eventually occupy different niches. As mentioned earlier in the chapter, a niche is the ecological role that an organism plays in nature. Organisms that occupy different niches exploit different resources in different ways. If differentiated populations with different niches ever migrate back into contact with one another, they may still remain distinct in their characteristics. The populations may occur in different habitats, have different feeding habits, or otherwise be separated by differences that arose in them while apart. For these or other reasons, the individuals of the differentiated populations may not hybridize with each other; if they do, functional hybrid individuals may not be formed, or the hybrids may be sterile. For animals that choose their mates, the differences between the populations that have originated in isolation may be so great that they may choose mates of their own kind, rather than those of the other, formerly isolated population. In other words, and for a variety of possible reasons, the populations may have become species.

> *Populations of organisms tend to become increasingly different from one another in all of their characteristics. If the process continues long enough, or the selective forces are strong enough, the populations may become so different that they are considered distinct species.*

Once species have formed, they are able to retain their identity because of two types of isolating mechanisms: (1) **prezygotic isolating mechanisms,** which prevent the formation of zygotes; and (2) **postzygotic isolating mechanisms,** which prevent the proper functioning of zygotes. Some of these isolating mechanisms also occur, although often in a less marked form, within species. When they do, they illustrate stages in the evolution of new species. The sections that follow discuss various prezygotic and postzygotic isolating mechanisms and offer examples of how these mechanisms operate to help species retain their identity.

Prezygotic Isolating Mechanisms

Prezygotic isolating mechanisms include geographical, ecological, temporal, behavioral, and mechanical isolation, as well as the prevention of gamete fusion.

Geographical Isolation

One prezygotic isolating mechanism is **geographical isolation.** Most species simply do not exist together in the same places. They are generally adapted to different climates or to different habitats, which means that natural hybridization between them is not possible. If brought together in zoos, parks, or botanical gardens, however, they may hybridize.

One example of geographical isolation involves the English oak, *Quercus robur,* which occurs in European areas with a relatively mild, oceanic climate. Its characteristics are quite similar to those of the valley oak, *Q. lobata,* of California, and quite different from those of the scrub oak, *Q. dumosa,* also of California and adjacent Baja California (figure 16.6). All of these species can hybridize with one another and form fertile hybrids. The English oak does not hybridize with the others in nature, however, simply because its geographical range does not overlap with theirs. Similarly, although lions, *Panthera leo,* and tigers, *P. tigris,* do not now occur together in nature, they do mate and produce hybrids in zoos. The hybrids in which the tiger is the father, called "tiglons," are viable and fertile; less is known about "ligers," hybrids in which the lion is the father.

Ecological Isolation

A second prezygotic isolating mechanism, called **ecological isolation,** is associated with species that occur in the same area, but in different habitats, and thus may not hybridize with each other. On the other hand, if they do hybridize with each another, the hybrids may not be well represented in the overall population because they may not be as *fit* in the habitat of either of their parents. **Fitness** is an evolutionary term that describes an organism's ability to successfully produce viable offspring. Thus, hybrids with parents from different habitats may not be compatible enough with their environment to successfully reproduce.

For example, in India, the ranges of lions and tigers overlapped until about 150 years ago. But even before this time, there were no records of natural hybrids. Lions stayed mainly in the open grassland and hunted in groups called prides; tigers tended to be solitary creatures of the forest. Because of their ecological and behavioral differences, lions and tigers rarely came into contact with one another, even though their ranges overlapped over thousands of square kilometers.

Similar situations occur among plants, like the two species of California oaks—the valley oak *Q. lobata* and the scrub oak *Q. dumosa*—described earlier. *Quercus lobata,* a graceful deciduous tree that can be as much as 35 meters tall, occurs in the fertile soils of open grassland on gentle slopes and valley floors in central and southern California. In contrast, *Q. dumosa* is an evergreen shrub, usually only 1 to 3 meters tall, that

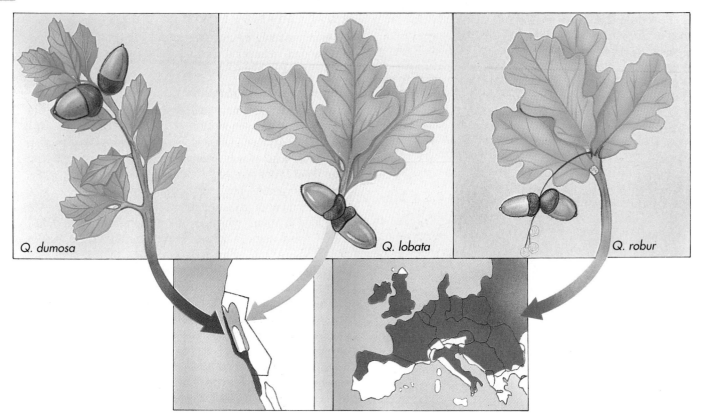

Figure 16.6 Geographical isolation.
*Three species of oaks—*Quercus robur, Q. lobata, *and* Q. dumosa *—have different leaf and acorn characteristics, yet all of them could hybridize with one another. Geographical distance,* *however, keeps the species separate:* Q. lobata *and* Q. dumosa *are found in different habitats in California, while* Q. robur *occurs in Europe.*

(a)

(b)

Figure 16.7 Ecological isolation.
*These closely related species of oak—*Quercus lobata (a) *and* Quercus dumosa (b)*—occur in different habitats and are thus adapted to different environmental pressures.*

often forms the kind of dense scrub known as chaparral and is found on steep slopes in less fertile soils. While these very different oaks can produce fully fertile hybrids, such hybrids are rare for two reasons: (1) The parents' sharply distinct habitats limit their occurrence together, and (2) there is no intermediate habitat where the hybrids might flourish (figure 16.7).

Temporal Isolation

Differences in the timing of breeding periods between two species creates another isolating mechanism, called **temporal isolation.** For example, *Lactuca graminifolia* and *L. canadensis,* two species of wild lettuce, grow together along roadsides throughout the southeastern United States. Hybrids between these two species can easily be made experimentally and are completely fertile. But such hybrids are rare in nature because *L. graminifolia* flowers in early spring and *L. canadensis* flowers in summer. When their blooming periods overlap, as they do occasionally, the two species do form hybrids, which may even be abundant locally.

Many closely related species of birds and amphibians also have different breeding seasons that prevent hybridization. For example, five species of frogs of the genus *Rana* occur together in most of the eastern United States. However, because each of their peak breeding times is different, hybrids are rare. In such insects as termites and ants, mating occurs when winged, reproductive individuals swarm from the nest. Species often differ in their swarming times, which eliminates hybridization between them.

Behavioral Isolation

Chapter 45 considers the often elaborate courtship and mating rituals of some groups of animals. Related species of organisms, such as birds, often differ in their mating rituals, which tends to keep these species distinct in nature, even if they do occur in the same places. Indeed, much animal communication is related to mate selection. When the mating rituals and other behavioral cues between two species differ to the degree that hybridization is prevented, this isolating mechanism is called **behavioral isolation.**

A striking example of behavioral isolation is found in the more than five hundred species of flies of the genus *Drosophila* that reside in the Hawaiian Islands. Many of these flies differ greatly from other species of *Drosophila,* exhibiting characteristics that can only be described as bizarre. The genus occurs throughout the world, but nowhere are the flies more remarkably diverse in their external appearance or behavior than in Hawaii. The Hawaiian species of *Drosophila* are long-lived and often very large, compared with their relatives on the mainland. The females are more uniform, while the males are often highly distinctive, displaying complex territorial behavior and elaborate courtship rituals (figure 16.8).

The patterns of mating behavior among the Hawaiian species of *Drosophila* are of great importance in maintaining the distinctiveness of the individual species. Despite the great differences between them, which are evident in figure 16.8, *D. heteroneura* and *D. silvestris* are very closely related and can produce fully fertile hybrids. They occur together over a wide area on the island of Hawaii, and yet hybridization has been observed at only one locality. These flies' very different and complex behavioral characteristics obviously play a major role in maintaining their distinctiveness.

Mechanical Isolation

Structural differences between some related species of animals also prevent mating. Aside from such obvious features as size, the structures of the male and female copulatory organs may be so incompatible as to prevent mating. In many insect and other arthropod groups, the sexual organs—particularly those of the male—are so diverse that they are used as a primary basis for classification. Evolutionary biologists generally presume that this structural diversity contributes to maintaining differences between species and call this barrier to hybridization **mechanical isolation.**

Similarly, the flowers of related species of plants often differ significantly in their proportions and structures, which can limit the transfer of pollen from one plant species to another. For example, bees may pick up the pollen of one species on one place on their bodies, and this area may not come into contact with the respective structures of the flowers of another plant species, so the pollen is not transferred. This difference then decreases the frequency of hybridization between the two species.

Prevention of Gamete Fusion

In animals that simply shed their gametes into water, such as frogs, the eggs and sperm derived from different species may not attract one another, thus preventing hybridization. Among land animals, the sperm of one species may function so poorly within the reproductive tract of another species that fertilization never takes place. In plants, growth of the pollen tubes, which are necessary for fertilization, may be impeded in hybrids between different species. Thus, in both plants and animals, such isolating mechanisms may prevent the union of gametes even after successful mating. Prevention of gamete fusion is the last kind of prezygotic isolating mechanism possible before hybrids are formed.

Prezygotic isolating mechanisms lead to reproductive isolation by preventing the formation of hybrid zygotes. The principal prezygotic isolating mechanisms are geographical, ecological, temporal, behavioral, and mechanical isolation and the prevention of gamete fusion.

Postzygotic Isolating Mechanisms

All of the factors discussed up to this point tend to prevent hybridization. But if hybridization does occur and zygotes are produced, many factors may still prevent those zygotes from developing into normal, functional, fertile individuals. Development in any species is a complex process. In hybrids, the genetic complements of two species may be so different that they cannot function together normally in embryonic development. For example, hybridization between sheep and goats usually produces embryos that die in the earliest developmental stages.

(a)

(b)

(c)

(d)

Figure 16.8 Behavioral isolation.

Males and females of closely related species of Hawaiian
Drosophila *engage in different forms of courtship and mating
rituals.* (a) Drosophila silvestris. *After approaching the female from
the rear, the male lunges forward while vibrating his wings. Then,
with his head under the wings of the female, he raises his forelegs
up and over the female's abdomen. Specialized hairs on the dorsal
surface of one of the leg segments are then "drummed" over the
dorsal surface of the female's abdomen.* (b) D. heteroneura. *A male
with extended wings approaches a female in a typical courtship
posture.* (c) D. heteroneura. *Two males interlock antennae as part
of the aggressive behavior involved in territorial defense.* (d) D.
davisetae. *In this species, the males raise their abdomens up and
over their backs and spray a chemical over the female. This
chemical acts as a mating signal to attract the female.*

For a long time, the leopard frogs (*Rana pipiens* complex) were believed to constitute a single species. Why, then, was it difficult or impossible, even in the laboratory, to produce viable hybrids between some of these frogs? Before the nature of these species was understood, biologists assumed that hybridization difficulties resulted from regional differentiation within one species. But when the mating calls of these species were demonstrated to differ substantially, the leopard frogs were recognized as a series of similar but distinct species. Such closely related species, often distinguished primarily by behavioral or other nonevident characteristics, are called **sibling species** (figure 16.9). Sibling species are very similar, but hybrids between them are usually rare.

Many examples of this kind, in which similar species initially have been distinguished only as a result of hybridization experiments, are known in plants. Sometimes, the hybrid embryos can be removed at an early stage and grown in an appropriate medium. When these hybrids are supplied with extra nutrients or other growth requirements that compensate for their weakness or inviability, they may complete their development normally.

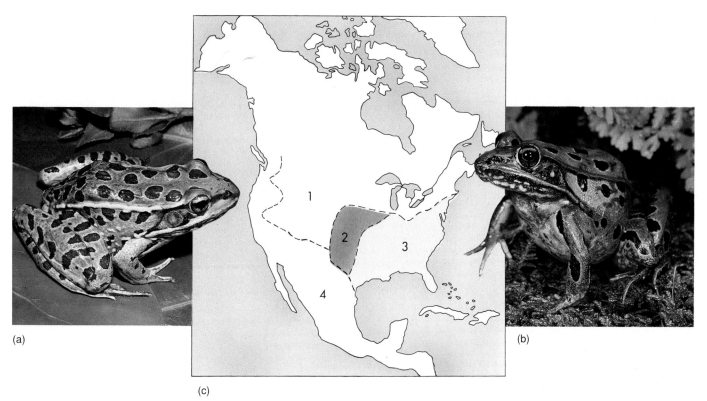

Figure 16.9 Sibling species of leopard frogs.
The different sibling species of the Rana pipiens *complex occupy different habitats in North America.* (a) *The leopard frog,* Rana pipiens, *in California.* (b) *The leopard frog,* R. pipiens sphenocephala *in Wisconsin.* (c) *Numbers indicate the following species in the geographical ranges shown:* (1) R. berlandieri, (2) R. blairi, (3) R. sphenocephala, *and* (4) R. pipiens.

Even if the hybrids survive the embryo stage, they may not develop normally. If they are weaker than their parents, they will almost certainly be eliminated in nature. Even if they are vigorous and strong, as in the case of the mule (a hybrid of the horse and the donkey) they may still be sterile and thus incapable of contributing to succeeding generations. Sex organ development in hybrids may be abnormal, the chromosomes derived from the respective parents may not pair properly, or their fertility may simply be lower than normal for other reasons. Table 16.1 summarizes the various reproductive isolating mechanisms.

> *Postzygotic isolating mechanisms lead to hybrid zygotes developing abnormally or failing to develop entirely or hybrids not being able to become established in nature.*

Combining Barriers to Hybridization: Reproductive Isolation

All of the kinds of reproductive isolation discussed thus far can arise in populations that are diverging from one another. These kinds of reproductive isolation all occur within certain species,

T a b l e 1 6 . 1 Reproductive Isolating Mechanics

1. Prezygotic mechanisms prevent the formation of zygotes.
a. Geographical isolation: The species occur in different areas.
b. Ecological isolation: The species live in different habitats and do not meet; there may be no habitat suitable for their hybrids.
c. Temporal isolation: The species reproduce at different seasons or different times of day.
d. Behavioral isolation: The behavior of the species may differ so that there is little or no attraction between them.
e. Mechanical isolation: Structural differences between species may prevent mating.
f. Prevention of gamete fusion: The gametes may not fuse, usually because of chemical factors.

2. Postzygotic mechanisms prevent the proper functioning of zygotes once they are formed. The inviability, sterility, and irregular development of hybrids are postzygotic mechanisms.

as would be expected from the way in which such differences evolve. Therefore, some populations of particular species cannot hybridize with one another, just as if they were distinct species.

Species formation is a continuous process, understandable because of the intermediate stages at all levels of differentiation. If populations that are partly differentiated have contact, they may still be able to interbreed freely, and the differences between them may then disappear over the course of time.

If their hybrids are partly sterile, however, or not as well adapted to the existing habitats as are their parents, the hybrids will be at a disadvantage. As a result, there may be selection for factors that limit the ability of the differentiated populations to hybridize. If hybrids are sterile or not as successful as their parents, individual plants or animals that do not hybridize may be more fit than those that do.

Most species are separated by combinations of the isolating mechanisms just discussed. For example, two related species may occur in different habitats, produce their gametes at different times of the year, have different behavioral patterns, and produce inviable embryos even if hybridization does take place. Such patterns, in which more than one factor functions in limiting the frequency of hybrids between two species, presumably arise for two reasons:

Figure 16.10 A view of the Galápagos Islands.
The isolation of the individual islands contributed to the diversity of the Galápagos finches.

1. The factors that limit hybridization arise primarily as by-products of adaptive change in populations. Consequently, several factors that limit hybridization often emerge simultaneously and may characterize the differentiated populations.

2. If differentiated populations have contact with one another, natural selection may strengthen the isolating mechanisms already present. For example, if the hybrids do not complete their development beyond the embryo stage, then any prezygotic isolating mechanism that limits the hybridization that produces the embryos would be advantageous. Individuals that form hybrids that do not function well in nature waste reproductive energy and are less fit than are individuals that do not form such hybrids.

Both kinds of forces that limit hybridization are part of the overall process of change in populations isolated from one another, although the forces may sometimes be strengthened when and if the differentiated populations migrate into contact with one another.

> *As a result of the way in which they originate, species are often separated by more than one factor. Some of these factors prevent the species from hybridizing at all. Others limit the success of hybrids once they are formed.*

Microevolution to Macroevolution: Examining Clusters of Species

One of the visible manifestations of species formation is the existence of groups of closely related species in certain locations. These species often have evolved relatively recently from a common ancestor. Such clusters are particularly impressive on groups of islands, in series of lakes, or in other sharply discontinuous habitats. The existence of these clusters makes sense only in the context of their arising by microevolutionary divergence from an ancestral form occupying diverse habitats. One of the best-known clusters of species is a group of finches found on the Galápagos Islands, where they were studied by Charles Darwin. Thirteen species of Darwin's finches occur on the Galápagos Islands. A fourteenth species lives on Cocos Island, which lies about 1,000 kilometers to the north.

The Galápagos Islands are a remarkable natural laboratory of evolution. The islands are all relatively young in geological terms, and they have never been connected with the adjacent mainland of South America or with any other source area. The lowlands of the Galápagos Islands are covered with thorn scrub. At higher elevations, which are attained only on the larger islands, are moist, dense forests. All of the Galápagos organisms have reached the islands by crossing the sea as a result of chance dispersal in the water or wind, or transport via another organism (figure 16.10).

Oceanic islands often show a disproportionate representation of certain groups of organisms. For example, in addition to the thirteen species of Darwin's finches on the Galápagos Islands, there are only fourteen other species of resident land birds. Perhaps the ancestor of Darwin's finches reached these islands earlier. In that case, all the habitats where birds occur on the mainland would have been unoccupied, and the ancestor of Darwin's finches would have been able to take advantage of them all. As new arrivals moved into these vacant habitats, adopting new lifestyles, they were subjected to a diverse set of selective pressures. Under these circumstances, the ancestral finches split into a series of diverse populations, and some of these eventually became species.

The descendants of the original finches that reached the Galápagos Islands now occupy many different habitats that

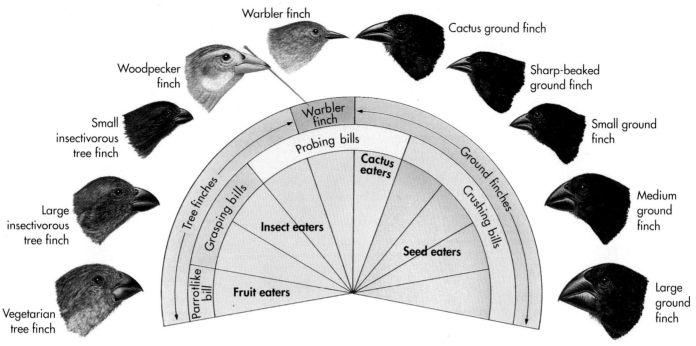

Figure 16.11 Darwin's finches.
Ten species of Darwin's finches from Indefatigable Island, one of the Galápagos Islands, show differences in bills and feeding habits. The bills of several of these species resemble those of different, distinct families of birds on the mainland. All of these birds are thought to have evolved from a single common ancestor.

encompass a distinct group of birds on the mainland. Among the thirteen species of Darwin's finches that inhabit the Galápagos, there are three main groups (figure 16.11):

1. *Ground finches* feed on seeds of different sizes. The size of their bills is related to the size of the seeds on which they feed.

2. *Tree finches* have bills suitable for feeding on insects. One of the tree finches has a parrotlike beak and feeds on buds and fruit in trees. Another has a chisel-like beak for carrying around a twig or cactus spine, which it uses to probe for insects in crevices. It is an extraordinary example of a bird that uses a tool.

3. The *warbler finch* is an unusual bird that plays the same ecological role in the Galápagos woods that warblers play on the mainland, searching continually over the leaves and branches for insects.

The evolution of Darwin's finches on the Galápagos Islands and Cocos Island provides a classic example of species formation and illustrates how adaptation to local conditions produces the divergence that is the heart of species formation. By the same sorts of processes, species are originating continuously in all groups of organisms.

The evolution of organisms on islands differs from that in mainland areas only in the degree of opportunity afforded for *rapid* evolution. Because evolution on islands is often rapid,

and therefore recent, Darwin and other investigators have found the study of plants and animals on islands especially informative.

> The clusters of species that have formed on the Galápagos Islands are a clear example of species formation arising by microevolutionary divergence from an ancestral form occupying different habitats—of microevolution leading to macroevolution.

The Evidence for Evolution: An Overview

This chapter and chapter 15 have considered the body of evidence that supports the three central tenets of evolution stated in this chapter's first paragraph. The evidence for macroevolution is so strong as to be beyond further question. Although scientists are currently engaged in lively discussions about the rate at which macroevolution proceeds, and about whether it proceeds gradually or in spurts, essentially all of them agree about the fact of macroevolutionary change.

The evidence for microevolution depends partly on observations of evolutionary change now in progress. In many cases, observing evolutionary change within populations of a single species has been possible, and such observations have shown selection to be a powerful agent of adaptive change, just as Darwin proposed in 1859. Many factors cause populations to change progressively over time, including both environmentally guided selection and such random factors as genetic drift.

The evidence that microevolutionary change leads to macroevolutionary change is extensive. Biologists have observed the stages of the species-forming process in many different plants, animals, and microorganisms. These stages are indicated by such features as genetic variation in individual populations, genetic differences between different populations, divergence of populations from one another in response to different adaptive pressures, ecological races, progressively stronger reproductive isolation, and clusters of related species, each species occupying an ecologically distinct, but often adjacent, habitat. These patterns are so frequently observed in nature that they leave no doubt of the way in which species normally originate. Microevolution drives macroevolution, just as Darwin proposed.

EVOLUTIONARY VIEWPOINT

Evolution is not simply a thing of the past involving dinosaurs, fossils, and extinct dodo birds. Evolution goes on all the time, and is going on right now, as local populations of species adapt to differing environmental demands and become increasingly distinct. Today's world only marks where we are now on an ongoing evolutionary journey.

SUMMARY

1. The theory of evolution is based on three lines of evidence: (1) that macroevolution occurs, (2) that natural selection is responsible for the adaptive nature of microevolution, and (3) that microevolutionary change (evolution within a species) leads to macroevolutionary change (evolution of new species).

2. Species are kinds of organisms that differ from one another in one or more characteristics and that do not normally hybridize freely with other species when they come into contact in nature. They often cannot hybridize with one another at all. Individuals within a given species, on the other hand, usually are able to interbreed freely.

3. Populations change as they adjust to the demands of their environments. Even populations of a given species that are close to one another geographically are normally effectively isolated. Such populations are free to diverge in ways that are responsive to the needs of their particular environment.

4. In general, if the selective forces bearing on population differ greatly, the populations will diverge rapidly; if these selective forces are similar, the populations will diverge slowly.

5. Ecological races and subspecies differentiate within species but often still intergrade with one another. The differences between races, in both plants and animals, are mostly determined by genetic factors.

6. Among the factors that separate populations, and species, from one another are geographical, ecological, temporal, behavioral, and mechanical isolation, as well as factors that inhibit the fusion of gametes or the normal development of the hybrid organism. Reproductive isolation between species arises as a normal by-product of the progressive differentiation of populations.

7. Clusters of species arise when the differentiation of a series of populations proceeds further, past race formation. On islands, such differentiation is often rapid because numerous open habitats are available. In many continental areas, differentiation is not as rapid.

8. The evidence that microevolutionary change leads to macroevolutionary change is that it is possible to directly observe the intermediate stages of the process.

REVIEWING THE CHAPTER

1. What is the species concept, and how did it originate?

2. What is the cornerstone of species formation?

3. How do populations of organisms diverge?

4. What is an ecological race or ecotype?

5. What are some barriers to hybridization?

6. What is a prezygotic isolating mechanism?

7. What is a postzygotic isolating mechanism?

8. What is reproductive isolation?

9. How do microevolution and macroevolution relate to clusters of species?

10. What is the evidence for evolution?

COMPLETING YOUR UNDERSTANDING

1. Darwin's observations that species intergrade was important evidence for which theory?

2. In Darwin's explanation of the relative constancy of a species, what is a niche?

3. Which of the following is the most realistic definition of species?

   a. A kind of organism.

   b. A kind of organism that consists of one or more populations within which the individuals are interfertile with one another and that do not interbreed with other species.

   c. A distinct kind of organism consisting of individuals that usually resemble one another more closely than they do other kinds of organisms and that do not normally interbreed with related kinds of organisms when they occur together in nature.

   d. A geographical race.

   e. The result of punctuated equilibrium.

4. What are the differences between clones and ecotypes, and how do they relate to plant biology?

5. Why does the English oak not hybridize with either the valley oak or scrub oak in nature?

6. What is the interrelationship between fitness and ecological isolation?

7. Why are hybrids rare between (*a*) lions and tigers and (*b*) the valley oak and scrub oak?

8. How is temporal isolation a barrier to hybridization?

9. What is behavioral isolation, and does it apply to both plants and animals?

10. What are some examples of mechanical isolation in plants and animals?

11. Which of the following is *not* a prezygotic isolating mechanism?
  a. Geographical isolation
  b. Ecological isolation
  c. Seasonal isolation
  d. Hybrid sterility
  e. Mechanical isolation

12. In both plants and animals, successful matings may occur between different species. In those matins, what are the mechanics that prevent gametes from fusing?

13. What are some of the factors that prevent hybrids between two species from developing viable offspring?

14. Why are species clusters particularly impressive on groups of islands or in a series of lakes?

15. Is the evolution of Darwin's finches on the Galápagos Islands an example of microevolution or macroevolution?

16. On the Galápagos Islands, the descendants of a South American finch have produced a cluster of species. Darwin's finches fill a variety of niches. Which niche is *not* filled by any of these finches?
  a. Seed eater
  b. Cactus eater
  c. Fish eater
  d. Insect eater
  e. Fruit eater

*17. In the fall of 1986, the U.S. Supreme Court heard arguments for and against a Louisiana law requiring that "creation science" be taught in public schools on an equal footing with evolution. One of the arguments advanced in support of the law was that evolution is as much a religion as creationism, reflecting the beliefs of scientists and their faith in a particular worldview, rather than the certain knowledge of objective reality that they claim. In June 1987, the decision of the Court was announced: By 7 to 2, the justices struck down the law, rejecting the "creation science" argument. How would you have voted?

*For discussion

FOR FURTHER READING

Goslow, G. E., Jr.; K. P. Dial; and F. A. Jenkins, Jr. 1990. Bird flight: Insights and considerations. *BioScience* 40 (2): 108–15. New techniques that show more than the wing participates in flight.

Gould, S. J. 1977. *Ever since Darwin*. New York: W. W. Norton. An entertaining and insightful collection of essays on evolution and Darwinism. In subsequent years, Gould published many additional volumes of essays, all well worth reading.

Hitching, F. H. 1982. *The neck of the giraffe, or where Darwin went wrong*. London: Chaucer Press (Pan). An entertaining and informal presentation of all the arguments currently being advanced against Darwin's theory.

May, R. 1992. How many species inhabit the earth? *Scientific American,* Oct., 42–48. How species form and how many exist today—questions that are relevant to efforts to conserve biological diversity and to manage our environment.

Ryan, M. J. 1990. Signals, species, and sexual selection. *American Scientist* 78 (1): 46–52. Studies of mate recognition in frogs and fishes that reveal preferences for individuals, populations, and even members of closely related species.

Waters, A. et al. 1991. *Plasmodium falciparum* appears to have arisen as a result of lateral transfer between avian and human hosts. *Proceedings of the National Academy of Sciences (USA)* 88 (April): 3140–144. The apparent discovery that humans caught the parasite that causes malaria from birds in the recent past, with the advent of agricultural society.

Weiner, J. 1994. *The beak of the finch*. New York: A. Knopf. A wonderful account of ongoing research on Darwin's finches that reviews evolution in action today. Few books give a better view of how research on evolution is really carried out.

EVOLUTION OF LIFE ON EARTH

A rich library of fossils in the earth's rocks bears witness to a diverse past, little of which survives today. These fossil crenoids have been extinct since the end of the Cretaceous, 65 million years ago.

All of the many kinds of organisms now living on earth represent only a small fraction of the total that have lived. Every living creature occupies a place once taken by a different kind of animal or plant. For example, mammals like ourselves play an ecological role once filled by dinosaurs. Dinosaurs were once the dominant land vertebrates, but today, no dinosaurs walk or fly above the earth (figure 17.1). The trilobites—sea organisms related to the living horseshoe crabs—dominated the ocean floor in the Cambrian period more than five hundred million years ago, but they are gone now (figure 17.2). So are the ammonites—octopuses with shells that were one of the dominant forms of ocean life in the Cretaceous period one hundred million years ago.

A great deal is known about dinosaurs and the many other forms of life that have preceded humans in the evolutionary parade of life on earth because of fossils. Fossils allow scientists to visualize the body of an organism that lived in the past because certain structures are mineralized and so preserved in rock. While the fossil record is incomplete, enough fossils have been found to piece together a fairly complete picture of the past, a picture that is reviewed in this chapter. First, a few words about fossils.

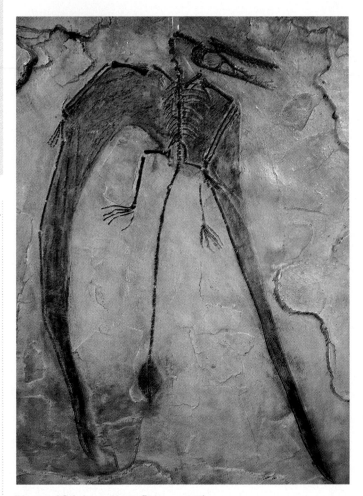

Figure 17.1 An extinct flying reptile.
Flight has evolved three separate times among vertebrates. Birds and bats are still with us, but pterosaurs, such as the one pictured, became extinct with the dinosaurs about sixty-five million years ago.

Fossils

A **fossil** is any record of an extinct organism; fossils may be nearly complete impressions of organisms, or merely burrows, tracks, molecules, or traces of their existence. Only fossils provide definite evidence about what extinct organisms, including the ancestors of those organisms living now, looked like. Fossils provide an actual record of organisms that once lived and an accurate understanding of where, when, and in what kind of environment they lived. If enough fossils of various ages are available, lines of evolutionary progression can be traced among individual groups of organisms.

Problems in Preservation

Only a minute fraction of the organisms living at any one time are preserved as fossils, most of them in **sedimentary rocks.** Sedimentary rocks are formed of particles of other rocks that are weathered off, deposited at the bottom of bodies of water or accumulated by wind, and then hardened into strata of the sort that Darwin found so filled with intriguing fossils in South America. During sedimentary rock formation, dead organisms are sometimes washed down along with inorganic debris, such as mud or sand, and eventually reach the bottom of some pond, lake, or ocean, where they may be incorporated into sedimentary rocks. In exceptional circumstances, fossils may be preserved in an organic substance, such as tar. The La Brea Tar Pits in Los Angeles illustrate this phenomenon (figure 17.3). In other instances, organisms themselves may form the whole deposit, as in the formation of coal or oil deposits. However, because many organisms do not live in places where their remains are likely to be preserved in sedimentary rocks, they are unlikely to be preserved as fossils.

For specific organisms to become well preserved as fossils, they must usually be buried before their decay is complete and before they are torn apart by scavengers. In addition, the process of decay, for at least some portions of the organism, must stop after the organism is buried. Because an organism's soft parts usually decay rapidly, only such structures as bones, teeth, shells, scales, leaves, and wood normally are available in

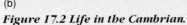

(b)

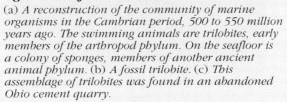

(a)

Figure 17.2 Life in the Cambrian.

(a) *A reconstruction of the community of marine organisms in the Cambrian period, 500 to 550 million years ago. The swimming animals are trilobites, early members of the arthropod phylum. On the seafloor is a colony of sponges, members of another ancient animal phylum. (b) A fossil trilobite. (c) This assemblage of trilobites was found in an abandoned Ohio cement quarry.*

(c)

(a)

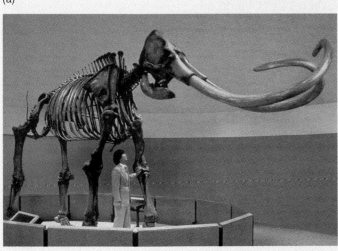

(b)

(c)

Figure 17.3 Life in downtown Los Angeles fifteen thousand years ago.

Trapped in large pools of tar, the skeletons of sabre-toothed cats (a), mammoths (b), and huge ground sloths (c) have been preserved in the La Brea Tar Pits.

Sidelight 17.1
THE DISCOVERY OF A LIVING COELACANTH

New discoveries relating to the history of life on earth are not always made in the fossil record. In 1938, scientists were surprised by the announcement that a trawler fishing in the Indian Ocean off the coast of South Africa had landed a large, very strange fish (Sidelight figure 17.1). The specimen, which was about 2 meters long, became rotten before it was seen by an ichthyologist (a scientist who specializes in the study of fish) and was skinned so that some of it could be saved. Nevertheless, as soon as J.L.B. Smith of the University of Grahamstown in South Africa saw it, he recognized it as a member of an ancient group of "lobe-finned" fishes (fishes in which the fins closely resemble the limbs of amphibians and other land vertebrates). These fishes had been described from fossils more than a century earlier and had been thought to have been extinct for about seventy million years! They were called coelacanths, and the newly discovered living fish was dubbed *Latimeria chalumnae*.

Coelacanths are well represented in the fossil record for more than 300 million years, from about 390 million years to about 70 million years ago. Until *Latimeria* was discovered alive, the group, which was somewhat similar to the ancestors of the terrestrial vertebrates, was assumed to have become extinct long ago. But here was indisputable evidence that one of their

Sidelight figure 17.1 A living fossil.

For many years, scientists thought that the coelacanth, well represented in the fossil record, was extinct. As you can see from this photograph, the report of their demise was premature.

descendants was still alive, swimming the warm waters of the western Indian Ocean.

Because the first specimen had been skinned, scientists had no opportunity at first to learn about its internal parts, features that were of great interest in terms of its relationship to terrestrial vertebrates and other fishes. Leaflets and posters were distributed among the fishing communities of southern and eastern Africa, offering rewards for another specimen of this remarkable fish. But it was not until 1952 that a second coelacanth was brought to the attention of scientists. It was landed in the Comoro Islands, about 3,000 kilometers northeast of where the first specimen was caught. Coelacanths were landed occasionally in the Comoros and were well known to local fishermen.

Living mostly at depths of 150 to 300 meters in the sea, *Latimeria* is a very strange animal. Its features mark it as a member of the evolutionary line that gave rise to the terrestrial tetrapods. Study of the dozens of specimens that have been landed since 1952 has shed additional light on the nature of this ancient and archaic group of vertebrates.

---

the fossil record. If fossils are exposed, they often disintegrate quickly; therefore, most of those that are found were exposed relatively recently.

When fossils are formed, the actual parts of the organisms are almost always replaced by minerals. A fossil rarely contains any of the original organic materials from the organism's body, but rather, mineralized replicas of those materials.

Interpreting the structure of fossil organisms normally requires using the hard parts to try to determine what the soft parts looked like. For organisms such as worms, which have no hard parts, fossils are rare. Even though soft-bodied animals undoubtedly evolved before their hard-bodied counterparts, there is little evidence of their history in the fossil record. Sometimes, soft-bodied animals are preserved in exceptionally fine-grained muds, where the supply of oxygen was poor while the muds were being deposited, thereby slowing deterioration.

Sidelight 17.1 describes the fascinating discovery of a "living fossil."

Fossils provide the concrete means for judging deductions about the history of particular groups of organisms. Fossils are found mainly in sedimentary rocks.

Dating Fossils

The direct methods of dating rocks and fossils mentioned in chapter 15 first became available in the late 1940s. Naturally occurring radioactive isotopes of certain elements are employed in this process. Such isotopes are unstable and decay over the course of time at a steady rate, producing other isotopes.

One of the most widely employed methods of dating—the **carbon-14 (^{14}C) method**—uses estimates of the different isotopes present in samples of carbon. Most carbon atoms have an atomic weight of 12; the symbol for this particular isotope of carbon is ^{12}C. But a fixed proportion of the atoms in a given sample of carbon consists of carbon with an atomic weight of 14 (^{14}C), an isotope with two more neutrons than ^{12}C. ^{14}C is produced from ^{12}C as a result of bombardment by particles from space. The carbon incorporated into the bodies of living organisms consists of this same fixed proportion of ^{14}C and ^{12}C. But after an organism dies and is no longer incorporating carbon, its ^{14}C gradually decays over time, by the loss of neutrons, to ^{12}C. It takes 5,600 years for half of the ^{14}C present in a sample to be converted to ^{12}C by this process; this length of time is called the **half-life** of the ^{14}C isotope. Given the relationships just outlined, a sample that had a quarter of its original proportion of ^{14}C remaining would be approximately 11,200 years old.

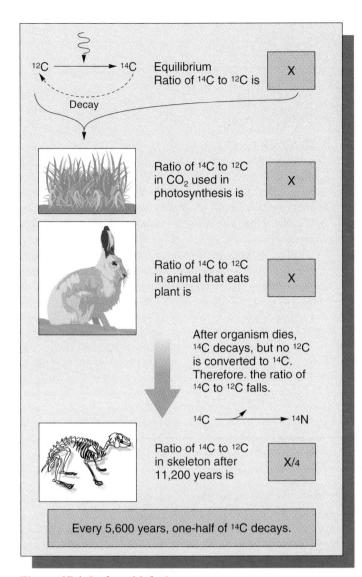

Figure 17.4 Carbon-14 dating.
Because no carbon is assimilated after death, the degree to which carbon-14 isotopes have decayed in a fossil provides a direct measure of the time since death occurred.

Thus, fossil material that contains carbon, as all organic material does, can be dated by measuring as accurately as possible the proportion of ^{14}C still present in the carbon, provided that the fossil material is less than about fifty thousand years old (figure 17.4). For older fossils, the amount of ^{14}C remaining is so small that it cannot be measured precisely enough to provide accurate age estimates.

The ages of older samples can sometimes be determined by studying the decay of other radioactive isotopes. For example, ^{40}K (potassium-40) decays very slowly into ^{40}Ar (argon-40) and can be used to date much older rocks. The half-life of ^{40}K is 1.3 billion years; other isotopes have even longer half-lives. These dating methods allow estimates of the ages of various rocks to be more precise.

Unfortunately, for sedimentary rocks, which incorporate pieces of other rocks, such dating methods can establish only the ages of the materials and not the time when they were incorporated into the sedimentary rocks—the time when the fossils were formed. But these methods do allow for estimations of the ages of even the oldest rocks on earth.

Fossil ages are estimated by determining the proportions of different isotopes of carbon in the organic material preserved in them or the proportions of different isotopes in the rocks where they are found. Radioactive isotopes change from one form to another over time, so the isotope proportions in a given sample provide an absolute date for the age of that sample.

Continental Drift

As discussed in chapter 3, the planets of the solar system, including the earth, began to coalesce approximately 4.6 billion years ago. The accretion (pulling together of fragments by gravity) of the early earth seems to have been completed about 4.5 billion years ago. After several hundred million years, the continents had formed. The oldest known rocks on earth are 3.8 billion years old. By the time they were consolidated, some portions of the earth's crust had begun to move. As these sections moved, they became thicker.

Discovery of Continental Movement

About two hundred million years ago, in the early Jurassic period, the major continents, following a long history of earlier movements, were all together in one great supercontinent. Alfred Wegener, the German scientist who in 1915 first proposed the idea of continental movement in his book *The Origin of Continents and Oceans*, called this giant land mass **Pangaea** (figure 17.5). Because the specific mechanism Wegener proposed to account for continental movement was not feasible, his theory fell into discredit with most geologists and biologists. Indeed, although it is now generally accepted, Wegener's theory of continental movement was denied by most scientists for nearly half a century.

In the early 1960s, new evidence provided a mechanism for continental movement. The theory that emerged visualized heavy, basaltic, ocean-floor rocks moving like a conveyor belt away from the midoceanic ridges where they were formed. In the process, they carried the lighter continental rocks along with them. For example, the Mid-Atlantic Ridge is an enormous zone of upwelling basaltic lava. It generates basaltic ocean-floor rocks that move out from the ridge both toward Europe and Africa and toward North and South America. As these rocks continue to be formed, both pairs of continents ride farther and farther apart from one another.

The earth's crust and associated upper mantle, which together are about 100 to 150 kilometers thick, are divided into

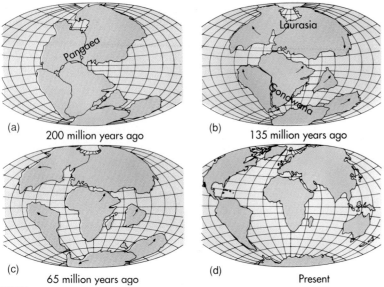

(a) 200 million years ago

(b) 135 million years ago

(c) 65 million years ago

(d) Present

▭ Figure 17.5 How the continents have moved.

The positions of the continents at (a) 200 million, (b) 135 million and (c) 65 million years ago, and (d) at present. Two hundred million years ago, when dinosaurs first flourished, all the continents were shoved together into one massive supercontinent—Pangaea. By 135 million years ago, they had separated into northern and southern clusters—Laurasia and Gondwana. When the dinosaurs disappeared sixty-five million years ago, sea levels were high, and the continents had vast inland seas.

plates—seven enormous ones and a series of smaller ones that lie between them. The theory that explains the movement of continents is thus called **plate tectonics** (figure 17.6).

Earthquakes are generally the result of the relative movement of these plates. Thus, an earthquake along the San Andreas Fault in California, such as the one that destroyed San Francisco in 1906 or the large ones that occurred in the San Francisco Bay region in October 1989 and Los Angeles in October 1987 and January 1994, results from the relative movement of two gigantic segments of the earth's crust and mantle. The far edges of the plates that caused these earthquakes are in Japan and Iceland.

Plate movements are also responsible for the formation of most mountains. The Himalayas have been thrust up to the highest elevations on earth as a result of the grinding, prolonged collision of the Indian subcontinent with Asia.

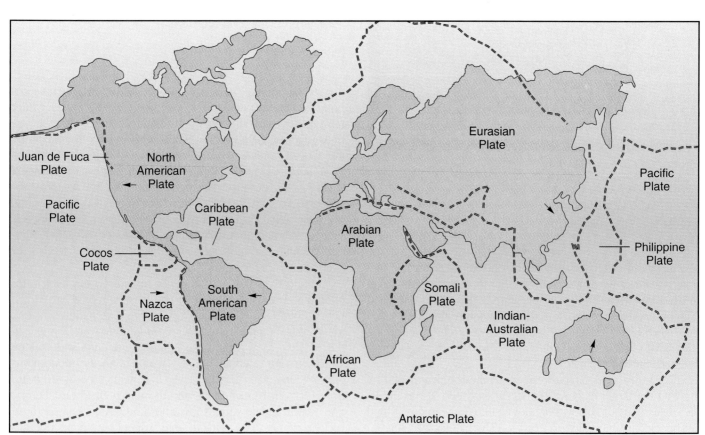

→ Direction of plate motion

▭ Figure 17.6 Plate tectonics.

The earth's crust and upper mantle are divided into seven enormous plates and a series of smaller plates that move relative to one another.

Lava flowing out of fissures in the earth's crust, such as the midoceanic ridges, form great plates of relatively heavy rocks 100 to 150 kilometers thick. These rocks move, carrying the lighter continents along with them. There are seven huge plates, all moving in relation to one another, and a number of smaller plates.

History of Continental Movement

The continents have gradually moved apart from their positions as parts of Pangaea two hundred million years ago. The opening of the South Atlantic Ocean about 125 to 130 million years ago separated Africa and South America, which had been directly connected earlier. Subsequently, South America moved slowly toward North America, with which it eventually became connected by a land bridge, the Isthmus of Panama, between 3.1 and 3.6 million years ago. The bridge allowed many South American plants and animals, such as the opossum and the armadillo, to migrate overland into North America. At the same time, numerous North American plants and animals, such as oaks, deer, and bears, moved into South America for the first time.

Changes on a greater or lesser scale in the positions of all the continents played a major role in the distribution patterns of organisms seen today. For example, the ratite birds—ostriches, emus, rheas, cassowaries, and kiwis—now thought to have diverged very early in the course of their evolution from all other living groups of birds, undoubtedly migrated between southern continents and islands that are now widely separated from one another, during the Mesozoic era, when direct overland migration was possible. Explaining present-day distributions from continental movements is a large part of **biogeography,** the study of where different kinds of organisms occur and how they got there.

Australia provides a striking example of the way in which continental movements have affected the nature and distribution of organisms. Until about fifty-three million years ago, Australia was joined with a much warmer Antarctica, as was South America.

A land connection between Australia and Antarctica, through the present position of Tasmania, seems to have persisted until about thirty-eight million years ago. On the other side of the world, the connection of southern South America with West Antarctica lasted until about twenty-three million years ago. With the warmer climates of the early Cenozoic era, more-or-less direct overland migration between Australia and South America, via Antarctica, was possible until about forty million years ago.

Marsupials, which are best represented in Australia and South America at present, seem clearly to have moved overland between these continents via Antarctica when this was still possible. Isolated from the placental mammals that had become abundant and diverse in other parts of the world, marsupials underwent a major episode of evolutionary spread in Australia.

As Australia moved northward, the Antarctic Ocean opened, and the cold Circumantarctic Current began to flow around the Southern Hemisphere. The formation of the Circumantarctic Current led directly to the development of the Antarctic ice sheet, which reached its full size by about ten million years ago. Ultimately, the cold temperatures associated with the formation of that enormous mass of ice also triggered the onset of widespread continental glaciation (the Ice Age) in the Northern Hemisphere during the past few million years.

The movements of continents over the past two hundred million years have profoundly affected the distribution of organisms on the earth. Some of the major events include the linkage of South America with North America about 3.1 to 3.6 million years ago and the separation of Australia and South America from Antarctica, which triggered the formation of the southern, and ultimately, the northern ice sheets.

A Time Line of Evolution

When scientists first began to study and date fossils, they had to find some way to organize the different time periods from which the fossils came. They divided the earth's past into large blocks of time called **eras.** Eras are further subdivided into smaller blocks of time called **periods,** and some periods, in turn, are subdivided into **epochs,** which can be divided into **ages.** The major geological eras, with their approximate dates in millions of years ago (abbreviated "MYA"), are as follows (table 17.1):

Archean era, 4600 to 2500 MYA. The earth formed during this earliest era, and the first cells appeared. Several kinds of bacteria fossils that date from this era have been found.

Proterozoic era, 2500 to 590 MYA. During this era, eukaryotes appeared, followed by the first multicellular organisms. Fossils of these early eukaryotes have been found.

Paleozoic era, 590 to 250 MYA. The name of this era is derived from the Greek words *paleos,* meaning "old," and *zoos,* meaning "life." The Paleozoic era is divided into six periods: the **Cambrian, Ordovician, Silurian, Devonian, Carboniferous,** and **Permian.** During the Cambrian period, all of the main invertebrate groups formed. Until very early fossils that date from the two previous eras were discovered recently, this was the oldest period from which fossils were known. Because of the Cambrian period's wealth of fossil information, the two eras that precede the Cambrian—the Archean and Proterozoic—are designated **Precambrian time.**

Mesozoic era, 250 to 65 MYA. The term *mesozoic* is taken from the Greek word *meso,* meaning "middle." The Mesozoic era is divided into three periods: the **Triassic, Jurassic,** and **Cretaceous.** This is the era of the dinosaurs, and many dinosaur fossils from this era have been recovered.

Cenozoic era, 65 MYA to present. The term *cenozoic* is derived from the Greek word *coenos,* meaning "recent." During this era, mammals and birds became diverse, and human evolution took place. The Cenozoic era is divided into two

T a b l e   1 7 . 1   Major Geological Eras

Era	Period	Epoch	MYA*	Life-Forms
Cenozoic	Quaternary	Pleistocene	1–2	
	Tertiary	Pliocene	7	First humans appear.
		Miocene	26	
		Oligocene	38	Monkeylike primates appear.
		Eocene	54	Eohippus appears. Small mammals undergo adaptive radiation.
		Paleocene	65	Major extinction event decimates the dinosaurs and many marine organisms.
Mesozoic	Cretaceous		136	Flowering plants appear. Insects become diverse.
	Jurassic		210	Large dinosaurs dominate the earth. First birds appear.
	Triassic		250	Small dinosaurs and first mammals appear.
Paleozoic	Permian		285	Major extinction event occurs. Most species disappear. Conifers appear.

T a b l e   1 7 . 1   continued

Era	Period	Epoch	MYA*		Life-Forms
	Carboniferous		370		First reptiles and arthropods appear. Coal deposits form. Horsetails, ferns, and seed-bearing plants are abundant.
	Devonian		410		"Age of the fishes"—fishes with bones and jaws appear. Amphibians also appear. Major extinction event affects marine invertebrates and fishes.
	Silurian		430		Notochord becomes flexible as single rod is replaced with separate pieces seen in the Ostracoderms (armored fish without bones, jaws, or teeth). Plants invade the land.
	Ordovician		500		First vertebrates appear. Major extinction event affects marine species.
	Cambrian		590		Major extinction event affects the trilobites. All of the main phyla appear.
Proterozoic			2500		First eukaryotic cells and multicellular eukaryotic animals appear. Oxygen-producing bacteria are present. Atmosphere and oceans are oxygenated. Chemical evolution results in formation of first cells. Stromatolites form. First rocks form. Earth is born.
Archean					

*Millions of years ago.

periods: the **Tertiary** and **Quaternary.** The Tertiary period is further divided into five epochs: the **Paleocene, Eocene, Oligocene, Miocene,** and **Pliocene.** The Quaternary period has two epochs: the **Pleistocene** and **Recent.** The first humans appeared during the Pliocene epoch of the Tertiary period.

Early History of Life on Earth

As mentioned earlier, the earth itself is about 4.6 billion years old, and the oldest rocks that have persisted in recognizable form are about 3.8 billion years old. For many years, scientists believed that such ancient rocks did not contain any fossils, but they now know that the fossils were simply too small to be seen clearly without an electron microscope. The earliest fossils found so far, all of them bacteria, are about 3.5 billion years old.

Massive limestone deposits called **stromatolites** became frequent in the fossil record about 2.8 billion years ago. Produced by cyanobacteria, stromatolites were abundant in virtually all freshwater and marine communities until about 1.6 billion years ago. Today, stromatolites are still being formed, but only under conditions of high salinity, aridity, and high light intensities.

About two billion years ago, many kinds of bacteria existed, including single, rounded cells; filaments apparently divided by cross-walls; tubular structures; branching filaments; and several unusual forms that do not fit well into any of the previous categories. For most of the time in which life has existed on earth, the only organisms in existence were bacteria. Although the most conspicuous structures they formed were the stromatolites, bacteria were clearly everywhere (figure 17.7). Bacteria dominated for at least two billion years.

The fossil record indicates that unicellular protists—the first eukaryotes—first appeared about 1.5 billion years ago. Multicellular animals that lived before the evolution of external skeletons in many cases were not well preserved. Traces of such organisms up to seven hundred million years old exist.

> *The oldest fossils are bacteria, dating from about 3.5 billion years ago. The oldest eukaryotic fossils—unicellular protists—are from about 1.5 billion years ago.*

With the evolution of hard external skeletons, shells, and other structures that were easily preserved, the nature of the fossil record changed dramatically (figure 17.8). The evolution of nearly all known major groups of organisms took place within a few tens of millions of years after the appearance of skeletons in their ancestors about 570 million years ago. Organisms' ability to manufacture such skeletons seems to have been a major evolutionary advance that made the further evolutionary radiation of these groups possible.

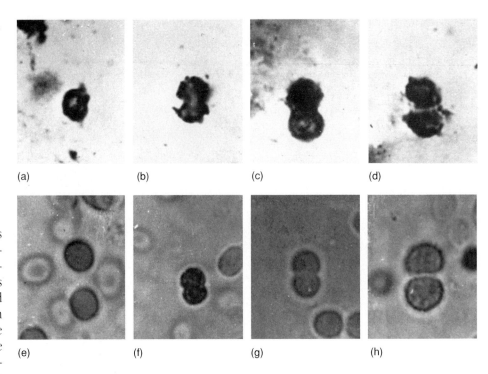

Figure 17.7 Bacteria have changed little in over three billion years.
The fossilized bacteria in (a) through (d) were found in South African stromatolites about 3.4 billion years old. Photos (e) through (h) show present-day bacteria that bear a striking resemblance to the fossilized bacteria.

> *Traces of multicellular animals are preserved in rocks up to seven hundred million years old. The evolution of hard skeletons, shells, and other structures may have been an advance that triggered the evolution of nearly all known major groups of organisms.*

The Paleozoic Era

Virtually all of the major groups of organisms that survive at the present time, except for the plants, originated and diversified during the Paleozoic era. During the Cambrian period (590 to 505 million years ago) of the Paleozoic, multicellular animals became more diverse more rapidly (in geological terms) than was ever the case later. **Phyla** (singular, *phylum*)—the major groups into which kingdoms are divided (such as mollusks, sponges, or flatworms)—appeared mainly at this time and exclusively in the sea. In contrast, plants originated on land, although they were derived from aquatic ancestors, the green algae. Thus, the diversification of animal life on earth is basically a marine record, and the fossils from the Paleozoic era all originated in the sea.

(a)

(b)

Figure 17.8 The Burgess Shale.
(a) *The Burgess Shale is a rock formation that formed from a fine-grained mud in the sea but has been uplifted to high elevations in the Rocky Mountains of British Columbia. Over one hundred kinds of early Cambrian fossils have been found there. Many of the fossils represent unique phyla. Only a few of the phyla present then still survive today.* (b) *Among the many extinct forms are tiny* Hallucigenia, *and the large predators,* Anomalocaris, *seen in this visualization of the early Cambrian seafloor.*

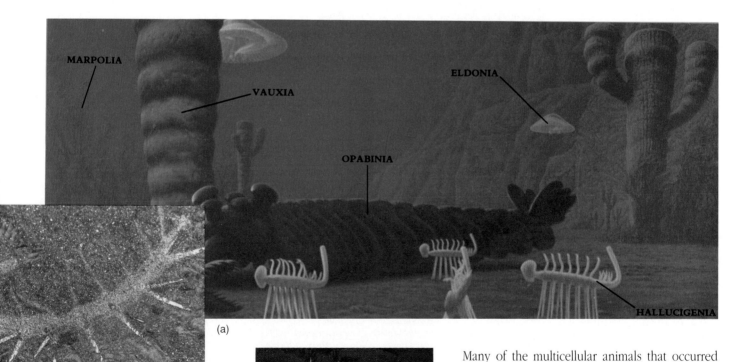

MARPOLIA

VAUXIA

ELDONIA

OPABINIA

HALLUCIGENIA

(a)

(b)

Figure 17.9 The puzzle of Hallucigenia.

(a) *This early fossil of* Hallucigenia, *12.5 mm long, was crushed flat in the Burgess Shale rock. Until recently, scientists thought it walked on its stiff spines, as portrayed in* (b) *The discovery of the new specimens in China* (c), *revealed a second set of legs hidden behind the animal.*

(c)

Many of the multicellular animals that occurred in the early Paleozoic era have no living relatives. Their fossils indicate that this was a period of experimentation with different body forms and ways of life, some of which ultimately led to the contemporary phyla of animals, and others to extinction (figure 17.9). For example, the trilobites (see figure 17.2) appear to be the ancestors of at least one living group, the horseshoe crabs, whereas the ammonites, which were abundant one hundred million years ago, have no surviving descendants.

Figure 17.10 Fossil from the Devonian period.
An insect (phylum Arthropoda) from the Devonian period, about 376 to 379 million years ago, found in what is now New York State. Land animals appeared about forty million years earlier, but they were initially very scarce.

The early Paleozoic era was a time of extensive diversification for marine animals. Many new kinds of animals appeared, some of which have persisted to the present.

Until the close of the Cambrian period of the Paleozoic, all of the animals in the sea fed on unicellular organisms that floated freely in the water. During the following Paleozoic period—the Ordovician—true predators appeared, and animals became even more diversified. The first corals also originated during the Ordovician and began to change the structure of marine communities permanently. By the end of this period, practically every mode of life that has ever existed—for example, bottom feeders, scavengers, carnivores, colonies, and driving multicellular forms—had already evolved. With all of the adaptive zones in which these animals occur filled, later evolution of novel forms was much more limited.

Invasion of the Land

Only a few phyla, or groups, of organisms have invaded the land successfully; most others have remained exclusively marine. The first organisms to colonize the land were the plants, about 410 million years ago. Plant features that favored land colonization, such as cuticle and bark, water-resistant outer coverings with specialized openings for gas exchange (see chapter 32), tissues for conducting water and nutrients, and drought-resistant pollen grains and spores, evolved. The ancestors of plants were specialized members of a group of photosynthetic protists known as the green algae. Although most algae are aquatic, the immediate ancestors of plants might themselves have been semiterrestrial. Nevertheless, the occupation of the land truly began with the plants.

The second major invasion of the land, and perhaps the most successful, was by the **arthropods,** a phylum of hard-shelled animals with jointed legs and a segmented body (figure 17.10). Arthropods were originally marine organisms; the trilobites in

(a)

(b)

Figure 17.11 Basic body patterns established long ago.

(a) *The fossil dragonfly, which is about 170 million years old (Jurassic period), closely resembles (b) its modern counterpart. This fossil dramatically illustrates the establishment of modern groups of organisms in the Mesozoic era. Many of these groups have persisted to the present day.*

figure 17.2 were members of this phylum. Among the descendants of the first arthropods to invade the land are winged insects (figure 17.11).

This second invasion of the land occurred at about the same time as the evolution of the plants, about 410 million years ago. The arthropod body plan has proven so well adapted to life on land that insects and other classes of this phylum now represent a large majority of all species of organisms. Among the features important to the success of arthropods on land are their drought-resistant cuticles and efficient structures for conserving water and exchanging gases with the atmosphere (see chapter 29). Plants undoubtedly colonized the land first because they would have been essential sources of food and shelter for the first arthropods that emerged from the sea.

Vertebrates initiated the third major invasion of the land (figure 17.12). Vertebrates are members of a phylum of animals called the **chordates** (see chapter 29) and include human ancestors. The first of the vertebrates were the amphibians, represented today by such animals as frogs, toads, and salamanders. The earliest amphibians known are from the end of the Devonian period, just over 360 million years ago (figure 17.13). Among their descendants on land are the reptiles. Different groups of reptiles, in turn, ultimately became the ancestors of the dinosaurs (figure 17.14), the birds, and the mammals, as discussed further in chapter 18. The air-breathing lungs of vertebrates; their scales, fur, and feathers, which regulate heat loss; their efficient circulatory and waste-removal systems; and their internal fertilization are all traits that admirably fit vertebrates to life on land.

That all three of these major groups of organisms—plants, arthropods, and vertebrates—colonized the land within a few tens of millions of years of one another is probably related to the development of suitable environmental conditions, such as the formation of a layer of ozone in the atmosphere, which blocked ultraviolet radiation. Ozone (O_3) forms in equilibrium

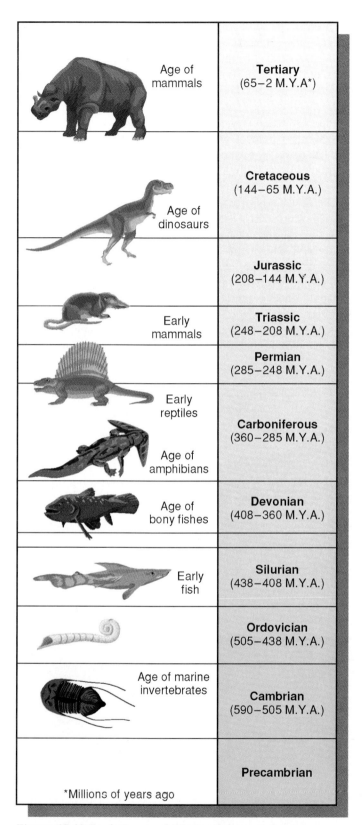

Age of mammals	**Tertiary** (65–2 M.Y.A*)
Age of dinosaurs	**Cretaceous** (144–65 M.Y.A.)
	Jurassic (208–144 M.Y.A.)
Early mammals	**Triassic** (248–208 M.Y.A.)
Early reptiles	**Permian** (285–248 M.Y.A.)
Age of amphibians	**Carboniferous** (360–285 M.Y.A.)
Age of bony fishes	**Devonian** (408–360 M.Y.A.)
Early fish	**Silurian** (438–408 M.Y.A.)
	Ordovician (505–438 M.Y.A.)
Age of marine invertebrates	**Cambrian** (590–505 M.Y.A.)
	Precambrian

*Millions of years ago

Figure 17.12 Evolutionary history of the vertebrates.
Successive groups of vertebrates have flourished for more than four hundred million years of the earth's history.

with oxygen (O_2) and thus was not abundant until the activities of photosynthetic bacteria had elevated the level of oxygen in the atmosphere sufficiently. These conditions have allowed multicellular organisms to exist in terrestrial habitats for more than four hundred million years, about a tenth of the age of the earth.

In addition to the plants, arthropods, and vertebrates, a fourth large group—the fungi, which constitute a distinct kingdom of organisms—also colonized the land, probably at about this same time. The success of the fungi on land might be related to the structure of their cell walls, which are rich in chitin, a water-impermeable substance that also forms the external skeletons of arthropods.

> *Plants and arthropods colonized the land about 410 million years ago. Amphibians, the first terrestrial vertebrates, arrived about fifty million years later. Fungi may also have colonized the land at about the same time as plants. Earlier, there were no terrestrial multicellular organisms.*

Mass Extinctions

One of the most prominent features of the history of life on the earth has been the periodic occurrence of major extinction episodes. Five such events have occurred during the course of geological time. In each of them, a large proportion of the organisms on the earth at that time became extinct. Four of these events occurred during the Paleozoic era, the first of them near the end of the Cambrian period (about 505 million years ago). At that time, most of the existing families of trilobites (see figure 17.2) became extinct. A second major extinction marked the close of the Ordovician period, about 438 million years ago, and a third occurred at the close of the Devonian period, about 370 million years ago.

The fourth and most drastic extinction event happened during the last ten million years of the Permian period, which ended the Paleozoic era, about 238 to 248 million years ago. Approximately 96 percent of all species of marine animals living at that time may have become extinct! The last of the trilobites and many other groups of organisms disappeared forever.

The fifth and most recent major extinction event occurred at the close of the Mesozoic era, sixty-five million years ago, and is discussed in more detail later in the chapter.

> *Five major extinction events have occurred in the history of life on the earth. The extinction event at the end of the Permian period decimated about 96 percent of the species of marine animals.*

Other large-scale extinction events, in addition to the five already noted, also have occurred. In 1983, John Sepkoski and David M. Raup of the University of Chicago reported the surprising correlation that such events appear to occur regularly every twenty-six to twenty-eight million years. The most recent such event occurred about eleven million years ago. The search for a cosmic event, such as periodic comet showers, that could cause major extinction events with such regularity is now under way, as other scientists attempt to test the validity of Sepkoski and Raup's correlation.

The major extinction event that ended the Mesozoic era may be correlated with the impact of a large meteorite, as discussed later in the chapter. Changes in the relative positions of the continents and oceans also have played an important role in extinctions on a regional scale. Numerous other hypotheses have been advanced to explain major extinction events, but scientists generally disagree as to their validity. The extinctions almost certainly had multiple causes. Currently, the activities of human beings are bringing about an episode of mass extinction fully comparable to anything that has occurred in the past (see chapter 24).

The Mesozoic Era

The Mesozoic era, which began about 248 million years ago and ended about 65 million years ago, was a time very different from the present and one of intensive evolution of terrestrial plants and animals.

The major evolutionary lines on land had been established during the mid-Paleozoic, but the evolutionary expansion of these lines—a radiation of novel forms that led to the establishment of the major groups of organisms living today—

occurred in the Mesozoic. Tracing the evolution of these lines requires first considering the events that happened just before the Mesozoic era, in the Permian period (286 to 248 million years ago), a time of drought and extensive glaciation that concluded the Paleozoic era.

As mentioned earlier, the Permian ended with the greatest wave of extinction in the history of life on the earth. In the sea, only about 4 percent of the species survived. The Mesozoic and Paleozoic eras were initially recognized as distinct from each other because the marine animals of the Paleozoic can be recognized instantly as different from those of the Mesozoic.

The few kinds of marine organisms that survived into the Mesozoic included gastropods (a group of mollusks that includes snails and their relatives), bivalves (mollusks such as oysters and clams), crustaceans (a group of arthropods that includes crabs, shrimp, and lobsters), fishes (aquatic chordates), and echinoderms (a phylum that includes starfish and sea urchins). These organisms began to evolve rapidly during the Mesozoic. Some of the new organisms produced had ways of living that were radically different from those of their ancestors. For example, the first efficient burrowers appeared among the

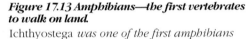

Figure 17.13 Amphibians—the first vertebrates to walk on land.

Ichthyostega *was one of the first amphibians with teleologically efficient limbs for crawling on land and a relatively advanced ear structure for picking up airborne sounds. Despite these features,* Ichthyostega, *which lived about 350 million years ago, was still quite fishlike in overall appearance.*

Figure 17.14 Dinosaurs.
Dinosaurs were remarkably diverse, as shown in this reconstruction from the Peabody Museum, Yale University.

echinoderms. Both on land and in the sea, the numbers of species of almost all groups of organisms have been climbing steadily since the Permian extinction 250 million years ago and are now at all-time highs.

Even though the evolutionary radiation of marine organisms during the Mesozoic was spectacular, the evolution of life on land during this era is of even greater interest, for humans are products of that history. Many of the significant events in the history of the vertebrates occurred during the Mesozoic era.

The History of Plants

The earliest known fossil plants are from about 410 million years ago. By the close of the Paleozoic, plants had become abundant and diverse. Shrubs and then trees evolved and came to form forests. These Carboniferous forests, in turn, formed many of the great coal deposits being consumed today. Much of the land was low and swampy at this time, which provided excellently preserved plant remains. The coal deposits provide a relatively complete record of the horsetails, ferns, and primitive seed-bearing plants that made up these ancient forests.

The last part of the Paleozoic era was cool and dry. The swamps of the Carboniferous forests, which existed at a time of worldwide moist and warm climates, largely disappeared.

The end of the Paleozoic seems to have been ecologically stressful in that many new life-forms originated. One of these groups was the conifers, a group of seed-bearing plants that is currently represented by pines, spruces, firs, and similar trees and shrubs. Today, the descendants of these conifers still form extensive forests in many temperate and subtropical areas. Seed-bearing plants with featherlike leaves, similar to the living group called cycads, became abundant in the Mesozoic era and helped to give that period its nickname—"the age of dinosaurs and cycads."

Ultimately, however, the flowering plants, which apparently originated during the second half of the Mesozoic, became the dominant land plants. The oldest fossils definitely known to be flowering plants are about 127 million years old. It seems likely that the group actually originated somewhat earlier, but no one is certain how much earlier. Like the mammals, the flowering plants were for a long time a minor group; however, they have been more abundant than any other group of plants now for about one hundred million years. The evolution of the flowering plants, which began in the Mesozoic era, has continued strongly to the present. Today, there are about 240,000 species of this large group, which greatly outnumbers all other kinds of plants.

As the flowering plants became more diverse, so did the insects, whose feeding habits were closely linked with the characteristics of the flowering plants; the two groups have evolved together (figure 17.15). Indeed, all groups of terrestrial organisms, including mammals, birds, and fungi, have characteristics that are largely related to those of the flowering plants. These groups now dominate life on land.

The earliest fossil plants, about 410 million years old, evolved into shrubs and then trees to form forests by the end of the Paleozoic era. Conifers evolved during the Permian period (286 to 248 million years ago). Flowering plants, which appeared in the fossil record about 127 million years ago, together with the conifers, have come to dominate the modern landscape.

Extinction of the Dinosaurs

Everyone is generally familiar with the disappearance of the dinosaurs, an event of global importance that took place about sixty-five million years ago, at the end of the Cretaceous period. Less discussed, but actually of more fundamental importance, was the disappearance of many other kinds of organisms at about the same time. Among the **plankton** (free-drifting protists) and other organisms that are still abundant in the sea, many of the larger (but still microscopic) forms suddenly disappeared about sixty-five million years ago, and a much lower number of smaller ones took their place. The same rapid changes occurred in at least some nonplanktonic marine animal groups, such as bivalves. The ammonites, a large and diverse group related to octopuses, but with shells, abruptly disappeared.

In 1980, a group of distinguished scientists, headed by physicist Luis W. Alvarez of the University of California, Berkeley, presented a controversial hypothesis for these drastic changes. Alvarez and his associates discovered that the usually rare element iridium was abundant in a thin layer that marked the end of the Cretaceous period in many parts of the world (figure 17.16). Iridium is rare on earth but common in meteorites. Alvarez and his colleagues proposed that, if a large meteorite, or asteroid, had struck the surface of the earth then, a dense cloud would have been thrown up. The cloud would have been rich in iridium, and as its particles settled, the iridium would have

Figure 17.15 The coevolution of plants and insects.
A blister beetle, Pyrota concinna, *eats the petals of a daisy in Zacatecas, Mexico. Beetles were among the first visitors to the early flowering plants, spreading pollen and thus bringing about cross-fertilization as they flew from flower to flower.*

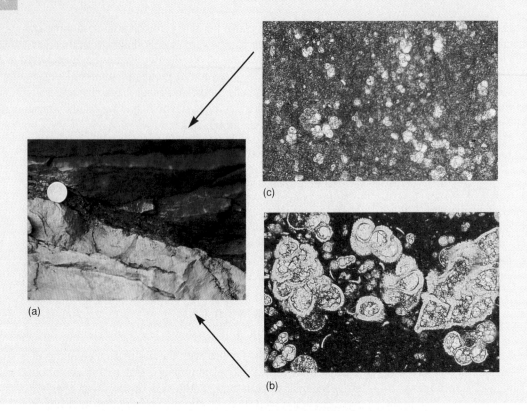

Figure 17.16 The end of the Cretaceous period seen in rocks.

(a) *White limestone (bottom layers) was deposited under the sea in the closing years of the Cretaceous period, and the much darker limestone (upper layers) was deposited during the first years of the Tertiary period. The two types of rock are separated by a layer of greenish clay about 1 centimeter thick (where the coin has been placed), in which iridium is abundant.* (b) *Large, ornamented, diverse foraminifera in sediments were deposited during the late Cretaceous period.* (c) *Small, relatively unornamented, much less diverse foraminifera were deposited a few million years later, during the early years of the Tertiary period. This drop in diversity in the sea was reflected on land in the extinction of all dinosaurs.*

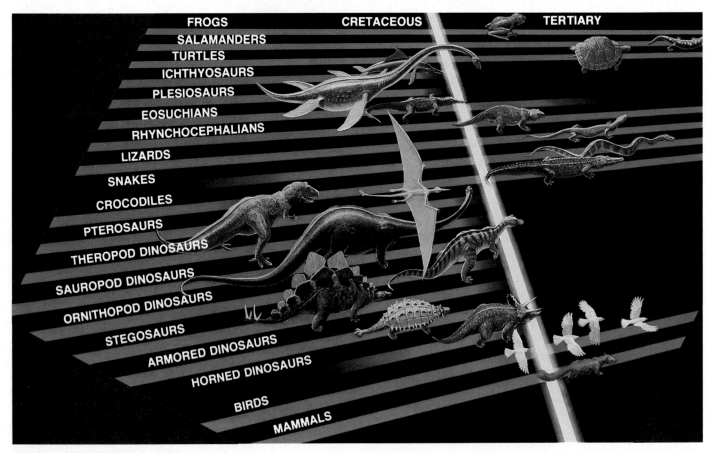

Figure 17.17 Extinction of the dinosaurs.

The dinosaurs became extinct sixty-five million years ago (yellow line) in a major extinction event that also eliminated all the great marine reptiles (plesiosaurs and ichthyosaurs), as well as the largest of the primitive land mammals. The birds and smaller mammals survived and went on to occupy the aerial and terrestrial niches left vacant by the dinosaurs. Crocodiles, small lizards, and turtles also survived, but reptiles never again achieved the diversity of the Cretaceous period.

been incorporated into the layers of sedimentary rock being deposited at that time. By darkening the world, the cloud would have greatly slowed or temporarily halted photosynthesis and driven many kinds of organisms to extinction (figure 17.17).

Calculations have shown that the cloud produced by a meteorite about 20 kilometers in diameter would certainly have caused these effects. Daytime conditions would have resembled those on a moonless night—below the amount of light required for photosynthesis—for several months. In biological communities such as the marine plankton, which are based directly on the continuous production of food by photosynthesis, this would have had serious disruptive effects and could have produced the sudden change seen in the fossil record. Disruption of photosynthesis may also have been responsible for the extinction of certain other kinds of organisms. But is it reasonable to assume that it would have eliminated the dinosaurs?

The Alvarez hypothesis is controversial. It is not clear that dinosaurs became extinct suddenly, as they would have by a meteorite collision. Also, it is not clear whether other kinds of animals and plants show the types of patterns that would have been predicted from the effects of a meteorite collision. Whether or not a meteorite impact caused widespread extinction sixty-five million years ago is a hypothesis that is still under active consideration.

> *The occurrence of a worldwide layer rich in iridium sixty-five million years old suggests that a giant meteorite struck the earth at that time and threw up a huge cloud. The role of this cloud in the extinction of organisms is being actively investigated.*

The Cenozoic Era: The World We Know

This chapter concludes with a brief account of some of the major evolutionary changes of the past sixty-five million years, changes that have resulted in the earth's current conditions.

The relatively warm and moist climates of the early Cenozoic era have gradually given way to today's colder and drier climates. The final separation of South America and Antarctica about twenty-seven million years ago established the Circumantarctic Current and set the stage for worldwide glaciation. The ice mass that formed as a result of this glaciation has made the climate cooler near the poles, warmer near the equator, and drier in the middle latitudes than ever before.

In general, forests covered most of the land area of the continents, except for Antarctica, until about fifteen million years

ago, when the forests began to recede rapidly and modern plant communities appeared. During this time, some of the continents were approaching one another again, after having been widely separated during most of the Cenozoic era.

Organisms in Australia and South America in particular evolved in isolation from all those in the rest of the world, mainly during the Cenozoic era. Evolution in such isolated regions has been responsible for the distinctive characteristics of the groups of plants and animals found in different regions of the world. During the past several million years, the formation of extensive deserts in northern Africa, the Middle East, and India made migration between Africa and Asia very difficult for the organisms of tropical forests. These desert barriers, in turn, provided further opportunities for evolution in isolation.

In general, the overall character of the Cenozoic era has been set by a deteriorating climate, sharp differences in habitats even within small areas, and the regional evolution of distinct groups of plants and animals. These factors have enhanced the opportunities for the rapid formation of many new species by the processes outlined in chapter 16.

> *Throughout the sixty-five million years of the Cenozoic era, the world climate has deteriorated (gotten colder and drier) steadily, and the distributions of organisms have become more and more regional in character.*

Looking Back

Most of the evolutionary processes discussed in this chapter actually have occurred relatively recently. If, as illustrated in figure 3.11, the history of life on earth is viewed as a single twenty-four-hour day, then eukaryotes originated at about 4 P.M. multicellular organisms appeared at about 8 P.M., and the land was invaded just before 10 P.M. Dinosaurs walked the earth from about 10:52 P.M. until about 11:40 P.M., and the whole age of mammals, of which humans are now the dominant product, has taken place in the day's last twenty minutes. Humans have been present on a twenty-four-hour earth for only forty seconds!

For most of the history of life on the earth, bacteria were the only organisms present. The atmospheric changes that they initiated and the symbiotic events in which they participated made possible the evolution of the protists—the first eukaryotes—and eventually, of the animals, plants, and fungi. Overall, the structural complexity and number of species of organisms on earth has greatly increased, but most of the metabolic diversity that exists today was already present in bacterial ancestors two billion years ago.

EVOLUTIONARY VIEWPOINT

Life first appeared on the earth more than 3.5 billion years ago, soon after the molten rock cooled enough to form a hard crust. All of life's early evolution took place in the sea. Life did not spread to land until very late, after photosynthesis by marine algae and bacteria had introduced oxygen to the earth's atmosphere, producing an ozone shield that protected the surface from intense ultraviolet radiation.

SUMMARY

1. Fossils provide a record of life in the past. They occur largely in sedimentary rocks, with the actual organic remains gradually being replaced by minerals. Fossil ages are estimated by determining the proportions of isotopes in the rocks where the fossil is found.

2. The earth originated about 4.6 billion years ago, and the oldest known rocks are about 3.8 billion years old.

3. About two hundred million years ago, the continents were clustered. They have been separating from one another ever since, thereby affecting the nature and distribution of organisms.

4. The major geological eras, with their approximate dates in millions of years ago (MYA), are: Archean (4600 to 2500 MYA), Proterozoic (2500 to 590 MYA), Paleozoic (590 to 250 MYA), Mesozoic (250 to 65 MYA), and Cenozoic (65 MYA to present).

5. Fossil bacteria about 3.5 billion years old are the oldest direct evidence of life on earth. Stromatolites—massive deposits of limestone formed by cyanobacteria—appear in the fossil record starting about 2.8 billion years ago.

6. The first unicellular eukaryotes appeared about 1.5 billion years ago; all earlier forms of life were bacteria. Multicellular animals appeared about seven hundred million years ago. The earliest soft-bodied forms are poorly represented in the fossil record.

7. The fossil record indicates that all extant (current or actually existing) phyla of organisms except plants appear to have evolved during the Cambrian period (590 to 500 million years ago). The evolution of hard skeletons, including shells and similar structures, about 570 million years ago seems to have been a fundamental evolutionary advance that made this evolutionary radiation possible.

8. Plants and terrestrial arthropods appeared about 410 million years ago; terrestrial vertebrates (amphibians) appeared about 360 million years ago. These groups of organisms, together with the fungi, have dominated life on land since then.

9. Five major episodes of mass extinction have occurred during the history of life on the earth. The most drastic was at the end of the Permian period, about 248 million years ago, when some 96 percent of all species of marine animals became extinct. In another extinction event at the end of the Mesozoic era, sixty-five million years ago, the dinosaurs and many other kinds of organisms disappeared.

10. During the Mesozoic, the outlines of life on the earth as it is currently known were established. Flowering plants were dominant by the end of the Mesozoic, and insects, mammals, birds, and other groups had begun to evolve in relation to the diversity of these plants.

11. The mass extinction event at the end of the Cretaceous period may have been caused by a giant meteorite striking the earth. The resulting cloud thrown up by this collision could potentially have darkened the world, slowed photosynthesis, and driven many kinds of organisms to extinction.

12. The Cenozoic era (sixty-five million years ago to the present) is characterized by a deteriorating world climate, sharp differences in habitats even within small areas, and the regional evolution of distinct groups of plants and animals. These factors have led to a steady increase in the number of species.

REVIEWING THE CHAPTER

1. What are fossils, how are they formed, and how are they dated?

2. What is the theory of plate tectonics, and how does it relate to evolution?

3. What were the major events that occurred in biological evolution in each of the five geological eras?

4. What were the first life-forms on the earth, and when did they first appear?

5. When during the Paleozoic was there more intense diversification of marine organisms than ever was the case later?

6. When did animals and plants invade the land? Which group was first and why?

7. In the history of the earth, when did mass extinctions occur?

8. What were the significant biological events during the Mesozoic era?

9. What were the significant events in the history of plants?

10. What is a current hypothesis about how dinosaurs became extinct?

11. What were the significant biological events during the Cenozoic era?

COMPLETING YOUR UNDERSTANDING

1. What kind of biological information do sedimentary rocks provide?

2. Why is the fossil record incomplete?

3. What are the coelacanths, and why are they of interest to evolutionists?

4. Who was Alfred Wegener, and what was Pangaea?

5. Which of the following continents approached one another most recently?
   a. Australia - Antarctica
   b. Europe - North America
   c. India - Asia
   d. North America - South America
   e. South America - Antarctica

6. How does biogeography support plate tectonics?

7. What triggered the onset of widespread continental glaciation (the Ice Age) in the Northern Hemisphere over the past few million years? What were the consequences of biological evolution from this event?

8. What do stromatolites tell us about the past history of life on the earth?

9. Before 1.5 billion years ago, the only forms of life on earth were

a. marine organisms.

b. cyanobacteria.

c. bacteria.

d. viruses.

e. soft-bodied.

10. What is significant about the Cambrian period with respect to biological evolution?

11. Which group—arthropods or mammals—were the most successful group of animals to invade the land and why?

12. What features were necessary in plants for them to survive on land?

13. What were the major groups of vertebrates to invade the land, and in what sequence did these invasions occur?

14. How does the ozone layer relate to biological evolution, and why should we be concerned about the ever-declining ozone shield around planet Earth?

15. When did fungi invade the land, and what is their significance to biological evolution?

16. What is the Alvarez hypothesis, and what does the rare element iridium tell us about the history of the earth?

17. Dinosaurs and mammals both lived throughout the Mesozoic era, a period of more than 150 million years. All this time, the dinosaurs were the dominant form, mammals being a minor group. Both mammals and small reptiles survived the Cretaceous extinction. Why do you suppose that reptiles did not go on to become dominant again, rather than mammals?

18. No iridium layer has been found preserved in the rocks representing four of the five major mass extinctions. How else could you account for these four extinction events?

19. How did the separation of Australia and South America lead to the formation of a huge sheet of ice over Antarctica? Do you think that a land bridge connecting Alaska to Siberia would have had a similar effect, forming a huge Arctic ice mass over the North Pole?

FOR FURTHER READING

Alvarez, W., and F. Asaro. 1990. What caused the mass extinction?—An extra-terrestrial impact. *Scientific American,* Oct., 78–84. A debate on the cause of the extinction of the dinosaurs. The discoverers of iridium deposited at the time argue that a giant meteorite collided with the earth.

Cortillot, V. 1990. What caused the mass extinction?—A volcanic eruption. *Scientific American,* Oct., 85–92. A companion article to the Alvarez/Asaro article. Geologists argue that the iridium came not from a meteorite but from within the earth, blown into the sky by enormous volcanoes produced when plumes in the earth's mantle reached the surface sixty-five million years ago.

Gore, R. 1989. Extinctions. *National Geographic,* June, 662–99. A superbly illustrated article about the events that have changed the nature of life in the past and are continuing at present.

Gould, S. J. 1989. *Wonderful life. The Burgess Shale and the nature of history.* New York: W. W. Norton. A marvelous account of the discovery, study, and interpretation of the world's oldest fossil animals.

Gray, J., and W. Shear. 1992. Early life on land. *American Scientist,* 80 (Sept.): 444–56. Article about minute fossils that offer evidence that life invaded the land millions of years earlier than previously thought.

Lessem, D. 1989. Secrets of the Gobi Desert. *Discover,* June, 40–46. New fossil discoveries that provide evidence about biological interchange between continents in ancient times.

Morell, V. 1993. How lethal was the K-T impact? *Science* 261 (Sept.): 1518–19. New information that the asteroid that hit the earth sixty-five million years ago appears bigger than previously thought. But did it kill the dinosaurs?

Richardson, J. 1986. Brachiopods. *Scientific American,* Sept., 100–106. A fascinating group of marine organisms, well represented in the fossil record, that illustrates many of the kinds of changes that have occurred during the history of life on the earth.

York, D. 1993. The earliest history of the earth. *Scientific American,* Jan., 90–96. Modern techniques of radioactive dating that enable geologists to pry the history of the planet's first billion years from ancient rock.

CHAPTER

18

THE STORY OF VERTEBRATE EVOLUTION

There are far fewer land mammals today than in the past. Two million years ago, many large mammals lived; almost all have disappeared as humans became common over the last twelve thousand years. The few that remain, like these African elephants, are in danger of extinction.

FOR REVIEW

Here are some important terms and concepts that have been discussed in previous chapters and that you will encounter again in this chapter. Review them before proceeding if necessary.

Theory of evolution (*chapter 15*)
How species evolve (*chapter 16*)
Major features of evolutionary history (*chapter 17*)
Origin of mammals (*chapter 17*)

The group of animals to which we belong—the vertebrates—originated in the sea as jawless fishes more than half a billion years ago.

None of these early fishes live today, but they represent a hallmark in evolutionary history, for from them evolved all of the large animals, such as lions and sharks and eagles. Most living vertebrates—over half of the 42,500 species—are fishes. The first vertebrates to walk on land were amphibians, ancestors of today's frogs and toads. Although amphibians evolved from bony fishes about 350 million years ago, they have never fully adapted to life away from water. The first vertebrates to conquer the land completely were amphibians' descendants—reptiles. In the milder climates of the Mesozoic, reptiles were the most successful of all terrestrial vertebrates, dominating the air, land, and sea for 250 million years. From reptiles, in turn, were derived the two most successful terrestrial groups of vertebrates today—the mammals and birds. This chapter reviews the evolutionary journey from jawless fishes to mammals such as ourselves (figure 18.1).

Figure 18.2 shows the evolutionary relationship between the seven major groups, or classes, of vertebrates that have living representatives. Phyla are divided into classes, which, in turn, are divided into smaller categories—called orders—such as rodents, carnivores, and bats among mammals. These orders, in turn, consist of one or more families, such as cats and weasels among the carnivores. Families, in turn, include genera and species. Such a system of classification, in which successively smaller units of classification are included within one another, like boxes within boxes, is called a hierarchical system. It is illustrated further in chapter 25.

General Characteristics of Vertebrates

Vertebrates Are Chordates

Vertebrates are members of the phylum Chordata, approximately 42,500 species whose members are characterized by a flexible rod that develops along the back in embryos. The simplest chordates are tunicates, marine animals that are anchored to one spot as adults and look a bit like jugs. Other chordates include birds, reptiles, amphibians, fishes, and mammals.

Chordates are distinguished by three principal features, not all necessarily present in adult animals, but present during the course of development: (1) a single, dorsal (along the back),

Figure 18.1 A close relative?
In evolutionary terms, this orangutan is thought to be closely related to humans. This possibility has disturbed many people since Darwin first established a logical explanation for evolution—namely, natural selection—in 1859. Although the theory of evolution by natural selection is now accepted by almost all biologists and indeed forms the foundation of modern biology, it still stirs heated debate in nonscientific circles.

hollow **nerve cord,** or main trunk, to which the nerves that reach the different parts of the body are connected; (2) a rod-shaped **notochord,** which forms between the nerve cord and the developing gut (which becomes the stomach and intestines) in the early embryo; and (3) **pharyngeal slits.** The **pharynx** is a muscular tube that connects the mouth cavity and the esophagus in chordates. It serves as the gateway to the digestive tract and to the windpipe, or **trachea.**

Vertebrates Have a Backbone

With the exception of tunicates and a small group of fishlike marine animals—the lancelets, which are considered in chapter 29—all chordates are vertebrates. In vertebrates, the notochord becomes surrounded and then replaced during the course of the embryo's development by a bony **vertebral column,** a tube of hollow bones called vertebrae, which encloses the dorsal nerve cord like a sleeve and protects it. In addition, vertebrates (except for the agnathans [jawless fishes]) have a distinct and well-differentiated head; as a result, they are sometimes called the **craniate chordates** (from the Greek word *kranion,* meaning

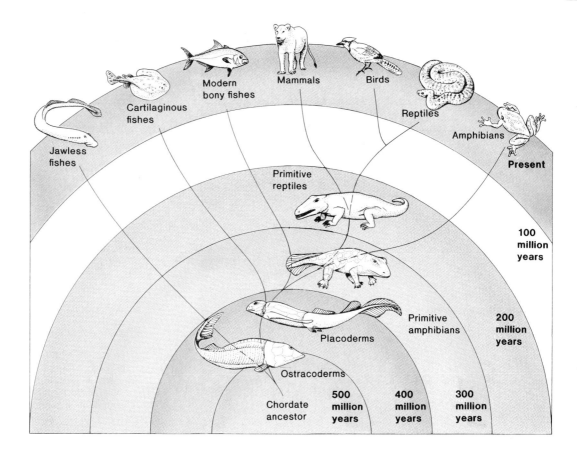

Figure 18.2 The seven classes of vertebrates.

Most of the kinds of primitive amphibians are now extinct, although two groups survive as frogs and salamanders. Primitive reptiles arose from amphibians and gave rise to mammals, to snakes and lizards, and to dinosaurs, which survive today as birds.

"skull"). Most vertebrates have a bony skeleton, although the living members of two of the classes of fishes—Agnatha (lampreys and hagfishes) and Chondrichthyes (sharks and rays)—have a cartilaginous one. In vertebrates, the notochord becomes surrounded and then replaced during the course of embryological development by the vertebral column. But all of the features of chordates, even among the most advanced members of the group, are evident in their embryos (figure 18.3).

> *Vertebrates are a group of chordates characterized by a vertebral column surrounding a dorsal nerve cord.*

Fishes: The First Vertebrates

The first vertebrates were fishes (table 18.1). They evolved in the ocean some 550 million years ago, when seas covered much of what is now land. Their fossils are found on every continent, and their descendants, the modern fishes, are the most numerous and diverse of the vertebrates. In fact, of the approximately 42,500 species of living chordates, about half are fishes. Fishes live in almost every imaginable aquatic habitat—in small puddles, in torrents, and in the perpetual dark thousands of meters beneath the ocean surface. They can endure temperatures varying from hot springs at nearly 40 degrees Celsius to near-freezing polar waters.

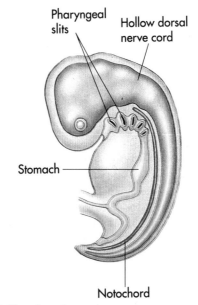

Figure 18.3 Chordate features.

Embryos reveal some of the principal distinguishing features of the chordates, even though these features may be lost during the course of development.

One of evolution's great successes, fishes have had a long history of change (figure 18.4). Ancestral fishes, which were jawless and lacked paired fins, eventually gave rise to five major evolutionary lines, all possessing jaws. Two of these groups are now extinct: the placoderms, which had extensive

Table 18.1 Major Groups of Fishes

Class	Typical Examples		Key Characteristics	Approximate Number of Living Species
Agnatha	Lampreys, hagfishes		Largely extinct group of jawless fishes with no paired appendages; the two surviving groups are parasites	63
Acanthodii	Spiny fishes		First fishes with jaws; all now extinct; paired fins supported by sharp spines	Extinct
Placodermi	Armored fishes		Jawed fishes with heavily armored heads; often quite large	Extinct
Chondrichthyes	Sharks, skates, rays		Streamlined hunters; skeleton of cartilage; no swim bladder; internal fertilization	850
Osteichthyes	Ray-finned fishes		Most diverse group of vertebrates; swim bladders and bony skeletons; paired fins supported by bony rays	18,000
	Lobe-finned fishes		Largely extinct group of bony fishes; ancestral to amphibians; paired lobed fins	7

body armor, and the spiny fishes, bizarre animals with spiny fins. The third group consists of sharks and rays, whose skeletons are made of cartilage. The remaining two groups—the ray-finned fishes and the lobe-finned fishes—have skeletons made of bone. Most living fishes are ray-finned fishes, but a few of the lobe-finned fishes also survive. Among the latter are the coelacanth discussed in chapter 17 and the lungfishes; they are members of the group from which amphibians were derived hundreds of millions of years ago, ultimately giving rise to the reptiles and to all other terrestrial vertebrates, living and extinct.

Jawless Fishes

Jawless fishes, members of the class Agnatha, which appeared about 550 million years ago in the mid-Cambrian period, were the only vertebrates for more than one hundred million years. Agnathans were small and are thought to have fed in a head-down position, their fins acting as stabilizers while their small mouths sucked up organic particles from the bottom.

Many of the major groups of agnathans are extinct. The living ones—the lampreys and hagfishes—belong to a single, fairly uniform group of scaleless, eel-like fishes. In these living agnathans, the notochord persists throughout the animal's life, and

the body is supported by an internal skeleton made of cartilage. But fossils show that the ancestral agnathans had bony skeletons and that such skeletons were lost during the course of evolution.

Lampreys have round mouths that function like suction cups. They use their specialized mouths to attach themselves to the outsides of bony fishes. Once attached, they rasp through the fish's skin with their tongues, which are covered with sharp spines, sucking out the fish's blood through the hole. Sometimes, lampreys are so abundant that they constitute a serious threat to commercial fisheries.

Jawless fishes, the agnathans, were the first vertebrates and for more than one hundred million years, the only vertebrates. Living representatives are the lampreys and hagfishes.

The Evolution of Jaws

Jaws first developed among vertebrates that lived about 410 million years ago, toward the close of the Silurian period. The biting jaws characteristic of modern vertebrates evolved by the modification of one or more of the gill arches, originally

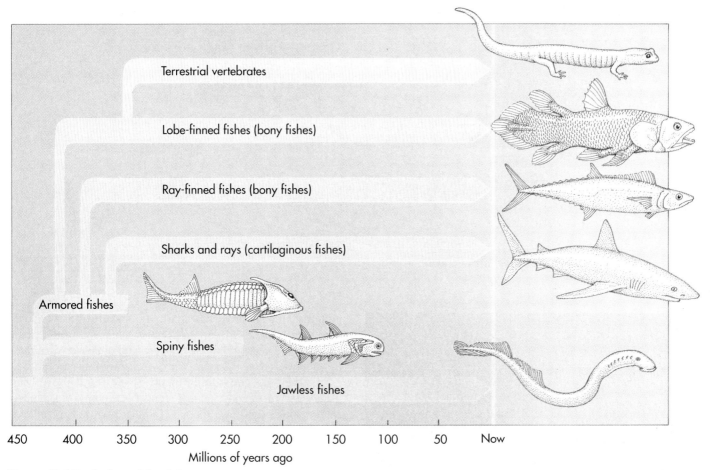

Figure 18.4 Evolution of the fishes.
The great lines of spiny and armored fishes that dominated the early seas are now extinct.

the areas between the gill slits (figure 18.5). Jaws allowed ancient fishes to become proficient predators, to defend themselves, and to collect food from places and in ways that had not been possible earlier. Jawed fishes were able to bite and chew their food instead of sucking it or filtering it in, as did all of the more primitive chordates.

Because of these features, the early jawed fishes largely replaced their jawless ancestors over the course of time. Thus, there are only about sixty-three living species of agnathans, whereas bony fishes have become dominant throughout the world's waters. Of the approximately 42,500 species of living chordates, about half are fishes.

The Early Jawed Fishes

The earliest jawed fishes, the placoderms, were heavily armored and much larger than agnathans—the largest approached 10 meters in length. Placoderms dominated the seas for fifty million years and then became extinct 345 million years ago, replaced by sharks and bony fishes. A second group that developed jaws were the spiny fishes. They had strong spines along the edges of their fins and a torpedo-shaped body with numerous paired fins. All are now extinct.

Sharks and Rays

Sharks appeared about 390 million years ago, probably evolving from placoderm ancestors. Sharks are members of the class Chondrichthyes, which also includes dogfish, skates, and rays (figure 18.6). Many hundreds of extinct species of Chondrichthyes are known from the fossil record, but there are only about 850 living ones. These fishes are mostly scavengers and carnivores. In the living species, the internal skeleton is composed of cartilage—a soft, light, and elastic material. Such skeletons are lighter and more buoyant than those of the kinds of fishes that the Chondrichthyes replaced. In addition to the evolutionary advance represented by such a skeleton, the Chondrichthyes were the first vertebrates to develop large, strong, mobile fins, which enabled them to pursue moving prey by quick adjustment of their motion through the water. With these advantages, they soon largely replaced placoderms throughout the world.

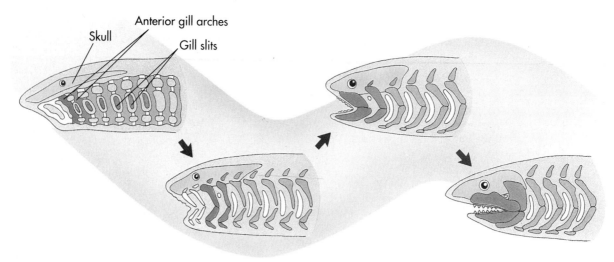

Figure 18.5 Evolution of the jaw.
Jaws evolved from the anterior gill arches of ancient, jawless fishes.

Skull

Anterior gill arches

Gill slits

Figure 18.6 A blue shark.
Sharks, fast-swimming predators, are the tigers of the sea. Like tigers, they are at the top of their food chain and have no natural enemies—except humans.

> *The sharks and other Chondrichthyes were among the first jawed vertebrates and the first to develop efficient fins. Another evolutionary advantage was their development of a cartilaginous skeleton.*

The skin of Chondrichthyes is covered everywhere with small, pointed denticles, similar in structure to the teeth of other vertebrates, which gives their skin a rough, sandpaper-like texture. Their teeth, which are abundant in the fossil record, are enlarged versions of these denticles.

Sharks drive themselves through the water by sinuous motions of the whole body and by their thrashing tails. Such

motions tend to drive the sharks downward, but their two spreading pectoral fins correct for this tendency. In the skates and rays, the pectoral fins are enlarged and undulate when these fishes move, giving the animals a very characteristic appearance.

For their oxygen supply, many sharks (and those bony fishes that swim constantly) depend on a constant stream of water that is forced over the gills and that brings dissolved oxygen with it. Others are able to pump water through their gills while they are stationary. Fishes that obtain their oxygen by moving constantly can literally drown if they are prevented from swimming. Drowned sharks are often found trapped in the nets used to protect Australian beaches.

Bony Fishes

The vast majority of the more than eighteen thousand known species of fishes are bony fishes (class Osteichthyes). The first bony fishes were lobe-finned. This group was once quite abundant, but today, only four genera survive, including the coelacanth (discussed in chapter 17) and the lungfishes. Amphibians evolved from this ancient group of fishes.

The earliest ray-finned fishes, which appeared more than three hundred million years ago, had a strong body armor consisting of heavy, overlapping scales. Fishes of this kind dominated for two hundred million years but were replaced one hundred million years ago by **teleost fishes,** which evolved a unique new way to stay stationary in the water with no effort. Teleost fishes possess a **swim bladder,** a gas-filled sac that allows fishes to regulate their buoyant density and thus offset the greater weight of a bony skeleton over a cartilaginous one (figure 18.7). Fishes with swim bladders can remain suspended at any depth in the water without moving their fins. Swim bladders apparently evolved as outpocketings of the pharynx, specialized for respiration, in fishes of the Devonian period.

Stationary bony fishes do not drown because they use muscles to draw water through their gills by way of their gill slits. These gill slits are simply the pharyngeal slits characteristic of all chordates.

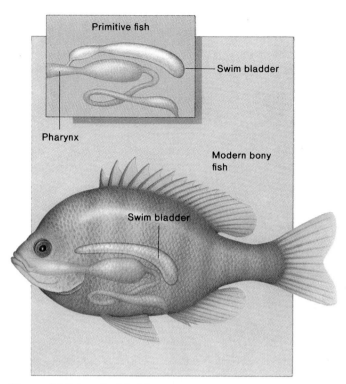

Figure 18.7 A swim bladder.
Bony fishes use swim bladders, which evolved from an outpocketing of the pharynx, to control their buoyancy in water.

Figure 18.8 The bony fishes.
The bony fishes (class Osteichthyes) are extremely diverse. The perch shown here is one of the most advanced—highly maneuverable and very adaptable.

Bony fishes are the most diverse of all groups of vertebrates and comprise about half of all species of living vertebrates (figure 18.8). The first bony fishes evolved in freshwater, as indicated by where early fossils are found and the characteristics of those fossils. Bony fishes apparently entered the oceans only after a number of their distinctive evolutionary lines had appeared. Today, they are abundant both in the modern seas and in freshwater, and many of them spend a portion of their lives in both environments. For example, Atlantic and Pacific salmon are born in the headwaters of rivers. They descend the rivers to the sea, where they lead the bulk of their lives, and then they return to the same rivers to spawn.

Bony fishes (class Osteichthyes) possess swim bladders, which enable them to increase their buoyancy (and thus offset the greater weight of a bony skeleton) and remain suspended in water without moving their fins. Bony fishes account for about half of all living vertebrate species.

Invasion of the Land: Key Adaptations for Survival

About 350 million years ago, one of the lobe-finned fishes took the first steps in the evolutionary journey onto land. This ancient, ancestral fish, which eventually became the ancestor of all **tetrapods** (terrestrial, basically four-limbed vertebrates), probably began to explore a land existence at the margins of freshwater ponds or swamps.

To survive on land, fishes needed several key adaptations. First, in the animals that ultimately became the earliest amphibians, primitive lungs gradually evolved into the efficient air-breathing organs seen in modern tetrapods. Originally, these lungs were supplementary organs that allowed animals to breathe air directly when this was necessary. Lungs of this sort still occur today in the lungfishes.

A second adaptation occurred in the circulation system, where a new blood vessel evolved to carry blood from the lung back to the heart, where the blood was pumped again to circulate it through the body. This adaptation was critical, since in fishes, passage of blood through fish gills slows its flow, so circulation to the rest of the fish body is sluggish (figure 18.9).

A third adaptation was the evolution of walking legs. In early terrestrial vertebrates, efficient locomotion on land was gradually made possible by the evolution of strong skeletal supports in the thoracic portion of the body, that part just behind the head. Such supports provided a more rigid base for the limbs, which were derived from the kinds of fins found in the lobe-finned fishes.

Three evolutionary advances—the abilities to use gaseous oxygen for respiration, to circulate it efficiently through the body, and to move from one place to another on land—became the basis for the success and later evolutionary radiation of vertebrates on land.

Amphibians

The first land vertebrates were amphibians, members of a class that first became abundant about 370 million years ago during the Carboniferous period (figure 18.10). They evolved pectoral

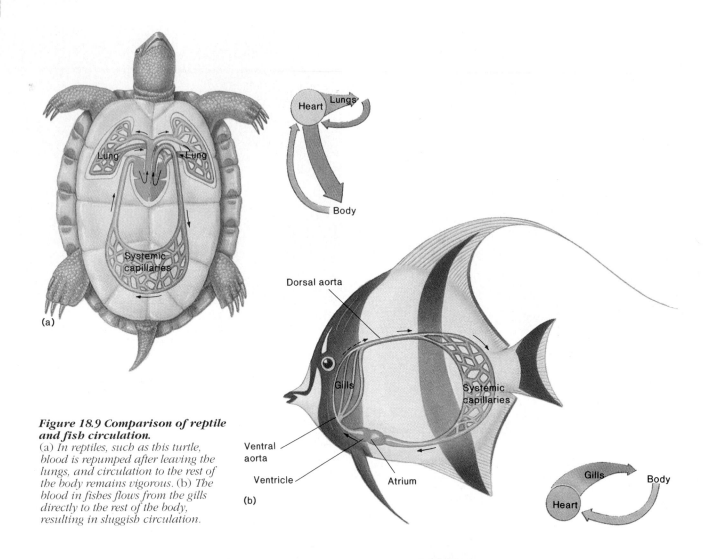

Figure 18.9 Comparison of reptile and fish circulation.
(a) *In reptiles, such as this turtle, blood is repumped after leaving the lungs, and circulation to the rest of the body remains vigorous.* (b) *The blood in fishes flows from the gills directly to the rest of the body, resulting in sluggish circulation.*

and pelvic legs to support their body weight on land (figure 18.11) and a pulmonary vein to send oxygenated blood from the lungs back to the heart for repumping. The early amphibians had rather fishlike bodies, short stubby legs, and lungs. They had scales and dry skin to avoid water loss (the wet skins of today's amphibians evolved much later). The early amphibians often grew to great size, and many had body armor (figure 18.12). These early forms died out completely 160 million years ago. The relatively few forms of amphibians that survive today depend on the availability of water during their early stages of development. Many of the species are quite common in moist places (table 18.2).

The two most familiar orders of living amphibians are: (1) those with tails—the salamanders, mud puppies, and newts, which form the order Caudata, with about 369 species; and (2) those that do not have tails as adults—the frogs and toads, which form the order Anura, with about 3,680 species (figure 18.13). The first true lungs evolved in amphibians, but these organs were relatively inefficient. Respiration also takes place through their moist, glandular skin and through the lining of their mouth. The constant water loss through the skins of amphibians is one reason why these animals generally must remain

Figure 18.10 An early aquatic amphibian.
Diplocaulus was an aquatic amphibian that is thought to have lived in the bottom of the swamps, lakes, and streams that covered Texas in the Permian. Diplocaulus had weak limbs and was one meter in length. Two of the bones at the back of its skull were greatly elongated, producing a triangular-shaped head that may have acted as a hydrofoil, allowing the animal to swim into the current.

in moist habitats. Amphibian larvae and adults of those species that remain permanently in the water respire by means of gills.

Amphibians lay their eggs, which lack water-retaining external membranes and shells and dry out rapidly, directly in water

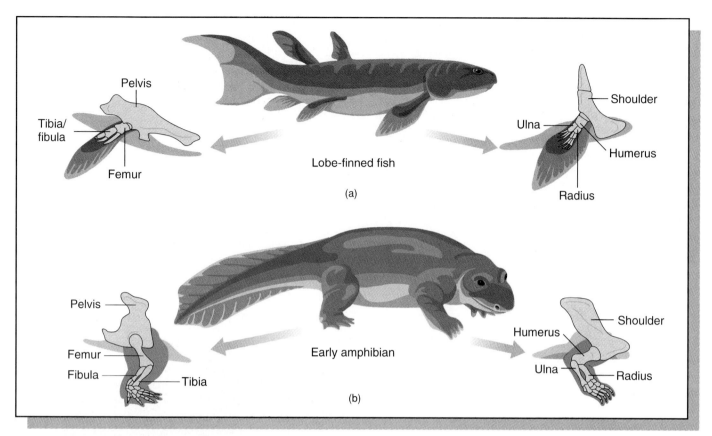

Figure 18.11 Legs evolved from fins.
(a) *The limbs of a lobe-finned fish. Some lobe-finned fishes could move out onto land.* (b) *The limbs of a primitive amphibian. As illustrated by their skeletal structure, the legs of primitive amphibians could clearly function on land better than could the fins of lobe-finned fishes.*

Cacops

Figure 18.12 Cacops, *a large extinct amphibian, had extensive body armor.*

or in moist places. Anuran larvae are tadpoles, which usually live in the water, where they feed on minute algae. The adults, which are highly specialized for jumping and very different from the larvae in appearance, are carnivorous. In contrast to the anurans, young salamanders are carnivorous like the adults and look like small versions of them. Many salamanders swim efficiently and return to water to breed.

> *Amphibians are terrestrial but still require a moist environment. Their eggs are laid in water, which is also where the larvae develop.*

Although amphibians appear primitive, they are, in fact, members of a successful group of animals, one that has survived over three hundred million years. Amphibians evolved long before the dinosaurs and, thus far, have outlasted them by sixty-five million years.

Reptiles

The first reptiles appeared during the age of amphibian dominance—the Carboniferous period, about three hundred million years ago (table 18.3). One of reptiles' most critical adaptations to life on land was the evolution of the **amniotic egg,** which protects the embryo from drying out, nourishes it, and enables

Table 18.2   Orders of Amphibians

Order	Typical Examples		Key Characteristics	Approximate Number of Living Species
Anura	Frogs, toads		Compact tailless body; large head fused to the trunk; rear limbs specialized for jumping	3,680
Caudata	Salamanders, newts		Slender body; long tail and limbs set out at right angles to the body	369
Gymnophiona	Caecilians		Tropical group with a snakelike body; no limbs; little or no tail	160

it to develop outside of water (figure 18.14). Amniotic eggs, characteristic of reptiles and birds (and of the very few egg-laying mammals), retain their own water. They contain a large **yolk,** the primary food supply for the embryo, and abundant albumen, or egg white, which provides additional nutrients and water. An egg membrane, called the **amnion,** surrounds the developing embryo, enclosing a liquid-filled space within which the embryo develops. Blood vessels grow out of the embryo through the egg membrane to the egg's surface, where they take in oxygen and release carbon dioxide. The egg is more easily permeable to these gases than to water. In most reptiles, the eggshell is leathery.

Another adaptation of reptiles to a terrestrial existence, in addition to the amniotic egg, is the presence of copulatory organs, which the male inserts into the female. Copulatory organs protect the eggs and sperm from drying out before their fusion.

> *A key innovation of reptiles for life on land was the water-tight amniotic egg, which contains embryonic nutrients and is permeable to gases but not water. In addition, copulatory organs protected reptiles' eggs and sperm from drying out.*

Body Temperature Regulation

The early reptiles, like amphibians and fishes but unlike birds and mammals today, were **ectothermic** (from the Greek words *ectos,* meaning "outside," and *thermos,* meaning "heat"), regulating their body temperatures by taking in heat from the environment. They developed a wide array of behavioral mechanisms for controlling their internal temperatures with remarkable precision. Even though ectothermic animals are called "cold-blooded," they often maintain bodily temperatures much warmer than their surroundings as a result of their behavior.

(a)

(b)

Figure 18.13 Two kinds of amphibians.
(a) *Red-eyed tree frog,* Agalychnic callidryas. (b) *An adult barred tiger salamander,* Ambystoma tigrinum.

Later reptiles appeared to evolve a means of generating their own body heat. For example, a growing body of evidence indicates that dinosaurs, like the birds and mammals of today, were **endothermic** (from the Greek word *endos,* meaning "inside"), capable of regulating their body temperatures internally. Endothermic animals, which are also sometimes called "warm-blooded" and **homeotherms** (from the Greek word *homios,*

Table 18.3 Orders of Reptiles

Order	Typical Examples		Key Characteristics	Approximate Number of Living Species
Ornithischia	Stegosaur		Dinosaurs with two pelvic bones facing backward, like a bird's pelvis; herbivores, with turtlelike upper beak; legs under body	Extinct
Saurischia	Tyrannosaur		Dinosaurs with one pelvic bone facing forward, the other back, like a lizard's pelvis; both plant-and flesh-eaters; legs under body	Extinct
Pterosauria	Pterodactyl		Flying reptiles; wings were made of skin stretched between fourth finger and body; wingspans of early forms typically 2 feet; later forms over 25 feet	Extinct
Plesiosauria	Plesiosaur		Barrel-shaped marine reptiles with sharp teeth and large paddle-shaped fins; some had snakelike necks twice as long as their body	Extinct
Ichthyosauria	Ichthyosaur		Streamlined marine reptiles with many body similarities to sharks and modern fishes	Extinct
Squamata, suborder Sauria	Gecko		Lizards; limbs set out at right angles to body; anus is in transverse (sideways) slit; most are terrestrial	3,800
Suborder Serpentes	Rattlesnake		Snakes; no legs; move by slithering; scaly skin is shed periodically; most are terrestrial	3,000
Testudines	Turtle		Ancient armored reptiles with shell of bony plates to which vertebrae and ribs are fused; sharp, horny beak without teeth	250
Crocodylia	Crocodile		Advanced reptiles with four-chambered heart and socketed teeth; anus is a longitudinal (lengthwise) slit; closest living relatives of birds	25
Rhynchocephalia	Tuatara		Sole survivors of a once successful group that largely disappeared before the dinosaurs; fused, wedgelike, socketless teeth; primitive third eye under skin of forehead	1

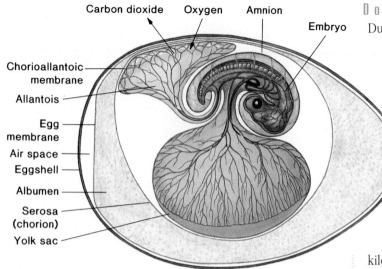

Carbon dioxide   Oxygen   Amnion

Embryo

Chorioallantoic
membrane

Allantois

Egg
membrane

Air space

Eggshell

Albumen

Serosa
(chorion)

Yolk sac

Figure 18.14 Evolution of a watertight egg.
The amniotic egg is perhaps the most important feature that allows reptiles to live in a wide variety of terrestrial habitats.

Figure 18.15 A common desert-dweller.
Lizards are among the most common vertebrates in desert environments. They are superbly adapted to retain water and require far less food than mammals.

meaning "similar"), are able to remain active at night, even if temperatures are cool. They are also able to live at higher elevations and farther north and south than cold-blooded reptiles and amphibians. Endotherms can maintain their internal organs within a very narrow range of temperature more or less independently of external conditions, an adaptation that makes endotherms' maintenance and functioning more efficient than that of ectotherms.

However, these advantages come at a considerable price: An endothermal system requires more energy to maintain. Eighty to ninety percent of the calories in endotherms' food goes to maintain body temperature. A cold-blooded reptile that is the same size as a mammal can therefore subsist on about a tenth of the food required by the mammal. For this reason, cold-blooded reptiles today survive in places like extreme deserts, where food would be too scarce to support many birds and mammals (figure 18.15).

Ectotherms regulate body temperature by taking in heat from the environment. Endotherms can regulate body temperature internally.

Dominant Reptile Groups

During the 250 million years that reptiles were the dominant large land vertebrates, a parade of changes occurred. Four major forms of reptiles took turns as the dominant type: pelycosaurs, therapsids, thecodonts, and dinosaurs (figure 18.16).

Pelycosaurs: Becoming a Better Predator

Early reptiles like **pelycosaurs** were better adapted to life on dry land than amphibians because they had evolved watertight eggs. They had powerful jaws because of an innovation in muscle arrangement: Their jaw muscles were anchored to holes in the skull so they could bite more powerfully (figure 18.17). An individual pelycosaur weighed in the neighborhood of 200 kilograms. With long, sharp, "steak-knife" teeth, pelycosaurs were the first land vertebrates able to kill beasts their own size. Dominant for 50 million years, pelycosaurs once made up some 70 percent of all land vertebrates (figure 18.18). They died out about 250 million years ago, replaced by their direct descendants—the therapsids.

Therapsids: Speeding Up Metabolism

Therapsids ate ten times more frequently than their pelycosaur ancestors, burning the extra food to produce body heat (figure 18.19). They were warm-blooded endotherms, maintaining a constant high body temperature. This permitted therapsids to be far more active than other vertebrates of that time, when winters were cold and long. For twenty million years, therapsids (also called "mammal-like reptiles") were the dominant land vertebrate, until largely replaced 230 million years ago by a cold-blooded reptile line—the thecodonts. Therapsids became extinct 170 million years ago, but not before giving rise to their descendants—the mammals.

Thecodonts: Wasting Less Energy

Thecodonts were cold-blooded—that is, they were ectotherms, like amphibians and early reptiles (figure 18.20). Thecodonts largely replaced therapsids when the world's climate warmed 230 million years ago. The therapsid's endothermy was no longer advantageous in the warmer climate, and cold-blooded thecodonts needed only a tenth as much food! Thecodonts were the first land vertebrates to be **bipedal**—to stand and walk on two feet. They were dominant for fifteen million years, until replaced by their direct descendants—the dinosaurs.

Dinosaurs: Learning to Run Upright

Dinosaurs evolved from thecodonts as bipedal predators about 220 million years ago (figure 18.21), with a significant improvement in body design: Their legs were positioned directly underneath their body. This placed the weight of the body directly over the legs, which allowed dinosaurs to run with great speed and agility. A dinosaur fossil can be distinguished from a thecodont fossil by the presence of a hole in the side of the hip socket. Because the dinosaur leg is

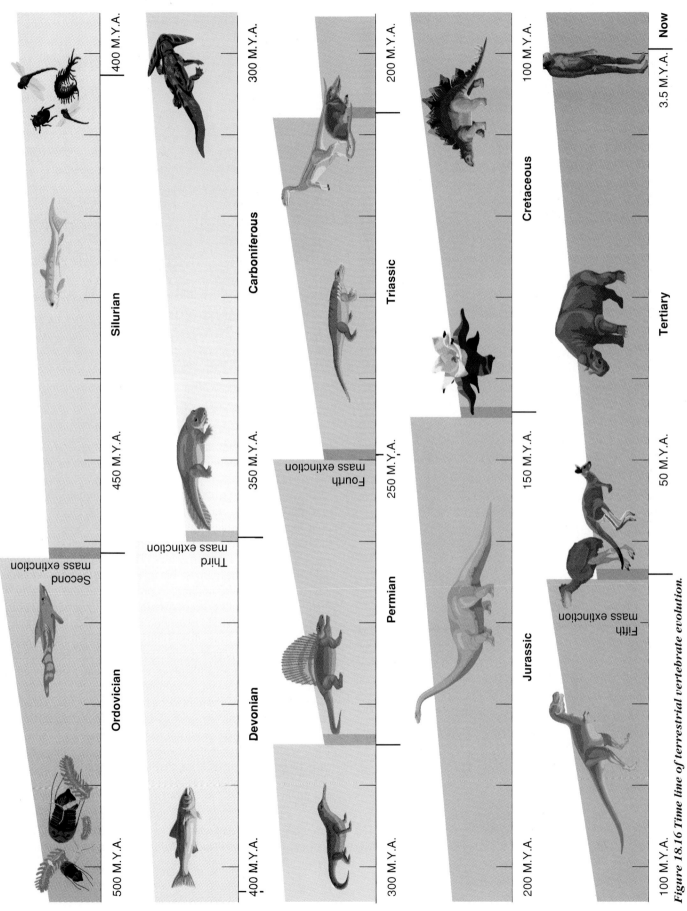

Figure 18.16 *Time line of terrestrial vertebrate evolution.*

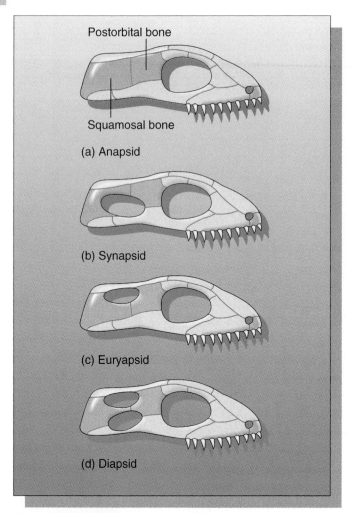

Figure 18.17 Four types of reptiles.
Here reptiles are put in four subclasses according to their skull design. (a) *Anapsids show no hole between postorbital and squamosal bones.* (b) *Synapsids show one hole between and below these bones.* (c) *Euryapsids show one hole between and above these bones.* (d) *Diapsids show two holes between these bones.*

Figure 18.18 Before the early dinosaurs.
The first dominant reptiles were synapsids, like mammals are today. Called pelycosaurs, some were herbivores; others were fierce predators. Many possessed heat-absorbing fins on their backs, leading to the nickname "sailbacks."

positioned underneath the socket, the force is directed upward, not inward, so there was no need for bone on the side of the socket (figure 18.22). Dinosaurs went on to become the most successful of all land vertebrates, dominating for 150 million years (figure 18.23). They disappeared abruptly 65 million years ago. Most scientists think that dinosaurs became extinct as the result of a meteor impact (see chapter 17). Perhaps, the warm-blooded but uninsulated dinosaurs could not survive the intense cold resulting from atmospheric debris blocking the sun.

Birds

Birds are evolutionary descendants of the dinosaurs (table 18.4). The earliest known bird, *Archaeopteryx*, is about 150 million years old. The seven known fossils of *Archaeopteryx*

were all found in a single quarry in southern Germany. In many ways, *Archaeopteryx* is little more than a reptile with feathers. Presumably, its ability to fly was somewhat limited, and it may have been a glider, assisted in gliding by its feathers (figure 18.24).

Birds like *Archaeopteryx* evolved from small, bipedal dinosaurs about 150 million years ago, but were not common until the flying reptiles, called pterosaurs, became extinct with the dinosaurs. Unlike pterosaurs, birds are insulated with feathers, evolved from reptilian scales. Their bones are hollow and light. In all but the earliest birds, the breastbone is enlarged with a keel down the midline, to provide a solid attachment for flight muscles. Unlike dinosaurs (and like thecodonts), birds have a well-developed collarbone (the so-called "wishbone"), which acts as a torque-absorbing strut linking the shoulders and is absolutely necessary for flight. Birds are so structurally similar to dinosaurs in all other respects that many scientists consider birds to be simply feathered dinosaurs.

The modern descendants of *Archaeopteryx* and other ancient birds are true masters of the air (figure 18.25). Their wings are forearms, modified over the course of evolution. Their scaly skin has feathers—flexible, strong, replaceable organs that form an excellent airfoil (a surface, such as a wing or rudder, that obtains resistance from the air through which it moves) for flying. The membranes that make up the flying surface of a bat or one of the extinct groups of flying reptiles can be severely damaged by a single rip. In contrast, the feathers of birds are individually replaceable. Even after the

Figure 18.19 Therapsids.
The direct descendants of pelycosaurs, these synapsids are the immediate ancestors of mammals. Many grew to be quite large. Therapsids walked on all fours, but adopted a more upright stance than pelycosaurs. Therapsids had complex teeth and a secondary palate so they could breathe and chew at the same time.

Figure 18.20 Thecodonts.
These diapsid reptiles were the first land vertebrates to be bipedal—to stand and walk on two feet. Most early thecodonts resembled crocodiles, but later forms stood upright. They were dominant for fifteen million years until they were replaced by their direct descendants, the dinosaurs.

loss of many individual feathers, a bird can keep flying. Several hundred microscopic hooks along the sides of the individual barbules of a feather attach the barbules to one another. These hooks unite the feather so that, in a marvelous way unique to birds, a feather is magnificently adapted for

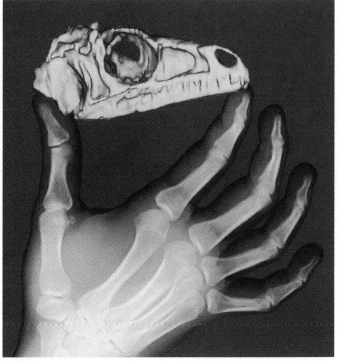

Figure 18.21 Eoraptor.
The earliest dinosaur known, Eoraptor, *was discovered in 1988 in 228-million-year-old Triassic rock in Argentina.*

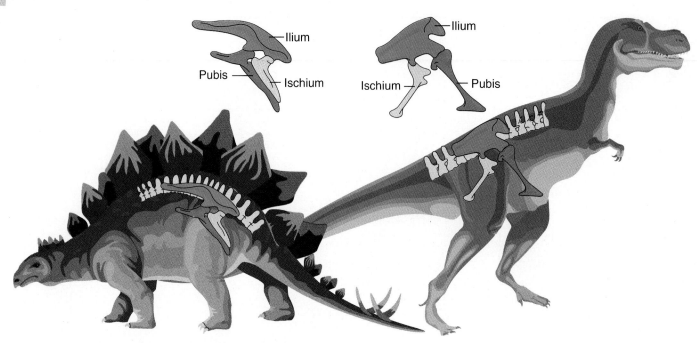

Figure 18.22 The two orders of dinosaurs differ in their hips.
Ornithischian ("bird-hipped") dinosaurs like Stegosaurus had a
rear-pointing pubis bone, while Saurischian ("lizard-hipped")
dinosaurs like the flesh-eating theropods had a forward-pointing
pubis. Birds evolved from lizard-hipped dinosaurs.

Figure 18.23 Many dinosaurs were very large.
These Chinese sauropods, Mamenchisaurus houchuanensis, were
relatives of the American diplodocus, and had a 33-foot-long neck!

Table 18.4   Major Orders of Birds

Order	Typical Examples	Key Characteristics	Approximate Number of Living Species
Passeriformes	Crows, mockingbirds, robins, sparrows, starlings, warblers	*Songbirds* Well-developed vocal organs; perching feet; dependent young	5,276 (Largest of all bird orders; contains over 60 percent of all bird species)
Apodiformes	Hummingbirds, swifts	*Fast fliers* Short legs; small bodies; rapid wing beat	428
Piciformes	Honeyguides, toucans, woodpeckers	*Woodpeckers or toucans* Grasping feet; chisel-like, sharp bills that can break down wood	383
Psittaciformes	Cockatoos, parrots	*Parrots* Large, powerful bills for crushing seeds; well-developed vocal organs	340
Charadriiformes	Auks, gulls, plovers, sandpipers, terns	*Shorebirds* Long, stiltlike legs; slender, probing bills	331
Columbiformes	Doves, pigeons	*Pigeons* Perching feet; rounded, stout bodies	303
Falconiformes	Eagles, falcons, hawks, vultures	*Birds of prey* Carnivorous; keen vision, sharp, pointed beaks for tearing flesh; active during the day	288
Galliformes	Chickens, grouse, pheasants, quail	*Gamebirds* Often limited flying ability; rounded bodies	268
Gruiformes	Bitterns, coots, cranes, rails	*Marsh birds* Long, stiltlike legs; diverse body shapes; marsh-dwellers	209
Anseriformes	Ducks, geese, swans	*Waterfowl* Webbed toes; broad bill with filtering ridges	150
Ciconiiformes	Herons, ibises, storks	*Waders* Long-legged; large bodies	114
Strigiformes	Barn owls, screech owl	*Owls* Nocturnal birds of prey; strong beaks; powerful feet	146
Procellariiformes	Albatrosses, petrels	*Seabirds* Tube-shaped bills; capable of flying for long periods of time	104
Sphenisciformes	Emperor penguins, crested penguins	*Penguins* Marine; modified wings for swimming; flightless; found only in Southern Hemisphere; thick coat of insulating feathers	18
Dinornithiformes	Kiwis	*Kiwis* Flightless; small; primitive; confined to New Zealand	2
Struthioniformes	Ostrich	*Ostrich* Powerful running legs; flightless; only two toes, very large	1

Figure 18.24 Archaeopteryx.
Artist's reconstruction of Archaeopteryx, *an early bird about the size of a crow. Closely related to its ancestors among the bipedal dinosaurs,* Archaeopteryx *lived in the forests of central Europe 150 million years ago. The teeth and long, jointed tail are features not found in any modern birds. Discovered in 1862,* Archaeopteryx *was cited by Darwin in support of his theory of evolution. The true feather colors of* Archaeopteryx *are not known.*

flight (figure 18.26). Overall, flight in birds is made possible by their light, hollow bones; the replacement of scales with feathers; and the development of highly efficient lungs that supply the large amounts of oxygen necessary to sustain muscle contraction during prolonged flight.

Flight has evolved four times since animals first invaded the land. Insects were the first to fly and remain by far the most numerous fliers today. They were followed by the pterosaurs—the diverse, flying relatives of the dinosaurs who flew for millions of years—longer than birds have been flying. Birds and bats are latecomers to the air and fill the ecological niche once occupied by pterosaurs. Birds began to become large and diverse about ninety to sixty-five million years ago. There are approximately nine thousand species of living birds.

> *Birds evolved from reptiles and, with the bats, are one of the two living vertebrate groups that have achieved full mastery of the air.*

Mammals

Mammals evolved from therapsids at the same time that the dinosaurs first appeared, about two hundred million years ago (table 18.5). Throughout dinosaurs' 150-million-year reign, all mammals were small, the largest no bigger than a cat. When the dinosaurs disappeared and world climates turned progressively colder, mammals became the world's dominant large land animal, as they still are.

Mammals enjoyed their maximum diversity fifteen million years ago. Since then, world climates have deteriorated, culminating in the great ice ages of the last several million years, during which many very large mammals evolved—all of which disappeared as humans became common over the last twelve thousand years. Most mammals living today are small, no larger today than during the reign of the dinosaurs. In fact, 71 percent of today's mammal species are rodents, bats, or insectivores, most of them small enough to hold in your hand.

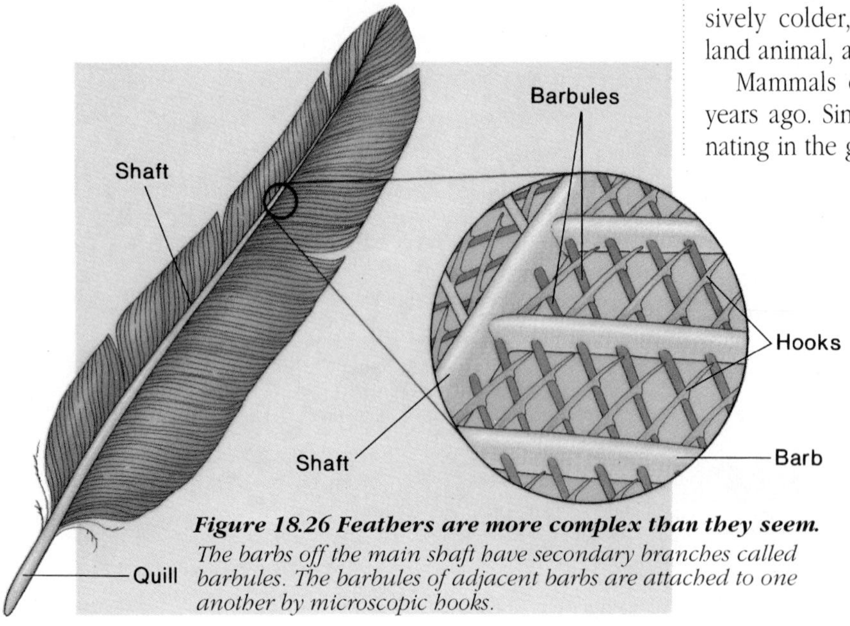

Figure 18.25 Masters of flight.
Most modern birds, such as this Caspian tern, are expert fliers. Their feathers, which provide an efficient and easily-controlled aerodynamic surface, evolved from reptilian scales.

Shaft

Barbules

Shaft

Hooks

Barb

Quill

Figure 18.26 Feathers are more complex than they seem.
The barbs off the main shaft have secondary branches called barbules. The barbules of adjacent barbs are attached to one another by microscopic hooks.

Table 18.5 Major Orders of Mammals

Class	Typical Examples		Key Characteristics	Approximate Number of Living Species
Rodentia	Beavers, mice, porcupines, rats, squirrels		*Small plant-eaters* Chisel-like incisor teeth	1,814
Chiroptera	Bats		*Flying mammals* Primarily fruit- or insect-eaters; elongated fingers; thin wing membrane; nocturnal; navigate by sonar	986
Insectivora	Moles, shrews		*Small, burrowing mammals* Insect-eaters; most primitive placental mammals; spend most of their time underground	390
Marsupialia	Kangaroos, koalas		*Pouched mammals* Have abdominal pouch for young	280
Carnivora	Bears, cats, dogs, weasels, wolves		*Carnivorous predators* Teeth adapted for shearing flesh; no native families in Australia	240
Primates	Apes, humans, lemurs, monkeys		*Tree-dwellers* Large brain size; binocular vision; opposable thumb; end product of a line that branched off early from other mammals.	233
Artiodactyla	Cattle, deer, giraffes, pigs, sheep		*Hoofed mammals* With two or four toes; herbivores	211
Cetacea	Dolphins, porpoises, whales		*Fully marine mammals* Streamlined bodies; front limbs modified into flippers; no hind limbs; blowholes on top of head; no hair except on muzzle	79
Lagomorpha	Rabbits, hares, pikas		*Rodentlike jumpers* Four upper incisors (rather than the two seen in rodents); hindlegs often longer than forelegs, an adaptation for jumping	69
Pinnipedia	Sea lions, seals, walruses		*Marine carnivores* Feed mainly on fish; limbs modified for swimming	34
Edentata	Anteaters, armadillos, sloths		*Toothless insect-eaters* Many are toothless, but have some degenerate, peglike teeth	30
Perissodactyla	Horses, rhinoceroses, zebras		*Hoofed mammals with one or three toes* Herbivorous teeth adapted for chewing	17
Proboscidea	Elephants		*Long-trunked herbivores* Two upper incisors elongated as tusks; largest living land animal	2

Characteristics of Mammals

There are about forty-five hundred species of living mammals (class Mammalia), including human beings. Like their therapsid ancestors, mammals are endotherms and devote 90 percent of their food energy to making heat. Their skin is covered with hair or fur during at least some stage of their life cycle. Like birds and crocodiles, mammals have a four-chambered heart with complete double circulation—that is, oxygen-rich and oxygen-poor blood circulate in separate systems. Mammals have copulatory organs like their reptilian ancestors, and fertilization is internal. Female mammals nourish their young with milk, a nutritious substance produced in the mother's mammary glands. Mammals' locomotion is advanced over that of reptiles because mammalian legs are positioned much farther under the body and are suspended from limb girdles, which permit greater leg mobility.

(a)

(b)

Figure 18.27 Monotremes.

(a) *Duck-billed platypus,* Ornithorhynchus anatinus, *at the edge of a stream in Australia.* (b) *Spiny echidna,* Tachyglossus aculeatus, *a desert-dweller unrelated to porcupines, which it resembles.*

Mammals nourish their young with milk, are covered with hair rather than scales or feathers, and have superior locomotion.

Monotremes

Most mammalian young are not enclosed in eggs when they are born. The only exceptions are the **monotremes,** or egg-laying mammals, which consist of the duck-billed platypus and the echidna (or spiny anteater) (figure 18.27). These animals occur only in Australia and New Guinea and are not known from the fossil record, although they must be an ancient group. In echidnas, the eggs are transferred to a special pouch, where they are held until the young hatch.

Marsupials

The **marsupials** are mammals in which the young are born early in their development, sometimes as soon as eight days after fertilization, and are retained in a pouch, or marsupium

(figure 18.28). Living marsupials are found only in Australia, where they are abundant and diverse, and in North and South America. Ancient fossil marsupials, forty to one hundred million years old, are found in North America, where they became extinct until their geologically recent reintroduction, which occurred after North and South America were linked three to four million years ago. The opossum, a familiar mammal in North America, arrived from the south in just this way.

Placental Mammals

Most modern mammals are **placental mammals.** The first organ to form during the course of their embryonic development is the placenta. The **placenta** is a specialized organ, held within the mother's womb, across which she supplies the offspring with food, water, and oxygen and through which she removes wastes. Both fetal and maternal blood vessels are abundant in the placenta, and substances can thus be exchanged efficiently between the bloodstreams of the mother and her offspring. In placental mammals, unlike marsupials, the young undergo a considerable period of development before they are born.

Some mammals (monotremes) lay eggs, others (marsupials) give birth to embryos that continue their development in pouches (as do monotreme eggs), and still others (placental mammals) nourish their developing embryos within the mother's body by means of a placenta until development is almost complete.

Placental mammals are extraordinarily diverse (figure 18.29). One evolutionary line of mammals—the bats—has joined insects and birds as the only group of living mammals that truly flies. Another group of mammals—the Cetacea (whales and porpoises)—has reverted to an aquatic habitat like that from which mammalian ancestors came hundreds of millions of years ago. Among the earliest of placental mammals to develop were the primates, from which we humans developed. The story of human evolution is examined in the next chapter.

Figure 18.28 A marsupial.
Kangaroo with young in its pouch.

Figure 18.29 A placental mammal.
This young Bengal tiger, Panthera tigris tigris, *will quickly grow to become an efficient and powerful predator. Tigers are not typical placental mammals because they are so large. More than half of all mammal species are small enough to fit in a person's hand.*

EVOLUTIONARY VIEWPOINT

The terrestrial vertebrates living today are only a dim reflection of yesterday's triumphs. Today's moist-skinned amphibians give little clue to their dry-skinned, armored ancestors. Who would guess, looking at today's reptiles, that warm-blooded reptiles once dominated the land, only to be driven extinct by cold-blooded ones? The history of terrestrial vertebrates is a long parade of change, driven by a changing environment. The "optimal" vertebrates changed as continents moved and conditions changed.

SUMMARY

1. Chordates are characterized by a single, dorsal, hollow nerve cord; a flexible rod, the notochord, that forms on the dorsal side of the gut in the early embryo; and pharyngeal slits.

2. Vertebrates compose most groups of the phylum Chordata. They differ from other chordates in that they usually possess a vertebral column, a distinct and well-differentiated head, and a bony skeleton.

3. Members of the class Agnatha differ from other vertebrates in that they lack jaws. Once abundant and diverse, they are represented among the living vertebrates only by the lampreys and the hagfishes.

4. The two living classes of fishes other than Agnatha have jaws, as do the members of the other four classes of vertebrates. Jawed fishes constitute about half of the estimated 42,500 species of vertebrates and are dominant in freshwater and saltwater everywhere. Of the two classes of jawed fishes, the Chondrichthyes, or cartilaginous fishes, consist of about 850 species of sharks, rays, and skates; the Osteichthyes, or bony fishes, consist of about 18,000 species.

5. Key adaptations to survival on land were the development of lungs, a more efficient circulation system, and walking legs.

6. The first land vertebrates were amphibians, one of the four classes of tetrapods (basically, four-limbed vertebrates). Amphibians depend on water and lay their eggs in moist places. In many species, the larvae also develop in water.

7. Reptiles were the first vertebrates fully adapted to terrestrial habitats. Amniotic eggs, which evolved in this group but are also characteristic of the birds and the very few egg-laying mammals, represent a significant adaptation to the widespread dry conditions on land.

8. Early reptiles were ectothermic—that is, their body temperatures were set by external influences. Later reptiles appear to have become endothermic, evolving means of regulating their body temperatures internally. Birds and mammals are endothermic. With few exceptions, all other living animals are ectothermic.

9. Birds evolved from dinosaurs. Birds' light, hollow bones, feathers, and highly efficient lungs have allowed birds to achieve full mastery of the air.

10. Mammals evolved from therapsids, and along with the birds, are the world's dominant land vertebrates today. Monotremes are egg-laying mammals, while marsupials are mammals that carry their young in pouches during development. Placental mammals nourish developing embryos within the mother's body.

REVIEWING THE CHAPTER

1. What are the general characteristics of vertebrates?

2. What were the first vertebrates?

3. When did jawed fishes appear?

4. What characterizes bony fishes?

5. What adaptations were necessary for animals to get from water to land?

6. How are amphibians distinguished?

7. How are reptiles characterized?

8. What two types of body temperature regulation have evolved?

9. How were therapsids different from pelycosaurs?

10. Why did cold-blooded thecodonts replace endothermic therapsids?

11. How and when did dinosaurs evolve?

12. How are birds characterized?

13. How are mammals distinguished?

14. How are monotremes identified?

15. What typifies marsupials?

16. How are placental mammals distinguished?

COMPLETING YOUR UNDERSTANDING

1. Of the living species of chordates, approximately what percent are fishes?

2. What is a lamprey, and why would a Great Lakes fisherman be concerned about them?

3. How and when did the biting jaw originate?

4. What were placoderms, when did they live, and what replaced them?

5. When did sharks first appear, and what characterizes them?

6. What prevents stationary bony fishes from drowning?

7. What is a tetrapod, and how do tetrapods relate to the evolution of primates?

8. Which animals belong to the orders Caudata and Anura?

9. Which order of dinosaurs is characterized by terrestrial flesh-eaters?

    a. Ornithischia   d. Pterosauria

    b. Saurischia   e. Plesiosauria

    c. Gymnophiona f. Ichthyosauria

10. Which of the following are not reptiles? [*two answers required*]

    a. Dinosaurs   d. Lampreys

    b. Turtles   e. Alligators

    c. Lizards   f. Salamanders

11. Which groups of animals are ectothermic, and which are endothermic?

12. When did therapsids become extinct, and what were their descendants?

13. What and when were the first bipedal animals?

14. How long did dinosaurs live, how did they originate, and do they have living descendants?

15. What is *Archaeopteryx,* and when did it live?

16. How do birds differ from pterosaurs and bats, and what is the origin of feathers?

17. What is the mechanical mechanism of flight in birds?

18. Why did it take so long for mammals to become the world's dominant animals?

19. Why do you think the majority of today's mammals are so small?

20. Which factors are responsible for a decline in the diversity of mammals over the past fifteen million years?

21. Which of the following groups of animals have (past or present) representatives with flight? [*four answers required*]

    a. Insects   d. Fishes

    b. Reptiles   e. Birds

    c. Amphibians f. Mammals

22. Which of the following orders are hoofed mammals? [*two answers required*]

    a. Artiodactyla   d. Primates

    b. Perissodactyla e. Chiroptera

    c. Rodentia   f. Pinnipedia

23. Why are monotremes and marsupials so restricted geographically?

24. How and when did the opossum arrive in North America?

25. What is the origin of whales and porpoises?

*26. Of the 42,500 species of living chordates, twice as many (about 19,000 fishes) live in the sea as on the surface of the land (about 6,000 reptiles and 4,500 mammals). Why do you think there are so many more species of chordates in the sea than on land?

*27. Flying reptiles were a very diverse group during the age of dinosaurs. However, although the present age might be considered the age of mammals, bats (the only flying mammals) exhibit far less diversity than did flying reptiles. Can you suggest an explanation for this?

*For discussion

FOR FURTHER READING

Bakker, R. T. 1986. *The dinosaur heresies*. New York: William Morrow. A fine discussion of the controversy about whether dinosaurs could control their own temperatures internally. Well-written and beautifully illustrated.

Czerkas, S., and S. Czerkas. 1991. *Dinosaurs—A global view*. New York: Mallard Press. A detailed, historical account of terrestrial vertebrate evolution. Highly recommended.

Eastman, J., and A. De Vries. 1986. Antarctic fishes. *Scientific American,* Nov., 106–14. Fascinating account of why fishes do not freeze in the Antarctic seas, even though ice forms in the oceans in which they swim. These biologists showed that the fishes produce their own potent antifreeze.

Horner, J., and J. Gorman. 1988. *Digging dinosaurs*. New York: Harper and Row. A highly readable account of how maternal care in dinosaurs was discovered. A good look at how research into dinosaurs is really done.

Norman, D. 1991. *Dinosaur!* New York: Prentice-Hall. A splendid introduction to the general topic of dinosaurs. Based on a popular television series, but far superior.

O'Brien, J. 1987. The ancestry of the giant panda. *Scientific American,* Nov., 102–7. A clear account of how modern molecular techniques are being used to solve a famous evolutionary puzzle: Is the panda a raccoon or a bear?

Spalding, D. 1993. *Dinosaur hunters*. Rocklin, Calif.: Prima Publishing. A lively account of the eccentric amateurs and obsessed professionals who "found" the dinosaurs.

Walker, A., and M. Teaford. 1989. The hunt for proconsul. *Scientific American,* Jan., 76–82. An exciting account of the last common ancestor of great apes and human beings.

Wallace, J. 1989. New discoveries about dinosaurs. *Science Year,* 42–76 (annual *World Book* supplement). New evidence that has convinced most scientists that many dinosaurs, once considered dull and plodding, were actually agile, fast-moving, and even social.

Wellnhofer, P. 1990. Archaeopteryx. *Scientific American,* May, 70–77. New information that is continuing to provide insight into the evolution of flight in birds.

HOW HUMANS EVOLVED

Humankind's closest living relatives are the African apes, chimpanzees, and gorillas, like this mountain gorilla from Rwanda. All of these African apes are rare, largely due to human encroachment of their habitat. Unless care is taken to preserve them from extinction, we will destroy this key link to our biological past.

FOR REVIEW

Here are some important terms and concepts that have been discussed in previous chapters and that you will encounter again in this chapter. Review them before proceeding if necessary.

Darwin's theory of evolution by natural selection (*chapters 1 and 15*)
DNA (*chapter 12*)
Creationism (*chapter 15*)
Natural selection (*chapter 15*)

In his book *The Descent of Man* (1871), Darwin suggested that humans evolved from ancestors of African apes—the gorilla and the chimpanzee. Although little fossil evidence existed in 1871 to support Darwin's case, numerous fossil discoveries since Darwin's death strongly support his hypothesis (figure 19.1). Human evolution is the part of the evolution story that most interests people and about which the most is known. This chapter describes the evolutionary journey that has led to humans. It is an exciting story, replete with controversy.

The idea that humans evolved from apelike ancestors has proven very controversial. Many people have refused to accept Darwin's theory because it views humanity as the product of a long evolutionary journey, rather than as the immediate product of Divine creation. In the United States, that controversy continues within the religious community. In the 1920s, several states passed laws making it illegal to teach Darwin's theory of evolution in public school classrooms, laws that were on the books and enforced till 1968, when the U.S. Supreme Court declared them unconstitutional. Even today, over twenty states have policies permitting local school districts to include creationism (the idea that humans are the direct product of Divine creation) as an alternative to evolution in biology courses, although laws actually requiring "equal time" were ruled unconstitutional in 1986. Human evolution thus remains controversial—not because of scientific doubt (the scientific evidence for human evolution is clear and compelling) but because of disagreement about where science stops and religion begins.

The Evolution of Primates

The story begins about eighty million years ago, during the age of dinosaurs, with the appearance of tree shrews—big-eyed mammals about the size of a fist that lived in trees and caught insects. Tree shrews were the ancestors of the first primates.

The Earliest Primates

The small, nimble tree shrews gave rise forty million years ago to larger animals called **primates.** Evolution, selecting for changes that made tree-living insect-catchers better and better at stalking their insect prey along slender branches, resulted in two distinct improvements in primates:

Figure 19.1 The trail of our ancestors.
These fossil footprints, made in Africa 3.7 million years ago, look like they might have been left by a mother and child walking on the beach. But these tracks are not human. Preserved in volcanic ash, they record the passage of two individuals of the genus Australopithecus, *the group from which our genus,* Homo, *evolved.*

1. *Grasping fingers and toes.* Unlike the clawed feet of tree shrews and squirrels, primates had grasping hands and feet that let them grip limbs, hang from branches, seize food, and use tools.

2. *Binocular vision.* Unlike the eyes of shrews and squirrels, which sit on each side of the head so that the two fields of vision do not overlap, the eyes of primates are shifted forward to the front of the face, producing overlapping **binocular vision** that lets the brain judge distance precisely.

Other mammals have binocular vision, but only primates have both binocular vision *and* grasping hands. These characteristics guided the evolution of increased intelligence that became the hallmark of primates.

The first primates were **prosimians** ("before monkeys"), which looked something like a cross between a squirrel and a cat (figure 19.2). By thirty-eight million years ago, prosimians were common in North America, Europe, Asia, and Africa. Only a few survive today, principally on the island of Madagascar about 300 miles off the east coast of Africa. Madagascar is about twice the size of the state of Arizona and is home to all of the roughly forty surviving species of lemurs, a prosimian the size of a tomcat with a long tail for balancing. Lemurs are **nocturnal**—that is, they hunt insects at night and sleep during the day. Today, the tropical rain forests of Madagascar are being rapidly destroyed by an expanding human population. As the lemur's forest home disappears, humans' oldest living link to their past may soon be extinct in the wild.

Monkeys: Day-Active Primates

About thirty-six million years ago, a revolution occurred in how primates lived: They became **diurnal**—that is, active during the day. These new day-active primates called **monkeys,** appear to have replaced prosimians rather rapidly.

Monkeys' diurnal activities had far-reaching effects. Because vision is much more important for daytime hunting, evolution favored many improvements in eye design, including the appearance for the first time of cones in the retina—color vision. The improved senses became governed by an expanded brain. Feeding mainly on fruits and leaves rather than on insects, monkeys were the first primates with an **opposable thumb**—one that stood out at an angle from the other digits and could be bent back against them to grasp an object or tool. Monkeys live in groups with complex social interactions and tend to care for their young for prolonged periods. This long childhood

Figure 19.2 Prosimians.
Ringtail lemur, Lemur catta. All living lemurs are found on the island of Madagascar, and all are in danger of extinction as the rain forests there are destroyed.

of learning seems to be a necessary part of the development of the large brains of higher primates.

Monkeys first evolved in Central Africa and are still common there. Some migrated early on to South America, where they developed in isolation. Their South American descendants today are easy to identify, since unlike African monkeys, they grasp objects with long prehensile tails.

Apes: The Path to Humanity

Humans did not evolve from monkeys, but from another kind of primate that evolved independently from prosimian ancestors about twenty-five million years ago—**apes** (figure 19.3). Apes have larger brains than monkeys, and none of them have tails. With the exception of the gibbon, which is small, all living apes are larger than any monkey. Apes exhibit the most adaptable behavior of any mammal except human beings. Together with the **hominids** (human beings and their direct ancestors), apes make up a group called the **hominoids.**

Figure 19.3 Apes.
(a) *Gorilla,* Gorilla gorilla.
(b) *Mueller gibbon,* Hylobates muelleri. (c) *Chimpanzee,* Pan troglodytes. (d) *Orangutan,* Pongo pygmaeus.

(a)

(b)

(c)

(d)

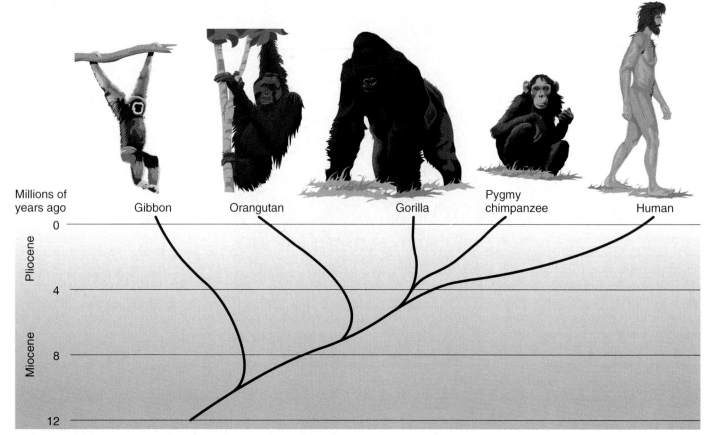

Figure 19.4 The evolution of living hominoids, the apes, and human beings.

Australopithecines may have split off from the ape-chimpanzee lineage as early as 5.5 million years ago. Humans are closely related to chimpanzees and gorillas; there is little doubt that an extraterrestrial taxonomist would place them in the same genus.

Once common, apes are rare today, living in relatively small areas in Africa and Asia. No apes ever occurred in North America.

Studies of ape DNA have explained a great deal about how apes evolved. The most primitive apes—the line leading to gibbons—diverged from other apes about ten million years ago, while orangutans split off about eight million years ago. The key split between the hominids and the line leading to gorillas and chimpanzees occurred less than five million years ago (figure 19.4). Because the split was so recent, the genes of humans and chimpanzees have not had time to evolve many differences—human and chimpanzee DNA differs in less than 3 percent of their nucleotide sequence. A human hemoglobin molecule and that of a chimpanzee differ in only a single amino acid!

The earliest primates were small, tree-dwelling insect-eaters that shared the world with dinosaurs. Monkeys and apes are their descendants.

Evolutionary Origins of Humans

Fifteen million years ago, the world's climate began to get cooler, and the great forests were largely replaced with open grassland savannas. In response to these changes, a new kind of ape was evolving, the direct ancestor of human beings.

Hominids

These new apes were classified as hominids—that is, of the human line (figure 19.5). They exhibited two critical early steps on the path leading to the evolution of humans:

1. *Bipedalism*—Walking upright on two feet. Apes have arms longer than their legs, a narrow pelvis, and a spinal cord that exits from the back of the skull. Early hominids, on the other hand, were build for **bipedalism** in that they had shorter arms (there are no trees to swing from in grassland savannas), a bowl-shaped pelvis (to center the weight of the body over the legs), and a spinal cord that exited from the bottom of the skull (the head was on *top* of the body, instead of in front of it).

Figure 19.5 Nearly human.
These four skulls, all photographed from the same angle, are among the best specimens available of the key Australopithecene species.

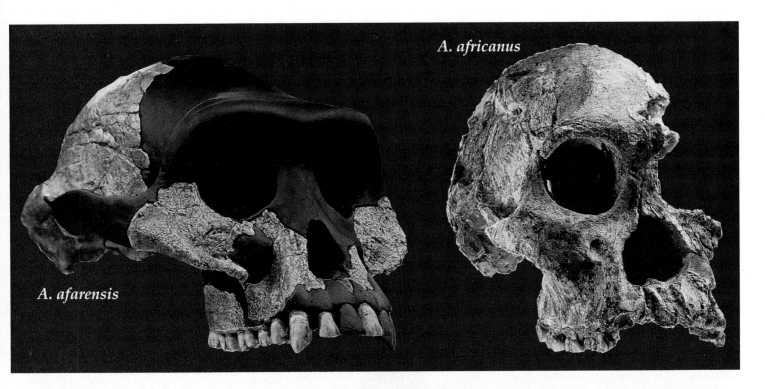

A. africanus

A. afarensis

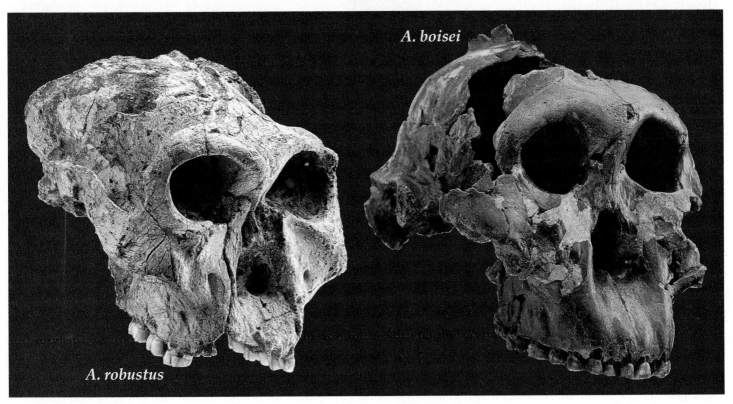

A. boisei

A. robustus

2. *Larger brains*—more brain per pound of body weight. Early hominids weighed about 40 pounds and were about 3½ feet tall, but their brains were typically far larger than an ape of similar size, occupying a volume of 400 to 500 cubic centimeters—bigger than a large gorilla's brain.

Discovery of *Australopithecus*

The first hominid was discovered in 1924 by Raymond Dart, an anatomy professor at Johannesburg in South Africa. One day, a mine worker brought into his office an unusual chunk of rock—actually, a rock-hard mixture of sand and soil. Picking away at it, Professor Dart uncovered a skull unlike any ape he had ever seen. Beautifully preserved, the skull was of a five-year-old individual, still with its milk teeth. It had a rounded jaw unlike the pointed jaw of apes, and the brain case indicated a brain far larger than any ape of this size. What riveted Dart's attention was that the rock in which the skull was embedded had been collected near other fossils that suggested that the rocks and their fossils were several million years old! At that time, the oldest reported fossils of hominids were less than five hundred thousand years old, so the ancientness of this skull was unexpecting and exciting. Nor was the suggestion of great age misleading—scientists now know Dart's skull to be fully 2.8 million years old. Dart called his find **Australopithecus africanus** (from the Latin *australo,* meaning "southern," and the Greek *pithecus,* meaning "ape"), the ape from South Africa—not human, but on the way.

An Evolutionary Tree with Many Branches

Dart argued that *Australopithecus* was the direct ancestor of humans, the long-sought "missing link" to the apes. At first, few believed he was right, but mounting evidence eventually convinced the scientific community. In 1938, a second, stockier kind of *Australopithecus* was unearthed in South Africa. Called *A. robustus,* it had massive teeth and jaws. In 1959 in East Africa, Mary Leakey discovered a third kind of *Australopithecus*—*A. boisei*—even more stockily built. Nicknamed "nutcracker man," *A. boisei* had a great bony ridge on the crest of the head to anchor immense jaw muscles—like a Mohawk haircut, but of bone. Like the other australopithecines, *A. boisei* was very old—almost two million years. In

1989, yet a fourth kind of australopithecine was reported—a massively-boned ancestor of *A. boisei*. The structure of their feet and pelvis indicates that all australopithecines walked upright. They all had big brains—from 400 to 500 cubic centimeters—and their teeth are more human than apelike. That they are humans' direct ancestors is no longer questioned.

Lucy: The Oldest Hominid

In 1974, anthropologist Don Johanson went to the remote Afar Desert of Ethiopia in search of early human fossils—and hit the jackpot. He found the most complete, best preserved skeleton of a prehuman hominid ever. Nicknamed "Lucy," the skeleton was 40 percent complete and over three million years old—the oldest reported australopithecine (figure 19.6). The skeleton was assigned the scientific name *A. afarensis.*

The shape of the pelvis indicated that Lucy was a female, and her leg bones proved she walked upright. Her teeth were distinctly humanlike, but the head shape resembled that of an ape, and the brain was not any larger than an ape's, about 400 cubic centimeters. For comparison, the average human brain today is about 1,350 cubic centimeters. Apparently, hominids walked upright before they acquired big brains.

Since Johanson's discovery, numerous other specimens of *A. afarensis* have been unearthed, and most researchers agree that these slightly built *A. afarensis* individuals represent the true base of the human family tree, the first members of the genus *Australopithecus,* ancestor to all the others. While researchers often disagree about which species goes on which branch of the human family tree, they all rec-

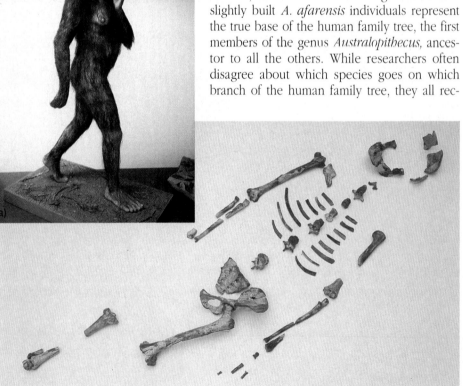

(a)

(b)

Figure 19.6 "Lucy."
(a) *This reconstruction was based on a careful study of muscle attachments to the skull and skeleton.* (b) *Lucy is the most complete skeleton of* Australopithecus *discovered so far.*

ognize the genus *Australopithecus* as being immediately ancestral to the human genus, **Homo,** our link to the past.

> *The australopithecines are the immediate ancestors of humans and are like us in many ways.*

The First Humans

The first humans evolved from australopithecine ancestors about two million years ago, only to be replaced, in turn, by a second "improved version" of humans that moved out of Africa and spread across the earth (figure 19.7). We are the third and only surviving species of humans.

Figure 19.7 Our own genus.
These four skulls illustrate the changes that have occurred during the evolution of the three species of humans. The Cro-Magnon skull is essentially the same as human skulls today. The skulls were photographed from the same angle.

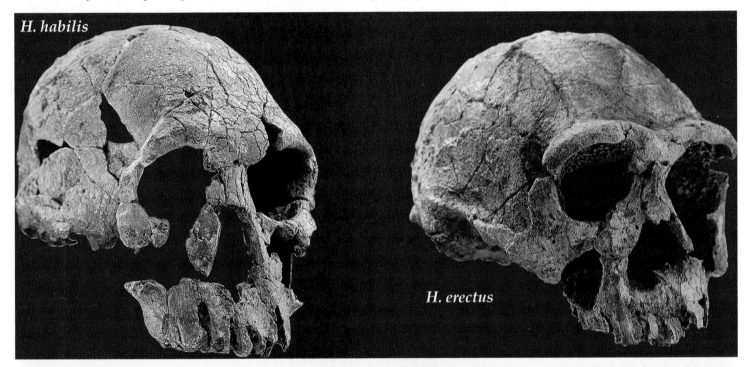

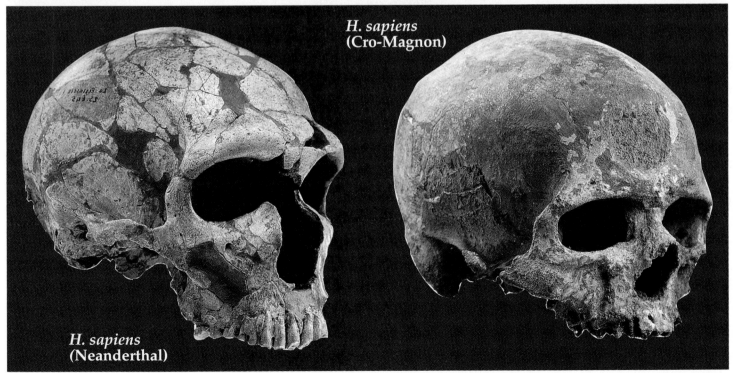

African Orgin: *Homo habilis*

In the early 1960s, stone tools were found scattered among hominid bones discovered close to the site where *A. boisei* had been unearthed. Although the fossils were badly crushed, painstaking reconstruction of the many pieces suggested a skull with a brain volume of about 640 cubic centimeters, much larger than the australopithecine range of 400 to 500 cubic centimeters. Many researchers questioned whether this fossil was human, but Richard Leakey's discovery in 1972 of a virtually complete skull silenced all critics. The skull, 1.6 million years old, had a brain volume of 775 cubic centimeters and many of the characteristics of human skulls—it was clearly human and not australopithecine. Because of its association with tools, this early human was called **Homo habilis,** meaning "handy man." Skeletons discovered in 1987 indicate that *H. habilis* was small in stature, like *Australopithecus. Homo habilis* lived in Africa for five hundred thousand years and then became extinct, replaced by a new kind of human with an even larger brain.

Out of Africa: *Homo erectus*

Details about *H. habilis* are sketchy, since only a few specimens are available for study. In fact, some scientists still dispute whether *H. habilis* qualified as a true human—it had not moved far from its australopithecine roots. About **Homo erectus,** the species that replaced *H. habilis,* however, there are no such doubts. The many specimens that have been found show that *H. erectus* was a true human.

Java Man

After the publication of Darwin's book, *On the Origin of Species* in 1859, there was much public discussion about "the missing link," the fossil ancestor common to both humans and apes. Puzzling over this, Dutch doctor and anatomist Eugene Dubois took a very simple approach to the problem: He went to the zoo. Observing the apes there, he was most drawn to the orangutans, the "old men of the forest" from Java and Borneo. Many of the orangutan's anatomical features seemed to resemble what a "missing link" should look like. So, Dubois closed his practice and sought out fossil evidence of the missing link in the orangutan's home country of Java.

Dubois set up practice in a river village in eastern Java. In 1891, while digging into a hill that villagers claimed had "dragon bones," he unearthed a skull cap and a thigh bone. He was very excited by his find for three reasons:

1. The structure of the thigh bone clearly indicated that the individual had long, straight legs and was an excellent walker.

2. The size of the skull cap suggested a *very* large brain, about 1,000 cubic centimeters—much larger than any ape.

3. Most surprising, the bones seemed very old—as much as five hundred thousand years old, judged by other fossils Dubois unearthed with them.

Dubois' fossil hominid was far older than any fossil hominid discovered up to that time, and at first, few scientist were willing to accept it as an ancient species of human. Finally, after

Figure 19.8 Peking man.
Workers in China excavate the fossil bed where numerous specimens of Peking man were unearthed in the 1920s and 1930s.

years of arguing to an unconvinced audience that the bones were indeed human, Dubois in disgust buried the **"Java man"** skull cap and thigh bone under the floorboards of his dining room and for thirty years refused to let anyone see them.

Peking Man

A generation later, scientists were forced to admit that Dubois had been right all along. In the 1920s, a skull was discovered near Peking (now Beijing), China, that closely resembled Java man. Continued excavation at the site eventually revealed fourteen skulls, many excellently preserved, together with lower jaws and other bones (figure 19.8). Crude tools were also found, and most important of all, the ashes of campfires. Casts of **"Peking man"** and the other fossils were distributed for study to laboratories around the world. The originals were loaded onto a truck and evacuated from Peking at the beginning of World War II, only to disappear into the confusion of history. No one knows what happened to the truck or its priceless cargo.

The Success of *Homo erectus*

Java man and Peking man are not recognized as belonging to the same species—*Homo erectus.* At a height of 5 feet, *H. erectus* was much taller than *H. habilis. Homo erectus* had a large brain (about 1,000 cubic centimeters) and walked erect. Its skull had prominent brow ridges and, like modern man, a rounded jaw. The shape of the skull interior suggests that *H. erectus* was able to talk.

Homo erectus has been shown to have come out of Africa. In 1976, a complete *H. erectus* skull discovered in East Africa was dated at 1.5 million years old, a million years older than the Java and Peking finds. The appearance of *H. erectus* in Africa 1.5 million years ago marked the beginning of the great human expansion. Far more successful than *H. habilis, H. erectus* quickly became widespread and abundant in Africa and within a million years had migrated into Asia and Europe. A social species, *H. erectus* lived in tribes of twenty to fifty people, often dwelling in caves (although evidence also indicates that they built crude, wooden shelters). They successfully hunted large animals, butchered them using flint and bone tools, and cooked them over fires. The site in China where Peking man was found contains the remains of horses, bears, elephants, deer, and rhinoceroses.

Homo erectus survived longer than any other species of human, for over a million years. These very adaptable humans only disappeared in Africa and Europe about five hundred thousand years ago, as modern humans were emerging. Interestingly, *H. erectus* survived much longer in Asia, until about 250,000 years ago.

Homo erectus was without doubt modern humans' immediate ancestor. From the neck down, *H. erectus* and modern humans are almost identical. Brain size is where the two differ.

> *Of the three species of humans that have evolved, two are now extinct. The second to evolve—Homo erectus—spread widely across Europe and Asia.*

Modern Humans: *Homo sapiens*

The evolutionary journey to modern humans ends with the appearance about five hundred thousand years ago of *Homo sapiens* ("wise man"), our own species. We are newcomers to the human family—*H. sapiens* has not been around nearly as long as *H. erectus* was. Still, humans have changed quite a bit since those first days.

Out of Africa—Again

The origin of human races is a much-debated point among scientists studying human evolution. Many argue that the different races each evolved from *H. erectus* independently and that each adapted to a different place—Orientals in Asia, Caucasians in Europe, Aborigines in Australia, and so on. Others believe that the same species would be unlikely to evolve more than once and argue that human races appeared *after H. sapiens* evolved from *H. erectus*. Recently, scientists studying **mitochondrial DNA** from living humans all over the world have argued that their research shows that all human races originated from one *H. sapiens* ancestor in Africa.

Scientists looked at mitochondrial DNA to study evolution because the DNA within mitochondria is transmitted only by females. Females' eggs carry many mitochondria within them that become part of a new baby, while sperm contribute no mitochondria to the new baby. Sperm carry their mitochondria wrapped around their tails and so do not inject them into the egg during fertilization. For that reason, particular versions of a mitochondrial gene can be traced back through a family tree, from mother to grandmother to great-grandmother.

Human races evolved only recently in the evolutionary scale of things, and there has not been enough time for many mitochondrial DNA differences to accumulate, so the exact human tree cannot be reliably traced using this approach. So far, however, the greatest number of different mitochondrial DNA sequences occur among modern Africans. Since DNA accumulates mutations over time, the oldest DNA should show the largest number of mutations. This result thus argues that humans have been living in Africa longer than on any other continent. While researchers are not in complete consensus, this line of investigation appears to suggest that *H. sapiens* evolved in Africa and that the human races evolved after that, and not independently from separate species of *H. erectus*. If this is correct, then *H. sapiens* was born in Africa and from there spread to all parts of the world, retracing the path taken by *H. erectus* half a million years before (figure 19.9).

Neanderthal Man

Homo sapiens first appeared in Europe as *H. erectus* was becoming rarer, about 130,000 years ago. Because the first *H. sapiens* fossils to be found, in 1856, were from the Neander Valley of Germany, these early European humans were called **Neanderthals** (*thal* means valley in old German). Compared with modern humans, European Neanderthals were short, stocky, and powerfully built. Their skulls were massive, with protruding faces and heavy, bony ridges over the brows. Their brains were even larger than those of modern humans!

The Neanderthals made diverse tools, including scrapers, spearheads, and hand axes. They lived in huts or caves. Rare at first outside of Africa, they became progressively more abundant in Europe and Asia, and by seventy thousand years ago had become common. Neanderthals took care of their injured and sick and commonly buried their dead, often placing food, weapons, and even flowers with the dead bodies. Such attention to the dead strongly suggests that they believed in a life after death. This is the first evidence of the kinds of symbolic thinking characteristic of modern humans.

Cro-Magnon Man

About thirty-four thousand years ago, the European Neanderthals were abruptly replaced by people of essentially modern character, the **Cro-Magnons** (named after the valley in France where they were discovered). Scientists can only speculate on the reasons for this sudden replacement, but it was complete all over Europe in a short period. Some evidence indicates that the Cro-Magnons came from Africa: Fossils of essentially modern aspect but as much as one hundred thousand years old have been found there. Cro-Magnons appear to have replaced the Neanderthals completely in the Middle East by forty thousand years ago and then spread across Europe, coexisting and possibly even interbreeding with the Neanderthals for several thousand years.

Cro-Magnons used sophisticated stone tools and also made tools out of bones and horns. They had a complex social organization and are thought to have had fully modern language capabilities. They lived by hunting. The world was cooler than it is now—the time of the last great Ice Age—and Europe was covered with grasslands inhabited by large herds of grazing animals. These herds are pictured in elaborate and often beautiful cave paintings throughout Europe (figure 19.10).

Humans of modern appearance eventually spread across Siberia to North America, which they reached at least thirteen thousand years ago, after the ice had begun to retreat and while a land bridge still connected Siberia and Alaska. By ten thousand years ago, about five million people populated the entire world (compared with the more than 5.6 billion now).

Homo sapiens Are Unique

We humans are animals and the product of an evolution marked by a progressive increase in brain size. We are the only animal that knows how to make and use tools effectively—a capability that, more than any other, has been responsible for our dominant position in the animal kingdom. While not the only animal capable of conceptual thought, we have refined and extended this ability until it has become the hallmark of our species. We use symbolic language, and with words, can shape concepts out of experience. This has allowed the accumulation of experience that can be transmitted from one generation to another. Thus, we have what no other animal has ever had: cultural evolution. Through culture, we have found ways to change and mold our environment to our needs, rather than evolving in response to environmental demands. We control our biological future in a way never before possible—an exciting potential and a frightening responsibility.

> *Our species may have evolved in Africa and then retraced the path taken by* Homo erectus, *replacing it in an outward wave of migration, although researchers have not yet reached a consensus on this.*

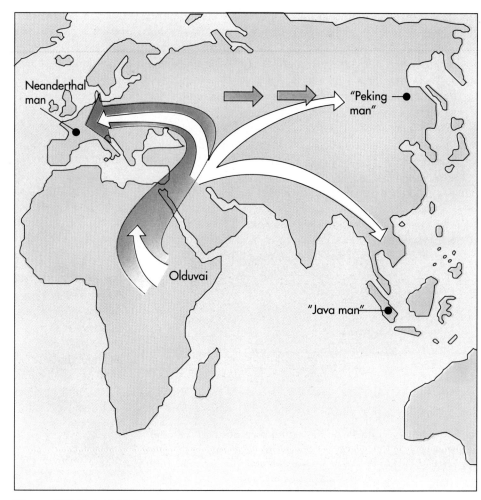

Figure 19.9 Out of Africa—twice.
Evidence now supports the theory that the human species first evolved in Africa. Homo *thus spread from Africa to Europe and Asia twice. First,* Homo erectus (white arrow) *spread as far as Java and China. Later,* H. erectus *was replaced by* Homo sapiens (red arrow) *in a second wave of migration.*

Figure 19.10 Cave painting.
Cro-Magnons, our immediate ancestors, made paintings of animals and sometimes hunters in caves, especially in Europe. The paintings were done during a period of about twenty thousand years that ended about eight to ten thousand years ago.

EVOLUTIONARY VIEWPOINT

Biologists may know more about the evolution of humans than about any other genus. Despite the sometimes spirited arguments among researchers about human origins, the detailed hominid fossil record clearly shows that members of the genus *Homo* have changed over time. Their probable ancestors—the australopithecines—had small brains, but were already bipedal. What might the next species in the genus *Homo* be like (*H. nextus*), if indeed there is a fourth species of humans?

SUMMARY

1. Primates first appeared about forty million years ago, evolving from tree shrews. The first primates were prosimians. Their grasping fingers and toes and binocular vision were distinct improvements that guided the evolution of primates' increased intelligence.

2. Diurnal (day-active) monkeys replaced the nocturnal prosimians about thirty-six million years ago. Monkeys exhibit complex social interactions and were the first primates with an opposable thumb.

3. The apes, which evolved independently from prosimian ancestors about twenty-five million years ago, gave rise to the gibbons, orangutans, and then to the gorillas, chimpanzees, and hominids (human beings and their direct ancestors). Apes and hominids collectively are termed hominoids.

4. The earliest hominids belonged to the genus *Australopithecus* and were the direct ancestors of humans. They exhibited bipedalism (walking upright on two feet) and larger brains (400 to 500 cubic centimeters). The oldest australopithecine fossil, nicknamed "Lucy," is over three million years old.

5. With an enlarged brain that was perhaps associated with the increased use of tools, the genus *Australopithecus* gave rise to humans belonging to the genus *Homo*. The first species of this genus—*Homo habilis*—appeared in Africa about two million years ago. It became extinct about 1.5 million years ago.

6. The second species of *Homo*—*Homo erectus*—appeared in Africa at least 1.5 million years ago. Within a million years, it migrated from Africa to Europe and Asia, as evidenced by the Java and Peking fossils. *Homo erectus* walked erect, had a large brain, (1,000 cubic centimeters), and is believed to have been able to talk.

7. The third (and current) species of *Homo*—*Homo sapiens*—replaced *H. erectus* about 500,000 years ago in Africa and Europe and about 250,000 years ago in Asia. The Neanderthals were early *H. sapiens* in Europe about 130,000 years ago. The Cro-Magnons (essentially modern) began to replace the Neanderthals by forty thousand years ago.

8. Studies of mitochondrial DNA suggest (but do not yet prove) that all of today's human races originated from *H. sapiens* in Africa.

REVIEWING THE CHAPTER

1. How did primates evolve?

2. What were the earliest primates?

3. How did monkeys differ from earlier primates?

4. How did apes lay the pathway to humanity?

5. What are the evolutionary origins of humans?

6. What advanced features are found in hominids?

7. What is the significance of *Australopithecus?*

8. Who is "Lucy"?

9. Who were the first humans?

10. What were the characteristics of *Homo habilis,* and where did this species live?

11. What were the characteristics of *Homo erectus,* and where did this species live?

12. Why was *Homo erectus* a very successful species?

13. How does *Homo sapiens* differ from *Homo erectus?*

14. What is the history of *Homo sapiens?*

15. What were the characteristics of Neanderthal man?

16. What were the characteristics of Cro-Magnon man?

17. Why are *Homo sapiens* unique?

COMPLETING YOUR UNDERSTANDING

1. What was Darwin's hypothesis on human evolution published in *The Descent of Man* (1871)?

2. What controversies arose in the United States between the religious community and the public school classroom as a result of Darwin's hypothesis on human evolution?

3. Which characteristics were selected for in the earliest primates to allow them to become successful?

4. What allowed for an increased intelligence in primates?

5. Create a hypothesis for why all living lemurs on the earth are on the island of Madagascar.

6. What is the future status of lemurs on planet Earth?

7. Why did monkeys replace prosimians?

8. How are apes distinguished from monkeys?

9. What is the best explanation for why the nucleotide sequence in the DNA of humans and chimpanzees is more than 97 percent similar?

10. When did the first hominids appear, and how did they originate?

11. What was the relationship between bipedalism and brain size in the earliest hominids?

12. Who was Raymond Dart, and what information did he provide about human evolution?

13. How many species of humans have there been, and where and when did they originate?

14. Why is there some doubt in the scientific community that *Homo habilis* was a true human?

15. Why did Eugene Dubois go to Java to seek fossil ancestors of modern humans? What were the frustrations associated with his discoveries there?

16. What is the bizarre ending to the discovery of "Peking man"?

17. How did *Homo erectus* differ from *Homo habilis?*

18. Why has there been so much interest in mitochondrial DNA with respect to tracing human evolution?

19. The greatest number of different mitochondrial DNA sequences in humans occurs in Africa. What does this information tell us about humans?

20. Where did Neanderthals originate, and where did they migrate to?

21. What evidence is there that Neanderthals may have believed in life after death?

22. How did Cro-Magnons differ from Neanderthals?

23. Is there any evidence that Cro-Magnons coexisted with Neanderthals? If so, where and when?

*24. Modern humans, *Homo sapiens,* evolved from *Homo erectus* less than one million years ago. Do you think that evolution of the genus *Homo* is over, or might another species of humans evolve within the next million years? Do you think this would involve the extinction of *H. sapiens?*

*25. Studies of the DNA of primates have revealed that humans differ from gorillas in only 1.4 percent of the DNA nucleotide sequences, and from chimpanzees in only 1.2 percent. This degree of genetic similarity (about 1 percent) is the same as is usually seen among "sibling" species (that is, species that have only recently evolved from a common ancestor). Yet humans are assigned not only to a different genus but to a different family! Do you think this is legitimate, or are humans just a rather unusual kind of African ape?

*For discussion

FOR FURTHER READING

Blumenschine, R., and J. Cavallo. 1992. Scavenging and human evolution. *Scientific American,* Oct., 90–96. An interesting if controversial argument that our early ancestors were better scavengers than hunters.

Gibbons, A. 1992. Mitochondrial Eve—Wounded, but not dead yet. *Science* 257 (Aug.): 873–75. An account of the disputed mitochondrial DNA analysis pointing to an African origin of modern *Homo sapiens.*

Milton, K. 1993. Diet and primate evolution. *Scientific American,* Aug., 86–93. The nature of food available in the early angiosperm forests of the Cretaceous and how that has shaped how the human evolutionary line has developed.

Simons, E. 1989. Human origins. *Science* 245 (Sept.): 1343–50. A thorough review of the skeletal evidence that indicates that all major steps in human evolution took place in Africa.

Thorne, A., and M. Wolpoff. 1992. The multiregional evolution of humans. *Scientific American,* April, 76–83. The argument against African origin of modern *H. sapiens.*

Walker, A., and M. Teaford. 1989. The hunt for proconsul. *Scientific American,* Jan., 76–82. An exciting account of the last common ancestor of great apes and human beings.

Wilson, A., and R. Cann. 1992. The recent African genesis of humans. *Scientific American,* April, 68–73. The original argument proposing a recent African origin of modern humans, employing a much-criticized phylogenetic analysis of mitochondrial DNA.

BIOLOGY 1040   ZOOLOGY   SPRING 2000

Instructor: Al Craig   Office room 1240   phone 333-7295
email: alcraigxx@aol.com
Text: Understanding Biology, Raven and Johnson

	TOPIC	READ
1	Introduction, Course overview	Ch. 1 ✓
2	Atoms and molecules	Ch. 2 ✓
3	Macromolecules	
4	Origin of life	Ch. 3 ✓
5	The cell	Ch. 4 ✓
6	Membranes separate compartments	Ch. 5 ✓
7	Bioenergetics	
8	Cell division	Ch. 9 ✓
9	Meiosis	
10	Genetics	Ch. 10 ✓
11	Gene expression	Ch. 12,13 ✓

Review for Exam #1
EXAM #1   FEBRUARY 17

12	Evolution	Ch. 15 ✓
13	Evolution- continued	Ch. 16,17 ✓
14	Kingdoms of life	Ch. 25 ✓
15	Viruses and bacteria	Ch. 26 ✓
16	Protista	Ch. 27 ✓
17	Animal body plans	Ch. 29
18	Simple animals	Ch. 29 (cont.)
19	Complex animals	Ch. 30
20	Tissues	Ch. 30(cont.)

Review for Exam #2
EXAM #2   MARCH 30

21	Organ systems- digestion	Ch. 38
22	Cellular respiration	Ch. 7
23	Organ systems- respiration	Ch. 39
24	Organ systems- circulation	Ch. 40
25	Nature of disease	Ch. 41
26	Organ systems- excretion	Ch. 42
27	Organ systems- nervous system	Ch. 34,35
28	Homeostasis	Ch. 33
29	Ecology/ populations	Ch. 20,21
30	Ecosystems	Ch. 22
31	Biosphere	Ch. 23,24

Review for Exam #3
EXAM #3

Office hours: I can be found in my office during the following :

Tuesday	10:00 am- 1:00 pm, 3:00-4:00 pm
Wednesday	12:00- 1:00 pm
Thursday	12:00- 1:00 pm

I AM ALSO AVAILABLE BY APPOINTMENT IF THESE HOURS ARE NOT CONVENIENT FOR YOU.

Grading :   There will be three hour exams, each is worth 25% of the final grade. The laboratory is worth the remaining 25% of the final grade.

The dates for the first two-hour exams are stated on the syllabus. The third hour exam is held during the final exam period, which will be announced during the semester.

Note: Exam #3 is the final exam. The final exam period is two and a half hours. You do not have the option of showing up to take the third (final) exam at any time during the exam period. You must be there at the beginning of the scheduled time. No exams will be distributed 30 minutes after the beginning of the final exam period. Therefore be on time for Exam #3. If you are late for exam #3 you will receive an incomplete and will have to make up the grade at a later date.

If for any reason you do poorly or not as well as you may have liked on either exam #1 or exam #2 you may redeem yourself at the time of exam # 3. At that time you may take another (ONE ONLY) exam on the same material as the exam that you previously were not satisfied with.

This fire in Brazil, and countless others like it, was set deliberately to clear the forest for livestock grazing. Soon much of Brazil's natural forest will be gone.

POPULATION DYNAMICS

*A satellite view of Los Angeles in 1975.
Much of the earth's human population is
crowded into large cities, many of them
even larger than Los Angeles.*

FOR REVIEW

Here are some important terms and concepts that have been discussed in previous chapters and that you will encounter again in this chapter. Review them before proceeding if necessary.

Adaptation (*chapter 15*)
How species originate (*chapter 16*)
Major features of evolution (*chapter 17*)

Stepping back and viewing the broad panorama of evolution, as we did in part 5, reveals the unity that underlies the diversity of life. All of us—camels, elephants, and humans—share a common history of adaptation and change. But our view should not be limited to looking backward. Right now, all the organisms on earth are involved in a day-to-day struggle to survive and reproduce. Each living thing on earth is interacting with its physical environment and with the organisms around it in a journey toward the future we all share (figure 20.1).

The House We Live In

In 1866, German biologist Ernst Haeckel first gave a name to the study of how organisms interact with their environment, both with temperature, moisture, and soil, and also with other organisms. He called this study of how organisms fit into their environment **ecology,** from the Greek words *oikos,* meaning "house, place where one lives," and *logos,* meaning "study of." Ecology, then is the study of the house in which we live.

Ecology is a complex but fascinating area of biology with many important implications for each of us. Every ecological principle would hold true even if there were no human beings in the world, but human activities are now profoundly affecting the operation of these principles. A rapidly growing human population—currently at an unprecedented level of more than 5.6 billion people—is severely straining the earth's sustaining capacity. In the face of this situation, ecological principles may enable humans to chart a sound future.

Ecology is concerned with the most complex level of biological integration, attempting to explain why particular kinds of organisms can be found living in one place and not another. The physical and biological variables that govern organisms' distribution, the factors that control the numbers of particular kinds of organisms and maintain them at certain levels, and the principles that allow prediction of the future behavior of assemblages of organisms are all areas of ecological interest.

Ecology is the study of the relationships of organisms with one another and with their environment.

Ecologists consider groups of different organisms at four progressively more inclusive levels of organization:

Figure 20.1 Competition in the forest.
Each plant in this forest competes with all the individuals around it for light, soil, nutrients, and moisture. The competition for available resources is fierce, especially during the relatively brief period in the spring before trees have so many leaves that photosynthesis is limited.

1. Individuals of a species that live together are called a **population.** The place in which these individuals live is called their **habitat.** The limits or boundaries of a population's habitat is called the **range** of the population.

2. Populations of different species that live together are called **communities.**

3. A community, together with the nonliving factors with which it interacts, is called an **ecosystem.** An ecosystem regulates the flow of energy, ultimately derived from the sun, and the cycling of the essential elements on which the lives of its plants, animals, and other organisms depend. Groups of ecosystems that include major assemblages of terrestrial plants, animals, and microorganisms and that occur over wide geographical areas are called **biomes.** Deserts, tropical forests, and grasslands are examples of biomes.

4. That part of the earth where life exists or can be supported is called the **biosphere.** The biosphere is, in effect, a collection of complex ecosystems.

The exploration of ecological principles begins in this chapter with a consideration of the properties of populations, emphasizing their dynamics. Chapter 21 discusses communities and the sorts of interactions that occur in them. Chapter 22 moves on to the dynamics of ecosystems, while chapter 23 considers the properties of biomes. Finally, chapter 24 examines the future of the biosphere and attempts to assess the impact of the

(a)

(b)

(c)

Figure 20.2 Highly endangered populations.
These animals exist in very small populations and are in danger of extinction unless the activities threatening them are brought under control. (a) The Sumatran rhinoceros, Didermoceros sumatrensis, *has been reduced to a few hundred individuals on the island of Sumatra. (b) The Mediterranean monk seal,* Monachus monachus. *Fewer than five hundred individuals of this species are believed to exist, and they are confined to remote, cliffbound coasts and islands in the Mediterranean. (c) The Kagu,* Rhynochetus jubatus, *the only member of a family of birds that is restricted to the island of New Caledonia in the southwestern Pacific Ocean. It has been brought to the brink of extinction by introduced dogs that have run wild on the island.*

many problems that plague it. The biosphere's future will depend on the ecological principles presented in this chapter and the four chapters that follow.

Characteristic Properties of Populations

A population consists of the individuals of a given species occurring together at one place and at one time. This flexible definition allows the world's human population, the population of protozoa in the gut of an individual termite, and the deer that inhabit a forest all to be described in similar terms.

All specific populations have characteristic properties, such as size, density and dispersion. Each population occupies a particular place and plays a particular role in its ecosystem; its role is defined as its **niche** in that ecosystem. Although the properties of populations are ultimately determined by the genetic properties of the individuals they include, a population is fundamentally an operating system with its own distinct properties. Understanding these properties is crucial for understanding the nature of life on earth.

Population Size

One of the critical properties of any population is **population size**—the number of individuals in a population. Population size has a direct bearing on a given population's ability to survive. Random events or natural disturbances may endanger a small population more than they endanger a large population. Inbreeding can also be a negative factor in population survival if the population is small and few potential mates are available. Inbreeding often leads directly to a lowering of vigor—that is, of hardiness and the ability to meet the everyday challenges of living. In addition, the reduced levels of variability that result from inbreeding are likely to detract from the population's ability to adjust to changing future conditions. Chapter 15 examined the risks of a small, nonvariable population in relation to cheetahs. If an entire species consists of only one or a few small populations, that

Population Density

In addition to population size, **population density**—the number of individuals that occur in a given area—is also highly significant. If the individuals of a population are widely spaced, they may rarely, if ever, encounter one another, and the reproductive capabilities—and therefore the future—of the population may be very limited, even if the absolute number of individuals over a wide area is relatively high.

Population Dispersion

A third significant property of populations is the **population dispersion** of individual organisms throughout the population's range. Put another way, dispersion describes the different population densities throughout a population's range. Populations may be spaced in a random manner, they may be clumped, or their distribution throughout the range may be uniform.

The rarest type of dispersion is **random dispersion** (figure 20.3*a*). Random dispersion occurs when few or no environmental features attract or repel the organisms in a population. As a result, organisms disperse to different parts of the range in an unpredictable way. Trees in a forest are sometimes ran-

...ies is likely to become ...nct, especially if it occurs in areas that have been or are being changed radically (figure 20.2).

(a) Random

(b) Clumped

(c) Uniform

Figure 20.3 Dispersion patterns in populations.
Random dispersion (a) *is rare, while clumped dispersion* (b) *is more common. Clumped patterns are indicative of an environment in which resources are unevenly distributed. Uniform dispersion* (c) *reflects a population that is evenly spaced throughout the range.*

tains the most nutrients. In an animal population of plant-eaters, individual animals may gather where the most food plants exist. Clumped dispersion illustrates the need for organisms to seek out and take advantage of the best resources needed for survival.

In a **uniform dispersion,** the individuals of a population are evenly spaced throughout the range (figure 20.3*c*). Uniform dispersions are often associated with environmental conditions that result in competition among population individuals for resources. For example, a plant population may become uniformly dispersed if environmental conditions, such as shading, are detrimental to plant growth. The plants in this population will compete with one another for light, and those that receive the best light will stay in their position. Animal populations may be uniformly dispersed as well because of competition among individuals or because of social interactions that involve territoriality.

A population is a group of individuals of a particular species that live together and can be described in terms of size, density, and dispersion.

Population Growth

In addition to size, density, and dispersion, another key characteristic of any population is its

domly dispersed. Since random dispersion occurs so rarely in nature, however, ecologists use it as a gauge to measure the other types of dispersion. Clumped and uniformly dispersed populations are identified by comparisons with a hypothetical, randomly dispersed population.

More common in nature is the **clumped dispersion** (figure 20.3*b*). In these populations, individuals seek out those areas of the range where resources are concentrated. In a plant population, individuals may cluster in an area where the soil con-

capacity to grow. Most populations tend to remain relatively constant in number, regardless of how many offspring the individuals produce. As mentioned in chapter 1, Darwin partly based his theory of natural selection on this seeming contradiction. Many calculations have shown that houseflies, bacteria, or even elephants would soon cover the world if their reproduction went unchecked. Under certain circumstances, population size can increase rapidly for a time (figure 20.4). Nature, however, operates to limit population growth (see Sidelight 20.1).

Actual Rate of Population Increase

The rate at which a population will increase when there are no limits of any sort on its rate of growth is called its **innate capacity for increase**, or **biotic potential.** This theoretical rate is almost always impossible to calculate, however, because there usually are limits to growth. What biologists actually tend to calculate is the **actual rate of population increase** (abbreviated as *r*), which is defined as the birth rate plus the number of immigrants into the population minus the death rate plus the number of emigrants out of the population:

$$r = (\text{birth rate} + \text{immigration}) - (\text{death rate} + \text{emigration})$$

A population's innate capacity for growth is constant, determined largely by the organism's physiology. The actual growth rate, on the other hand, depends on both the birth rate and the death rate and so changes as the population increases in size. In general, as a population increases and begins to exhaust its resources, its death rate rises. The number of individuals in a population from its beginning grows rapidly at first, increasing exponentially (2, 4, 8, 16, 32, 64, 128, . . .) (figure 20.5). Soon, however, the rate of increase slows as the death rate begins to rise. Eventually, just as many individuals are dying as are being born. The early, rapid phase of population growth lasts only for a short period, usually when an organism reaches a new habitat in which it has abundant resources. Examples of this phenomenon include the marsh weed called purple loosestrife reaching the wetlands of North America from Europe for the first time (see figure 20.4); algae colonizing a newly formed pond; and the first terrestrial organisms that arrive on an island recently thrust up from the sea.

Figure 20.4 Establishment of a new population.
When an organism reaches a new habitat in which it has abundant resources, it may experience rapid population growth for a relatively short period. For example, European purple loosestrife (Lythrum salicaria) *was introduced to North America some time before 1860 and has since become naturalized over thousands of square miles of marshes and other wet places.*

Time (hours)	Number of bacteria	Growth curve
10.0	1,048,576	
9.5	524,288	
9.0	262,144	
8.5	131,072	
8.0	65,536	
7.5	32,768	
7.0	16,384	
6.5	8,192	
6.0	4,096	
5.5	2,048	
5.0	1,024	
4.5	512	
4.0	256	
3.5	128	
3.0	64	
2.5	32	
2.0	16	
1.5	8	
1.0	4	
0.5	2	
0.0	1	

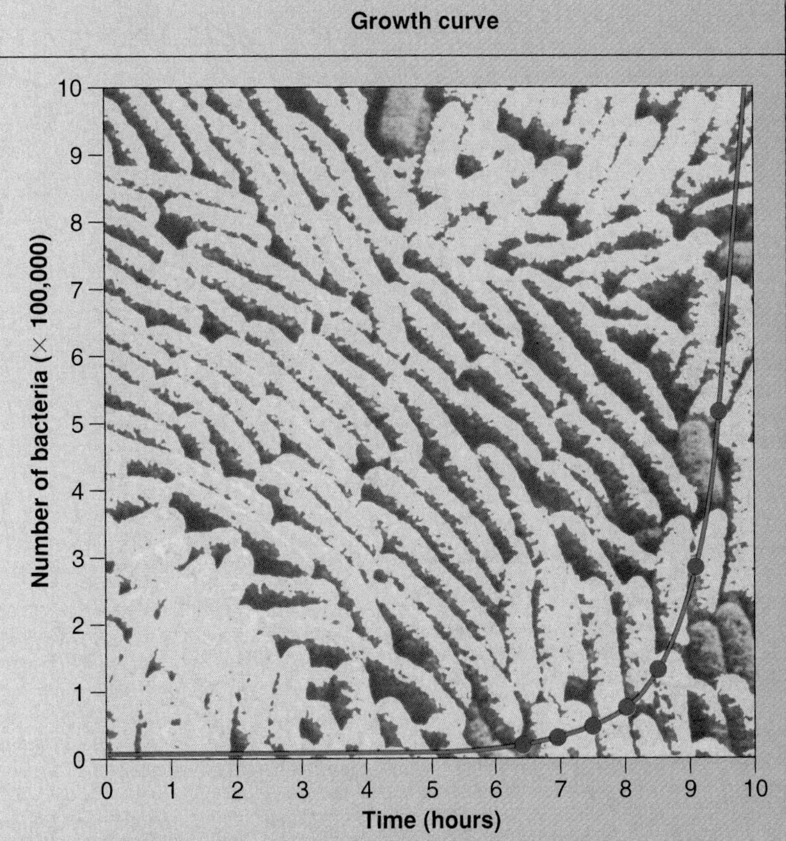

Figure 20.5 Exponential growth in a population of bacteria.
In just ten hours, this population grew from just one individual to over one million individuals!

Sidelight 20.1

THE BLACK DEATH: COMPARISONS WITH AIDS

One of the greatest disasters to ever befall the human population was the worldwide epidemic of bubonic plague that raged in the mid-fourteenth century. Called the "Black Death," this epidemic not only affected world population numbers but also caused important social and economic changes in areas hardest hit by the disease. In this modern era, what parallels can be drawn between the fear and anguish the Black Death engendered six hundred years ago and current feelings about AIDS and its victims?

Like AIDS, bubonic plague had no cure in the fourteenth century. The disease is caused by a bacterium called *Pasteurella pestis* that lives in rats. These rats in turn, are hosts to fleas that ingest the bacteria when they bite an infected rat. Fourteenth-century notions of sanitation did not extend to flea extermination, and flea infestation was a common nuisance in medieval households. Experts believe that the disease first started in Asia and that trading ships brought infected rats to eastern Europe in the mid-fourteenth century. Spreading west, the disease ravaged all parts of Europe, including France, Italy, and England. Based on medieval chronicles of the plague, demographers have noted that entire cities were felled by the disease, and some estimate that as many as a third of Europe's total population perished in the epidemic (Sidelight figure 20.1).

Bubonic plague spread so rapidly and with such deadly force because it is highly contagious. Unlike AIDS, bubonic plague can be transmitted even from casual contact. The fleas and rats that carried the bacteria were very common in the fourteenth century and provided another route for infection. In addition, the disease strikes quickly and kills quickly: Chroniclers noted that a person could seem entirely well upon waking in the morning and be dead by nightfall. Finally, the large numbers of the dead overwhelmed the cemeteries; sometimes, the dead were simply left on doorsteps, where they would be collected and dumped into mass graves, resulting in a center of contagion that could infect even more people.

The medieval medical establishment, based as it was on equal parts of superstition and ancient Greek medical practices, could only guess about the epidemic's cause and grope for ways to contend with it. With no knowledge of microscopic bacteria or carriers of disease (such as rats and fleas), doctors believed that the Black Death was caused by foul air, evil spirits, or God's edict against the sin of the world. The most accepted cause came from doctors in Paris, who proclaimed that a triple conjunction of Saturn, Jupiter, and Mars in March 1348 unleashed the epidemic on the world. Efforts to contain the contagion focused on quarantine, exotic treatments, and prayer. Most of the treatments were designed to pull the infection out of the body. Bleeding, the application of hot plasters, laxatives, and enemas

Sidelight figure 20.1
A medieval tapestry of the plague.

were thought to be effective in causing the infection to leave the body; however, none of these treatments were of much use. Doctors know today that the only cure for bubonic plague is the quick administration of antibiotics and quarantine; without the "magic bullet" that antibiotics provide, sufferers from this disease inevitably die.

There is no "magic bullet" where AIDS is concerned, but scientists do know that AIDS is caused by a virus and that virus transmittal from one person to another is, fortunately, far from casual. For infection to take place, the

AIDS virus, which is present in body fluids, must be introduced into the body of another person through body fluids. Activities that allow this exchange of body fluids include sex and the sharing of hypodermic needles.

Because large numbers of the world population, especially in Europe, died, the Black Death resulted in sweeping economic and social changes, the effects of which were felt for hundreds of years. The shortage of labor led to laborers' demands for higher wages and fairer treatment; in this early labor movement, the sentiments that triggered the peasant revolts of the late fourteenth century were born. Outbreaks of labor movements continued until the nineteenth century, and the Black Death, which struck poor and rich alike, has been said to have jolted European peasants and serfs out of their complacency and prompted them to question the feudal system of class and wealth. On an emotional level, the remaining population, afraid and ravaged by grief, looked for answers. Upon examining the Catholic church more closely, many found that the church was corrupt and did not offer solace and comfort for the average person. This feeling continued for many years, steadily building until Martin Luther, Henry VIII, and others broke with the Catholic church altogether in the sixteenth century.

Today, AIDS has already resulted in some social changes. At first, because the epidemic began in the U.S. homosexual population, some government leaders reviled the homosexual lifestyle and pointed to AIDS as their justification. Now, with AIDS moving into the heterosexual cohort of the population, this type of rhetoric is rarely heard. The homosexual segment of society has been instrumental in bringing the epidemic to the public eye and has continually pushed for more explicit education and prevention measures. Many feel that AIDS is responsible for promoting the "new monogamy" that some see making inroads into American society. A new emphasis on safer sex and ways to prevent infection are now also part of American culture. It remains to be seen what other effects the AIDS epidemic will have on society, but these effects will surely be related to how many people succumb to the disease before a cure is found.

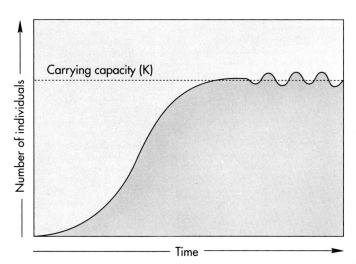

Figure 20.6 The sigmoid growth curve.
The sigmoid growth curve begins with a period of exponential growth like that shown in figure 20.5. When the population approaches its environmental limits (K), growth slows and finally stabilizes, fluctuating around the maximum number of individuals that the environment will hold.

The actual rate of population increase equals (the birth rate plus the number of immigrants into the population) minus (the death rate plus the number of emigrants out of the population).

Carrying Capacity

No matter how rapidly new populations grow, they eventually reach some environmental limit imposed by shortages of an important factor, such as space, light, water, or nutrients. A population ultimately stabilizes at a certain size, called the **carrying capacity** of the particular place where the population lives. The carrying capacity, symbolized by *K*, is the number of individuals that can be supported at that place indefinitely. As carrying capacity is approached, the population's rate of growth slows greatly because, in effect, there is less "room" for each individual. Graphically, this relationship is the S-shaped **sigmoid growth curve,** characteristic of biological populations (figure 20.6).

The size at which a population stabilizes in a particular place is the carrying capacity of that place for that species.

Density-Dependent and Density-Independent Effects

As a population approaches its carrying capacity for a particular habitat, competition for resources, emigration, and accumulation of toxic waste products all increase. The resources for which population members must increasingly compete may include food, shelter, light, mating sites, or any other factor necessary for them to carry out their life cycle and reproduce.

Effects such as these occur as a result of population density and are called **density-dependent effects.** Among animals, density-dependent effects are often accompanied by hormonal changes that may bring about alterations in behavior that directly affect ultimate population size. One striking example occurs in migratory locusts (actually a kind of "short-horned" grasshopper), which, when they are crowded, produce hormones that cause them to enter a new, migratory phase: The locusts take off as a swarm and fly long distances to new habitats. In contrast, **density-independent effects** are caused by such factors as the weather and physical disruption of the habitat—factors that operate regardless of population size.

Density-dependent effects are controlled by factors that come into play particularly when the population size is larger; density-independent effects are controlled by factors that operate regardless of population size.

Optimal Yield

Modern agricultural practice depends to some extent on the characteristics of the sigmoid growth curve. Early in a population's history, resources are not yet limiting the growth of individuals. For best yields, this is the best time to "harvest" a population. Commercial fisheries also attempt to operate so that fish populations are always harvested in the steep, rapidly growing parts of the growth curve. The point of **optimal yield** (maximum sustainable catch from the population) lies partway up the sigmoid curve. Harvesting the population of an economically desirable species near this point results in much better yields than can be obtained either when the population approaches the carrying capacity of its habitat or when it is very small. Overharvesting a population that is smaller than this critical size can destroy the population's productivity for many years, as apparently happened to the Peruvian anchovy fishery after the population had been depressed by the 1972 El Niño weather disturbance. In forestry, the harvesting of mature trees for plywood, pulp, or similar products tends to exploit the r part of the sigmoid curve; proper management [ ]res an understanding of the characteristics of the populations being harvested. Although determining the population levels of commercially valuable species is often difficult, such estimates are the subject of much study because they are critical from both a commercial and ecological point of view.

> *In natural systems exploited by humans, the aim is optimal yield, which involves exploiting the population of the most productive part of the rising portion of the sigmoid growth curve.*

r Strategists and *K* Strategists

Many species, such as annual plants, a number of insects, and bacteria, have very fast rates of population growth. In some species of bacteria, an individual bacterium can reproduce in less than twenty minutes. Growth this rapid cannot be controlled effectively by reducing population sizes. In such species, small surviving populations begin an exponential pattern of growth and soon regain their original sizes. In contrast, a comparable reduction in population size among relatively slow-breeding organisms, such as whales, rhinoceroses, California redwoods, or most tropical rain forest trees, can lead directly to their extinction (see figure 20.2).

Slow-growing populations of organisms tend to be limited in number by the environment's carrying capacity and are sometimes called **K strategists.** Such organisms, including the whales and redwoods just mentioned, usually live in fairly stable, predictable habitats. In contrast, species whose populations are characterized by very rapid growth followed by sudden crashes in population size tend to live in unpredictable and rapidly changing environments. These organisms have a high intrinsic rate of increase, or *r,* and are called **r strategists.** The growth curve for bacteria shown in figure 20.5 demonstrates the characteristic high rate of increase for *r* strategists. Many organisms are neither "pure" *r* strategists nor "pure" *K* strategists. Rather, their reproductive strategies lie somewhere between these two extremes or change from one extreme to the other under certain environmental circumstances.

> *Some populations (K strategists) are limited by the environment's carrying capacity; others (r strategists) are not.*

Figure 20.7 An r strategist: The cockroach.
Cockroaches, a major household pest, produce eighty young about every six months. If every cockroach that hatched survived, kitchens might look like this. One pair of cockroaches could produce this horde of over 130,000 individuals in just three generations. The exhibit shown here is in the Smithsonian Institution's National Museum of Natural History.

In general, *r* strategists produce many offspring as a result of each reproductive event. Their offspring are small, mature rapidly, and receive little or no parental care. Such organisms include cockroaches and mice (figure 20.7).

In contrast, *K* strategists produce few offspring. These offspring are large, mature slowly, and often receive parental care. This group consists of such organisms as coconut palms, whales, and humans (figure 20.8). Many *K* strategists are in danger of extinction.

The Effect of the Human Population on *r* and *K* Strategists

Human beings use their culture to avoid or postpone the effects of their population size, which is currently increasing at the rate of about 1.6 percent per year. Some ninety-five million people are added annually—three people every second—to a total population that already numbers more than 5.6 billion people. At this rate, the world population will double in about forty-three years. As discussed in chapter 24, such a rate of growth has potentially grave consequences. By reducing the population sizes of most other species as our own enormous populations grow so rapidly, we are driving to extinction those relatively large and slow-breeding species that we view as desirable, while favoring those weed and pest species that compete directly with us for food and other critical resources.

Figure 20.8 A K strategist: The humpbacked whale.
Most whales have one calf at a time. Roger Paynes, who has contributed to scientific understanding of whales, wrote: "In our long, sad history, we have brought hundreds of species to extinction. Unless we change our priorities, we will soon eradicate hundreds of thousands more. There is a curious fact, however; among all of the species we have destroyed, there is not one that occurred worldwide. . . the closest we have come to taking that fatally insane step is with the (humpbacked) whale—a species that once occurred off the western and eastern shores of every continent. We came so close to destroying that species that it really is something of a miracle that it survived."

The rapidly growing human population is tending to exterminate K strategists, such as whales, elephants, and forest trees, while favoring r strategists, including houseflies, cockroaches, and dandelions.

Mortality and Survivorship

A population's intrinsic rate of increase depends on the ages of the organisms in it and the reproductive performance of the individuals in the various age groups. Very young and very old individuals obviously are not as reproductively active as the rest of the population. The speed with which a population can grow depends on how many of its members are reproducing, and thus, on the population's **age distribution.** A population with many reproducing individuals will grow much faster than one with few reproducing individuals. Age distributions differ greatly from species to species and even, to some extent, from place to place within a given species.

One way to express a population's age distribution is the **survivorship curve. Survivorship** is defined as the percentage of an original population that survives to a given age. Survivorship curves are based on data collected in a summary called a **life table.** Life tables originated when insurance companies began studying the human population to find out when individuals in the population were expected to die (it is certainly ironic that these studies are called *life* tables!). Ecologists discovered that this method of recording mortality rates could be applied to other populations. A life table shows how many individuals are alive at any one time and expresses this information in a variety of different ways: the number of individuals dying during an interval of time, the **mortality rate** (the percentage of an original population that are dead at any given age), and an estimation of the remaining life span of an individual at any one time (figure 20.9). Life tables can be used to examine mortality patterns in a given population, and when graphed in a survivorship curve, these individual patterns of populations can be grouped into types.

Figure 20.10 shows samples of different kinds of survivorship curves. In the hydra, individuals are equally likely to die at any age, as indicated by the straight-line survivorship curve. Oysters, on the other hand, produce vast numbers of offspring, only a few of which live to reproduce. Once oyster offspring become established and grow into reproductive individuals, however, their mortality is extremely low. Even though human infants are susceptible to death at relatively high rates, humans' highest mortality rates occur later in life, in their postreproductive years.

By convention, there are three types of survivorship curves: Type I represents the kind of life cycle found in humans, where a large proportion of the individuals appear to reach their physiologically determined maximum age; type II reflects the situation found in hydra, where the mortality rate remains more or less constant at all ages; and type III represents the kind of life cycle found in oysters, where mortality is high in the early stages but then declines. Many animal and protist populations in nature probably have survivorship curves that lie somewhere between those characteristic of type II and type III, and many plant populations, with high mortality at the seed and seedling stages, are probably closer to type III. Humans have probably approached type I more and more closely through the years, with the birth rate remaining relatively constant or declining somewhat, but the death rate dropping markedly.

In type I survivorship curves, a large number of individuals reach their maximum theoretical age. In type II curves, mortality is largely independent of age. In type III curves, mortality is highest during the young stages. These curves are convenient ways to describe the different kinds of life histories.

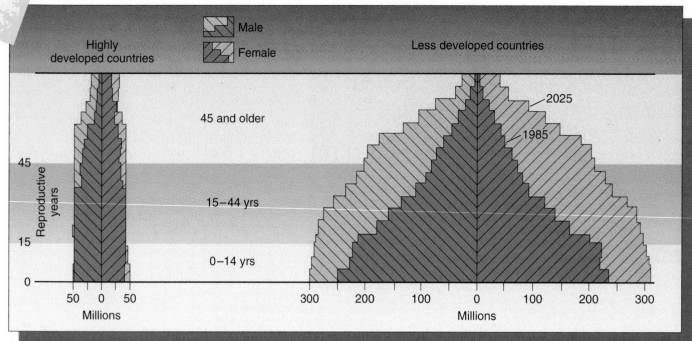

Male

Female

Highly developed countries

Less developed countries

45 and older

2025

1985

45

15

0

Reproductive years

15–44 yrs

0–14 yrs

50   0   50

Millions

300   200   100   0   100   200   300

Millions

Figure 20.9 The coming population explosion will hit less-developed countries hardest.

Less-developed countries have triangular age distribution profiles, with much of their population yet to enter childbearing age. When all of the young people begin to bear children, the population will experience rapid growth.

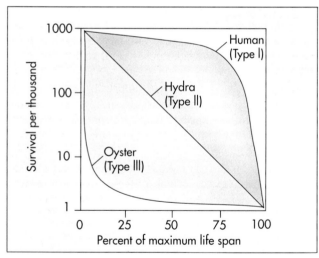

Figure 20.10 Survivorship curves.

The shapes of the curves are determined by the percentages of individuals who die at specific ages. Type I curves represent populations in which most individuals reach their physiologically determined maximum age. Type II curves represent populations in which the mortality rate is fairly constant at all ages. Type III curves indicate populations in which mortality is high in the early stages but then declines.

Human Population Growth

As mentioned earlier, the human population numbers more than 5.6 billion people. The rapid growth of the human population over the last three hundred years has widespread implications. If the human population continues to grow, competition for resources will increase. Currently available resources will be exploited, and the potential for environmental harm will be high. Why is the human population growing so rapidly, and how will this rapid growth affect future generations? How can the growth be contained? The first step in answering these questions is to understand the problem.

Demography and Population Pyramids

Demography, or the statistical study of populations, is the tool that scientists employ to obtain accurate numbers of population size. The term *demography* comes from two Greek words: *demos,* meaning "the people" (the same root as for the word *democracy*), and *graphos,* meaning "measurement." Demography allows predictions of how population sizes will fluctuate in the future. It takes into account a population's age distribution and its changing size through time.

A population whose size remains the same through time is called a **stable population.** In stable populations, births plus immigration exactly balance deaths plus emigration. Population size and age distribution remain constant.

Characteristics of the human population can be illustrated graphically with a **population pyramid**—a bar graph using five-year age categories (figure 20.11). In a population pyramid, males are conventionally placed to the left of the vertical age axis, females to the right. A population pyramid thus shows the composition of a population by age and sex.

Figure 20.11 shows population pyramids for Kenya, the United States, and Austria in 1990. In the U.S. population pyramid, the number of cohorts (groups of individuals) fifty-five to fifty-nine years old is smaller than those preceding and following since these people were born during the depression years. The cohort from the years 1945 to 1964 represents the "baby boom."

The shape of the Kenyan population pyramid in figure 20.11 is characteristic of developing countries, where the number of children less than fifteen years old often exceeds 40 percent of the total population. When these individuals reach childbearing age, the population will grow explosively. Even if the Kenyan people limited family size to two children (half of the present average), effective immediately, the country's population growth would not level off until the middle of the next century.

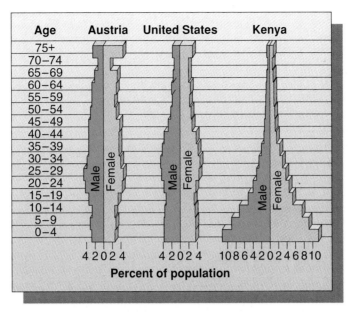

Figure 20.11 Three 1990 population pyramids.
Population pyramids are graphed according to a population's age distribution. Kenya's pyramid has a broad base because of the great number of individuals at or below childbearing age. The U.S. pyramid clearly demonstrates the larger number of individuals in the "baby boom" cohort—the pyramid bulges from the years 1945 to 1964. Austria's pyramid has a narrow base, and thus Austria is in a negative growth phase—that is, fewer people in the future will be at childbearing age.

History of Human Population Growth

How did we get to our present situation, in which the rapidly growing human population is now threatening our survival? An understanding of this rapid growth requires going back to the beginning. The earliest fossils that are clearly *Homo sapiens* (modern humans) come from western Europe and are about a hundred thousand years old. Human beings reached North America at least twelve to thirteen thousand years ago, crossing the narrow straits between Siberia and Alaska and moving swiftly to the southern tip of South America. By ten thousand years ago, when the continental ice sheets withdrew and agriculture first developed, there were about five million people on earth, distributed over all the continents except Antarctica. With the new and much more dependable food sources that became available as a result of agriculture, the human population began to grow more rapidly. Towns and then cities developed in the areas where agriculture was practiced, and food was sufficiently abundant; such settlements had become widespread by five thousand years ago. About two thousand years ago, an estimated 130 million people lived on the earth—less than half the population of the United States and Canada today.

The ability to live together on a long-term basis in relatively large settlements made possible, in turn, the specialization of professions in these centers. This was a necessary condition for the development of modern culture. Such advances as metal tools and utensils could not develop until after towns had been formed.

The blip at the bend in the curve of figure 20.12 reflects the outbreak of bubonic plague in 1348. Within a generation, seventy-five million people, about 30 percent of Europe's population, died. But the human population recovered quickly from this disaster and rapidly replaced the lost numbers. By 1650, the world population had doubled, and doubled again, reaching five hundred million, with many people living in substantial urban centers where many important innovations were occurring. The Renaissance in Europe, with its renewed interest in science, ultimately led to the establishment of industry in the seventeenth century and to the Industrial Revolution of the late eighteenth and early nineteenth centuries.

The Present Population Situation

For the past three hundred years, and probably for much longer, the human birth rate (as a global average) has remained nearly constant, at about thirty births per year per thousand people. It may be lower now—about twenty-five births per year per thousand people—but this difference may not be significant. However, with better sanitation and improved medical techniques, the death rate has fallen steadily, to an estimated 1994 level of about nine deaths per thousand people per year. The difference between these two figures amounts to an annual worldwide increase in human population of approximately 1.6 percent. While such a rate of increase may seem

392

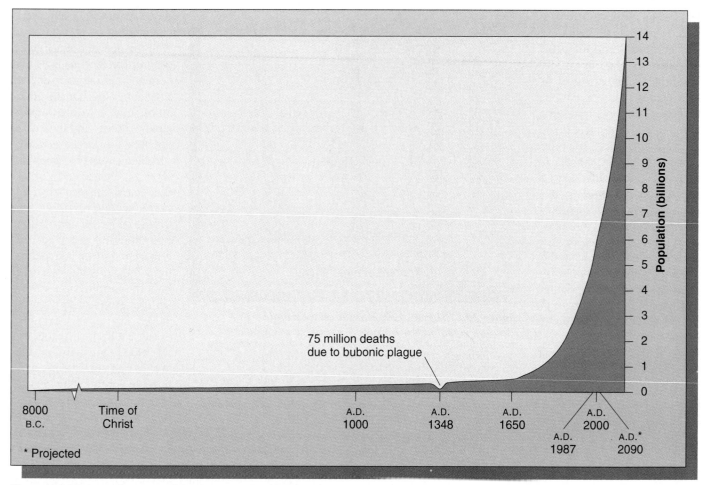

Figure 20.12 Growth curve of the human population.
Over the past three hundred years, the world population has been growing steadily.

relatively small, it could potentially lead to a doubling of the world population from its present level of more than 5.6 billion people in only forty-three years!

The world population reached an unprecedented five billion people by early 1987, and the *annual* increase in population amounts to about ninety million people, a number substantially larger than the total population of Britain or Germany. At this rate, more than 246,000 people are added to the world population each day, or more than 170 every minute! The world population is expected to rise to well over six billion people by the end of the century, and then perhaps to a stable level as high as thirteen billion by the year 2090, according to United Nations estimates.

In view of the limited resources available to the human population and the need to learn how to manage those resources well, the first and most necessary step toward global prosperity is to stabilize the human population. One of the surest signs of the pressure being placed on the environment is human utilization of about 40 percent of the total net global photosynthetic productivity on land. Given that statistic, a doubling of the human population in forty-three years poses extraordinarily se-

vere problems. The facts virtually demand restraint in population growth. If and when technology is developed that would allow greater numbers of people to inhabit the earth in a stable condition, the human population can be increased to whatever level might be appropriate.

> *In the late 1990s, the global human population of more than 5.6 billion people will grow at a rate of approximately 1.6 percent annually. At that rate, it would reach about 6.2 billion people by the year 2000 and about 8.4 billion by the year 2025.*

By the year 2000, about 60 percent of the people in the world will be living in countries that are at least partly tropical or subtropical. An additional 20 percent will be living in China, and the remaining 20 percent—one in five—in the so-called developed, or industrialized, countries: Europe, Russia, Japan, the United States, Canada, Australia, and New Zealand. Whereas the populations of the developed countries are growing at an annual rate of only about 0.3 percent, those of the less developed, mostly tropical countries (excluding China) are growing at an annual rate estimated in 1994 to be about 2.2 percent. For every person living in an industrialized country in 1950, there were two people living elsewhere; by 2020, just seventy years later, there will be five.

Because people in industrialized countries control about 85 percent of the world's wealth and material goods and enjoy standards of living perhaps twenty times higher than those found in many developing countries, the changing ratios of total population in each of these sectors should be a matter of special concern to everyone. For example, the infant mortality rate in industrial countries in 1994 was ten per thousand, whereas in developing countries (excluding China), it was seventy per thousand; the respective life expectancies at birth were seventy-five years versus sixty years, the latter number being even lower in those developing countries where many people are malnourished.

As mentioned earlier in the chapter, a population's age structure determines how fast the population will grow. For this reason, predicting the future growth patterns of a population requires knowing what proportion of individuals have not yet reached childbearing age. In developed countries, such as the United States, about a fifth of the population consists of people under fifteen years of age; in developing countries, such as Mexico, the proportion is typically about twice as high. Thus, even if the policies that most tropical and subtropical countries have established to limit population growth are carried out consistently for decades, the populations of these countries will continue to grow well into the next century, and individuals in the developed countries will constitute a smaller and smaller proportion of the world's population. For example, even if India, with a mid-1994 population level of about 912 million people (with 36 percent under fifteen years old), managed to reach a simple replacement reproductive rate by the year 2000, its population would still not stop growing until the middle of the next century. At present rates and patterns of growth and given the country's age structure, India will have a population of nearly 1.4 billion people by the year 2025 and will still be growing rapidly.

Most countries are devoting considerable attention to slowing the growth rate of their populations, and there are genuine signs of progress. Some countries have provided government-sponsored birth control facilities to aid families in reducing the number of children and in spacing the births of children. These

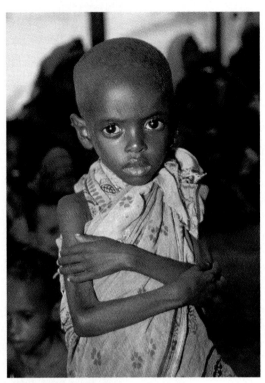

Figure 20.13 Death by deprivation.
Starvation and disease are major killers in many parts of the world, particularly among the young. When large numbers of people fled Rwanda during the civil unrest in 1994, over a hundred thousand people died of cholera and dysentery in crowded refugee camps.

programs are also accompanied by public service announcements and advertising that encourage birth control and limiting the number of births. In addition, studies have shown that the education of women acts to lower the birth rate; some countries are experimenting with programs focusing on female education. Depending on the success of these programs, the world population may stabilize by the close of the next century at about thirteen billion people. No one knows whether the world can support so many people indefinitely. Finding a way to do so is the greatest task facing humanity's future. The quality of life available for our children in the next century will depend to a large extent on our success.

Food and Population: An Example of Dwindling Resources

Even though experts estimate that enough food is produced in the world to provide an adequate diet for everyone in it, food distribution is so unequal that large numbers of people live in hunger. In the United States, Russia, Japan, and Europe, there is, on the average, a quarter to a third *more* food available per person than the United Nations Food and Agriculture Organization (FAO) regards as the caloric intake necessary to maintain moderate physical activity. In contrast, such countries as Bangladesh, Ecuador, and Kenya have 10 to 15 percent *less* than this minimum available per person and, for the most part, little cash with which to purchase more. Worldwide, over five hundred million people each consume less than 80 percent of the U.N.-recommended minimum standards for caloric intake, a diet insufficient to prevent stunted growth and serious health risks. Only the United States, Canada, Argentina, and a few European countries have emerged as consistent food exporters.

Of the approximately 4.5 billion people living in the tropics and China in 1994, the World Bank estimated that some 1.5 billion—nearly a quarter of the total world population—were living in extreme poverty, not having adequate food, clothing, and shelter for themselves and their families from day to day. Among these malnourished people, UNICEF (United Nations International Children's Emergency Fund) estimates that about thirteen million children under the age of five starve to death every year or die of diseases complicated by malnutrition (figure 20.13). Worse, many millions of additional children exist only in a state of lethargy, their mental capacities often permanently damaged by their lack of access to adequate quantities of food.

> *About 1.5 billion people live in a state of extreme poverty; many of them are malnourished.*

One of the most alarming trends in developing countries is the massive movement to urban centers. Mexico City, the largest city in the world today, is plagued by smog, traffic, waste disposal, and other problems, all worsened by the incredible congestion of over twenty million inhabitants (figure 20.14)! The prospects of supplying adequate food, water, and sanitation to a city whose population will increase to over thirty million by the end of the century are almost unimaginable.

Nor is Mexico City's astonishing urban growth unique. Only seven cities had a population larger than five million in 1950; by the year 2000, there will be fifty-seven such megacities, forty-two of them in the Third World. New York City has almost sixteen million inhabitants, and Mexico City, Tokyo, and São Paulo are bigger than New York! These four cities alone contain over seventy million inhabitants—half as many as the entire human population two thousand years ago. The earth is peppered with gigantic cities like New York, growing rapidly. In 1992, there were sixteen cities with populations greater than ten million. By the year 2000, there will be twenty-six such cities—and all but three of these will be in the South (figure 20.15).

Figure 20.14 Mexico City, the world's largest city.
Mexico City has serious problems related to its large population, such as sanitation difficulties, air pollution, and poverty.

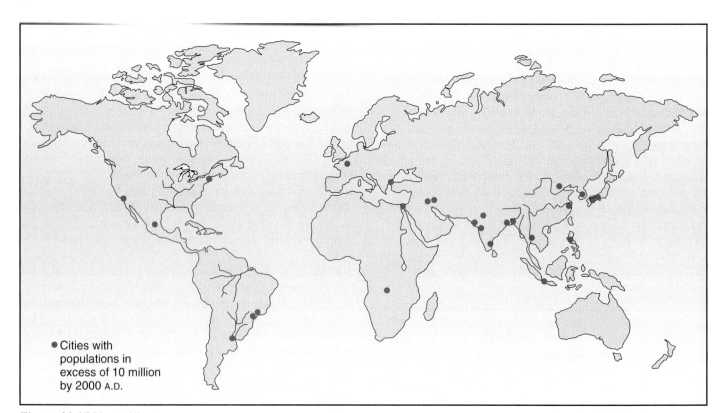

● Cities with
populations in
excess of 10 million
by 2000 A.D.

Figure 20.15 Megacities.
In 1920, the world's urban population amounted to 360 million; by the end of the century, it will be nearly three billion. Only seven cities had a population larger than five million in 1950; by 2000, there will be fifty-seven such megacities, forty-two of them in the Third World. Twenty-six cities are likely to have a population greater than ten million.

The proportion of poor, hungry people in the less-developed countries, now amounting to a third of their populations, will grow rapidly unless successful solutions to their problems are developed in the near future.

One of the greatest and most immediate challenges facing today's world is to produce enough food to feed the expanding human population. Even though world food production has expanded 2.6 times since 1950—more rapidly than the human population—virtually all the lands that can be cultivated are already under cultivation. A quarter of the world's topsoil has been lost from agricultural lands during this period, and the human population continues to grow explosively. Much of the world is populated by very large numbers of hungry people who are rapidly destroying the sustainable productivity of the lands they inhabit. At the same time, consumption in industrialized countries such as the United States, running at twenty to thirty times the rate in developing countries, is having an even greater adverse effect on the future of humanity. Chapter 24 examines the prospects for increased agricultural productivity in the future and how those prospects might be improved.

EVOLUTIONARY VIEWPOINT

One of the most important events in the history of life on the earth has been the decline in the human death rate since 1600, resulting in explosive population growth. Because humans are capable of impacting the environment so seriously, no species is immune to the consequences of the swelling population of people. Millions of years of adaptive evolution can be wiped out by a parking lot. No amount of conservation can long preserve our diverse biological world if human population growth continues unchecked.

SUMMARY

1. Ecology is the study of the relationships of organisms with one another and their environment. Ecologists study groups of organisms at four progressively more inclusive levels of organization: populations, communities, ecosystems, and the biosphere.

2. A population is a group of individuals of a particular species that live together. Specific populations have characteristic properties, such as size (the number of individuals in a population), density (the number of population individuals that occur in a given area), and dispersion (random, clumped, or uniform).

3. The actual rate of population increase is defined as the birth rate plus the number of immigrants into the population minus the death rate plus the number of emigrants out of the population.

4. Most populations exhibit a sigmoid growth curve, which implies a relatively slow start in growth, a rapid increase, and then a leveling off as the environment's carrying capacity is approached. Optimal yields are obtained when populations are harvested in the rapid growth phase.

5. *K* strategists are slow-growing, few-offspring populations, whose size tends to be limited by the carrying capacity of their environment. In contrast, *r* strategists have large broods and rapid rates of population growth.

6. Survivorship curves describe the characteristics of growth in different populations. In type I populations, a large proportion of the individuals approach their physiologically determined age limits. Type II populations have a constant mortality throughout their lives. Type III populations have very high mortality in their early stages of growth, but an individual surviving beyond that point is likely to live a very long time.

7. Demography is the statistical study of populations. Demographers use population pyramids to graph a population's age structure. In the human population, demography has revealed that, in developing countries such as Kenya, as much as 40 percent of the population is under the age of fifteen. When these individuals reach childbearing age, the population will grow explosively.

8. The human population has grown rapidly over the last three hundred years because of a constant birth rate and a declining death rate. By the year 2090, the world population could reach thirteen billion, which is more than double the current human population size. Most of the human population will live in nonindustrial, developing countries.

9. One of the most alarming problems associated with the rapidly growing human population is how to supply food to the large populations living in developing countries. Providing enough food for the expanding human population is one of the greatest challenges facing today's world.

REVIEWING THE CHAPTER

1. What is the definition of ecology?

2. What are populations?

3. What are three characteristic properties of specific populations?

4. How is population growth defined?

5. What is the formula for the actual rate of population increase?

6. What is the carrying capacity of a population?

7. How do plagues affect population growth?

8. What are density-dependent and density-independent effects?

9. How are *r* strategist and *K* strategist organisms defined?

10. What is the interrelationship between mortality and survivorship?

11. Should there be concern over human population growth?

12. What is demography?

13. What is the history of human population growth?

14. What is the present situation in human population growth?

15. What is the current status of food supply with regard to human population density?

16. What does the future hold for the human population and food supply?

COMPLETING YOUR UNDERSTANDING

1. How do the terms *habitat, range, community, biome,* and *biosphere* relate to populations of organisms?

2. What is the effect of density on future population growth?

3. In plants, what are the mechanisms that allow for populations to be distributed in a uniform, random, or clumped manner?

4. In populations, how is the actual rate of increase (*r*) defined?

5. How is AIDS not like the bubonic plague with respect to transmission?

6. What positive social effects, if any, came from the bubonic plague and currently are coming from the AIDS epidemic?

7. What is the significance of the S-shaped sigmoid growth curve in biological populations?

8. Which of the following organisms are *K* strategists? [*three answers required*]
  a. Bacteria   d. Humans
  b. Redwoods   e. Dandelions
  c. Cockroaches   f. Whales

9. What are the differences between type I, type II, and type III survivorship curves? Give an example of an organism that would fit each type.

10. From figure 20.11, which country—Kenya, the United States, or Austria—should be most concerned about its future population and why? Can you think of one or more factors not considered in the figure that could alter the future U.S. population?

11. What is the best explanation for the increase in the estimated human population of five million (ten thousand years ago) to 130 million (two thousand years ago)?

12. According to current projections, when will the human population double again? What will the human population be at that time?

13. From current projections, how will the human population be distributed in the year 2025?

14. What are the current differences in infant mortality and life expectancy between industrial and developing countries?

15. What are some of the governmental and educational efforts currently underway in developing countries to deal with human population growth?

16. What percentage of the current human population lives in extreme poverty?

17. What are some of the quality-of-life problems facing megacities in both industrial and developing countries?

18. As the human population increases, what are some of the major obstacles to increasing the food supply worldwide?

FOR FURTHER READING

Beddington, J. R., and R. M. May. 1982. The harvesting of interacting species in a natural ecosystem. *Scientific American,* Nov., 62–69. An analysis of the changing populations of whales and other animals that feed on the krill (shrimp) populations in the Antarctic Ocean to exemplify the problems of using a biological resource without destroying it.

Daily, G. C., and P. R. Ehrlich. 1992. Population, sustainability, and earth's carrying capacity. *BioScience,* Sept., 761–70. An argument for finding ways to sustain the growing human population without undermining the planet's potential for supporting future generations.

Population Reference Bureau. 1995. *World population data sheet–1995.* Washington, D.C.: Population Reference Bureau. A data sheet, published every year and free to anyone who requests it, that lists the current population statistics for every country in the world. For information, contact the Population Reference Bureau at (202) 483–1100.

World Resources Institute. *World resources, 1992–93.* 1993. New York: Oxford University Press. An up-to-date accounting of the world's growing population and the prospects for sustainable development of the world's resources.

HOW SPECIES INTERACT WITHIN COMMUNITIES

Alchemy at its best—this wild cucumber vine (Echinocystis lobata) converts light energy into chemical energy and interacts closely with its ecosystem by contributing to cycling of nutrients.

FOR REVIEW

Here are some important terms and concepts that have been discussed in previous chapters and that you will encounter again in this chapter. Review them before proceeding if necessary.

Natural selection (*chapter 15*)
Energy flow (*chapter 7*)
Adaptation (*chapter 15*)
Communities (*chapter 20*)
Populations (*chapter 20*)
Ecosystems (*chapter 20*)
Niche (*chapter 20*)

How an ecosystem functions is largely determined by the interactions of the organisms living within it. This set of interacting organisms is called a community. The relationships among the organisms of an ecosystem's community are the keys to understanding that ecosystem.

The magnificent redwood forest that extends along the coast of central and northern California and into the southwestern corner of Oregon is an example of a community. Within it, the most obvious organisms are redwood trees, *Sequoia sempervirens*. These trees are the sole survivors of a genus that once was distributed throughout much of the Northern Hemisphere. With the redwoods are regularly associated a number of other plants and animals (figure 21.1). Their coexistence is partly made possible because of the special conditions created by the redwoods themselves: shade, water (dripping from the branches), and relatively cool temperatures. This particular, distinctive assemblage of organisms is called the redwood community.

This community is recognized largely because of the redwoods themselves. The overall distributions of the other kinds of organisms that occur together with the redwoods differ greatly from one another: Some are smaller than the distribution of the redwoods, and some are larger. In the redwood community, their **niches** (the roles organisms play in an ecosystem) overlap, which is why, in this community, they occur together, a recognizable group. (Niches are discussed in more detail later in the chapter.) However, viewing the community as a kind of "superorganism" (the term that some ecologists in the past used to refer to communities) is incorrect. Rather, it is a collection of distinct organisms with overlapping requirements in close proximity to one another.

The redwood community also has a historical dimension: The organisms now characteristic of this community have each had a complex and unique evolutionary history. At different times in the past, they evolved and then came to be associated with the redwoods. The present-day redwood community has come into existence only during the past few million years. During this period, the redwood's range was reduced as climatic change eliminated the tree from many of the areas in which it grew previously. In a historical sense, then, the redwood community recognized today represents the concordance of the individual histories of the plants, animals, and microorganisms that compose it.

Figure 21.1 The redwood community.
The forests of coastal California and southwestern Oregon are dominated by redwoods, although many other kinds of plants live beneath them.

A redwood community is recognized largely because of the presence of the redwood trees, but many other kinds of organisms exist in this community as well. The community exists because the niches of these species overlap in it.

Although they are not superorganisms, many communities are very similar in their species composition and appearance over wide areas. For example, the open savanna grassland that stretches across much of Africa includes many plant and animal species that coexist over thousands of square kilometers. Interactions among these organisms, some of which have evolved over millions of years, are similar throughout these communities. This chapter explores some of these interactions and then considers the ways in which the distributions of organisms are limited—why particular kinds of organisms occur in some places, and not in others.

Coevolution

The interactions among organisms that characterize particular communities and ecosystems have arisen as a result of the organisms' evolutionary history. **Primary producers** (photosynthetic organisms, including plants, algae, and photosynthetic bacteria), **herbivores** (organisms that consume photosynthetic organisms), and second and higher level predators (organisms that consume other predators) have changed and adjusted to one another continually over millions of years. For example, many of the features of flowering plants have evolved in relation to the way their pollen is dispersed by animals, especially insects (discussed further in chapter 31). These animals, in turn, have evolved a number of special traits that enable them to obtain food or other resources efficiently from the plants they visit. In addition, the seeds of many flowering plants have features that make them more likely to be dispersed to new areas of favorable habitat.

Such interactions, which involve the long-term, mutual evolutionary adjustment of the characteristics of the members of biological communities in relation to one another, are examples of **coevolution.** This chapter examines some examples of coevolution, grouping them under the general headings of competition, predator-prey interactions, plant and animal defenses, and symbiosis.

> Coevolution *is a term that describes the long-term evolutionary adjustment of one group of organisms to another.*

Figure 21.2 The flour beetle Tribolium confusum, showing adults, larvae, and pupae.
This species, and the closely similar T. castaneum, *were used in experiments that tested what happens when two species of organisms compete with one another.*

> The principle of competitive exclusion states that, if two species are competing with one another for the same limited resource in the same place, one of them will be able to use that resource more efficiently than the other and eventually will drive the second species to extinction locally.

The principle of competitive exclusion has been demonstrated in a number of laboratory experiments, but not always with predictable results. For example, when Thomas Park and his colleagues at the University of Chicago grew two species of flour beetle, *Tribolium,* together in the same container of flour, one species always became extinct, but the species that survived varied. *Tribolium castaneum* usually won under relatively hot and damp conditions, whereas *T. confusum* won under cooler, drier conditions. Subsequent experiments with these species demonstrated that a genetic component was also involved in the unpredictability of the outcome. Some strains of one species would win over some—but not all—strains of the other under a given set of conditions (figure 21.2).

Competition

Competition between individuals of two or more species for the same resources is a factor that may limit population size. When two or more species compete in this way, it is called **interspecific competition.** Interspecific competition is not the same as competition among individuals of a single species, which is called **intraspecific competition.**

The Principles of Competitive Exclusion

Two species that both use a scarce or **limiting resource** are said to compete with one another. Interspecific competition is greatest among organisms that occupy the same **trophic level.** A trophic level is a step in the flow of energy through an ecosystem, such as the step at which plants manufacture food or the step at which carnivores feed on other animals. Thus, plants compete mainly with other plants, herbivores with other herbivores, and carnivores with other carnivores. In addition, competition is more acute among organisms that closely resemble one another than among organisms that are less similar.

More than fifty years ago, Soviet ecologist G. F. Gause formulated what he called the **principle of competitive exclusion.** This principle states that, if two species are competing with one another for the same limited resource, then one of the species will be able to use that resource more efficiently than the other, and the former will therefore eventually eliminate the latter locally.

Competition in Nature

The principle of competitive exclusion can also be studied through observation and experiment in nature. J. H. Connell of the University of California, Santa Barbara, investigated the competitive interactions between two species of barnacles (marine crustaceans) that grew together on the same rocks along the coast of Scotland. Free-swimming barnacle larvae settle down, cement themselves to rocks, and then remain permanently attached to that point. Of the two species Connell studied, *Chthamalus stellatus* lives in shallower water, where it is often exposed to air by tidal action, and *Balanus balanoides* occurs lower down, where it is rarely exposed to the atmosphere. Connell found that, in this deeper zone, *Balanus* could always outcompete *Chthamalus* by crowding it off the rocks, undercutting it, and replacing it even where it had begun to grow. But when Connell removed *Balanus* from the area, *Chthamalus* was easily able to occupy the deeper zone, indicating that no physiological or other general obstacles prevented it from becoming established in that area. In contrast, *Balanus* cannot survive in the shallow-water habitats of *Chthamalus*. *Balanus* evidently does not have the special physiological and morphological adaptations that allow *Chthamalus* to occupy this zone.

In another set of field observations, the late Princeton ecologist Robert MacArthur studied five species of warblers—small, insect-eating birds that coexist during part of the year in the forests of the northeastern United States and adjacent Canada.

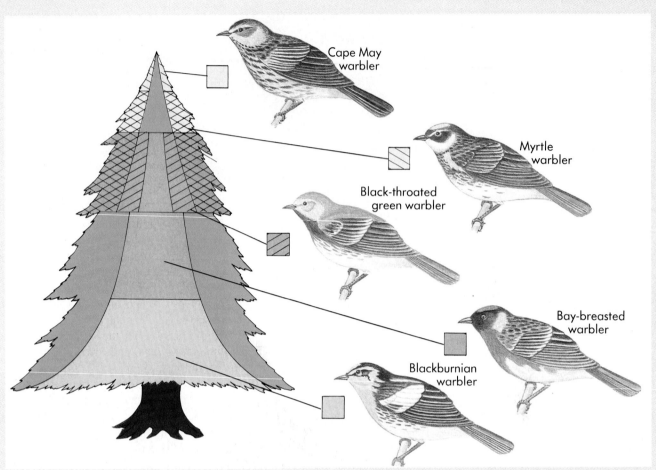

Figure 21.3 Using environmental resources differently avoids competition.
Although all five species of warbler shown here feed on insects in the same spruce trees at the same time, they mainly feed in different parts of the tree and in different ways. In this way, competition is avoided.

Although they all appeared to be competing for the same resources, MacArthur found that each species actually spent most of its time feeding in different parts of the trees, so each ate different subsets of insects in those trees. Some of the warbler species fed on insects near the ends of the branches, while others regularly penetrated well into the foliage, and some stayed high on the trees, while others fed on the lower branches. These patterns were recombined in different ways characteristic of each of the warbler species. As a result of these different feeding habits, each species of warbler actually occupied a different niche; in other words they had different ways of utilizing the environment's resources and thus did not directly compete with one another for limited resources (figure 21.3).

> *Species of barnacles, warblers, and many other kinds of organisms can coexist because their niches differ—in other words, they use environmental resources differently. Thus, competitive exclusion does not occur.*

During the past decade, there has been a lively debate about the real importance of competition in nature in relation to community structure. Environmental changes greatly affect the outcome of competitive situations. Furthermore, in the kinds of highly diverse situations that occur in nature, there are opportunities for many species to coexist in the same environment, so evaluating the competitive interactions that may be occurring is often very difficult.

The Niche

Studies of the principle of competitive exclusion led to the development of the concept of the niche. As mentioned earlier, a niche may be defined most simply as the role an organism plays in an ecosystem. It may be described in terms of space, food, temperature, appropriate conditions for mating, requirements for moisture, and so on. A full portrait of an organism's niche also includes the organism's behavior and how this behavior changes at different times of the day and during different seasons. In other words, *niche* is not synonymous with *habitat*. The habitat of an organism—the place where it lives—is defined by some, but by no means all, of the factors that make up its niche. The factors that make up an organism's niche determine whether the organism can exist in a given ecosystem and also how many species can exist there together.

The **actual niche** of an organism—the role the organism actually plays in a particular ecosystem—is distinguished from the

(a)

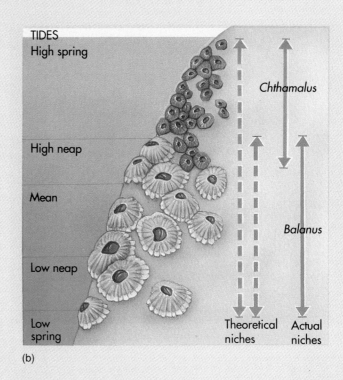

(b)

Figure 21.4 Competition that limits niche use.
(a) Chthamalus *(the smaller, smooth barnacles) and* Balanus *(the larger, ridged barnacles) grow together on a rock along the coast of Scotland. (b) The distribution of the two species with respect to different water levels. Investigators found that* Chthamalus *could live in both deep and shallow zones (its theoretical niche), but that* Balanus *forced* Chthamalus *out of that part of its theoretical niche that overlapped the actual niche of* Balanus.

niche the organism might occupy if competitors were not present, called the **theoretical niche.** Thus, the theoretical niche of the barnacle *Chthamalus* in Connell's experiments in Scotland included that of *Balanus,* but *Chthamalus's* actual niche was much narrower because of competition from *Balanus* in *Balanus's* actual niche (figure 21.4).

> *A niche may be defined as the role that an organism plays in the environment. An organism's theoretical niche is the niche the organism would occupy if competitors were not present. An organism's actual niche is the niche the organism actually occupies under natural circumstances.*

The flour beetles of genus *Tribolium* discussed earlier also demonstrate some interesting properties of the niche. If *Tribolium* is grown in pure flour along with beetles of a second genus, *Oryzaephilus, Oryzaephilus* is driven to extinction by means only partly understood. But if small pieces of glass tubing are added to the flour, providing refuges for *Oryzaephilus,* then both kinds of flour beetles will coexist indefinitely. In a sense, this experiment suggests why so many kinds of organisms can coexist in a really complex ecosystem, such as a tropical rain forest, and why competition is more direct in an ecosystem with fewer species, such as agricultural fields.

Gause's principle of competitive exclusion can be restated, in terms of niches, as follows: *No two species can occupy the same niche.* While species can and do coexist while competing for the same resources, Gause's theory predicts that, when two species coexist on a long-term basis, one or more features of their niches will always differ; otherwise, one species will inevitably become extinct.

Predator-Prey Interactions

Predation, like competition, is a factor that may limit population size. In this sense, predation includes everything from one kind of animal capturing and eating another to **parasitism,** the condition of an organism living in or on another organism, at whose expense the parasite is maintained. Predation and parasitism are two ends of a biological spectrum between which there is no clearly marked distinction. They are governed by similar principles.

When experimental populations are set up under very simple circumstances in the laboratory—as Gause did with *Paramecium* and its predator protozoan *Didinium*—the predator often exterminates its prey and then becomes extinct itself, having nothing to eat (figure 21.5). If refuges are provided for the prey, prey populations will be driven to low levels but can recover. The populations of predators and prey then tend to follow a cyclical pattern: The low population levels of the prey species provide scant food for the predators, who, in turn, become scarce, which allows the prey to recover and again become abundant. Population cycles are characteristic of some species of small mammals, such as lemmings, and may be stimulated, at least in some situations, by their predators (figure 21.6).

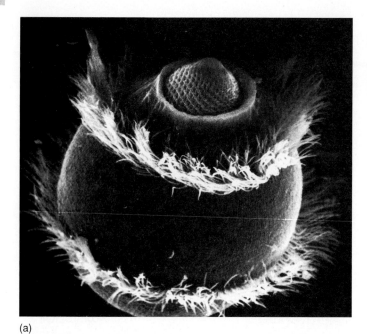

(a)

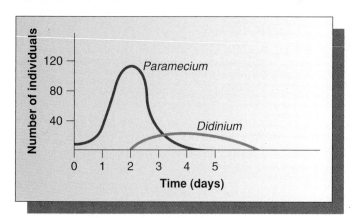

(b)

Figure 21.5 Gause's experiments on predation.
(a) *Egg-shaped* Didinium, *the predator, has almost completely ingested its prey—the smaller protist* Paramecium. (b) *The graph demonstrates that, when* Didinium *is added to a population of* Paramecium, *the numbers of* Didinium *initially rise, while the numbers of* Paramecium *steadily fall. As the* Paramecium *are depleted, however, the* Didinium *also die off.*

(a)

(b)

Figure 21.6 A predator-prey cycle.
(a) *A snowshoe hare being chased by a lynx.* (b) *The numbers of lynxes and snowshoe hares oscillate in tune with each other in northern Canada. The data are based on numbers of animal pelts from 1845 to 1935. As the number of snowshoe hares grows, so does the number of lynxes, and the cycle repeats about every nine years. The number of snowshoe hares is not controlled by predators, but by available resources. The number of lynxes, on the other hand, is controlled by the availability of prey, or snowshoe hares.*

Relationships of this sort are important for biological control. If the prey becomes so rare that the predator species only encounters it infrequently, the predator itself may become extinct. Ideally, in the case of deliberate biological control, the prey will survive in small numbers, thus making it possible for small populations of the predator to also survive and for the prey species to be controlled indefinitely.

For example, in Australia, prickly pear cacti (*Opuntia*), introduced from Latin America, once overran the ranges and became so abundant that vast areas were closed to cattle grazing. The situation was changed dramatically with the introduction of the moth *Cactoblastis*, from the areas where the cacti were native. The moth larvae feed on the pads of the cacti and rapidly destroy the plants. Within relatively few years, the moth reduced the formerly abundant cactus to the status of a rare species in many regions. Today, finding an individual of *Cactoblastis* is rare, but the moth is still present and evidently keeps the cactus in check (figure 21.7).

The future is more problematic for the American chestnut (*Castanea americana*), a species of tree that has virtually been driven to extinction by the accidental introduction of the Asiatic fungus, *Cryptonectria* (*Endothia*) *parasitica*. The chestnut used to be a dominant or codominant tree throughout much of the forests of eastern and central temperate North America. Chestnut blight, the disease caused by the fungus, was first seen in North America in New York State in 1904. It killed most North American chestnut trees in the next thirty years. Today, the American chestnut survives largely as sprouts that grow every year from

(a)

(b)

Figure 21.7 Biological control of pests.
(a) *In 1926, prickly pear cacti* (Opuntia), *introduced from Latin America, choked many of the pastures of Australia with their rampant growth.* (b) *The cacti were eventually controlled by the introduction of* Cactoblastis, *a cactus-feeding moth from the areas where the cacti were native.* (c) *Three years later, the cacti had been almost completely cleared.*

(c)

the trunks of trees that were killed decades ago. A few populations of chestnuts appear to be isolated from chestnut blight in remote areas, whereas others apparently are genetically resistant.

Predator-prey relationships operate in many other familiar ways. Organisms that cause diseases that completely kill their host species are not successful because this eliminates the disease-causing organism's own source of food. Thus, less virulent strains of the disease-causing organism are favored by natural selection and will survive. For example, rabbits were introduced to Australia and New Zealand as a convenient source of meat, but they soon ran wild, with devastating effects on the countryside. When the virus causing the disease myxomatosis was introduced to control the rabbits, most of the rabbits soon died. The most virulent strains of the virus disappeared with the dead rabbits, and less lethal strains became apparent in the remaining rabbit populations. At the same time, strains of rabbits that were resistant to the disease began to appear. Now, the populations of both organisms have achieved an equilibrium relationship in which they can coexist indefinitely (figure 21.8).

> *An organism that causes a disease that always kills its host will also die. An organism that causes a disease with sub-lethal effects will have the chance to spread to another host.*

The predator-prey relationships between large carnivores and grazing mammals are sometimes deceiving. For example, on Isle Royale in Lake Superior, moose multiplied freely in isolation. When wolves later reached the island by crossing over the ice in winter, as the moose had done earlier, biologists assumed that the wolves were playing the determining role in controlling the moose population. More careful studies, however, have demonstrated that this is not the case (figure 21.9). The moose that the wolves eat are mostly old and diseased and would not have survived long anyway. In general, the moose are controlled by the amount of food available to them, their diseases, and many factors other than the wolves.

Wolves have been returning slowly to remnants of their former habitat in the United States outside of Alaska, where fewer than thirteen hundred wolves live. Only two of the twenty-four

Figure 21.8 Introduced species often explode in numbers.
This photo shows rabbits killed by Australian hunters. Imported from Europe as a source of meat, rabbits ran wild in Australia, until they were eventually controlled by a lethal virus that causes the disease myxomatosis.

Figure 21.9 Wolves chasing a moose—what will be the outcome?

On Isle Royale, in Michigan, a large pack of hungry wolves chased a moose for 2 kilometers. The moose finally turned and faced the wolves, who by that time were exhausted from running through chest-deep snow. The wolves laid down, and the moose walked away. Thus, predator-prey relationships are not always clear-cut. Isle Royale wolves only prey on moose who are old and infirm; rarely are they able to run down a healthy moose.

subspecies that originally roamed North America currently survive in this area. Because wolves kill a small amount of livestock, their presence in farming areas is controversial. Their reintroduction into their former range in Yellowstone National Park, where biologists have set a goal of ten breeding pairs, has likewise met with opposition from neighboring ranchers. On the other hand, there is no known case in North America of injury to a human being by a wild, nonrabid wolf, sensational stories to the contrary.

The intricate interactions between predators and prey are essential factors in the maintenance of groups of organisms occurring together that are rich and diverse in species. By controlling the levels of some species, predators make possible the continued existence of others in that same community. In other words, by keeping the numbers of individuals of some of the competing species low, the predators prevent or greatly reduce competitive exclusion. Such patterns are particularly characteristic of biological communities in intertidal marine habitats. For example, in preying selectively on bivalves, sea stars prevent bivalves from monopolizing all the space in intertidal marine habitats, thereby opening the habitats to many other kinds of organisms. In many instances, predators appear to play the key role in allowing species to coexist because situations that appear to violate the principle of competitive exclusion are, in fact, frequent in nature.

Plant and Animal Defenses

Both plants and animals have devised complex mechanisms with which to ward off predators. Plants, for example, can contain chemical substances that are poisonous to other organisms. Other defenses include physical structures, such as spines and thorns, that effectively shield plants from attack. Some animals also employ chemical defenses to ward off predators and herald their poisonous nature with dramatic coloration. Still other animals mimic poisonous species by adopting their coloration. The sections that follow examine these defenses in more detail.

Plant-Herbivore Relationships

The plant-herbivore interface is the point where the greatest energy transfer in the world's ecosystems occurs, with the exception of the interface between organic matter and decomposers. In plant-herbivore interactions, some 300,000 species of autotrophic organisms (organisms that make their own food), including about 250,000 species of plants, use about 1 percent of the sunlight energy that falls on their leaves and green stems. Then at least two million species of herbivores feed on these autotrophs. The herbivores integrate the materials in the plants, algae, or photosynthetic bacteria into their own bodies, actually converting about 10 percent of the energy stored in these organisms. The nature of life on earth has been largely determined by the ways in which plants avoid being eaten and the ways in which herbivores succeed in finding plants to eat.

Plant Defenses

The most obvious ways plants have for limiting herbivore activities are physical defenses. Thorns, spines, and prickles discourage browsers, although some animals have learned to overcome these obstacles (figure 21.10). The defensive role of plant hairs, especially those with a glandular, sticky tip, is less obvious. The enormous abundance of insects and the scale at which they operate as herbivores provide some indication of

Figure 21.10 Many herbivores eat only particular food plants.

The gerenuk is an antelope that grazes successfully on some of the spiny shrubs and trees of the African savanna that other browsers do not use as food.

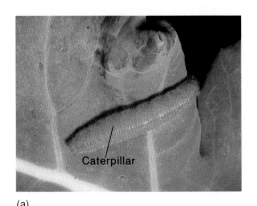

(a)

(b)

Figure 21.11 Some insect herbivores can tolerate toxins.
The green caterpillars (a) of the cabbage butterfly (b) are able to break down the toxic mustard oils that prevent most insects from eating cabbage.

why hairs of various kinds are so important to plants. Simple strengthening of plant parts, as by the deposition of silica in the leaves, is also a protective element in some plants, such as grasses. If enough silica is present in their cells, the plants may simply be too tough to eat.

Significant as these physical adaptations are, plants' chemical defenses are even more numerous and diverse. Since most animals can manufacture only eight of the twenty required amino acids, they must obtain all of the others from plants. By restricting the amount of one or more of the twelve amino acids that animals require from them, plants can limit their nutritional suitability for animals. As a result, they may be protected from herbivores.

Perhaps the most important element in plants' defenses against herbivores are the **secondary chemical compounds,** so called to distinguish them from the **primary chemical compounds** (compounds that are regularly formed as components of the major metabolic pathways, such as respiration). Virtually all plants, and apparently many algae as well, contain structurally diverse secondary chemical compounds that play a defensive role. Estimates suggest the existence of fifty to one hundred thousand secondary plant compounds; some fifteen thousand of these have already been analyzed chemically.

> *Secondary compounds, chemicals not involved in primary metabolic processes, play the dominant role in protecting plants from being eaten by herbivores or predators.*

Much is now known about the roles of secondary compounds in nature. As a rule, different compounds are characteristic of particular groups of plants. For example, the mustard family (Brassicaceae) is characterized by a group of chemicals known as mustard oils, the substances that give pungent aromas and tastes to such plants as mustard, cabbage, watercress, radish, and horseradish. These same tastes signal the presence of chemicals that are toxic to many, but not all, groups of insects (figure 21.11).

Another group of plants, the potato and tomato family (Solanaceae), is rich in alkaloids and steroids, complex molecules that exhibit an enormous range of structural diversity.

Such compounds occur very widely among the flowering plants, but the particular kinds found in different groups vary greatly. Sometimes, the compounds are involved in more complex defensive systems. For example, plants of the milkweed family (Asclepiadaceae) and the closely related dogbane family (Apocynaceae) tend to produce a milky sap that deters herbivores from eating them. In addition, these plants usually also contain cardiac glycosides, molecules named for their drastic effect on vertebrate heart function.

One of the best-known plant groups with toxic effects comprises the related poison ivy, poison oak, and poison sumac (*Toxicodendron* species). All contain a gummy oil called urushiol, which contains a substance that causes a severe rash in susceptible people. Urushiol can persist for years on clothes or other objects with which it has come into contact. At least 140,000 cases cause absence from work in the United States annually, and many more are not reported. Approximately one out of two people is at least moderately sensitive to these plants. Immediate washing with soap removes the excess urushiol and prevents its further spread, but the reaction of exposed body parts appears to be immediate. Urushiol almost certainly functions to protect the plants in which it occurs from herbivores, but this relationship has not been demonstrated experimentally.

As another example, the seeds of castor beans (*Ricinus communis*) produce the protein ricin, which attacks ribosomes. Ricin is similar to a toxin produced by the bacterium that causes bacterial dysentery (*Shigella dysenteriae*). It removes an adenine nucleotide from a specific position in the RNA chain within a ribosome; this area of the ribosome is the same in all animals, which explains the wide toxicity of castor bean seeds. Why such an unusual enzyme in a bacterium should resemble so closely one found in the seeds of a plant is unclear. Perhaps, the bacterium acquired the gene from the plant in the course of its evolution.

Many plant parts are toxic to humans. One familiar example is the houseplant, *Dieffenbachia* (also called dumb cane), a member of the same plant family as philodendrons. Touching a leaf of this plant with the tongue results in intense pain and eventually produces ulcers and corrosive burns, because dumb cane leaves have special ejector cells that expel needlelike crystals of calcium oxalate that penetrate the skin. Particularly in families with children, people should take care to know the properties of the plants they grow.

These few examples demonstrate that many plants are protected by a rich and varied chemical arsenal. Mustard oils, alkaloids, steroids, and other classes of secondary compounds are toxic to most herbivores. Consequently, most herbivores tend to avoid plants with these compounds. Therefore, the pattern of occurrence of such chemicals has greatly affected the evolution of the plants themselves and of the herbivores, especially the insects, that feed on the plants.

Producing Defenses When They Are Needed

Some of the secondary compounds that plants use to defend themselves from herbivores are not normally present in their tissues. Rather, the plants produce the compounds as needed, in much the same way that immune systems operate in animals. When a leaf of a tomato, potato, or alfalfa plant is injured, a chemical message derived from fragments of ruptured cell walls travels rapidly through the plant. This induces the synthesis and accumulation of proteins that inhibit digestion in an animal's digestive tract. These inhibitory enzymes currently are being identified in several laboratories.

Similarly, the infection of a plant by a fungus or bacterium may cause the plant to synthesize a molecule that retards the spread of the infection. A familiar example of such a process is provided by the chemicals that color apples brown when they are cut; interacting with proteins, these chemicals make the apple less attractive to caterpillars, fungi, and other organisms.

Such defensive strategies are metabolically efficient since the plants are able to avoid producing the chemical until it is actually needed, using the energy instead for their growth and maintenance. Parasites, such as fungi, that regularly infect particular kinds of plants are often tolerant of plants' particular defensive substances.

How Herbivores Evolve to Neutralize Plants' Defenses

Associated with each family or other group of plants that is protected by a particular kind of secondary compound are certain groups of herbivores that are able to feed on these plants, often as their exclusive food source. How do these animals manage to avoid the plants' chemical defenses, and what are the evolutionary antecedents and ecological consequences of such patterns of specialization?

Some herbivore groups feed on plants of many families, others on plants of just a few families. In general, the herbivores that feed on a restricted array of plant families, perhaps only one, feed on plant groups in which certain kinds of secondary compounds are well represented. For example, the larvae of cabbage butterflies (subfamily Pierinae) feed almost exclusively on plants of the mustard and caper families, as well as on a few other small plant families characterized by the presence of mustard oils. Similarly, the caterpillars of the monarch butterflies and their relatives (subfamily Danainae) feed on plants of the milkweed and dogbane families. No other groups of butterflies have larvae that feed on these particular plants, which indicates the efficiency of the chemicals in retard-

ing feeding by most kinds of butterflies. Other kinds of insects that feed on these plants are usually specialized within their groups, as are the cabbage and monarch butterflies.

The evolution of these particular patterns can be explained as follows: Once the ancestors of the caper and mustard families acquired the ability to manufacture mustard oils, they were protected for a time against most or all herbivores. The mustard oil plants apparently succeeded so well that they and their descendants evolved and migrated, eventually giving rise to the thousands of species of mustards and capers that now grow worldwide.

At some point, certain groups of insects—for example, the cabbage butterflies—developed the ability to break down the mustard oils and thus feed on these mustards and capers without harming themselves. Having acquired this ability, the butterflies were able to use a new resource without competition from other herbivores. Often, then, in groups of insects such as the cabbage butterflies, sense organs have evolved that are particularly sensitive to the secondary compounds that their food plants produce. Clearly, the cabbage butterflies and the plants of the mustard and caper families have a coevolutionary relationship.

> *Many groups of plants are protected from most herbivores by their secondary chemical compounds. But once the members of a particular herbivore group acquire the ability to feed on these plants, the herbivores have gained access to a new resource, which they can exploit without competition from other herbivores.*

Although the role of secondary chemical compounds in protecting plants from herbivores was first discovered in flowering plants, other groups of photosynthetic organisms have similar defenses. Thus, many red, brown, and green algae produce compounds that deter feeding by such herbivores as fishes and that also inhibit the growth of bacteria in culture. Some evidence suggests that the group of shrimplike crustaceans known as amphipods, which are often abundant on algae, may be as important in feeding on algae in marine environments as insects are in feeding on plants on land. Both marine and terrestrial animals may also manufacture toxic chemicals that deter their predators. Therefore, similar systems seem to operate both in the sea and on land that deter the dominant herbivores from consuming the primary producers of each community and that deter carnivores from consuming herbivores.

Chemical Defenses in Animals

Some groups of animals that feed on plants rich in secondary chemical compounds receive an extra benefit, one of great ecological importance. For example, when the caterpillars of monarch butterflies feed on plants of the milkweed family, they do not break down the cardiac glycosides that protect these plants from most herbivores. Instead, they concentrate, store, and pass the cardiac glycosides through the chrysalis stage to the adult butterfly and even to the eggs; all stages are then

(a)   (b)

(c)   (d)

Figure 21.12 Monarch butterflies make themselves poisonous.
All stages of the monarch butterfly's life cycle are protected from birds and other predators by the poisonous chemicals in the milkweeds and dogbanes on which the butterflies feed as larvae. (a) A cage-reared bluejay, which has never seen a monarch butterfly before, eats one. (b) The same bird a few minutes later, regurgitating the butterfly. This bird is not likely to eat an orange and black insect again! Both caterpillars (c) and adult butterflies (d) "advertise" their poisonous nature with warning coloration.

vae of cabbage butterflies, that feed on plants with well-marked chemical defenses, also usually are not brightly colored because they are able to break down the molecules involved, rather than store them.

Some marine animals, such as certain nudibranchs (sea slugs; see chapter 29), acquire defensive chemicals or defensive cells from their prey. For example, hydroids often provide such stinging cells to animals that graze on them. Off the coast of Panama, the large nudibranch *Aplysia* grazes selectively on red algae of the genus *Laurencia,* which is protected by elatol, a powerful inhibitor of cell division. This may be why few fish feed on *Aplysia.* Marine animals, algae, and flowering plants are currently being investigated because of the enormous diversity of chemical compounds found in these organisms that might provide new drugs to fight cancer and other diseases or that might be sources of antibiotics.

Animals also manufacture many of the startling array of substances that they use to perform an incredible variety of defensive functions. Bees, wasps, predatory bugs, scorpions, spiders, and many other arthropods manufacture chemicals that they use to defend themselves and to hunt for prey.

A variety of chemical defenses is also found among the vertebrates. Frogs of the family Dendrobatidae, for example, produce toxic alkaloids in the mucus that covers their skin. Some of these toxins are so powerful that a few micrograms are deadly if injected into the human bloodstream. Some South American native tribes tip their darts with this poison; for this reason, these frogs are sometimes called "dart-poison frogs." Most of the more than one hundred species of these frogs are brightly colored; they are favorites in zoos and aquariums for that reason (figure 21.14).

themselves protected from predators! A bird that eats a monarch butterfly quickly regurgitates it and thenceforth avoids the conspicuous orange and black pattern that characterizes the adult monarch (figure 21.12*a,b*). Locally, however, some birds have acquired the ability to tolerate the protective chemicals and are able to eat the monarchs.

Insects that feed regularly on plants of the milkweed family are generally brightly colored. Among them are brightly colored cerambycid beetles, whose larvae feed on the roots of the milkweed plants; bright blue or green chrysomelid beetles; and bright red bugs (order Hemiptera). In some parts of the world, there are also bright red grasshoppers and other very obvious insects. These herbivores "advertise" their poisonous nature by their bright colors, using an ecological strategy known as **warning coloration** (figure 21.12*c,d*).

Insects that eat plants whose chemical defenses are less obvious than those of the milkweeds are seldom brightly colored. In fact, many of these insects are **cryptically colored**— colored so as to blend in with their surroundings and thus be hidden from predators (figure 21.13). Insects, such as the lar-

Like plants, animals defend themselves chemically from potential predators. Some animals acquire distasteful or poisonous chemicals from plants they eat; others manufacture their own defensive chemicals.

Warning Coloration

Warning coloration (also known as **aposematic coloration**) is characteristic of animals with effective defense systems, including not only poisons, but also stings, bites, and other means of

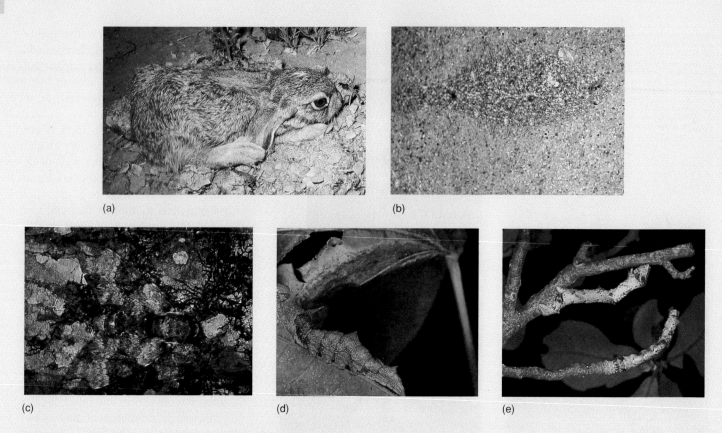

(a)

(b)

(c)

(d)

(e)

Figure 21.13 Striking examples of cryptic coloration.
(a) *A young jackrabbit in the desert near Tucson, Arizona, displaying both adaptive stillness and camouflage.* (b) *Protective coloration in a flatfish on the seafloor.* (c) *A tropical hawkmoth at rest, blending in perfectly with the moss-and-lichen-covered tree* bark. (d) *The brown-green-brown coloring of the unicorn caterpillar, which gives it the appearance of a dried leaf edge.* (e) *An inchworm caterpillar* (Necophora quernaria), *which closely resembles a twig.*

Figure 21.14 Warning: poisonous frog!
Frogs of the family Dendrobatidae, such as this individual, are abundant in the forests of Latin America and are extremely poisonous to vertebrates. More than two hundred different alkaloids have been isolated and identified in these frogs, and some of them are playing important roles in neuromuscular research. Two groups of Indians in western Columbia obtain a potent poison for blow gun darts from extremely toxic species of these frogs that live in their region.

repelling predators. Such organisms benefit by clearly advertising their defenses—for example, by exhibiting colors not normally found in that particular habitat (see figures 21.12 *c,d* and 21.14). Otherwise, the distasteful or poisonous individual runs the risk of being killed while protecting itself.

> *Warning coloration serves to keep potential predators away from poisonous or otherwise dangerous prey.*

Of course, the animals that display warning coloration must also remain together if the system is to be effective. A lone individual that exhibits its warning coloration but is eaten will not deliver a message useful for the survival of other animals with a similar appearance. But if genetically related individuals are similarly colored and live in the same vicinity, the selective advantage is obvious. Such animals tend to live together in family groups, unlike those that are cryptically colored. If camouflaged animals lived together in groups, one might be discovered by a potential predator, offering a valuable clue to the presence of others.

Models

Monarch   Pipevine swallowtail

Mimics

(a)   Viceroy   (b)   Red-spotted purple

Figure 21.15 Batesian mimicry: Two examples.

(a) *The viceroy butterfly is a North American mimic of the poisonous monarch. Although the viceroy is not related to the monarch, it looks a lot like it, and so predators that have learned to avoid distasteful monarchs avoid viceroys, too. (b) The red-spotted purple butterfly is another member of the same genus as the viceroy and thus is more closely related to it than is the monarch. However, it does not look at all like the viceroy. Instead, it resembles another poisonous butterfly, the pipevine swallowtail. The very different appearances of the viceroy and red-spotted purple butterflies, vividly illustrate how selection can drastically change the appearances of mimics.*

Mimicry

Many unprotected species have come, during the course of their evolution, to resemble distasteful ones that exhibit warning coloration. Provided that the unprotected animals are present in numbers that are low relative to those of the species that they resemble, they, too, will be avoided by predators. Such a pattern of resemblance is called **Batesian mimicry,** after British naturalist H. W. Bates, who first brought it to general attention in the 1860s. Bates also carried out the first scientific studies of this phenomenon, which has been of great interest to naturalists ever since.

> *In Batesian mimicry, unprotected species resemble other species that are distasteful. Both species exhibit warning coloration. Predators avoid unprotected mimics because of the mimics' coloring.*

Many of the best-known examples of Batesian mimicry occur among butterflies and moths. Predators in systems of this kind apparently use visual cues to hunt for their prey. Other-wise, similar patterns of coloration would not offer any protection to species without chemical defenses.

The groups of butterflies that are the **models** in Batesian mimicry are, not surprisingly, members of groups whose larvae feed on only one or a few closely related plant families that are strongly protected chemically. The model butterflies take poisonous molecules from these plants and retain them in their own bodies. The mimic butterflies, in contrast, belong to groups in which larvae feeding habits are not so restricted. As caterpillars, these butterflies feed on a number of different plant families, although not those protected by toxic chemicals.

One well-known mimic among North American butterflies is the viceroy butterfly, *Limenitis archippus*. This butterfly, which resembles the poisonous monarch, ranges from central Canada south through much of the United States east of the Sierra Nevada and Cascade Range into Mexico. The larvae feed on willows and cottonwoods, and neither they nor the adults are distasteful to birds (figure 21.15). Interestingly, the viceroy larvae are hidden from predators on the leaves of their host plants because they resemble bird droppings, whereas the distasteful monarch larvae are very conspicuous.

Another kind of mimicry, **Muellerian mimicry,** was named for German biologist Fritz Mueller, who first described it in 1878. In Muellerian mimicry, several unrelated but protected animal species come to resemble one another. Thus, a number of different kinds of stinging wasps have black-and-yellow striped abdomens, but they may not all be descended from a common ancestor with similar coloration. In general, yellow-and-black and bright red tend to be common color patterns that presumably warn predators relying on vision to avoid such animals. Proving that a resemblance between two protected animals actually represents Muellerian mimicry is more difficult than demonstrating Batesian mimicry experimentally.

In both Batesian and Muellerian mimicry, mimic and model must not only look alike but also act alike to deceive predators. For example, the members of several families of beetles that resemble wasps behave surprisingly like the wasps they mimic, flying often and actively from place to place. Mimics must also spend most of their time in the same habitats as their models. If they did not, predators would discover that all of those conspicuous animals are not only easily seen but also quite tasty! If

Models

Batesian mimics

Figure 21.16 More mimics: Batesian and Muellerian mimicry.
The familiar yellow-and-black stripes of wasps (the model, top photo) form the basis for large Batesian and Muellerian mimicry complexes. Shown are Batesian mimics representing three separate orders of insects, all rarer than wasps and with behavior patterns similar to those of the dangerous wasps they resemble. None of these mimics sting, yet all are conspicuous members of the communities where they occur, flying about actively and remaining in full view at all times. Because these insects all resemble each other, *they are Muellerian mimics of each other.*

the animals that resemble one another are all poisonous, or dangerous, they still gain an advantage by resembling one another, thus achieving collective protection (figure 21.16).

> *Muellerian mimicry is a phenomenon in which two or more unrelated but protected species resemble one another, thus achieving a kind of group defense.*

Symbiosis

Symbiotic relationships are those in which two kinds of organisms live together. All symbiotic relationships provide the potential for coevolution between the organisms involved, and in many instances, the results of this coevolution are fascinating. The major kinds of symbiotic relationships are: (1) **commensalism,** in which one species benefits while the other neither benefits nor is harmed; (2) **mutualism,** in which

both participating species benefit; and (3) **parasitism,** in which one species benefits but the other is harmed. Parasitism, as mentioned earlier, can also be viewed as a form of predation.

Examples of symbiosis include lichens, which are associations of certain fungi and green algae or cyanobacteria (discussed in more detail in chapter 27). Another important example are the mycorrhizae, associations between fungi and the roots of plants in which the fungi expedite the absorption of certain nutrients by the plants, which in turn provide the fungi with carbohydrates. Similarly, the association between bacteria and the root nodules that occur in legumes and certain other plants enables the host plants to convert atmospheric nitrogen to a form usable by organisms and allows the bacteria to obtain carbohydrates. A coral reef is a highly complex symbiotic system, involving not only the coral animals but also algae and other autotrophic organisms that are intermingled with the coral animals and contribute greatly to the reef's net productivity. More broadly, a coral reef is an entire biological community, one in which symbiotic relationships are especially prominent.

> Symbiotic relationships are those in which two or more kinds of organisms live together in often elaborate, more or less permanent relationships. In commensalism, one species benefits while the other is unaffected. In mutualism, both participating species benefit. In parasitism, one species benefits while the other is harmed.

Commensalism

In nature, the individuals of one species often grow attached to those of another. Two examples are: (1) birds nesting in trees and (2) smaller plants, called **epiphytes,** growing on the branches of other plants. In both of these cases, the host plant is generally unharmed, while the organism that nests or grows on it benefits.

Similarly, various marine animals, such as barnacles, grow on other, often actively moving sea animals and thus are carried passively from place to place. These "passengers" presumably gain more protection from predation than they would if they were fixed in one place, and they also reach new sources of food. The increased water circulation that such animals receive as their host moves around may be of great importance, particularly if the passengers are filter-feeders. The passengers' gametes are also more widely dispersed than would be the case otherwise.

The best-known examples of commensalism—those that virtually define the concept—involve the relationships between certain small tropical fishes and sea anemones, marine animals with stinging tentacles (see chapter 29). These fishes have evolved the ability to live among the tentacles of the sea anemones, even though the tentacles would quickly paralyze other fishes that touched them. The anemone fishes gain protection from other predators by remaining among the anemones' tentacles and also feed on the leftovers from the meals of the host anemone, remaining uninjured under remarkable circumstances (figure 21.17).

In the examples mentioned, it is difficult to be certain whether the host partner receives a benefit or not, and the boundary between commensalism and mutualism is not clearcut. For instance, having particles of food removed from its tentacles may make the sea anemone better able to catch other prey. The same holds true for the sea animals on which barnacles grow: The sea animal may benefit from having the barnacles remove other parasites, such as lice, from its skin. Thus, the factors involved in a particular association can be learned only by patient observation and experimentation.

Mutualism

Examples of mutualism are of fundamental importance in determining the structure of biological communities. In the tropics, for example, leafcutter ants are often so abundant that they can remove a quarter or more of the total leaf surface of the plants in a given area. They do not eat these leaves directly; rather, they take them to their underground nests, where they chew them up and inoculate them with the spores of particular

Figure 21.17 Commensalism: Sea anemones and clownfish.
These clownfish often form commensal relationships with sea anemones, gaining protection by remaining among the anemones' tentacles and gleaning scraps from the anemones' food.

Figure 21.18 Mutualism: Ants and aphids.
These ants are tending to willow aphids, obtaining the "honeydew" that the aphids excrete continuously, moving the aphids from place to place, and protecting them from potential predators.

fungi. These fungi are cultivated by the ants and brought from one specially prepared bed to another, where they grow and reproduce. In turn, the fungi constitute the primary food of the ants and their larvae. The relationship between the leafcutter ants and these fungi, fueled by material cut from the leaves of plants, is an excellent example of mutualism.

Another relationship of this kind involves ants and aphids. Aphids, also called greenflies, are small insects that suck fluids from living plants with their piercing mouthparts. They extract a certain amount of the sucrose and other nutrients from this fluid, but much runs out, in an altered form, through their anus. Certain ants have taken advantage of this habit—in effect domesticating the aphids—by protecting the aphids from predators, carrying the aphids to new plants, and using the "honeydew" that the aphids excrete as food (figure 21.18).

(a)

(b)

(c)

(d)

Figure 21.19 Ants and acacias, a classic story of coevolution and mutualism.
(a) *Ants of the genus* Pseudomyrmex *live within the spinelike stipules of the bull's horn acacias.* (b) *The ants harvest the sugar-rich fluid found in the nectaries at the base of the acacia leaves and* (c) *the Beltian bodies found at the ends of the acacia leaflets.* (d) *The ants' activities benefit the acacias by clearing away much of the vegetation that crowds the acacias, and the ants also attack most kinds of herbivores that threaten the acacias.*

A particularly striking example of mutualism involving ants concerns certain Latin American species of the plant genus *Acacia*. In these species, the leaf parts called stipules are modified as paired, hollow thorns; consequently, these particular species are called "bull's horn acacias." The thorns are inhabited by stinging ants of the genus *Pseudomyrmex,* which do not nest anywhere else.

At the tip of the leaflets of these acacias are unique, protein-rich bodies called Beltian bodies—after Thomas Belt, a nineteenth-century British naturalist who first wrote about them based on his experiences in Nicaragua. Beltian bodies do not occur in species of *Acacia* that are not inhabited by ants, and their role is clear: They serve as the ants' primary food. The plants also secrete nectar from glands near the bases of their leaves. The ants consume this nectar and feed it and the Beltian bodies to their larvae as well (figure 21.19*a–c*).

Obviously, this association is beneficial to the ants: The ants and their larvae are protected within the swollen thorns, and the acacias provide a balanced diet of sugar-rich nectar and protein-rich Beltian bodies. What, if anything, do the ants do for the plants? This question fascinated observers for nearly a century, until 1967, when Daniel Janzen, then a graduate student at the University of California, Berkeley, discovered the answer in a beautifully conceived and executed series of field experiments. Janzen's discoveries were based first on very careful observations. Once he had formulated his hypotheses about the nature of the ant/acacia system, he poisoned the ant colonies in some of the plants and studied the fates of those particular plants, which were growing in the same habitats as others that had healthy ant colonies.

Janzen's experiments demonstrated that whenever any herbivore lands on the branches or leaves of an ant-inhabited acacia, the ants immediately attack and devour it. Thus, the ants protect the acacias from being eaten, and the herbivores also provide additional food for the ants, which continually patrol the branches.

Janzen also found that the ants that live in the bull's horn acacias help their hosts to compete with other plants. The ants cut away any branches of other plants that touch the bull's horn acacia in which they are living—creating, in effect, a tunnel of light through which the acacia can grow, even in the lush deciduous forests of lowland Central America (figure 21.19*d*). Without the ants, the acacia is unable to compete successfully in this habitat. Finally, Janzen discovered that the ants bring organic material into their nests and that the part that they do not consume, together with ant excretions, provide the acacias with an abundant source of nitrogen, which is an essential nutrient.

Related species of acacias that do not have the special features of the bull's horn acacias and are not protected by ants often have bitter-tasting substances in their leaves that the bull's horn acacias lack. Evidently, these bitter-tasting substances protect the acacias in which they occur in much the same way that the ants protect the acacias that they inhabit.

Parasitism

The concept of parasitism seems obvious, but individual instances are often surprisingly difficult to distinguish from predation, as mentioned earlier in the chapter, and from other kinds of symbiosis. Many instances of parasitism are well known. For example, vertebrates are parasitized by members of many different groups of animals and protists. Invertebrates also have a variety of parasites that live within their bodies. However, bacteria and viruses are often not considered parasites, even though they fit the definition precisely. Lice, which live on the bodies of vertebrates—mainly birds and mammals—are normally considered parasites, but mosquitoes are not, even though they draw food from the same birds and mammals in a similar manner. Mosquitoes are not considered parasites because their interaction with their host is short-lived: The mosquitoes fly away when they have finished feeding. However, mosquitoes are closely associated ecologically with the animals from which they draw blood, and they also synchronize their diurnal and seasonal activities closely with those of their hosts. The interrelationship between mosquito and host, although technically not parasitic, is very close.

Internal parasitism is generally marked by much more extreme specialization than is external parasitism, as shown by the many protist and invertebrate parasites that infect humans. The more closely the life of the parasite is linked with that of its host, the more its physical characteristics and behavior are likely to have been modified during the course of its evolution (figure 21.20). The same, of course, is true of symbiotic relationships of all sorts. Conditions within the body of another organism are very different from those encountered outside and are apt to be much more constant in every way. Consequently, the structure of the parasite is often simplified, and unnecessary armaments and structures are lost as it evolves.

Despite these qualifications and inconsistencies in terminology, the general meaning of the term *parasite* is clear. Parasitism may be regarded as a special form of predation in which the predator is much smaller than the prey and remains closely associated with it. Parasitism is harmful to the prey organism and beneficial to the parasite.

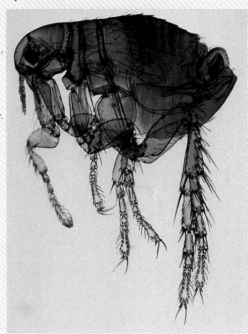

Figure 21.20 Parasitism.

The human flea, Pulex irritans. Fleas' structural and behavioral modifications are related to their parasitic way of life. These parasites are flattened from side to side and slip easily through hair. Their ancestors were larger, more brightly colored, and had wings.

EVOLUTIONARY VIEWPOINT

While many discussions of evolution focus on competition, cooperation has been—and is now—at least as important an evolutionary factor. Symbiosis can be seen at every level of biological organization. Consider a tree: Its leaves carry out photosynthesis with symbiotic chloroplasts; its roots assimilate nutrients with the aid of mycorrhizae; its bark is spattered with symbiotic lichens; and its branches are patrolled by ants in mutualistic cooperation. Without symbiosis, the tree would be inconceivable.

SUMMARY

1. Communities are populations of different organisms that live together. They exist because the organisms' ecological requirements and environmental tolerances overlap in the areas where the organisms occur together.

2. Even in the redwood community, which is dominated by one very obvious kind of organism, the different organisms function individually and somewhat independently, and they display genetic variability from place to place. Communities are continually evolving and change dynamically through time.

3. Coevolution is the process by which different kinds of organisms adjust to one another by genetic change over long periods of time. It is a stepwise process that ultimately involves adjustment of both groups of organisms.

4. Interspecific competition is competition between two species for the same resources. G. F. Gause formulated the principle of competitive exclusion, which states that, if two species are competing for the same limited resource, one of the species will be able to use that resource more efficiently than the other and will eventually eliminate the other species.

5. A niche is the role that an organism plays in the environment. The theoretical niche of an organism is the niche that an organism might occupy if no competitors were present. The actual niche is the niche that an organism occupies, with all competitors in place.

6. Predation includes the killing of one organism by another for food, as well as parasitism. Predation cycles between predator and prey predict that, when a predator has killed a large portion of the prey population, the predator population numbers will also decline for a time until the prey population is built up again.

7. Plants are often protected from herbivores, fungi, and other agents by chemicals that they manufacture. Such chemicals, which are not part of the plant's primary metabolism, are called secondary compounds.

8. Particular classes of secondary compounds are usually characteristic of individual plant families or groups of closely related families. The herbivores that can feed on such plants either break the secondary compounds down or store them in their

bodies. If they do the latter, the chemicals may, in turn, protect the herbivores from their predators.

9. Warning, or aposematic, coloration makes the organisms that possess it obvious and is characteristic of organisms that are poisonous, have stings, or are otherwise harmful. In contrast, cryptic coloration, or camouflage, is characteristic of organisms that are not specially protected.

10. In Batesian mimicry, a palatable or nontoxic organism resembles another kind of organism that is distasteful or toxic. Muellerian mimicry occurs when several toxic or dangerous kinds of organisms resemble one another.

11. In Batesian mimicry, the models are generally herbivores that feed on plants that contain toxic secondary compounds. The models retain such chemicals in their bodies. The mimics, in contrast, usually feed on plants that are not protected in this way.

12. Symbiotic relationships are those in which two kinds of organisms live together. The three major kinds of symbiotic relationships are: (1) commensalism, in which one species benefits and the other is unaffected; (2) mutualism, in which both participating organisms benefit; and (3) parasitism, in which one species benefits but the other is harmed.

REVIEWING THE CHAPTER

1. How does coevolution work?
2. What is interspecific competition?
3. What is the principle of competitive exclusion?
4. How does competition work in nature?
5. In ecology, what is a niche?
6. How do predators and prey interact?

7. How do plants and animals defend themselves?
8. What is a plant-herbivore relationship?
9. How do plants produce defenses when needed?
10. How did some herbivores evolve to neutralize plants' chemical defenses?
11. How do animals obtain and make chemicals for defense?

12. How did animals evolve warning coloration?
13. What is mimicry?
14. What is symbiosis?
15. What is a commensal symbiotic relationship?
16. What are features of mutualism?
17. What is the concept of parasitism?

COMPLETING YOUR UNDERSTANDING

1. What is responsible for the current reduced size of the redwood community?
2. What are primary producers and herbivores?
3. How are interspecific and intraspecific competition different?
4. What is a trophic level, and how does it relate to interspecific competition?
5. In J. H. Connell's investigations of two species of barnacles living on the coast of Scotland, why was the shallower water species uncommon in deeper water, even though it could live there?
6. How do actual and theoretical niches differ?
7. How can Gause's principle of competitive exclusion be restated in terms of niches?
8. What are the commonalities between predation and parasitism?
9. From figure 21.6, what controls the number of hares? Can you give an example of a situation similar to that of the hare in the human population at the present time?

10. What are some examples, both beneficial and harmful, of introducing species into biomes where they did not exist previously?
11. Autotrophic organisms utilize about _____ percent of the energy contained in the sunlight that falls on them.
  a. 1
  b. 10
  c. 25
  d. 50
  e. 75
  f. 100
12. How do primary and secondary chemical compounds function in plants?
13. Adult monarch butterflies are primarily protected from birds by
  a. flying rapidly.
  b. resembling the poisonous viceroy butterfly.
  c. manufacturing poisonous molecules that are toxic to birds.
  d. obtaining poisons from the plants that their larvae eat and retaining them.
  e. Muellerian mimicry.

14. Why do species of marine animals, algae, and flowering plants offer encouragement for finding new drugs to fight cancer and other diseases? What are some of the blockades that could stifle this type of research?
15. A species of poisonous animal is best protected if it
  a. is cryptically colored.
  b. is aposematically colored.
  c. is camouflaged.
  d. moves slowly.
  e. moves rapidly.
16. In Batesian mimicry, why are the mimics rare?
17. How do Batesian and Muellerian mimicries differ from one another?
18. Yellowjacket wasps are protected because
  a. they obtain poisonous chemicals from the plants on which they feed.
  b. they sting.
  c. they are Muellerian mimics.
  d. they are Batesian mimics.
  e. they occur in large colonies.

19. Which of the following are examples of commensalism? [*two answers required*]

    a. Birds nesting in trees

    b. Ants/fungi/leaves

    c. Mosquitoes and mammals

    d. Lice

    e. Ants and aphids

    f. Epiphytes

20. How do you think the intensity of competition would differ between an Arctic tundra and a tropical rain forest? How would you test your answer?

21. Which kinds of ecosystems are most resistant to introduced species? Why?

22. Some kinds of insects are more resistant to insecticides than are others. Given the information in this chapter, can you think of a reason why?

FOR FURTHER READING

Barrett, S. C. H. 1987. Mimicry in plants. *Scientific American* 257: 76–83. A discussion of how mimicry in plants attracts pollinators or deters herbivores.

Davies, N., and M. Brooke. 1991. Coevolution of the cuckoo and its hosts. *Scientific American,* Jan., 92–98. The classic story of an evolutionary "arms race" between a parasite and its host.

De Vries, P. 1992. Stinging caterpillars, ants and symbiosis. *Scientific American,* Oct., 76–82. Article about the intricate symbiosis shared by ants and caterpillars, a symbiosis that says a lot about interdependence within ecosystems and how it arises.

Duffy, J. E., and M. E. Hay. 1990. Seaweed adaptations to herbivory. *BioScience* 40 (5): 368–85. A discussion of the chemical, structural, and morphological defenses that are abundant in the seaweeds and that sometimes change from season to season.

Fautin, D. G. 1987. Who are those little orange fish and why are they living in a sea anemone? *Pacific Discovery* 40 (2): 18–29. A beautifully illustrated account of clownfishes and the sea anemones they inhabit.

Handel, S., and A. Beattie. 1990. Seed dispersal by ants. *Scientific American,* Aug., 76–83. How plant species induce ants to spread the plants' seeds with special food lures and other adaptations.

Morse, D. H. 1985. Milkweeds and their visitors. *Scientific American,* July, 112–19. Detailed study of the nectar-feeders, herbivores, predators, and parasites that gather on milkweed plants, forming a model ecological community.

Pietsch, T., and D. Grobecker. 1990. Frogfishes. *Scientific American,* June, 96–103. Masters of aggressive mimicry—voracious carnivores that can gulp prey faster than any other vertebrate predator.

Rosenthal, G. A. 1986. The chemical defenses of higher plants. *Scientific American* 254: 94–99. Interesting account of the array of chemical defenses produced by plants.

Walker, T. 1991. Butterflies and bad taste. *Science News,* June, 348–49. Evidence that viceroy butterflies, the classic example of tasty creatures mimicking an unpalatable one, do not taste good after all!

YNAMICS OF ECOSYSTEMS

The idyllic dairy farm, carved out of a Vermont hillside, is an example of a natural biome—temperate deciduous forest—that has been converted into a much less diverse artificial community. Agricultural land is a biological community, but is simpler and less stable than a natural biological community.

FOR REVIEW

Here are some important terms and concepts that have been discussed in previous chapters and that you will encounter again in this chapter. Review them before proceeding if necessary.

Nitrogen (*chapter 3*)
Metabolic energy (*chapter 7*)
Carbon fixation (*chapter 8*)
Communities (*chapters 20 and 21*)

The earth provides living organisms with much more than a place to stand or swim. Many chemicals cycle between our bodies and the physical environment around us. We live in a delicate balance with our physical environment, a balance easily disturbed by human activities.

Ecosystems

In ecological terms, populations of different organisms that live together in a particular place are called **communities.** The community of organisms that live in a place, together with the nonliving factors with which the organisms interact, is called an **ecosystem.** Within ecosystems, the essential elements on which the lives of constituent plants, animals, and other organisms depend are cycled, and energy flow is regulated. Great differences in climate across the face of the globe over billions of years have resulted in the creation of diverse terrestrial and marine ecosystems (figure 22.1).

The individual organisms and populations of organisms in an ecosystem act as part of an integrated whole, adjust over time to their role in the ecosystem, and relate to one another in complex ways that are only partly understood. Despite their differences, all ecosystems are governed by the same principles and restricted by the same limitations. The earth is a closed system with respect to nutrients, but an open one in terms of energy. That is, no new nutrients are being added from outside, but energy from the sun is constantly being added. Ecosystems function to regulate the cycling of those nutrients and the capture and expenditure of that energy (figure 22.2). As you will see in this chapter, all organisms, including human beings, depend on the activities of a few other organisms—plants, algae, different kinds of bacteria in the cases of carbon and nitrogen, for example—for the basic components of life.

> *A community is the interacting set of different kinds of organisms that occur together at a particular place. An ecosystem is that set of organisms, together with the nonliving factors with which it interacts. Within ecosystems, nutrients are cycled throughout, and energy is regulated and transferred.*

Figure 22.1 Example of an ecosystem—a desert at night.
Unlike the moon, all of the earth's surface, even its deserts, is teeming with life, although it may not always appear so. The same ecological principles apply to the organization of all of the earth's ecosystems, both on land and in the sea, although the details differ greatly.

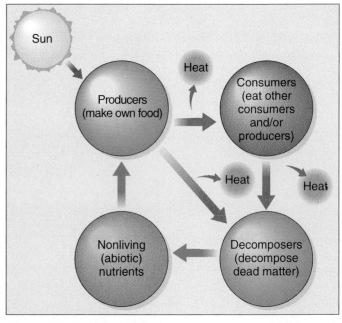

Figure 22.2 Cycling within an ecosystem.
Energy enters the ecosystem from the sun and leaves the ecosystem in the form of heat. (Energy, however, is not created or destroyed; when energy leaves an ecosystem, it is in a converted form.) Nutrients do not leave the ecosystem but, instead, cycle through it.

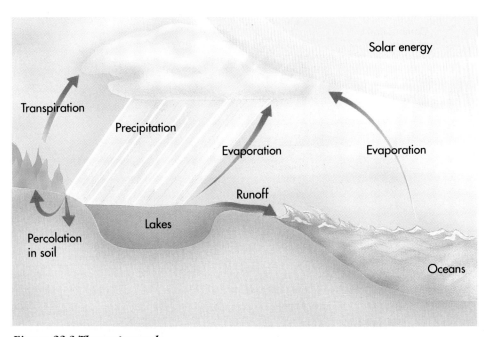

Figure 22.3 The water cycle.
Most of the water present on the earth is held in oceans and the atmosphere. Only a small amount is present in the soil or in organisms.

Nutrient Cycling in Ecosystems

All of the substances that occur in organisms, including water, carbon, nitrogen, and oxygen, as well as a number of other substances that are ultimately derived from the weathering of rocks, cycle through ecosystems. These cycles are geological ones that involve the biologically controlled cycling of chemicals and therefore are called **biogeochemical cycles.** Carbon (in the form of carbon dioxide), nitrogen, and oxygen primarily enter the bodies of organisms from the atmosphere, whereas phosphorus, potassium, sulfur, magnesium, calcium, sodium, iron, and cobalt, all of which are essential for plant growth (see chapter 32), come from rocks. All organisms require carbon, hydrogen, oxygen, nitrogen, phosphorus, and sulfur in relatively large quantities; the other elements are required in smaller amounts.

Nutrients in ecosystems are said to be *cycled:* They are incorporated from the atmosphere or from weathered rock first into the bodies of organisms. They then sometimes pass from these organisms into the bodies of other organisms that feed on the primary ones. Ultimately, through decomposition, they are returned to the nonliving world, but are often incorporated again into the bodies of other organisms. Some examples will help to clarify the ways in which different cycles function.

The Water Cycle

All life directly depends on the presence of water, since the bodies of most organisms consist mainly of this substance. Water is the source of the hydrogen ions whose movements generate ATP in organisms, and for that reason alone, it is indispensable to their functioning. Thus, the **water cycle** is the most familiar of all biogeochemical cycles.

Oceans cover nearly three-fourths of the earth. From their surfaces, water evaporates into the atmosphere, a process powered by energy from the sun. Plants also add water to the atmosphere through **transpiration** (discussed in detail in chapter 32). This water eventually precipitates back to the earth in the form of rain or snow. Most of it falls directly into the oceans, but some falls onto the land, where it passes into surface and subsurface bodies of freshwater. Only about 2 percent of all the water on earth is found frozen, held in the soil, or incorporated into the bodies of organisms. The rest is free water, circulating between the atmosphere and the earth. Regardless of where this water is held temporarily, it eventually returns to the atmosphere and the oceans (figure 22.3).

Much less obvious than the surface waters, seen in streams, lakes, and ponds, is the **groundwater,** which occurs in permeable, saturated, underground layers of rock, sand, and gravel called **aquifers.** In many areas, groundwater is the most important water reservoir; for example, in the United States, more than 96 percent of all freshwater is groundwater. Groundwater flows much more slowly than surface water, anywhere from a few millimeters to as much as a meter or so per day. In the United States, groundwater is the source of about a fourth of the water used for all purposes and provides about half of the population with drinking water. Three-fourths of American cities and most U.S. rural areas rely, at least in part, on groundwater reserves. Throughout the world, groundwater use is growing much more rapidly than surface water use.

> *Some 96 percent of the freshwater in the United States consists of groundwater. This groundwater, which already provides a fourth of all the water used in the United States, will be used even more extensively in the future.*

The Carbon Cycle

The **carbon cycle** is based on carbon dioxide, which makes up only about 0.03 percent of the atmosphere. Worldwide synthesis of organic compounds from carbon dioxide and water uses about a tenth of the roughly 700 billion metric tons of carbon dioxide in the atmosphere each year. This enormous amount of biological activity is the result of the combined activities of photosynthetic bacteria, algae, and plants. All heterotrophic organisms—including the nonphotosynthetic bacteria and protists, the animals, and the relatively few plants that have lost the ability to photosynthesize—obtain their carbon indirectly from the organisms that fix it (see chapter 8 for a discussion of carbon fixation). When their bodies decompose, organisms release carbon dioxide to the atmosphere again. Once there, it can be reincorporated into the bodies of other organisms (figure 22.4).

About a tenth of the estimated 700 billion metric tons of carbon dioxide in the atmosphere is fixed annually by the process of photosynthesis.

In addition to the roughly 700 billion metric tons of carbon dioxide in the atmosphere, approximately 1 trillion metric tons of carbon are dissolved in the oceans; more than half of this quantity is in the upper layers, where photosynthesis takes place. The fossil fuels, primarily oil and coal, contain more than 5 trillion additional metric tons of carbon, and between 600 billion and 1 trillion metric tons are locked up in living organisms at any one time.

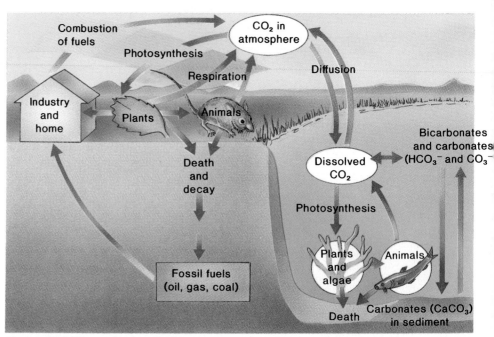

Figure 22.4 The carbon cycle.
All organisms depend on photosynthesizing organisms that fix the carbon dioxide (CO_2) in the air into a usable form. When a nonphotosynthetic animal eats a plant, it is incorporating carbon compounds into its body. Carbon is returned to the atmosphere when the animal breathes out carbon dioxide and when plants and animals decay after death.

Carbon dioxide and certain other gases, such as methane, that occur in the earth's atmosphere transmit radiant energy from the sun but trap the longer wavelengths of infrared light, or heat, and prevent them from radiating into space. This creates what is known as the **greenhouse effect.** Planets that lack this type of "trapping" atmosphere are much colder than those that possess one. If the earth did not have a "trapping" atmosphere, the average earth temperature would be about –20 degrees Celsius, instead of the actual +15 degrees Celsius.

The earth's greenhouse effect is intensifying because of human activities involving, for example, fossil fuel combustion and certain types of waste disposal. These activities are increasing the amounts of carbon dioxide, chlorofluorocarbons (CFCs), nitrogen oxides, and methane—all "greenhouse gases"—in the atmosphere. Before widespread industrialization, carbon dioxide concentration in the atmosphere was about 260 to 280 parts per million (ppm). During the twenty-five-year period starting in 1958, this concentration increased from 315 ppm to more than 340 ppm and is continuing to rise rapidly. The rise in average global temperatures of about 0.5 degrees Celsius during the past century is consistent with these increased carbon dioxide concentrations in the atmosphere, but scientists have not yet been able to demonstrate with certainty that the temperature increase was caused by the atmospheric changes.

Projected increases in the amounts of atmospheric greenhouse gases over the next fifty years could increase average global temperatures from 1 to 4 degrees Celsius, a matter of serious concern because of such associated effects as shifts in prime agricultural lands, changes in sea levels, and alterations in rainfall patterns. The projected changes concern *average* temperatures, and as such, cannot be connected with particular hot or cold days or seasons. In 1992 and 1993, *below-average* temperatures were recorded worldwide, but scientists believe that these cooler temperatures were a result of recent volcanic eruptions, such as the Mount Pinatubo 1992 eruption in the Philippines. Volcanic eruptions send large amounts of dust and debris high into the atmosphere and act as a shield against the sun's rays. Despite these recent temperature readings, experts feel that the earth is warming now more rapidly than at any period in the past, including the times when glaciers were melting during the ice ages.

The Nitrogen Cycle

Nitrogen gas constitutes nearly 80 percent of the earth's atmosphere by volume, but the total amount of fixed nitrogen in the soil, oceans, and the bodies of organisms is only about 0.03 percent of that figure. Most organisms cannot use atmospheric nitrogen because few of them possess the special enzyme system necessary to break the very strong triple bond of nitrogen gas (N_2). Those that *do* have the enzymes (several genera of bacteria) carry out a process called **nitrogen fixa-tion.** Once nitrogen has been fixed, it cycles within biological systems. All living organisms depend on nitrogen fixation to synthesize proteins, nucleic acids, and other necessary nitrogen-containing compounds. Nitrogen fixation is the means by which a very small fraction of the enormous reservoir of nitrogen that exists in the earth's atmosphere is made available for biological processes (figure 22.5).

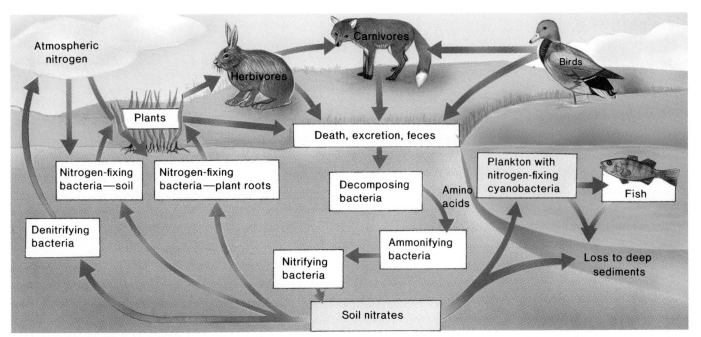

Figure 22.5 The nitrogen cycle.
Like the carbon cycle, all organisms depend on those organisms that can fix nitrogen into a usable form.

Although dozens of genera of bacteria have the ability to fix nitrogen, only a few that form symbiotic associations with plants usually fix enough nitrogen to be of major significance. Plants such as the legumes (members of the plant family Fabaceae, which includes peas, beans, alfalfa, mesquite, and many other well-known plants) that form symbiotic associations with nitrogen-fixing bacteria can grow in soils with very low amounts of nitrogen (figure 22.6). A legume crop may add as much as 300 to 350 kilograms of nitrogen per hectare per year, whereas all other sources may add only about 15 kilograms. The importance of these symbiotic associations is obvious.

> *Although nitrogen gas constitutes nearly 80 percent of the earth's atmosphere, it becomes available to organisms almost entirely through the activities of a few genera of bacteria.*

Figure 22.6 A nitrogen-fixing plant.
Beans are able to fix nitrogen into a form that all organisms can use. Bean plants have symbiotic relationships with nitrogen-fixing bacteria that occur in their root nodules.

The Phosphorus Cycle

In all biogeochemical cycles other than those involving water, carbon, oxygen, and nitrogen, the reservoir of the nutrient exists in mineral form, rather than in the atmosphere. The **phosphorus cycle** is presented here as a representative example of all other mineral cycles because of the critical role phosphorus plays in plant nutrition worldwide (figure 22.7).

Phosphates—charged phosphorus ions—exist in the soil only in small amounts because they are relatively insoluble and are present only in certain kinds of rocks. If the phosphates are not lost to the sea by way of rivers and streams, they may be absorbed by plants and incorporated into compounds, such as ATP, nucleic acids, and membrane phospholipids. Fungi associ-

ated with plant roots—the associations called **mycorrhizae**—facilitate the transfer of phosphates from the soil to plants (see chapter 27). Plants that grow in very poor soils and do not form mycorrhizae may instead have clusters of fine roots that perform a similar function.

When animals or plants die, shed parts, or, in the case of animals, excrete waste products, phosphorus may be returned to the soil or water and then cycle again through other organisms. When phosphates are lost to the deep sea, they may be recycled through the activities of seabirds that eat fishes and other animals that feed in deep waters (figure 22.8).

Energy Flow in Ecosystems

An ecosystem includes two different kinds of living components: autotrophs and heterotrophs. Autotrophs ("self-feeders"), consisting of plants, algae, and some bacteria, are able to capture light energy and to manufacture their own food. To support themselves, heterotrophs ("other feeders"), including animals, fungi, most protists and bacteria, and nongreen plants, must obtain organic molecules synthesized by autotrophs.

Once energy enters an ecosystem—mainly after it is captured through photosynthesis—it is slowly released as metabolic processes proceed. The autotrophs that first acquire this energy provide all of the energy that heterotrophs use. The organisms that make up an ecosystem have become adapted through time to delay the release of the energy obtained from the sun back into space. Energy flows one way through an ecosystem, whereas nutrients are continually recycled.

Productivity

Primary productivity is the total amount of light energy converted to organic compounds in a given area per unit of time. An ecosystem's **net primary productivity** is the total amount of energy fixed by photosynthesis per unit of time, minus that which is expended by the metabolic activities of ecosystem organisms. The total weight of all ecosystem organisms, called the ecosystem's **biomass,** increases as a result of the ecosystem's net production.

Some ecosystems, such as cornfields or cattail swamps, have a high net primary productivity. Others, such as tropical rain forests, also have a relatively high net primary productivity, but a rain forest has a much larger biomass than a cornfield. Consequently, a rain forest's net primary productivity is much lower in relation to its total biomass. In ecosystems such as sugarcane fields, coral reefs, and estuaries, the net primary productivity may range from roughly 3,500 to 9,000 grams of organic material per square meter per year. The productivity of marshlands and tropical forests is somewhat less; that of deserts is about 200 grams.

Trophic Levels

Green plants are the **primary producers** (photosynthesizers) of an ecosystem. They generally convert about 1 percent of the energy that falls on their leaves into food energy, although in especially productive systems, this percentage may be a little

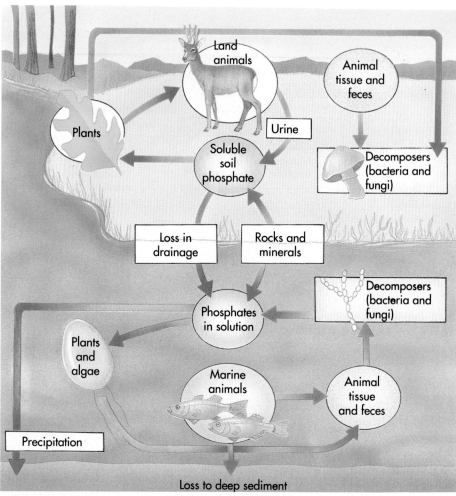

Figure 22.7 The phosphorus cycle.
Phosphorus is a necessary nutrient in ecosystems, but phosphorus amounts are critical. If too much phosphorus is added to soil or water, the delicate ecological balance can be upset, and organisms can die.

Figure 22.8 Living the phosphorus cycle.
These pelicans roosting on a small island in the Gulf of California, Mexico, bring up phosphorus from the deepest layers of the sea by eating fish and other marine animals. The phosphorus from these fish and marine organisms is deposited on land when the pelicans excrete their waste material, called guano.

higher. When a green plant is consumed by another organism, usually only about 10 percent of the plant's accumulated energy is actually converted into the body of the consumer. Primary producers are characterized as autotrophs because they make their own food.

Several levels of heterotrophs—organisms that eat other organisms—are recognized. The **primary consumers,** or herbivores, feed directly on the green plants. **Secondary consumers**—carnivores and the parasites of animals—feed directly or indirectly on the herbivores. **Decomposers** break down the organic matter accumulated in the bodies of other organisms. Another, more general, term that includes decomposers is **detritivores.** Detritivores are organisms that live on the refuse of an ecosystem—not only on dead organisms but also on the cast-off parts of organisms. Detritivores include large scavengers, such as crabs, vultures, and jackals, as well as decomposers, such as bacteria and fungi.

All of these levels, and usually additional ones, are represented in any fairly complicated ecosystem. They are called **trophic levels,** from the Greek word *trophos,* meaning "feeder." Organisms from each of these levels, feeding on one another, make up a series called a **food chain.** The length and complexity of food chains vary greatly. In real life, a given kind of organism feeding on only one other kind of organism is rather rare. Usually, each will feed on two or more other kinds and, in turn, will be fed on by several other kinds of organisms. When diagrammed, the relationship appears as a series of branching lines, rather than as one straight line; it is called a **food web** (figure 22.9).

A certain amount of the energy that organisms ingest and retain at a given trophic level goes toward heat production. A great deal of the energy is used for digestion and work, and usually 40 percent or less goes toward growth and reproduction. An invertebrate such as a worm or insect typically uses about a quarter of this 40 percent for growth. In other words, about 10 percent of the food that an invertebrate eats is turned into its own body and thus into potential food for its predators. Although the comparable figure varies from approximately 5 percent in carnivores to nearly 20 percent for herbivores, 10 percent is a good average value for the amount of organic matter present at each step in a food chain, or each successive trophic level, and the amount that reaches the next level.

A plant fixes about 1 percent of the sun's energy that falls on its green parts. The successive members of a food chain, in turn, process about 10 percent of the available energy in the organisms on which they feed into their own bodies.

Lamont Cole of Cornell University studied the flow of energy in a freshwater ecosystem in Lake Cayuga in upstate New York. He calculated that about 150 calories of each 1,000 calories of potential energy that are fixed by algae and cyanobacteria are transferred into the bodies of small heterotrophs (figure 22.10). Of these, about 30 calories are incorporated into the bodies of smelt, the principal secondary consumers of the system. Humans who eat the smelt gain about 6 calories from each 1,000 calories that originally entered the system. If, on the other hand, trout eat the smelt and humans eat the trout, humans gain only about 1.2 calories from each original 1,000.

Relationships of this kind make it clear that organisms, including people, that subsist on an all-plant diet obviously have more food and energy available to them than do carnivores. Such considerations will become increasingly important in the future, not only for the efficient management of fisheries, but also in an effort to maximize food yield for a hungry and increasingly overcrowded world (see Sidelight figure 22.1).

Food chains generally consist of only three or four steps. The loss of energy at each step is so great that very little of the original energy is still usable after having been incorporated successively into the bodies of organisms at four trophic levels. Generally, far more individuals exist at the lower trophic levels of any ecosystem than at the higher ones. Similarly, the biomass of the primary producers in a given ecosystem is greater than that of the primary consumers, with successive trophic levels having a lower and lower biomass and correspondingly less potential energy. Larger animals characteristically are members of the higher trophic levels; to some extent, they *must* be larger in order to capture enough prey to support themselves.

Ecological Pyramids

An ecosystem's trophic structure determines the energy flow through the ecosystem and is influenced both by the numbers and sizes of the organisms (that is, how much energy they can supply and how much they consume) and by the amount of energy lost at each transfer. These relationships, if shown diagrammatically, appear as **ecological pyramids.**

Pyramid of Numbers: Counting Individuals

Counting all the individuals in an ecosystem and assigning each to a trophic level would result in a **pyramid of numbers** (figure 22.11*a*). Such a pyramid is not a good representation of the flow of energy through the ecosystem because it does not take into account that some organisms are bigger than others. Oak trees, for example, contribute more than grass plants, and bears more than fleas. In addition, such pyramids rarely count the decomposers, who number in the billions.

Pyramid of Biomass: Weighing Biomass

Weighing all the individuals at each trophic level of an ecosystem would result in a **pyramid of biomass** that would not be influenced by differences in size (figure 22.11*b*). Such pyramids give a good if indirect representation of an ecosystem's energy flow. For most ecosystems on land, the pyramid of biomass points "up," with smaller trophic levels on top. But some aquatic ecosystems have an inverted pyramid of biomass because the biomass of the heterotrophic zooplankton is larger than that of the autotrophic phytoplankton. This is only possible because the phytoplankton are reproducing at a prodigious rate. Generally, this is a very unusual situation among ecosystems.

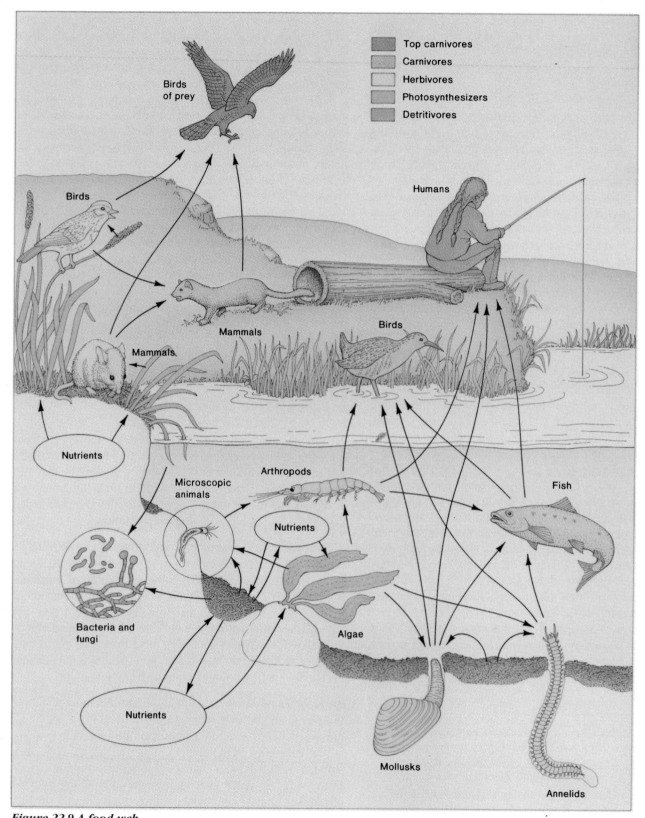

Top carnivores
Carnivores
Herbivores
Photosynthesizers
Detritivores

Birds of prey

Birds

Humans

Mammals

Mammals

Birds

Nutrients

Arthropods

Fish

Microscopic animals

Nutrients

Bacteria and fungi

Nutrients

Algae

Mollusks

Annelids

Figure 22.9 A food web.
This food web is much more complicated than a food chain. The arrows show the complex relationships between the different-colored trophic levels.

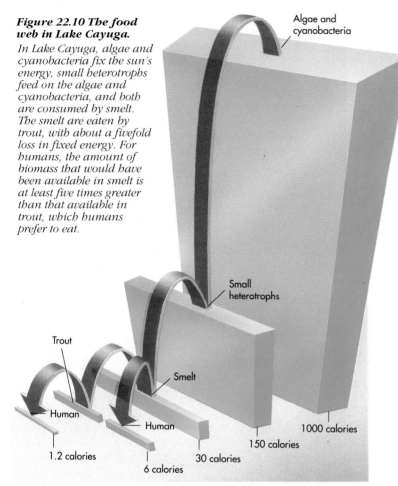

Figure 22.10 The food web in Lake Cayuga.

In Lake Cayuga, algae and cyanobacteria fix the sun's energy, small heterotrophs feed on the algae and cyanobacteria, and both are consumed by smelt. The smelt are eaten by trout, with about a fivefold loss in fixed energy. For humans, the amount of biomass that would have been available in smelt is at least five times greater than that available in trout, which humans prefer to eat.

Algae and cyanobacteria

Small heterotrophs

Trout

Smelt

Human

Human

1.2 calories

6 calories

30 calories

150 calories

1000 calories

Pyramid of Energy: Measuring Energy Loss

A really accurate picture of the flow of energy through an ecosystem requires measuring energy flow directly at each point of transfer. The resulting **pyramid of energy** is not skewed by size or metabolic rate differences, and it is always "right side up" (pyramids of energy cannot be inverted because of the necessary loss of energy at each step) (figure 22.11*c*). Constructing such a pyramid is difficult in that the actual amounts of energy individuals take in, how much they burn up during metabolism, how much they invest in growth, and how much they lose as waste must be accurately measured. In the few cases where this has been done carefully, an important point has emerged: *Less than 17 percent of the energy entering one trophic level becomes available to organisms at the next level.*

Ecological Succession

Even when the climate of a given area remains stable year after year, ecosystems have a tendency to change from simple to complex in a process known as **succession.** This process is familiar to anyone who has seen a vacant lot or cleared woods slowly but surely become occupied with larger and larger plants and more and more different kinds of them, or a pond become filled with vegetation that encroaches from the sides and gradually turns the pond into dry land.

Succession is continuous and worldwide in scope. If a wooded area is cleared, and the clearing is left alone, plants will slowly reclaim the area. Eventually, the traces of the clearing will disappear, and the whole area will again be woods. This kind of succession, which occurs in areas that have been disturbed and that were originally occupied by living organisms, is called **secondary succession.** Humans are often responsible for initiating secondary succession throughout the regions of the world that they inhabit. Secondary succession may also take place after fire has burned off an area, for example, or after a volcanic eruption such as occurred at Mount St. Helens in 1980 (figure 22.12).

Primary succession, in contrast to secondary succession, occurs on some bare, lifeless substrate, such as rocks or open water, where organisms gradually occupy the area and change its nature. New islands formed in the ocean by undersea volcanoes undergo primary succession. On bare rocks, cyanobacteria, algae, or lichens (associations of fungi and algae discussed in chapter 27) may grow first. Acidic secretions from the lichens help to break down the stone and form small pockets of soil. Mosses may then colonize these pockets of soil, eventually followed by ferns and the seedlings of other plants. Over many thousands of years, or even longer, the rocks may be completely broken down, and the vegetation over an area where there was once a rock outcrop may be just like that of the surrounding grassland or forest.

Similarly, a lake may gradually accumulate organic matter and fill in. Plants standing along the edges of the lake, such as cattails and rushes, and those growing submerged, such as pondweeds, together with other organisms, may contribute to the formation of a rich organic soil. As this process continues, the lake may increasingly be filled in with terrestrial vegetation. Eventually, the area where the lake once stood, like the rock outcrop just described, may become an indistinguishable part of the surrounding vegetation.

> *Primary succession takes place in areas that are originally open, such as dry rock faces or new ponds. Secondary succession, in contrast, takes place in areas that have been disturbed after having been occupied by living things earlier.*

Over a very long period, ponds and bare rocks in the same region may come to feature the same kind of vegetation as one another—the vegetation characteristic of the region as a whole. This relationship led American ecologist F. E. Clements, in the early twentieth century, to propose the concept of **climax vegetation.** The term refers to the belief that this characteristic vegetation type is thought to be controlled by the region's *climate.* However, with an increasing realization that (1) the climate keeps changing, (2) the process of

Pyramid of numbers

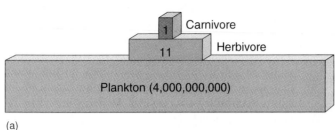

(a)

Pyramid of biomass

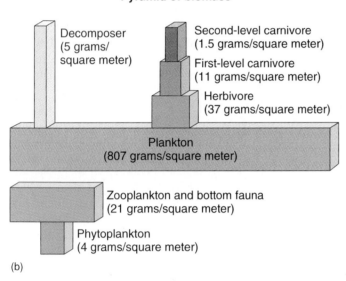

(b)

Pyramid of energy

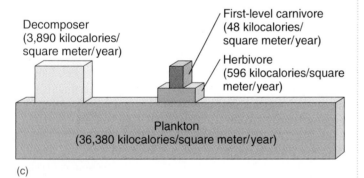

(c)

Figure 22.11 Ecological pyramids.
Ecological pyramids measure different characteristics of each trophic level. (a) *Pyramid of numbers.* (b) *Pyramid of biomass.* (c) *Pyramid of energy.*

succession is often very slow, and (3) the nature of a region's vegetation is determined to a greater extent by human activities, ecologists no longer consider the term *climax vegetation* as useful as they once did.

One of the most interesting features of succession is that the organisms involved in its early stages often are participants in symbiotic systems. Lichens, so important in the early colonization of rocks, are symbiotic associations of cyanobacteria or algae and fungi. This association appears to enable lichens to withstand harsh conditions of moisture and temper-

(a)

(b)

Figure 22.12 Succession.
Mount St. Helens in the state of Washington erupted violently on 18 May 1980. The lateral blast devastated more than 600 square kilometers of forest and recreation lands within fifteen minutes. (a) *This is how an area near Clearwater Creek looked four months after the eruption.* (b) *Five years later, succession was underway at the same spot, with shrubs, blueberries, and dogwoods following the first plants that became established immediately after the blast.*

ature. Legumes and other plants that harbor nitrogen-fixing bacteria in nodules on their roots are often among the first colonists of relatively infertile soils. Trees with mycorrhizae are apparently more resistant to conditions of moisture stress and perhaps other environmental extremes than are trees without mycorrhizae. The first land plants are believed to have had mycorrhizae that were highly beneficial to them in the sterile, open habitats that they encountered.

Symbiotic associations, such as lichens, legumes with their associated nitrogen-fixing bacteria, and plants with mycorrhizal fungi, appear to play an unusually important role in early successional communities.

Sidelight 22.1

EATING LOW ON THE FOOD CHAIN: GOOD FOR YOU AND GOOD FOR THE ENVIRONMENT

These days, what we eat has become a popular subject. Experts debate the merits of oat bran versus wheat bran, and of fats found in fish versus fats found in beef, and stress the importance of lowering blood cholesterol. All of these debates are aimed at designing a human diet that prevents disease and maintains the body at its optimal weight. But what about the environment? In the 1960s, ecologists and environmentalists coined the phrase "eating low on the food chain" as the best way to maximize the world's food resources and to plan a healthy diet. In the 1990s, as food resources dwindle and environmental concerns mount, more and more people are following this advice.

"Eating low on the food chain" is based on the ecological principles discussed in this chapter. As mentioned in the chapter discussion, approximately 10 percent of the energy in one trophic level is made available to the consumer in the next trophic level. The energy that does not make it to the next trophic level is not lost; rather, it is either expended by the metabolism of the organisms in the higher trophic level and is returned to the ecosystem in the form of heat, or it is used in such processes as reproduction. However this energy is expended, it is not available to the organisms in the next trophic level as food. This "loss" of energy is evident in the Lake Cayuga ecosystem (figure 22.10): From a total pool of 1,000 calories that are available in the algae and cyanobac-

teria trophic level, humans receive only 1.2 calories when they eat from the trophic level occupied by trout.

These numbers become even more exaggerated at the trophic levels occupied by grains, cattle, and humans. Using 1979 statistics, ecologists determined that each cow destined for the meat market consumes about 2,500 pounds of grain. It takes 16 pounds of grain to produce 1 pound of beef. Analysis of the nutrient composition of these amounts re-

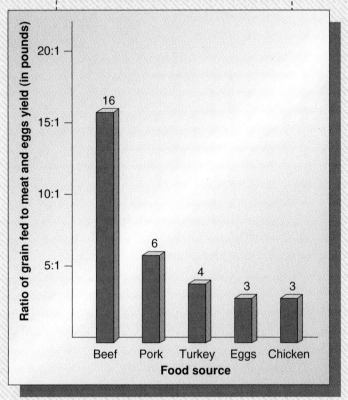

Sidelight figure 22.1 Energy loss from livestock production.
As this graph shows, the ratios of energy (grain) required to produce meat or eggs suggest that plant foods are a more energy-efficient choice for human diets.

veals that the 16 pounds of grain contain twenty-one times the calories in the 1 pound of beef, eight times the protein, and three times

the fat (Sidelight figure 22.1). Investigating further, ecologists found that, in 1979, 145 million tons of grain were fed to all livestock animals (including cattle, hogs, and chickens), and the meat and eggs yield was 21 million tons. The energy "loss" of this endeavor was the amount of energy contained in 124 million tons of grain, a large amount indeed.

In addition, the raising of livestock requires huge amounts of water, fertilizers, and equipment, which together contribute to the pollution of the environment and to the higher price per pound of meat as compared to vegetable foods. Furthermore, nutrition experts have recently warned that a diet high in animal fat can be detrimental to human health and lead to cardiovascular disease and some cancers. As noted earlier, the 16 pounds of grain fed to cattle to produce 1 pound of beef contain only three times the fat that the beef does. This is good news to those who wish to curtail their fat intake. Not only are plant foods cheaper and kinder to the environment, their lower fat content make them healthier food choices than meat.

Not everyone has the temperament or the motivation to become a complete vegetarian and eat exclusively from the lower rungs of the food chain. Like any lifestyle change, becoming a vegetarian requires modifying ingrained tastes and habits. Also, vegetarians need to make sure that they receive adequate amounts of protein, a nutrient that meat contains in abundance and that plant foods supply adequately only in certain combinations. However, everyone could reduce their meat consumption, perhaps by eating one vegetarian meal a month and then slowly increasing the number of meatless meals over time. The food choices we make have an impact not only on our bodies, but on the environment as well.

The following general characteristics appear to hold for succession of all sorts: As any ecosystem matures, total biomass increases, while net productivity decreases. The earlier successional stages are more productive than the later ones. Agricultural systems are examples of early successional stages in which the process is intentionally not allowed to go to completion and the net productivity is high. There are many more species in mature ecosystems than in immature ones, and the number of heterotrophic species increases even more rapidly than the number of autotrophic species. This progression is related to the decreasing net productivity of increasingly mature ecosystems and to the fact that mature ecosystems have a greater ability to regulate nu-

trient cycling. The plants and animals that appear in the later stages of succession may be more specialized, in general, than those that exist in the earlier stages. The late-successional species fit together into more complex communities and have much narrower ecological requirements, or niches.

Communities at early successional stages have a lower total biomass, higher net productivity, fewer species, many fewer heterotrophic species, and less capacity to regulate the cycling of nutrients than do communities at later successional stages.

EVOLUTIONARY VIEWPOINT

Energy flows into and through an ecosystem in one direction, while nutrients cycle within an ecosystem. As an ecosystem's biological members experience competition on the one hand and cooperation via symbiosis on the other hand, they enter into increasingly complex associations, all of which depend critically on continued energy flow and nutrient cycling. Evolution within ecosystems promotes both stability and interdependence.

SUMMARY

1. Populations of different organisms that live together in a particular place are called communities. A community, together with the nonliving components of its environment, is called an ecosystem. Ecosystems regulate energy flow and nutrient cycling.

2. Only 2 percent of the water on earth is fixed in any way; the rest is free. In the United States, 96 percent of the freshwater is groundwater.

3. About 10 percent of the roughly 700 billion metric tons of free carbon dioxide in the atmosphere is fixed each year through photosynthesis. An additional trillion metric tons of carbon is dissolved in the ocean, and five times that amount is locked up as coal, oil, and gas. About as much carbon exists in living organisms at any one time as there is in the atmosphere.

4. Carbon dioxide and certain other heat-absorbing gases in the earth's atmosphere are responsible for the earth's greenhouse effect, which is intensifying because of the production of these gases associated with industrialization.

5. A very small fraction of the enormous reservoir of nitrogen that exists in the earth's atmosphere is made available to organisms through the activities of a few genera of bacteria.

6. Carbon, nitrogen, and oxygen have gaseous or liquid reservoirs, as does water. All of the other nutrients, such as phosphorus, have solid reservoirs.

7. An ecosystem's net primary productivity is the total amount of energy fixed by photosynthesis per unit of time, minus that which is expended by the metabolic activities of ecosystem organisms.

8. Plants convert about 1 percent of the energy that falls on their leaves to food energy. Herbivores that eat the plants and the other animals that eat the herbivores constitute a set of trophic levels. At each of these levels, only about 10 percent of the energy fixed in the food is fixed in the body of the animal that eats that food. For this reason, food chains are always relatively short.

9. Ecological pyramids of numbers, biomass, and energy measure the passage of resources through each trophic level.

10. Ecosystems tend to become more complex over time through a process known as succession. Primary succession takes place in areas that are originally bare, such as rocks or open water. Secondary succession takes place in areas where the initial communities of organisms have been disturbed.

11. Both types of succession lead ultimately to the formation of climax communities, whose nature is controlled primarily by the climate of the area concerned, although the human influence on many of these communities is increasing. Such communities have more total biomass, less net productivity, more species, many more heterotrophic species, and a higher capability of regulating nutrient cycling than do earlier successional stages.

REVIEWING THE CHAPTER

1. What is an ecosystem?
2. How are nutrients cycled through ecosystems?
3. How is water cycled?
4. How is carbon cycled?
5. How is nitrogen cycled?
6. How is phosphorus cycled?
7. How does energy flow through ecosystems?
8. What are trophic levels?
9. What is a pyramid of numbers?
10. What is a pyramid of biomass?
11. What is a pyramid of energy?
12. How does ecological succession occur?

COMPLETING YOUR UNDERSTANDING

1. What does a biogeochemical cycle do?
2. Plants add water to the atmosphere in a process called
  a. evaporation.
  b. groundwater flow.
  c. transpiration.
  d. biogeochemical cycling.
3. The greenhouse effect
  a. may have begun already, but experts disagree.
  b. is responsible for the carbon dioxide in the atmosphere.
  c. keeps the earth about 35 degrees Celsius warmer than it would be without an atmosphere.
  d. is being decreased as a result of human activities.
  e. has not been demonstrated.
4. What is the greenhouse effect, and how does it relate to human activity? Since 1992, what appears to have temporarily slowed the increase in global temperature from the greenhouse effect?

5. Which gas makes up most of the earth's atmosphere?
  a. Carbon dioxide
  b. Hydrogen
  c. Nitrogen
  d. Oxygen
  e. Helium
  f. Sulfur dioxide
6. What is the process of nitrogen fixation, and how could it become more beneficial to agriculture?
7. What are mycorrhizae, and how do they relate to the phosphorus cycle?

8. How do primary productivity and net primary productivity differ?

9. Arrange the communities that follow in descending order of their net productivity.

    a.   Surface of the Greenland ice cap

    b.   Tropical rain forest

    c.   Temperate woodland

    d.   Desert scrub

    e.   Cornfield

10. How could you increase the net primary productivity of a desert?

11. What is the difference between a food chain and a food web?

12. When a green plant is consumed by another organism, how much of the plant's energy is converted into the body of the consumer?

    a.   None of it

    b.   All of it

    c.   1 percent of it

    d.   10 percent of it

    e.   One-third of it

13. Are grain-fed cattle a good enterprise? Why or why not?

14. What are the benefits and cautions associated with becoming a complete vegetarian?

15. Why is a pyramid of numbers not a good representation of energy flow through an ecosystem?

16. What important point has emerged from carefully measured pyramids of energy?

17. Why do current-day ecologists no longer consider the term *climax vegetation* as useful as they once did?

18. Can you think of relatively recent examples of volcanic activity that would demonstrate both primary and secondary succession?

19. How do symbiotic associations relate to ecological succession?

20. Why does the productivity of an ecosystem change as it becomes more mature?

21. At what successional stage would you characterize a field of wheat? What does this imply as to its stability and productivity?

FOR FURTHER READING

Colinvaux, P. A. 1989. The past and future Amazon. *Scientific American,* May, 102–8. A very interesting article, based on current research, about historical changes in the Amazon rain forest ecosystem.

Houghton, R. A., and G. M. Woodwell. 1989. Global climatic change. *Scientific American,* April, 36–44. A good summary of recent research on global warming.

Paul, W. et al. 1990. The global carbon cycle. *American Scientist* 78: 310–26. The dynamic responses of natural systems to carbon dioxide that may determine the future of the earth's climate.

Pimm, S. et al. 1991. Food web patterns and their consequences. *Nature* 350 (April): 669–74. A review of current ecological knowledge of food webs.

Sisson, R. 1986. Tide pools: Windows between land and sea. *National Geographic* 169 (2): 252–59. A beautifully illustrated tour of a California tide pool, alive with organisms.

Spencer, C. et al. 1991. Shrimp stocking, salmon collapse, and eagle displacement. *BioScience* 41: 14–21. Cascading interactions in the food web of a large aquatic ecosystem.

CHAPTER
23

ATMOSPHERE, OCEANS, AND BIOMES

An exceptionally clear view of the earth as seen from space makes it readily evident that water covers much of the earth's surface and also swirls around the atmosphere.

FOR REVIEW

Here are some important terms and concepts that have been discussed in previous chapters and that you will encounter again in this chapter. Review them before proceeding if necessary.

Major features of evolution (*chapter 17*)
Communities (*chapters 20 and 21*)
Biogeochemical cycles (*chapter 22*)
Ecosystems (*chapter 22*)

Evolution shapes natural communities to suit their environments. Over billions of years, great differences in climate across the face of the globe have resulted in the creation of diverse terrestrial and marine communities. These are major assemblages of plants, animals, and microorganisms that occur over wide areas and have distinctive characteristics that separate them from others. The climate differences that have led to the evolution of these different communities are caused ultimately by the major atmospheric and oceanic circulation patterns, which are driven by the unequal distribution of heat from the sun.

Groups of terrestrial communities that occur over wide areas and that are easily recognized by their overall appearance and characteristic climates are called **biomes.** Biomes are reviewed in this chapter, as are analogous assemblages that occur in the sea. Each biome is similar in its structure and appearance wherever it occurs on earth and differs significantly from other biomes. In this text, biomes provide a convenient means for discussing the properties of life on earth from an ecological perspective.

Biome distribution results from the interaction of earth features, such as different soil types or the occurrence of mountains and valleys, with two key physical factors: (1) the amount of solar heat that reaches different parts of the earth and the seasonal variation in that heat, and (2) global atmospheric circulation and the resulting patterns of oceanic circulation. Together, these factors determine the local climate, and in particular, the amounts and distribution of precipitation.

The General Circulation of the Atmosphere

The world contains a great diversity of ecosystems because climates vary a great deal from place to place. On any given day, Miami, Florida, and Bangor, Maine, often have very different weather. There is no mystery about this. The tropics are warmer than the temperate regions because the sun's rays arrive almost perpendicular to regions near the equator, whereas their angle of incidence spreads them out over a much greater area near the poles, providing less energy per unit area. Because the earth is a sphere, some parts of the earth receive more energy from the sun than other parts. This is responsible for many of the earth's major climatic differences and thus, indirectly, for much of ecosystem diversity (figure 23.1).

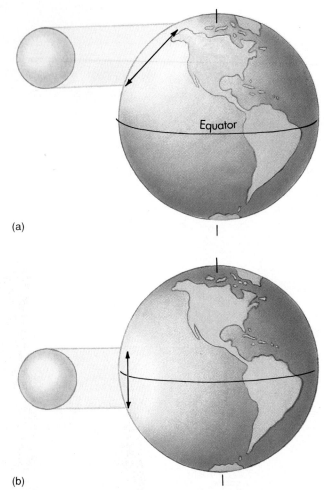

(a)

(b)

Figure 23.1 Relationship between the earth and the sun critical in determining the nature and distribution of life on earth.
A beam of solar energy striking the earth in middle latitudes (a) is spread over a wider area of the earth's surface than a similar beam striking the earth near the equator (b).

The earth's annual orbit around the sun and its daily rotation on its own axis are also both important in determining world climate. Because of the daily cycle, the climate at a given latitude is relatively constant. Because of the annual cycle and the inclination of the earth's axis at approximately 23.5 degrees from its plane of revolution around the sun, all parts of the earth away from the equator experience a progression of seasons. One of the poles is closer to the sun than the other at all times because the angle and direction of the earth's inclination are maintained as the earth rotates around the sun.

How Major Atmospheric Circulation Patterns Affect Precipitation and Climate

The major atmospheric circulation patterns result from the interactions between six large air masses. These great air masses occur in pairs, with one air mass of the pair occurring in the northern latitudes and the other occurring in the southern latitudes. These air masses affect climate because the rising and falling of an air mass influences its temperature, which, in turn, influences its moisture-holding capacity.

Cold air is more dense than warm air, so warm air rises and cold air falls. When air falls from a higher level, where it is cool, to a lower level, where it is warmer, the air is warmed. Warm air holds more moisture than cold air, so falling air that is also warming tends to absorb moisture like a sponge. In regions of the earth where air is falling and being warmed, there is thus little precipitation. On the other hand, when warm air at a lower level rises, it is cooled, and as this cooling occurs, moisture leaves the air in the form of precipitation. These phenomena are the keys to understanding climate.

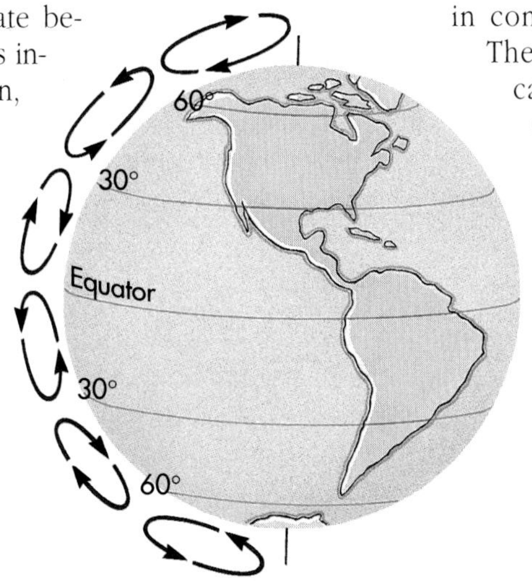

Figure 23.2 Atmospheric circulation patterns.
These result from air movement out from and back to the earth's surface within six large air masses.

The two air masses that sit on either side of the equator, heated by the sun, are rising air masses (figure 23.2). These warm air masses rise and flow in the direction of the two poles. As this warm air rises, it cools and loses most of its moisture, which explains why the tropics receive so much rain. When the air masses that have risen reach about 20 to 30 degrees north or south latitude, the air, now cooled, sinks and becomes reheated, producing a zone of decreased precipitation. Consequently, most of the great deserts of the world lie near latitudes 20 to 30 degrees north and south of the equator. Air at these latitudes is still warmer than it is in the polar regions, and thus it continues to flow toward the poles. Air rises and cools again at about 60 degrees north and south latitude and flows back toward the equator. The rising and cooling air sheds its moisture in the form of precipitation, and thus the great temperate forests of the world are located between 30 and 60 degrees north and south latitude. Finally, another air mass rises at 60 degrees north and south latitude, producing zones of very low temperature and precipitation known as the polar regions. The complex interactions of all of these air masses result in the major wind patterns that blow over the earth.

Warm air rises near the equator, descends and produces arid zones at about 30 degrees north and south latitude, flows toward the poles again, and then rises at about 60 degrees north and south latitude and moves back toward the equator. Air masses also rise at 60 degrees north and south latitude and move toward the poles, where they produce zones of low temperature and precipitation.

Although, as noted earlier, most of the major deserts of the earth occur at latitudes 20 to 30 degrees north and south due to the interaction of air masses, some deserts also form

in continental interiors for a different reason. These deserts have limited precipitation because of their distance from the sea and sometimes because mountain ranges intercept the moisture-laden winds from the sea, a phenomenon called the **rain shadow effect.** In the rain shadow effect, as moisture-laden air from the sea blows toward a mountain range, the air's moisture-holding capacity decreases, resulting in increased precipitation on the windward side of the mountains—the side from which the wind is blowing. As the air descends the other side of the mountain—the leeward side—it is warmed, and its moisture-holding capacity increases, tending to block precipitation. Leeward sides of mountains are thus much drier than windward sides, and the vegetation is often very different (figure 23.3).

Most of the great deserts and associated arid areas of the world lie at about 30 degrees north and south latitude. Other major deserts occur in continental interiors and may have formed as a result of the rain shadow effect.

Circulation Patterns in the Ocean

Oceanic circulation patterns are determined by the patterns of atmospheric circulation just discussed, but they are modified by the location of the landmasses around which and against which the ocean currents must flow. Oceanic circulation is dominated by huge surface **gyrals,** circular patterns that move around the subtropical oceans at about 30 degrees north latitude and 30 degrees south latitude. Gyrals move clockwise in the Northern Hemisphere and counterclockwise in the Southern Hemisphere.

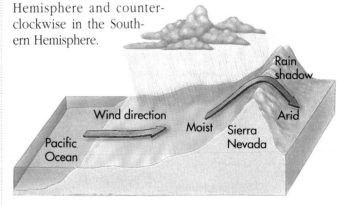

Figure 23.3 The rain shadow effect.
Moisture-laden winds from the Pacific Ocean rise and are cooled when they encounter the Sierra Nevada. As they cool, their moisture-holding capacity decreases, resulting in precipitation. As the air descends on the east side of the range, it warms, its moisture-holding capacity increases, and it picks up moisture from its surroundings. As a result, desert conditions prevail on the east side of the mountains.

They profoundly affect life not only in the oceans but also on coastal lands by the ways in which they redistribute heat. For example, the Gulf Stream, in the North Atlantic, swings away from North America near Cape Hatteras, North Carolina, and reaches Europe near the southern British Isles. Because of the Gulf Stream, western Europe is much warmer and thus more temperate than is eastern North America at the same latitudes.

In South America, the Humboldt Current carries cold water northward up the west coast and helps to make possible an abundance of marine life that supports the fisheries of Peru and northern Chile. Marine birds, which feed on these organisms, are responsible for the commercially important guano deposits of these countries. These deposits are rich in phosphorus, which is brought up from the ocean depths by the upwelling of cold water that occurs in the Pacific Ocean. When the coastal waters are not as cold as usual, the devastating phenomenon called "El Niño" affects the vitality of both marine and terrestrial populations of animals and plants worldwide (see Sidelight figure 23.1).

In the ocean, huge surface gyrals move around the subtropical zones between approximately 30 degrees north and south latitude, clockwise in the Northern Hemisphere and counterclockwise in the Southern Hemisphere.

Sidelight 23.1
EL NIÑO–SOUTHERN OSCILLATION EVENTS

The west coast of South America is normally a highly productive area, cooled by the Humboldt Current, which sweeps up from the south and even makes it possible for penguins, which are cold-water birds, to live on the Galápagos Islands near the equator. The productivity of the surface waters depends on this current and on the cool, nutrient-rich waters that well up from the depths to replace the depleted waters at the surface. In a normal January, however, warm water flows down the coast from the tropics to southern Peru and northern Chile. Local fishermen named this current El Niño ("The Christ Child") because the current occurs near Christmas and is normally benevolent. But scientists now reserve this name for a catastrophic version of the same phenomenon, one that is felt not only locally but on a global scale.

Such events, which are generally known as El Niño–Southern Oscillation or ENSO events, are now known to be of global significance in perturbing the general circulation of the atmosphere, as do episodes of enhanced cold-water upwelling. ENSOs occur every 2 to 6 years. The term "southern oscillation" is used to describe the atmospheric changes that accompany the ocean warming of an El Niño event. The factors that lead to an ENSO are unknown, although they seem to involve major changes in atmospheric pressure systems, and there is a net movement of warm water from the western Pacific to the central and eastern Pacific. When this occurs, commercial fish stocks virtually disappear from the waters of Peru and northern Chile and the plankton drop to a twentieth of their normal abundance. The commercially valuable anchovy fisheries of Peru were essentially permanently destroyed by the 1972 ENSO event. Violent winter storms lashed the coast of California, accompanied by flooding and a colder and wetter winter than normal in Florida and along the Gulf Coast. A particularly severe ENSO seems to occur about every 25 years or so. One such oscillation began in October 1986 and extended through early 1988; it was apparently correlated with unusual drought in the states of Washington and Oregon in the summer of 1987.

In the ENSO of 1982 to 1983, many birds starved to death on islands throughout the central Pacific, and a huge number of bird colonies simply gave up breeding. On Christmas Island, an important breeding site about 2,000 kilometers south of Honolulu, 95 percent of the 14 million birds that normally nest on the island, representing eighteen species, simply left, abandoning their eggs, young, and nests; by June 1983, there were only 150,000 birds on the island. Even for the Farallon Islands, which lie about 45 kilometers off San Francisco, many birds starved, and the large breeding ground was tragically disrupted. Many eggs failed to hatch, and thousands of chicks starved (Sidelight figure 23.1). On the other hand, the moist, rainy conditions that accompany an ENSO event are favorable for crops on land; the Peruvian rice crop, for example, is strongly favored by such a climatic regimen. Land birds, such as Darwin's finches on the Galápagos Islands, breed abundantly during an ENSO, and their population sizes increased remarkably during the ENSO of 1982 to 1983.

Sidelight figure 23.1 Double-edged effects of ENSO events.

(a) *Daphne Crater, Galápagos Islands, containing about eight hundred pairs of the nesting tropical seabirds known as blue-footed boobies in 1975, a normal year.* (b) *The same crater in 1983, an El Niño year. The crater is ringed with abundant vegetation, due to El Niño's moist, rainy conditions, but no boobies are nesting there because El Niño conditions have sharply reduced the birds' food supply of fish and plankton.*

(a)   (b)

The Oceans

Nearly three-fourths of the earth's surface is covered by ocean. The seas have an *average* depth of more than 3 kilometers and are, for the most part, cold and dark. Heterotrophic organisms are found even at the greatest ocean depths, which reach nearly 11 kilometers in the Marianas Trench of the western Pacific Ocean, but photosynthetic organisms, which rely on energy from the sun, are confined to the upper few hundred meters of water, where light rays can penetrate. Organisms that live below this level obtain almost all of their food indirectly, as a result of photosynthetic activities that occur above. These activities result in organic detritus that drifts downward.

Many fewer species live in the sea than on land. Probably more than 85 percent of all species of organisms occur on land, including the great majority of members of a few large groups—especially insects, mites, fungi, and plants. Each of these groups has marine representatives, but these comprise only a very small fraction of the total number of species. On land, the barriers between habitats are sharper, and variations in elevation, parent rock, degree of exposure, and other factors have all been crucial to the evolution of the millions of species of terrestrial organisms. On the other hand, most phyla originated in the sea, and almost all are represented there now, whereas only a few phyla occur on land.

> *Although representatives of almost every phylum occur in the sea, relatively few phyla occur on land. However, the few terrestrial phyla have many species: An estimated 85 percent of living species of organisms are terrestrial. The enormous evolutionary success of the few land phyla is due to the boundaries between different habitats being sharper on land than they are in the sea.*

The marine environment consists of three major kinds of habitats: (1) the **neritic zone,** the zone of shallow waters along the coasts of the continents; (2) the **surface layers** of the open sea; and (3) the **abyssal zone,** the deep-water areas of the oceans.

The Neritic Zone

The neritic zone of shallow water is small in area, but it is inhabited by very large numbers of species compared with other parts of the ocean (figure 23.4). The intense and sometimes violent interaction between sea and land in this zone gives a selective advantage to well-secured organisms that can withstand being washed away by the continual beating of the waves. Part of this zone, the **intertidal,** or **littoral,** region is exposed to the air whenever the tides recede.

Because of the way in which it gives access to the land, the intertidal zone must have been home for the ancestors of the first land organisms. The complex structures necessary to anchor them and protect them from drying out in this turbulent zone seem in some cases to have made possible their success on land. Some of the organisms that occur in intertidal areas, such as fiddler crabs, have internal biological clocks that allow them to attune their feeding activities to the daily ebb and flow of the tides. The world's great fisheries also occur on banks in the coastal zones (figure 23.5), where nutrients, derived from the land, are often more abundant than in the open ocean.

Partly enclosed bodies of water, such as those which often form at river mouths and in coastal bays, where the salinity is intermediate between that of salt- and freshwater, are called **estuaries.** Estuaries are among the most naturally fertile areas in the world, often with rich stands of submerged and emergent plants, algae, and microscopic organisms. They provide the breeding grounds for most coastal shellfish and fish that are harvested both in the estuaries and in open water.

The Surface Zone

Drifting freely in the upper, better-illuminated waters of the ocean is a diverse biological community, primarily consisting of microscopic organisms called the **plankton.** Fishes and other larger organisms that swim in these same waters constitute the **nekton,** whose members feed mainly on plankton. Together, the organisms that make up the plankton and the nekton provide all of the food for those that live below. Some of the members of the plankton, including the algae and some bacteria, are photosynthetic. Collectively these organisms account for about 40 percent of all the photosynthesis that takes place on earth, and even more by some calculations. Most of the plankton occurs in the top 100 meters of the sea, the zone into which light from the surface penetrates freely. Perhaps half of the total photosynthesis in this zone is carried out by organisms less than 10 micrometers in diameter, including cyanobacteria and the smallest algae.

The Abyssal Zone

In the deep waters of the sea, below the top 300 meters, occur some of the most bizarre organisms found on earth. Many of these animals have some form of bioluminescence, by means of which they communicate with one another or attract their prey. In the mud of the ocean floor, or along rifts from which warm water issues, live similar assemblages of peculiar creatures. Bacteria are apparently rather frequent in the deeper layers of the sea and as decomposers are as important in this zone as they are on land and in freshwater habitats.

> *Most of the life in the world's oceans occurs near its shores. In the open ocean, the food chain is based on extensive photosynthesis carried out by phytoplankton near the surface.*

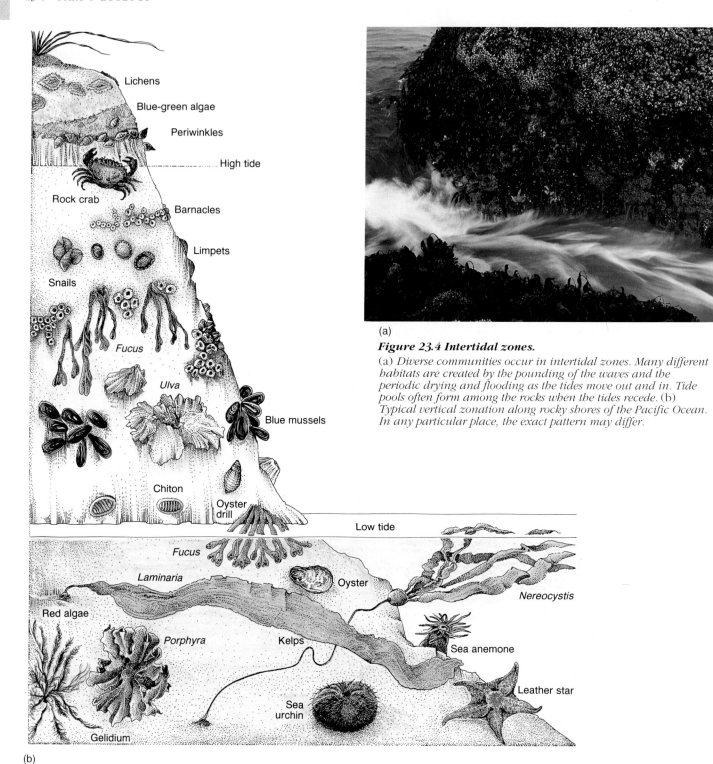

(a)

Figure 23.4 Intertidal zones.
(a) *Diverse communities occur in intertidal zones. Many different habitats are created by the pounding of the waves and the periodic drying and flooding as the tides move out and in. Tide pools often form among the rocks when the tides recede.* (b) *Typical vertical zonation along rocky shores of the Pacific Ocean. In any particular place, the exact pattern may differ.*

(a)

Figure 23.5 An endangered estuary: Chesapeake Bay.

Chesapeake Bay, which has more than 11,300 kilometers of shoreline, is an estuary and drains more than 166,000 square kilometers in one of the most densely populated and heavily industrialized areas in North America. (a) The body of open water is about 320 kilometers long and, at some points, nearly 50 kilometers wide. (b) Large metropolitan areas and shipping facilities make Chesapeake Bay one of the busiest natural harbors anywhere. (c) One of the most biologically productive bodies of water in the world, Chesapeake Bay yielded an annual average of about 275,000 kilograms of fish in the 1960s, but only a tenth as much in the 1980s. The human population of the area grew 50 percent during the same period. (d) More than 290 oil spills were reported in the bay in one year (1983), with oil transport and commercial shipping expected to double by the year 2020. This grebe is coated from an oil spill off the mouth of the Potomac River. (e) Pesticides, increases in nutrients, and uncontrolled erosion from certain agricultural practices block the light needed for photosynthesis and upset the delicate ecological balance on which the bay's productivity depends. The states that border the bay are cooperating, with the assistance of the Environmental Protection Agency, to try to restore Chesapeake Bay's former productivity.

(b)

(c)

(d)

(e)

Freshwater Communities

Freshwater habitats are distinct from both marine and terrestrial ones, but they are very limited in area. Inland lakes cover about 1.8 percent of the earth's surface, and running water covers about 0.3 percent. All freshwater habitats are strongly connected with terrestrial ones, with marshes and swamps constituting intermediate habitats. In addition, a large amount of organic and inorganic material continually enters bodies of freshwater from communities growing on the land nearby (figure 23.6). Many kinds of organisms are restricted to freshwater habitats. When they occur in rivers and streams, they must be able to attach themselves in such a way as to resist or avoid the effects of current, or risk being swept away. In bodies of standing water, such matters are of much less importance (figure 23.7).

Ponds and lakes have three zones in which organisms occur: (1) a **littoral** (or shore) **zone;** (2) a **limnetic zone,** inhabited by plankton and other organisms that live in open water; and (3) a **profundal zone,** below the limits of effective light penetration. **Thermal stratification** is characteristic of the larger lakes in temperate regions and is the process whereby water at a temperature of 4 degrees Celsius (which is when water is densest), sinks beneath water that is either

Figure 23.6 A nutrient-rich stream.
In this stream in the North Coast Ranges of California in summer, as in all streams, much organic material falls or seeps into the water from the communities along the edges. This input is responsible for much of the stream's biological productivity.

(a) (b) (c)

(d) (e) (f)

Figure 23.7 Freshwater organisms.
(a) *Speckled darter.* (b) *Green frog.* (c) *Freshwater snail.*
(d) *Giant waterbug with eggs on its back.* (e) *Damselfly nymph.*
(f) *Bladderwort.*

(a)

(b)

Figure 23.8 Eutrophic and oligotrophic bodies of water.

(a) *In this eutrophic farm pond, the surface bloom of green algae reflects the plentiful supply of nutrients in the water.*
(b) *Lake Tahoe, an oligotrophic lake, lies high in the Sierra*

Nevada on the border between California and Nevada. The drainage of fertilizers applied to the plantings around residences, business concerns, and recreational facilities bordering the lake poses an everpresent threat to the maintenance of the water's deep blue color.

warmer or cooler. In winter, water at 4 degrees Celsius sinks beneath cooler water that freezes at the surface at 0 degrees Celsius. Below the ice, the water remains between 0 and 4 degrees Celsius, and plants and animals survive there. In spring, as the ice melts, the surface water is warmed to 4 degrees Celsius and sinks below the cooler water, bringing the cooler water to the top with nutrients from the lake's lower regions. This process is known as the **spring overturn.**

In summer, warmer water forms a layer over the cooler waters (about 4 degrees Celsius) that lie below. In the area between these two layers—the **thermocline**—temperature changes abruptly. Depending on the climate of the particular area, the warm upper layer may become as much as 20 meters thick during the summer. In the autumn, its temperature drops until it reaches that of the cooler layer underneath—4 degrees Celsius. When this occurs, the upper and lower layers mix—a process called the **fall overturn.** Therefore, colder waters reach the surfaces of lakes in the spring and fall, bringing up fresh supplies of dissolved nutrients.

Lakes can be divided into two categories, based on their production of organic matter. **Eutrophic lakes** have an abundant supply of minerals and organic matter (figure 23.8*a*). Oxygen is depleted below the thermocline in the summer because of the abundant organic material and the high rate at which aerobic decomposers in the lower layer use oxygen. These stagnant waters again reach the surface after the fall overturn. In **oligotrophic lakes,** on the other hand, organic matter and nutrients are relatively scarce. Such lakes are often deeper than eutrophic ones, and their deep waters are always rich in oxygen (figure 23.8*b*). Oligotrophic lakes are highly susceptible to pollution from excess phosphorus from such sources as fertilizer runoff, sewage, and detergents.

Freshwater communities have a distinct vertical organization that is strongly influenced by seasonal changes at latitudes where lakes and ponds experience freezing winter temperatures.

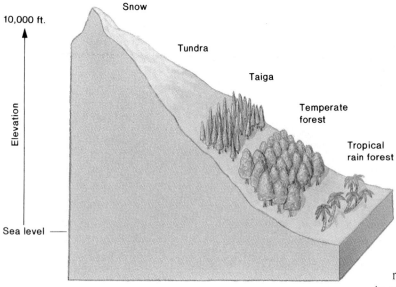

10,000 ft.

Snow

Tundra

Taiga

Temperate forest

Tropical rain forest

Elevation

Sea level

Figure 23.9 Altitude influences biomes in same way as latitude.

Tropical mountainsides have biomes that, as elevation increases, mirror biomes normally found far north and far south of the equator.

BIOMES

Biomes are climatically defined assemblages of organisms that have a characteristic appearance and are distributed over a wide land area. Latitude and altitude have similar influences on biomes—that is, biomes found far north and far south of the equator at sea level also occur in the tropics, but at high mountain elevations (figure 23.9).

Biomes are classified in several ways, but in this text, they are grouped into fourteen categories. Figure 23.10 shows the geographical distribution of these fourteen biomes. The biomes differ remarkably from one another because they have evolved in regions with very different climates. Distinctive features of the six major biomes—tropical rain forest, savanna, desert, temperate grassland, temperate deciduous forest, and taiga—are now discussed in more detail.

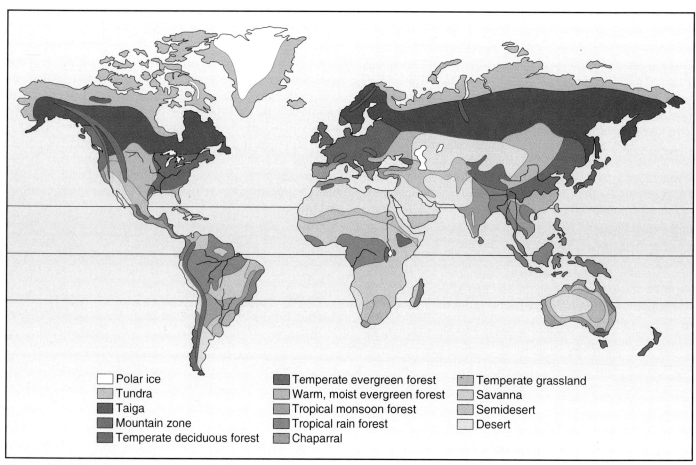

☐ Polar ice
☐ Tundra
■ Taiga
■ Mountain zone
■ Temperate deciduous forest

■ Temperate evergreen forest
☐ Warm, moist evergreen forest
☐ Tropical monsoon forest
☐ Tropical rain forest
☐ Chaparral

■ Temperate grassland
☐ Savanna
☐ Semidesert
☐ Desert

Figure 23.10 Distribution of the earth's biomes.

Biomes vary remarkably because they have evolved in regions with very different climates.

Tropical Rain Forest: Lush and Equatorial

The **tropical rain forest** is the richest biome in terms of number of species, probably containing at least half of the species of terrestrial organisms—at least several million! The communities that make up tropical rain forests are diverse in that each kind of animal, plant, or microorganism is often represented in a given area by very few individuals. For example, seldom are there fewer than forty species of tree per hectare; this is four or five times as many as are typical of temperate forests. In a single square mile of tropical rain forest in Peru or Brazil, there may be fourteen hundred or more species of butterflies—more than twice the total number found in the United States and Canada combined! The ways of life of tropical organisms are often specialized and highly unusual.

Rainfall in tropical rain forests is generally 200 to 450 centimeters per year, with little difference in the rain's distribution from season to season. Tropical rain forests have a high net productivity, even though they exist mainly on quite infertile soils. Most of the nutrients are held within the plants themselves and are rapidly recycled when the plants die or when parts, such as leaves, are lost. Most of the roots of the tall trees spread out in a thin layer of soil, often no more than a few centimeters thick. These roots transfer nutrients from leaves and other fallen or-ganic debris quickly and efficiently back to the trees. When people clear tropical rain forests by what are called slash-and-burn methods, an abundant supply of nutrients runs out into the soil. This temporary fertilization makes crop cultivation possible for a few years, but then, as nutrients are depleted, the people must move on. If too many people attempt to cultivate a given region, the forest is destroyed (see chapter 24).

There are extensive tropical rain forests in South America, particularly in and around the Amazon Basin; in Africa, particularly in portions of West Africa; and in Southeast Asia. Human populations in these areas are rapidly destroying the rain forests. Consequently, few rain forests will be left in an undisturbed condition anywhere in the world by the first part of the next century. The disturbance and destruction of tropical rain forests will be accompanied by the extinction of as many as a quarter of the total species on earth during the next fifty years.

Savanna: Dry, Tropical Grassland

In areas of reduced annual precipitation, or prolonged annual dry seasons, open tropical and subtropical **deciduous forests**—forests in which most of the trees and shrubs lose their leaves at some season of the year—give way to the kind of open grassland called **savannas.** Huge herds of grazing mammals, with their associated predators, inhabit the savannas of Africa. Such animal communities occurred in North America during the Pleistocene epoch but have persisted mainly in Africa. On a global scale, the savanna biome is transitional between tropical rain forest and desert. Generally, 90 to 150 centimeters of rain fall each year in savannas. Temperatures annually fluctuate more here than in the tropical rain forests, and there is seasonal drought. These factors have led to the evolution of an open landscape, often with widely spaced trees; many of the animals and plants are active only during the rainy season. Savannas have often been converted to agricultural purposes throughout the world and provide most of the agricultural products for many tropical and subtropical countries.

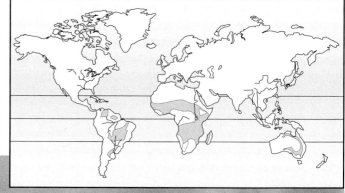

Desert: Arid and Burning Hot

Less than 25 centimeters of annual precipitation usually falls in the world's **desert** areas—so low an amount that water is the predominant controlling variable for most biological processes. In desert regions, the vegetation is characteristically sparse. As mentioned earlier in the chapter, such regions occur mostly around 20 to 30 degrees north and south latitude, where the warm air that rises near the equator falls and precipitation is limited. Deserts are most extensive in continental interiors, especially in Africa (the Sahara Desert), Eurasia, and Australia. Less than 5 percent of North America is desert. Desert organisms are often bizarre in appearance or way of life.

Desert survival depends on water conservation by structural, behavioral, or physiological adaptations. Plants and animals may restrict their activity to favorable times of the year, when water is present. To avoid high temperatures, most desert vertebrates live in deep, cool, and sometimes even somewhat moist burrows. Those that are active over a greater portion of the year emerge from these burrows only at night, when temperatures are relatively cool. Some, such as camels, can drink large quantities of water when it is available and can then safely withstand the loss of much of it. Many animals simply migrate to or through the desert, where they exploit food that may be abundant seasonally. When the food disappears, the animals move on to more favorable areas.

Temperate Grassland: Sea of Grass

Temperate grasslands once covered much of the interior of North America, and they were widespread in Eurasia and South America as well. Such grasslands are often highly productive when converted to agriculture. Many of the rich agricultural lands in the United States and southern Canada were originally occupied by **prairies,** another name for temperate grasslands. In eastern Europe and Central Asia, they are called **steppes.** The roots of perennial grasses characteristically penetrate far into the soil, and grassland soils tend to be deep and fertile.

Temperate grasslands are often populated by herds of grazing mammals. In North America, the prairies were once inhabited by huge herds of bison and pronghorns. Wolves, bears, and other predators, as well as various groups of American Indians, hunted the herds for food and clothing. The herds are almost all gone now, with most of the prairies having been converted to the richest agricultural region on earth, stretching across a wide region of the north-central United States and adjacent Canada.

Temperate Deciduous Forest: Rich with Hardwood

In areas of the Northern Hemisphere with relatively warm summers, relatively cold winters, and sufficient precipitation, **temperate deciduous forest** occurs. This biome covers very large areas, particularly over much of the eastern United States and Canada and an extensive region in Eurasia. Deer, bears, beavers, and raccoons are the familiar animals of the temperate regions. Many of the plants flower before the trees form their leaves in the early spring. In temperate deciduous forests, annual precipitation generally ranges from 75 to 250 centimeters, well distributed throughout the year but generally unavailable to animals and plants in the winter because it is frozen. Because the temperate deciduous forests represent the remnants of more extensive forests that stretched across North America and Eurasia several million years ago, these remaining areas—especially those in eastern Asia and eastern North America—share animals and plants that were once more widespread. Alligators, for example, are found only in China and in the southeastern United States. The temperate deciduous forest is much richer in species in eastern Asia than in either North America or Europe because climatic conditions have remained more constant and favorable for survival there than in the other two regions.

Taiga: Great Evergreen Forest of the North

Taiga is the northern forest of coniferous trees, primarily spruce, hemlock, and fir, that extends across vast areas of Eurasia and North America. Here, the winters are long and cold, and most of the limited amount of precipitation falls in the summer. Many large mammals, including elk, moose, deer, and such carnivores as wolves, bears, lynx, and wolverines, live in the taiga. Traditionally, fur trapping has been extensive in this region, which is also important in lumber production. Because of the latitude where taiga occurs, the days are short in winter (as little as six hours) and correspondingly long in summer. During the summer, plants may grow rapidly, and crops often attain a large size in a surprisingly short time. Marshes, lakes, and ponds are common and are often fringed by willows or birches. Most of the trees in the taiga tend to occur in dense stands of one or a few species.

The world's terrestrial communities can be divided into a series of widespread types, called biomes, that are found wherever similar patterns of seasonal moisture and temperature occur.

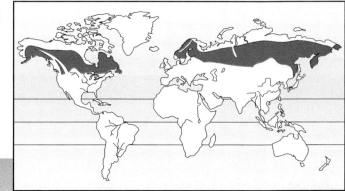

Other Biomes

Chaparral

The **chaparral** of California and adjacent regions is historically derived from deciduous forests. It consists of evergreen, often spiny shrubs and low trees that form extensive communities in dry-summer regions. Because of its relatively dry conditions, chaparral is frequently subjected to periodic fires. Plant species found in chaparral have adapted to these periodic fires in several ways. For example, some chaparral plant seeds can germinate only when they have been exposed to the hot temperatures generated during a fire. California fires are life threatening to humans that live in these regions, but paradoxically, they are life sustaining to many chaparral plants.

Tundra

Farthest north in Eurasia, North America, and their associated islands, between the taiga and the permanent ice, lies the open, often boggy biome known as the **tundra.** This enormous biome, extremely uniform in appearance, covers a fifth of the earth's land surface. Trees are small and are mostly confined to the margins of streams and lakes. In general, the tundra resembles some parts of the prairies.

Annual precipitation in the tundra is very low, usually less than 25 centimeters, and the water is unavailable for most of the year because it is frozen. During the brief Arctic summers, water often sits on frozen ground, making the tundra surface extremely boggy. **Permafrost,** or permanent ice, usually exists within a meter of the surface.

Polar Ice

Ice caps lie over the Arctic Ocean in the north and Antarctica in the south. The poles receive almost no precipitation, so although ice is abundant, freshwater is scarce. The sun barely rises in the winter months. Life in Antarctica is largely limited to the coasts. Because the Antarctic ice cap lies over a landmass, it is not warmed by the latent heat of circulating ocean water and becomes very cold. As a result, only bacteria, algae, and some small insects inhabit the vast Antarctic interior.

Mountain Zone (Alpine)

Because increasing altitude produces many of the same changes in temperature and moisture as increasing latitude, the tops of mountains have a typical windswept vegetation similar in many respects to tundra. Few if any trees are able to grow in this alpine zone, which, like polar regions, is alive with life in the warm summer months. During the harsh winter, little grows.

Temperate Evergreen Forest

Temperate evergreen forests occur in regions where winters are mild and there is a strong, seasonal dry period. The pine forests of the western United States, the California oak woodlands, and the Australian eucalyptus forests are typical temperate evergreen forests.

Warm, Moist Evergreen Forest

Extensive evergreen forests occur in temperate regions where winters are mild and moisture is plentiful. These can be seen in coastal China, in the pine forests covering much of the southeastern United States, and in the coastal redwood forests of northern California.

Tropical Monsoon Forest

Tropical monsoon forests occur in the tropics and semitropics at slightly higher latitudes than rain forests or where local climates are drier. Most trees in these forests are deciduous, losing many of their leaves during the dry season. Rainfall is typically very seasonal, measuring several inches daily in the monsoon season and approaching drought conditions in the dry season, particularly in locations far from oceans, such as central India.

Semidesert (Tropical Dry Forest)

Semidesert or tropical dry forests occur in tropical regions with less rain than monsoon forests but more rain than savannas. Vegetation is dominated by bushes and trees with thorns and spikes, which is why these regions are also known as thornwood forests. Plants survive on one or a few short rainy periods each year, growing intensively in response to the moisture. The brief rain is followed by a long dry season in which leaves fall from plants and there is little or no growth.

The Fate of the Earth

This chapter's survey of biomes and the corresponding communities that occur in the sea should provide insight into the different kinds of climate under which these areas evolved over tens or hundreds of millions of years. In just the last several hundred years, in contrast, human populations have grown over most of the land surface of the globe. Having already converted many biomes to agriculture, urban areas, or other uses, human populations now are attacking those biomes—such as the tropical rain forests—that are less suitable for exploitation and about which less is known. The human population doubled from 1950 to 1987, and most of that growth occurred in the warmer portions of the globe. There are now so many of us that the whole earth must be managed as a single system if our descendants are to find a stable ecological situation, one in which they can live out their lives in peace and relative prosperity. The next chapter examines some of the factors that will determine success or failure in this great enterprise.

EVOLUTIONARY VIEWPOINT

A relatively small number of terrestrial communities (biomes) occupy the earth's surface, each type of community being much the same wherever it is encountered on the earth. The similarity over such large areas reflects similar underlying evolutionary pressures. Knowledge of annual rainfall and mean annual temperature permits a surprisingly accurate prediction of which biome will occur in any given part of the world.

SUMMARY

1. Biomes and the corresponding communities that exist in the sea are ecosystems that occur over wide areas and that are easily recognized by their overall appearance and characteristic climates.

2. Biome distribution on the earth's surface is largely determined by two key physical factors: (1) how much solar heat reaches a particular place and (2) global atmospheric circulation. Together, these factors determine the local climate and the amounts and distribution of precipitation.

3. The major atmospheric circulation patterns result from interactions between six large air masses. Warm air rises near the equator and flows toward the poles, descending at about 30 degrees north and south latitude. Because the air is falling in these regions, it is warmed, and its moisture-holding capacity is therefore increased. The great deserts of the world are formed in these latitudes. Air rises and cools again at about 60 degrees north and south latitude and flows back toward the equator. Air masses also rise at 60 degrees north and south latitude and move toward the poles, creating zones of low temperature and precipitation.

4. Huge oceanic circulation patterns are driven by surface gyrals that move around subtropical zones between approximately 30 degrees north and south latitude. These gyrals spin clockwise in the Northern Hemisphere and counterclockwise in the Southern Hemisphere.

5. Ocean communities occur in three major kinds of environments: (1) the neritic zone, (2) the surface layers, and (3) the abyssal zone. The neritic zone, which lies along the coasts, is small in area but very productive and rich in species. The surface layers are the habitat of plankton (drifting organisms) and nekton (actively swimming organisms). The abyssal zone is the deep-water areas of the oceans.

6. Freshwater habitats comprise only about 2.1 percent of the earth's surface; most of them are ponds and lakes. As autumn changes to winter, the warmer water near the surface cools and sinks (because water is most dense at 4 degrees Celsius), thereby mixing the water of the lake. When winter changes to spring, the surface water warms to 4 degrees Celsius and sinks, resulting in similar mixing. Eutrophic lakes have an abundant supply of minerals and organic matter, while in oligotrophic lakes, organic matter and nutrients are relatively scarce.

7. Biomes are major terrestrial assemblages of plants, animals, and microorganisms that occur together in similar habitats. Biomes are defined largely by climate. This text recognizes six primary biomes: (1) tropical rain forest, (2) savanna, (3) desert, (4) temperate grassland, (5) temperate deciduous forest, and (6) taiga. Eight minor biomes are also discussed: (1) chaparral; (2) tundra; (3) polar ice; (4) mountain zone; (5) temperate evergreen forest; (6) warm, moist evergreen forest; (7) tropical monsoon forest; and (8) semidesert.

REVIEWING THE CHAPTER

1. Why does circulation occur in the earth's atmosphere?

2. What are the major atmospheric circulation patterns?

3. What are the circulation patterns in the oceans?

4. What is an ENSO, and why does it occur?

5. What are the communities in the oceans?

6. What is the neritic zone, and what occupies it?

7. What are the surface and abyssal zones, and what occupies them?

8. What are the communities of freshwater?

9. How are biomes defined?

10. How are tropical rain forests characterized?

11. What are savannas?

12. How are deserts distinguished?

13. What characterizes the temperate grasslands?

14. What are temperate deciduous forests?

15. How is the taiga characterized?

16. What distinguishes the chaparral?

17. What are the major features of the tundra?

18. How is the polar ice biome characterized?

19. How are mountain zones distinguished?

20. What is a temperate evergreen forest?

21. What are the major features of the warm, moist evergreen forest?

22. What distinguishes the tropical monsoon forest?

23. What is a semidesert?

24. With an increasing human population, what is the fate of the earth?

COMPLETING YOUR UNDERSTANDING

1. Seasons occur because
   a. the sun revolves slowly around the earth.
   b. the earth revolves slowly around the sun.
   c. warm air is concentrated near the equator.
   d. the earth is tilted relative to the sun's rays.
   e. the moon creates a pull on the earth's atmosphere.

2. Why are the great deserts of the world located near 20 to 30 degrees north and south latitude?

3. What is a rain shadow effect, and what causes it?

4. What kinds of biological communities would you expect to find on the windward and leeward sides of a mountain range in an area where the annual precipitation ranges between 20 and 100 centimeters per year and is distributed mainly in one rainy season? How would these differences affect human existence in each area?

5. What is the correlation between the major wind currents across the surface of the earth and ocean circulation? What are gyrals?

6. Why are biologists interested in monitoring ENSOs?

7. Why do more species live on land than in the sea? Why are there fewer phyla on land than in the sea?

8. What is vertical zonation, and where is it found?

9. Why are estuaries among the most naturally fertile areas in the world? Should there be a concern for "saving" estuaries? Why?

10. Phytoplankton are found in the _____ zone in oceans.
    a. neritic
    b. surface
    c. abyssal

11. Why have recent discoveries of life in the abyssal zone surprised scientists?

12. What are littoral, limnetic, and profundal zones, and how do they relate to freshwater communities?

13. Why would thermal stratification not be likely to occur in a tropical freshwater lake?

14. How do spring and fall overturns benefit lake ecosystems?

15. How do eutrophic and oligotrophic lakes differ? In the United States, where would you locate examples of them?

16. List the biomes that follow in order from north to south as they occur in the Northern Hemisphere.
    a. Taiga
    b. Savanna
    c. Temperate deciduous forest
    d. Tundra
    e. Desert
    f. Tropical rain forest

17. Why do you think most terrestrial species are located in tropical rain forests? What concerns should we have for the extensive deforestation occurring there?

18. Large herds of grazing mammals, along with their associated predators, are characteristic of the _____ biome.
    a. tundra
    b. tropical rain forest
    c. savanna
    d. temperate deciduous forest
    e. taiga
    f. tropical rain forest

19. Which of the following features are characteristic of some desert organisms? [*three answers required*]
    a. Additional leaves in the summer
    b. Seed germination related to moisture availability
    c. Deep burrows
    d. Ability to retain large quantities of water
    e. Nocturnal activity

20. Prairies and steppes are characteristic of the
    a. desert.
    b. temperate grasslands.
    c. savanna.
    d. taiga.
    e. tundra.
    f. tropical rain forest.

21. What is the problem with humans building homes and communities in the chaparral?

22. Which biome covers 20 percent of the earth's surface?
    a. Tundra
    b. Taiga
    c. Temperate deciduous forest
    d. Tropical rain forest
    e. Desert
    f. Chaparral

23. Near the coast in Southern California, there are two major plant communities. Right along the ocean is the coastal sage community, which is dominated by low shrubs that often wither or lose their leaves in the summer. Higher up, on the ridges, occurs the chaparral, a community dominated by evergreen shrubs. Which of these communities would you think grows in the area that receives more rainfall? Why?

FOR FURTHER READING

Attenborough, D. 1984. *The living planet: A portrait of the earth.* London, England: William Collins Sons and British Broadcasting Corporation. A beautifully written and illustrated account of the biomes.

Forsyth, A. 1990. *Portraits of the rain forest.* Ontario, Canada: Camden House Publications, Camden East. A magnificently illustrated and balanced account of the rain forest.

Gore, R. 1990. Between Monterey tides. *National Geographic,* Feb., 2–43. Beautifully illustrated account of life along the California coast.

McNaughton, S. J. 1984. Grazing lawns: Animals in herds, plant form, and coevolution. *American Naturalist* 124: 863–86. The fascinating story of how differences in grasses and in the populations of animals that depend on grasses help to determine the structure of the savanna biome.

Norse, E. 1990. *Ancient forests of the Pacific Northwest.* Washington, D.C.: Island Press. A thorough and interesting account of the controversy surrounding the harvest of ancient forests in the U.S. Pacific Northwest.

Perry, D. R. 1984. The canopy of the tropical rain forest. *Scientific American,* Nov., 138–47. Article that describes efforts to explore the richest and least understood biological community on earth.

Whitehead, J. A. 1989. Giant ocean cataracts. *Scientific American,* Feb., 50–57. Enormous undersea cataracts that play a crucial role in determining the chemistry and climate of the deep ocean.

OUR CHANGING ENVIRONMENT

Not visible in this satellite image of North America are the very significant changes produced since humans first began to live here some 12,000 years ago. Most of these changes have occurred in the last 150 years.

Figure 24.1 is a satellite photograph of the city of New York in the spring of 1985. As such, it is a portrait of 15.3 million people. The wakes of ships in the harbor can be seen, as can the wharves that line the shores of the Hudson River. The Verrazano Bridge is visible in the lower left, joining Brooklyn to New Jersey where the Hudson River empties into the Atlantic Ocean. Individual buildings cannot be seen—the scale is too large for that—and from the satellite, it is not obvious that the ground below teems with people. At the moment this picture was taken, millions of people within the satellite's view were talking, while hundreds of thousands of cars struggled through traffic, hearts were broken, babies were born, and dead people were buried. Were the satellite lens but sharp enough, we would see frozen in time all of this and more—15.3 million people busy at life, a panorama of modern industrial society. All of human history led to this photograph, and in it lies the future of humanity.

Our futures, and those of all of the people on the earth, are linked to the unseen millions in this photograph, for we share the earth with them. The wisdom with which they manage their environment has deep significance for us all, for their environment is also ours. As human numbers increase, so does the human impact on our common environment, posing new challenges for us all.

Holding hands, the 15.3 million people of New York City would form a chain that would stretch halfway around the world. And the earth is peppered with rapidly growing cities like New York. More than a dozen other cities have more than ten million inhabitants. The three cities that were larger than New York City in 1991 (Mexico City, Tokyo, and São Paulo) together have over fifty million inhabitants. As noted in chapter 20, in 1987, the human population of the earth reached—and passed—a significant milestone: five billion individuals.

The central challenge to the ingenuity of environmental scientists, politicians, and every citizen is this: There are a lot of people on the earth. And a lot of people consume huge amounts of food and water, use a great deal of energy and raw materials, and produce tremendous waste. They also have the potential to solve the problems that arise in an increasingly crowded world, to do what can be done to meet the challenge. This chapter delves into the details of how today's living affects the environment within which all future humans must live and examines the efforts being mounted to lessen the adverse impact and to increase potential benefits.

Figure 24.1 New York City, as seen from a satellite.
This city, one of the world's largest, is home to 15.3 million people.

Feeding a Hungry World

As noted in chapter 20, world population is expanding at an alarming rate. Most of this burgeoning human population is located in developing countries, which sometimes lack the agricultural expertise or technology to produce enough food to feed their starving numbers. In the past two decades, the agricultural strategies that the human population relies on to produce food have been shown to oftentimes cause great harm to the environment. The clear-cutting of tropical rain forests, for instance, to make room for cattle grazing and crop planting is one of the gravest ecological problems facing the world today. This section looks at possible solutions to feeding the world population—solutions that will also sustain human environments. These challenges are immediate, and the future of the world depends on meeting them.

The Inadequacy of Traditional Food Crops

As a beginning point in meeting the challenge of providing more food for more people, new kinds of crops need to be identified. Nearly all of the major crops now grown in the

(a)

(b)

Figure 24.2 New tropical tree crops.
(a) *Oil palms are grown throughout the tropical regions as a source of oil. On infertile soils, trees often are much more successful than cereal crops.* (b) *The roots of* Casuarina, *a fast-growing tree, are inhabited by nitrogen-fixing actinomycetes, which form nodules there.* Casuarina *is receiving increasing attention as a possible source of firewood and for watershed protection in the tropics.*

world have been cultivated for hundreds, or even thousands, of years. Only a few, including rubber and oil palms, have entered widespread cultivation since 1800 (figure 24.2). One key attribute for which nearly all important crops were first selected was ease of growth by relatively simple methods.

Currently, just three plant species—rice, wheat, and corn—supply more than half of all human energy requirements. Only about 150 kinds of plants are used extensively, and only about 5,000 plant species have ever been used for food. Estimates indicate that tens of thousands of additional kinds of plants, among the world total of some 250,000 species, could be used for human food if their properties were fully explored and they were brought into cultivation. Plants also have many additional uses: For example, oral contraceptives were produced from Mexican yams for many years; the muscle relaxants used in surgery worldwide came from curare, an Amazonian vine used traditionally to poison hunting darts; and the cure for Hodgkin's disease was developed in the early 1970s from the rosy periwinkle, a widely cultivated plant native to Madagascar. More recently, extracts from the bark of yew trees have been used effectively to treat breast cancer. Sidelight 24.1 examines additional recent triumphs in the search for economically important plants.

The reasons for which new crops are selected now are often very different from those that appealed to the humans

who developed agriculture, living as they did in small groups around the foothills of the Near East or on the temperate slopes of Mexican mountains. Standards of cultivation have changed, and many plant products—for example, oils, drugs, and other chemicals—are used now that would not have led to the plants' cultivation earlier. In a time when human activities threaten to drive many of the world's plants and other organisms to extinction in the very near future, new and useful crops that fit the multiple needs of modern society must be sought before they are gone forever.

Only about 150 kinds of plants, out of the roughly 250,000 known, play an important role in international trade at present. Many more could be developed by a careful search for new crops.

The Prospects for More Food

The most promising strategy for improving the world food supply is to increase the productivity of crops already being grown. Much of the improvement in food production must take place in the tropics and subtropics, where a rapidly growing majority of the world's people already live, including most of those enduring a life of extreme poverty. These people cannot be fed by exports from the industrial nations alone, which contribute only about 8 percent of their total food at present and where agricultural lands are already heavily exploited.

During the 1950s and 1960s, the so-called **Green Revolution** took place as a result of the development of new, improved strains of wheat and rice. Wheat production in Mexico increased nearly tenfold between 1950 and 1970, and Mexico temporarily became a wheat exporter rather than a wheat importer (figure 24.3). During the same decades, food production in India was largely able to outstrip even a population growth of approximately 2.3 percent annually, and China became self-sufficient in food.

Despite the Green Revolution's apparent success, improvements were limited. The agricultural techniques employed required adequate machinery, as well as large expenditures of energy and abundant supplies of fertilizers, pesticides, and herbicides. For example, in the United States, about a thousand times as much energy is needed to produce the same amount of wheat that results from traditional farming methods in India. Industrial-country methods can produce large volumes of wheat, but the greater volume comes at a greater energy cost

Figure 24.3 A positive result of the Green Revolution: Dwarf wheat.
Development of such improved strains as dwarf wheat helped to make Mexico a wheat exporter during part of the 1960s and 1970s.

Sidelight 24.1
THE SEARCH FOR NEW CROPS

With their vast and diverse arsenal of natural insecticides, plants offer a reservoir of useful products that have barely been tapped. A number of these plants have proven useful already, and many more will do so in the future.

One of the most fertile areas for biological exploration is the study of plant use by people who live in direct contact with natural plant communities. These people know a great deal about the plants with which they come into contact, and the screening that they have performed over many generations can, if its results are understood and catalogued, help scientists to recognize new, useful plants. For example, the herbal curer in the jungles of southern Surinam has a profound knowledge of uses of the plants that grow in his region, but that knowledge is being lost as civilization and modern medicine move into the area.

Recent triumphs in the search for economically important plants include:

1. Guayule (*Parthenium argenteum*) (Sidelight figure 24.1a). This widespread shrub of northern Mexico and the southwestern United States is a rich source of rubber, with rubber yields from some cultivated strains up to 20 percent. Guayule is now being cultivated in more that thirty countries, and efforts are being made to bring it into even wider cultivation.

2. Periwinkle (*Catharanthus roseus*) (Sidelight figure 24.1b). Periwinkle is a garden plant, now widespread in cultivation; as a native plant, it is found only in Madagascar. Two drugs, vinblastine and vincristine, which have been developed from periwinkle by Eli Lilly & Company, are effective in the treatment of certain forms of leukemia. For example, a child who develops Hodgkin's disease, a form of leukemia, now has a 90 percent chance of survival if treated with these drugs. In 1950, such a child would have had only a 20 percent chance to live.

3. Grain amaranths (*Amaranthus*) (Sidelight figure 24.1c). The grain amaranths, fast-growing plants that produce abundant grain rich in lysine—an amino acid that is rare in most plant proteins but essential for animal nutrition—are now being investigated seriously for widespread development. Important grain crops of the Latin American highlands in the days of the Incas and Aztecs, the grain amaranths are now minor crops in Latin America. Their use was suppressed because they played a role in pagan ceremonies of which the Spaniards disapproved.

4. Winged bean (*Psophocarpus tetragonolobus*) (Sidelight figure 24.1d). The winged bean is a nitrogen-fixing tropical vine that produces highly nutritious seeds, pods, and leaves. It tubers are eaten like potatoes, and its seeds produce large quantities of edible oil. First cultivated locally in New Guinea and Southeast Asia, the winged bean has spread since the 1970s throughout the tropics, where it holds great promise as a productive source of food.

(a)

(b)

(c)

(d)

Sidelight figure 24.1 Recent triumphs in the search for economically important plants.

(a) *Guayule* (Parthenium argenteum).
(b) *Periwinkle* (Catharanthus roseus).
(c) *Grain amaranths* (Amaranthus).
(d) *Winged bean* (Psophocarpus tetragonolobus).

per unit volume. In developing countries, energy prices are often held at artificially low levels, so a poor rural farmer may actually find it more expensive to grow an equivalent amount of grain than a large-scale farmer using developed-world technology. In this respect, the introduction of Green Revolution methods in some regions has actually worsened poverty for many of the people and lessened their access to food, fuel, and other commodities on which they depend.

The Green Revolution has resulted in increased food production in many parts of the world through the use of improved crop strains. However, these strains often depend on increased inputs of fertilizers, water, pesticides, and herbicides, as well as the greater use of machinery—means not normally available to the poor.

Certain nutritional problems have also arisen in connection with the Green Revolution. Overconcentration on cereal crops has tended to lower the production of other nutritionally important plants, including legumes, oilseeds, and vegetables of all kinds. Legumes and cereals are often nutritionally combined because they provide a balanced set of amino acids required by human beings for proper growth. The varied strains of crops that are grown on small farms may also be driven out by fewer kinds of modern strains and fewer crops, which produce a better yield if large-scale inputs of chemicals and the use of machinery are possible (figure 24.4). Despite these short-term advantages, the loss of the unique, traditional strains of crops presently cultivated by small, rural farmers throughout the world may, in the long run, ultimately prevent these particular crop plants from growing in less favorable habitats or withstanding important diseases (figure 24.5). **Monoculture**—the exclusive cultivation of a single crop over wide areas—is an efficient way to use certain kinds of soils, but it carries the risk of an

Figure 24.4 A traditional farm in Thailand.

On it grow coconuts, sugar palms, and litchi nuts, and the nitrogen-fixing water fern Azolla *grows in ponds. Land crabs are frequent and are harvested as a source of protein. Farms of this type are extremely suitable for many sites in the tropics.*

entire crop being destroyed with the appearance of a single pest species or disease.

Biologists play a crucial role in the improvement of crops and in the development of new ones by applying traditional methods of plant breeding and selection to many crops of importance in the tropics and subtropics, in addition to wheat, corn, and rice (figure 24.6). Genetic engineering techniques (see chapter 14) will make it possible to produce plants that are resistant to specific herbicides. These herbicides can then be applied much more effectively for weed control and without harming the crop plants. Genetic engineers are also developing new strains of plants that will grow successfully in areas where those particular plants could not grow before. Eventually, the introduction of desirable plant characteristics, such as the ability to fix nitrogen, to carry out C_4 photosynthesis (see chapter 8), or to produce substances that deter pests and diseases in abundance, will be

Figure 24.5 Development of new strains in the tropics.

The discovery of hearty new strains of crops in the tropics is sometimes a lucky accident.
(a) *Most corn (*Zea mays*) is genetically uniform and thus difficult to improve by selection.*
(b) *This field in Mexico is dominated by a wild perennial relative,* Zea diploperennis, *discovered in 1977.* Zea diploperennis *is resistant to five of the seven main types of virus diseases that damage Z. mays. The two species can be crossed to produce fertile hybrids.*
(c) *Spikelets of Z. diploperennis. (d) This great improvement was almost lost. Zea diploperennis might never have been found at all because it occurs in only one field. Its single population could easily have been destroyed by the rapidly increasing human population pressures of the area.*

(a)

(b)

Figure 24.6 From annoying weed to useful crop.
(a) *Water hyacinth, despite its beautiful flowers, is generally regarded as a noxious weed that clogs waterways throughout the tropics. (b) The rapid growth that makes hyacinth such a harmful weed allows it to produce large amounts of biomass rapidly. Here it is being harvested in India as animal feed. It can be readily dried and burned as a source of energy as well.*

possible. The ability to transfer genes between organisms, which became a practical technique in 1973, has become immensely important in the improvement of crop plants.

The oceans were once regarded as an inexhaustible food source, but overexploitation of their resources is actually limiting the world catch from year to year, and these catches are costing more in terms of energy. The mismanagement of fisheries, mainly through overfishing, local pollution, and the destruction of fish breeding and feeding grounds, has already lowered the catch of fish in the sea by about 20 percent from its maximum levels. The decline in the numbers of whales in world oceans is a tragic and well-known example of the way fisheries have been and continue to be destroyed.

New kinds of food, such as microorganisms grown in culture in nutrient solutions, are also being developed. For example, the photosynthetic, nitrogen-fixing cyanobacterium *Spirulina* is being investigated in several countries as a possible commercial food source. It is a traditional food in Africa, Mexico, and other regions. *Spirulina* thrives in very alkaline water and has a higher protein content than soybeans; the ponds in which it grows are ten times more productive, on average, than wheat fields. Such protein-rich concentrates of microorganisms could provide important nutritional supplements to human diets. But psychological barriers must be overcome to persuade people to eat such foods, and the processes required to produce these foods tend to be energy expensive (figure 24.7).

Tropical Rain Forest Destruction

More than half of the world's people live in the tropics, and this percentage is increasing rapidly. For global stability, and for the sustainable management of the biosphere, problems of food production and regional stability must be solved in the areas where most people live. World trade, political and economic stability, and the future of most species of plants, animals, fungi, and microorganisms depend on addressing these problems in an adequate fashion.

Many people in the tropics engage in **shifting agriculture**—clearing and cultivating a patch of tropical rain forest, growing crops for a few years, and then moving on (figure 24.8). The soil's fertility has by then returned to its original, very low level—the temporary enrichment that resulted from cutting and burning the forest has been exhausted—and the cultivator must move on and clear another patch of forest. Such agricultural systems work well where human populations are relatively low, but as these numbers grow, there is little opportunity even for the cultivation of traditional crops such as manioc (tapioca, cassava). Firewood gathering is also hastening the demise of many tropical forests. About 1.5 billion people worldwide—a third of the global population—depend on firewood as their major fuel source and are cutting local supplies faster than the trees can regenerate themselves.

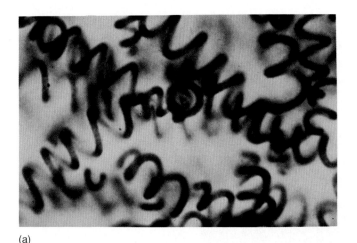

(a)

(b)

(c)

Figure 24.7 **Spirulina, *a* *cyanobacterium that is being grown increasingly as a source of protein.***

(a) *Individuals of* Spirulina, *each about 250 micrometers long.* (b) *A mat of* Spirulina *on a conveyer belt in a processor plant in Mexico.* (c) *Ponds for the production of* Spirulina *in Thailand.*

Figure 24.8 Shifting agriculture.

A family has cleared a small patch of forest in the northern Amazon Basin of Brazil. They will be able to grow crops on it for a few years and then will need to move on to another part of the original forest. Such activities can be sustained indefinitely if population levels are low enough, but that will rarely be the case in the closing years of the twentieth century.

Figure 24.9 On the edge of destruction.
The enormous destruction of these forests has caused erosion, mass extinction, and harmful shifts in the cycling of nutrients in the tropics.

Extinction in the Tropical Rain Forest

As a result of such overexploitation, experts predict that little undisturbed tropical forest will remain anywhere in the world by early in the next century. Many areas now occupied by forest will still be tree-covered, but those trees will represent only a small percentage of those that currently grow in these areas. Many species of plants, animals, fungi, and microorganisms can reproduce only under the conditions present in undisturbed tropical forest. Consequently, they are threatened with extinction or, at the very least, exclusion from large areas. In fact, a fifth or even more of the species on earth may become extinct during the next thirty or forty years, amounting to a million or more species. From another perspective, several species per day are probably becoming extinct now. Many of these inhabit ecologically devastated islands, such as St. Helena in the South Atlantic and Rodrigues in the western Indian Ocean, or similarly devastated areas on continents, such as the Atlantic forests of Brazil or the lowlands of western Ecuador (figure 24.10). By early in the next century, the rate could easily reach several species per *hour* and continue to climb for at least another fifty years. Overall, this would amount to an extinction event that has been unparalleled for at least sixty-five million years, since the end of the age of dinosaurs. The number of species in danger of extinction during our lifetimes is far greater than the number that became extinct at that distant time.

The prospect of such an extinction event is of major concern. Only one out of every six tropical organisms has even been given a scientific name, so many species that are about to become extinct will never have been seen by any human being. As these species disappear, so does the opportunity to learn about them, not only scientifically, but also in terms of their possible benefits for humanity. The plants, animals, and

Most other kinds of tropical forests have already been largely destroyed; they tend to grow in more fertile soils and were exploited by human beings earlier than were the tropical rain forests now under massive attack. About 6 million square kilometers of tropical rain forest are estimated to still exist in a relatively undisturbed form. This area, about three-fourths the size of the United States exclusive of Alaska, represents about half of the original extent of tropical rain forest. From it, about 160,000 square kilometers are being clear-cut per year (the rate now exceeds 1.5 acres per second), with perhaps an equivalent amount severely disturbed by shifting agriculture, firewood gathering, and other practices. The total area of tropical rain forest destroyed—and therefore permanently removed from the world total—amounts to an area greater than the size of Indiana each year (figure 24.9). At this rate, all of the tropical rain forests in the world will be gone in about thirty years, but in many regions, the rate of destruction is much more rapid.

(a)

(b)

Figure 24.10 Escape from extinction.

(a) *This palm is the last individual of its species,* Hyophorbe verschaffeltii, *known in nature. It is shown here on the island of Rodrigues, in the Indian Ocean, the same island where that famous flightless pigeon, the dodo, became extinct more than three centuries ago.* (b) Hyophorbe verschaffeltii *is not likely to become* extinct because it is preserved on a university campus on the island of Mauritius. The palms are the second most important family of plants in the world economically, being surpassed only by the grasses.

microorganisms that share this planet are all unique, and with the loss of any species, humanity loses forever the chance to use it for any purpose. That 80 percent of the entire human food supply is based primarily on twenty kinds of plants, out of the quarter million kinds available, should give everyone pause to consider what it means to be living in a generation during which a high proportion of the remainder are being permanently lost. Many of the lost species would surely be of great use if their properties were known.

What Can Biologists Do?

In the face of this crisis, biologists can help to design plans for finding those organisms most likely to be of use and for saving them from extinction. They must also participate in sound, globally based schemes to preserve as much as possible of the biological diversity of life so that later generations have as many options as possible. With the loss of tropical rain forest, and of biological communities throughout the world, humanity is permanently losing many opportunities not only for knowledge, but also for increased prosperity. Biologists must understand this message and inform their fellow citizens of its importance.

The consequences of uncontrolled deforestation in the tropics and subtropics can be viewed in another way: Tropical rain forests are complex, productive biomes that work well in the areas where they have evolved. Unfortunately, biologists do not know how to replace tropical rain forests with other productive biomes that will support human beings. Cutting a forest or opening a prairie in the North Temperate Zone provides the basis for a farm that can be worked for generations. In the tropics, for the most part, biologists simply do not know how to engage in continuous agriculture in areas not now under cultivation. When humans clear a tropical forest, they engage in a one-time consumption of natural resources that will not be available again (figure 24.11). The complex ecosystems that have been built up over millions of years are now being dismantled, in almost complete ignorance, by the human species.

Biologists must learn more about the construction of sustainable agricultural ecosystems that will meet human needs in tropical and subtropical regions. The ecological principles reviewed in the previous three chapters are universal principles. The undisturbed tropical rain forest has one of the highest rates of net primary productivity of any plant community

(a)

(b)

(c)

(d)

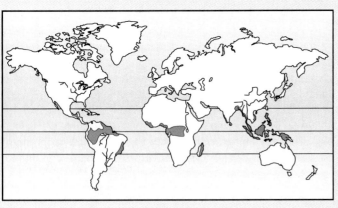

(e)

Figure 24.11 Destroying the tropical rain forests.
(a) *These fires are destroying the rain forest in Brazil, which is being cleared for cattle pasture.* (b) *The flames are so widespread and so high that they can be viewed from space. This 1988 satellite photo shows a plume of smoke generated from the burning of the rain forest.* (c) *The consequences of deforestation can be seen on these elevation slopes in Ecuador, which now support only low-grade pastures where highly productive forest grew in the 1970s.* (d) *As time passes, the consequences of tropical deforestation may become even more severe, as shown here by the extensive erosion in this area of Tanzania from which forest has been removed.* (e) *Map showing world locations of rain forests.*

on earth, and thus, logically, could be harvested for human purposes in a sustainable, intelligent way. Standing by passively while tropical rain forests are consumed is unthinkable. Sound development of the tropics—an urgent matter in view of the very large numbers of humans who live in tropical countries—must be based on sound biology as well as on achieving stable human population levels and alleviating the problems of poverty and widespread malnutrition. Biologists must increasingly address problems that have traditionally been the concern of agronomists, animal breeders, and other agriculturists, and must apply the results of their research to the creation of a stable global ecosystem.

> *Many of the world's tropical rain forests are being rapidly destroyed, leading to widespread extinction of many species that might have potentially benefited humans. Those with knowledge of biology must help in reaching sound decisions that will slow this destruction.*

Human Impact on the Environment: A Selection of Concerns

The simplest way to gain a feeling for some of the other dimensions of humanity's impact on the environment is simply to scan the top stories of any newspaper or news magazine or to watch television news shows. Although only a sampling, the "front page" issues that follow say a great deal about the scale and complexity of the current challenge.

Nuclear Power

The Chernobyl Incident

At 1:24 A.M. on 26 April, 1986, one of the four reactors of the Chernobyl nuclear power plant blew up. Located in Ukraine, 100 kilometers north of Kiev, Chernobyl was one of the largest nuclear power plants in Europe, producing a thousand megawatts of electricity—enough to light a medium-sized city. Early in the morning of April 26th, workers at the plant were hurrying to complete a series of tests to determine how the generator of reactor #4 performed during a power reduction and took a foolish shortcut: They shut off all the safety systems. The Chernobyl reactors were graphite reactors designed with a series of emergency systems that shut the reactors down at low power, because the core is unstable then—and the workers turned these emergency systems off. A power surge occurred during the test, and there was nothing to dampen it. Power zoomed to hundreds of times the safe maximum, and a

Table 24.1
Radiation Vocabulary

Radioactivity—The emission of nuclear particles and rays by unstable atoms as they decay into more stable forms. It is measured in curies. One curie equals thirty-seven billion disintegrations a second.

Dose—A measure of the amount of radiation absorbed by the body, usually stated in rem units. An acute dose of 600 rem will usually result in death within sixty days.

Background radiation—Radiation from natural sources, such as cosmic rays and radon gas, as well as from man-made sources, such as atomic testing. In the United States, this averages about 100 millirem (one millionth of a rem) a year. For comparison, a chest X ray exposes an individual to 20 millirem.

Figure 24.12 Three Mile Island nuclear power plant.
This plant was the site of a nuclear accident in March 1979.

white-hot blast with the force of a ton of dynamite partially melted the fuel rods and heated a vast head of steam that blew the reactor apart.

The explosion and heat sent up a plume 5 kilometers high, carrying several tons of uranium dioxide fuel and fission products. By Russian accounts, over 100 megacuries of radioactivity were released, making it the largest nuclear accident ever reported (table 24.1). By way of comparison, the Three Mile Island accident in Pennsylvania in 1979 released 17 curies, millions of times less (figure 24.12). The cloud from Chernobyl

...st northwest, then southeast, spreading the radioac-...band across central Europe from Scandinavia to ...ece. Within a 30-kilometer radius of the reactor, at least one-fifth of the population, some twenty-four thousand people, received serious radiation doses (greater than 45 rem). Thirty-one individuals died as a direct result of radiation poisoning—most of them firefighters who succeeded in preventing the fire from spreading to nearby reactors. Today, almost ten years after the accident, the area surrounding Chernobyl is still a ghost town.

For western Ukraine and the rest of Europe, the radiation dose was much lower but still significant. Data provided by the Russians indicate that radiation outside of the immediate Chernobyl area will be expected to be responsible for from five thousand to seventy-five thousand cancer deaths because of the large numbers of people exposed.

The Promise and Challenges of Nuclear Power

Industrial society grew for over 150 years on a diet of cheap energy. Until recently, much of this energy was derived from burning fossil fuels—wood, coal, and oil. But as these fuel sources became increasingly scarce and the cost of locating and extracting new deposits ever more expensive, modern society began to look elsewhere for energy. The great promise of nuclear power is that it provides an alternative source of plentiful energy. Although with current technology nuclear power is not cheap (power plants are expensive to build and operate), its raw material—uranium ore—is so common in the earth's crust that the supply is unlikely to run out.

For all its promise of plentiful energy, nuclear power presents several new problems that must be mastered before its full potential can be realized. One serious challenge is the need to ensure the safety of the approximately 390 nuclear reactors now in operation. A second challenge is safe disposal of the radioactive wastes produced by nuclear power plants and safe decommissioning of plants that have reached the end of their useful lives (about twenty-five years). By 1990, approximately thirty-five plants were more than twenty-five years old, and not one had been safely decommissioned. A third challenge is the need to guard against terrorism and sabotage because the technology of nuclear power generation is closely linked to that of nuclear weapons.

Because of these challenges associated with nuclear power, other energy sources, such as the sun and the wind, should continue to be investigated and developed. Energy conservation, however, is the most immediate and cost-effective way to address energy needs. As much as 75 percent of the electricity produced in the United States and Canada currently is wasted through the use of inefficient appliances, according to scientists at the Lawrence Berkeley Laboratory. The use of highly efficient motors, lights, heaters, air conditioners, refrigerators, and other items that are currently available could lead to large energy savings and greatly alleviate the problem of greenhouse gas emission. For example, a new, compact, fluorescent light bulb uses a fifth of the electricity required by conventional lighting, provides equal or better lighting, lasts up to thirteen times longer than incandescent bulbs, and provides substantial cost savings.

> *Nuclear power has the potential to provide plentiful energy to the world, but it poses severe unsolved problems of waste disposal and safety.*

Fossil Fuels and Global Warming

The burning of coal and oil to obtain energy produces two very undesirable chemical by-products: sulfur and carbon dioxide. As discussed later in this chapter, the sulfur emitted from burning coal is a principal cause of acid rain. Of perhaps even greater importance, the carbon dioxide produced from the burning of all fossil fuels is a major greenhouse gas—it absorbs solar energy, trapping heat within the atmosphere. As mentioned in chapter 22, levels of carbon dioxide in the earth's atmosphere have risen alarmingly, and scientists warn that the climatic warming that will result may have catastrophic consequences. For these reasons, and also because fossil fuel supplies are limited, replacements for fossil fuels need to be found.

Pollution

The Day the Rhine Almost Died

The river in figure 24.13 is the Rhine, a broad ribbon of water running through the heart of Europe. From high in the Alps that separate Italy and Switzerland, the Rhine flows north across the industrial regions of Germany before reaching Holland and the sea. Given the sheer amount of goods that are produced and shipped on or near its shores, the Rhine is one of the world's most commercially important rivers, far exceeding the Mississippi. The Rhine is also, where it crosses the mountains between Mainz and Coblenz, one of the most beautiful rivers on earth. On the first day of November 1986, the Rhine almost died.

Figure 24.13 The picturesque Rhine river.
The site of a dangerous chemical spill in November 1986, the Rhine miraculously recovered.

The blow that struck at the life of the Rhine did not at first seem so deadly. Firefighters were fighting a blaze that morning in Basel, Switzerland. The fire was gutting a huge warehouse, into which the firefighters poured streams of water to dampen the flames. The warehouse was that of a giant chemical company—Sandoz. In the rush to contain the fire, no one thought to ask what chemicals were stored in the warehouse. By the time the fire was out, the streams of water had washed about 30 tons of mercury and pesticides into the Rhine.

Flowing down the river, the deadly wall of poison killed everything it passed. For hundreds of kilometers, the river surface was blanketed with dead fish. Many cities that use the Rhine's waters for drinking had little time to make other arrangements. Even the plants in the river began to die. All across Germany, from Switzerland to the sea, the river reeked of rotting fish, and not one drop of the water was safe to drink or even touch.

Six months later, Swiss and German environmental scientists monitoring the effects of the accident were able to report that the blow to the Rhine was not mortal. Enough small aquatic invertebrates and plants had survived to provide a basis for the eventual return of fish and other water life, and the river was rapidly washing out the remaining residues from the spill. A lesson difficult to ignore, the spill on the Rhine caused the gov-

ernments of Germany and Switzerland to intensify efforts to protect the river from future industrial accidents and to regulate the creation of chemical and industrial plants on its shores.

The Threat of Pollution

The pollution of the Rhine is a story that can be told countless times in different places in the industrial world, from Love Canal in New York, to the James River in Virginia, to Times Beach in Missouri. In the private sector, many people do not think about where the garbage they leave out by the curb for collection actually ends up. The truth would shock many people: Landfills that are used as dumping grounds for garbage are rapidly filling up, especially in heavily populated U.S. areas. "Garbage barges" heaped with trash from New York have tried in vain to find places to dispose of their burdens. Garbage barges and closed landfills are a reality of modern existence, where pollution and its effects are being felt in all corners of the globe. The sections that follow examine different kinds of pollution and offer some tentative solutions. The good news is that many of these solutions depend on you, the citizen, and your motivation to clean up the environment.

Solid-Waste Pollution

Solid waste can be defined as the everyday garbage disposed of by both private citizens and industry. Ninety percent of all landfilled municipal waste is paper. Efforts are underway to employ genetically engineered bacteria to convert this waste paper to ethanol for fuel. Some solid wastes, such as paper products, break down easily in landfills, but others do not. For example, the polymers known as plastics, which industry produces in abundance, break down slowly if at all in nature. Scientists are attempting to develop strains of bacteria that can decompose plastics, but so far, their efforts have been unsuccessful. Recently, however, some breakthroughs have been made with some types of plastics: The plastic bags used in supermarkets can be recycled to make other plastic items, and plastic soft drink containers have also been made from recycled plastics.

Recycling is the key to solid-waste management. In addition to paper being converted to ethanol and recycled plastic being made into other plastics, glass, tin, paper, and other materials can also be used to manufacture new materials. But these efforts depend solely on the willingness of individuals and industries to collect, sort, and deliver their recyclable garbage to recycling centers. The good news is that more people than ever before are recycling their trash. In areas where recycling is prohibitive (such as in some large cities that do not have amounts of many new kinds of chemicals, such as pesticides, herbicides, and fertilizers. Developed countries such as the United States now attempt to carefully monitor the side effects of these chemicals. Unfortunately, however, large quantities of many toxic chemicals that were manufactured in the past still circulate in the ecosystems of these nations.

For example, the chlorinated hydrocarbons, a class of compounds that includes DDT, chlordane, lindane, and dieldrin, have all been banned for normal use in the United States, where they were once used very widely. These molecules break down very slowly and accumulate in animal fat. Furthermore, as they pass through a food chain, they are increasingly concentrated, a process called **biological magnification** (figure 24.14). DDT caused serious problems by leading to the production of very thin, fragile eggshells in many bird species in the United States and elsewhere until the late 1960s, when it was banned in time to save the birds from extinction. Chlorinated hydrocarbons have many other undesirable side effects, several of which are still poorly understood.

In the Air Air pollution is a major problem in the world's large cities. In Mexico City, for example, oxygen is sold routinely on street corners for patrons to inhale, and schoolchildren are frequently kept at home by governmental edict when the air is considered too dangerous for young lungs to bear. The culprits in air pollution are carbon monoxide, hydrocarbons, and sulfur oxides. Carbon monoxide and hydrocarbons are emitted by cars, trucks, and buses; sulfur oxides are a by-product of the burning of fossil fuels to generate power for industry.

Climate and the pollutants present in the air contribute to the type of pollution that is characteristic of a region. Cities such as New York, Boston, and Philadelphia are known as **brown air** cities because the pollutants in the air are usually sulfur oxides emitted by industry. Cities such as Los Angeles, however, are called **gray air** cities because the pollutants in the air undergo chemical reactions in the strong sunlight. Solar energy converts the carbon monoxide and hydrocarbons abundant in the air around Los Angeles into **ozone.** While ozone is a protective substance when high in the atmosphere (the ozone layer around the earth protects against harmful ultraviolet rays), when present in lower atmospheric layers, it can be deadly.

Efforts to control air pollution have focused on persuading people to leave their cars at home. New mass transit systems in some cities, such as Washington, D.C., have been successful, but mass transit has not met with favor in other cities. Another solution has been to design cars that run "cleaner" than previously built automobiles and to institute emissions testing for all vehicles. Finally, the enforcement of strict air pollution standards should lessen the amount of industrial pollution.

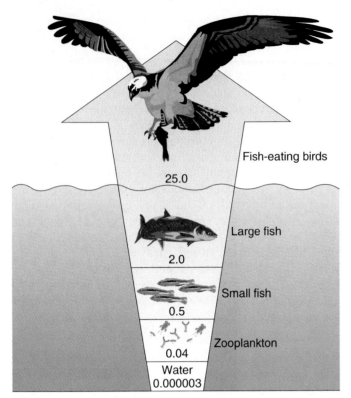

Fish-eating birds
25.0

Large fish
2.0

Small fish
0.5

Zooplankton
0.04

Water
0.000003

Amount of DDT (parts per million)

Figure 24.14 Biological magnification.
The amount of DDT increases (is "magnified") as you progress up a food chain.

recycling centers), concerned citizens are buying only those products that are packaged in recycled materials and are asking manufacturers to cut down on the amount of packaging. All of these efforts will help to reduce the amount of solid waste in the environment and lead to a cleaner earth.

Water Pollution

Water pollution also exists on a global scale. Not enough water is available to dispose of the diverse substances that today's enormous human population produces continuously. Despite the implementation of ever-improving methods of sewage treatment throughout the world, lakes, streams, and groundwater are becoming increasingly polluted. Household detergents, which contain phosphates, may flow into oligotrophic lakes and lead to their eutrophication, as discussed in chapter 23. This leads to an overgrowth of algae and a rapid deterioration of water quality.

Chemical Pollutants

On Land Widespread agriculture, increasingly carried out by modern methods, introduces into the global ecosystem huge

Figure 24.15 Four Corners power plant in New Mexico.
The tall smokestacks spew acidic smoke high into the atmosphere. It falls back to earth in the form of acid rain.

"Green" Chemistry Obviously, a "back-to-nature" approach—one that ignores the important contributions made to the current standard of living by the intelligent use of chemicals—will not care adequately for the needs of today's world population. Nor will such a backward approach feed the additional billions of people expected over the next few decades. On the other hand, technology must be used as intelligently as possible and with due regard for the protection of the productive capacity of all parts of the earth. Sidelight 2.1 in chapter 2 discusses current efforts to develop more environmentally safe ("green") chemicals.

> *Many ecological problems are created by society's tendency to dump chemicals as if the environment had an unlimited capacity to absorb them. As these chemicals accumulate, sensitive components of ecosystems are damaged, sometimes with disastrous consequences.*

Acid Rain

The smokestacks in figure 24.15 are those of the Four Corners power plant in New Mexico. This facility burns coal, sending the smoke high into the atmosphere through these stacks, each of which is over 65 meters tall. The smoke that the stacks belch out contains high concentrations of sulfur, which smells bad (like rotten eggs) and produces acid when it combines with the water vapor in air. The intent of those who designed the plant was to release the sulfur-rich smoke high in the atmosphere, where the winds would disperse and dilute it. This sort of solution to the problem posed by burning high-sulfur coal was first introduced in Britain in the mid-1950s and rapidly became popular in the United States and Europe: About eight hundred such stacks were built in the United States alone. The problem is that the stacks only carry the acid produced by the fuels away from the areas where the acid is produced. Thus, while London no longer suffers from acid fogs, the forests and lakes of Sweden are being destroyed.

The environmental effects of this acidity are serious. The sulfur introduced into the upper atmosphere combines with water vapor to produce sulfuric acid, and when the water later falls as rain or snow, the precipitation is **acid rain.** Natural rainwater rarely has a pH lower than 5.6; in the northeastern United States, rain and snow now have a pH of about 3.8, about a hundred times as acidic as the usual limit. Because the prevailing winds in the temperate latitudes, where most industries are concentrated, are westerlies, the sulfur emissions released by U.S. midwestern plants primarily return to the earth in rain and snow that falls in the eastern United States and Canada, and similar patterns occur in Europe.

Acid precipitation destroys life. Thousands of the lakes of northern Sweden and Norway are now eerily clear and no longer support fish. In the northeastern United States and eastern Canada, tens of thousands of lakes are dying biologically as a result of acid precipitation (figure 24.16*a*). At pH levels below 5.0, many fish species and other aquatic animals die, unable to reproduce under these conditions. In southern Sweden and elsewhere, groundwater regularly has a pH between 4.0 and 6.0, its acidity resulting from the acid precipitation that is slowly filtering down into the underground reservoirs, thus threatening the water supplies of future generations.

(a)

Figure 24.16 Acid rain kills.
(a) *Twin Pond in the Adirondacks of upstate New York is one of the many lakes in the region in which acidity levels have killed the fishes, amphibians, and most of the other kinds of animals and plants that once lived there.*
(b)–(d) *Camel's Hump Mountain in Vermont.* (b) *The healthy forest as it was in 1963.* (c) *A view from the same spot twenty years later, in 1983, showing many trees diseased and dying.* (d) *A closer view of some of the dying trees in this area.*

(b)

(c)

(d)

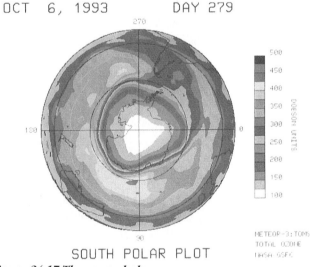

OCT 6, 1993   DAY 279

SOUTH POLAR PLOT

Figure 24.17 The ozone hole.
Satellite picture of the Antarctic ozone hole, taken on 6 October, 1993.

Trees also suffer from acid rain. The Black Forest in Germany and forests in the eastern United States and Canada have sustained monumental damage. At least 3.5 million hectares of forest in the Northern Hemisphere are believed to have been affected by acid precipitation, and the problem is clearly growing (figure 24.16*b–d*).

The solution at first seems obvious: Capture and remove the emissions instead of releasing them into the atmosphere. But the execution of this solution poses serious difficulties. First, it is expensive. Reliable estimates of the cost of installing and maintaining the necessary "scrubbers" in the United States approach $4 to $5 billion a year. Although this is no more than a hundredth of the amount that will ultimately be spent to "bail out" failed savings and loans associations, national priorities are not yet clearly focused on a healthy environment. An additional difficulty is that the polluter and the recipient of the pollution are far from one another, and neither wants to pay so much for what they view as someone else's problem. The U.S. Clean Air Act revisions of 1990 significantly addressed this problem for the first time.

The Ozone Hole

The swirling colors of figure 24.17 are a satellite view of the South Pole in 1993. This is not the view your eye would see, but rather, a computer reconstruction in which the colors represent different concentrations of ozone (O_3, a form of oxygen gas [O_2]). As easily seen in the satellite picture, over Antarctica is an "ozone hole," an area about the size of the United States within which the ozone concentration is very much less than it is elsewhere. The hole is not a permanent feature, but rather one that becomes evident each year for a few months at the onset of the Antarctic spring. Every September since 1975, the ozone hole has reappeared, and each year, the layer of ozone is thinner and the hole is larger. In 1985, the minimum ozone concentration in the hole was 30 percent lower than it had been five years earlier. Recent photographs show the ozone layer to be thinning over North America as well.

The major cause of the ozone depletion was suggested in the early 1970s but was only accepted slowly. **Chlorofluorocarbons (CFCs),** chemicals that have been manufactured in large amounts since their invention in the 1920s for use in cooling systems, fire extinguishers, and Styrofoam containers, percolate up through the atmosphere and reduce O_3 molecules to O_2 molecules. Although other factors have also been implicated in ozone depletion, the CFCs' role is so predominant that worldwide agreements to phase out their production by the year 2000 have been signed. Production of CFCs and other ozone-destroying chemicals is banned in the United States after 1995. Nonetheless, large amounts of CFCs that were manufactured earlier, and that are still being manufactured, continue to move slowly upward through the atmosphere. The problem will grow worse before the ozone layer that protects all life is stabilized once again.

The thinning of the ozone layer in the stratosphere, 25 to 40 kilometers above the earth's surface, is a matter of serious biological concern. This layer protects key biological molecules, especially proteins and nucleic acids, from the harmful ultraviolet solar rays that continuously bombard the earth. Life on land, in fact, may have become possible only when the oxygen layer was sufficiently thick that it generated enough ozone, in chemical equilibrium with the oxygen, that the earth's surface was sufficiently shielded from these destructive rays. This factor may account for the three billion years in which all life was aquatic.

Ultraviolet radiation is a serious human health concern: Every 1 percent drop in the atmospheric ozone content is estimated to lead to a 6 percent increase in the incidence of skin cancers. The drop of approximately 3 percent that has occurred worldwide, therefore, is estimated to have led to as much as a 20 percent increase in skin cancers, which are one of the more lethal diseases afflicting human beings. Humans are not the only organisms affected by ultraviolet radiation; other organisms, such as the photosynthetic plankton species that are so important to global productivity, are apparently very susceptible to ultraviolet radiation.

Environmental Science

Environmental scientists attempt to find solutions to environmental problems, considering them in a broad context. Unlike biology or ecology, which are sciences that seek to learn general principles about how life functions, environmental science is an applied science, dedicated to solving problems. Its basic tools are derived from the sciences of ecology, geology, meteorology (the study of climate), the social sciences, and the many other areas of knowledge that bear on the functioning of the environment and human management of it. Environmental science addresses the problems created by rapid human population growth: an increasing need for energy, a depletion of resources, and a growing level of pollution.

Components of Environmental Science

The problems faced by planet earth are not insurmountable. A combination of scientific investigation and public action, when brought to bear effectively, can solve environmental problems that seem intractable. Viewed simply, there are five components to solving any environmental problem:

1. Assessment. The first stage in addressing any environmental problem is scientific analysis, the gathering of information. Data must be collected and experiments performed in order to construct a "model" that describes the situation. Such a model can be used to make predictions about the future course of events.

2. Risk analysis. The **environmental impact**—the potential effects of environmental intervention (what could be expected to happen if a particular course of action were followed)—is then assessed, using the results of scientific analysis. Evaluation of not only the potential for solving the environmental problem, but also any adverse effects that a plan of action might create, is also necessary.

3. Public education. When a clear choice can be made among alternative courses of action, the public must be informed. This involves explaining the problem in terms the public can understand, presenting the alternative actions available, and disclosing the probable costs and results of the different choices.

4. Political action. The public, through its elected officials, chooses a course of action and implements it. The choices are particularly difficult to implement when environmental problems transcend national boundaries.

5. Follow-through. The result of any action taken should be carefully monitored, both to see if the environmental problem is being solved and, more basically, to evaluate and improve the initial evaluation and modeling of the problem. Every environmental intervention is an experiment, and the knowledge gained from each one is needed for future applications.

What Biologists Have to Contribute

The development of appropriate solutions to the world's environmental problems must rest partly on the shoulders of politicians, economists, bankers, engineers—many different kinds of people. However, the application of basic biological principles is vital to finding permanent solutions to these problems and achieving a stable, productive world. This chapter has reviewed a number of biological solutions to environmental problems.

The energy that bombards the earth comes in essentially inexhaustible amounts from the sun; living systems capture that energy and fix it in molecules that can be utilized by all organisms to sustain their life processes. Humans and all other living beings depend on the proper utilization of that renewable energy and of the other materials on which civilization depends. The processes involved need to be better understood and new systems developed to work efficiently in areas where they are not available now. That is the challenge for the future—a challenge that must be met soon. In many parts of the world, the future is happening right now (figure 24.18).

Everyone should have a basic scientific education so that they can understand the basis for their continued existence on earth and the steps that must be taken to improve the quality of living. Biology should play a major part in that education and is of critical importance in improving the standard of living

Figure 24.18 The future is now.
The girl gazing from this page faces an uncertain future. She is a refugee: The whims of war have destroyed her home, her family, all that is familiar to her. Her expression carries a message about our own future. The message is that our future, like hers, is now on us, and that the problems humanity faces in living on an increasingly unstable, overcrowded, and polluted earth are no longer hypothetical, a dilemma for our children, but are with us today and demanding a solution.

for fellow human beings. Biological literacy is no longer a luxury for intelligent human beings who want to play a constructive role in improving the world; it has become a necessity.

Environmental problems are so central to the earth's future that all of us need to become informed and active in seeking solutions to these problems while there is still time to act.

EVOLUTIONARY VIEWPOINT

In natural populations, the numbers of individuals increase until the environment's carrying capacity is reached. Numbers then stabilize around that maximum sustainable number. No one knows if the world's human population has exceeded the earth's carrying capacity. We know only that human population is still growing at a very rapid rate.

SUMMARY

1. An explosively growing human population is placing considerable stress on the environment, consuming ever-increasing quantities of food and water, using a great deal of energy and raw materials, and producing enormous amounts of wastes and pollution.

2. Possible solutions to feeding the world population include identifying new kinds of food crops, increasing the productivity of crops already being grown, and developing new kinds of food, such as microorganisms grown in culture in nutrient solutions.

3. The cutting and burning of the tropical rain forests of the world to make pasture and cropland is producing a massive wave of biological extinction.

4. Among the many challenges to the environment posed by human activities is the release of harmful materials into the environment: The release of carbon dioxide from burning fossil fuels may increase the world's temperature and so alter weather and ocean levels; the escape of radioactive materials into the atmosphere may increase the incidence of cancer; the release of pollutants into the air and water and onto the land may harm ecosystems and in some instances even make some areas unfit for human habitation; the release of industrial smoke into the upper atmosphere leads to acid rain that kills forests and lakes; and the release of such chemicals as chlorofluorocarbons destroys the atmosphere's ozone and so exposes the world to more ultraviolet radiation.

5. All of these environmental challenges can and must be addressed. Today, environmental scientists and concerned citizens are actively searching for constructive solutions to these problems through assessment, risk analysis, public education, political action, and follow-through.

6. Application of biological principles to the human condition has never been more necessary. Only when society and many talented individuals work to solve these grave problems will the future Planet Earth be preserved in all its magnificence for our children and grandchildren to enjoy.

REVIEWING THE CHAPTER

1. How is a hungry world going to be fed?
2. Why are traditional food crops inadequate?
3. What are the prospects for more food and new foods?
4. What are the ramifications of tropical rain forest destruction?
5. How serious is extinction in the tropical rain forest?
6. What can be done to curtail extinctions?

7. What is the human impact on the environment?
8. What are the threats from nuclear power?
9. What are the promises and challenges of nuclear power?
10. What problems are associated with the burning of fossil fuels?
11. In what ways do we contribute to pollution?
12. How is pollution threatening?
13. What is solid-waste pollution?
14. How do we contribute to water pollution?

15. What are chemical pollutants?
16. What are the consequences of acid rain?
17. How are we breaking down the ozone shield?
18. Why is environmental science an applied science?
19. What are the components of environmental science?
20. How can biologists further contribute to the resolution of environmental problems?

COMPLETING YOUR UNDERSTANDING

1. We obtain more than half of our food energy from how many different kinds of plants?

    a. 200

    b. 1,800

    c. 3

    d. 12

    e. 20

2. What is the evidence that untested plants could provide human food or other beneficial products?

3. What was the Green Revolution, and what were some of its side effects?

4. What can be learned from indigenous peoples living in close contact with their environment?

5. How did the Green Revolution create certain nutritional problems?

6. Why has the ocean not materialized as our answer to the food supply?

7. In today's world, what are the drawbacks of engaging in shifting agriculture in the tropics?

8. The most promising source of additional food for the near future is

    a. culturing microorganisms.

    b. hydroponic culture.

    c. improving existing cultivation.

    d. developing new farmlands.

    e. farming the sea.

9. When are all tropical rain forests, except for a few preserves, expected to disappear?

10. During the next thirty to forty years, what percentage of the world's species are expected to become extinct?

11. How can we curtail tropical deforestation?

12. How will the Chernobyl incident "live on" for those people who were close to the accident?

13. What are some of the alternatives to fossil fuels and nuclear power as energy sources? Why are these alternatives not being used today on a large scale?

14. The most important way to alleviate global warming in the near future involves

    a. nuclear energy.

    b. solar energy.

    c. wind energy.

    d. energy conservation.

    e. installing scrubbers in coal-burning plants.

15. Are "garbage barges" something of the past? Why or why not?

16. What is one of the most important basics in solid-waste management?

17. What is biological magnification, and what is an example of it?

18. What is the difference between a "brown air" and a "gray air" city?

19. What is the main contributing factor to acid precipitation, and what is the solution to the problem?

20. Should residents of North America have health-related concerns about ozone depletion, or is this just a problem for inhabitants of the Southern Hemisphere?

21. Which component of environmental science analyzes the environmental impact?

22. Decreased birth rates have ultimately followed decreased death rates in many countries—for example, Germany and Great Britain. Do you think this will eventually occur worldwide and solve our population problems?

23. Some have argued that U.S. attempts to promote lowering of the birth rate in the underdeveloped countries of the tropics is no more than economic imperialism and that it is in the best interests of these countries that their populations grow as rapidly as possible. How would you respond to this argument?

FOR FURTHER READING

Ausubel, J. 1991. A second look at the impacts of climate change. *American Scientist*, May, 210–21. Article that suggests that we may be worrying too much about crops and coastlines, and too little about water and wildlife.

Brown, L., ed. 1994. *State of the world*. New York: W. W. Norton. A highly recommended, easily read summary of the ecological problems faced by an overcrowded and hungry world, with an excellent chapter of suggested solutions. A new edition appears every year.

Ellis, W. S. 1990. The Aral. A Soviet sea lies dying. *National Geographic*, Feb., 73–92. Overproduction of cotton that led to overuse of water and to fertilizer and pesticide pollution that has almost completely destroyed the Aral Sea.

The endless cycle. 1990. *Natural History*, May. An entire issue devoted to recycling. Unless we follow nature's example and return used goods to the production cycle, we face a bleak future.

The global warming debate. 1993. *National Geographic*, Spring. An entire issue of *Research and Exploration* devoted to a detailed examination of this very controversial issue.

Managing Planet Earth. 1989. *Scientific American*, September. An entire issue devoted to many aspects of the management of the earth, with a number of excellent articles.

Matthews S., and J. Sugar. 1990. Is our world warming? *National Geographic*, Oct., 66–99. A vivid, well-illustrated discussion of the greenhouse effect, its causes, and the possible consequences of global warming.

Reganold, J. et al. 1990. Sustainable agriculture. *National Geographic*, June, 112–20. Traditional conservation methods combined with modern technology that can reduce farmers' dependence on possibly dangerous chemicals.

Reganold, J. P., R. I. Papendick, and J. F. Parr. 1990. Sustainable agriculture. *Scientific American*, June, 112–20. Nontraditional approaches to agriculture that offer both financial and environmental rewards.

Repetto, R. 1990. Deforestation in the tropics. *National Geographic*, April, 36–42. Government policies that encourage excessive logging and clearing for ranches and farms and are to blame for accelerating the destruction of tropical forests.

Toon, O., and R. Turco. 1991. Polar stratospheric clouds and ozone depletion. *Scientific American*, June, 68–74. The story of how the "ozone hole" forms every spring in the high skies over Antarctica.

Vietmeyer, N. D. 1986. Lesser-known plants of potential use in agriculture and forestry. *Science* 232: 1379–84. An excellent, brief survey of little-known but extremely useful plants that humans could use to improve food and firewood supplies in the future.

Weiner, J. 1990. *The next one hundred years: Shaping the future of our living earth.* New York: Bantam Books. A thorough review of the ecological problems that confront us.

What on earth are we doing? 1990. *National Wildlife* 28(2). An excellent issue of this outstanding magazine, entirely devoted to the state of the environment, that presents many actions that you can take to help solve the problems discussed in this chapter.

White, R. 1990. The great climate debate. *National Geographic*, July, 36–43. A thoughtful article that asks whether we should take steps now to avoid consequences we cannot foresee and also discusses the scientific and political controversy surrounding the greenhouse effect and the prospect of global warming.

Wilson, E. O. 1989. Threats to biodiversity. *Scientific American*, Sept., 108–16. Habitat destruction, mainly in the tropics, that is driving thousands of species to extinction.

EXPLORATIONS

Interactive Software: **Pollution of a Freshwater Lake**

This interactive exercise allows students to explore how the addition of certain "harmless" chemicals can pollute a lake by allowing algae to grow. Bacteria feeding on dead algae use up all the oxygen dissolved in the water, killing the lake. The exercise presents a map of Lake Washington, indicating the location of sewage treatment plants and the chemical composition of their effluent. Students can investigate the nature of pollution by altering the chemicals in the effluent. They will discover that phosphates in the effluent lead to algal growth, which is followed by bacterial growth (they feed on the dead algae), and a precipitous drop in levels of dissolved oxygen. Students can then attempt to "clean up" the lake by altering the nature and amount of effluent. Their success will depend largely upon how early in the pollution process they initiate their recovery efforts.

Questions to Explore:

1. How does the growth of algae, which are photosynthetic and make oxygen, lead to oxygen depletion of the lake?

2. Why not simply poison the algae in the lake?

3. How long can you wait before cleaning up the lake and still succeed?

4. Can you envision any way to successfully avoid oxygen depletion of the lake without stopping the discharge of treated sewage into the lake?

These hyacinth macaws (Andorhynchus hyacinthinus) are but one of many endangered species. Much of the world's biological diversity is threatened as modern society alters and often destroys natural habitats.

THE SIX KINGDOMS OF LIFE

Theses Brazilian jaguars (Panthera onca) are settling a disagreement force-fully. Centuries of competition within and between species have shaped the bi-ological world as we know it today. Each species is in its own way a survivor.

FOR REVIEW

Here are some important terms and concepts that have been discussed in previous chapters and that you will encounter again in this chapter. Review them before proceeding if necessary.

The six kingdoms of life (*chapter 1*)
Prokaryotic versus eukaryotic cells (*chapter 4*)
Definition of a species (*chapter 16*)
How species form (*chapter 16*)
Biological classification (*chapter 18*)

So far, this text has stressed common themes and general principles that apply to all organisms. Now, however, the focus shifts to the *differences* among organisms, to the diversity of organisms that make up the biological world. The remainder of this text examines the parade of life on the earth, from the simplest microbes to jellyfish, insects, elephants, bears, and redwood trees. Like going to the zoo, examining biological diversity is interesting, challenging, and a great deal of fun (figure 25.1).

The Classification of Organisms

At least ten million different kinds of organisms live on the earth. For biologists to discuss and study them, the organisms must have names. From earliest recorded history, organisms have been grouped into basic units, such as oaks, cats, and horses. Eventually, these units, which were often given the same names used by the Romans and Greeks, began to be called **genera** (singular, **genus,** from the Latin, meaning "race," "stock," or "kind"). Starting in the Middle Ages, these names were written in Latin, the language of scholarship at that time, or given a Latin form. Thus, oaks were assigned to the genus *Quercus,* cats to *Felis,* and horses to *Equus*—names that the Romans applied to these groups of organisms—and new Latin names were invented for genera not known in antiquity.

> *Genera names were the basic points of reference in classification systems, and these names eventually came to be written in Latin.*

The Polynomial System: The Need to Classify Species

Following the Renaissance in Europe, scientists began to see the need for a system that would more precisely define organisms in terms of species. As mentioned in chapter 16, English clergyman and scientist John Ray (1627–1705) proposed that a species be defined as consisting of individual organisms that could breed with one another and produce offspring that were still of that species. In other words, species provided the most precise description of individual organisms, and a system of naming species was needed that would take this precision into account.

Before the 1750s, scholars usually added a series of additional descriptive terms to the name of the genus when they wanted to designate a particular species. (Both the singular and plural of the word *species* are the same.) These phrases, starting with the name of the genus, made up what came to be known as **polynomials,** strings of Latin words and phrases consisting of up to twelve, fifteen, or even more words. But this zeal for precision had its drawbacks. Not only were such polynomials cumbersome, they also could be altered at will by later authors. As a result, a given organism really did not have a single name that was its alone. Instead, polynomial names were a series of different descriptive phrases that could be related to one another only by scholars. This system of naming was burdensome and imprecise.

Figure 25.1 Two kingdoms meet in life and death.
This weevil, lying on a leaf in the tropical rain forest of Costa Rica, is a kind of beetle and belongs to kingdom Animalia. It has been killed by the fungus whose fruiting structure now rises above it, scattering spores to the four winds; the fungus is a member of kingdom Fungi. The filaments of the fungus grew through the body of the weevil, killed it, and converted its body into the body of the fungus.

The Binomial System

The simplified system of naming plants, animals, and microorganisms that has been standard for more than two centuries stems from the works of Swedish biologist Carl Linnaeus (1707–1778; figure 25.2). Linnaeus's ambition, like that of some of his predecessors, was to catalog all the kinds of organisms and minerals. In the 1750s, he produced several major works that, like his earlier books, employed the polynomial system, which had been well established for nearly a century as the basic system for naming organisms.

But as a kind of shorthand, Linnaeus also included a two-part name, or **binomial,** for each species that became the standard method of species designation. The first part of the name is the genus, and the second part is the epithet. By convention, the genus

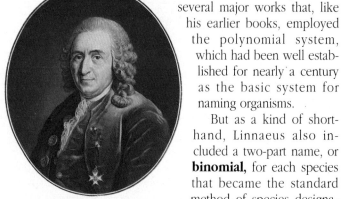

Figure 25.2 Carl Linnaeus (1707–1778).
This Swedish biologist devised the system of naming organisms that is still in use today.

and epithet are Latinized (that is, they follow Latin rules of prefixes and suffixes) and are always italicized; only the first (genus) word of the binomial name is capitalized. For example, Linnaeus designated the willow oak *Quercus phellos* and the red oak *Quercus rubra,* even though he also included the polynomial names for these species (figure 25.3). The short names were so convenient that Linnaeus's binomial system was immediately accepted.

Quercus phellos
(Willow oak)

Quercus rubra
(Red oak)

Figure 25.3 Two species of oak.
Although they are clearly oaks (Quercus), these two species differ sharply in the shapes and sizes of their leaves and in many other features, including their overall geographical distributions.

In Linnaeus's binomial system, a species name consists of two parts: the first part is the genus; the second part is the epithet.

By agreement, the scientific names of organisms are the same throughout the world and thus provide a standard and precise way of communicating about organisms, whether a particular scholar speaks Chinese, Arabic, Russian, or English. In contrast, locally given names often differ greatly from place to place; thus, a bear in Australia may be either a koala bear (a marsupial) or one of the large carnivores (placental mammals) more commonly known by this name. For general communication, however, one standard set of names is important. Linnaeus's binomial shorthand system has served the science of biology well for nearly 250 years.

What Is a Species? A Brief Review

As explained in chapter 16, no absolute criteria can be applied to the definition of a species. Individuals that belong to a given species—for example, dogs—may look very unlike one another (figure 25.4). Nevertheless, they are generally capable of breeding with one another, and the characteristics of the different forms can appear in the progeny of a single mated pair. Species remain relatively constant in their characteristics, can be distinguished from other species, and do not normally interbreed when they come together with other species in nature. For example, dogs are not capable of interbreeding with foxes, which, although they are generally similar, are members of another, completely distinct, group of mammals. In contrast, dogs

can and do form fully or partly fertile hybrids with related species, such as wolves and coyotes, which are also members of the genus *Canis*. The transfer of characteristics between these species has, in some areas, changed the characteristics of both of the interbreeding units.

The criteria just mentioned for species apply primarily to those that regularly **outcross**—interbreed with individuals other than those like themselves. In some groups of organisms, including bacteria and many eukaryotes, **asexual reproduction**—reproduction without sex—predominates. The species of these organisms clearly cannot be characterized in the same way as are the species of outcrossing animals and plants; they do not interbreed with one another, much less with the individuals of other species. Despite these difficulties, biologists generally agree on the kinds of units classified as species, although these units share no uniform biological characteristics.

Species differ from one another in at least one characteristic and generally do not interbreed freely with one another where their ranges overlap in nature.

Since the time of Linnaeus, about 1.4 million species have been named. This is a far greater number of organisms than Linnaeus suspected to exist when he was developing his system of classification in the eighteenth century. But the actual number of species in the world is undoubtedly much greater, judging from the very large number of species still being discovered. **Taxonomists**—scientists who study and classify different kinds of organisms—estimate that there are at least ten million species of organisms on the earth. While at least two-thirds of these—more than six million species—occur in the tropics, no more than five hundred thousand tropical species have been named.

The Taxonomic Hierarchy

Taxonomy is defined as the systematic classification of organisms based on their similarities. Biological systems of classification are **hierarchical,** with the basic units, called **taxa,**

The ancestor: *Canis lupus*

The wolf is still common in the wild in northern Canada and Alaska

Dogs of ancient lineage: *Canis familiaris*

Afghan hound. Swift, with strong jaws, bred as a hunter to run down quarry

Shih-tzu. Docile and intelligent, bred as pets at the imperial court of ancient China

Greyhound. Often used for racing, bred for speed by European aristocrats

Great Dane. One of the largest dogs, bred in Germany for size and bravery as boar hounds

Newer breeds

Bulldog. Strong and tenacious biters, bred in England for the now-outlawed sport of bull-baiting

Chihuahua. One of the smallest dogs, bred as tiny and lively lap dogs from ancient Mexican lines

Retriever. The most popular pet today, good-natured and docile, bred to hunt land and water birds

Poodle. Perhaps the most intelligent dog, bred in France for companionship from a line of water dogs

Figure 25.4 All dogs are members of the same species.
Domestic dogs are descended from the wolf, but hundreds of different breeds have been produced by controlled breeding and selection. All of them are fully able to interbreed.

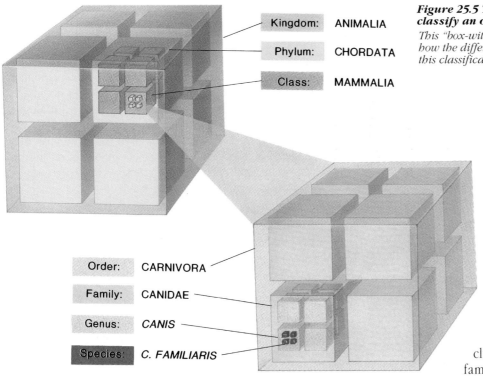

Figure 25.5 The hierarchical system used to classify an organism.
This "box-within-a-box" diagram clearly demonstrates how the different taxa are inclusive of one another in this classification of a dog.

Kingdom: ANIMALIA
Phylum: CHORDATA
Class: MAMMALIA
Order: CARNIVORA
Family: CANIDAE
Genus: *CANIS*
Species: *C. FAMILIARIS*

arranged like boxes within boxes (figure 25.5). In such a system, grouping genera into the larger, more inclusive taxa known as **families** is an effective way to remember and study their characteristics. Families of organisms, such as "finches," "legumes," and "squirrels," include many genera and have characteristics of their own. Thus, the oaks (*Quercus*), beeches (*Fagus*), and chestnuts (*Castanea*) are grouped, along with other genera, into the beech family Fagaceae because of the many features they have in common. Similarly, the tree squirrels (*Sciurus*), Siberian and western North American chipmunks (*Eutamias*), and marmots (*Marmota*) are grouped with other related mammals in the family Sciuridae (figure 25.6). Knowing which family a genus belongs to immediately communicates many of the genus's characteristics.

The **taxonomic system** also includes several more inclusive taxa than families, making possible more efficient communication about organisms. Families are grouped into **orders,** orders into **classes,** and classes into **phyla** (singular, *phylum*). The phyla, in turn, are grouped into **kingdoms,** the most inclusive units of classification (see figure 25.5). A mnemonic device for remembering the order of the different taxa is the phrase: "**K**indly **P**ay **C**ash **O**r **F**urnish **G**ood **S**ecurity" (kingdom—phylum—class—order—family—genus—species).

(a)

(b)

(c)

Figure 25.6 Diversity within the family Sciuridae.
All of the organisms here have the same classification up until their genus. This classification is kingdom Animalia, phylum Chordata, class Mammalia, order Rodentia, and family Sciuridae. (a) Tamiasciurus hudsonicus, or red squirrel, is an avid tree-dweller. (b) Eutamias townsendii, or Townsend's chipmunk, is a member of a genus of diurnal, brightly colored, very active ground-dwelling squirrels. (c) Marmota flaviventris, the yellow-bodied marmot, is a large squirrel that lives in burrows. Marmots are social, tolerant, and playful. They live in harems that consist of a single territorial male together with numerous females and their offspring.

> *Species are grouped into genera, genera into families, families into orders, orders into classes, and classes into phyla. Phyla are the basic taxa within kingdoms. Such a system is hierarchical.*

Table 25.1   Sample Classifications of Three Representative Organisms

	Human Being	**Honeybee**	**Red Oak**
Kingdom	Animalia	Animalia	Plantae
Phylum	Chordata	Arthropoda	Anthophyta
Class	Mammalia	Insecta	Dicotyledones
Order	Primates	Hymenoptera	Fagales
Family	Hominidae	Apidae	Fagaceae
Genus	*Homo*	*Apis*	*Quercus*
Species	*Homo sapiens*	*Apis mellifera*	*Quercus rubra*

Table 25.1 shows how the human species, the honeybee, and the red oak are placed in taxa at the seven hierarchical levels. The taxa at the different levels may include many, a few, or only one entry, depending on the nature of the relationships in the particular groups involved. Thus, there is only one living genus of the family Hominidae, but, as mentioned earlier, several living genera of the family Fagaceae. Each taxon at every level implies to someone familiar with the system, or with access to the appropriate books, both a series of characteristics that pertain to that group and also a series of organisms that belong to it.

How Do Taxonomists Classify?

While taxonomy is often thought of as a boring science, it is, in fact, one of the more lively branches of biology, often controversial and rarely dull. Perhaps the oldest and most fundamental disagreement within taxonomy concerns what its role in biology should be, what it should be trying to accomplish. There are two fundamental viewpoints, one the Linnaean approach of classifying and naming, and the other the Darwinian approach of tracing evolutionary history. Each viewpoint has led to extreme schools within taxonomy, and in practice, both viewpoints influence how taxonomy is done today.

Phenetics: Classifying by Morphological Similarity

Ever since Linnaeus, biologists have been naming new species by carefully noting differences among organisms. The essence of this process lies in deciding what differences between species are important. The shape, size, and appearance of an individual is called its morphology, and differences in morphology have always formed the backbone of taxonomy.

In the 1950s, some taxonomists began to use computers to compare many morphological traits at once—not only their presence or absence, but also quantitative information on size and color. This approach, called **numerical taxonomy** or **phenetics,** judges taxonomic affinities entirely on the basis of measurable similarities and differences. As many characteristics as possible are compared, in order to minimize the contribution of a few characters that might resemble one another because of parallel evolution, such as discussed in chapter 15 **(analogy),** rather than because of taxonomic affinity **(homology).** Strictly phenetic approaches are not widely used today because they make no attempt to judge the phylogenetic affinity of organisms.

Cladistics: Classifying by Evolutionary Relationships

At the opposite end of the taxonomic spectrum are biologists who consider only evolutionary relatedness in assigning taxonomic affinity, ignoring the degree of morphological similarity or difference. This school of taxonomy is called **cladistics.** Cladistics (from the Greek word *clados,* meaning "branch") classifies organisms according to the historical order in which branches arise along a phylogenetic tree. Only the order of branching is considered in assigning position on the phylogenetic tree.

Cladistics is ideally suited to molecular data, particularly data on DNA sequence divergence, such as discussed in Sidelight 25.1. All that is necessary is to ascertain which differences arose after a branch diverged from the evolutionary tree. These differences, called **derived characteristics,** are shared by all members of a branch but are not present before the branch. On the phylogenetic tree of plants, for example, vascular tissue is a derived characteristic, shared by all vascular plants but not present in plants like mosses that evolved before vascular plants. Among the vascular plants, all the plants that possess seeds are placed on one branch of the evolutionary tree, distinct from plants like ferns that evolved before the advent of seeds. Among the seed plants, the plants on the branch occupied by angiosperms possess flowers, while the gymnosperms do not. Further along the angiosperm branch is another branch containing a group of plants whose seedlings have two leaves, called dicots, while no other angiosperms do. A phylogenetic tree constructed in this fashion is called a **cladogram.** A cladogram shows the *order* of evolutionary descent, not the extent of divergence.

Taxonomy Today

In practice, taxonomy today utilizes information from both phenetics and cladistics. A clear example of the conflict that can arise between *order* of divergence and *magnitude* of divergence is provided by taxonomists' assignment of birds to the separate class Aves, while retaining crocodiles in the class Reptilia—even though crocodiles are more closely related to birds than to other reptiles. It is clear that birds and crocodiles share many characters, such as a four-chambered heart, and a cladogram would group birds and crocodiles together, since the traits are viewed as derived rather than as having evolved independently (figure 25.7). Taxonomists today, however, assign birds to their own unique class Aves because birds have diverged so fundamentally since they separated from the

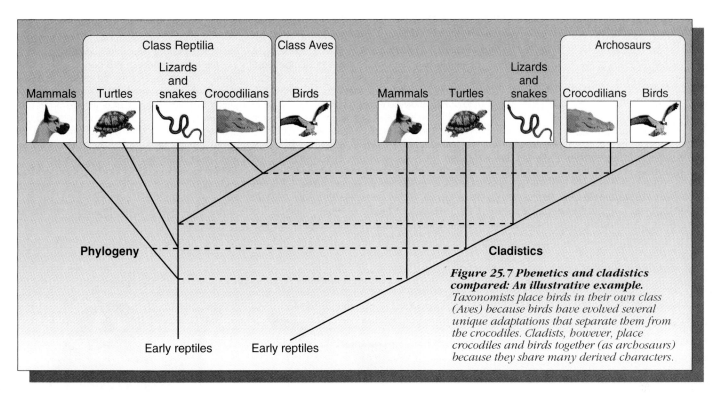

Figure 25.7 Phenetics and cladistics compared: An illustrative example.
Taxonomists place birds in their own class (Aves) because birds have evolved several unique adaptations that separate them from the crocodiles. Cladists, however, place crocodiles and birds together (as archosaurs) because they share many derived characters.

crocodiles, evolving feathers, hollow bones, and a host of other adaptations to flight. Ultimately, the taxonomic decision to classify crocodiles with reptiles is a practical one. The extent of divergence since birds branched from crocodiles is, in this case, too important biologically to ignore.

> *Both information about the order of evolutionary divergence (cladistics) and the magnitude of the changes which have occurred (phenetics) are used by taxonomists today.*

The Six Kingdoms of Organisms

The kingdom is the most inclusive category in which organisms are grouped. Until about ten years ago, scientists generally categorized organisms into five kingdoms. Recently, however, many scientists have added a sixth kingdom to this taxonomic category by dividing the bacteria into two separate kingdoms.

Eukaryotes and Prokaryotes: A Fundamental Division

During the past fifty years, biologists have recognized that the most fundamental distinction among groups of organisms is not that between animals and plants, but rather that which separates prokaryotes and eukaryotes. The prokaryotes, or bacteria, differ from all other organisms in their lack of membrane-bound organelles and microtubules, as well as in their possession of simple flagella. Their DNA is not associated with histone proteins and is not located in a nucleus. Bacteria do not undergo sexual recombination in the same sense that eukaryotic organisms do,

although forms of genetic recombination occur occasionally in some bacteria. The cell walls of most bacteria contain muramic acid, which is only one of the many biochemical peculiarities by which they differ from all eukaryotes. Bacteria existed for at least 2.5 billion years before the appearance of eukaryotes. In recognition of their highly distinctive features, the bacteria traditionally have been assigned to a kingdom of their own, **Monera.**

As noted earlier, biologists have recently divided the Monera into two separate kingdoms: In kingdom **Archaebacteria,** the bacteria possess unique genetic machinery and metabolism. (Many, for example, use sulfur compounds as an energy source to make food, and some can withstand high temperatures.) Bacteria of the other kingdom, the **Eubacteria,** do not have these unique characteristics. Eubacteria can be either photosynthetic or heterotrophic (figure 25.8). The current scheme of the evolution of the eukaryotic kingdoms from these two prokaryotic kingdoms is diagrammed in figure 25.9.

The features of eukaryotes contrast sharply with those of all prokaryotes. All eukaryotes have microtubules in their cytoplasm, and their cells are characterized by classes of membrane-bound organelles. They have a clearly defined nucleus that is enclosed within a double membrane, the nuclear envelope. Their chromosomes are complex structures in which the DNA is characteristically associated with histone proteins; the chromosomes divide and are distributed in a regular manner to the daughter cells as a result of the process of mitosis. When flagella and cilia are present in eukaryotes, they have a characteristic 9 + 2 structure that consists of microtubules. Mitochondria are among the complex organelles that occur in the cells of all eukaryotes. In addition, eukaryotes exhibit integrated multicellularity of a kind not found in bacteria, as well as true sexual reproduction.

Sidelight 25.1
WHAT ARE THE RELATIVES OF THE GIANT PANDA? A MOLECULAR SOLUTION

Ever since the giant panda, a unique mammal of the mountain forests of southwestern China, was brought to the attention of European scientists, scientists worldwide have speculated about its true relationships. Although widespread throughout most of central China as recently as 150 to 200 years ago, the giant panda is now confined to six widely separated areas, amounting in total to 29,500 square kilometers. These areas are mostly in central Sichuan Province, with an outlying area in the Qinling Mountains of Shaanxi Province to the northeast. The total wild population of giant pandas at present probably numbers as few as 1,150 to 1,200 individuals. In addition, twelve giant pandas are still alive in zoos outside of China, and sixty to seventy are in zoos within China.

Although Pere David, who in 1869 provided the first scientific description of the giant panda for European scientists, considered the panda to be a bear, others pointed out that its bones and teeth resemble those of the raccoons, a family of mammals from the New World. While scientists accepted that pandas resemble both bears and raccoons, they could not agree about which group of mammals pandas resemble more closely.

The lesser panda, or red panda (*Ailurus fulgens*), which is another fascinating Asian mammal, seems similar to giant pandas in many respects, yet it is also similar to the raccoons. Partly because of these relationships, the red panda has often been grouped with the giant panda in a subfamily of the raccoon family, as if they both were kinds of raccoons. While the red panda lacks the giant panda's remarkable false thumb (an "extra" structure with which the giant panda grasps bamboo stalks while eating them), the two have similar

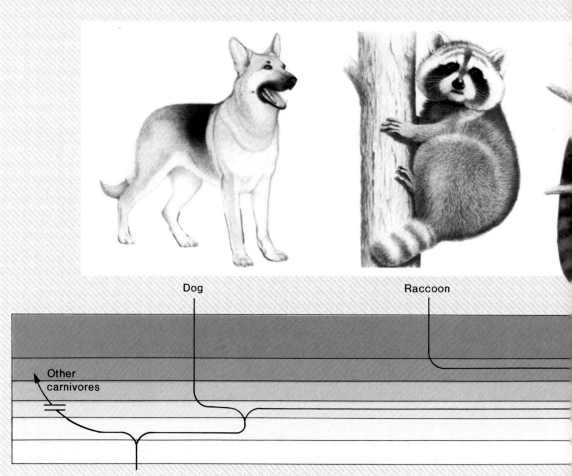

Dog   Raccoon

Other carnivores

Sidelight figure 25.1 Is the panda a bear or a raccoon?
Modern DNA analysis has provided a clear answer to this old puzzle: Both! The giant panda is a bear; the red panda is a raccoon.

behavior: Both feed on bamboos and occur in the same general area. Scientists now know the resemblance to be misleading.

By matching reactions between blood serum samples from these animals, a comparison that was first made in the 1950s, scientists soon came to the conclusion that the giant panda was indeed related to the bears. By the same methods, the red panda showed up as a separate evolutionary line, independent from, but related to, both bears and raccoons. The famous field naturalist George Schaller, of the New York Zoological Society's Wildlife Conservation International Division, who has probably spent more time studying the behavior of pandas in the wild than anyone else, con-cluded on the basis of their behavioral traits that red pandas and giant pandas are closely related to one another and that they both more closely resemble bears than raccoons.

In the 1980s, however, Stephen O'Brien of the U.S. National Cancer Institute and his coworkers at the U.S. National Zoological Park neatly determined the relatives of the pandas using molecular methods of analysis—DNA hybridization, isozyme similarities, immunological comparisons, and the study of chromosomal morphology after the application of special staining techniques that reveal satellite DNA and other differentiated regions. Their conclusions, illustrated in Sidelight figure 25.1, clearly indicate that the red panda represents an evolutionary line that diverged from the raccoons fairly soon after they had separated from the bears and that the giant panda diverged from the bears much more recently. As to timing, the original split between raccoons and bears probably occurred thirty to fifty million years ago, that between the red panda and the raccoons about ten million years later, or about twenty to thirty million years ago. Although the red panda and the giant panda share a number of structural and behavioral features, they do not share a common ancestor closer than the common ancestor of the bears and the raccoons. In these studies, modern molecular information has led to the solution of a long-standing evolutionary problem: The red panda is a raccoon, not a bear.

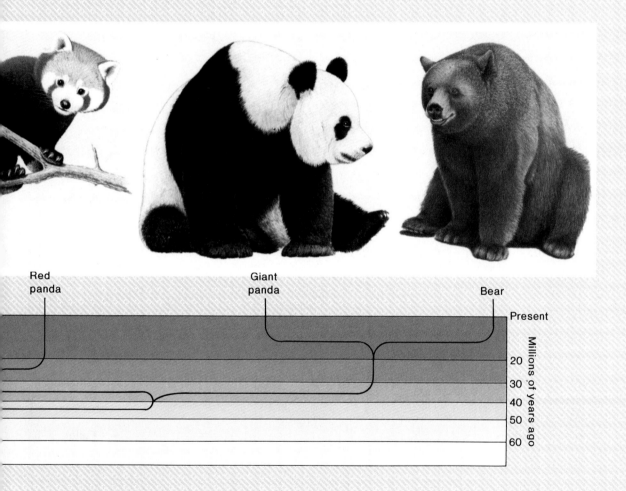

Red panda   Giant panda   Bear

Present

Millions of years ago

20
30
40
50
60

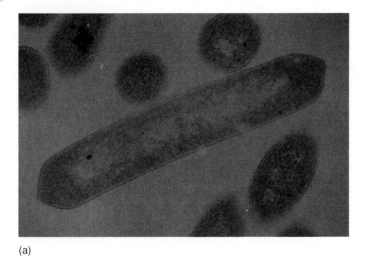

(a)

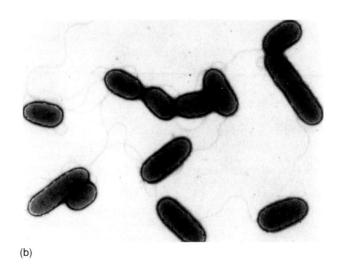

(b)

Figure 25.8 The two bacterial kingdoms, Archaebacteria and Eubacteria.
(a) Archaebacteria, *represented here by* Methanobacterium, *one of the methanogenic bacteria.* (b) Eubacteria, *represented by* Pseudomonas, *a soil bacteria responsible for many plant diseases.*

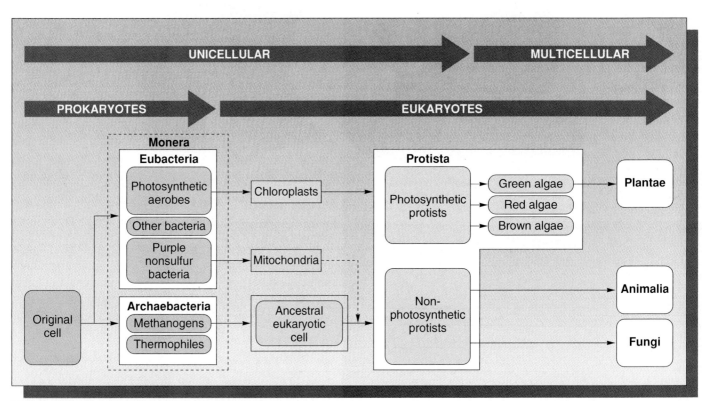

Figure 25.9 Origin of the eukaryotic kingdoms.
The solid lines *indicate evolutionary relationships, while the* dotted lines *indicate symbiotic events.*

Despite their often bewildering diversity of form, eukaryotes are much less diverse metabolically than are bacteria. All eukaryotes are believed to have evolved from a common eubacterial ancestor. Eukaryotes consist primarily of groups of single-celled organisms, together with several multicellular groups that were derived from single-celled ancestral eukaryotes.

The Kingdoms of Eukaryotic Organisms

The classification of organisms, especially eukaryotes, into kingdoms is clearly somewhat arbitrary. In the **six-kingdom system,** four of the kingdoms are eukaryotic organisms, while the fifth and sixth kingdoms—Archaebacteria and Eubacteria—contain

Figure 25.10 Kingdom Protista: Diverse, mostly unicellular organisms.
Volvox *lives in highly organized colonies in which some cells become specialized for reproduction. The colonies are so complex that some researchers consider them multicellular organisms.*

Figure 25.11 Kingdom Plantae: Multicellular, terrestrial, photosynthesizing organisms.
A blue columbine (Aquilegia caerulea) *among subalpine fir trees in Colorado.*

the prokaryotes, as mentioned earlier. In the six-kingdom system of classification, the array of diverse eukaryotic phyla consisting predominantly of single-celled organisms is assigned to kingdom Protista (see chapter 27). Three major multicellular lines, each with a great many species, are assigned to separate kingdoms: Animalia, Plantae, and Fungi. Each of these three multicellular kingdoms evolved from a different ancestor among the Protista. The characteristics of *all* the kingdoms of life, including the bacterial kingdoms, are summarized in table 25.2.

Kingdom Protista

Kingdom Protista has sometimes been referred to as the "grab-bag" kingdom because the protists contain all eukaryotes that do not seem to fit in the other eukaryotic kingdoms (figure 25.10). As a result, kingdom Protista contains a wide variety of organisms that display different physical characteristics, modes of nutrition, and lifestyles.

In older systems of classification, all photosynthetic protists were considered plants, even if they were unicellular. Protists that ingested their food like the animals were considered small, very simple animals. The photosynthetic protists were called **algae** (singular, *alga*), and the traditional name for heterotrophic protists is **protozoa** or **protozoans.** Although today these names are still in use, most scientists no longer consider algae plants or protozoa animals because this classification scheme does not reflect the actual relationships between the different groups of protists. This text considers algae and protozoa protists.

Kingdom Plantae

The ancestor of plants was a protist that could photosynthesize—a green alga. Plants are multicellular and have several common characteristics: They can photosynthetically manufacture their own food; their cells possess cell walls made of cellulose and other polysaccharides; they are generally immobile

(although they have mobile sperm); and they have a characteristic life cycle (discussed later in the chapter) that is found only in plants (figure 25.11).

Kingdom Fungi

Fungi are predominantly multicellular, although kingdom Fungi does contain a unicellular group, the yeasts. Fungi are generally filamentous in form, and they exhibit a particular mode of nutrition: Fungi secrete digestive enzymes that break down food particles into smaller pieces, which the fungi then absorb. Fungi are important ecologically because they are decomposers; that is, they break down dead organisms and return organism nutrients to the ecosystem for reuse (figure 25.12).

Kingdom Animalia

Animals are all multicellular, and their cells lack cell walls. Of all the kingdoms, animals are set apart by the possession of a nervous system, which can often be complex. Their mode of nutrition is ingestion: Animals take food into their bodies, and digestive enzymes break the food into smaller particles internally. Most animals are highly motile and can move around easily from place to place (figure 25.13).

Figure 25.12 Kingdom Fungi: Ecologically important, primarily multicellular organisms with absorptive nutrition.
The "shaggy mane" mushroom (Coprinus comatus).

Table 25.2   Characteristics of the Six Kingdoms

Kingdom	Cell Type	Nuclear Envelope	Mitochondria	Chloroplasts	Cell Wall
Archaebacteria	Prokaryotic	Absent	Absent	None	Noncellulose (polysaccharide plus amino acids)
Eubacteria	Prokaryotic	Absent	Absent	None (photosynthetic membranes in some types)	Noncellulose (polysaccharide plus amino acids)
Protista	Eukaryotic	Present	Present	Present (some groups)	Present in some forms, various types
Fungi	Eukaryotic	Present	Present	Absent	Chitin and other noncellulose polysaccharides
Plantae	Eukaryotic	Present	Present	Present	Cellulose and other polysaccharides
Animalia	Eukaryotic	Present	Present	Absent	Absent

Figure 25.13 Kingdom Animalia: Multicellular, motile organisms with ingestive nutrition.
Pika (Ochotona princeps) *collecting winter food supply.*

Six kingdoms of organisms are recognized. Three of them—Archaebacteria, Eubacteria, and Protista—are predominantly unicellular. The other three kingdoms—Fungi, Animalia, and Plantae—are multicellular and are derived from individual groups of protists.

Features of Eukaryotic Evolution

The features that set eukaryotes apart from the prokaryotes—organelles, multicellularity, sexuality, and their life cycles—evolved over billions of years. The sections that follow examine how these features evolved and their effects on the organisms that developed them.

Endosymbiosis and the Origin of Eukaryotes

As discussed in chapter 4, two of the principal organelles of eukaryotic cells—mitochondria and chloroplasts—share a number of unusual characteristics with each other and with the bacteria from which they were apparently derived.

As mentioned in chapter 17, the first eukaryotes appeared about 1.5 billion years ago. Their descendants soon became abundant and diverse. With few exceptions, all modern eukaryotes possess mitochondria, which clearly indicates that these organelles were acquired early in the group's history. Mitochondria are most similar to nonsulfur purple bacteria, the group that probably gave rise to them originally (see chapter 4).

In contrast to mitochondria, which are relatively uniform from group to group, chloroplasts fall into three classes that are distinct in their biochemistry. Each of these three classes seems to have been derived from a different bacterial ancestor.

Even today, so many bacteria and unicellular protists are symbiotic that the incorporation of smaller organisms with desirable features into cells does not appear to be a difficult process. The **endosymbiosis** that gave rise to the chloroplasts of different groups of protists seems to have taken place independently, and the same groups of bacteria appear to have been involved more than once.

Eukaryotic cells acquired chloroplasts by endosymbiosis at least several times during the evolution of different groups of protists.

Evolution of Multicellularity

True **multicellularity,** a condition in which the activities of individual cells are coordinated and the cells themselves are in contact, is unique to eukaryotes and one of their primary

Means of Genetic Recombination, If Present	Mode of Nutrition	Motility	Multicellularity	Nervous System
Conjugation, transduction, transformation	Mostly autotrophic (chemosynthetic)	Bacterial flagella, gliding or nonmotile	Absent	None
Conjugation, transduction, transformation	Photosynthetic or heterotrophic	Bacterial flagella, gliding or nonmotile	Absent	None
Fertilization and meiosis	Photosynthetic or heterotrophic, or combination of these	9 + 2 cilia and flagella amoeboid, contractile fibrils	Absent in most forms	Simple mechanisms for conducting stimuli in some forms
Fertilization and meiosis	Absorption	Nonmotile	Present in most forms	None
Fertilization and meiosis	Photosynthetic, chlorophylls *a* and *b*	9 + 2 cilia and flagella in gametes of some forms; none in most forms	Present in all forms	None
Fertilization and meiosis	Digestion	9 + 2 cilia and flagella, contractile fibrils	Present in all forms	Present, often complex

characteristics. The cell walls of bacteria occasionally adhere to one another, and bacterial cells may also be held together within a common sheath. Consequently, some bacteria form filaments, sheets, or three-dimensional bodies, but the degree of integration of individual bacteria within these cell groups is much more limited than is the case in multicellular eukaryotes.

Multicellularity evolved numerous times among the protists. The brown, green, and red algae all evolved multicellularity independently, as did animals, plants, and fungi. The evolution of multicellularity allowed organisms to deal with their environments in novel ways.

Distinct types of cells, tissues, and organs can be differentiated within the complex bodies of multicellular organisms. With such functional division within its body, a multicellular organism can protect itself, move about, seek mates and prey, and carry out other activities on a scale and with a complexity that would have been impossible for its unicellular ancestors. With all of these advantages, it is not surprising that multicellularity has arisen independently so many times (figure 25.14).

Multicellularity evolved numerous times among the protists. All of the complex differentiations associated with advanced life-forms depend on multicellularity.

Evolution of Sexuality

The third major characteristic of eukaryotic organisms as a group is sexuality. Although some genetic material is interchanged in bacteria (see chapter 26), it is certainly not a regular, predictable mechanism in the same sense that sex is in eukaryotes. The regular alternation between **syngamy**—the union of male and female gametes—and meiosis constitutes the **life cycle** characteristic of eukaryotes. It differs sharply from anything found in bacteria.

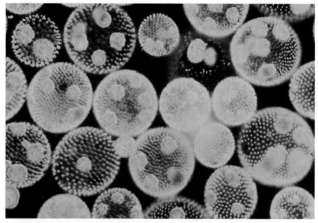

Figure 25.14 Multicellularity in a protist **(Volvox).**
Individual, motile, unicellular, green algae are united in Volvox *as a hollow ball of cells that moves by means of the beating of the flagella of its individual cells. A few cells near the posterior end of the colony are reproductive cells, but most are relatively undifferentiated. Some species of* Volvox *have cytoplasmic connections between the cells that function in the coordination of colony activities.* Volvox *represents an unusual form of multicellularity that evolved independently of all other multicellular organisms.*

Some eukaryotes are always haploid, but animals and plants are diploid during most of their lives. Plant and animal cells have two sets of chromosomes, derived, respectively, from their male and female parents. These chromosomes segregate regularly by the process of meiosis. Because meiosis involves crossing-over, no two products of a single meiotic event are ever identical. As a result, the offspring of sexual, eukaryotic organisms vary widely, thus providing the raw material for evolution.

In many of the unicellular phyla of protists, sexual reproduction occurs only during times of stress. Meiosis may have evolved originally as a means of producing new, well-adapted forms that would increase survival chances during such times.

Key:  Haploid

Diploid

Figure 25.15 Three types of eukaryotic life cycles.
(a) *Zygotic meiosis, a life cycle found in most protists.* (b) *Gametic meiosis, a life cycle typical of animals.* (c) *Sporic meiosis, a life cycle found in plants and fungi.*

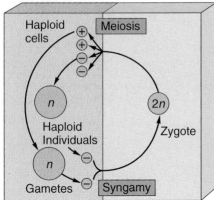

(a) Zygotic meiosis

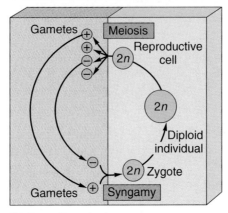

(b) Gametic meiosis

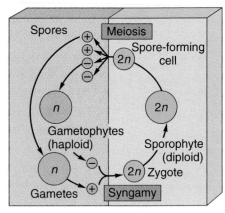

(c) Sporic meiosis

The first eukaryotes were probably haploid, and diploids seem to have arisen on a number of separate occasions by the fusion of haploid cells, which then eventually divided by meiosis.

Sexual organisms are able to evolve rapidly in relation to environmental demands because they produce variable progeny. Sexual reproduction, involving the regular alternation between syngamy and meiosis, makes evolutionary adjustment possible.

Life Cycles

Eukaryotes are characterized by three major types of life cycles. In the simplest of these, the zygote is the only diploid cell. Such a life cycle is said to be characterized by **zygotic meiosis** because the zygote immediately undergoes meiosis. A second type of life cycle is found in most animals. The gametes are the only haploid cells, and they undergo **gametic meiosis.** Meiosis produces more gametes, which fuse, giving rise to a zygote. With the third type of life cycle, found in plants, there is a regular alternation between a multicellular haploid phase and a multicellular diploid phase. The diploid phase produces **spores** that give rise to the haploid phase, and the haploid phase produces gametes that fuse to form the zygote. The zygote is the first cell of the multicellular diploid phase. This kind of life cycle is characterized by **alternation of generations** and **sporic meiosis.** These three major types of eukaryotic life cycles are diagrammed in figure 25.15.

In a life cycle characterized by zygotic meiosis, the zygote is the only diploid cell. In a life cycle characterized by gametic meiosis, the gametes are the only haploid cells. In plants, there is an alternation of generations, in which both diploid and haploid cells divide by mitosis. This type of life cycle is characterized by sporic meiosis.

Viruses: A Special Case

All of the groups discussed so far are clearly organisms, but another group—the **viruses**—lies on the borderline between living and nonliving. Viruses present a special classification problem in that they are not organisms. Nonliving but capable of replication, viruses are basically fragments of nucleic acids probably derived from cells. They occur in both eukaryotes and bacteria. Most of them have the capacity to organize protein coats around themselves. Viruses are able to direct the machinery of their host cells to manufacture more virus material, but they cannot exist on their own. For that reason, they cannot logically be placed in any of the kingdoms (figure 25.16). Viruses are discussed in more detail in chapter 26.

A Look Ahead

The remaining chapters in part 7 discuss in detail the diversity of the major groups of organisms, stressing their evolutionary relationships with one another. Multicellularity and sexuality, as discussed in this chapter, are major themes that underlie the evolutionary advances to be examined. The colonization of the land, which was carried out by multicellular organisms complex enough to function in this harsh habitat, is the feature that contributed most to the diversification of the major groups of organisms seen today.

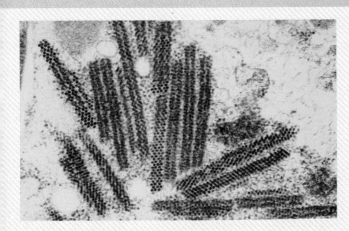

Figure 25.16 Viruses are not alive.
A kind of virus called GD7 is shown here replicating itself in a human white blood cell. Viruses are nonliving segments of DNA and RNA, ultimately derived from the genomes of either bacteria or eukaryotes, most with protein coats, or capsids. Viruses have the ability to take over the protein- and nucleic-acid-synthesizing machinery of their host cells to make more viruses, but they are not biologically active unless inside a host cell. For this reason, viruses are not considered organisms, and so are not included in the six kingdoms of life.

EVOLUTIONARY VIEWPOINT

Taxonomy serves two purposes in biology: It gives names to kinds of organisms and suggests how organisms are related to one another. In its second role, taxonomy has become a very active area of modern biology. The advent of DNA-level taxonomic data has lent impetus to cladistic analyses of evolutionary phylogeny, which are well suited to data in which all characters bear equal weight. However one views the value of cladistics, the new molecular data clearly are having a major impact on current knowledge of how organisms are related.

SUMMARY

1. Carl Linnaeus, an eighteenth-century Swedish scientist, developed the system of naming organisms that is still used today. In it, every species of organism has a binomial name, which consists of two words, the first of which is the genus name and the second of which, when combined with the genus name, designates the species.

2. Species remain relatively constant in their characteristics, can be distinguished from other species, and do not normally interbreed freely with other species in nature.

3. There are probably at least ten million species of plants, animals, and microorganisms in the world. Only about 1.4 million of them have been named thus far, and only about five hundred thousand of the estimated six million or more species that occur in the tropics have been identified.

4. In the hierarchical system of classification used to describe organisms, genera are grouped into families, families into orders, orders into classes, classes into phyla, and phyla into kingdoms.

5. Taxonomy involves a study of both the order in which groups evolve (cladistics) and the degree to which they diverge from one another (phenetics).

6. All organisms can be categorized into six kingdoms. The fundamental distinction among living organisms is that which separates the bacteria, which have a prokaryotic form of cellular organization, from the eukaryotes.

7. Bacteria have recently been divided into two separate kingdoms: the Archaebacteria and Eubacteria.

8. Many diverse phyla of eukaryotes are placed in the kingdom Protista. These organisms display different physical characteristics, modes of nutrition, and lifestyles. Virtually all phyla of Protista have at least some multicellular representatives, and three—red algae, brown algae, and green algae—are exclusively or largely multicellular.

9. Three groups of multicellular eukaryotes are so large and differ so sharply from their ancestors among the Protista that each is classified as a different kingdom. These are the animals (Animalia), plants (Plantae), and fungi (Fungi).

10. The features that set eukaryotes apart from prokaryotes include organelles like mitochondria and chloroplasts acquired by symbiosis. True multicellularity and sexuality are also both exclusive properties of the eukaryotes. Multicellularity confers a degree of protection from the environment and the ability to carry out a wider range of activities than is available to unicellular organisms. Sexuality permits extensive, orderly, genetic recombination. Eukaryotes are also characterized by three major types of life cycles: zygotic meiosis, gametic meiosis, and sporic meiosis.

11. Viruses are not organisms and are not included in the classification of organisms. They are fragments of the genomes of organisms.

REVIEWING THE CHAPTER

1. What is the significance of Linnaeus's binomial system of classification, and why was it revolutionary for its time?

2. Why is it difficult to list specific criteria for species?

3. What are the different categories used in the taxonomic hierarchy, from the most inclusive to the least inclusive?

4. How does phenetics differ from cladistics?

5. What are the six kingdoms of life, and how do they differ from one another?

6. Why was kingdom Monera divided into two separate kingdoms?

7. What is endosymbiosis, and what does it have to do with eukaryotic organelles?

8. What are some of the advantages of multicellularity?

9. How does sexuality make evolutionary adjustment among eukaryotes possible?

10. What are the three different life cycles found in eukaryotes, and how do they differ from one another?

11. Why are viruses not included in the six kingdoms of life?

COMPLETING YOUR UNDERSTANDING

1. Each species has a two-part name or _____. The first part is called the _____ , and the second part is called the _____ . As an example, the human species name is _____ _____.

2. How is a species best defined? Would this definition apply to all organisms in all kingdoms? If not, how would you amend your definition to include the other organisms?

3. What is the current assessment of the number of living species on our planet? What criticisms might you have for this projected number and why? What problems might be encountered in the immediate future to come up with a more accurate estimate of the number of living species than is currently suggested?

4. What was Linnaeus's rationale for dividing all species into just two groupings—plants and animals? What is today's rationale for placing all species into six groupings or kingdoms? Two hundred years from now, would you expect the number of groupings to be six? Why or why not?

5. What are the major differences between prokaryotes and eukaryotes? Do they have any similarities? If so, what are they?

6. In what ways are eukaryotes less diverse than prokaryotes?

7. Which kingdom has a combination of unicellular and multicellular organisms? Would there be justification to remove these multicellular members and transfer them to another kingdom? Why or why not? Are classification systems natural or artificial?

8. Which kingdoms are heterotrophic, and how are they distinguished from one another?

9. How is endosymbiosis related to the evolution of a eukaryotic cell?

10. What are the relationships between mitochondria, chloroplasts, and bacteria?

11. What kinds of life cycles are characteristic of animals, plants, and protists?

12. Why are viruses considered nonliving?

FOR FURTHER READING

Gibbons, A. 1991. Systematics goes molecular. *Science* 251:872–74. A brief account of how DNA sequence comparisons are revolutionizing modern taxonomy.

Gould, S. J. 1986. Evolution and the triumph of homology, or why history matters. *American Scientist* 74:60–69. An excellent account of Gould's influential views on the roles of chance and causation in evolution.

Gould, S. J. 1992. What is a species? *Discover* 12:40–44, One famous evolutionist's view on a question that biologists are still debating a century after Darwin.

Margulis, L., and K.V. Schwartz. 1982. *Five kingdoms: An illustrated guide to the phyla of life on earth*. San Francisco: W. H. Freeman. A marvelous account of the diversity of organisms.

Beautifully illustrated and presented in a logical fashion throughout.

Pääbo, S. 1993. Ancient DNA. *Scientific American* 11:86–92. An account of the work that led to the writing of *Jurassic Park*—the search for DNA information from long-extinct organisms.

Thomas, R. et al. 1989. The phylogeny of the extinct marsupial wolf. *Nature* 340:465–67. An attempt to unravel the evolutionary history of one of the most fascinating of recently extinct animals.

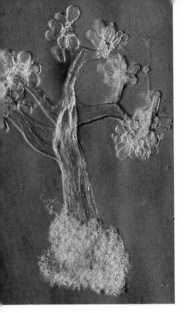

Figure 26.2 Bacteria are not multicellular, but . . .

Although no bacteria are truly multicellular, some adhere to one another. In Chondromyces crocatus, *one of the so-called "gliding" bacteria, the rod-shaped individuals move together, forming the composite spore-bearing structures shown here. Millions of spores, which basically are individual bacteria, are released from these structures.*

Bacterial Forms

Bacteria are mostly simple in form, usually being straight and rod-shaped **(bacilli),** spherical **(cocci),** or long and spirally coiled **(spirilla).** Some bacteria change into stalked structures; grow long, branched filaments; or form erect structures that release the **spores,** single-celled bodies that grow into new bacteria (figure 26.2). Among the rod-shaped bacteria, some adhere end-to-end after they have divided, forming chains. Still other bacteria form different kinds of cell groups.

Binary Fission

Bacteria, like all other living cells, divide. In bacterial division, called **binary fission,** an individual cell simply increases in size and divides in two (see chapter 9). The cell membrane and cell wall grow inward and eventually divide the cell by forming a new wall from the outside toward the center of the old cell.

> *Bacterial cells have a very simple structure. The cytoplasm contains no internal compartments or organelles and is bounded by a membrane encased within a cell wall composed of one or more layers of polysaccharide. Bacilli, cocci, and spirilla are common bacterial forms. Bacteria divide by binary fission.*

Bacterial Ecology and Metabolic Diversity

Bacteria occur in the widest possible range of habitats and play key ecological roles in virtually all of them. Some bacteria thrive in hot springs, where the usual temperatures may range as high as 78 degrees Celsius; others have been recovered living beneath 430 meters of ice in Antarctica. Bacteria are abundant in groundwater, where they were once thought to be absent. Still other bacteria, capable of dividing only under high pressures, exist around deep-sea vents, where the water temperatures run as high as 360 degrees Celsius. These bacteria use the energy they get from converting hydrogen sulfide to sulfur to power their growth and reproduction, and all other deep-sea vent organisms live directly or indirectly by consuming these bacteria.

Bacteria are able to play such a varied ecological role in virtually all habitats on the earth because of their metabolic diversity. Different kinds of bacteria vary extensively from one another in almost all aspects of their metabolism, in marked contrast to eukaryotes, which are relatively uniform. All eukaryotes follow the same general patterns of respiration, glucose breakdown, photosynthesis, and synthesis of nucleic acids and proteins. Bacteria are far more diverse in their ways of carrying out each of these vital functions (see table 26.1).

Archaebacteria

As noted in chapter 25, bacteria are divided into two kingdoms, the Archaebacteria and the Eubacteria. Of the two kingdoms, the **archaebacteria** are the more ancient. Most archaebacteria are autotrophs; only a few, however, photosynthesize. Instead, archaebacteria derive the energy they use for their metabolic activities from the oxidation of chemical energy sources, such as the reduced gases—ammonia (NH_3), methane (CH_4), or hydrogen sulfide (H_2S). In the presence of one of these chemicals, archaebacteria can manufacture their own amino acids and proteins.

The environments where these chemical energy sources are found are often harsh and forbidding to most other kinds of organisms. Archaebacteria can be divided into three groups: methanogens, thermoacidophiles, and halophiles. The names of the groups are descriptive of the types of environments in which these bacteria thrive or of the endproducts of their metabolism.

Methanogens, as their name implies, manufacture methane as a result of their metabolic activities. Methanogens die in the presence of oxygen. Thus, they are found in swamps and marshes in which all the oxygen has been consumed by the other organisms that live in these environments. The methane smell characteristic of swamps and marshes is courtesy of the methanogens.

Thermoacidophiles favor extremely hot and acidic environments, such as hot springs (figure 26.3). Many thermoacidophiles use hydrogen sulfide as their energy source.

Figure 26.3 Thermoacidophiles live in hot springs.

These archaebacteria growing in Sulfide Spring, Yellowstone National Park, Wyoming, are able to tolerate high acid levels and very hot temperatures.

Table 26.1   Bacteria

Major Group	Typical Examples	Key Characteristics
		Archaebacteria
Archaebacteria	Methanogens, halobacteria	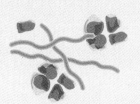 Bacteria that are not members of the kingdom Eubacteria; probably the oldest form of life on earth; anaerobic, with unusual cell walls; some produce methane; others reduce sulfur
		Eubacteria
Actinomyces	*Streptomyces, Actinomyces*	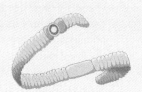 Gram-positive bacteria; form branching filaments and produce spores; often mistaken for fungi; produce many commonly-used antibiotics, including streptomycin and tetracycline; one of the most common types of soil bacteria; also common in dental plaque
Chemoautotrophs	sulfur bacteria, *Nitrobacter, Nitrosomonas*	 Bacteria able to obtain their energy from inorganic chemicals; most extract chemical energy from reduced gases such as SH_2 (hydrogen sulfide), NH_3 (ammonia), and CH_4 (methane); play a key role in the nitrogen cycle
Cyanobacteria	*Oscillatoria, Spirulina*	 A form of photosynthetic bacteria common in both marine and freshwater environments; deeply pigmented; often responsible for "blooms" in polluted waters
Enterobacteria	*E. coli, Salmonella, Vibrio cholerae*	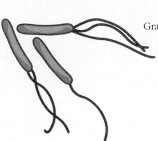 Gram-negative rod-shaped bacteria; do not form spores; usually aerobic heterotrophy; many important diseases are caused by enterics, including bubonic plague and cholera

Major Group	Typical Examples		Key Characteristics
			Eubacteria
Gliding and Budding Bacteria	Myxobacteria, *Chondromyces*		Gram-negative bacteria; exhibit gliding mobility by secreting slimy polysaccharides over which masses of cells glide; some groups form upright multicellular structures carrying spores called fruiting bodies
Pseudomonads	*Pseudomonas*		Gram-negative heterotrophic rods with polar flagella; very common form of soil bacteria; also contains many important plant pathogens
Rickettsias and chlamydias	*Rickettsia, chlamydia*		Small gram-negative intracellular parasites; the rickettsia life cycle involves both mammals and arthropods such as fleas and ticks; Rickettsia are responsibile for many fatal human diseases, including typhus (*Rickettsia prowazekii*) and Rocky Mountain spotted fever; chlamydial infections are one of the most common sexually transmitted diseases
Spirochaetes	*Treponema*		Long, coil-shaped cells with flagella at both ends; common in aquatic environments; a parasitic form is responsible for the disease syphilis

Halophiles thrive in very salty environments, such as the Great Salt Lake and the Dead Sea. These environments are extremely basic and are rich in hydrogen ions. Halophiles use the pumping of hydrogen ions out of their cells to drive ATP synthesis.

Archaebacteria, the most ancient group of living organisms are largely autotrophic and obtain energy by oxidizing chemicals. Their chemical energy sources are often found in hot, acid, salty, or anaerobic environments, all of which are inhospitable to most other organisms.

Eubacteria

Eubacteria comprise a diverse group of bacteria that differ from the archaebacteria in that they cannot withstand the harsh environments of archaebacteria, and in an evolutionary sense, they are younger than the archaebacteria. The many phyla in kingdom Eubacteria cannot be covered in detail in this text. Instead, the discussion will offer a glimpse of eubacteria diversity and concentrate especially on the so-called pathogenic bacteria—bacteria that cause harm to other organisms.

Photosynthetic Bacteria

Like plants and algae, the photosynthetic bacteria contain chlorophyll, but it is not held within chloroplasts. Instead, photosynthesis takes place on a system of internal membranes. Bacterial photosynthetic processes are diverse and reflect the long evolutionary path to photosynthesis as it occurs in plants. One group of photosynthetic bacteria—the **cyanobacteria**—was discussed in chapters 3 and 17 in connection with the evolution of life on the earth. The activities of the cyanobacteria appear to have been decisive in bringing about the increase of free oxygen in the earth's atmosphere from less than 1 percent to about 20 percent. This change, in turn, appears to have been crucial to the evolution of eukaryotic life based on aerobic respiration. The few hundred living species of cyanobacteria are important relics of earth's history and still of great interest and importance today (figure 26.4).

Figure 26.4 A cyanobacterium.
This filamentous cyanobacteria, Anabaena, exhibits some of the closest approaches to multicellularity among the bacteria.

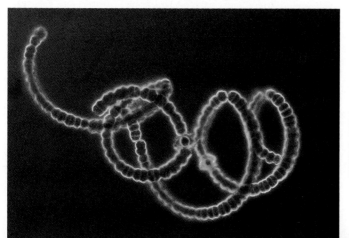

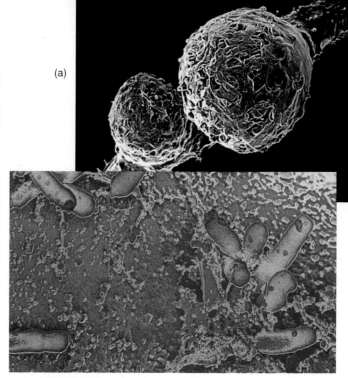

(a)

(b)
Figure 26.5 Root nodules.
(a) *Scanning electron micrograph of root nodules on the pea plant,* Pisum sativum, *caused by nitrogen-fixing bacteria* (Rhizobium leguminoserum) *that infect the root hairs.* (b) *The orange cylinders are individual* Rhizobium *bacteria on the root hairs of* Pisum sativum.

Nitrogen-Fixing Bacteria

Nitrogen-fixing bacteria convert atmospheric nitrogen to a form in which it can be used by living organisms as part of the nitrogen cycle (see figure 22.5). *Rhizobium,* which lives in nodules on the roots of legumes, and *Frankia,* which forms nitrogen-fixing nodules on the roots of a few kinds of nonleguminous plants, are examples of this type of bacteria (figure 26.5). A few other genera of nitrogen-fixing bacteria live free in the soil or in freshwater aquatic or marine environments, but *Rhizobium* is by far the most important of all nitrogen-fixing organisms. *Rhizobium* is a member of a group of bacteria that are aerobic (require oxygen for growth), gram-negative, and mostly motile by means of flagella. These bacteria are heterotrophic, meaning that they do not make their own food, but obtain it from other organisms.

In addition to these heterotrophic bacteria, many genera of cyanobacteria, discussed in the previous section on photosynthetic bacteria, have the ability to fix nitrogen. In Asia, rice can often be grown continuously in the same fields without the addition of nitrogen fertilizer because of the presence of nitrogen-fixing cyanobacteria in the paddies. One of the most important nitrogen-fixing genera of cyanobacteria is *Anabaena* (see figure 26.4).

Other Heterotrophic Bacteria

In addition to the nitrogen-fixing heterotrophic bacteria are numerous other phyla of heterotrophic bacteria. Most bacteria are heterotrophs, obtaining their energy from organic material formed by other organisms. Organic material must be broken down if it is to be used again by organisms. Bacteria and fungi play the leading role in breaking down organic molecules formed by biological processes, thereby making the nutrients in these molecules available once more for recycling. Decomposition is just as indispensable to the continuation of life on the earth as is photosynthesis.

Disease-causing bacteria are also heterotrophic, obtaining their nutrients from living tissues. Disease-causing bacteria are discussed in detail in the next section.

> *Eubacteria are a metabolically diverse group that obtain their energy from sunlight (photosynthetic bacteria) or from organic material formed by other organisms (heterotrophs), rather than from chemicals. Some bacteria are able to fix atmospheric nitrogen, an ability upon which all other organisms depend for supplies of usable nitrogen.*

Pathogenic Bacteria

Pathogenic (disease-causing) **bacteria** are a scourge of both plants and animals. Almost all plants are susceptible to one or more kinds of bacterial disease. Symptoms vary, but bacterial diseases are commonly manifested as spots of various sizes on the stems, flowers, or fruits, or by wilting or local rotting. A familiar bacterial disease is citrus canker, which broke out in Florida in August 1984 and led to the destruction of over four million citrus seedlings in the first four months. The ongoing effects of this disease, combined with freezes, threatened the future of Florida's $2.5 billion citrus industry.

Many human diseases are also caused by bacteria, including typhoid fever, dysentery, plague, cholera, typhus, tetanus, bacterial pneumonia, whooping cough, and diphtheria. Enormous sums are spent annually to reduce the likelihood of these infections and their devastating effects. In industrialized countries, bacterial infections can usually be controlled with antibiotics, but most people in the world do not have access to such drugs.

Several genera of pathogenic bacteria are particularly dangerous to humans. Members of the genus *Streptococcus* are associated with strep throat, scarlet fever, rheumatic fever, and other infections. Interestingly, the scarlet fever bacterium produces its characteristic and deadly **toxin** (poison) only if it is infected with the appropriate bacterial virus. Tuberculosis, another bacterial disease, is still a leading cause of death in humans (see Sidelight 26.1). These diseases are mainly spread through the air.

Another important airborne microbe is the pathogenic genus *Staphylococcus,* which causes widespread hospital infections. Toxic shock syndrome, caused by certain strains of *S. aureus,* is characterized by fever, lowered blood pressure, vomiting, diarrhea, and a rash in which the skin peels. About 85 percent of the cases of toxic shock syndrome have occurred in menstruating women who are using tampons at the same time the disease occurs, but both men and women can contract the disease. Superabsorbent tampons of a kind no longer manufactured contained fibers that provided an environment that enhanced production of the bacterium's disease-causing toxin.

One of the more recently detected human bacterial diseases is legionellosis (Legionnaire's disease), which is believed to affect about 125,000 people in the United States annually. The disease develops into a severe form of pneumonia that proves fatal to 15 to 20 percent of its victims if left untreated. Discovered in 1976, when a mysterious lung ailment proved fatal to thirty-five American Legion members attending a convention in Philadelphia, legionellosis is caused by small, flagellated, rod-shaped bacteria with pointed ends that have been given the name *Legionella*. These bacteria are common in water, preferring warm water at about 40 to 50 degrees Celsius. They can be spread through air-conditioning units that are not cleaned regularly enough. In the human body, they destroy the monocytes, a type of white blood cell that normally plays a major defensive role against most microorganisms. The bacteria are gram-negative and can be destroyed by treatment with the antibiotic erythromycin.

A number of important bacterial diseases are sexually transmitted. Among the most common are gonorrhea, caused by the bacterium *Neisseria gonorrhoeae*, and syphilis, caused by *Treponema pallidum*, a corkscrew-shaped bacterium of the group called the spirochaetes. As discussed in chapter 4, some scientists believe that, following their incorporation into eukaryotic cells, spirochaetes gave rise to centrioles. At any rate, both gonorrhea and syphilis were easily controlled until recently by such antibiotics as penicillin. But the appearance of penicillin-resistant strains of gonorrhea has made treatment of this disease much more difficult.

Gonorrhea is much more common and less serious than syphilis, which can be fatal. Gonorrhea infected about five hundred people per one hundred thousand in the United States each year during the 1980s, syphilis fewer than twenty in one hundred thousand. Men can detect gonorrhea easily because of a discharge of pus from the penis and burning sensations during urination. In women, gonorrhea is more difficult to detect, producing, at most, very mild symptoms. If untreated, however, it can lead to infection and inflammation of the oviducts, which can cause their blockage and lead to sterility. Within three weeks of infection, syphilis generally produces a hard, painful ulcer called a **chancre** that soon disappears. A generalized skin rash appears two to four months later, after which the disease may become inactive or may ultimately damage the nervous system or circulatory system. With syphilis, as with gonorrhea, antibiotic resistance has recently become a serious problem.

More common than either syphilis or gonorrhea are **chlamydial infections** caused by the bacterium *Chlamydia trachomatis*. These infections, which are usually relatively mild, are controllable with the antibiotic tetracycline, but if left untreated, can cause serious complications.

A particularly troublesome bacterial disease that is now spreading throughout the United States is Lyme disease, caused by the spirochaete *Borrelia burgdorferi*. The disease, named for the village of Old Lyme, Connecticut, where it was first recognized, is an inflammatory ailment of humans that is spread by ticks. The symptoms of Lyme disease somewhat resemble those of arthritis. The juvenile stages of the tick *Ixodes*

Sidelight 26.1
THE RISE OF TUBERCULOSIS

In 1992, the Centers for Disease Control reported an alarming statistic: The number of tuberculosis cases nationwide had increased 18 percent since 1980. The frightening aspect of this statistic is that, unlike AIDS, tuberculosis is spread by casual contact. The bacterium that causes tuberculosis, *Mycobacterium tuberculosis,* is present in the lungs of affected individuals and is transmitted through the air by a cough or a sneeze (Sidelight figure 26.1). Standing close, hugging, and shaking hands are all nonrisk behaviors for contracting AIDS, but are high-risk in the case of tuberculosis.

The cause of this rise in tuberculosis is complex and involves several social, medical, and even genetic factors. The medical establishment in the 1940s thought that it had vanquished this longtime human scourge with the discovery of antibiotics that attacked and killed the tuberculosis-causing bacteria. Since that time, tuberculosis research has been relegated to the back burner. *Mycobacterium tuberculosis* bacteria are difficult and dangerous to work with due to their ease of transmission, and few pharmaceutical companies have been willing to pour their research dollars into designing new antibiotics for a bacterium that seemed to be on the wane. But in the 1980s, the AIDS epidemic provided the bacteria with a new niche to plunder—the immune-compromised bodies of those suffering from AIDS. Because the bacteria is so easily transmitted, the numbers of tuberculosis cases began to rise, particularly among the homeless and other individuals who live on the fringe of society, out of reach of the medical establishment. A new epidemic was born.

Compounding the problem is the ability of the *M. tuberculosis* bacteria to spontaneously mutate in response to the antibiotic treatment used to eradicate the disease. This is allowing new strains of these bacteria to elude destruction by antibiotics, such as Isohiazia, rifampin, and streptomycin. To combat this tendency, doctors usually prescribe two or three different antibiotics to tuberculosis patients. The theory is that one of the antibiotics will kill most of the bacteria, and the other two will take care of those bacteria that have mutated into a different resistant strain in response to the first antibiotic.

While the theory makes logical sense, in practice, it sometimes fails. To kill all of the bacteria, the tuberculosis patient must take these antibiotics for six months. Usually, the symptoms (fever, cough, and general malaise) disappear after three weeks, and some people simply stop taking their antibiotics when the symptoms abate. Other patients, such as the homeless, cannot maintain the treatment in the face of their daily struggles to find food or shelter. Meanwhile, these half-treated patients still harbor live *M. tuberculosis* bacteria that have been exposed to low levels of the antibiotic. In these individuals, there is strong selection favoring antibiotic-resistant mutant strains of the bacterium. Not surprisingly, strains of *M. tuberculosis* highly resistant to many antibiotics are now appearing and are able to transmit the disease.

What can be done to alleviate this growing threat? Experts contend that research into the *M. tuberculosis* bacterium must focus on its mutational activity and that antibiotic research should focus on the development of drugs that target highly conserved parts of the cell that are not readily altered by mutation and that are common to all strains of the bacteria.

In a general sense, however, a program needs to be instituted that will lessen the number of mutations seen in *all* bacteria, not just those that cause tuberculosis. Many scientists think that overprescription of antibiotics has caused the rise in resistant bacterial strains. Some experts think that half of all antibiotics prescribed in the United States are unnecessary and that some physicians simply prescribe an antibiotic to pacify their patients with a sore throat or runny nose. Antibiotics do nothing against viral infections, and prescribing antibiotics in these cases is not only medically useless but can contribute to the development of resistant strains of bacteria.

The next time you are prescribed an antibiotic, make sure that its use is warranted—that you have a true bacterial infection. Otherwise, you might be risking both your money and the health of others.

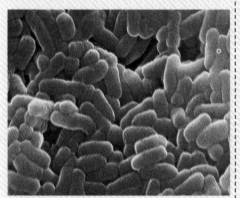

Sidelight figure 26.1 Mycobacterium tuberculosis.

False-color scanning electron micrograph of Mycobacterium tuberculosis, *the rodlike bacteria that is the causative agent of tuberculosis in humans.*

dammini, which is the primary vector of the disease, attach themselves to white-footed mice and occasionally to humans. For many victims, the first sign of infection is a rash that resembles a bull's-eye, which may appear up to a month after the tick's bite. At the end of the summer, when the ticks are adults, they move to different mammals, especially white-tailed deer (in the eastern United States), and mate on the host. The ticks also infect birds and are presumably spread over wide distances in this way. Fortunately, Lyme disease, if properly diagnosed, can be treated effectively with penicillin or tetracycline.

One human bacterial disease—dental caries, or cavities—affects almost everyone. The disease arises in the film on our teeth, which is called **dental plaque.** This film consists largely of bacterial cells surrounded by a polysaccharide layer. Most of the bacteria are filamentous cells, classified as *Leptotrichia buccalis,* which extend out perpendicular to the surface of the tooth, but many other bacterial species are also present. Tooth decay is caused by the bacteria present in the plaque, which persists especially in places that are difficult to reach with a toothbrush (figure 26.6). Diets high in sugars are especially harmful to the teeth because lactic-acid bacteria (especially *Streptococcus sanguis* and *S. mutans*) ferment the sugars to lactic acid, a substance that causes the local loss of calcium from the teeth. Once the hard tissue has started to break down, the breakdown of proteins in the tooth enamel starts, and tooth decay begins in earnest. Fluoride makes teeth more resistant to decay because it retards the loss of calcium. Because tooth decay is an infectious disease caused by bacteria, it can, in principle, be controlled by antibiotics. A number of studies regarding this are now underway.

> *Bacteria are important disease-causing organisms in plants and animals, including humans. Among the many human diseases that bacteria cause are scarlet fever, rheumatic fever, tuberculosis, typhoid, and legionellosis.*

(a)

Figure 26.6 Tooth decay is a bacterial disease.
(a) *Bacteria grow readily in the human mouth, and their waste products can cause caries (tooth decay) and gum disease.*

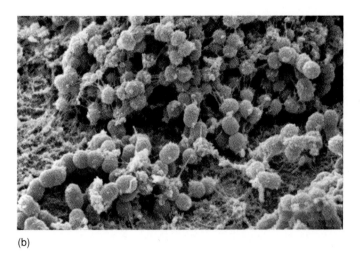

(b)

(b) *A colony of bacteria growing on a tooth is called plaque. This plaque of* Streptococcus mutans *is magnified 31,000 times.*

Viruses

The simplest organisms living on the earth today are bacteria, and scientists believe that bacteria closely resemble the first living organisms to evolve. Even simpler than the bacteria, however, are the viruses. The earliest indirect observations of viruses, other than simple observations of their effects, were made near the end of the nineteenth century. At that time, several groups of European scientists, working independently, concluded that the infectious agents associated with a plant disease known as tobacco mosaic and those associated with hoof-and-mouth disease in cattle were not bacteria. They reached this conclusion because these infectious agents were not filtered out of solutions by the kinds of fine-pored porcelain filters routinely used to remove bacteria from various mixtures. Upon investigating the properties of the filtered material, the scientists found that, not only were viruses much smaller than any known bacteria, they also could reproduce themselves only within living host cells and therefore lacked some of the critical machinery by which living cells are able to reproduce themselves.

The Structure of Viruses

For many years after their discovery, viruses were regarded as very primitive forms of life, perhaps the ancestors of bacteria. This view is now known to be incorrect. The true nature of viruses first became evident in 1933, when Wendell Stanley, then of the Rockefeller Institute, prepared an extract of tobacco mosaic virus and purified it. Surprisingly, the purified virus precipitated (separated out of solution) in the form of crystals (figure 26.7a). Stanley was able to show by this method that viruses are more accurately described as chemical matter than as living organisms, at least in any normal sense of the word *living*. The purified crystals still retained the ability to infect healthy tobacco plants and so clearly *were* the virus itself, not merely a chemical derived from it. With Stanley's experiments, scientists began to understand the nature of viruses for the first time.

Within a few years, other scientists were able to follow up on Stanley's discovery and demonstrate that tobacco mosaic virus consists simply of an RNA molecule surrounded by a coat of protein molecules (figure 26.7b). Many plant viruses have a similar composition, but most other viruses have DNA in place of RNA. Nearly all viruses form a protein sheath, or **capsid,** around their nucleic acid core. In addition, many viruses have an **envelope,** rich in proteins, lipids, and glycoprotein molecules, around the capsid (figure 26.8).

Most viruses have an overall structure that is usually **helical** or **isometrical.** Helical viruses, such as the tobacco mosaic virus, have a rodlike or threadlike appearance. Isometric ones have a roughly spherical shape.

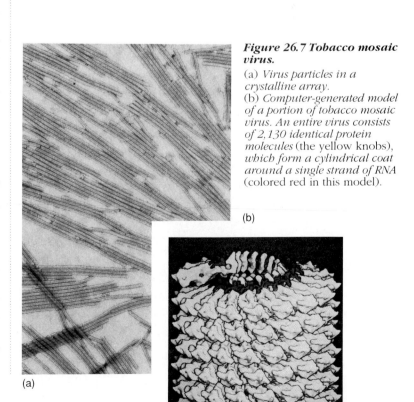

Figure 26.7 Tobacco mosaic virus.

(a) *Virus particles in a crystalline array.*
(b) *Computer-generated model of a portion of tobacco mosaic virus. An entire virus consists of 2,130 identical protein molecules* (the yellow knobs), *which form a cylindrical coat around a single strand of RNA* (colored red in this model).

(b)

(a)

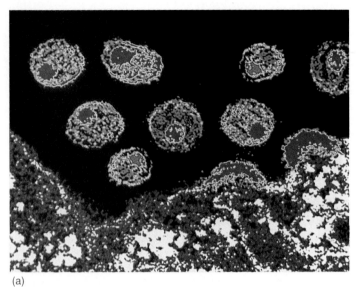

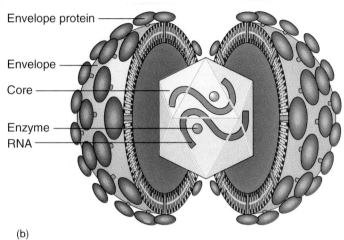

Envelope protein

Envelope

Core

Enzyme
RNA

(a)   (b)

Figure 26.8 Structure of a typical plant or animal virus.
(a) *HIV virus particles budding from a lymphocyte.* (b) *Structure of the AIDS virus.*

Another type of virus, called a **bacteriophage,** has a different type of structure. Bacteriophages are discussed in chapter 12, with regard to Hershey and Chase's experiment that led to the conclusion that the genetic material was DNA, not protein. Bacteriophages exclusively infect bacteria. Like other viruses, bacteriophages contain nucleic acid surrounded by a capsule of protein. However, bacteriophages have "legs," called tail fibers, that allow them to dock with a bacterial cell. The tail fibers are attached to a tail. Bacteriophages are much more complex than viruses that infect plants or animals; the different components of the bacteriophage—the tail fibers and the tail—are composed of different types of protein (figure 26.9).

A third type of infectious agent that resembles viruses but is not actually a virus is a **viroid.** Viroids are circular molecules of RNA that contain from 250 to 400 nucleotides. Viroids lack capsids, and no proteins are associated with them. Viroids are very small—about one-tenth the size of the next smallest known agents of infectious disease. Currently, research indicates that all viroids infect plants. Various viroids are associated with a dozen plant diseases. One of them is cadang-cadang, a disease of coconuts that has caused millions of dollars of damage in the Philippines.

> *Most viruses form a protein sheath, or capsid, around their nucleic acid core and a lipid-rich protein envelope around the capsid.*

How Do Viruses Replicate?

Viruses do not have the ability to grow or replicate on their own. Viruses reproduce only when they enter cells and utilize the cellular machinery of their hosts. They are able to reproduce in this way because they carry genes that are translated into proteins by the cell's genetic machinery, leading to the production of more viruses. Outside of its host cell, a virus is simply a fragment of nucleic acid, usually encased in protein. The fact that individual kinds of viruses contain only a single type of nucleic acid, either DNA or RNA, is one of the major reasons why they can reproduce only within living cells. All true organisms contain both DNA and RNA, and both are essential components of their genetic machinery. Viruses also lack ribosomes and all of the enzymes necessary for protein synthesis and energy production.

In light of their simple chemical nature and total inability to exist independently from other organisms, earlier theories that viruses represent a kind of halfway stage between life and

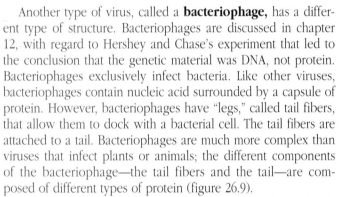

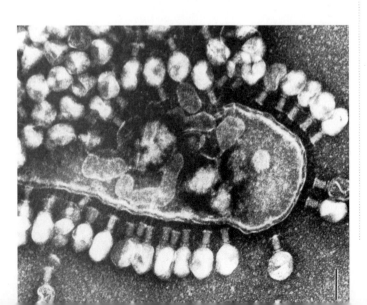

Figure 26.9 Bacteriophages surrounding a bacterium.
Bacteriophages use their tail fibers to dock with bacterial cells and then inject their genetic material inside the bacterial cell, where it takes over the bacterial cell's genetic machinery to make more bacteriophage.

nonlife have now largely been abandoned. Instead, viruses are now viewed as fragments of the genomes of other organisms; they could not have existed independently of preexisting organisms.

A computer's operation is directed by a set of instructions in a program, just as a cell is directed by DNA-encoded instructions. A new program can be introduced into the computer that will cause the computer to cease what it is doing and instead devote all of its energies to making copies of the introduced program. But the new program is not itself a computer and cannot make copies of itself when outside the computer, lying on the desk. The introduced program, like a virus, is simply a set of instructions.

> *Viruses are fragments of DNA or RNA that have become detached from the genomes of bacteria or eukaryotes and have the ability to replicate themselves within cells. Viruses are nonliving and are not organisms. Their genetic material consists of DNA or RNA, but not both.*

The Diversity of Viruses

Viruses occur in virtually every kind of organism that has been investigated for their presence. Viruses are almost always highly specific in the hosts they infect and do not reproduce anywhere else. In light of this observation, there should be nearly as many kinds of viruses as there are kinds of organisms—perhaps millions of them. Because a given organism often has more than one kind of virus, the actual number of kinds of viruses might even be much greater.

Viruses vary greatly in appearance. The smallest viruses are only about 17 nanometers in diameter, the largest ones up to 1,000 nanometers (1 micrometer) in their greatest dimension (figure 26.10). The largest viruses, therefore, are just visible with a light microscope. Most viruses can be detected only by using the higher resolution of an electron microscope. Viruses are directly comparable with molecules in size, a hydrogen atom being about 0.1 nanometer in diameter and a large protein molecule being several hundred nanometers in its greatest dimension.

Viruses and Disease

Depending on which genes a virus genomic fragment carries, it can often seriously disrupt the normal functioning of the cells that it infects. For thousands of years, diseases caused by viruses have been known and feared. Among them are smallpox (see Sidelight 26.2), chicken pox, measles, German measles (rubella), mumps, influenza, colds, infectious hepatitis, yellow fever, polio, rabies, and AIDS, as well as many other diseases not as well known. One series of viral diseases that has been discussed a great deal in recent years is herpes, which includes one category of viruses (herpesvirus 1) associated with cold sores and fever blisters, another (herpesvirus 2) that causes venereal disease, a third (herpesvirus 3) that is responsible for chicken pox and sometimes shingles, and several others.

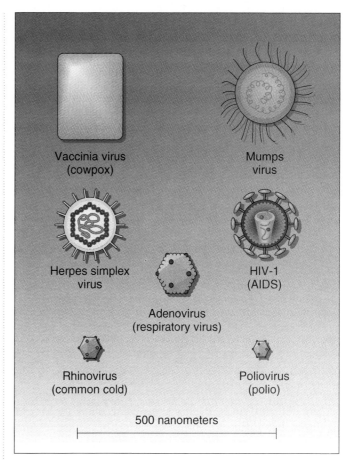

Figure 26.10 Virus sizes.
At the scale shown here, a human hair would be 8 meters thick. Thus, as you can see, viruses are extremely small particles.

A Viral Case Study: HIV

A new and particularly vicious viral disease, **acquired immunodeficiency syndrome (AIDS),** was first reported in the United States in 1981. AIDS is caused by a virus called **human immunodeficiency virus (HIV).** Affected individuals have no resistance to infection, and all of them eventually die of diseases that noninfected individuals easily ward off. Few who contract AIDS survive more than a few years.

AIDS is infectious and thus very dangerous. The risk of AIDS transmission from an infected individual to a healthy one in the course of day-to-day contact is essentially nonexistent. The transfer of body fluids, such as blood, semen, or vaginal fluid, or the use of nonsterile needles, between infected and healthy individuals, however, poses a severe risk.

As a result, the incidence of AIDS is growing very rapidly in the United States. Over one million people were estimated to have been infected by HIV by the late 1980s. Many—perhaps all of them—will eventually come down with AIDS. AIDS incidence is already very high in many African countries. Implications of the AIDS epidemic for college students are discussed in chapter 43.

Clinical symptoms typically do not begin to develop until after a long latency period—generally eight to ten years—after the initial infection with HIV. During this long interval, carriers of HIV have no clinical symptoms but are apparently fully infectious, which makes the spread of AIDS very difficult to control.

502

Sidelight 26.2

THE ERADICATION OF SMALLPOX

One of the greatest triumphs of modern medicine has been the eradication of smallpox everywhere in the world. This dread disease killed millions of people, and when first introduced to new populations, as in the Spanish conquest of Mexico, it sometimes eliminated half or more of the total population. Humans were the only identified hosts of the smallpox virus. Thus, if all susceptible people could be inoculated, the disease could be eradicated.

Officials of the World Health Organization, established in 1948 as a specialized agency of the United Nations, noted that smallpox had already been eliminated through vaccinations in North America and Europe. By 1959, the disease had been eliminated throughout the Western Hemisphere, except for five South American countries, and an intensive worldwide campaign was initiated. As late as 1967, smallpox was still endemic in thirty-three

countries, and the campaign appeared to be faltering. Countries throughout the world, including the Soviet Union and the United States, manufactured and donated large quantities of vaccine, and improved methods of vaccination were developed. Although reporting was poor, probably ten to fifteen million cases of smallpox occurred worldwide in 1967.

Attention was then focused on areas where cases had actually occurred. The last case of smallpox on the Indian subcontinent was contracted by a three-year-old girl on 16 October 1975. By 1977, Somalia was the last country of the world in which the scourge of smallpox persisted. Ali Maow Maalin, a twenty-three-year-old resident of Merka, Somalia, contracted the last known case of smallpox reported anywhere in the world in 1977 (Sidelight figure 26.2).

Because the smallpox virus requires humans to spread, its total absence anywhere in the world since 1977 ensures that it is extinct—except in government research laboratories in the United States and Russia. The destruction of these last samples of the virus is being actively discussed as this edition is published.

Sidelight figure 26.2 The last smallpox victim.

Ali Maow Maalin of Somalia was the last known individual in the world to have smallpox.

Many different kinds of efforts are being made to combat AIDS. Sexual abstinence by infected individuals, the use of condoms, and strictly hygienic procedures in the use of needles and in certain medical practices are all important.

How HIV Compromises the Immune System

In normal individuals, an army of specialized cells patrols the bloodstream, attacking and destroying any invading bacteria or viruses. But in AIDS patients, this army of defenders is vanquished. One special kind of white blood cell, called a T4 cell, which is discussed further in chapter 41, is required to rouse the defending cells to action, but in AIDS patients, T4 cells are few or nonexistent. Without these crucial immune system cells, the body cannot mount a defense against invading bacteria or viruses. AIDS patients inevitably die of infections that a healthy individual can easily stave off.

In the early 1980s, biologists all over the world began working to determine the cause of AIDS. It was not long before the infectious agent, HIV, was identified by laboratories in France and the United States. Study of HIV revealed it to be closely related to an African vervet, or green monkey, virus, suggesting that it might have been introduced to humans in Central Africa by a monkey bite. The virus homes in on T4 cells, infecting and killing them until none are left (figure 26.11).

The HIV Infection Cycle

Viruses infect bacteria, animals, plants, and probably all other organisms. The way in which HIV infects humans provides an example of how viral infections proceed. Most other viral infections follow a similar course, although the details of entry and replication differ in individual cases. The HIV infection cycle is outlined in figure 26.12.

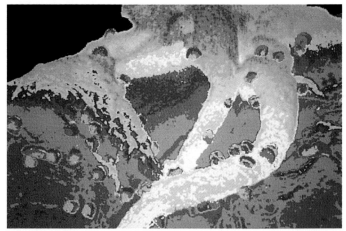

Figure 26.11 HIV attacking T4 lymphocytes.

On its outer surface, HIV virus possesses a glycoprotein called gp160 that is able to recognize and bind to the CD4 cell-surface receptor proteins present on T4 lymphocytes, triggering entry of the virus into the cell.

Attachment When HIV is introduced into the human bloodstream, the virus particle circulates throughout the entire body but only infects T4 cells that it encounters. Most other animal viruses are similarly narrow in their requirements. Polio goes only to certain spinal nerve cells, hepatitis to the liver, and rabies to the brain.

How does a virus such as AIDS recognize a specific kind of target cell, such as a T4 cell? Recall from chapter 5 that every kind of cell in the human body has a specific array of cell-surface receptors, which are designed to bind particular hormones

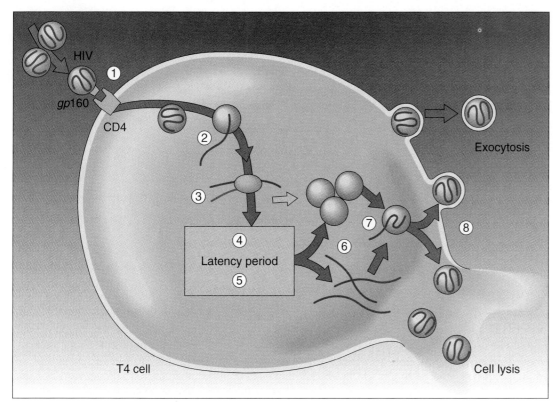

Figure 26.12 How HIV infects T4 cells.
(1) *The gp160 glycoprotein of HIV attaches to the CD4 receptor protein on the surface of a T4 lymphocyte, initiating endocytosis. (2) The viral RNA is released into the cell's cytoplasm. (3) A DNA copy is made of the virus RNA. (4) The HIV DNA undergoes a long latency period. It is not clear whether the HIV DNA becomes incorporated into the host chromosome or not. (5) After a period of some eight to ten years, the virus initiates active replication. (6) Both HIV RNA and HIV proteins are made. (7) Complete HIV particles are assembled. (8) Some T4 cells lyse, releasing free HIV, whereas other HIV particles exit the T4 cell by exocytosis.*

Replication Once within the host cell, the HIV particle sheds its protective coat. This leaves a single strand of virus RNA floating in the cytoplasm, along with a virus enzyme that was also within the virus shell. This enzyme, called **reverse transcriptase,** synthesizes a double strand of DNA complementary to the virus RNA. This double-stranded DNA then inserts itself into the chromosomes of the host T4 cell, where one of two things then happens: Either the copy remains quiet ("latent"), its genes not transcribed into virus proteins, or the copy becomes active, taking over part of the host cell machinery and directing it to produce many copies of the virus. In the latter case, the cell eventually dies, releasing thousands of new virus particles, which infect other T4 cells.

Not all viruses contain RNA and the enzyme reverse transcriptase. In fact, HIV is a member of a group of relatively rare viruses called **retroviruses.** Retroviruses are so named because they contain reverse transcriptase. Other viruses contain RNA but no reverse transcriptase; they are able to replicate because the RNA acts directly as mRNA once inside the host cell, attaching to the host's ribosomes. The common cold virus is one such virus. Other viruses contain DNA; these viruses include the herpesvirus, which causes cold sores and genital herpes, and the papillomaviruses, which cause warts in animals, including humans.

> *Some viruses have the ability to incorporate themselves into the chromosomes of their hosts. In this form, they are said to be latent if they do not begin replicating themselves immediately.*

The conditions that determine when the latent HIV will become active are not well understood. Infections by other microbes seem to trigger AIDS outbreaks, perhaps because the infected T4 cells are involved in the immune response. Many latent human viruses are known to be activated by external stimuli, such as ultraviolet radiation. This is precisely what happens when a fever blister develops on your lip because of the activation of a latent herpesvirus, or when a cell bearing a latent retrovirus is suddenly converted into a rapidly dividing cancerous cell by a carcinogen, a process described in chapter 13.

and growth factors. Cells also possess one or more kinds of cell-surface markers that they use to identify themselves to other, similar cells. These cell-surface markers are glycoproteins. HIV recognizes T4 cells because each HIV particle possesses a glycoprotein (called *gp160*) on its surface that precisely fits a protein called CD4 on the T4 cell surface. Other viruses possess other glycoproteins that key in on other cell types. Because most human cells lack the CD4 surface marker, they are immune to HIV infection. But when HIV encounters T4 cells, it is able to attach to the T4 cell surfaces, as these cells contain the proper CD4 marker on their surfaces.

Entry After docking with a T4 cell, HIV penetrates the cell membrane. Like many other animal viruses, HIV enters the cell by endocytosis, with the cell membrane folding inward to form a deep cavity around the virus particle. Plant viruses normally enter the cells of their host at points of injury, whereas bacteriophages shed their coats outside their host cell, injecting their nucleic acids through the host cell wall and making new coats as they reproduce themselves within.

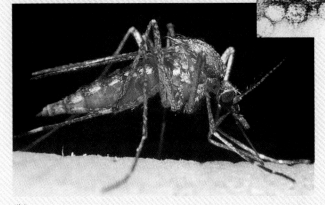

Figure 26.13 Viral diseases often linked to animal carriers.
(a) *The virus that causes yellow fever is found in monkeys.*
(b) *The* Aedes *mosquito spread yellow fever from monkeys to human beings. The building of the Panama Canal became possible only after the mosquito was eradicated.*

(a)

(b)

One of the most important viruses in human history has been the yellow fever virus (figure 26.13). This virus is one of the arboviruses, viruses spread by arthropods—*Aedes* mosquitoes in the case of yellow fever.

In their modes of dispersal, as in their manner of infection, viruses are amazingly versatile. As indicated by the recent history of HIV, and by scientists' increasing understanding of the role of viruses and viruslike particles in causing cancer, many new viral diseases will appear among the increasingly crowded human populations of the future. The very best efforts of biological scientists will be necessary to deal with them effectively.

HIV, the virus that causes AIDS, is an RNA virus that replicates inside human cells by first making a DNA copy of itself. It is only able to gain entrance to a few types of cells—those possessing a particular cell-surface marker recognized by the glycoproteins on its own surface.

Bacteria and Viruses: Simple But Versatile

For at least two billion years—more than half of the history of life on the earth—bacteria were the only organisms in existence. They have survived to the present in rich diversity by exploiting an amazingly diverse set of habitats, some of them unchanged since the beginnings of evolution. Ecologically, their metabolic activities are of fundamental importance. In a broader sense, bacteria also contribute directly to the functioning of all but a very few eukaryotes because they are the ancestors of mitochondria. Given that all chloroplasts originated as symbiotic bacteria, it may be said that even photosynthesis, like chemosynthesis and nitrogen fixation, is exclusively a property of bacteria.

Although bacteria are clearly very simple living organisms, viruses are best thought of as segments of genomes. Scientists have learned a great deal about virus structure, but much remains to be learned, especially since the nature of viruses suggests that new forms are evolving constantly. Unknown particles whose existence is yet to be even suspected are likely discoveries in the future.

EVOLUTIONARY VIEWPOINT

The concept of evolution often conjures up images of dinosaurs and fossils, but in fact, evolution goes on all around us every day, and in no group more extensively than the bacteria. Bacteria have populated the earth for 3.5 billion years, and if success is measured by numbers, bacteria are easily the most successful of all life-forms. More bacteria exist in a spoonful of dirt than there are vertebrates on the earth. Because they reproduce so quickly, often producing a new generation in less than an hour, bacteria can adapt to changing circumstances more quickly than any other kind of organism.

SUMMARY

1. Bacteria are the only organisms with prokaryotic cellular organization. They are the oldest and simplest organisms but are metabolically much more diverse than all of the other forms of life on earth combined.

2. Most bacteria have cell walls consisting of a network of polysaccharide molecules connected by short, peptide cross-links. In gram-negative bacteria, a full membrane, which contains lipopolysaccharide, is deposited over this layer. Bacteria reproduce by binary fission.

3. Bacteria are rod-shaped (bacilli), spherical (cocci), or spirally coiled (spirilla). Bacilli or cocci may adhere in small groups or chains.

4. Bacteria are divided into two kingdoms: kingdom Archaebacteria and kingdom Eubacteria. Archaebacteria use energy derived from chemical sources to power their metabolic activities. Three groups are: Methanogens produce methane as a result of their metabolism; thermoacidophiles inhabit hot springs and use sulfur compounds as an energy source; and halophiles live in extremely salty environments and use the pumping of hydrogen ions out of their cells to power ATP production.

5. Some eubacteria are photosynthetic and are called cyanobacteria. Cyanobacteria are responsible for the increase in free oxygen in the atmosphere that was crucial to eukaryotic evolution.

6. Some cyanobacteria and heterotrophic bacteria have the ability to fix nitrogen. Some of these bacteria live in nodules on the roots of plants, whereas others are free-living, found in the soil or in freshwater aquatic or marine environments.

7. Heterotrophic bacteria include many groups of pathogenic (disease-causing) bacteria. Bacterial diseases include typhoid fever, strep throat, toxic shock syndrome, Legionnaire's disease, tuberculosis, a number of sexually transmitted diseases, and Lyme disease, among many others.

8. Viruses are fragments of bacterial or eukaryotic genomes that are able to replicate within cells by using the genetic machinery of those cells. They are not alive and are not organisms.

9. Viruses consist of a nucleic acid surrounded by a protein coat or capsid. Viruses that infect animals and plants are usually either helical or isometrical (spherical). Bacteriophages are viruses that infect bacteria. They typically have tail fibers and tails.

10. The cells to which a virus will attach are determined by the proteins that make up the coats and envelopes of the virus.

11. AIDS, or acquired immunodeficiency syndrome, is caused by human immunodeficiency virus (HIV). HIV attacks the immune system and is transferred between individuals through bodily fluids, such as blood or semen.

REVIEWING THE CHAPTER

1. What are the structural differences between gram-positive and gram-negative cell walls in bacteria?

2. What are bacilli, cocci, and spirilla bacteria?

3. What is the justification for two kingdoms of prokaryotes?

4. What are three types of archaebacteria, and how do they differ metabolically?

5. How do cyanobacteria differ from other prokaryotes?

6. What role do nitrogen-fixing bacteria and cyanobacteria play in ecosystems?

7. What are the names of at least three bacterial diseases that you could contract?

8. What are the differences between plant and animal viruses, bacteriophages, and viroids?

9. Why are viruses not considered to be living?

10. What is the infectious agent that causes AIDS, and how do humans acquire the AIDS virus?

11. What are T4 cells, and what do they have to do with AIDS?

12. Do all viruses replicate in the same way?

COMPLETING YOUR UNDERSTANDING

1. How old are the oldest bacteria, and what is the evidence for this?

2. How do human diseases relate to gram-negative and gram-positive bacteria and antibiotics?

3. How does a bacterial cell divide, and how does this differ from cell divisions in the human body?

4. How do chemoautotrophic bacteria acquire energy for growth?

5. What are some of the beneficial implications of bacteria and cyanobacteria in agriculture?

6. When was Legionnaire's disease first reported, and what were the circumstances that brought it to national attention?

7. Why is it worthless to take antibiotics for the common cold?

8. What is the most common sexually transmitted disease?

9. What explanation can you give for the increase of Lyme disease and tuberculosis? How are these diseases transmitted, and what treatments are currently available?

10. What is dental plaque, and how can it be minimized?

11. The true nature of viruses first became apparent in 1933 from an examination of
    a. elephants.
    b. tobacco.
    c. humans.
    d. monkeys.
    e. horses.
    f. none of the above.

12. What are some of the structural differences in viruses?

13. The genetic information in viruses is encoded in a molecule of
    a. RNA.
    b. DNA.
    c. proteins.
    d. either RNA or DNA, but not both.
    e. either proteins or lipids, but not both.
    f. lipids.

14. Which appeared first in evolution—bacteria or viruses? Why?

15. How big is a virus? (Use the metric system.)

16. Of the sexually transmitted diseases—gonorrhea, syphilis, chlamydia, and herpes—which ones can and cannot be treated with antibiotics? Why?

17. If smallpox has been eradicated worldwide, then why do you think the U.S. and Russian governments still keep the virus in restricted laboratories?

18. What does HIV do to the human immune system?

19. Which of the following are *not* involved in AIDS transmission? [*two answers required*]
    a. Semen
    b. Kissing
    c. Nonsterile needles
    d. Vaginal fluids
    e. Insect bites
    f. Blood transfusions

20. What is a retrovirus?

FOR FURTHER READING

Aral, S., and K. Holmes. 1991. Sexually transmitted diseases in the AIDS era. *Scientific American*, Feb., 62–69. Gonorrhea, syphilis, and other infections are still common and need increased attention.

Donoghue, H. 1987. A mouthful of microbial ecology. *New Scientist*, February 5, 61–65. Excellent and well-illustrated account of the communities of bacteria that inhabit human mouths and their effects.

Fischetti, V. 1991. Streptococcal M protein. *Scientific American,* June, 58–65. How the bacteria that cause strep throat and rheumatic fever depend on this cell-surface protein to evade the body's defense.

Habicht, G. S., G. Beck, and J. L. Benach. 1987. Lyme disease. *Scientific American*, July, 78–83. A discussion of the many fascinating properties of this rapidly spreading disease.

Kiester, E. 1990. A curiosity turned into the first silver bullet against death. *Smithsonian*, Nov., 72–76. A very engaging account of the discovery of penicillin.

Mee, C. L., Jr. 1990. How a mysterious disease laid low Europe's masses. *Smithsonian,* Feb., 66–79. The story of bubonic plague, the Black Death that changed the character of Europe in the fourteenth century.

Read, R. 1989. Bacteria with a sticky touch. *New Scientist* 124 (Oct.): 38–41. Article about how a bacterium's first step when it enters the body is often to anchor itself to a cell. Understanding how it does this may make it possible to design new drugs to combat infection.

Resenberg, Z., and A. Fauci. 1990. Inside the AIDS virus. *New Scientist,* February 10, 51–54. A look at the mechanisms by which the AIDS virus destroys the immune system, information of central importance in developing a cure.

Shapiro, J. A. 1988. Bacteria as multicellular organisms. *Scientific American,* June, 82–89. Explores the sophisticated temporal and spatial control systems that govern colonies of bacteria.

Weiss, R. 1992. On the track of "killer" TB. *Science* 255 (January): 148–50. The appearance of drug-resistant strains of the disease in the wake of the AIDS epidemic that has led to vigorous efforts to find weapons against it.

Woese, C. 1981. Archaebacteria. *Scientific American,* June, 98–122. A now-classic presentation of the arguments that the Archaebacteria represent an independent kingdom of organisms.

EXPLORATIONS

Interactive Software: **AIDS**

This interactive exercise allows students to explore the risk factors associated with the AIDS epidemic by varying key aspects of behavior and policy. The exercise presents a world map on which appear charts of the incidence of AIDS in that geographical region by year and projections of the future state of the epidemic. Students can investigate the past history of the epidemic and explore the potential future consequences of altering patterns of sexual behavior and drug use, use of safe sex, and level of public education. The potential for explosive future growth of the epidemic in Asia will become evident as future projections are explored in different world regions.

Questions to Explore:

1. Why do so many more people have HIV than AIDS?

2. Can you think of a way to have HIV and never get AIDS?

3. What would be the effect of a mutation in *gp*160 that allowed it to recognize a mosquito blood cell surface protein as well as CD4?

4. What is the role of public education in the AIDS epidemic?

5. When is the AIDS epidemic going to be over?

6. What do you imagine a successful HIV vaccine might be like?

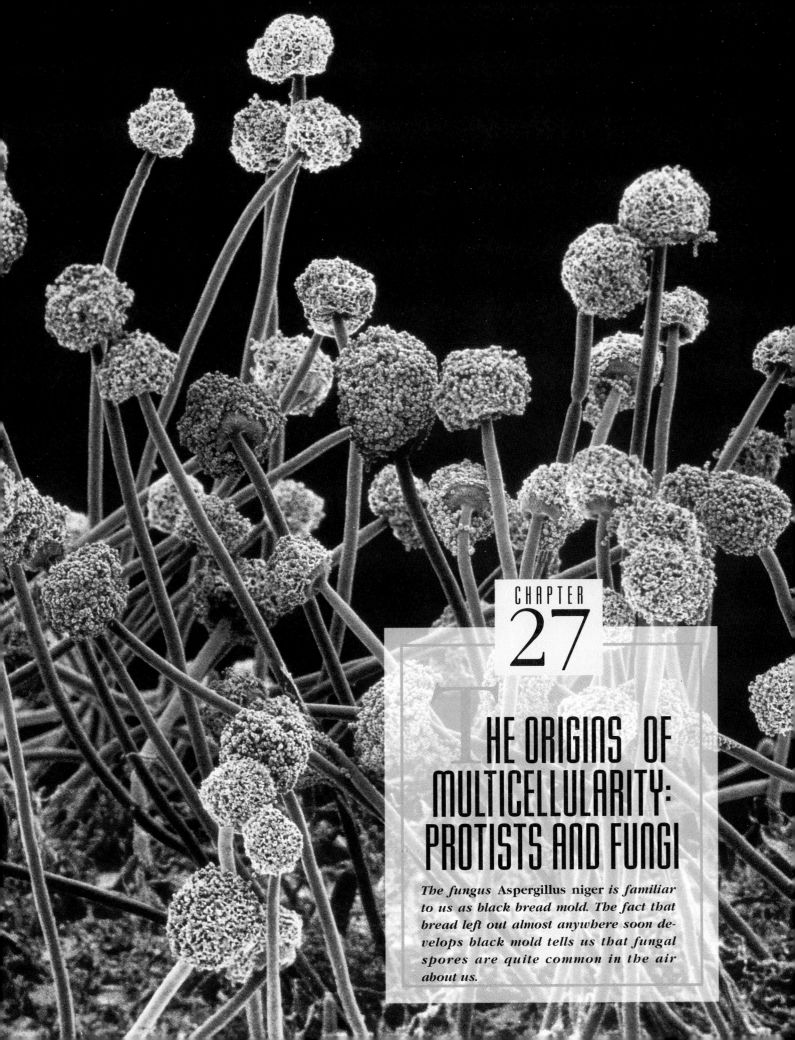

HE ORIGINS OF MULTICELLULARITY: PROTISTS AND FUNGI

The fungus Aspergillus niger *is familiar to us as black bread mold. The fact that bread left out almost anywhere soon develops black mold tells us that fungal spores are quite common in the air about us.*

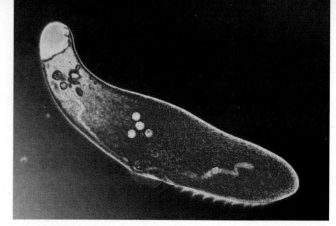

Figure 27.1 Protists.
Although multicellularity has evolved many times among the protists, most protists are microscopic, single-celled organisms. This one is typical in that it has a complex interior organization. Protists are far more diverse than any other kingdom of life.

Most kinds of eukaryotes are protists, members of the very diverse kingdom known as Protista (figure 27.1). Protists are typically unicellular organisms, but two phyla of algae—the red and brown algae—consist almost entirely of multicellular ones. A third phylum—the green algae—has many multicellular representatives. Multicellularity originated in each of these groups independently. Other protists gave rise to three of the six kingdoms recognized as distinct: plants, animals, and fungi (figure 27.2). Plants manufacture their own food, animals ingest it, and fungi absorb it after releasing enzymes into their surroundings. Later chapters discuss the plants and animals; this chapter includes an examination of the fungi. The fungi are a kingdom of diverse organisms that have a great impact on the functioning of the biosphere, decomposing organic materials and returning them to ecosystems.

The Protists: The Most Diverse Kingdom of Eukaryotes

While all six kingdoms of life contain great diversity, no other kingdom approaches the vast array of form and function found among the protists. Among the protists are the simplest eukaryotes, but even single-celled protists have a structure far more complex than that of bacteria (figure 27.3). The phyla considered in this chapter consist mainly of microscopic, unicellular organisms, yet they also include massive, multicellular kelps that may exceed 100 meters in length. In their ways of life, their habitats, and the details of their life cycles, the protists are extraordinarily diverse.

Among the phyla that make up this great group are those that apparently gave rise to the other multicellular evolutionary lines known as animals, plants, and fungi (figure 27.4). The choanoflagellates, one of the groups of zoomastigotes, appear certainly to be the ancestors of the sponges, and according to many biologists, probably of all animals; the fundamental similarity between their unique structure and that of the collar cells of sponges (see chapter 29) is simply too great to be explained in any other way. Green algae, which share with plants the same photosynthetic pigments, cell wall structure, and chief storage product (starch), certainly include the ancestors of plants, a kingdom that achieved its distinctive

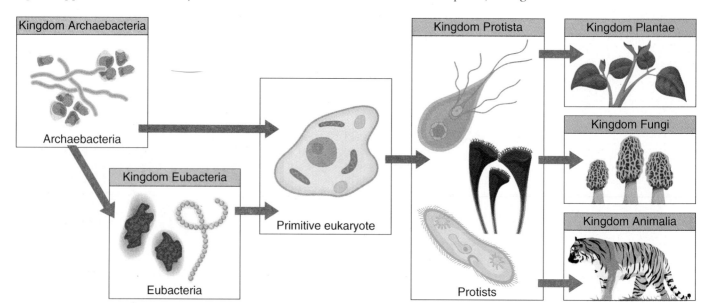

Figure 27.2 The evolutionary radiation of the eukaryotes.
The oldest surviving organisms are the members of kingdom Archaebacteria. From them arose both the Eubacteria and, it is now thought, the primitive eukaryotes. The Eubacteria gave rise to the mitochondria present within eukaryotic cells.

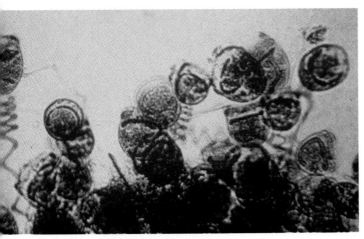

Figure 27.3 Single-celled protist.
Vorticella, *which is heterotrophic, feeds largely on bacteria and has a retractable stalk. This is an amazing degree of complexity for a single cell, going far beyond the complexity found in any bacterium.*

features when it invaded the land. The ancestors of the fungi are not known because their features are distinct from those of all living protists.

Symbiosis and the Origin of Eukaryotes

Unraveling the evolutionary relationships of the protists requires understanding the history of the major symbiotic organelles: mitochondria and chloroplasts (see chapter 4). Because these organelles were derived from different groups of bacteria, their relationship to each other does not necessarily reflect the overall relationships of the cells within which they now occur. Both kinds of relationships must be kept in mind to appreciate the protists properly.

Mitochondria are thought to have originated from bacterial ancestors with characteristics similar to the purple nonsulfur bacteria. Mitochondria are characteristic of all eukaryotes except for several zoomastigotes (discussed later in the chapter) and *Pelomyxa palustris,* an unusual amoeba-like organism found on the muddy bottoms of freshwater ponds. *Pelomyxa,* a unique organism that is placed in its own phylum, lacks mitosis, and its nuclei divide somewhat like those of bacteria. The existence of a few protists that lack mitochondria probably indicates that the initial stages of this group's evolution took place before mitochondria were acquired. But this event must have taken place at about the same time that eukaryotes originated—approximately 1.5 billion years ago.

> *Mitochondria, which probably originated from symbiotic purple nonsulfur bacteria, are absent in two unusual groups of protists. These organisms may represent a stage of evolution before ancestral eukaryotes acquired mitochondria.*

The mitochondria that occur in all but a very few eukaryotes generally possess similar features, which makes it difficult to determine whether these mitochondria all originated from a single symbiotic event, or whether similar bacteria became symbiotic independently in different groups of early eukaryotes.

For chloroplasts, however, the story is very different. Three biochemically distinct classes of chloroplasts exist, each resembling a different bacterial ancestor. Thus, the chloroplasts of red algae (Rhodophyta) possess chlorophyll *a,* carotenoids, and an unusual class of accessory pigments called phycobilins. These chloroplasts were almost certainly derived from symbiotic cyanobacteria. In contrast, the green chloroplasts of plants and

Figure 27.4 Evolutionary relationships of the six kingdoms.

The solid-color arrows indicate evolutionary relationships; the dotted lines *indicate symbiotic events.*

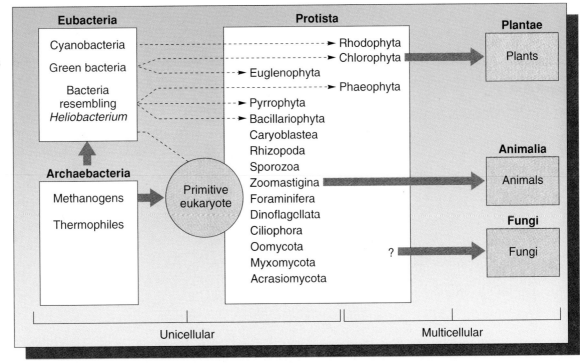

green algae seem to have been derived from bacterial ancestors similar to *Prochloron* (see figure 4.7). Finally, the chloroplasts of several other groups (brown algae, diatoms, and dinoflagellates), which have chlorophylls *a* and *c*, carotenoids, and distinctive, yellowish-brown pigments, probably were derived from a third group of bacteria.

Eukaryotic cells acquired chloroplasts by symbiosis not once, but at least three times during the evolution of different groups of protists.

Major Groups of Protists

Over thirty-five thousand different living species of protists have been described, and no doubt thousands more are yet to be discovered. Another forty thousand species are known from the fossil record. Protists display a bewildering array of shapes. Some are microscopic; others are bigger than a human. Some are photosynthetic; others are efficient hunters; still others are parasites. Some live in the oceans; others live in freshwater; still others live within other organisms. No other kingdom is nearly as diverse as Protista.

Table 27.1   The Protists

Group	Phyla	Typical Examples		Key Characteristics	Approximate Number of Living Species
Sarcodina				No permanent locomotor apparatus; heterotrophs	13,500
Amoebas	Rhizopoda	Amoeba		Move by pseudopodia	
Forams	Foraminifera	Forams		Rigid shells; move by protoplasmic streaming	
Algae				Photosynthetic; multicellular or largely multicellular	12,500
Red	Rhodophyta			Chlorophyll *a* + red pigment	
Brown	Phaeophyta	Kelp		Chlorophyll *a* + chlorophyll *c*	
Green	Chlorophyta	*Chlamydomonas*		Chlorophyll *a* + chlorophyll *b*	
Diatoms	Bacillariophyta	Radiolarians		Double shell of silica; photosynthetic; unicellular	11,500
Flagellates				Locomotor flagellae	9,500
Dinoflagellates	Dinoflagellata	Red tides		Photosynthetic; unicellular; two flagellae	
Zoomastigotes	Zoomastigina	Trypanosomes		Heterotrophic; unicellular	
Euglenoids	Euglenophyta	*Euglena*		Some photosynthetic, others heterotrophic; unicellular	

Given this awesome diversity, biologists have not found it easy to sort the Protista into sensible groupings. Years ago, when only four kingdoms were commonly recognized (monera, plants, animals, and fungi), biologists lumped the photosynthetic protists, called algae, in with the plants, and called the hunting protists "protozoa" or little animals; rusts, mildews, molds, and related protists were called fungi. Plants, animals, and fungi are now defined as strictly multicellular groups that originated from different ancestors among the protists. All other eukaryotic organisms, with their incredible diversity, are retained within the Kingdom Protista.

There are fourteen major phyla of protists. Some biologists choose to group them in the old traditional way (algae, protists, molds), while others group them by how members of the phyla obtain energy (photosynthetic or heterotrophic). In this text, the fourteen major phyla are grouped into seven categories reflecting key shared characteristics (table 27.1).

Group	Phyla	Typical Examples		Key Characteristics	Approximate Number of Living Species
Ciliates	Ciliophora	*Paramecium*		Many cilia; two nuclei per cell; fixed cell shape; heterotrophic; unicellular	8,000
Sporozoans	Sporozoa	*Plasmodium*		Nonmotile; spore-forming; unicellular parasites	3,900
Molds				Heterotrophs with restricted mobility; cell walls made of carbohydrate	1,150
Cellular slime molds	Acrasiomycota			Colonial aggregations of individual cells; most closely related to amoebas	
Plasmodial slime molds	Myxomycota			Stream along as a multinucleate mass of cytoplasm	
Water molds	Oomycota	Rusts and mildew		Terrestrial and freshwater	

Sarcodina: Amoeba-Like Protists

Two major phyla of protists, called the Sarcodina, have no permanent locomotor apparatus—no legs or appendages to help them move around. The amoebas (phylum Rhizopoda) are heterotrophic protists that move from place to place by means of their **pseudopodia,** from the Greek words meaning "false" and "feet" (figure 27.5). Amoebas are abundant throughout the world in freshwater and saltwater, as well as in the soil. Many species are parasites of animals. Amoebas are estimated to infect more than ten million U.S. citizens, about two million of whom show symptoms of amoebic dysentery. Amoebas lack cell walls, flagella, meiosis, and any form of sexuality, but they do carry out mitosis and possess mitochondria. The second phyla without a permanent locomotor apparatus—the forams (phylum Foraminifera)—are a very diverse group of marine heterotrophs with rigid shells that move by protoplasmic streaming.

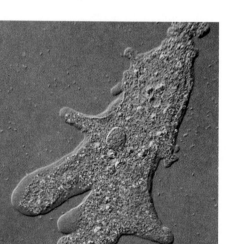

Figure 27.5 An amoeba.
Amoeba proteus *is a relatively large amoeba commonly used in teaching and for research in cell biology. The projections are pseudopodia; an amoeba moves simply by flowing into them. The amoeba's nucleus is plainly visible.*

Algae

Three phyla of photosynthetic protists—Chlorophyta, Rhodophyta, and Phaeophyta—consist largely or entirely of multicellular organisms. Phylum Chlorophyta—the green algae—consists of about seven thousand species. Most are aquatic, but some are semiterrestrial in damp places. They are biochemically similar to the plants, which are believed to have been derived from multicellular green algae. The growth form of the marine genus *Ulva* is platelike (the individuals are two cells thick); that of the freshwater genus *Volvox* is a globular sphere of flagellated cells (see figure 25.14). Other green algae are filamentous, such as *Spirogyra,* or unicellular and sometimes flagellated. Flagellated unicellular organisms, such as *Chlamydomonas* (see Life Cycle 27.1), apparently gave rise to colonial algae such as *Volvox,* in which the individual cells are so integrated with one another that *Volvox* can be considered truly multicellular. *Acetabularia,* which was discussed in Key Experiment 12.1, is a member of this phylum. Individual cells of *Acetabularia* are very large. Many of the green algae have cell walls composed totally or partly of cellulose.

Life Cycle 27.1

CHLAMYDOMONAS

Individual cells of *Chlamydomonas* (a microscopic, biflagellated alga) are haploid and divide asexually, producing identical copies of themselves. At times, such haploid cells act as gametes, fusing to produce a zygote, as shown in the lower right-hand side of the diagram. The zygote develops a thick, resistant wall, becoming a zygospore. Meiosis takes place, ultimately resulting in the release of four haploid individuals. Because of the segregation during meiosis, two of these individuals are what is called the + strain; the other two, the − strain. Only + and − individuals are capable of mating with one another, although both may also divide asexually and reproduce themselves.

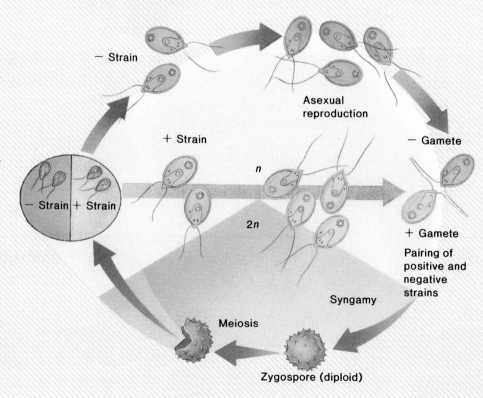

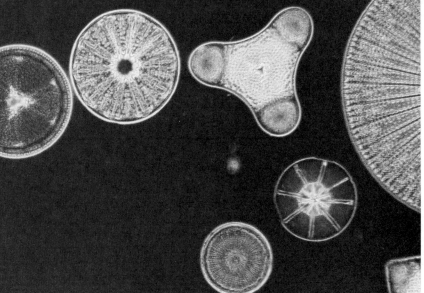

Figure 27.6 Several different kinds of diatoms.
The diverse shapes of the silica shells of some members of this phylum are only hinted at in the forms shown here.

The other two groups of multicellular photosynthetic algae are primarily marine. The red algae (phylum Rhodophyta) comprise about four thousand species whose chloroplasts are similar to those of the cyanobacteria. An unusual feature of the red algae, some of which grow at greater depths in the sea than any other photosynthetic organism, is that they completely lack flagellated cells at any stage in their life cycle. Some members of this phylum, as well as some green and brown algae, are harvested as important food sources in the Orient. Agar, a mucilaginous material extracted from the cell walls of red algae, is used as a culture medium for bacteria and other microorganisms and for a wide variety of commercial applications, including making the capsules that contain vitamins and drugs.

The brown algae (phylum Phaeophyta) are unrelated to either the red or green algae. They have flagellated reproductive cells and biochemically distinct chloroplasts. About 1,500 species of brown algae exist, some of which attain great size (up to 100 meters long) and complexity. The kelps and rockweeds are one large group of brown algae that are often commercially harvested for fertilizer or for alginates, a group of substances widely used as thickening agents and colloid stabilizers in food, textile, cosmetic, and pharmaceutical industries.

Diatoms

Diatoms (phylum Bacillariophyta), of which there are about 11,500 living species and many others known only as fossils, have a very characteristic "shell" made of two parts, like the halves of a box. Chemically, the shell is opaline silica (figure 27.6). Huge masses of diatoms have been deposited in certain locations, such as near Lompoc, California, where the strata are hundreds of meters thick.

Flagellates

Two major and one smaller phyla of protists are characterized by locomotor flagellae. Photosynthetic flagellates are the dinoflagellates (phylum Dinoflagellata). The approximately one thousand species of dinoflagellates are important as the causal agent of "red tides," during which millions of dinoflagellates, secreting toxic substances, color the sea red and poison fish and shellfish by the thousands. Free-living dinoflagellates have two flagella, one lying in a shallow girdle around the center of the cell, which is armed with stiff cellulose plates, and the other directed backward. Other dinoflagellates are symbiotic with marine animals such as coral. Symbiotic dinoflagellates allow a coral colony to produce its own food by photosynthesis and to flourish in nutrient-poor tropical waters. "Bleached" coral—coral from which the dinoflagellates have been lost—is unhealthy and perhaps dying.

Heterotrophic flagellates are called zoomastigotes (phylum Zoomastigina), with cells propelled by the beating of one or more long flagella. Many of the thousands of species of zoomastigotes are parasitic. The serious and widespread diseases trypanosomiasis, or "sleeping sickness" (figure 27.7), East Coast fever, and Chagas' disease, which are especially prevalent in the tropics, are caused by trypanosomes, one of the groups of zoomastigotes. Sleeping sickness has kept cattle out of much of Africa, thus complicating the problems of feeding the people.

Trypanosomes have complicated life cycles. When they are ready to spread to a mammal from the flies that carry them around, the trypanosomes acquire a thick coat of glycoprotein antigens that protects them from the host's antibodies. Trypanosomes can change this coat so quickly that the immune systems of cattle and humans cannot mount a defense against it. A major effort is currently underway to develop vaccines against trypanosomes, and prototype vaccines are being tested in the field.

Other groups of zoomastigotes live in the guts of wood roaches and termites

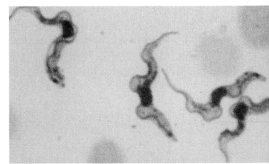

(a)

(b)

Figure 27.7 Sleeping sickness— A protist's disease.

(a) *Trypanosoma,* the protists that cause sleeping sickness, among red blood cells. The nuclei (dark-staining bodies), the anterior (forward-projecting) flagella, and the undulating, changeable shape of the trypanosomes are visible in this photograph. (b) A tsetse fly, shown here sucking blood from a human arm, in Tanzania, East Africa. Tsetse flies transmit the trypanosomes that cause sleeping sickness.

Sidelight 27.1
HIKER'S DIARRHEA

A common complaint of unsuspecting hikers who assume that the water in cold mountain streams is fine for drinking is "hiker's diarrhea." This inconvenient malady is caused by *Giardia lamblia*, a ciliate found throughout the world, including all parts of the United States and Canada (Sidelight figure 27.1). It occurs in water, including the clear water of mountain streams and the water supplies of some cities, infecting at least forty species of wild and domesticated animals in addition to humans. In 1984 in Philadelphia, 175,000 people had to boil their drinking water for several days following the appearance of *Giardia* in the city's water system. Although most individuals exhibit no symptoms if they drink water infested with *Giardia*, many suffer nausea, cramps, bloating, vomiting, and diarrhea. Only thirty-five years ago, *Giardia* was thought to be harmless; today, at least sixteen million U.S. residents alone are infected by it.

Giardia lives in colonies in the upper small intestine of its host, where it occurs in a motile form that cannot survive outside the host's body for long. It is spread in the feces of infected individuals in the form of dormant, football-shaped cysts—sometimes at levels as high as three hundred million cysts per gram of stool. These cysts can survive at least two months, and they last longer in cool water, such as that of mountain streams. They are relatively resistant to the usual water-treatment agents, such as chlorine and iodine, but are killed at temperatures greater than about 65 degrees Celsius. Apparently, infected wild animals, such as beavers, can release *Giardia* cysts that lead to infection in humans. Pollution by humans or domestic animals therefore is not necessary for stream water to be dangerous, although pollution can be an important contributing factor. There are at least three species of *Giardia* and many distinct strains; how many of them attack humans and under what circumstances are not known with certainty.

In the wilderness, good sanitation is important in preventing the spread of *Giardia*. Dogs, which readily contract and spread the disease,

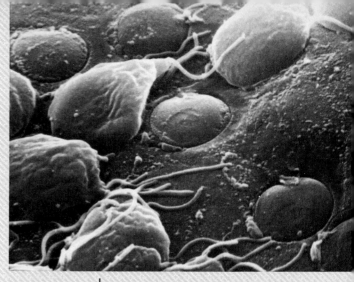

***Sidelight figure 27.1* Giardia lamblia.** *This ciliate, found in water, can induce nausea, cramps, and diarrhea in humans.*

should not be taken into pristine wilderness areas. Drinking water should be filtered—the filter must be capable of eliminating particles as small as 1 micrometer in diameter—or boiled for at least one minute. Water from natural streams or lakes should never be consumed directly, regardless of how clean it looks. In other regions, good sanitation methods are important to prevent not only *Giardia* but also other diseases.

---

and possess cellulase enzymes that allow these insects to digest cellulose and thus live on a diet of wood. The insects are unable to manufacture these enzymes themselves, but the presence of the zoomastigotes, which are always in their guts, enables them to live as if they did have this ability. Members of this phylum also include the choanoflagellates, which have a characteristic collar at one end of the cell, from which a single flagellum protrudes. Choanoflagellates are clearly the ancestors of the sponges and probably the ancestors of all other animals as well.

The third smaller phylum of flagellates, called euglenoids (phylum Euglenophyta), contains some members that are photosynthetic and others that are heterotrophic. Euglenoids are closely related to zoomastigotes, and some biologists include them within them.

Ciliates

The ciliates (phylum Ciliophora) are a large phylum of heterotrophic, unicellular protists, comprising at least eight thousand structurally complex, primarily aquatic species (see Sidelight 27.1). As their name indicates, most ciliates possess large numbers of cilia (fine flagella). Perhaps the best-known member of the phylum is *Paramecium*, which has been the subject of comprehensive genetic and life-history studies, as has its relative *Tetrahymena*. Many ciliates move about by the beating of their cilia, which occur in lines on their cells. *Vorticella* remains anchored in one place, using its cilia to sweep food into its gullet (see figure 27.3). Ciliates are characterized

by an unusual mode of reproduction known as **conjugation.** During conjugation, individuals exchange nuclei through connecting tubes.

Despite their unicellularity, ciliates are extremely complex organisms, inspiring some biologists to consider them organisms without cell boundaries, rather than single cells. Indeed, based on recent comparisons of DNA sequences, molecular taxonomists are beginning to argue that ciliates are so different from all other protists that they should be placed in a kingdom of their own.

Sporozoans

The well-known disease malaria is caused by the sporozoans (phylum Sporozoa), which are nonmotile, spore-forming parasites of animals. Nearly four thousand species are known. Approximately 250 million people are afflicted by malaria at any one time, and two to four million of them die each year. The symptoms include chills, fever, sweating, an enlarged and tender spleen, confusion, and great thirst. Malaria kills most children under five years of age who contract it.

The disease is caused by sporozoans of the genus *Plasmodium*, which are carried by mosquitoes of the genus *Anopheles* (see Life Cycle 27.2). Once the *Plasmodium* organisms reach the bloodstreams of humans or other mammals, they move to the liver, where they begin to divide. They then pass back into the bloodstream and invade the red blood cells, dividing rapidly within them and causing them to become enlarged and, ultimately, to rupture. This releases toxic substances throughout

Life Cycle 27.2

PLASMODIUM

Plasmodium is the sporozoan that causes malaria. The life cycle stages—called sporozoites, merozoites, and gametocytes—each produce different antigens, and they are sensitive to different antibodies. The gene encoding the sporozoite antigen was cloned in 1984, but it is not certain how effective a vaccine against sporozoites might be. When a mosquito inserts its proboscis into a human blood vessel, it injects about a thousand sporozoites. They travel to the liver within a few minutes, where they are no longer exposed to antibodies circulating in the blood. If even one sporozoite reaches the liver, it will multiply rapidly there and still cause malaria. The number of malaria parasites increases roughly eightfold

every twenty-four hours after they enter the host's body. A compound vaccination against sporozoites, merozoites, and gametocytes would probably be the most effective preventive measure, and such a vaccine is now under development. Human trials are under way and promise eventually to control the disease, even though the complexity of the system is great.

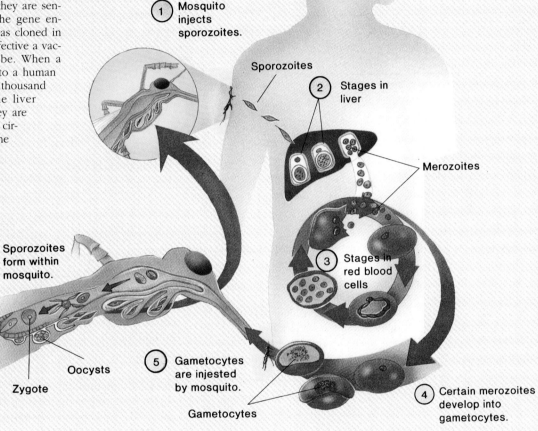

1. Mosquito injects sporozoites.

Sporozoites

2. Stages in liver

Merozoites

3. Stages in red blood cells

4. Certain merozoites develop into gametocytes.

5. Gametocytes are injested by mosquito.

Gametocytes

6. Sporozoites form within mosquito.

Oocysts

Zygote

the body of the host, bringing about the well-known cycle of fever and chills that is characteristic of malaria. Malaria has proven difficult to control both because the mosquitoes have become resistant to insecticides and because the parasites have developed resistance to the chemicals, such as quinine, that are used to kill them.

Molds

Among the unrelated phyla of protists that have historically, and incorrectly, been considered fungi are two that have amoeba-like stages in their life cycles. The plasmodial slime molds (phylum Myxomycota) exist mainly as flowing, multinucleate masses that feed on bacteria and other small bits of organic matter. The more than five hundred species of this phylum differentiate into often ornate spore-forming bodies during reproduction. In contrast, the cellular slime molds

(phylum Acrasiomycota), of which there are only a few dozen species, exist chiefly as single "amoebas" for most of their life cycle but swarm together at certain stages to form resistant spores that are dispersed (see Life Cycle 27.3). Despite the similarity of their names, these two phyla are clearly unrelated to one another. A third group of protists—the water molds (phylum Oomycota)—were also regarded as fungi in the past but are most closely related to amoebas. Water molds have two unequal flagellae and motile spores. Rusts and mildew are common examples of water molds.

The protists, the most diverse of all the phyla, are so different that taxonomists are constantly proposing new ways to classify them. The seven groups proposed in this chapter offer one convenient approach.

Life Cycle 27.3

CELLULAR SLIME MOLD

The individual organisms of the cellular slime molds behave as separate "amoebas," moving through the soil or other substrate and ingesting bacteria and other smaller organisms. At a certain phase of their life cycle, the individual organisms aggregate and form a moving mass, the slug, that eventually transforms itself into a spore-containing body, in which the amoebas become encysted as spores.

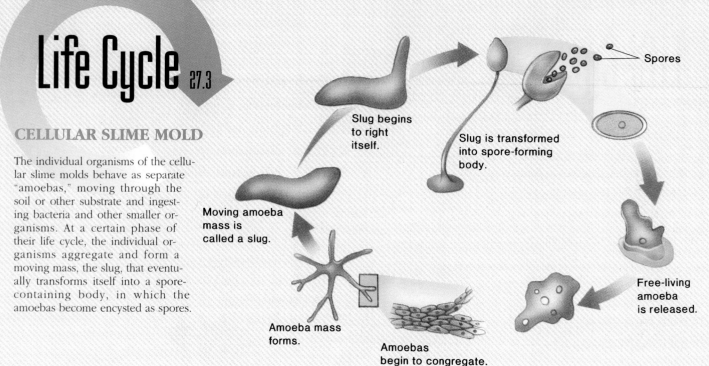

Slug begins to right itself.

Slug is transformed into spore-forming body.

Spores

Moving amoeba mass is called a slug.

Free-living amoeba is released.

Amoeba mass forms.

Amoebas begin to congregate.

The Fungi

Fungi are an ancient group of organisms at least four hundred million, and perhaps eight hundred million, years old. This distinct kingdom of organisms comprises approximately seventy-seven thousand described species, and many more await discovery. Scientists who study fungi are called **mycologists.** Although fungi have traditionally been included in the plant kingdom (which is why the major groups are called divisions and not phyla), they have no chlorophyll and resemble plants only in lacking mobility and in growing from the ends of somewhat linear bodies. But even these similarities prove misleading when fungi are examined closely. Some plants have motile sperm with flagella; therefore, plants certainly originated from ancestors that had flagella. In contrast, no fungi ever have flagella, and no evidence indicates that their ancestors possessed flagella. Fungi are basically filamentous in their growth form, consisting of slender filaments, whereas plants are multicellular in three dimensions. Unlike plants, fungi obtain their food by secreting enzymes out of their bodies and onto or into the substrate. They then absorb into their bodies the materials that these enzymes make available.

Fungi absorb their food after digesting it externally by secreting enzymes. This unique mode of nutrition, combined with their filamentous growth and complete lack of flagella, make the members of this kingdom highly distinctive.

Fungal Ecology

Fungi, along with bacteria, play an essential role as decomposers in the biosphere. They break down organic materials and return the substances locked up in these molecules to circulation in the ecosystem. In this way, critical biological building blocks, such as compounds of carbon, nitrogen, and phosphorus, that have been incorporated into the bodies of living organisms are released and made available for other organisms.

Many fungi are harmful because they decay, rot, and spoil many different materials as they obtain food. They also can cause serious diseases in plants and animals, including human beings. But other fungi are extremely useful. The manufacture of both bread and beer depends on the biochemical activities of yeasts, single-celled fungi that produce abundant quantities of ethanol and carbon dioxide. Both cheese and wine achieve their delicate flavors because of the metabolic processes of certain fungi, and other fungi make possible the manufacture of such oriental delicacies as soy sauce and tofu. Vast industries depend on the biochemical manufacture of organic substances such as citric acid by using fungi in culture, and yeasts are now employed on a large scale to produce protein for the enrichment of animal food. Many antibiotics, including the first one widely used—penicillin—are derived from fungi. Other fungi are used to convert complex organic molecules into other molecules, such as in the synthesis of many commercially important steroids.

Fungal Structure

Fungi exist mainly as slender filaments, or **hyphae** (singular, *hypha*), that are barely visible to the naked eye (figure 27.8). These hyphae may be divided into cells by cross-walls called **septa** (singular, *septum*). But the septa rarely form a complete barrier, except for those separating the reproductive cells. Cytoplasm characteristically flows freely throughout the hyphae, passing through the major pores in the septa. Because of this cytoplasmic streaming, proteins, which are synthesized

(a)

(b)

(c)

(d)

Figure 27.8 Representatives of two phyla of fungi: Ascomycetes and Basidiomycetes.
All visible structures of fleshy fungi, such as the ones shown here, arise from an extensive network of filaments (hyphae) that penetrate and interweave with the substrate on which they grow. (a) A morel, Morchella esculenta, a delicious, edible ascomycete (phylum Ascomycota) that appears in early spring in the northern,
temperate woods (especially under oaks). (b) A cup fungus (phylum Ascomycota) in the rain forest of the Amazon Basin. (c) Amanita muscaria, the fly agaric, a poisonous basidiomycete (phylum Basidiomycota). (d) A shelf fungus, Coriolus versicolor (phylum Basidiomycota), growing on a tree trunk. Basidia, the reproductive structures, line the underside of this fungus.

throughout the hyphae, may be carried to the actively growing tips of the hyphae. As a result, fungal hyphae may grow rapidly when abundant food and water are available and the temperature is high enough.

A mass of hyphae is called a **mycelium** (plural, *mycelia*). This term, like *mycologist,* is derived from the Greek word *myketos,* meaning "fungus." Fungal mycelia constitute a system that may be many kilometers long, although it is concentrated in a much smaller area. This system grows through and penetrates the environment of the fungus, resulting in a unique relationship between the fungus and its environment. All parts of a fungus are metabolically active, continually interacting with the soil, wood, or other material in which the mycelia are growing (figure 27.9).

Fungi exist primarily in the form of filamentous hyphae, which are completely divided by septa only when reproductive organs are formed. These hyphae surround and penetrate the substrate within which the fungi are growing.

In three of the four divisions of fungi, structures composed of interwoven hyphae, such as mushrooms, puffballs, and morels, are formed at certain stages of the life cycle. These structures may expand rapidly because of cytoplasmic streaming and growth in the kilometers of hyphae from which they arise. For this reason, mushrooms can force their way through tennis court surfaces or appear suddenly in a lawn.

The cell walls of fungi are not formed of cellulose, as are those of plants and some groups of protists. Other polysaccharides, such as chitin, which occurs especially frequently, are typical constituents of fungal cell walls. Chitin is the same material that makes up the major portion of the hard shells, or exoskeletons, of arthropods, a phylum of animals that includes insects

and crustaceans (see chapter 29). Chitin is far more resistant to microbial degradation than is cellulose.

Mitosis in fungi differs from that found in other organisms. The nuclear envelope does not break down and re-form, and the spindle apparatus is formed within it. In addition, all fungi lack centrioles. Overall, fungal features suggest that the kingdom originated from some unknown group of single-celled eukaryotes that lacked flagella. Certainly, the fungi differ sharply from all other groups of living organisms.

Spores, always nonmotile, constitute a common means of reproduction among the fungi. They may be formed through either sexual or asexual processes. When the spores land in a suitable place, they germinate, giving rise to a new fungal hypha. Because the spores are very small, they may remain suspended in the air for long periods and may be blown great distances from their place of origin, which explains the extremely wide distributions of many kinds of fungi. Unfortunately, many fungi that cause plant and animal diseases are spread rapidly and widely by such means.

Major Groups of Fungi

As shown in table 27.2, fungi comprise four phyla, representatives of two of which are shown in figure 27.8. The evolutionary relationships between these groups are not clear, although the zygomycetes, which have the fewest number of species, are the simplest in structure.

Figure 27.9 Mycelia.
A fungal mycelium growing through leaves on the forest floor in Maryland.

Table 27.2   Fungi

Phylum	Typical Examples		Key Characteristics	Approximate Number of Living Species
Ascomycota	yeasts, truffels, morels		Develop by sexual means; spores are formed inside a sac called an ascus; asexual reproduction is also common	30,000
Deuteromycota	*Aspergillus, Penicillium*		Sexual life cycle has not been observed; most are thought to be ascomycetes that have lost the ability to reproduce sexually	17,000
Basidiomycota	mushrooms, toadstools, rusts		Develop by sexual means; spores are born on club-shaped structures called basidia; only the terminal cell of a basidium produces spores (all cells of an ascus do); asexual reproduction is rare	16,000
Zygomycota	*Rhizopus* (black bread mold)		Develop sexually and asexually; multinucleate hyphae lack septa except for reproductive structures; fusion of hyphae leads directly to formation of a zygote, which divides by meiosis when it germinates	665

Ascomycetes

The ascomycetes (phylum Ascomycota) comprise about thirty thousand described species of fungi, including such familiar and economically important fungi as cup fungi and morels (see figure 27.8*a, b*), truffles, and yeasts (the only single-celled fungi). Ascomycetes also include the organisms that cause many of the most serious plant diseases. The chestnut blight, *Cryphonectria parasitica,* has almost exterminated the American chestnut throughout its native range, and Dutch elm disease, *Ceratocystis ulmi,* has decimated elms around the world.

The characteristic reproductive structure of the ascomycetes is the **ascus** (plural, *asci*), a club-shaped element that is formed within a structure consisting of densely interwoven hyphae, the **ascocarp.** The hyphae of ascomycetes are haploid. Syngamy, which occurs in the young asci, is immediately followed by meiosis, producing four, eight, or more spores. Asexual reproduction by means of **conidia** (singular, *conidium*), which are multinucleate spores in fungi cut off at the ends of the hyphae, is characteristic of most ascomycetes.

Deuteromycetes

Asexual reproduction is also a common feature among the approximately seventeen thousand species of deuteromycetes (Fungi Imperfecti), a group of fungi in which sexual reproduction is not known. The deuteromycetes do not comprise a true phylum, but rather, a grouping of all fungi whose sexual

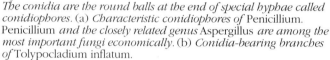

(a)

(b)

Figure 27.10
Deuteromycetes (Fungi Imperfecti).

Scanning electron micrographs of conidia (spores) of deuteromycetes, ascomycetes in which sexual reproduction is unknown. The conidia are the round balls at the end of special hyphae called conidiophores. (a) Characteristic conidiophores of Penicillium. Penicillium *and the closely related genus* Aspergillus *are among the most important fungi economically. (b) Conidia-bearing branches of* Tolypocladium inflatum.

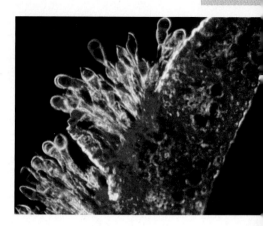

Figure 27.11 Rusts.
Wheat rust, Puccinia graminis, *one of about seven thousand species of rusts, all of them plant pathogens. This species causes enormous economic losses to wheat wherever it grows and is combated largely by breeding resistant wheat varieties. Mutation and recombination in wheat rust constantly produce new virulent strains, thus making it necessary to continuously replace the existing wheat varieties. Wheat rust alternates between two different hosts—wheat and barberries—and needs both to complete its life cycle. The sexual stages of wheat rust take place only on barberries, and the eradication of these plants helps to control this disease.*

reproductive stages have not been observed. Given the features of their hyphae, most deuteromycetes would be classified as ascomycetes if their sexual structures were found.

Because of the great economic importance of some deuteromycetes, they must be classified separately so that individual species can be identified. Among the economically important genera are *Penicillium* and *Aspergillus*. Some species of *Penicillium* are sources of the well-known antibiotic penicillin, and others give the characteristic flavor to such cheeses as Roquefort and Camembert. *Aspergillus* is used to ferment soy sauce (figure 27.10).

Basidiomycetes

Basidiomycetes, the third most diverse phylum of fungi (phylum Basidiomycota), with about sixteen thousand described species, are the most familiar fungi. These include not only the mushrooms, toadstools, puffballs, jelly fungi, and shelf fungi, but also many important plant pathogens among the groups called rusts and smuts (figure 27.11). In place of the asci of ascomycetes, basidiomycetes form **basidia** (singular, *basidium*). At the apex of the basidia, the spores that result from meiosis are elevated; otherwise, the details of the life cycle are generally similar to those of the ascomycetes (see Life Cycle 27.4). Asexual reproduction, however, is relatively uncommon.

Zygomycetes

Zygomycetes (phylum Zygomycota) are the least diverse of the four fungal phyla, comprising about 665 described species. The name of this phylum refers to its chief characteristic, the production of sexual structures called **zygosporangia** (singular, *zygosporangium*). (In these fungi, the zygosporangium often has been called a zygospore, but because it contains several to many zygotes, this text refers to it as a zygosporangium.) A zygosporangium is formed following fusion of two of the simple reproductive organs of these fungi, which are called **gametangia** (singular, *gametangium*). Within a zygosporangium, the gametes—which are simply nuclei—fuse, forming one or more diploid nuclei, or zygotes.

The zygosporangia of members of this phylum differ greatly and are often highly ornate. Spores are also formed without sexual reproduction; these asexual spores are produced in bodies called **sporangia** (singular, *sporangium*). The life cycle of the black bread mold, *Rhizopus*, is shown in Life Cycle 27.5.

Lichens

A **lichen** is a symbiotic association between an ascomycete and a photosynthetic partner (table 27.3). (About a dozen species of basidiomycetes also form associations with algae, but they are closely related to free-living basidiomycetes and do not resemble any other lichens.)

Table 27.3 Symbiotic Associations

Group	Key Characteristics	Approximate Number of Species
Lichens	Fungi (almost always ascomycetes) in the cells of which are cyanobacteria or green algae or both; derives its energy from its photosynthetic partner and cannot survive without it; able to survive freezing or drying out; can invade the harshest of habitats; breaks down rocks and sets the stage for invasion by other organisms	13,500
Mycorrhizae	Symbiotic associations between fungi and plants; 80 percent of all plants have mycorrhizae in their roots; by far the most common are endomycorrhizae, in which zygomycete fungal hyphae penetrate the cells of the plant root; some plants of temperate regions have ectomycorrhizae, in which basidiomycete hyphae (or, rarely, ascomycete) surround but do not penetrate the root	5,000

Life Cycle 27.4

BASIDIOMYCETE

In primary mycelia, there is only one nucleus in each cell; in secondary mycelia, which are formed by the fusion of primary mycelia (plasmogamy), there are two nuclei, one derived from each of the strains that gave rise to the secondary mycelia, within each cell. Secondary mycelia ultimately may become massed and interwoven, forming the basidiocarp, within which basidia line the gills. Meiosis immediately follows syngamy in these basidia.

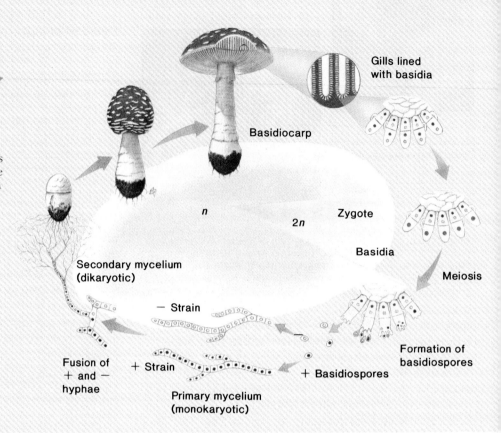

Life Cycle 27.5

RHIZOPUS

The hyphae grow over the surface of the bread or other material on which the fungus feeds, producing erect, sporangium-bearing stalks in clumps. If both + and − strains are present in a colony, they may grow together, and their nuclei may fuse, producing a zygote. This zygote, which is the only diploid cell of the life cycle, acquires a thick, black coat and is then called a zygosporangium. Meiosis occurs during its germination, and normal, haploid hyphae grow from the resulting haploid cells.

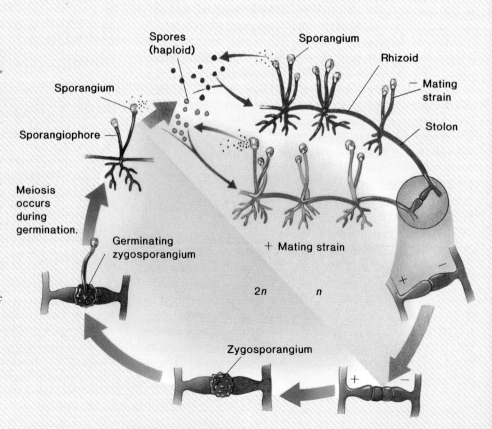

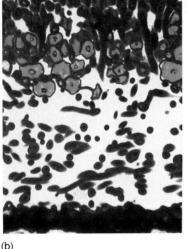

(a)   (b)

Figure 27.12 Lichens. (a) *Lichens on a fog-swept rock in coastal California.* (b) *Structure of a lichen. This transverse section of a lichen shows the fungal hyphae (more densely packed into a protective layer on the top) and, especially, the bottom layer of the lichen. The green cells near the lichen's upper surface are those of a green alga. Penetrated by fungal hyphae, the green alga supplies carbohydrates to the fungus.*

There are about 13,500 described species of lichens (figure 27.12*a*). Most of the visible body of a lichen consists of its fungus, but within the tissues of that fungus are found either cyanobacteria or green algae, or both. Specialized fungal hyphae penetrate the photosynthetic cells held within them and transfer nutrients directly to the fungal partner, which, in turn, protects its bacterial or algal cells from drying out (figure 27.12*b*).

The durable construction of the fungus, linked with the photosynthetic properties of its partner, have enabled lichens to invade the harshest of habitats: the tops of mountains, the farthest north and south latitudes, and dry, bare rock faces in the desert. In such desolate, exposed areas, lichens are often the first colonists, breaking down the rocks and setting the stage for the invasion of other organisms. Lichens with a cyanobacterium for a photosynthetic partner have a particular advantage because they are able to fix atmospheric nitrogen for themselves; they also contribute nitrogen to their habitat, where it is used by other pioneering organisms.

Lichens are often strikingly colored because of their pigments, which probably play a role in protecting the photosynthetic partner from the destructive action of the sun's rays. These same pigments may be extracted from the lichens and used as natural dyes, as they were in the traditional method of manufacturing Harris tweed (now, however, it is colored with synthetic dyes).

Lichens can survive in inhospitable habitats partly because they are able to dry or freeze to a condition that could be called suspended animation. Once the drought or cold has passed, the lichens recover quickly and resume their normal metabolic activities, including photosynthesis. Lichen growth may be extremely slow in harsh environments; many relatively small ones appear to be thousands of years old and therefore are among the oldest living things on the earth. Lichens are extremely sensitive to atmospheric pollutants and thus can be used as bioindicators of air quality. Many species are characteristically absent near major cities and other sources of pollution.

Lichens are symbiotic associations between ascomycetes and a photosynthetic partner (green algae, cyanobacteria, or both).

Mycorrhizae

The roots of about 80 percent of all kinds of plants normally are involved in symbiotic relationships with certain specific kinds of fungi; these fungi have been estimated to account for as much as 15 percent of the weight of the plant's roots in many cases. Associations of this kind are termed **mycorrhizae,** from the Greek words meaning "fungus" and "roots" (see table 27.3). To a certain extent, the fungi involved in mycorrhizal associations replace and have the same function as root hairs, fine projections from the epidermis (outermost cell layer) of the terminal portions of roots. When mycorrhizae are present, they aid in the direct transfer of phosphorus, zinc, copper, and probably other nutrients from the soil to plant roots. The plant, on the other hand, supplies organic carbon to the symbiotic fungus (figure 27.13).

Mycorrhizae are symbiotic associations between plants and fungi.

The earliest fossil plants often are found to have mycorrhizal roots. Such associations may have played an important role in plants' initial invasion of the land. The soils available at that time would have been relatively infertile and completely lacking in organic matter. Plants that form mycorrhizal associations are particularly successful in similar situations today. Given this fact and the fossil evidence, the suggestion that mycorrhizal associations were characteristic of the earliest plants seems reasonable.

Figure 27.13 Mycorrhizae aid plant growth.
Soybeans without mycorrhizae (left) *and with different strains of mycorrhizae* (center and right)*.*

EVOLUTIONARY VIEWPOINT

Protists are such a diverse group that all other kingdoms pale by comparison. Indeed, the major groups of protists are as different from one another as animals are from fungi. Many biologists are beginning to argue that the ciliates are so different that they should be declared a separate kingdom! Fungi are easily the most foreign of the multicellular kingdoms. Multinucleate, with shared cytoplasm, represent a radically different evolutionary road than the one taken by animals.

SUMMARY

1. Kingdom Protista consists of the exclusively or predominantly unicellular phyla of eukaryotes, together with three phyla that include many multicellular organisms: the red algae, brown algae, and green algae. No other kingdom approaches the highly diverse array of form and function found among the protists.

2. *Pelomyxa palustris,* an amoeba-like organism, lacks both mitosis and mitochondria and apparently represents a very early stage in the evolution of eukaryotes.

3. The three major multicellular groups of eukaryotes—plants, animals, and fungi—all originated from protists. But because the three groups are so large and important, each is considered a separate kingdom. The three are not related directly to one another. The plants originated from green algae; the sponges, and probably all animals, originated from the choanoflagellates, a group of zoomastigotes; and the ancestors of fungi are unknown.

4. Chloroplasts originated independently several times in the protists. The process apparently involved the members of at least three different groups of bacteria: cyanobacteria (in the red algae); *Prochloron*-like organisms (in the green algae and euglenoids); and bacteria of a third, different group (in the brown algae, diatoms, and dinoflagellates).

5. In this text, fourteen phyla of protists are grouped into seven categories that reflect key shared characteristics.

6. The Sarcodina consist of two phyla: The Rhizopoda and Foraminifera. These protists have no permanent locomotor apparatus and either move by pseudopodia or protoplasmic streaming.

7. The algae consist of three phyla of photosynthetic protists. Green algae (phylum Chlorophyta) are a highly diverse group of organisms that are abundant in the sea, freshwater, and damp semiterrestrial habitats. Plants were derived from a multicellular green algae. Red algae (phylum Rhodophyta) lack flagellated cells. Brown algae (phylum Phaeophyta) may be quite large and include the kelp.

8. Diatoms (phylum Bacillariophyta) have a characteristic double shell of opaline silica.

9. The flagellates consist of three phyla of protists with locomotor flagellae. Dinoflagellates (phylum Dinoflagellata) are photosynthetic. Among the zoomastigotes (phylum Zoomastigina) are the organisms that cause sleeping sickness and several other primarily tropical diseases. Another group of Zoomastigina, the choanoflagellates, are the ancestors of sponges and probably all animals. A third group, the euglenoids (phylum Euglenophyta), includes a number of genera in which chloroplasts occur.

10. There are about eight thousand named species of ciliates (phylum Ciliophora). These protists are extremely complex with numerous cilia.

11. The malaria parasite, *Plasmodium,* is a member of the phylum Sporozoa. Carried by mosquitoes, it multiplies rapidly in the liver of humans and other primates. Cyclical fevers are characteristic of malaria because of the release of toxins into the bloodstream of the host.

12. The molds consist of three unrelated phyla of protists: plasmodial slime molds (phylum Myxomycota), cellular slime molds (phylum Acrasiomycota), and water molds (phylum Oomycota).

13. The fungi are a distinct kingdom of eukaryotic organisms characterized by their filamentous growth form, lack of chlorophyll and motile cells, chitin-rich cell walls, characteristic form of mitosis, and external digestion of food by the secretion of enzymes. Along with the bacteria, fungi are the decomposers of the biosphere.

14. Fungal filaments, called hyphae, collectively make up a mass called the mycelium. Mitosis in fungi occurs within the nuclear envelope.

15. Fungi comprise four phyla: Ascomycetes (phylum Ascomycota) have reproductive structures called asci, and asexual reproduction is characteristic. Deuteromycetes (Fungi Imperfecta) are fungi whose sexual reproductive stages have not been observed. In basidiomycetes (phylum Basidiomycota), reproduction is typically sexual. Zygomycetes (phylum Zygomycota) produce sexual structures called zygosporangia, within which gametes fuse, forming zygotes.

16. Symbiotic systems involving fungi include lichens and mycorrhizae. The fungal partners in lichens are ascomycetes, which derive their nutrients from green algae, cyanobacteria, or both. Mycorrhizae are symbiotic associations between plants and fungi and are characteristic of about 80 percent of all plant species.

REVIEWING THE CHAPTER

1. What are the distinguishing features of kingdom Protista?

2. How does symbiosis relate to the origin of eukaryotic cells?

3. How do the major groups of protists differ from one another?

4. What are some of the major diseases caused by protists?

5. What are the distinguishing features of kingdom Fungi?

6. What is the structure of a fungus, and how do fungi reproduce?

7. How do the major groups of fungi differ from one another?

8. What are some of the major diseases caused by fungi?

9. What are lichens and mycorrhizae?

COMPLETING YOUR UNDERSTANDING

1. What is the significance of purple nonsulfur bacteria and the evolution of eukaryotic cells?

2. Did chloroplasts originate in eukaryotes from one or more ancestral stocks? What is the evidence for your answer?

3. What role would photosynthetic protists such as diatoms and dinoflagellates play in the earth's biosphere?

4. What characteristic of green, red, and brown algae makes including these groups in the kingdom Protista unusual?

5. What are alginates and agars? Where do they come from? How do we use these substances in everyday life?

6. If you were on a camping trip in a remote area away from human populations, would it be necessary to boil or filter your drinking water acquired from a nearby stream? Why or why not?

7. What is conjugation, and what benefit does it give to those protists where it occurs?

8. What causes sleeping sickness, amoebic dysentery, and malaria? What steps are being taken to reduce these diseases? What are some of the problems with eradication?

9. What are slime molds and water molds? Why were they transferred from kingdom Fungi to kingdom Protista?

10. What is the major role of fungi in the biosphere? Do fungi act alone in this role?

11. A mass of fungal filaments is called a
a. hyphae.
b. mycelium.
c. septa.
d. colony.
e. pseudopodia.
f. kelp.

12. What type of reproductive structures would be associated with Dutch elm disease?

13. To which group of fungi do mushrooms belong?

14. Which group of fungi is characterized by a thick-walled zygosporangium?

15. Which group of fungi lacks sexual reproduction? Name one important fungus that belongs to this group.

16. What is penicillin? Where does it come from? What is it used for?

17. How are lichens used as bioindicators?

18. How do mycorrhizae function? What role did mycorrhizae probably play in plants' invasion of land hundreds of millions of years ago?

FOR FURTHER READING

Ahmadjian, V., and S. Paracer. 1986. *Symbiosis: An introduction to biological associations.* Hanover, N.H.: University Press of New England. An outstanding account of lichens and other symbiotic systems.

Brusca, R., and G. Brusca. 1990. *Invertebrates.* Sunderland, Mass.: Sinauer. A very thoughtful book on animal diversity with an excellent chapter on heterotrophic protists.

Cochran, M. F. 1990. Back from the brink. Chestnuts. *National Geographic,* Feb., 128–40. How the American chestnut, nearly destroyed by the chestnut blight, is being rescued by dedicated scientists and volunteers.

Kosikowski, F. V. 1985. Cheese. *Scientific American,* May, 52–59. A fascinating account of how more than two thousand varieties of cheese are made and how bacteria and fungi participate in the process.

Lambrecht, F. 1985. Trypanosomes and hominid evolution. *BioScience* 35: 640–46. A fascinating article that charts the probable effects of sleeping sickness in determining the course of human history.

McKnight, K. H., and V. McKnight. 1987. *A field guide to mushrooms of North America.* Peterson Field Guide Series. New York: Houghton Mifflin. An excellent guide that includes most of the common and edible species of North America.

Oaks, S. et al. 1991. *Malaria—Obstacles and opportunities.* Washington, D.C.: National Academy Press. The report of a high-level committee of scientists on the recent worldwide resurgence of malaria, which is staging a dramatic comeback in many countries where it was thought to be under control. Malaria already kills more humans each year than any other communicable disease, and the numbers are rising.

Saffo, M. B. 1987. New light on seaweeds. *BioScience* 37: 654–64. A discussion of how seaweeds occur at different depths in the ocean, their photosynthetic pigments being efficient in harvesting energy for the particular light waves that reach that depth.

Waters, A. et al. 1991. *Plasmodium falciparum* appears to have arisen as a result of lateral transfer between avian and human hosts. *Proceedings of the National Academy of Sciences U.S.A.* 88 (April): 3140–44. The suggestion that humans may have caught malaria from chickens!

CHAPTER
28

PLANT DIVERSITY

These decorative Dutch tulips, a blanket of color, stretch off into the horizon. Many familiar plants have been domesticated—deliberately bred to exhibit particular traits.

FOR REVIEW

Here are some important terms and concepts that have been discussed in previous chapters and that you will encounter again in this chapter. Review them before proceeding if necessary.

Carbon fixation (*chapter 8*)
Evolutionary history of plants (*chapter 17*)
Classification of organisms (*chapter 25*)
Mycorrhizae (*chapter 27*)

Of the six kingdoms of living organisms, the three discussed in chapters 26 and 27—bacteria, protists, and fungi—consist of organisms that are mostly small relative to humans. Bacteria and most protists are unicellular, and fungi, although multicellular, are rarely as large as a fist. But the plant kingdom consists of many species that consist of large individuals. A tree can be 30 meters or more tall, with a mass of many tons. Plants are the dominant organisms of the terrestrial landscape and are believed to have been the first organisms to invade the land successfully from the sea, where life first evolved and flourished.

Despite all of the diversity of life on the earth, only three groups are able to fix carbon; by doing so, they provide for the sustenance of not only themselves, but of all other living organisms. The algae and photosynthetic bacteria carry out most photosynthesis in the sea, whereas plants are the dominant photosynthetic organisms on land. Plants are multicellular eukaryotic organisms that have: (1) cellulose-rich cell walls; (2) chloroplasts that contain chlorophylls *a* and *b*, together with carotenoids; and (3) starch as their primary carbohydrate food reserve. This chapter discusses the characteristics of the major plant groups (figure 28.1). The major plant phyla used to be called "divisions" (and still are by many botanists), but to maintain consistency with the other kingdoms, this text will adopt the newer approach of referring to major plant groups as phyla (table 28.1). Chapters 30 to 32 treat specific aspects of plant biology in more detail.

Figure 28.1 Representatives of four phyla of plants.
(a) *A moss gametophyte (phylum Bryophyta).*
(b) *A Norway spruce,* Picea nigra *(phylum Coniferophyta), in the Alps. Seeds are produced in the large cones, and pollen is produced in the smaller ones.* (c) *Oak tree,* Quercus robur *(phylum Anthophyta).*
(d) *Maidenhair fern,* Adiantum pedatum *(phylum Pterophyta).*

(a)

(b)

(c)

(d)

The Green Invasion of the Land

Plants, fungi, and insects are the only major groups of organisms that occur almost exclusively on land; several other phyla, including chordates and mollusks (chapter 29), are also well represented on land. The groups that occur almost exclusively in terrestrial habitats—the plants, fungi, and insects—all probably evolved there, whereas the chordates and mollusks clearly originated in the water. Of the groups that evolved on land, the ancestors of the plants were almost certainly the first to become terrestrial.

A major evolutionary challenge in the transition from an aquatic to a terrestrial habitat is **desiccation,** the tendency of organisms to lose water to the air. In the organisms that were most successful on land, therefore, various structures evolved to conserve water. In plants, **vascular systems**—strands of specialized cells that conduct water and carbohydrates—evolved to supply water to different plant parts.

Relatively efficient vascular systems evolved in some of the earliest plants, and only two phyla that have living representatives—the liverworts (phylum Hepaticophyta) and the hornworts (phylum Anthocerophyta)—lack them completely. These two phyla, along with the mosses (phylum Bryophyta), which have reduced vascular systems, have been grouped as **bryophytes.** Most scientists, however, have concluded that the "bryophytes," like the "algae," are a group of organisms not directly related to one another. The remaining nine plant phyla, which have evident and efficient conducting systems, are called the **vascular plants;** they probably share a common ancestor with the mosses (see table 28.1).

Vascular plants are so named because they have vascular tissue. The word *vascular* comes from the Latin word *vasculum,* meaning "a vessel or duct," and refers, in the case of plants, to their conducting systems. Vascular tissue consists of specialized strands of elongated cells that run from near the tip of a plant's roots through its stems and into its leaves. Different kinds of vascular tissue conduct both water with dissolved minerals (nutrients), which comes in mainly through the roots, and carbohydrates, which are manufactured in the green parts of the plant (figures 28.2 and 28.3). Therefore, water and nutrients

reach all parts of the plant, as do the carbohydrates that provide energy for synthesizing the plant's different structures. The vascular systems of those mosses that possess them may be less efficient than those of the vascular plants, but perhaps mosses, being much smaller on average than vascular plants, have less need for efficient conducting systems.

Most plants also are well protected from drying out by the **cuticle,** an outer covering formed from a waxy substance called cutin. The cuticles that cover the exposed surfaces of plants are impermeable to water and thus provide a key barrier to water loss. Passages do exist through the cuticle, however, in the form of specialized pores called **stomata** (singular, *stoma*) in the leaves and sometimes the green portions of the stems. Stomata, which occur on at least some portions of all plants except liverworts, allow carbon dioxide to pass into the plant bodies and allow water and oxygen to pass out of them.

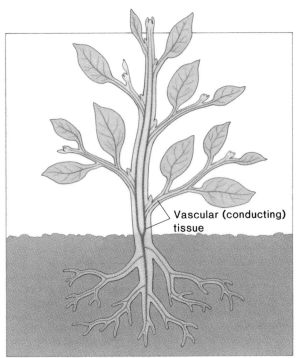

Figure 28.3 The architecture of a plant.
A vascular plant's body design is determined largely by the need to transport water through the plant. As shown in this diagram of a plant's vascular (conducting) tissues, water passes continuously in through the roots, up through the stems, and out through the leaves. At the same time, sugar (carbohydrate) molecules, manufactured as a result of photosynthesis in the leaves and green stems, pass through a parallel conducting system to all parts of the plant, where they are used for growth.

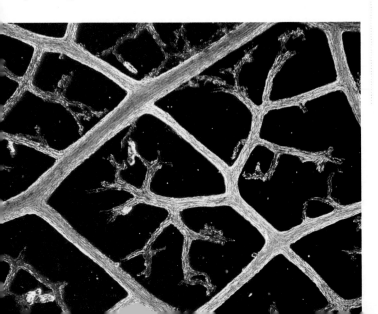

Figure 28.2 Vascular tissues.
The veins of a vascular plant contain strands of specialized cells for conducting carbohydrates, as well as water with its dissolved minerals. The veins run from the tips of the roots to the tips of the shoots and throughout the leaves, as shown here in this greatly enlarged photomicrograph of a cleared leaf.

Table 28.1   Plants

Phylum	Approximate Number of Species	Key Characteristics	Typical Examples
Anthophyta (flowering plants)	235,000	Flowering; also called angiosperms; characterized by ovules that are fully enclosed by the carpel; fertilization involves two sperm nuclei; one forms the gamete, the other fuses with polar bodies to form endosperm for the seed; after fertilization, carpels and the fertilized ovules (now seeds) mature to become fruit	Oak trees, corn, wheat, roses
Bryophyta (mosses)	16,600	Without vascular tissues; lack true roots, and leaves; live in moist habitats and obtain nutrients by osmosis and diffusion; two other phyla, Hepaticophyta (liverworts) and Anthocerophyta (hornworts), are also considered bryophytes, and make up 40 percent of all bryophyte species	*Polytrichum, Sphagnum* (peat moss)
Pterophyta (ferns)	12,000	Seedless vascular plants; haploid spores germinate into free-living haploid individuals; two minor phyla, Sphenophyta (horsetails) and Psilophyta (whisk ferns) contain twenty-one additional species	*Azolla, Sphaeropteris* (tree ferns)
Lycophyta (lycopods)	1,000	Seedless vascular plants similar in appearance to mosses, but diploid; found in moist habitats	*Lycopodium,* club mosses

Phylum	Approximate Number of Species		Key Characteristics	Typical Examples
Coniferophyta (conifers)	550		Gymnosperms; ovules within a carpel but partially exposed at time of pollination; flowerless; seeds are dispersed by the wind; sperm lack flagella; leaves are needlelike or scalelike; most species are evergreens, and live in dense stands; among the most common trees on earth	Pines, spruce, fir, redwood, cedar
Cycadophyta (cycads)	100		Gymnosperms; very slow growing, palmlike trees; sperm have flagella, but reach vicinity of egg by a pollen tube	Cycads, sago palms
Gnetophyta (shrub teas)	70		Gymnosperms; nonmotile sperm; shrubs and vines	Mormon tea, *Welwitschia*
Ginkgophyta (ginkgo)	1		Gymnosperms; fanlike leaves that are dropped in winter (deciduous); seeds fleshy and ill-scented; motile sperm	Ginkgo trees

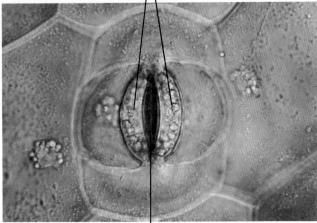

Guard
cells

Stoma

Figure 28.4 A stoma.
The guard cells flanking the stoma contain chloroplasts, unlike other epidermal cells. Water passes out through the stoma, and carbon dioxide enters by the same portal. The mechanism for opening and closing a stoma is described in chapter 32.

The cells that border stomata expand and contract, thus controlling water loss while allowing the entrance of carbon dioxide (figure 28.4). Most plants depend on the constant flow of water in through the roots or lower portions and out through the stomata.

> *Vascular plants and some mosses have specialized plumbing, called the vascular system. The vascular system involves strands of elongated, specialized cells that transport water and dissolved nutrients and other strands that transport carbohydrate molecules.*

In addition to their structural features, another key factor in plants' successful occupation of terrestrial habitats was the special relationship they developed with fungi. Mycorrhizae are characteristic of 80 percent of all plants and are frequently seen in early plant fossils (see chapter 27). They were probably critical to the success of these early plants in the harsh habitats available at the time of the first land invasion.

Many other features developed gradually and aided plants' evolutionary success on land. For example, above-ground and below-ground parts of the first plants were fundamentally the same. Later, roots and shoots with specialized structures evolved, each suited to its particular environment. **Leaves,** expanded areas of photosynthetically active tissue, evolved and diversified in relationship to the varied land habitats. Specializations in key reproductive features improved the methods by which plants protected their embryos and were dispersed from place to place. Flowers and seeds were of particular importance.

The more specialized roots, stems, leaves, and reproductive features that evolved in plants were important factors in the group's overwhelming success on land. An estimated 266,000 species of this kingdom currently dominate every part of the terrestrial landscape, except the extreme polar regions and the highest mountaintops.

The Plant Life Cycle

Alternation of Generations

The different kinds of plant life cycles provide an important key to understanding plants' evolutionary relationships. All plants exhibit **alternation of generations,** in which a diploid generation, or **sporophyte** ("spore plant"), alternates with a haploid generation, or **gametophyte** ("gamete plant"). The terms *sporophyte* and *gametophyte* indicate the kinds of reproductive structures that the respective generations produce.

Most adult animals are diploid, and in this respect, they resemble the sporophyte generation of a plant. Such animals, however, produce eggs and sperm, which fuse directly to form a zygote. In contrast, the sporophyte generation of a plant does not produce gametes as a result of meiosis. Instead, meiosis takes place in specialized cells, called **spore mother cells,** and results in the production of haploid **spores,** the first cells of the gametophyte generation. Spores do not fuse with one another, as gametes do; instead, they divide by meiosis, producing a multicellular haploid individual, the gametophyte.

In turn, the gametes—eggs and sperm—eventually are produced by the gametophyte as a result of mitosis. They are haploid, like the gametophyte that produces them. When they fuse to form a zygote, the first cell of the next sporophyte generation has come into existence. The zygote grows into a sporophyte in which meiosis ultimately occurs (figure 28.5).

> *Plant life cycles are marked by an alternation of generations of diploid sporophytes with haploid gametophytes. As a result of meiosis, sporophytes produce haploid spores, which grow into gametophytes. Gametophytes produce gametes, as a result of mitosis. The fusion of gametes produces a zygote, the first cell of the sporophyte generation.*

In ferns, mosses, and liverworts, the gametophyte is green and free-living; in most other plants, the gametophyte is not green and is nutritionally dependent on the sporophyte. A moss or liverwort exhibits largely gametophyte tissue; the sporophytes are smaller brown or yellow structures attached to or enclosed within gametophyte tissues. In most other plants, the gametophytes are always much smaller than the sporophytes; in many, they are nutritionally dependent on the sporophytes and enclosed within their tissues. When looking at a vascular plant, what one sees is a sporophyte, with rare exceptions.

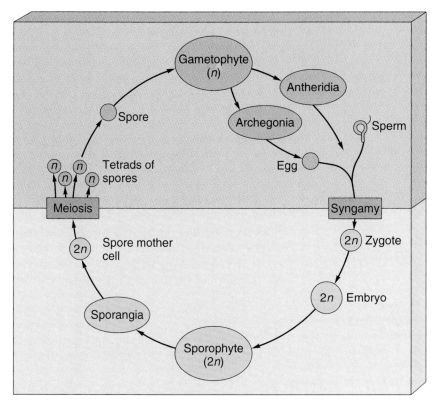

Figure 28.5 A generalized plant life cycle.
In a life cycle of this kind, gametophytes, which are haploid (n), alternate with sporophytes, which are diploid (2n). Antheridia (male) and archegonia (female), which are the sex organs (gametangia), are produced by the gametophyte, and they, in turn, respectively produce sperm and eggs. The sperm and egg ultimately come together in the process of syngamy to produce the first diploid cell of the sporophyte generation, the zygote. Meiosis takes place within the sporangia, the spore-producing organs of the sporophyte, resulting in the production of the spores, which are haploid and are the first cells of the gametophyte generation.

The Specialization of Gametophytes

Gametes of the first plants differentiated within specialized organs called gametangia (singular, *gametangium*), in which the eggs and sperm were surrounded by a jacket of cells. Complex, multicellular gametangia of this sort are still found in all members of the plant kingdom that do not form seeds and in some seed-forming plants as well. Eggs and sperm are formed within different kinds of gametangia. Those in which eggs are formed are called **archegonia** (singular, *archegonium*), and those in which sperm are formed are called **antheridia** (singular, *antheridium*). An archegonium produces only one egg; an antheridium produces many sperm. These structures usually look very different from one another (figure 28.6).

In some plants, including the ferns, antheridia and archegonia occur together on the same gametophyte. In others, and typically in mosses, the two kinds of gametangia are on separate gametophytes. In the more advanced vascular plants, including all but a very few of the vascular plants that form seeds, the gametangia have been lost

during the course of evolution. The eggs or sperm develop directly from individual cells of the respective gametophyte. Some of the gametophytes bear only eggs, and others bear only sperm.

When one kind of gametophyte bears antheridia and another kind bears archegonia, the two kinds of gametophytes may look different from one another. If they do, the gametophytes that form antheridia are called **microgametophytes** ("small gametophytes"), and those that form archegonia are called **megagametophytes** ("large gametophytes"). In nearly all plants with two kinds of gametophytes, the gametophytes arise from two kinds of spores: **microspores** and **megaspores**. Plants that produce two different-looking spores are called **heterosporous**. Those that produce only one kind of spore are called **homosporous**.

Spores are formed as a result of meiosis in the sporophyte generation. Their differentiation occurs within specialized multicellular structures called sporangia (singular, *sporangium*). If a plant forms both megaspores and microspores, each of these will be formed in a different kind of sporangium called, respectively, **megasporangia** and **microsporangia** (figure 28.7).

The descriptions of the principal plant phyla that follow note a progressive reduction of the gametophyte from group to group and increasing specialization for life on land, culminating with the remarkable structural adaptations of flowering plants.

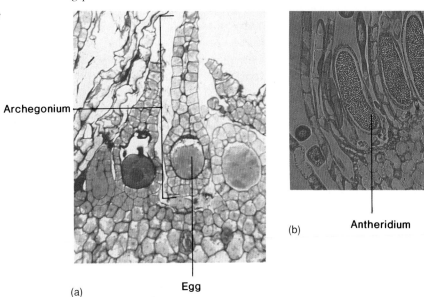

Figure 28.6 Gametangia in plants.
(a)*Transection through the archegonium of the liverwort* Marchantia. *A single egg differentiates within the lower, swollen portion of the archegonium.* (b) *Transection through a group of moss antheridia. The smaller cells in each of these elongate structures will give rise to sperm that, when liberated by the rupturing antheridium, swim through free water to the mouth of the archegonium.*

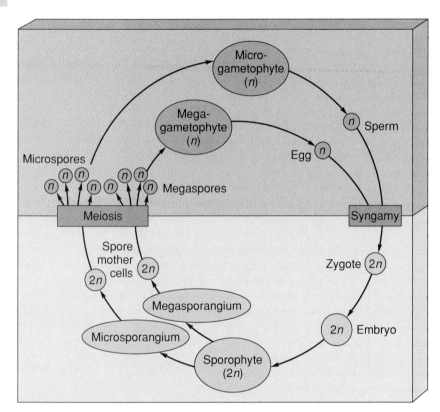

Figure 28.7 The life cycle of a heterosporous vascular plant.
In any heterosporous plant, the sporophyte generation (often diploid) produces two kinds of gametophytes by meiosis, each with half the sporophyte's number of chromosomes. The union of two such gametophytes (egg and sperm) by syngamy produces the next sporophyte generation.

Mosses, Liverworts, and Hornworts

Now regarded as separate phyla, not directly related to one another, the mosses (phylum Bryophyta), liverworts (phylum Hepaticophyta), and hornworts (phylum Anthocerophyta) are simple plants that retain many primitive features (figure 28.8). As mentioned earlier, they were formerly collectively called the "bryophytes," but that term is no longer believed to reflect a direct relationship between these groups. Mosses and liverworts are especially common in relatively moist places, both in the tropics and in temperate regions. In the Arctic and Antarctic, mosses are the most abundant plants and also boast the greatest number of species. Regardless of where they grow, mosses, liverworts, and hornworts

Figure 28.8 A simple plant.
A liverwort, Marchantia. *The sporophytes are borne within the tissues of the umbrella-shaped structures that arise from the surface of the flat, green, creeping gametophyte. These particular structures develop archegonia within their tissues; another kind of similar structure, borne on different plants of* Marchantia, *produces the antheridia.*

require free water at some time of the year to reproduce sexually because their sperm swim through the water to the mouth of the archegonium.

> *The term bryophyte has traditionally referred to the mosses, liverworts, and hornworts. These, however, are now understood to be phyla of relatively simple plants that are not directly related to one another.*

Mosses

Mosses are a diverse group of about ten thousand species of small plants. Many other small, tufted plants are mistakenly called "mosses"; for example, Spanish moss is actually a flowering plant, a relative of the pineapple. In many mosses, the stems of the gametophytes have a central strand of water-conducting cells. Unlike the similar conducting cells that occur in vascular plants, the conducting cells of mosses lack specialized wall thickenings. In some mosses, these conducting cells are surrounded by carbohydrate-conducting cells resembling those of the vascular plants. Mosses also have a cuticle and stomata. Although all of these structures are less complex than those of the vascular plants, they probably had a common origin with them.

In mosses, as in liverworts and hornworts, the sporophytes are borne on the gametophytes, from which the sporophytes derive their food. Moss sporophytes take six to eighteen months to develop and are generally elevated on a stalk (see Life Cycle 28.1).

Liverworts and Hornworts

Liverworts constitute a phylum of about sixty-five hundred species of plants that are generally inconspicuous individually but that may form masses in moist places. Their name dates from the Middle Ages and relates to the liver-shaped outline of the gametophyte in some genera. Liverworts lack conducting tissue, cuticle, and stomata and are the simplest of all living plants. Their sporophytes often remain enclosed in gametophyte tissue until they are mature; in outline, their life cycle resembles that of the mosses.

The six genera and approximately one hundred species of hornworts do possess stomata, together with a number of unique features that separate them from both mosses and liverworts and strongly suggest that there is no direct evolutionary connection between the members of these groups.

> *Mosses often have specialized conducting strands, a cuticle, and stomata, whereas liverworts, which are the most primitive living plants, lack all three features. In the course of the life cycles of both groups, the sporophyte remains attached to, and nutritionally dependent on, the gametophyte.*

Life Cycle 28.1

MOSS

On the gametophytes, which are haploid, the sperm are released from antheridia. They then swim through free water to the archegonia and down their neck to the egg. Fertilization takes place there; the resulting zygote develops into a sporophyte, which is diploid. The sporophyte grows out of the archegonium and differentiates into a slender, basal stalk, or seta, which has a swollen capsule at its apex. The capsule is covered, at least at first, with a cap, or calyptra, formed from the swollen archegonium. The sporophyte grows on the gametophyte and eventually produces spores as a result of meiosis. The spores are shed from the capsule after a specialized lid—the operculum—drops off. The spores germinate, giving rise to gametophytes. The gametophytes initially are threadlike; they grow along the ground. Ultimately, buds form on them, from which leafy gametophytes arise.

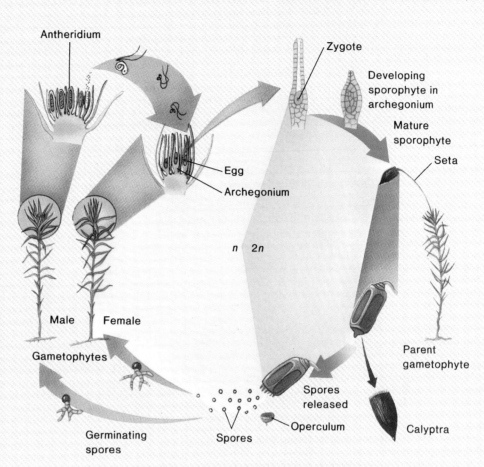

Vascular Plants

The first vascular plants appeared approximately 430 million years ago. Fossils of these early plants are rare. As mentioned earlier, fungi (mycorrhizae) are found associated with the roots of many of these fossils, suggesting that this symbiosis may have played a key role in plants' successful invasion of the land.

Vascular plants are characterized by: (1) their large, dominant, and nutritionally independent sporophytes; (2) efficient conducting tissues; (3) specialized leaves, stems, and roots; (4) cuticles and stomata; and (5) the evolution of seeds in some. Vascular plants appear to be derived from an immediate common ancestor, which is apparently closely related to the ancestor of the mosses.

> *Vascular plants are characterized by their dominant sporophytes; efficient conducting tissues; specialized stems, leaves, and roots; cuticles and stomata; and in some of them, seeds.*

Growth of Vascular Plants

Early vascular plants were characterized by **primary growth,** growth that results from cell division at the tips of the stems and roots. Primary growth is also characteristic of herbaceous plants and of the young stems of all plants today (see chapter 30). This sort of growth is like erecting a smokestack by continuing to add bricks to the top; the smokestack gets taller without the part that is already there becoming larger. Within stems that have resulted from primary growth are **vascular bundles** of elongated cells that conduct water with dissolved minerals (nutrients) and carbohydrates throughout the plant body. Vascular bundles are a key characteristic of vascular plants; they are not present in any other group.

Secondary growth was an important early development in the evolution of vascular plants. In secondary growth, a cylinder of cells around the plant's periphery actively divides. Secondary growth causes a plant to grow in diameter. Only after the evolution of secondary growth could vascular plants develop thick trunks and therefore grow tall. This evolutionary advance made possible the development of forests and, consequently, the domination of the land by plants. According to the fossil record, secondary growth had evolved independently in several different groups of vascular plants by the middle of the Devonian period, approximately 380 million years ago.

> *Primary growth results from cell division at the tips of stems and roots, whereas secondary growth results from the division of a cylinder of cells around the plant's periphery.*

Conducting Systems of Vascular Plants

Two types of conducting elements in the earliest plants are characteristic of the vascular plants as a group: **Sieve elements** are soft-walled cells that conduct carbohydrates away from the areas where the carbohydrates are manufactured. **Tracheary elements** are hard-walled cells that transport water and dissolved nutrients up from the roots. Both kinds of cells are elongated, and both occur in strands. More specialized versions of each occur in the flowering plants, or angiosperms, than in other vascular plants. Sieve elements are the characteristic cell types of a tissue called **phloem;** tracheary elements are characteristic of a tissue called **xylem.** In primary tissues, which result from primary growth, these two types of tissue are often associated with one another in the same vascular strands.

> *Water and nutrients are carried in the xylem, which consists primarily of hard-walled cells called tracheary elements. Carbohydrates, in contrast, are carried in the phloem, which consists of soft-walled cells called sieve elements.*

What Is a Seed?

Seeds are a characteristic feature of some groups of vascular plants, collectively known as **seed plants.** A **seed** contains an embryo surrounded by a protective coat. The embryo's development has been temporarily arrested (figure 28.9). Seeds are the means by which plants, being rooted in the ground, are dispersed to new places. Many seeds have devices, such as the wings on the seeds of pines or maples or the plumes on the seeds of a dandelion, that help them to travel efficiently.

The seed protects the embryonic plant from drying out or being eaten when it is at its most vulnerable stage. Most seeds have abundant food stored in them, either inside the embryo or in specialized storage tissue. The rapidly growing young plant uses the seed as a ready source of energy; the seed thus plays the same role as the yolk of an egg. The evolution of the seed was clearly a critical step in the domination of the land by plants.

Seedless Vascular Plants

The members of four phyla of vascular plants with living representatives do not form seeds. The ferns are members of the most familiar phylum of seedless vascular plants (phylum Pterophyta), which includes about twelve thousand living species. The club mosses and their relatives (phylum Lycophyta) include four living genera with a total of about a thousand species. The horsetails, with a single genus (*Equisetum*) and about fifteen species, make up the phylum Sphenophyta. The fourth phylum, the Psilophyta, consists of two genera, the whisk ferns.

The life cycle of a fern differs from that of a moss primarily in the much more complex development, independence, and dominance of the fern's sporophyte (see Life Cycle 28.2).

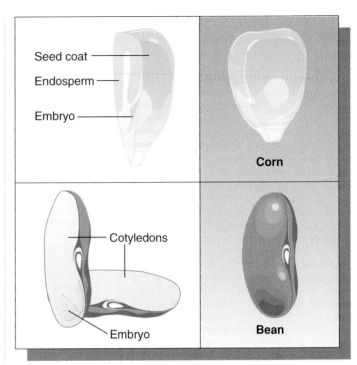

Figure 28.9 Basic structure of seeds.
The seed coat, formed of sporophytic tissue from the parent, protects the embryo—the dormant young plant of the next sporophyte generation—within. The first leaves, specialized in a bean for food storage, are the cotyledons. Seeds are drought resistant and readily dispersed. Most seeds have abundant food for the developing embryo stored in the endosperm.

Gymnosperms

In four of the phyla of living seed plants, the ovules—the structures that eventually become the seeds after fertilization—are not completely enclosed by the tissues of the parent individual on which they are carried at the time of pollination. Members of these four phyla—the conifers, cycads, ginkgoes, and gnetophytes—are called **gymnosperms** (from the Greek *gymnos*, meaning "naked," and *sperma*, meaning "seed"; in other words, naked-seeded plants). In fact, the seeds of gymnosperms often become enclosed by the tissues of their parents by the time the seeds are mature, but their ovules are naked at the time of pollination.

The Cycads, Ginkgo, and Gnetophytes

True seeds apparently evolved only once in the history of plants, and the four living phyla of gymnosperms, together with the flowering plants, or angiosperms, are thought to have descended from a seed-bearing common ancestor. But the gymnosperms are clearly an artificial group, including all seed-bearing plants that do not possess angiosperms' special features.

The four phyla with living species differ greatly from one another. For example, the cycads and ginkgo still have motile sperm, as do all seedless plants. The cycads (phylum Cycadophyta) comprise ten genera and about a hundred species, widespread throughout the warmer portions of the world. One

Life Cycle 28.2

FERN

The gametophytes, which are haploid, grow in moist places. Rhizoids (anchoring structures) project from their lower surface. Eggs and sperm develop in archegonia and antheridia, respectively, on gametophytes' lower surface near the apical notch, the region of the most rapid cell division. The sperm, when released, swim through free water to the mouth of the archegonium, entering and fertilizing the single egg. Following the fusion of egg and sperm to form a zygote—the first cell of the diploid sporophyte generation—the zygote starts to grow within the archegonium. Eventually, the sporophyte becomes much larger than the gametophyte—it is what is known as a fern plant. Most ferns have more or less horizontal stems, called rhizomes, that creep along below the ground. On the sporophyte's leaves, called fronds, occur clusters of sporangia (called sori; singular, *sorus*), within which meiosis occurs and spores are formed. The release of these spores, which is explosive in many ferns, and their germination, lead to the development of new gametophytes.

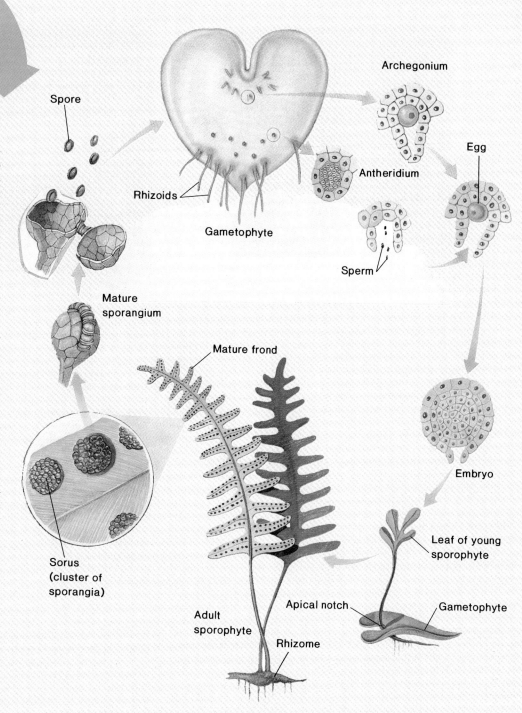

of the most familiar cycads is the sago palm. This is not a true palm, however, despite the palmlike appearance of its leaves; palms are flowering plants. The ginkgo (*Ginkgo biloba*) is the only living species of the phylum Ginkgophyta. Its seeds resemble small plums, with a fleshy, unpleasantly scented outer covering. The ginkgo is not known anywhere as a wild plant, but it was preserved in cultivation around the temples and in the gardens of Japan and China. Phylum Gnetophyta, comprising three genera, including Ephedra, or Mormon tea, and about seventy species, is the group most closely related to the flowering plants.

> *Seed plants are apparently derived from a common ancestor and consist of four phyla of gymnosperms, together with the angiosperms, or flowering plants.*

The Conifers

The most familiar phylum of gymnosperms is Coniferophyta, the conifers (figure 28.10). With about 550 living species, the conifers include such tree groups as pines, spruces, firs, yews,

Figure 28.10 A conifer.
Engelmann spruce, Picea engelmannii, *in the Rocky Mountains.*

(a)

(b)

redwoods, bald cypress, junipers, cedars, and others. Most conifers are evergreens that form vast forests in many of the temperate and cooler regions of the globe. Many are very important commercially as major sources of timber and pulp. In the conifer life cycle, pollen grains, which contain the male gametes, are carried by the wind to the vicinity of the ovules, the structures holding the female gametes. The ovules are carried on cone scales (in pines and their relatives) or similar structures. Within the seeds of conifers and other gymnosperms, stored food is provided for the embryo within tissue derived from the megagametophyte. The life cycle of the pine is presented in Life Cycle 28.3 as an example of the kinds of life cycles found in conifers.

Angiosperms: The Flowering Plants

Flowering plants, or **angiosperms** (phylum Anthophyta), are the dominant photosynthetic organisms nearly everywhere on land. This great group of some 235,000 species includes familiar trees (except conifers), shrubs, herbs, grasses, vegetables, and grains—in short, nearly all of the plants commonly seen every day. Virtually all human food is derived, directly or indirectly, from the flowering plants; in fact, more than half of the calories humans consume come from just three species: rice, corn (maize), and wheat. In addition, angiosperms are valuable sources of timber, pulp, textiles, medicines, waxes, and resins—a wide array of important products, with many more awaiting discovery.

In a sense, angiosperms' remarkable evolutionary success is the culmination of the plant line of evolution. The earliest record of flowering plants is from approximately 123 million years ago, in the early Cretaceous period. Angiosperms, therefore, are by far the youngest of all the plant phyla, despite their world dominance for most of the last hundred million years.

Angiosperm Structure and Reproduction

Angiosperms differ from other seed plants in their possession of flowers and fruits, the characteristic structures of phylum An-

Figure 28.11 The remarkable diversity of angiosperm flowers.
(a) *Wild geranium,* Geranium, *a woodland plant, showing the five free petals, ten stamens, and fused carpel.* (b) *Tiger lily,* Lilium canadense, *with six free, colored, attractive flower parts (they cannot be separated in lilies into sepals and petals) and six free anthers (the parts of the stamens where pollen are formed). The carpels are fused together.* (c) *Fragrant water lily,* Nymphaea odorata, *a flower with numerous free, spirally arranged parts that intergrade with one another in form.*

(c)

thophyta (figure 28.11). A flower basically consists of four **whorls;** a whorl is a circle of parts present at a single level along an axis. The two outer whorls of the angiosperm flower are: (1) the outer **calyx,** the individual parts of which (the **sepals**) are often green and leaflike, surrounding the flower; and (2) the inner **corolla,** the individual parts of which are called **petals** (figure 28.12). The petals are often colored and attractive to insects and other animals that visit the flowers and spread the flowers' pollen.

The inner whorls contain the male and/or female gametophytes. The third whorl, called the **androecium,** consists of the **stamens,** which produce pollen grains. Stamens usually consist of an **anther** (the pollen-bearing portion) and a **filament.**

The fourth and innermost flower whorl, called the **gynoecium,** consists of the **carpels.** In angiosperms, the ovules are enclosed within the carpels, and pollination is indirect. This contrasts with the other phyla of seed plants—the gymnosperms—in which the ovules are carried on the surface of scales, and pollen is often blown to them by the wind.

Life Cycle 28.3

PINE

In all seed plants, the gametophyte generation is greatly reduced. Pine microsporangia are borne in pairs on the surface of the thin scales of the relatively delicate pollen-bearing cones. A germinating pollen grain is the mature microgametophyte of a pine. Megagametophytes, in contrast, develop within the tissues of the ovule. The familiar seed-bearing cones of pines are much heavier and more substantial structures than the pollen-bearing cones. Two ovules, and ultimately two seeds, are borne on the upper surface of each scale. In the spring, when the seed-bearing cones are small and young, their scales are slightly separated. Drops of sticky fluid, to which the airborne pollen grains adhere, form between these scales. After a pollen grain has reached such a drop, it germinates, and a slender pollen tube grows toward the egg. When the pollen tube grows to the vicinity of the megagametophyte, sperm are released, fertilizing the egg and producing a zygote there. The development of the zygote into an embryo takes place within the ovule, which matures into a seed. Eventually, the seed falls from the cone and germinates, the embryo resuming growth and becoming a new pine tree.

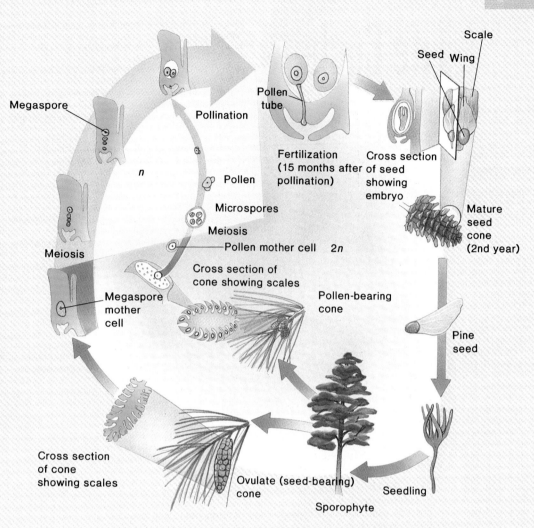

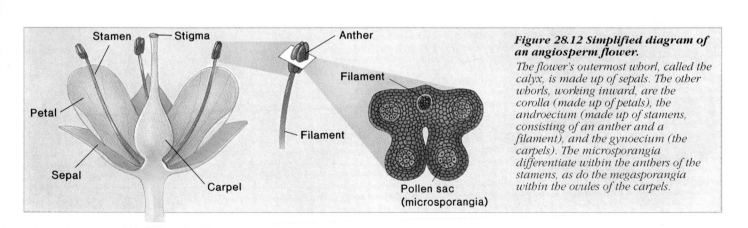

Figure 28.12 Simplified diagram of an angiosperm flower.

The flower's outermost whorl, called the calyx, is made up of sepals. The other whorls, working inward, are the corolla (made up of petals), the androecium (made up of stamens, consisting of an anther and a filament), and the gynoecium (the carpels). The microsporangia differentiate within the anthers of the stamens, as do the megasporangia within the ovules of the carpels.

In many angiosperms, pollen grains are carried from flower to flower by insects; in others, the pollen is transported passively in the wind, as it is in most other seed plants. Chapter 31 discusses the details of angiosperm reproduction. Briefly, however, the pollen reaches a specialized area of the carpel, called the **stigma,** where each pollen grain then forms a **pollen tube,** which grows through the tissue of the carpel and ultimately reaches the egg. Within each pollen grain are two sperm, which travel down the tube. One sperm fuses with the egg to form a zygote.

During meiosis, most angiosperms produce eight haploid nuclei within an **embryo sac.** One of these nuclei becomes an egg. Two others, held together in a large cell at the center of the embryo sac, are called **polar nuclei.** The second sperm from the pollen grain fuses with these two polar nuclei, forming the **primary endosperm nucleus.** Because the primary endosperm nucleus is formed from the fusion of three haploid (n) nuclei, it is triploid ($3n$), whereas the zygote is diploid ($2n$). The primary endosperm nucleus divides rapidly by mitosis, giving rise to a

Life Cycle 28.4

ANGIOSPERM

As in the pine, the sporophyte is the dominant generation. Eggs form within the megagametophyte, or embryo sac, inside the ovules, which, in turn, are enclosed in the carpels—members of the inner whorl of the flower. The carpel is differentiated in most angiosperms into a slender portion, or style, ending in a stigma, the surface on which the pollen grains germinate. The pollen grains, meanwhile, are formed within the sporangia of the anthers and complete their differentiation to their mature, three-celled stage either before or after grains are shed. Fertilization is distinctive in angiosperms, being a double process. A sperm and an egg come together, producing a zygote; at the same time, another sperm fuses with the two polar nuclei, producing the primary endosperm nucleus, which is triploid. Both the zygote and the primary endosperm nucleus divide mitotically, giving rise, respectively, to the embryo and the endosperm. The endosperm is the tissue, unique to angiosperms, that nourishes the embryo and young plant.

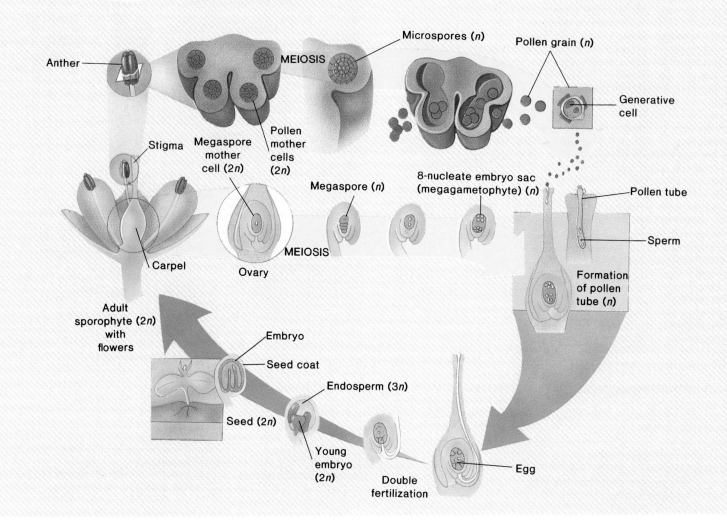

specialized kind of triploid nutritive tissue called the **endosperm,** one of the distinctive features of angiosperms. In different groups of angiosperms, the endosperm either is digested by the growing embryo or is retained in the mature seed to nourish the germinating seedling. The unique process whereby one sperm fuses with the egg while the other fuses with the polar nuclei is called **double fertilization.** This process is characteristic of angiosperms and is found nowhere else. Life Cycle 28.4 outlines the angiosperm life cycle.

Double fertilization, a process unique to the angiosperms, occurs when one sperm fertilizes the egg and the second one fuses with the polar nuclei. These two events result in the formation of the zygote and the primary endosperm nucleus, respectively. The latter divides to produce the endosperm, the nutritive tissue that occurs in the seeds of angiosperms.

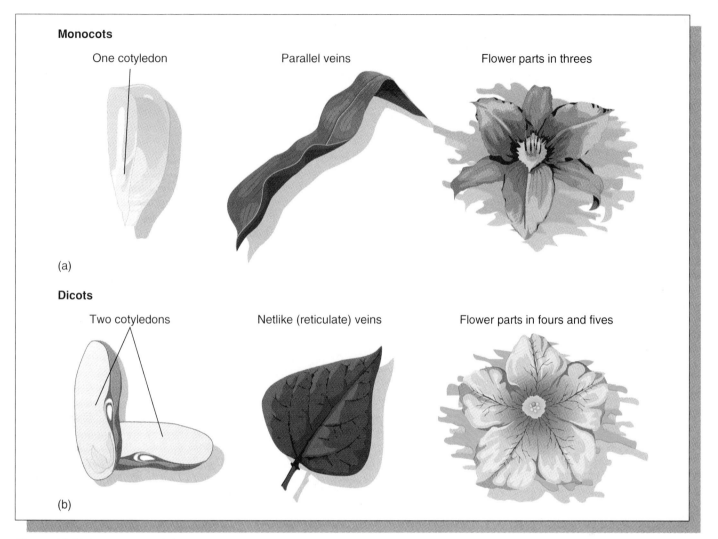

Figure 28.13 Monocots and dicots.
(a) *Monocots are characterized by one cotyledon, parallel veins, and the occurrence of flower parts in threes.* (b) *Dicots have two cotyledons and netlike (reticulate) veins. Their flower parts occur in fours and fives.*

Monocots and Dicots

The two classes of angiosperms (phylum Anthophyta) are: (1) the Monocotyledones, or **monocots** (about 65,000 species), and (2) the Dicotyledones, or **dicots** (about 170,000 species). The monocots include lilies, grasses, cattails, palms, agaves, yuccas, pondweeds, orchids, and irises. The dicots include the great majority of familiar angiosperms of all kinds.

Monocots and dicots differ from one another in several features. For example, the venation of monocot leaves usually consists of parallel veins, whereas dicot leaves generally have netlike (reticulate) veins. In monocot flowers, the members of a given whorl generally occur in threes; in dicots, they are usually in fours and fives. Monocot embryos, as the name of the class implies, generally have one seedling leaf, or **cotyledon;** dicot embryos have two cotyledons (figure 28.13). The arrangement of cotyledons in monocots has been derived from that in

dicots through the suppression of one of the cotyledons. Monocots, which were evidently derived from primitive dicots, share with them a similar kind of single-pored pollen.

The two classes of angiosperms are Monocotyledones (monocots) and Dicotyledones (dicots). Monocots have one cotyledon (seedling leaf), usually parallel venation in their leaves, and flower parts often in threes. Dicots have two cotyledons, usually netlike (reticulate) venation, and flower parts in fours or fives.

A Very Successful Group

Plants have come to dominate the land all over the world, with vascular plants, and especially angiosperms, playing the leading role in most habitats. Many angiosperm features seem to be correlated with successful growth under arid and semiarid conditions, which have become more widespread during the history of this phylum. Flowers effectively ensure the transfer of gametes over substantial distances and therefore promote outcrossing because insects, birds, and other animals are more precise in carrying pollen from one plant to another than is the wind. Fruits, often carried from place to place by animals, were probably especially effective in assisting the spread of some of these early angiosperms from one patch of suitable habitat to another. The tough, often leathery leaves of the flowering plants, their efficient cuticles and stomata, and their specialized conducting elements were all important factors in survival and growth under arid conditions, just as they are in stressful conditions today. The many natural insecticides produced by these plants were essential for their survival as well. As the early angiosperms evolved, all of the features that contributed to their success continued to evolve rapidly.

EVOLUTIONARY VIEWPOINT

Because of their great diversity and rapid rise to evolutionary dominance among the plants, it has long been supposed that flowering plants co-evolved with their pollinators, the insects, themselves the most diverse of animals. Interestingly, recent evidence suggests that the evolutionary interaction may have been more one-way: A number of insect fossils that are older than flowering plants have been found.

SUMMARY

1. The evolution of conducting tissues, cuticle, stomata, leaves, and seeds made plants less dependent on free water and were important factors in the overwhelming success of the group on land.

2. Plants all have an alternation of generations, in which gametophytes, which are haploid, alternate with sporophytes, which are diploid. The spores that sporophytes form as a result of meiosis grow into gametophytes, which produce gametes (sperm and eggs) as a result of meiosis.

3. In plants that do not form seeds and in a few seed-forming plants as well, gametes are differentiated within specialized organs called gametangia. The great majority of vascular plants do not have gametangia; the eggs or sperm develop directly from individual cells of the respective gametophyte.

4. In all seed plants and a few ferns, the gametophytes are either female (megagametophytes) or male (microgametophytes). Megagametophytes produce only eggs; microgametophytes produce only sperm. The megagametophytes and microgametophytes, in turn, are produced from distinct kinds of spores—megaspores and microspores.

5. There are twelve phyla of plants with living representatives, of which seven are seedless and five form seeds. Liverworts (phylum Hepaticophyta) are the simplest living plants. Traditionally, they have been grouped with other simple plants—the mosses (phylum Bryophyta) and hornworts (phylum Anthocerophyta)—as the "bryophytes," but that term is no longer believed to reflect a direct relationship between these groups.

6. Liverworts, mosses, hornworts, and ferns have green, nutritionally independent gametophytes. In the first three phyla, the sporophytes are nutritionally dependent on the gametophytes.

7. Vascular plants, consisting of nine living phyla, have well-defined conducting tissues. Xylem is specialized to conduct water and dissolved materials, while phloem is specialized to conduct carbohydrate molecules that the plants manufacture. Seeds are characteristic of some groups of vascular plants. Four phyla of vascular plants, however, are seedless.

8. The gymnosperms have exposed ovules at the time of pollination, and pollination usually does not involve animals. In the other phyla of seed plants—the angiosperms—the ovules are enclosed at the time of pollination, which is indirect and often mediated by animals.

9. Angiosperms evolved at least 123 million years ago, in the early Cretaceous period, and have been the dominant plants on land for about one hundred million years.

10. Angiosperm flowers typically consist of four whorls. From the inside outward, these whorls are called the gynoecium (consisting of the carpels), androecium (stamens), corolla (petals), and calyx (sepals).

11. Angiosperm seeds have a unique kind of nutritive tissue—the endosperm—derived as part of the process of double fertilization, which is unique to the division. In other seed plants, the role of the endosperm is played by the tissue of the megagametophyte.

12. There are two classes of angiosperms. Monocotyledones, or monocots, include about 65,000 species, largely with parallel venation, flower parts in threes, and one cotyledon. Dicotyledones, or dicots, include approximately 170,000 species, usually with reticulate venation, flower parts in fours or fives, and two cotyledons.

13. Angiosperms were successful on land because of their relatively drought-resistant vegetative features, including their specialized vascular systems, cuticles, and stomata. In addition, flowers facilitated the precise transfer of pollen and therefore outcrossing, even when the stationary individual plants were widely separated. Fruits enabled angiosperms to disperse widely from one favorable patch of habitat to another.

REVIEWING THE CHAPTER

1. What characteristics evolved in plants that allowed them to live on land?

2. What are the main characteristics that distinguish the different phyla of plants?

3. What are the major phases of a plant life cycle?

4. How would you distinguish mosses, liverworts, and hornworts from one another?

5. What are the prominent characteristics of vascular plants?

6. How do vascular plants grow?

7. How do vascular plants conduct food and water?

8. What is a seed?

9. Which vascular plants are seedless?

10. What characterizes a gymnosperm?

11. How are angiosperms separated from all other plants?

12. What are monocots and dicots, and how do they differ from each other?

COMPLETING YOUR UNDERSTANDING

1. What explanation can you give for some 80 percent of plants having mycorrhizal associations?

2. What advantage did plants gain from the evolution of the leaf?

3. What is the fundamental difference between animals and plants when comparing their life cycles?

4. What are the functions of spores and gametes, and when in the life cycle of a plant are they produced?

5. Why would gametangia in plants produce a single egg per archegonium and many sperm per antheridium?

6. What are megasporangia and microsporangia, and what role do they play in a plant life cycle?

7. Why should mosses, liverworts, and hornworts no longer be collectively called bryophytes?

8. What is primary and secondary growth, and which plants exhibit these growth patterns?

9. What are xylem and phloem, and in which plants are they found?

10. In the evolution of land plants, what is the significance of seeds? How do seeds relate to plant survival?

11. Ferns, club mosses, horsetails, and whisk ferns are collectively called

    a. angiosperms.

    b. bryophytes.

    c. seed plants.

    d. gymnosperms.

    e. kelps.

    f. seedless vascular plants.

12. Which groups of plants belong to the gymnosperms? Which of these groups has the greatest number of species?

13. What are some of the commercial uses of gymnosperms and angiosperms?

14. How old is the earliest record of flowering plants? How does this compare to other land plants for origin and duration?

15. What are the parts of a flower, and how do each of these parts function?

16. What is double fertilization, and how does this relate to the success of angiosperms?

17. What is a cotyledon, and in which plants is it found?

18. Which plant phylum has the greatest number of species? Why?

19. What roles have animals played in the success of flowering plants? How do flowering plants relate to the success of animal evolution? Be sure to understand the significance of co-evolution.

FOR FURTHER READING

Gensel, P. G., and H. N. Andrews. 1987. Evolution of early land plants. *American Scientist* 75:478–89. Excellent account of what scientists know about the first plants to invade the land.

Heywood, V. H., ed. 1978. *Flowering plants of the world*. New York: Mayflower Books. An excellent account of the diversity of angiosperms.

Nicklas, K. J. 1987. Aerodynamics of wind pollination. *Scientific American,* July, 90–95. Why the capture of pollen blown about by the wind is not as random as it seems.

Norstog, K. 1987. Cycads and the origin of insect pollination. *American Scientist* 75: 270–79. A discussion of why angiosperms are not the only insect-pollinated plants.

Paolillo, D. J. 1981. The swimming sperms of land plants. *BioScience* 31: 367–73. Excellent account of the sperm in those groups of plants that retain them.

Schofield, W. B. 1985. *Introduction to bryology.* New York: Macmillan. An outstanding analysis of the liverworts, mosses, and hornworts, with many general chapters.

ANIMAL DIVERSITY

The great diversity of animals we see today is the result of millions of years of evolution. This cardinal is more closely related to dinosaurs than to you!

FOR REVIEW

Here are some important terms and concepts that have been discussed in previous chapters and that you will encounter again in this chapter. Review them before proceeding if necessary.

Major features of evolutionary history (*chapter 17*)
Vertebrate evolution (*chapter 18*)
Classification (*chapter 25*)
The six kingdoms of life (*chapter 25*)
Choanoflagellates (*chapter 27*)

One kingdom of life remains to discuss: Kingdom Animalia. Animals are multicellular heterotrophs that ingest their food; there are no photosynthetic animals and no unicellular ones. We are animals, and so are fleas and worms and jellyfish. The animal kingdom evolved from protists in water, and much of its evolution since then has been in the sea. Of the thirty-five phyla in the animal kingdom, only three have been overwhelmingly successful as land-dwellers: the arthropod phylum of spiders and insects (figure 29.1), the mollusks (there are more species of terrestrial mollusks than terrestrial vertebrates), and our own chordate phylum, which includes the vertebrates.

This chapter traces the long evolutionary history of the animals. All of the major animal phyla first evolved in the sea during the Cambrian period. All but three of them are still water-dwellers. Only the phyla containing vertebrates, mollusks, and arthropods (spiders and insects) contain members that are fully terrestrial. The earliest animals to evolve in the sea had no distinct tissues or organs. Later, a succession of animals with well-defined tissues evolved—first, solid worms, and then worms and mollusks with progressively more complex body cavities. The next innovation in body design was segmentation, in which bodies were assembled from similar subunits, like the

cars of a train. In the arthropods, jointed appendages evolved. A radical change in the organization of the embryo accompanied the evolution of the sea stars and their relatives—the echinoderms—and of our own phylum, the chordates. The evolutionary relationships among the major groups of animals are presented visually in figure 29.2.

General Features of Animals

Animals are extraordinarily diverse in form. They range in size from a few that are smaller than many protists to others that are huge, like the truly enormous whales and giant squids. Animal cells are also exceedingly diverse in form and function and are of fundamental importance in making up animals' complex bodies, which function under a wide variety of circumstances. Except for the sponges, animal cells are organized into **tissues,** which are groups of cells combined into a structural and functional unit. In most animals, two or more kinds of tissues are organized into complex **organs.**

Most animals reproduce sexually. Their gametes—eggs and sperm—do not divide by mitosis. With few exceptions, animals are diploid; their gametes are the only haploid cells in their life cycles. The complex form of a given animal develops from a zygote formed from the union of male and female gametes. In a characteristic process of embryonic development, discussed in chapter 44, the zygote first undergoes a series of mitotic divisions and becomes a hollow ball of cells called the **blastula.** This developmental stage occurs in all animals. In most animals, the blastula folds inward at one point to form a hollow sac with an opening at one end called the **blastopore.** An embryo with a blastopore is called a **gastrula.** The subsequent growth and movement of the cells of the gastrula produce the digestive system. The details of early embryonic development differ widely from one phylum of animals to another and often provide important clues to the evolutionary relationships among the phyla.

Animals possess a complex array of cell types. Most are diploid and reproduce sexually. All animals start development by forming a hollow ball of cells, but the details of the embryonic development that follows differ widely.

Sponges: Animals without Tissues

Sponges are the simplest of animals. Sponge cells are not organized into tissues. Sponges lack organs, and most sponges completely lack symmetry. Sponge bodies consist of little more than masses of cells embedded in a gelatinous matrix. There is relatively little coordination among the cells: A sponge can pass through a fine silk mesh, with individual clumps of cells separating, and then reaggregate on the other side.

The body of a young sponge is shaped like a sac or vase. The outside body wall is covered by a layer of flattened cells

Figure 29.1 Desert tarantula (Aphonopelma chalcodes) on sandstone.
This spider belongs to the arthropod phylum. The arthropod phylum and our own phylum, the chordates, are the two most successful land-dwelling phyla.

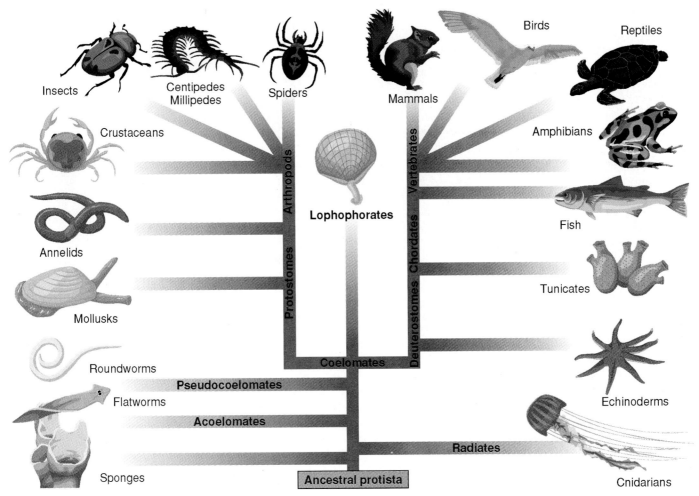

Figure 29.2 The animal ancestral tree.

called the **epithelial wall.** Facing into the internal cavity are specialized, flagellated cells called **choanocytes,** or collar cells. Between the choanocytes and epithelial wall is a gelatinous, protein-rich matrix in which occur various types of amoeboid cells, minute needles of calcium carbonate or silica called **spicules,** and fibers of a tough protein called **spongin.** The spicules and spongin may occur together, or only one of the two may be present. These elements not only strengthen the sponge body, but also deter predators.

The body of a sponge is perforated by tiny holes. The name of the phylum—Porifera—refers to this system of pores. The beating of the flagella of the many choanocytes that line the body cavity draws water in through the pores and drives it through the sponge, thus providing the means by which the sponge acquires food and oxygen and expels wastes. The flagellum of each choanocyte beats independently. In some sponges, 1 cubic centimeter of tissue can propel more than 20 liters of water a day in and out of the sponge body! Water movement through sponge pores and channels is a primitive form of the circulatory systems that occur in other, more complex animals.

Sponges are unique in the animal kingdom because they possess choanocytes, which are special flagellated cells whose beating drives water through the body cavity.

Microscopic examination of individual choanocytes reveals an important substructure: The base of each flagellum is surrounded by a collar of small, hairlike projections that resemble a picket fence. The choanocyte's beating flagellum draws water through the openings in the collar. Any food particles in the water are also drawn in and trapped. The trapped particles pass directly through the plasma membrane and are later digested either by the choanocyte itself or by a neighboring amoeboid cell (figure 29.3).

Each choanocyte closely resembles a protist with a single flagellum and is exactly like the group of unicellular zoomastigotes called choanoflagellates (see chapter 27). This close resemblance indicates that choanoflagellates are almost certainly the ancestors of sponges. They may also be the ancestors of the other animals as well, but a direct relationship between sponges and other groups has not been shown with certainty.

Figure 29.3 Anatomy of a sponge.
The sponge draws water into its central cavity, or spongocoel, through thousands of tiny pores, beating the flagella of its choanocytes to pull the water inward. As the water passes in, the collars of the choanocytes act as filters to capture microscopic food from the seawater.

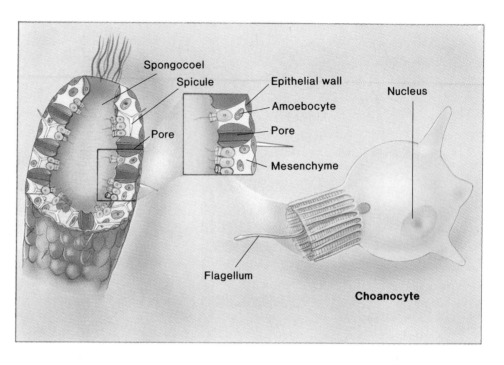

There are about five thousand species of marine sponges and about 150 additional freshwater species (figure 29.4). In the sea, sponges are abundant at all depths. Although some sponges are tiny, no more than a few millimeters across, others, such as the loggerhead sponges, may reach 2 meters or more in diameter. Although larval sponges are free-swimming, the adults are **sessile,** or anchored in place.

Eumetazoans

All animals other than sponges are called **eumetazoans,** or "true animals" because they have a definite shape and symmetry and nearly always distinct tissues. Three distinct cell layers form in the embryos of all eumetazoans: an outer **ectoderm,** an inner **endoderm,** and an in-between **mesoderm.** These layers ultimately differentiate into the tissues of the adult animal. In general, the nervous system and outer covering layers, called **integuments,** develop from the ectoderm, the muscles and skeletal elements develop from the mesoderm, and the intestine and digestive organs develop from the endoderm.

> *All eumetazoans possess three embryonic tissues: ectoderm, mesoderm, and endoderm.*

Cnidarians: Radially Symmetrical Animals

Only the two most primitive of the eumetazoan phyla are **radially symmetrical,** with parts arranged around a central axis like the petals of a daisy. These two phyla are Cnidaria (pronounced nī-DAH-rē-ah), which includes jellyfish, hydra, sea anemones, and corals, and Ctenophora (pronounced tē-NŌ-fō-rah), a minor phylum that includes the comb jellies. The bodies of all other eumetazoans are marked by a fundamental bilateral symmetry.

All cnidarians are carnivores that capture their prey, such as fishes and crustaceans, with the tentacles that ring their mouth. Their bodies, like those of all eumetazoans, consist of three layers and differ greatly from those of sponges. Cnidarians are predominantly marine and basically gelatinous in construction. More than nine thousand species are known (figure 29.5).

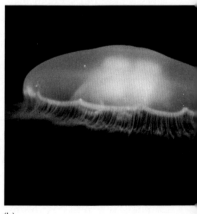

(a)   (b)

Figure 29.4 A barrel sponge.
Barrel sponges are large, and their forms are somewhat organized.

Figure 29.5 Representative cnidarians.
(a) *Yellow cup coral.* (b) *A jellyfish,* Aurelia aurita.

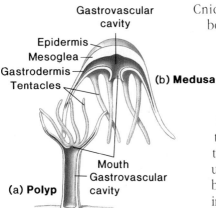

Figure 29.6 Body forms of cnidarians.

(a) *Polyp body form.* (b) *Medusa body form. These two phases alternate in many cnidarian life cycles, but many cnidarians— including the corals and sea anemones, for example—exist only as polyps, whereas others exist only as medusae.*

Cnidarians have two basic body forms: **polyps** and **medusae.** Polyps are cylindrical, pipe-shaped animals that are usually attached to a rock. The polyp mouth faces away from the rock on which the animal is growing and therefore is often directed upward. Many polyps build up a hard shell, an internal skeleton, or both. In contrast, most medusae are free-floating and are often umbrella-shaped. Their mouths usually point downward, and the tentacles hang down around the mouth. Medusae are commonly known as jellyfish because of their thick, gelatinous interior, called the mesoglea. Many cnidarians occur only as polyps, whereas others exist only as medusae; still others alternate between these two phases during the course of their life cycles (figure 29.6).

Individual cnidarian species may either be medusae, which are floating, bell-shaped animals with the mouth directed downward, or polyps, which are anchored animals with the mouth directed upward. These two forms alternate during the life cycle of some cnidarians.

Nematocysts

Cnidarian tentacles bear stinging cells called **cnidocytes.** The name of the phylum Cnidaria refers to these cells, which are highly distinctive and occur in no other group of organisms. Within each cnidocyte is a **nematocyst,** which is best thought of as a small but very powerful harpoon. Cnidarians use nematocysts to spear their prey and then draw the harpooned prey back to the tentacle containing the cnidocyte. The cnidocyte uses water pressure to propel the harpoon. Using transmembrane channels, each cnidocyte builds up a very high internal concentration of ions. Because the cnidocyte's membrane is not permeable to water, this creates an intense osmotic pressure. If a flagellum-like trigger on the cnidocyte is touched, other transmembrane channels open, permitting water to rush in. The resulting hydrostatic pressure pushes the barbed filament of the nematocyst violently outward. Because the filament is shot from the nematocyst so forcefully, the barb can penetrate even the hard shell of crustaceans. Nematocyst discharge is one of the fastest cellular processes in nature. The entire process takes place in about 3 milliseconds, with a maximum velocity of 2 meters per second (figure 29.7).

Cnidarians characteristically possess a specialized kind of cell called a cnidocyte. Each cnidocyte contains a nematocyst, which is like a harpoon and is used to attack prey. Nematocysts are found in no phylum other than Cnidaria.

Extracellular Digestion

A major evolutionary innovation in cnidarians, as compared with sponges, is extracellular digestion of food—that is, digestion within a gut cavity rather than within individual cells. This evolutionary advance has been retained by all of the more advanced groups of animals. Cnidarians have a digestive cavity with only one opening—the mouth (see figure 29.6). Digestive enzymes, primarily proteases, are released from cells lining the walls of the cavity and partially break down the food. But unlike the process in more advanced invertebrates, cnidarian digestion is not completely extracellular: Food is fragmented into small bits, which are then engulfed by the cells lining the gut by the process of phagocytosis.

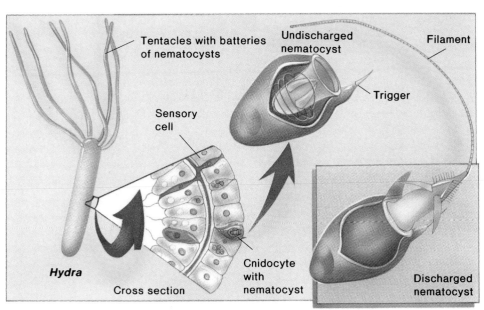

Figure 29.7 Structure of cnidocyte with nematocyst in Hydra.

Hydra *spear their prey with tiny harpoons called nematocysts. Present on both the body and particularly on the tentacles, nematocysts are the "sting" of jellyfish.*

The evolution of extracellular digestion that precedes the phagocytosis and intracellular digestion in cnidarians is important because, for the first time, an organism could digest an animal larger than itself. Cnidarians tackle the job a little at a time.

Cnidarians are the most primitive animals that exhibit extracellular digestion.

Figure 29.8 How radial and bilateral symmetry differ.
(a) *Radial symmetry is the regular arrangement of parts around a central axis, so that any plane passing through the central axis divides the organism into halves that are approximate mirror images. (b) Bilateral symmetry is reflected in a body form in which the right and left halves of an organism are approximate mirror images.*

Evolution of Bilateral Symmetry

Unlike a radially symmetrical animal, a **bilaterally symmetrical** animal has a right half and a left half that are mirror images of each other. There are a top and a bottom, better known, respectively, as the **dorsal** and **ventral** portions of the animal. There is also a front, or **anterior,** end and a back, or **posterior** end, and therefore, right and left sides (figure 29.8). All eumetazoans other than cnidarians and ctenophores are bilaterally symmetrical (even sea stars). Bilaterally symmetrical animals constitute a major advance because this symmetry allows different parts of the body to become specialized in different ways.

Bilaterally symmetrical animals exhibit one of three basic body plans (figure 29.9):

1. Some bilaterally symmetrical animals, called **acoelomates,** have no body cavity at all, other than the digestive system.

2. In another group of phyla, the body cavity develops between the mesoderm and the endoderm, rather than within the mesoderm. Because of the way it originates, the body cavity of these animals lacks the characteristic lining derived from mesoderm that is found in a true coelom. For that reason, this kind of cavity is called a **pseudocoel,** and the animals in which it occurs are called **pseudocoelomates.**

(a) Radial symmetry

Dorsal

Medial

Lateral

Ventral

(b) Bilateral symmetry

Dorsal

Anterior

Posterior

Ventral

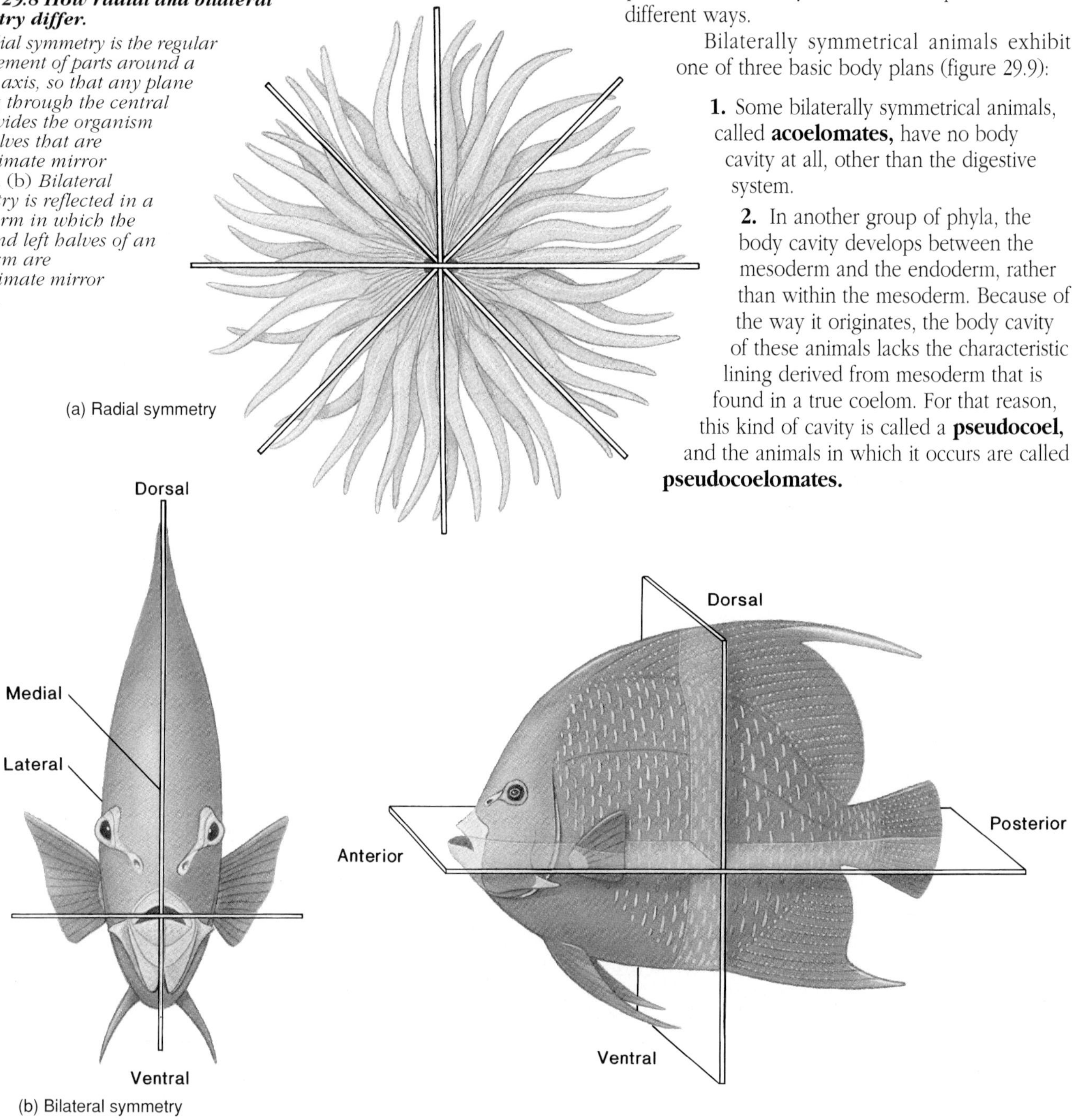

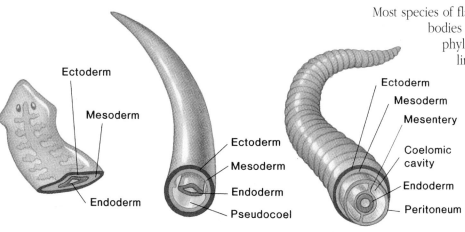

(a) Acoelomate   **(b) Pseudocoelomate**   **(c) Coelomate**

Figure 29.9 Three body plans for bilaterally symmetrical animals.
(a) *Acoelomates have no body cavity.* (b) *Pseudocoelomates develop a body cavity between the mesoderm and endoderm.* (c) *Coelomates have a body cavity bounded by mesoderm.*

3. In the more advanced phyla, the mesoderm opens during development, forming a particular kind of body cavity called the **coelom.** The digestive, reproductive, and other internal organs develop within or around the margins of this coelom and are suspended within it by double layers of mesoderm known as **mesenteries.** Animals in which a coelom develops are called **coelomates.**

The body architecture of bilaterally symmetrical animals follows one of three patterns: (1) acoelomate, possessing no body cavity; (2) pseudocoelomate, possessing a cavity between the endoderm and the mesoderm; and (3) coelomate, possessing a cavity bounded by mesoderm. All vertebrates are coelomates.

Solid Worms: The Acoelomate Phyla

Among bilaterally symmetrical animals, the acoelomates have the simplest body plan: They are solid worms that lack any internal cavity other than the digestive tract (see figure 29.9a). By far the largest phylum of acoelomates, with about fifteen thousand species, is Platyhelminthes, which includes the flatworms. These ribbon-shaped, soft-bodied animals are flattened from top to bottom, like a piece of tape or ribbon (figure 29.10).

Although in structure they are among the simplest of all the bilaterally symmetrical animals, flatworms exhibit traces of many of the evolutionary trends that are so highly developed among members of more advanced phyla. For example, flatworms have distinct bilateral symmetry and a definite head at the anterior end. They also are the simplest animals in which organs occur.

Most species of flatworms are parasitic and occur within the bodies of members of almost every other animal phylum. Flatworms range in size from 1 millimeter or less to many meters long, as in some of the tapeworms.

The acoelomates, typified by the flatworms, are the most primitive bilaterally symmetrical animals and are the simplest animals in which true organs occur.

Flatworms lack circulatory systems, and most have a gut with only one opening. Therefore, they excrete wastes directly into the gut and out through the mouth. To a lesser extent, they also excrete wastes by means of specialized, bulblike cells lined with cilia that function primarily to regulate the organism's water balance. The nervous systems of flatworms are simple; only the tiny swellings that occur near the leading (anterior) end of some flatworms resemble brains.

The free-living flatworms belong to the class Turbellaria, a large group with about 121 families. There are two classes of parasitic flatworms: flukes (class Trematoda) and tapeworms (class Cestoda). Among the best-known flukes are the blood flukes of the genus *Schistosoma*, which afflict some two to three hundred million people (about one in twenty of the world's population) throughout tropical Asia, Africa, Latin America, and the Middle East. The eggs of *Schistosoma* leave

Figure 29.10 Flatworms (phylum Platyhelminthes).
(a) *A common flatworm,* Planaria. (b) *A free-living marine flatworm.*

***Figure 29.11 Structure of the beef tapeworm,* Taenia saginata.**

The scolex—the attachment organ—usually bears several suckers. Each proglottid contains both male and female reproductive organs. Embryos emerge from a gravid (pregnant) proglottid through a genital pore, leave their host with the feces, and are deposited on leaves, in water, and in other places, from which they may be picked up by another animal.

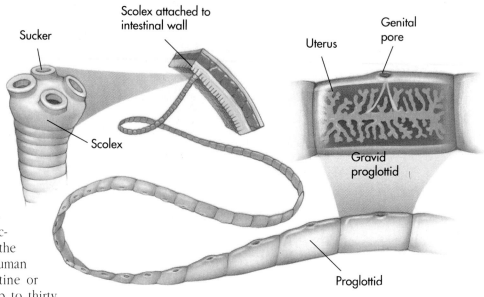

the human body through urine and feces, and their larvae develop within the bodies of freshwater snails. Eventually, these snails release a highly infectious stage of the worm's life cycle; the individuals of this stage burrow into human skin and eventually reach the intestine or bladder, where they may live for up to thirty years, causing inflammation, discomfort, and sometimes death.

Tapeworms also alternate between various hosts: The juvenile beef tapeworm occurs in cattle, but the mature form of the tapeworm occurs in humans. In a tapeworm, the attachment organ, or **scolex,** attaches to the intestinal wall; the body consists mainly of a series of repetitive segments, the **proglottids.** A mature beef tapeworm may reach 10 meters or more in length (figure 29.11). The proglottids that are shed pass out of humans in the feces, scattering embryos that may be ingested by cattle. About 1 percent of all U.S. cattle are infected.

Evolution of a Body Cavity

All other bilaterally symmetrical animals have a body organization that differs from that of the solid worms in an important way: They possess an internal body cavity. A solid body is limited in that there is no internal circulatory or digestive system, and all internal organs are pressed on by muscles and are thus deformed by muscular activity. Even though flatworms have digestive systems, they are subject to the same type of problems.

An internal body cavity circumvents these limitations. Perhaps its most important advantage is that body organs are located within a fluid-filled enclosure, where they can function without having to resist pressures from the surrounding muscles. In addition, the fluid that fills the cavity may act as a circulatory system, freely transporting food, water, waste, and gases throughout the body. Without the free circulation made possible by such a system, every cell of an animal must be within a short distance of oxygen, water, and all of the other substances that it requires.

An animal's digestive system, like its circulatory system, also functions much more efficiently within an internal body cavity, such as a coelom, than when embedded in other tissues. When the gut is suspended within such a cavity, food can pass through the gut freely and at a rate controlled by the animal because the gut does not open and close when the animal moves. By controlling the rate of food passage, the body allows the food to be digested much more efficiently than would

be possible otherwise. Waste removal is also much more efficient under such circumstances.

In addition, an internal body cavity provides space within which the gonads (ovaries and testes) can expand, allowing the accumulation of large numbers of eggs and sperm. Such accumulation helps to make possible all of the diverse modifications of breeding strategy that characterize the phyla of higher animals. Furthermore, large numbers of gametes can be released when the conditions are as favorable as possible for offspring survival.

Nematodes

Nematodes (phylum Nematoda) are bilaterally symmetrical, cylindrical, unsegmented worms that possess a pseudocoelomate body plan (see figure 29.9*b*). They lack a defined circulatory system; the circulatory role is performed by the fluids that move within the pseudocoel. Nematodes, like all coelomates, have a complete, one-way digestive tract that functions like an assembly line, the food being acted on in different ways in each section: First, the food is broken down and absorbed, then the wastes are treated and stored, and so on.

The nematode body is covered by a flexible, thick cuticle that is shed periodically as the individual grows. A layer of muscles extends beneath the epidermis along the length of the worm. These longitudinal muscles push both against the cuticle and against the pseudocoel, whipping the body from side to side as the nematode moves.

Twelve thousand species are recognized in phylum Nematoda, most of them microscopic animals that live in soil (figure 29.12). Nematodes occur practically everywhere. A spadeful of fertile soil has been estimated to contain, on average, one million nematodes. So many new kinds of nematodes are found when environments are sampled that some scientists estimate that there actually may be a half million or more species of this phylum, the great majority of which have never been collected or studied.

Figure 29.12 Nematodes.
One square meter of ordinary garden, lawn, or forest soil teems with two to four million nematodes. Although most are similar in form, they range from about 0.2 millimeter to about 6 millimeters long. A trained nematologist (student of nematodes) must examine slide mounts with a compound microscope to determine which species are present.

Almost every species of plant and animal that has been studied has been found to have at least one parasitic species of nematode living on or in it, and nematodes are one of the most serious groups of pests on agricultural and horticultural crops throughout the world. Roundworms and hookworms are common nematode parasites of humans in the United States.

Rotifers

A second phylum consisting of animals with a pseudocoelomate body plan is Rotifera, the rotifers. Rotifers are common, small, basically aquatic animals that have a crown of cilia at their heads; they range from 0.04 to 2 millimeters long. There are about two thousand species throughout the world. Bilaterally symmetrical and covered with chitin, rotifers depend on their cilia for both locomotion and feeding, ingesting bacteria, protists, and small animals. They are often called "wheel animals" because the cilia, when they are beating together, resemble spokes radiating from a wheel. Each female rotifer lays between eight and twenty large eggs during the course of her life.

An internal body cavity is one of the most significant advances in the animal body plan. The pseudocoelom of nematodes and rotifers protects the internal organs from being deformed by body movements and permits internal circulation of materials.

Advent of the Coelomates

Two of the three bilaterally symmetrical body plans have already been discussed: the acoelomates, represented by the solid flatworms, and the pseudocoelomates, which are worms with a body cavity that develops between their mesoderm and endoderm. Even though both of these body plans have proven very successful, a third way of organizing the body evolved and occurs in the bulk of the animal kingdom—the "higher" invertebrates and vertebrates. This body organization involves the development of a coelom, which is a body cavity that originates within the mesoderm (see figure 29.9c).

Advantage of the Coelom

Both pseudocoelomate and coelomate animals possess a fluid-filled body cavity, a great improvement in body design when compared with the less advanced solid worms. What, then, is the functional difference between a pseudocoel and a coelom, and why has the coelom been so much more successful in evolutionary terms?

In coelomates, the body cavity develops, not between the endoderm and the mesoderm, but entirely within the mesoderm. This makes it easier for complex organ systems to develop. The evolutionary specialization of coelomates' internal organs has far exceeded that of the pseudocoelomates. For example, very few pseudocoelomates possess a true circulatory system, whereas many coelomates have such a system.

In addition, the presence of a coelom allows the digestive tract, by its coiling or folding within the coelom, to be longer than the animal itself. The longer passage allows for storage organs for undigested food, longer exposure to enzymes for more complete digestion, and even storage and final processing of food remnants. Such an arrangement allows an animal to eat a great deal when it is safe to do so and then to hide during the digestive process, thus limiting the animal's exposure to predators. The tube within the coelom architecture is also more flexible, thus allowing the animal greater freedom to move.

The evolution of the coelom was a major improvement in animal body architecture. It permitted the development of a closed circulatory system, provided a fluid environment within which digestive, sexual, and other organs could be suspended, and facilitated muscle-driven body movement.

An Embryonic Revolution

Two major branches of coelomate animals represent two distinct evolutionary lines. In the first, which includes the mollusks, annelids, and arthropods as well as some smaller phyla, the mouth (stoma) develops from or near the blastopore. This pattern of embryonic development also occurs in all noncoelomate animals. An animal whose mouth develops in this way is called a **protostome.** If the animal has a distinct anus or anal pore, it develops later, in another region of the embryo. This kind of developmental pattern is so widespread that it virtually ensures that this pattern is the original one for animals as a whole and that it was characteristic of the common ancestor of all eumetazoans.

A second pattern of embryological development occurs in the echinoderms, the chordates, and a few other small, related phyla. In these animals, the anus forms from or near the blastopore, and the mouth forms later, on another part of the blastula. Animals in this group of phyla are called the **deuterostomes.** They are clearly related to one another by their shared pattern of embryonic development.

The progressive division of cells during embryonic growth is called **cleavage.** The pattern of cell cleavage relative to the embryo's polar axis determines the way in which the cells are arrayed. In nearly all protostomes, each new cell buds off at an angle oblique to the polar axis. As a result, a new cell nestles into the space between the adjacent older ones, resulting in a closely packed array. This kind of pattern is called **spiral cleavage** because a line drawn through a sequence of dividing cells spirals outward from the polar axis (figure 29.13*a*). In deuterostomes, however, the cells divide parallel to and at right angles to the polar axis. As a result, the pairs of cells that result from each division are positioned directly above and below one another; this process gives rise to a loosely packed array of cells. Such a pattern is called **radial cleavage** because a line drawn through a sequence of dividing cells describes a radius outward from the polar axis (figure 29.13*b*).

In protostomes, each embryonic cell's developmental fate is fixed when that cell first appears. Even at the four-celled stage, each cell is different, and not one of them, if separated from the others, can develop into a complete animal because the chemicals that act as developmental signals are localized in different parts of the egg. Consequently, the cleavage divisions that occur after fertilization separate different signals into different daughter cells. On the other hand, in deuterostomes, the first cleavage divisions of the fertilized embryo result in identical daughter cells, any one of which can, if separated, develop into a complete organism. The commitment of individual cells to developmental pathways occurs later.

Discussion of the coelomate animals begins with the three major phyla of protostomes: the mollusks, annelids, and arthropods. These three phyla include such familiar animals as clams, snails, octopuses, earthworms, lobsters, spiders, and insects. The members of these phyla exhibit all of the major advances associated with the evolution of the coelom. An examination of the two largest phyla of deuterostomes—the echinoderms and the chordates—concludes the chapter.

Echinoderms and chordates are deuterostomes, with radial cleavage of cells during embryonic development. Almost all other animals are protostomes, with a spiral pattern of cell cleavage.

Mollusks

The **mollusks** (phylum Mollusca) include the snails, clams, scallops, oysters, cuttlefishes, octopuses, and slugs. Three classes of mollusks are representative of the phylum: (1) Gastropoda—snails, slugs, limpets, and their relatives; (2) Bivalvia—clams, oysters, scallops, and their relatives; and (3) Cephalopoda—squids, octopuses, cuttlefishes, and nautiluses (figure 29.14).

In terms of named species, the mollusks are the largest animal phylum, except for the arthropods. At least 110,000 species of mollusks have been identified, and probably at least that many remain to be discovered. Mollusks are abundant in marine, freshwater, and terrestrial habitats. With about thirty-five thousand species, the terrestrial mollusks far outnumber the roughly twenty thousand species of terrestrial vertebrates.

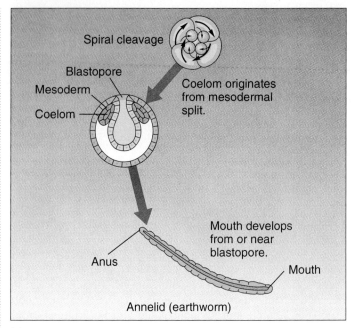

(a)

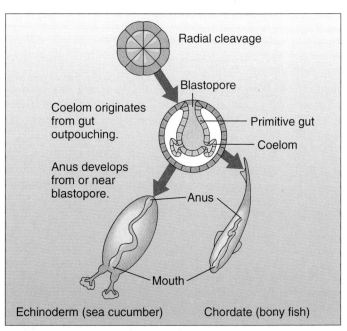

(b)

Figure 29.13 The two basic patterns of embryonic development.

The progressive division of cells during embryonic growth is called cleavage. (a) In spiral cleavage, each new cell buds off at an angle oblique to the axis of the embryo; this kind of cleavage is characteristic of nearly all protostomes. (b) In radial cleavage, which is characteristic of deuterostomes, the cells divide parallel to and at right angles to the polar axis. As a result, the pairs of cells that result from each division are positioned directly above and below one another. These differences are correlated with the other features of embryonic development illustrated here.

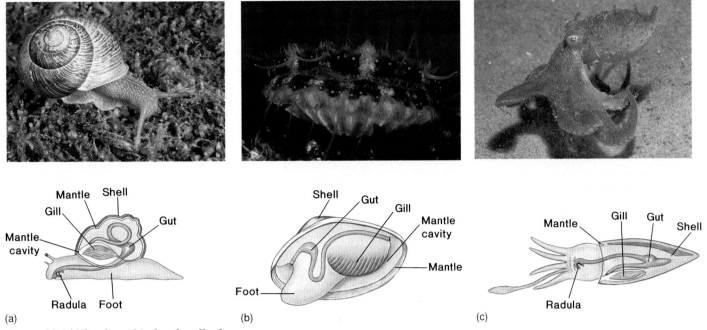

Figure 29.14 The three kinds of mollusks.
(a) *Gastropod.* (b) *Bivalve.* (c) *Cephalopod.*

The mollusk body is composed of a head, a **visceral mass** that contains the organs, a muscular foot used in locomotion, and a heavy fold of tissue, called a **mantle,** wrapped around the visceral mass like a cape. Organs of digestion, excretion, and reproduction are located in the visceral mass. The folds of the mantle enclose a cavity between the dorsal wall and the visceral mass. Within this mantle cavity are the mollusk's gills, which consist of a specialized respiratory system of filamentous projections, rich in blood vessels, that capture oxygen and release carbon dioxide. Mollusk gills are very efficient: Many gilled mollusks extract 50 percent or more of the dissolved oxygen from the water that passes through the mantle cavity. In most members of this phylum, the other surface of the mantle also secretes a protective **shell.** Many mollusks can withdraw for protection into their mantle cavity, which lies within the shell. In the squids and octopuses, the mantle cavity has been modified to create the jet-propulsion system that enables the animals to move rapidly through the water.

The foot of a mollusk is muscular and may be adapted for locomotion, for attachment, for food capture (in squids and octopuses), or for various combinations of these functions. Some mollusks secrete mucus that forms a path along which they glide on their foot. In the cephalopods (squids and octopuses), the foot is divided into arms, called tentacles.

A characteristic feature of mollusks is the **radula,** a rasping tonguelike organ. All members of the phylum have radulas except bivalves, in which the radula was almost certainly lost during evolution. The radula consists primarily of chitin and is covered with rows of pointed, backward-curving teeth. Some of the snails and their relatives—members of a group known as the gastropods—use the radula to scrape algae and other food materials off their substrates and to then convey this food to the digestive tract. Other gastropods are active predators, using their radulas to puncture and extract food from their prey.

> *Mollusks are the second largest phylum of animals in terms of named species. They characteristically have bodies with three distinct sections: a head, a visceral mass enclosed by a mantle, and a foot. All mollusks except the bivalves also possess a unique rasping tongue called a radula.*

The Rise of Segmentation

Unlike mollusks, the other major phyla of protostomes—annelids and arthropods—have segmented bodies. Just as it is efficient for workers to construct a tunnel from a series of identical prefabricated parts, so also are these advanced protostome coelomates "assembled" as a chain of identical segments, like the boxcars of a train. Segmentation underlies the organization of all advanced animals. In some adult arthropods, the segments are fused, making the underlying segmentation difficult to perceive, but segmentation is usually apparent in embryological development. Even among vertebrates, the backbone and muscular areas are segmented.

> *Segmentation is a feature of the advanced coelomate phyla, notably the annelids and the arthropods, although it is not obvious in some of them.*

(a)

(b)

Figure 29.15 Annelids.
(a) *Earthworms are the terrestrial annelids. This night crawler,*
Lumbricus terrestris, *is in its burrow.* (b) *Shiny bristle worm,*
Oenone fulgida, *a polychaete.*

Annelids

The **annelids** (phylum Annelida), one of the major animal phyla, are segmented worms (figure 29.15). They are abundant in marine, freshwater, and terrestrial habitats throughout the world. Their segments are visible externally as a series of ring-like structures running the length of the body. Internally, the segments are divided from one another by partitions. In each of the cylindrical segments of these animals, the digestive and excretory organs are repeated in tandem.

Leeches are one of three classes of annelid worms (class Hirudinea). They are freshwater predators or bloodsuckers and were formerly used in medicine for bloodletting, which was thought to be beneficial. Currently, the enzymes that leeches use to prevent blood clotting are extracted for treating certain conditions in which blood clots are a serious risk.

The two other classes of annelids are Polychaeta, a largely marine group of about eight thousand species, and Oligochaeta, the earthworms, with about thirty-one hundred species. Earthworms are often present in very large numbers and are extremely important in soil aeration and enrichment.

Annelids' basic body plan is a tube within a tube: The internal digestive tract is a tube suspended within the coelom, which is a tube running from mouth to anus (figure 29.16). The anterior segments have become modified, and a well-developed **cerebral ganglion,** or **brain,** is contained in one of them. The sensory organs are mainly concentrated near the worm's anterior end. Some of these are sensitive to light, and elaborate eyes with lenses and retinas have evolved in certain members of the phylum. Separate nerve centers, or ganglia, are located in each segment but are interconnected by nerve cords. These nerve cords are responsible for coordinating the worm's activities.

Annelids use their muscles to crawl, burrow, and swim. In each segment, the muscles play against the fluid in the coelom. This fluid creates a hydrostatic (liquid-supported) skeleton that gives the segment rigidity, like an inflated balloon. Because each segment is separate, each is able to contract or expand independently. Therefore, a long body can move in quite complex ways. When an earthworm crawls on a flat surface, it lengthens some parts of its body while shortening others.

Each segment of an annelid typically possesses **setae** (singular, *seta*), which are bristles of chitin that help to anchor the worms during locomotion or when they are in their burrow. Because of the setae, annelids are often called "bristle worms." The setae are absent in all but one species of the leeches.

The annelids are characterized by serial segmentation. The body is composed of numerous similar segments, each with its own circulatory, excretory, and neural elements and each with its own array of setae.

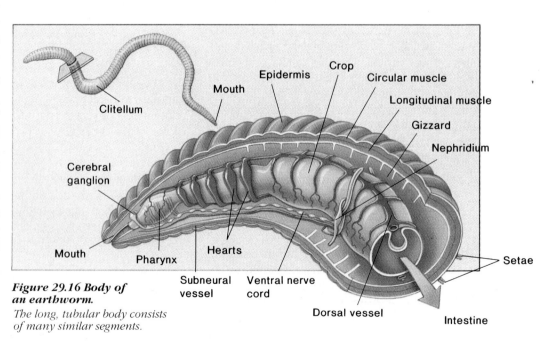

Figure 29.16 Body of an earthworm.
The long, tubular body consists of many similar segments.

Labels: Clitellum, Mouth, Epidermis, Crop, Circular muscle, Longitudinal muscle, Gizzard, Nephridium, Cerebral ganglion, Mouth, Pharynx, Hearts, Subneural vessel, Ventral nerve cord, Dorsal vessel, Intestine, Setae

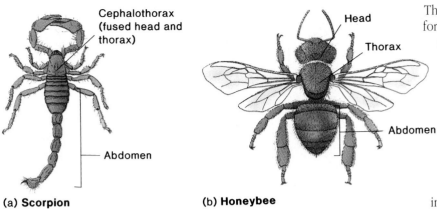

Cephalothorax
(fused head and
thorax)

Abdomen

(a) Scorpion

Head

Thorax

Abdomen

(b) Honeybee

*Figure 29.17 Evolution from
many to few body segments among the arthropods.*
*The scorpion (a) and the honeybee (b) are examples of arthropods
with different numbers of body segments.*

Arthropods

With the evolution of the first annelids, many of the major innovations of animal structure had already appeared: the division of tissues into three primary types (endoderm, mesoderm, and ectoderm), bilateral symmetry, coelomic body architecture, and segmentation. However, one further innovation remained: jointed appendages. This development marks the origin of the body plan characteristic of the most successful of all animal groups, the arthropods.

The name *arthropod* comes from the two Greek words *arthros,* meaning "jointed," and *podes,* meaning "feet." All arthropods have jointed appendages. The numbers of these appendages are progressively reduced in the more advanced members of the phylum, and the nature of the appendages differs greatly in different subgroups. Thus, individual appendages may be modified into antennae, mouthparts of various kinds, or legs.

Arthropod bodies are segmented like those of annelids, a phylum to which at least some of the arthropods are clearly related. Some classes of arthropods have many body segments. In others, the segments have become fused together into functional groups, such as the head or thorax of an insect (figure 29.17). But even in these, the original segments can be distinguished during larval development.

> *Arthropods, like annelid worms, are segmented, but in many
> arthropods, the individual segments are fused into func-
> tional groups, such as the head or thorax of an insect.*

The arthropods have a rigid external skeleton, or **exoskeleton,** in which chitin is an important element (figure 29.18). The exoskeleton provides places for muscle attachment, protects the animal from predators and injury, and most importantly, impedes water loss. As an individual outgrows its exoskeleton, the exoskeleton splits open and is shed. This process, called **ecdysis,** is controlled by hormones.

The eggs of arthropods develop into immature forms that may bear little or no resemblance to the adults of the same species. Most members of this phylum change their characteristics as they develop from stage to stage, a process called **metamorphosis.**

Arthropods are by far the most successful of all life-forms. Approximately a million animal species—about two-thirds of all the named species on the earth—are members of this gigantic phylum, and many millions more are believed to be awaiting discovery. There are several times as many species of arthropods as there are of all other plants, animals, and microorganisms put together. A hectare of lowland tropical forest is estimated to be inhabited, on average, by forty-one thousand insect species. Many suburban gardens have fifteen hundred or more species of this gigantic class of organisms.

The insects and other arthropods are abundant in all of the habitats on this planet, but they especially dominate the land, where along with the flowering plants and the vertebrates, they determine the very structure of life. In terms of individuals, approximately a billion billion insects are estimated to be alive at any one time!

Figure 29.18 Arthropod exoskeletons.
(a) *Some arthropods have a very tough exoskeleton, like this beetle.*
(b) *Others have a fragile exoskeleton, like this spot dragonfly.*

(a)

(b)

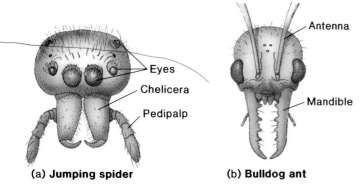

(a) Jumping spider   **(b) Bulldog ant**

Figure 29.19 Differentiating between chelicerates and mandibulates.
(a) *In the chelicerates, such as the jumping spider, the chelicerae are the foremost body appendages, followed by the pedipalps (which resemble legs but have one less segment). (b) In the mandibulates, in contrast, such as the bulldog ant, the antennae are the foremost appendages, and the mandibles are next.*

Diversity of the Arthropods

Many arthropods have jaws, or **mandibles,** formed by the modification of one of the pairs of anterior appendages. (The appendages nearest the anterior end are one or more pairs of sensory antennae, and the next are the mandibles.) These arthropods, called **mandibulates,** include the crustaceans, insects, centipedes, millipedes, and a few other small groups. The remaining arthropods, which include the spiders, mites, scorpions, and a few others, lack mandibles and are called **chelicerates.** Their mouthparts, known as **chelicerae** (singular, *chelicera*), evolved from the appendages nearest the animal's anterior end. They often take the form of pincers or fangs (figure 29.19). The crustaceans seem to have evolved mandibles separately and are not thought to be directly related to the insects and other mandibulates.

Chelicerates

The chelicerate fossil record goes back as far as that of any multicellular animals, about 630 million years. A major group of extinct arthropods, the trilobites, was also abundant then, and the living horseshoe crabs seem to be directly descended from them. By far the largest of the three classes of chelicerates is the largely terrestrial class Arachnida, with some fifty-seven thousand named species, including the spiders, ticks, mites, scorpions, and daddy longlegs (figure 29.20). Scorpions are probably the most ancient group of terrestrial arthropods; they are known from as early as the Silurian period, some 425 million years ago. There are about thirty-five thousand named species of spiders (order Araneae). The order Acari, the mites, is the largest in terms of number of species and the most diverse of the arachnids.

Mandibulates: Crustaceans

The crustaceans (subphylum Crustacea, with a single class) are a large, diverse group of primarily aquatic organisms,

(a)

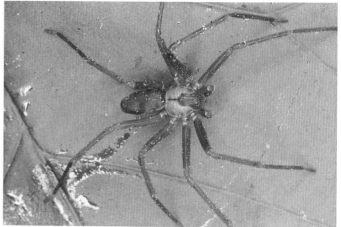

(b)

Figure 29.20 Arachnids.
(a) *One of the two poisonous spiders in the United States and Canada, the black widow spider,* Latrodectus mactans.
(b) *The other genus of poisonous spiders of this area, the brown recluse,* Loxosceles reclusa. *Both species are common throughout temperate and subtropical North America, but bites are rare in humans.*

including some thirty-five thousand species of crabs, shrimps, lobsters, crayfish, barnacles, water fleas, pillbugs, and related groups (figure 29.21). Often incredibly abundant in marine and freshwater habitats and playing a role of critical importance in virtually all aquatic ecosystems, crustaceans have been called "the insects of the water." Most crustaceans have two pairs of antennae, three pairs of chewing appendages, and various numbers of pairs of legs. Crustaceans differ from the insects—but resemble the centipedes and millipedes—in that they have legs on their abdomen as well as on their thorax. They are the only arthropods with two pairs of antennae.

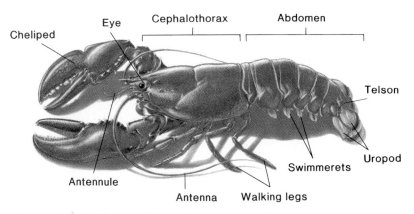

Figure 29.21 Body of a lobster, Homarus americanus.

Some of the specialized terms used to describe crustaceans are indicated. For example, the head and thorax are fused together into a cephalothorax. Appendages called swimmerets occur in lines along the sides of the abdomen and are used in reproduction and also for swimming. Flattened appendages known as uropods form a kind of compound "paddle" at the end of the abdomen. Lobsters may also have a telson, or tail spine.

Mandibulates: Insects and Their Relatives

The insects, class Insecta, are by far the largest group of arthropods, whether measured in terms of numbers of species or numbers of individuals; as such, they are the most abundant group of eukaryotes on earth (figure 29.22). Insects live in every conceivable habitat on land and in freshwater, and a few have even invaded the sea. One scientist calculated that about two hundred million insects are probably alive at any one time for each person on the earth! More than 70 percent of all the named animal species are insects, and the actual proportion is undoubtedly much higher because millions of additional forms await detection, classification, and naming. About 90,000 described species are found in the United States and Canada, but the actual number of species in this region probably approaches 125,000. Table 29.1 presents a glimpse of the enormous diversity of insects.

(a)

(b)

(c)

(d)

(e)

(f)

Figure 29.22 Insects and some relatives.

(a) *Copulating grasshoppers (order Orthoptera).* (b) *Often called the "praying mantis" because of the way its upper arms are folded, the mantid is a powerful predator.* (c) *A butterfly,* Enodia anthedon. *As larvae called caterpillars, butterflies have no wings and are herbivores. After complete metamorphosis, the winged adults feed on flower nectar.* (d) *An African blister beetle, which has spread its two tough, leathery forewings, exposing its delicate flying wings. More than 350,000 species of beetles—a quarter of all species of animals—have been described.* (e) *A millipede. Millipedes are sedentary herbivores, whereas centipedes are active predators. Millipedes and centipedes are closely related to insects; all of the other animals illustrated here are insects.* (f) *A mayfly,* Hexagenia limbata, *immediately after molting to reach the adult form. Adult mayflies, the reproductive stage, do not feed, and they live for only a few hours or a few days. Their nymphs, which are aquatic, live underwater for months or even years.*

Table 29.1   Major Orders of Insects

Order	Typical Examples	Key Characteristics	Approximate Number of Named Species
Coleoptera	Beetles	The most diverse animal order; two pairs of wings; the front pair of wings is a hard cover that partially protects the transparent rear pair of flying wings; heavily armored exoskeleton; biting and chewing mouthparts; complete metamorphosis	350,000
Diptera	Flies	Some that bite people and other mammals are considered to be pests; the front flying wings are transparent; the hind wings are reduced to knobby balancing organs; sucking, piercing, and lapping mouthparts; complete metamorphosis	120,000
Lepidoptera	Butterflies	Often collected for their beauty; two pairs of broad, scaly flying wings, often brightly colored; hairy body; tubelike, sucking mouthparts; complete metamorphosis	120,000
Hymenoptera	Bees	Often social, known to many people because of their sting; two pairs of transparent flying wings; mobile head and well-developed eyes; often possess stingers; chewing and sucking mouthparts; complete metamorphosis	100,000
Hemiptera	Bedbugs	Many so tiny, they are difficult to see; often live on blood; two pairs of wings, or wingless; piercing, sucking mouthparts; complete metamorphosis	60,000

Insects have three body sections (head, thorax, and abdomen); three pairs of legs, all attached to the thorax; and one pair of antennae (figure 29.23). Most insects have compound eyes, which are composed of many independent visual units, among them the ocelli (simple eyes). The insect thorax consists of three segments, each of which has a pair of legs, although occasionally, one or more of these pairs of legs is absent. The thorax is almost entirely filled with muscles that operate the legs and wings. Insects have two pairs of wings, which are attached to segments of the thorax, although in the flies and some other groups, one or both of these pairs have been lost during the course of evolution. The wings of adult insects are solid sheets of chitin with strengthening veins made of chitin tubules. Like many other arthropods, insects have no single major respiratory organ. Their respiratory system consists of small, branched, cuticle-lined air ducts called **tracheae** (singular, *trachea*), which are a series of tubes that transmit oxygen throughout the body. The tracheae ultimately branch into very small **tracheoles.** Air passes into the tracheae through specialized openings called **spiracles,** which occur on many of the segments of the thorax and abdomen.

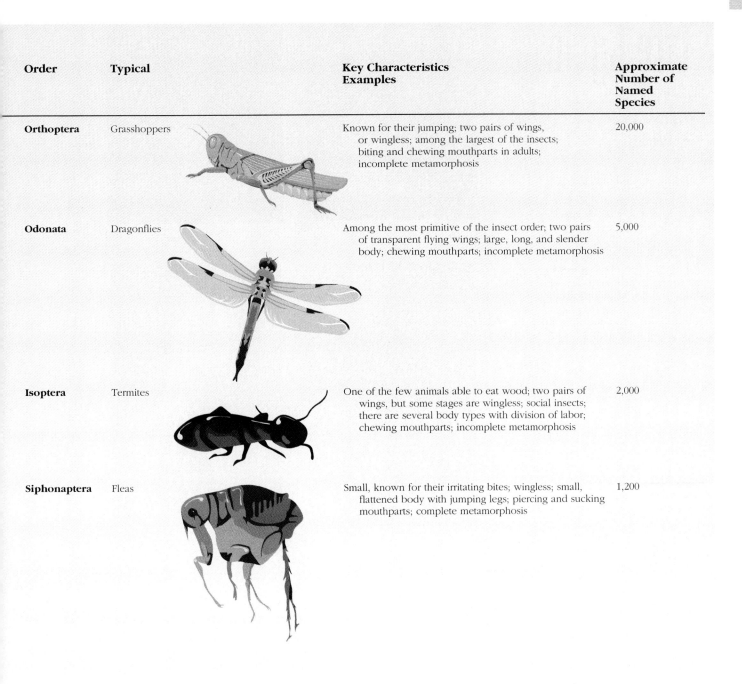

Order	Typical	Key Characteristics Examples	Approximate Number of Named Species
Orthoptera	Grasshoppers	Known for their jumping; two pairs of wings, or wingless; among the largest of the insects; biting and chewing mouthparts in adults; incomplete metamorphosis	20,000
Odonata	Dragonflies	Among the most primitive of the insect order; two pairs of transparent flying wings; large, long, and slender body; chewing mouthparts; incomplete metamorphosis	5,000
Isoptera	Termites	One of the few animals able to eat wood; two pairs of wings, but some stages are wingless; social insects; there are several body types with division of labor; chewing mouthparts; incomplete metamorphosis	2,000
Siphonaptera	Fleas	Small, known for their irritating bites; wingless; small, flattened body with jumping legs; piercing and sucking mouthparts; complete metamorphosis	1,200

Also characteristic of insects are complex patterns of metamorphosis, in which certain stages give way to others. In some insects, such as grasshoppers, cockroaches, and termites, the stages grade from one to the other without abrupt changes; in others, such as butterflies, moths, flies, and beetles, the changes are abrupt (Life Cycle 29.1).

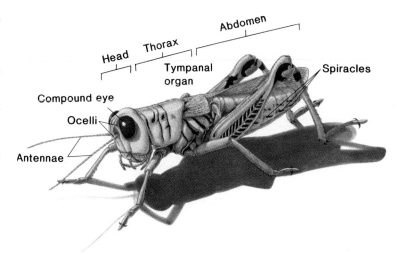

Figure 29.23 Body of a grasshopper.
The diagram illustrates the major structural features of insects, the most numerous group of arthropods. The left wing has been removed to show the features of the abdomen.

Life Cycle 29.1

Monarch Butterfly, *Danaus plexippus*

The life cycle of the monarch butterfly illustrates complete metamorphosis:

(1) The life cycle of *D. plexippus* begins as a tiny, translucent fertilized egg, laid on a leaf.

(2) The egg hatches a larva (caterpillar), which feeds on the leaf. The pseudopods, or false legs, that occur on the abdomen of larval butterflies and moths are not related to the true legs that occur in the adults.

(3) After growing and shedding its tough skin several times, the larva undergoes metamorphosis within a tough case called a chrysalis.

(4) A chrysalis.

(5) The colors of the wings of the adult butterfly can be seen through the thin outer skin of the chrysalis. The adult is nearly ready to emerge.

(6–8) After several weeks, an adult butterfly emerges from the chrysalis and is ready within a few hours to fly.

(9) An adult monarch butterfly lives for several months and may migrate a considerable distance during its lifetime.

(1)

(2)

(5)

(6)

Deuterostomes

Two outwardly dissimilar large phyla—Echinodermata and Chordata—have a series of key embryological features that are very different from those of other animal phyla. Because these features are extremely unlikely to have evolved more than once, both echinoderms and chordates are believed to share a common ancestry. They are the members of the group called the deuterostomes. Deuterostomes diverged from protostome ancestors more than 630 million years ago.

Deuterostomes, like protostomes, are coelomates. But they differ fundamentally from protostomes in the way the embryo grows (see figure 29.13). The blastopore of a deuterostome becomes the animal's anus, and the mouth develops at the other end. Deuterostomes have radial, rather than spiral, cleavage. Daughter cells are identical for a brief period of development. Each daughter cell, if separated from the others at an early

enough stage, can develop into a complete organism. Also, whole groups of cells move around during embryonic development to form new tissue associations.

Echinoderms

Echinoderms (phylum Echinodermata) are an ancient group of marine animals that are very well represented in the fossil record. The term *echinoderm* means "spine-skin," an apt description of many members of this phylum, which comprises about six thousand living species. Many of the most familiar seashore animals, such as the sea stars (starfish), brittle stars, sea urchins, sand dollars and sea cucumbers, are echinoderms (figure 29.24). All are bilaterally symmetrical as larvae but radially symmetrical as adults; thus, they are said to be fundamentally bilaterally symmetrical and are believed to have evolved from bilaterally symmetrical ancestors. Echinoderms are well represented not only in the shallow waters of the sea but also

(3)   (4)

(7)   (8)   (9)

Figure 29.24 An echinoderm.
A brittle star, Ophiothrix.

in its abyssal depths. All but a few of them are bottom-dwellers. The adults range from a few millimeters to more than 1 meter in diameter or length.

Echinoderms have a five-part body plan, corresponding to the arms of a sea star. As adults, these animals have no head or brain. Their nervous systems consist of central nerve rings, from which branches arise. The animals are capable of complex response patterns, but function is not centralized. Centralization of the nervous system apparently is not feasible in animals with radial symmetry.

The **water vascular system** of an echinoderm radiates from a ring canal that encircles the animal's esophagus. Five radial canals extend into each of the five parts of the body and determine its basic symmetry. Water enters the water vascular system through a sievelike plate on the animal's surface, from which it flows through a tube to the ring canal. The five radial canals, in turn, extend out of the animal through short side branches into the many hollow tube feet. In some echinoderms,

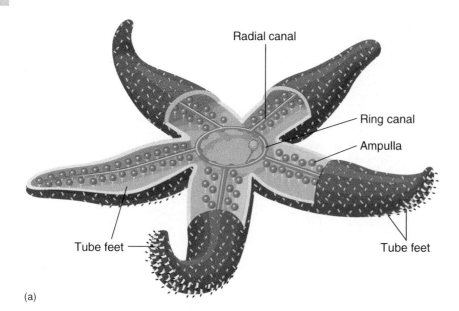

(a)

Radial canal

Ring canal

Ampulla

Tube feet

Tube feet

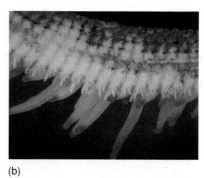

(b)

Figure 29.25 Structure of an echinoderm.
(a) *The echinoderm body plan, emphasizing the water vascular system of a sea star.* (b) *The extended tube feet of a sea star,* Ludia magnifica.

each tube foot has a sucker at its end. At the base of each tube foot is a fluid-containing muscular sac, the **ampulla.** When the sac contracts, the fluid is prevented from entering the radial canal and is forced into the tube foot, thus extending it. When extended, the foot attaches itself to the substrate. Longitudinal muscles can then shorten the foot, and the animal is pulled forward to a new position as water is forced back into the foot by the ampulla. Such action, repeated in many tube feet, allows sea stars, sand dollars, and sea urchins to slowly move along (figure 29.25).

As adults, echinoderms are radially symmetrical animals whose bodies typically have a pattern that repeats itself five times. They are able to move by means of a unique water vascular system connected to their tube feet.

Chordates

The chordates (phylum Chordata) are the best understood and most familiar group of animals. There are some 42,500 species of chordates, a phylum that includes the birds, reptiles, amphibians, fishes, and mammals (and therefore the human species).

The three principal features of chordates were presented in chapter 18, where the evolution of vertebrates was discussed, but are reviewed here briefly: All chordates have (1) a single, hollow nerve cord, (2) a notochord, and (3) pharyngeal slits (see figure 18.3), and each of these features has played an important role in the phylum's evolution. In the more advanced vertebrates, the dorsal nerve cord becomes differentiated into the brain and spinal cord. The notochord, which persists throughout the life cycle of some of the invertebrate chordates, becomes surrounded and then is replaced by the vertebral column during vertebrates' embryological development.

Pharyngeal slits are present in all vertebrate embryos but are lost later in the development of the terrestrial vertebrates. However, they provide a clue to the group's aquatic ancestry.

Chordates are characterized by a single, hollow dorsal nerve cord. At some point in the embryonic development of all chordates, the notochord, which is a flexible rod, forms dorsal to the gut, and slits are present in the pharynx. The notochord persists into the adult stage in the less advanced chordates.

In addition to these three principal features, many other characteristics distinguish the chordates. In their body plan, chordates are more or less segmented, and distinct blocks of muscles, called somites, can be seen clearly in many less specialized forms (figure 29.26). Chordates have an internal skeleton against which the muscles work, and either this skeleton or the notochord makes possible chordates' extraordinary powers of locomotion. Finally, chordates have a tail that extends beyond the anus, at least during their embryonic development; nearly all other animals have a terminal anus.

The chordates are divided into three major groups. Two of them—the tunicates and the lancelets—are **acraniates;** that is, they lack a brain. The **craniate** chordates are the **vertebrates.**

Tunicates

The tunicates (subphylum Tunicata) are a group of about 1,250 species of marine animals. As adults, most of them are sessile (fixed in one spot) and lack visible signs of segmentation or of a body cavity (figure 29.27).

Discerning the evolutionary relationships of an adult tunicate by examining its features would be difficult. But the tadpolelike larvae of tunicates plainly exhibit all basic chordate

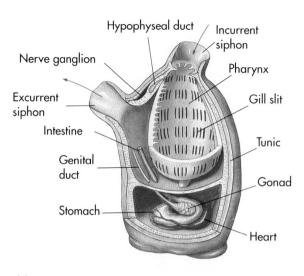

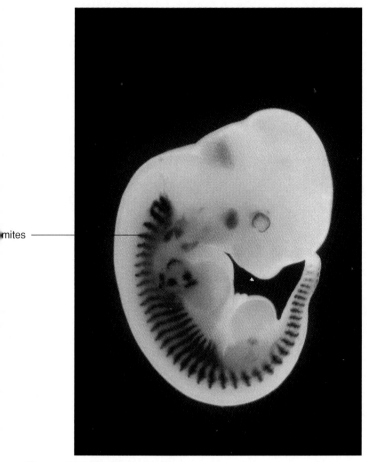

Figure 29.26 A mouse embryo.
At 11.5 days of development, the muscle is already divided into segments called somites (stained dark in this photo), *reflecting the fundamentally segmented nature of all chordates.*

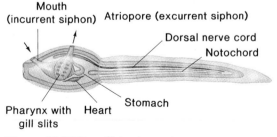

(b)

Figure 29.27 Tunicates.
(a) *Diagram of tunicate structure.* (b) *A beautiful blue and gold tunicate.*

Figure 29.28 Our distant cousins.
The structure of a larval tunicate is much like that of the postulated common ancestor of the chordates.

characteristics (notochord, nerve cord, and pharyngeal slits) and mark tunicates as having the most primitive combination of features found in any chordate (figure 29.28). Tunicate larvae do not feed and have a poorly developed gut. They remain free-swimming for no more than a few days. Then they settle to the bottom and attach themselves to a suitable substrate by means of a sucker.

Lancelets

The lancelets (subphylum Cephalochordata) are scaleless, fishlike marine chordates a few centimeters long. They occur widely in shallow waters throughout the world's oceans. The approximately twenty-three species of lancelets were given their English name because of their similarity to a lancet, a small, two-edged surgical knife. The notochord persists throughout the lancelet life cycle and runs the entire length of the dorsal nerve cord (figure 29.29).

Vertebrates

Vertebrates (subphylum Vertebrata) differ from other chordates in that they usually possess a vertebral column, which replaces the notochord to a greater or lesser extent in adult individuals. In addition, the vertebrates, or craniate chordates, have a distinct head with a skull and brain. The hollow dorsal nerve cord of most vertebrates is protected with a U-shaped groove formed by paired projections from the vertebral column (figure 29.30).

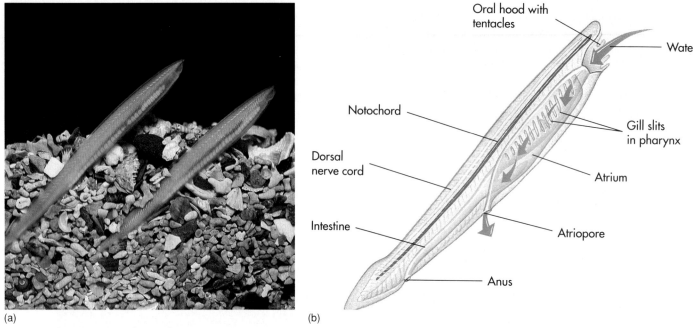

(a)   (b)

Figure 29.29 Body of a lancelet.

(a) *Two lancelets,* Branchiostoma lanceolatum, *partly buried in shell gravel, with their anterior ends protruding. The muscle segments are clearly visible in this photograph. The numerous square, pale yellow objects along the side of the body are gonads, indicating that these are male lancelets.*

(b) *The structure of a lancelet, showing the path by which water is pulled through the animal. The oral hood projects beyond the mouth and bears sensory tentacles, which also ring the mouth itself.*

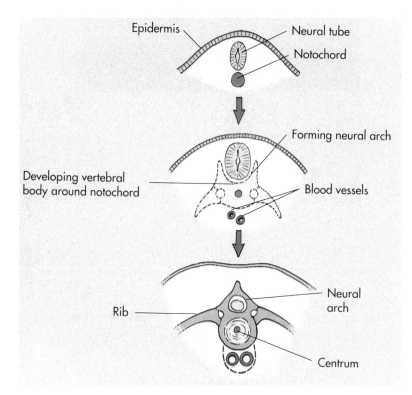

Figure 29.30 Embryonic development of a vertebrate.

During the course of development, the flexible notochord is surrounded and eventually replaced by a cartilaginous or bony covering, the centrum. The neural tube is protected by an arch above the centrum, and the vertebrae may also have a hemal arch, which protects the major blood vessels below the centrum. The vertebral column is a strong, flexible rod against which the muscles pull when the animal swims or moves.

The seven classes of living vertebrates were discussed and illustrated in chapter 18; three of them are fishes, and four are tetrapods. The classes of fishes are: (1) Agnatha, the lampreys and hagfishes; (2) Chondrichthyes, the cartilaginous fishes, sharks, skates, and rays; and (3) Osteichthyes, the bony fishes, the dominant group of fishes today. The four classes of tetrapods are: (1) Amphibia, the amphibians, including salamanders, frogs, and toads; (2) Reptilia, the reptiles; (3) Aves, the birds; and (4) Mammalia, the mammals.

A Look Back

This chapter has examined ten animal phyla—Porifera, Cnidaria, Platyhelminthes, Nematoda, Rotifera, Mollusca, Annelida, Arthropoda, Echinodermata, and Chordata—whose diversity in body structure provides an overview of the evolution of the animal body plan. Table 29.2 briefly reviews these ten phyla, as well as several additional animal phyla of significance.

Table 29.2 Animals

Phylum	Typical Examples	Key Characteristics	Approximate Number of Named Species
Arthropoda (arthropods)	Beetles, other insects, crabs, spiders	Most successful of all animal phyla; chitinous exoskeleton covering segmented bodies with paired, jointed appendages; many insect groups have wings	1,000,000
Mollusca (mollusks)	Snails, oysters, octopuses, nudibranchs	Soft-bodied coelomates whose bodies are divided into three parts: head-foot, visceral mass, and mantle; many have shells; almost all possess a unique rasping tongue called a radula; 35,000 species are terrestrial	110,000
Chordata (chordates)	Mammals, fish, dinosaurs, birds	Segmented coelomates with a notochord; possess a dorsal nerve cord, pharyngeal slits, and a tail at some stage of life; in vertebrates, the notochord is replaced during development by the spinal column; 20,000 species are terrestrial	42,500
Platyhelminthes (flatworms)	*Planaria*, tapeworms, liver flukes	Solid, unsegmented, bilaterally symmetrical worms; no body cavity; digestive cavity has only one opening	15,000
Nematoda (roundworms)	*Ascaris*, pinworms, hookworms, *Filaria*	Pseudocoelomate, unsegmented, bilaterally symmetrical worms; tubular digestive tract passing from mouth to anus; tiny; without cilia; live in great numbers in soil; some are important animal parasites	12,000+

(continued)

Table 29.2   Animals (continued)

Phylum	Typical Examples		Key Characteristics	Approximate Number of Named Species
Annelida (segmented worms)	Earthworms, polychaetes, leeches, beach tube worms		Coelomate, serially segmented, bilaterally symmetrical worms; complete digestive tract; terrestrial forms have bristles called setae on each segment that anchor them during crawling	12,000
Cnidaria (jellyfish)	Jellyfish, hydra, corals, sea anemones		Soft, gelatinous, radially symmetrical bodies whose digestive cavity has a single opening; possess tentacles armed with stinging cells called cnidocytes that shoot sharp harpoons called nematocysts; marine	10,100
Echinodermata (starfish)	Sea stars, sea urchins, sand dollars, sea cucumbers		Deuterostomes with radially symmetrical adult bodies; endoskeleton of calcium plates; five-part body plan and unique water vascular system with tube feet; able to regenerate lost body parts; marine	6,000
Porifera (sponges)	Barrel sponges, boring sponges, basket sponges, vase sponges		Asymmetrical bodies; without distinct tissues or organs; saclike body consists of two layers breeched by many pores; internal cavity lined with food-filtering cells called choanocytes; mostly marine (150 species live in freshwater)	5,150
Bryozoa (moss animals)	*Bowerbankia, Plumatella,* sea mats, sea moss		Microscopic, aquatic deuterostomes that form branching colonies, possess U-shaped row of ciliated tentacles for feeding called a lophophore that usually protrudes through pores in a hard exoskeleton; also called "Ectoprocta" because the anus or proct is external to the lophophore	4,000

Phylum	Typical Examples		Key Characteristics	Approximate Number of Named Species
Rotifera (wheel animals)	Rotifers		Small, aquatic pseudocoelomates with a crown of cilia around mouth resembling a wheel; almost all live in freshwater	2,000
Minor worms	Velvet worms, acorn worms, arrowworms, giant tube worms		**Chaetognatha** (arrowworms): coelomate deuterostomes; bilaterally symmetrical; large eyes and powerful jaws **Hemichordata** (acorn worms): marine worms with dorsal *and* ventral nerve cords **Onychophora** (velvet worms): protostomes with a chitinous exoskeleton; evolutionary relicts **Pognophora** (tube worms): sessile deep-sea worms with long tentacles; live within chitinous tubes attached to the ocean floor **Rhynchocoela** (ribbon worms): Acoelomate, bilaterally symmetrical marine worms with long, extendable proboscis	980
Brachiopoda (lamp shells)	*Lingula*		Like bryozoans, possess a lophophore, but within two clamlike shells; more than 30,000 species known as fossils	250
Ctenophora (sea walnuts)	Comb jellies, sea walnuts		Gelatinous, almost transparent, often bioluminescent marine animals; eight bands of cilia; largest animals that use cilia for locomotion; complete digestive tract with anal pore	100
Phoronida (lophophores)	*Phoronis*		Lophophorate, protostome tube worms; often live in dense populations; unique U-shaped gut, instead of the straight digestive tube of other tube worms	12

(continued)

Table 29.2   Animals (continued)

Phylum	Typical Examples		Key Characteristics	Approximate Number of Named Species
Loricifera (sand animals)	*Nanoloricus mysticus*		Tiny, bilaterally symmetrical pseudocoelomates that live in spaces between grains of sand; mouthparts possess a unique flexible tube; most recently-discovered animal phylum (1983)	6

EVOLUTIONARY VIEWPOINT

The evolution of the animal body plan was not a simple linear progression of "improvements," and it certainly involved more than the few stages highlighted in this chapter. However, focusing in this chapter on some of the major phyla that characterized the main variations in body architecture helps us to appreciate the central lesson of animal diversity, which is that the different animal phyla did not evolve all at once, but rather, represent a long history of evolution and adaptation.

SUMMARY

1. The animals, kingdom Animalia, comprise some thirty-five phyla and at least four million species. Animals are heterotrophic, multicellular organisms that ingest their food.

2. The sponges (phylum Porifera) are characterized by specialized, flagellated cells called choanocytes; a lack of symmetry in body organization; and a lack of tissues and organs.

3. All animals other than sponges are called eumetazoans and have radially or bilaterally symmetrical body plans and distinct tissues. The cnidarians (phylum Cnidaria) are predominantly marine, radially symmetrical animals with unique stinging cells called cnidocytes, each of which contains a specialized harpoon apparatus, or nematocyst.

4. Bilaterally symmetrical animals include three major evolutionary lines: acoelomates, which lack a body cavity; pseudocoelomates, which develop a body cavity between the mesoderm and the endoderm; and coelomates, which develop a body cavity (coelom) within the mesoderm.

5. Acoelomates are the most primitive bilaterally symmetrical animals. They lack an internal cavity, except for the digestive system, and are the simplest animals that have organs, which are structures made up of two or more tissues. The most prominent phylum of acoelomates is Platyhelminthes, which includes the free-living flatworms and the parasitic flukes and tapeworms.

6. Pseudocoelomates, exemplified by the nematodes (phylum Nematoda) and rotifers (phylum Rotifera), have a body cavity that develops between the mesoderm and the endoderm. This pseudocoel provides a place to which muscles can attach, giving these worms enhanced powers of movement.

7. The two major branches of coelomate animals—the protostomes and the deuterostomes—represent two distinct evolutionary lines. The major phyla of protostomes are Mollusca, Annelida, and Arthropoda. The major phyla of deuterostomes are Echinodermata and Chordata.

8. In the protostomes, the mouth develops from or near the blastopore, and early divisions of the embryo are spiral. At early stages of development, the fate of the individual cells is already determined, and they cannot develop individually into a whole animal.

9. In the deuterostomes, the anus develops from or near the blastopore, and the mouth forms later, on another part of the blastula. At early stages of development, each cell of the embryo can differentiate into a whole animal.

10. Arthropods are the most successful of all animals in terms of numbers of individuals and numbers of species, as well as in terms of ecological diversification. Like the annelids, arthropods have segmented bodies, but in arthropods, some of the segments have become fused during the course of evolution.

11. Insects, which have six legs, are the largest class of arthropods, with an estimated three hundred million insects alive at any one time for each person on the earth.

12. Echinoderms are marine deuterostomes that are radially symmetrical as adults. They have a unique water vascular system that includes tube feet, by means of which some echinoderms move.

13. Chordates are characterized by a single, hollow dorsal nerve cord; a notochord; and pharyngeal slits, at least as embryos. There are three chordate subphyla: tunicates, lancelets, and vertebrates.

REVIEWING THE CHAPTER

1. What distinguishes animals from life-forms in other kingdoms?

2. Why are sponges the simplest of animals?

3. What are the distinguishing features of eumetazoans?

4. Which animal phyla are radially symmetrical?

5. What are some of the distinguishing features of cnidarians?

6. What characterizes bilateral symmetry?

7. In bilaterally symmetrical animals, what are the differences in body cavities between phyla?

8. What distinguishes solid worms from other animal phyla?

9. How did the body cavity evolve?

10. What distinguishes nematodes from other animals?

11. What are the distinguishing features of rotifers?

12. What are the advantages of a true coelom?

13. What are the two distinct evolutionary lines in coelomate animals?

14. What characteristics distinguish mollusks from other animals?

15. Which animals are segmented?

16. How are annelids different from other animals?

17. What are arthropods?

18. How do arthropods differ from one another?

19. What are chelicerates, and how far back do they go in the fossil record?

20. Which animals are mandibulates?

21. How do deuterostomes differ from protostomes?

22. How are echinoderms distinguished from other animals?

23. What is a chordate, and how are the three subphyla in this group separated from one another?

COMPLETING YOUR UNDERSTANDING

1. All major groups of animal phyla first evolved in the sea during the _____ geological time period, or approximately _____ millions of years ago.

2. What are the differences between tissues and organs?

3. What is the blastula, and where does it originate?

4. What is the evidence that kingdom Animalia probably originated from protists called choanoflagellates?

5. Which animals are not eumetazoans?

6. From an evolutionary point of view, which is the most primitive: asymmetry, radial symmetry, or bilateral symmetry? Which is the most advanced?

7. How are polyps different from medusae? Which animals have these body forms?

8. What are cnidocytes and nematocysts? How do these structures function? In which phylum are these structures found?

9. What distinguishes dorsal, ventral, anterior, and posterior? To which groups of animals do these terms apply?

10. What are the origins of and differences between a coelom and pseudocoelom? What is an acoelomate animal?

11. What are tapeworms and blood flukes? How do humans acquire them?

12. In which habitats would you find nematodes and rotifers?

13. What is the significance of embryology as it relates to animal evolution?

14. How are arachnids and insects both similar and different?

15. What is the best explanation for the tremendous success of insects?

16. What role do arthropods play in the earth's biosphere? What role do chordates play?

17. By far the largest number of animal species belong to the

    a. mammals.   d. mollusks.

    b. arthropods.   e. cnidarians.

    c. fishes.   f. birds.

18. What are tracheae and spiracles? How do they function, and where are they found?

19. Which group of insects has by far the largest number of known species?

20. Why are chordates and echinoderms, which are very dissimilar, believed to be members of the same evolutionary line?

21. Why are echinoderms considered to be fundamentally bilaterally symmetrical when the adult forms are radially symmetrical?

22. What characteristics are found in chordates but not in other animals?

23. To which phylum do tunicates belong, and in which habitat do they live?

24. What is the fundamental difference between animal and plant life cycles? (Review chapter 28 if necessary.)

25. Which kingdom other than Animalia exhibits extracellular digestion? (Review a previous chapter to answer this, if necessary.)

FOR FURTHER READING

Alexander, R. 1990. *Animals.* Cambridge, England: Cambridge University Press. An exciting and lucid account of the ways in which various groups of animals have adjusted to their modes of life.

Bavendam, F. 1989. Even for ethereal phantasms, it's a dog-eat-dog world. *Smithsonian,* Aug., 94–101. Aspects of the lives of some of the incredibly beautiful sea slugs, or nudibranchs.

Bieri, R., and E. Thuesen. 1990. The strange worm, *Bethybelos. American Scientist* 78: 542–49. The discovery of a single specimen of an unusual arrowworm that suggests a new way of looking at the evolution of animal nervous systems.

Evans, H., and K. O'Neill. 1991. Beewolves. *Scientific American,* Aug., 70–76. Article describing the female beewolves, which are voracious predators, each year capturing many bees to feed their young.

Hadley, N. 1986. The arthropod cuticle. *Scientific American,* July, 104–12. A major innovation aiding the success of insects as they conquered the terrestrial environment: the cuticle, far more than a simple waxy covering.

Holldobler, B., and E. O. Wilson. 1990. *The ants.* Cambridge, Mass.: Harvard University Press. A wonderful book, filled with exciting and informative insights into this incredibly diverse group of insects.

Martin, J. 1990. The engaging habits of chameleons suggest mirth more than menace. *Smithsonian,* June, 44–53. Wonderful variety in a single group of tropical lizards.

Ricketts, E. F., J. Calvin, and J. W. Hedgpeth. 1985. *Between Pacific tides.* 5th ed. Stanford, Calif.: Stanford University Press. An outstanding, ecologically oriented treatment of the marine organisms of the U.S. Pacific Coast.

Tuttle, R. 1990. Apes of the world. *American Scientist* 78 (March): 115–25. A survey of our closest living relatives that reveals rich prospects for further study and an urgent need for conservation.

Wilson, E. O. 1990. Empire of the ants. *Discover,* March, 45–50. Outstanding essay on the fascinating ways of ants.

Wootton, R. 1990. The mechanical design of insect wings. *Scientific American,* Nov., 114–20. Subtle details of engineering and design that reveal how insect wings are remarkably adapted to the acrobatics of flight.

A forest is more than a field of trees. This German beech forest is a complex, thriving community. When such forests are clear-cut or destroyed by pollution, they cannot simply be replanted like a cornfield. Their diversity is the product of countless delicate interactions, few of which we understand in any detail.

THE STRUCTURE AND FUNCTION OF PLANT TISSUES

Plants are not only biologically interesting and commercially important, but they are also breathtakingly beautiful. This wild lupine leaf (Lupinus sp.), covered with dew, adds delicate artwork to a dawning day.

The plants that cover the earth's surface in such bewildering variety all possess the same fundamental architecture. This chapter is devoted to outlining the basic structure of plants and to analyzing how roots and shoots differ in growth and form. Although roots and shoots differ in their basic structure, growth at the tips throughout the life of the individual is characteristic of both. All parts of plants have an outer covering, called dermal tissue, and ground tissue, within which is embedded vascular tissue that conducts water, nutrients, and food throughout the plant. Plants cannot move, but they adjust to their environment by growing and changing their form. Although the similarities between a cactus, an orchid, and a pine tree may not at first be obvious, plants have a fundamental unity of structure that is reflected in the construction plan of their respective bodies; in the way they grow, produce, and transport their food; and in how they regulate their development.

Organization of a Plant

A vascular plant is organized along a vertical axis, like a pipe. The part below ground is called the **root;** the part above ground is called the **shoot** (figure 30.1). The root penetrates the soil and absorbs water and various ions, which are crucial for plant nutrition. It also anchors the plant. The shoot consists of stem and leaves. The **stem** serves as a framework for the positioning of the **leaves,** where most photosynthesis takes place. The arrangement, size, and other characteristics of the leaves are critically important in the plant's production of food. Flowers, other reproductive organs, and ultimately, fruits and seeds are formed on the shoot as well.

An embryo formed within a seed remains dormant for a time. Such an embryo consists of an axis, usually with one or two cotyledons, or embryonic leaves. As discussed in chapter 28, monocot embryos usually have a single cotyledon, whereas dicot embryos usually have two. The food in a mature seed may be stored either in endosperm (a common condition in monocots) or in the cotyledons (as in many dicots) (figure 30.2).

In an embryo, the **apical meristems**—regions of active cell division that occur at the tips of roots and shoots—differentiate

early, thus establishing the growth pattern that will persist throughout the plant's life (figure 30.3). In the germination of a seed, the embryonic root may emerge first, anchoring the seedling in the soil, or the shoot may emerge at the same time or even earlier.

An apical meristem is a region of active cell division that occurs at or near the tips of the roots and shoots of plants.

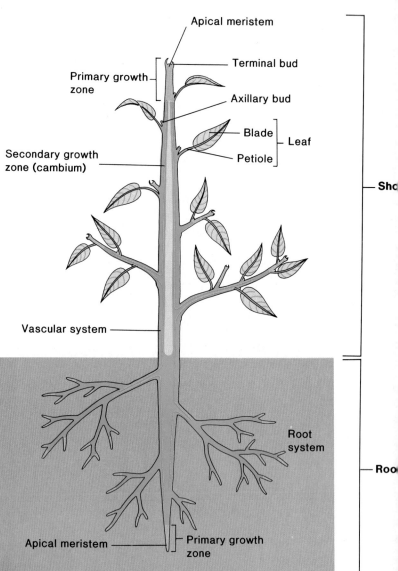

Figure 30.1 The body of a plant.

The terms in this illustration are explained in text discussions of different parts of the plant body, which consists of a shoot (stems and leaves) and root. Primary growth takes place as a result of the division of clusters of cells, the apical meristems, which are located at the ends of the roots and the stems. Secondary growth takes place laterally, allowing the plant to enlarge in girth. The gray areas are zones of active elongation.

Figure 30.2 Stages of germination in the common bean, Phaseolus vulgaris.

(a) *In the bean, as in most dicots, the food that is initially produced in the endosperm is absorbed by the embryo during the course of its development and is primarily located in the cotyledons by the time the seed is mature.* (b) *Stages of germination in the common bean.*

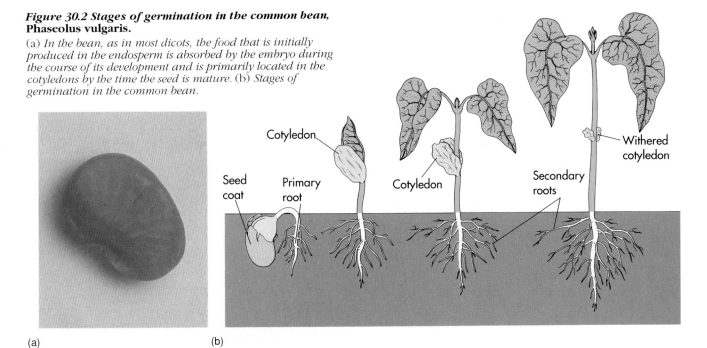

Cotyledon

Seed coat

Primary root

Cotyledon

Secondary roots

Withered cotyledon

(a)

(b)

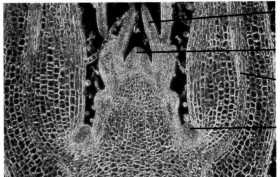

Young leaf primordium

Apical meristem

Older leaf primordium

Lateral bud

Figure 30.3 Where leaves originate.

The leaves of Coleus *are a familiar sight in greenhouses and as bedding plants. The leaves are borne in opposite pairs, and each pair alternates in direction. The transection of a shoot apex in* Coleus *shown here indicates how this kind of leaf arrangement is initiated. (A primordium is a cell or organ in its earliest stages of differentiation.)*

Tissue Types in Plants

The organs of a plant—the leaves, roots, and stem—are composed of different combinations of tissues, just as your legs are composed of bone, muscle, and connective tissue. A tissue is a group of similar cells—cells that are specialized in the same way—organized into a structural and functional unit. Plants have three major tissue types: (1) **vascular tissue,** which conducts water and dissolved minerals up the plant and conducts the products of photosynthesis throughout; (2) **ground tissue,** in which the vascular tissue is embedded; and (3) **dermal tissue,** the outer protective covering of the plant.

Each tissue type is composed of distinctive kinds of cells, whose structures are related to the functions of the tissues in which they occur. For example, as shown in chapter 28, vascular tissue is composed of xylem, which conducts water and dissolved minerals, and phloem, which conducts carbohydrates (mostly sucrose) that the plant uses as food. Together, xylem and phloem are the two principal types of conducting tissue in vascular plants.

> *The three major types of tissues in plants are: (1) vascular tissue, which conducts water, minerals, carbohydrates, and other substances throughout the plant; (2) ground tissue, which fills the plant's interior; and (3) dermal tissue, which covers the outside of the plant.*

Types of Meristems

Animals grow all over. As children grow into adults, their torsos grow at the same time their legs do. If, instead, children grew in only one place, with their legs getting longer and longer, they would be growing in a way similar to the way plants grow.

Plants contain zones of unspecialized cells called **meristems,** whose only function is to divide. Every time one of these cells divides, one of its two offspring remains in the meristem, whereas the other differentiates into one of the three kinds of plant tissue and ultimately becomes part of the plant body.

Primary growth in plants is initiated by the apical meristems. The growth of these meristems results primarily in the extension of the plant body (see figure 30.1). As it elongates, it forms what is known as the **primary plant body,** which is

576

made up of the **primary tissues.** The primary plant body comprises the young, soft shoots and roots of a tree or shrub, or the entire plant in some short-lived plants.

Secondary growth involves the activity of the **lateral meristems,** which are cylinders of meristematic tissue. The continued division of their cells results primarily in the thickening of the plant body (see figure 30.1). There are two kinds of lateral meristems: the **vascular cambium,** which gives rise to ultimately thick accumulations of secondary xylem and phloem, and the **cork cambium,** from which arise the outer layers of bark on both roots and shoots. The tissues formed from the lateral meristems, constituting most of the bulk of trees and shrubs, are known individually as **secondary tissues** and collectively as the **secondary plant body.**

The ways in which meristems function in the production of a mature plant determine the plant's nature. A **woody plant,** such as a tree or shrub, has experienced extensive secondary growth. An **herbaceous plant** is one in which secondary growth has been limited. Herbaceous plants produce new shoots each year, either from underground portions of the plant or from seeds. If they complete their entire life cycle within a year, they are called **annuals;** if shoots are produced year after year, they are called **perennials.** Members of a less common class of plants—**biennials**—form a leafy shoot the first year and then go on to flower the second year.

> *The primary plant body, which includes the young, soft shoots and roots, arises from the apical meristems. Once the lateral meristems begin to function, they produce the secondary plant body, which is characterized by thick accumulations of conducting tissue and the other cell types associated with it.*

Plant Cell Types

Ground Tissue

Parenchyma cells are the least specialized and the most common of all plant cell types; they form masses in leaves, stems, and roots. Parenchyma cells, unlike some other cell types, are characteristically alive at maturity, with fully functional cytoplasm and a nucleus. They are capable, therefore, of further division (figure 30.4). Most parenchyma cells have only **primary cell walls,** which are mostly cellulose that is laid down while the cells are still growing. **Secondary cell walls,** in contrast, are deposited between the cytoplasm and primary wall of a fully expanded cell.

Collenchyma cells, which are also living at maturity, form strands or continuous cylinders beneath the epidermis of stems or leaf stalks and along veins in leaves. They are usually elongated, with unevenly thickened primary walls, which are their distinguishing feature. Strands of collenchyma provide much of the support for plant organs in which secondary growth has not occurred (figure 30.5).

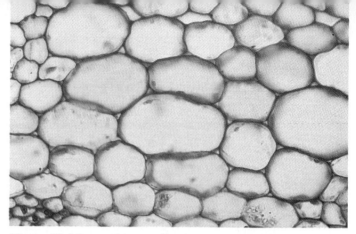

Figure 30.4 Parenchyma cells.
Cross section of parenchyma cells from grass. Only primary cell walls are seen in this living tissue.

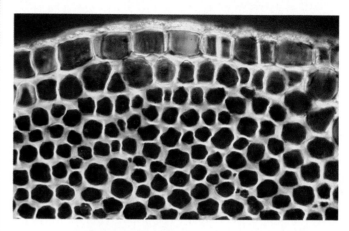

Figure 30.5 Collenchyma cells.
Cross section of collenchyma cells, with thickened side walls, from a young branch of elderberry (Sambucus). *In other kinds of collenchyma cells, the thickened areas may occur at the corners of the cells or in other kinds of strips.*

> *Parenchyma cells, which are usually living at maturity, are the most common type of cells in the primary plant body. They lack secondary cell walls. Collenchyma cells, which are also living at maturity, are elongated cells with unevenly thickened primary walls. They provide much of the support for plants in which secondary growth has not occurred.*

In contrast to parenchyma and collenchyma cells, **sclerenchyma cells** have tough, thick secondary walls; they usually do not contain living protoplasts when mature. There are two types of sclerenchyma: **fibers,** which are long, slender cells that usually form strands, and **sclereids,** which are variable in shape but often branched. Sclereids are sometimes called stone cells because they make up the bulk of the stones of peaches and other "stone" fruits, as well as that of nut shells (figure 30.6). Both fibers and sclereids are tough and thick-walled and strengthen the tissues in which they occur.

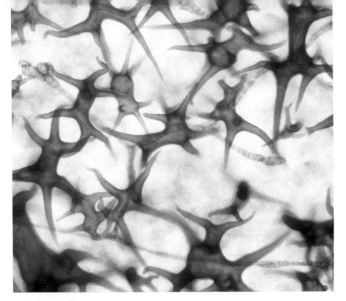

Figure 30.6 Sclereids.
*Clusters of sclereids ("stone cells"), stained blue in this
preparation, in the pulp of a pear. Such clusters of sclereids give
pears their gritty texture.*

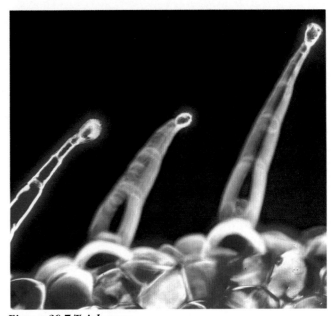

Figure 30.7 Trichomes.
*Three of the multicellular trichomes that cover the leaves of
African violets,* Saintpaulia. *A covering of trichomes creates a layer
of more humid air near the leaf surface, enabling the plant to
conserve available water supplies.*

Dermal Tissue

Flattened **epidermal cells,** which are often covered with a
thick, waxy layer called the **cuticle,** cover all parts of the primary plant body. These are the most abundant cells in the
plant epidermis, or skin. They protect the plant and provide an
effective barrier against water loss. A number of types of specialized cells occur among the epidermal cells, including guard
cells, trichomes, and root hairs.

Guard cells are paired cells. With the opening that lies between them, they make up the **stomata** (singular, *stoma*),
which occur frequently in the epidermis of leaves and occasionally on outer parts of the shoot, such as on stems or fruits.
Oxygen and carbon dioxide passage into and out of the leaves,
as well as water loss, take place almost exclusively through the
stomata, which open and shut in response to such external
factors as supply of moisture (see chapter 32).

Trichomes are superficial outgrowths of the epidermis,
like the hairs of mammals, whose forms vary greatly in different kinds of plants. They play a major role in controlling water
loss from leaves and other plant parts and in regulating the
temperature of plant parts (figure 30.7).

Similar to trichomes, but actually extensions of single epidermal cells, are root hairs, which occur in masses just behind
the very ends of the roots. They keep the roots in intimate
contact with soil particles and are soon worn off as the root
continues to grow (figure 30.8).

Vascular Tissue

Vascular plants contain two kinds of conducting or vascular
tissue: the xylem and the phloem.

Xylem

Xylem is plants' principal water-conducting tissue, forming a
continuous system that runs throughout the plant body. Within
this system, water (and the minerals dissolved in the water)
passes from the roots up through the shoot in an unbroken
stream; chapter 32 discusses how this stream is maintained.

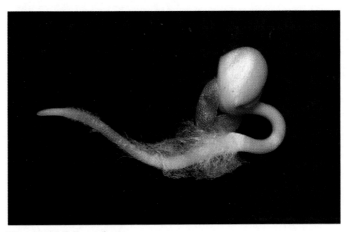

Figure 30.8 Root hairs.
A germinating seedling of radish, Raphanus sativus, *showing the
abundant fine root hairs that form in the back of the root apex.*

When water reaches the leaves, much of it passes into the air
as water vapor, mainly through the stomata.

The two principal types of conducting elements in the
xylem are **tracheids** and **vessel elements,** both of which
have thick secondary walls, are elongated, and have no living
protoplast at maturity. In conducting elements composed of
tracheids, water flows from tracheid to tracheid through openings called **pits** in the secondary walls. In contrast, vessel elements have not only pits but also definite openings, or
perforations, in their end walls by which they are linked together and through which water flows. A linked row of vessel
elements forms a vessel (figure 30.9). Primitive angiosperms
have only tracheids, but the majority of living angiosperms
have vessels. Vessels conduct water much more efficiently
than do strands of tracheids. In addition to conducting cells,
xylem likewise includes fibers and parenchyma cells.

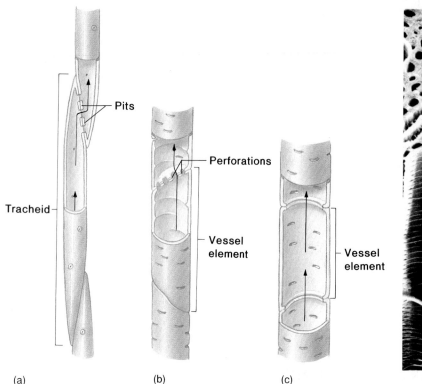

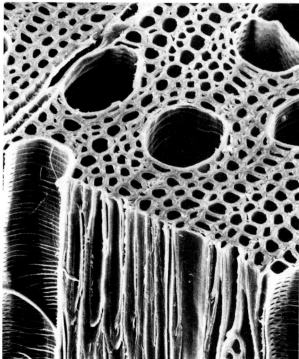

Figure 30.9 Comparison of vessel elements and tracheids.
(a) *In tracheids, the water passes from cell to cell by means of pits.*
(b) *In vessel elements, water moves by way of perforations, which may be simple or interrupted by bars.* (c) *Open-ended vessel* elements. *The photo is a scanning electron micrograph of the wood of a red maple* (Acer rubrum *(×350).*

The major types of conducting cells of the xylem are tracheids and vessel elements. Fibers and parenchyma cells are also incorporated into the xylem.

Phloem

Phloem is the principal food-conducting tissue in vascular plants. Different kinds of plants have one of two different kinds of phloem cells: **sieve cells** or **sieve-tube members.** Clusters of pores known as **sieve areas** occur on both kinds of cells and connect the protoplasts of adjoining sieve cells and sieve-tube members. Both cell types are living, but their nuclei are lost during maturation.

Angiosperms contain sieve-tube members. In sieve-tube members, the pores in some of the sieve areas are larger than those in others; such sieve areas are called **sieve plates.** Sieve-tube members occur end to end, forming longitudinal series called **sieve tubes.** Other vascular plants contain sieve cells. In these, the pores in all of the sieve areas are roughly the same diameter. Specialized parenchyma cells known as **companion cells** occur regularly in association with sieve-tube members, but are absent in plants with sieve cells (figure 30.10). In an evolutionary sense, sieve-tube members clearly are advanced over sieve cells because they are more specialized and presumably more efficient.

The principal cell types in phloem are sieve cells or sieve-tube members, both of which lack a nucleus at maturity. Sieve-tube members are associated with specialized parenchyma cells called companion cells.

Shoots

Technically, a plant shoot is that portion that lies above the cotyledons, and the plant root is the portion below the cotyledons. In most plants, all of the aboveground parts are portions of the shoot. Leaves are usually the most prominent shoot organs and determine the shoot's appearance.

Leaves

Leaves, outgrowths of the shoot apex, are the light-capturing organs of most plants (figure 30.11). The only exceptions to this are found in some plants, such as cacti, whose green stems have largely taken over the function of photosynthesis for the plant.

The apical meristems of stems and roots are capable of growing indefinitely under appropriate conditions. Leaves, in contrast, grow by means of **marginal meristems,** which flank their thick central portions. These marginal meristems grow outward and ultimately form the **blade** (flattened portion) of the leaf, while the central portion becomes the midrib. Once a

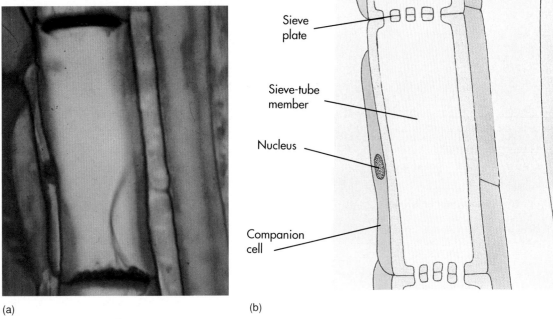

(a)   (b)

Figure 30.10 Sieve tubes.

(a) *Sieve-tube member from the phloem of squash* (Cucurbita), *connected with the cells above and below to form a sieve tube.* (b) *In this diagram of* (a), *note the thickened end walls, which are at right angles to the sieve tube. The narrow cell with the nucleus at the left of the sieve-tube member is a companion cell.*

(a)   (d)

(b)   (c)   (e)

Figure 30.11 Leaves.

Angiosperm leaves are stunningly variable. They are the primary way plants capture light and regulate their water loss. (a) *Diverse leaves in the herb layer of a Costa Rican rain forest.* (b) *A compound leaf: marijuana* (Cannabis sativa). *Such a compound leaf, in which the leaflets join the petiole (the leaf's slender stalk) at one point, is said to be palmately compound. A compound leaf is associated with a single lateral bud, located where the petiole is attached to the stem.* (c) *A simple leaf, its margin deeply lobed, from the tulip tree* (Liriodendron tulipifera). (d) *Another*

compound leaf from a member of the legume family in the lowland forest of Peru. Such a compound leaf, in which the leaflets are attached all along the main axis, is said to be pinnately compound. (e) *Many unusual arrangements of leaves occur in different kinds of plants. For example, in this miner's lettuce* (Claytonia perfoliata), *an herb of the Pacific states, two leaves are completely fused below each of the clusters of flowers, which seem, therefore, to arise from the center of a single leaf.*

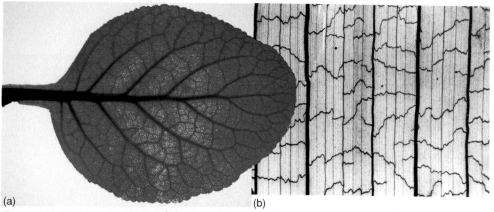

(a) (b)

Figure 30.12 A comparison of dicot and monocot leaves.
(a) *The leaves of dicots, such as this African violet relative from Sri Lanka, have net, or reticulate, venation.* (b) *Those of monocots, such as this Latin American palm, have parallel venation. Both leaves have been cleared with chemicals, and* (a) *has been stained with a red dye to make its veins show up more clearly.*

leaf is fully expanded, its marginal meristems cease to function. Their growth is called **determinate,** whereas that of apical meristems is **indeterminate.**

In addition to the flattened blade, most leaves have a slender stalk, the **petiole.** Two leaflike organs, the **stipules,** may flank the base of the petiole where it joins the stem. Veins, usually consisting of both xylem and phloem, run through the leaves. As mentioned in chapter 28, in many monocots, the veins are parallel; in most dicots, the pattern is net or reticulate venation (figure 30.12).

A typical leaf contains masses of parenchyma, called **mesophyll** ("middle leaf"), through which the vascular bundles, or veins, run. Beneath the upper epidermis of a leaf are one or more layers of closely packed, columnlike parenchyma cells called **palisade parenchyma.** The rest of the leaf interior, except for the veins, consists of a tissue called **spongy parenchyma.** Between the spongy parenchyma cells are large intercellular spaces that function in gas exchange and in the passage of carbon dioxide from the atmosphere to the mesophyll cells. These intercellular spaces are connected, directly or indirectly, with the stomata (figure 30.13).

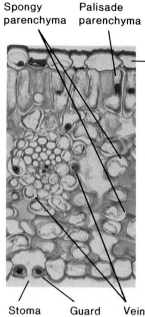

Spongy parenchyma   Palisade parenchyma

—— Epidermis

Stoma   Guard cell   Vein

Figure 30.13 A leaf in cross section.
Transection of a lily leaf, showing palisade and spongy parenchyma; a vascular bundle, or vein; and the epidermis, with paired guard cells flanking the stoma.

The cells of the mesophyll, especially those near the leaf surface, are packed with chloroplasts. These cells constitute the plant's primary site of photosynthesis. Water and minerals are brought from the roots to the leaves in the xylem strands of the veins. Once the water reaches the ends of the veins, it passes into the photosynthetic mesophyll cells. Because the surfaces of these cells border on the intercellular spaces of the leaf interior, and because water can pass through cell membranes and cell walls easily, much of the water evaporates into the intercellular space and can then escape from the leaf through the open stomata. Thus, stomata, which allow absorption of essential carbon dioxide, also allow the escape of water. At the same time that water is leaving through the stomata, the products of photosynthesis are being transported from the leaves to all other parts of the plant through the phloem.

> *The mesophyll in a leaf consists of two types of parenchyma cells, both packed with chloroplasts. Palisade parenchyma cells are columnar and closely packed together, whereas spongy parenchyma cells are loosely packed and separated by large intercellular spaces.*

Stems

As mentioned earlier, the stem is that part of the shoot that serves as the framework for the positioning of the leaves. It experiences both primary and secondary growth and also is the source of an economically important plant product—wood.

Primary Growth

In the primary growth of a shoot, leaves first appear as **leaf primordia** (singular, *primordium*), or rudimentary young leaves, which cluster around the apical meristem, unfolding and growing as the stem itself elongates (figure 30.14). The places on the stem at which leaves form are called **nodes.** The portions of the stem between these attachment points are called the **internodes.** As the leaves expand to maturity, a **bud,** a tiny, undeveloped side-shoot, develops in the **axil** of each leaf, the angle between a branch or leaf and the stem from which it arises (see figure 30.3, where the buds are visible as densely staining masses at the base of the larger pair of leaf primordia). These buds, which have their own leaves, may elongate and form lateral branches, or they may remain small and dormant. A hormone diffusing downward from the terminal bud of the shoot continuously suppresses the expansion of the lateral buds (see chapter 32, where plant hormones are discussed in more detail). These buds begin to expand when the terminal bud is removed. Therefore, gardeners who wish to produce bushy plants or dense hedges crop off the tops of the plants, thus removing their terminal buds.

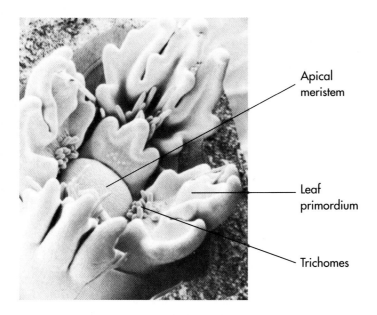

Apical
meristem

Leaf
primordium

Trichomes

Figure 30.14 A shoot apex.
Scanning electron micrograph of the shoot apex of a silver maple,
Acer saccharinum, *showing a developing shoot during summer,*
the season of active growth. The apical meristem, leaf primordia,
and trichomes are plainly visible.

Within the soft, young stems, the strands of vascular tissue, xylem and phloem, either occur as a cylinder in the outer portion of the stem, as is common in dicots, or are scattered through it, as is common in moncots (figure 30.15). The vascular bundles contain both primary xylem and primary phloem (figure 30.16). At the stage when only primary growth has occurred, the inner portion of the ground tissue of a stem is called the **pith,** and the outer portion is the **cortex.**

Secondary Growth

In stems, secondary growth is initiated by the differentiation of the vascular cambium, which consists of a thin cylinder of actively dividing cells that is located between the bark and the main stem in mature woody plants. The vascular cambium differentiates from parenchyma cells within the vascular bundles of the stem, between the primary xylem and the primary phloem (figure 30.17). The cylindrical form of the vascular cambium is completed by the differentiation of some of the parenchyma cells that lie between the bundles. Once established, the vascular cambium consists of elongated, somewhat flattened cells with large vacuoles. The cells that divide from the vascular cambium outwardly, toward the bark, become secondary phloem; those that divide from it inwardly become secondary xylem. The cells of the vascular cambium also divide laterally (side by side), allowing the stem to thicken as the tree or shrub becomes more mature.

While the vascular cambium is becoming established, a second kind of lateral cambium, the cork cambium, normally also develops in the stem's outer layers. The cork cambium usually consists of plates of dividing cells that move deeper and deeper into the stem as they divide. Outwardly, the cork cambium splits off densely packed **cork cells;** they contain a fatty

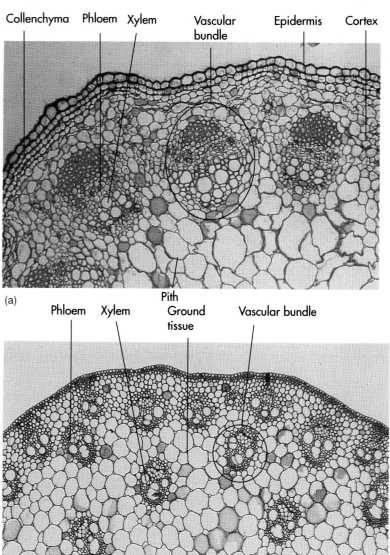

Collenchyma   Phloem   Xylem   Vascular bundle   Epidermis   Cortex

(a)

Phloem   Xylem   Pith Ground tissue   Vascular bundle

(b)

Figure 30.15 A comparison of dicot and monocot stems.
(a) *Transection of a young stem of the common sunflower,*
Helianthus annuus, *a dicot, in which the vascular bundles are*
arranged around the outside of the stem. (b) *Transection of corn,*
Zea mays, *a monocot, with the scattered vascular bundles*
characteristic of the class.

Figure 30.16
A vascular bundle.
Transection of a
vascular bundle from a
buttercup, Ranunculus
acris, *showing the xylem*
and phloem.

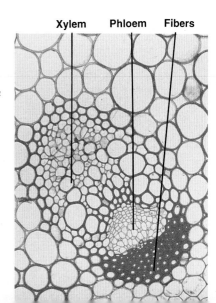

Xylem   Phloem   Fibers

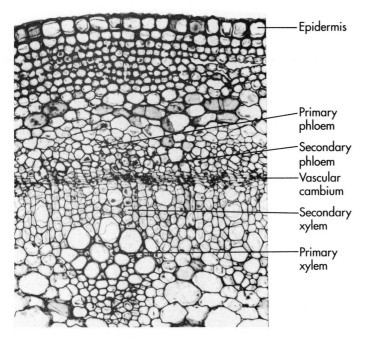

- Epidermis
- Primary phloem
- Secondary phloem
- Vascular cambium
- Secondary xylem
- Primary xylem

Figure 30.17 Vascular cambium.
Early stage in the differentiation of the vascular cambium in an elderberry (Sambucus canadensis) *stem. The outer part of the cortex consists of collenchyma and the inner part of parenchyma.*

substance and are nearly impermeable to water. Cork cells are dead at maturity. Inwardly, the cork cambium divides to produce a dense layer of parenchyma cells. The cork, the cork cambium that produces it, and this dense layer of parenchyma cells make up a layer called the **periderm,** which is the plant's outer protective covering. Oxygen reaches the layers of living cells under the periderm through areas of loosely organized cells called **lenticels,** which are often easily identifiable on the bark's outer surface (figure 30.18).

> *The periderm consists of cork to the outside, the cork cambium in the middle, and layers of parenchyma cells to the inside of a stem or root. Penetrated by areas of more loosely packed tissue known as lenticels, the periderm retards water loss from the secondary plant body.*

Cork, which covers the surfaces of mature stems or roots, takes the place of the epidermis, which performs a similar function in the younger parts of the plant. The term **bark** refers to all of the tissues of a mature stem or root outside of the vascular cambium. Because the vascular cambium has the thinnest-walled cells that occur anywhere in a secondary plant body, it is the layer at which bark breaks away from the accumulated secondary xylem. The inner layers of the bark are primarily secondary phloem, with the remains of the primary phloem crushed among them. Bark's outer layers consist of the periderm, and the very outermost ones are cork (figure 30.19).

Wood

Wood is one of the most useful, economically important, and beautiful products obtained from plants. Anatomically, **wood** is

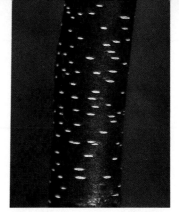
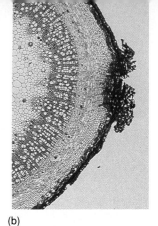

(a) (b)

Figure 30.18 Lenticels.
(a) *Lenticels are the numerous, small, pale, raised areas visible on the bark of paper birch,* Betula papyrifera. *Lenticels allow oxygen to readily diffuse into the living tissues immediately below the bark of woody plants. Highly variable in form in different species, lenticels are a valuable aid to the identification of deciduous trees and shrubs in winter.* (b) *Transection through a lenticel in a stem of elderberry,* Sambucus canadensis.

(a) (b)

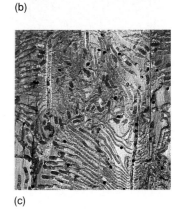

(c)

Figure 30.19 Bark.
(a) *Bark of valley oak,* Quercus lobata, *in California.* (b) *Bark is highly characteristic of individual trees and shrubs, which often can be identified from the bark, even when the trees are leafless, as in this madrone,* Arbutus menziesii. (c) *Galleries constructed by elm bark beetles, which spread Dutch elm disease from tree to tree, in the cambium of an elm tree. Such beetles bore through the thin layer of living cells that separates bark from the trunk of a tree, thus gaining access to the carbohydrates passing through the phloem.*

accumulated secondary xylem. As the secondary xylem ages, its cells become infiltrated with gums and resins, and the wood becomes darker. For this reason, the wood located nearer the central regions of a given trunk, called **heartwood,** is often darker and denser than the wood nearer the vascular cambium, called **sapwood,** which is still actively involved in transport within the plant (figure 30.20). Commercially, wood is divided into hardwoods and softwoods. **Hardwoods** are the woods of dicots, regardless of how hard or soft they actually may be; **softwoods** are the woods of conifers.

Figure 30.20 Heartwood and sapwood.

The distinction between heartwood (dark central portion) and sapwood (light outer portion) is evident in this sawed-off limb of ponderosa pine (Pinus ponderosa) in the mountains of California.

Because of the way it is accumulated, wood often displays rings. In temperate regions, these rings are **annual rings:** They reflect the fact that the vascular cambium divides more actively when water is plentiful and temperatures are suitable for growth than when water is scarce and the weather is cold. The abrupt discontinuity between the layers of larger cells, with proportionately thinner walls, that form in the growing season (in most temperate regions, during the spring and early summer) and those that form later is often very evident. For this reason, the annual rings in a tree trunk can be used to calculate the tree's age (figure 30.21).

Roots

Roots have simpler patterns of organization and development than do stems (figure 30.22). Although different patterns exist, the kind of root described here is found in many dicots. There

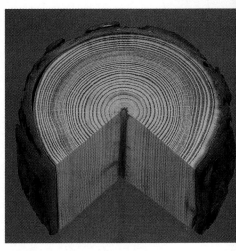

Figure 30.21 Annual rings in a section of pine (Pinus).
These rings can be used to calculate the tree's age.

is no pith in the center of the vascular tissue in most dicot roots. Instead, these roots have a central column of xylem with radiating arms. Alternating with the radiating arms of xylem are strands of primary phloem. Surrounding the column of vascular tissue and forming its outer boundary is a cylinder of cells one or more cell layers thick called the **pericycle.** Branch, or lateral, roots are formed from cells of the pericycle. The outer layer of the root, as in the shoot, is the **epidermis.** The mass of parenchyma in which the root's vascular tissue is located is the cortex. Its innermost layer—the **endodermis**—consists of specialized cells that regulate the flow of water between the vascular tissues and the root's outer portion. The endodermis lies just outside of the pericycle. Endodermis cells are surrounded by a thickened, waxy band called the **Casparian strip.** It is by the differential passage of minerals and nutrients through endodermis cells that the plant regulates its supply of minerals (figure 30.23).

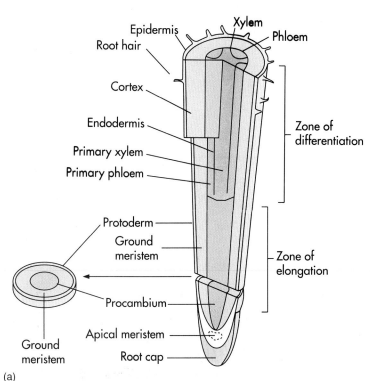

(a)

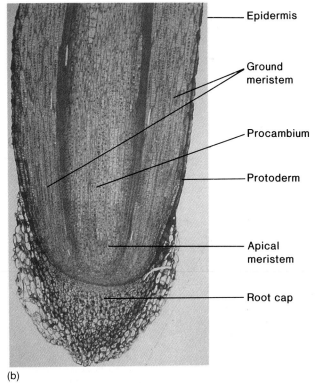

(b)

Figure 30.22 Root structure.

(a) *Diagram of primary meristems in the root, showing their relation to the apical meristem. The three primary meristems are the protoderm, which differentiates further into epidermis; the procambium, which differentiates further into primary vascular* strands; and the ground meristem, which differentiates further into ground tissue. (b) *Median longitudinal section of a root tip in corn,* Zea mays, *showing the differentiation of epidermis, cortex, and column of vascular tissues.*

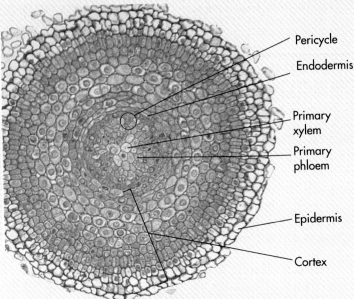

Figure 30.23 A root cross section.
Cross section through a root of a buttercup, Ranunculus californicus. *The primary xylem and phloem, cortex, and epidermis are visible.*

Most dicot roots have a central column of primary xylem with radiating arms and strands of primary phloem that alternate between these arms, surrounded by a layer one or more cells thick called the pericycle. The innermost layer of the cortex, or endodermis, consists of cells surrounded by a thickened, waxy band called the Casparian strip.

The apical meristem of the root divides and produces cells both inwardly, back toward the body of the plant, and outwardly. Outward cell division results in the formation of a

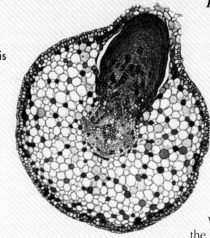

Figure 30.24 Lateral roots.
A lateral root growing out through the cortex of the black willow, Salix nigra. *Lateral roots originate beneath the surface of the main root, whereas lateral stems originate at the surface.*

thimblelike mass of relatively unorganized cells, the **root cap,** which covers and protects the root's apical meristem as it grows through the soil.

The root elongates relatively rapidly just behind its tip. Abundant root hairs form above that zone. Virtually all water and minerals are absorbed from the soil through the root hairs, which greatly increase the root's surface area and absorptive powers. In plants with mycorrhizae (chapter 27), the root hairs are often greatly reduced in number, and the fungal filaments of the mycorrhizae play a role similar to that of the root hairs.

One of the fundamental differences between roots and shoots has to do with the nature of their branching. In stems, branching occurs from buds on the stem surface; in roots, branching is initiated well back of the root apex as a result of cell divisions in the pericycle. The lateral root primordia grow out through the cortex toward the surface of the root, eventually breaking through and becoming established as lateral roots (figure 30.24). Secondary growth in roots, both main roots and laterals, is similar to that in stems, with the vascular cambium being initiated after the division of cells located between the primary xylem and the primary phloem.

EVOLUTIONARY VIEWPOINT

The first plants appeared about 440 million years ago, but they were relatively small until vascular plants evolved secondary growth 380 million years ago. Secondary growth makes it possible for a plant to increase in diameter and thus become thick-trunked and tall. This advance made possible the evolution of forests.

SUMMARY

1. A plant body is basically a vertical axis with two parts: a root and a shoot. Within it are three principal tissue types: vascular tissue, ground tissue, and dermal tissue. Vascular tissue conducts substances through the plant, ground tissue is the matrix in which the vascular tissue is embedded, and dermal tissue covers the outside of the plant.

2. Plants grow in length by means of their apical meristems, zones of active cell division at the ends of the roots and the shoots. Plants grow in diameter by means of their lateral meristems, which produce the secondary plant body.

3. Plants that complete their entire life cycle within a year are called annuals. Those that require two years to reach maturity and then flower just once are called biennials. Plants that flower year after year once they have reached maturity are perennials.

4. Parenchyma cells are the most common type of plant cell. They usually have only primary cell walls, which are composed mainly of cellulose that is laid down while the cells are still growing. Parenchyma cells are alive at maturity.

5. Collenchyma cells often form strands or continuous cylinders in the plant body, for which they provide the chief source of strength. They are recognized by the uneven thickenings of their primary cell walls.

6. Sclerenchyma cells have tough, thick secondary walls. Fibers, which are elongated, and sclereids, which are variable in shape but often branched, are sclerenchyma cells that strengthen the tissues in which they occur.

7. Flattened epidermal cells protect the plant and prevent water loss. Guard cells, trichomes, and root hairs are specialized types of epidermal cells.

8. Tracheids and vessel elements are the principal conducting elements of xylem. They have thick cell walls and lack protoplasts at maturity. Water reaches the leaves after entering the plant through the roots and passing upward via the xylem. Water vapor passes out of leaves by evaporating into the intercellular spaces and then escaping through the stomata. Carbon dioxide enters the plant by the same route in reverse.

9. Carbohydrates are conducted through the plant primarily in the phloem, whose elongated conducting cells are living but lack a nucleus. These conducting cells are called sieve cells and sieve-tube members.

10. Leaf growth is determinate, while growth of shoots and roots is indeterminate. Leaves,

which are mainly flattened organs of the shoot specialized for photosynthesis, expand in size by means of marginal meristems.

11. Secondary growth in both stems and roots takes place following the formation of lateral meristems known as vascular cambia. These cylinders of dividing cells form xylem internally and phloem externally. As a result of their activity, the girth of a plant increases.

12. The cork cambium forms in both roots and stems during the initial stages of establishment of the vascular cambium. It produces cork externally and a dense layer of parenchyma internally. The cork, cork

cambium, and underlying parenchyma layers collectively are called the periderm. In the stem, the periderm is perforated by areas of loosely organized tissue called lenticels.

13. Wood is accumulated secondary xylem. It often displays rings because it grows at different rates during different seasons.

14. Stems branch because of the growth of buds that form externally at the point where the leaves join the stem. Roots branch by forming centers of cell division within their cortex. Young roots grow out through the cortex, eventually breaking through the surface of the root.

REVIEWING THE CHAPTER

1. How is the plant body organized?

2. What are the major tissue types in plants?

3. How does primary growth in plants differ from secondary growth, and what types of meristems are involved in each?

4. What are parenchyma cells, and how do they function?

5. How do collenchyma cells function?

6. What is sclerenchyma, and how does it function?

7. What is the makeup of the epidermis, and how do epidermal cells function?

8. What is xylem, what are its primary conducting elements, and how do they function?

9. What is phloem, what are its primary cell types, and how do they function?

10. What is the shoot in a plant?

11. How are leaves structured?

12. How do leaves develop?

13. How are stems structured?

14. How do stems develop?

15. Anatomically speaking, what is wood?

16. How are roots structured?

17. How do roots develop?

COMPLETING YOUR UNDERSTANDING

1. If you made a garden salad of lettuce, cucumber, tomatoes, celery, red bell peppers, carrots, and onions, you would be eating mostly which plant tissue?

2. Of primary and secondary tissues, which is responsible for increases in plant length, and which is responsible for increases in plant girth?

3. Where are the apical and lateral meristems located?

4. If you wanted the same plants to come up year after year in your flower garden, you would plant

    a. annuals.   c. perennials.

    b. biennials.   d. all of the above.

5. Roots become thicker

    a. because of the accumulation of branch roots.

    b. as they take in water.

    c. when they form lateral meristems that become active.

    d. because they lack a cork cambium.

    e. because of the activity of the periderm.

6. The gritty texture in the outer core of pears is caused by

    a. xylem.   d. collenchyma.

    b. phloem.   e. cork.

    c. sclerenchyma. f. endodermis.

7. Which of the following structures are predominantly found in leaves? [*two answers required*]

    a. Guard cells   d. Mesophyll

    b. Xylem   e. Endodermis

    c. Root hairs   f. Pericycle

8. Which of the following cell types are nonliving at maturity? [*two answers required*]

    a. Collenchyma d. Tracheids

    b. Parenchyma e. Companion cells

    c. Sclerenchyma f. Sieve cells

9. Why are vessels more efficient than tracheids in water conduction?

10. From an evolutionary point of view, why are sieve-tube members more advanced than sieve cells?

11. How can you distinguish between a monocot leaf and a dicot leaf?

12. What are the similarities and differences in function between leaves' palisade and spongy parenchyma cells?

13. If you were closely examining a houseplant, how would you distinguish nodes from internodes?

14. Which leaf cells are the primary site of photosynthesis?

15. Branch stems arise

    a. from lateral meristems.

    b. from the axils of leaf primordia.

    c. internally.

    d. from the endodermis.

    e. from primary phloem strands.

16. In your garden, how would you make your plants more bushy? Why does this work?

17. How are the pith and cortex similar and different?

18. How do monocot and dicot stems differ from one another?

19. What is the difference between cork, bark, and periderm? Where would you find these structures in plants?

20. What happens at the lenticels?

    a. Water absorption

    b. Gas exchange

    c. Photosynthesis

    d. Production of the periderm

    e. *a* and *c*

21. Are the terms *hardwoods* and *softwoods* clearly defined? If not, why continue to use them?

22. Would you expect to find annual rings in trees in the tropical rain forest? Why or why not?

23. Cells of the _____ regulate the flow of water laterally between the vascular tissues and the cell layers in the outer portions of the root.

    a. Periderm   d. Pith

    b. Endodermis   e. Xylem

    c. Pericycle

24. Which of the following structures would be absent from roots growing in soil?

    a. Guard cells   d. Chloroplasts

    b. Endodermis   e. Cork

    c. Root hairs   f. Xylem

25. If you went to the grocery store and bought onions, carrots, white potatoes, celery, green peppers, tomatoes, lettuce, and radishes, which ones would be roots and why?

26. Branch roots arise

    a. in the axils of other branch roots.

    b. in the axils of root hairs.

    c. from the endodermis.

    d. from the pericycle.

    e. from the cork cambium.

27. Why is celery a healthy food choice?

*28. If you hammer a nail into the trunk of a tree 2 meters above the ground when the tree is 6 meters tall, how far above the ground will the nail be when the tree is 12 meters tall?

*For discussion

FOR FURTHER READING

Galston, A. W., P. J. Davis, and R. L. Satter. 1980. *The life of the green plant*. 3d ed. Englewood Cliffs, N. J.: Prentice-Hall. Very well-written, physiologically oriented treatment of the structure and functioning of plants.

Raven, P. H., R. F. Evert, and H. Curtis. 1981. *The biology of plants*. 4th ed. New York: Worth Publishers. A comprehensive treatment of general botany, emphasizing structural botany.

Sandved, K. B., and G. T. Prance. 1984. *Leaves*. New York: Crown Publishers. A beautiful introduction to the diversity of leaves.

Swain, R. B. 1988. Notes from the radical underground. *Discover*, Nov., 16–18. Cooperation and competition between tree roots—a new view of the relationships among plants.

Woodward, I. 1989. Plants, water and climate. Part One. *New Scientist*, February 18, 1–4. Excellent account of the way water moves through plants.

FLOWERING PLANT REPRODUCTION

A field of California poppies (Eschscholtzia californica). Each of these seemingly countless flowers must be visited by an insect, often several times, to ensure the pollination occurs that is required to produce a new generation.

FOR REVIEW

Flowering plants, or angiosperms, dominate the earth except for the great northern forests, the polar regions, the high mountains, and the driest deserts. Among the features that have contributed to their success are their unique reproductive structures, which include the flower and the fruit. Flowers bring about the precise transfer of pollen by insects and other animals, which allows plants to exchange gametes with one another, even though each plant is rooted in one place. Fruits play an important role in the dispersal of angiosperms from place to place. Not only were both flowers and fruits key elements in angiosperms' early success, but their evolution produced most of the striking differences seen among different angiosperms today (figure 31.1).

Each of us deals with plant reproduction every day without thinking about it: The bread we eat is made from the ground-up fruit of a grass—wheat; the roses given as a symbol of love evolved as structures attractive to insects; and the honey spread on our toast is produced by bees from nectar, the bribe a flower uses to induce the bees to carry the flower's gametes to another plant. This chapter examines how flowering plants reproduce.

Figure 31.1 Flowers are often beautiful.
The rich colors and textures of this flower head of a South African species of Gazania, *a member of the sunflower family (Asteraceae), have evolved in response to the sensory perceptions and activities of insects. There are two types of flowers in the head; the outermost are extended into rays, and the inner ones, called disc flowers, are symmetrical.*

Formation of Angiosperm Gametes

As mentioned in chapter 28, all plant life cycles are characterized by an alternation of generations, in which a diploid sporophyte generation gives rise to a haploid gametophyte generation. In angiosperms, the gametophyte generation is very small and is completely enclosed within the tissues of the parent sporophyte. The male gametophytes, or microgametophytes, are **pollen grains.** The female gametophyte, or megagametophyte, is the **embryo sac.** Pollen grains and the embryo sac both are produced in separate, specialized structures of the angiosperm flower.

Like animals, therefore, angiosperms have separate structures for producing male and female gametes. The reproductive organs of angiosperms and animals, however, are different in two ways: First, in angiosperms, both male and female structures usually occur together in the same individual flower, while animals require two sexes of individuals. Second, angiosperm reproductive structures are not permanent parts of the adult individual. Angiosperm flowers and reproductive organs develop seasonally; these flowering seasons correspond to times of the year most favorable for pollination.

Structure of the Angiosperm Flower

As noted in chapter 28, a typical angiosperm flower is composed of whorls, a circle of parts present at a single level along an axis. The outermost whorl consists of structures called the **sepals.** The sepals protect the other flower whorls and serve as the flower's attachment point to the stalk, or **peduncle.** The sepals are usually green, although in some angiosperm species, they are brightly colored. All of the sepals together are called the **calyx.**

The second whorl of a flower is composed of **petals.** Most angiosperm petals are vibrantly colored, and in those plants pollinated by animals (such as insects and birds), the petals may have characteristic shapes and features that attract the pollinating animal. All of the petals together are called the **corolla.**

The third whorl of a flower is composed of the male and female reproductive structures. The male reproductive structures are the **stamens.** Each stamen consists of a **filament,** a slender stalk, to which is attached the **anther.** At the end of each anther are two **pollen sacs,** in which the **pollen grains** develop. All of the stamens of an angiosperm flower are called the **androecium,** which means "male household" (figure 31.2).

The female reproductive structures are the **pistils.** Each pistil is composed of a **stigma, style,** and **ovary.** The stigma, the top part of the pistil, is covered with a sticky, sugary liquid to which pollen grains adhere during pollination. The liquid also nourishes the pollen grain as it makes its way to the ovary. The style is the elongated portion of the pistil that leads to the ovary. The ovary encloses the **carpel,** which contains the **ovules.** Within the ovules, the haploid megaspores develop into **embryo sacs** that will contain the egg. After fertilization, the ovules develop into seeds that give rise to the sporophyte

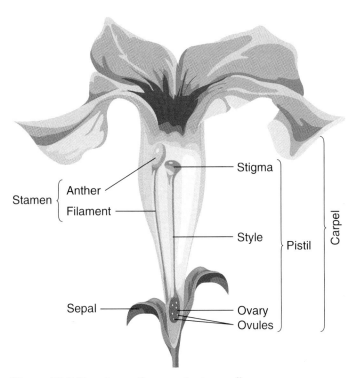

Figure 31.2 Structure of an angiosperm flower.
The male reproductive structure is called the stamen and consists of the anther and filament. The female reproductive structure is called the pistil and consists of the stigma, style, and ovary. The ovary contains the carpels, which enclose the ovules. Some ovaries have one carpel; others have several. The ovary shown here contains one carpel.

generation. All of the pistils of an angiosperm flower are called the **gynoecium,** meaning "female household" (see figure 31.2).

> *Angiosperm flowers consist of whorls. The first whorl contains the sepals, and the second whorl contains the petals. The third whorl contains the reproductive structures. The male reproductive structure, called the stamen, consists of the filament and the anther. The female reproductive structure, called the pistil, consists of the stigma, style, and ovary. The ovary contains the carpel, which in turn, encloses the ovules, where eggs develop.*

Pollen Formation

Pollen grains form in the two pollen sacs located in the anther. Each pollen sac contains specialized chambers (the microsporangia, see chapter 27) in which the microspore mother cells are enclosed and protected. The microspore mother cells undergo meiosis to form four haploid microspores. Each microspore then undergoes a mitotic division to form four pollen grains.

Pollen grain shapes are specialized for specific flower species. As discussed in more detail later in the chapter, fertilization requires that the pollen grain grow a tube that penetrates

the style until the ovary is encountered. Most pollen grains have a furrow from which this pollen tube emerges. Some pollen grains have three furrows.

> *Haploid microspores are formed from the meiotic division of the microspore mother cells contained in the microsporangia of the pollen sacs. The haploid microspores divide mitotically to form haploid pollen grains.*

Egg Formation

Eggs develop in the carpel of the angiosperm flower. Within the carpel are chambers (the megasporangia, see chapter 27) where the ovules develop. Within each ovule are megaspore mother cells. Each megaspore mother cell undergoes meiosis to produce four haploid megaspores. Only one of these megaspores, however, survives; the rest are absorbed by the ovule. The lone remaining megaspore undergoes repeated mitotic divisions to produce eight haploid nuclei that are enclosed within an embryo sac, or megagametophyte. Within the embryo sac, the eight nuclei are arranged in precise positions. One nucleus is located near the opening of the embryo sac; this nucleus is the egg cell. The other seven are arranged around the embryo sac: Two are located in the middle of the embryo sac and are called **polar nuclei;** two nuclei flank the egg cell; and the other three nuclei are located at the end of the embryo sac, opposite the egg cell.

> *Megaspores are formed from the meiotic division of the megaspore mother cell contained in the megasporangia of the carpel. Three of the megaspores formed are absorbed; the fourth divides mitotically to produce eight haploid nuclei. One of these nuclei becomes the egg cell. The other seven nuclei are arranged in precise positions in the embryo sac.*

Pollination

Pollination is the process by which pollen is placed on the stigma, initiating fertilization. The pollen may be carried to the flower by wind or by animals, or it may originate within the individual flower itself. When pollen from a flower's anther pollinates the same flower's stigma, the process is called **self-pollination.**

Pollination by Animals

In many angiosperms, the pollen grains are carried from flower to flower by insects and other animals that visit the flowers for food or other rewards, or are deceived into doing so because the flower's characteristics suggest such rewards (figure 31.3). A liquid called **nectar,** which is rich in sugar as well as amino acids and other substances, is often the reward sought by animals. Successful pollination depends on the plants attracting insects and other animals regularly enough that the pollen is carried from one flower of that particular species to another.

Figure 31.3 A bumblebee, Bombus, *covered with pollen while visiting a flower.*
This bee will transfer large quantities of pollen to the next flower it visits.

The relationship between such animals, known as **pollinators,** and the flowering plants has been important to the evolution of both groups. By using insects to transfer pollen, the flowering plants can disperse their gametes on a regular and more or less controlled basis, despite their being anchored to substrate (figure 31.4).

For pollination by animals to be effective, a particular insect or other animal must visit numerous plant individuals of the same species. Flowers' color and form have been shaped by evolution to promote such specialization. Yellow flowers are particularly attractive to bees, whereas red flowers attract birds but are not particularly noticed by insects. Some flowers have very long floral tubes with the nectar produced deep within them; only the long, slender beaks of hummingbirds or the long, coiled tongues of moths or butterflies can reach such nectar supplies.

The most numerous insect-pollinated angiosperms are those pollinated by bees, a large group of insects consisting of some twenty thousand species. Bees are the most frequent, characteristic, and constant visitors to particular kinds of flowers today, but they were not abundant and may not even have existed when angiosperms first appeared. The diversity of flowering plants seen today is closely related to later specialization of angiosperms in relation to bees.

Pollination by Wind

In certain angiosperms, pollen is blown about by the wind and reaches the stigmas passively, as it does in most gymnosperms. For such a system to operate efficiently, the individuals of a given plant species must grow relatively close together because wind does not carry pollen very far or very precisely, as compared with insects or other animals. Because gymnosperms, such as spruces or pines, grow in dense stands, wind pollination is very effective. Wind-pollinated angiosperms, such as birches, alders, and ragweed, also tend to grow in dense stands. The flowers of wind-pollinated angiosperms are usually small, greenish, and odorless, and their petals are reduced in size or absent. They typically produce large quantities of pollen.

Whether the first angiosperms were pollinated by insects or by the wind cannot be determined with certainty. The association with insects, however, is an ancient one for the flowering plants, and wind pollination seems almost certainly to have evolved secondarily in all angiosperms in which it occurs today.

Self-Pollination

In some angiosperms, the pollen does not reach other individuals at all: Instead, it is shed directly onto the stigma of the same flower, sometimes in bud. This results in self-pollination and inbreeding, with the evolutionary consequences discussed in chapter 15.

Pollination can be accomplished by the action of animals or wind or by self-pollination.

Figure 31.4 Pollination by an insect.
A clouded yellow butterfly, native to Europe, prepares to land on a thistle. Pollen adhering to the butterfly from previous visits to other thistles may soon be transferred to this plant.

Fertilization, Seed Development, and Fruit Development

Fertilization in angiosperms is a complex, somewhat unusual process in which two sperm cells are utilized. This unique process is called **double fertilization** and results in the fertilization of the egg and the formation of a nutrient substance that nourishes the embryo. Once fertilization is complete, the embryo develops by dividing numerous times. Meanwhile, protective tissues enclose the embryo, resulting in the formation of the **seed.** The seed, in turn, is enclosed in another structure called the **fruit.** These typical angiosperm structures evolved in response to the need for seeds to be dispersed over long distances to ensure genetic variability. Seeds and fruits are tied to the activity of animals, who carry the seeds to new habitats.

Fertilization

Once a pollen grain, spread by the wind, the action of an animal, or self-pollination, adheres to the sticky, sugary substance that covers the stigma, it begins to grow a **pollen tube** that pierces the style. The pollen tube, nourished by the sugary substance, grows until it reaches the ovule in the ovary. Meanwhile, the pollen grain inside the tube divides to form two sperm cells (figure 31.5).

The pollen tube eventually reaches the embryo sac in the ovule. At the entry to the embryo sac, the tip of the pollen tube bursts and releases the two sperm cells. Simultaneously, the two nuclei that flank the egg cell disintegrate, and one of the sperm cells fertilizes the egg cell, forming a zygote. The other sperm cell fuses with the two polar nuclei located at the center of the embryo sac, forming the triploid (3n) **primary endosperm nucleus.** The primary endosperm nucleus eventually develops into the **endosperm,** which nourishes the embryo.

Fertilization in angiosperms involves double fertilization. After pollination, a pollen tube grows from the pollen grain to the ovary, depositing two sperm cells in the embryo sac. One sperm cell fertilizes the egg cell. The other sperm cell fuses with two of the nuclei in the embryo sac to form the triploid primary endosperm nucleus.

Seed Formation

The entire series of events that occur between fertilization and maturity is called **development.** During development, cells become progressively more specialized, or **differentiated.** The first stage in the development of a plant zygote is active cell division to form an organized mass of cells, the **embryo.** In angiosperms, the differentiation of cell types within the embryo begins almost immediately after fertilization. By the fifth day, the principal tissue systems can be detected within the embryo mass, and within another day, the root and shoot apical meristems can be detected. Cells do not move during embryonic plant development, as they do in animals. In animals, specific cell movements play an important

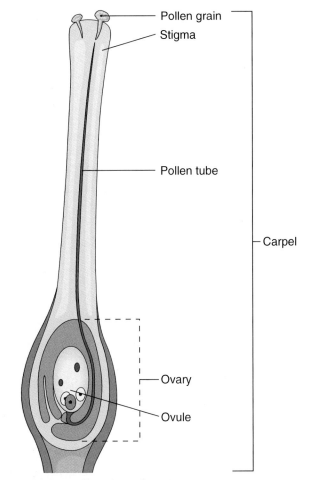

Figure 31.5 A pollination tube.
This pollen grain has successfully grown down the long stigma to reach and fertilize the ovule.

role in development. In plants, however, cells differentiate where they are formed, their positions determining in large measure their future developmental fates.

The primary endosperm nucleus develops at the same time as the embryo. Immediately after fertilization, the primary endosperm nucleus divides repeatedly to form many triploid, free-floating nuclei. In dicots, walls develop around the free-floating endosperm nuclei, forming the endosperm, which contains such nutrients as starches, oils, and fats. The endosperm does not develop to the same extent in monocots: Most of the nutrient material is contained in the cotyledons, and the endosperm contains only a small amount of protein.

Early in the development of an angiosperm embryo, a profoundly significant event occurs: The embryo simply stops developing and becomes dormant. In many plants, embryo development is arrested soon after apical meristems and the first leaves, or cotyledons, are differentiated. The **integuments**—the coats surrounding the embryo—develop into a relatively impermeable **seed coat,** which encloses the quiescent embryo within the seed, together with a source of stored food (see Sidelight 31.1).

Sidelight 31.1

THE FLOWERING OF BAMBOOS AND THE STARVATION OF PANDAS

In many Asian bamboo species, all the individuals of the species flower and set seed simultaneously. These cycles tend to occur at very long intervals, ranging from three years upward. Most of the cycles are between fifteen and sixty years long. The most extreme example is the Chinese species *Phyllostachys bambusoides*, which seeded massively and throughout its range in the years 919, 1114, between 1716 and 1735, in 1833 to 1847, and in the late 1960s. The last event involved cultivated plants in widely separated areas, including England, Russia, and Alabama. *P. bambusoides*, therefore, has a flowering cycle of about 120 years, set by an internal clock that runs more or less independently of environmental circumstances.

The bamboos with cycles of this kind spread mainly by the production of rhizomes, horizontal underground stems. When they do flower, nearly all individuals flower at once, set large quantities of seed, and die. Huge numbers of animals, including rats, pigs, and pheasants, often migrate into areas where bamboos are fruiting to feed on the seeds; humans also gather them for food. Apparently, the plants put all of their energy into producing such vast amounts of seeds that large numbers of seedlings survive, even though most of the seeds are eaten.

A particular conservation problem that attracted widespread notice in the 1980s is the relationship between the flowering of bamboos (primarily of the genus *Fargesia*) and the survival of the giant panda, a spectacular animal that still lives in a few mountain ranges in southwestern China. Humans have occupied so much of the former range of the panda that the wild population now consists of only a few thousand individuals, which live only in several widely separated areas. In these places, only one or a few species of bamboo provide virtually all of the pandas' food. The mass flowering of some of these bamboo species in 1982 to 1984, which led to their death over large areas, drove many of the pandas to the brink of starvation, and only a massive international effort made it possible to rescue many of the animals (Sidelight figure 31.1). The predictable episodes of mass flowering in bamboos will need to be considered carefully in planning for the pandas' survival.

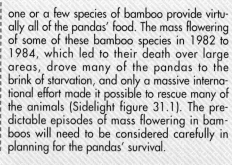

(a) (b)

Sidelight figure 31.1

(a) *A panda in a large enclosure of a stand of bamboo,* Fargesia spathacea. *Bamboos of the genus* Fargesia *are the most important panda food in the Min Mountains of northern Sichuan Province, China, where this photograph was taken.* (b) *A Chinese scientist examining a stand of the bamboo* Fargesia nitida *that has flowered and dropped its leaves; soon, the whole plant will be dead. This bamboo covers large areas of the Min Mountains and other mountainous areas where pandas live. It flowered extensively in the Min Mountains between 1974 and 1976, and again between 1982 and 1984; as a result, many pandas needed to be rescued and fed in captivity.*

Figure 31.6 How a seed germinates.

The germination of a seed, such as this garden pea, involves the fracture of the seed coat, from which the young shoot and root emerge.

Figure 31.7 Seeds can remain dormant for long periods.

This seedling was grown from seeds of lotus recovered from the mud of a dry lake bed in Manchuria, northern China. The radiocarbon age of this seed indicates that it was formed around the year 1515; another seed that was germinated was estimated to be at least a century older. The coin is included in the photo to give some idea of the size.

Once a seed coat forms around the embryo, most of the embryo's metabolic activities cease; a mature seed contains only about 10 percent water. Under these conditions, the seed and the young plant within it are very stable. **Germination,** or the resumption of metabolic activities which leads to the growth of a mature plant, cannot take place until water and oxygen reach the embryo, a process that sometimes involves cracking the seed (figure 31.6). Seeds of some plants have been known to remain viable for hundreds of years (figure 31.7).

Environmental factors help to ensure that the plant will germinate only under appropriate conditions. Sometimes, seeds are held within tough fruits that will not crack until exposed to the heat of a fire, a strategy that clearly results in the germination of a plant in an open habitat. Other seeds germinate only when inhibitory chemicals have been washed out of their seed coats, guaranteeing germination when sufficient water is available. Still other seeds germinate only after they pass through the intestines of birds or mammals or are regurgitated by them. This process both weakens the seed coats and ensures dispersal.

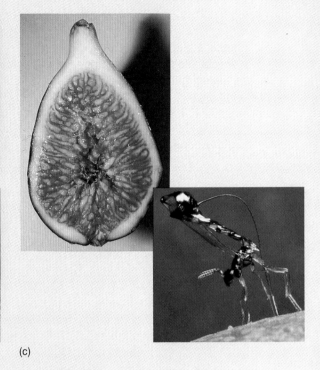

(a)

(b)

(c)

Figure 31.8 Animal-dispersed fruits.
(a) *The bright red berries of this honeysuckle,* Lonicera hispidula, *are highly attractive to birds, just as are red flowers. Birds may carry the berry seeds either internally or stuck to their feet for great distances.*

(b) *The spiny fruits of this burgrass,* Cenchrus incertus, *adhere readily to any passing animal, as you will know if you have stepped on them.*

(c) *The fruits of figs,* Ficus carica, *are flower clusters turned inside out. They are pollinated by tiny, specialized wasps, which require figs to complete their life cycle. The wasps enter through the hole in the end of the fig. When mature, figs are consumed by various animals, and the seeds of the individual tiny flowers are scattered about by birds and mammals.*

While an embryo is embedded in its seed, water and oxygen are largely excluded and the metabolism in the seed is greatly slowed. To reactivate the metabolism, the embryo has only to be provided with water, oxygen, and a source of metabolic energy. The final release of arrested development—germination—is then cued to specific signals from the environment.

Seeds are clearly important adaptively in at least three respects:

1. Seeds permit plants to postpone development until conditions are favorable. Under conditions in which young plants might or might not become established, a plant can "afford" to have some seeds germinate because others remain dormant.

2. By tying the reinitiation of development to environmental factors, seeds permit embryo development to be synchronized with critical aspects of the plant's habitat.

3. Perhaps most important, seed dispersal facilitates the migration and dispersal of genotypes into new habitats. The seed also offers maximum protection to the young plant at its most vulnerable developmental stage.

The developing embryo becomes enclosed in a seed. Within the seed, the embryo stops developing until conditions are conducive to its germination.

Fruits

During seed formation, the carpel of the flower ovary begins to develop into fruit. Paralleling the evolution of angiosperm flowers, and of equal importance to angiosperm success, has been the evolution of these fruits. Fruits form in many ways and exhibit a wide array of modes of specialization in relation to their dispersal.

Fruits that have fleshy coverings, often black, bright blue, or red, are normally dispersed by birds and other vertebrates. Like the red flowers discussed in relation to pollination by birds, the red fruits signal an abundant food supply (figure 31.8a). By feeding on these fruits, the birds and other animals carry seeds from place to place before excreting the seeds as solid waste. The seeds, not harmed by the animal digestive system, thus are transferred from one suitable habitat to another. Other fruits are dispersed by attaching themselves to the fur of mammals or the feathers of birds (figure 31.8*b*).

Specialized fruit and seed dispersal has evolved many different times in the flowering plants (figure 31.8*c*). For instance, mammals such as squirrels disperse and bury seeds and often do not find them again. Other fruits or seeds have "wings" and are blown about by the wind. The wings on the seeds of pines and those on the fruits of ashes or maples play identical ecological roles (figure 31.9*a*). Dandelions and related plants provide a familiar example of a kind of fruit that is dispersed by the wind (figure 31.9*b*). Seed dispersal of such plants as milkweeds, willows, and cottonwoods is similar. In tumbleweeds, the whole plant scatters seeds as it is blown about by the wind (figure 31.9*c*).

Still other fruits, such as those of mangroves, coconuts, and certain other plants that characteristically occur on or near beaches, swamps, or other bodies of water, are regularly spread from place to place in the water (figure 31.10). Dispersal of this sort is especially important in the colonization of distant island groups, such as the Hawaiian Islands. The seeds

(a)   (b)

Figure 31.9 Wind-dispersed fruits.
(a) *The double fruits of the maple,* Acer, *when mature, are blown considerable distances from their parent trees.*
(b) *False dandelion,* Pyropappus caroliniana. *The "parachutes" disperse the dandelion fruits widely in the wind, much to the gardener's displeasure.*
(c) *Tumbleweed,* Salsola, *in which the whole dead plant becomes a light, windblown structure that scatters seeds as it rolls about.*

(c)

Figure 31.11 Development of angiosperms.
(a) **Dicot development in a soybean.** *The two cotyledons of the dicot are pulled up through the soil along with the hypocotyl (the stem below the cotyledons). The cotyledons are the first leaves that perform photosynthesis. As other leaves develop, they take over photosynthesis entirely, and the cotyledons shrivel and fall off the stem. Flowers develop in buds at the nodes.*
(b) **Monocot development in wheat.** *Monocots have one, single, underdeveloped cotyledon, which does not appear in the development of the mature plant. The coleoptile is a primordial stem; it encloses and protects the shoot and leaves as it pushes its way up through the soil.*

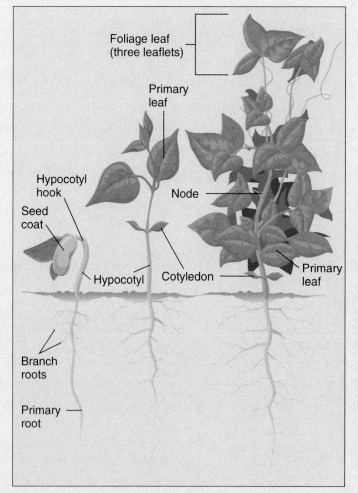

Foliage leaf
(three leaflets)

Primary
leaf

Hypocotyl
hook   Node

Seed
coat

Hypocotyl   Cotyledon   Primary
leaf

Branch
roots

Primary
root

(a)

Figure 31.10 Water-dispersed fruits.
A fruit of the coconut, Cocos nucifera, *sprouts on a sandy beach. One of the most useful plants for humans in the tropics, coconuts have become established even on the most distant islands by drifting in the waves.*

of about 280 original flowering plant species must have reached Hawaii to have evolved into the minimum of 956 native species found there today.

Fruits have evolved alongside flowers and are important to seed dispersal.

Germination

What happens to a seed when it encounters conditions suitable for its germination? First, it imbibes water. Seed tissues are so dry at the start of germination that the seed takes up water with great force, after which metabolism resumes. Initially, the metabolism may be anaerobic, but when the seed coat ruptures, aerobic metabolism takes over. At this point, oxygen must be available to the developing embryo because plants, which drown for the same reason people do, require oxygen for active growth. Few plants produce seeds that germinate successfully underwater, although some, such as rice, have evolved a tolerance of anaerobic conditions. Figure 31.11 shows the development of a dicot and monocot from germination to maturity.

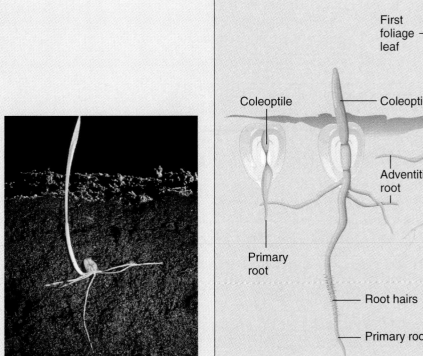

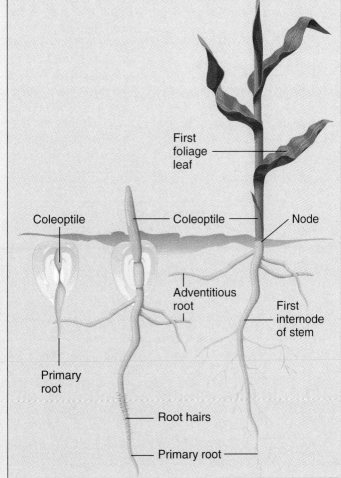

First foliage leaf

Coleoptile — Coleoptile — Node

Adventitious root

First internode of stem

Primary root

Root hairs

Primary root

(b)

Germination and early seedling growth require the mobilization of metabolic reserves stored in the starch grains of **amyloplasts** (organelles specialized to store starch) and in protein bodies. Fats and oils are also important food reserves in some kinds of seeds. They can readily be digested during germination to produce glycerol and fatty acids, which yield energy through aerobic respiration and can also be converted to glucose. Any of these reserves may be stored in the embryo itself or in the endosperm, depending on the kind of plant.

How are the genes that transcribe the enzymes involved in the mobilization of food resources activated? Experimental studies have shown that, in the endosperm of the cereal grains at least, this occurs when the embryo synthesizes hormones called gibberellins (see chapter 32). These hormones initiate a burst of mRNA and protein synthesis. Whether the gibberellins act directly on the DNA or through chemical intermediates in the cytoplasm is not known. DNA synthesis apparently does not occur during the early stages of seed germination but becomes important when the **radicle,** or embryonic root, has grown out of the seed coats.

> *During early germination and seedling establishment, the vital mobilization of the food reserves stored in the embryo or the endosperm is mediated by hormones.*

Growth and Differentiation

Once a seed has germinated, the plant's further development depends on the activities of the meristematic tissues, which interact with the environment. As described in chapter 30, the shoot and root apical meristems give rise to all of the other cells of the adult plant.

Differentiation, or the formation of specialized tissues, occurs in five stages in plants (figure 31.12):

Stage 1 is the formation of the embryo by cell division of the zygote formed by the sperm's fertilization of the egg within the ovule.

Stage 2 is the differentiation within the embryo of the apical meristems, which begins almost immediately in angiosperms. The apical meristems can be detected when there are as few as forty cells in the growing embryo. Apical meristems are largely responsible for primary growth.

Stage 3 is the differentiation from apical meristems of the vascular and cork cambia (see chapter 30), which are largely responsible for secondary growth.

Stage 4 is the production of **primordia,** which are cells fully committed to becoming leaves, shoots, or roots. Leaf and shoot primordia develop directly from apical meristem cells, whereas root primordia develop from the root cambium, called the pericycle (see chapter 30).

Stage 5 is the production of fully differentiated tissues and structures, including xylem, phloem, leaves, shoots and roots.

After a seed germinates, the pattern of growth and differentiation that was established in the embryo is repeated indefinitely until the plant dies. But differentiation in plants, unlike that in animals, is largely reversible. Botanists first demonstrated in the 1950s that individual differentiated cells isolated from mature individuals could give rise to entire individuals. F. C. Steward was able to induce isolated bits of phloem taken from carrots to form new plants, plants that were normal in appearance and fully fertile (see Key Experiment 31.1). Regeneration of entire plants from differentiated tissue has since been carried out in many plants, including cotton, tomatoes, and cherries.

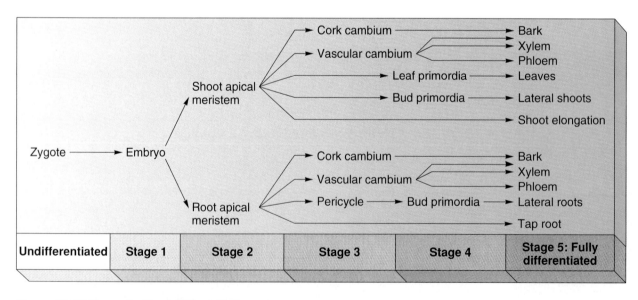

Figure 31.12 Stages in plant differentiation.
As this diagram shows, the different cells and tissues in a plant all originate from the shoot and root apical meristems.

KEY EXPERIMENT 31.1

2000

STEWARD'S EXPERIMENT TO GROW A NEW PLANT FROM PLANT TISSUE

◀ 1958

Phloem tissue was isolated from carrots in the laboratory of F. C. Steward at Cornell University. The discs of tissue were grown in a flask in which the medium was constantly agitated so as to bring a fresh supply of nutrients to the masses of callus (undifferentiated cells) that soon formed. From the phloem tissue, Steward was able to induce the formation of normal, fully fertile carrot plants.

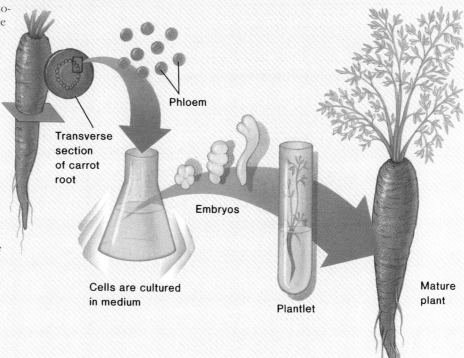

Transverse section of carrot root

Phloem

Cells are cultured in medium

Embryos

Plantlet

Mature plant

1900

These experiments clearly demonstrate that the original differentiated phloem tissue still contains all of the genetic potential needed for the differentiation of entire plants. No information is lost during plant tissue differentiation, and no irreversible steps are taken.

In plants, regeneration is not confined to the laboratory. Nature does it, too. Asexual reproduction is a regeneration process in which differentiated root or stem cells form a new shoot with its own set of roots. As a result, entire plants may form from horizontal roots, as well as from aboveground and underground runners, called **stolons** and **rhizomes**, respectively. Colonies of aspen, *Populus tremuloides,* often consist of a single individual that has given rise to a colony of genetically identical trees by producing new stems from its horizontal roots (figure 31.13).

Cell differentiation in plants is reversible; that is, any cell of the plant contains all the genetic information necessary for the development of a whole individual.

Figure 31.13 Clones of aspen.
This grove of aspen trees beneath the Grand Teton mountains of Wyoming is a single large colony produced by individuals spreading underground and periodically sending up new shoots.

Reproductive Strategies of Plants

Flowers, seeds, fruits, and asexual reproduction in plants can all be viewed as ways in which the plants increase or decrease the rate of recombination, a genetic process that has been one of the most powerful forces in the evolution of life on the earth. By rapidly generating new combinations of alleles, recombination gives natural selection the raw material from which to select new and better-adapted phenotypes. But recombination is not always advantageous. Species that are well adapted to their environments have little to gain by shuffling their genes about because most new combinations are less well-attuned to the plants' present environments. In plants, reproductive strategies that favor **outcrossing** (hybridization with another individual of the same species that is not a close relative), thereby promoting recombination, occur in some species, whereas self-pollination strategies that favor inbreeding, thereby minimizing recombination, occur in other species.

Factors Promoting Outcrossing

Outcrossing is critical for the adaptation and evolution of all eukaryotic organisms. Outcrossing is enhanced in certain plant species by the possession of separate male and female flowers. A flower that has only ovules, not pollen, is called **pistillate;** functionally, it is female. A flower that produces only pollen is called **staminate;** functionally, it is male (figure 31.14).

Staminate and pistillate flowers may occur on separate individuals in a given plant species, as in willows. Such plants, whose sporophytes produce either only ovules or only pollen, are called **dioecious** (from the Greek words meaning "two houses"). In other kinds of plants—oaks, birches, and ragweed—the two types of flowers are produced on the same plant; such plants are called **monoecious** ("one house").

The separation of ovule-producing and pollen-producing flowers that occurs in dioecious and monoecious plants makes outcrossing even more likely than it would be otherwise. But most angiosperms are neither dioecious nor

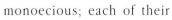

Figure 31.14 Staminate and pistillate flowers.
Birches have two different kinds of flowers, pollen-producing (staminate) ones, which hang down in long, yellowish tassels, and ovule-bearing (pistillate) ones, which mature into characteristic conelike structures and occur in small, reddish-brown clusters. Both are shown here.

monoecious; each of their flowers includes both pollen-producing and ovule-producing structures—that is, stamens and carpels.

An even simpler way that many flowers promote outcrossing is through the physical separation of the anthers and the stigmas. If the flower is constructed in such a way that these organs do not come into contact with one another, the pollen tends to be transferred to the stigma of another flower, rather than to the stigma of its own flower.

Another device that occurs widely in flowering plants and increases outcrossing is **genetic self-incompatibility.** In self-incompatible plants, the pollen from a given individual does not function on the stigmas of that individual, or the embryos resulting from self-fertilization do not function. Self-incompatibility is a mechanism that increases outcrossing, even though the flowers of such plants may produce both pollen and ovules, and their stamens and stigmas may be mature at the same time.

Factors Promoting Self-Pollination

Self-pollination is also very frequent among angiosperms. In fact, probably more than half of the angiosperms that occur in temperate regions self-pollinate regularly. Most of these have small, relatively inconspicuous flowers in which the pollen is shed directly onto the stigma, sometimes even before the bud opens (figure 31.15).

Pistil —
Anther —

(a)

— Unopened pistil

— Anther

(b)

— Anther

— Open pistil

(c)

Figure 31.15 Sex in the genus **Epilobium.**
(a) Epilobium ciliatum. *In this species, the anthers shed their pollen directly onto the large, creamy, undivided stigmas of the same flower, resulting in self-pollination.* (b and c) Epilobium angustifolium *is strongly outcrossing and is one of the first plants in which the process of pollination was studied. In it, the flowers are at first only staminate, and the anthers shed pollen. Then, immediately following pollen shedding, the stigma swings up so* that it is level with the anthers and its four lobes open, as shown in (c). The flower at this point is now fully pistillate. The flowers open progressively up the stem, so that the bees visit the lowest ones first. Working up the stem, the bees encounter the pollen-shedding, staminate-phase flowers, become covered with pollen, and carry it passively to the lower, functionally pistillate flowers of another plant.*

If outcrossing is just as important for plants, in genetic terms, as it is for animals, then why are so many plant species self-pollinating? There are two basic reasons for the frequency of self-pollinated angiosperms:

1. Self-pollination is advantageous in harsh climates. Because self-pollinating plants do not need to be visited by animals to produce seed, they can grow in areas where the kinds of insects or other animals that might visit them are absent or very scarce, as in the Arctic or at high elevations.

2. In genetic terms, self-pollination produces offspring that are more uniform than those that result from outcrossing. Such offspring may contain high proportions of individuals that are well adapted to particular habitats. Where the habitat occurs regularly, inbreeding is advantageous because it produces a greater proportion of well-adapted progeny than does outcrossing. Many successful weeds are self-pollinating because the habitat of weeds has been made uniform and spread all over the world by human beings.

Whether a plant is outcrossing or self-pollinated seems to depend largely on the nature of its environment, with harsh climates favoring the more uniform types produced by self-pollination.

EVOLUTIONARY VIEWPOINT

Ninety percent of all living plants are angiosperms (flowering plants). They comprise over 235,000 species, including hardwood trees, shrubs, grasses, vegetables, and grains—in short, nearly all everyday plants except conifers. Angiosperms have found a better solution to a key problem posed by terrestrial living: the inherent conflict between the need to obtain nutrients (solved by roots, which anchor the plant to one place) and the need to find mates (solved by making the male gametes tiny so that they can be transported to other plants). The wind pollination of gymnosperms is very inefficient. Insect pollination is far more efficient and allows a plant species to exist in widely dispersed populations. Angiosperms that live in dense populations, however, often dispense with insect pollination. For example, grasses, which are wind-pollinated, are among the most successful angiosperms.

SUMMARY

1. The flowers of angiosperms contain the plant's male and female reproductive structures. The male structures are called stamens, and each stamen consists of a filament and an anther. The female structures are called pistils, and each pistil consists of a stigma, style, and ovary. The ovary contains the carpels, which enclose the ovules in which the megaspores develop.

2. Pollen grain formation begins with the meiotic division of the microspore mother cells contained in the microsporangia. This meiosis produces four haploid microspores, which divide mitotically to form four pollen grains.

3. Egg cell formation begins with the meiotic division of the megaspore mother cells contained in the megasporangia. This meiosis produces four haploid megaspores. One of these megaspores divides mitotically to produce eight haploid nuclei within an embryo sac. One of these nuclei becomes the egg cell. The other nuclei are arranged in positions around the embryo sac; two of these nuclei, called polar nuclei, will become the endosperm.

4. Angiosperm flowers make possible the precise transfer of pollen and thus enable even widely separated plants to outcross effectively. Pollen is transferred by animals, particularly insects, and by the wind. Self-pollination is also possible.

5. Fertilization occurs when a pollen grain lands on the stigma. The pollen grain grows a pollen tube that eventually reaches the embryo. Within the pollen tube, the pollen grain divides to form two sperm cells. One sperm cell fertilizes the egg cell; the other sperm cell fuses with the two polar nuclei to form the primary endosperm nucleus.

6. The transformation of a zygote into a mature individual, initiated immediately after fertilization, is called development. Development is a process of progressive specialization that results in differentiation, the production of highly individualized tissues and structures.

7. Differentiation of angiosperm embryos ceases after the major organs have developed, and each now-dormant embryo is encased within a dry casing, becoming a seed that maintains a state of suspended development until the seed is broken or moistened. Resumption of metabolic activity after the period of dormancy is called germination.

8. Angiosperm seeds remain within the carpel, which develops into a fruit. Animals that consume fruit may carry the seeds for long distances before excreting them as solid waste. The seeds, not harmed by the animal digestive system, can then germinate in the new location.

9. Because of their rigid, relatively impermeable seed coat, seeds germinate only when they receive water and appropriate environmental cues. In seed germination, mobilization of the food reserves stored in the cotyledons and in the endosperm is critical. In cereal grains, this process is mediated by hormones known as gibberellins.

10. In plants, differentiation occurs in five stages and is largely reversible. Whole plants have been regenerated from cultures of single cells.

11. Outcrossing in different angiosperms is promoted by the separation of the pollen-producing and ovule-producing structures into different flowers, or even into different individuals; by the separation of the pollen and the stigmas within a given flower with respect to position or time of maturation; and by genetic self-incompatibility.

12. Self-pollination occurs in more than half of the angiosperm species of temperate regions. It is most common among plants in harsh climates or those that occur in widespread, uniform habitats, such as weeds.

REVIEWING THE CHAPTER

1. How are gametes formed in angiosperms?
2. What is the structure of an angiosperm flower?
3. How is pollen formed?
4. How do angiosperms produce egg cells?
5. What is pollination?
6. Why do animals pollinate angiosperms?
7. Which plants are wind-pollinated?
8. How does self-pollination work?
9. What is the unique process of double fertilization?
10. What is the mechanism of fertilization in plants?
11. How are seeds formed?
12. Exactly what is a fruit?
13. How do seeds germinate?
14. How do plants grow and differentiate?
15. What are the reproductive strategies of plants?
16. What factors promote outcrossing?
17. What factors promote self-pollination?

COMPLETING YOUR UNDERSTANDING

1. In which biomes do flowering plants *not* dominate the vegetation?
2. What is the difference between a pollen grain and a male gamete?
3. In most cases, which flower part would attract animals?

    a. Sepal   c. Stamen
    b. Petal   d. Carpel

4. What are the functions of the anther and filament in the male flower part?
5. How do the stigma and style function in the female flower part?
6. What is the difference between a microspore and a pollen grain?
7. Which type of cell division—mitosis or meiosis—produces gametes in plants?
8. What are megaspores, where are they found, and what do they give rise to?
9. Flowers with very long tubes with the nectar produced deep within them are visited by [*three answers required*]

    a. humming-   d. moths.
       birds.   e. beetles.
    b. bees.   f. butterflies.
    c. ants.

10. If you were planting a flower garden and wanted to attract hummingbirds, you would plant more flowers that are

    a. yellow.   d. red.
    b. blue.   e. of any color.
    c. white.   f. gold.

11. Why do most wind-pollinated angiosperm trees produce flowers before they produce leaves?
12. The most numerous insect-pollinated angiosperms are those pollinated by

    a. hummingbirds.
    b. beetles.
    c. bees.
    d. butterflies.
    e. the wind.

13. What is the function of a pollen tube, and how does it grow?
14. What is the endosperm, what is its function, and how is it derived?
15. How do integuments function in seeds?
16. Why is plant development so much more closely linked with environmental cues than animal development?
17. How are angiosperm seeds and fruits dispersed?
18. What is imbibition, and where does it occur in plants?
19. What is an amyloplast, and how does it relate to seed germination?
20. Leaf and bud primordia develop from

    a. the xylem.
    b. the epidermis.
    c. the apical meristem.
    d. the vascular cambium.
    e. none of the above.

21. In plant differentiation, which of the following are lacking in roots (see figure 31.12)? [*two answers required*]

    a. Bark   d. Vascular
                   cambium
    b. Bud   e. Lateral shoots
    c. Leaf   f. Lateral roots
      primordia

22. How can entire plants—for example, carrots, cotton, tomatoes, and cherries—be regenerated from differentiated tissue?
23. If you were skiing in Colorado and you noted entire mountainsides populated with uniform patches of aspens, how would you interpret these stands of aspens?
24. What is a pistillate flower? What is a staminate flower?
25. What is the advantage of a plant being genetically self-incompatible?
26. Why does self-pollination occur in plants?
27. Self-pollination occurs within at least some species of a wide variety of different kinds of plants, usually those living in harsh environments or among "weedy" species. Why do you suppose that animals coping with similar environments never developed similar reproductive strategies? Can you think of any animals that *do* self-fertilize?
28. Why have grasses been so successful, even though they are not insect-pollinated?
29. Why did the wild population of the giant panda bear have to rely on humans for its survival in China in the mid-1980s?
*30. When you eat a tossed salad, what plant parts are you consuming?

*For discussion

FOR FURTHER READING

Batra, S. W. T. 1984. Solitary bees. *Scientific American,* Feb., 120–27. Excellent article on these diverse and fascinating pollinators, some of which are of great commercial importance.

Cook, R. E. 1983. Clonal plant populations. *American Scientist* 71: 244–53. Excellent discussion on the role of asexual reproduction among plants in natural populations.

Evans, M. et al. 1986. How roots respond to gravity. *Scientific American,* Dec., 112–19. Modern studies of this fascinating problem.

Janzen, D. H. 1976. Why bamboos wait so long to flower. *Annual Reviews of Ecology and Systematics* 7: 347–91. A now classic essay about the natural history of bamboos.

Johnson, W., and C. Adkisson. 1986. Airlifting the oaks. *Natural History,* Oct., 40–46. A familiar example of animal dispersal of plant fruit and seeds: Bluejays spread acorns widely to new habitats.

Robacker, D. et al. 1988. Floral aroma. *BioScience* 38: 390–96. An interesting discussion of the odors that plants utilize in attracting their pollinators.

Schaller, G. B. 1993. *The last panda.* Chicago: University of Chicago Press. The famous naturalist's personal account of his long struggle against nature and bureaucracy to understand and save pandas.

Schaller, G. B., H. Jinchu, P. Wenshi, and Z. Jing. 1985. *The giant pandas of Wolong.* Chicago, Ill.: University of Chicago Press. A fascinating account of the mutual adaptations of pandas and bamboo. This book offers the first glimpse of the life of pandas in their remote mountain home in western China.

THE LIVING PLANT: TRANSPORT AND GROWTH

In order to hold its head up toward the sun, each of these sunflowers must have proper hydraulic support, an efficient plumbing system that collects and transports water, minerals, and organic plant products.

FOR REVIEW

Here are some important terms and concepts that have been discussed in previous chapters and that you will encounter again in this chapter. Review them before proceeding if necessary.

Osmotic pressure (*chapter 5*)
Transport of ions across membranes (*chapter 5*)
Photosynthesis (*chapter 8*)
Apical meristems (*chapter 30*)
Xylem (*chapter 30*)
Phloem (*chapter 30*)
Seed germination (*chapter 31*)

At first glance, a plant may not appear vibrantly alive. It does not bound from one place to another like a gazelle; nor does it growl, or respond to a caress. But its appearance is deceiving, and its internal structure is complex. Plants have a conducting system, as humans do, for transporting fluids and nutrients from one part to another and special organs for reproduction and the gathering of energy. And like humans, plants regulate their growth and organ functioning with hormones, chemicals that act as messengers to coordinate the many activities of the body.

This chapter focuses on the internal activities of living plants, first discussing the movement of fluids and nutrients within plants. Then, the hormones that control how plants grow are examined in more detail. Finally, the chapter looks at plant responses to environmental stimuli, such as light, gravity, and touch.

Water Movement

Functionally, a plant is essentially a tube with its base embedded in the ground. At the base of the tube are roots, and at its top are leaves. For a plant to function, two kinds of transport processes must occur: First, the carbohydrate molecules produced in the leaves by photosynthesis must be carried to all of the other living plant cells. To accomplish this, liquid, with these carbohydrate molecules dissolved in it, must move both up and down the tube. Second, nutrients and water in the ground must be taken up by the roots and ferried to the leaves and other plant cells. In this process, liquid moves up the tube. Plants accomplish these two processes by using chains of specialized cells: Those of the phloem transport photosynthetically produced carbohydrates up and down the tube, and those of the xylem carry water and minerals upward (figure 32.1).

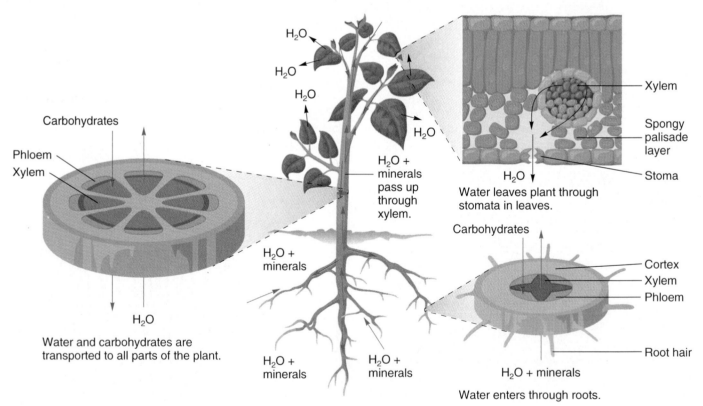

Figure 32.1 The flow of materials into, out of, and within a plant.
Water and minerals enter through the roots of a plant and are transported through the xylem to all parts of the plant body (blue arrows). Water leaves the plant through the stomata in the leaves (black arrows). Carbohydrates synthesized in the leaves are circulated throughout the plant by the phloem (red arrows).

Cohesion-Adhesion-Tension Theory

Many of the leaves of a large tree may be more than ten stories off the ground. How does a tree manage to raise water so high? If a long, hollow tube, closed at one end, is filled with water and placed, open end down, in a full bucket of water, gravity acts (pushes) on the column of air over the bucket: The weight of the air (at sea level) exerts an amount of pressure that is defined as 1 atmosphere downward on the water in the bucket and thus presses the water up into the tube. But gravity also acts to pull the water within the tube down. The interaction of these two forces determines the water level in the tube. At sea level, the water rises to about 10.4 meters. If the tube is any higher, a vacuum forms in the upper, closed end of the tube and fills with water vapor.

Opening the tube and blowing air across the upper end demonstrates how water rises higher than 10.4 meters in a plant. The stream of relatively dry air causes water molecules to evaporate from the water surface in the tube. The water level in the tube does not fall because, as water molecules are drawn from the top, they are replenished by new water molecules forced up from the bottom. This, in essence, is what happens in plants. The passage of air across leaf surfaces results in the loss of water by evaporation, creating a "pull" at the open upper end of the "tube." Meanwhile, new water molecules entering through the roots of the plant are pushed up by atmospheric pressure. In addition to these pushing and pulling forces, the **adhesion** of water molecules to the walls of the very narrow tubes that occur in plants also helps to maintain water flow to the tops of plants.

Water rises in a plant beyond the point at which it would be supported by atmospheric pressure (10.4 meters at sea level) because evaporation from its leaves produces a "pull" on the entire water column all the way down to the roots.

A column of water in a tall tree does not collapse simply because of its weight because water molecules have an inherent strength that arises from their tendency to form hydrogen bonds with one another. These hydrogen bonds cause **cohesion** of the water molecules; in other words, a column of water resists separation. This resistance, called **tensile strength,** varies inversely with the diameter of the column; that is, the smaller the diameter of the column, the greater the tensile strength. Therefore, plants must have very narrow transporting vessels to take advantage of tensile strength.

How the combination of the forces of gravity, tensile strength, and cohesion affect water movement in plants is called the **cohesion-adhesion-tension theory.**

Transpiration

The cohesion-adhesion-tension theory explains the process by which water leaves a plant, a process called **transpiration.**

More than 90 percent of the water taken in by plant roots is ultimately lost to the atmosphere, almost all of it from the leaves. It passes out primarily through the stomata in the form of water vapor. On its journey from the plant's interior to the outside, a molecule of water first passes into the pockets of air within the leaf by evaporating from the walls of the spongy mesophyll that lines the intercellular spaces. These intercellular spaces open to the outside of the leaf by way of the stomata. The water that evaporates from these surfaces of the spongy mesophyll cells is continuously replenished from the tips of the veinlets in the leaves. Because the strands of xylem conduct water within the plant in an unbroken stream all the way from the roots to the leaves, when a portion of the water vapor in the intercellular spaces passes out through the stomata, the supply of water vapor in these spaces is continually renewed (figure 32.2).

The humidity level of these intercellular spaces is 100 percent, meaning that the air in the spaces is saturated with water. Outside the leaf, the humidity is much lower. Water within the intercellular spaces thus passes out of the stomata from an area of high water concentration (the intercellular spaces) to an area of low water concentration (the outside). Transpiration is therefore dependent on osmosis, the movement of water from areas of high concentration to areas of low concentration.

Because plants are constantly losing water to the atmosphere, and because the presence of this water is essential to photosynthesis and other vital plant activities, growing plants depend on a continuous stream of water entering and leaving their bodies at all times. For this reason, water must always be available to plant roots.

Such structural features as the stomata, the cuticle, and the intercellular spaces in leaves have evolved in response to one or both of two contradictory requirements: minimizing the loss of water to the atmosphere, on the one hand, and admitting carbon dioxide, which is essential for photosynthesis, on the other. How plants resolve this problem is discussed a little later. A consideration of how roots absorb water must come first.

Water Absorption by Roots

Most of the water absorbed by plants comes in through the root hairs, which collectively have an enormous surface area. The root hairs are always **turgid**—plump and swollen with water—because they have a higher concentration of dissolved minerals than does the water in the soil solution; water, therefore, tends to move into them steadily. Once inside the roots, water passes inward to the conducting elements of the xylem.

Water is not the only substance that enters the roots by passing into the cells of root hairs. Membranes of root hair cells contain a variety of ion transport channels that actively pump specific ions into the plant, even against large concentration gradients. These ions, many of which are plant nutrients, are then transported throughout the plant as a component of the water flowing through the xylem.

At night, when the relative humidity may approach 100 percent at the leaf surface, there may be no transpiration from the leaves. Under these circumstances, the upward pull exerted on the column of water becomes very small or nonexistent. Active transport of ions into the roots, however, continues. This results in an increasingly high ion concentration within the cells,

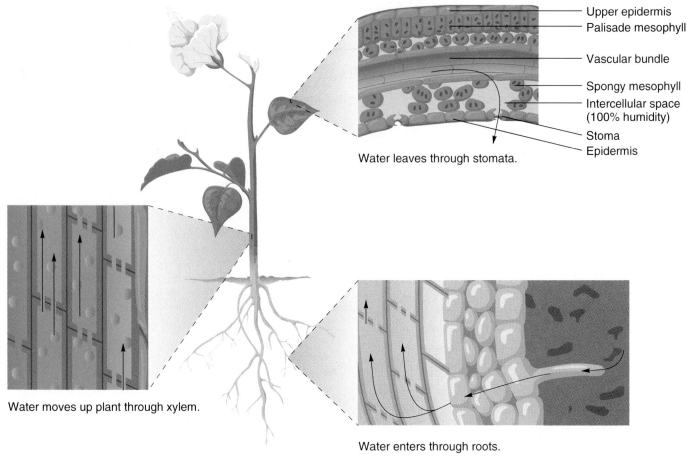

Upper epidermis
Palisade mesophyll

Vascular bundle

Spongy mesophyll
Intercellular space
(100% humidity)
Stoma
Epidermis

Water leaves through stomata.

Water moves up plant through xylem.

Water enters through roots.

Figure 32.2 Transpiration.
The movement of water upward in the xylem, combined with the entrance of water through the roots, causes water to leave through the stomata in the leaves.

a concentration that causes water to be drawn into the root hair cells by osmosis. The result is the movement of water into the plant and up the xylem columns because of **root pressure.**

Evidence of root pressure can be seen in the process called **guttation,** in which droplets of water are forced out of the leaves through special openings (*not* the stomata) at the tips and sides of the leaves. Although guttation droplets resemble droplets of dew, they are not water condensed by air, as dew droplets are. Instead, guttation droplets are water that has been literally forced out of the leaves by root pressure.

Root pressure, which is active primarily at night, is caused by the continued, active accumulation of ions by plant roots at times when transpiration from the leaves is very low or absent.

Regulation of Transpiration: Open and Closed Stomata

The only way plants can control water loss on a short-term basis is to close their stomata. Many plants can do this when subjected to water stress. But the stomata must be open at least part of the time so that carbon dioxide, which is necessary for

photosynthesis, can enter the plant. In its pattern of opening or closing its stomata, a plant must respond to both the need to conserve water and the need to admit carbon dioxide.

The stomata open and close because of changes in the water pressure of their guard cells. Stomatal guard cells are the only epidermis cells with chloroplasts and have a distinctive shape—they are thicker on the side next to the stomatal opening and thinner on their other sides and ends. When the guard cells are turgid (plump and swollen with water), they become bowed in shape, as do their thick inner walls, thus opening the stomata as wide as possible (figure 32.3).

Guard Cells and Ion Transport Channels

The guard cells use ATP-powered ion transport channels in their plasma membranes to concentrate ions actively, which causes water to enter the guard cells osmotically. As the guard cells accumulate water, they become turgid, opening the stomata. The guard cells remain turgid only as long as the active transport channels pump ions, chiefly potassium (K^+), into the guard cells and so maintain the higher solute concentration. Thus, keeping the stomata open requires a constant expenditure of ATP. When the active transport of ions into the guard cells ceases, the higher ion concentration within the guard cells

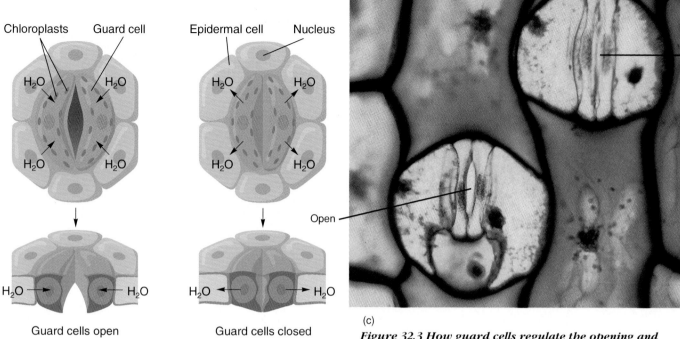

Figure 32.3 How guard cells regulate the opening and closing of stomata.
(a) *When guard cells contain a high level of solutes, water enters the guard cells, causing them to bow outward. This bowing opens the stoma. (b) When guard cells contain a low level of solutes, water leaves the guard cells, causing them to become flaccid. This flaccidity closes the stoma (c) In this* Tradescantia *leaf, both open and closed stomata can be seen.*

causes ions to move by diffusion into surrounding cells. Ultimately, water leaves the guard cells also, which then become somewhat "limp" or deflated, and the stomata between them close.

> *Stomata open when their guard cells become turgid. Guard cells' inner surfaces are thickest, and they bow inward when pressure within the cells is high. Keeping the guard cells turgid requires a constant expenditure of ATP.*

Environmental Factors That Influence the Opening and Closing of Stomata

A number of environmental factors affect the opening and closing of stomata. The most important is water loss. The stomata of plants that are wilted because of a lack of water tend to close. An increase in carbon dioxide concentration also causes the stomata of most species to close. In most plant species, stomata open in the light and close in the dark. Very high temperatures (above 30 to 35 degrees Celsius) also tend to cause stomata to close. Finally, stomata exhibit daily rhythms of opening and closing that appear to be controlled within the plant by factors that are only poorly understood.

Carbohydrate Transport

Most of the carbohydrates manufactured in plant leaves and other green parts are moved through the phloem to other parts of the plant. This process, known as **translocation,** makes suitable carbohydrate building blocks available at the plant's actively growing regions. The carbohydrates concentrated in storage organs such as tubers, often in the form of starch, are also converted into transportable molecules, such as sucrose, and moved through the phloem. The liquid in the phloem contains 10 to 25 percent dissolved solid matter, almost all of which is sucrose.

Movement of substances in the phloem can be remarkably fast—rates of 50 to 100 centimeters per hour have been measured—and is a passive process that does not require the expenditure of energy. The **mass flow** of materials transported in the phloem occurs because of water pressure, which develops as a result of osmosis. First, sucrose produced as a result of photosynthesis is actively "loaded" into the sieve tubes of the vascular bundles. This loading increases the solute concentration of the sieve tubes, so water passes into them by osmosis. An area where the sucrose is made is called a **source;** an area where sucrose is delivered from the sieve tubes is called a

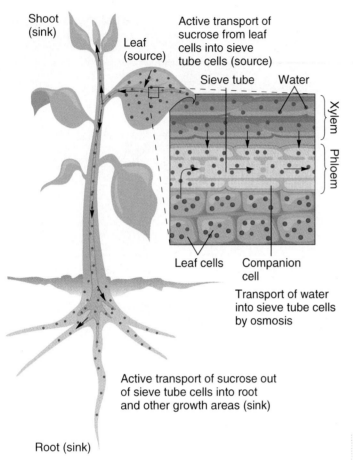

Figure 32.4 Mass flow.
Carbohydrates made in the leaves (the "source") are loaded into sieve tubes, and water follows by osmosis. At points where carbohydrates are needed (the "sinks," such as roots and shoots), the carbohydrates are unloaded, and water again follows by osmosis.

sink. Sinks include the roots and other regions where the sucrose is being unloaded. There the solute concentration of the sieve tubes is decreased as the sucrose is removed. As a result of these processes, water moves in the sieve tubes from the areas where sucrose is being taken in to those areas where it is being withdrawn, and the sucrose moves passively with the water (figure 32.4).

Essential Plant Nutrients

Just as human beings need certain nutrients, such as carbohydrates, amino acids, and vitamins, to survive, so also do plants need various nutrients to remain alive and healthy. Lack of an important nutrient can cause a plant to develop conditions related to the nutrient deficit, which may make the plant more susceptible to disease or even death.

Plants require a number of inorganic nutrients. Some of these are **macronutrients,** which plants need in relatively large amounts, and others are **micronutrients,** which are required in trace amounts (table 32.1). The nine macronutrients are: carbon, hydrogen, and oxygen (the three elements found in all organic compounds); and nitrogen, potassium, calcium, phosphorus, magnesium, and sulfur. Each of these nutrients

Table 32.1 Plant Nutrients	
Nutrient	**Relative Abundance in Plant Tissue (ppm)**
Macronutrients (Nutrients Required in Greater Concentration)	
Hydrogen	60,000,000
Carbon	35,000,000
Oxygen	30,000,000
Nitrogen	1,000,000
Potassium	250,000
Calcium	125,000
Magnesium	80,000
Phosphorus	60,000
Sulfur	30,000
Micronutrients (Nutrients Required in Minute Quantities)	
Chlorine	3,000
Iron	2,000
Boron	2,000
Manganese	1,000
Zinc	300
Copper	100
Molybdenum	1

Note: ppm = parts per million. Parts per million equals units of an element by weight per million units of oven-dried plant material.

approaches or exceeds 1 percent of a healthy plant's dry weight. The seven micronutrient elements, which in most plants constitute from less than one to several hundred parts per million by dry weight, are iron, chlorine, copper, manganese, zinc, molybdenum, and boron. Nutrients move primarily through plants as solutes in the water column in the xylem.

> *Plants need macronutrients in relatively large amounts (approaching or exceeding 1 percent of a plant's dry weight). Plants need micronutrients in only trace amounts of one to several hundred parts per million.*

The six macronutrients in addition to carbon, hydrogen, and oxygen are involved in plant metabolism in many ways. Potassium ions regulate the **turgor pressure** (the pressure within a cell that results from water moving into the cell) of guard cells and therefore the rate at which the plant loses water and takes in carbon dioxide. Calcium is an essential component of the **middle lamellae,** the structural elements laid down between plant cell walls, and also helps to maintain the physical integrity of membranes. Magnesium is a part of the chlorophyll molecule. The presence of phosphorus in many key biological molecules such as nucleic acids and ATP has been explored in detail in earlier chapters. Nitrogen is an essential part of amino acids and proteins, chlorophyll, and the nucleotides that make up nucleic acids. Sulfur is a key component of an amino acid (cysteine) essential in building proteins. Sidelight 32.1 describes how some plants—the carnivorous plants—obtain their nitrogen in a way very different from other plants.

Sidelight 32.1
CARNIVOROUS PLANTS

Some plants are able to use other organisms directly as sources of nitrogen just as animals normally do. These are the carnivorous plants. Carnivorous plants often grow in acidic soils, such as bogs—habitats not favorable for the growth of most legumes or of nitrifying bacteria. By capturing and digesting small animals directly, such plants obtain adequate nitrogen supplies and thus are able to grow in these seemingly unfavorable environments.

Carnivorous plants have adaptations used to lure and trap insects and other small animals. The plants digest their prey with enzymes secreted from various kinds of glands.

The Venus flytrap (*Dionaea muscipula*), which grows in the bogs of coastal North and South Carolina, has three sensitive hairs on each side of each leaf, which, when touched, trigger the two halves of the leaf to snap together. Once enfolded by a leaf, the prey of a Venus flytrap is digested by enzymes secreted from the leaf surfaces (Sidelight figure 32.1).

Pitcher plants attract insects by the bright, flowerlike colors within their pitcher-shaped leaves and perhaps also by sugar-rich secretions. Once inside the pitchers, insects slide down into the cavity of the leaf, which is filled with water, digestive enzymes, and half-digested prey.

Bladderworts, *Utricularia*, are aquatic. They sweep small animals into their bladderlike leaves by the rapid action of a springlike trapdoor and then digest these animals. In the sundews, the glandular trichomes secrete both sticky mucilage, which traps small animals, and digestive enzymes.

Sidelight figure 32.1 Carnivorous plants.
(a) *Venus flytrap, which inhabits low, boggy ground in North and South Carolina.* (b) *A Venus flytrap that has snapped together, imprisoning a fly.*

A number of ions are components of enzyme systems and serve as cofactors (see chapter 6) in essential biochemical reactions. Potassium, which reaches higher local concentrations in plants than any other elements except carbon and oxygen, affects the conformation of many proteins and probably affects at least sixty different enzymes while they are functioning. Zinc appears to play a similar role in the synthesis of the important plant growth hormone auxin. Plants with an inadequate supply of zinc display symptoms that derive mainly from a lack of cell elongation, apparently reflecting a shortage of auxin.

As in animals, when an essential nutrient is in short supply, plants display characteristic deficiency symptoms. For this reason, a trained observer can often tell what chemicals should be supplied simply by observing a plant's appearance.

Some plants have specific nutritional requirements not shared by others. For example, while grasses require silica because it helps to retard their complete destruction by herbivores, silica is not required by plants in general. Cobalt is essential for the normal growth of the nitrogen-fixing bacteria associated with the nodules of legumes. Nickel seems to be essential for soybeans, and its role in the nutrition of other plants requires further investigation.

In general, animals acquire the elements they need through plants. Plants therefore form an indispensable link between animals and the reservoirs of chemicals in nature. Some elements that animals require, such as iodine, come by way of plants but are not required by the plants. Iodine is very rare in soils; a shortage of it in the human diet can lead to the condition known as goiter.

Regulating Plant Growth: Plant Hormones

Hormones are chemical substances produced in small, often minute, quantities in one part of an organism and then transported to another part of the organism, where they stimulate certain physiological processes and inhibit others. How they act in a particular instance is influenced both by what the hormones themselves are and by how they affect the particular tissue that receives their message.

In animals, hormones are usually produced at definite sites, normally in organs that are solely concerned with hormone production. In plants, on the other hand, hormones are produced in tissues that are not specialized for that purpose but that carry out other, usually more obvious, functions. Nor do plant hormones have definite target areas.

At least five major kinds of hormones are found in plants: **auxin, cytokinins, gibberellins, ethylene,** and **abscisic acid.** Other kinds of plant hormones certainly exist but are less well understood. The study of plant hormones, especially attempts to understand how hormones produce their effects, is an active and important field of current research.

Five major kinds of plant hormones are reasonably well understood: auxin, cytokinins, gibberellins, ethylene, and abscisic acid.

Discovery of the First Plant Hormone

In his later years, the great evolutionist Charles Darwin became increasingly devoted to the study of plants. In 1881, he and his son Francis published a book called *The Power of Movement in Plants*, in which they reported their systematic experiments concerning the way in which growing plants bend toward light, a phenomenon known as **phototropism.**

The Darwins found that young grass seedlings normally bent strongly toward a source of light if the light came primarily from one side. However, if the upper part of a seedling was covered with a cylinder of metal foil so that no light reached its tip, the shoot would not bend. The Darwins obtained this result even though direct light reached the region where the bending normally occurred. However, if the end of a shoot was covered with a gelatin cap, which transmitted light, the shoot would bend as if it were not covered at all (see Key Experiment 32.1).

KEY EXPERIMENT 32.1

2000

THE DARWINS' DISCOVERY OF AUXIN

Charles and Francis Darwin found that a young grass seedling normally bends toward the light (1). If they covered the tip of the seedling with a lightproof collar, however, the seedling did not bend toward the light (2). When the tip of the seedling was covered with a transparent collar, the bending did occur (3). When the Darwins placed the collar below the tip, the seedling again bent toward the light (4). From these experiments, the Darwins concluded that, in response to light, an "influence" that causes bending was transmitted from the tip of the seedling to the area below the tip, where bending usually occurs.

◄ 1881

1800

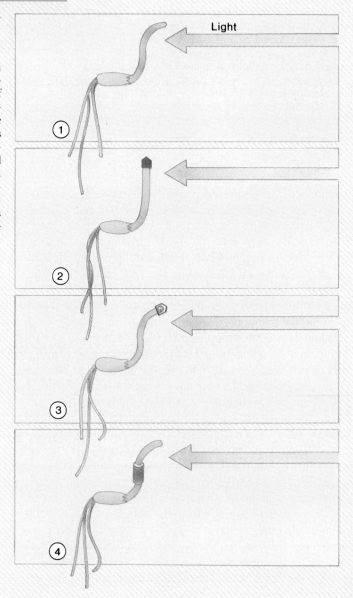

To explain this unexpected finding, the Darwins hypothesized that, when the shoots were illuminated from one side, an "influence" that arose in the uppermost part of the shoot was then transmitted downward, causing the shoot to bend. For some thirty years, the Darwins' perceptive experiments remained the sole source of information about this interesting phenomenon. Then, several botanists conducted a series of experiments that demonstrated that the substance causing the shoots to bend was a chemical. The botanists cut off the tip of a grass seedling and then replaced it, but separated it from the rest of the seedling by a block of agar, a gelatinous medium often used in biological experiments. The seedling reacted as if there had been no change. Something was evidently passing from the tip of the seedling through the agar into the region where the bending occurred. The "something" was a plant hormone called **auxin.** Auxin is now known to regulate cell growth in plants.

How Auxin Controls Growth

How auxin controls plant growth was discovered in 1926 by Frits Went, a Dutch plant physiologist, in the course of studies for his doctoral dissertation. Carrying the earlier experiments an important step farther, Went cut off the tips of grass seedlings that had been illuminated normally and set these tips on agar. He then took grass seedlings that had been grown in the dark and cut off their tips in a similar way. Finally, Went cut tiny blocks from the agar on which the tips of the light-grown seedlings had been placed and put them on the tops of the decapitated dark-grown seedlings, but set off to one side. Even though these seedlings had not been exposed to the light themselves, they bent *away* from the side on which the agar blocks were placed (see Key Experiment 32.2).

From his experiments, Went was able to show that the substance that flowed into the agar from the tips of the light-grown grass seedlings enhanced cell elongation. This chemical messenger caused the tissues on the side of the seedling into which it flowed to grow more than those on the opposite side. He named the substance that he had discovered auxin, from the Greek word *auxein*, meaning "to increase."

Went's experiments provided a basis for understanding the responses that the Darwins had obtained some forty-five years earlier: Grass seedlings bend toward the light because the auxin contents on the two sides of the shoot differ. The side of the shoot that is in the shade has more auxin; therefore, its cells elongate more than those on the lighted side, bending the plant toward the light. Later experiments by other investigators showed that auxin in normal plants migrates from the illuminated side to the dark side in response to light and thus causes the plant to bend toward the light.

KEY EXPERIMENT 32.2

2000

WENT'S DEMONSTRATION OF HOW AUXIN AFFECTS PLANT GROWTH

Frits Went, a Dutch plant physiologist, discovered how auxin controls plant growth. Went removed the tips of grass seedlings and put them on agar. Auxin flowed from the tips of the seedlings into the agar blocks (1). Went then placed these blocks of agar on one side of the ends of grass seedlings that had been grown in the dark and from which the tips had been removed (2). The seedlings bent away from the side on which the auxin-filled agar block was placed (3). Went concluded that auxin promoted cell elongation and that it accumulated on the side of a grass seedling bent away from the light.

◀ **1926**

1900

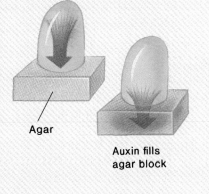

Auxin in tip of seedling

Agar

Auxin fills agar block

(1)

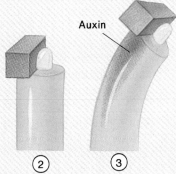

Auxin

(2) (3)

The Major Plant Hormones

Since the discovery of auxin, other experiments conducted on plants have revealed the presence and activities of other plant hormones. These hormones play different roles in plant growth and maintenance.

Auxin

Only one form of auxin—indoleacetic acid (IAA)—occurs in nature. IAA is produced at the shoot tip in the region of the apical meristem and diffuses continuously downward. The term *auxin* is now used to refer both to the naturally occurring substance and to those related synthetic molecules that produce similar effects.

Auxin acts to increase the plasticity of the plant cell wall, allowing it to stretch more during active cell growth. Because very low concentrations of auxin promote cell wall plasticity, auxin must be broken down rapidly to prevent its accumulation. Plants break auxin down by means of the enzyme **indoleacetic acid oxidase.** By controlling the levels of both IAA and IAA oxidase, plants can precisely regulate their growth.

Auxin controls various plant responses in addition to those involved in phototropism. One of these is the suppression of lateral bud growth, as discussed in chapter 30. How can auxin, a growth promoter, also inhibit growth? Apparently, the cells around lateral buds produce the chemical ethylene under the influence of auxin. The ethylene, in turn, inhibits the growth of the lateral buds. When the terminal bud is removed, removing the surface of auxin, the lateral buds grow, producing bushy plants. The number of flowers on an individual plant also increases in this situation.

> *The only known naturally occurring auxin—indoleacetic acid (IAA)—is produced in the apical meristems of shoots and diffuses downward, suppressing the growth of lateral buds. In young grass seedlings and other herbs, auxin plays a major role in stem elongation, migrating from the illuminated portions of the stem to the dark portions, thereby causing the stems to grow toward the light.*

Synthetic Auxins

Synthetic auxins are routinely used to control weeds. When applied as herbicides, they are applied in higher concentrations than those at which IAA normally occurs in plants. One of the most important of the synthetic auxins used in this way is 2,4-dichlorophenoxyacetic acid, usually known as 2,4-D. It kills weeds in lawns without harming the grass because 2,4-D affects only broad-leaved dicots. When treated, the weeds literally "grow to death," rapidly depleting all metabolic reserves so that no source of energy remains for transport or other essential functions.

Closely related to 2,4-D is the herbicide 2,4,5-trichlorophenoxyacetic acid (2,4,5-T), which is widely used to kill woody seedlings and weeds. Notorious as the "Agent Orange" of the Vietnam War, 2,4,5-T is easily contaminated with a by-product of its manufacture, **dioxin.** Dioxin, which is believed to be harmful to people, is the subject of great environmental concern.

Cytokinins

A **cytokinin** is a plant hormone that, in combination with auxin, stimulates cell division in plants and determines the course of differentiation. Substances with these properties are widespread, both in bacteria and in eukaryotes. In vascular plants, most cytokinins seem to be produced in the roots, from which they are then transported throughout the rest of the plant. Cytokinins apparently stimulate cell division by influencing the synthesis or activation of proteins specifically required for mitosis.

The naturally occurring cytokinins all appear to be derivatives of the purine base adenine (interestingly, adenine is a component of DNA). In contrast to auxin, cytokinins *promote* the growth of lateral buds and *inhibit* the formation of lateral roots. Consequently, the balance between cytokinins and auxin determines, along with other factors, the appearance of a mature plant.

> *Cytokinins are plant hormones that, in combination with auxin, stimulate cell division and determine the course of differentiation. Most are produced in the roots and transported to the rest of the plant.*

Gibberellins

Gibberellins are named for the fungus genus *Gibberella*, which causes a disease of rice in which the plants grow to be abnormally tall. This "foolish seedling disease" of rice was investigated in the 1920s by Japanese scientists, who found that, by growing the fungus in culture, they could obtain a chemical completely free of the fungus itself that affected the rice plants in a way similar to the fungus. This substance, isolated in 1939 and chemically characterized in 1954, was the first of what proved to be a large class of naturally occurring plant hormones called the gibberellins.

Synthesized in the apical portions of both shoots and roots, gibberellins have important effects on stem elongation in plants and play the leading role in controlling this process in mature trees and shrubs. In these plants, the application of gibberellins characteristically promotes internode (the spaces between leaf nodes on stems) elongation, and this effect is enhanced if auxin is also present. Gibberellins are also involved with many other aspects of plant growth, such as flowering inducement and the hastening of seed germination (figure 32.5).

> *Gibberellins, a very common and important class of plant hormones, are produced in the apical regions of shoots and roots and play the major role in controlling stem elongation for most plants, acting in concert with auxin and other plant hormones.*

Figure 32.5 Gibberellin.
Although more than sixty gibberellins have been isolated from natural sources, apparently only one is active in shoot elongation. As shown in this photo, cabbage will produce tall flowering shoots when treated with this form of gibberellin.

Ethylene

Long before its role as a plant hormone was appreciated, the simple hydrocarbon **ethylene** ($H_2C{=}CH_2$) was known to defoliate plants when it leaked from gas lights in street lamps. But ethylene is a natural product of plant metabolism and appears to be the main factor in the formation of specialized cell layers that precedes the dropping off of leaves, flowers, and fruit from plants. As mentioned earlier, auxin, diffusing down from the apical meri-stem of the stem, may stimulate the production of ethylene in the tissues around the lateral buds and thus retard bud growth. Ethylene also suppresses the stem and root elongation, probably for similar reasons.

Ethylene is also produced in large quantities during a certain phase of fruit ripening, when fruits' respiration is proceeding at its most rapid rate. At this phase, complex carbohydrates are broken down into simple sugars, cell walls become soft, and the volatile compounds associated with flavor and scent in the ripe fruits are produced. When applied to fruits, ethylene hastens their ripening.

One of the first lines of evidence that led to the recognition of ethylene as a plant hormone was the observation that gases from oranges caused premature ripening in bananas. Such relationships have led to major commercial uses. Tomatoes are often picked green and then artificially ripened as desired by the application of ethylene. Ethylene is widely used to speed the ripening of lemons and oranges as well. Carbon dioxide produces effects opposite to those of ethylene in fruits, and fruits that are being shipped are often kept in an atmosphere of carbon dioxide if they are not intended to ripen yet.

> *Ethylene, a simple, gaseous hydrocarbon, is a naturally occurring plant hormone. It plays the key role in controlling the falling off of leaves, flowers, and fruit from plants. Ethylene also hastens the ripening of fruit.*

Abscisic Acid

Abscisic acid is a naturally occurring plant hormone that is synthesized mainly in mature green leaves, fruits, and root caps. The hormone was given its name because applications of

the hormone stimulate leaves to age rapidly and fall off (the process of *abscission*), but evidence that abscisic acid plays an important natural role in this process is scant. Abscisic acid suppresses the growth and elongation of buds and promotes aging, counteracting some of the effects of the gibberellins (which stimulate bud growth and elongation) and auxin (which tends to retard aging).

Plant Responses to Environmental Stimuli

Plants are living organisms, and as such, respond to different environmental stimuli in a variety of ways. As discussed earlier in the chapter, plants bend toward light as they grow in response to an environmental stimulus. A host of other plant responses, including flowering, dropping of leaves, and the yellowing of leaves due to loss of chlorophyll, are also prompted by various environmental stimuli. This section explores some of these plant responses and the environmental factors that cause them.

Tropisms

Tropisms, or movement responses to external stimuli, control patterns of plant growth and thus plant appearance. Plants adjust to environmental conditions by tropism responses. With **positive tropisms,** the movement or reaction is in the direction of the stimulus source, whereas with **negative tropisms,** movement or growth is in the opposite direction. Three major classes of plant tropisms are considered here: phototropism, gravitropism, and thigmotropism.

Phototropism

Phototropism, the bending of plants toward unidirectional sources of light, was introduced in the discussion of auxin. In general, stems are positively phototropic, growing toward the light, whereas roots are negatively phototropic, growing away from it. The phototropic reactions of stems are clearly of adaptive value because they allow plants to capture greater amounts of light than would otherwise be possible. Auxin is involved in most, if not all, of plants' phototropic growth responses.

> *Phototropisms are plants' growth responses to a unidirectional source of light. They are mostly, if not entirely, mediated by auxin.*

Gravitropism

Another familiar plant response is **gravitropism,** formerly known as geotropism, which causes stems to tend to grow upward and roots downward (figure 32.6). Both of these responses clearly have adaptive significance: Stems that grow upward are apt to receive more light than those that do not; roots that grow downward are more apt to encounter a more favorable environment than those that do not. The phenomenon is now called gravitropism because it is clearly a response to gravity and not to the earth (prefix "geo") as such.

Figure 32.6 Tropism guides plant growth. *The branches of this fallen tree are growing straight up because they are negatively gravitropic and also positively phototropic.*

Amyloplasts, starch-containing plastids found in plant cells, probably play an important role in plants' perception of gravity. The amyloplasts are heavy capsules with large amounts of calcium and starch. In roots, amyloplasts apparently occur in the central cells of the root cap; removing the root cap stops the roots' responses to gravity in most cases. In shoots, on the other hand, gravity is clearly sensed along the whole length of the stem, probably by the functioning of similar amyloplasts in certain cells.

Gravity causes amyloplasts to fall to the lower side of a given cell. There, they apparently set in motion a series of reactions that eventually causes the shoots and roots to bend. Amyloplasts reach the lower side of their cells within a minute, and the bending of the root or shoot may occur within as little as ten minutes.

Gravitropism, a plant's response to gravity, generally causes shoots to grow up and roots to grow down. The force of gravity apparently is sensed in special cells with amyloplasts, starch-containing plastids.

Thigmotropism

Still another commonly observed response of plants is **thigmotropism,** a name derived from the Greek root *thigma,* meaning "touch." Thigmotropism is defined as the response of plants to touch. Examples include plant tendrils that rapidly curl around and cling to stems or other objects, and twining plants, such as bindweed, that coil around objects. These behaviors are the result of rapid growth responses to touch. Specialized groups of cells in the epidermis appear to be concerned with thigmotropic reactions, but again, their exact mode of action is not well understood.

Thigmotropisms are plants' growth responses to touch.

Turgor Movements

As defined earlier in the chapter, turgor pressure is the pressure within a cell that results from water moving into the cell. Some kinds of plant movements are based on reversible changes in the turgor pressure of specific cells, rather than on differential growth or cell enlargement. One of the most familiar examples involves the changing leaf positions that certain plants exhibit at night and during the day. For example, the attractively spotted leaves of the prayer plant (*Maranta*) spread horizontally during the day due to high turgor pressure but become more or less vertical at night as a result of low turgor pressure. The leaves' daytime horizontal orientation ensures that light energy spreads over a large portion of the leaves, thereby providing more light energy for photosynthesis.

Turgor movements of plants are reversible and involve changes in the turgor pressure of specific cells that allow plants to orient their leaves and flowers in different positions.

Photoperiodism

Essentially all eukaryotic organisms are affected by the cycle of night and day, and many features of plant growth and development are keyed to changes in the proportions of light and dark in the daily twenty-four-hour cycle. Such responses constitute **photoperiodism,** a mechanism by which organisms measure seasonal changes in relative day and night length. One of the most obvious of these photoperiodic reactions concerns angiosperm flower production.

Day length changes with the seasons; the farther from the equator one is, the greater the variation. Plants' flowering responses fall into three basic categories in relation to day length: **Short-day plants** begin to form flowers when days become shorter than a critical length and nights become longer. **Long-day plants,** on the other hand, initiate flowers when days become longer than a certain length and nights become shorter. Thus, many fall flowers are short-day plants, and many spring and early-summer flowers are long-day plants (figure 32.7). Commercial plant-growers use these responses to day length to time flower blooming for specific holidays or occasions.

In addition to long-day and short-day plants, a number of plants are described as **day-neutral.** Day-neutral plants produce flowers whenever environmental conditions are suitable, without regard to day length.

Short-day plants form flowers when days become shorter than a certain critical length. Long-day plants form flowers when days become longer than a certain length. Day-neutral plants do not have specific day-length requirements for flowering.

Phytochromes

Flowering responses to daylight are controlled by several chemicals that interact in complex ways. Although the nature

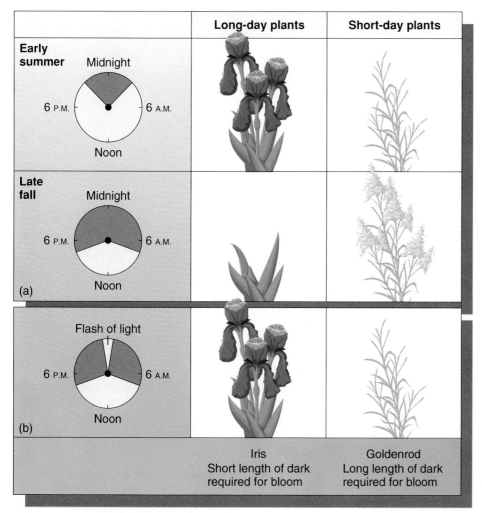

	Long-day plants	Short-day plants
Early summer (a)		
Late fall (a)		
(b)		
	Iris Short length of dark required for bloom	Goldenrod Long length of dark required for bloom

Figure 32.7 Photoperiodism in plants.
(a) *The iris is a long-day plant that is stimulated by short spring nights to bloom in the spring. The goldenrod is a short-day plant that is stimulated by the long nights of fall to bloom in the fall.*
(b) *If the long night of winter is artificially interrupted by a flash of light, the goldenrod will not bloom and the iris will. In each case, the duration of uninterrupted darkness determines when flowering will occur.*

of some of these chemicals has been deduced, how the various chemicals work together to promote or inhibit flowering responses is still being debated.

One class of chemicals—the **phytochromes**—has been shown to influence plants' flowering responses. Phytochromes are pigment molecules, meaning that they absorb light energy and undergo changes in their chemical structure as a result of this energy absorption. In plants, when light in the red wavelengths (about 660 nanometers) strikes the phytochrome molecule P_r, the molecule is converted to the phytochrome molecule P_{fr}. The reaction is reversed when light in the far-red wavelengths (about 730 nanometers) strikes P_{fr}: P_{fr} is converted back to P_r. Daylight contains both red and far-red wavelengths, but night light contains only far-red wavelengths. Thus, during nighttime, any P_{fr} that remains in the plant is converted to P_r. Of the two pigments, only P_{fr} is biologically active or able to exert a biological response on a plant.

Scientists hypothesize that, in short-day plants, P_{fr} acts as a flowering inhibitor, while in long-day plants, it acts as a flowering promoter. During seasons of long days and short nights, P_{fr}

accumulates in short-day plants during the day and is destroyed at night. However, enough P_{fr} is left over at daybreak to inhibit flowering. As the days get shorter and the nights get longer, however, short-day plants have a chance to "catch up" and destroy larger amounts of the flowering inhibitor P_{fr} during the longer nights. Thus, short-day plants flower during seasons with longer nights and shorter days.

In contrast, during seasons of long days and short nights, P_{fr} accumulates in long-day plants during the day and is destroyed at night, but enough P_{fr} remains at daybreak to promote flowering. As the days shorten and the nights lengthen, however, the long nights provide a longer opportunity for P_{fr} destruction, and flowering in the long-day plants is inhibited.

Flower-Inhibiting and Flower-Inducing Hormones

The conversion of P_r to P_{fr} and the influence of P_{fr} on flowering in different types of plants is not the whole story of the flowering response in plants. When scientists tested the phytochrome hypothesis in plants, they found that P_{fr} disappears in plants within three to four hours of darkness. They concluded that phytochrome conversion alone cannot control the flowering response but that some other chemical that interacted with P_{fr} is involved.

After many experiments, scientists hypothesized that a hormone that they hypothetically called **florigen** also exerted a flower-inducing response in plants. Another hormone, not yet isolated or named, seemed to induce a flower-inhibiting response in plants. These hormones are produced in the leaves and are transmitted to the bud, where they exert their different effects. In long-day plants, florigen is produced during long days, and the flower-inhibiting hormone is produced during short days. In short-day plants, the flower-inhibiting hormone is produced during long days and florigen is produced during short days. Although much experimental evidence points to the existence of these hormones, scientists have not yet been able to isolate and study these particular chemical messengers.

The interaction of phytochromes P_r and P_{fr}, flower-inhibiting hormone, and flower-inducing hormone (florigen) controls flowering responses in plants.

Dormancy

Plants respond to their external environment largely by changes in growth rate. Plants' ability to stop growing altogether when conditions are not favorable—to become dormant—is critical to their survival.

In temperate regions, dormancy is generally associated with winter, when low temperatures and the unavailability of water because of freezing make it impossible for plants to grow. During this season, the buds of deciduous trees and shrubs remain dormant, and the apical meristems remain well protected inside enfolding scales. Perennial herbs spend the winter underground as stout stems or roots packed with stored food. Many other kinds of plants, including most annuals, pass the winter as seeds.

In climates that are seasonally dry, dormancy occurs primarily during the dry season, whenever in the year it falls. In dry conditions, plants remain dormant by using strategies similar to those of temperate-zone plants.

Annual plants occur frequently only in areas of seasonal drought. Seeds are ideal mechanisms for allowing annual plants to bypass the dry season, when water supplies are insufficient for growth. When it rains, the seeds germinate, and the plants grow rapidly to take advantage of the relatively short period of water availability.

EVOLUTIONARY VIEWPOINT

Cutting and burning the world's tropical rain forests is having a devastating effect on local climates, converting wet forests into arid semidesert where rain is infrequent. Why? The trees of the rain forest act like great pipes, sucking water from the ground and releasing it by transpiration high in the sky, where it can fall back as rain. Cutting the tree breaks this cycle; the water simply runs off the land in streams and rivers. Once broken, this cycle is not easy to restore. Once gone, the forests will not soon return.

SUMMARY

1. Water flows through plants in a continuous column, driven mainly by transpiration through the stomata. Plants can control water loss primarily by closing stomata. The cohesion of water molecules and their adhesion to the walls of the very narrow cell columns through which they pass are additional important factors in maintaining the flow of water to the tops of plants.

2. Water enters the plant through the roots. When the pull on the column of water is small, such as at night, the amount of water entering the xylem increases. Guttation is a result of this high root pressure. When their guard cells are turgid, stomata open and bow out, thus causing the thickened inner walls of these cells to bow away from the opening.

3. The movement of dissolved sucrose and other carbohydrates in the phloem does not require energy. Sucrose is loaded into the phloem near sites of photosynthesis and unloaded at the places where it is required.

4. Nutrients move primarily through plants as solutes in the water column in the xylem.

5. The nine macronutrients—substances that are each present at concentrations approaching or exceeding 1 percent or more of a plant's dry weight—are carbon, hydrogen, oxygen, nitrogen, potassium, calcium, phosphorus, magnesium, and sulfur. Plants also need seven known micronutrients, each present at concentrations of from one to several hundred parts per million of dry weight.

6. Hormones are chemical substances produced in small quantities in one part of an organism and then transported to another part of the organism, where they bring about physiological responses. The tissues in which plant hormones are produced are not specialized particularly for that purpose; nor are there usually clearly defined receptor tissues or organs.

7. The five major classes of naturally occurring plant hormones are: auxin, cytokinins, gibberellins, ethylene, and abscisic acid. They often interact with one another in bringing about growth responses.

8. Auxin is produced at the tips of shoots and diffuses downward, suppressing the growth of lateral buds. In young grass seedlings and other herbs, auxin plays a major role in promoting stem elongation.

9. Cytokinins are necessary for mitosis and cell division in plants. They promote the growth of lateral buds and inhibit the formation of lateral roots.

10. Gibberellins play the major role in controlling stem elongation in most plants.

11. Ethylene is a gas that functions as a plant hormone and is widely used to hasten fruit ripening.

12. Abscisic acid is a naturally occurring plant hormone that suppresses the growth and elongation of buds and promotes aging.

13. Tropisms in plants are growth responses to external stimuli. A phototropism is a response to light; a gravitropism is a response to gravity; and a thigmotropism is a response to touch.

14. Turgor movements are reversible but important elements in the adaptation of plants to their environments in that they allow plants to orient their leaves and flowers in different positions.

15. The flowering responses of plants fall into two basic categories in relation to day length: Short-day plants begin to form flowers when the days become shorter than a given critical length; long-day plants do so when the days become longer than a certain length. Phytochromes and flowering-response hormones interact to promote or inhibit the flowering response.

16. Dormancy is a necessary part of plant adaptation that allows a plant to bypass unfavorable seasons, such as winter, when water may be frozen, or periods of drought. Dormancy also allows plants to survive in many areas where they would be unable to grow otherwise.

REVIEWING THE CHAPTER

1. How does the cohesion-adhesion-tension theory explain the movement of water in plants?

2. What is the process of transpiration?

3. How is water absorbed by roots?

4. How is the transpiration rate in plants regulated?

5. How do ion transport channels regulate the activity of guard cells?

6. How do environmental factors influence the opening and closing of stomata?

7. How are carbohydrates transported in plants?

8. What macronutrients and micronutrients are essential for plants?

9. How do hormones regulate plant growth?

10. How did the Darwins' experiments help lead to the discovery of auxin?

11. How did Frits Went discover that the shoot tip uses auxin to control growth?

12. What are the major plant hormones?

13. What is auxin, and how does it function in plants?

14. For what purpose are synthetic auxins used?

15. What are cytokinins, and how do they function in plants?

16. How do gibberellins function?

17. What is ethylene, and how does it function in plants?

18. How does abscisic acid function in plants?

19. How do plants respond to environmental stimuli?

20. What are tropisms?

21. What is phototropism?

22. How does gravitropism affect plants?

23. What is thigmotropism?

24. In plants, what are turgor movements?

25. How do plants respond to photoperiodism?

26. How do phytochromes influence the flowering process?

27. How do flower-inhibiting and flower-inducing hormones control flowering responses in plants?

28. Why do plants become dormant?

COMPLETING YOUR UNDERSTANDING

1. Why are the water-transporting vessels in plants very narrow?

2. Why does the water column in the world's tallest redwoods not collapse?

3. What is the relationship between root pressure and guttation?

4. What essential ingredients for photosynthesis pass through the stomata?

5. Why are the inner surfaces of stomatal guard cells the thickest?

6. The liquid in the phloem contains 10 to 25 percent dissolved solid matter, primarily

    a. starch.   d. K ions.

    b. protein.   e. lipids.

    c. ATP.   f. sucrose.

7. In mass flow, what plant parts constitute sources and sinks?

8. Which macronutrients are found in all organic compounds?

9. In plants, how are macronutrients and micronutrients defined?

10. How do plants use calcium and magnesium?

11. Why do plants need zinc?

12. Why are carnivorous plants often found growing in acidic soils such as bogs?

13. If you place a houseplant near a window, why should you rotate it periodically?

14. In plants, what is the function of IAA oxidase?

15. What is dioxin, and why is it an environmental concern?

16. How do auxin and cytokinins differ in their functions? How are they similar in function?

17. How were gibberellins discovered in plants?

18. When you go to the supermarket, what evidence is there that some produce is artificially ripened?

19. How does abscisic acid counteract some of the effects of gibberellins and auxin?

20. In plants, what causes phototropism, gravitropism, and thigmotropism?

21. If you wanted to increase the turgor pressure in one of your houseplants, you would

    a. move the plant to a dark room.

    b. add essential macronutrients to the soil.

    c. add water to the soil.

    d. add saltwater to the soil.

    e. none of the above.

22. Why do gardeners plant bulbs such as tulips and daffodils in the fall rather than in the spring?

23. Would the flowering in a long-day plant be affected if it received a flash of light in the middle of its night cycle? Why or why not? (See figure 32.7.)

24. What type of plant would you purchase from a nursery if you were looking for something that would flower throughout the growing season?

25. Why is florigen an elusive plant hormone?

26. Why are annual plants frequently found in areas of seasonal drought?

*__27.__ Why do gardeners often remove many of a plant's leaves after transplanting it?

*__28.__ If you grew a plant that initially weighed 200 grams but eventually weighed 50 kilograms in a pot, would you expect the soil in the pot to change weight? If so, how much, and why?

*__29.__ When poinsettias are kept inside a house following the holiday season, they rarely bloom again. Why do you think this might be, and what might you do to get them to produce flowers a second time?

*For discussion

FOR FURTHER READING

Evans, M., R. Moore, and K. Hasenstein. 1986. How roots respond to gravity. *Scientific American,* Dec., 112–20. Speculation that roots grow down and not up because calcium ions settle in the roots, where they activate proteins that promote growth.

Hardwick, R. 1986. Construction kits for modular plants. *New Scientist,* April 10, 39–42. An introduction to the ways in which plant growth follows repetitive patterns.

Mansfield, T. A., and W. J. Davies. 1985. Mechanisms for leaf control of gas exchange. *BioScience* 35: 158–68. Excellent review of some of the factors involved in stomatal opening and the ways in which they are integrated.

Mooney, H. A. et al. 1988. Plant physiological ecology today. *BioScience* 37: 18–67. Nearly an entire issue of *BioScience* devoted to the ways in which plants cope with their environments and in which scientists study them. Highly recommended.

Sandved, K., and G. Prance. 1984. *Leaves.* New York: Crown Publishers. An incredibly beautiful introduction to the diversity of leaves.

Sisler, E. C., and S. F. Yang. 1984. Ethylene, the gaseous plant hormone. *BioScience* 33: 233–38. An up-to-date review of this important plant hormone.

Stewart, D. 1990. Green giants. *Discover,* April, 61–64. More than majestic trees, sequoias are triumphs of hydraulics, wind resistance, and architecture.

Grey wolves during the breeding season. The male is lighter in color, while the female is the black-phase individual. Like other mammals and birds, wolves devote considerable time and effort to raising their young. Most reptiles and amphibians, in contrast, often leave their offspring to fend for themselves.

THE VERTEBRATE BODY

Vertebrate tissues can be soft, hard, or even liquid. This is "spongy bone" from a vertebra in the lower back, magnified ten times.

FOR REVIEW

Here are some important terms and concepts that have been discussed in previous chapters and that you will encounter again in this chapter. Review them before proceeding if necessary.

Sodium-potassium pump (*chapter 5*)
Evolution of vertebrates (*chapter 18*)
Basic structure of chordates (*chapters 18 and 29*)
Coelom (*chapter 29*)

The vertebrate body is a biological machine of exquisite complexity. Bears and goldfish, rattlesnakes and hummingbirds, the different kinds of vertebrates at first glance seem very different from one another. But the differences are not nearly as striking as the similarities. All vertebrates share the same basic body plan, with the same sorts of organs operating in much the same way.

This chapter begins a detailed consideration of vertebrate biology and of the often intricate and fascinating structure of vertebrate bodies (figure 33.1). The focus is on human biology because the architecture of the human body provides a good focal point for discussing vertebrate bodies in general and because human biology is of particular importance to all of us. We all want to know how our body works and why it functions the way it does.

Not all vertebrates are the same, of course, and some of the differences are important. For example, a fish does not breathe the same way you do. Thus, the discussion is not limited to human beings, but rather, describes the human animal in the context of vertebrate diversity. The differences reflect human evolutionary history and offer important lessons in why we function the way we do.

Figure 33.1 The interior of bone.
While bone is often thought of as hard and solid, the interiors of many bones are composed of a surprisingly delicate latticework. Bone, like most tissues present in the body, is a dynamic structure, constantly renewing itself.

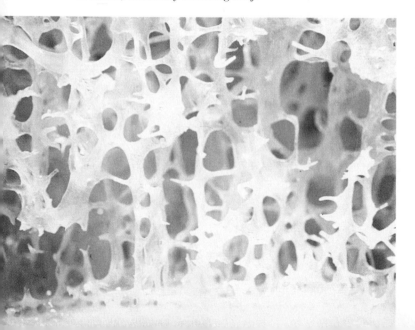

The Human Animal

An incredible machine of great beauty, the human body has the same general architecture of all vertebrates.

Body Architecture

The architecture of the human body includes a long tube that travels from one end of the body to the other, from mouth to anus. This tube is suspended within an internal body cavity called the **coelom** (see chapter 29). In human beings, the coelom is divided into three parts: (1) the **thoracic cavity,** which contains the heart and lungs; (2) the **abdominal cavity,** which contains the stomach, intestines, and liver; and (3) the **pelvic cavity,** which contains the reproductive organs. All vertebrate bodies are supported by an internal scaffold or skeleton made up of jointed bones that grow as the body grows. A bony skull, or **cranial cavity,** surrounds the brain; and a column of bones, the **vertebrae,** surrounds the **dorsal nerve cord,** or **spinal cord** (figure 33.2).

Human beings are mammals and, like all other mammals, are **homeotherms,** which means that they regulate their internal temperature at a relatively constant value. Humans keep their temperature at about 37 degrees Celsius (98.6 degrees

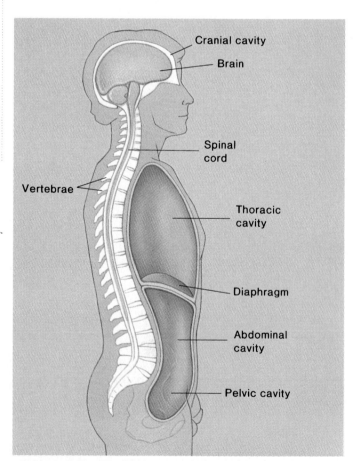

Figure 33.2 Architecture of the human body.
The human body contains four cavities (the cranial, thoracic, abdominal, and pelvic cavities), and a muscular diaphragm divides the coelom into the thoracic and abdominal cavities. The human body also has a dorsal central nervous system, consisting of a spinal cord and brain, enclosed in vertebrae and the cranial cavity.

Fahrenheit). Like all other mammals, human beings have hair, rather than scales or feathers; and like all other mammals except monotremes (for example, the duck-billed platypus), humans do not lay eggs but give birth to young who require parental nurturing. Human development is a lengthy process compared with other mammals; infants are nursed for a long time and mature slowly.

Body Organization

Vertebrate bodies, like those of all other multicellular animals, are composed of different cell types. The bodies of adult vertebrates consist of between fifty and several hundred different kinds of cells, depending on the kind of vertebrate and the degree of differentiation identified between cell types. Groups of similar cells are organized into **tissues,** which are structural and functional units.

Organs are body structures composed of several different tissues grouped together to carry out a specific function. The heart is an organ. It contains cardiac muscle tissue wrapped in connective tissue and is the target of many nerves. All these tissues work together to pump blood through the body.

An **organ system** is a group of organs that function together to carry out principal body activities. For example, the digestive organ system is composed of individual organs concerned with the breaking up of food (teeth), the passage of food to the stomach (esophagus), food storage (stomach), food digestion and absorption (intestine), and the expulsion of solid residue (rectum). The liver, a chemical reprocessing plant, is also a component of the digestive system. The human body contains eleven principal organ systems. Their components and functions are described in table 33.1.

The architecture of the animal body is basically that of a tube suspended in a cavity. The animal's tissues are made of many different cell types, and the tissues combine in various ways to form the organs that carry out specific functions.

Tissues

Early in the development of any vertebrate, the growing mass of cells differentiates into three fundamental cell layers: **endoderm, mesoderm,** and **ectoderm.** These three kinds of embryonic cell layers differentiate, in turn, into the hundreds of different cell types characteristic of the adult vertebrate body. As discussed in chapter 28, where they were reviewed in an evolutionary context, these diverse cell types are traditionally grouped on a functional basis into four basic types of tissues: **epithelial tissue, connective tissue, muscle tissue,** and **nervous tissue** (figure 33.3). Of these, connective tissues are particularly diverse. Blood cells, for example, are classified as one kind of connective tissue, whereas bone is another.

Epithelial Tissue: Guards and Protects

Epithelial cells are the body's guards and protectors. These cells cover the body surface and determine which substances enter it and which do not. The organization of the vertebrate body is fundamentally tubular, with developmental derivatives of ectodermal cells covering the outside (skin), those of endodermal cells lining the hollow inner core (the digestive canal and gut), and those of mesodermal cells lining the body cavity

Table 33.1   The Major Vertebrate Organ Systems

System	Functions	Components	Detailed Treatment
Integumentary	Covers the body and protects it	Skin, hair, nails, and sweat glands	Chapter 33
Nervous	Receives stimuli, integrates information, and directs the body	Nerves, sense organs, brain, and spinal cord	Chapters 34 and 35
Skeletal	Protects the body and provides support for locomotion and movement	Bones, cartilage, and ligaments	Chapter 36
Muscular	Produces body movement	Skeletal muscle, cardiac muscle, and smooth muscle	Chapter 36
Endocrine	Coordinates and integrates body activities	Pituitary, adrenal, thyroid, and other ductless glands	Chapter 37
Digestive	Captures soluble nutrients from ingested food	Mouth, esophagus, stomach, intestines, liver, and pancreas	Chapter 38
Respiratory	Captures oxygen and exchanges gases	Lungs, trachea, and other air passageways	Chapter 39
Circulatory	Transports cells and materials throughout the body	Heart, blood vessels, blood, lymph, and lymph structures	Chapter 40
Immune	Removes foreign bodies from the bloodstream	Lymphocytes, macrophages, and antibodies	Chapter 41
Urinary	Removes metabolic wastes from the bloodstream	Kidney, bladder, and associated ducts	Chapter 42
Reproductive	Carries out reproduction	Testes, ovaries, and associated reproductive structures	Chapter 43

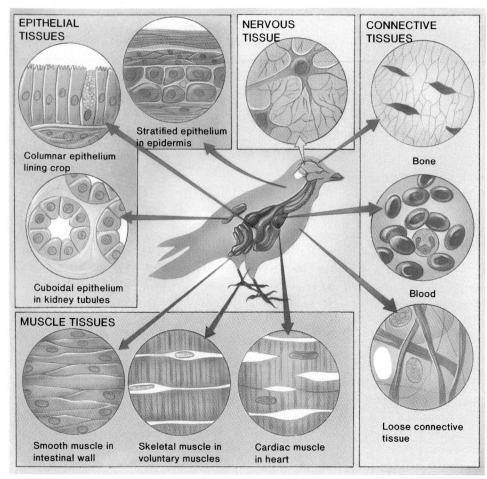

EPITHELIAL TISSUES

Stratified epithelium in epidermis

Columnar epithelium lining crop

Cuboidal epithelium in kidney tubules

NERVOUS TISSUE

CONNECTIVE TISSUES

Bone

Blood

Loose connective tissue

MUSCLE TISSUES

Smooth muscle in intestinal wall

Skeletal muscle in voluntary muscles

Cardiac muscle in heart

Figure 33.3 Vertebrate tissue types.
Epithelial tissues are indicated by blue arrows, connective tissue by green arrows, muscle tissue by red arrows, and nervous tissue by a yellow arrow.

(coelom). All of these kinds of cells are broadly similar in form and function and are collectively called the **epithelium.** The body's epithelial layers function in four different ways:

1. They protect the tissues beneath them from dehydration and mechanical damage, a particularly important function in land-dwelling vertebrates.

2. They provide a selectively permeable barrier that can facilitate or impede the passage of materials into the tissues beneath. Because epithelium encases all body surfaces, every substance that enters or leaves the body must cross an epithelial layer.

3. They provide sensory surfaces. Many of the body's sense organs are, in fact, modified epithelial cells.

4. They secrete materials. Most secretory glands are derived from invaginations of layers of epithelial cells that occur during embryonic development.

Layers of epithelial tissue are usually only one or a few cells thick. Individual epithelial cells possess a small amount of cytoplasm and have a relatively low metabolic rate. Few blood vessels pass through the epithelium; instead, nutrients, gases, and wastes in epithelial tissue circulate by diffusion from the capillaries of neighboring tissue.

Epithelial tissues possess remarkable regenerative abilities, with the cells of epithelial layers constantly being replaced throughout the organism's life. For example, the liver, which is a gland formed of epithelial tissue, can readily regenerate substantial portions of tissue that have been surgically removed from it. The cells lining the digestive tract are replaced every few days.

Epithelial tissues are classified by the number of cell layers in the tissue and by the shapes of the cell themselves. For instance, **simple epithelial tissues** consist of one cell layer, while **stratified epithelial tissues** consist of more than one cell layer. The cell shapes of epithelial cells are **squamous** (roughly square-shaped), **cuboidal** (cube-shaped), **columnar** (column-shaped), and **transitional** (composed of different cell shapes). In addition, the cells may be ciliated or nonciliated. Thus, **simple squamous epithelium** is composed of one layer of roughly square-shaped, nonciliated cells. **Simple cuboidal epithelium** is composed of one layer of cube-shaped, nonciliated cells. The various epithelium types are found in different places in the body and perform different functions, as shown in table 33.2.

> *Epithelial tissue is composed of different shapes of cells arranged in either one layer (simple) or more than one layer (stratified). Epithelial tissues protect the tissues beneath them from dehydration and other damage, provide a selectively permeable barrier for materials, provide sensory surfaces, and secrete materials.*

Connective Tissue: Supports the Body

Connective tissue cells provide the body with its structural building blocks and also with its most potent defenses. Connective tissue is derived from the mesoderm and consists of three elements: cells; **ground substance,** a material deposited between the connective tissue cells that holds cells together; and fibers. Together, the ground substance and fibers are called the **matrix,** and connective tissue cells are embedded in this matrix.

Table 33.2 Some Types of Epithelial Tissue

Type of Epithelial Tissue			Location	Function
Simple squamous			Lines major organs (heart, air sacs of lungs, Bowman's capsule of kidney); lines body cavity	Absorption, exchange of materials, filtration, secretion
	Cytoplasm Nucleus			
Simple cuboidal			Lines tubules and ducts of glands; covers surface of ovary; lines interior of eye	Absorption and secretion
	Cytoplasm Interior of kidney tubules Nucleus			
Simple columnar			Lines gastrointestinal tract	Secretion from special goblet cells of materials, absorption
	Globular cell Cytoplasm Nucleus			
Stratified squamous			Lines interior of mouth, tongue, esophagus, vagina	Protection
Transitional			Lines urinary bladder	Permits stretching

All three elements in the different connective tissues can vary. For example, the matrix may be liquid, as in blood, or it can be very hard and inflexible, as in bone. Connective tissue cells are just as various and perform a myriad of functions. In blood, for instance, there are red blood cells that are efficient oxygen carriers, macrophages that function in the immune system to engulf foreign cells, and plasma cells, which also function in the immune system to produce antibodies.

Connective Tissue Cells

Although space does not permit a listing of all the cells found in connective tissue, several are noted here:

Fibroblasts are large, branching, flat cells that produce matrix (figure 33.4*a*). They are active in wound healing and are the main components of scar tissue. **Plasma cells,** as already mentioned, function in the immune system to make the different antibodies that the body needs to ward off infection. Plasma cells are small and white, and are found in abundance in the intestinal tract. **Fixed macrophages** are irregular in shape and function in the immune system to engulf foreign cells, such as invading bacteria or viruses. **Mast cells** secrete chemicals in response to **trauma,** or injury. Mast cells produce antihistamine, which dilates blood vessels and causes the inflammation response, and they also secrete a chemical called heparin that prevents blood from clotting in the blood vessels. **Erythrocytes** are red blood cells (figure 33.4*b*). **Leukocytes** are white blood cells, and **adipocytes** are fat cells. There are also **osteocytes** (bone cells), **chondrocytes** (cartilage cells), and **reticular cells** (cells that contain collagen, a fibrous protein).

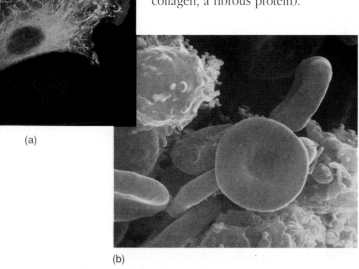

(a)

(b)

Figure 33.4 Connective tissue cells.
(a) *Fibroblast. The different proteins in this fibroblast are made visible by fluorescence microscopy.* (b) *Red blood cells, or erythrocytes, are flattened spheres, typically with a depressed center.*

Connective Tissue Matrix: Fibers and Ground Substance

Connective tissue is distinguished by the type of matrix laid down between the individual cells. As mentioned earlier, the matrix consists of the ground substance and fibers.

Ground substance can be a fluid, a gel, or a solid. It supports the connective tissue cells and provides a medium through which substances are exchanged between the connective tissue cells and the blood.

The three types of fibers found in connective tissue are **collagenous fibers, elastic fibers,** and **reticular fibers.** Collagenous fibers often occur in bundles, which makes them very tough and resistant. Composed of the protein collagen, they are found in many connective tissues, such as bone, cartilage, tendons, and ligaments. Elastic fibers are composed of the protein elastin, which makes them very stretchable. Reticular fibers are also composed of collagen but are augmented with glycoprotein. These fibers usually provide the internal scaffold for such soft organs as the spleen and the liver.

This text recognizes eight types of connective tissue: **loose connective tissue, dense connective tissue, elastic connective tissue, reticular connective tissue, cartilage, bone, vascular connective tissue,** and **adipose tissue.** Table 33.3 shows examples of these different connective tissues and lists their locations and functions in the body.

> *Connective tissue provides support, structure, and protection for the body. Connective tissue is composed of connective tissue cells embedded in a matrix, which is composed of ground substance and fibers. The type of matrix determines the nature and function of the connective tissue.*

Muscle Tissue: Provides for Movement

Muscle cells, formed from the mesoderm early in development, are the "workhorses" of the vertebrate animal. Muscle cells' distinguishing characteristic is their relative abundance of actin and myosin microfilaments. (Microfilaments are microscopic fibers.) These protein microfilaments are present as a fine network in all eukaryotic cells (see chapter 4) but are far more abundant in muscle cells.

Vertebrates possess three different kinds of muscle cells (table 33.4): smooth muscle, skeletal muscle, and cardiac muscle. The last two types are called striated muscle to distinguish them from nonstriated, smooth muscle, in which microfilaments are only loosely organized. In striated muscle cells, the actin microfilaments are bunched together with thicker microfilaments of myosin into many thousands of fibers called **myofibrils.** Myofibrils shorten when the actin and myosin filaments slide past each other.

Table 33.3   Types of Connective Tissue

Type of Connective Tissue		Location	Function
Loose connective tissue			
		Deep layers of skin, blood vessels, nerves, body organs	Support, elasticity
Dense connective tissue			
		Tendons, ligaments	Attaches structures to one another; provides great strength
Elastic connective tissue			
		Lungs, arteries, trachea, vocal chords	Provides elasticity
Reticular connective tissue			
		Spleen, liver, lymph nodes	Provides internal scaffold for soft organs
Cartilage			
	Chondrocyte Matrix	Ends of long bones; tip of nose; parts of larynx, trachea	Provides flexibility and support

(continued)

Table 33.3   (continued)

Type of Connective Tissue		Location	Function
Bone		Bones	Protection, support, muscle attachment
Haversian canal — Osteocyte			
Vascular connective tissue		Within blood vessels	Transport of oxygen and carbon dioxide; immune response; blood clotting
Plasma cell — Leukocyte — Red blood cell			
Adipose tissue		Deep layers of skin; surrounds heart and kidneys; padding around joints	Support, protection, heat conservation, energy source
Adipocytes			

Interactions of actin and myosin microfilaments can cause muscle cells to change shape, as in smooth muscle. Because muscle cells have so many microfilaments all aligned the same way, a considerable force is generated when the actin and myosin microfilaments interact.

Muscle cells contain microscopic fibers of actin and myosin called microfilaments. By pulling against one another, actin and myosin cause muscle cells to change shape, often with considerable force.

Nervous Tissue: Conducts Signals Rapidly

Nervous tissue, the fourth major class of vertebrate tissue, is composed of two kinds of cells: (1) **neurons,** which are specialized for the rapid transmission of nerve impulses from one organ to another; and (2) **supporting glial cells,** which support and insulate the neurons. The supporting cells also assist in the propagation of the nerve impulse by playing an important part in the maintenance of the ionic composition of nervous tissue. In addition, supporting cells are believed to supply neurons with nutrients and other molecules.

Table 33.4 Types of Muscle Tissue

Type of Muscle Tissue	Location	Function
Smooth muscle tissue  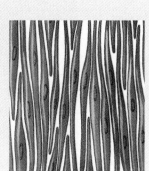	Gastrointestinal tract, uterus, urinary bladder, blood vessels	Propulsion of materials
Cardiac muscle tissue 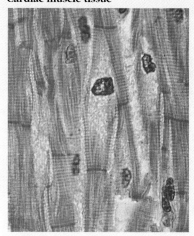 	Heart	Contraction
Skeletal muscle tissue  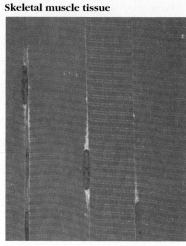 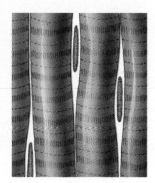	Attached to bones	Movement

Neurons are specialized to conduct signals rapidly throughout the body. Their membranes are rich in ion-selective channels, water-filled pores that control ion movement into and out of the cell. This maintains a voltage difference between the cell's interior and exterior. When ion channels in a local area of the membrane open, ions enter from the exterior. This flow of ions, called a current, temporarily wipes out the voltage difference—a process called **depolarization.** Depolarization in a local area of the membrane tends to open nearby ion channels in the neuron membrane, resulting in a wave of electrical activity that travels, or propagates, down the entire nerve length as a **nerve impulse.** The nature of nerve impulses varies, as described in more detail in chapter 34, but all nerve impulses propagate along nerves as waves of membrane depolarization.

The neuron cell body contains the cell nucleus. From the cell body project two kinds of cytoplasmic extensions that carry out the neuron's transmission functions. The first of these extensions consists of **dendrites,** threadlike protrusions that act as antennae for the reception of nerve impulses from other cells or sensory systems. The second kind of cytoplasmic extension is the **axon,** a long, tubular extension of the cell that carries the nerve impulse away from the cell body, often for considerable distances (figure 33.5).

Axons can make nerve cells very long. For example, a single neuron that activates the muscles in the human thumb may have its cell body in the spinal cord and an axon that extends all the way across the shoulder and down the arm to the thumb. Single neuron cells more than a meter in length are common.

Nerves of the vertebrate body appear as fine, white threads when viewed with the naked eye but are actually bundles of axons. Like a telephone trunk cable, these nerves include large numbers of independent communications channels—bundles of hundreds of axons, each connecting a nerve cell to a muscle fiber. In addition, the nerve contains numerous supporting glial cells bunched around the axons.

Neurons have plasma membranes rich in ion channels that pump ions out of the cell, creating a voltage difference across the membrane. A nerve impulse occurs when the ions re-enter the cell, creating a current that then propagates down the neuron's surface.

Organs: Different Tissues Working Together

The four classes of vertebrate tissues discussed in this chapter (epithelium, connective, muscle, and nervous) are the building blocks of the vertebrate body. Each body organ, a structure that carries out a specific function, is composed of these tissues, organized and assembled in various ways. Different tissue combinations are found in different organs.

The many organs that, working together, carry out the principal body activities are referred to as organ systems. Much of vertebrate biology is concerned with the functioning of organ

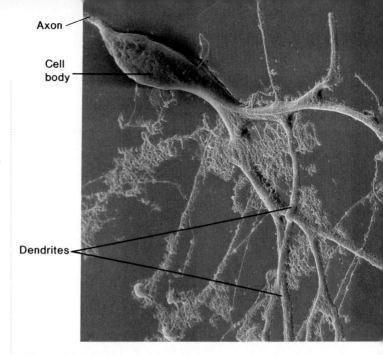

Figure 33.5 Anatomy of a neuron.
Neurons are specialized to transmit nerve impulses. A neuron axon receives nerve impulses from other neurons, and the dendrites transmit the nerve impulse to subsequent neurons.

systems. The eleven major organ systems of the human body all work together (see table 33.1). The remainder of the text is devoted to describing them in more detail.

Homeostasis

As the animal body has evolved, specialization has increased, and nowhere is this more evident than in the many cell types that make up the body of a vertebrate. Each cell type is a sophisticated machine, finely tuned to carry out a precise role within the body. Such specialization of cell function is possible only when extracellular conditions, such as temperature, pH, and concentrations of glucose and oxygen, remain constant. Maintenance of constant extracellular conditions within the body is called **homeostasis.** Homeostasis allows cells to function efficiently and to relate to one another properly.

The process of homeostasis acts in a similar way to a residential thermostat. If a thermostat is set at 67 degrees Fahrenheit and the temperature drops below this set point, an electrical signal is sent through a wire that runs from the thermostat to the furnace and switches the furnace "on" automatically. When the temperature reaches the set point of 67 degrees, another signal is sent to the furnace to switch the furnace "off." The human body has many mechanisms that act like a thermostat to ensure that internal conditions do not vary outside of a narrow range. These internal conditions include the optimal temperature at which the body functions, the correct levels of glucose in the blood, and normal levels of blood pH.

Components of Homeostasis

Several elements work together to maintain homeostasis. Scattered throughout every body organ and tissue are **sensory receptors,** whose sole function is to continually test the body's temperature, pH, glucose levels, blood pressure, and a host of

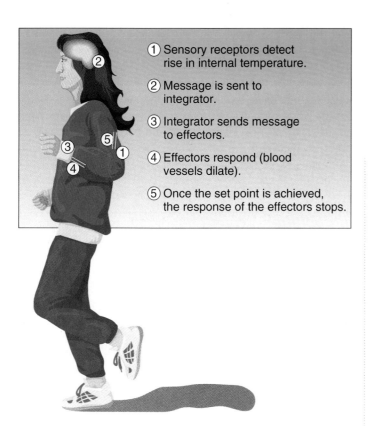

① Sensory receptors detect rise in internal temperature.

② Message is sent to integrator.

③ Integrator sends message to effectors.

④ Effectors respond (blood vessels dilate).

⑤ Once the set point is achieved, the response of the effectors stops.

Figure 33.6 A negative feedback loop: Temperature regulation.

When you exercise, your body's internal temperature rises. Sensory receptors detect this change and send a message to the brain (the integrator), which, in turn, sends a message to the blood vessels (the effectors) to dilate. This dilation accounts for your red face after strenuous exercise: The dilation causes more blood to course through the blood vessels. The double line in the diagram indicates that, once the set point is achieved, the effectors' response stops.

other internal conditions. All of these conditions must be maintained within very narrow limits, or set points, for the body to function properly.

When a particular condition deviates from a set point, a sensory receptor sends a signal to an **integrator.** In the human body, integrators are organs that are able to bring about changes in internal conditions. The **hypothalamus,** the part of the brain that regulates temperature and the amounts of different hormone levels secreted into the blood, is the body's most important integrator. Once it receives a signal from a sensory receptor, the integrator, in turn, sends a message to an **effector,** a specific tissue or organ that changes its function in response to the integrator's message. For example, when body temperature rises, the hypothalamus signals blood vessels to dilate, thereby releasing trapped heat. When the internal condition returns to the set point, the integrator stops sending its signal, and homeostasis is restored (figure 33.6). Sidelight 33.1 offers a discussion of how set points affect weight loss.

Sidelight 33.1
ADIPOCYTES AND HOMEOSTASIS: THE SET-POINT THEORY

You constantly diet but can never seem to keep weight off. Some of your friends can eat anything they want and not gain weight, but it seems that if you even *look* at food, you put on pounds. Sound familiar? If so, you might be one of the millions of people with a genetic predisposition to store fat. Your failure to keep weight off is not a willpower problem but a problem of the number and size of your fat cells, or adipocytes. In an evolutionary sense, this condition is not a problem at all: In fact, in another type of culture, one in which food is only sporadically available, the possession of numerous and larger adipocytes is an evolutionary advantage.

Nutritional experts explain the mechanism by which some persons store fat more quickly than others with the **set-point theory.** This theory states that at certain periods of infancy and childhood, the number and size of the body's adipocytes are influenced and can be changed. At age two, for example, if the child eats large amounts of food and most of this food is stored as fat, the number and size of the adipocytes used for this fat storage increases.

Once this change occurs, reducing the number and size of the adipocytes is very difficult. In fact, the body maintains the rate of metabolism, or the speed with which food is converted into the building blocks of the body and into fat, at the optimal rate needed to maintain the adipocytes at their set number and size. This rate is called the body's set point. The set point is a homeostatic phenomenon; that is, it changes in response to outside factors, such as greatly diminished food intake, and keeps the number of adipocytes at their preset levels.

For example, if a person with a large number of adipocytes goes on a "crash" diet, the weight loss may be slow or nonexistent because the rate of metabolism adjusts to this starvation diet. The body clings to every calorie it is given and converts it quickly into fat. Any weight loss is probably from lean muscle or water. This type of metabolism was an obvious advantage for our hunter-gatherer ancestors.

Food was not plentiful—it was either "feast or famine." During feast times, the body stored most of the calories consumed in adipocytes to ward off the long periods of starvation when food was scarce.

The body's set point, however, can be revved up through exercise. Exercise need not be vigorous or sustained for long periods to initiate this change; many physicians recommend a brisk walk at least three or four times a week as sufficient to change the set point. And because quick-weight-loss schemes play into the exact mechanism that the set point seems to protect—the risk of sudden starvation—nutritionists recommend sensible diets that result in a weight loss of about a pound a week. If weight loss is slow, the body adjusts gradually to the change in calories, and the set-point mechanism is not triggered.

The diet should also be one that the dieter will stick to for life. Drastic dietary changes at any time upset the body's metabolism. Finally, a diet that is low in fat, low in protein, and high in complex carbohydrates is the best way to lose and maintain weight. With these tips in mind, the set point can be changed and a healthier outlook achieved.

A Negative Feedback Loop: Regulation of Blood Glucose Levels

The temperature regulation example just described pertains to a type of homeostasis called **negative feedback.** The whole process from beginning to end is called a **negative feedback loop** because the end of the process—restoration of the set point—is the same as the beginning of the process. In negative feedback, a disturbance triggers processes that change the body's condition. These changes are then detected and reversed.

Regulation of blood glucose levels provides another of the many examples of negative feedback in the human body: After a big dinner, digestion introduces a large amount of glucose into the blood within a short period. High glucose levels are a disturbance from the normal levels the body requires to function properly. Thus, when sensory receptors in the blood vessels detect that glucose levels are exceeding normal values, the brain (the integrator) sends a message to the liver cells (the effectors). The liver cells convert this excess glucose to a storage form called glycogen, thereby causing blood glucose levels to fall back within the normal range (figure 33.7). Similarly, when glucose levels in the blood fall below the normal range, liver cells break down the liver's glycogen stores to add more glucose to the bloodstream. Thus, glucose levels in the fluid surrounding body cells change very little over the course of a day,

even though the body's glucose intake may be concentrated within a short period. This is the essence of negative feedback.

A Positive Feedback Loop: Labor

Negative feedback loops detect and seek to reverse disturbances from normal levels. However, certain deviations from normal limits are necessary for certain body functions, and in **positive feedback,** these disturbances are accentuated and prolonged. Labor, for example, involves strong contractions of the uterus. These contractions are definitely disturbances in the body's normal functioning, but they are maintained and prolonged by a hormone called oxytocin. Oxytocin is secreted from the hypothalamus as long as the fetus's head is pushing against the opening of the uterus. Once the fetus is expelled, the flow of oxytocin slows, and uterine contractions cease. Positive feedback loops are rare, however, in the human body and are usually associated with disease states and abnormal functioning.

Homeostasis is the maintenance of internal body conditions within very narrow limits. Sensory receptors detect disturbances and send messages to integrators, which, in turn, send signals to effectors, which initiate the changes that will bring conditions back within the narrow limits. Homeostasis that operates in this way is called negative feedback. In positive feedback, a disturbance is accentuated and prolonged, as in labor. The human body has many negative feedback loops but few positive ones.

Homeostasis: Central to the Operation of the Vertebrate Body

The next eleven chapters examine the functioning of vertebrates' organ systems in detail. These are shown in overview in figure 33.8. In every case, homeostatic mechanisms coordinate the functioning of these organ systems within the narrow limits dictated by the complexity of these highly specialized animals.

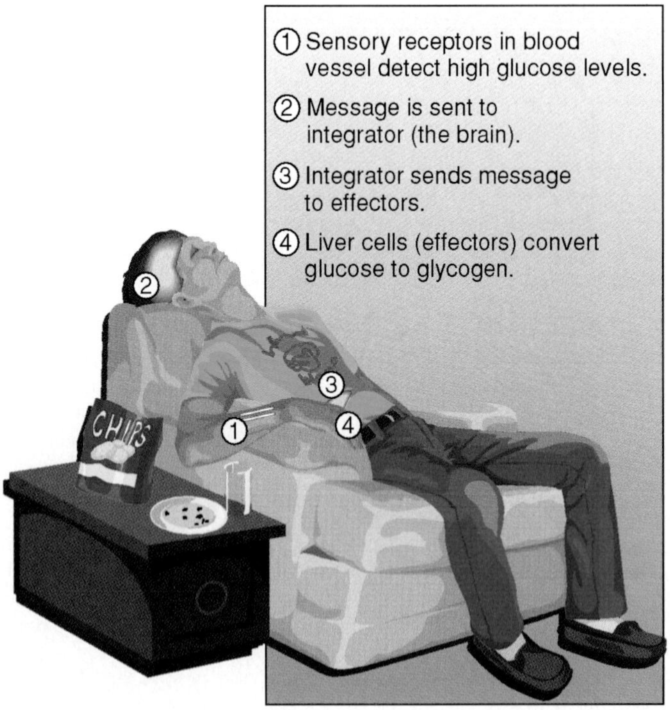

1. Sensory receptors in blood vessel detect high glucose levels.
2. Message is sent to integrator (the brain).
3. Integrator sends message to effectors.
4. Liver cells (effectors) convert glucose to glycogen.

Figure 33.7 A negative feedback loop: Regulation of blood glucose levels.

After a heavy meal, glucose levels in the blood reach very high levels. Sensory receptors detect this change and send a message to the brain (the integrator), which, in turn, sends a message to liver cells (the effectors). Liver cells then convert the excess glucose in the blood to glycogen, a storage form of glucose. The double line in the diagram indicates that, once the set point is achieved, the effectors' response stops.

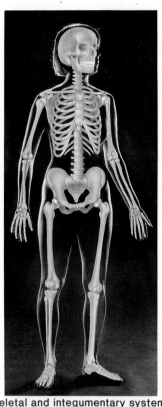

Skeletal and integumentary systems

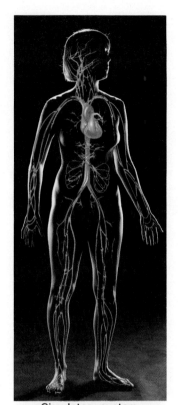

Circulatory system

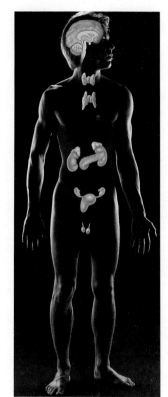

Endocrine system

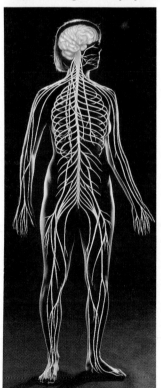

Nervous system

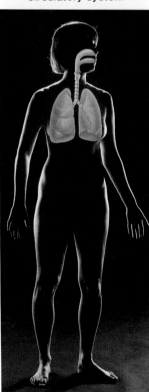

Respiratory system

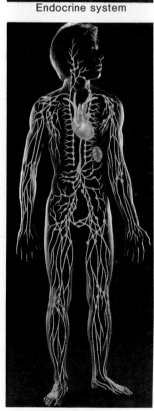

Lymphatic system

Figure 33.8 Eleven vertebrate body systems.
In each of these systems, organs work together to perform specific functions. For example, the circulatory system contains the heart and blood vessels, which work together to pump the blood through the body. Some of the systems themselves also work together. The circulatory system works in concert with the respiratory system to circulate oxygen through the body and to dispose of carbon dioxide that accumulates in body tissues.

(continued)

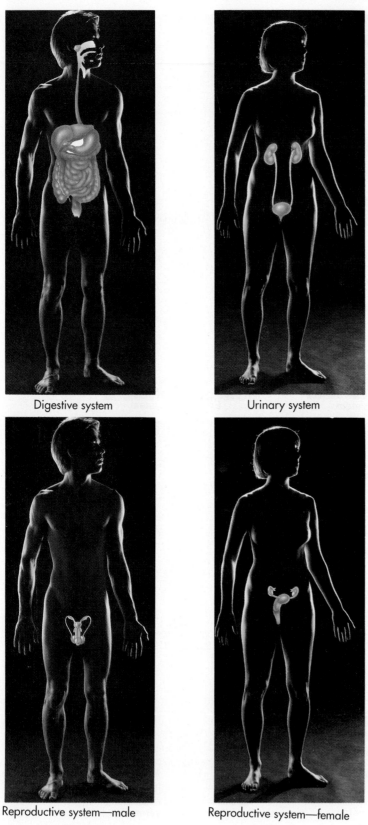

Digestive system

Urinary system

Muscular system

Reproductive system—male

Reproductive system—female

Figure 33.8 (continued)

EVOLUTIONARY VIEWPOINT

While the same basic tissue types occur in all vertebrates, each phylum of animals has a distinct array of tissues and organs, often unique to it alone. For example, there are no counterparts among vertebrate tissues and organs to the pneumatocysts of cnidarians, the flame cells of Platyhelminthes, the radulas of mollusks, or the Malpighian tubules of insects.

SUMMARY

1. The human body has the same general architecture of all vertebrates: a hollow digestive tract suspended within the coelom and supported by an internal scaffold or skeleton. A variety of different cell types are organized into tissues, different tissues are grouped into organs, and different organs are grouped into organ systems.

2. The eleven principal vertebrate organ systems are: integumentary, nervous, skeletal, muscular, endocrine, digestive, respiratory, circulatory, immune, urinary, and reproductive.

3. The four basic types of tissue are epithelial tissue, connective tissue, muscle tissue, and nervous tissue.

4. Epithelium covers body surfaces. The lining of the major body cavities is composed of epithelium, as is the exterior skin. Epithelial tissues protect the tissues beneath them from dehydration and other damage, provide a selectively permeable barrier for materials, provide sensory surfaces, and secrete materials.

5. Connective tissue supports the body mechanically and defensively. It consists of connective tissue cells embedded within a matrix. Connective tissue cells include fibroblasts, plasma cells, macrophages, mast cells, leukocytes, adipocytes, and osteocytes, to name a few. The different kinds of connective tissue are distinguished primarily by the kind of matrix laid down between individual cells.

6. Contraction of cells within muscle tissue provides the force for mechanical movement of the body. Vertebrates have three different

kinds of muscle cells: smooth muscle, skeletal muscle, and cardiac muscle.

7. Nervous tissue is composed of two kinds of cells: neurons, which are specialized for the rapid transmission of nerve impulses, and supporting glial cells, which support and insulate the neurons, help to maintain the ionic composition of nervous tissue, and are believed to supply neurons with nutrients and other molecules.

8. Organs are body structures composed of several different tissues grouped together into a structural and functional unit. Organ systems are groups of organs that function together to carry out the principal activities of the body.

9. The organs of the body employ feedback loops to maintain constant extracellular conditions around the many cells of the body, a condition called homeostasis. In a negative feedback loop, sensory receptors detect a change in set point and then send a message to an integrator, an organ able to initiate changes in internal conditions. An integrator, in turn, sends a signal to effectors, specific tissues or organs that change their function in response to the integrator's message. When the condition returns to set point, the integrator stops sending its signal. Positive feedback loops are rare.

REVIEWING THE CHAPTER

1. What is the body architecture of the human animal?

2. How is the vertebrate body organized?

3. What are the basic tissue types in vertebrates?

4. How does epithelium function as a protective tissue?

5. What are the cell types of epithelial tissues?

6. How does connective tissue support the body?

7. What are the cells of connective tissue?

8. How do fibers and ground substance function in the matrix of connective tissue?

9. How does muscle tissue provide for movement?

10. How does nervous tissue conduct signals rapidly?

11. How do organs work together?

12. How is homeostasis achieved?

13. What are the components of homeostasis?

14. How does a negative feedback loop work?

15. What is a positive feedback loop?

COMPLETING YOUR UNDERSTANDING

1. What does it mean for a human to be a homeotherm?

2. What is the function of an organ system?

3. What is the origin and function of the endoderm, mesoderm, and ectoderm?

4. Which of the following is *not* one of the four basic types of tissues of the adult vertebrate body?

  a. Muscle   d. Mesoderm
  b. Nervous   e. Epithelium
  c. Connective

5. Which of the following is *not* a function of epithelial layers?

  a. To secrete material
  b. To store and distribute substances throughout the body
  c. To protect the tissues beneath from dehydration and mechanical damage
  d. To provide a selectively permeable barrier
  e. To provide sensory surfaces

6. The type of cell in the epithelial tissue that lines the gastrointestinal tract and functions in secretion and absorption is

  a. simple squamous.
  b. simple cuboidal.
  c. simple columnar.

  d. stratified squamous.
  e. transitional.
  f. none of the above.

7. Which of the following is *not* a cell type of connective tissue?

  a. Leukocytes
  b. Fat cells
  c. Columnar cells
  d. Erythrocytes
  e. Macrophages

8. Which cell type in the connective tissue would function in the immune system to engulf foreign cells?

  a. Fibroblasts   d. Mast
  b. Plasma   e. Fixed macrophages
  c. Adipocytes   f. None of the above

9. Which type of connective tissue would be located in tendons and ligaments and why?

a. Loose   d. Reticular

b. Dense   e. Cartilage

c. Elastic   f. Vascular

10. Which of the following is a function of cartilage?

a. Insulation

b. Flexible support

c. Strong connections

d. Shock absorption

e. Immune defenses

f. Protection of internal organs

11. Which type of tissue is responsible for voluntary motion, such as walking and lifting?

12. What is depolarization, and why does it occur?

13. In the human body, why are some neurons more than a meter in length?

14. If the human body was not properly regulating its temperature, what might you suggest as a malfunction?

15. Which cell type is responsible for hormone secretion?

16. What is the set-point theory, and to whom does it apply?

17. What is usually associated with positive feedback loops?

18. Land was successfully invaded four times—by plants, fungi, arthropods, and vertebrates. Because bodies are far less buoyant in air than in water, each of these four groups evolved a characteristic hard substance to lend mechanical support. Describe and contrast these four substances, discussing their advantages and disadvantages. Can you imagine any other substance (such as plastic) that would have been superior to any of these?

19. Your body contains 206 bones. As you grow, all of these bones must increase in size and maintain proper proportions with one another. How is the growth of these bones coordinated?

20. What are the basic elements of feedback control that must be present for an organism to regulate its internal environment? Where would you expect each to be located in the case of the cardiovascular system?

FOR FURTHER READING

Caplan, A. 1984. Cartilage. *Scientific American,* Oct., 84–97. An interesting account of the many roles played by cartilage in the vertebrate body.

Currey, J. 1984. *The mechanical adaptations of bones.* Princeton, N.J.: Princeton University Press. A functional analysis of why different bones are structured the way they are, with an unusually well-integrated evolutionary perspective.

Golde, D. 1991. The stem cell. *Scientific American,* Dec., 86–93. How isolating and storing stem cells (prolific bone marrow tissue that produces both red and white blood cells) is leading to improved treatments for many blood diseases.

Houk, J. C. 1988. Control strategies in physiological systems. *FASEB Journal,* Feb., 97. General discussion of control systems written for a general audience.

National Geographic Society. 1986. *The incredible machine.* Washington, D.C.: National Geographic Society. A series of outstanding articles on the human body, focusing on its major organ systems. Beautifully illustrated and fun to read.

Rosenfeld, A. 1988. There's more to skin than meets the eye. *Smithsonian,* May, 159–80. An entertaining and informative account of the many tasks carried out by the body's largest organ.

Seeley, R., T. Stephens, and P. Tate. 1992. *Anatomy and physiology.* 2d ed. St. Louis, Mo.: Mosby. A student-oriented introduction to human anatomy, with an excellent and visually pleasing treatment of tissues in chapter 4.

EXPLORATIONS

Interactive Software:
Life Span and Lifestyle

This interactive exercise allows students to explore how their lifestyle decisions influence their life expectancy. The exercise presents a diagram of a typical human life span, shown as an age distribution of U.S. life expectancies (how long a U.S. citizen of a certain age can expect to live). The student explores the change in probability of survival when certain activities are initiated. The student may begin or stop smoking at any age (cigarettes induce cancer as well as lung and heart disease); eat or stop eating a diet high in animal fat (which induces heart disease and stroke and is highly correlated with breast cancer in women); eat or stop eating a high-protein diet (which seems to induce colon cancer); and initiate or cease regular exercise (which counteracts cardiovascular disease). By varying patterns of behavior, the student soon learns that survival reflects lifestyle.

Questions to Explore:

1. Can exercise counteract the lowered life expectancy due to smoking?

2. Is it more beneficial to give up smoking or a red-meat diet?

3. Is smoking more harmful at one age than another?

4. Is there any age when diet does not matter?

5. What lifestyle is associated with the longest life expectancy? The shortest?

HOW ANIMALS TRANSMIT INFORMATION

A Purkinje neuron (nerve cell) from the cerebellum of a human brain, magnified 400 times. Purkinje neurons, a key building block of the CNS, make many connections within the brain.

FOR REVIEW

Here are some important terms and concepts that have been discussed in previous chapters and that you will encounter again in this chapter. Review them before proceeding if necessary.

Gap junctions (*chapter 5*)
Sodium-potassium pump (*chapter 5*)
Neuron (*chapter 33*)
Depolarization (*chapter 33*)

Successful performance of any complex machine requires proper communication among its parts, and the vertebrate body is no exception. Body cells can communicate with one another in several ways. One simple way is by direct cell contact, with an open cytoplasmic connection between two cells that permits the passage of ions and small molecules. Communication of this kind is provided by gap junctions, discussed in chapter 5. While direct cell contact through gap junctions provides a ready means of communication between adjacent cells, it does not allow for rapid and efficient communication between distant tissues.

For distant communication, the body sometimes sends chemical instructions to various tissues. The instructions are in the form of hormones, small chemical molecules that act as messengers circulating within the body's bloodstream. This kind of command system, however, can be too slow. For example, if the message to be delivered to the leg muscles is, "Contract quickly, we are being pursued by a leopard," a quicker means of communication than hormones is desirable.

All complex animals possess rapid means of tissue communication in the form of specialized cells called **neurons,** which, as mentioned in chapter 33, are cells specialized for the transmission of nerve impulses. A **nerve impulse,** also called an **action potential,** is an electrical excitation of the neuron membrane. Nerve impulses are *not* true electrical currents. The nature of a nerve impulse, however, is electrical because it is based on the separation of positive and negative charges across the neuron membrane. A battery has a positive charge at one end and a negative charge at the other, and this charge separation is what generates the battery's power. This same type of charge separation is found in a neuron, but it is the disruption of the charge separation that causes the nerve impulse.

Without neurons and their special properties, we would not be able to eat, run, study, or think. Many elements work together to allow us, for example, to escape from a pouncing leopard, and neurons provide the connections for these elements. Our eyes, which are sensory organs, detect the approach of the dangerous animal. Neurons located in the eye send a nerve impulse to the brain, which quickly assesses the situation and sends a nerve impulse along neurons to our leg muscles, which are signalled to contract.

Neurons are clearly the chief functional units of the nervous system, connecting the various body organs, muscles, and tissues to the body's "control center," the central nervous system. This chapter focuses on the neuron and examines in detail the specializations that tailor it to the transmission of nerve impulses. The chapter concludes with an in-depth examination of how drugs affect the nervous system.

The Neuron

The body contains many different neurons, some of them tiny with a few projections, others bushy with projections, and still others with extensions several meters long. **Sensory neurons** collect information and transmit nerve impulses to the brain. They are found in sensory organs, such as the eyes, ears, and nasal epithelium. **Motor neurons** signal muscles or organs to act. They receive instructions from **brain neurons,** which integrate incoming information from sensory neurons and send information to motor neurons (figure 34.1).

Despite differences in appearance and function, all neurons have the same basic architecture (figure 34.2). Extending from the body of all but the simplest neurons are one or more

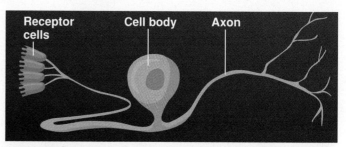

(a) **Sensory Neuron**

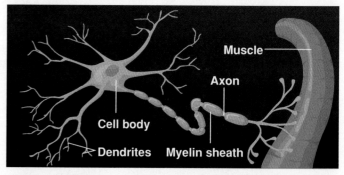

(b) **Motor Neuron**

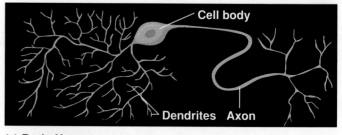

(c) **Brain Neuron**

Figure 34.1 Types of vertebrate neurons.
(a) *Sensory neurons, which carry nerve impulses from sense organs to the brain, typically have dendrites only in specific receptor cells.* (b) *The axons of many motor neurons, which carry commands from the brain to muscles or glands, are encased at intervals by Schwann cells.* (c) *Brain neurons often possess highly branched, extensive dendrites.*

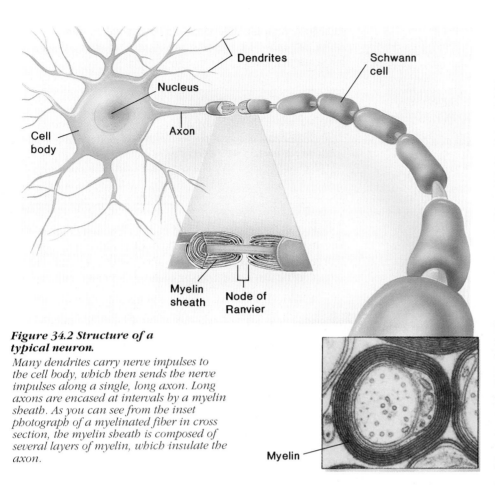

Figure 34.2 Structure of a typical neuron.
Many dendrites carry nerve impulses to the cell body, which then sends the nerve impulses along a single, long axon. Long axons are encased at intervals by a myelin sheath. As you can see from the inset photograph of a myelinated fiber in cross section, the myelin sheath is composed of several layers of myelin, which insulate the axon.

Figure 34.3 Some neurons are quite long.
From the toe of this giraffe, a single axon can extend all the way up to the pelvis.

cytoplasmic extensions called **dendrites.** Most nerve cells possess a profusion of dendrites, many of which are branched. The dendrites enable the **cell body** of the neuron to receive inputs from many different sources simultaneously. The surface of the neuron cell body integrates the nerve impulses arriving from the many different dendrites. If the resulting membrane excitation is large enough, it travels outward from the cell body as an electrical nerve impulse along an **axon** (see figure 34.2). Most nerve cells possess a single axon, which may be quite long. The axons controlling muscles in your legs are more than a meter long, and even longer ones occur in larger mammals. In a giraffe, a single axon travels from the hoof all the way up to the pelvis (figure 34.3).

Most neurons are unable to survive alone for long; they require the nutritional support provided by companion **neuroglial cells.** More than half the volume of vertebrate nervous systems is composed of supporting neuroglial cells. In many neurons, including the motor neurons that extend from brain to muscles, impulse transmission along the very long axons is facilitated by neuroglial **Schwann cells,** which envelop the axon at intervals and act as electrical insulators. The Schwann cells form a **myelin sheath,** a flattened sheath of fatty material found in many, but not all, vertebrate neurons. The myelin sheath is interrupted at the spaces between Schwann cells to form gaps called **nodes of Ranvier,** where the axon is in direct contact with the surrounding intercellular fluid. An axon and its associated Schwann cells, or cells with similar properties, form a **myelinated fiber** (see figure 34.2). Bundles of nonmyelinated and myelinated neurons, which in cross section

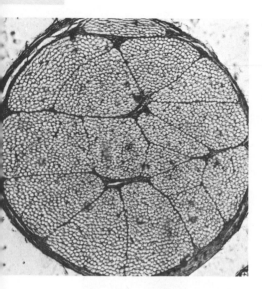

Figure 34.4 A nerve.
Nerves are a bundle of axons bound together by connective tissue. In this micrograph of a nerve cross section, many myelinated neuron axons are visible, each looking something like a severed hose.

look somewhat like telephone cables, are called **nerves** (figure 34.4). Sidelight 34.1 presents an overview of multiple sclerosis, a disease that ravages the myelin sheaths.

Nerve cells specialized for electrical signal transmission are called neurons. Typically, a nerve impulse is received by a dendrite branch, which passes it to the cell body, where its influence is merged with that of other incoming nerve impulses. The resulting nerve impulse is passed outward along a single, long axon.

The Nerve Impulse

The neuron maintains a separation of positive and negative charges across its cell membrane, and disruption of this charge separation causes the transmission of a nerve impulse. An

---

Sidelight 34.1
MULTIPLE SCLEROSIS: A WAR OF NERVES

Multiple sclerosis is a debilitating disease that afflicts about 350,000 Americans (Sidelight figure 34.1). The disease typically strikes young adults in their twenties and thirties with the onset of unsettling and seemingly unrelated symptoms: visual impairments, such as blurriness or double vision; clumsiness; vertigo; "pins and needles" sensations; and sometimes, urinary incontinence or frequent urinary infections. Despite their variety, these symptoms have one root cause: multiple sclerosis, in which the myelin sheaths covering neurons are progressively destroyed and replaced by hard scars, or **plaques.** The destruction of the myelin sheaths can have

devastating consequences for the individual because the transmission of nerve impulses along these neurons is impaired. Neurons may not be able to transmit the information, for example, that the bladder needs to be voided, that the feet are touching the ground, or that an object has been detected in the retina.

Most people afflicted with multiple sclerosis experience steady attacks after the initial attack for the first four or five years. Then, the disease may go into remission for as many as ten to twenty years, only to resurface again with a vengeance. With each attack, more and more myelin sheaths are destroyed, and more plaques develop. The result is a steady loss of function, and many multiple sclerosis patients become confined to wheelchairs. Almost all patients who suffer from the disease eventually die from it.

A variety of therapies are used to treat the disease, but none have been very successful. However, a new, promising treatment recently approved by the U.S. Food and Drug Administration has been shown to slow the course of the disease in some patients. The new treatment is based on the theory that multiple sclerosis is a disease of the immune system.

Chapter 26 noted that HIV, the virus that causes AIDS, wreaks its destruction on the human body by infiltrating the body's T-cells, the immune cells responsible for alerting the body's immune response. In multiple sclerosis, these same T-cells are not killed but are *overactive.* Unfortunately, the activity is misdirected, causing the T-cells to unleash their destruction on the myelin sheaths. When activated, T-cells release a chemical called gamma interferon. Researchers were tipped off that T-cells are involved in multiple sclerosis when they found that the level of gamma interferon in the cerebrospinal fluid (the fluid that cushions the brain and circulates into the spinal cord) of multiple sclerosis patients was extremely high just before and after an attack. The researchers theorized that injecting patients with a chemical that counteracts gamma interferon would slow the course of the disease. This chemical, called beta interferon, does just that: It lessens the activity of T-cells and curtails their production of gamma interferon.

Researchers are hopeful that, for the first time, they have found a therapy that reduces the number and severity of attacks in a significant number of patients. They stress, however, that beta interferon therapy is not a cure—some patients participating in the drug trials showed no improvement, and some even got worse. But beta interferon still offers a glimmer of hope to the thousands afflicted with multiple sclerosis and has opened the door to more experimental treatments based on the immune system.

Sidelight figure 34.1 Multiple sclerosis.
Annette Funicello, famous as a mouseketeer on a popular 1950s TV show, is a victim of multiple sclerosis.

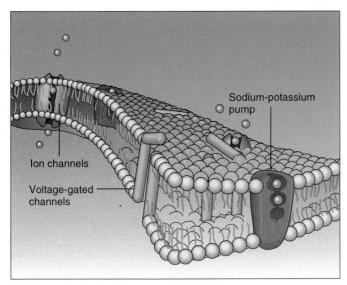

Figure 34.5 Landscape of a neuronal membrane.
Embedded in the neuronal membrane are the sodium-potassium pump, ion channels, and voltage-gated channels. Each of these protein channels plays a role in sustaining the charge separation across the neuronal membrane.

understanding of nerve impulse transmission first requires familiarity with the landscape of the neuronal membrane and the protein channels embedded in this membrane. Hundreds of different protein channels probably traverse plasma membranes, but only three kinds are essential to an understanding of nerve impulse transmission (figure 34.5):

1. *The sodium-potassium pump.* This protein channel, already discussed in chapter 5, pumps three sodium (Na^+) ions *out* of the neuron while pumping two potassium (K^+) ions *into* the neuron. The result is that more K^+ ions are inside the neuron than outside the neuron, and more Na^+ ions are outside the neuron than inside the neuron.

2. *Ion channels.* Ion channels are specifically designed for the movement of different ions; for example, there are K^+ ion channels and Na^+ ion channels. These channels are always "open" and allow the diffusion, or "leakage," of ions into and out of the neuron according to concentration gradients. For example, as noted earlier, because of the sodium-potassium pump, the concentration of K^+ ions is higher inside the neuron than outside. Therefore, K^+ ions tend to diffuse out of the neuron through the K^+ ion channels. A neuronal membrane has many K^+ ion channels but very few Na^+ ion channels. This difference in number is significant to maintaining a neuron's charge separation.

3. *Voltage-gated channels.* A neuron has two types of voltage-gated channels: One type only transmits Na^+ ions, and one type only transmits K^+ ions. However, both

types of voltage-gated channels work in the same general way. The term *voltage* is defined as the measurement of a charge separation. As used to describe these two channels, *voltage* means simply that, when these channels are employed in the neuron membrane, the charge separation across the membrane changes. The term *gated* means that these channels are usually closed but that they can open when stimulated.

The Resting Potential: The Neuron at Rest

When the charge separation across the neuronal membrane is maintained, the neuron is said to be at rest. This condition is called the **resting potential.** A resting neuron is negatively charged on the inside and positively charged on the outside. Physicists call such a charge separation a **polarization,** and thus, a resting neuron is said to be polarized. The sodium-potassium pump, combined with the action of the K^+ ion channels, maintain the polarization across the neuronal membrane.

As already explained, the action of the sodium-potassium pump results in a concentration difference across the neuronal membrane, with more Na^+ ions outside the neuron than inside, and more K^+ ions inside the neuron than outside. This concentration difference, however, does not result in the polarization of the neuron. Polarization occurs due to the action of the K^+ ion channels. As the K^+ ions are pumped into the neuron, they also simultaneously leak out of the neuron through the K^+ ion channels because of the K^+ concentration gradient (more K^+ ions inside the neuron than outside).

Leakage of K^+ ions out of the neuron results in the neuron's interior becoming more negative, relative to its exterior. The interior is more negative because many fewer positively charged ions are on the inside of the neuron than on the outside. Outside the neuron are Na^+ ions that have been pumped out by the sodium-potassium pump and K^+ ions that have leaked out through the open K^+ ion channels.

While the Na^+ ion channels present in a neuron allow Na^+ ions to leak into the neuron, this leakage has little effect for two reasons: (1) There are very few Na^+ ion channels, and (2) the small number of Na^+ ions that leak into the neuron are quickly swept out by the sodium-potassium pump.

Using sophisticated instruments, scientists have been able to measure the voltage of the neuronal interior as -70 millivolts. The next sections explain how this voltage changes when a nerve impulse is initiated along the neuron.

> *In a resting neuron, the neuronal interior is more negatively charged than its exterior. The sodium-potassium pump maintains a concentration gradient of Na^+ and K^+, with a high concentration of Na^+ outside the neuron and a high concentration of K^+ inside the neuron. Leakage of K^+ ions through K^+ ion channels—from inside the neuron to outside—is what causes neuron polarization.*

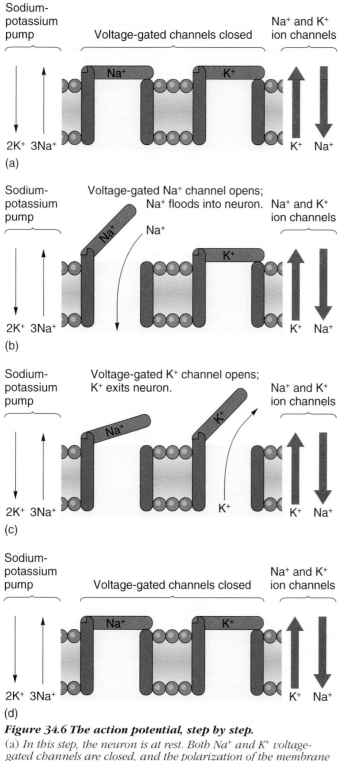

Figure 34.6 The action potential, step by step.
(a) *In this step, the neuron is at rest. Both Na⁺ and K⁺ voltage-gated channels are closed, and the polarization of the membrane is maintained by the ion channels and the sodium-potassium pump.* (b) *When a stimulus is applied at a site on the neuronal membrane, the voltage-gated Na⁺ channels open, allowing Na⁺ ions to enter the neuron. This entry wipes out the membrane's polarization.* (c) *Almost instantaneously, the voltage-gated K⁺ channels open, allowing K⁺ ions to leave the neuron.* (d) *The exit of the K⁺ ions quickly restores the resting potential.*

The Action Potential: Initiating a Nerve Impulse

The transmission of a nerve impulse, called an **action potential** to differentiate it from the resting potential, occurs in four phases: (1) initiation of an action potential by a sufficiently large stimulus, called a **threshold stimulus;** (2) its transmission along a neuron; (3) its transfer to a **target cell,** which can be a neuron or a muscle cell; and (4) its effect on the target cell.

Sensory receptors located all over the body are designed to detect different stimuli. Once detected, the stimuli are passed along to neurons located within or close to the sensory receptor itself. A host of different stimuli, such as light (as in the neurons of the eye), chemical changes, touch, heat, and cold can trigger an action potential, but only if the stimulus is large enough. Action potentials are all-or-nothing phenomena—that is, they do not occur without sufficient stimulus, but once they start, they cannot be stopped. All action potentials are also alike: Every action potential causes the same amount of voltage change in the neuron's interior. There are no large or small action potentials, only large and small stimuli.

Action potentials temporarily disrupt the polarization of a neuron, resulting in **depolarization.** When a stimulus is applied to a site on the neuron, the closed, voltage-gated Na⁺ channels suddenly open, allowing Na⁺ ions to flood into the neuron (figure 34.6*a,b*). This sudden movement of positive ions into the neuron causes the neuron's interior to develop a positive charge relative to its exterior. Next, the voltage-gated Na⁺ channels close and the voltage-gated K⁺ channels suddenly open and assist the K⁺ ion channels in sweeping K⁺ ions out of the neuron, thus restoring the resting potential (figure 34.6*c,d*).

The depolarization and restoration of the resting potential takes only about 5 milliseconds. Fully one hundred such cycles could occur, one after another, in the time it takes to say the word *nerve* (figure 34.7).

An action potential is initiated by a threshold stimulus. The stimulus causes the voltage-gated Na⁺ channels to open, allowing a flood of Na⁺ ions into the neuron's interior. This flood of positive ions temporarily wipes out the neuron's polarization. The action of the voltage-gated K⁺ channels quickly restores the neuron to its resting potential.

Transmission of an Action Potential along a Neuron

An action potential is transmitted along a neuron when the depolarization of one site of the neuron stimulates the depolarization of adjacent neuron sites. The fleeting depolarization of one site is enough to "ignite" nearby voltage-gated Na⁺ channels, causing them to open and permit Na⁺ ions to enter the neuron,

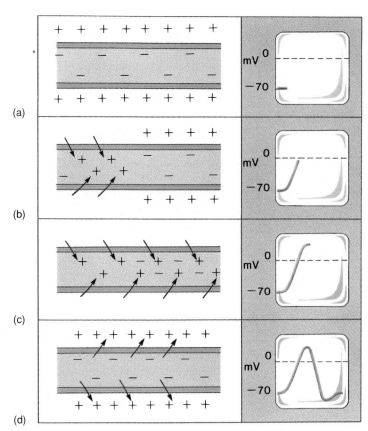

Figure 34.7 Schematic diagram of the action potential.
(a) *At rest, there is an excess of exterior positive charge and a net interior negative charge of about −70 millivolts (mV). (b) When a neuron is stimulated, Na⁺ ions enter, abolishing the voltage difference. (c) Enough positive ions enter to establish a net interior positive charge. (d) The exit of K⁺ ions then returns the interior to a net negative charge.*

thus depolarizing these sites as well. In this way, the action potential is propagated along the neuron (figure 34.8). Like a burning fuse, an action potential is usually initiated at one end and travels in one direction, but it can also be initiated in the middle of a neuron and travel out from both directions, as if "lit" in the middle.

Measurements taken from the neuronal interior during an action potential show that the neuron achieves a voltage of 40 millivolts. As mentioned earlier, during the resting potential, the neuronal interior has a voltage of −70 millivolts. Action potentials thus cause the voltage of the neuronal interior to jump a total potential difference of 110 millivolts (figure 34.9)!

The time required for voltage-gated Na⁺ channels to "recover" from their active state and close again is called the **refractory period.** During the refractory period, which lasts only fractions of a second, no action potentials can be transmitted.

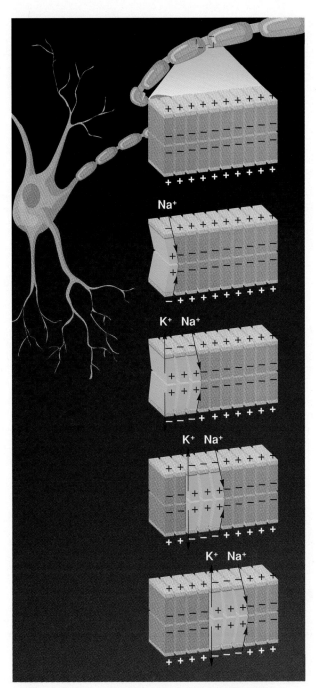

Figure 34.8 Transmission of an action potential.
The transmission of an action potential resembles the ignition of a fuse. The depolarization at one site of the neuronal membrane "ignites" the adjacent site, until the action potential is propagated along the length of the axon.

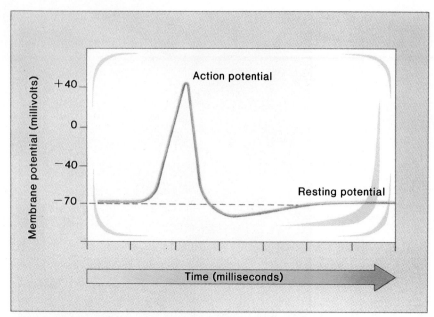

Figure 34.9 Graph of an action potential.

An action potential generates 40 millivolts of electricity within the neuronal interior; the resting potential generates −70 millivolts. This change is always the same for any given neuron, and each neuron requires the same length of time to recover.

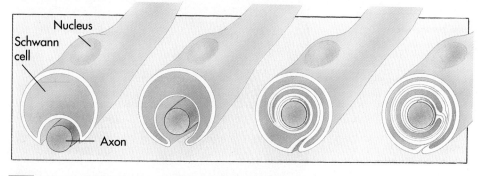

Figure 34.10 How a myelin sheath is made.

Development of the myelin sheath that surrounds many neuron axons involves the envelopment of the axon by a Schwann cell. The progressive growth of the Schwann cell membrane around the axon contributes to the many layers characteristic of myelin sheaths. These layers act as an excellent electrical insulator.

An action potential is propagated along a neuron as one area of depolarization "ignites" nearby voltage-gated Na^+ channels to open. The time it takes for the voltage-gated Na^+ channels to close again is called the refractory period. During this time, no action potentials can be initiated in the neuron.

Saltatory Conduction

The speed at which the action potential is transmitted along the neuron and out along the axon depends on several factors, one of which is the axon diameter. The larger the axon diameter, the faster the action potential will be conducted. Some species of squid and other marine invertebrates have giant axons that ensure the rapid transmission of action potentials.

Some vertebrates have increased the speed of action potential conduction by insulating their axons. These vertebrate neurons possess axons sheathed at intervals by neuroglial Schwann cells. Schwann cells envelop the axon, wrapping their cell membrane around it many times to produce a series of layers. This lipid-rich envelope of membrane layers, called a myelin sheath, prevents the transport of ions across the neuron membrane beneath it and thus acts as an electrical insulator, creating a region of high electrical resistance on the axon (figure 34.10).

Schwann cells are spaced one after the other along an axon, with nodes of Ranvier separating each Schwann cell from the next. These nodes are critical to nerve impulse propagation in these cells. Within the small gap represented by each node, the axon surface is exposed to the fluid surrounding the nerve. Voltage-gated ion channels are concentrated in these zones. The direct fluid contact permits ion transport through these channels and the generation of an action potential. The action potential does not continuously spread down the axon because it cannot flow across the insulating Schwann cells. Instead, it "jumps" from one node to the next, a very fast form of nerve impulse conduction known as **saltatory conduction** (from the Latin word *saltare*, meaning "to jump"). An impulse conducted in this fashion moves very fast, up to 120 meters per second for large-diameter neurons (figure 34.11). Saltatory conduction is also very inexpensive metabolically because the movements of Na^+ and K^+ ions are reduced. The sodium-potassium pump does not have to be as active when only the nodes undergo depolarization, not the entire neuron.

Not all nerve impulses propagate as a wave of depolarization spreading along the neuronal membrane. On some vertebrate neurons, impulses travel via saltatory conduction, rapidly leaping from node to node over insulated portions.

Crossing the Gap: The Synaptic Cleft

An action potential passing down the axon of a neuron eventually reaches the end of the axon and is passed along either to another neuron or to a target tissue, such as a muscle. Axons, however, do not actually make direct contact with other neurons or with target tissue. Instead, a narrow intercellular gap, 10 to 20 nanometers across, separates the axon

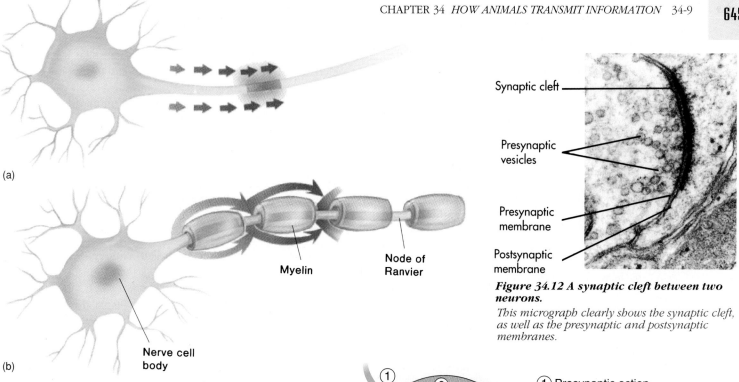

(a)

(b)

Nerve cell body

Myelin

Node of Ranvier

Figure 34.11 Saltatory conduction.
(a) *In an unmyelinated fiber, each portion of the membrane becomes depolarized in turn, like a row of falling dominoes.* (b) *The nerve impulse moves faster along a myelinated fiber because the wave of depolarization jumps from node to node, without ever depolarizing the insulated membrane segments between the nodes.*

Synaptic cleft

Presynaptic vesicles

Presynaptic membrane

Postsynaptic membrane

Figure 34.12 A synaptic cleft between two neurons.
This micrograph clearly shows the synaptic cleft, as well as the presynaptic and postsynaptic membranes.

tip and the target neuron or tissue. This gap is called the **synaptic cleft.** The membrane of the neuron that passes the action potential along to the target neuron or tissue is called the **presynaptic membrane;** the membrane of the receiving neuron or muscle cell is called the **postsynaptic membrane** (figure 34.12). A synaptic cleft between two neurons is called a **synapse;** a synaptic cleft between a neuron and a muscle cell is called a **neuromuscular junction.**

The gap between a neuron and another neuron or target cell is called a synaptic cleft. The membrane of the transmitting neuron is called the presynaptic membrane; the membrane of the receiving neuron or tissue is called the postsynaptic membrane. A synaptic cleft between two neurons is called a synapse; between a neuron and a muscle cell, the cleft is called a neuromuscular junction.

How Action Potentials Are Transmitted: An Overview

If presynaptic and postsynaptic membranes never actually touch one another, how is the action potential transmitted? In neural synapses, the action potential is transmitted either electrically or chemically across the synaptic cleft. In electrical transmission, special channels in the presynaptic and postsynaptic membranes allow the depolarization to be transmitted across the cleft. In chemical transmission, special chemicals called **neurotransmitters** are released from the presynaptic

ACh

ACh

(1) Presynaptic action potential arrives at synapse.

(2) Acetylcholine is released from vesicles.

(3) Acetylcholine binds to Na^+ gated channel, causing postsynaptic membrane excitation.

(4) Postsynaptic action potential.

(5) Acetylcholine is broken down into acetic acid and choline by acetylcholinesterase.

(6) Acetylcholine is resynthesized.

Na^+ chemically gated channel

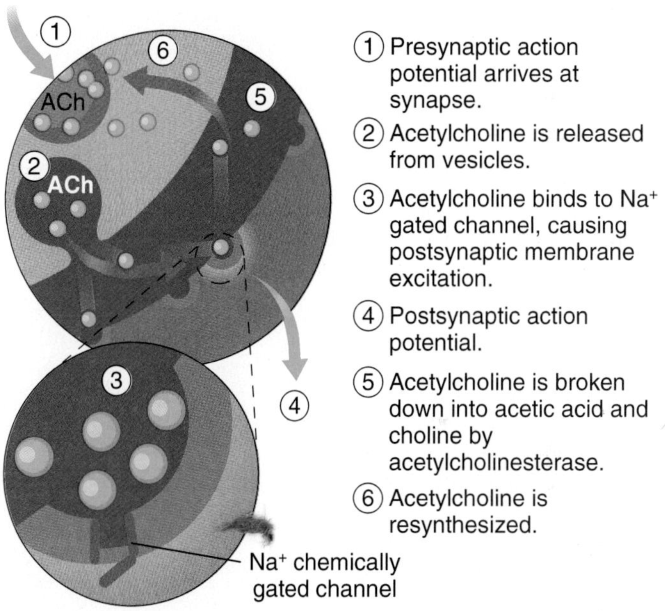

Figure 34.13 Chemical transmission.
Chemical transmission involves the release of neurotransmitters from vesicles in the presynaptic membrane. The neurotransmitters bind to receptors in the postsynaptic membranes. This binding induces chemically gated ion channels to open, which causes the postsynaptic membrane to become depolarized. The release of neurotransmitters from the presynaptic membrane causes calcium (Ca^{++}) to be released into the cytoplasm of the muscle cell. This release causes the muscle to contract. These six steps include the breakdown of acetylcholine into acetic acid and choline. These constituent parts will be reused in subsequent transmissions.

membrane when an action potential reaches it; these neurotransmitters cross the cleft and bind to special receptors in the postsynaptic membrane (figure 34.13). This binding causes special ion channels to open, initiating a wave of depolarization. Although action potentials are transmitted both electrically and chemically in neural synapses, only chemical transmission is found in neuromuscular junctions.

The ion channels located in the postsynaptic membrane are different from the voltage-gated ion channels discussed earlier. These postsynaptic channels are chemically gated, meaning that they open when stimulated by a chemical (in this case, a neurotransmitter). These chemically gated channels, like the voltage-gated channels, are specific for certain ions. There are chemically gated Na^+ channels, as well as chemically gated channels for such ions as chloride (Cl), potassium (K), and a host of other ions. This variety has implications in the type of neurotransmitter used in chemical transmission and in the effects that the binding of the neurotransmitter has in the postsynaptic cell.

Action potentials can be transmitted electrically or chemically. In chemical transmission, special chemicals called neurotransmitters are released from the presynaptic membrane by the stimulus of the action potential, cross the cleft, and bind to receptors in the postsynaptic membrane. This binding induces special ion channels to open, initiating a wave of depolarization.

Chemical Neurotransmitters

The advantages of chemical synaptic clefts, compared with those that utilize electrical transmission, are that the specific chemical neurotransmitter can be different in different junctions, and the nature of the chemically gated ion channels activated by a particular neurotransmitter can also be different. Over sixty different chemicals that act as specific neurotransmitters

or that can modify the activity of neurotransmitters have been identified. The result is a great diversity in the response of postsynaptic cells. The events that occur within the synaptic cleft when an action potential arrives depend on the identity of the particular neurotransmitter released into the cleft.

An understanding of what happens requires looking first at the junction between a nerve and a muscle cell, where the situation is relatively simple, and then examining nerve-nerve junctions, where the situation is more complex. A brief discussion of the categories of chemical neurotransmitters concludes this section.

Neuromuscular Junctions

In synapses with muscle cells, called neuromuscular junctions, the neurotransmitter used to transmit the action potential from the presynaptic membrane to the postsynaptic membrane is **acetylcholine.** Depolarization of the presynaptic membrane causes acetylcholine to be released from storage vesicles in the presynaptic membrane. Crossing the gap, the acetylcholine molecules bind to receptors in the postsynaptic muscle membrane, opening chemically gated Na^+ channels. During the millisecond that these Na^+ channels are open, some ten thousand Na^+ ions flow inward (see figure 34.13 and Sidelight 34.2). This ion flow depolarizes the adjacent postsynaptic muscle cell membrane, which contains voltage-gated Na^+ channels. In this way, acetylcholine initiates a wave of depolarization that passes down the muscle cell. This wave of depolarization, in turn, causes calcium ions that have been stored in the endoplasmic reticulum (called the sarcoplasmic reticulum) of the muscle cell to be released into the cytoplasm, and this calcium release triggers muscle contraction.

Sidelight 34.2
THE POSTSYNAPTIC MEMBRANE: A CLOSER LOOK

Scientists have recently succeeded in reconstructing the receptor protein that is the target of the neurotransmitter acetylcholine. Protruding through postsynaptic membranes, these proteins respond to acetylcholine by opening a water-filled channel through the membrane for Na^+ ions to diffuse. The protein has five membrane-spanning subunits arrayed in a circle that opens to create the water channel. Sidelight figure 34.2 shows a cross section through the receptor protein, illustrating the long, vertical wall made by the subunits as they traverse the membrane.

Sidelight figure 34.2 Receptor protein.
(a) *Cross section of receptor protein channel.* (b) *Electron density map of cross section of channel shown in* (a).

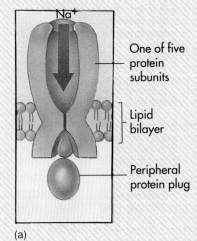

(a)

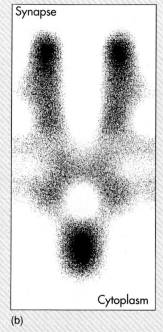

(b)

> *At a neuromuscular junction, acetylcholine released from an axon tip depolarizes the muscle cell membrane, permitting the release of calcium ions into the cytoplasm of the muscle cell, which triggers muscle contraction.*

At the neuromuscular junction, the residual neurotransmitter remaining in the synaptic cleft after the last nerve impulse must be broken down. Otherwise, the postsynaptic membrane simply remains depolarized because acetylcholine cannot easily diffuse away. Leftover acetylcholine is broken into its two constituent parts—acetic acid and choline—by the enzyme **acetylcholinesterase,** which is present in the synaptic cleft. The acetic acid and choline molecules are then transported back to the presynaptic cell so that they can be reused in subsequent transmissions.

Interestingly, acetylcholinesterase is one of the fastest-acting enzymes in the vertebrate body, cleaving one acetylcholine molecule every 40 microseconds. This rapid breakdown permits as many as a thousand nerve impulses per second to be transmitted across the neuromuscular junction. Many organic phosphate compounds, such as the nerve gases tabun and sarin and the agricultural insecticide parathion, are potent inhibitors of acetylcholinesterase. Because they produce continuous neuromuscular transmission, such compounds can be lethal to vertebrates. Breathing, for example, requires muscular contraction, as does blood circulation.

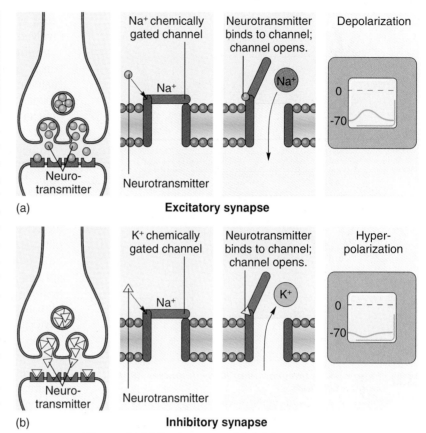

Figure 34.14 Excitatory and inhibitory synapses.
(a) *In excitatory synapses, the binding of the neurotransmitter causes Na$^+$ chemically gated channels to open, and depolarization results.* (b) *In inhibitory synapses, the binding of the neurotransmitter causes K$^+$ chemically gated channels to open, and hyperpolarization results.*

Neural Synapses

When the axon connection is with another neuron, rather than with a muscle, the resulting response can be either depolarizing or **hyperpolarizing,** which is the opposite of depolarizing. In hyperpolarization, the polarization of the neuron is *continued.* The neurons in the eyes that transmit action potentials to the brain are hyperpolarizing.

Neural synapses that utilize depolarization are called **excitatory synapses.** In excitatory synapses, the receptor that binds the neurotransmitter on the postsynaptic membrane is a chemically gated channel. When a neurotransmitter such as acetylcholine binds to the receptor, the channel opens and allows diffusion of all small ions into and out of the postsynaptic membrane according to their concentration gradients. Because there is more Na$^+$ outside the membrane than inside, the result of this binding is primarily the entry of Na$^+$ ions into the postsynaptic membrane. This entry of Na$^+$ ions causes depolarization, which is sometimes called an **excitatory postsynaptic potential** (figure 34.14*a*).

On the other hand, neural synapses that use hyperpolarization are called **inhibitory synapses.** In inhibitory synapses, the receptor on the postsynaptic membrane is a chemically gated potassium (K$^+$) channel (or in some cases, a chloride channel).

The binding of a neurotransmitter (**gamma-aminobutyric acid, or GABA,** is a typical example) causes the channel to open and allows K$^+$ ions to exit out of the postsynaptic membrane. This exit of K$^+$ ions causes the membrane interior to become more negative, which reduces the ability of the postsynaptic membrane to depolarize. This hyperpolarizing response is sometimes called an **inhibitory postsynaptic potential** (figure 34.14*b*).

An individual neuron can possess both kinds of synaptic connections to other neurons. When signals from both excitatory and inhibitory synapses reach the neuron's cell body, the depolarizing and hyperpolarizing effects interact. This integration of effects leads to a summation in which the various excitatory and inhibitory effects taken together either cancel each other out or reinforce each other (figure 34.15).

> *There are two types of neural synapses: excitatory, which results in depolarization; and inhibitory, which results in hyperpolarization. Integration of the effects from excitatory and inhibitory synapses results in a summation in which the effects either reinforce each other or cancel each other out.*

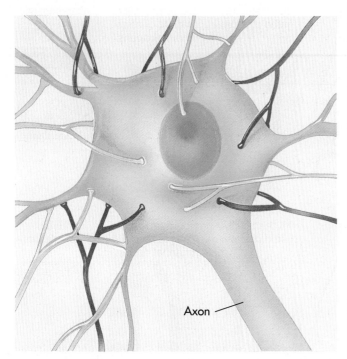

■□ *Figure 34.15 Integration of nerve impulses on the neuron cell body.*
The cell body receives both inhibitory (red) *and excitatory potentials* (blue). *The summed influences of all these impulses determines whether the axonal membrane will be sufficiently depolarized to initiate a propagating action potential.*

Categories of Chemical Neurotransmitters

The effect of a chemical neurotransmitter depends on the nature of the postsynaptic ion channel activated by that neurotransmitter. Therefore, a particular compound may be excitatory or inhibitory in its effects. Neurotransmitters fall into several chemical categories. In most cases, they have been "assigned" a particular role, based on studies showing that drugs known to interfere with or augment the effect of a particular neurotransmitter have consistent behavioral effects.

One category of neurotransmitters is the **monoamines.** Acetylcholine is a monoamine widely used in the brain. A deficiency of acetylcholine has been suggested as a cause of the type of dementia known as **Alzheimer's disease.** Another monoamine—norepinephrine—probably affects arousal and some emotional responses. Other monoamine neurotransmitters are dopamine and serotonin. Dopamine appears to be involved in emotional behavior, with excess dopamine causing schizophrenia. Dopamine inhibitors are used in the treatment of this disease. Dopamine is also important in the motor systems of the body. Dopamine deficiencies cause the uncontrolled shaking movements seen in Parkinson's disease, which can be effectively treated by administration of the precursor to dopamine, L-dopa. Overproduction of dopamine may be the

cause of some depressive illnesses. Serotonin appears to be involved in the regulation of the sleep cycle.

Amino acids are another category of neurotransmitters. Glutamate is probably the main excitatory neurotransmitter in the central nervous system, whereas glycine is one of the major inhibitory neurotransmitters. GABA is also inhibitory, and GABA deficits seem to cause certain types of anxiety. Antianxiety drugs such as diazepam (Valium) increase GABA levels.

A third category of neurotransmitters are **neuropeptides** that act as **neuromodulators.** Learning and memory have begun to be understood in terms of long-term synaptic changes, some of which appear to be mediated by neuromodulators. The effects of neuromodulators on neuronal activity are more subtle than the brief changes in ion permeability that result in postsynaptic potentials. In many cases, these substances seem to cause structural modifications of ion channel proteins, so that their response to a particular neurotransmitter is altered. Many neuromodulators are small proteins called neuropeptides. Examples include vasopressin (antidiuretic hormone), somatostatin, oxytocin, the enkephalins, and the endorphins.

The distinction between neuromodulators and neurotransmitters is not a matter of chemical identity. The difference lies in the time scale and nature of the effect of the substance on information processing. Neurotransmitters mediate effects that are immediate and quickly over. Neuromodulators mediate effects that are slow and longer lasting and that typically involve second messengers within the cell. These messengers modify the behavior of membrane receptors for a neurotransmitter or the operation of voltage-gated ion channels. Neuromodulators are discussed in more detail in the next section.

Most of the many kinds of neurotransmitters fall into one of three chemical categories: (1) monoamines like acetylcholine and dopamine, (2) amino acids like GABA and glutamate, and (3) neuropeptides that act as neuromodulators.

Drugs and the Nervous System

What is an **addiction?** Why is it so difficult for a person addicted to a drug such as cocaine to stop using the drug? What causes addiction, and can addiction be overcome?

Scientists searching for the answers to these questions have discovered that addiction is not only a psychological phenomenon, but a physiological one as well. While many people have the mistaken belief that addicts are weak individuals who lack the willpower to relinquish their addictions, experts now know that the effects of some drugs cause structural changes in certain neurons in the body. These changes are not permanent; in fact, the neurons return to their preaddictive state once the consumption of the drug is stopped. But often, the sudden stoppage of a drug causes unpleasant and sometimes fatal symptoms that result from the neurons readjusting to their normal state. These symptoms, called **withdrawal symptoms,** prevent many addicts from kicking their habits.

An understanding of addiction requires first an understanding of the role of neuromodulators (discussed in the previous section) in the transmission of nerve impulses. To review: Nerve impulses are transmitted from one neuron to another across synaptic clefts—gaps between neurons—through the action of neurotransmitters. When a nerve impulse arrives at the presynaptic membrane, the depolarization causes neurotransmitters to be released from the presynaptic membrane into the synaptic cleft. The neurotransmitters then bind to receptors on the postsynaptic membrane. This binding causes chemically gated channels in the postsynaptic membrane to open, and the flooding of ions into the neuronal membrane initiates the depolarization of the postsynaptic neuron.

Neuromodulators interact with neurotransmitters during nerve impulse transmission. Some neuromodulators slow down the destruction of the neurotransmitter in the synaptic cleft; the result is that the neurotransmitter is present for longer periods of time in the cleft and can continually bind to the receptors. Other neuromodulators aid in the release of the neurotransmitter from the presynaptic membrane, and still others inhibit the reabsorption of the neurotransmitter. Thus, neuromodulators act to keep the neurotransmitter active longer in the synaptic cleft (figure 34.16).

The effects of neuromodulators are important to the understanding of addiction because many addictive drugs simulate the action of neuromodulators in the body. A drug that mimics a neuromodulator increases the amount of neurotransmitter in the synaptic cleft. To compensate for the increased amount of neurotransmitter in the synaptic cleft and thus the repeated binding of the neurotransmitter to receptors in the postsynaptic membrane, *the postsynaptic membrane gradually reduces its number of neurotransmitter receptors*. This gradual reduction accounts for the addiction that many individuals develop for some drugs (figure 34.17). Without the drug, transmission across the synapse now fails. With continued and increasing drug use, the effect becomes even more pronounced, until even high levels of the drug cannot induce transmission. This buildup of tolerance for the drug is the direct result of the decrease in the number of neurotransmitter receptors.

When the addict stops taking the drug for a long period of time, the constant stimulation of the postsynaptic membrane ceases, and the body eventually readjusts its sensitivity to neurotransmitters by *adding* receptors to the postsynaptic membrane until the number again approaches that before the addiction. Thus, it *is* possible to overcome an addiction; however, the addict experiences many unpleasant symptoms while the neurons adjust to the stoppage of the drug and receptors are added.

Many substances cause these changes in neurons. Some of these substances are illegal; others are legal and can be purchased at the nearest grocery store. What follows is a summary of the effects and dangers of some of the most familiar drugs.

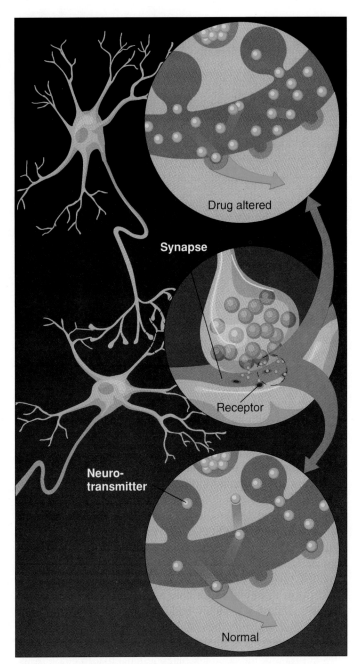

Figure 34.16 How a nerve impulse crosses a synapse.
When a nerve impulse reaches the tip of an axon, the electrical signal passes across the synaptic gap as a chemical signal. Chemical signals called neurotransmitters are released from the axon tip into the synapse. They then pass across the gap and bind to specific receptor proteins on the postsynaptic neuron's membrane, opening ion channels that fire the postsynaptic neuron. Then the neurotransmitters are quickly reabsorbed or broken down.

Addiction to drugs occurs when the body makes adjustments to compensate for the effects of the drug. Often, this involves reducing the number of receptors on postsynaptic membranes of particular central nervous system pathways. After this change, the pathway cannot operate without the drug.

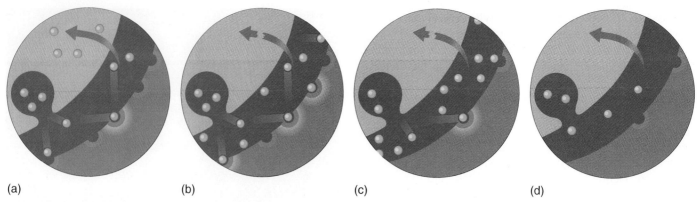

(a)   (b)   (c)   (d)

Figure 34.17 The chemical nature of addiction.

(a) *In a normal synapse, neurotransmitters are rapidly reabsorbed by the presynaptic neuron.* (b) *When a drug blocks removal of a neurotransmitter, receptors across the synapse are flooded with excess neurotransmitters.* (c) *The postsynaptic neuron responds to this overload by lowering the number of receptors in a synapse.* (d) *Now if the drug is withdrawn, allowing the excess neurotransmitters to be removed so that firing of the presynaptic neuron produces only normal levels of neurotransmitter in the synapse, there are too few postsynaptic receptors to achieve a threshold and fire the postsynaptic nerve.*

Cocaine

Cocaine is found in the leaves of the coca plant that grows in the Andes Mountains of South America. The Spanish conquerors who colonized the area noticed that the natives chewed coca leaves and that the coca leaves reduced both fatigue and appetite. For these reasons, the Spanish encouraged the natives to cultivate and use the coca plant.

Cocaine was first isolated from coca leaves in 1844. Hailed as a "wonder drug," cocaine was used in a variety of homeopathic medicines in the United States and even found its way into a new soda fountain drink, Coca Cola (the cocaine was replaced with caffeine in 1903). Cocaine's addictive effects were noted with alarm in the late nineteenth century, and by 1914, the U.S. government had declared cocaine an illegal substance.

In the 1970s, however, recreational use of cocaine became popular. Users either sniff a powdery form of cocaine or inject a treated form of cocaine, called freebase, directly into the bloodstream. A new form of cocaine, called crack, surfaced in the 1980s. Crack is a highly addictive form of cocaine that is smoked. Because the cocaine is absorbed by the lungs, this form of cocaine acts quickly, is relatively inexpensive, and does not require the complex paraphernalia that freebase does. Crack use has become an epidemic in some cities and has contributed to a growing crime rate and gang warfare.

Cocaine is a stimulant that acts on the brain's limbic system, the body's "pleasure center." Cocaine mimics a neuromodulator that blocks the reabsorption of dopamine by neurons in the limbic system. The result of this blockage is that dopamine stays in the synaptic clefts of these neurons and continually binds to the receptors in the postsynaptic membrane, causing intense pleasure, increased energy, and feelings of power (figure 34.18). The neurons respond to this continual stimulation by reducing the number of dopamine receptors in the postsynaptic membranes, and thus more and more of the drug is needed for the addict to experience the pleasurable effects that the dopamine binding elicits.

As the addiction builds, cocaine addicts find that their pleasure centers cannot function at all without the stimulation of

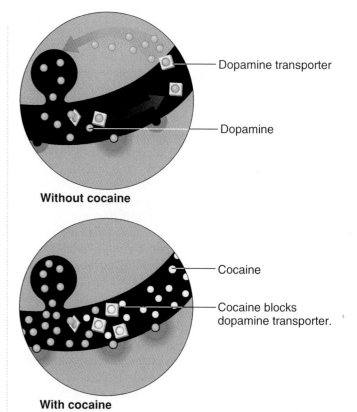

Without cocaine

With cocaine

Dopamine transporter

Dopamine

Cocaine

Cocaine blocks dopamine transporter.

Figure 34.18 Cocaine blocks neurotransmitter reabsorption.

Cocaine binds to the dopamine transporter protein, blocking the dopamine building site. Without the ability to bind the transporter, dopamine levels build up in the synapse, firing the postsynaptic membrane more often.

the drug. The "crash" that is experienced as the drug's effects wear off becomes more pronounced, and the addict begins to suffer from severe mood swings, from deep depression during the crash to euphoria when the drug is once again introduced into the body. Cocaine is an extremely difficult addiction to overcome, but with appropriate counseling, it can be done.

Cocaine acts as a neuromodulator to block reabsorption of dopamine in the synapses of pleasure pathways in the brain. The body eventually responds to the increased firing of these synapses by "turning down the volume"—reducing the number of dopamine receptors in the affected synapses.

Narcotics

Narcotics are pain-killing drugs and include heroin, morphine, and methadone. All of these drugs are derived from opium, a gummy, resinlike substance found in the unripe pods of the poppy plant. Ancient Egyptians understood the effects of this drug and used it to kill pain, and eventually, to give pleasure and to lessen anxiety.

In the United States, a derivative of opium called morphine was discovered around the year 1800. Morphine was used widely by the medical profession, and like cocaine, found its way into a variety of over-the-counter medications. Once its addictive qualities were documented during the 1870s, however, its used was banned in the legal market.

In 1898, however, chemists came up with heroin, a new form of opium that its inventors claimed was nonaddictive. Like morphine, heroin was added to many nonprescription medications, even cough medicine. In the early twentieth century, heroin, too, became illegal, as lawmakers and physicians discovered its addictive potential. Heroin use declined until the 1960s, when its illegal use became rampant. Part of the heroin problem during this decade was a result of Vietnam veterans returning from the war. Since heroin production is a cottage industry in Asia, heroin was plentiful during the war, and many soldiers became addicted. In the 1970s, heroin use again declined, but recently, there has been another upsurge in its abuse.

Narcotics such as heroin and morphine act in a different way than cocaine and other addictive drugs. Instead of decreasing the number of neurotransmitter receptors in the postsynaptic membranes of neurons, narcotics *increase* the number of receptors. Narcotics mimic a neuromodulator in the body's pain response. When a tissue in the body is damaged, the nerve endings in the damaged tissue release neuromodulators, such as prostaglandins. These neuromodulators assist the nerve impulses in traveling along neurons in the spinal cord to the brain. The neurotransmitter used in this transmission is called **substance P.** Once the brain receives the pain message, it automatically acts to turn off the pain response by signaling the neurons leading to the spine to release substances called **enkephalins.** Enkephalins are neuromodulators that shut off the pain signals being sent to the brain. Thus, enkephalins are chemicals that act to inhibit the pain response. Narcotics mimic the chemical structure of enkephalins. They bind to the same receptors used by enkephalins and therefore block pain signals from reaching the brain. With continued narcotic use, the neurons in the pain pathway increase the number of enkephalin receptors. The neurons, in effect, become desensitized to pain signals.

Narcotic addiction has extremely unpleasant withdrawal effects. When the drug is withdrawn from the body, the pain pathway neurons become extremely sensitive. Because the number of enkephalin receptors has been increased, more receptors are left unbound by enkephalins, and pain impulses are not blocked. Thus, the withdrawing addict experiences *more* pain than normal until the number of receptors reaches its preaddiction level.

Many narcotic addicts who wish to overcome their addiction resort to methadone treatments. Methadone is a narcotic that also mimics enkephalin; it is given orally and so is absorbed across the small intestine. This mode of absorption is slower and does not cause the intense, pleasurable "rush" of heroin. During the course of methadone treatment, the dose is *gradually* decreased to lessen withdrawal symptoms. However, as with all drugs, the addict eventually reaches a point at which the drug must be stopped altogether if recovery is to be complete.

Heroin and other narcotics mimic neuromodulators called enkephalins that normally function in the central nervous system's pain detection pathway to dampen pain signals.

LSD and Marijuana

LSD, or lysergic acid diethylamide, is derived from a fungus that grows on rye plants. LSD is a **hallucinogen;** that is, it produces euphoria and hallucinations. LSD acts on the neurons of the brain's cerebral cortex, the body's "thinking center," by inhibiting the action of the neurotransmitter serotonin. In the body, serotonin inhibits the brain's tendency to free-associate; in effect, serotonin keeps the brain "on the right track." When serotonin is prevented from being released in the synaptic clefts, it cannot bind to receptors and produce its inhibitory effects. LSD's hallucinatory effects are a direct result of its action on these inhibitor neurons.

Marijuana is a drug that is derived from the leaves of the plant *Cannabis sativa*. The leaves are dried and then smoked. The active ingredient in marijuana is tetrahydrocannabinol, or THC. THC has a variety of effects, some of which are similar to the effects of alcohol and some of which are similar to the serotonin-inhibiting effects of LSD. Researchers have also found specific receptors in the brain that THC might bind to. Marijuana has sometimes been called a nonaddictive drug, but researchers studying the drug are beginning to dispute this claim.

Barbiturates and Tranquilizers

Barbiturates, like Seconal and Nembutal, are prescription drugs for reducing anxiety and inducing sleep. They act on the brain's cerebral cortex and can be highly addictive.

Tranquilizers, such as Valium and Librium, do not induce sleep, but like barbiturates, they are highly addictive. Alcohol increases the effects of both these drugs. The combination of these drugs and alcohol, however, can depress the body's respiratory system, leading to coma and sometimes death.

Amphetamines

Also known as "speed" or "uppers," amphetamines are stimulants. Like cocaine and opium, these drugs were first considered panaceas for a wide variety of ailments, including asthma, narcolepsy, and even weight control. The legal and illegal production of amphetamines increased, and soon people who needed to stay awake for long periods of time, such as shift workers, were using amphetamines. The long-term use of amphetamines, however, causes bizarre, often violent behavior, and for this reason, the U.S. government enacted laws that moderate the prescription of these drugs. For example, amphetamines can no longer be prescribed solely for weight loss or sleep prevention.

Amphetamines stimulate the body's fight-or-flight response, which readies the body for battle. Appetite is decreased, and large amounts of glucose flood the system, supplying increased energy. Amphetamines tap this response because they resemble a group of neuromodulators, discussed earlier in the chapter, called monoamines. Monoamines mediate the body's fight-or-flight response. Addiction to amphetamines lowers the number of monoamine receptors in the body, and thus, more of the drug is needed to experience the drug's effects. Withdrawal from amphetamines involves the readjustment of these receptors to their normal levels, and therefore, withdrawal symptoms include fatigue and anxiety.

Alcohol

Ethyl alcohol is a product of the fermentation of sugars in certain plants, such as barley, corn, and grapes. Ancient cultures probably made alcoholic drinks for practical reasons: The juice made from grapes soon spoiled, but when this juice was fermented, the alcohol in the juice prevented spoilage. Alcoholic beverages were probably safer than the water that these cultures drank; as a result, those who drank alcoholic drinks were sick less often than those who drank water. Alcoholic beverages were imbibed at all hours of the day: for breakfast, lunch, and dinner. By the sixteenth century, water purification had become more developed, and alcoholic drinks became less a matter of survival and more of a popular social pastime. Today, alcohol is a means of unwinding with friends or celebrating special occasions.

Alcohol does not directly affect the receptors in neurons; instead, it initiates changes in the neuron's plasma membrane that are similar to the changes caused by general anesthetics. The result of this alteration is a general inhibition of all nerve signals, particularly in those neurons that inhibit the limbic system. The limbic system contains both a "pleasure center" and a "punishment center"; some experts believe that alcohol knocks out the neurons that stimulate the punishment center, leading to the loss of inhibition associated with the effects of alcohol.

Prolonged use of alcohol can have destructive effects on the body. For example, alcohol does not cross the small intestine into the bloodstream; it enters the bloodstream by crossing the stomach walls. Ethyl alcohol irritates the stomach wall and can erode the protective mucus lining the stomach, leading to ulcers. Alcohol can damage the liver as well: The blood is filtered

Figure 34.19 Fetal alcohol syndrome.
Even small amounts of alcohol early in pregnancy can have devastating effects on a fetus. Because many women do not know they are pregnant for several months, it is important that sexually active women moderate their alcohol use.

through the liver to remove impurities; if large amounts of alcohol are present in the liver, the liver must work overtime to remove the alcohol from the bloodstream. An overworked liver becomes swollen and can become riddled with scar tissue, a condition called **cirrhosis.** Alcohol can have particularly damaging effects on a growing fetus. If a pregnant woman drinks alcohol, the alcohol can cross the placenta and cause a variety of detrimental effects, such as **fetal alcohol syndrome,** which is characterized by low birth weight, learning disabilities, and poor impulse control (figure 34.19).

Alcohol addiction can be overcome, but without careful monitoring, severe alcoholics can die from the effects of withdrawal. Known as **delirium tremens,** these withdrawal symptoms can cause hallucinations, respiratory distress, and paralysis. However, with medical attention and appropriate counseling, alcoholics can beat their addiction.

Alcohol affects the nervous system by altering the properties of neuronal membranes, producing a general inhibition of all nerve signals.

Nicotine

Nicotine is the active ingredient found in cigarettes and chewing tobacco. It is highly addictive, and the U.S. surgeon general has likened nicotine addiction to that of cocaine. Nicotine acts as a neuromodulator and binds to a variety of receptors in the

Figure 34.20 Smoking, the most common addiction.
Cigarette smoke contains high levels of nicotine, a highly addictive drug. When evidence surfaced that cigarette manufacturers manipulate tobacco to maintain high nicotine levels, Congress held public hearings on whether cigarettes should be a controlled substance like many other addictive drugs.

brain, including receptors meant for acetylcholine. This binding causes a transient, pleasant effect and a lessening of anxiety. Some nicotine users claim that, when they are agitated, smoking helps to calm them down, and that, when they are depressed, smoking energizes them and allows them to think more clearly. These effects are short-lived, however; to keep

these effects going, the smoker or chewer must keep blood nicotine levels steady by smoking or chewing at regular intervals during the day (figure 34.20).

Because nicotine addiction lowers the number of neurotransmitter receptors in certain neurons, stopping nicotine intake results in withdrawal symptoms, such as depression, anxiety, and an intensive nicotine craving. Quitting, however, also causes some positive side effects: Since smoke and "chew" deadens taste buds, quitting reactivates the sense of taste, and many ex-smokers and chewers are able to taste food for the first time in years. Their breath smells better, and their respiratory tracts quickly clear, making breathing easier. After about ten years, the risk of cancer for those who have quit smoking is about the same as for those who have never smoked (see chapter 13 for a complete discussion of the link between cancer and smoking).

Many programs designed to help people kick the nicotine habit are available. The best programs are those that encourage the complete stoppage of nicotine with support from others who are attempting to quit. The new nicotine patch also offers hope to quitters by providing a gradually reduced dose of nicotine over a period of several weeks. As with methadone treatments, however, at some point, the quitter must be weaned completely from the drug to restore the normal number of neuron receptors. Quitting a nicotine addiction is difficult, for nicotine is among the most addictive of drugs, but every year thousands succeed.

Nicotine is a highly addictive neuromodulator that acts at many levels in the brain. Nicotine addiction, like cocaine addiction, results from changes in the number of receptors in affected synapses. The hard-to-break cigarette habit, still common among Americans, is a direct consequence of nicotine addiction.

EVOLUTIONARY VIEWPOINT

Nerves are by no means limited to vertebrates. All animals except sponges have some form of nervous system. The squid, a kind of mollusk, has a giant axon used in its escape response that has one of the most rapid speeds of transmission of any neuron, due to its large diameter. Often, nerves are hooked up in simple monosynaptic reflex arcs that permit very speedy responses. That is why swatting an insect with your hand is so difficult: The approaching air wave deflects sensory hairs that fire the fly's flight reflex before your hand can reach the fly.

SUMMARY

1. A neuron is a cell specialized for the transmission of nerve impulses. A neuron consists of dendrites, which receive nerve impulses, a cell body, and an axon, which transmits nerve impulses.

2. A neuron is at rest when the neuron's interior is more negatively charged than its exterior; a condition called polarization. This resting potential is created by the active pumping of the sodium-potassium pump of sodium (Na^+) ions across the membrane out of the neuron and the pumping of potassium (K^+) ions across the membrane into the neuron. In addition, numerous K^+ ion channels located in the neuronal membrane allow K^+ ions to diffuse out of the neuron. A neuronal membrane has many K^+ ion channels but very few open Na^+ ion channels. This difference in number is significant to the maintenance of the charge separation of a neuron.

3. An action potential, or nerve impulse, is initiated by a threshold stimulus, which means that the stimulus must be large enough to cause the voltage-gated Na^+ channels to open. However, all action potentials are the same "strength." This is called the all-or-nothing response.

4. An action potential causes the depolarization of the neuronal membrane. Depolarization causes the opening of voltage-gated Na^+ channels, allowing a flood of Na^+ ions to enter the neuron. Polarization is almost instantaneously restored by the opening of voltage-gated K^+ channels, which allow K^+ ions to leave the neuron.

5. An action potential is propagated along a neuron as one area of depolarization "ignites" nearby voltage-gated Na^+ channels to open. On some vertebrate neurons, however, impulses travel via saltatory conduction, rapidly leaping from node to node over insulated portions.

6. Action potentials are transmitted to receiving neurons or target tissue across synaptic clefts. The membrane of the transmitting neuron is the presynaptic membrane; the membrane of the receiving neuron or tissue is the postsynaptic membrane. Action potentials can be transmitted chemically or electrically. Chemical transmission involves chemically gated ion channels in the postsynaptic membrane that open as a result of the binding of neurotransmitters to receptors in the postsynaptic membrane.

7. A neuromuscular junction is a synaptic cleft between a neuron and a muscle cell. The neurotransmitter acetylcholine passes from the presynaptic membrane into the cleft and binds to receptors on the postsynaptic membrane. This binding opens Na^+ ion channels in the postsynaptic cleft, allowing Na^+ ions to enter the muscle cell. This entry causes a depolarization, which, in turn, stimulates the release of calcium into the cytoplasm of the muscle cell. This calcium release triggers muscle contraction.

8. A neural synapse can be excitatory or inhibitory. An excitatory synapse causes a depolarization of the receiving neuron. An inhibitory synapse causes a hyperpolarization (a continuation of the polarization) of the receiving neuron.

9. The cell body of a neuron integrates both incoming excitatory and inhibitory nerve impulses. This integration leads to a summation of the effects of these impulses in which the effects are either reinforced or canceled out. The final outcome is dependent on the mix of signals received.

10. Three categories of chemical neurotransmitters are monoamines, amino acids, and neuropeptides that act as neuromodulators.

11. Addiction to drugs is a consequence of adjustments the central nervous system makes to compensate for inappropriate levels of activity induced by drugs. Often, these adjustments involve changes in the number of receptors in postsynaptic membranes, as evidenced by the chapter discussions of cocaine, amphetamines, and nicotine.

REVIEWING THE CHAPTER

1. What are the types, structure, and function of neurons?

2. How are nerve impulses transmitted?

3. How does a neuron maintain its resting potential?

4. How is a nerve impulse (action potential) initiated?

5. How is the action potential transmitted along a neuron?

6. What is a saltatory conduction, and how does it work?

7. How is the action potential transmitted to target cells?

8. How do neuromuscular junctions work?

9. How do the different neural synapses work?

10. What are three different categories of chemical transmitters?

11. How do drugs affect the nervous system?

12. What is the origin of cocaine, and what are the effects of using it?

13. What are narcotics, and what are the effects if addiction occurs?

14. How do LSD and marijuana affect the nervous system?

15. What are the effects of using barbiturates, tranquilizers, or amphetamines?

16. What effects does alcohol have on the nervous system?

17. How does nicotine affect the nervous system?

COMPLETING YOUR UNDERSTANDING

1. Why are nerve impulses not true electrical currents?

2. Neurons receive nutritional support from
   a. Schwann cells.
   b. nodes of Ranvier.
   c. myelinated fibers.
   d. axons.
   e. neuroglial cells.
   f. dendrites.

3. What causes multiple sclerosis, and can anything be done to assist patients with this disease?

4. How is the term *voltage* used in nerve impulse transmissions?

5. Why is a neuron's exterior more positively charged than its interior?

6. Why does a small stimulus not produce a small action potential and a large one not produce a large action potential?

7. After a threshold stimulus, how is a neuron restored to its resting potential?

8. Why does the refractory period occur?

9. Why is saltatory conduction "inexpensive" metabolically?

10. The synaptic cleft, an intercellular gap separating the axon tip and the target neuron, is _____ across.
   a. 10–20 nanometers
   b. 1–10 millimeters
   c. 10–20 micrometers
   d. 1–10 picometers
   e. 1 meter
   f. 1 decimeter

11. Why is there such a great diversity in the response of postsynaptic cells?

12. Why is it necessary to break down the residual neurotransmitter remaining in the synaptic cleft after the last nerve impulse?

13. What are polarization, hyperpolarization, and depolarization, and when do they occur?

14. What is GABA, and how does it function?

15. What has been suggested as a cause for Alzheimer's disease?

16. What are some of the emotional behaviors caused by an excess or deficiency in the neurotransmitter dopamine?

17. What distinguishes neuromodulators from neurotransmitters?

18. Many addictive drugs simulate the action of _____ in the human body.

19. In a drug addict, what are the causes of withdrawal symptoms?

20. How does an individual build up a tolerance for an addictive drug?

21. What is the difference between freebase and crack, two forms of cocaine?

22. Why does a withdrawing narcotic addict experience more pain than normal?

23. What is the origin of LSD and marijuana?

24. What are the consequences of consuming alcohol when also taking barbiturates and tranquilizers?

25. What is "speed" or "uppers," and how do people obtain these drugs?

26. In past cultures, why were alcoholic beverages probably consumed at all hours of the day?

27. Why has the U.S. surgeon general compared the addiction of nicotine to that of cocaine?

28. Local anesthetics used by dentists to deaden pain are blockers of Na^+ channels. Why?

FOR FURTHER READING

Dunant, Y., and M. Israel. 1985. The release of acetylcholine. *Scientific American*, April, 58–66. A challenge to the accepted theory that acetylcholine is emitted by synaptic vesicles.

Hoffman, M. 1991. A new role for gases—neurotransmission. *Science* 252 (June): 1788–89. How the remarkable discovery that nitric oxide carries nerve impulses is revolutionizing ideas of nerve transmission.

Holloway, M. 1991. Rx for addiction. *Scientific American*, March, 94–104. An up-to-date look at the molecular mechanisms underlying drug addiction.

Katz, B. 1966. *Nerve, muscle, and synapse*. New York: McGraw-Hill. A classic description of how a nerve impulse arises and is propagated. Concise and unusually lucid.

Kuffler, S. W., and J. G. Nicholls. 1984. *From neuron to brain: A cellular approach to the function of the nervous system*. 2d ed. Sunderland, Mass.: Sinauer Associates. A superb overview of the mechanisms of nerve excitation and transmission.

Marx, J. 1990. Marijuana receptor gene cloned. *Science* 249 (August): 624–26. The isolation of the gene encoding the marijuana receptor that may lead to a better understanding of how the brain controls pain.

Melzack, R. 1990. The tragedy of needless pain. *Scientific American*, Feb., 27–34. An account of why morphine, when administered to control pain, may not be addictive.

Musto, D. 1991. Opium, cocaine, and marijuana in American history. *Scientific American*, July, 40–47. A brief history of drug use by the general public over the last two hundred years.

Neher, E., and B. Sakmann. 1992. The patch clamp technique. *Scientific American*, March, 44–51. Nobel-Prize-winning work (1991), in which a tiny section of membrane is isolated to manipulate the ion channels.

Stevens, C. F. 1979. The neuron. *Scientific American*, Sept., 54–65. A description of the structure and functioning of a typical nerve cell.

Veca, A., and J. Dreisbach. 1988. Classical neurotransmitters and their significance within the nervous system. *Journal of Chemical Education* 65: 108–20. Written for the non-neuroscientist, provides an up-to-date survey of the role of acetylcholine, norepinephrine, and amino acid neurotransmitters.

EXPLORATIONS

Interactive Software:
Nerve Conduction

This interactive exercise allows students to explore how voltage-gated channels cause a nerve impulse to pass down a motor axon by enabling them to alter the architecture of the neuron. The exercise presents a diagram of a motor neuron axon, showing the series of voltage-gated channels in the membrane. Students can investigate the consequences of extending or reducing the zones covered by the myelin sheath, measuring the speed of conduction along the axon. By altering the diameter of the axon, students can explore the surprisingly great influence of axon diameter upon the speed with which the impulse travels down the axon.

Questions to Explore:

1. If you were to stimulate the tip of an axon, would a nerve impulse move back toward the cell body?

2. Could you generate an action potential by opening voltage-gated potassium channels?

3. Can you increase the speed a nerve impulse travels by increasing the magnitude of the action potential?

4. Can you increase the speed of conduction by shortening the intervals between nodes of Ranvier?

Interactive Software:
Synaptic Transmission

This interactive exercise allows students to explore the ways neurons employ chemicals to pass nerve impulses from one cell to another by enabling them to vary the nature of the chemicals. The exercise presents a diagram of a junction between a neuron and a target cell, containing a variety of chemically gated and voltage-gated channels. Students can investigate the difference between excitatory and inhibitory synapses by varying the nature of the chemical released into the synapse and seeing what kind of channels are opened in response. Students will also observe the consequences of the channel opening to transmission of the nerve impulse.

Questions to Explore:

1. What happens if a neuron releases both stimulatory and inhibitory neurotransmitters into a synapse simultaneously?

2. Does increasing the amount of neurotransmitter alter the speed of transmission across the synapse?

3. Can a neurotransmitter be excitatory in one synapse and inhibitory in another?

4. What is the advantage of a synapse over a direct physical connection between two nerves?

5. Why can't the nerve impulse leap across the synaptic cleft electrically, like it does when jumping from one node of Ranvier to another in saltatory conduction?

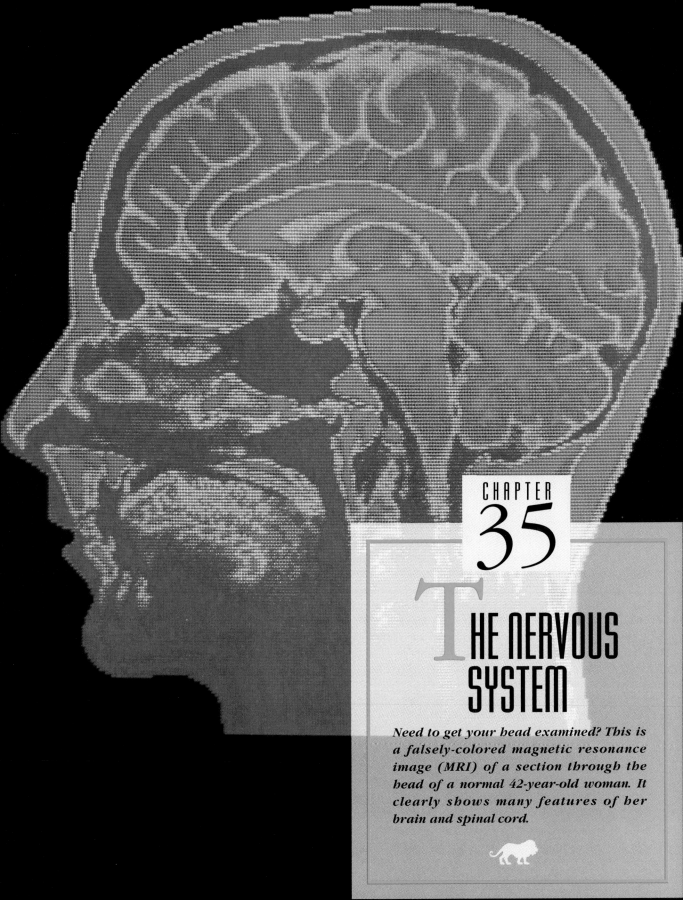

35

THE NERVOUS SYSTEM

Need to get your head examined? This is a falsely-colored magnetic resonance image (MRI) of a section through the head of a normal 42-year-old woman. It clearly shows many features of her brain and spinal cord.

FOR REVIEW

Here are some important terms and concepts that have been discussed in previous chapters and that you will encounter again in this chapter. Review them before proceeding if necessary.

Neuron (*chapters 33 and 34*)
Depolarization (*chapters 33 and 34*)
Muscle contraction (*chapter 34*)
Myelination (*chapter 34*)
Chemically gated ion channels (*chapter 34*)

Nerves extend throughout the vertebrate body, a highway system for information. This network of neurons—the nervous system—provides a highly organized communication network that allows the body to gather information about the body's condition and the external environment, to process and integrate that information, and to issue commands to the body's muscles and glands. The basic architecture of the nervous system is similar throughout the animal kingdom.

Organization of the Vertebrate Nervous System

All vertebrate nervous systems can be said to have one underlying mechanism and three basic elements. The underlying mechanism is the nerve impulse, or action potential. The three basic elements are: (1) a central processing region, or **brain;** (2) nerves that bring information to the brain; and (3) nerves that transmit commands from the brain. The vertebrate nervous system consists of two distinct functional components: the central nervous system and the peripheral nervous system (figure 35.1).

The Central Nervous System

The **central nervous system** is composed of the brain and the **spinal cord** and is the site where information relayed from one set of nerves scattered throughout the body is processed. Using this information, the central nervous system then issues commands through a second set of nerves that make muscles contract or glands secrete. (These two separate sets of nerves are described in the next section.) The central nervous system, then, functions as both a processing center and a command center: All action potentials generated by sensory receptors are processed and interpreted in the central nervous system; then the central nervous system sends commands to the muscles and glands of the body that are based on the information relayed by the action potentials.

The central nervous system relies on the information detected by the body's sensory receptors. The many types of receptors can be divided into two primary groups: those that sense internal stimuli and those that detect external stimuli.

This chapter examines both types of receptors and how the information is relayed to the central nervous system.

Within the central nervous system are individual neurons, axons, and long dendrites bundled together like the strands of a telephone cable. These bundles of nerve fibers are called **tracts.** The cell bodies from which these tracts extend are often clustered into groups, which in the central nervous system are called **nuclei.** The spinal cord is made up of many such tracts, and the brain contains many nuclei with characteristic functions.

The Peripheral Nervous System

Working in concert with the central nervous system is the **peripheral nervous system** (see figure 35.1*b*). The peripheral nervous system is so named because it consists of all nervous structures that are "outside" of, or peripheral to, the central nervous system. The peripheral nervous system is composed of all the nerve pathways of the body. These pathways are commonly divided into two groups: **sensory,** or **afferent, pathways,** which transmit information from the body to the central nervous system; and **motor,** or **efferent, pathways,** which transmit commands to the body from the central nervous system. The motor pathways can be further subdivided into the **voluntary,** or **somatic, nervous system,** which relays commands to skeletal muscles; and the **involuntary,** or **autonomic** (from the Greek words *auto,* meaning "self," and *nomos,* meaning "law"; therefore, "self-controlling") **nervous system,** whose nerves stimulate secretion from glands and excite or inhibit the smooth muscles of the body. The voluntary nervous system can be controlled by conscious thought; we can, for instance, command our hands to move when we are fixing dinner. The involuntary nervous system, in contrast, cannot be controlled by conscious thought. We cannot, for example, tell the smooth muscles in our digestive tract to speed up their action.

Within the involuntary peripheral nervous system are two types of nerves: the **sympathetic** and **parasympathetic** nerves. Both types perform different functions, as described late in the chapter.

In the peripheral nervous system, nerve fibers are bundled to form cables called **nerves.** The cell bodies from which the nerves extend are called **ganglia.**

Along with the voluntary and involuntary nervous systems is a **neuroendocrine system,** which consists of a network of **endocrine glands** that secrete **hormones.** Hormones are chemical messengers that are sent through the body's bloodstream from the endocrine glands and that act on specific organs and perform a variety of functions. The neuroendocrine system is, along with the nerve pathways, a means by which the central nervous system exerts control over body functions. The neuroendocrine system's response to commands from the central nervous system is much slower than that of the nerve pathways, but it is more long-lasting. The neuroendocrine system is discussed in more detail in chapter 37.

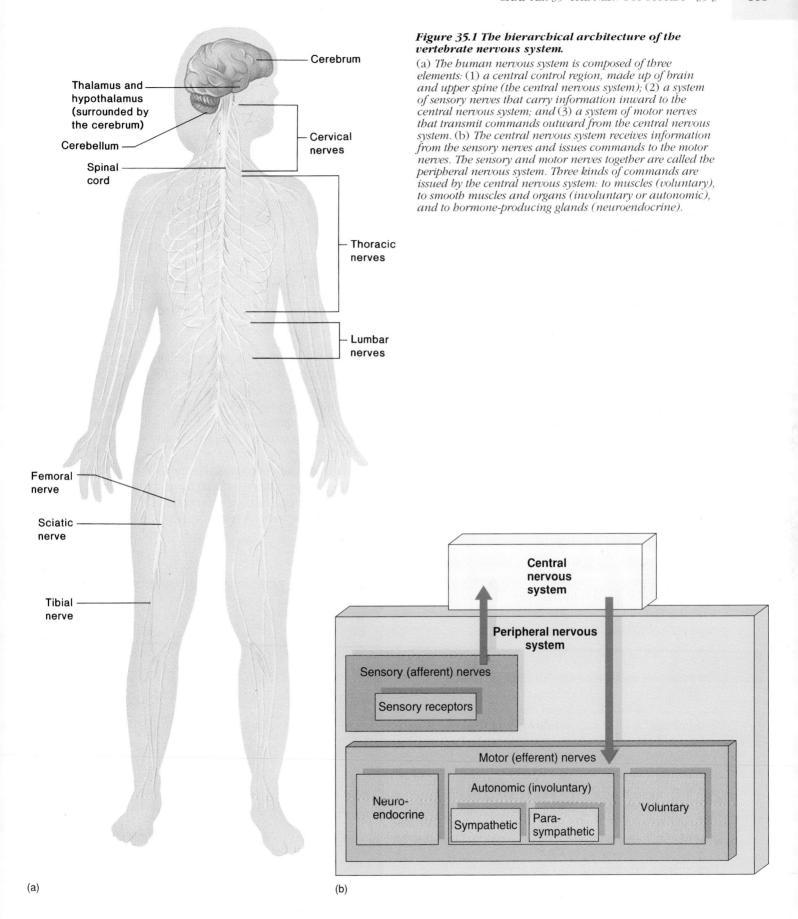

Figure 35.1 The hierarchical architecture of the vertebrate nervous system.
(a) *The human nervous system is composed of three elements: (1) a central control region, made up of brain and upper spine (the central nervous system); (2) a system of sensory nerves that carry information inward to the central nervous system; and (3) a system of motor nerves that transmit commands outward from the central nervous system. (b) The central nervous system receives information from the sensory nerves and issues commands to the motor nerves. The sensory and motor nerves together are called the peripheral nervous system. Three kinds of commands are issued by the central nervous system: to muscles (voluntary), to smooth muscles and organs (involuntary or autonomic), and to hormone-producing glands (neuroendocrine).*

(a)

(b)

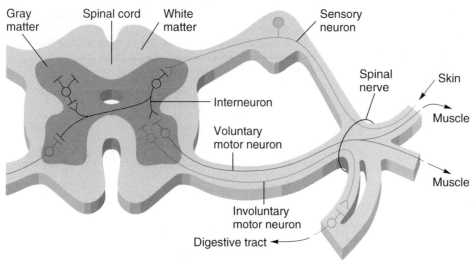

Figure 35.2 Structure of a spinal nerve.
The interior of a spinal nerve is composed of gray matter, which is surrounded by white matter. Sensory neurons enter the spinal nerve and synapse with interneurons. These interneurons, in turn, synapse with involuntary and voluntary motor neurons. A nerve impulse traveling along the incoming sensory nerve can either be shunted to the brain for further processing, or it can travel out again to the motor nerves that command skeletal muscles (voluntary) or smooth muscles (involuntary).

The vertebrate nervous system consists of a brain, spinal cord, and a large number of nerves that communicate with the brain. Nerves either bring information to the brain in the form of senses or relay information from the brain, which results in movement or glandular secretion.

Structure and Function of the Central Nervous System

As noted earlier, the vertebrate central nervous system is composed of the spinal cord and brain. These two structures work together to process information and to issue commands to all parts of the body.

The Spinal Cord

The vertebrate spinal cord extends from the brain and is encased inside vertebrae, special bones that have a hole in the middle, through which the spinal cord runs. The spinal cord is hollow and is filled with **cerebrospinal fluid,** a fluid that bathes and protects the brain against shock. Because this fluid circulates around the brain, doctors can test for a brain infection by "tapping" the spinal cord for a small amount of cerebrospinal fluid, which can then be analyzed.

The spinal cord is made of many **spinal nerves;** a group of spinal nerves that cluster in particular regions is called a **plexus.** An individual spinal nerve is composed of a central region containing neuron cell bodies that are not myelinated (see chapter 34). Because they are unmyelinated, these cell bodies are called the **gray matter.** The gray matter is surrounded by an area called the **white matter,** which consists of two types of myelinated neurons: sensory and motor. Within the spinal cord, the sensory neurons are organized

into the **ascending tracts,** and the motor neurons are organized into the **descending tracts.**

The spinal cord has two functions: First, the spinal cord acts as a relay center for the information sent to the brain and for the commands that the brain issues. Second, the spinal cord can itself issue commands to motor neurons. In fact, the spinal cord controls most of the muscles in the body.

Figure 35.2 shows that the gray matter, which is centrally located in the spinal nerve, contains neurons called **interneurons.** An interneuron is any neuron that receives or sends information. In the spinal nerves, interneurons receive information from the body through sensory neurons and can either (1) pass this information along to the brain for further processing, or (2) send commands along motor neurons that control muscles. These muscles can be under voluntary control, as in the muscles of the legs (figure 35.3). Or they can be under involuntary control, as in the muscles of the digestive tract.

Later in the chapter, the discussion of the peripheral nervous system will show how the spinal cord controls these muscles through the motor nerves. For now, it is only necessary to know that the spinal cord plays a dual function in the central nervous system: relay of information to the brain and direct control of the body's musculature.

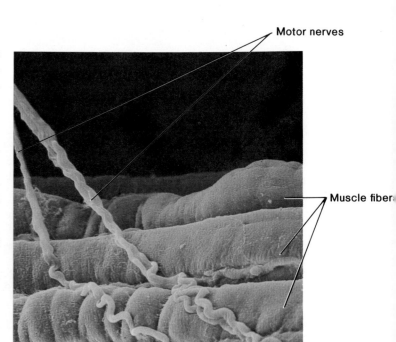

Figure 35.3 Motor nerves attached to muscle.
The slender, twisted threads are motor nerves, and the long, thick strands at the bottom are muscle fibers. Often, a nerve will carry many independent fibers that branch off to establish contact with different target muscles.

The spinal cord is made up of spinal nerves. Each spinal nerve consists of centrally located gray matter, which integrates information sent from sensory neurons. The gray matter either sends the information to the brain for further processing or issues commands through motor neurons to the muscles of the body.

Evolution of the Vertebrate Brain

The structure and function of the vertebrate brain has long been a subject of scientific inquiry. Egyptian hieroglyphics sketch the rough contours of the human brain, and even now, scientists continue to probe its mysteries. Despite ongoing research, scientists are still not sure how the brain performs many of its functions. For instance, scientists continue to look for the mechanism by which the brain stores memories and do not understand how some memories can be seemingly "locked away," only to surface during times of stress. The brain is the most complex vertebrate organ ever to evolve and can perform a bewildering variety of complex functions.

The earliest vertebrates had far more complex brains than their ancestors. Casts of the interior braincases of fossil agnathans (fish that swam five hundred million years ago) show that, although very small, these brains already had the three principal divisions that characterize the brains of all contemporary vertebrates: (1) the **hindbrain,** or **rhombencephalon;** (2) the **midbrain,** or **mesencephalon;** and (3) the **forebrain,** or **prosencephalon** (figure 35.4).

The Hindbrain

The hindbrain is the principal component of these early brains, as it still is in fishes today. Composed of the **pons,** the **medulla oblongata,** and the **cerebellum,** the hindbrain may be considered an extension of the spinal cord devoted primarily to coordinating certain motor responses, called **reflexes.** Tracts run up and down the spinal cord to the hindbrain. The hindbrain, in turn, integrates the sensory signals that come in from the body and determines the pattern of motor response.

Much of this coordination is carried out by the cerebellum ("little cerebrum"), a small extension of the hindbrain. In advanced vertebrates, the cerebellum plays an increasingly important role as a coordinating center and is correspondingly larger than it is in the fishes. In all vertebrates, the cerebellum processes data on the current position and movement of each limb, the state of relaxation or contraction of the muscles affecting that limb, and the general position of the body and its relation to the outside world. These data are gathered in the cerebellum and synthesized, and the resulting orders are issued to motor pathways.

The Midbrain and Forebrain

In fishes, the remainder of the brain is devoted to the reception and processing of sensory information. The second major division of the fish brain, the midbrain, is composed primarily of the **optic lobes,** which receive and process visual information. The third major division, the **forebrain,** processes olfactory (smell) information. Together, the midbrain and the pons and medulla oblongata of the hindbrain make up the **brain stem** of the vertebrate brain.

In early vertebrates, the principal brain component was the hindbrain, devoted largely to coordinating motor reflexes.

Advent of a Dominant Forebrain

Starting with the amphibians and exhibited much more prominently in the reptiles is a pattern that became a dominant evolutionary trend in the development of the vertebrate brain: The processing of sensory information became increasingly centered in the forebrain (figure 35.5).

The forebrain in reptiles, amphibians, birds, and mammals is composed of two elements with distinct functions. The **diencephalon** (from the Greek word *dia,* meaning "between") mainly consists of the **thalamus** and **hypothalamus,** important centers for sensory information processing and internal homeostasis. The **telencephalon,** or "end-brain" (from the Greek word *telos,* meaning "end"), includes the **cerebrum** and all those higher functions associated with it.

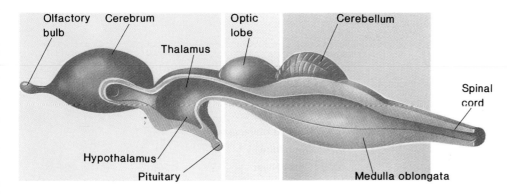

Forebrain (prosencephalon) **Midbrain (mesencephalon)** **Hindbrain (rhombencephalon)**

Figure 35.4 Basic organization of the vertebrate brain seen in the brains of primitive fishes.

Primitive fish brains are divided into the same regions seen in differing proportions in all vertebrate brains: the hindbrain, which is the largest portion of the brain in fishes; the midbrain, which in fishes is a small zone devoted to processing visual information; and the forebrain, which in fishes is devoted primarily to processing olfactory (smell) information. In the brains of terrestrial vertebrates, the forebrain plays a far more dominant role than it does in fishes.

In land-dwelling vertebrates, information processing is increasingly centered in the forebrain.

The Diencephalon The thalamus is an intermediate processing center for sensory information received from auditory and optical sensory receptors. Information from other sensory receptors, such as information about posture derived from the muscles located throughout the body, and information about orientation, derived from sensors in the ear, is also passed along to the thalamus, but this type of information is first combined with information from the cerebellum in the hindbrain before it is shunted to the thalamus. From the thalamus, this sensory information is passed along to the cerebrum for further processing and integration.

The hypothalamus, which like the thalamus is located in the diencephalon, integrates all visceral activities. It controls areas in the medulla oblongata, which, in turn, regulate body temperature, respiration, and heartbeat. It also directs the secretions of the brain's major hormone-producing gland, the **pituitary gland.** The hypothalamus is linked by a network of neurons to some areas of the cerebral cortex. This network, which along with the hypothalamus is called the **limbic system,** is responsible for many of vertebrates' most deep-seated drives and emotions.

The Telencephalon: Stereotyped Behavior

The telencephalon of reptiles and birds consists largely of a layer of nerve tissue called the **corpus striatum,** which controls complicated stereotyped behavior. No other group has anything like the corpus striatum. Much of what is considered "behavior" in birds and reptiles is really composed of predetermined neural reactions. For example, the complex mating rituals of some birds are often choreographed by nerve paths within the corpus striatum.

Expansion of the Cerebrum

Fishes and reptiles have brains that are small compared with the size of their bodies; mammals and birds have brains that are proportionally large (figure 35.6). The increase in brain size in the mammals largely reflects the great expansion of the cerebrum, the dominant part of the mammalian brain.

The cerebrum is the center for correlation, association, and learning in the mammalian brain—the so-called "higher" functions that set vertebrates apart from invertebrates. As mentioned earlier, the cerebrum receives sensory information from the thalamus. Commands issued from the cerebrum travel along motor nerves that pass straight through the brain to the motor tracts located in the spinal cord. These motor nerves are called the **pyramidal (corticospinal) tract,** a structure that is particularly prominent in primate brains. Motor neurons, which are attached to these motor tracts, serve the muscles of the body.

The brains of mammals and birds are large relative to their body size. This reflects great enlargement of the cerebrum, the center for correlation, association, and learning.

The Human Brain

The cerebrum, located at the very front of the human brain, is so large compared with the rest of the brain that it appears to envelop it (figure 35.7). In the brains of humans and other placental mammals, the cerebrum is split into two halves, or hemispheres,

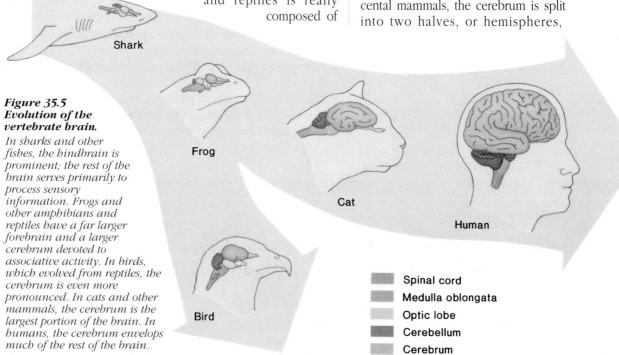

Figure 35.5
Evolution of the vertebrate brain.

In sharks and other fishes, the hindbrain is prominent; the rest of the brain serves primarily to process sensory information. Frogs and other amphibians and reptiles have a far larger forebrain and a larger cerebrum devoted to associative activity. In birds, which evolved from reptiles, the cerebrum is even more pronounced. In cats and other mammals, the cerebrum is the largest portion of the brain. In humans, the cerebrum envelops much of the rest of the brain.

Shark

Frog

Bird

Cat

Human

■ Spinal cord
■ Medulla oblongata
■ Optic lobe
■ Cerebellum
■ Cerebrum

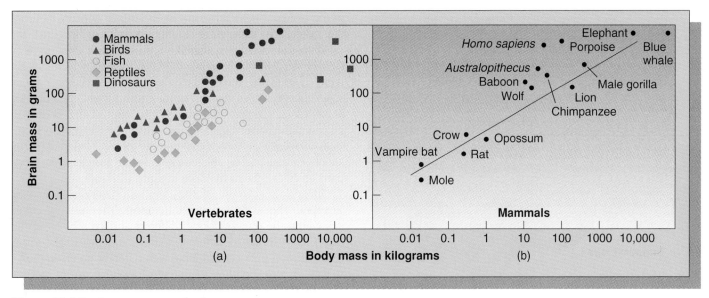

Figure 35.6 Brain mass versus body mass.
Among most vertebrates, brain weight is a relatively constant proportion of body weight, so that a plot of brain mass versus body mass results in a straight line. (a) However, as this graph shows, the proportion of brain mass to body mass is much greater in birds than in reptiles, and the proportion is even greater in mammals. (b) Among mammals, humans have the greatest brain mass per unit of body mass. In second place are porpoises.

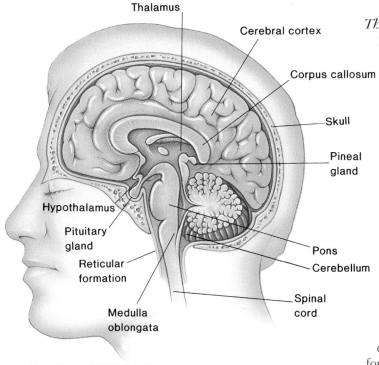

Figure 35.7 The human brain.
The cerebrum envelopes the rest of the brain. Only the cerebral cortex, part of the cerebrum, is visible in surface view.

which are connected only by a nerve tract called the **corpus callosum.** Each hemisphere of the brain is divided further by two deep grooves into four lobes, designated the **frontal, parietal, temporal,** and **occipital lobes** (figure 35.8). Thus, the brain has two hemispheres with four lobes in each, for a total of eight lobes.

The Cerebral Cortex

Much of the cerebrum's neural activity occurs within a thin, gray layer only a few millimeters thick on its outer surface, overlying a solid white region that consists of myelinated nerve fibers. This layer, called the **cerebral cortex,** is densely packed with neuron cell bodies. The human cerebral cortex contains more than ten billion nerve cells, roughly 10 percent of all the neurons in the brain. The surface of the cerebral cortex is highly convoluted, particularly in human brains, a property that increases its surface area (and number of cell bodies) threefold.

Sensory and Motor Cortices The cerebral cortex is divided into four different regions, or cortices. Three cortices—the **auditory cortex,** the **visual cortex,** and the **somatosensory cortex**—receive sensory information from various parts of the body; the fourth is the **motor cortex,** which coordinates motor commands sent from the cerebral cortex to motor nerves.

Scientists ascertained the locations of these different cortices by studying the effect of injuries to particular parts of the cerebrum. An injury in one part of the cerebral cortex, for example, might hamper a person's ability to see; thus, the injured region controls vision. Other studies have involved electrically stimulating the motor and somatosensory cortices and then noting the effect on the human body. For example, if a leg muscle twitches when a particular region of the motor cortex is stimulated, scientists infer that the stimulated area controls the muscle that twitched. Figure 35.9 demonstrates the results of another study, which involved the use of positron-emission tomography (PET).

Studies of this kind have produced a rough map of the human brain. The motor cortex straddles the rearmost portion

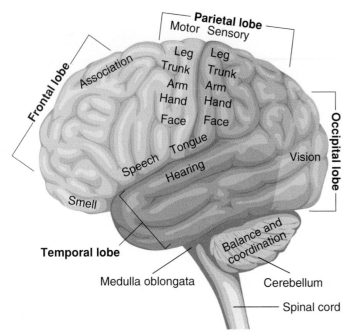

Figure 35.8 Major functional regions of the human brain.
Both hemispheres of the human brain are divided into four lobes: the parietal, frontal, temporal, and occipital.

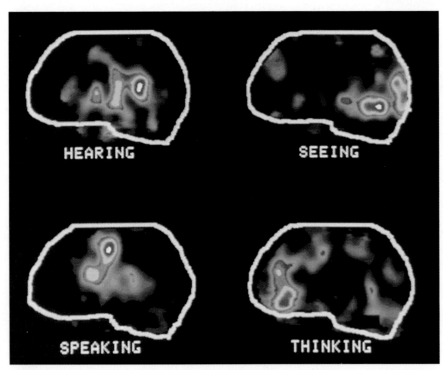

Figure 35.9 Brain experiments.
With PET (positron-emission tomography) scans, scientists have been able to peer into the brain and see which areas are most active when experimental subjects perform different tasks. As these scans show, different areas of the brain are active when we hear a spoken word, see and read that same word silently, speak the word out loud, and think of and say a word related to the first.

of the frontal lobe. Each point on its surface contains neurons that control the movement of different muscles. In effect, the motor cortex contains a maplike representation of neurons of the human body. The somatosensory cortex, which lies directly behind the motor cortex on the edge of the parietal lobe, contains a similar map of the human body, the contours of which are shown in figure 35.10. Some of the body part representations are larger in this illustration than others; this size difference connotes the number of neurons that correspond to that part of the body. In other words, some body parts are represented by more neurons than others.

The other cortices of the cerebral cortex can also be "mapped," but with essential differences. The auditory cortex lies within the temporal lobe; different surface regions of this cortex correspond to different sound frequencies. The visual cortex lies on the occipital lobe, with different points in the visual field corresponding to different positions on the **retina,** the area at the back of the eye where images are projected. The visual cortex is also specialized to detect shapes and the movement of objects.

As mentioned earlier, the cerebrum is split into two hemispheres connected by the corpus callosum. This splitting has implications for where the sensory information is received in the cerebrum. Somewhere along the way, the ascending (sensory) nerve tracts cross the descending (motor) tracts. Thus, information from the right side of the body is detected in the sensory areas of the left cerebral hemisphere, and motor commands issued from the left hemisphere affect muscles and organs on the body's right side.

> *The cerebral cortex is divided into the auditory, visual, somatosensory, and motor cortices. Information from the right side of the body is detected in the left cerebral hemisphere; motor commands from the left cerebral hemisphere affect the right side of the body.*

Associative Cortex Only a small portion of the total surface of the cerebral cortex is occupied by the motor and sensory cortices. The remainder of the cerebral cortex is referred to as **associative cortex.** This appears to be the site of sensory information integration from several senses and of higher mental activities, such as planning and contemplation. The associative cortex represents a far greater portion of the total cortex in primates than it does in any other mammals and reaches its greatest extent in human beings. In a mouse, for example, 95 percent of the surface of the cerebral cortex is occupied by motor and sensory areas. In

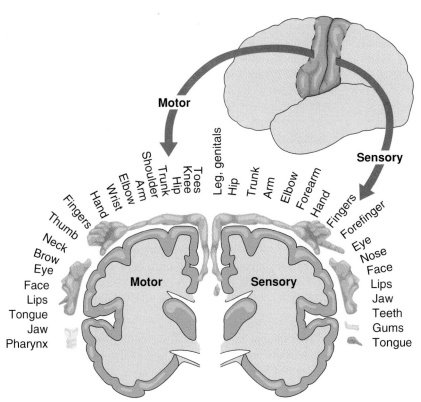

Figure 35.10 Motor and somatosensory cortices of the cerebral cortex.
Each region of the motor and somatosensory cortices is associated with a different part of the human body, as indicated in this stylized map.

humans, only 5 percent of the surface is devoted to motor and sensory functions; the remainder is associative cortex.

As mentioned earlier, the cerebrum consists of two hemispheres, identical in form but differing in function. Although each hemisphere contains similar sensory and motor regions, the hemispheres seem to be responsible for different associative activities. Injury to the left hemisphere of the cerebrum often results in the partial or total loss of speech, but a similar injury to the right side does not. Several **speech centers** control different aspects of speech; two of them are **Broca's area,** which controls the muscles necessary for speech, and **Wernicke's area,** which controls the form, or syntax, of speech. These areas are almost always located on the left hemisphere of right-handed people; in about half of left-handed people, they are located on the right hemisphere. Injury to Wernicke's area produces fluent, grammatical, but meaningless speech; injury to Broca's area destroys the ability to speak altogether, but not the ability to write or read.

Injuries to other sites on the surface of the brain's left hemisphere result in impairment of the ability to read, write, or do arithmetic. Comparable injuries to the right hemisphere have very different effects, resulting in impairment of three-dimensional vision, musical ability, or the ability to recognize patterns and to solve inductive problems.

Some have speculated that the human brain is composed of a "rational" left hemisphere and an "intuitive" right hemisphere because many of the more imaginative associative activities

seem to be carried out by the right hemisphere. Some people even argue that the brain actually has two consciousnesses, one dominant over the other, or that the right brain hemisphere is the more ancient, with language and reasoning having evolved much later. Why associative activities cluster in different areas of the brain remains a mystery.

The Limbic System

Moving down from the cerebral cortex is the forebrain, which was introduced earlier in the chapter. Within the forebrain are the thalamus and hypothalamus, as well as structures called the **amygdala** and the **hippocampus.** Parts of the thalamus and hypothalamus and the amygdala and hippocampus make up the limbic system, which in vertebrates controls deep-seated, unconscious drives and emotions (figure 35.11).

Scientists studying this fascinating part of the brain are not sure of the exact mechanism that controls these drives, but they hypothesize that the limbic system contains a pleasure center and a punishment center. When the pleasure center is stimulated, a person is motivated to sustain the activity that continues the stimulation. If the punishment center is stimulated, a person is motivated to stop the activity that continues the stimulation. Scientists further hypothesize that, when the limbic system's control of these centers becomes abnormal, abnormal behaviors, such as overeating, result.

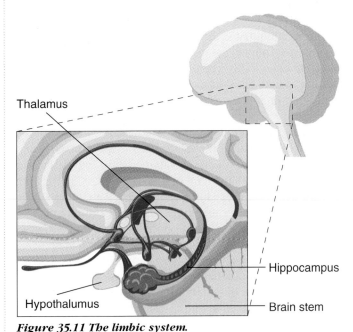

Figure 35.11 The limbic system.
The limbic system, shaded in blue, *controls humans' most deep-seated drives and emotions.*

The Reticular Formation

The **reticular formation** is composed of neurons that extend from the thalamus to the beginning of the spinal cord. These neurons receive sensory information from all parts of the cerebral cortex and then enhance or inhibit the information. When a person is awake, the reticular formation enhances sensory information that is necessary for survival, while inhibiting the information that is merely "background"—for example, how the chair that a person is sitting on feels, the intensity of light in the room, and so on.

When the individual is asleep, however, the reticular formation filters the sensory information that continues to reach the cerebral cortex, which allows the individual to ignore this input. Sometimes, the reticular formation is aided in its filtering duties by people darkening the room and reducing the level of noise when they are ready to go to sleep, but as everyone who has had an early class knows, the reticular formation sometimes does a good enough job even without assistance!

> *The limbic system in the forebrain is responsible for interpretation and control of "pleasure" and "punishment," while the reticular system performs a regulatory function, accenting sensory stimuli during awake times and dampening them during sleep.*

Exploring the Brain: Memory and Sleep

Two areas of great interest to scientists and laypeople alike are the ways in which the brain produces memory and sleep. Brief overviews of these phenomena and of how the brain functions in bringing about these two different experiences follow.

Memory

There are two types of memory: **short-term memory** and **long-term memory.** Short-term memory is the type used for looking up a phone number and then remembering it long enough to dial it. This type of memory is not meant to be permanently stored. Short-term memory is also very limited; only about ten words or numbers can be committed to short-term memory.

New studies indicate that these transient, "working"-type memories are stored in the hippocampus, the structure mentioned earlier that is associated with the limbic system. Further studies indicate that short-term memories are stored in the form of short-term neural excitation; experiments with animals have shown that short-term memories can be removed from the brain by application of electrical shock.

Long-term memories, researchers believe, start out in the same location as short-term memories—that is, the hippocampus. However, if an action or experience occurs repeatedly, or is profound, or involves learning or conditioning, the memory stored in the hippocampus is shunted to one or many long-term memory sites in the human brain. Long-term storage may involve long-term modifications of transmission at individual synapses, perhaps by altering the type of chemically gated ion channel present or the channel's response (see chapter 34 for the role of chemically gated ion channels in synapse transmission).

A very interesting phenomenon involves the sudden recall of a "forgotten" memory, sometimes called a flashback. Flashbacks of extremely stressful experiences have been reported by Vietnam veterans and victims of child sexual abuse. Scientists believe that these flashbacks are a result of the fracturing of experiences into long-term memories. When a person suffers an extremely stressful situation, the memory can become fractured in the hippocampus before being shunted to different sites for long-term storage; for example, the visual part of the memory is carried to the visual cortex, the auditory part is carried to the auditory cortex, and so on. Later, when the person experiences a similar stressful situation, the neurons that lead to each of the memory sites are stimulated at once, and a flashback is experienced. Research in this area is ongoing, and scientists are now searching for a chemical basis that affects long-term memory.

Sleep

The refreshment that vertebrates receive from a good night's sleep remains unexplained by scientists. Despite much research on this most basic of vertebrate activities, the "why" behind the recovery aspect of sleep remains a mystery.

Human beings exhibit five distinct sleep stages. These stages have been measured by instruments that detect the brain's electrical activity. These measurements, called electroencephalograms, or EEGs, show the height and frequency of the brain's electrical rhythms, or waves. Stages 1 through 4 are characterized by brain waves that show increasing height and less frequency. During these stages, the body relaxes, breathing becomes slower and deeper, and the heart rate slows. In stage 4, the period of deepest sleep, the waves are slow and very high. A person in this stage of sleep is very difficult to awaken.

In stage 5, however, the brain-wave patterns change dramatically, becoming irregular and fast. Despite the high rate of brain activity, the muscles of the body remain deeply relaxed, but heart rate and respiration increase. This stage of sleep, also called **rapid-eye-movement (REM) sleep,** is so named because individuals in this stage of sleep move their eyes rapidly back and forth under the eyelids, as if watching a moving picture. Not surprisingly, this is the stage in which most dreaming takes place. Sleepers at this stage can be awakened easily. Some researchers have suggested that, during REM sleep, an individual processes the experiences of the day into long-term memory. The often-suggested activity of reviewing notes for a test right before going to sleep may, therefore, have some basis in fact.

> *Memory, whether short-term or long-term, is a function of storing electrical impulse patterns in different regions of the brain. Sleep, while its unquestionably necessary refreshing aspect still remains mysterious, may contribute to the processing of long-term memory during the REM stage.*

Detecting Stimuli: The Body's Sensory Receptors

All input to the central nervous system arrives in the same form, as nerve impulses in sensory neurons. Every arriving nerve impulse is identical to every other one. The information that the brain derives from sensory input is based on the frequency and pattern with which these impulses arrive and on the identity of the specific neuron that transmits it. To the brain, sunsets, the music of a symphony, and searing pain are all the same, differing only in the source of the impulse and its frequency and pattern. Thus, if the auditory nerve is artificially stimulated, the central nervous system perceives the stimulation as a noise. If the optic nerve is artificially stimulated (for example, by pressing on the eyes), the stimulation is perceived as a flash of light.

Except for visual system photoreceptors, all sensory receptors are able to initiate nerve impulses by opening **stimulus-gated Na$^+$ ion channels** within sensory neuron membranes, thereby depolarizing the membranes. As their name implies, these channels are opened by a mechanical stimulation—a disturbance such as touch, heat, or cold. The receptors differ from one another with respect to the nature of the environmental input that triggers this event. Many kinds of receptors have evolved among vertebrates, with each receptor sensitive to a different aspect of the environment. These receptors and their location, structure, and process are summarized in table 35.1.

Sensing Internal Information

Traditionally, the sensing of information that relates to the body itself—its internal condition and position—is known as **interoception,** or inner perception (figure 35.12*a*). In contrast, the sensing of the exterior is called **exteroception** (figure 35.12*b*). Many of the neurons and receptors that monitor body

Table 35.1 Sensory Transduction Among the Vertebrates

Stimulus	Receptor	Location	Structure	Underlying Process
Temperature	Heat receptors and cold receptors	Skin, hypothalamus	Simple nerve endings	Temperature change alters activity of ion channels in membrane.
Blood chemistry	Carotid bodies	Arterial walls	Simple nerve endings	Change in O$_2$ and H$^+$ concentration alters membrane charge.
Pain	Nociceptors	Body surfaces	Simple nerve endings	Changes in pressure or temperature open membrane channels.
Muscle contraction	Stretch receptors	Within muscles	Spiral nerve endings wrapped around muscle spindle	Stretch of spindle deforms nerve.
Blood pressure	Baroreceptors	Arterial branches	Nerve that extends over thin part of arterial wall	Stretch at arterial wall deforms nerve.
Touch	Meissner's corpuscles, Merkel cells	Surface of skin	Nerve ending within elastic capsule	Rapid or extended change in pressure deforms nerve.
Vibration	Pacinian corpuscles	Deep within skin	Nerve ending within elastic capsule	Severe change in pressure deforms nerve.
Balance	Statocysts	Outer chambers of inner ear	Pebble and cilia	Pebble presses against cilia.
Motion	Cupula	Semicircular canals of inner ear	Collection of cilia	Fluid movement deforms cilia.
	Lateral-line organ	Within grooves on body surface of fish	Collection of cilia	Fluid movement deforms cilia.
Taste	Taste bud cells	Mouth; skin of fish	Chemoreceptors	Particular molecules bind to specific receptors in membrane.
Smell	Olfactory neurons	Nasal passage	Chemoreceptors	Particular molecules bind to specific receptors in membrane.
Hearing	Organ of Corti	Cochlea of inner ear	Cilia between membranes	Sound waves in fluid deform membrane.
Vision	Rod and cone cells	Retina of eye	Array of photosensitive pigment	Light initiates process that closes ion channels.
Heat	Pit organ	Face of snake	Temperature receptors in two chambers	Temperature of surface and interior chambers is compared.
Electricity	Ampullae of Lorenzini	Within skin of fish	Closed vesicles with asymmetrical ion channel distribution	Electric field alters ion distribution on membranes.
Magnetism	Unknown	Unknown	Unknown	Deflection at magnetic field initiates nerve impulse?

(a)   (b)

Figure 35.12 Two classes of sensory receptors.
(a) *Interoception senses internal stimuli. Dancers, for example, are able to maintain their proper stance by using their internal sense of balance.* (b) *Exteroception perceives external stimuli. This leaf frog learns about the world around it by using eyes to perceive patterns of light, just as you are doing in reading this page.*

functions are simpler than those that monitor the external environment. The simplest sensory receptors are free nerve endings that depolarize in response to direct physical stimulation, to temperature, to chemicals like oxygen diffusing into the nerve cell, or to a binding or stretching of the neuron cell membrane. The vertebrate body uses a variety of such **interoceptors** to obtain information about its internal conditioning.

Temperature Change

Two kinds of nerve endings in the skin are sensitive to temperature changes: Cold receptors are stimulated by a low temperature, whereas warm receptors are stimulated by a high temperature. Comparisons of information from the two allow detection of absolute temperature and temperature changes.

Blood Chemistry

Special receptors called **carotid bodies** are embedded in artery walls at several locations in the circulatory system. By sensing carbon dioxide levels, the carotid bodies provide the sensory input that the body uses to regulate its rate of respiration. When carbon dioxide rises above normal levels, this information is conveyed to the central nervous system, which reacts by increasing respiration to eliminate the excess carbon dioxide.

Pain

A stimulus that causes or is about to cause tissue damage is perceived as **pain.** Such a stimulus elicits an array of central nervous system responses, including reflexive withdrawal of a body segment from a source of stimulus and changes in heartbeat and blood pressure. The receptors that produce these effects are called **nociceptors.** They consist of the ends of small, sparsely myelinated nerve fibers located within tissue, usually near the surfaces where damage is most likely to occur. Physical deformation of the ends of the nerve initiates depolarization.

Muscle Contraction

Buried deep within the muscles of all vertebrates, except the bony fishes, are specialized muscle fibers called **muscle spindles.** Wrapped around each muscle spindle is the end of a sensory neuron, called a **stretch receptor.** When a muscle is stretched, the spindle fiber elongates, stretching the spiral nerve ending of the stretch receptor and stimulating it repeatedly to "fire" or initiate an action potential (figure 35.13). When the muscle contracts, the tension on the spindle lessens, and the stretch receptor ceases to fire. Thus, the frequency of stretch receptor discharges along the sensory nerve fiber indicates muscle length and the rate of change of muscle length at any given moment.

Other stretch receptors located in the tendons that attach the muscles to the skeleton monitor the tension produced by skeletal muscles. The central nervous system uses this information to control movements that involve the combined action of several muscles, such as those involved in breathing or walking.

Blood Pressure

Blood pressure is sensed by a highly branched network of nerve endings called **baroreceptors,** or pressure receptors, within the walls of major arteries in several locations in the circulatory system. When blood pressure increases, the arterial wall is stretched most where it is thinnest—in the region of the baroreceptor—thereby increasing the rate of firing of the sensory neuron. A decrease in blood pressure causes the arterial wall to move inward and lowers the rate of neuron

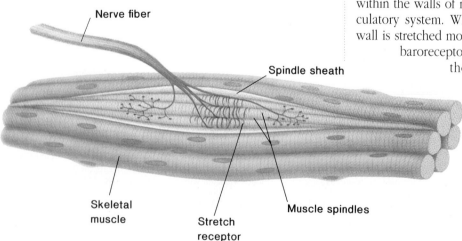

Nerve fiber

Spindle sheath

Skeletal muscle

Stretch receptor

Muscle spindles

Figure 35.13 Stretch receptors embedded within skeletal muscle.
Stretching of the muscle elongates the muscle spindles, which deforms the stretch receptors. This deformation stimulates the nerve endings on the stretch receptors to fire.

depolarization. Thus, the frequency of impulses arriving from baroreceptors allows the central nervous system to continuously monitor blood pressure: the higher the frequency of nerve impulses, the higher the blood pressure.

Touch and Vibration

Touch is sensed by many different types of pressure receptors below the skin surface. **Meissner's corpuscles** and **Merkel cells** are responsible for the great sensitivity of the fingertips. Meissner's corpuscles fire in response to rapid changes in pressure, and Merkel cells measure the duration and extent to which pressure is applied.

Deep below the skin of vertebrates lie vibration-sensitive receptors, called **Pacinian corpuscles,** that are stimulated by deep pressure. Other receptors produce the sensation referred to as itching (figure 35.14).

> *Mechanical stimuli of different types, such as touch and temperature, initiate nerve impulses by deforming the sensory neuron membrane. Such deformation opens stimulus-gated Na⁺ channels in the membrane and initiates depolarization.*

Balance

All vertebrates possess gravity receptors known as **statocysts** or **otoliths.** In humans, these are located in two hollow chambers, called the **utricle** and the **saccule,** within the inner ear and provide the information the brain uses to perceive balance (figure 35.15). How these receptors work can be demonstrated with a pencil standing in a glass. No matter which way the glass is tipped, the pencil rolls along the rim, applying pressure to the lip of the glass. Determining the direction in which the glass is tipped only requires knowing where on the rim the pressure is being applied. The body uses receptors of this sort, broadly called **proprioceptors** (from the Greek word *proprios,* meaning "self"), to sense its position in space. Gravity serves as a reference point; changes in the pressure applied to a nerve ending by a heavy object within the receptor provide the stimulus.

Motion

Vertebrates sense motion in a way similar to that used to detect vertical position: by employing a receptor in which fluid deflects cilia in a direction opposite to that of the motion. Within the inner ear are three fluid-filled **semicircular canals,** each oriented in a different plane at right angles to the other two so that motion in any direction can be detected. Protruding into the canals are sensory cells called **hair cells,** which are connected to sensory nerves. As figure 35.15 shows, hair cells are topped by two types of cilia: short **stereocilia** and one, longer **kinocilium.** The movement of the stereocilia and the kinocilium is transmitted through the hair cell, and this information is then conducted through the sensory nerve to the brain. Because the three canals are oriented in all three

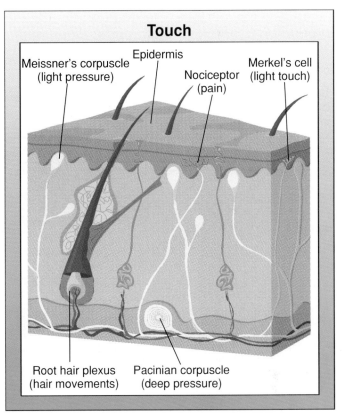

Figure 35.14 Touch receptors in human skin.
These receptors detect different types of touch, from deep pressure to light touch.

planes, movement in any plane is sensed by at least one of them. Complex movements are analyzed by comparisons of the sensory input from each canal.

> *Complex mechanical receptors can respond to pressure, to gravity, or to motion. In each case, the receptors employ mechanical devices like levers to convert the information to a mechanical stimulus, which then initiates the depolarization of a sensory membrane by deformation.*

Sensing the External Environment

The sensory receptors examined thus far have been directed at the body's internal environment or at determining the body's position in space. Sensing the nature of objects in the external environment—for example, an object standing some distance away—is the challenge for **exteroceptors.** The amount of information that any sensory exteroceptor can obtain about a distant object is limited both by the nature of the stimulus sensed by the exteroceptor and by the medium, either air or water, through which the stimulus must move to reach the exteroceptor.

Balance and movement

Figure 35.15 Receptors for balance and motion.

(a) *The semicircular canals contain special receptors that help the body detect motion in three different planes of orientation.* (b) *As you can see in the electron micrograph, groups of long cilia form tentlike assemblies called cupulae that project out into passages within the semicircular canals of the ear. Movement causes fluid in the canal to press against the cupula. Because there are three semicircular canals at right angles to one another, movement in any plane can be detected.*

Vestibule
Semicircular canals
Cochlea
Oval window
Round window
Vestibular nerve
Anterior canal
Lateral canal
Utricle
Saccule
Posterior canal
Kinocilium
Stereocilia
Support cell
Nucleus
Sensory nerve ending
Motor nerve ending
Gelatinous layer
Cupula
Hair cell
Nerve fibers
(a)

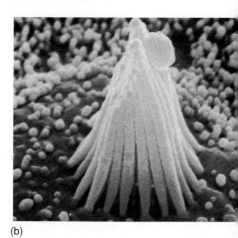

(b)

The four primary senses that detect objects at a distance use different classes of exteroceptors. Taste and smell use chemical receptors; hearing uses mechanical receptors; and vision uses electromagnetic receptors that sense photons of light. Some vertebrates also employ other sensory systems, such as heat, electricity, and magnetism.

Taste and Smell

The simplest exteroceptors, like the simplest interoceptors, are chemical ones. Embedded within the membranes of sensory nerve endings or of sensory cells associated with sensory neurons are specific chemical receptors. These chemical receptors control chemically gated ion channels and thus induce depolarization when they bind particular molecules.

Two chemical sensory systems utilize different receptors and process information at different locations in the brain: (1) taste, in which the receptors are specialized sensory cells, and (2) smell, in which the receptors are neurons. Despite differences in how the stimuli are received and the information processed, there is little difference in the chemical stimuli to which the two chemical senses respond. In taste, the receptors are **taste buds** located in the mouth (figure 35.16). In smell, the receptors are olfactory neurons whose cell bodies are embedded in the epithelium of the upper portion of the nasal passage (figure 35.17).

Hearing

Terrestrial vertebrates detect vibration in air by means of mechanical receptors located within the ear. In the ears of mammals, sound waves beat against the large **tympanic membrane** (eardrum) of the **outer ear,** causing corresponding vibrations in three small bones: the **hammer, anvil,** and **stirrup** (figure 35.18). These bones act together as a lever system, increasing the force of the vibration. The third in line of the levers—the stirrup—pushes against another membrane, the **oval window.** Because the oval window is smaller than the tympanic membrane, vibration against it produces more force per unit area of membrane. The combination of the lever system and area difference amplifies sound.

The chamber in which all these events occur, called the **middle ear,** is connected to the pharynx by the **eustachian tube.** The eustachian tube ensures equal air pressure in both the middle and outer ears. The familiar "ear popping" associated with landing in an aircraft or with the rapid descent of an

Taste

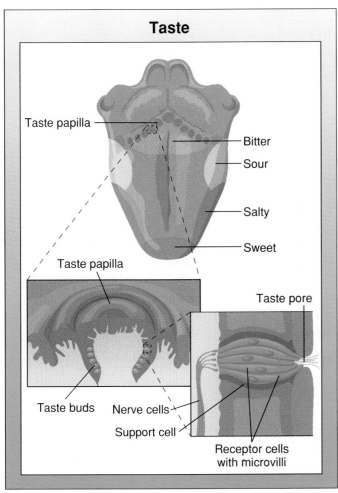

Figure 35.16 Taste buds.
Four types of taste buds that detect four types of taste (salty, sweet, sour, and bitter) are located on different areas of the tongue. These taste buds are grouped in small projections called papillae. Individual taste buds are composed of receptor cells that open into the mouth through a taste pore.

Smell

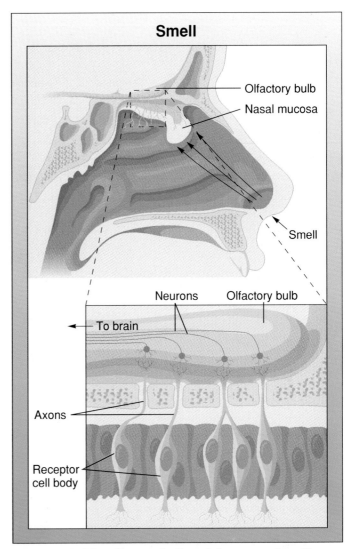

Figure 35.17 The olfactory bulb and the sense of smell.
The olfactory bulb contains the olfactory neurons that lead to the brain, and these neurons are connected by their actions to receptor cells located in the nasal mucosa. The chemicals that make up a particular smell stimulate these receptor cells to fire their axons. The nerve impulse is carried up to the neurons and from there proceeds to the brain.

elevator in a tall building is a result of pressure equalization between the two sides of the eardrum. The oval window functions as the door to the **inner ear,** where hearing actually takes place. The chamber of the inner ear is shaped like a tightly coiled snail shell and is called the **cochlea,** from the Latin word meaning "snail."

The actual organ of hearing is called the **organ of Corti.** Hair cells located within the organ of Corti function as the auditory receptors. The hair cells rest on the **basilar membrane,** which bisects the cochlea. The hair cells do not project into the fluid filling the cochlea. Instead, they are covered with another membrane, called the **tectorial membrane.**

The bending of the basilar membrane as it vibrates causes the hair cells pressed against the tectorial membrane to bend,

depolarizing their associated sensory neurons. Sounds of different frequencies cause different portions of the basilar membrane to vibrate and thus to fire different sensory neurons. The central nervous system perceives sound intensity in terms of the frequency of discharge in particular nerve fibers. Sound frequency is coded in terms of the pattern of neurons on the basilar membrane that are firing sensory impulses. High frequencies produce the greatest deflection near the oval window. Low frequencies primarily affect the apex of the cochlea.

Hearing

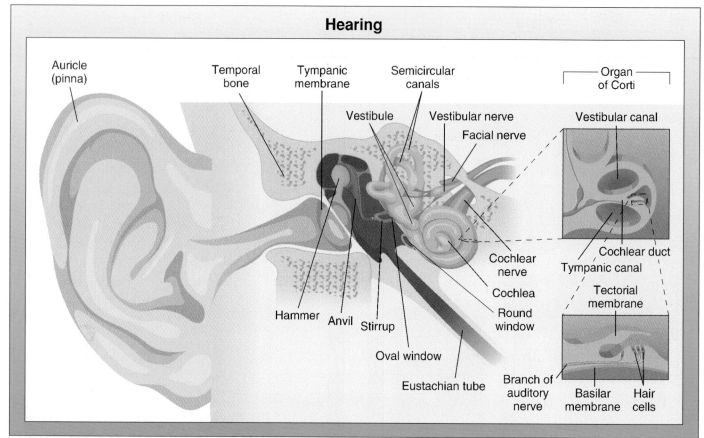

Figure 35.18 The various structures of the human ear associated with hearing.

The organ of Corti is the actual organ of hearing and consists of the vestibular canal, the cochlear ducts, the tympanic canal, and its tectorial membrane. Hair cells attached to the underside of the tectorial membrane receive sound waves and conduct the nerve impulse stimulated by these sound waves to the brain via the auditory nerve.

Ability to hear depends on the flexibility of the basilar membrane, a flexibility that changes with age. Humans are not able to hear low-pitched sounds—below 20 vibrations or cycles per second—although some other vertebrates can. As children, human beings can hear high-pitched sounds, up to 20,000 cycles per second, but this ability decreases progressively through middle age. Other vertebrates can hear sounds at far higher frequencies. Dogs readily detect sounds of 40,000 cycles per second. Thus, dogs can hear a high-pitched dog whistle, while the person blowing the whistle hears nothing.

Vision

Light is perhaps the most useful stimulus for learning about the external environment. Because light travels in a straight line and arrives virtually instantaneously, visual information can be used to determine both the direction and the distance of an object. No other stimulus provides as much detailed information.

The type of problem associated with perceiving light is not one encountered before in this discussion of sensory systems. All receptors described to this point have been either chemical or mechanical. None of them is able to respond to the electromagnetic energy provided by photons of light. Vertebrates have vision, the perception of light, because of a specialized sensory apparatus called an eye. Eyes have evolved many times independently in different groups of animals.

Eyes contain sensory receptors that detect photons of light. These receptors are located in the back of the eye, called the retina, which is organized something like a camera. Light that falls on the eye is focused by a lens on the receptors on the retina, just as the lens of a camera focuses light on film.

How does a photon-detecting sensory receptor work? Just as is the case in photosynthesis, the primary photoevent of vision is the absorption of a photon of light by a pigment. The visual pigment is called *cis*-**retinal,** which is the cleavage product of beta-carotene, a photosynthetic pigment in plants (see figure 8.3). In vertebrates, the *cis*-retinal pigment is coupled to a transmembrane protein called **opsin** to form **rhodopsin** and **iodopsin.**

The visual pigments in vertebrate eyes are located in the tips of specialized sensory cells called **rod** and **cone cells** (figure 35.19). Rod cells, which contain rhodopsin, are responsible for black-and-white vision. Cone cells, which contain iodopsin, function in color vision. There are three different kinds of cone cells, each with a different kind of iodopsin

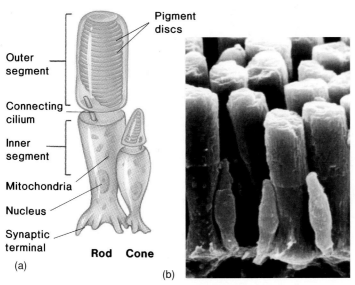

Figure 35.19 Rods and cones.
(a) *The broad, tubular cell on the left is a rod; the shorter, tapered cell next to it, a cone. (b) Although not obvious in this electron micrograph of rods and cones, the pigment-containing outer segments of these cells are separated from the rest of the cells by a partition, through which there is only a narrow passage, called the connecting cilium.*

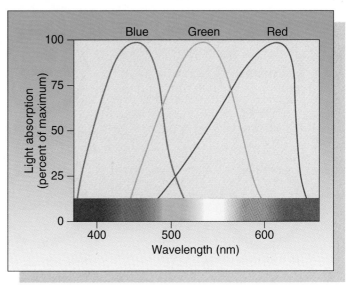

Figure 35.20 The absorption spectrum of human vision.
The wavelength of light absorbed by the visual cone cells is shifted, depending on the protein to which the pigment is bound. There are three such proteins, producing cones that absorb at 455 nanometers (blue), 530 nanometers (green), and 625 nanometers (red).

molecule that preferentially absorbs light of different wavelengths (figure 35.20). In each case, photons of light hyperpolarize the photoreceptors via a specific second-messenger system.

Vertebrate eyes are lens-focused (figure 35.21). Light first passes through a transparent layer, the **cornea,** which begins to focus the light onto the retina. Light then passes through the **lens,** a structure that completes the focusing. The lens is a thick disk filled with transparent jelly, somewhat resembling a flattened balloon. In mammals, the lens is attached by suspensor ligaments to **ciliary muscles.** When these muscles

contract, they change the shape of the lens and thus the point of focus on the rear of the eye (figure 35.22). In amphibians and fishes, the lens does not change shape. These animals instead focus their images by moving the lens in and out, thus operating the same way that a camera does. In all vertebrates, the amount of light entering the eye is controlled by a shutter, called the **iris,** between the cornea and the lens. The iris reduces the size of the transparent zone, or **pupil,** of the eye through which the light passes. Surrounding the entire eye is a tough, protective membrane called the **sclera.**

As mentioned earlier, the field of receptor cells—rods and cones—that lines the back of the eye is the retina. The retina contains about three million cones, most of them located in the central region of the retina called the **fovea,** and approximately one billion rods. The eye forms a sharp image in the central, or foveal, region of the retina, which is a region composed almost entirely of cone cells.

Each foveal cone cell makes a one-to-one connection with a special neuron called a **bipolar cell** (figure 35.23). Each bipolar cell is connected, in turn, to an individual visual **ganglion cell,** whose axon is part of the **optic nerve.** The optic nerve transmits visual impulses more or less directly to the brain. The frequency of pulses transmitted by any one receptor provides information about light intensity. The pattern of

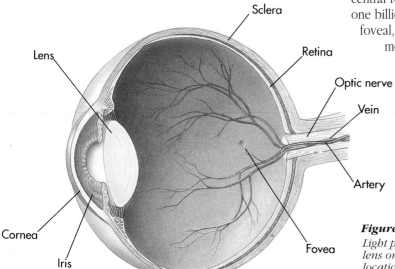

Figure 35.21 Structure of the human eye.
Light passes through the transparent cornea and is focused by the lens on the rear surface of the eye—the retina—at a particular location called the fovea. The retina is rich in rods and cones.

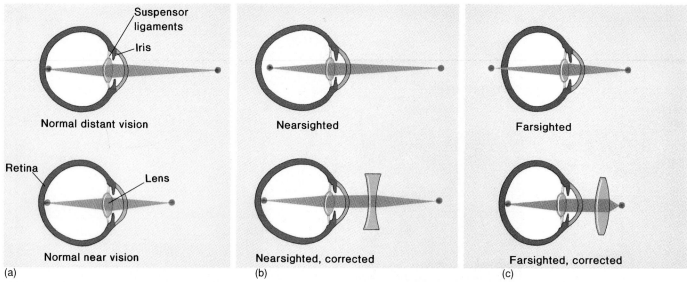

(a)   (b)   (c)

Figure 35.22 Focusing the human eye.
(a) *Contraction of ciliary muscles pulls on suspensor ligaments and changes the shape of the lens, which alters its point of focus forward or backward.* (b) *In nearsighted people, the ciliary muscles place the point of focus in front of the fovea rather than on it. The problem can be corrected with glasses or contact lenses,* *which extend the focal point back to where it should be.* (c) *In farsighted people, the ciliary muscles make the opposite error, placing the point behind the retina. Corrective lenses can shorten the focal point.*

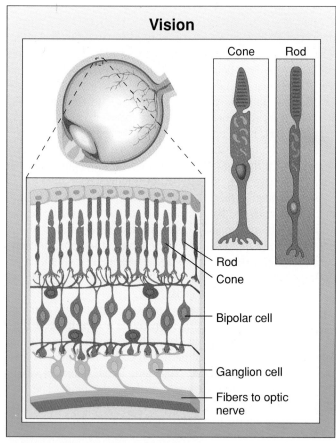

Vision

Figure 35.23 Structure of the retina.
Note that the rods and cones are at the rear of the retina, not the front. Light passes through several layers of ganglion and bipolar cells before it reaches the rods and cones.

action potentials present in different foveal axons sends a point-by-point image of the visual field to the visual cortex. The pattern also contains information related to an object's shape, orientation, and movement. Three different types of cone cells (red, blue, and green) provide information about the color of the image.

Most vertebrates have two eyes, one on each side of the head. When both are trained on the same object, the image that each sees is slightly different because each eye views the object from a different angle. The bilateral location of eyes, with its slight displacement of images (an effect called **parallax**), permits sensitive depth perception. By comparing the differences between the images provided by the two eyes with the physical distance to specific objects, vertebrates learn to interpret different degrees of disparity between the two images as representing different distances. This is called **stereoscopic vision** (figure 35.24). Humans are not born with the ability to perceive distance; they learn it. Stereoscopic vision develops in babies within only a period of months.

Other Environmental Senses in Vertebrates

Vision is the primary sense used by all vertebrates that live in a light-filled environment, but the wavelength and intensity of visible light are by no means the only stimuli available for assessing this environment. Animals are known that sense polarized light (the octopus), ultrasound (bats—figure 35.25), heat (snakes), electricity (catfish), and magnetism (birds). Most vertebrate sensory systems evolved early, while vertebrates were still confined to the sea. Some, such as sensors that detect electric fields, are not used by land vertebrates, presumably because air is a very poor conductor of electric currents. Others,

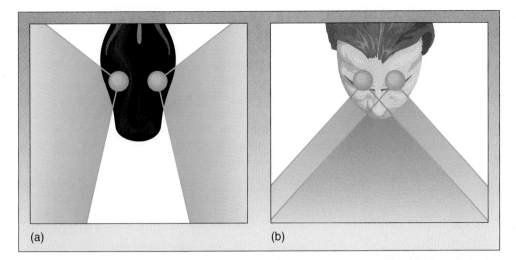

Figure 35.24 Stereoscopic vision.
(a) *When eyes are located on the sides of the head, the two vision fields do not overlap and stereoscopic vision does not occur.* (b) *Vertebrates have eyes located toward the front of the head, so the two fields of vision overlap, resulting in stereoscopic vision.*

Figure 35.25 The bat-moth game: using ultrasound to detect prey.
This bat is emitting high-frequency "chirps" as it flies. It then listens for the sound's reflection against the moth. By timing how long it takes for a sound to return, the bat can effectively catch its prey even in total darkness.

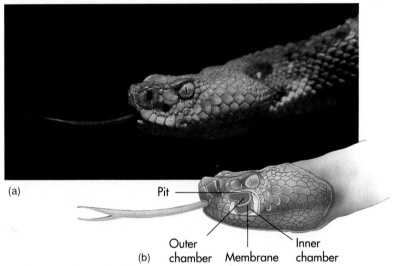

(a)

Pit

Outer chamber Membrane Inner chamber

(b)

Figure 35.26 Sensing heat.
(a) *The depression between the nostril and eye of this rattlesnake opens into the pit organ. Rattlesnakes possess a pair of pit organs, one located on each side of the head between the eye and the nostril.* (b) *The pit organ is composed of two chambers separated by a membrane. A snake with such organs can locate and strike a motionless warm animal in total darkness by perceiving the thermal radiation emanating from the body of the prey.*

such as sensors that respond to liquid pressure waves on fish, are used by land vertebrates to detect pressure waves in air—to hear. Because the terrestrial environment is so physically different from the sea, it has presented some new sensory opportunities. Air, for example, transmits heat much better than water does, and snakes have evolved heat sensors that detect heat sources, and in so doing, provide quite detailed three-dimensional images (figure 35.26).

No one vertebrate has a perfect repertoire of sensory systems that allows it to discriminate all potential cues. Rather, vertebrates have evolved sensory nervous systems that meet particular evolutionary challenges. Thus, some hunting mammals, such as dogs, have evolved highly capable olfactory systems, but primates have not. Interestingly, the human environment contains detailed sensory information that could potentially be utilized with the proper sensory receptors. For example, humans cannot utilize thermal radiation, as snakes do, or ultrasound, as bats do, to construct three-dimensional images.

Exteroreceptors are receptors sensitive to external stimuli. The most common examples of these are the chemoreceptors (smell and taste), photoreceptors (vision), and mechanoreceptors (hearing and touch).

Structure and Function of the Peripheral Nervous System

As described in detail earlier in the chapter, sensory information from receptors is sent along sensory nerves to the central nervous system. The hair cells in the semicircular canals of the ear, the ganglion cells of the eye, and the sensory neurons in the nasal epithelium are all attached to sensory nerves that wind their way to the brain and spinal cord. These sensory nerves are part of the peripheral nervous system, the part of the nervous system that is outside the central nervous system and that feeds information to and receives information from the central nervous system.

Once the sensory information is processed, the central nervous system issues commands to muscles, organs, and glands along motor nerves (figure 35.27). These motor nerves are also part of the peripheral nervous system. Some motor nerves affect skeletal muscles and are subject to conscious control; thus, they are part of the voluntary peripheral nervous system. These are the pathways that coordinate fingers to grasp a pencil, that spin the body during a dance, and that put one foot ahead of the other for walking.

Other motor nerves affect the smooth muscles, glands, and organs. These effects are not subject to conscious control and thus are part of the involuntary or autonomic peripheral nervous system. These pathways are responsible for maintaining the body's internal environment: They monitor the body's temperature, digest food, and regulate blood pH. Thus, the involuntary pathways of the peripheral nervous system are basically concerned with homeostasis, the maintenance of constant conditions within the body.

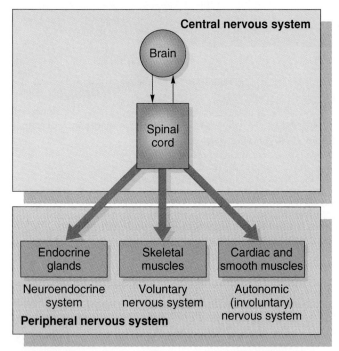

Figure 35.27 How the central nervous system issues commands.
The central nervous system issues commands via three different systems: The neuroendocrine system is a network of endocrine glands whose hormone production is controlled by commands from the central nervous system. The voluntary nervous system is a network of motor nerves that extend to the skeletal muscles. The involuntary (autonomic) nervous system is a network of motor nerves that extend to cardiac and smooth muscles and some glands.

The neuroendocrine system is also considered part of the peripheral nervous system. The neuroendocrine system receives commands from the central nervous system, but it does not carry out its effects by the action of motor nerves. Instead, the neuroendocrine system acts on organs by the secretion of hormones from glands. Hormones, as opposed to nerve signals, can affect changes to organs at greater distances. Neuroendocrine control in the vertebrate body is also associated with long-term changes in activity level. Chapter 37 presents a more detailed look at the neuroendocrine system.

Neuromuscular Control: How Motor Nerves Control Muscles

As described earlier in the chapter, the voluntary peripheral nervous system directs the skeletal muscles of the body and regulates their contractions by means of stretch receptors. When a muscle extends or contracts, receptors in the muscle spindle are depolarized, initiating a nerve impulse that travels along sensory nerve fibers to a spinal nerve in the spinal cord. The sensory neuron in the spinal nerve shunts the nerve impulse to the centrally located gray matter in the spinal nerve and then "drops off" the nerve impulse to an interneuron located in the gray matter. To be more specific, the sensory neuron in the spinal nerve makes a synaptic connection with the interneuron, thereby transmitting the nerve impulse. The interneuron either sends the nerve impulse to the brain for further processing, or it initiates a response.

In vertebrate animals, the motion that results from a nerve impulse that is initiated in the spinal nerve interneuron is called a **reflex.** In some animals of relatively simple construction, reflexes are the only muscle responses available. The pathway that the nerve impulse takes through the spinal nerve is called a **reflex arc.**

Some reflex arcs are called **monosynaptic reflex arcs** because the sensory neuron makes direct synaptic contact with a motor neuron in the spinal nerve—there is no interneuron "middleman" between them. In these reflex arcs, the motor neuron in the spinal nerve is connected directly to a muscle. Monosynaptic reflex arcs are relatively stable but are usually associated with more complex motor pathways. It is through these more complex paths that voluntary control is established.

In humans, the most familiar reflex arc that exists on its own, and not as part of another complex pathway is the knee-jerk response. This reflex arc is monosynaptic because it does not involve interneurons. If the ligament directly below the kneecap is struck lightly by the edge of a hand or by a doctor's rubber hammer, the sudden pull that results stretches the muscles of the upper leg, which are attached to the ligament. Stretch receptors in these muscles immediately send an impulse along sensory nerve fibers to the spinal cord, where these fibers synapse directly with a motor neuron. This motor neuron extends back to the upper leg muscles, stimulating them to contract and the leg to jerk upward (figure 35.28). Such reflexes play an important role in maintaining posture.

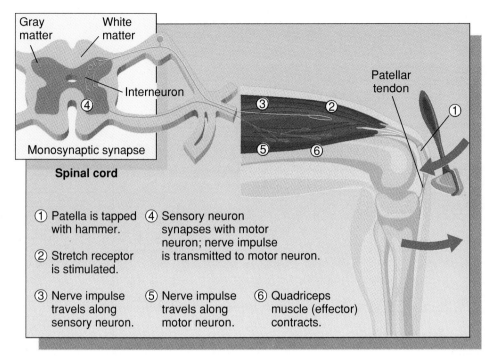

- ① Patella is tapped with hammer.
- ② Stretch receptor is stimulated.
- ③ Nerve impulse travels along sensory neuron.
- ④ Sensory neuron synapses with motor neuron; nerve impulse is transmitted to motor neuron.
- ⑤ Nerve impulse travels along motor neuron.
- ⑥ Quadriceps muscle (effector) contracts.

Figure 35.28 The knee-jerk reflex.
This reflex arc does not contain an interneuron, so it is monosynaptic.

A reflex is a simple behavior produced by a neural information-response circuit called a reflex arc.

Neurovisceral Control

The glands, smooth muscle, and cardiac muscle of the body (the "viscera") respond to a second network of motor neurons—the involuntary peripheral nervous system. The involuntary peripheral nervous system plays a major role in controlling the body's internal environment and thus in maintaining homeostasis.

Components of the Involuntary Peripheral Nervous System

The involuntary peripheral nervous system controls the body's glands and involuntary muscles. In this regard, two elements of the involuntary peripheral nervous system usually act antagonistically (in opposition to each other) (figure 35.29):

1. The **parasympathetic nervous system** consists of a network of (*a*) long motor axons that make synaptic contact with ganglia (the cell bodies from which nerves extend) in the immediate vicinity of an organ and (*b*) short motor neurons extending from the ganglia to the organ.

2. The **sympathetic nervous system** is composed of a network of (*a*) short motor axons that extend to ganglia located near the spine and (*b*) long motor neurons extending from the ganglia directly to each target organ.

Control of the Involuntary Peripheral Nervous System

Involuntary nerve impulses must cross two synapses in traveling from the central nervous system out to a target organ, whether they travel along sympathetic or parasympathetic nerves. The first synapse is in the ganglion, between the axon of a neuron extending from the central nervous system and the dendrites of the autonomic neuron's cell body. The second synapse is between the autonomic neuron's axon and the target organ. The neurotransmitter in the ganglion is acetylcholine for both sympathetic and parasympathetic nerves. However, the neurotransmitter between the terminal autonomic neuron axon and the target organ is different in the two antagonistic elements of the autonomic nervous systems. In the parasympathetic system, the neurotransmitter at the terminal synapse is acetylcholine, just as it is in the ganglion. In the sympathetic system, the neurotransmitter at the terminal synapse is either adrenaline (epinephrine) or noradrenaline (norepinephrine), both of which have an effect *opposite* to that of acetylcholine.

Most glands (except the adrenal gland), smooth muscles, and cardiac muscle controlled by the involuntary peripheral nervous system have inputs from *both* the sympathetic and parasympathetic systems. Usually, these two systems antagonize one another, but either may be excitatory. Thus, depending on which of the two components of the involuntary nervous system is selected by the central nervous system, an arriving signal will either stimulate or inhibit the organ (figure 35.30).

Each gland, smooth muscle, and cardiac muscle constantly receives stimulatory signals through one nerve and inhibitory signals by way of the other nerve. The central nervous system controls activity in each case by varying the ratio of the two signals.

The glands and involuntary muscles of the body are innervated by two opposing sets of motor nerves.

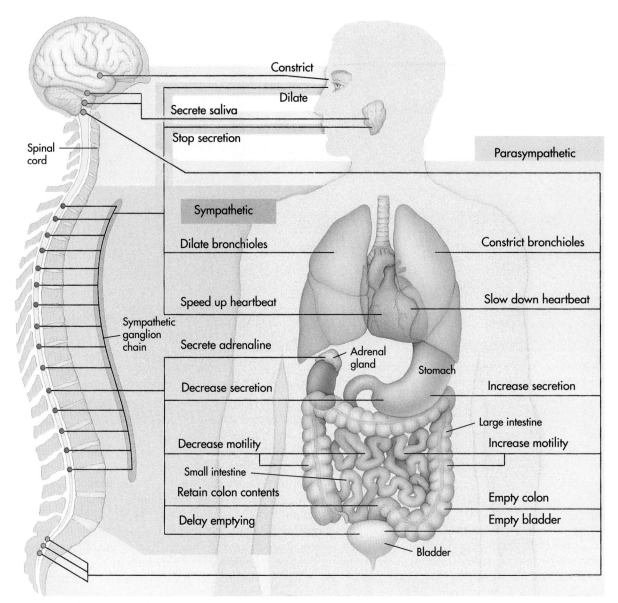

***Figure 35.29 The sympathetic and parasympathetic
nervous systems.***

*The ganglia of sympathetic nerves are located near the spine, and
the ganglia of parasympathetic nerves are located far from the
spine, near the organs they affect. As you can see, a nerve path
runs from both of the systems to every organ indicated, except the
adrenal gland.*

Thus, an organ receiving nerves from both components of
the involuntary nervous system is subject to the effects of two
opposing neurotransmitters. If the sympathetic nerve ending
excites a particular organ, the parasympathetic synapse usually
inhibits it. For example, the sympathetic system speeds up the
heart and inhibits gastrointestinal motility and secretion,
thereby inhibiting digestion, whereas the parasympathetic
system slows down the heart and increases gastrointestinal
activity. In many cases, the opposing systems are organized so
that the parasympathetic system stimulates the activity of normal body functions—for example, the churning of the stomach, the contractions of the intestine, and the secretions of the
salivary glands. The sympathetic system, on the other hand,
generally mobilizes the body for greater activity, as in increased respiration or a faster heartbeat.

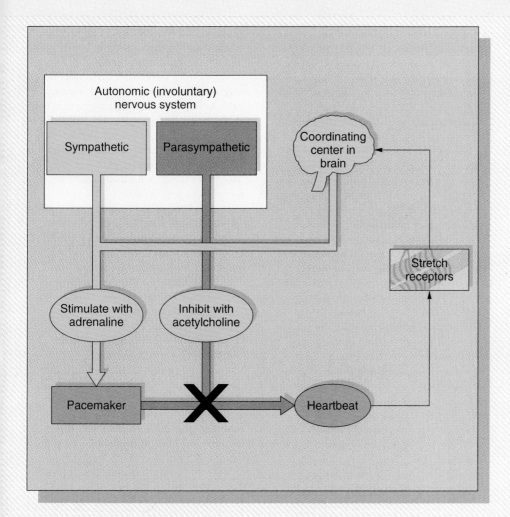

Figure 35.30 The control of heartbeat.

Heartbeat is controlled at two levels: (1) Each heartbeat is initiated by the pacemaker, which is stimulated or inhibited by signals from the involuntary nervous system. (2) Stretch receptors in the heart muscle provide information to a coordinating center in the brain, which modifies the involuntary commands.

EVOLUTIONARY VIEWPOINT

The brains and sensory systems of vertebrates exhibit considerable diversity. Birds, which evolved from reptiles long after mammals did, have a brain in which the corpus striatum, rather than the cerebrum, is the dominant element in the forebrain; much of birds' "behavior" is composed of completely "hard-wired" neural reactions. Nor are the so-called five senses of humans the only vertebrate senses: Rattlesnakes can "see" heat; bats can use a form of sonar to form detailed images with sound; and birds can use terrestrial magnetism to navigate.

SUMMARY

1. The vertebrate nervous system is made up of the central nervous system, consisting of the brain and the spinal cord, and the peripheral nervous system. Within the peripheral nervous system, sensory pathways transmit information *to* the central nervous system, while motor pathways transmit commands *from* the central nervous system.

2. The motor pathways of the peripheral nervous system are divided into somatic, or voluntary, pathways, which relay commands to skeletal muscles, and autonomic, or involuntary, pathways, which stimulate the glands and smooth muscles of the body.

3. The spinal cord is made up of spinal nerves and has two functions: (1) It is a relay center for information sent to the brain and for the commands that the brain issues, and (2) it can issue commands through motor neurons to the muscles of the body.

4. A vertebrate brain consists of a hindbrain, or rhombencephalon; a midbrain, or mesencephalon; and a forebrain, or prosencephalon. The hindbrain was the principal brain component of early vertebrates, with the forebrain becoming increasingly dominant in reptiles, amphibians, birds, and mammals.

5. In birds and mammals, the brain is much larger in proportion to the body than is the case in other vertebrates, reflecting a great increase in the size of the cerebrum. The cerebrum is the center for correlation, association, and learning in the mammalian brain.

6. In humans and other primates, the cerebrum is split into two hemispheres, connected by a nerve tract called the corpus callosum. Each hemisphere is divided further by deep grooves into four lobes: the frontal, parietal, temporal, and occipital lobes. The lobes have different functions.

7. The cerebral cortex is divided into the auditory, visual, somatosensory, and motor cortices. Studies of the effects of injuries to particular parts of the cerebrum have produced "maps" of each cortex that show what parts of the body are represented by what parts of the different cortices.

8. Only a small part of the cerebral cortex is occupied by the motor and sensory cortices. The remainder is referred to as associative cortex, which appears to be where sensory information is integrated and higher mental activities occur.

9. The limbic system in humans is responsible for deep-seated drives and emotions. It probably consists of a pleasure center and a punishment center.

10. The reticular formation acts as a screening device in awake animals by enhancing important sensory information or inhibiting unnecessary sensory information. In sleeping animals, the reticular formation acts as a filter that screens out much sensory information.

11. There are two types of memory: Short-term memory is very limited and is only temporarily stored in the hippocampus. Long-term memory is initially stored in the hippocampus but may then be shunted to one or many long-term memory sites in the human brain.

12. Five stages of sleep are categorized by the height and frequency of associated brain waves.

13. Almost all sensory receptors are able to initiate nerve impulses by opening stimulus-gated Na$^+$ ion channels within sensory neuron membranes. No matter what type of stimulus—whether mechanical, chemical, or electromagnetic—the response is usually depolarization of a sensory neuron membrane.

14. Many of the body's internal receptors (interoceptors) are simple receptors in which a nerve ending becomes depolarized in response to direct physical stimulation, deformation of a nerve membrane, or chemical- or temperature-induced opening of ion channels in the nerve membrane.

15. Complex mechanical receptors are required for balance and motion. The receptors use mechanical devices like levers to convert pressure, gravity, or motion information to a mechanical stimulus, which then initiates depolarization of a sensory membrane.

16. Exteroceptors sense the nature of external objects. The four primary senses use different classes of exteroceptors. The information obtained is limited by the nature of the stimulus and by the medium through which the stimulus must move.

17. The senses of taste and smell use chemical sensory systems that utilize different receptors and process information at different brain locations.

18. Hearing requires mechanical receptors. Hearing organs in terrestrial vertebrates amplify airborne sound waves and direct them at a fluid-containing chamber—the cochlea—within the ear. Mechanical receptors in the chamber are then deformed by the fluid-borne sound waves.

19. Vision, the perception of light, requires receptors that can respond to the electromagnetic energy provided by photons of light. As in photosynthesis, a pigment is used as a primary photoreceptor. The vertebrate eye is designed like a lens-focused camera. The fovea at the center of the retina transmits a point-by-point image to the brain.

20. Other stimuli sensed by vertebrate receptors include polarized light, ultrasound, heat, electricity, and magnetism.

21. The motor neurons of the peripheral nervous system can be divided into two groups: the voluntary peripheral pathways and the involuntary peripheral pathways. These motor neurons carry out commands issued by the central nervous system.

22. In the voluntary peripheral nervous system, the contraction of skeletal muscle can be controlled by reflex arcs, which are coordinated by spinal nerves. A monosynaptic reflex arc is a pathway that does not use an interneuron.

23. In the involuntary peripheral nervous system, smooth muscles and glands are directed by antagonistic nerve pairs; one stimulates and the other inhibits activity. In general, the parasympathetic nerves stimulate the activity of normal internal body functions and inhibit alarm responses, while the sympathetic nerves perform the opposite functions.

REVIEWING THE CHAPTER

1. How is the nervous system organized in vertebrates?

2. What two body parts compose the central nervous system?

3. What is the peripheral nervous system?

4. What is the structure and function of the central nervous system?

5. What is the structure of the spinal cord, and how does it function?

6. How does the brain function, and what is its structure?

7. How did the vertebrate brain evolve?

8. What is the function and structure of the hindbrain?

9. How do the midbrain and forebrain function, and what are their structures?

10. How did the forebrain become dominant in vertebrates?

11. Where does stereotyped behavior occur?

12. How did the cerebrum expand?

13. How is the human brain characterized?

14. What is the function and structure of the cerebral cortex?

15. How does the limbic system function, and what is its structure?

16. What is the reticular formation, and how does it function?

17. What are two types of memory, and how do they work?

18. What are the distinct stages of sleep in humans?

19. What are the body's sensory receptors?

20. How is internal information sensed?

21. How is temperature sensed?

22. Which receptors detect changes in blood chemistry?

23. How is pain detected?

24. How and why do muscles contract?

25. How is blood pressure monitored?

26. How are touch and vibration sensed?

27. How is balance achieved?

28. How do vertebrates sense motion?

29. How is the external environment sensed?

30. What is the basis of taste and smell?

31. How is hearing detected?

32. How does vision work, and why is it a learning stimulus?

33. What other environmental senses occur in vertebrates?

34. What is the structure and function of the peripheral nervous system?

35. What is the process of neuromuscular control?

36. What is neurovisceral control?

37. What are the components of the involuntary peripheral nervous system?

38. How is the involuntary peripheral nervous system controlled?

COMPLETING YOUR UNDERSTANDING

1. Which of the following are part of the peripheral nervous system? [*two answers required*]

a. Tracts   c. Nuclei

b. Nerves   d. Ganglia

2. How can doctors tell if there is an infection in the brain?

3. Why is there any interest in examining a five-hundred-million-year-old fish with respect to vertebrate evolution?

4. Many writers have stated that the hindbrain of the human brain is its most primitive element. How would you respond to this proposition?

5. If in an accident the brain stem were twisted, what parts of the brain would be affected, and what would be the consequences?

6. What is responsible for emotion in vertebrates?

7. How does the pyramidal tract function, and in which animals is it prominent?

8. Which of the following animals has the largest brain mass per unit of body mass?

a. Whale   d. Chimpanzee

b. Human   e. Wolf

c. Gorilla   f. Porpoise

9. How are the different parts of the human brain mapped?

10. What is the correlation between control of muscles and organs and the left and right cerebral hemispheres?

11. What is the associative cortex, and in which animals is it prominent?

12. Why has it been hypothesized that the right-brain hemisphere is more ancient than the left-brain hemisphere in humans?

13. What is the basis for the hypothesis that, when the limbic system is not operating normally abnormal behavior, such as overeating, can be the result?

14. When you are driving your car, how would you detect that your driver's seat feels uncomfortable?

15. If one of your friends verbally gave you a telephone number for a "hot date"

during a telephone conversation, and you failed to write it down, what scientific explanation could you give for not remembering the phone number two hours later?

16. Do you or your family know anyone who served in a war who has frequent flashbacks of stressful war experiences? What is the scientific explanation for these flashbacks?

17. How are sleep patterns measured in humans?

18. What is the association between stimulus-gated Na^+ ion channels and the body's sensory receptors?

19. Which of the following stimuli would alter ion distribution on membranes?

a. Magnetism   d. Taste

b. Motion   e. Electricity

c. Heat   f. Muscle contraction

20. Which receptors are associated with the sensing of blood pressure?

a. Stretch receptors

b. Baroreceptors

c. Merkel cells

d. Nociceptors

e. Otoliths

f. Oval windows

21. If one of your older relatives who lived alone fell to the kitchen floor while cooking lunch, what scientific explanation might be given for the sudden change? If this relative were to orient himself or herself to crawl to the telephone to get help, would this alter your explanation?

22. The ability to hear often decreases with age because

a. the cilia degenerate.

b. the hair cells stiffen.

c. the flexibility of the basilar membrane changes.

d. the tympanic membrane breaks.

e. the tympanic canal straightens.

23. How does the eye lens differ between mammals and amphibians?

24. Which sensory systems are more applicable to land animals and why?

25. Why is the neuroendocrine system considered part of the peripheral nervous system and not the central nervous system?

26. Why does the knee-jerk response occur if a physician lightly strikes the ligament below the kneecap with a rubber hammer?

27. Why do the parasympathetic and sympathetic systems frequently oppose one another?

28. The autonomic nervous system is composed of which two of the following elements?

a. The sympathetic nervous system

b. The sympatric nervous system

c. The voluntary nervous system

d. The peripheral nervous system

e. The parasympathetic nervous system

29. The transmitter at autonomic ganglia is

a. norepinephrine.

b. acetylcholine.

c. serotonin.

d. glycine.

e. glutamic acid.

30. Most of us have sensed at one time or another an oncoming storm by detecting the increase in humidity in the air. What sort of receptors detect humidity? Why do you suppose that hot days seem so much hotter when it is humid?

31. The literature of science fiction is awash with stories of extrasensory perception, and some research laboratories are actively engaged in attempts to demonstrate it. Can you describe a sensory receptor that might function in "extrasensory" perception?

32. The heat-detecting pit receptor of snakes is a very effective means of "seeing" at night. It is the same sort of sensory system employed by soldiers in "snooper-scopes" and in heat-seeking missiles. Why do you suppose that other night-active vertebrates, such as bats, have not evolved this sort of sensory system?

FOR FURTHER READING

Allport, S. 1986. *Explorers of the black box: The search for the cellular basis of memory*. New York: W. W. Norton. A vivid account of the pioneering studies of Eric Kandel and others in their efforts to demonstrate how we remember. Easy to read, this book shows scientists in action, gathering data and disputing among themselves about what the data mean.

Barlow, R. 1990. What the brain tells the eye. *Scientific American*, April, 90–95. How the brain may exercise substantial control over just what the eye can detect.

Freeman, W. 1991. The physiology of perception. *Scientific American*, Feb., 78–85. Why the chaotic collective activity of millions of neurons may be necessary for rapid recognition of patterns.

Gibbons, B. 1986. The intimate sense of smell. *National Geographic,* Sept., 324–61. A detailed account of the human sense of smell. Interesting and well-illustrated.

Masland, R. 1986. The functional architecture of the retina. *Scientific American,* Dec., 102–11. A close look at how neurobiologists are beginning to study the paths of individual neurons by selectively injecting single cells with fluorescent dyes.

Miller, J. A. 1990. A matter of taste. *BioScience* 40(2): 78–82. Innovative methods and materials that have made possible new insights into the way our sense of taste works.

Selkoe, D. 1991. Amyloid protein and Alzheimer's disease. *Scientific American,* Nov., 68–79. The story of how, in 1907, Alois Alzheimer observed deposits of "a peculiar substance" in brains of patients who had suffered from senile dementia. Called amyloid plaques, these deposits now seem to be the cause of at least some forms of Alzheimer's disease.

Suga, N. 1990. Biosonar and neural computation in bats. *Scientific American,* June, 60–68. How the bat brain is organized to extract information from biosonar signals. Biology at its best—highly recommended.

EXPLORATIONS

Interactive Software:
Drug Addiction

This interactive exercise allows students to learn about the physical basis of drug addiction by exploring the direct consequences of the addictive drug cocaine on a nerve. The exercise presents an animated diagram of a single nerve synapse within the limbic system of the human brain. The neurotransmitter dopamine crosses this synapse to produce feelings of pleasure in the central nervous system. The student can explore the consequences of introducing cocaine into the synapse, watching it bind up the transporter that normally removes dopamine from the synapse. The result is that dopamine levels stay high and fire the synapse repeatedly, producing euphoria. Students can then watch the synapse as it adjusts to this higher neurotransmitter level by lowering the number of its postsynaptic receptor channels. Now there is no pleasure without the drug, which is how addiction starts. By varying the amount and frequency of cocaine use, the student can explore how patterns of drug use reinforce addiction.

Questions to Explore:

1. How do neurotransmitters pass a nerve impulse across a synapse?

2. How are neurotransmitters removed from a synapse after a nerve impulse has passed?

3. What would happen if neurotransmitters were *not* removed from a synapse?

4. How do nerves respond to prolonged high levels of neurotransmitters?

5. How can addiction be reversed?

6. Can you discover a way to use cocaine and avoid addiction?

HOW ANIMALS MOVE

Scanning electron micrograph (magnified 200 times) of the house dust mite, who crawls through the particles of dust on your mattress while you sleep, hunting for the shed human cells that are its food. Because its feces are a potent antigen, many people are strongly allergic to house dust (really to the feces of the mites).

FOR REVIEW

Here are some important terms and concepts that have been discussed in previous chapters and that you will encounter again in this chapter. Review them before proceeding if necessary.

Actin filaments (*chapter 4*)
Cytoskeleton (*chapter 4*)
ATP (*chapter 6*)
Vertebrate muscle tissue (*chapter 33*)
Vertebrate connective tissue (*chapter 33*)
Depolarization (*chapters 33 and 34*)

One of the most obvious differences between plants and animals is that plants remain in one spot, while most animals move about from one place to another—swimming, flying, or walking. Of the three multicellular kingdoms that evolved from protists, only animals explore their environment in this active way; plants and fungi move only by growing or by becoming the passive passengers of wind and water. To move, all animals use the same basic mechanism: muscle contraction. This chapter examines how animals, especially vertebrates, use muscles to achieve movement.

Mechanical Problems Posed by Movement

The Need for Force

The basic problem posed by movement is that gravity tends to hold objects in one place. A large boulder cannot be lifted with one hand because gravity is pulling down on the boulder far harder than the hand can push up. Moving the boulder requires lifting with a greater force than gravity is exerting. All motion must meet this simple requirement.

Organisms use the chemical energy of ATP to supply that force. When an ATP molecule is split into ADP and P_i (inorganic phosphate), 7.3 kilocalories of energy are made available for the work of movement. Organisms apply this energy to alter the length of structural elements within muscle cells, causing the cells to shorten. When many muscle cells shorten all at once, they can exert great force.

The Need for a Skeleton

If this were all there is to movement, organisms would not move. Instead, they would simply pulsate as their muscles contracted and relaxed, contracted and relaxed, in futile cycles. For a muscle to produce movement, it must direct its force against another object. Some soft-bodied invertebrates like slugs, which live on land but have no shell or internal skeleton, move by attaching themselves to the material over which they move.

Figure 36.1 Jellyfish with hydroskeleton.
Movement of the muscle tissues to expel water causes the jellyfish to move forward. The small cnidarian Polyorchis penicillatus *is common in early summer in the Santa Barbara Channel. Each tentacle has a specialized sense organ known as an ocellus, marked by red pigment at its base.*

Hydroskeletons

Most soft-bodied invertebrates, which have neither an internal nor an external skeleton, solve the problem of what to direct muscle force against by using the relative incompressibility of the water within their bodies as a kind of skeleton, called a **hydroskeleton.** They simply direct the force of their muscles against the water. Both earthworms and jellyfish have hydroskeletons. In jellyfish, the action of the muscles expelling the water produces an opposite reaction, which causes the animal to move (figure 36.1).

Exoskeletons

Most animals are able to move because the opposite ends of their muscles are attached to hard parts of their bodies, so that, for example, muscle contraction results in the beating of a wing or the lifting of a leg. When the hard body part to which the muscles are attached is a shell that encases or surrounds the body, the shell is called an **exoskeleton.** Arthropods, for example, have muscles that are attached to a rigid chitin exoskeleton, which enables them to swim, walk, and fly.

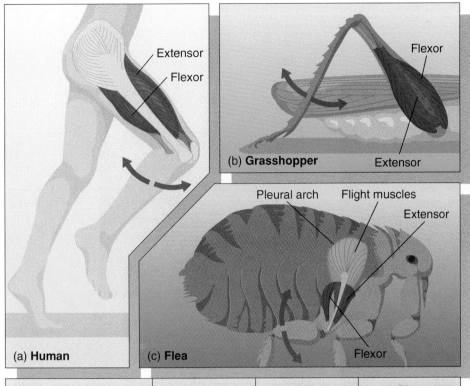

(a) **Human**

(b) **Grasshopper**

(c) **Flea**

(d)	Human	Grasshopper	Flea
Body mass	70 kg	3 g	0.5 mg
Typical jump height	60 cm	59 cm	20 cm
Acceleration (gravities)	1.5	15	245

Figure 36.2 Muscles in action.
In humans, grasshoppers, and fleas, antagonistic muscle pairs control leg movement. Muscles can exert force only by becoming shorter. All three animals shown here have double sets of muscles that work in opposite directions. In the human (a), the muscles are attached to bone, whereas in the flea and grasshopper (b and c), the muscles are attached to the inside of the skeleton. In each case, the flexor muscles move the lower leg closer to the body, and the extensor muscles move it away. (d) Fleas, grasshoppers, and humans all jump to similar heights relative to their size.

These muscles usually occur in pairs that operate antagonistically to each other. In grasshoppers' legs, for example, a **flexor muscle** controls forward leg movement, and an **extensor muscle** controls backward movement. Both muscles are attached to the same beginning and end points so that forward and backward movements can be coordinated (figure 36.2). In flying insects, antagonistic muscle pairs are called **elevators** and **depressors,** and the antagonistic movements of these muscles cause the wings to beat.

As long as an individual is small enough, attachment to an exoskeleton is an effective strategy. However, the rules of mechanics require that the exoskeleton must be much thicker in large insects than in small ones to bear the pull of the muscles. In an insect the size of a human being, the exoskeleton would need to be so thick that the animal could hardly move. This relationship limits insect size.

Endoskeletons

In vertebrates, muscles are attached to an internal scaffold of bone, an **endoskeleton,** that is both rigid and flexible and yet able to bear far more weight than chitin. Instead of a rigid exterior skeleton, vertebrates have a soft, flexible exterior—skin—that stretches to accommodate body movements (figure 36.3). If vertebrate skin did not stretch, it would tear whenever an arm or leg was bent. Skin flexibility is a necessary component of vertebrate movement, which otherwise is determined largely by muscles and their attachments to bones. Vertebrate skin is described in detail in chapter 41.

Movement, especially on land, is complicated by gravity. Muscles have great contractility, but for this to be useful, the muscles must exert a force against something. Force can be exerted against anything providing resistance: water, external skeletons, or internal skeletons.

The Human Skeleton

The endoskeleton of humans is made up of 206 individual bones (figure 36.4). These can be grouped according to their functions. The 80 bones of the **axial skeleton** support the main body axis, and the 126 bones of the **appendicular**

Figure 36.3 On the move.
The different movements of animals are the result of muscle contractions.

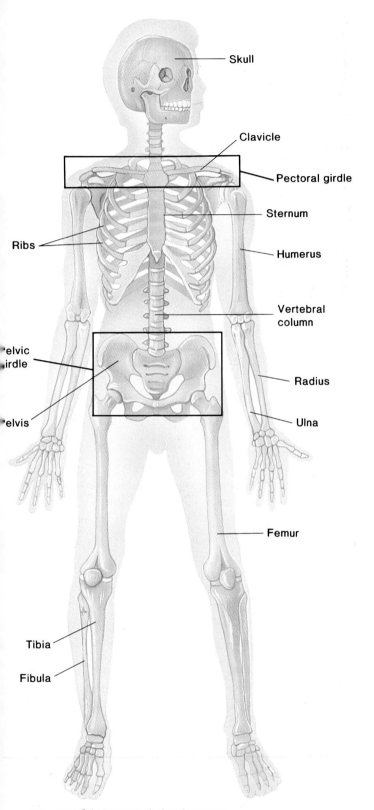

Figure 36.4 Human skeletal system.
The human skeleton consists of 206 individual bones, only a few of which are labeled here.

skeleton support the arms and legs (figure 36.5). Interestingly, the body's motor control systems have evolved so that the muscles of the axial skeleton (postural muscle) and the appendages (manipulatory muscle) are controlled more or less independently.

The Axial Skeleton

The axial skeleton is made up of the skull, backbone, and rib cage. Of the skull's twenty-eight bones, eight form the **cranium,** which encases the brain; the rest are facial bones and middle-ear bones. The skull also contains the **hyoid bone.** It is suspended at the back of the jaw by muscles and a form of connective tissue called a **ligament** and supports the base of the tongue.

The skull is attached to the anterior (upper) end of the backbone, which is also called the **spine** or **vertebral column.** The spine is made up of thirty-three **vertebrae,** stacked to provide a flexible column that surrounds and protects the spinal cord. Curving forward from the vertebrae are twelve pairs of ribs, which are attached at the front to the breastbone, or **sternum,** forming a protective cage around the heart and lungs.

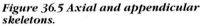

Figure 36.5 Axial and appendicular skeletons.
The axial skeleton (shown in red) provides the body's vertical support, and the appendicular skeleton (shown in yellow) is composed of the body's appendages (arms and legs).

The Appendicular Skeleton

The 126 bones of the appendicular skeleton are attached to the axial skeleton at the shoulders and hips. The shoulder or **pectoral girdle** is composed of two large, flat shoulder blades (scapulae), each connected to the breastbone by a slender, curved collarbone (clavicle). The arms are attached to the pectoral girdle. Each arm and hand contains thirty-two bones. The clavicle is the most frequently broken bone of the body because a fall on an outstretched arm results in a large component of the force being transmitted to the clavicle.

The **pelvic girdle** forms a bowl that provides strong connections for the legs, which carry the body weight. Each leg and foot contains thirty bones.

The human skeleton consists of 206 bones arranged into an axial, or central, skeleton, and an appendicular skeleton containing the bones of the arms and legs. Bone serves not only as an attachment site for muscles, but also as a protective covering for those organs encased by it, such as the brain.

Bones: What Vertebrate Skeletons Are Made Of

Bone is a special form of connective tissue in which collagen fibers (see chapter 33) are coated with a calcium phosphate salt. The great advantage of bone over chitin as a structural material is that it is strong without being brittle.

The properties of bone can be compared to those of fiberglass. Fiberglass is composed of glass fibers embedded in epoxy glue. The individual fibers are rigid, giving great strength, but they are brittle. The epoxy component of fiberglass, on the other hand, is flexible but weak. The composite—fiberglass—is both rigid and strong. When a fiber breaks due to stress and a crack begins to form, the crack runs into glue before it reaches another fiber. The glue distorts and reduces the concentration of the stress, and consequently, the adjacent fibers are not exposed to the same high stress. In effect, the glue acts to spread the stress over many fibers.

Bone's internal structure is similar to that of fiberglass because the collagen fibers in bone run in various directions. Small, needle-shaped crystals of a calcium-containing mineral—**hydroxyapatite**—surround and impregnate bone collagen fibers. The fibers run parallel to the axes of long bones and also to the curved ends of bones in joints. Because of the placement of these fibers, no crack can penetrate far into bone without encountering a hard mass of hydroxyapatite crystals embedded in a collagenous matrix. Bone is more rigid than collagen, and it is also more flexible and resistant to fracture than either hydroxyapatite or chitin.

Bone consists of many concentric layers called **lamellae,** like so many layers of paint on an old pipe. Each layer is composed of repeating units called **Haversian canals,** narrow channels that run parallel to the length of bone. Haversian canals contain a central canal that houses blood vessels and nerve cells. The blood vessels provide a lifeline to bone-forming cells called **osteoblasts,** and the nerve cells in the central canal control the diameter of the blood vessels and thus the flow of blood within them. Spaces in the periphery of the Haversian canals are called **lacunae.** The lacunae contain **osteocytes,** which can be thought of as "mature" osteoblasts that no longer form bone (figure 36.6).

Bone is a dynamic, living tissue that is constantly being renewed. New bone is formed by osteoblasts located in the layer of bone that has not yet organized itself into Haversian canals.

As an osteoblast secretes collagen fibers and calcium crystals, it becomes trapped in the new bone that it is forming. At this point, the osteoblast is then called an osteocyte. Osteocytes do not form new bone; instead, they function to maintain the daily cellular activities of the bone tissue.

Bone is a connective tissue in which collagen fibers are surrounded by hydroxyapatite. The internal structure of bone is organized into Haversian canals. New bone is formed by osteoblasts, which eventually become trapped in the new bone that they lay down. These mature, trapped osteoblasts are called osteocytes, and they maintain the daily cellular activities of bone tissue.

The bones of the vertebrate skeleton have two structural elements: The ends and interiors of long bones are composed of an open lattice of bone called **spongy bone tissue,** or **marrow.** Most of the body's red blood cells are formed within this lattice framework. Surrounding the spongy bone tissue at the core of bones are concentric layers of **compact bone tissue,** in which the collagen fibers are laid down in a far denser pattern than in the marrow (see figure 36.6). Compact bone tissue gives bones the strength to withstand mechanical stress.

Joints: How Bones of the Skeleton Are Attached to One Another

Mechanical body movement occurs when bones move relative to one another at intersection points called **joints.** There are three kinds of joints (figure 36.7):

1. Sutures are nearly immobile joints connected by a thin layer of fibrous connective tissue. The cranial bones of the skull are joined by sutures. In a fetus, these bones are not fully formed, and open areas of connective tissue between the bones ("soft spots," or **fontanels**) allow the bones to slide over one another slightly as the fetus makes its journey down the birth canal during childbirth. Later, bone replaces this connective tissue.

2. Cartilaginous joints are slightly movable joints bridged by articular cartilage. The vertebral bones of the spine are separated by pads of cartilage called **intervertebral disks.** The disks allow some movement while acting as efficient shock absorbers.

3. Freely movable joints are swinging joints bridged by pads of articular cartilage and held together by tough strips of connective tissue called **ligaments.** The pads of cartilage that tip the two bones do not touch. Fluid-filled membranes, called **synovial membranes,** lubricate the gliding of the bones across one another. **Rheumatoid arthritis** is a degenerative and very painful disorder in which the immune system attacks these synovial membranes; white blood cells degrade the cartilage and other connective tissue, and bone is deposited in the joint.

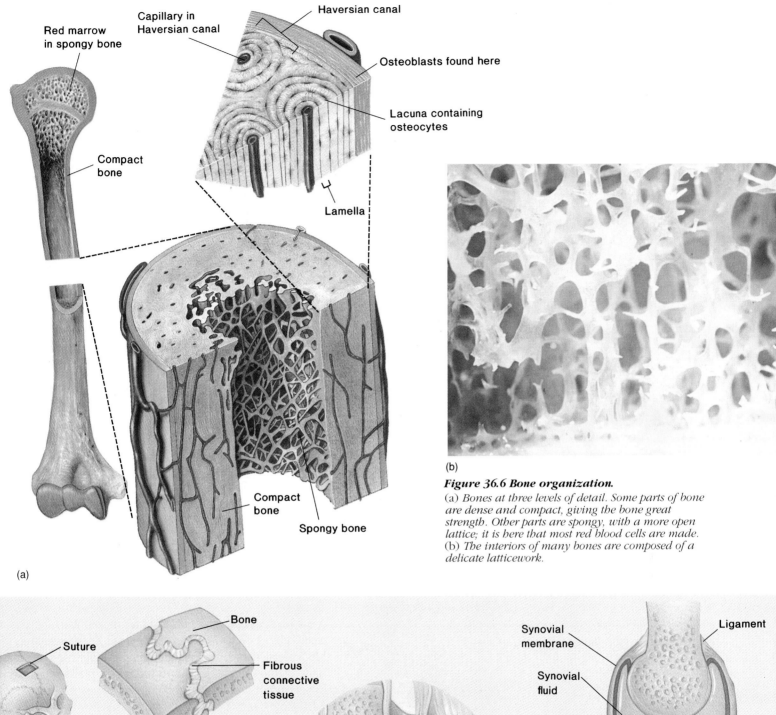

Red marrow in spongy bone

Compact bone

Capillary in Haversian canal

Haversian canal

Osteoblasts found here

Lacuna containing osteocytes

Lamella

Compact bone

Spongy bone

(a)

(b)

Figure 36.6 Bone organization.
(a) *Bones at three levels of detail. Some parts of bone are dense and compact, giving the bone great strength. Other parts are spongy, with a more open lattice; it is here that most red blood cells are made.* (b) *The interiors of many bones are composed of a delicate latticework.*

Suture

Bone

Fibrous connective tissue

(a) Suture

Figure 36.7 Three types of joints.
(a) *Sutures are immovable joints that are held together by fibrous connective tissue.* (b) *Cartilaginous joints contain articular cartilage that cushions the two bones of the joint.* (c) *In freely movable joints, a synovial membrane filled with synovial fluid provides lubrication.*

Body of vertebra

Articular cartilage

Intervertebral disk

(b) Cartilaginous joint

Synovial membrane

Synovial fluid

Fibrous capsule

Ligament

Articular cartilage

(c) Freely movable (synovial) joint

Tendons: How Bones Are Connected to Muscles

Muscles are attached to bones by straps of dense, collagenous connective tissue called **tendons.** Bones pivot about joints because of where the tendons are attached to the bones. Each muscle acts to pull on a specific bone. One end of the muscle—the **origin**—attaches to a bone that remains stationary during a contraction. This provides an object against which the muscle can pull. The other end of the muscle—the **insertion**—is attached to a bone that moves if the muscle contracts.

In the freely movable joints of vertebrates, muscles are attached in opposing pairs: flexors and extensors. When the flexor muscle of the leg contracts, the lower leg is moved closer to the thigh. When the extensor muscle of the leg contracts, the lower leg is moved in the opposing direction, further away (figure 36.8). Efficient use of body energy requires that antagonistic muscle pairs be controlled. If, for example, the triceps contracted at the same time as the biceps, the arm would not move.

A joint can be immovable (suture, skull), slightly movable (cartilaginous, spine), or freely movable (knee). Ligaments are tough strips of connective tissue that hold bones to bones, while tendons are similar in composition but attach muscles to bones.

Muscle: How the Body Moves

Even though the cells of almost all eukaryotic organisms appear capable of shape changes, many multicellular animals have evolved specialized cells devoted almost exclusively to this purpose. These cells contain numerous filaments of the proteins **actin** and **myosin.** Such specialized animal cells are called **muscle cells.**

As mentioned in chapter 33, vertebrates possess three different kinds of muscle cells: smooth muscle, skeletal muscle, and cardiac muscle (see table 33.4). Of these, skeletal and cardiac muscle show clear patterns of striations caused by the geometric arrangement of actin and myosin filaments. Taken together, all the muscle cells make up the body's muscular system (figure 36.9).

Skeletal muscles are the muscles associated with the skeleton. They, along with cardiac muscle, are called **striated muscles** because they are marked with obvious lines. Skeletal muscle cells are produced during development by the fusion of several cells at their ends to form a very long fiber. Each muscle cell or muscle fiber still contains all the original nuclei pushed out to the periphery of the cytoplasm.

Each skeletal muscle is a tissue made up of numerous individual muscle cells that act as a unit and that are specialized for rapid contractions and large forces. In contrast, smooth muscles are specialized for slow, maintained contraction with minimal energy utilization. Contraction of smooth muscle is analogous to a large raft being towed upstream by many small canoes, with each canoe bound to the raft by its own towline. In other words, each smooth muscle cell participates individually in muscle contraction. The contraction of skeletal muscle, on the other hand, is analogous to placing all the rowers (muscle cells) in one galley, where they row in concert, pulling the raft far more effectively.

How Muscles Work

Muscle cells move by expanding and contracting portions of their surfaces. This ability to alter surface relationships arises from dynamic changes in their cytoskeletons (supporting matrices of protein within eukaryotic cells; see chapter 4). The key elements driving these changes are tiny cables within muscle cells called **myofilaments,** which are long chains of the proteins actin and myosin.

The Microscopic Structure of Muscle

A micrograph of a muscle cell shows several rectangular-shaped blocks traversed by parallel lines. These repeating blocks, called **sarcomeres,** are the individual contracting units of the muscle cell. A **myofibril** is made up of a long chain of sarcomeres, lined up like the cars in a train. All the sarcomeres of a muscle cell contracting together produce the characteristic movement of a muscle cell.

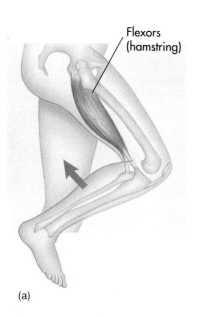
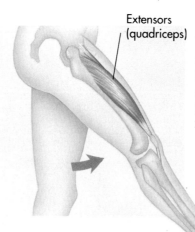

Flexors (hamstring)

Extensors (quadriceps)

(a)

(b)

Figure 36.8 Flexor and extensor muscles.
(a) *Flexors move the bone to which they are attached back toward the body.* (b) *Extensors move the bone away from the body. Antagonistic muscle pairs, such as hamstrings and quadriceps, allow humans to walk and run.*

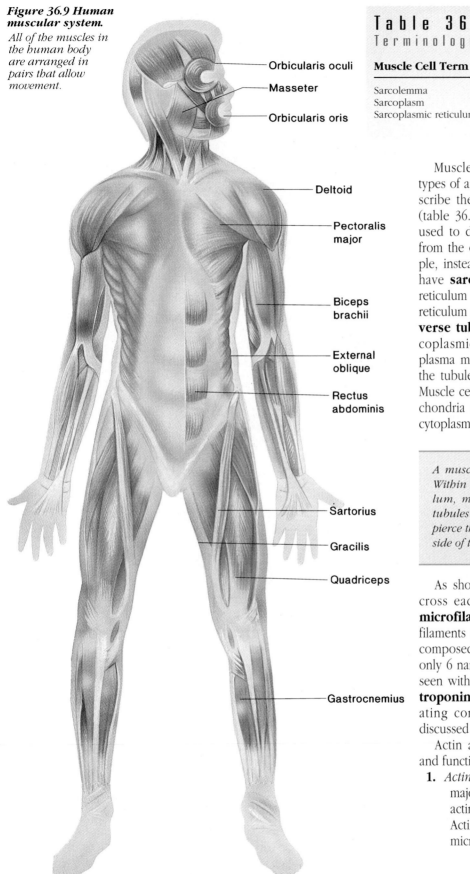

Figure 36.9 Human muscular system.

All of the muscles in the human body are arranged in pairs that allow movement.

Orbicularis oculi
Masseter
Orbicularis oris
Deltoid
Pectoralis major
Biceps brachii
External oblique
Rectus abdominis
Sartorius
Gracilis
Quadriceps
Gastrocnemius

Table 36.1 Muscle Cell Terminology

Muscle Cell Term	General Animal Cell Term
Sarcolemma	Plasmalemma (plasma membrane)
Sarcoplasm	Cytoplasm
Sarcoplasmic reticulum	Endoplasmic reticulum

Muscle cells have the same anatomy as other types of animal cells, but the terminology used to describe their internal structures is slightly different (table 36.1). The prefix *sarco,* meaning "flesh," is used to differentiate certain muscle cell organelles from the organelles of other animal cells. For example, instead of endoplasmic reticulums, muscle cells have **sarcoplasmic reticulums;** the sarcoplasmic reticulum is comparable to the smooth endoplasmic reticulum of other cells. Hollow tubules called **transverse tubules (T-tubules)** wrap around the sarcoplasmic reticulum and pierce the muscle cell plasma membrane (called the **sarcolemma**) so that the tubules open to the outside of the muscle cell. Muscle cells also contain a nucleus and several mitochondria scattered throughout the **sarcoplasm,** or cytoplasm, of the muscle cell (figure 36.10).

A muscle cell is surrounded by the sarcolemma. Within the sarcoplasm are the sarcoplasmic reticulum, mitochondria, and a nucleus. Transverse tubules traverse the sarcoplasmic reticulum and pierce the sarcolemma so that they open to the outside of the muscle cell.

As shown in figure 36.10, the parallel lines that cross each sarcomere are the actin and myosin **microfilaments.** The prefix *micro* connotes that the filaments are microscopic. The myofilaments that are composed of microfilaments are also very small—only 6 nanometers thick—which is far too fine to be seen with the naked eye. Two additional proteins—**troponin** and **tropomyosin**—are important in initiating contractions in skeletal muscle; they are discussed in more detail later in the chapter.

Actin and myosin have characteristic structures and functions in the sarcomere:

1. *Actin.* Actin microfilaments are one of the two major components of myofilaments. Individual actin proteins are the size of a small enzyme. Actin molecules polymerize to form thin microfilaments. The microfilaments consist of

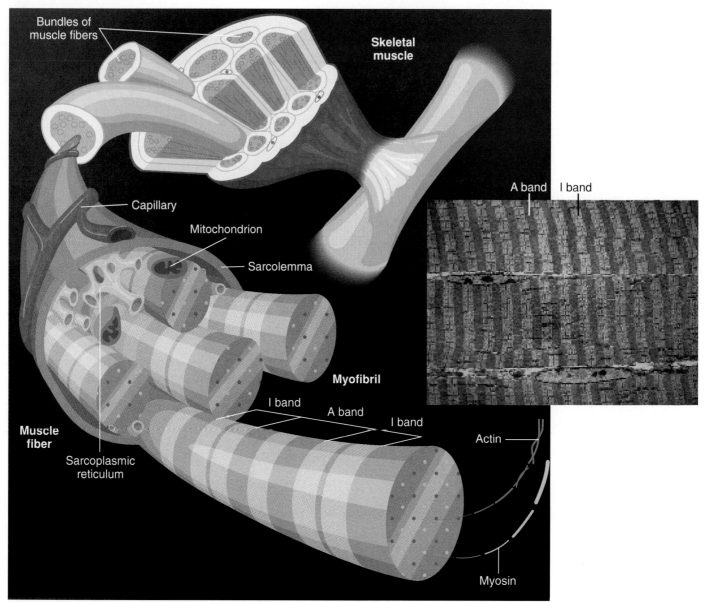

Figure 36.10 Arrangement of myofilaments in a muscle fiber.

This diagram shows a large to small view of the muscle fiber. As you can see, individual myofilaments are composed of actin and myosin microfilaments.

two strings of monomers wrapped around one another, like two strands of pearls loosely wound together. The result is a long, thin, helical microfilament with a diameter of about 6 nanometers (figure 36.11a).

2. *Myosin.* The other major protein associated with myofilaments—myosin—is composed of several myosin protein molecules that are more than ten times longer than an individual actin microfilament. Myosin has an unusual shape: One end of the molecule consists of a very long rod, whereas the other end consists of a double-headed globular region. In electron micrographs, a myosin molecule looks like a two-headed snake (figure

36.11b). A myosin microfilament is composed of several myosin molecules (figure 36.11c).

The secret of muscle contraction lies in the interactions of the actin and myosin microfilaments that form the myofilaments of a muscle cell. Figure 36.12 shows a myofilament composed of a myosin microfilament interposed between two pairs of actin microfilaments. The heads of the myosin microfilament jut out toward the actin microfilaments on each side, making contact with the actin microfilaments for brief intervals. As explained in more detail later in the chapter, these brief contacts cause the myofilament to shorten, and the muscle cell contracts.

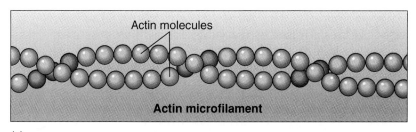

(a)

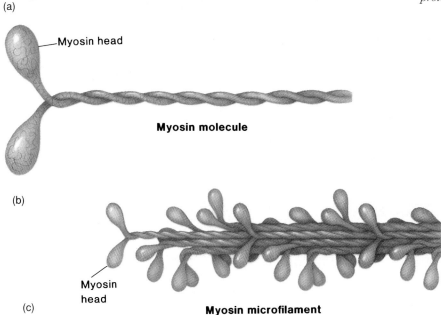

(b)

(c)

Figure 36.11 Structure of actin and myosin microfilaments.

(a) *Actin microfilaments are composed of two actin molecules wrapped around each other.* (b) *Each myosin molecule is a coil of two chains wrapped around one another. At the end of each chain is a globular region referred to as the "head."* (c) *Myosin microfilaments are composed of several myosin molecules bunched together. The myosin heads protrude from the microfilament.*

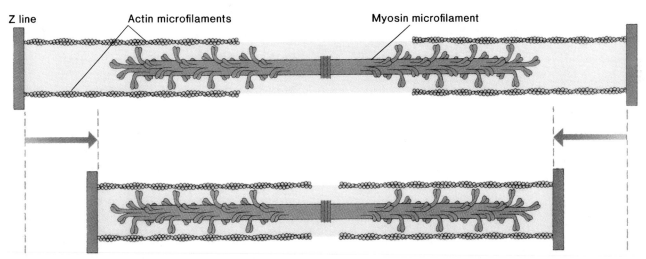

Figure 36.12 Interaction of actin and myosin microfilaments in vertebrate muscle.

As the right-hand end of the myosin microfilament "walks" along the actin microfilaments, pulling them toward the center, the left-hand end of the same myosin microfilament "walks" in a leftward direction, pulling its actin microfilaments toward the center. The outcome is that both ends of the actin microfilaments move toward the center, resulting in contraction.

The myofilaments are arranged in a pattern within the sarcomere. The parallel lines within the sarcomere are denser and darker in the middle of the sarcomere and fainter at either side (figure 36.13*a*). The denser portion of the lines are the myofilaments, which appear denser because of the overlapping of the actin and myosin microfilaments. This portion of the sarcomere is called the **A band.** In the middle of the A band is an area composed of only myosin microfilaments; this area is called the **H zone.** The lighter lines out to the sides of the A band are composed of actin microfilaments only; these are called the **I bands.** Finally, the dense lines that separate the individual sarcomeres are called the **Z lines.** Z lines are composed of a protein called **actinin,** which is another essential protein (besides actin and myosin) necessary for muscle contraction (figure 36.13*b, c*).

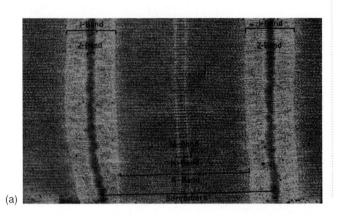

(a)

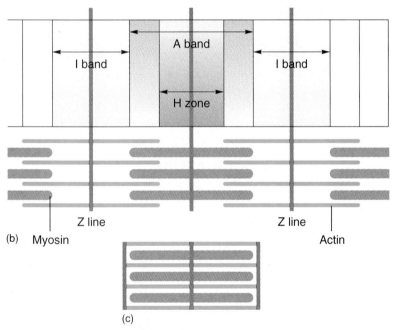

(b)   Myosin   Z line   Z line   Actin

(c)

Figure 36.13 Structure of a sarcomere.
(a) *Micrograph of a sarcomere. The muscle is at rest.* (b) *Compare (b) to (a). A center portion is darker than the other portions; these lines are the A bands, which are composed of myofilaments. In the center of the A bands, a lighter band represents bands of myosin microfilaments, the H zone. On either side of the A bands are lighter lines composed of actin microfilaments; these are I bands. The Z lines at the ends of the sarcomere are composed of actinin and anchor the actin microfilaments.* (c) *The A band in its contracted state. In contraction, the actin and myosin microfilaments move toward the center, and the I band "disappears."*

The microfilaments are arranged in patterns in the sarcomere. The dense A band contains myofilaments. The H zone in the middle of the A bands is composed of myosin microfilaments only; the I bands on either side of the A bands are composed of actin microfilaments only. The Z lines are composed of actinin and separate the sarcomeres.

How Myofilaments Contract: The Sliding Filament Theory of Muscle Contraction

As mentioned earlier, the myosin heads of the myosin microfilaments jut outward toward the actin microfilaments on either side. During muscle contraction, the myosin heads change their tilt so that they can briefly connect to the actin microfilaments. These connections are called **cross-bridges.** Almost immediately after a cross-bridge is made, the myosin head returns to its original position. This action is repeated many times so that, in effect, the myosin "walks" step by step along the actin microfilament. The actin microfilaments are pulled inward and thus slide past the myosin filaments (see figure 36.12). Each step that the myosin heads take along the actin microfilament requires one molecule of ATP.

How does the formation of cross-bridges lead to the contraction of the muscle cell? Because the actin microfilaments are anchored at both ends to the actinin found in the Z lines, the zones between the actinin anchors shorten when the actin microfilaments slide past the myosin. This sliding of the actin microfilaments inward as a result of the formation of myosin cross-bridges is called the **sliding filament theory** of muscle contraction (figure 36.14).

The sliding filament theory of muscle contraction states that the formation of cross-bridges (contacts between the myosin heads and the actin microfilaments) causes the actin microfilaments to slide inward past the myosin. The actin microfilaments shorten, or contract.

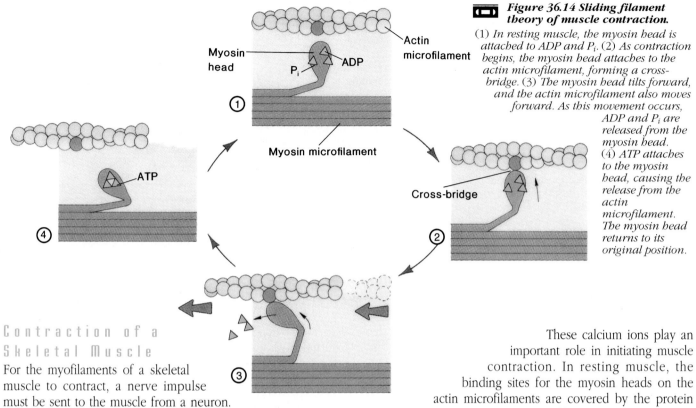

Figure 36.14 Sliding filament theory of muscle contraction.
(1) *In resting muscle, the myosin head is attached to ADP and P_i. (2) As contraction begins, the myosin head attaches to the actin microfilament, forming a cross-bridge. (3) The myosin head tilts forward, and the actin microfilament also moves forward. As this movement occurs, ADP and P_i are released from the myosin head. (4) ATP attaches to the myosin head, causing the release from the actin microfilament. The myosin head returns to its original position.*

Contraction of a Skeletal Muscle

For the myofilaments of a skeletal muscle to contract, a nerve impulse must be sent to the muscle from a neuron. This nerve impulse initiates contraction by causing calcium ions to be released into the sarcoplasm of the muscle cell. These calcium ions interact with two other proteins mentioned earlier—troponin and tropomyosin—to uncover myosin binding sites on the actin microfilaments.

How Nerves Signal Muscles to Contract

In vertebrate striated skeletal muscle, contraction is initiated by a nerve impulse. The nerve impulse arrives as a wave of depolarization along the neuronal membrane. As explained in chapter 34, the axon of the neuron that transmits the nerve impulse is embedded in the surface of the muscle fiber, forming a **neuromuscular junction.** When a wave of depolarization reaches the axon, at the point where the axon attaches to a muscle (the **motor endplate**), it causes the neuronal membrane to release the chemical acetylcholine into the junction. The acetylcholine passes across to the sarcolemma and opens the ion channels of that membrane, depolarizing it (figure 36.15).

The Calcium "Switch"

How does the depolarization of the muscle membrane cause the muscle fibers to contract? The sarcoplasmic reticulum of the muscle cell is embedded with numerous ion channels that permit calcium ions to pass through. Thus, in resting muscle, calcium ions enter the sarcoplasmic reticulum and are stored there. When the sarcolemma is depolarized, the nerve impulse is carried from the T-tubules located within the sarcolemma deep into the muscle cell. When the T-tubules contact the sarcoplasmic reticulum, the nerve impulse depolarizes the sarcoplasmic reticulum and causes calcium ions to be released from its interior into the sarcoplasm.

These calcium ions play an important role in initiating muscle contraction. In resting muscle, the binding sites for the myosin heads on the actin microfilaments are covered by the protein tropomyosin. Tropomyosin is a microfilament protein like actin, but interspersed along the tropomyosin are molecules of the globular protein troponin. The troponin anchors the tropomyosin to the actin microfilament by binding both the tropomyosin and actin. When calcium ions are released into the sarcoplasm, they bind to the troponin molecules, which causes the shape of the troponin molecules to change. As a result of this change, the tropomyosin is repositioned out of the

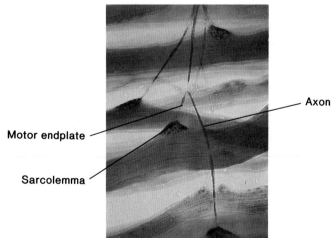

Figure 36.15 A neuromuscular junction.
The axon of a neuron contacts the motor endplate of the muscle cell. Acetylcholine released from the axon enters the synaptic cleft. The acetylcholine opens ion channels in the sarcolemma, depolarizing the membrane.

way of the myosin binding sites on the actin microfilament. Thus, the binding of calcium ions to troponin molecules uncovers the myosin binding sites on the actin. The myosin heads are now free to form cross-bridges with the actin microfilaments and, with ATP expenditure, move along the actin in a stepwise fashion (figure 36.16).

Calcium ions are stored in the sarcoplasmic reticulum of a resting muscle cell. When the sarcolemma is depolarized, the nerve impulse is conducted into the muscle cell by the T-tubules. The nerve impulse depolarizes the sarcoplasmic reticulum, and calcium ions are released into the sarcoplasm. These calcium ions bind to the troponin that, with tropomyosin, covers the myosin binding sites on the actin. The calcium binding causes the uncovering of the myosin binding sites on the actin so that contraction can take place.

The Motor Unit: Strength and Duration of Muscle Contraction

Muscle contraction is used to perform all kinds of functions, but these functions do not all require the same amount of muscle contraction. Walking, for instance, does not require the same kind of muscle contraction as running. The amount of muscle contraction is controlled by regulating the number of muscle fibers that contract.

A **motor unit** is defined as a single motor neuron and all the muscle myofilaments it controls. A single motor neuron can control as many as 150 myofilaments; some motor neurons control many fewer than this number, depending on the location in the body. When a motor neuron is stimulated, all the myofilaments within the motor unit contract. Thus, a muscle cell's strength of contraction is controlled by the number of motor units that contract at one time.

Increasing the strength of contraction is called **motor unit recruitment** or **summation.** All of the motor units in a muscle do not contract at one time. The term *summation* refers to the fact that, as more and more motor neurons in a muscle are stimulated, more motor units are activated to contract.

This type of contraction prevents all of the myofilaments in the motor units of a muscle from tiring too easily. In addition, the asynchronous contraction of the motor units produces smooth muscle movement, as opposed to the jerky movement that would result if all the motor units contracted simultaneously. Finally, recruitment allows some motor units to be in a contracted state at all times, a condition that contributes to **muscle tone.** Muscle tone is essential in maintaining posture.

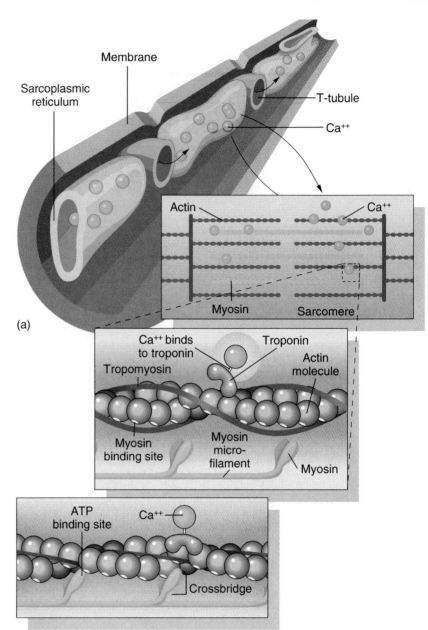

(a)

(b)

Figure 36.16 The calcium "switch" that initiates muscle contraction.

(a) *In resting muscle, calcium ions are stored in the sarcoplasmic reticulum. When an action potential reaches the muscle cell, the T-tubules carry the action potential deep into the sarcoplasm. The action potential causes the sarcoplasmic reticulum to release its store of calcium ions.* (b) *In resting muscle, the myosin binding sites are covered by troponin and tropomyosin. The calcium ions released into the sarcoplasm as a result of an action potential bind to the troponin. This binding causes the tropomyosin and troponin to move out of the way of the myosin binding sites, leaving the myosin heads free to bind to the actin microfilament.*

A single motor neuron and all the muscle myofilaments it controls is called a motor unit. As more motor units are stimulated, more and more of a muscle's complement of myofibrils are brought into the contraction, increasing its force.

EVOLUTIONARY VIEWPOINT

The basic structure of skeletal muscle has changed little during vertebrate evolution (although the nature of motor units is quite flexible). All vertebrate skeletal muscle is fired by a calcium ion (Ca^{++}) trigger, the calcium ion influx into the muscle cell initiated by a nerve impulse. The universal necessity of a properly functioning calcium ion trigger has made the maintenance of proper calcium levels in the vertebrate body a key regulatory goal. The gland that secretes the calcium-ion-regulating hormone is one of only two hormone-producing glands whose removal is fatal to any vertebrate.

SUMMARY

1. In vertebrates, movement results from the contraction of muscles anchored to bones. When the limb containing the bone is pulled to a new position, the skin stretches to accommodate the change.

2. The human skeleton is organized into the axial and appendicular skeletons. The axial skeleton provides the body's main vertical support; the appendicular skeleton contains the bones of the appendages, such as the legs and arms.

3. Bone is a form of connective tissue in which collagen fibers are impregnated with calcium salts. The salts act like the glass fibers in fiberglass, creating a material that is strong but not brittle.

4. Bones move relative to one another at intersection points called joints. The three types of joints are: (1) sutures, (2) cartilaginous joints, and (3) freely movable joints.

5. Tendons are straps of dense, collagenous connective tissue attached to bones that allow bones to pivot about joints.

6. Muscle cells contain numerous filaments of the proteins actin and myosin. Skeletal muscles are striated muscles associated with the skeleton that are specialized for rapid contractions and large forces.

7. A muscle cell is made up of individual sarcomeres separated by Z lines. The plasma membrane of a muscle cell is the sarcolemma; the sarcoplasmic reticulum (endoplasmic reticulum) is bound by transverse tubules (T-tubules) that open to the outside of the cell. The cytoplasm of a muscle cell is called the sarcoplasm.

8. The myofilaments in the sarcomere are arranged in a characteristic pattern. The A bands are myofilaments. The H zone in the middle of the A lines consists of myosin microfilaments only. The faint I bands on either side of the A bands consist of actin microfilaments only.

9. Muscle cells contract as a result of shortening of myofilaments within the cytoskeleton. The myofilaments are composed of the proteins actin and myosin, together with two other proteins that together control the contraction.

10. In a myofilament, the myosin is located between adjacent actin microfilaments. Changes in the shape of the heads of the myosin molecule, driven by the splitting of ATP molecules, cause the myosin molecule to form cross-bridges with the actin microfilament. Repeated cross-bridge formations lead to the myosin "walking" step by step along the actin microfilament, producing contraction of the myofilament. This is called the sliding filament theory of muscle contraction.

11. In vertebrate skeletal muscle, contraction is initiated by a nerve impulse. Acetylcholine passes across the neuromuscular junction from the nerve to the muscle, initiating the process that causes the muscle to contract.

12. Muscle contraction is controlled by a calcium "switch." Depolarization of the sarcolemma causes calcium ions to be released from the sarcoplasmic reticulum into the sarcoplasm. The calcium ions bind to troponin, which, together with tropomyosin, is bound to actin microfilaments in resting muscle and covers the myosin binding sites on actin. The binding of calcium to troponin causes the tropomyosin to move out of the way of the myosin binding sites. Myosin is now free to form cross-bridges with the actin microfilaments.

13. A motor unit is one motor nerve and all the muscle myofilaments it controls. Contraction strength is controlled by the number of motor units activated. Motor unit summation or recruitment is the sum of the force exerted by the nonsimultaneous contractions of the motor units.

REVIEWING THE CHAPTER

1. What mechanical problems are posed by movement?

2. Why is there a need for a skeleton?

3. Which animals have hydroskeletons?

4. In which animals are exoskeletons found?

5. What is an endoskeleton, and which animals have them?

6. How is the human skeleton structured?

7. What is the structure and function of the axial skeleton?

8. What is the structure and function of the appendicular skeleton?

9. What is the composition of the vertebrate skeleton?

10. What is the internal structure of bone?

11. How are the bones of the skeleton attached to one another?

12. How do tendons connect bone to muscle?

13. How do muscles provide movement for the body?

14. What is the microscopic structure of muscle?

15. How do myofilaments contract?

16. How does skeletal muscle contract?

17. How do nerves signal muscles to contract?

18. What is the calcium "switch," and how does it work?

19. How is the motor unit the strength and duration of muscle contraction?

COMPLETING YOUR UNDERSTANDING

1. Why do many people not think of plants as being alive in the same way as animals?

2. From where is the energy for lifting an object derived?

3. What are elevator and depressor muscles, where are they found, and how do they function?

4. Why are insects not the size of dogs, cats, or humans?

5. Why must vertebrates have flexible skin?

6. The two bones that compose the forearm are the _____ and _____.

    a. Tibia and humerus
    b. Humerus and fibula
    c. Fibula and femur
    d. Ulna and radius
    e. Radius and fibula
    f. Femur and tibula

7. Where do the appendicular and axial skeletons connect?

8. What advantages does bone have over chitin as a structural material?

9. How is the structure of bone similar to that of fiberglass?

10. Bone-forming cells are called
    a. osteoblasts.
    b. osteocytes.
    c. lamellae.
    d. lacunae.
    e. hydroxyapatites.
    f. joints.

11. Haversian canals
    a. have bone lamellae laid down around them in concentric rings.
    b. run parallel to the long axes of bones.
    c. are interconnected.
    d. have blood vessels and nerves running through them.
    e. all of the above.

12. Which is more likely to be found at the core of bones?
    a. Compact bone tissue
    b. Spongy bone tissue (marrow)
    c. Osteoblasts
    d. Myofilaments
    e. Nothing; the interior is hollow

13. In a human fetus, why are the cranial bones not fully formed at the time of childbirth?

14. What is the function of cartilaginous joints?

15. What is the basis of rheumatoid arthritis?

16. What allows bones to pivot about joints in the arms and legs?

17. When you lift an object, why do the triceps and biceps muscles in your arm contract at different times?

18. Striated muscles include _____ and _____ muscles.
    a. Skeletal and smooth
    b. Cardiac and skeletal
    c. Smooth and cardiac
    d. Only smooth

19. Why is the terminology for describing the internal structure of muscle cells different from that used to describe other animal cells?

20. What are troponin and tropomyosin, and how do they function?

21. How do myosin and actin filaments differ, and how do they interact with one another?

22. Which of the following is *not* a major component of muscle?
    a. Albumin
    b. Actin
    c. Myosin
    d. Troponin
    e. Tropomyosin

23. When a sarcomere contracts, which of the following disappear?
    a. I band
    b. A band
    c. H zone
    d. Z line
    e. Actin
    f. Myosin

24. What is muscle tone, and how is it achieved?

25. Myofilaments can contract forcefully, pulling membranes attached to the two ends toward one another. Myofilaments cannot expand, however, pushing membranes attached to the two ends of a myofilament apart from one another. Why can myofilaments pull but not push?

26. Among long-distance runners and committed joggers, the long bones of the leg often develop "stress fractures," numerous fine cracks running parallel to one another along the lines of stress. In most instances, stress fractures occur when runners push themselves much farther than they are accustomed to running. Runners who train by gradually increasing the distances they run develop stress fractures rarely, if ever. What protects this second kind of runner?

FOR FURTHER READING

Alexander, R. 1991. How dinosaurs ran. *Scientific American*, April, 130–36. Takes the approach of a structural engineer to argue that many giant dinosaurs were formidable running machines.

Carafoli, E., and J. Penniston. 1985. The calcium signal. *Scientific American*, Nov., 70–78. Discusses how the release of calcium ions is the only known way in which the electricity of the nervous system is able to produce changes in the body. Nerves regulate all muscle contractions and hormone secretions by controlling the level of calcium ions.

Cohen, C. 1975. The protein switch of muscle contraction. *Scientific American*, Nov., 36–45. Explains how proteins associated with myofilaments interact with calcium ions to trigger contraction.

Gordon, K. 1989. Adaptive nature of skeleton design. *BioScience*, Dec., 784–90. Plasticity that allows changes in strength and locomotion.

Goslow, G. et al. 1990. Bird flight—Insights and complications. *BioScience*, Feb., 108–15. Discussion of how flapping flight evolved. While this evolution is still a puzzle, key questions have been answered.

Hildebrand, M. 1987. The mechanics of horse legs. *American Scientist*, Nov., 594–601. A delightful analysis of the horse leg as a lever system, from the perspective of physics.

Schmidt-Nielsen, K. 1983. *Animal physiology: Adaptation and environment*. 3d ed. New York: Cambridge University Press. Chapter 11 presents an outstanding treatment of muscles and bones from an evolutionary perspective.

Weeks, O. 1989. Vertebrate skeletal muscle: Power source for locomotion. *BioScience*, Dec., 791–99. How the organization of muscle fibers is highly adaptable.

EXPLORATIONS

Interactive Software:
Muscle Contraction

This interactive exercise allows students to explore how the proteins in muscle cells interact with one another to produce muscle contraction. The exercise presents a diagram of a myofilament, with myosin "walking" along actin. Students can explore how changing the ATP concentration affects the speed of muscle contraction, learning that diminished ATP slows rather than weakens the force of an individual myofibril's contraction. Students can then investigate a muscle fiber made of many myofilaments, and explore how the force of the fiber's contraction depends upon changes in calcium ion levels and why a more forceful muscle contraction results from repeated firing of the motor neuron.

Questions to Explore:

1. Why are skeletal muscles able to pull but not push?

2. How can the force exerted by an individual sarcomere be changed?

3. What would be the expected effect of punching a tiny hole in the sarcoplasmic reticulum?

4. What is the effect of calcium deficiency upon muscle contraction?

HORMONES

*Humans do not have a monopoly on af-
fection. These Japanese macaques live
in a close-knit community whose members cooperate to ensure successful
breeding and raising of offspring. Not
everybody has a family at the same
time—hormones control when particular
animals breed.*

FOR REVIEW

Here are some important terms and concepts that have been discussed in previous chapters and that you will encounter again in this chapter. Review them before proceeding if necessary.

Receptor proteins (*chapter 5*)
Neurotransmitters (*chapter 34*)
Operation of neurons (*chapter 34*)
Central nervous system (*chapter 35*)
Sympathetic nervous system (*chapter 35*)

The tissues and organs of an adult mammal participate in a multitude of activities: capturing oxygen and digesting food, walking and singing, and seeing far into the distance. All of these activities must be coordinated to avoid conflict and to maximize interaction (figure 37.1). This is why we think of ourselves as *organisms,* rather than as smoothly functioning collections of organs. Integration of the many activities of the vertebrate body is the primary function of the central nervous system. Some central nervous system control of body functions is directed through the motor nerves of the voluntary and autonomic components of the peripheral nervous system. This chapter examines a second control system—the **neuroendocrine system**—which differs in its speed of expression, duration of response, and narrowness of application.

The Importance of Chemical Messengers

To control the vertebrate body, the central nervous system employs a battery of specific chemical messengers called **hormones.** Why use a chemical messenger rather than an electrical one, the kind that neurons transmit? For electrical signals to be effective, they must be transmitted to individual cells. This requires many neurons and is wasteful when the desired result is to affect the metabolic or other activity of a group of tissues, organs, or organ systems. The advantages of chemical messengers over electrical signals are twofold: First, chemicals can spread to all tissues via the blood.

Figure 37.1 Organisms carry out a host of activities.
These two people, very much in love, are probably not aware of many of their bodies' complex activities, which include capturing oxygen, digesting food, pumping blood, maintaining body temperature, and carrying out countless other tasks.

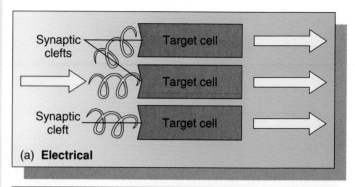

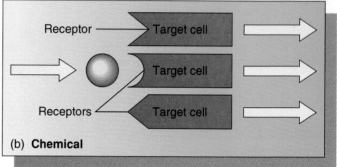

Figure 37.2 Chemical signals are specific.
(a) *An electrical signal might excite many adjacent target cells to fire.* (b) *A chemical signal, however, can be directed at a specific target cell because of the specificity of the receptor.*

Second, each kind of hormone molecule has a unique shape, much as every human face is unique. The mechanism that links a hormone to a specific intracellular process is based on the specific binding of a hormone to a particular receptor. Because hormones have unique shapes, they require uniquely shaped receptors to bind to. This specificity is not seen in the electrical transmission of neurons, where every action potential is alike (figure 37.2).

The Role of Receptor Proteins

The body recognizes a chemical messenger with a particular shape by designing a template that exactly matches the shape of a potential signal molecule—like a glove fits a hand. With such a template, the body can recognize a chemical messenger with exquisite precision, selecting one individual molecule from billions of others. These marvelous templates are the **receptor proteins** discussed in chapter 5. The AIDS virus, for example, infects certain cells of the immune system, and not the cells of the lungs or foot, because the immune cells possess a particular cell-surface receptor that the virus recognizes.

Chapters 34 and 35 discussed how receptor proteins play a critical role in the nervous system as the targets of neurotransmitters. Nerve cells have highly specific cell-surface receptors embedded in their membranes. Each receptor is tuned to respond to a different neurotransmitter molecule. How the neuron responds to stimulation depends on which of its receptors encounters its particular neurotransmitter. In the neuroendocrine system, receptor proteins are located in the membranes of **target cells,** the cells on which the chemical messengers exerts its effect.

The great advantage of a chemical messenger is that it can be directed at a particular protein receptor on its target cells. In each case, the body's operating principle is the same: Only cells whose membranes contain an appropriate receptor protein will respond to a molecular message.

> *Chemical communication within the vertebrate body involves two elements: chemical messengers and receptor proteins on target cells. The system is highly specific because each receptor protein has a shape that only its particular chemical messenger molecule fits.*

How Chemical Signals Are Sent

The path of communication within the vertebrate body can be visualized as a series of simple steps:

1. *Issuing the command.* The central nervous system is the most important regulator of body activities, although other organs can also independently control the release of chemical messengers. Because the central nervous system plays such an active role in controlling hormone release, the various hormones, the glands that secrete them, and the central nervous system can be thought of as acting in concert in the neuroendocrine system. For many hormones, an area of the brain called the **hypothalamus** controls the release of chemical messengers from the **pituitary gland.** The pituitary gland is discussed later in the chapter.

2. *Transporting the signal.* Hormones may act on an adjacent cell, may be carried throughout the body by the bloodstream, or may even pass to a different organism.

3. *Hitting the target.* When a hormone encounters a target cell with a matching receptor protein, the hormone binds to that receptor.

4. *Having an effect.* When a hormone binds to a receptor protein, the receptor protein responds by changing shape, which triggers a change in cell activity.

Types of Chemical Messengers

Table 37.1 shows four types of chemical messengers: (1) hormones released from endocrine glands; (2) **neurohormones** released by neurons; (3) **second messengers** that work only within cells; and (4) neurotransmitters released by axons.

Hormones are stable enough to be transported in active form far from where they are produced and typically act at a distant site. Thus, hormones are very different from neurotransmitters, which tend to have effects that last only for a short time. Hormones are designed to have a more lasting effect.

Most hormones are produced in and secreted from ductless **endocrine glands.** Endocrine glands need to be distinguished from **exocrine glands,** which are glands with ducts that secrete sweat, milk, digestive enzymes, and other materials from the body (figure 37.3).

Another type of chemical messenger—the neurohormones—is secreted by neurons but act like "true" hormones: Once

Table 37.1 Types of Chemical Messengers

Type	Release Site	Function
Hormone	Endocrine glands	Causes change in target cell activity
Neurohormone	Neurons	Causes change in target cell activity
Second messenger	Synthesized within cell in response to hormone binding to receptor	Causes change in target cell activity
Neurotransmitter	Presynaptic membrane	Opens ion channels in postsynaptic membrane and permits transmission of action potential

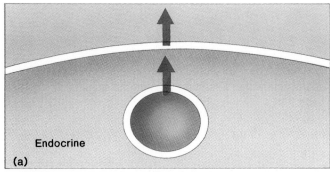

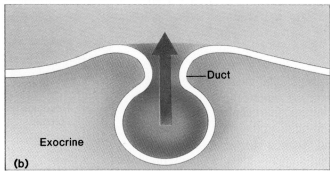

Figure 37.3 Endocrine and exocrine glands.
(a) *Endocrine glands produce and secrete hormones but do not have ducts.* (b) *Exocrine glands have ducts and secrete sweat, milk, digestive enzymes, and other materials.*

secreted into the bloodstream, neurohormones travel through the body until they encounter a target cell with the appropriate receptor protein. The neurohormone's binding to the receptor induces changes in the target cell's cellular activity.

Second messengers are a third type of chemical messenger. While not true hormones because they are not secreted by an endocrine gland, second messengers are found inside cells and respond to the binding of a hormone to the receptor protein on the cell membrane. The hormone itself does not enter the cell. The *binding* of the hormone, however, induces the second messenger inside the cell to initiate changes in cellular activity. Some second messengers that play important roles in hormone action are **cyclic adenosine monophosphate (cAMP), cyclic guanosine monophosphate (cGMP),** and **inositol triphosphate (IP$_3$).**

Neurotransmitters such as acetylcholine are also considered chemical messengers. As discussed in chapter 34, neurotransmitters are released at the synaptic cleft between a neuron and a target cell—either another neuron or a muscle cell. Neurotransmitters bind to receptors on the target cell membrane, thereby allowing the nerve impulse from the presynaptic cell membrane to be transmitted across the synaptic cleft.

> *Chemical signals in animals are sent by either the nervous or endocrine systems. Hormones, neurohormones, and second messengers are chemicals that effect a change in a target cell. Neurotransmitters are responsible for the transmission of a nerve impulse.*

How Hormones Work: Getting the Message Across

Neurohormones and endocrine hormones act in one of two fundamental ways: Either they enter the target cell, or they do not.

Steroid Hormones Enter Cells

Some receptor proteins designed to recognize hormones are located in the target cell cytoplasm. The hormones in these cases are lipid-soluble molecules, typically steroids, that pass across the cell membrane and bind to receptors within the cytoplasm. This complex of receptor and hormone then binds to the DNA in the nucleus and causes a change in the pattern of gene activity—initiating transcription

of some genes while repressing the transcription of others (figure 37.4). The change in gene activity is responsible for the hormone's effect. The steroids that weight lifters and other athletes sometimes use turn on genes and thus trick muscle cells into added growth (see Sidelight 37.1).

All **steroid hormones** are derived from cholesterol, a complex molecule composed of three six-membered carbon rings and one five-membered carbon ring that resembles a fragment of chain-link fence. The hormones that promote the development of secondary sexual characteristics are steroids. They include cortisone and testosterone, as well as the hormones estrogen and progesterone, which control the female reproductive system.

> *Steroid hormones enter a target cell, bind to a cytoplasmic receptor protein, and penetrate the nucleus, where they initiate the transcription of some genes while repressing the transcription of others.*

Peptide Hormones Do Not Enter Cells

Other hormone receptor proteins are embedded within the cell membrane, with their recognition region directed outward from the cell surface. Hormones, typically peptides, bind to these

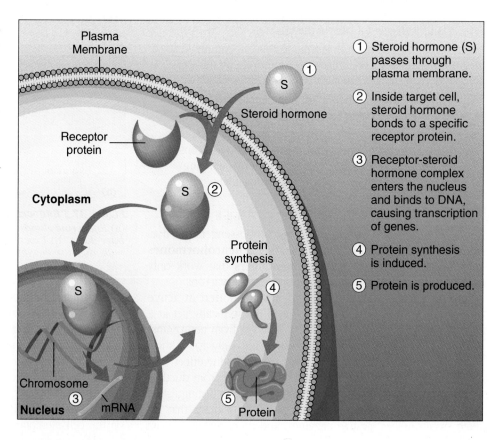

Figure 37.4 How steroid hormones work.
Steroid hormones are lipid soluble and thus readily pass through the plasma membrane of cells into the cytoplasm. There, they bind to specific receptor proteins, forming a complex that enters the nucleus and binds to specific regulatory sites on chromosomes. The binding initiates transcription of the genes regulated by the site and thus results in the production of specific proteins.

Plasma Membrane

Receptor protein

Cytoplasm

Steroid hormone

Protein synthesis

Chromosome

Nucleus   mRNA

Protein

1. Steroid hormone (S) passes through plasma membrane.
2. Inside target cell, steroid hormone bonds to a specific receptor protein.
3. Receptor-steroid hormone complex enters the nucleus and binds to DNA, causing transcription of genes.
4. Protein synthesis is induced.
5. Protein is produced.

SIDELIGHT 37.1
THE ABUSE OF ANABOLIC STEROIDS

Athletes' use of anabolic steroids has been well documented by the media. Lyle Alzado, who died of brain cancer in 1992, was once a talented football player, who, upon his cancer diagnosis, went public with his former steroid use. Alzado believed that steroid abuse had caused his brain cancer. While the cancer connection is tenuous, athletes' abuse of anabolic steroids causes serious medical problems, some of them irreversible.

Anabolic steroids are synthetic compounds that resemble the male sex hormone testosterone. The "anabolic" descriptor refers to *anabolism*, which is any process that results in the synthesis of a biological molecule. Injection of anabolic steroids into the muscles causes the muscle cells to produce more protein, and protein is the building block of muscle fibers. The more muscle fibers there are in a muscle, the bulkier the muscle becomes. Athletes have found that anabolic steroid injections result in bigger muscles and increased strength, and in the competitive arena of the sports world, these attributes are desirable (Sidelight figure 37.1).

Anabolic steroids, however, have many dangerous side effects. In both men and women, steroids can cause liver damage, heart disease, acne, and psychological disorders. Some experts also believe that steroid use can

Sidelight figure 37.1 Many athletes have succumbed to steroid abuse. *Although steroids stimulate muscle growth, their abuse carries many serious health risks.*

be linked to cancer. In men, side effects include a shriveling of the testes and the development of breasts. Impotence and a reduced sperm count can also result. In women, the body becomes "masculinized," and there may be a cessation of menstruation, development of facial and body hair, and a deepening of the voice.

Anabolic steroids are now illegal, but a bustling black market continues to supply these drugs to athletes. For many sports events, blood screens have been implemented to detect athletes' use of anabolic steroids. Fortunately, many athletes have stopped using anabolic steroids because of the side effects and the negative publicity afforded to the abuse of these compounds. Younger athletes are less likely to abuse anabolic steroids for these reasons.

---

receptors on the cell surface. This binding then triggers events within the cell cytoplasm, usually through chemical intermediates known as second messengers (figure 37.5).

Some **peptide hormones,** such as epinephrine (also called adrenaline), are small molecules derived from the amino acid tyrosine; others are short polypeptide chains. Most hormones that circulate within the brain belong to this second class. Still other hormones, such as insulin, are large proteins, consisting of very long polypeptide chains.

> *Peptide hormones do not enter their target cells. Instead, they interact with a receptor on the cell surface and initiate changes within the cell that are mediated by the actions of second messengers.*

Figure 37.5 How peptide hormones work.
This simplified diagram shows that the binding of peptide hormones (P) to receptors in the plasma membranes of target cells induces changes inside the cell. These changes are mediated by the action of chemicals known as second messengers.

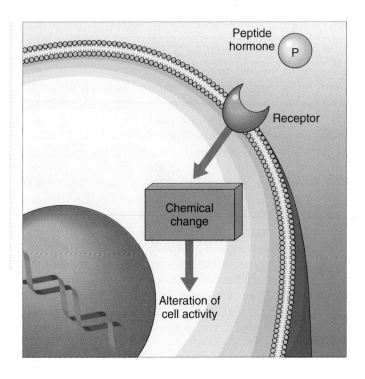

Peptide hormone P

Receptor

Chemical change

Alteration of cell activity

How does the binding of a peptide hormone to the *surface* of a cell produce changes within? The peptide hormone insulin provides a well-studied example of how peptide hormones achieve their effect within target cells.

Most vertebrate cells have insulin receptors in their membranes, the number ranging from fewer than a hundred to more than a hundred thousand in some liver cells. The receptors are glycoproteins (see chapter 4). One part protrudes from the cell surface and binds insulin, whereas another part spans the membrane and extends into the cytoplasm. Two stages of activity follow the binding of insulin to a receptor:

Stage 1: Activation by Insulin

The binding of insulin to the receptor causes the shape of the insulin receptor to change. Within the cell, phosphate groups are actively added to the tyrosine amino acid side groups of proteins. This phosphorylation of proteins activates stage 2 of the insulin response.

Stage 2: Amplification of Signal

As a result of insulin receptor-induced enzyme activity, small molecules of cyclic AMP are formed from ATP. Cyclic AMP is a second messenger and activates a variety of cell enzymes that stimulates the uptake of glucose from the blood and the formation of glycogen, causing blood glucose levels to fall. In addition to cyclic AMP, insulin binding also promotes the production of inositol triphosphate, another second messenger, which is cleaved from the plasma membrane. Thus, a single insulin molecule binding to its receptor releases many second messenger molecules into the cell, and each second messenger activates many enzyme molecules in an expanding cascade of response.

Second Messengers

As previously described, second messengers are intermediary compounds that couple extracellular signals to intracellular processes and also amplify a hormonal signal. Second messengers set into action a variety of events within the affected cell, depending on the cell's enzymatic profile.

Earl Sutherland first described and isolated a second messenger—cyclic adenosine monophosphate (cAMP)—in the early 1960s. When hormones such as epinephrine bind receptors on liver cells, the receptor changes shape and binds a cell protein called **G protein,** causing it, in turn, to bind the nucleotide GTP and activate another membrane protein, **adenylate cyclase.** The result of these complex interactions is the production of large amounts of cAMP by the activated adenylate cyclase. cAMP is the amplified form of the epinephrine hormonal message. As mentioned earlier, another common second messenger is inositol phosphate.

A variety of peptide hormones use cAMP as a second messenger. cAMP has different effects in various target cells because different enzymes are present in different target cells and tissues. In muscle cells, cAMP is induced by epinephrine and activates the enzyme protein kinase-A, which, in turn, activates the

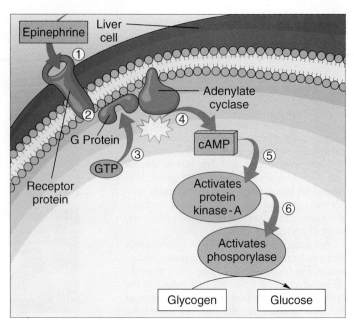

Figure 37.6 How second messengers work.
The synthesis of second messengers in the cell is activated by the binding of peptide hormones to the receptors located in the plasma membranes. In this diagram, (1) a peptide hormone (epinephrine) binds to a specific receptor protein. (2) The hormone binding causes changes in the shape of the receptor protein.
(3) The receptor protein binds to a protein called G. (The 1994 Nobel Prize in Medicine was awarded for the discovery of G proteins.) (4) The G protein then binds to GTP and activates a membrane protein called adenylate cyclase. Adenylate cyclase produces cAMP. (5) cAMP activates protein kinase-A, an enzyme. (6) Protein kinase-A activates phosphorylase, which breaks glycogen into glucose.

enzyme that breaks down glycogen into glucose (figure 37.6). In cells of the ovary, cAMP is induced by a luteinizing hormone and stimulates the cells to produce a specific cell enzyme.

> *Second messengers are cytoplasmic signal molecules produced in response to a peptide hormone. They often greatly amplify the original signal.*

Unique Thyroxine: Goes Directly into the Nucleus

One important endocrine hormone is neither steroid nor peptide. The hormone **thyroxine,** produced by the thyroid gland, is a modified form of the amino acid tyrosine and contains four iodine atoms. Thyroxine acts on target cells in a unique way: It diffuses directly into the cell cytoplasm and then into the cell nucleus, where it interacts with a receptor protein attached to DNA to initiate production of particular growth-promoting messenger RNAs. Thyroxine is the only hormone known to go directly into a target cell nucleus without first binding to receptors on the cell membrane or in the cytoplasm.

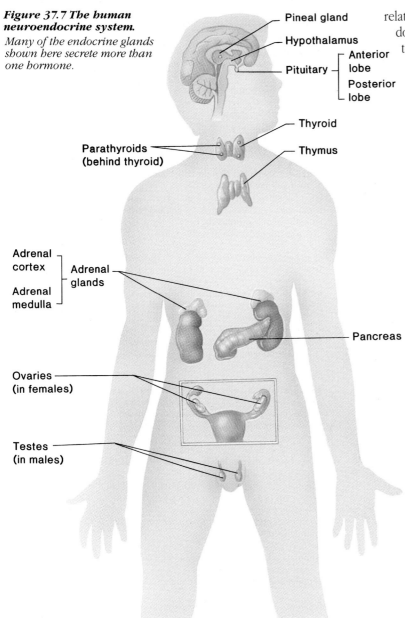

Figure 37.7 The human neuroendocrine system.

Many of the endocrine glands shown here secrete more than one hormone.

Pineal gland
Hypothalamus
Pituitary — Anterior lobe / Posterior lobe
Thyroid
Parathyroids (behind thyroid)
Thymus
Adrenal cortex
Adrenal medulla
Adrenal glands
Pancreas
Ovaries (in females)
Testes (in males)

The Major Endocrine Glands and Their Hormones

Vertebrates have about a dozen major endocrine glands that together make up the neuroendocrine system (figure 37.7). This section briefly examines the principal endocrine glands and the hormones they produce (table 37.2).

The Pituitary: Many Hormones, Many Effects

The pituitary, located in a bony recess in the brain below the hypothalamus, is the production site of nine major hormones. Because many of these hormones act principally to influence other endocrine glands, the pituitary was regarded until relatively recently as a "master gland," orchestrating the endocrine system. In fact, as is discussed later in the chapter, that role is reserved for the hypothalamus because the hypothalamus controls the pituitary. The pituitary, however, remains one of the most important endocrine glands.

The pituitary is actually two glands. The back (*posterior*) end regulates water conservation, milk letdown, and uterine contractions in women; the front (*anterior*) end regulates other endocrine glands (figure 37.8).

The Posterior Pituitary

The role of the **posterior pituitary** first became evident in 1912, when a remarkable medical case was reported: A man who had been shot in the head developed a surprising disorder—he began to urinate every thirty minutes, unceasingly. The bullet had lodged in the pituitary gland, and subsequent research demonstrated that removal of the pituitary produces these unusual symptoms. Pituitary extracts were shown to contain a substance that makes the kidneys conserve water, and eventually, in the early 1950s, the peptide hormone vasopressin (called **antidiuretic hormone, ADH**) was isolated. ADH regulates the kidneys' water retention. When ADH is missing, the kidneys cannot retain water, which is why the bullet led to excessive urination (and why excessive alcohol, which inhibits ADH secretion, has the same effect).

The posterior pituitary also produces a second hormone of very similar structure to ADH—both are short peptides composed of nine amino acids—but very different function, called **oxytocin.** Sensory receptors in the nipples send messages to the hypothalamus, causing oxytocin release. Oxytocin stimulates muscle contraction around the ducts into which mammary glands secrete milk, thereby initiating milk release. Oxytocin also stimulates uterine contractions in women during childbirth, which is why the uterus of a nursing mother returns to normal size more quickly than does the uterus of a mother who does not nurse her baby.

ADH and oxytocin are both neurohormones because they are synthesized inside neurons within the hypothalamus. They are then transported down nerve cell axons from the hypothalamus to synapses located within the posterior pituitary, where they are stored in the axon terminals. The neurohormones are released into the bloodstream when a nerve impulse from the hypothalamus reaches the terminals.

The posterior lobe of the pituitary, a distinct gland, has neural connections with the hypothalamus. It secretes the important hormones ADH and oxytocin.

Table 37.2   Principal Endocrine Glands and their Hormones

Endocrine Gland and Hormone	Target Tissue	Principal Action
1. Posterior lobe of pituitary		
Antidiuretic hormone (ADH)	Kidney tubules	Stimulates reabsorption of water; conserves water
Oxytocin	Uterus	Stimulates contraction of uterus
	Mammary glands	Stimulates milk letdown
2. Anterior lobe of pituitary		
Growth hormone (GH)	General	Stimulates growth
Thyroid-stimulating hormone (TSH)	Thyroid gland	Stimulates secretion of thyroid hormones
Gonadotropic hormones		
Luteinizing hormone (LH)	Sex organs	Stimulates ovulation (females); stimulates secretion of testosterone (males)
Follicle-stimulating hormone (FSH)	Sex organs	Stimulates ovarian follicle (females) and sperm production (males)
Adrenocorticotropic hormone (ACTH)	Adrenal cortex	Stimulates secretion of adrenal cortical hormones
Prolactin (PRL)	Mammary glands	Stimulates milk production
Melanocyte-stimulating hormone (MSH)	Melanin-producing cells	Controls pigmentation in some animals
3. Thyroid gland		
Thyroid hormone (thyroxine)	General	Stimulates metabolic rate; essential to normal growth and development
Calcitonin	Bone	Lowers blood calcium levels by inhibiting loss of calcium from bone
4. Parathyroid glands		
Parathyroid hormone (PTH)	Bone, kidneys, digestive tract	Increases blood calcium levels by stimulating bone breakdown; stimulates calcium reabsorption from in kidneys; activates vitamin D
5. Adrenal medulla		
Epinephrine and norepinephrine	Muscle, cardiac muscle, blood vessels	Initiate stress response; increase heart rate, blood pressure, metabolic rate; dilate blood vessels; mobilize fat stores; raise blood sugar levels
6. Adrenal cortex		
Aldosterone	Kidney tubules	Maintains proper balance of sodium and potassium ions in blood
Cortisol	General	Aids in adaptation to long-term stress; raises blood glucose levels; mobilizes fat stores; stimulates carbohydrate metabolism; acts to reduce inflammation
7. Pancreas (islets of Langerhans)		
Insulin	General	Lowers blood glucose levels; increases storage of glycogen
Glucagon	Liver, fat tissue	Raises blood glucose levels; stimulates breakdown of glycogen in liver
8. Ovary		
Estrogens	General female reproductive structures	Stimulate development of secondary sexual characteristics and growth of sex organs at puberty; prompt monthly preparation of uterus for pregnancy
Progesterone	Uterus	Completes preparation of uterus for pregnancy
	Mammary glands	Stimulates development
9. Testis		
Testosterone	General	Stimulates development of secondary sexual characteristics and growth spurt at puberty
	Male reproductive structures	Stimulates development of sex organs; stimulates production of sperm
10. Pineal gland		
Melatonin	Sex organs (?), pigment cells	Function not well understood; influences pigmentation in some animals; may control biorhythms in some animals; may help control onset of puberty in humans
11. Thymus		
Thymosin	White blood cells	Stimulates maturation and production of white blood cells

The Anterior Pituitary

The key role of the **anterior pituitary** first became understood in 1909, when a thirty-eight-year-old South Dakota farmer was cured of the growth disorder acromegaly by the surgical removal of a pituitary tumor. Acromegaly, a form of giantism in which the jaw begins to protrude and features thicken, is almost always associated with pituitary tumors (figure 37.9). Robert Wadlow, born in Alton, Illinois, in 1928, grew to a height of 8 feet, 11 inches—the tallest human being ever recorded—and weighed 475 pounds before he died from infection at age 22

(figure 37.10). Skull X rays showed that he had a pituitary tumor. So did the 8-foot, 2-inch Irish giant Charles Byrne, born in 1761; his skeleton, preserved in the Royal College of Surgeons, London, also shows the effects of a pituitary tumor.

Pituitary tumors produce giants because the tumor cells produce prodigious amounts of a growth-promoting hormone. This **growth hormone (GH),** a peptide of 191 amino acids, is normally produced in only minute amounts by the anterior pituitary gland and usually only during periods of body growth, such as infancy and puberty.

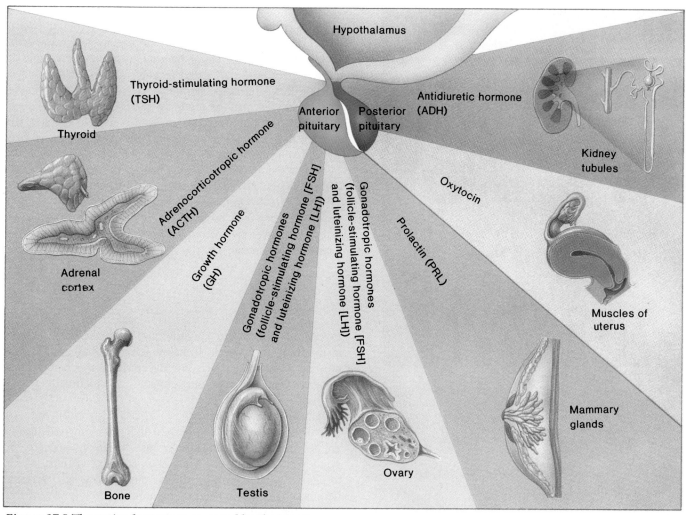

Figure 37.8 The major hormones secreted by the pituitary gland.

The pituitary gland is divided into two glands: the anterior pituitary and the posterior pituitary. Each side of the pituitary secretes different hormones.

The vertebrate anterior pituitary gland is now known to produce seven major peptide hormones, each very similar in structure and controlled by a **releasing factor** secreted from cells in the hypothalamus:

1. *Thyroid-stimulating hormone (TSH).* TSH stimulates the thyroid gland to produce thyroxine, which, in turn, simulates oxidative respiration.

2. *Luteinizing hormone (LH).* LH plays an important role in the female menstrual cycle (see chapter 43). It also stimulates the male sex organs to produce testosterone, which initiates and maintains the development of male secondary sexual characteristics, those external features not involved in reproduction.

Figure 37.9 Acromegaly.

This ancient carving of the Egyptian pharaoh Akhenaton, who ruled from 1379–1362 B.C., exhibits many characteristics of acromegaly. It may be the oldest known case.

Figure 37.10 The Alton giant.
This photo, taken when Robert Wadlow of Alton, Illinois, was thirteen years old, shows him towering over his father and nine-year-old brother. Born a normal size, he developed a growth-hormone-secreting pituitary tumor as a young child and never stopped growing during his twenty-two years of life.

3. *Follicle-stimulating hormone (FSH).* FSH is significant in the female menstrual cycle (see chapter 43). In males, it stimulates certain cells in the testes to produce a hormone that regulates sperm development.

4. *Adrenocorticotropic hormone (ACTH).* ACTH stimulates the adrenal cortex to produce corticosteroid hormones. Some of these hormones regulate the production of glucose from fat. Others regulate the balance of sodium and potassium ions in the blood. Still others contribute to the development of male secondary sexual characteristics.

5. *Growth hormone (GH).* GH stimulates muscle and bone growth throughout the body.

6. *Prolactin (PRL).* PRL stimulates the breasts to produce milk.

7. *Melanocyte-stimulating hormone (MSH).* In reptiles and amphibians, MSH stimulates color changes in the epidermis. This hormone has no known function in mammals.

The anterior lobe of the pituitary secretes seven major peptide hormones. Very similar in structure, they have a wide variety of functions.

In addition to their endocrine functions, these and many other hormones have been shown to be associated with particular cell populations within the central nervous system. Biologists are actively researching the as yet unknown role of these hormones in the central nervous system. Whatever their function there, the same hormones clearly may play different roles in different parts of the body because cells have different enzymatic profiles. Hormones are signals used in different tissues for different reasons, just as a raised hand is a signal that has different meanings in different contexts: In class, it indicates a question; in a football game, it signals a "fair catch"; and when a police officer does it on a street, it means "stop."

Vertebrate hormone evolution has been a conservative process. Rather than produce a new hormone for every use, vertebrates have often adapted a hormone already on hand for a new use, when this new use does not cause confusion.

The Thyroid: A Metabolic Thermostat

The **thyroid gland** is shaped like a shield (from the Greek *thyros,* meaning "shield") and lies just below the Adam's apple in the front of the neck. It makes several hormones, the two most important of which are **thyroxine,** which increases metabolic rate and promotes growth, and **calcitonin,** which stimulates calcium uptake.

Without adequate thyroxine (also called thyroid hormone), growth is retarded. Children with underactive thyroid glands are not able to carry out carbohydrate breakdown and protein synthesis at normal rates, a condition called **cretinism,** which results in stunted growth. Mental retardation is also a result because thyroxine is needed for normal development of the central nervous system. Adults with too little of this hormone also have slowed metabolism, which affects their mental performance.

As mentioned earlier, the hormone thyroxine is made from the amino acid tyrosine, with four iodine atoms added to it. If the iodine concentration in a person's diet is too low, the thyroid cannot make adequate amounts of thyroxine and will grow larger in a futile attempt to manufacture more of the hormone. The greatly enlarged thyroid gland that results is called a **goiter** (figure 37.11). This need for iodine in the diet is why iodine is added to table salt.

The Parathyroids: Bone Builders

The **parathyroid glands** are four small glands attached to the thyroid. Small and unobtrusive, they were ignored by researchers until well into the twentieth century. The first suggestions that the parathyroids produce a hormone came from experiments in which the parathyroids were removed from dogs: The concentration of calcium in the dogs' blood plummeted to less than half the normal level. However, if an extract of parathyroid gland was administered, calcium levels returned to normal. If an excess was administered, calcium levels became *too* high,

Figure 37.11 A goiter.
This condition is caused by a lack of iodine in the diet.

and the dogs' bones literally were dismantled by the extract. The parathyroid glands clearly were producing a hormone that acted on calcium uptake into, and release from, bone.

The hormone produced by the parathyroids is called **parathyroid hormone (PTH).** It is one of only two hormones in the human body that are absolutely essential for survival. (The other, discussed in the next section, is aldosterone, produced by the adrenal glands.) PTH regulates calcium levels in the blood. As mentioned in chapter 36, calcium ions are the key actors in vertebrate muscle contraction; by altering calcium release, nerve impulses cause muscles to contract. Life is not possible without the muscles that pump the heart and drive the body, and these muscles cannot function if calcium levels are not kept within narrow limits. Calcium is also important for normal nerve activity.

PTH acts as a fail-safe to ensure that calcium levels never fall too low. Released into the bloodstream, PTH travels to the bones and acts on the bone-producing cells, stimulating them to dismantle bone tissue and release calcium into the blood-stream. PTH also acts on the kidneys to resorb calcium ions from the urine and leads to the activation of vitamin D, necessary for calcium absorption by the intestine. A diet deficient in vitamin D leads to poor bone formation, a condition called **rickets.**

An adequate intake of calcium is also necessary to keep the body running smoothly. PTH is synthesized by the parathyroids in response to falling levels of calcium ions in the blood; the body essentially sacrifices bone to keep calcium levels within the narrow limits necessary for proper functioning of muscle and nerve. **Osteoporosis,** a condition that causes the bones to become fragile and susceptible to breakage, has been linked to low blood calcium levels. Without adequate calcium intake, the body continually "raids" the bones to keep calcium levels high. If this raiding continues, the bones become brittle and can sometimes even shrink in size.

The thyroid also plays a role in maintaining proper calcium levels. If calcium levels in the blood become too high, the thyroid hormone calcitonin, mentioned earlier, stimulates calcium deposition in bone. Calcitonin and PTH thus work together to maintain proper blood calcium levels (figure 37.12).

The Adrenals: Two Glands in One

There are two **adrenal glands,** one located just above each kidney. Each adrenal gland is composed of two parts: an inner core called the **medulla,** which produces the peptide hormones epinephrine and norepinephrine, and an outer layer called the **cortex,** which produces the steroid hormones **cortisol** and **aldosterone.**

Adrenal Medulla: Emergency Warning Siren

The medulla releases epinephrine and norepinephrine in times of stress. Epinephrine and norepinephrine act in the body as emergency signals that stimulate rapid deployment of body fuel. The "alarm" response throughout the body is identical to the individual effects achieved by the sympathetic nervous system (see chapter 35) but is longer lasting. Among the effects of these hormones are an accelerated heartbeat, increased blood pressure, higher levels of blood sugar, dilated blood vessels, and increased blood flow to the heart and lungs. These hormones thus can be thought of as extensions of the sympathetic nervous system.

Adrenal Cortex: Maintaining the Proper Amount of Salt

The adrenal cortex releases two hormones: cortisol and aldosterone. Cortisol (also called hydrocortisone) acts on many different cells in the body to maintain nutritional well-being. It stimulates carbohydrate metabolism and acts to reduce inflammation. Derivatives of this hormone, such as prednisone, have widespread medical use as anti-inflammatory agents. The ability of many cortisol-derived steroids to stimulate muscle growth has also led to the abuse of so-called anabolic steroids by athletes, as discussed earlier in Sidelight 37.1.

Aldosterone acts primarily at the kidney to promote the uptake of sodium and other salts from the urine. Sodium ions play critical roles in nerve conduction and many other body functions. Their concentration also has a critical influence on blood pressure. Without aldosterone, sodium ions are not retrieved from body fluids and are lost in the urine. Loss of salt in the blood causes water to leave the bloodstream and enter cells; thus, blood pressure falls.

Aldosterone also acts in the opposite way to promote the export of potassium out of the body, stimulating the kidneys to secrete potassium ions into the urine. When aldosterone levels are too low, potassium levels in the blood may rise to dangerous levels.

Aldosterone works with ADH, secreted by the posterior pituitary, to maintain the proper levels of salts in the blood. Figure 37.13 shows how these hormones work in concert. As mentioned earlier, aldosterone is, with PTH, one of the two endocrine hormones essential for survival. Removal of the adrenal glands is invariably fatal.

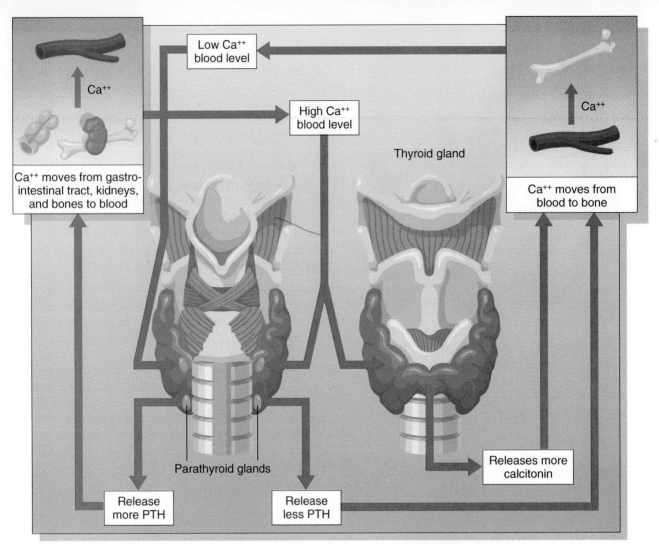

Low Ca⁺⁺ blood level

High Ca⁺⁺ blood level

Ca⁺⁺

Ca⁺⁺ moves from gastro-intestinal tract, kidneys, and bones to blood

Thyroid gland

Ca⁺⁺

Ca⁺⁺ moves from blood to bone

Parathyroid glands

Release more PTH

Release less PTH

Releases more calcitonin

Figure 37.12 The interaction of PTH and calcitonin to regulate calcium (Ca⁺⁺) levels in the blood.

PTH acts to remove calcium from bone and deposit it in the blood. Calcitonin acts to remove calcium from the blood and deposit it in bone.

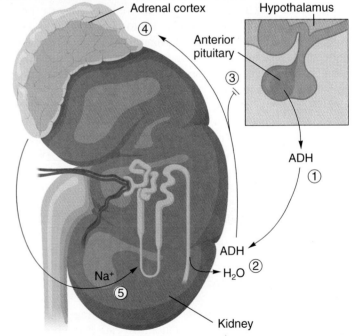

Adrenal cortex   Hypothalamus

Anterior pituitary

ADH

ADH

→ H₂O

Na⁺

Kidney

Figure 37.13 How hormones control water and salt levels in the blood.

Control of water (H₂O) and salt (Na⁺) balance within the kidney is centered in the hypothalamus. The posterior pituitary produces antidiuretic hormone (ADH), which renders the kidney's collecting ducts freely permeable to water. As a result, water leaves the ducts and flows into the blood, increasing water retention. When water retention is too high, blood pressure rises. Pressure-sensitive receptors in the hypothalamus detect this and cause ADH production to shut down. If the level of salts in the blood falls, the adrenal cortex initiates production of the hormone aldosterone, which stimulates salt reabsorption by the kidney ducts.

① ADH is released from anterior pituitary.

② ADH causes kidney tubules to release water into blood.

③ High water pressure shuts down ADH.

④ High water pressure dilutes salts in blood; aldosterone is released.

⑤ Aldosterone causes kidney ducts to reabsorb sodium and other salts.

The Pancreas: Regulation of Glucose Levels

The **pancreas gland** is located behind the stomach and is connected to the front end of the small intestine by a small tube. It secretes a variety of digestive enzymes into the gut through this tube and for a long time was thought to be solely an exocrine gland. In 1869, however, a German medical student named Paul Langerhans described some unusual cell clusters scattered throughout the pancreas.

By the end of the nineteenth century, doctors had begun to notice that patients with injuries to the pancreas often develop **diabetes mellitus,** a common and serious disorder involving elevated blood glucose levels. In 1893, it was suggested that the clusters of cells in the pancreas, which came to be called **islets of Langerhans,** produced something that prevented diabetes mellitus. That substance is now known to be the peptide hormone **insulin.**

Individuals with diabetes mellitus are unable to take up glucose from the blood, even though the level of blood glucose is high. In Type I diabetes mellitus, insulin secretion is abnormally low, a problem caused by a malfunction of insulin-producing cells. In Type II diabetes mellitus, the number of insulin receptors on the target tissue is abnormally low, but the level of insulin in the blood is normal. Individuals afflicted with both types of diabetes literally starve: They lose weight and may eventually suffer brain damage and even death if their condition is untreated.

The peptide hormone insulin was not actually isolated until 1922, when two young doctors working in a Toronto hospital succeeded where many others had not. On 11 January 1922 they injected an extract purified from beef pancreas glands into a thirteen-year-old boy, a diabetic whose weight had fallen to 65 pounds and who was not expected to survive. The hospital record note gives no indication of the historic importance of the trial, only stating, "15 cc of MacLeod's serum. 7 1/2 cc into each buttock." With this single injection, the glucose level in the boy's blood fell 25 percent. A more potent extract soon brought levels down to near normal. This was the first instance of successful insulin therapy. Today, individuals with diabetes can be treated with daily supplies of insulin. Others who suffer from the disease, especially Type II, are treated by a combination of exercise and diet. Active research on the possibility of transplanting islets of Langerhans holds much promise of a lasting treatment for diabetes.

> *Diabetes is a condition in which individuals are unable to obtain glucose from their blood because they either lack insulin (Type I) or have an abnormally low number of insulin receptors (Type II). It is a serious disease and can be fatal if untreated.*

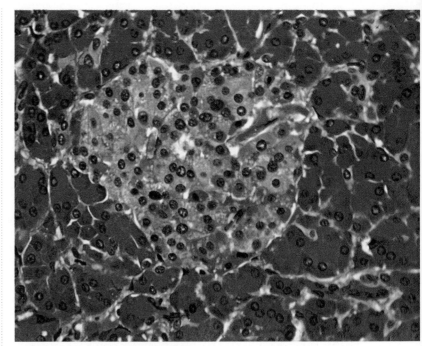

Figure 37.14 Islets of Langerhans.
Glucagon and insulin are produced by clumps of cells within the pancreas called islets of Langerhans (stained dark in this preparation).

The islets of Langerhans in the pancreas are now known to produce *two* hormones—insulin and **glucagon**—that interact to govern blood glucose levels (figure 37.14). Insulin is a storage hormone, designed to put away nutrients for leaner times. It promotes the accumulation of glycogen in the liver and triglycerides in fat cells. After a meal, **beta-cells** in the islets of Langerhans secrete insulin, storing glucose to be used later. Later, when body activity causes blood glucose levels to fall, other cells in the islets of Langerhans, called **alpha-cells,** secrete glucagon, which causes liver cells to release stored glucose and fat cells to break down triglycerides. Insulin and glucagon thus work together to keep blood glucose levels within narrow bounds (figure 37.15).

Other Endocrine Glands

The **ovaries** and **testes** are important endocrine glands, producing the sex hormones estrogen, progesterone, and testosterone, which are described in detail in chapter 43. Surprisingly, the gut is also a major endocrine gland, secreting hormones that regulate the release of digestive enzymes, which, in turn, play a key role in food digestion (see chapter 38). Hormones secreted by the same structure on which they act are called **endogenous hormones.** The **thymus,** a gland located in the neck below the thyroid gland, produces a hormone called **thymosin,** which promotes the production and maturation of white blood cells.

Figure 37.15 Hormonal control of blood glucose levels.

(a) *When blood glucose levels are low, cells within the pancreas release the hormone glucagon into the bloodstream; other cells within the adrenal gland, situated on top of the kidneys, release the hormone adrenaline into the bloodstream. When they reach the liver, glucagon and adrenaline both act to increase the liver's breakdown of glycogen to glucose.* (b) *When blood glucose levels are high, other cells within the pancreas produce the hormone insulin, which stimulates the liver and muscles to convert glucose into glycogen. Blood glucose levels determine the levels of insulin and glycogen in the blood via feedback loops to the pancreas and adrenal gland.*

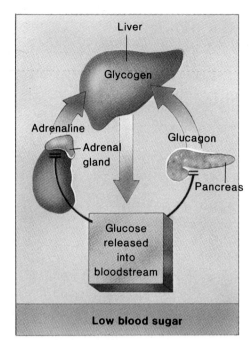

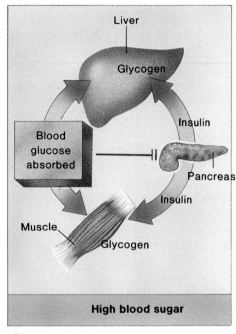

(a)　　　　(b)

The only other major endocrine gland is the **pineal gland,** which is small—about the size of a pea—is shaped like a pinecone (hence, its name), and sits in the center of the brain. It was the last endocrine gland discovered in the human body, and its function is still not completely understood. In 1958, scientists discovered that the pineal gland secretes the hormone **melatonin,** a derivative of the amino acid tryptophan. In hamsters, melatonin regulates reproductive biology; in frogs, it influences pigmentation; but in humans, its function is a topic of ongoing research.

The pineal gland in reptiles is located closer to the surface and is called the "third eye" because it is structurally similar to the retina and responds directly to light. In humans, the pineal gland is not connected to the central nervous system directly, but the gland *is* connected, via the sympathetic nervous system, to the eyes. The human pineal gland seems to release melatonin as a response to darkness, in a cyclical biological rhythm keyed to daylight. Perhaps, the pineal gland is involved in establishing daily biorhythms. It has also been implicated in such mood disorders as winter depression, also called SADS (seasonal affective disorder syndrome), and in a variety of roles concerning sexual development.

How the Brain Controls the Endocrine System

The eleven major endocrine glands (see table 37.2) do not function independently. Their activities are coordinated at two levels: (1) The six anterior pituitary hormones found in humans regulate many of the activities of the other endocrine glands, and (2) the pituitary gland is controlled by the central nervous system via the hypothalamus.

Releasing Hormones: The Brain's Chemical Messengers

How the brain regulates the pituitary gland was, until recently, one of the great medical mysteries. The pituitary is suspended by a short stalk to the hypothalamus, which is part of the diencephalon region of the forebrain, located at the base of the brain (figure 37.16). Within the hypothalamus, information about the body's many internal functions is processed, and regulatory commands are issued. For example, some of these commands involve regulating body temperature, food and water intake, reproductive behavior, and response to pain and emotion. These commands are issued to the pituitary gland, which, in turn, sends chemical signals to the body's various hormone-producing glands. Injury to the hypothalamus causes a decrease in the anterior pituitary's production of hormones. The hypothalamus thus regulates the body through a chain of command in the same way that a general gives orders to a chief of staff, who relays them to lower-ranking commanders.

The hypothalamus processes information about the body's internal functions and issues regulatory commands that direct the pituitary gland to send further chemical signals to the various hormone-producing glands of the body.

No nerves connect the anterior pituitary gland to the hypothalamus or to any other part of the brain. How, then, are the commands sent from the hypothalamus to the pituitary? In the 1930s, researchers discovered a network of tiny blood vessels that spans the short distance between the hypothalamus and anterior pituitary, and they wondered if, perhaps, the hypothalamus employed chemical messengers. The experimental difficulties in isolating such hypothalamic hormones were great because very little of the hormones are present in any one brain. After concentrating the hypothalamus glands from one million pigs, however, researchers were able to isolate the first of these hormones, a short peptide

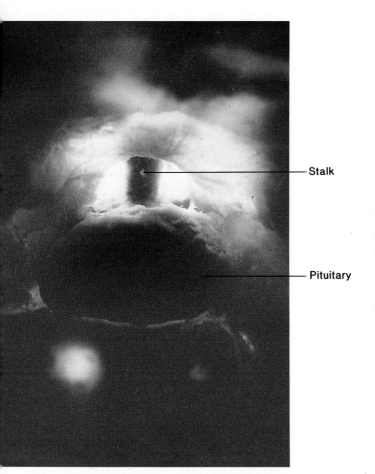

Figure 37.16 The pituitary gland hanging by a short stalk from the hypothalamus.
The pituitary gland regulates the hormone production of many of the body's endocrine glands. Here, it is enlarged fifteen times.

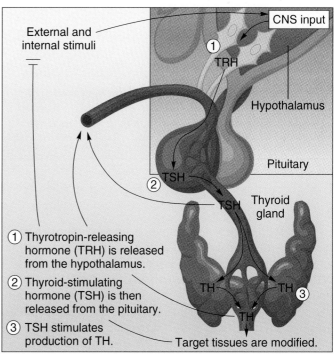

① Thyrotropin-releasing hormone (TRH) is released from the hypothalamus.

② Thyroid-stimulating hormone (TSH) is then released from the pituitary.

③ TSH stimulates production of TH.

Figure 37.17 The release of thyroid hormone.
Thyroid hormone (TH) from the thyroid gland is the end result of a long series of events controlled by the central nervous system (CNS).

called **thyrotropin-releasing hormone (TRH),** in 1969. The release of TRH from the hypothalamus triggers secretion of thyrotropin (thyroid-stimulating hormone, TSH) from the anterior pituitary, and TSH, in turn, stimulates the thyroid to make thyroid hormone (TH) (figure 37.17).

Six other hypothalamic regulatory hormones have since been isolated, which together govern all the hormones secreted by the anterior pituitary. For each releasing hormone secreted by the hypothalamus, the anterior pituitary synthesizes a corresponding hormone. When the pituitary gland receives a releasing hormone from the hypothalamus, the anterior lobe responds by secreting the corresponding pituitary hormone, which initiates the release of specific hormones in other endocrine glands.

The anterior pituitary is connected to the hypothalamus by special blood vessels only a few millimeters long. Through these vessels passes a group of hypothalamic releasing hormones, which command the anterior pituitary to initiate the production (release) of specific hormones in distant endocrine glands.

How the Hypothalamus Regulates Hormone Production

Hormone production in the anterior pituitary gland is controlled by the hypothalamus in two ways: (1) central nervous system control and (2) feedback control.

Central Nervous System Control

Production of growth hormone (GH), prolactin (PRL), and melanocyte-stimulating hormone (MSH) is controlled by both releasing and inhibitory signals produced in the central nervous system by the hypothalamus. For example, the releasing signal for GH is the **growth-hormone-releasing hormone (GHRH)** produced by the hypothalamus. GHRH stimulates the anterior pituitary to produce GH. The inhibiting signal, which the hypothalamus produces at the same time, is **somatostatin.** Somatostatin inhibits the anterior pituitary from producing GH. The hypothalamus thus regulates growth by mediating the relative production rates of GHRH and somatostatin. It regulates PRL and MSH production in a similar way. Thus, the release and inhibition of hormones is controlled by the release of *other* hormones.

Feedback Control

The levels of all other hormones produced by the anterior pituitary are controlled by negative feedback from the target glands. For example, when luteinizing hormone (LH) stimulates the gonads to release testosterone into the bloodstream, that testosterone, in turn, inhibits the hypothalamus. The hypothalamus then ceases to transmit LH-releasing hormone to the pituitary.

> *Pituitary hormone production is regulated by (1) pairs of hypothalamic hormones with opposite effects and (2) feedback loops, through which the hypothalamus is sensitive to blood hormone levels.*

Nonendocrine Hormones

Neuropeptides

As discussed earlier in the chapter, neurons produce chemical messengers called neurohormones that act on targets located far away from the site of neurohormone production. Oxytocin is a neurohormone, as is antidiuretic hormone (ADH). Both neurohormones act on target cells located far from the pituitary, where they are produced. Other neurohormones, called **neuropeptides,** are produced in the brain *and* act on target cells in the brain.

In 1974, Swedish researchers were first to isolate two small neuropeptides, called **enkephalins,** that were only five amino acid units long and that acted as powerful narcotics. The enkephalins appear to play a role in integrating sensory impulses from pain receptors. Morphine has a potent analgesic (pain-relieving) effect on the central nervous system because it mimics the effects of the enkephalins.

A second type of neuropeptide hormone—**endorphins,** which are polypeptides thirty-two amino acid units long—has since been found. Endorphins appear to regulate emotional responses in the brain.

More than twenty neuropeptides have now been identified (and some investigators believe that the number will eventually exceed one hundred). Neuropeptides act as neurotransmitters in the brain. Their study is revolutionizing how scientists view the brain's internal activity. For example, scientists are currently investigating the role of these neuropeptides in long-term memory storage. The role of neuropeptides within the brain is one of the most active areas of biological research.

Prostaglandins

Prostaglandins are modified lipids produced from membrane phospholipids by virtually all cells. They do not circulate in the blood, but rather, accumulate in regions of tissue disturbance or injury. They stimulate smooth muscle contraction and expansion, as well as blood vessel contraction. Overproduction of prostaglandins, however, causes pain due to increased blood vessel and muscle contraction, as in a headache (in which the blood vessels in the head forcefully constrict) and menstrual cramps (in which the smooth muscles of the uterus contract at very high levels). Aspirin and other antiprostaglandin medication relieve headache pain and menstrual cramps because they inhibit prostaglandin production.

Atrial Peptides

Another nonendocrine hormone of considerable interest is **atrial natriuretic hormone (ANH),** a small peptide that is manufactured in the heart and that circulates throughout the body. This hormone is another example of an endogenous hormone. Receptors for ANH have been identified in cells of blood vessels, kidneys, and adrenal glands. Because the atrial peptides apparently help to regulate blood pressure and volume, they are being investigated as a potential treatment for high blood pressure.

> *Some hormonelike substances are produced by nonendocrine organs. These include neuropeptides, such as oxytocin, ADH, and endorphins; prostaglandins, which cause localized pain such as headaches; and atrial peptides such as ANH, which contribute to the regulation of blood pressure.*

Hormones: The Body's Messengers

Hormonal communication is essential to the body's proper functioning. The many kinds of hormones encountered in this chapter form a network of communication that reaches to all body tissues (figure 37.18). The hormones are all ultimately under the control of the central nervous system, of which they are a chemical extension. Hormones are mentioned repeatedly in the chapters that follow. They are one of the principal integrating elements of the vertebrate body.

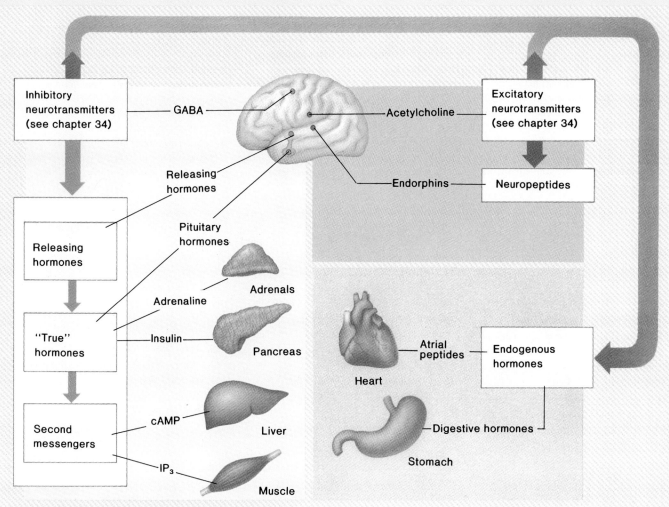

Figure 37.18 Overview of neuroendocrine control.
This diagram shows the different types of chemical messengers, their origin, and their target tissues.

SUMMARY

1. Chemical messengers are highly specific, affecting only particular target cells. This exquisite sensitivity is possible because each messenger is keyed to a particular receptor protein present in the membranes of target cells and not in other cells.

2. Hormones are stable chemical messengers that act on cells located far from where the hormone is produced. Neurohormones are produced by neurons but act in the same fashion as "true" hormones. Second messengers are molecules within cells that initiate changes in cellular activity in response to the binding of a hormone to a receptor.

3. Most hormones are either peptides that interact with receptors on the target cell surface, thereby activating enzymes within, or steroids, which enter into cells and alter the cells' transcription patterns.

4. The posterior lobe of the pituitary, a distinct gland, secretes oxytocin and antidiuretic hormone (ADH) into the bloodstream. This gland is linked directly to the hypothalamus by neural connections.

5. The anterior lobe of the pituitary, another distinct gland, is connected to the hypothalamus by special blood vessels a few millimeters long. It secretes seven principal kinds of hormones, each corresponding to a specific releasing hormone from the hypothalamus.

6. The thyroid gland is noted for the hormones thyroxine, which increases metabolic rate and promotes growth, and calcitonin, which stimulates calcium uptake.

7. The parathyroids, four small glands attached to the thyroid, produce parathyroid hormone (PTH), which acts as a fail-safe to make sure calcium levels never fall too low.

EVOLUTIONARY VIEWPOINT

In the animal kingdom, hormones often play important roles in communication between organisms. Many insects are attracted to potential mates by sex hormones called pheromones. Vertebrates also commonly employ such hormones to signal sexual receptiveness. Humans and other apes lack an estrus cycle; humans also have a less acute sense of smell than other mammals. Perhaps for these reasons, humans appear to lack sexual pheromones—although some scientists claim to have identified a sex-pheromone receptor in the human nose.

8. The adrenal medulla releases epinephrine and norepinephrine in times of stress to produce an "alarm" response throughout the body. The adrenal cortex releases cortisol, which helps to maintain nutritional well-being, and aldosterone, which maintains a proper balance of sodium and potassium ions in the blood.

9. Clusters of cells in the pancreas, called the islets of Langerhans, produce two hormones—insulin and glucagon—that work together to regulate blood glucose levels.

10. The brain maintains long-term control over physiological processes by synthesizing releasing hormones in the hypothalamus. These hormones direct the synthesis of specific circulating hormones by the pituitary gland. Pituitary hormones travel out into the body and initiate the synthesis of particular hormones in target tissues.

11. Pituitary hormone production is regulated by pairs of hypothalamic hormones that have opposite effects and by negative feedback control.

12. Neuropeptides are produced by hormones in the brain and act as neurotransmitters. Prostaglandins are nonendocrine chemical messengers that cause muscle and blood vessel contraction. Atrial natriuretic hormone (ANH) is manufactured by the heart and regulates blood pressure and volume.

13. Most hormones circulating in the bloodstream are produced by endocrine glands, whose activity is under the direct control of the central nervous system.

REVIEWING THE CHAPTER

1. What is the importance of chemical messengers?

2. What role do receptor proteins play?

3. How are chemical signals sent?

4. Where are chemical messengers made?

5. What are the types and functions of chemical messengers?

6. How do hormones work?

7. How do steroid hormones enter cells?

8. How and where do peptide hormones function?

9. What is insulin, and how does it work?

10. What are second messengers, and how do they function?

11. Why is the hormone thyroxine unique?

12. What are the major endocrine glands, and where are they found in the body?

13. What are the effects of the many hormones associated with the pituitary glands?

14. How is the thyroid gland a metabolic thermostat?

15. How do parathyroids function?

16. What are the functions of the adrenal glands?

17. How does the pancreas regulate glucose levels?

18. How do the other endocrine glands function?

19. How does the brain control the endocrine system?

20. What are the brain's chemical messengers, and how do they function?

21. How is the central nervous system regulated?

22. How does feedback control work?

23. What are the nonendocrine hormones?

24. How do neuropeptides function?

25. What are the effects of prostaglandins?

26. What is the function of atrial peptides?

27. How do the body's messengers communicate?

COMPLETING YOUR UNDERSTANDING

1. Why does each type of hormone molecule have a unique shape?

2. What are target cells, and how do they function?

3. What is the difference in release site between a neurohormone and a hormone?

4. What is the origin of steroid hormones?

5. Why do some athletes use anabolic steroids? What are some of the dangerous side effects of using anabolic steroids?

6. What is adrenaline, and how does it function?

7. The hormone that regulates the kidneys' retention of water is

    a. oxytocin.   d. LH.

    b. TSH.   e. ACTH.

    c. GH.   f. ADH.

8. How are pituitary tumors responsible for producing giant humans?

9. Why is it a good idea to occasionally consume iodized salt?

10. What causes rickets? What is a simple remedy for rickets?

11. Which two hormones does the adrenal cortex release? [*two answers required*]

    a. Cortisol   d. Calcitonin

    b. Epinephrine   e. Norepinephrine

    c. Aldosterone   f. Thyroxine

12. Where are the islets of Langerhans found, and how do they function?

13. What is the difference between Type I and Type II diabetes?

14. What is SADS, and how is it correlated with the human endocrine system?

15. Which of the following endocrine glands are found in the brain?

    a. Pineal   d. Pituitary

    b. Thymus   e. Parathyroids

    c. Pancreas   f. Adrenal cortex

16. Why are biologists so actively researching neuropeptides?

17. Why would taking aspirin relieve menstrual cramps?

18. Why are atrial peptides being investigated as a potential treatment for high blood pressure?

19. If you were lost in the desert with a case of liquor and were desperately thirsty, would you drink the liquor? Explain your answer.

20. Why do you suppose that the brain goes to the trouble of synthesizing releasing hormones, rather than simply directing the production of the pituitary hormones immediately?

*For discussion

FOR FURTHER READING

Atkinson, M., and M. MacLaren. 1990. What causes diabetes? *Scientific American,* July, 62–71. Explains why, for insulin-dependent diabetic patients, the answer is an autoimmune ambush of the body's insulin-producing cells. Why the attack begins and persists is now becoming clear.

Baulieu, E., and P. Kelly. 1990. *Hormones.* Paris: Hermann Press. A comprehensive text on hormones, with chapters written by individual experts.

Berridge, M. 1985. The molecular basis of communication within the cell. *Scientific American,* Oct., 142–52. An up-to-date account of what is known about second messengers in the cell.

Bloom, F. E. 1981. Neuropeptides. *Scientific American,* Oct., 148–68. An account of recent advances in the study of endorphins and other brain hormones.

Cantin, M., and J. Genest. 1986. The heart as an endocrine gland. *Scientific American,* Feb., 76–81. A good example of how the body self-regulates its activities. In addition to pumping blood, the heart secretes a hormone that fine-tunes blood pressure control.

Davis, J. 1984. *Endorphins: New waves in brain chemistry.* New York: Doubleday. A popular account of current research on brain hormones.

Rassmussen, H. 1989. The cycling of calcium as an intercellular messenger. *Scientific American,* Oct., 66–73. How calcium is involved in a variety of prolonged responses within cells.

Rosen, O. 1987. After insulin binds. *Science,* Sept., 1452–58. A review of how the insulin receptor works—it is much like growth factor receptors and some cancer-inducing genes.

EXPLORATIONS

Interactive Software:
Hormone Action

This interactive exercise allows students to explore how the hormone insulin regulates levels of blood sugar. The exercise presents a diagram of a human body with liver and muscles highlighted and levels of circulating blood glucose indicated. An insert shows a section of plasma membrane with insulin receptors in cross section. By varying diet and amount of exercise, students can investigate how the interaction of insulin and glucagon keeps levels of blood glucose constant, despite wide fluctuations in dietary intake of calories and utilization of calories for exercise.

Questions to Explore:

1. How does insulin know what cells to affect?

2. Why does the body use *two* hormones to regulate blood glucose levels?

3. Can you envision a way that obese individuals could avoid contracting Type II diabetes?

4. Why cannot a diabetic person simply eat more to compensate for the lack of insulin?

HOW ANIMALS DIGEST FOOD

What's for dinner? Hopefully not the photographer! These sharp teeth and powerful jaws are only the start to the digestive process of this Florida crocodile (Crocodylus americanus).

FOR REVIEW

Here are some important terms and concepts that have been discussed in previous chapters and that you will encounter again in this chapter. Review them before proceeding if necessary.

Acid (*chapter 2*)
Lysosomes (*chapter 4*)
Phagocytosis (*chapter 5*)
Oxidative respiration (*chapter 7*)
Insulin and glucagon (*chapter 37*)

A vertebrate body is a complex organization of many cells. Like a city, it contains many individuals with specialized functions. It has its own police (macrophages), its own construction workers (fibroblasts), and its own telephone company (the nervous system). Just as in a city, the vertebrate body does not have any "farmers" (that is, photosynthetic cells), and food must be trucked in from elsewhere.

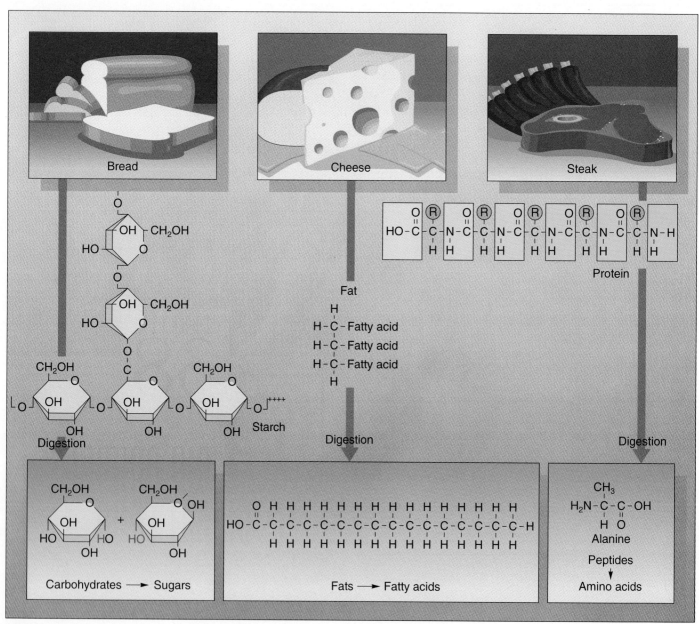

Figure 38.1 Complex molecules broken down into simpler compounds during digestion.

The body cannot incorporate large, complex molecules, such as starch, saturated fats, and proteins. The process of digestion breaks these molecules into the smaller sugars, fatty acids, and amino acids that can then be absorbed for use in the body.

All vertebrate cells are nourished with food that the animal obtains outside itself and transports to individual cells. Many of vertebrates' major organ systems are involved in this energy acquisition: The digestive system processes organic foodstuffs; the respiratory system acquires the oxygen necessary to metabolize these organic foodstuffs; the circulatory system transports both food and oxygen to the individual body cells; and the excretory system rids the body of waste produced by the metabolism of organic compounds. This chapter examines the first of these activities—digestion.

The Nature of Digestion

Animals are thermodynamic machines—expending energy to maintain order throughout their bodies—and energy sources are therefore essential for survival. Animals obtain the metabolic energy needed for growth and activity by degrading the chemical bonds of organic molecules. These breakdown processes, referred to as **catabolism,** are considered in detail in chapters 6 and 7. They include the breakdown of sugar molecules in glycolysis and the oxidation of pyruvate in the citric acid cycle. What catabolic processes have in common is that they act on amino acids, lipids, sugars, and fragments of these molecules to produce energy in the form of ATP, together with water and carbon dioxide (CO_2) as waste products.

But few organisms contain significant concentrations of free sugars and amino acids. Instead, the simple molecules are incorporated into long chains—into starches, fats, and proteins. Thus, eating another organism does not in itself provide a rich source of such molecules to an animal. Before an animal can obtain energy from its food, it must degrade these molecules into the simple compounds from which the molecules were built. This process is called **digestion** (figure 38.1).

Types of Digestion

Like most other body systems, digestion has become increasingly complex during the evolution of animals. Animals' protist ancestors simply incorporated food particles within their bodies by phagocytosis, combining these particles with enzymes in a process of **intracellular digestion.** Cnidarians break down the bodies of their prey by secreting enzymes into their central cavity and then incorporating prey fragments into their cells by phagocytosis. This form of digestion is an intermediate point between true intracellular digestion and true **extracellular digestion** (figure 38.2).

Flatworms exhibit true extracellular digestion because they have a highly branched digestive system with only one opening. Because of its branching, such a digestive system has a greatly increased absorptive surface. Although some of the food particles are broken down by the extracellular secretion of enzymes, as in cnidarians, most are simply incorporated into the cells that line the flatworm's digestive tract and are digested there (figure 38.3*a*).

Other, more complex animals have a more complete digestive tract, with a mouth and an anus, and various methods for breaking down the food mechanically, as by grinding in

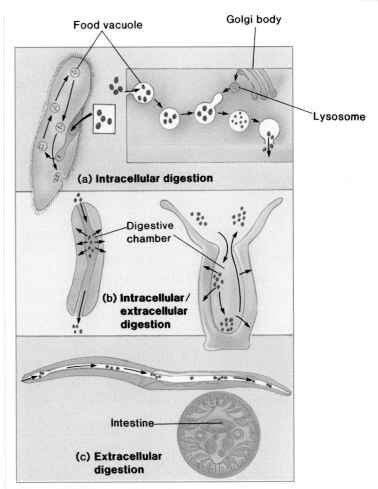

Figure 38.2 Types of digestion.
(a) *Intracellular digestion takes place within an organism's individual cells. These cells take in food particles through the process of phagocytosis.* (b) *Cnidarians were the first animals to use a digestive cavity, in which the food particles are broken down extracellularly (outside of cells). Final digestion, however, is carried out intracellularly, through the process of phagocytosis.* (c) *Roundworms practice true extracellular digestion, in which food passes through a "digestive tube" in one direction. Digestive products are absorbed into the organism during passage, and residual matter is eliminated from the tube's far end.*

the gizzard of an earthworm or bird, or enzymatically. Many such animals also have a crop for storing food for later use (figure 38.3*b, c*).

Agents of Digestion

Digestion is carried out in two ways. One involves a variety of highly specific enzymes (table 38.1). Enzymes that break up proteins into amino acids are called **proteases;** enzymes that break up starches and other carbohydrates into sugars are called **amylases;** and enzymes that break up lipids and fats into small segments are called **lipases.** Other enzymes, known as **DNase** and **RNase,** break up DNA and RNA, respectively.

A second agent of digestion is **hydrochloric acid (HCl),** which is secreted by the stomach when food is introduced.

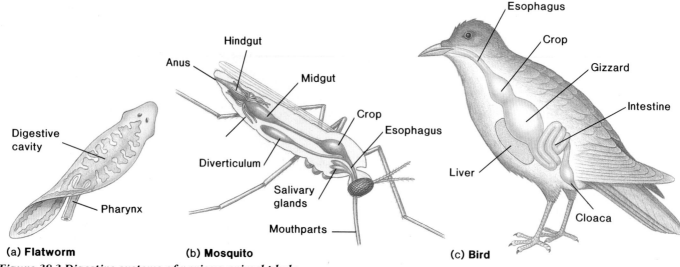

Figure 38.3 Digestive systems of various animal phyla.
(a) *The digestive system of a flatworm, in which there is only one opening to the outside. Most flatworm digestion takes place after food particles are incorporated into the cells that line the digestive cavity.* (b) *The digestive system of an adult female mosquito. Most of the blood that the mosquito sucks from another animal is stored* in the diverticulum for later use. (c) *The digestive system of a bird. A bird's crop plays a role similar to that of the diverticulum of the mosquito. In the gizzard, hard food is mixed and crushed with rocks and pebbles, often called grit. In this way, food is prepared for more efficient digestion in the intestine.*

Hydrochloric acid plays a very small role in breaking up proteins into smaller pieces, but its primary function is to convert the enzyme **pepsinogen** into **pepsin,** a protease. The action of pepsin on proteins is the primary means by which proteins are broken down into smaller peptide chains.

Proteins are the only nutrients degraded in the stomach. Most digestive enzymes, with the exception of pepsin, do not function in the stomach's high acidity. Starches and fats are broken down in the more basic environment of the small intestine.

Table 38.1		Digestive Enzymes	
Location	**Enzyme**	**Substrate**	**Digestion Product**
Salivary gland	Amylase	Starch, glycogen	Disaccharides
Stomach	Pepsin	Proteins	Short peptides
Small intestine	Peptidases	Short peptides	Amino acids
	Nucleases	DNA, RNA	Sugars, nucleic acid bases
	Lactase		
	Maltase	Disaccharides	Glucose, monosaccharides
	Sucrase		
Pancreas	Lipase	Triglycerides	Fatty acids, glycerol
	Trypsin	Proteins	Peptides
	Chymotrypsin		Peptides
	DNase	DNA	Nucleotides
	RNase	RNA	Nucleotides

In the stomach, pepsin breaks down proteins into smaller peptides. Starches are degraded by amylases, and lipids are degraded by lipases in the more basic environment of the small intestine.

Organization of the Vertebrate Digestive System

The digestive tract's general organization is the same in all vertebrates, although different elements are emphasized in different groups. In all vertebrates, proteins are broken down in the stomach, after which food passes to the upper part of the small intestine, called the **duodenum,** where a battery of digestive enzymes continues the digestive process. The products of digestion then pass across the wall of the small intestine into the bloodstream. Figure 38.4 illustrates the organization of the human digestive system, which is typical of the kinds of digestive systems found in vertebrates.

Specializations among the digestive systems of different kinds of vertebrates reflect differences in the way these animals live:

1. The initial components of the gastrointestinal (that is, passing through the stomach and intestine) tract are the **mouth** and **pharynx.** The pharynx is the gateway to the **esophagus.** Fishes have a large pharynx with gill slits, not unlike that of the lancelets, whereas air-breathing vertebrates have a greatly reduced pharynx.

2. Adult amphibians, which are carnivores, have a short intestine. The food they ingest is readily digested, including the soluble carbohydrate glycogen. Many birds, in contrast, subsist on plant material. The primary structural component of plants is cellulose, a rigid carbohydrate that resists digestion. Birds

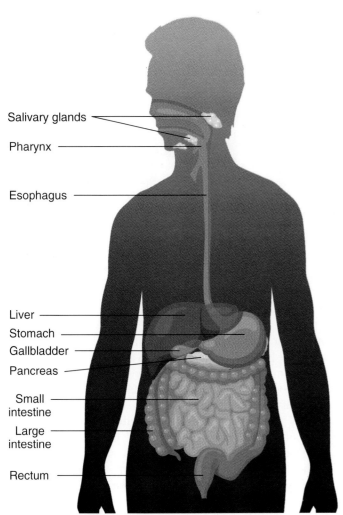

Salivary glands

Pharynx

Esophagus

Liver

Stomach

Gallbladder

Pancreas

Small intestine

Large intestine

Rectum

Figure 38.4 The human digestive system.
The human digestive system is shaped somewhat like a tube with openings at both ends.

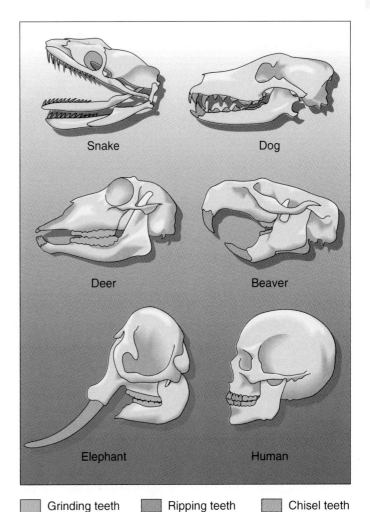

Snake   Dog

Deer   Beaver

Elephant   Human

☐ Grinding teeth   ☐ Ripping teeth   ☐ Chisel teeth

Figure 38.5 Vertebrate teeth—tools to fit many functions.
Vertebrate teeth are specialized for particular tasks. In the snake, the teeth slope backward to aid in the retention of prey during swallowing. In carnivores, such as dogs, "canine" teeth specialized for ripping food predominate. In herbivores, such as deer, flat, grinding teeth predominate. In the beaver, specialized teeth at the front of the mouth act as chisels. In the elephant, two of the upper teeth are specialized as weapons. Humans are omnivores, consuming both plant and animal food, and so have both canine and grinding teeth.

have a convoluted small intestine, by means of which they prolong the process of digestion and aid absorption of digestion products.

3. Many animals have teeth, and chewing (mastication) breaks up food into small particles and mixes food with fluid secretions. Birds, which lack teeth to masticate their food, break up food in their stomachs, which have two chambers. In one of these chambers—the gizzard—small pebbles ingested by the bird are churned together with the food by muscular action; this churning serves to grind up the seeds and other hard plant material into smaller chunks before their digestion in the stomach's second chamber.

4. Mammals that digest grass and other vegetation often have stomachs with multiple chambers, where bacteria aid cellulose digestion.

The Digestive Journey

As food makes its way through the vertebrate digestive system, it is first broken down into smaller fragments, which are then absorbed across the wall of the small intestine. Any material

that cannot be digested, such as cellulose, continues to move through the gastrointestinal tract and is eventually excreted.

The Mouth: Where It All Begins

All vertebrates take in food through the mouth, which in all groups except birds typically contains teeth. The teeth are specialized in different ways, depending on whether the vertebrates usually feed on animals or plants and how they obtain what they eat. Human beings are **omnivores,** eating both plant and animal food regularly. As a result, human teeth are structurally intermediate between the pointed, cutting teeth characteristic of carnivores and the flat, grinding teeth characteristic of herbivores (figure 38.5).

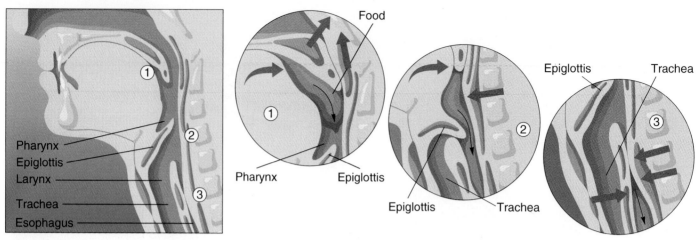

Figure 38.6 How humans swallow.
(**1**) *As food moves past the opening at the back of the mouth, muscles push the back part of the roof of the mouth (the soft palate) up against the nasal passage, sealing it off* (blue arrows).

(**2**) *In addition, the epiglottis folds back and seals off the respiratory passage.* (**3**) *The epiglottis returns to its original position as the food slips farther down the esophagus.*

Within the vertebrate mouth, the tongue mixes the food with a mucous solution, the **saliva.** In humans, saliva is secreted into the mouth by three pairs of **salivary glands** located above and below the jaw and in the connective tissue of the tongue. The saliva moistens and lubricates the food so that it can be swallowed more readily and does not abrade the tissue it passes on its way down through the esophagus. The saliva also contains the enzyme **amylase,** which initiates the breakdown of starch and other complex polysaccharides into smaller fragments. This action by amylase is the first of the many digestive processes that occur as the food passes through the digestive tract. However, salivary amylase is not essential because similar enzymes found in the pancreas can digest carbohydrates very effectively.

> *Vertebrate teeth serve to shred animal tissue and to grind plant material. In humans, the ripping and tearing teeth are in front, the grinding teeth in the rear. Saliva secreted into the mouth moistens the food, which aids its journey into the digestive system and begins the enzyme-catalyzed process of degradation.*

Food's Journey to the Stomach

After passing through the opening at the back of the mouth, food passes through the upper **esophageal sphincter** and enters the esophagus, which connects the pharynx to the stomach. No further digestion takes place in the esophagus. Its role is to move food down toward the stomach. In adult humans, the esophagus is about 25 centimeters long, and its lower end opens into the stomach proper. The lower two-thirds of the esophagus is enveloped in smooth muscle. Successive waves of contraction of these muscles move food down through the esophagus to the stomach. Such rhythmic sequences of waves

of muscular contraction in the walls of a tube are called **peristalsis.** Because the movement of food through the esophagus is primarily caused by these peristaltic contractions, humans can swallow even if they are upside down (figure 38.6).

The exit of food from the esophagus to the stomach is controlled by the **lower esophageal (cardiac) sphincter.** Contraction of this sphincter prevents food in the stomach from moving back up the esophagus.

If the sphincter does not close properly, the food backs up into the esophagus, and the acid coating the food causes a burning sensation called "heartburn" (even though it actually has nothing to do with the heart). This problem often arises during pregnancy, when the digestive organs are displaced far upward.

The Stomach: Preliminary Digestion

The stomach is a saclike portion of the digestive tract and has several functions: temporary food storage, mechanical breakdown of food, and the chemical digestion of proteins (figure 38.7). The digestive process is organized within the stomach as ingested food is collected, proteins are partially hydrolyzed, and the stomach contents are fed in a controlled fashion into the primary digestive organ, the small intestine.

The interior of the stomach, like that of the rest of the digestive tract, is continuous with the outside of the body. The epithelium lies under a deep layer of connective tissue, called **mucosa,** below which is located a complex array of muscles, blood vessels, and nerves. Covering the mucosa is the **submucosa,** which also contains blood vessels and nerves. Two sets of smooth muscles that lie over the submucosa move partially digested food through the digestive tract. The **serosa**—the top layer of the tract—cushions and protects the layers underneath. Holding the digestive tract in place is a thin layer of tissue called the **mesentery** (figure 38.8).

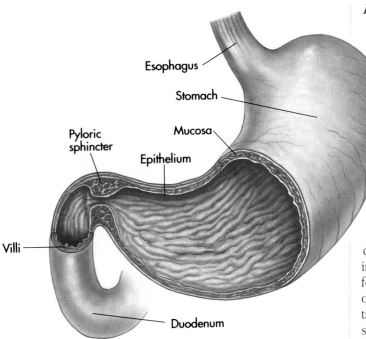

Figure 38.7 The stomach.
The stomach stores food for short periods of time and begins digestion of proteins. The stomach also breaks food down mechanically by its contractions

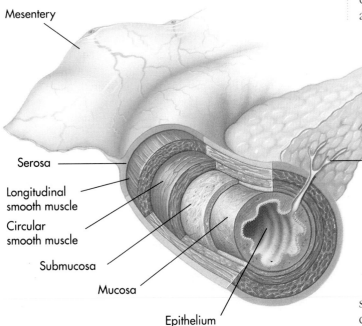

Figure 38.8 Organization of the vertebrate digestive tract.
The many layers that are laid over the epithelium of the digestive tract protect the epithelium and supply it with blood. The smooth muscles also allow food to pass through the digestive tract.

How the Stomach Digests Proteins

The epithelium of the stomach is the source of the "digestive juices," a combination of hydrochloric acid (HCl) and digestive enzymes that, when mixed with food, forms a semisolid material called **chyme.** The upper epithelial surface of the stomach is dotted with deep depressions called **gastric pits.** In them, the epithelial membrane is invaginated, forming exocrine glands within the mucosa. These exocrine glands contain two kinds of secreting cells: (1) **parietal cells,** which secrete HCl, and (2) **chief cells,** which secrete pepsinogen. Within the stomach, HCl cleaves a terminal fragment from the pepsinogen, converting pepsinogen into the protease pepsin (figure 38.9).

Many of the epithelial cells that line the stomach are specialized for mucus secretion. This mucus, which is produced in large quantities, lubricates the stomach wall and facilitates food movement within the stomach. It also protects the cells of the stomach wall from abrasion by the food. Most important, however, the mucus protects the stomach walls from the stomach's own digestive juices—the **gastric fluid**—which otherwise would eat the stomach lining away (for example, causing ulcers).

The human stomach secretes about 2 liters of HCl every day, creating a very concentrated acid solution—about three million times more acidic than the blood. HCl breaks up connective tissue. The very low pH values, between 1.5 and 2.5, created by HCl disrupt the attraction between carboxyl and amino side groups of proteins that is responsible for proteins' tertiary structure. This process causes the folded proteins of connective tissue to open out and disrupts their associations with one another.

The action of HCl disintegrates proteins into smaller fragments and thus is an essential prelude to digestion. However, HCl's ability to break proteins down into amino acids is very limited. Protein is further digested by the enzyme pepsin, which cleaves proteins into short polypeptides. Because the fragments produced by pepsin are large and frequently charged, they cannot pass across the epithelial membrane. Except for water, some vitamins, and alcohol, no absorption takes place through the stomach wall. Although pepsin aids protein digestion, it is not essential because the pancreas secretes enough proteases to digest proteins completely.

Stomach Secretion of the Hormone Gastrin

Overproduction of acid in the stomach makes it impossible for the body to neutralize the acid later in the small intestine, a

Table 38.2   Hormones of Digestion

Hormone	Source	Stimulus	Action	Note
Gastrin	Pyloric portion of stomach	Entry of food into stomach	Secretion of HCl	Unusual in that it acts on same organ that secretes it
Cholecystokinin (CCK)	Duodenum	Arrival of food in small intestine	Stimulates gallbladder contraction, and so the release of bile into intestine; stimulates secretion of digestive enzymes by pancreas	CCK bears a striking structural resemblance to gastrin
Secretin	Duodenum	HCl in duodenum	Stimulates pancreas to secrete bicarbonate, which neutralizes stomach acid	The first hormone to be discovered (1902)

step that is essential for the terminal stages of digestion. The stomach controls acid production by means of hormones produced by endocrine cells scattered throughout its epithelial layer (table 38.2). The hormone **gastrin** regulates HCl synthesis by the parietal cells of the gastric pits, permitting such synthesis to occur only when the pH of the stomach contents is higher than about 1.5 (see figure 38.9).

Some stomachs greatly overproduce gastrin, which results in excessive acid production. The excessive acid may attack and burn holes, called **duodenal ulcers,** in the walls of the small intestine. The contents of the small intestine are not normally acidic, and this organ is much less able to withstand the disruptive actions of stomach acids than is the stomach wall. For this reason, over 90 percent of all ulcers are duodenal, although other ulcers sometimes occur in the stomach when its mucous barrier is damaged by, for example, aspirin or alcohol.

In the stomach, concentrated acid breaks up connective tissue and protein into molecular fragments, which are further digested by pepsin into short polypeptides. Carbohydrates and fats are not digested in the stomach.

Because the stomach's inner surface is highly convoluted, it can fold up when empty and open out like an expanding balloon as it fills with food. The stomach's capacity, of course, is limited. The human stomach has a volume of about 50 milliliters when empty; when full, it may have a volume fifty times larger, from 2 to 4 liters. Carnivores, which often consume large meals at infrequent intervals, possess stomachs that are able to distend much more than can human stomachs or those of most other mammals.

The Pyloric Sphincter: The Traffic Light for Food Leaving the Stomach

The digestive tract exits from the stomach at a muscular constriction known as the **pyloric sphincter** (see figure 38.7). The pyloric sphincter is the gate to the small intestine, the organ within which the final stages of digestion occur. The pyloric sphincter, therefore, is the traffic light of the digestive system. The small intestine's capacity is limited, and its digestive processes take time. Consequently, for efficient digestion, only relatively small portions of food can be introduced from the stomach into the small intestine at any one time. When a small volume of food, which is by now the semisolid, highly acidic substance chyme, passes into the small intestine, the acid introduced with the chyme acts as a signal and prompts the closing of the pyloric sphincter. Over time, the food is digested, and the acid that entered the small intestine with the food is neutralized. At a certain point in the process, the pH of the small intestine reaches a level that signals the pyloric sphincter to open once again. Another small portion of food is introduced from the stomach into the small intestine, and the process continues.

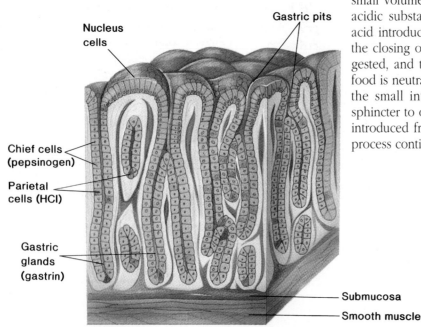

Nucleus cells

Gastric pits

Chief cells (pepsinogen)

Parietal cells (HCl)

Gastric glands (gastrin)

Submucosa

Smooth muscle

Figure 38.9 Gastric pits.
Gastric pits are deep invaginations of the stomach epithelium down into the underlying mucosa. Within the gastric pits are chief cells, which secrete pepsinogen, and parietal cells, which secrete hydrochloric acid (HCl). Hydrochloric acid cleaves pepsinogen into pepsin, which is used to degrade proteins into shorter peptides.

The Small Intestine: Terminal Digestion and Absorption

The small intestine is the true digestive vat of the vertebrate body. Within it, carbohydrates, proteins, and fats are broken down into sugars, amino acids, and fatty acids. Once these small molecules have been produced, they all pass across the epithelial wall of the small intestine into the bloodstream. Some of the enzymes necessary for these digestive processes are secreted by the cells of the intestinal wall. Most, however, are introduced into a short initial segment of the small intestine, the duodenum, through a duct from a gland called the pancreas.

The small intestine is approximately 6 meters long. The first 25 centimeters, about 4 percent of the total length, is the duodenum, where pancreatic enzymes and bile from the liver (discussed later in the chapter) enter the intestine, initiating digestion. The bloodstream absorbs water and the products of digestion in the later sections of the intestine, the **jejunum** and the **ileum.**

The epithelial wall of the small intestine is covered with fine, microscopic, fingerlike projections called **villi.** In turn, each of the epithelial cells covering the villi is covered on its outer surface by a field of cytoplasmic projections called **microvilli.** Both kinds of projections greatly increase the absorptive surface of the epithelium lining the small intestine. The average surface area of the small intestine of an adult human being is about 300 square meters. The membranes of the epithelial cells contain enzymes that complete digestion, as well as transmembrane protein carriers that actively transport sugars and amino acids across the membrane. Fatty acids cross passively by diffusion. After being absorbed by the small intestine, the sugar and amino acid end products of digestion enter capillaries in the villi, while the triglyceride and fatty acid end products enter lymph vessels known as **lacteals.**

> *Most digestion occurs in the first 25 centimeters of the 6-meter length of the small intestine, in a zone called the duodenum. The rest of the small intestine is devoted to the absorption of the products of digestion.*

A startling amount of material passes through the small intestine. An average human consumes about 800 grams of solid food and 1,200 milliliters of water each day, for a total volume of about 2 liters. To this amount is added about 1.5 liters of fluid from the salivary glands, 2 liters from the gastric secretions of the stomach, 1.5 liters from the pancreas, 0.5 liter from the liver, and 1.5 liters of intestinal secretions, for a total of a remarkable 9 liters.

However, although the flux is great, the *net* passage is small. Almost all of these fluids and solids are reabsorbed during their passage through the small and large intestine, with about 8.5 liters passing across the wall of the small intestine and an additional 350 milliliters through the wall of the large intestine. Of the 800 grams of solid and 9 liters of liquid that enter the digestive tract in one day, only about 50 grams of solid and 100 milliliters of liquid leave the body as **feces.** The normal fluid absorption efficiency of the digestive tract thus approaches a phenomenal 99 percent.

Organs That Support the Digestive System

In addition to the stomach and intestines, two other organs—the pancreas and liver—play key roles in the digestive process.

The Pancreas: Making Digestive Enzymes

The **pancreas** is a large gland situated near the junction of the stomach and the small intestine. It is one of the body's major exocrine glands, secreting a host of different enzymes that act in the duodenum to break down carbohydrates, proteins, and fats. These enzymes include proteases for breaking down proteins, lipases for digesting fats, and enzymes for breaking down carbohydrates. The pancreas also functions as an endocrine gland.

The pancreas has two types of exocrine cells. The first type secretes digestive enzymes. The second type functions exactly opposite to the parietal cells of the gastric pits. Instead of secreting an acid (HCl), these specialized cells secrete a base—**bicarbonate.** The alkaline bicarbonate is critical to successful digestion, since most of the enzymes secreted by the pancreas will not work in acid solution. The introduction of bicarbonate into the duodenum neutralizes the acid derived from the stomach and thus permits digestive enzymes to function. Since acid is secreted in the stomach, and bicarbonate is secreted in the intestine, digestion has no net effect on the body's acid-base balance (see chapter 42).

Bicarbonate secretion from the pancreas is controlled by the hormone **secretin,** just as the secretion of HCl in the stomach is controlled by gastrin. Interestingly, secretin was the first hormone to be discovered. Secretin is secreted in response to the presence of HCl in the stomach.

In addition to enzyme and bicarbonate secretion, the pancreas has yet a third function critical to metabolism. As noted in chapter 37, the islets of Langerhans, distributed throughout the exocrine regions of the pancreas, function as endocrine glands, producing the hormones that act in the liver and elsewhere to regulate blood glucose levels.

The Liver: The Body's Principal Metabolic Factory

Because fats are insoluble in water, they tend to enter the small intestine as small globules that are not attacked readily by the enzymes secreted by the pancreas. Before fats can be digested by pancreatic lipases, they must be made soluble. This process is carried out by a collection of detergent molecules secreted by a second gland, the **liver.**

Functions of the Liver The liver is the body's principal metabolic factory, turning foodstuffs arriving from the digestive tract in the bloodstream into substances utilized by the different body cells. It is the body's largest internal organ. In an adult human, the liver weighs about 1.5 kilograms and is the size of a football.

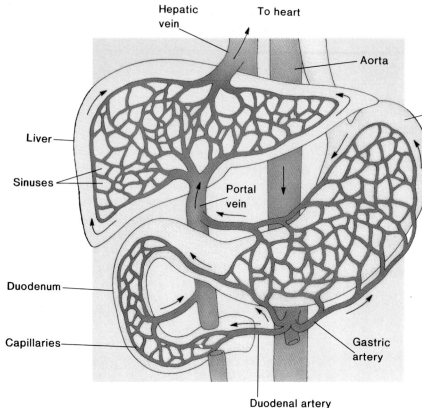

Hepatic vein

To heart

Aorta

Stomach

Liver

Sinuses

Portal vein

Duodenum

Capillaries

Gastric artery

Duodenal artery

Maintaining a constant level of metabolites in the blood requires active control by the body's organs. A moment's reflection shows why. Most vertebrates eat sporadically. In the United States, most people eat three meals a day. Food enters the digestive system at intervals separated by long periods of fasting. Much of the food is digested relatively quickly, with the metabolites, such as glucose and amino acids, passing through the lining of the small intestine into the bloodstream. Without active control, the levels of these and other metabolites would quickly rise in the blood right after a meal and then would fall rapidly during a period of starvation, when the metabolites are being removed from the bloodstream by metabolizing cells and are not being replenished.

Metabolite levels in the blood are controlled by establishing a metabolic reservoir, or bank, in the liver. The liver regulates the blood's metabolite levels by means of a special shunt in the circulatory system. Blood from the stomach and small intestine, rich with the products of digestion, flows into the **portal vein,** which carries it not to the heart but to the liver. Within the liver, the blood passes through a network of fine passages called **sinuses.** Only then is it collected into the **hepatic vein** and delivered to the vena cava and the heart (figure 38.10).

When excessive amounts of glucose are present in the blood passing through the liver, a situation that occurs soon after a meal, the liver converts the excess glucose into the starchlike glucose polymer glycogen, stimulated by the hormone insulin. The liver stores this glycogen. Some glycogen is also stored in the muscles, where it is easily available to fuel muscle contraction. When blood glucose levels decrease, as they do in a period of fasting or between meals, the glucose deficit in the blood plasma is made up from the glycogen reservoir, stimulated by the hormone glucagon. The human liver stores enough glycogen to supply glucose to the bloodstream for about ten hours of fasting. If fasting continues, the liver begins to convert other molecules, such as amino acids, into glucose to maintain the blood glucose level. The liver, the body's metabolic reservoir, thus acts much like a bank, making deposits and withdrawals in the "currency" of glucose molecules.

The liver carries out a wide variety of metabolic functions, many of which are discussed in subsequent chapters. It supplies quick energy, metabolizes alcohol, makes proteins, stores vitamins and minerals, regulates blood clotting, regulates cholesterol production, and detoxifies poisons.

The liver also produces **bile,** which contains the detergent molecules already mentioned, known as **bile salts.** The liver secretes the bile through a duct into the duodenum. Bile acts as a superdetergent. It combines with fats to form microscopic droplets called **micelles** in a process known as **emulsification.** The components of bile include bile salts, cholesterol, and the phospholipid lecithin, all of which combine to render fat soluble. Human beings concentrate and store bile manufactured in the liver in the **gallbladder.** When chyme enters the small intestine, a hormone known as **cholecystokinin (CCK)** stimulates gallbladder contraction to release bile into the duodenum.

How the Liver Regulates Blood Glucose Levels Vertebrate blood requires a relatively constant composition, since the different body tissues need specific compounds as energy sources. Brain cells, for example, can store very little glucose and lack the enzymes to convert fat or amino acids into glucose. Brain cells are thus very sensitive to the level of available glucose. They are totally dependent on blood plasma for glucose and cease to function if blood glucose levels fall much below normal values.

> *In addition to its many other roles, the liver acts to maintain blood glucose levels within narrow bounds.*

Also like a bank, the liver exchanges currencies, converting other molecules, such as amino acids and fats, to glucose and subsequently glycogen for storage. Excess amino acids in the blood are converted to glucose by liver enzymes, and the glucose is stored as glycogen. The first step in this conversion is the removal of the amino group ($-NH_3$) from the amino acid, a process called deamination. Unlike plants, animals cannot reuse the nitrogen from these amino groups and must excrete it as nitrogenous waste. The product of amino acid deamination—ammonia (NH_4^+)—forms a complex with carbon dioxide to form urea. In some vertebrates, amino acids are not deaminated directly, but instead are converted to uric acid. The liver releases the uric acid into the bloodstream, where the kidneys subsequently remove it.

The liver has a limited storage capacity. When its glycogen reservoir is full, it continues to remove excess glucose molecules from the blood by converting them to fat, which is stored elsewhere in the body. In humans, for example, long periods of overeating and the resulting chronic oversupply of glucose frequently result in the deposition of fat around the stomach or on the hips.

The Large Intestine:
Concentration of Solid Wastes

The large intestine, or **colon,** is much shorter than the small intestine, occupying approximately the last meter of the intestinal tract. No digestion takes place within the large intestine, and only about 4 percent of body fluids are absorbed there. The large intestine is not convoluted, lying instead in three relatively straight segments, and its inner surface does not possess villi. Consequently, the large intestine has less than one-thirtieth the absorptive surface area of the small intestine. Although sodium, vitamin K, and some other products of bacterial metabolism are absorbed across its wall, the large intestine's primary function is to act as a refuse dump. Within it, undigested material, including large amounts of plant fiber and cellulose, is compacted and stored. Many bacteria live and actively divide within the large intestine, where they play a role in the processing of undigested material into the final excretory product feces. Bacterial fermentation produces gas within the colon at a rate of about 500 milliliters per day. This rate increases greatly after the consumption of beans or other vegetable matter because undigested plant matter provides material for fermentation.

> *The large intestine serves primarily to compact the solid refuse remaining after digestion, thereby facilitating elimination.*

The final segment of the gastrointestinal tract is a short extension of the large intestine called the **rectum.** Compacted solids within the colon pass into the rectum as a result of the peristaltic contractions of the muscles encasing the large intestine. From the rectum, the solid material passes out of the anus through two **anal sphincters.** The first is composed of smooth muscle; it opens involuntarily in response to a pressure-generated nerve signal from the rectum. The second sphincter, in contrast, is composed of striated muscle. It is subject to voluntary control from the brain, thus permitting a conscious decision to delay defecation.

Figure 38.11 summarizes the events of the digestive process.

Nutrition

Vertebrate food ingestion serves two ends: It provides a source of energy, and it also provides raw materials that the animal is not able to manufacture for itself (figure 38.12). There are two basic nutritional states: (1) the **absorptive state,** which immediately follows meals, and (2) the **postabsorptive state,** during which reserve stores of energy must be tapped. The digestion process is matched to the needs of the body by both neural and hormonal control systems. Neural control is mediated by external parasympathetic input and local reflexes, whereas hormonal control primarily involves insulin and glucagon. The liver maintains a very constant level of glucose in the blood and stores several hours' reserve of glucose in the form of glycogen. Any intake of food in excess of that required to maintain the glycogen reserve results in one of two consequences: Either the excess glucose is metabolized by the muscles and other body cells, or it is converted to fat and stored within fat cells. Fats have a much higher energy content per gram than carbohydrates and proteins and represent a very efficient way to store energy. This is represented by the simple equation

$$FOOD - EXERCISE = FAT$$

In wealthy countries such as those of North America and Europe, obesity results from chronic overeating and from unbalanced diets high in fat and is a significant human health problem. In the United States, about 30 percent of middle-aged women and 15 percent of middle-aged men are classified as overweight, weighing at least 20 percent more than the average weight for their height (figure 38.13). Being overweight is strongly correlated with coronary heart disease and many other disorders.

Classes of Nutrients

To maintain itself and sustain daily cellular activities, the body needs to take in a variety of **nutrients.** Nutrients are substances that the body requires as cellular components or enzymes. These include proteins, fats, carbohydrates, and vitamins and minerals. The trick to avoiding obesity and critical imbalances in any of these nutrients is understanding how much of each nutrient is needed to maintain a healthy body.

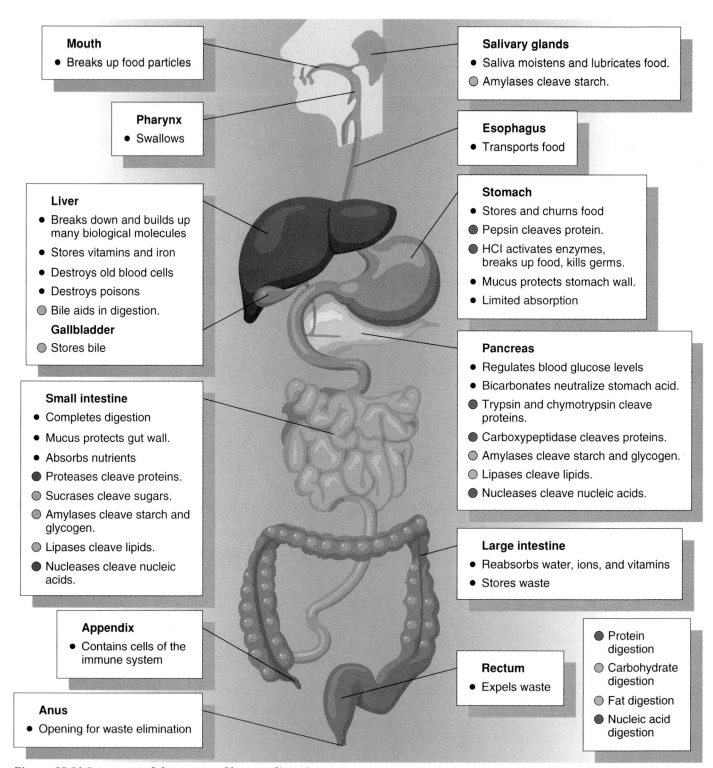

Mouth
- Breaks up food particles

Pharynx
- Swallows

Liver
- Breaks down and builds up many biological molecules
- Stores vitamins and iron
- Destroys old blood cells
- Destroys poisons
- Bile aids in digestion.

Gallbladder
- Stores bile

Small intestine
- Completes digestion
- Mucus protects gut wall.
- Absorbs nutrients
- Proteases cleave proteins.
- Sucrases cleave sugars.
- Amylases cleave starch and glycogen.
- Lipases cleave lipids.
- Nucleases cleave nucleic acids.

Appendix
- Contains cells of the immune system

Anus
- Opening for waste elimination

Salivary glands
- Saliva moistens and lubricates food.
- Amylases cleave starch.

Esophagus
- Transports food

Stomach
- Stores and churns food
- Pepsin cleaves protein.
- HCl activates enzymes, breaks up food, kills germs.
- Mucus protects stomach wall.
- Limited absorption

Pancreas
- Regulates blood glucose levels
- Bicarbonates neutralize stomach acid.
- Trypsin and chymotrypsin cleave proteins.
- Carboxypeptidase cleaves proteins.
- Amylases cleave starch and glycogen.
- Lipases cleave lipids.
- Nucleases cleave nucleic acids.

Large intestine
- Reabsorbs water, ions, and vitamins
- Stores waste

Rectum
- Expels waste

- Protein digestion
- Carbohydrate digestion
- Fat digestion
- Nucleic acid digestion

Figure 38.11 Summary of the events of human digestion.
This diagram shows the different sites of digestion throughout the digestive tract.

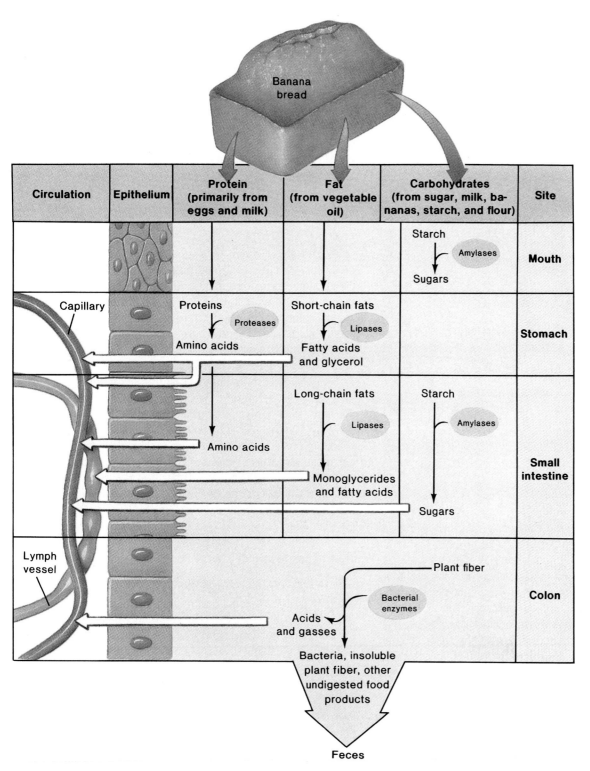

Circulation	Epithelium	Protein (primarily from eggs and milk)	Fat (from vegetable oil)	Carbohydrates (from sugar, milk, bananas, starch, and flour)	Site
				Starch → Amylases → Sugars	**Mouth**
Capillary		Proteins → Proteases → Amino acids	Short-chain fats → Lipases → Fatty acids and glycerol		**Stomach**
		Amino acids	Long-chain fats → Lipases → Monoglycerides and fatty acids	Starch → Amylases → Sugars	**Small intestine**
Lymph vessel			Acids and gasses ← Bacterial enzymes ← Plant fiber	Bacteria, insoluble plant fiber, other undigested food products	**Colon**

Feces

Figure 38.12 What happens to what you eat—a molecular journey.

This diagram shows the fates of the different nutrients found in a familiar food—banana bread. Proteins are degraded in the stomach and absorbed across the small intestine. Fats are also degraded to a certain extent in the stomach, but most breakdown takes places in the small intestine. The products of this breakdown, monoglycerides and fatty acids, are absorbed by the lacteals of the lymph vessels. Carbohydrates are initially broken down in the mouth, but the bulk of their breakdown takes place in the small intestine. The sugars that result from this breakdown are also absorbed by the small intestine. Any indigestible material, such as fiber, is excreted.

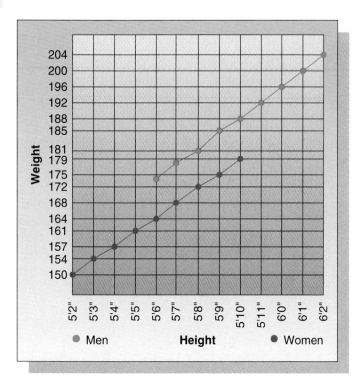

Figure 38.13 Obesity is a significant human health problem.

Obesity is usually characterized as the state of being more than 20 percent heavier than the average person of the same sex and height. These are the weight values at which obesity begins in Americans for a variety of heights, using average weights compiled in 1985.

Proteins

Proteins are components of the enzymes that facilitate chemical reactions of metabolism. Structural proteins, such as actin, myosin, and collagen, make up the cells of muscles and connective tissues. As explained in chapter 2, proteins are composed of chains of peptides, which, in turn, are composed of strings of amino acids. The human body cannot make eight of the twenty amino acids that exist in organisms and thus must acquire them from food sources. These eight are called the **essential amino acids.** A **complete protein** source contains all eight of the essential amino acids, while an **incomplete protein** source contains only some. Regular intake of complete proteins is essential for good health.

Fats

The problem with eating proteins is that most foods that are high in these essential amino acids are also high in fats. While essential in the human diet—fats provide quick energy and are important components of cell membranes—only about a tablespoon of fat per day is necessary for the body to run smoothly. Most people, in their effort to take in enough protein, take in too much fat. An excess of fat in the diet has been shown to contribute to heart disease, obesity, stroke, and some types of cancers, including colon and breast cancer.

Meats, cheeses, and other dairy products are excellent sources of complete protein but are very high in fat content.

Experts now recommend that these usual protein sources be replaced with incomplete plant proteins, which can yield all eight of the essential amino acids if combined properly. For example, the combination of lentils and beans yields a full complement of the essential eight amino acids, but with much less fat than does a steak or hamburger. Experts also recommend the avoidance of saturated fats, in which every carbon atom in the fat molecule is bound to a hydrogen molecule, in favor of unsaturated fats. Unsaturated fats are more fluid than saturated fats and do not build up in arteries as much as saturated fats do. Substitutions such as these, along with a conscious decision to reduce fat intake, are now considered worthwhile goals for everyone.

Carbohydrates

Carbohydrates, which are found in starchy foods, such as breads, cereals, potatoes, and pasta, are another important nutrient. Once considered fattening, carbohydrates are now known to be usually low in fat and calories (unless smothered in butter and fatty sauces). Carbohydrates are quickly used up as an energy source, and so are not converted as readily into fat. Experts are now recommending that carbohydrate intake be increased while protein and fat intake be lowered.

Sometimes, carbohydrates contain dietary **fiber** that adds bulk to foods and ensures their quick passage through the large intestine. Diets low in fiber have been linked to colon cancer, possibly because such a diet slows down the movement of wastes in the large intestine, exposing the large intestine to various toxins and waste products that can eventually cause cancer.

Because carbohydrates can be broken down into glucose, a simple sugar that is the essential fuel for body activities, the body does not need excess sugar intake. However, many people eat sugary foods and overload their bodies with this nutrient. Continued excess sugar intake can lead to diabetes and other ailments, including tooth decay and general malnutrition.

Vitamins and Minerals

Along with the essential amino acids, humans have also lost the ability to manufacture other substances that are essential for maintaining cellular processes. **Vitamins** are substances that cannot be synthesized by humans but that are required cofactors for intracellular enzyme systems. Human beings, monkeys, and guinea pigs, for example, have lost the ability to synthesize ascorbic acid (vitamin C) and will develop the disease **scurvy**—characterized by weakness, spongy gums, and bleeding of the skin and mucous membranes—if vitamin C is not supplied in sufficient quantities in their diets. As shown in table 38.3, humans require at least thirteen different vitamins.

Minerals are inorganic substances that are also required by humans in small, or **trace,** amounts. Magnesium, iron, and folic acid are examples of the essential minerals that humans must obtain in their food.

Proteins, fats, carbohydrates, and vitamins and minerals are nutrients required by the body in different amounts.

Table 38.3   Major Vitamins

Vitamin	Function	Dietary Source	Recommended Daily Allowance (milligrams)	Deficiency Symptoms	Solubility
Vitamin A (retinol)	Used in making visual pigments, maintenance of epithelial tissues	Green vegetables, milk products, liver	1	Night blindness, flaky skin	Fat
B-Complex Vitamins					
B_1	Coenzyme in carbon dioxide removal during cellular respiration	Meat, grains, legumes	1.5	Beriberi, weakening of heart, edema	Water
B_2 (riboflavin)	Part of coenzymes FAD and FMN, which play metabolic roles	In many different kinds of foods	1.8	Inflammation and breakdown of skin, eye irritation	Water
B_3 (niacin)	Part of coenzymes NAD^+ and $NADP^+$	Liver, lean meats, grains	20	Pellagra, inflammation of nerves, mental disorders	Water
B_5 (pantothenic acid)	Part of coenzyme-A, a key connection between carbohydrate and fat metabolism	In many different kinds of foods	5 to 10	Rare: fatigue, loss of coordination	Water
B_6 (pyridoxine)	Coenzyme in many phases of amino acid metabolism	Cereals, vegetables, meats	2	Anemia, convulsions, irritability	Water
B_{12} (cyanocobalamin)	Coenzyme in the production of nucleic acids	Red meat, dairy products	0.003	Pernicious anemia	Water
Biotin	Coenzyme in fat synthesis and amino acid metabolism	Meat, vegetables	Minute	Rare: depression, nausea	Water
Folic acid	Coenzyme in amino acid and nucleic acid metabolism	Green vegetables	0.4	Anemia, diarrhea	Water
Vitamin C	Important in forming collagen, cement of bone, teeth, connective tissue of blood vessels; may help maintain resistance to infection	Fruits, green leafy vegetables	45	Scurvy, breakdown of skin, blood vessels	Water
Vitamin D (calciferol)	Increases absorption of calcium and promotes bone formation	Dairy products, cod-liver oil	0.01	Rickets, bone deformities	Fat
Vitamin E (tocopherol)	Protects fatty acids and cell membranes from oxidation	Margarine, seeds, green leafy vegetables	15	Rare	Fat
Vitamin K	Essential to blood clotting	Green leafy vegetables	0.03	Severe bleeding	Fat

Dietary Guidelines

Recently, the U.S. Department of Agriculture (USDA) issued a new set of dietary guidelines designed to help people choose foods for a healthy diet. These guidelines are in the form of a pyramid (figure 38.14). At the base of the pyramid, and comprising the largest section of the pyramid, are carbohydrates. The USDA recommends eating six to eleven servings of these foods daily. The next tier of the pyramid is occupied by vegetables and fruits. These foods are high in essential vitamins and minerals, and also contain some protein while being low in fat. The USDA recommends eating three to five servings of vegetables and two to four servings of fruits daily. Protein foods occupy the next highest tier, and because these foods tend to be high in fat, the recommendation for daily intake is lower than the other tiers: two to three servings of dairy products and two to three servings of meat, fish, eggs, and beans per day. At the very top of the pyramid, occupying the smallest section, are fats, oils, and sweets. These should be eaten very sparingly.

In addition to following the guidelines suggested by the food pyramid, a healthy diet also requires:

1. Eating a variety of foods so that all the required nutrients will be represented.

2. Maintaining a healthy, although realistic, weight. If weight loss is desired, a loss of about 1 to 2 pounds per week is considered healthy (see Sidelight 38.1). A lower calorie intake

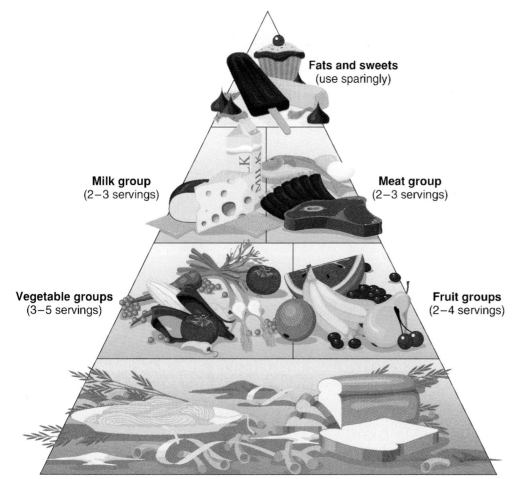

Fats and sweets
(use sparingly)

Milk group
(2–3 servings)

Meat group
(2–3 servings)

Vegetable groups
(3–5 servings)

Fruit groups
(2–4 servings)

Bread and cereal group
(6–11 servings)

Figure 38.14 The food pyramid.
This food guide allows diet planning according to current nutritional theories. A healthy diet uses more foods at the base of the pyramid and fewer at the top.

is important for weight loss, but exercise can speed up the process.

3. Avoiding cholesterol-laden foods. Cholesterol is a component of some fatty foods, and studies have shown that cholesterol contributes to the buildup of fat plugs, or plaques, in the arteries. These plaques can block blood flow and lead to heart attack and stroke.

4. Reducing salt intake. In excess amounts, salt can cause high blood pressure and lead to heart disease.

The food pyramid specifies the amount of each class of nutrient that is required on a daily basis. Additional dietary guidelines include eating a variety of foods, maintaining a healthy weight, avoiding cholesterol-laden food, and reducing salt intake.

Sidelight 38.1
DANGEROUS EATING HABITS

In the United States, serious eating disorders have become much more common since the mid-1970s. The most frequent of these are *anorexia nervosa*, a condition in which afflicted persons literally starve themselves, and *bulimia*, a condition in which individuals gorge themselves and then cause themselves to vomit so that their weight stays constant. For reasons not yet understood, 90 to 95 percent of those suffering from these eating disorders are female. Researchers estimate that 5 to 10 percent of American adolescent girls and young women suffer from eating disorders. As many as one in five female high school and college students may suffer from bulimia.

Those who suffer from anorexia are typically shy, well-behaved young women who feel embarrassed about their bodies. Eventually, they often begin to show signs of severe malnutrition and abnormally low temperatures and pulse rates. Although malfunction of the pituitary gland may produce similar symptoms, anorexia nervosa is now understood to be a psychiatric disturbance with severe psychic and psychological consequences. Its medical consequences may be severe enough to result in death.

Bulimia is usually accompanied by severe feelings of guilt, depression, anxiety, and helplessness. The eating binges characteristic of bulimia often involve large quantities of carbohydrate-rich junk food. The accompanying medical consequences result from the frequent purging, which reduces the body's supply of potassium. Potassium plays an important role in regulating body fluids; its loss may lead to muscle weakness or paralysis, an irregular heartbeat, and kidney disease. Like anorexia nervosa, bulimia often leads to decreased resistance to infection.

Although both anorexia nervosa and bulimia are now believed to be psychiatric disorders, scientists continue to search for physiological explanations for these serious conditions. Changes in the levels of certain hormones are associated with these conditions, but whether these changes cause or result from the conditions has not been demonstrated clearly.

The shame often associated with these conditions must be recognized as a problem, and professional help should be sought promptly. Both a psychiatrist or psychologist and a physician should be involved, and the family must play a strong, supportive role for the treatment to be effective. Even with effective treatment, both conditions may persist for years. Parents, teachers, and others must be alert to the existence of anorexia and bulimia and provide the kind of education about food and diet that, coupled with a supportive environment, will lead to early detection and cure of these eating disorders.

EVOLUTIONARY VIEWPOINT

The strong association between high-fat, low-fiber diets and cancers of the intestinal tract strongly suggests that the human gastrointestinal tract evolved to handle a very different set of food-stuffs than those currently consumed by most Americans. One of the most common life-threatening cancers among Americans—cancer of the colon—is rare among Chinese people, who eat one-third the amount of fat.

SUMMARY

1. Digestion is the rendering of parts of organisms into amino acids and sugars, which organisms can metabolize.

2. Vertebrates have a one-way digestive tract. The initial portion leads from a mouth through an esophagus to a stomach.

3. In most mammals, the stomach juices are concentrated acid, in which the protein-digesting enzyme pepsin is active.

4. Food passes from the stomach to the small intestine, where the pH is neutralized and a variety of enzymes, many synthesized in the pancreas, act to complete digestion. Most digestion occurs in the first 25 centimeters of the small intestine, in a zone called the duodenum.

5. The products of digestion are absorbed across the walls of the small intestine, which possess numerous villi and so achieve a very great surface area. Amino acids and sugars are transported by specific transmembrane channels, whereas fatty acids, which are lipid soluble, cross passively by diffusion. In the process of digestion, fats need to be made soluble by detergent molecules (bile salts) secreted by the liver.

6. Glucose and other metabolic products of digestion do not enter the general circulation directly but instead flow to the liver. The liver removes and stores any excess metabolic products and maintains glucose levels within narrow bounds.

7. The large intestine has little digestive or absorptive activity. It functions principally to compact the refuse left over from digestion for easier elimination.

8. The classes of nutrients are: proteins, fats, carbohydrates, and vitamins and minerals. A complete protein source is one that contains all eight of the amino acids that the body cannot synthesize.

9. Vertebrates lack the enzymes necessary to synthesize many necessary compounds and must obtain these enzymes, called vitamins, from their diet. A number of trace minerals must also be present in the diet.

10. The food pyramid provides dietary guidelines for daily food intake of carbohydrates, fruits and vegetables, protein foods, and fats. Other guidelines for a healthy diet include eating a variety of foods, maintaining a healthy weight, avoiding cholesterol, and reducing salt intake.

REVIEWING THE CHAPTER

1. What is the nature of digestion?

2. What are the types of digestion?

3. What are the agents of digestion?

4. How is the vertebrate digestive system organized?

5. What is the digestive pathway?

6. What happens in the mouth?

7. How is food processed in the stomach?

8. How does preliminary digestion occur in the stomach?

9. How does the stomach digest proteins?

10. Why does the stomach secrete the hormone gastrin?

11. What happens when food leaves the stomach?

12. How does the small intestine function?

13. What are the organs that support the digestive system?

14. How does the pancreas make digestive enzymes?

15. Why is the liver a major component of the digestive system?

16. What are the functions of the liver?

17. How does the liver regulate blood glucose levels?

18. How does the large intestine function?

19. What are the nutritional states?

20. How are nutrients classified?

21. How are proteins used?

22. Why are fats essential in the human diet?

23. Why are carbohydrates good for you?

24. Why are vitamins and minerals essential?

25. What are the dietary guidelines recommended by the U.S. Department of Agriculture?

26. What are some dangerous eating disorders and their consequences?

COMPLETING YOUR UNDERSTANDING

1. What is phagocytosis, and how does it work in the digestive system?

2. Where are the digestive enzymes located in the human body?

3. Why are most of the digestive enzymes not found in the stomach?

4. Why do birds need to ingest small stones or pebbles?

5. Why do humans have teeth that are intermediate between pointed, cutting teeth and flat, grinding teeth?

6. Starchy foods such as potatoes begin being digested in the
    a. mouth   d. duodenum.
    b. esophagus. e. small intestine.
    c. stomach.

7. In the stomach, why is the epithelium membrane invaginated in the gastric pits?

8. Why must the stomach be so acidic?

9. Protein-rich foods such as steak are broken down in the stomach by the combined action of [*two answers required*]
    a. low pH.   d. pepsin.
    b. high pH.   e. bile.
    c. amylase.

10. What are the cause and consequences of duodenal ulcers?

11. From what you have read about digestion, would it be more efficient to eat large amounts of food a few times per day or small amounts of food several times a day?

12. Where does most digestion and absorption occur in the body?

    a. Mouth   d. Large intestine

    b. Stomach   e. Small intestine

    c. Esophagus   f. Rectum

13. Which is longer in the human body—the esophagus, the small intestine or the large intestine? Why?

14. How efficient is fluid absorption in the human digestive tract?

15. Why must the pancreas secrete bicarbonate, an alkaline (basic) substance?

16. What might contribute to a gallbladder attack?

17. Under what conditions would the liver convert amino acids, the components of proteins, into glucose, a carbohydrate?

18. How is the liver involved in the formation of the waste product uric acid?

19. What is the basis for the deposition of fat around the stomach or on the hips?

20. Why and how does fermentation occur within the large intestine?

21. Why are the two anal sphincters composed of different muscle types?

22. Why are so many Americans overweight? What is the solution to this problem? Is this a concern to the college-age population? Why or why not?

23. What is an essential amino acid?

24. Many of us add sugar to our cereal or coffee and salt to our food prior to consumption. Why is this not a healthy practice?

25. Vitamin C deficiency causes the disease

    a. measles.   d. rickets.

    b. anemia.   e. beriberi.

    c. scurvy.   f. pellagra.

26. Which vitamin is a good source of an antioxidant (protects from oxidation)?

    a. Vitamin C   d. Vitamin K

    b. Vitamin E   e. Niacin

    c. Biotin   f. Vitamin D

27. Why is it important to read labels concerning protein, cholesterol, salt, and fat content when grocery shopping?

28. Eating disorders such as anorexia nervosa and bulimia are associated with changes in levels of

    a. vitamins.   d. nucleic acids.

    b. minerals.   e. fatty acids.

    c. carbohydrates.   f. hormones.

29. What correlation can be made between high rates of colon cancer in the United States and the fact that colon cancer is a rarity in the Chinese population?

*30. Human beings obtain vitamin K from symbiotic bacteria living in their gastrointestinal tract. Many bacteria also produce ascorbic acid (vitamin C). Can you suggest a reason why people have not evolved a symbiotic relationship with bacteria that would result in their obtaining bacterial vitamin C?

*For discussion

FOR FURTHER READING

Abraham, S., and D. Llewellyn-Jones. 1984. *Eating disorders—the facts.* Oxford, England: Oxford University Press. Covers the physiological and psychological factors leading to obesity, anorexia, and bulimia.

Cohen, L. 1987. Diet and cancer. *Scientific American,* Nov., 42–51. A summary of the evidence for an association between high-fat, low-fiber diets and certain cancers.

DeGabriele, R. 1980. The physiology of the koala. *Scientific American,* July, 110–17. Australian marsupials that are adapted to a highly specific and unusual diet.

Diesendorf, M. 1986. The mystery of declining tooth decay. *Nature,* July, 125–29. Article that argues that fluoridation does no good. Large reductions in tooth decay are being reported in both fluoridated and unfluoridated countries.

Harris, A. R. 1988. Why do more women get ulcers? *Chatelaine* 61 (March): 32. Discusses the link between ulcers in women and such risk factors as smoking, alcohol, and stress.

Moog, F. 1981. The lining of the small intestine. *Scientific American,* Nov., 154–76. A clear description of the most important absorptive surface in the human body.

Scrimshaw, N. S., and V. R. Young. 1976. The requirements of human nutrition. *Scientific American,* Sept., 50–64. Explains what you should eat and what, perhaps, you should not. A particularly good treatment of the important roles of trace elements in the human diet.

Wardlaw, G., and P. Insel. 1990. *Perspectives in nutrition.* St. Louis, Mo.: Mosby. Provides a comprehensive, up-to-date treatment of all aspects of nutrition.

Winick, M. 1988. *Control of appetite, current concepts in nutrition.* New York: John Wiley & Sons. Describes appetite regulation by control centers in the brain and receptors in the gastrointestinal tract. Discusses that the central nervous system pathways involved in satiety involve endogenous opiates and mediate responses to stress and reward.

EXPLORATIONS

Interactive Software:
Diet and Weight Loss

This interactive exercise allows students to explore how diet and exercise interact to determine whether we gain or lose weight by altering both the kinds of food eaten and the amounts of each consumed, as well as the amount of exercise. The exercise presents a diagram of central metabolism, tracing the path of food to ATP production and/or fat. Students will learn the power of the basic equation "food minus energy used equals fat." By varying the proportions of carbohydrates, proteins, and fats, students can investigate how different diets influence the "burn fat/make fat" decision, learning that the key is the level of blood glucose, not ATP. For example, a high-calorie diet can actually cause you to lose weight by burning fat if that diet is very low in fat.

Questions to Explore:

1. Can you invent a diet that reduces weight without reducing overall calorie consumption?

2. In what ways can you lose weight without exercise?

3. If you consume large but not unreasonable portions of foods that are high in calories (rich in fats) three times a day, how much exercise is required to *lose* weight?

How Animals Capture Oxygen

A normal chest X ray, enhanced to show the lungs clearly. The heart is the pear-shaped red object behind the vertical white column that is the esophagus.

FOR REVIEW

Here are some important terms and concepts that have been discussed in previous chapters and that you will encounter again in this chapter. Review them before proceeding if necessary.

Chemistry of carbon dioxide (*chapter 2*)
Diffusion (*chapter 5*)
Oxidative respiration (*chapter 7*)
Adaptation of vertebrates to terrestrial living (*chapters 18 and 29*)
Red blood cells (*chapter 33*)

The energy for growing, moving, and thinking comes from food. The body pries this energy out of food molecules using a biochemical process called cellular respiration (discussed in chapter 7). In cellular respiration, animals capture energy from food molecules by harvesting energetic electrons and then using these electrons to drive a series of proton-pumping channels in the mitochondrial membrane to generate ATP. Afterward, the electrons are donated to oxygen gas (O_2), which combines with hydrogen (H_2) to form water (H_2O). The carbon atoms left over after the electrons have been stripped off combine with oxygen and are released as carbon dioxide (CO_2). In most vertebrates, the water produced as a result of this metabolism is simply diluted into the much larger volume of the body's internal fluid. Cellular respiration is, in effect, a process that utilizes oxygen and produces carbon dioxide.

Any process that uses oxygen and produces carbon dioxide is a respiratory process. While the term *cellular respiration* pertains to respiration at the cellular level, the general term ***respiration*** describes the uptake of oxygen from the environment and the disposal of carbon dioxide at the body system level.

Respiration at the body system level involves a host of processes not found at the cellular level, like the mechanics of breathing and the exchange of oxygen and carbon dioxide in the arteries and veins. These processes are the subject of this chapter. Respiration is one of the principal physiological challenges facing all animals.

Where Oxygen Comes From: The Composition of Air

As discussed in chapters 3 and 8, each of the oxygen gas molecules that are the raw material of respiration has been produced by photosynthesis. Initially, photosynthetic organisms in the oceans released oxygen gas molecules into the water, and these molecules then diffused into the atmosphere. Later, terrestrial plants began to release additional oxygen directly into the air. Both processes are still occurring. The present atmosphere, called **air,** is rich in oxygen. Dry air has a constant composition: 78.09 percent nitrogen, 20.95 percent oxygen, 0.93 percent argon and other "inert" gases, and 0.03 percent carbon dioxide.

The amount of air present at a given altitude is usually expressed in terms that depend on its weight. A column of air standing on the ground and extending up into the sky as far as the atmosphere goes has many gas molecules in it, and they all experience the force of gravity. For this reason, the column, which some might consider "as light as air," actually weighs a lot. This weight cannot be sensed because it is equally distributed over the entire body.

All of this air weighs enough to push down on one end of a U-shaped column of mercury sufficiently to raise the other end of the column 760 millimeters at sea level under a set of specified, standard conditions. An apparatus that measures air pressure in this way is called a **barometer;** 760 millimeters of mercury (mm Hg) is therefore the average **barometric pressure** of air at sea level. This pressure is also defined as 1 atmosphere of pressure (figure 39.1).

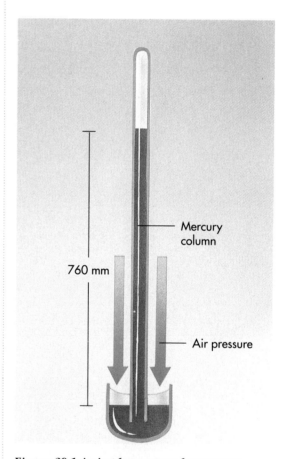

Figure 39.1 A simple mercury barometer.
The weight of air pressing down on the surface of the mercury in the open dish pushes the mercury down into the dish and up the tube. The greater the air pressure pushing down on the mercury surface, the farther up the tube the mercury is forced. At sea level, air pressure causes a standard column of mercury to rise 760 millimeters.

Within the column of air, each separate gas exerts a pressure as if it existed alone. The total air pressure is simply made up of the component **partial pressures (P)** of each of the individual gases. Thus, at sea level, the total of 760 millimeters of air pressure is composed of:

$$760 \times 78.09\% = 593.5 \text{ millimeters of nitrogen}$$

$$760 \times 20.95\% = 159.2 \text{ millimeters of oxygen}$$

$$760 \times 0.93\% = 7.1 \text{ millimeters of argon and other inert gases}$$

$$760 \times 0.03\% = 0.2 \text{ millimeter of carbon dioxide}$$

At altitudes above 6,000 meters, human beings do not survive long. The air still contains 20.95 percent oxygen, but the total air pressure is lower, as is the partial pressure of oxygen

(P_{O_2}). The atmospheric pressure at such a height is about 380 millimeters, so the partial pressure of oxygen is only:

$$380 \times 20.95\% = 79.6 \text{ millimeters of oxygen}$$

This figure is only half of the oxygen partial pressure at sea level. Since the oxygen content of the blood is related to the partial pressure, the amount of oxygen available to body tissues is reduced (see Sidelight 39.1).

> *The oxygen required for cellular respiration is a product of photosynthesis. The air humans breathe and need for survival is a mixture of oxygen (about 21 percent) and nitrogen (about 79 percent), each of which exert corresponding partial atmospheric pressures.*

Sidelight 39.1
HIGH-ALTITUDE SICKNESS

In the children's book *A Wrinkle in Time* by Madeleine L'Engle (1962), the children in the story hold magic flowers to their noses as they soar into the clouds on a unicorn. The children need those flowers, the book explains, because at high altitudes, there is not enough oxygen in the air for survival. In real life, mountain climbers experience this same phenomenon: At higher elevation levels, atmospheric pressure decreases, while the relative concentration of oxygen in the air stays the same. Thus, the absolute amount of oxygen in the air decreases at these high elevations. A lack of oxygen can have serious consequences for the human body, but surprisingly, some mountain climbers and other athletes can adjust to these conditions over time, a process called *acclimatization* (Sidelight figure 39.1).

The general term used to describe the effects of high-altitude sickness is *hypoxia*, or "low oxygen." Hypoxia causes a spectrum of symptoms along a continuum, and the severity of the hypoxia increases with the altitude. At altitudes between 2,500 and 3,000 meters, many people experience headaches, nausea, fatigue, shortness of breath, a decrease in appetite, and a concomitant increase in thirst. All of these symptoms are the outward signs of the body's attempts to cope with the lower oxygen levels. For example, the headaches are caused by the distention of blood vessels and increased blood flow to the brain to compensate for lower oxygen levels. In addition to headaches, the increased blood pressure on the brain can affect the higher brain centers, leading to some of the symptoms associated with alcohol consumption, such as poor judgment.

The next phase of high-altitude sickness, called high-altitude pulmonary edema, occurs at about 3,000 to 3,500 meters above sea level. Symptoms occur about thirty-six to seventy-two hours after reaching this altitude. Edema refers to the leakage of fluid from capillaries in the lungs into the alveoli. If too much fluid leaks into these air sacs, or if the body does not reabsorb it quickly enough, the afflicted individual can literally drown in his or her own secretions.

At 3,600 meters above sea level, cerebral edema may develop. In this type of edema, fluid leaks from capillaries into the brain. The brain becomes waterlogged, producing hallucinations and extreme disorientation. If left untreated, cerebral edema can cause death.

Above 3,600 meters, chronic high-altitude sickness develops. Most often seen in people who have failed to acclimatize to high altitudes, chronic high-altitude sickness is characterized by swelling of the feet and ankles, chest pains, and increased red blood cell count. In some cases, heart failure may result. The only cure for this type of high-altitude sickness is removal of the patient to lower altitudes.

Some mountain climbers are able to attain altitudes of up to 8,000 meters, where the level of oxygen is but one-third that at sea level, because they have acclimatized to the lower oxygen levels. Careful climbs always have scheduled days when camps are made at different altitudes along the climbing route, thereby allowing climbers' bodies to make some remarkable adjustments to the lower oxygen levels: for example, normally, 20 to 30 percent of the body's capillaries are inactive, but at high altitudes, these resting capillaries are suddenly thrust into action, recruited into circulating oxygenated blood in the body. More hemoglobin molecules are formed so that oxygen can be carried more efficiently on red blood cells, and

the number of red blood cells themselves increase. In addition, changes in enzyme function enhance cellular respiration, providing the climber with more energy.

Despite the body's incredible ability to adjust to high-altitude conditions, some body functions remain impaired at high altitudes. The body simply cannot function at the rate it does at sea level. Movements become slower, and coordination drops off. All of these changes, however, are completely reversible once lower altitudes are reached.

Sidelight figure 39.1
Sir Edmond Hillary and Sherpa Tenting Norgay set off on a final stage of their first ascent of Mount Everest (1953). At 8,848 meters (29,028 feet), it is the tallest mountain on earth, and presents difficulties to climbers because of a lack of oxygen.

The Evolution of Respiration

Animals do not actively capture oxygen from their environments; to do so would require expending more energy than the animals would obtain by utilizing the captured oxygen. In all animals, oxygen capture and carbon dioxide discharge are passive processes: The gases move into and out of cells by diffusion. The force that drives the movement of the gases is the difference in the oxygen partial pressure between the interior of the organism and its exterior environment. Oxygen will diffuse into a cell only if the oxygen partial pressure in the cell is less than the oxygen partial pressure in the air or water surrounding the cell. The greater the difference in partial pressures (the more oxygen outside the cell than inside), the more rapid is the diffusion.

Oxygen diffuses slowly. The oxygen levels required by cellular respiration in most organisms cannot be obtained by diffusion alone over distances greater than about 0.5 millimeter. This factor severely limits the size of organisms that obtain their oxygen entirely by diffusion from the environment into the cytoplasm. Single-celled protists are small enough that diffusion distance presents no problem, but as the size of an organism increases, the problem soon becomes significant.

The mechanism of respiration has changed significantly during the evolution of animals. In general, the changes have tended to optimize the rate of diffusion by (1) increasing the concentration difference in partial pressure between the organism and its environment, (2) increasing the surface area over which diffusion takes place, and (3) decreasing the thickness of tissue through which the gas must pass to reach the organism's interior (figure 39.2).

Creating a Water Current

Most of the more primitive animal phyla possess no special respiratory organs. The sponges (phylum Porifera), cnidarians (phylum Cnidaria), many flatworms (phylum Platyhelminthes) and roundworms (phylum Nematoda), and some annelids (phylum Annelida) all obtain their oxygen by diffusion directly from surrounding water. How do they overcome the limits imposed by diffusion? By beating with their cilia, these organisms create a water current, by means of which they continuously replace the water over the diffusion surface. Because of this continuous replenishment with water containing fresh oxygen, *the exterior oxygen partial pressure does not decrease as diffusion proceeds.* Although each oxygen gas molecule that passes into the organism has been removed from the surrounding volume of water, the exterior oxygen partial pressure does not fall

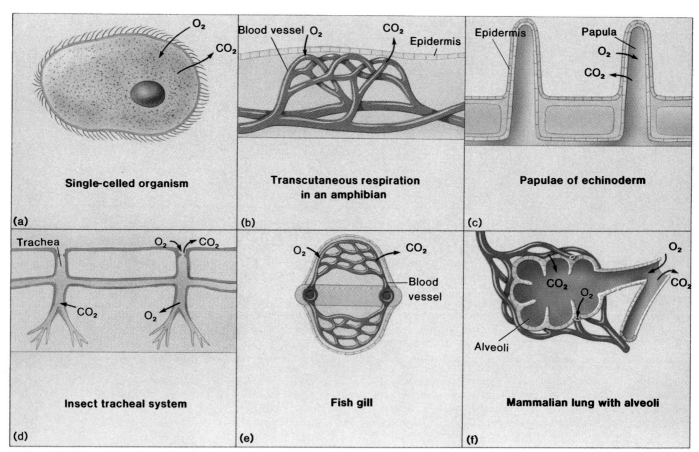

Figure 39.2 Gas exchange in animals.
(a) *Gases diffuse directly into single-celled organisms.*
(b) *Amphibians and many other multicellular organisms respire through their skin (transcutaneous respiration).* (c) *Echinoderms have protruding papulae, which provide an increased respiratory*
surface. (d) *Insects respire through tracheae, which open to the outside.* (e) *Fish gills provide a very large respiratory surface and employ countercurrent flow.* (f) *Mammalian lungs provide a large respiratory surface but do not permit countercurrent flow.*

because a new volume of water with a higher oxygen partial pressure is constantly replacing the one from which oxygen has been removed.

Increasing the Diffusion Surface

Most of the more advanced invertebrates (mollusks, arthropods, echinoderms), as well as the vertebrates, possess special respiratory organs that both increase the surface area available for diffusion and reduce the thickness of the tissue separating the internal fluid from the surroundings (see figure 39.2). These organs are of two kinds: (1) those that facilitate exchange with water and (2) those that facilitate exchange with air. As a rough rule of thumb, aquatic respiratory organs increase the diffusion surface by tissue extensions, called **gills,** that project from the body out into the water. Atmospheric respiratory organs, on the other hand, involve invaginations into the body; in terrestrial vertebrates, the respiratory organs are internal sacs called **lungs.**

Perhaps the simplest of the respiratory organs are the **tracheae** of arthropods (see chapter 29). Tracheae are extensive series of passages connecting the surface of the animal to all portions of its body (see figure 39.2*d*). Oxygen diffuses from these passages directly to the cells, without the intervention of an active circulatory system. Piping air directly to the cells in this manner works very well in small organisms, such as insects, since air must move only a relatively short distance within their bodies.

Animals acquire oxygen through diffusion. Diffusion is a passive process and occurs because of the pressure differentials in oxygen concentrations. As organisms increase in size, they require specialized structures that can support a membrane thin enough for diffusion to occur (hence, the evolution of gills and lungs).

The Gill: An Aqueous Respiratory Machine

The gill—by far the most successful aqueous respiratory organ—evolved among the bony fishes. In these animals, water passes through the mouth into two cavities situated behind the mouth on each side of the head. From these cavities, the water passes back out of the body. The gills hang like curtains between the mouth and the entrance to each of the cavities.

Many fishes that swim continuously, such as tuna, have practically immobile **gill covers** over their gill cavities. They swim with their mouths partly open, constantly forcing water over the gills. Most bony fishes, however, have flexible gill covers that permit a pumping motion using muscles. In such gills, there is an uninterrupted one-way flow of water over the gills, even when the fish is not swimming (figure 39.3).

In addition to maintaining a high diffusion rate by providing a continuous water flow, the gills of fishes are constructed in such a way that they actually maximize the oxygen partial pressure difference between tissue and environment. The structure of a gill shows how this is accomplished. Each gill is composed of thin, membranous gill filaments that project out into the flow of water. Within each filament are rows of thin, disklike **lamellae** arrayed parallel to the direction of water movement. Water flows over these lamellae from front to back. Within each lamella, blood circulation is arranged so that *blood is carried in the direction opposite to water movement, from the back of the lamella to the front* (figure 39.4).

Because the water flowing over lamellae and the blood flowing within lamellae run in opposite directions, the difference in oxygen partial pressures is maximized. At the back of the gill, the least-oxygenated blood meets the least-oxygenated water and is able to remove oxygen from the water. By the time the blood reaches the front of the gill, it has acquired abundant oxygen but acquires still more oxygen by diffusion from the water entering the gill. It can do this because the new water entering the front of the lamellae is richer in oxygen than the water that has already flowed past the gills and lost some of its oxygen. This kind of **countercurrent flow** ensures a continuous oxygen partial pressure gradient, and diffusion continues all along the gill.

Figure 39.3 How a fish breathes.
The gills are suspended under a hard gill cover between the mouth and the entrance to each gill cavity. Breathing occurs in two stages: (a) When the oral valve of the mouth is open, closing the gill cover increases the volume of the mouth cavity so that water is drawn in. (b) When the oral valve of the mouth is closed, opening the gill cover decreases the volume of the mouth cavity, forcing water out past the gills to the outside.

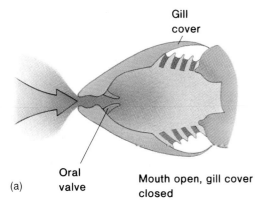

Gill cover

Oral valve

(a) Mouth open, gill cover closed

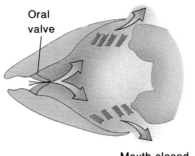

Oral valve

Mouth closed, gill cover open

(b)

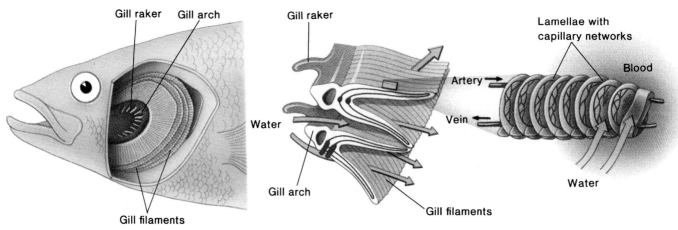

Figure 39.4 Structure of a fish gill.
Water passes from the gill arch over the filaments (from left to right in the diagram). Water always passes the lamellae in the same direction—which is opposite to the direction the blood *circulates across the lamellae. This opposite orientation of blood and water flow is critical to the success of the gill's operation.*

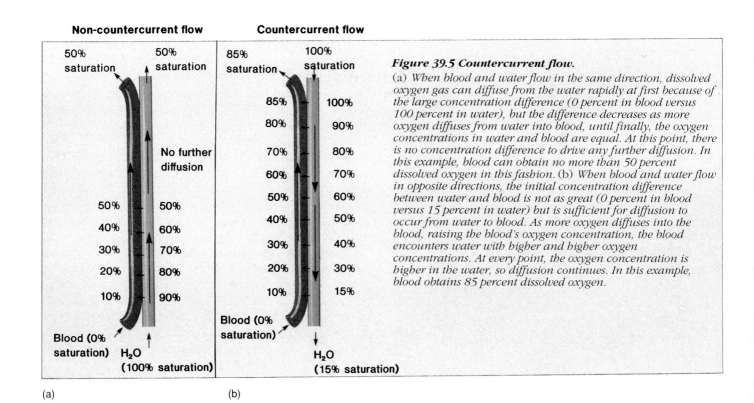

Figure 39.5 Countercurrent flow.
(a) *When blood and water flow in the same direction, dissolved oxygen gas can diffuse from the water rapidly at first because of the large concentration difference (0 percent in blood versus 100 percent in water), but the difference decreases as more oxygen diffuses from water into blood, until finally, the oxygen concentrations in water and blood are equal. At this point, there is no concentration difference to drive any further diffusion. In this example, blood can obtain no more than 50 percent dissolved oxygen in this fashion.* (b) *When blood and water flow in opposite directions, the initial concentration difference between water and blood is not as great (0 percent in blood versus 15 percent in water) but is sufficient for diffusion to occur from water to blood. As more oxygen diffuses into the blood, raising the blood's oxygen concentration, the blood encounters water with higher and higher oxygen concentrations. At every point, the oxygen concentration is higher in the water, so diffusion continues. In this example, blood obtains 85 percent dissolved oxygen.*

If water and blood flowed in the same direction, the oxygen partial pressure difference would be high initially, as the oxygen-free blood met the new water that was entering. The difference would fall rapidly, however, as the water lost oxygen to the blood. Since the oxygen partial pressure of the blood rises as the water oxygen falls, much of the oxygen in the water would remain there when the blood and water oxygen partial pressures became equal and diffusion ceased (figure 39.5*a*).

In countercurrent flow, in contrast, the blood oxygen level becomes lower and lower as the water oxygen level falls. The result is that, in countercurrent flow, the blood can attain oxygen partial pressures as high as those that exist in the water entering the gills (figure 39.5*b*). Fish gills are the most efficient respiratory machines among large organisms. Gills are able to maximize the rate of diffusion in an oxygen-poor medium, obtaining as much as 85 percent of the available oxygen.

The Lung: From Aquatic to Atmospheric Breathing

Although aquatic animals were the first to evolve, their descendants have successfully invaded land many times. The respiratory challenge on land is very different from that in water.

Water is relatively poor in dissolved oxygen, containing only 5 to 10 milliliters of oxygen per liter of water. Air, in contrast, is rich in oxygen, containing about 210 milliliters of oxygen per liter of air. Not surprisingly, many members of otherwise aquatic groups—including many mollusks, crustaceans, and fishes—use atmospheric air as an oxygen source.

When organisms first became fully terrestrial, air became their oxygen source. An entirely new respiratory apparatus evolved, one based on internal passages rather than on gills. Even though they were superb oxygen-capturing mechanisms, gills were not maintained in terrestrial organisms for two principal reasons:

1. Air is less buoyant than water. Because gills' fine, membranous lamellae lack structural strength, they must be supported by water to avoid collapsing. A fish out of water, although awash in oxygen, soon suffocates because its gills collapse into a mass of tissue. This collapse greatly reduces the gills' diffusion surface.

2. Water diffuses into air through the process of **evaporation.** Atmospheric air is rarely saturated with water vapor, except immediately after a rainstorm. Consequently, organisms that live in air are constantly losing water to the atmosphere. Gills would have provided an enormous surface for water loss.

Two main systems of internal oxygen exchange evolved among terrestrial organisms. One was the tracheae of insects, mentioned earlier, and the other was the lung. Both systems sacrifice respiratory efficiency to maximize water retention. Insects prevent excessive water loss by closing the external openings of the tracheae whenever body carbon dioxide levels are below a certain level.

Lungs, in contrast, minimize the effects of drying out by eliminating the one-way flow of oxygen that so increased the efficiency of aquatic respiratory systems. In organisms with lungs, the air moves into the lung through a tubular passage and then back out again via the same passage. When each breath is completed, the lung still contains a volume of air, the **residual volume.** In human beings, this volume is about 1,200 milliliters. Each inhalation adds from 500 milliliters (resting) to 3,000 milliliters (exercising) of additional air. Each exhalation removes approximately the same volume as inhalation added, reducing the air volume in the lung once more to about 1,200

milliliters. Because the diffusion surfaces of lungs are not exposed to fully oxygenated air, but rather to a mixture of fresh and partially depleted air, the difference in oxygen partial pressures is far from maximal, and lungs' respiratory efficiency is much less than that of gills. Oxygen capture is lessened by this two-way flow of air—but so is water loss.

Evolution of the Lung: Amphibians to Birds

Oxygen is so much more plentiful in air than in water that low respiratory efficiency apparently did not present a critical problem to early land-dwellers. The amphibian lung, which evolved from the fish swim bladder, an important regulator of buoyancy, is essentially a sac with a convoluted internal membrane (figure 39.6a). The convoluted inner membrane surface of the amphibian lung allows increased diffusion, but not enough to provide all the needed oxygen. Consequently, amphibians obtain much of their oxygen by diffusion through their moist skin.

Reptiles are far more active than amphibians and have significantly greater metabolic demands for oxygen. The early reptiles could not rely on their skins for respiration. Living fully on land, reptiles are "watertight," avoiding desiccation by possessing a dry, scaly skin. Again, as in aquatic organisms, the respiratory apparatus changed in ways that tended to optimize respiratory efficiency. The lungs of reptiles possess many small chambers called **alveoli,** clustered together like a bunch of grapes (figure 39.6b). The alveoli greatly increase the lung's diffusion surface.

Metabolic demands for oxygen became even greater with the evolution of birds and mammals, which, unlike reptiles and amphibians, maintain a constant body temperature by heating their bodies metabolically. The lungs of mammals contain many clusters of alveoli (figure 39.6c). Humans, for example, have about three hundred million alveoli in their two lungs. The increased number of alveoli enlarged yet again the lung's total diffusion surface. In humans, the total surface devoted to diffusion can be as much as 80 square meters, an area about forty-two times the body's surface area.

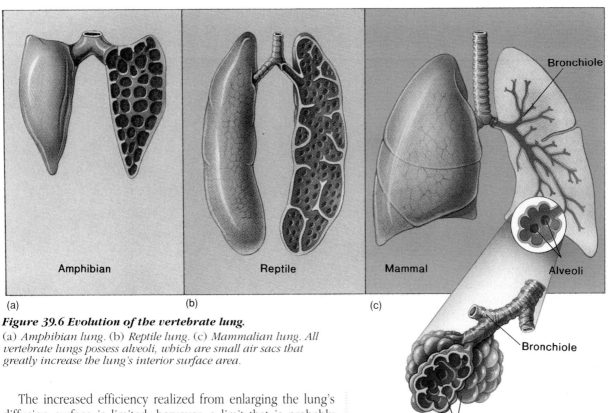

Figure 39.6 Evolution of the vertebrate lung.
(a) *Amphibian lung.* (b) *Reptile lung.* (c) *Mammalian lung. All vertebrate lungs possess alveoli, which are small air sacs that greatly increase the lung's interior surface area.*

The increased efficiency realized from enlarging the lung's diffusion surface is limited, however, a limit that is probably approached by the more active mammals. The advent of birds and flying introduced respiratory demands that exceeded the capacity of a saclike lung. Many birds rapidly beat their wings for prolonged periods during flight. Such rapid wing movement depends on frequent contraction of wing muscles and uses up a lot of energy quickly. Flying birds must carry out intensive cellular respiration to replenish the ATP expended to contract flight muscles. They thus require a great deal of oxygen, more oxygen than a saclike lung—even one with a large surface, such as a mammalian lung—is capable of delivering. The lungs of birds cope with the demands of flight by employing a new respiratory mechanism, one that significantly improves respiratory efficiency.

An avian lung works like a two-cycle pump. When a bird inhales air, the air passes directly to a nondiffusing chamber called a **posterior air sac.** When the bird exhales, the air flows into a lung. On the following inhalation, the air passes from the lung to a second air sac, the **anterior air sac.** Finally, on the second exhalation, the air flows from the anterior air sac out of the body. This complicated passage has the advantage of creating a unidirectional flow of air through the lungs! Thus, there is no dead volume as in the mammalian lung, and the air passing through a bird lung is always fully oxygenated (figure 39.7).

In fish gills, blood and water flow in opposite directions, 180 degrees apart, whereas in bird lungs, the latticework of capillaries is arranged *across* the air flow, at a 90-degree angle. This **crosscurrent flow** is not as efficient at extracting oxygen as a fish gill, but the oxygenated blood leaving the lung can still contain more oxygen than exhaled air, a capacity not achievable by mammalian lungs.

Birds have evolved a complex system of air sacs and two-cycle breathing that ultimately provides for a very efficient one-way flow of air through the lungs. There is no "dead air space," as there is in the mammalian lung.

The Human Respiratory System

The human respiratory system, as already noted, is not as efficient as that of birds, but it does provide a means of acquiring oxygen and disposing of carbon dioxide with a minimum of water loss. The organs associated with the human respiratory system include openings that allow air to be taken into the body (the mouth [oral cavity] and nostrils), a one-cycle pump that facilitates this air intake (the lungs), and structures that provide surfaces for gas exchange (the alveoli). Working in concert, these structures provide an efficient, land-adapted mechanism of oxygen intake and circulation and carbon dioxide release.

Structure of the Human Respiratory System

Air normally enters the **nostrils,** where it is warmed and cleaned by small hairs that act as filters to remove dust and other particles. As air passes through the **nasal cavity,** an extensive array of cilia on the nasal cavity's epithelial lining further filter and moisten the air. The air then passes through the

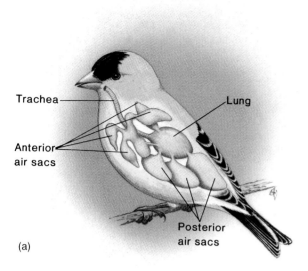

Figure 39.7 How a bird breathes.
(a) *A bird's respiratory system is composed of anterior air sacs, lungs, and posterior air sacs. (b) Breathing occurs in two cycles: In Cycle 1, air is drawn from the trachea into the posterior air sac and then is exhaled into a lung. In Cycle 2, the air is drawn from the lung into the anterior air sac and then is exhaled through the* trachea. *Passage of air through the lungs is always in the same direction, from posterior to anterior (right to left here). Because blood circulates in the lung from anterior to posterior, the lung achieves a type of countercurrent flow that is very efficient in picking up oxygen from the air.*

back of the mouth, crossing the path of food as it enters first the **larynx,** or voice box, and then the **trachea.** The air passes down the trachea, through the two **bronchi** (each of which is attached to a lung) and then to the lungs (figure 39.8). As discussed in chapter 38, the epiglottis is a flap of tissue that covers the trachea whenever food is swallowed; otherwise, food might lodge in the trachea and prevent air from entering the lungs (see figure 38.6).

The bronchi not only provide the passageway for air to the lungs, they also serve a cleaning function. Ciliated cells in the bronchi secrete mucus that entraps foreign particles (figure 39.9). These cells then carry the particles upward to the back of the mouth, where they can be swallowed. An individual is estimated to swallow, on average, about a quart of this mucus daily!

Human lungs are located in the chest, or **thoracic cavity.** The lungs contain millions of small sacs—the alveoli mentioned earlier. So many capillaries surround these alveoli that it is as if blood was flowing over them in a constant sheet. The alveoli are connected to the bronchi by a branching network of tubules called **bronchioles,** some of which are surrounded by smooth muscle and are sensitive to metabolites and oxygen.

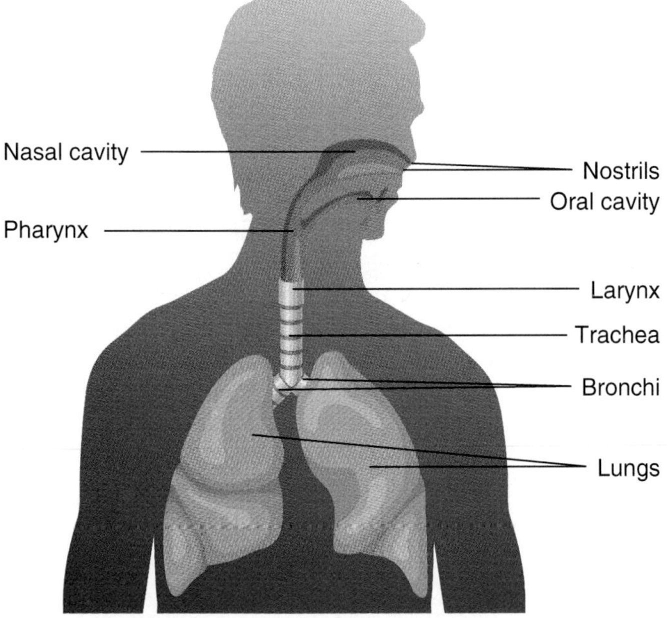

Figure 39.8 The human respiratory system.
The human respiratory system contains openings for the intake of air (nostrils and oral cavity), a passageway for the air to the lungs (nasal cavity, larynx, trachea, and bronchi), and an organ that facilitates gas exchange (lungs). Within the lungs, but not shown here, are alveoli (air sacs) and bronchioles (small passages that connect the alveoli).

Figure 39.9 "Cleaner" cells in the bronchi.
Respiratory, ciliated cells line each bronchus. These cells secrete a mucus that traps foreign particles. The cilia sweep these particles upward into the area at the back of the mouth, where they can then be swallowed.

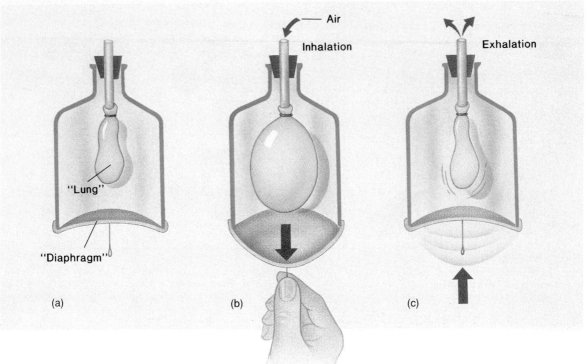

Figure 39.10 A simple experiment that shows how you breathe.

In the jar is a balloon (a). When the diaphragm is pulled down, as shown in (b), the balloon expands; when it is relaxed (c), the balloon contracts. In the same way, air is taken into the lungs when the diaphragm pulls down, expanding the volume of the lung cavity. When the diaphragm pushes back, the volume decreases, and air is expelled.

How Humans Breathe

The human respiratory apparatus is simple in structure, functioning as a one-cycle pump. The thoracic cavity is bounded on its sides by the ribs and on the bottom by a thick layer of muscle, the **diaphragm,** which separates the thoracic cavity from the abdominal cavity. Each lung is covered by a very thin, smooth membrane called the **pleural membrane.** A second pleural membrane marks the interior boundary of the thoracic cavity, into which the lungs hang. Within the cavity, lungs are supported by water—the **intrapleural fluid**—which also permits an even application of pressure to all parts of the lungs.

The pleural membranes can be visualized as a system of two balloons of different sizes, one nested inside the other, with the space between them completely filled with a very thin film of water. The inner balloon opens out to the atmosphere. Thus, two forces act on this inner balloon: Air pressure from the atmosphere pushes it outward, and water pressure from the intrapleural fluid pushes it inward (figure 39.10).

The active pumping of air in and out through the lungs is called **breathing.** During **inhalation,** contraction of three sets of muscles attached to the ribs (the sternocleidomastoids, the pectoralis major, and the external intercostals) causes the walls of the chest cavity to expand so that the rib cage moves outward and upward (figure 39.11*a*). The diaphragm, which is dome-shaped when relaxed, contracts and moves downward. In effect, the outer pleural membrane (outer balloon) has been enlarged by pulling it in all directions. This expansion causes the fluid pressure to decrease to a level less than that of the

air pressure within the lung (the inner balloon). Since fluids are incompressible, the wall of the inner balloon is pulled out. (If you have difficulty visualizing this, think about how much force must be applied to pull apart a wet pair of glass slides, compared to moving them horizontally.) As the inner balloon expands, its internal air pressure tends to decrease, and since the inner balloon is connected to the atmosphere, air rapidly moves in from the atmosphere to equalize pressure.

During **exhalation,** the ribs and diaphragm return to their original resting positions. In doing so, they exert pressure on the fluid. This pressure is transmitted uniformly over the entire surface of the lung (the inner balloon), forcing air from the inner cavity back out to the atmosphere (figure 39.11*b*). Intrapleural pressure is always less than atmospheric pressure and pressure within the alveoli.

The active pumping of air in and out through the lungs is called breathing. Expansion of the chest cavity draws air into the lungs, and the return of the diaphragm and ribs to their resting positions drives air from the lungs.

Gas Transport and Exchange

When oxygen diffuses from the air into the moist cells lining the lung's inner surface, its journey has just begun. Passing from these cells into the bloodstream, the oxygen is carried throughout the body by the circulatory system (described in

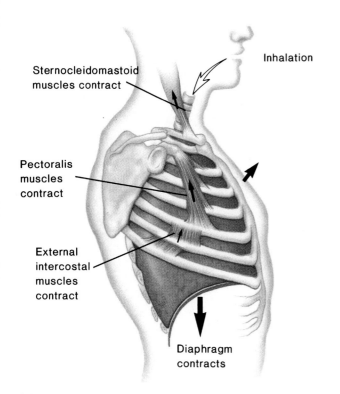

Inhalation

Sternocleidomastoid muscles contract

Pectoralis muscles contract

External intercostal muscles contract

Diaphragm contracts

(a) Inhalation

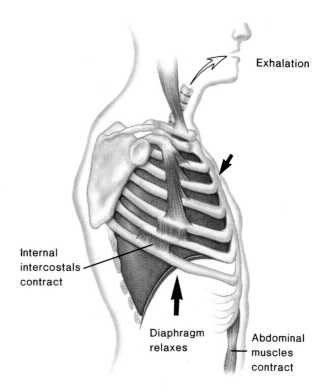

Exhalation

Internal intercostals contract

Diaphragm relaxes

Abdominal muscles contract

(b) Exhalation

Figure 39.11 Movement of the rib cage and diaphragm during inhalation and exhalation.

(a) *During inhalation, the diaphragm contracts and moves downward, and the chest cavity expands, increasing its volume. As a result of the larger volume, air is sucked in through the*

trachea. (b) *During exhalation, the diaphragm and rib cage return to their original positions, reducing the volume of the chest cavity and forcing air outward through the trachea.*

chapter 40). If transport depended only on diffusion, unassisted by a circulatory system, a molecule of oxygen would require an estimated three years to diffuse from the lung to the toe.

Oxygen moves within the circulatory system on carrier proteins that bind dissolved molecules of oxygen in the capillaries surrounding the lung alveoli. The carrier proteins later release their oxygen molecules to metabolizing cells at distant body locations.

All vertebrates use the oxygen carrier protein hemoglobin. **Hemoglobin** is composed of four polypeptide subunits; each of the four polypeptides is combined with iron in such a way that oxygen can be bound reversibly to the iron. Hemoglobin is synthesized by erythrocytes (red blood cells) and remains within these cells, which circulate in the bloodstream like ships bearing cargo.

Hemoglobin has a high affinity for oxygen, but it also has a high affinity for other molecules, such as carbon monoxide. When carbon monoxide is in the air, hemoglobin will bind to these molecules before it will bind to oxygen. This attraction is why carbon monoxide is so deadly to humans.

Oxygen Transport

Only a small amount (5 percent) of oxygen gas diffuses from the alveoli of the lung directly into the blood plasma and is transported to body tissues dissolved in the fluid. The remaining

95 percent is bound to hemoglobin within red blood cells. The higher the oxygen partial pressure in the air within the lungs, the more oxygen will dissolve in the blood and combine with hemoglobin.

Hemoglobin molecules act like little oxygen sponges, soaking oxygen up within red blood cells and causing more to diffuse in from the blood plasma. At the oxygen partial pressure encountered in the blood supply of the lung, most hemoglobin molecules are saturated with oxygen. In the tissues, oxygen partial pressure is much lower, and hemoglobin gives up its bound oxygen.

In tissue, the presence of carbon dioxide (CO_2) causes the hemoglobin molecule to assume a shape that allows it to give up its oxygen more easily. This augments the unloading of oxygen from hemoglobin. The effect of carbon dioxide on oxygen binding, called the **Bohr effect,** is significant because carbon dioxide is produced by tissues at the site of cell metabolism. For this reason, blood unloads oxygen more readily to those tissues undergoing metabolism and generating carbon dioxide.

Most oxygen is transported to body tissues by being bound to hemoglobin on red blood cells.

Carbon Dioxide Transport

At the same time that red blood cells are unloading oxygen, they are also absorbing carbon dioxide from the tissue. Perhaps 20 percent of the carbon dioxide that the blood absorbs is bound to hemoglobin. Another 8 percent diffuses from body tissues back into the blood plasma. The remaining 72 percent of the carbon dioxide diffuses from the plasma into the cytoplasm of red blood cells. There, an enzyme—**carbonic anhydrase**—catalyzes the combination of carbon dioxide with water to form carbonic acid (H_2CO_3), which dissociates into bicarbonate (HCO_3^-) and hydrogen (H^+) ions. This process removes large amounts of carbon dioxide from the blood plasma, facilitating the diffusion of more carbon dioxide into the plasma from the surrounding tissue. The facilitation is critical to carbon dioxide removal, since the difference in carbon dioxide concentration between blood and tissue is not large (only 5 percent).

The red blood cells carry their cargo of bicarbonate ions back to the lungs. The lower carbon dioxide concentration in the air inside the lungs causes the carbonic anhydrase reaction to reverse, releasing gaseous carbon dioxide, which diffuses outward from the blood into the alveoli. With the next exhalation, this carbon dioxide leaves the body. The 20 percent of the carbon dioxide that is bound to hemoglobin also leaves because hemoglobin has a greater affinity for oxygen than for carbon dioxide at low carbon dioxide concentrations. The diffusion of carbon dioxide outward from the red blood cells causes the hemoglobin within these cells to release its bound carbon dioxide and to take up oxygen instead. The red blood cells, with their newly bound oxygen, then start the next respiratory journey (figure 39.12).

> *Most of the carbon dioxide (CO_2) in body tissues enters red blood cells and dissociates to form hydrogen (H^+) ions and bicarbonate. The red blood cells carry this cargo to the lungs, where the bicarbonate and hydrogen ions recombine to form carbon dioxide, which is then exhaled.*

How the Brain Controls Breathing

Breathing rates fluctuate with the activities being performed. For example, breathing rates increase during heavy exercise and decrease during sleep. During exercise, more oxygen is needed to sustain the process of cellular respiration that provides energy to the muscles. During sleep, muscles are not used, and so the levels of oxygen needed in cellular respiration decrease. The brain plays an important role in adjusting these breathing rates to the needs of the body.

The part of the brain that controls respiration is called the **respiratory center,** located in the pons of the brain stem. The respiratory center is actually composed of two areas: One controls inspiration, and one controls expiration. Located near the respiratory center, in the medulla oblongata of the brain stem, is a chemoreceptor area that monitors blood carbon dioxide levels and blood pH. When blood carbon dioxide levels increase, or blood pH decreases, the chemoreceptor area sends a nerve impulse to the respiratory center, which, in turn, sends a

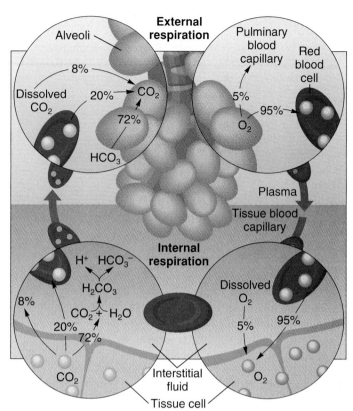

Figure 39.12 Summary of the transport of oxygen and carbon dioxide in the respiratory system.
Oxygen: *As the diagram shows, 5 percent of oxygen simply diffuses from lung alveoli into the blood plasma. The bulk of the oxygen taken in, however, is bound to hemoglobin on red blood cells. These oxygenated red blood cells then circulate throughout the body.* Carbon dioxide: *Only 8 percent of carbon dioxide diffuses from body tissues back into the blood plasma. Another 20 percent is bound to hemoglobin on red blood cells for transport back to the lungs. Seventy-two percent of the carbon dioxide in body tissues is combined with water to produce hydrogen (H^+) ions and bicarbonate. These two products are carried in the red blood cells back to the lungs, where they are recombined into gaseous carbon dioxide, which is then exhaled.*

nerve impulse to the muscles of the diaphragm and rib cage, signaling inspiration (figure 39.13).

Why does a decrease in blood pH signal inspiration? As mentioned earlier, most carbon dioxide is removed from body tissues by being broken down into bicarbonate and hydrogen (H^+) ions. When blood carbon dioxide levels are high, more hydrogen ions are produced as the carbon dioxide is broken down into its constituent parts. A high level of hydrogen ions causes blood pH to decrease. Thus, a falling blood pH indicates that more oxygen is needed in the respiratory system.

In addition to the chemoreceptors in the medulla oblongata are chemoreceptors in the **aortic** and **carotid bodies.** The aorta is a major artery that is the first "leg" that the blood takes in its circulation throughout the body; the aortic bodies are small, vascular structures located on the aorta. Carotid bodies are small structures located near the aorta. When these chemoreceptors detect falling oxygen levels in the blood, they send a nerve impulse to the respiratory center, which then sends a nerve impulse to the diaphragm and muscles of the rib

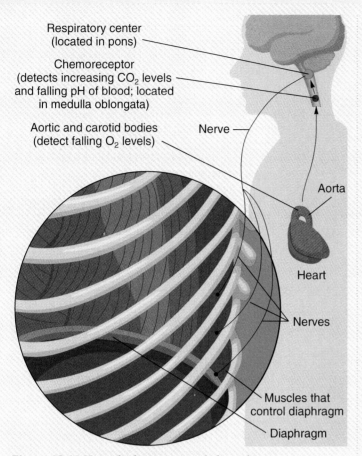

Respiratory center
(located in pons)

Chemoreceptor
(detects increasing CO_2 levels
and falling pH of blood; located
in medulla oblongata)

Aortic and carotid bodies
(detect falling O_2 levels)

Nerve

Aorta

Heart

Nerves

Muscles that
control diaphragm

Diaphragm

Figure 39.13 How the brain controls breathing.
A respiratory center that is attached by nerves to the muscles of the rib cage and diaphragm controls vertebrate breathing rates. The respiratory center receives input from chemoreceptors in the medulla oblongata, which detect increased carbon dioxide levels and falling pH levels in the blood. Other chemoreceptors, located in the aortic and carotid bodies, detect falling oxygen levels.

cage to contract (see figure 39.13). These chemoreceptors are not very sensitive; they are only activated when oxygen levels fall to about half their normal level. Nevertheless, these chemoreceptors can be instrumental when the body is under extreme stress, as during shock or at high altitudes.

> *The respiratory center of the brain controls the rate of breathing. When chemoreceptors located near the respiratory center detect an increase in blood carbon dioxide levels or a decrease in blood pH, they send a nerve impulse to the respiratory center, which, in turn, stimulates the diaphragm and rib cage to initiate inhalation. Chemoreceptors in the aortic and carotid bodies detect falling oxygen levels.*

Looking Ahead to Circulation

Chapter 40 considers the journey of red blood cells in more detail. Traveling between the lungs (where they acquire oxygen and release carbon dioxide) and tissues (where they release oxygen and acquire carbon dioxide), the body's red blood cells traverse a complex highway that passes to all parts of the body. The red blood cells are not the only traffic on this highway. Just as roads and sidewalks transport all the commerce of a city, so the vertebrate circulatory system transports all the material that moves from one part of the body to another.

EVOLUTIONARY VIEWPOINT

The evolution of the vertebrate lung is a particularly clear example of the progressive nature of adaptation, each alteration achieving the improvement in function necessary for the altered lifestyle of the organism in question. For example, bird lungs are far more efficient than those of mammals so that birds can fly. Mammals are unlikely to ever evolve such superefficient lungs because current mammalian lungs are getting the job done. One cannot help but wonder, however, about the lungs of bats. Perhaps evolution will favor novelty there, improving the flight capabilities of these very successful mammals.

SUMMARY

1. The oxygen required for cellular respiration is a product of photosynthesis. The air humans breathe, however, contains mostly nitrogen and is primarily an oxygen-nitrogen mix at about a 1:4 ratio. The partial pressure of these gases (about 600 millimeters of mercury for nitrogen and 159 millimeters of mercury for oxygen, at sea level) allows for the passive diffusion of oxygen into and carbon dioxide out of the assorted animal respiratory systems.

2. All animals obtain oxygen by diffusion. The evolution of respiratory mechanisms among animals has tended to favor changes that improve the rate of diffusion. These changes include decreasing the length of the path over which diffusion occurs, increasing the surface area over which diffusion occurs, and maximizing the difference in oxygen partial pressures between environment and tissue.

3. The most efficient aquatic respiratory organ is the gill of bony fishes. Fishes take water in through their mouths, move it past their gills, and pass it out of their bodies. This one-way flow is the secret of high respiratory efficiency, since it permits fishes to establish a countercurrent flow of blood: Blood vessels are located within the gills in such a way that blood flows in a direction opposite that of water.

4. Gills will not work in air, since air is not buoyant enough to support their fine latticework of passages. That is why fishes drown in air, even though much more oxygen is available in air than in water.

5. The amphibian lung is a simple sac with two-way flow in and out. Gaseous diffusion across the lung's small internal surface area is supplemented by diffusion across amphibians' moist skin.

6. The further evolution of the lung in reptiles and mammals has involved a progressive increase in the lung's internal surface area, achieved by partitioning the inner surface into increasingly numerous chambers called alveoli. A human lung possesses about three hundred million alveoli, with a combined surface area that is about forty-two times the body's surface area.

7. A fundamental change in the atmospheric lung is characteristic of birds, which use a series of air chambers and two-cycle breathing to effect a one-way flow of air. A one-way flow of the diffusing medium permits a form of countercurrent flow, which is by far the most efficient diffusion mechanism.

8. Humans, like other terrestrial vertebrates, breathe by expanding and contracting the cavity within which the lungs hang. These actions expand the lungs, sucking air inward, and then compress the air, forcing it out of the lungs.

9. In the respiration of terrestrial vertebrates, hemoglobin within the red blood cells binds to the oxygen, which diffuses across the lung capillaries into the blood plasma. The circulating system carries these red blood cells to the respiring tissues of the body.

10. In tissues, there is much less oxygen than in the blood and much more carbon dioxide because of the consumption of oxygen and generation of carbon dioxide by respiring cells. Hemoglobin responds to the higher carbon dioxide concentration by unloading its oxygen, which diffuses out of the red blood cells into the tissue. The carbon dioxide is then absorbed into the red blood cells and dissociates into hydrogen (H^+) ions and bicarbonate. The red blood cells carry this cargo back to the lungs, where the hydrogen ions and bicarbonate recombine to form gaseous carbon dioxide, which is then exhaled.

11. A respiratory center in the brain controls the rate of breathing. When chemoreceptors located near the respiratory center detect increased blood carbon dioxide levels or decreased blood pH, they send a nerve impulse to the respiratory center, which, in turn, stimulates the muscles of the rib cage and diaphragm to contract. Chemoreceptors in the aortic and carotid bodies send a nerve impulse to the respiratory center when they detect falling oxygen levels. These chemoreceptors, however, are not as sensitive as the chemoreceptors that detect pH and carbon dioxide levels.

REVIEWING THE CHAPTER

1. What is the source of the oxygen we breathe?

2. What is high-altitude sickness?

3. How did respiration evolve in animals?

4. In animals, how is water current created?

5. How is the diffusion surface increased?

6. Why is the gill an aqueous respiratory machine?

7. How did the transition in breathing from water to land occur?

8. How did the lung evolve?

9. What is the structure of the human respiratory system?

10. How do humans breathe?

11. How is oxygen transported and exchanged?

12. How is carbon dioxide transported and exchanged?

13. How does the brain control breathing?

COMPLETING YOUR UNDERSTANDING

1. How do respiration and cellular respiration differ?

2. Approximately what percent of the earth's atmosphere is molecular oxygen?

3. How did great civilizations, such as that of the Incas in Peru, survive at high altitudes?

4. Why would humans not survive long at altitudes above 6,000 meters?

5. Why are tracheae such successful respiratory organs in insects?

6. How have fishes, such as tuna, survived having mostly immobile gill covers?

7. Within the lamellae of gills, the blood circulation is arranged so that blood is carried in the opposite direction to the movement of water. The functional significance of this arrangement is that

 a. it helps to maintain the temperature of the organism equal to the water temperature, thus enhancing diffusion.

 b. it results from a developmental constraint.

 c. it ensures a continuous gradient of concentration difference between the blood and the water, so that diffusion continues all along the gill.

 d. it increases the surface area for diffusion.

 e. it allows some kinds of fishes to continue to get oxygen even if they are not moving.

8. What is a fundamental difference in the way air moves through tracheae in insects and lungs in other animals?

9. Why is the respiratory efficiency of lungs much less than that of gills?

10. Why must amphibians obtain much of their oxygen by diffusion through their skin?

11. In reptiles, birds, and mammals, what purpose do the alveoli serve?

12. Why do birds need a unidirectional flow of air through their lungs?

13. Arrange the following in the order in which air contacts them during breathing, starting from the mouth and nose.

 a. Bronchus   d. Hemoglobin

 b. Alveoli   e. Trachea

 c. Larynx

14. Air normally enters the body through the nostrils, which are filled with hairs. What is the function of these hairs?

   a. They filter out dust and other particles.

   b. They are a noise-making device.

   c. They slow and regulate the passage of air.

   d. They moisten the air.

   e. They allow you to breathe while you eat.

15. What is the epiglottis, and why is it necessary?

16. Why is carbon monoxide so deadly to humans?

17. What is the Bohr effect, and how does it work?

18. Most oxygen transported in the blood

   a. is dissolved in the plasma.

   b. is carried by lymphocytes.

   c. is bound to hemoglobin.

   d. reacts with water to form bicarbonate.

   e. is carried by myoglobin.

19. How does carbonic anhydrase relate to the removal of large amounts of carbon dioxide from the blood plasma, and why is this critical?

20. An increase in the carbon dioxide concentration of blood causes the same effect on hemoglobin's ability to combine with oxygen as does a decrease in the partial pressure of oxygen within the lungs. True or false?

21. How do chemoreceptors in the aortic and carotid bodies become activated?

22. How would smoking cigarettes interfere with the human breathing process?

23. Why could being hit in the stomach cause a rapid exit of air from the lungs?

*24. Often, people who appear to have drowned can be revived, in some cases after being underwater for as long as half an hour. In every case of full recovery after extended submergence, however, the person has been in very cold water. Is this observation consistent with the fact that oxygen is twice as soluble in water at 0 degrees Celsius as it is at 30 degrees Celsius?

*25. Can you think of a reason why a respiratory system has not evolved in which oxygen is actively transported across respiratory membranes, in place of the passive process of diffusion across these membranes?

*26. If, by accident, your pleural membrane were punctured, would you be able to breathe?

*For discussion

FOR FURTHER READING

Feder, M., and W. Burggren. 1985. Skin breathing in vertebrates. *Scientific American*, Nov., 126–42. Discusses how many vertebrates do a significant portion of their breathing through their skin.

Perutz, M. F. 1978. Hemoglobin structure and respiratory transport. *Scientific American*, Dec., 92–125. A classic account of how hemoglobin changes its shape to facilitate oxygen binding and unloading, by the man who won a Nobel Prize for unraveling hemoglobin structure.

Raloff, J. 1987. New clues to smog's effect on lungs. *Science News* 132 (8 August): 86. Describes how the lung's defense mechanisms can be damaged by environmental contamination.

Schmidt-Nielsen, K. 1971. How birds breathe. *Scientific American*, Dec., 73–79. A fascinating account of the discovery of unidirectional flow in avian lungs, by a great comparative physiologist.

Zapol, W. 1987. Diving adaptations of the Weddell seal. *Scientific American*, June, 100–105. A delightful account of how Antarctic seals are able to hold their breath longer than most mammals.

CIRCULATION

Each milliliter of your blood contains about five billion oxygen-carrying red blood cells, here magnified 1,132 times.

FOR REVIEW

Here are some important terms and concepts that have been discussed in previous chapters and that you will encounter again in this chapter. Review them before proceeding if necessary.

Glycogen (*chapter 2*)
Gills (*chapters 29 and 39*)
Erythrocytes (*chapter 33*)
Depolarization (*chapters 33 and 34*)
How hemoglobin carries oxygen (*chapter 39*)

Of the many tissues of the vertebrate body, few have the emotive impact of blood. In movies and in real life, the sight of blood connotes violence and bodily harm. In literature, blood symbolizes the life force, which is not a bad analogy in real life either. This chapter is about blood, the highways that circulate it through the body, and the heart that drives it along its journey. Although vertebrates possess many other organ systems that are necessary for life, the activities of blood bind these systems into a functioning whole.

Evolution of Circulatory Systems

The capture of nutrients and gases from the environment is an essential task for all living organisms. In animals, with their relatively large bodies and many organs and organ systems, this function has become complex.

In less complicated animals, such as roundworms, the fluid within the body cavity constitutes a primitive kind of circulatory system, one that permits materials to pass from one cell to another without leaving the organism. Within such simple systems, called **open circulatory systems,** circulating fluid and body fluid generally are indistinguishable. Arthropods (see chapter 29) have open circulatory systems. A muscled tube within the central body cavity forces the cavity fluid out through a network of interior channels and spaces. The fluid then flows back into the central cavity.

Most animals, however, have **closed circulatory systems,** in which the circulatory system fluid travels throughout the body within closed tubes or vessels, separated from the rest of body fluids. In annelids, for example, two major tubes or vessels extend the length of the worm, with branches extending out from each tube to the muscles, skin, and digestive organs (figure 40.1). A great advantage of closed circulatory systems is that they regulate fluid flow by means of muscle-driven changes in vessel diameters. In other words, with a closed circulatory system, different parts of the body can maintain different circulation rates.

The closed circulatory system characteristic of all vertebrates has five principal functions: (1) nutrient and waste transport, (2) oxygen and carbon dioxide transport, (3) temperature maintenance, (4) hormone circulation, and (5) immune defense. Temperature maintenance and hormone circulation are associated with the passage of the blood throughout the body.

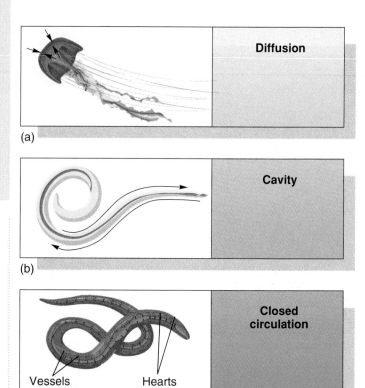

(a)

Diffusion

(b)

Cavity

(c) Vessels   Hearts

Closed circulation

(d) Tubular heart

Open circulation

Figure 40.1 Evolution of circulatory systems.
(a) *In simple diffusion, as in this jellyfish, the organism's internal fluid reaches all parts of the body. This diffusion is a passive process and does not require the organism to move or to expend energy.* (b) *As a nematode moves, its contracting muscles push fluid back and forth within the body cavity. This cavity type of circulation is a form of open circulation.* (c) *The closed circulatory system of an annelid contains several hearts that pump blood through blood vessels to the body tissues, and a second set of blood vessels that pump the blood back from the tissues to the heart.* (d) *Insects have an open circulatory system. The tubular heart pumps blood out to body tissues, from which the blood seeps back rather than traveling in vessels.*

Immune defense is associated with several types of blood cells that compose the circulatory fluid. This chapter looks in detail at the structure of vertebrate circulatory systems and how these structures perform these different functions.

Open circulatory systems involve the free movement of blood within the body cavity. In a closed circulatory system, blood is contained within closed vessels and transports such materials as nutrients, wastes, dissolved gases, hormones, and cells of the immune system. A closed circulatory system is also instrumental in maintaining body temperature.

The Vertebrate Circulatory System

The vertebrate circulatory system has three elements: (1) the **heart,** a muscular pump considered later in this chapter; (2) the **blood vessels,** a network of tubes that pass through the body; and (3) the **blood** that circulates within these vessels. The plumbing of the closed circuit—the heart and vessels—is known collectively as the **cardiovascular system** (figure 40.2).

Blood moves within the cardiovascular system, leaving the heart through vessels known as **arteries.** From the arteries, the blood passes into a network of **arterioles,** or smaller arteries. From these, it eventually is forced through the **capillaries,** a fine latticework of very narrow tubes (from the Latin *capillus,* meaning "a hair"). While passing through these capillaries, the blood exchanges gases and metabolites with body cells. After traversing the capillaries, the blood passes into a third kind of vessel—the **venules,** or small **veins.** A network of venules and larger veins collects the circulated blood and carries it back to the heart.

The Blood

About 8 percent of the body mass of most vertebrates is the blood circulating through their bodies. Blood is composed of a fluid plasma, together with three principal kinds of cells— erythrocytes, leukocytes, and platelets—that circulate within that fluid. These cells account for approximately 40 percent of blood's volume.

Blood Plasma: The Blood's Fluid

Blood plasma is a complex solution of water with three very different components:

1. *Metabolites and wastes.* If the circulatory system is thought of as the "highway" of the vertebrate body, the blood contains the "traffic" traveling on that highway. Dissolved within the plasma are glucose, lipids, and all the other metabolites, vitamins, hormones, and wastes that circulate among body cells.

2. *Salts and ions.* Like the seas in which life arose, plasma is a dilute salt solution. The chief plasma ions are sodium, chloride, and bicarbonate. Trace amounts of other salts, such as calcium and magnesium, as well as of metallic ions, including copper, potassium, and zinc, are also present. Plasma composition, therefore, is not unlike that of seawater.

3. *Proteins.* Blood plasma is 90 percent water. Passing by all the cells of the body, blood would soon lose most of its water to them by osmosis if it did not contain as high a concentration of proteins as the cells it passes. Water does not move from the blood vessels into the surrounding cells because blood plasma contains proteins. Antibody and globulin proteins that are active in the immune system are found in blood plasma, as is a small amount of fibrinogen, a protein that plays a role in blood clotting. Taken together, however, these proteins make up less than half of the amount of protein necessary to balance the protein content of other body

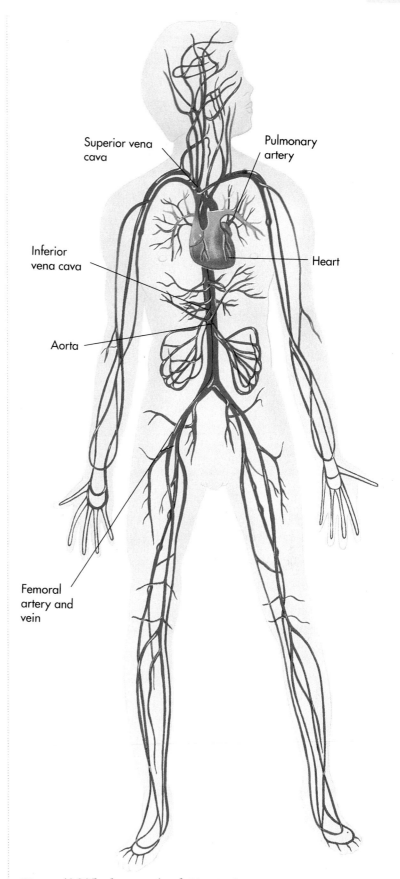

Figure 40.2 The human circulatory system.
This system is composed of the heart, a muscular pump; a system of blood vessels; and blood.

cells. The rest consists of the protein **serum albumin,** which circulates in the blood plasma as an osmotic counterforce. Human blood contains 46 grams of serum albumin per liter.

Blood plasma contains metabolites, wastes, and a variety of ions and salts. It also contains high concentrations of the protein serum albumin, which functions to keep the blood plasma in osmotic equilibrium with body cells.

Erythrocytes: Hemoglobin Carriers

Each milliliter of blood contains about five billion oxygen-carrying **erythrocytes,** or **red blood cells** (table 40.1). Each erythrocyte is a flat disk with a central depression on both sides, something like a doughnut with a hole that does not go all the way through (figure 40.3). Attached to the outer membranes of erythrocytes is a collection of polysaccharides, which determines the blood group of the individual (see chapter 11). Almost the entire interior of each cell is packed with hemoglobin.

Table 40.1 Types of Blood Cells

Blood Cell	Life Span in Blood	Function
Erythrocyte	120 days	Oxygen and carbon dioxide transport
Neutrophil	7 hours	Immune defenses
Eosinophil	Unknown	Defense against parasites
Basophil	Unknown	Inflammatory response
Monocyte	3 days	Immune surveillance (precursor of macrophage)
B-lymphocyte	Unknown	Antibody production (precursor of plasma cells)
T-lymphocyte	Unknown	Cellular immune response
Platelets	7–8 days	Blood clotting

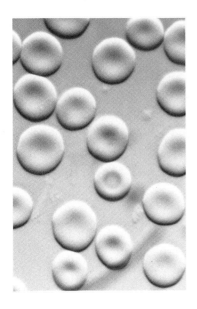

Figure 40.3 Human erythrocytes, magnified 1,000 times.

Human erythrocytes lack nuclei, which gives them a characteristic collapsed appearance, rather like a pillow on which someone has sat.

Mature erythrocytes in mammals contain neither a nucleus nor protein-synthesizing machinery. Because they lack a nucleus, these cells are unable to repair themselves and therefore have a rather short life; any one erythrocyte lives only about four months. New erythrocytes are constantly being synthesized and released into the blood by cells within the soft, interior marrow of bones.

Leukocytes: Body Defense

Fewer than 1 percent of the cells in human blood are **leukocytes,** or **white blood cells;** there are one or two leukocytes for every thousand red blood cells. Leukocytes are larger than red blood cells, contain no hemoglobin, and are essentially colorless. Each of the several kinds of leukocytes has a different function. All of these functions, however, are related to the body's immune defense against invading microorganisms and other foreign substances, as discussed in chapter 41. Leukocytes are not confined to the bloodstream; they also migrate out into the **interstitial fluid,** or the fluid that bathes all body cells.

If the skin is pricked or cut, some of the injured cells release chemicals that cause nearby capillaries to expand. The resulting increase in blood flow is one component of the **inflammatory response,** which makes the wound look red and feel warm. In inflammation, **granulocytes,** which are circulating leukocytes, push out through the walls of the distended capillaries to the injury site. The granulocytes are classified into three groups by their staining properties. Fifty to seventy percent of them are **neutrophils,** which begin to stick to the interior walls of the blood vessels at the injury site. They then form projections that enable them to push their way into infected tissues, where they engulf microorganisms and other foreign particles. **Basophils,** a second kind of granulocyte, contain granules that rupture and release chemicals that enhance the inflammatory response; they are important in causing allergic responses. The function of the third kind of granulocyte—the **eosinophils**—is not clear, but they may play a role in defending against parasitic infections.

Another kind of circulating leukocyte—the **monocyte**—is also attracted to inflammation sites, where it is converted into a **macrophage**—enlarged, amoeba-like cells that entrap microorganisms and particles of foreign matter. Monocytes usually arrive at the inflammation site after the neutrophils and are important in the immune response because they play a key role in antibody production (see chapter 41).

Platelets: Blood Clotting

Certain large cells within the bone marrow, called **megakaryocytes,** regularly pinch off bits of their cytoplasm. These cell fragments, called **platelets,** contain no nuclei. Platelets enter the bloodstream, where they play an important role in controlling blood clotting. A blood clot is a seal of a ruptured blood vessel. The ruptured vessel seals itself by generating a matrix of long fibers and trapped cells that fills the gap from components present in the plasma. In a clot, the gluey substance is a protein called **fibrin,** which is derived from the protein fibrinogen. Fibrin sticks platelets together to form a strong, tight seal.

Scientists have discovered that the fibrin that forms blood clots is generated in a spreading cascade of molecular events. The clotting process is initiated by injury to blood vessel cells. The injury attracts platelets, which release a protein factor that starts the cascade. At each stage in the cascade that follows, proteins from cells and blood combine in fast-rising waves, involving many more molecules in each progressive step. Billions of molecules of fibrin can be formed from a single clot-initiating event (figure 40.4).

Erythrocytes are the red blood cells, responsible for the transport of oxygen. Leukocytes contain the cells of the body's immune system, such as lymphocytes, eosinophils, basophils, and neutrophils. Platelets are cell fragments that are instrumental in the clotting process.

Blood Vessels: The Circulatory Network

Vertebrates have many types of blood vessels, which comprise the body's circulatory network. Blood vessels ferry nutrients and wastes to their appropriate locations, circulate oxygen to tissues, and pick up carbon dioxide for disposal by the lungs. Blood vessels also play a major role in temperature regulation in vertebrates.

Arteries: Highways from the Heart

Arteries carry blood away from the heart. Artery walls consist of three layers of tissue. The innermost one is composed of a thin layer of **endothelial cells.** Surrounding these cells is a thick layer of smooth muscle and elastic fibers, which, in turn, is encased within an envelope of protective connective tissue (figure 40.5*a*). Because this sheath is elastic, the artery is able to expand its volume considerably in response to a pulse of fluid pressure, much as a tubular balloon might respond to air blown into it. Steady contraction of the muscle layer strengthens the wall of the vessel against overexpansion.

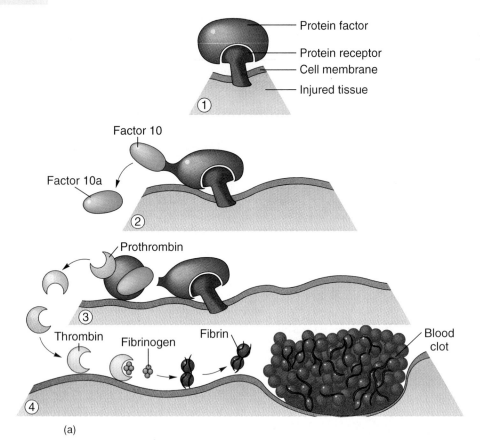

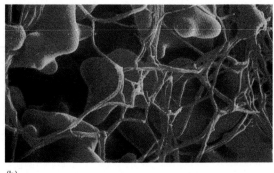

(b)

Figure 40.4 Blood clotting: A cascade of molecular events.

(a) *Blood clotting depends on the availability of fibrin, a fibrous protein that entraps platelets to form a blood clot. Fibrin production is the result of a complex series of molecular events, which in this diagram is summarized in four steps:*
(1) *A protein receptor on the cell membrane of injured tissue binds a protein factor.* (2) *This complex of protein receptor and protein factor binds another protein factor called factor 10. Factor 10 immediately changes to its "active" form, factor 10a.* (3) *Each molecule of factor 10a binds to another protein factor, which initiates the conversion of prothrombin to thrombin.*
(4) *Thrombin interacts with fibrinogen, promoting its conversion to fibrin. Finally, fibrin entraps platelets that seal the wound in the injured tissue.*
(b) *A micrograph of clotted blood, showing platelets within a network of fibrin strands.*

Arterioles: Little Arteries

Arterioles differ from arteries in two ways: They are smaller in diameter, and the muscle layer that surrounds an arteriole can be relaxed under the influence of hormones and metabolites to enlarge the diameter. When the diameter increases, blood flow also increases, an advantage during times of high metabolic activity. Conversely, most arterioles are in contact with many nerve fibers that, when stimulated, cause the arteriole's muscular lining to contract, thereby constricting the arteriole's diameter. Such contraction limits blood flow to the extremities during periods of low temperature or stress. Humans turn pale when they are scared or need to conserve heat because contractions of this kind constrict the arterioles in the skin. Humans blush for the opposite reason. When they overheat or are embarrassed, the nerve fibers connected to muscles surrounding the arterioles are inhibited, which relaxes the smooth muscle and causes the arterioles in the skin to dilate. This brings heat to the surface for escape.

Capillaries: Where Exchange Takes Place

Blood is a medium of exchange, carrying materials to and from the many cells of the body. All of this exchange occurs in the capillaries, which have narrow walls suitable for the passage of gases and metabolites. Capillaries have the simplest structure of any element in the cardiovascular system (figure 40.5b). They are little more than tubes one cell thick and, on the average, about 1 millimeter long; they connect the arterioles with the venules (discussed in the next section). The internal diameter of capillaries is, on the average, about 8 micrometers. Surprisingly, this is little more than the diameter of a red blood cell (5 to 7 micrometers). However, red blood cells squeeze through these fine tubes without difficulty (figure 40.6).

The intimate contact between capillary walls and the membranes of red blood cells facilitates the diffusion of gases and metabolites between them. No body cell is more than 100 micrometers from a capillary. At any one moment, about 5 percent of human blood is in the capillaries.

Some capillaries, called **thoroughfare channels,** connect arterioles and venules directly. From these channels, loops of true capillaries leave and return. Almost all exchange between the blood and body cells occurs through these loops. Entry to each loop is guarded by a ring of muscle called a **precapillary sphincter,** which, when closed, blocks flow through the capillary (figure 40.7). Such restriction of entry to the capillaries in surface tissue is another powerful means of limiting heat loss from an animal's body during periods of cold.

The entire body is permeated with a fine mesh of capillaries, a network that amounts to several thousand miles in overall length. If all the capillaries in a human body were laid end to end, they would extend across the United States!

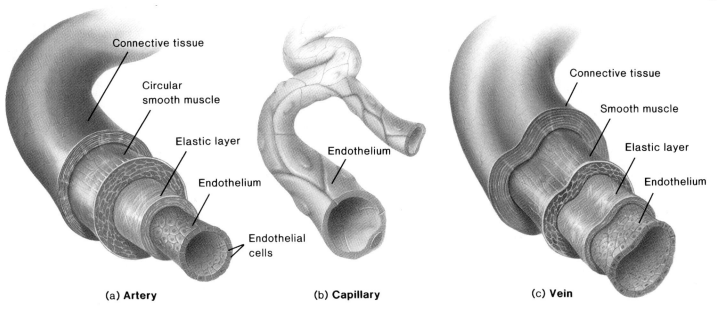

Figure 40.5 Structure of important blood vessels.
(a) *Artery.* (b) *Capillary.* (c) *Vein. Note that the artery has a thicker smooth muscle layer than does the vein.*

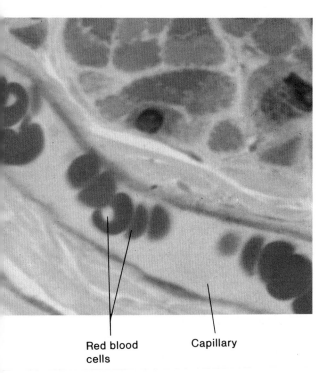

Figure 40.6 Red blood cells in single file in a capillary.

Many capillaries are even smaller than that shown here from the bladder of a monkey. Red blood cells will even pass through capillaries narrower than their own diameters, pushed along by the pressure generated by a pumping heart.

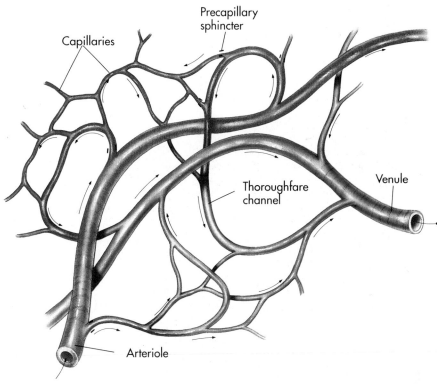

Figure 40.7 Thoroughfare channels: Where tissues meet red blood cells.

Capillary networks connect arterioles with venules. The most direct connection is via thoroughfare channels that connect arterioles directly to venules. Branching from these thoroughfare channels is a network of finer channels—the capillary network. Most of the exchange between tissues and red blood cells occurs while the red blood cells are in this capillary network.

Although individual capillaries have high resistance to flow because of their small diameters, the cross-sectional area of the extensive capillary network is greater than that of the arteries leading to it. Thus, blood pressure is actually lower in the capillaries than in the arteries.

Veins: Returning Blood to the Heart

Veins are vessels that return blood to the heart. Veins do not have to accommodate the pulsing pressures that arteries do because much of the heartbeat's force is weakened by the high resistance and great cross-sectional area of the capillary network. The walls of veins, although similar in structure to those of arteries, have much thinner layers of muscle and elastic fiber (figure 40.5c). An empty artery is still a hollow tube, like a pipe, but when a vein is empty, its walls collapse like an empty balloon (figure 40.8).

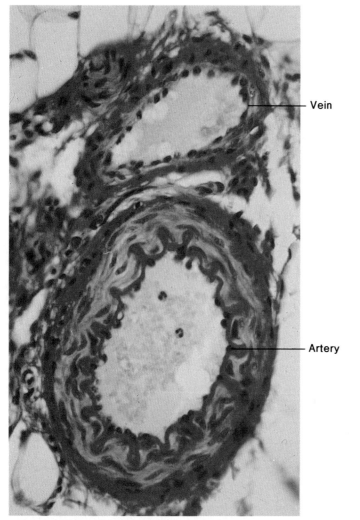

— Vein

— Artery

Figure 40.8 A closer look at blood vessels.
The vein (top) *has the same general structure as the artery* (bottom), *but much thinner layers of muscle and elastic fibers. An artery will retain its shape when empty, but a vein will collapse.*

The internal passageway of veins is often quite large. The diameter of the largest veins in the human body—the **venae cavae,** which lead into the heart—is fully 3 centimeters. Veins are so much larger than arteries because the blood pressure within veins is lower; the larger tube minimizes any further resistance to blood flow. Thus, only a small pressure difference is required to return blood to the heart. Veins also have unidirectional valves that aid blood's return to the heart.

Venules: Little Veins

Venules are smaller veins, just like arterioles are smaller arteries. As already mentioned, venules connect with arterioles to facilitate the exchange of materials between the blood and body tissues. The "connectors" between venules and arterioles are the capillaries.

> *Vertebrates have five types of blood vessels: (1) Arteries carry blood away from the heart; (2) arterioles are smaller arteries that respond to hormones and other metabolites; (3) capillaries facilitate exchange between tissues and blood; (4) veins return blood to the heart; and (5) venules are smaller veins that are connected to arterioles by the capillaries.*

Role of Blood Vessels in Blood Pressure

Everyone at one time or another has their **blood pressure** measured. What the doctor measures during this procedure is the force of the blood against the walls of the blood vessels. As discussed later in the chapter, in vertebrates, the heart pumps blood forcibly into the blood vessels. In the blood vessels that lead directly from the heart, blood pressure is high, due to this forcible heart contraction. As blood travels throughout the body in blood vessels, however, the farther the vessels are from the heart, the lower the blood pressure. In healthy organisms, the body maintains a constant blood pressure over the entire network of blood vessels because, in areas where blood pressure tends to fall, the blood moves from areas of high pressure to low pressure, and blood pressure is equalized.

> *Blood pressure is the force of blood against the walls of the blood vessels. Blood pressure is maintained evenly throughout all the blood vessels of the body.*

Sometimes, however, blood pressure cannot be maintained because blood vessels may be constricted due to **cardiovascular disease,** in which blood vessels are narrowed by cholesterol deposits called **plaque.** When blood vessels are narrowed, blood moves through them under greater force, and blood pressure rises, a condition called **hypertension.** If the force is high enough, the blood vessel may rupture. Sidelight 40.1 discusses hypertension in more detail and its relation to heart disease, as does a later section in this chapter.

At the turn of the century, the great killers were the infectious diseases. Now, however, the leading causes of death in the United States are reflected in the following 1992 statistics from the U.S. National Center for Health Statistics:

1. Heart disease   720,480   deaths
2. Cancer   521,090   deaths
3. Stroke   143,640   deaths
4. Chronic lung   91,440   deaths
   diseases
5. Accidents   86,310   deaths

Unlike infectious disease, chronic illness does not strike at random, threatening everyone equally. On the contrary, deaths from chronic illnesses are so greatly influenced by individual living habits that it is possible to predict in large measure how one will die from how one lives. As many as 70 percent of the deaths from these top killers could be prevented by altered living habits.

Heart disease is the greatest killer in the United States today. It is not, however, an inevitable result of being alive. A few simple changes in lifestyle can drastically lower the risk of heart disease. Strokes, which are caused by stoppage of blood flow to part of the brain, share many of the same risk factors as heart disease. The three principal risk factors are:

Smoking.

About a fourth of Americans (fifty-one million) smoke, leading to almost a third of heart-disease deaths because smoking elevates cholesterol levels and thus atherosclerosis, according to the U.S. surgeon general. Smokers also run a 20 to 60 percent greater chance of stroke. It is not easy to quit smoking—nicotine has been declared an addictive drug—but the best way for smokers to avoid death by heart disease and stroke is to throw their cigarettes away.

High cholesterol levels.

High blood cholesterol levels—over 200 for the average adult—can be lowered through diet. Fewer saturated (animal) fats (such as eggs and meat) and more plant fats should be eaten. If blood cholesterol is over 240, many physicians recommend cholesterol-lowering drugs.

High blood pressure.

Hypertension (high blood pressure) is a killer. High blood pressure causes up to half of all strokes. Blood pressure exceeding 140/90 can be lowered with a low-salt, low-fat diet.

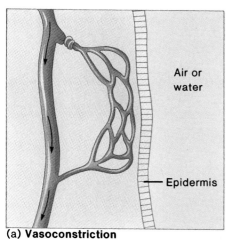

(a) Vasoconstriction

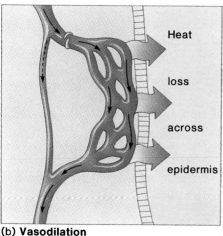

(b) Vasodilation

Figure 40.9 Regulation of body temperature.
The amount of heat lost at the body's surface can be regulated by controlling the flow of blood to the surface. (a) Constriction of surface blood vessels limits flow and lessens heat loss. (b) Dilation of surface blood vessels increases blood flow and thus increases heat loss.

Role of Blood Vessels in the Control of Body Temperature

Mammals and birds differ from most other organisms in that they maintain a constant body temperature by expending metabolic energy. As discussed in chapter 35, mammals closely monitor their internal temperature, using a battery of temperature-sensitive sensory receptors to inform the central nervous system when body temperature strays beyond normal bounds. If the body overheats, the central nervous system initiates a variety of actions to lower temperature, including sweating, panting, and dilation of arterioles in the skin. If the body becomes too cool, the central nervous system initiates constriction of surface arterioles to limit heat loss. Control of heat loss from the body surface through changes in skin blood flow is one of the most important aspects of thermoregulation (figure 40.9).

In all vertebrates, regardless of how they maintain their body temperature, blood circulation distributes heat more or less uniformly throughout the body. This heat circulation is accomplished first by a network of fine blood vessels that pass immediately beneath the body's external surfaces—the surfaces in contact with the environment. The blood that circulates within this network absorbs heat from a hot environment and releases heat to a cold one. The further circulation of this heated or cooled blood to the interior of the animal's body tends to adjust the animal's internal temperature in the direction of the external environment. In many marine mammals, heat loss to cold ocean waters is limited by countercurrent flow in the limbs, which removes heat from arterial blood before it reaches the surface (figure 40.10).

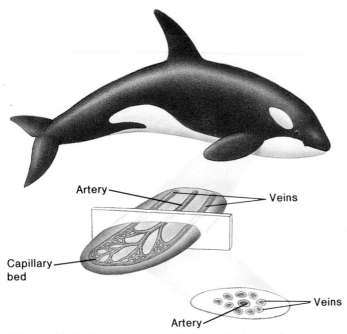

Figure 40.10 Countercurrent flow can limit heat loss.
The killer whale utilizes countercurrent flow in its fins to limit heat loss. Heat is removed from the arterial blood by nearby veins before the blood circulates to the body's surface.

The Lymphatic System: Recovering Lost Fluid

The cardiovascular system is said to be closed because all of its tubes are connected with one another, and none is simply open-ended. In another sense, however, the system is open—open to passage through capillary walls. Fluids are forced out across the thin walls of capillaries on the arteriole side by the pressure that builds up because of the great reduction in vessel diameter. Although this loss is an unavoidable consequence of a closed circulatory system (which could not function properly without tiny vessels with thin walls), it makes maintaining the integrity of that system difficult.

Difficulties arise because of how *much* liquid is lost from the cardiovascular system as blood passes through the capillaries. In a human being, about 3 liters of fluid leave the cardiovascular system in this way each day, a quantity amounting to more than half the body's total supply of about 5.6 liters of blood.

To counteract the effects of this process, the body utilizes a second *open* circulatory system called the **lymphatic system.** The elements of the lymphatic system gather liquid from the interstitial fluid and return it to the cardiovascular circulation. Open-ended **lymph capillaries** gather up fluids by diffusion and carry them through a series of progressively larger vessels to two large **lymphatic vessels,** which resemble veins. These lymphatic vessels drain into veins in the lower part of the neck through one-way valves. Once within the lymphatic system, this fluid is called **lymph** (figure 40.11).

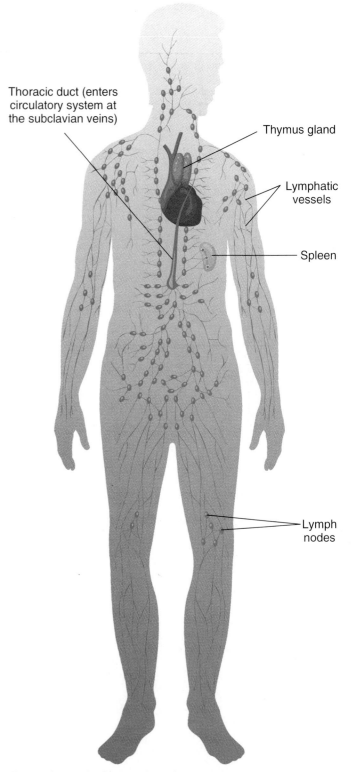

Figure 40.11 The human lymphatic system.
The thoracic duct is the main duct that collects lymph from the lymph vessels. The thoracic duct returns lymph to the general circulation by connecting to the subclavian vein, which is located in the vicinity of the shoulder.

Before being drained into the veins for recirculation in the body, lymph passes through structures called **lymph nodes.** Lymph nodes act as filters and remove bacteria, wastes, and other debris from the lymph so that these materials do not circulate through the body. Lymph nodes also contain many leukocytes, which engulf invading cells that might cause harm to the body. Lymph nodes that are working overtime to remove these foreign invaders may swell because of their overactivity. Humans might experience this swelling when ill with a particularly severe flu or other illness.

Another organ of the lymphatic system is the **spleen.** The spleen acts as a filtering device for blood cells, removing wastes and other impurities. Doctors are able to remove the spleen with no dangerous effects, suggesting that the spleen plays no essential function in the human body. However, sudden rupturing of the spleen, such as during a car accident, can release the filtered impurities into the body, leading to serious complications.

The heart does not pump fluid through the lymphatic system; instead, fluid is propelled through the lymphatic system when lymphatic vessels are squeezed by the movements of the body's muscles. Lymphatic vessels contain a series of one-way valves, which permit movement only in the direction of the neck (figure 40.12). When body muscles are inactive, lymph can build up in body tissues, a condition known as edema. Persons who are confined to bed or who are for other reasons immobile may experience edema as a result of low muscle activity.

Much of the fluid within blood plasma is forced out during passage through the capillaries. This fluid is collected by an open circulatory system—the lymphatic system—and is returned to the bloodstream.

The Heart

Any closed circulatory system requires both a system of passageways through which fluid can circulate and a pump to force the fluid through them. This pump—the vertebrate heart—is the key to vertebrate circulation.

Evolution of the Vertebrate Heart

Early chordates, such as the lancelets, had simple tubular hearts, which amounted to little more than a specialized, muscular zone of the ventral artery, which beat in simple waves of contraction. When gills evolved in the early fishes, a more efficient pump was necessary to force blood through the fine capillary network. The fish heart is shaped somewhat like a continuous tube with four chambers arrayed one after the other. The first two chambers (the **sinus venosus** and the **atrium**) are collection chambers; the second two (the **ventricle** and the **conus arteriosus**) are pumping chambers. When a fish heart beats, contraction starts in the rear chamber (the sinus venosus) and spreads progressively forward to the conus arteriosus (figures 40.13 and 40.14*a*). The same heartbeat sequence is characteristic of all vertebrate hearts.

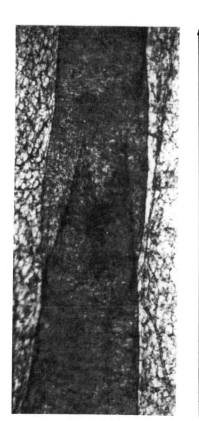

Figure 40.12 A lymphatic vessel, magnified 25 times.
Flow from bottom to top is not impeded because such flow tends to force the inner cone, or valve, open. Flow from top to bottom is prevented because such flow tends to force the inner valve closed.

Flow impeded

Flow not impeded

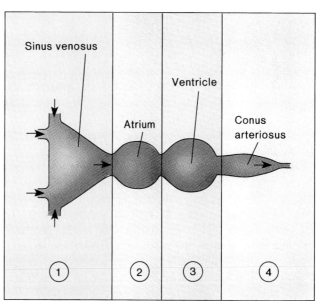

Figure 40.13 Schematic diagram of the fish heart.
(1) *The sinus venosus is a large collection chamber that contributes the least possible resistance to venous flow.* (2) *The atrium quickly delivers blood to the pump in increments of suitable volume.* (3) *The ventricle is a thick-walled pumping chamber that propels the blood through the gills.* (4) *Blood leaving the heart passes through the elongated conus arteriosus to smooth the pulsations and to add still more thrust.*

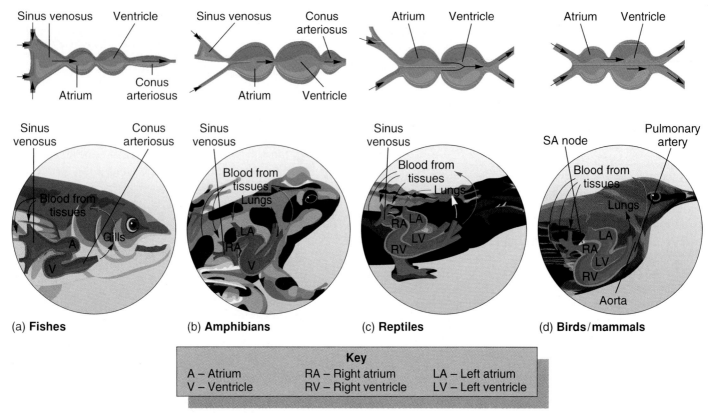

(a) **Fishes**   (b) **Amphibians**   (c) **Reptiles**   (d) **Birds/mammals**

Key		
A – Atrium	RA – Right atrium	LA – Left atrium
V – Ventricle	RV – Right ventricle	LV – Left ventricle

Figure 40.14 Evolution of the vertebrate heart.
(a) *In fishes, the heart is shaped somewhat like a continuous tube, with no separations between the different chambers.*
(b) *Amphibians have the beginning of a separation between the right and left halves of the heart.* (c) *In reptiles, this separation is even more complete, and three distinct chambers are evident.*
(d) *In birds and mammals, the separation is complete, and four distinct chambers are visible.*

After fish blood leaves the gills, it circulates through the rest of the body. In terrestrial vertebrates, however, the blood that receives oxygen and discharges carbon dioxide in the lungs is returned to the heart by two large veins—the **pulmonary veins**—and then pumped forcefully out through the body. This results in much-improved rates of circulation.

To achieve this double circulation, the arrangement of the amphibian heart differs from that found in fishes in two ways. First, the atrium is divided into two chambers: a right atrium, which receives unoxygenated blood from the sinus venosus for circulation to the lungs, and a left atrium, which receives oxygenated blood from the lungs through the pulmonary vein for circulation throughout the body. This division extends partway into the ventricle. Second, the conus arteriosus is partly separated by a dividing wall, which directs oxygenated blood into the aorta (which leads to the body's network of arteries) and unoxygenated blood into the pulmonary arteries (which lead to the lungs) (figure 40.14b). Reptilian hearts show even more complete separation (figure 40.14c).

Taken together, these modifications divide the circulatory system into two separate pathways: (1) the **pulmonary circulation,** between the heart and the lungs; and (2) the **systemic circulation,** between the heart and the rest of the body (figure 40.15). However, since the divisions in the heart ventricles and conus arteriosus are not complete, oxygenated and unoxygenated blood are mixed, somewhat reducing the oxygen level of the blood pumped to the body.

Further evolution among the vertebrates has resulted in the complete closing of the dividing wall in the ventricle, which results in a total division of the pumping chamber into two parts in birds and mammals (figure 40.14d). In these groups, the four-chambered heart acts as a double pump: The left side pumps oxygenated blood to the general body circulation, and the right side pumps unoxygenated blood from the veins to the lungs. Such efficient hearts function well in helping to maintain constant internal temperatures in birds and mammals. They also circulate blood through the lungs much more rapidly and efficiently than in other vertebrates, thus greatly increasing the efficiency with which oxygen is captured by the bloodstream.

Evolution of the vertebrate heart has occurred in discrete stages. An early, unspecialized muscular tube (seen in lancelets) changed to a series of sequential chambers in fish. In the amphibians and reptiles, a septum dividing the chambers in half became more prominent, forming separate systemic and pulmonary circulation systems. Mammals and birds have a complete septum, and therefore, total separation of oxygenated and deoxygenated blood.

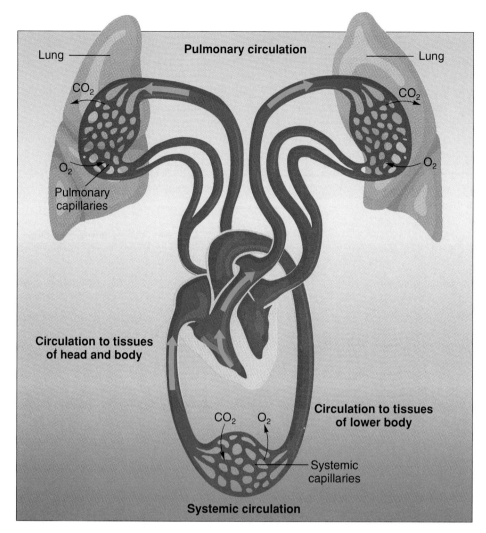

Pulmonary circulation

Lung

CO_2

O_2

Pulmonary capillaries

Circulation to tissues of head and body

Circulation to tissues of lower body

CO_2 O_2

Systemic capillaries

Systemic circulation

Lung

CO_2

O_2

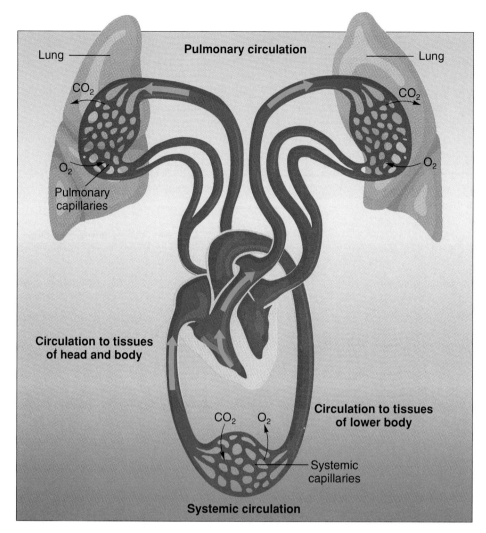

⑤

Superior vena cava

Aorta

Pulmonary valve

Right atrium

Tricuspid valve

Inferior vena cava

④

⑩

⑨

⑥

③

①

⑧

⑦

②

Right ventricle

● Deoxygenated blood
● Oxygenated blood

Pulmonary artery

Pulmonary veins

Left atrium

Mitral valve

Aortic valve

Left ventricle

Figure 40.15 Pulmonary circulation and systemic circulation.

The circulation of blood in mammals has two circuits: In pulmonary circulation, blood circulates from the heart to the lungs and back again for oxygenation and disposal of carbon dioxide. In systemic circulation, the oxygenated blood is circulated throughout the body, and the deoxygenated blood returns to the heart.

Structure and Function of the Human Heart

The human heart, like that of all mammals and birds, is a double pump; a frontal section through the heart clearly shows its organization (figure 40.16). The left side has two connected chambers, and so does the right, but the two sides are not connected with one another.

The journey of blood through the human heart begins with oxygenated blood entering the heart from the lungs (see figure 40.16). Oxygenated blood from the lungs enters the left side of the heart, emptying directly into the **left atrium** through large vessels called the **pulmonary veins.**

From the left atrium, blood flows through an opening into the adjoining chamber, the **left ventricle.** Most of this flow, roughly 80 percent, occurs while the heart is relaxed. When the heart starts to contract, the atrium contracts first, pushing the remaining 20 percent of its blood into the ventricle.

Figure 40.16 Path of blood through the human heart.

(1) Oxygenated blood from the lungs enters via the pulmonary veins into the left atrium. (2) Blood flows into the left ventricle. (3) Blood is pumped from the left ventricle through the aortic valve. (4) Blood leaves the heart through the aorta. (5) Deoxygenated blood enters the heart through the superior vena cava. (6) Blood flows into the right atrium. (7) Deoxygenated blood flows from the inferior vena cava into the right atrium. (8) Blood from both venae cavae flows into the right ventricle. (9) The right ventricle pumps the blood out through the pulmonary valve. (10) The deoxygenated blood leaves the heart for the lungs through the pulmonary artery.

After a slight delay, the left ventricle contracts. The walls of the ventricle are far more muscular than those of the atrium, and as a result, this contraction is much stronger. It forces most of the blood out of the left ventricle in a single strong pulse. The blood is prevented from going back into the atrium by a large one-way valve—the **mitral valve**—whose flaps are pushed shut as the ventricle contracts. Strong fibers that prevent the flaps from moving too far when closing are attached to the flap edges. If the flaps moved too far, they would project out into the atrium. The fibers that prevent this operate in much the same way as a chain on a screen door—the door can be opened only as far as the slack in the chain permits.

Prevented from reentering the left atrium, the blood takes the only other passage out of the contracting left ventricle: It moves through a second opening that leads into a large vessel called the **aorta.** The aorta is separated from the left ventricle by a one-way valve, the **aortic valve.** Unlike the mitral valve, the aortic valve is oriented to permit the flow of blood *out of* the ventricle. Once this outward flow has occurred, the aortic valve closes, thus preventing the reentry of blood from the aorta into the heart.

The aorta and all the other blood vessels that carry blood away from the heart are arteries. Many of these arteries branch from the aorta, carrying oxygen-rich blood to all parts of the body. The first to branch are the coronary arteries, which carry freshly oxygenated blood to the heart itself; heart muscles do not obtain their supply of blood from within the heart.

The blood that flows into the arterial system eventually returns to the heart after delivering its cargo of oxygen to body cells. In returning, it passes through a series of veins and eventually enters the right side of the heart. Two large veins collect blood from the systemic circulation. The **superior vena cava** drains the upper body, the **inferior vena cava** the lower body. These veins empty deoxygenated blood into the **right atrium.**

The right side of the heart is similar in organization to the left side. Blood passes from the right atrium into the **right ventricle** through a one-way valve, the **tricuspid valve.** It passes out of the contracting right ventricle through a second valve—the **pulmonary valve**—into the **pulmonary arteries,** which carry the deoxygenated blood to the lungs. The blood then returns from the lungs to the left side of the heart with a new cargo of oxygen, which is pumped to the rest of the body.

> *The human heart is a double pump, consisting of four discrete chambers. The atria receive blood from the venae cavae and the lungs, while the ventricles pump the blood out to the body and lungs. Four one-way valves keep blood flowing in one direction. A complete septum keeps oxygenated blood from mixing with deoxygenated blood.*

How the Heart Contracts

The overall contraction of the heart consists of a carefully orchestrated series of muscle contractions. First, the atria contract together, followed by the ventricles. Contraction is initiated by the **sinoatrial (SA) node,** the small cluster of excitatory cardiac muscle cells derived from the sinus venosus and embedded in the upper wall of the right atrium. The cells of the SA node act as a pacemaker for the rest of the heart. Their membranes spontaneously depolarize with a regular rhythm that determines the rhythm of the heart's beating. Each depolarization initiated within this pacemaker region passes quickly from one cardiac muscle cell to another in a wave that envelops both the left and the right atria almost instantaneously.

> *Contraction of the heart is initiated by the periodic spontaneous depolarization of cells of the SA node. The resulting wave of depolarization passes over both the left and right atria and causes their muscle cells to contract.*

The wave of depolarization does not immediately spread to the ventricles. Almost 0.1 second passes before the lower half of the heart starts to contract. The reason for the delay is that the atria of the heart are separated from the ventricles by connective tissue, and connective tissue cannot propagate a depolarization wave. The depolarization would not pass to the ventricles at all except for a slender connection of cardiac muscle cells known as the **atrioventricular (AV) node,** which connects to a strand of specialized muscle in the ventricular septum known as the **bundle of His.** Bundle branches divide to the right and left. On reaching the apex of the heart, each branch further divides into **Purkinje fibers,** which initiate contraction there. The passage of the wave triggers the almost simultaneous contraction of all the cells of the right and left ventricles because the bundle branches conduct very rapidly.

The cells involved in the passage of the depolarization wave from the atria to the ventricles have small diameters. Thus, they propagate the depolarization slowly, causing the delay mentioned. This delay permits the atria to finish emptying their contents into the corresponding ventricles before those ventricles start to contract.

> *The passage of the wave of depolarization from the AV node via the bundle of His and bundle branches to the Purkinje fibers causes the almost simultaneous contraction of all the cells of the right and left ventricles.*

Monitoring the Heart's Performance

The heartbeat is not simply a squeeze-release, squeeze-release cycle, but rather, a series of events in a predictable order. The simplest way to monitor heartbeat is to listen with a stethoscope to the heart at work. The first sound—a low-pitched *lub*—is the closing of the mitral and tricuspid valves at the start of ventricular contraction. The higher-pitched *dub* heard a little later is the closing of the pulmonary and aortic valves at the end of ventricular contraction. If the valves are not closing fully, or if they open too narrowly, turbulence is created within the heart. This turbulence can be heard as a **heart murmur.** It often sounds like liquid sloshing.

A second way to examine the events of the heartbeat is to monitor the blood pressure. During the first part of the heartbeat, the atria are filling. At this time, the pressure in the arteries leading from the left side of the heart out to the tissues of the body decreases slightly as the blood moves out of the arteries, through the vascular system, and into the atria. This period is referred to as the resting period, or **diastolic period,** and the lowest arterial pressure is called the diastolic pressure. During contraction of the left ventricle, a pulse of blood is forced into the systemic arterial system, immediately raising the blood pressure within these vessels. This pushing period, which ends with the closing of the aortic valve, is referred to as the pumping period, or **systolic period.** The highest arterial pressure is called the systolic pressure.

Blood pressure values are measured in millimeters of mercury; they reflect the height to which a column of mercury would be raised in a tube by an equivalent pressure. By the conventional system of measurement, normal blood pressure values are 70 to 90 diastolic and 110 to 130 systolic. These would be abbreviated as 110/70 or 130/90, respectively. When the inner walls of the arteries accumulate fats, as they do in the condition known as **atherosclerosis,** the diameters of the passageways are narrowed, and systolic blood pressure is elevated.

A third way to monitor the progress of events during a heartbeat is to measure the waves of depolarization. Because the human body basically consists of water, it conducts electrical currents rather well. A wave of membrane depolarization passing over the surface of the heart generates an electrical current that passes in a wave throughout the body. The magnitude of this electrical pulse is tiny, but it can be detected with sensors placed on the skin. A recording made of these impulses is called an **electrocardiogram.**

In a normal heartbeat, three successive electrical pulses are recorded. First, there is an atrial excitation, caused by the depolarization associated with atrial contraction (P wave). A 0.1 second later, there is a much stronger ventricular excitation, reflecting both the depolarization of the ventricles and the relaxation of the atria (QRS wave). Finally, perhaps 0.2 second later, there is a third pulse, caused by the relaxation of the ventricles (T wave) (figure 40.17).

The heart's performance can be evaluated in one of three principal ways: listening to it with a stethoscope, monitoring blood pressure, or performing an electrocardiogram.

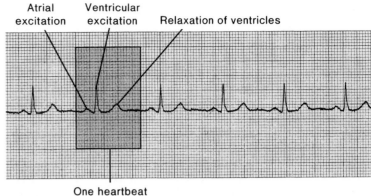

Atrial excitation   Ventricular excitation   Relaxation of ventricles

One heartbeat

Figure 40.17 An electrocardiogram.
An electrocardiogram measures the electrical activity of the depolarizations that cause the heart to contract.

Cardiovascular Disease

Cardiovascular diseases are those conditions that affect the heart, blood vessels, and to some extent, the blood. Even with the AIDS epidemic and the rise of infectious diseases, such as tuberculosis, the current leading cause of death in the United States (and the leading cause for the last thirty years) is cardiovascular disease. Like AIDS, but unlike most other infectious diseases, steps can be taken to lessen the chances of developing cardiovascular disease.

What Is Cardiovascular Disease?

Understanding the causes and the course that cardiovascular disease takes is a preliminary step toward prevention. Researchers believe that the main cause of most cardiovascular disease is the condition mentioned previously known as atherosclerosis. Atherosclerosis is a buildup of cholesterol deposits, fatty substances, and cellular debris within the blood vessels. Using a microscopic probe, scientists have been able to view these deposits, called plaque, in artery walls. When plaque buildup is severe, the blood vessels can no longer expand and contract properly, and blood moves through them with difficulty (figure 40.18). This slowdown in blood movement can have serious consequences, including angina pectoris, heart attack, stroke, and hypertension.

Angina Pectoris

Angina pectoris literally means "chest pain." The pain may occur in the heart and can often occur in the left arm or shoulder as well. Angina pectoris is a warning sign that the blood supply to the heart is inadequate and it is often a prelude to a heart attack.

Heart Attack

Known in the medical field as a *myocardial infarction* (which literally means "heart muscle death"), a heart attack is a severe condition in which part of the heart muscle dies as a result of

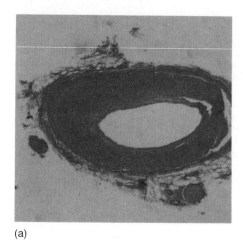

(a)

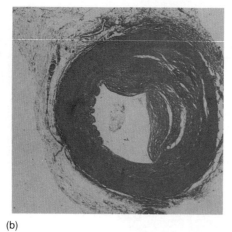

(b)

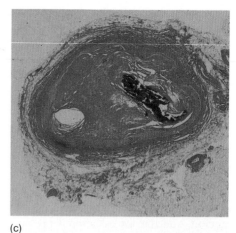

(c)

Figure 40.18 The path to a heart attack.
(a) *This coronary artery shows only minor blockage.* (b) *This artery exhibits severe atherosclerosis—much of the passage is blocked by buildup on the artery's interior walls.* (c) *This coronary artery is essentially completely blocked.*

insufficient blood supply to the heart. Heart attacks are the main cause of cardiovascular deaths in the United States and account for a fifth of all deaths. Recovery from a heart attack is possible if the segment of damaged heart tissue is small enough that the other blood vessels in the heart can enlarge their capacity and resupply the damaged tissue.

Heart attacks are often seen in persons with severe atherosclerosis or a related condition called **arteriosclerosis,** in which the plaques in the arteries leading away from the heart have become calcified. Arteriosclerosis is sometimes called "hardening of the arteries" because the calcium deposits within the arteries are hard and inflexible. Not only is blood flow through these arteries restricted, but the arteries also lack the ability to expand, as normal arteries do, to accommodate the volume of blood pumped out by the heart. This forces the heart to work harder and thus leads to a heart attack.

Strokes

A stroke is a condition in which a blood vessel bursts in the brain. Like heart attacks, some strokes are caused by a narrowing of blood vessels by plaques and thus can be considered a cardiovascular disease (figure 40.19). Other strokes are caused by a blood clot not related to cardiovascular disease; for example, some blood clots form in the brain as a result of cancer.

Although their causes may be different, strokes can have lasting consequences for the sufferer. Strokes damage brain tissue, and depending on the part of the brain in which the stroke occurs, the results can be either mild or severe. Some people suffer little or no aftereffects from a stroke. Others are paralyzed, lose the ability to speak, or die. Strokes are the third leading cause of death in the United States.

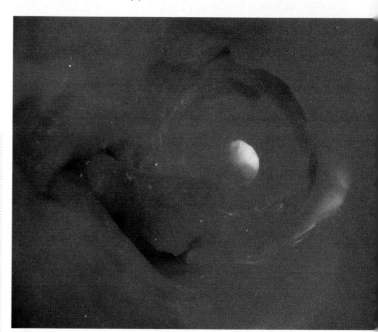

Figure 40.19 A look down an artery.
The interior of an artery caked with fatty deposits. Such deposits retard blood circulation. Fragments that break off can block arterial branches and produce strokes or heart attacks.

Hypertension

High blood pressure, or hypertension, has been linked to atherosclerosis, although the connection is by no means direct. The direct cause of hypertension involves high levels of a hormone called renin in the blood. Renin is produced by the kidneys and causes salt and water to be retained in body tissues. This retention, in turn, leads to high blood pressure. High blood pressure increases the heart's work to pump blood, which may cause the heart to fail. Hypertension also increases the pressure on the walls of blood vessels, greatly increasing the chances of stroke and heart attack. Atherosclerosis increases hypertension's detrimental effects on the heart and blood vessels, and because of this connection, hypertension plays a major role in the development of cardiovascular disease.

Preventing Cardiovascular Disease

Researchers studying the problem of cardiovascular disease have come up with a set of risk factors that contribute to the development of cardiovascular disease. These risk factors are of two types: (1) risky behaviors that, if indulged in, greatly increase the probability that an individual will develop cardiovascular disease; and (2) risk factors that an individual cannot control, such as a father who died of a heart attack or a genetic predisposition toward a high blood cholesterol level. The two types of risk factors complement each other in interesting ways. For example, the fact that a father died of a heart attack does not necessarily mean that his children will die of a heart attack, only that their chances for such a fate may be increased. However, they can lessen this probability by eliminating many of the risky behaviors that are within their control. The more risky behaviors that they eliminate, the less likely that they will have a heart attack.

Behaviors that put individuals at risk for cardiovascular disease include high blood cholesterol levels, hypertension, smoking, lack of physical activity, obesity, and abuse of certain drugs (cocaine is one). The risk factors that are beyond the control of individuals but that should alert persons that they are at higher risk for cardiovascular disease include early onset of a heart attack in a parent, being male, and being over age fifty.

An "antirisk" factor for cardiovascular disease that is currently under investigation is aspirin intake. A study involving twenty-two thousand male physicians who took 325 milligrams of aspirin every day over a six-year period showed that the men who took aspirin significantly reduced their risk of suffering a first heart attack. A similar study involving female nurses showed the same preventive effect of aspirin. However, before rushing out and buying a bottle of aspirin, individuals should be aware that aspirin intake can cause a host of physical problems, such as ulcers, and can interfere with blood clotting. Aspirin intake should be changed only under a physician's supervision.

The most prudent course of action to prevent cardiovascular disease is to eliminate the risky behaviors shown to contribute to the disease. A low-fat, low-cholesterol diet, a cessation of smoking, and the maintenance of a healthy weight through adequate exercise are the best ways to ensure a healthy heart. In addition, regular testing of blood cholesterol levels and blood pressure is a good idea, especially if cardiovascular disease runs in the family.

Cardiovascular disease is caused by anything that interferes with the function of the heart. Family genetic history or (usually) consequences of unhealthy life habits may result in atherosclerosis, angina, myocardial infarction, arteriosclerosis, stroke, or hypertension. Most heart-related problems can be prevented by a good diet, exercise, and avoidance of drugs (including nicotine).

The Central Importance of Circulation

The evolution of multicellular organisms has depended on the ability to circulate nutrients and other materials to the various body cells and to carry metabolic wastes away from them. The digestive processes described in chapter 38, by which vertebrates obtain metabolizable foodstuffs, and the respiratory processes described in chapter 39, by which vertebrates obtain the oxygen necessary for aerobic metabolism, both depend on the transport of food and oxygen to cells and the removal of the end products of their metabolism. Vertebrates carefully regulate the operation of their circulatory systems. By doing so, they are able to integrate their body activities. This regulation is carried out by the nervous system, the subject of chapter 35.

EVOLUTIONARY VIEWPOINT

The reason why our circulatory systems do not operate at considerably higher pressures than they do is the need to maintain a very thin epithelial wall between air and blood in the lungs so that diffusion can readily take place. Higher pressures would necessitate strengthening the one-cell-thick walls of these capillaries, which would render them useless as gas exchange devices. Higher pressures without strengthened blood capillaries, such as result from atherosclerosis, could cause the thin capillary walls to burst. This is why strokes are associated with heart disease and diet.

SUMMARY

1. Vertebrates have a closed circulatory system, which permits better control of circulation rates. Closed circulatory systems have five principal functions: (1) nutrient and waste transport, (2) oxygen and carbon dioxide transport, (3) temperature maintenance, (4) hormone circulation, and (5) immune defense.

2. The vertebrate circulatory system has three elements: (1) the blood, (2) the blood vessels, and (3) the heart. The general flow of blood circulation through the body is a circuit starting from the heart, which pumps blood out via muscled arteries to the capillary networks that interlace body tissues. The blood returns to the heart from these capillaries via the veins.

3. The plasma of the circulating blood contains metabolites, wastes, and a variety of salts and ions. It also contains high concentrations of the protein serum albumin, which functions to maintain the blood's osmotic equilibrium with surrounding tissues. Blood also contains red blood cells (erythrocytes), which circulate oxygen; white blood cells (leukocytes), which function in the immune response; and platelets, which play an important role in blood clotting.

4. Vertebrates have five types of blood vessels. Arteries carry blood away from the heart. Arterioles are smaller in diameter than arteries but can dilate or contract their diameters to control blood flow to extremities. Capillaries facilitate the exchange of gases and metabolites between tissues and blood. Veins return blood to the heart. Venules are smaller veins and connect with arterioles to facilitate exchanges between the blood and body tissues.

5. Blood pressure is the measure of the force of blood against the walls of blood vessels. High blood pressure is a dangerous medical condition that can result in the rupture of a blood vessel.

6. Blood vessels also function in temperature regulation. A network of blood vessels near the surface of the skin constrict or dilate to retain or release heat from the body.

7. A second open circulatory system—the lymphatic system—gathers liquid from the body that has been lost from the circulatory system by diffusion and returns it via a system of lymph capillaries, lymph vessels, and lymphatic ducts to veins in the lower part of the neck.

8. The four-chambered mammalian heart evolved from two chambers of the fish heart by the creation of dividing walls within the two central chambers. The chambers at the ends of the fish heart were gradually lost, although its pacemaker cells have been retained in their original location in the mammalian heart.

9. The heart of mammals and birds is a double pump, pushing both pulmonary (lung) circulation and systemic (general body) circulation. Because the two circulations are kept separate within the heart, the systemic circulation receives only fully oxygenated blood.

10. Contraction of the heart is initiated at the sinoatrial (SA) node, or pacemaker, by the periodic spontaneous depolarization of these cells. The resulting wave of depolarization spreads across the surfaces of the two atrial chambers, causing all these cells to contract.

11. Passage of the wave of depolarization to the ventricles is briefly delayed because of connective tissue that insulates the atria from the ventricles and does not propagate a depolarization wave. Only a narrow channel of cardiac muscle cells (called the atrioventricular [AV] node) that connects the atria and the ventricles allows the depolarization to pass. The delay in the passage of the wave of depolarization permits the atria to empty completely into the ventricles before ventricular contraction.

12. The heartbeat can be heard, or monitored, by tracking changes in blood pressure through the period of filling and contracting of the atria (the diastolic period) and contraction of the left ventricle (the systolic period). The waves of depolarization can also be measured directly; a recording of these pulses is called an electrocardiogram.

13. The primary cause of most cardiovascular disease is atherosclerosis, which is a buildup of cholesterol deposits, fatty substances, and cellular debris within blood vessels. The consequences of the resulting slowdown in blood movement can include angina pectoris, heart attacks, strokes, and hypertension. Prevention of cardiovascular disease involves eliminating the risky behaviors, such as smoking and overeating, that contribute to its development.

REVIEWING THE CHAPTER

1. How did the circulatory system evolve?

2. What is the composition of the circulatory system?

3. What is the makeup of blood plasma?

4. How do erythrocytes carry hemoglobin?

5. How do leukocytes defend the body?

6. How do platelets help to clot blood?

7. How are blood vessels the circulatory network of the body?

8. How do arteries function?

9. How do arterioles differ from arteries?

10. How do capillaries function?

11. How do veins and venules function?

12. What is the role of blood vessels in blood pressure?

13. How do blood vessels control body temperature?

14. How does the lymphatic system recover lost fluid?

15. How did the vertebrate heart evolve?

16. What is the structure and function of the human heart?

17. How does the heart contract?

18. How is the heart's performance monitored?

19. What is cardiovascular disease?

20. What is angina pectoris?

21. Why do heart attacks occur?

22. What is a stroke?

23. What is the cause of hypertension?

24. What can be done to prevent cardiovascular disease?

25. Why is circulation important?

COMPLETING YOUR UNDERSTANDING

1. What is the advantage of a closed circulatory system over that of an open circulatory system?

2. What are the major functions of the closed circulatory system?

3. Blood exchanges gases and metabolites with the cells of the body while passing through

    a. capillaries.   c. arteries.

    b. venules.   d. arterioles.

4. Blood would lose most of its water to body cells if it did not contain a high concentration of

    a. lipids.   d. proteins.

    b. glucose.   e. fatty acids.

    c. glycogen. f. nucleic acids.

5. Which of the following leukocytes is important in the immune response?

    a. Eosinophils c. Macrophages

    b. Basophils   d. Neutrophils

6. The walls of arteries, from the inside to the outside, are composed of

    a. endothelial cells, smooth muscle and elastic fibers, and protective connective tissue.

    b. smooth muscle and elastic fibers, protective connective tissue, and endothelial cells.

    c. protective connective tissue, smooth muscle and elastic fibers, and endothelial cells.

    d. endothelial cells, protective connective tissue, and smooth muscle and elastic fibers.

7. What is the scientific explanation for turning pale and blushing?

8. What is the correlation between the diameters of capillaries and erythrocytes?

9. In general, veins are larger in diameter than arteries. Why?

    a. To reduce resistance to flow

    b. To increase resistance to flow

    c. Because cholesterol builds up only in arteries

    d. Because arteries are more elastic

    e. None of the above

10. How do animals limit heat loss from their bodies during cold periods?

11. Why are there so many leukocytes in the lymph nodes?

12. What is the spleen, and how does it function?

13. A condition known as edema is associated with
   a. a ruptured spleen.
   b. kidney failure.
   c. hypertension.
   d. lymph buildup.
   e. pulmonary veins.
   f. serum albumin.

14. Which chambers in the heart of a fish are pumping chambers? [*two answers required*]
   a. Atrium
   b. Conus arteriosus
   c. Ventricle
   d. Sinus venosus

15. What is the difference between pulmonary and systemic circulation?

16. Blood pumped out of the left ventricle moves into the
   a. left atrium.
   b. pulmonary veins.
   c. right ventricle.
   d. aorta.
   e. superior vena cava.

17. In what order does blood flow through the four chambers of your heart, starting from the lungs?

18. Which structures in the human heart receive and carry deoxygenated blood? [*three answers required*]
   a. Right atrium
   b. Left atrium
   c. Aorta
   d. Left ventricle
   e. Right ventricle
   f. Both venae cavae

19. Which of the following events occurs first in the cardiac cycle?
   a. Atrial contraction
   b. An action potential in the SA node
   c. Arterial systolic pressure
   d. The QRS wave of the electrocardiogram
   e. Ventricular contraction

20. What is a heart murmur, and what causes it?

21. Given the blood pressure 104/65, can you explain what these two numbers mean and how they were determined? What health condition does this blood pressure indicate and why?

22. In a normal heartbeat monitored by an electrocardiogram, the T wave measures a pulse caused by
   a. relaxation of the ventricles.
   b. relaxation of the atria.
   c. atrial excitation.
   d. ventricular excitation.

23. What are the consequences of plaque buildup on arterial walls?

24. What causes strokes, and what are the consequences of having one?

25. How do kidneys relate to hypertension?

26. Why should you have regular testing of blood cholesterol levels and blood pressure? Do you know what your readings are?

*27. Instead of evolving an entire second open circulatory system—the lymphatic system—to collect water lost from the blood plasma during passage through the capillaries, why have not vertebrates simply increased the level of serum albumin in their blood?

*28. The hearts of the more advanced vertebrates pump blood entirely by pushing action. Why do you suppose that a heart that acts like a suction pump, drawing blood into the heart as it expands, rather than pushing it out as the heart contracts, has not evolved?

*For discussion

FOR FURTHER READING

Eisenberg, M. et al. 1986. Sudden cardiac death. *Scientific American*, May, 37–93. An account of the role of rapid medical intervention in coping with serious heart attacks.

Goldstein, G., and A. L. Belz. 1986. The blood-brain barrier. *Scientific American*, Sept., 74–83. Explains why many chemicals will not pass from the bloodstream to the cells of the brain because of a "barrier" resulting from the structure of brain capillaries, which possess an unusual array of membrane channels.

Lawn, R., and G. Vehar. 1986. The molecular genetics of hemophilia. *Scientific American*, March, 48–56. Describes inherited clotting factor defects responsible for hemophilia.

Lillywhite, H. 1988. Snakes, blood circulation, and gravity. *Scientific American*, Dec., 92–98. Explains why, when a snake climbs a tree, all the blood does not rush to its tail.

Robinson, T., S. Factor, and E. Sonneblink. 1986. The heart as a suction pump. *Scientific American*, June, 84–91. The argument that the heart is aided greatly in its Herculean task of beating millions of times between birth and death by a very clever trick: Contraction compresses elastic elements within the heart muscles, which then bounce back to expand the ventricles.

Uterman, G. 1989. The mysteries of lipoprotein (a). *Science* 246 (Nov.): 904-10. A description of how lipoprotein (a), when present in high concentrations in the bloodstream, signals the likelihood of heart attack and stroke.

Vines, G. 1989. Diet, drugs, and heart disease. *New Scientist*, Feb., 44–49. Affirmation that high blood cholesterol levels give people heart disease. Now the question is what to do about it.

Zucker, M. B. 1980. The functioning of the blood platelets. *Scientific American*, June, 86–103. A description of the many roles of platelets in human health, with emphasis on their role in blood clotting.

EXPLORATIONS

Interactive Software:
Evolution of the Heart

This interactive exercise allows students to learn how the human heart works by exploring the consequences of the changes that took place during its evolution. The exercise presents a cross-sectional diagram of a heart, as it would appear in a fish, at the beginning of the vertebrate heart's evolution. The fish heart has four sequential chambers and no internal septum. The students can explore the consequences of the evolutionary changes that have taken place by forming a septum and extending it through the heart while watching the effects on blood pressure and oxygen delivery to tissues. In the fish, blood pressure is low while oxygenation is high. In land vertebrates, the evolutionary goal is to maximize both.

Questions to Explore:
   1. Why do lungs decrease blood pressure?
   2. What is the advantage of high blood pressure?
   3. Why does a fish drown in air?
   4. Why does a bird's heart work "better" than a frog's?

How The Body Defends Itself

HIV viruses (artificially colored orange) being released by exocytosis from an infected lymphocyte. Currently, over 17 million people worldwide have been infected by this fatal immune-system-destroying virus that was first detected in 1982.

FOR REVIEW

Here are some important terms and concepts that have been discussed in previous chapters and that you will encounter again in this chapter. Review them before proceeding if necessary.

Phagocytosis (*chapter 5*)
Cell-surface markers (*chapter 5*)
Recombination (*chapter 14*)
AIDS (*chapter 26*)
Macrophage (*chapters 33 and 40*)
Lymphocyte (*chapters 33 and 40*)

The most dangerous enemies that vertebrates face are their distant relatives, the **microbes,** single-celled creatures too tiny to be seen with the naked eye. Every vertebrate body offers microbes a feast in nutrients, as well as a warm, sheltered environment in which to grow and reproduce. Like Europeans first discovering the New World, a microbe entering a vertebrate body is faced with a rich ecosystem ripe for plundering. The earth is awash with microbes, and no vertebrate body could long withstand their onslaught unprotected.

Humans survive because of the evolution of a variety of effective defenses against constant microbial attack. As these defenses are reviewed in this chapter, it will become apparent that they are far from perfect—microbial infection is still a major cause of death among humans. Some twenty-two million Americans and Europeans died of flu within eighteen months in 1918–1919 (figure 41.1). More than one million people will die

of malaria *this year.* Attempts to improve defenses against infection are among the most active areas of scientific research today.

The Body's Defenses

A bank protects its assets from robbery in two ways: First, the money is kept in a sturdy safe, and alarm and camera systems alert the police if the safe is breached or if a robber enters the bank. Second, bank employees must identify themselves with a code to open the safe. Thus, only employees who know the correct code are allowed to enter the safe and to come in contact with the bank's assets.

The human body has a similar, two-tiered approach to defense against **pathogens,** microbes or viruses with the potential for harming an organism. The first tier of defenses, called the **nonspecific defenses,** are mechanisms that protect the body from invasion by *any* foreign cell or virus. Nonspecific defenses include barriers, such as the skin and mucous membranes, as well as other mechanisms, such as the inflammatory response and chemicals that kill foreign invaders. Like the bank's safe and alarms, the body's nonspecific defenses comprise the very rapid initial response to infection.

The human body's second tier of defenses are called the **specific defenses.** Specific defenses are the cells of the **immune system.** These cells roam about the vertebrate body and scan all body cells for proper "identification." If a particular cell displays an ID that shows it to be a foreign cell, the specific defenses "mark" the foreign cell for destruction. These marked cells are then destroyed by the nonspecific defenses. Like a bank's ID system, the specific defenses are very sensitive. They are able to detect the presence of a single foreign cell among the billions of cells in the body and then to take steps to ensure its elimination.

Nonspecific defenses protect the body from any potential pathogens, using such mechanisms as barriers, killer chemicals or cells, and inflammation. Specific defenses, by contrast, are cells of the immune system that target a particular pathogen and systematically act to destroy it.

Nonspecific Defenses

Nonspecific defenses can be divided into two broad categories: (1) barriers and (2) mechanisms (including chemicals) that kill foreign cells (table 41.1). Nonspecific defenses all have one property in common: They respond to *any* microbial infection. They do not pause to inquire as to the invader's identity but swing into action immediately.

Barriers: A Layer of Protection Against Invasion

Barriers include skin, mucous membranes, and ciliated membranes. Skin and mucous membranes actually prevent microbes and viruses from invading the body. Ciliated membranes act as "sweepers" that remove any foreign cells that manage to breach the other barriers.

Figure 41.1 Flu epidemic of 1918 killed twenty-two million people worldwide in eighteen months.
With twenty-five million Americans infected during the influenza epidemic, it was hard to provide care for everyone. The Red Cross often worked around the clock.

Table 41.1   Nonspecific and Specific Defenses

Defense	Components	Function
Nonspecific Defenses		
Barriers	Skin, mucous membranes, ciliated membranes	Provide a physical barrier against invading microbes and viruses
Chemical defenses	Lysozyme	Found on skin and in cells of tissues; lyses (ruptures) foreign cells
	Histamine	Dilates blood vessels; attracts white blood cells to site of injury
Cells	Macrophages	Engulf microbes
	Neutrophils	Engulf microbes; release chemicals that kill foreign cell and itself
	Natural killer cells	Punch hole in foreign cell, causing it to lyse
Complement system	Proteins	Form a membrane attack complex that punches a hole in foreign cell, causing it to lyse
Inflammatory response	Histamines, prostaglandins	Dilate blood vessels and increase their permeability; attract white blood cells to site of injury
	Macrophages, neutrophils	Engulf microbes
	Monocytes	Release chemicals that kill microbes
Temperature response	Pyrogens	Raise internal body temperature, which stimulates phagocytosis and inhibits microbe growth
Specific Defenses		
Cell-mediated response	Helper T-cells	Initiate response; activate cytotoxic T-cells
	Cytotoxic T-cells	Destroy infected cells
	Memory cells	Linger in bloodstream to attack same pathogen if second infection occurs
Humoral (antibody) response	Helper T-cells	Initiate response; activate B-cells
	B-cells	Bind to antigen; proliferate
	Plasma cells	Are secreting B-cells; secrete antibody protein
Memory cells		Linger in bloodstream to attack same pathogen if second infection occurs

Skin

Skin is the outermost layer of the vertebrate body and provides the body's first defense against microbial invasion. It also keeps the body watertight so that excessive water is not lost to the air by evaporation. Skin is the largest organ of the vertebrate body, accounting for 15 percent of an adult human's total weight. Many other specialized cells are crammed in among skin cells: One square centimeter of human skin, about what a dime covers, contains two hundred nerve endings, ten hairs and muscles, one hundred sweat glands, fifteen oil glands, three blood vessels, twelve heat-sensing organs, two cold-sensing organs, and twenty-five pressure-sensing organs.

Vertebrate skin is composed of three layers: an outer **epidermis,** a lower **dermis,** and an underlying layer of **subcutaneous tissue** (figure 41.2).

Epidermis: Skin of the Vertebrate Body The epidermis of skin is from ten to thirty cells thick, about as thick as this page. The outer layer, called the stratum corneum, is the one seen when looking at, for example, an arm or face. Epidermal cells are continuously abraded, injured, and worn by friction and stress during the body's many activities. They also lose moisture and dry out. The body deals with this damage by continuously shedding stratum corneum cells and replacing them with new cells produced deep within the epidermis.

The cells of the innermost layer of epidermis, called the **stratum basal layer,** are among the most actively dividing cells of the vertebrate body. New cells migrate upward, forming keratin protein, which makes the skin tough, as they move. Each cell eventually arrives at the outer surface and takes its turn in the stratum corneum, ready to be shed and replaced by a newer cell. A cell normally lives in the stratum corneum for about a month. **Psoriasis,** familiar to some four million Americans as persistent dandruff, is a chronic skin disorder in which new cells reach the epidermal surface every three or four days, about seven times faster than normal.

Dermis: Support and Insulation The dermis of skin is from fifteen to forty times thicker than the epidermis and has a fine network of blood vessels passing through it. The thick dermis provides structural support for the epidermis and a matrix for the many nerve endings, muscles, and specialized cells residing within skin. The wrinkling that occurs with age takes place here (figure 41.3). The leather used to manufacture belts is derived from very thick animal dermis.

Subcutaneous Tissue: Shock Absorption The layer of subcutaneous tissue below the dermis is composed primarily of fat-rich cells. They act as shock absorbers and provide insulation, which conserves body heat. This tissue varies greatly in thickness in different parts of the body. The eyelids have none of it, whereas the buttocks and thighs may have a lot of it. The subcutaneous tissue of the skin on the soles of the feet may be a quarter of an inch thick or more.

Figure 41.2 Human skin.

Human skin contains many structures that perform a variety of functions. One of these functions is to provide a physical barrier that protects the body against microbial infection.

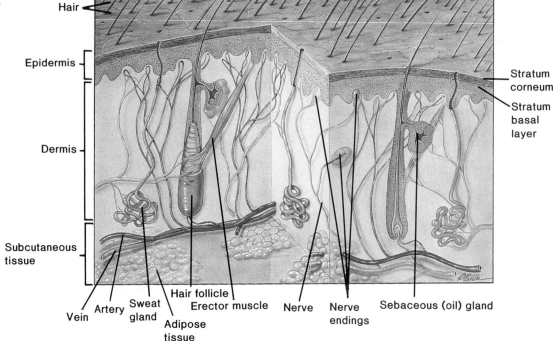

Hair
Epidermis
Dermis
Subcutaneous tissue
Stratum corneum
Stratum basal layer
Vein   Artery   Sweat gland   Hair follicle   Erector muscle   Nerve   Nerve endings   Sebaceous (oil) gland
Adipose tissue

(a)

(b)

Figure 41.3 The skin on the face ages more dramatically than skin elsewhere on the body.

(a) *At nineteen, this woman's face appears youthful and smooth.* (b) *Forty years later, her production of skin oil is much less; her face appears less smooth and elastic and begins to exhibit wrinkles.*

The Battle on the Skin Surface The skin not only defends the body by providing a nearly impenetrable barrier but also reinforces this defense with chemical weapons on the surface. The oil and sweat glands within the human epidermis, for example, lower the pH at the skin's surface to 3 to 5, an acid level that inhibits the growth of many microorganisms. Sweat also contains the enzyme **lysozyme,** which attacks and digests the cell walls of many bacteria.

Mucous Membranes

In addition to skin, mucous membranes located at various entry points in the body provide a physical barrier against invasion. The mucus that these membranes secrete entraps foreign invaders so that the invaders can be removed from the body. For instance, dust and other particles become entrapped in the mucus secreted from membranes lining the nasal cavity; a sneeze or cough then ejects these particles. Other mucous membranes line the urinary, reproductive, and gastrointestinal tracts—anywhere the body opens to the outside.

Ciliated Membranes

The vertebrate respiratory tract—specifically, the smaller bronchi and bronchioles—are lined with mucous membranes comprised of ciliated cells. These cells secrete a sticky mucus that traps microbes present in inhaled air before the microbes reach the warm, moist lungs, which are ideal microbial breeding grounds. The cilia on the cells lining these passages continually sweep the mucus upward, where it can be swallowed. Once swallowed, the microbes are destroyed in the stomach's highly acidic environment.

> *The barriers of the nonspecific defenses include skin, mucous membranes, and ciliated membranes.*

Chemical Defenses

The body secretes several types of chemicals that lyse (rupture) microbes. For instance, as already mentioned, the sweat of vertebrates contains the enzyme lysozyme, which attacks and kills microbes. Lysozyme is also found in tears, saliva, and nasal secretions.

Many bacteria are present in food, so a particularly strong mechanism is needed to protect the body at this point of entry. The hydrochloric acid found in the stomach and the protein-digesting enzymes found in the small intestine kill these invading bacteria.

Another chemical, called **histamine,** is released from basophils (a type of white blood cell, see chapter 40) when a tissue is injured. Histamine causes blood vessels around the injury site to dilate and also increases their permeability. These effects, in turn, attract more immune cells to the injury site so that healing can begin.

Cells That Kill Invading Microbes

Perhaps the most important of the vertebrate body's nonspecific defenses are the cells that patrol the bloodstream and attack invading microbes. Each of the three basic kinds of patrolling cells—macrophages, neutrophils, and natural killer cells—kills invading organisms differently:

1. *Foot soldiers.* **Macrophages** ("big eaters") kill bacteria one at a time by ingesting them, much as an amoeba ingests a food particle. Flowing cytoplasmic extensions stick to the invading bacterium and pull it inside the macrophage by means of endocytosis (figure 41.4). Once inside the macrophage, the bacterium is killed very efficiently: The membrane-lined vacuole containing the bacterium is fused with the enzyme, which is a powerful lysosome. Although some macrophages are fixed within particular organs, including the lungs, liver sinusoids, spleen, and brain, most macrophages patrol the body's byways, circulating in the blood, lymph, and interstitial fluid between cells.

2. *Kamikazes.* **Neutrophils** are white blood cells that ingest bacteria in the same way macrophages do. Both of these defending cells are **phagocytes** (cells that kill invading cells by engulfing them). Neutrophils, however, are kamikazes: They also release chemicals (identical to household bleach) to "neutralize" the entire area, killing any other bacteria in the neighborhood and themselves in the process. Macrophages, on the other hand, kill only one invading cell at a time, but live to keep on doing it.

3. *Internal security patrol.* **Natural killer cells** do not kill invading microbes but, rather, the cells infected by the microbes. They are particularly effective at detecting and attacking body cells with viral infections. Natural killer cells are not phagocytes; instead, they kill by puncturing

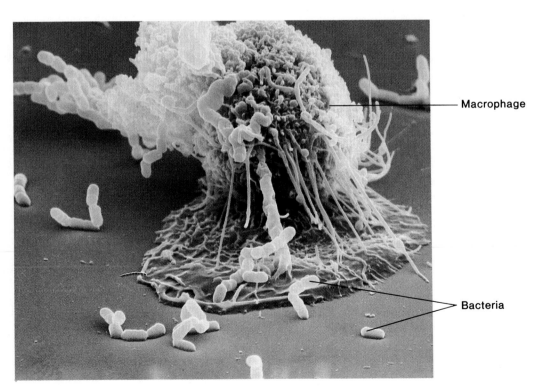

Figure 41.4 A macrophage in action.
In this scanning electron micrograph, a macrophage is "fishing" for a bacterium with long, sticky, cytoplasmic extensions. Bacterial cells unfortunate enough to come in contact with the extensions are drawn back to the macrophage and engulfed.

Macrophage

Bacteria

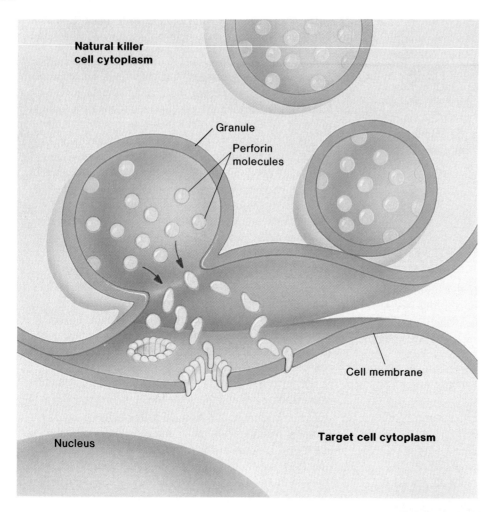

Natural killer cell cytoplasm

Granule

Perforin molecules

Cell membrane

Nucleus

Target cell cytoplasm

Figure 41.5 How natural killer cells kill their target cells.
Binding of the natural killer cell to the target cell initiates a chain of events. Granules loaded with perforin molecules move to the natural killer cell's outer cell membrane and disgorge their contents over the target cell. Perforin molecules insert themselves into the target cell membrane like the staves of a barrel, forming a pore that admits water and ruptures the target cell.

The "nonself" MHC marker on a foreign cell is called an **antigen.** Because an antigen so clearly identifies a nonself cell, foreign cells *themselves* are sometimes called antigens.

Proteins That Kill Invading Microbes: The Complement System

The vertebrate body employs a very effective chemical defense that "complements" its cellular defenses. The **complement system** consists of a battery of more than a dozen different proteins that normally circulate in the blood plasma in an inactive state. The defensive activity is triggered by the cell walls

the membrane of the target cell (figure 41.5). The puncture allows water to rush into the target cell, which swells and bursts (figure 41.6). Natural killer cells are also able to detect and kill cancer cells before the cancer cells develop into tumors. The vigilant surveillance by natural killer cells is one of the body's most potent defenses against cancer.

If these three kinds of cells turn on the body itself, as in **autoimmune diseases,** the results can be fatal. How, then, do these defending cells distinguish *self* (the body's own cells) from *nonself* (foreign cells)? The patrolling cells do not attack their own body because all body cells contain a cell-surface protein marker that identifies them (see chapter 5). This protein protrudes from the cell surface and is called a **MHC (major histocompatibility complex) marker.** The MHC marker acts like a "dogtag"—each person has a different version, although all the cells within a particular person contain the same one. Macrophages, neutrophils, and natural killer cells simply ignore any cells that have only "self" MHC markers. This type of defense is called "nonspecific" because these three types of defending cells will attack *any* cell that lacks the "self" MHC marker.

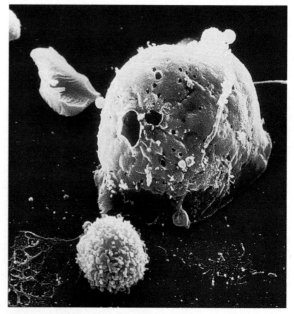

Figure 41.6 Death of a tumor cell.
A natural killer cell has attacked this cancer cell, punching a hole in its cell membrane. Water has rushed in, making it balloon out. Soon, it will burst.

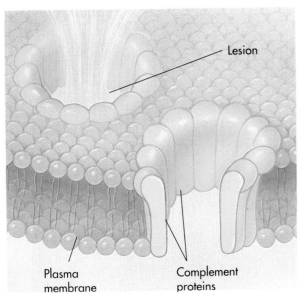

Figure 41.7 The complement system.
The complement system is composed of an array of proteins, which when activated, form a channel within a foreign cell's membrane that allows water to rush in, lysing the cell. In this way, complement proteins resemble the natural killer cells that also attack and lyse foreign cells.

of bacteria and fungi and also by the binding of antibodies to invading microbes (discussed later in this chapter). Upon detection of a bacterial cell wall, complement system proteins interact to form a **membrane attack complex (MAC)** like that produced by natural killer cells. The MAC inserts itself into the pathogen's cell membrane, forming a hole. Like a dagger through the heart, this wound is fatal to the invading cell: Water pulled in by osmosis causes the cell to swell and burst (figure 41.7).

Complement system proteins also act to amplify the effect of other body defenses. Some amplify the inflammatory response (discussed next) by stimulating histamine release, others attract phagocytes to infection sites, and still others coat invading microbes, roughening their surfaces so that macrophages can more readily stick to them.

Other proteins that play a key role in body defenses are **interferons.** Released by virus-infected cells, they diffuse out to other cells and inhibit viruses' ability to infect those cells. As discussed later in the chapter, their principal role is to sound the alert for the immune system.

The Inflammatory Response

One of the most generalized nonspecific responses to infection is the **inflammatory response,** which is initiated by infected or injured immune cells called **mast cells.** These infected or injured mast cells release chemical alarm signals, most notably histamines (discussed earlier in the chapter) and prostaglandins (discussed in chapter 37). These alarm signals promote the dilation of blood vessels around the injured area, which both

increases the blood flow to the site of infection or injury and, by stretching their thin walls, makes the capillaries more permeable. This is what produces the redness and swelling so often associated with infection. The larger, leakier capillaries promote the migration of phagocytes (macrophages and neutrophils) from the blood to the interstitial fluid surrounding cells, where the phagocytes can engulf bacteria. Neutrophils arrive first, spilling out chemicals that kill the bacteria in the vicinity (as well as tissue cells and themselves), followed by macrophages that clean up the remains of all the dead cells (figure 41.8). This counterattack by phagocytes can take a considerable toll; the pus associated with some infections is a mixture of dead or dying neutrophils, broken down tissue cells, and dead microbes.

The Temperature Response

When macrophages encounter invading microbes, they release chemical substances called **pyrogens** (from the Greek *pyr,* meaning "fire"), which pass through the bloodstream to the brain. When the pyrogens reach the cluster of neurons in the hypothalamus that serves as the body's thermostat, they act to boost the body's temperature several degrees above the normal value of 37 degrees Celsius (98.6 degrees Fahrenheit). The higher-than-normal temperature that results is called a **fever.** Fever contributes to the body's defense by stimulating phagocytosis, by inhibiting microbial growth, and by causing the body to reduce blood levels of iron, which bacteria need in large amounts to grow. Very high fevers, however, are dangerous because excessive heat may inactivate critical enzymes. In general, temperatures greater than 103 degrees Fahrenheit are considered dangerous; those greater than 105 degrees Fahrenheit are often fatal.

> *In addition to barriers, the nonspecific defenses include cells that kill invading microbes, the complement system, the inflammatory response, and the temperature response.*

Only occasionally are bacteria or viruses able to overwhelm the body's nonspecific defenses. When this happens, a second line of defense—the specific defenses of the immune system—take over. Unlike other defenses, the immune system remembers previous encounters with potential invaders and is ready for them if they reappear.

Specific Defenses: The Immune System

Few pass through childhood without being infected by a microbe. Measles, chickenpox, mumps—these are rites of passage, childhood illnesses that most people experience before their teens *and never catch again.* Once a person has had measles, he or she is immune. This immunity to childhood diseases is provided by the specific defenses of the immune system, the

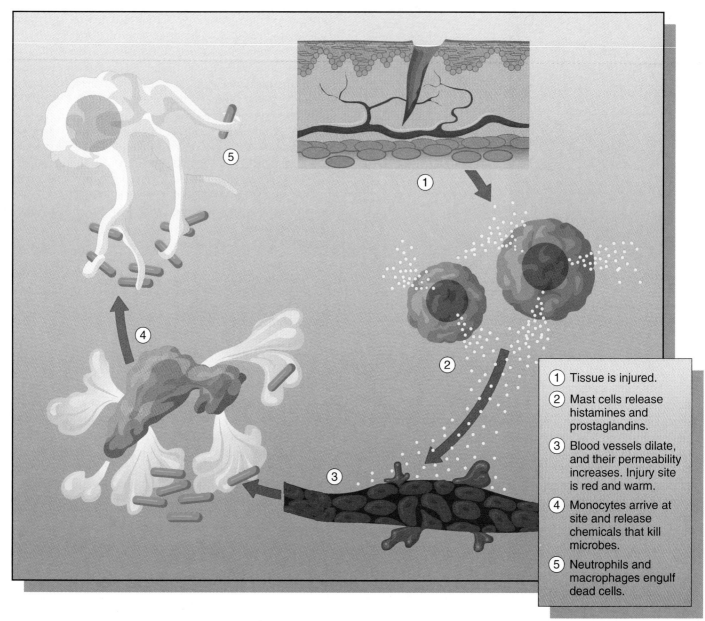

Figure 41.8 The sequence of events of the inflammatory response.

This response can be likened to a cleaning procedure in which bacteria are first killed and then are "mopped up" by phagocytes (neutrophils and macrophages).

1. Tissue is injured.
2. Mast cells release histamines and prostaglandins.
3. Blood vessels dilate, and their permeability increases. Injury site is red and warm.
4. Monocytes arrive at site and release chemicals that kill microbes.
5. Neutrophils and macrophages engulf dead cells.

body's most powerful means of resisting infection. The specific defenses are the backbone of health, protecting the individual not only from childhood diseases, but also from many other far more serious diseases. It is only in the last few years, as the result of an explosion of interest and knowledge, that biologists have begun to get a clear idea of how these defenses work.

Jenner's Discovery of the Immune Response

In 1796, an English country doctor named Edward Jenner conducted an experiment that marked the beginning of the study of immunology. Smallpox was a common and deadly disease in those days, and only those who had previously had the disease and survived it were immune from the infection—except, Jenner observed, milkmaids. Milkmaids who had caught another, much milder form of "the pox," called cowpox (it was caught by people who worked with cows), rarely caught smallpox. It was as if they had already had the disease. Jenner set out to test the idea that cowpox conferred protection against smallpox. He deliberately infected people with material that induced cowpox, after which many of these individuals were immune to smallpox, just as Jenner had predicted (figure 41.9).

Jenner's work demonstrated that the human body can effectively protect itself against disease when it is able to make

Figure 41.9 Edward Jenner and vaccination.
This famous painting shows Edward Jenner inoculating patients with cowpox in the 1790s and thus protecting them from smallpox. The underlying principles of vaccination were not understood until more than a century later.

suitable preparations. Smallpox is now known to be caused by a virus called variola; cowpox is caused by a different, although similar, virus. Jenner's patients injected with cowpox virus mounted a defense against the cowpox infection, a defense that was also effective against later infection by the similar smallpox virus.

Jenner's procedure of injecting a harmless microbe into a person or animal to confer resistance to a dangerous one is called **vaccination.** Modern attempts to develop resistance to herpes and other diseases are focusing on the virus **vaccinia,** which is related to the cowpox virus used by Jenner. Such attempts are employing genetic engineering methods (see chapter 14) to incorporate genes encoding the protein-polysaccharide coats of the other viruses into the chromosome of vaccinia. Since vaccinia does not cause disease, the body is exposed to the protein coats of the other viruses in a harmless way, which enables the body to build up resistance to them.

Pasteur's Cholera Experiment

It was a long time before people learned how one microbe can confer resistance to another, however. Further important information was added almost a century after Jenner by Louis Pasteur of France. In 1879, Pasteur was studying fowl cholera, a serious disease in chickens that is now known to be caused by a bacterium. From diseased chickens, Pasteur could isolate a culture of bacteria that would elicit the disease if injected into other healthy birds. One day, Pasteur accidentally left his bacterial culture out on a shelf at the end of the day and went on vacation. Two weeks later, he returned and injected this extract

into healthy birds. The extract had been weakened; the injected birds became only slightly ill and then recovered. Surprisingly, however, the vaccinated birds stayed healthy even if injected with massive doses of active fowl cholera bacteria, whereas control chickens receiving the same injections all died. Clearly, something about the bacteria could elicit immunity, if only the bacteria did not kill the bird first.

Just how is this immunity conferred? What happened in Pasteur's experiments is that the weakened cholera extract marshaled the chickens' specific immune defenses. The injected antigen was recognized as a "nonself" cell by the chickens' immune systems, and in response, the chickens' immune systems produced proteins called **antibodies.** Individual antibodies are made to the exact specifications of all the possible antigens that can invade the body. The antibodies produced in response to the cholera antigen performed two functions: They marked the antigen for destruction by the nonspecific defenses and then remained in the bloodstream to protect the chickens against future infection *by the same antigen.* Thus, the antibodies were able to recognize any future cholera invaders, weakened or normal, and to prevent them from causing disease (figure 41.10).

The specific defenses are aptly named because they are directed against specific antigens. There are two kinds of specific defenses: the cell-mediated immune response and the humoral immune response (see table 41.1). Both are examined in detail later in the chapter. First, however, is a discussion of the types of cells active in both responses and a look at how these responses are roused—how the alarm is sounded—in the body when a pathogen invades.

An immune response takes place when foreign proteins, called antigens, cause the production of other proteins, called antibodies, which mark the antigens for destruction by the nonspecific defenses and recognize any antigens of the same sort with which they may come in contact in the future.

Cells of the Immune System

The human immune system is neither localized in one place in the body nor controlled by any central organ, such as the brain. Rather, it is composed of a host of individual cells, an army of defenders that rush to an infection site to combat invading microorganisms. These cells—the white blood cells, or leukocytes, mentioned in chapter 40—are produced in the bone marrow and circulate in blood and lymph. Of the one hundred trillion cells in an adult human being, two in every hundred, or two trillion (2×10^{12}), are white blood cells. Although not bound together, the body's white blood cells exchange information and act in concert as a functional, integrated system. They are found not only in blood and lymph, but also in lymph nodes, the spleen, the liver, the thymus, and bone marrow.

White blood cells are larger than the red blood cells that ferry oxygen to the body's tissues, but they are formed from

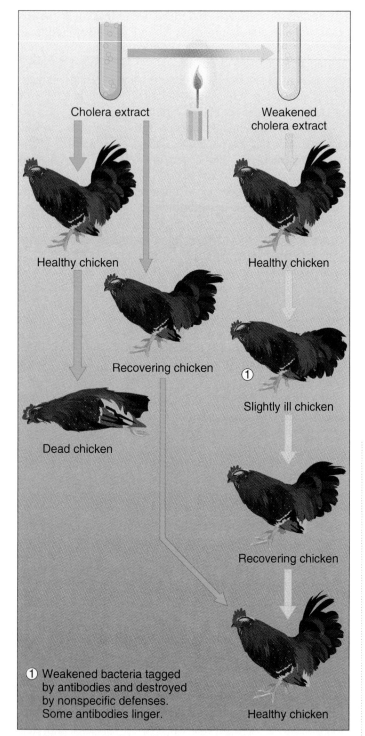

Figure 41.10 Pasteur's cholera experiment.

Pasteur's discovery of the antibody immune response was the result of a lucky accident: He forgot that he left an extract of cholera on a laboratory shelf before going on vacation. The weakened cholera extract that he injected his chickens with upon his return elicited an immune response that protected the chickens from infection in subsequent exposures to "normal" cholera extracts.

Cholera extract

Weakened cholera extract

Healthy chicken

Healthy chicken

Recovering chicken

① Slightly ill chicken

Dead chicken

Recovering chicken

① Weakened bacteria tagged by antibodies and destroyed by nonspecific defenses. Some antibodies linger.

Healthy chicken

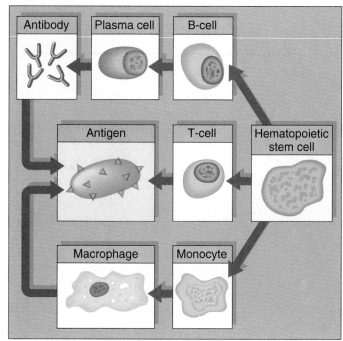

Antibody   Plasma cell   B-cell

Antigen   T-cell   Hematopoietic stem cell

Macrophage   Monocyte

Figure 41.11 Players in the immune response.

Three different kinds of cells mount attacks against the antigen-studded foreign cell in the center of the diagram. As described in the text, each line of attack actually involves the cooperative action of many cell types.

the same cells in bone marrow, called **hematopoietic stem cells** (figure 41.11). Unlike mature red blood cells, every white blood cell has a nucleus. Four kinds of white blood cells are involved in the immune system: phagocytes (including macrophages), **T-cells, B-cells,** and the nonspecific natural killer cells discussed earlier in the chapter. T- and B-cells are a subclass of leukocytes that are referred to as **lymphocytes.** Table 41.2 lists all the cells that participate in the specific defenses, as well as their functions.

The immune system is composed of white blood cells. Four principal classes are involved: phagocytes (including macrophages), natural killer cells, and two kinds of lympho-cytes (T-cells and B-cells).

Architecture of the Immune Defense: Specific Defenses

The specific defenses use a patrolling army of cells to attack and destroy invading microorganisms and to eliminate infected cells. Before the specific defenses can swing into action, however, they must be activated. This activation is akin to sounding an alarm. In response to this alarm, one or both specific defenses—the humoral or cell-mediated response—are marshaled against the invader.

Table 41.2   Cells of the Immune System

Cell Type	Function
Helper T-cells	Commanders of the immune responses, helper T-cells detect infection and sound the alarm, initiating both T-cell and B-cell responses.
Inducer T-cells	Not involved in the immediate response to infection, these cells mediate the maturation of T-cells that are involved.
Cytotoxic T-cells	Recruited by helper T-cells, these are the foot soldiers of the immune response, detecting and killing bacteria and infected body cells.
Suppressor T-cells	These cells dampen the activity of T- and B-cells, scaling back the defense after the infection has been checked.
B-cells	Precursors of plasma cells, these cells are specialized to recognize particular foreign antigens.
Plasma cells	Biochemical factories, these cells are devoted to the production of antibody directed against a particular foreign antigen.
Mast cells	These cells are initiators of the inflammatory response, which aids the arrival of white blood cells at an infection site.
Monocytes	These are the precursors of macrophages.
Macrophages	The body's first line of defense, they also serve as antigen-presenting cells to B-cells; later, they engulf antibody-covered cells.
Natural killer cells	These leukocytes recognize and kill foreign cells: Natural killer (NK) cells detect and kill a broad range of foreign cells; killer (K) cells attack only antibody-coated cells.

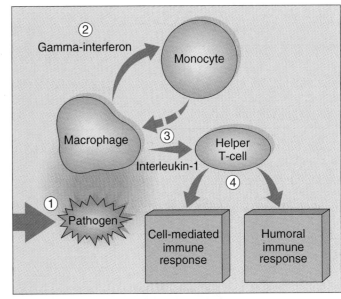

Figure 41.12 Sounding the alarm.
Invasion by a pathogen (a microbe or virus) initiates an alarm-sounding response that activates the cell-mediated and humoral responses. (1) The antigen-studded pathogen attracts macrophages. (2) The macrophage activity causes gamma-interferon to be released from the macrophages, which, in turn, causes monocytes to be converted into more macrophages. (3) This conversion releases interleukin-1 from the monocytes, which activates helper T-cells. (4) The helper T-cells, in turn, activate the cell-mediated and humoral immune responses.

Sounding the Alarm

Macrophages respond to a pathogen encounter by secreting a chemical alarm signal—proteins that initiate the immune response. Among these alarm signals are **gamma-interferon,** which activates monocytes to mature into macrophages, and **interleukin-1,** which activates a class of white blood cells called **helper T-cells.** Helper T-cells simultaneously initiate two different parallel immune responses: the cell-mediated immune response and the humoral immune response (figure 41.12).

The Cell-Mediated Immune Response

The activation of helper T-cells by interleukin-1 unleashes a chain of events known as the **cell-mediated immune response** (see table 41.1). In this response, special **cytotoxic T-cells** (*cytotoxic* means "cell poisoning") recognize and destroy infected body cells. Their mechanism of killing is the same as that of natural killer cells—they puncture the membranes of target cells. Helper T-cells initiate the response, activating both cytotoxic T-cells and other elements of the system. Once the pathogen is destroyed, **memory cells** linger in the bloodstream to attack the same pathogen if it invades in the future. Figure 41.13 provides a more detailed look at the

cell-mediated immune response. Cell-mediated immunity is essential for the destruction of host cells that have been infected by viruses or that have become abnormal (for example, in some cancers).

> *Macrophages alert the body to the presence of an intruder through the release of a chemical alarm that mobilizes the helper T-cells of the immune system. Helper T-cells then go on to activate both the cell-mediated and humoral immune responses. The cell-mediated response is immediate, attacking and disabling invading pathogens.*

The Humoral Immune Response

When a helper T-cell is stimulated to respond to a foreign antigen, it not only activates the T-cell-mediated immune response, as just described, but also simultaneously activates a second, longer-range defense, referred to as the **humoral,** or **antibody, immune response** (see table 41.1). The key player in this stage of defense against infection is another kind of lymphocyte, the B-cell. B-cells recognize invading pathogens much as T-cells do, but unlike T-cells, they do not attack pathogens directly. Rather, they mark them for destruction

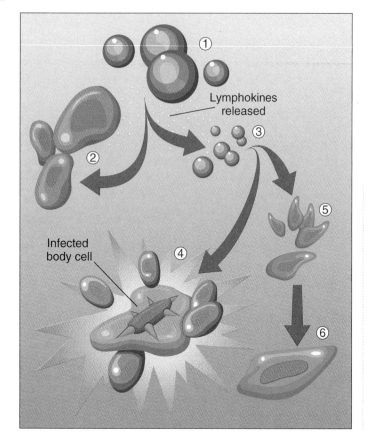

▭▭ *Figure 41.13 The cell-mediated immune response.*
(1) *When the helper T-cells are activated by the alarm-sounding response (diagrammed in figure 41.12), they release chemicals called lymphokines. (2) Lymphokines attract macrophages to the infection site. (3) Lymphokines also stimulate the multiplication of T-cells that have bound to the antigen. (4) Cytotoxic T-cells then swing into action and destroy the body cells that have been infected by the antigen. (5) As the infection subsides, suppressor T-cells multiply rapidly and "turn off" the immune response. (6) A small population of suppressor T-cells persists as memory cells that will be activated if the same antigen invades the body again.*

by the nonspecific body defenses. The humoral immune response is specialized to destroy invading bacteria and viruses and to inactivate foreign molecules that otherwise would be toxic.

How the Antibody Defense Works

Each B-cell has on its surface about one hundred thousand copies of a protein called an antibody that is designed to bind to foreign particles. Because each B-cell bears a different version of the antibody on its surface, each B-cell is specialized to recognize a different foreign antigen. The body contains many different B-cells, so there is almost always at least one B-cell that will bind to the surface of *any* microbe. At the onset of an infection, the antibody on the surface of one or more B-cells binds to antigens on the surface of the microbe. This induces the B-cell to proliferate (figure 41.14).

After about five days and numerous cell divisions, each B-cell that was stimulated by the antigen to proliferate has produced a large clone of cells. Some of the proliferating B-cells then stop reproducing and dedicate all of their resources to producing and secreting more copies of the antibody protein that responded to the antigen. The secreting B-cells are then called **plasma cells.** Plasma B-cells live only a few days but secrete a great deal of antibody protein during that time. One cell will typically secrete more than two thousand molecules per second (figure 41.15). Antibodies constitute about 20 percent by weight of the total protein in blood plasma, forming a plasma protein fraction known as

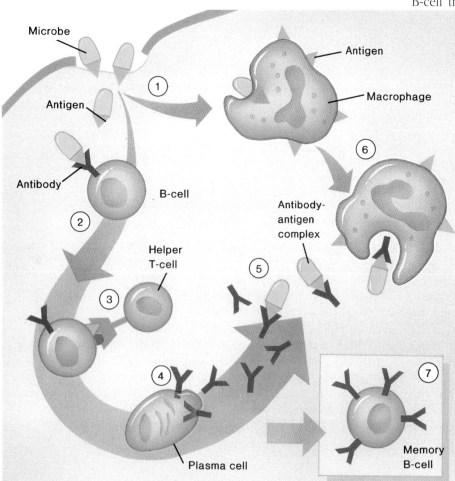

▭▭ *Figure 41.14 The humoral immune response.*

(1) *The invasion by a microbe triggers an immediate response by macrophages, and those macrophages that ingest bacterial cells display the bacterial antigens on their surfaces. (2) Other bacterial cells are bound by an antibody that protrudes from the surface of a B-cell. (3) The B-cell then interacts with a helper T-cell and is thus activated to divide. (4) Some of the B-cell progeny differentiate into plasma cells, which secrete antibody. (5) The bacteria are bound by free antibody molecules released by the plasma B-cells. (6) The antibody-bound bacteria are thus marked for destruction by macrophages. (7) Other B-cell progeny differentiate into memory cells, able to respond to the antigen at a future date.*

gamma-globulins. Antibodies bind to molecules on the surface of a virus or bacterium, thereby marking the microbe for destruction by macrophages, natural killer cells, or complement system proteins.

In the antibody defense, one or more B-cells recognize a foreign antigen and are stimulated to divide repeatedly, creating a large number of cells with the same antibody. Some of them proceed to produce large quantities of antibody molecules directed against the antigen. The antibodies bind to any antigen they encounter and mark for destruction cells or viruses bearing the antigen.

How Antibodies Recognize Antigens

The cell-surface receptors of lymphocytes (T-cell receptors and B-cell antibodies) can recognize specific antigens with great precision. Even single amino acid differences between proteins can often be discriminated, with a receptor recognizing one form and not the other. This high degree of precision is a necessary property of the immune system, since without it, foreign antigens often would not be identified. The differences between "self" and foreign ("nonself") molecules can be very subtle.

Antibody molecules (also called immunoglobulins) consist of four polypeptide chains. There are two identical short strands, called **light chains,** and two identical long strands, called **heavy chains** (figure 41.16*a*). The amino-acid sequences of the two kinds of chains suggest that the chains evolved from a single ancestral sequence of about 110 amino acids. Modern light chains contain 2 of these basic 110 amino acid units or domains; heavy chains contain 3, or in some cases, 4. The four chains are held together by disulfide (—S—S—) bonds, forming a Y-shaped molecule.

The specificity of antibodies resides in the two arms of the Y.

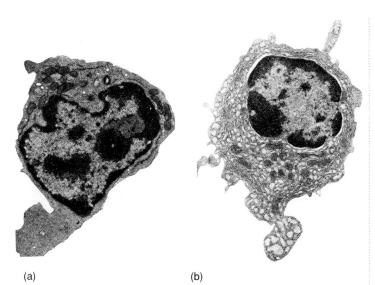

(a)   (b)

Figure 41.15 A B-cell and a plasma B-cell.
The B-cell (a) has a less prominent network of endoplasmic reticulum than the plasma B-cell. (b) Antibodies are made on the endoplasmic reticulum.

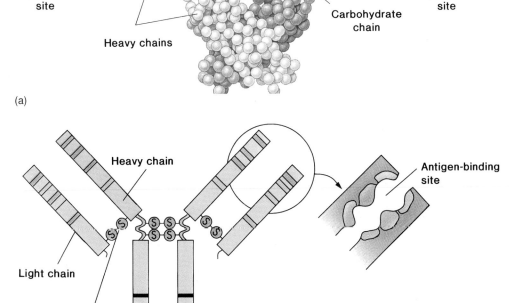

Light chains

Antigen-binding site

Heavy chains

Antigen-binding site

Carbohydrate chain

(a)

Heavy chain

Antigen-binding site

Light chain

Disulfide bond

(b)

Figure 41.16 The structure of an antibody molecule.
(a) *In this molecular model of an antibody, each amino acid is represented by a small sphere. The heavy chains are colored* blue, *the light chains,* red. *The four chains wind around one another to form a* Y *shape, with two identical antigen-binding sites at the arms of the* Y. (b) *In this schematic drawing of an antibody molecule, the antigen-binding sites on the arms are enhanced.*

Three small segments at the end of each arm come together to form a cleft that acts as the binding site for the antigen (figure 41.16*b*). Both arms always have exactly the same cleft. The specificity of the antibody molecule for an antigen depends on the precise shape of these clefts. An antigen fits into one of the clefts like a hand into a glove. Changes in the amino acid sequence of an antibody can alter the shape of the antibody's clefts and thereby change the antigen that can bind to that antibody, just as changing the size of a glove alters whose hand can fit it.

> *An antibody molecule recognizes a specific antigen because it possesses two clefts or depressions into which an antigen can fit, much as a substrate fits into an enzyme's active site. Changes in the amino-acid sequence at the position of these clefts alter the shape of the clefts and thus change the identity of the antigen able to fit into them.*

How the Immune System Responds to So Many Different Foreign Antigens

The vertebrate immune response is capable of recognizing as foreign practically any of the literally millions of different antigens. A human being is able to make between 10^6 and 10^9 different antibody molecules. How this is done has long been a puzzle, since research has demonstrated that only a few hundred antibody-encoding genes are present on vertebrate chromosomes, not millions. What process, then, is responsible for generating antibodies' great diversity? Two ideas were proposed:

1. The **instructional theory** suggested that the antigen elicited the appropriate antibody, like a shopper ordering a custom-made suit.

2. The **clonal theory** proposed that millions of different kinds of stem cells existed in the bone marrow and that an antigen caused those few encoding an appropriate antibody to proliferate, creating a clone of descendants expressing the appropriate antibody.

The clonal theory is now known to be correct. Within human bone marrow, the stem cells destined to form B-cells express an incredible diversity of antibody-encoding genes. Each cell encodes only one form of antibody, but almost every cell is different.

Vertebrates are able to generate millions of different stem cells, each producing a unique antibody, by rearranging parts of the antibody-encoding genes as each B-cell matures. Antibody genes do not exist as single sequences of nucleotides, as do the genes encoding all other proteins, but rather, are first *assembled* by joining together three or four DNA segments. Each segment, corresponding to a region of the antibody molecule, is encoded at a different site on the chromosome. These chromosomal sites are composed of a cluster of similar sequences, each sequence varying from others in its cluster by small degrees. When an antibody is assembled, one sequence is selected at random from each cluster, and the DNA sequences selected from the various clusters are brought together by DNA recombination to form a composite gene

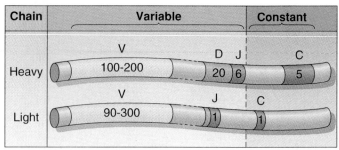

Chain	Variable				Constant
Heavy	V 100-200		D 20	J 6	C 5
Light	V 90-300			J 1	C 1

Figure 41.17 The antibody library.
Shown here are gene segments for the heavy and light chains of an antibody. The gene segments that specify both chains are divided into variable regions, in which there are many, many possible gene combinations, and constant regions, which do not change from one antibody to another. The variable region of the heavy chain contains a V segment, a D segment, and a J segment. The variable region of the light chain contains a V segment and J segment only. The diagram shows how many different copies of each segment are possible. Thus, one to two hundred segments of the V segment, twenty D segments, and six J segments are possible for a heavy chain. Because of this high number of possible different segments, millions of different segment combinations are possible.

(figure 41.17). The process is not unlike going into a large department store and picking at random one coat, one shirt or blouse, one pair of pants, one pair of socks, and one pair of shoes—few people would come out of the store wearing the same outfit.

Because a cell may end up with any heavy-chain gene and any light-chain gene during its maturation, the total number of different antibodies possible is staggering. It is:

Heavy (16,000 combinations) × Light (1,200 combinations)
= 19 million different possible antibodies

Every mature stem cell divides to produce a clone of descendant lymphocytes, and each cell of the clone carries the particular rearranged gene assembled earlier, when that stem cell was undergoing maturation. As a result, all the cells of a clone produce the specific immune receptor encoded by that stem cell, and no other. Because an adult vertebrate contains many millions of stem cells and each stem cell undergoes the maturation process independently, millions of different cells exist in each of us, each specializing in a different antibody.

> *Antibodies are encoded by genes assembled during stem-cell maturation by rearrangement of the DNA. Because each component is selected at random from many possibilities, a vast array of different antibodies is produced.*

The Nature of Immunity

When a particular B-cell is stimulated by an invading microbe to begin dividing, producing a clone of proliferating cells with the same antibody, all of these identical cells do not go on to become plasma cells. Instead, many persist as circulating lymphocytes called memory cells. These provide an accelerated response to any later encounter with the stimulating antigen because many cells can now respond to it, rather than a few.

Primary and Secondary Immune Responses

The first time a particular kind of pathogen invades the vertebrate body, only a few B-cells may have the antibody that can recognize it. This first encounter sets off what is called a **primary immune response.** It takes several days for these few cells to form a clone of cells that will produce antibodies. The *next time* the body is invaded by the same pathogen, however, the immune system is ready. As a result of the first infection, a small army of B-cells—the memory cells—can now recognize that pathogen. As mentioned earlier, only some of the clones of dividing B-cells became plasma cells in the first response; all the others are still there—a host of memory cells patrolling the bloodstream. Because each of these memory cells is well along the road to becoming a plasma cell, the **secondary immune response** is swifter, and because there are so many more of them, the response is much stronger. With each succeeding encounter, the bank of memory cells carrying that antibody becomes larger, so that the immune response grows even quicker and stronger (figure 41.18).

Memory cells can survive for several decades, which is why most people rarely contract mumps, chickenpox, or measles a second time. Memory cells are also why vaccination against measles, polio, and smallpox are effective against these diseases. The microbes causing these childhood diseases have a surface that changes little from year to year, so the same antibody is effective decades later. Other diseases, such as flu, are caused by microbes whose surface-specifying genes mutate rapidly. Thus, new strains appear every year or so that are not recognized by memory cells from previous infections. That is why immunity to flu lasts only a few years—the memory cells

persist, but the antigen continually shifts to new forms. The ability of flu to defeat human immune defenses by constantly changing its surface is but one of several strategies pathogens employ to defeat the vertebrate immune system.

Active and Passive Immunity

Two principal types of immunity are conferred upon the body. **Active immunity,** as the name suggests, depends on responses by a person's own immune system; it often confers lifelong immunity. **Passive immunity,** by contrast, does not involve the person's own immune system and is temporary. Both active and passive immunities can be naturally or artificially acquired.

An example of naturally acquired active immunity is the protection from disease acquired after becoming ill with a specific pathogen (such as chickenpox). After contracting chickenpox once in life, the individual is thereafter immune. Artificially acquired active immunity is essentially the same, except that, instead of actually contracting the disease, a physician administers a vaccine, with the same results. Naturally acquired passive immunity is conferred to a newborn by its mother. The baby is born temporarily immune to certain diseases and continues to receive helpful antibodies in mother's milk. This does not last for more than three or four months. Some individuals want a temporary "booster" to a failing or weakened immune system. This is where artificially acquired passive immunity comes into play: The patient is given an injection of gamma-globulin, antibodies produced by the immune system. The effect is temporary, and some wonder if it really serves any purpose at all.

> *The first time the body encounters a pathogen, either naturally (by getting sick) or artificially (by getting a vaccination), it becomes actively immunized and mounts a primary immune response. Any subsequent exposure to the same pathogen triggers a secondary immune response, faster and stronger than the primary response. Active immunity involves the individual's own immune system and is often lifelong, while passive immunity does not involve the person's own immune system and is temporary.*

Disorders of the Immune System

As is probably apparent from reading this chapter, disorders of the immune system can have serious consequences. When the body's defenses are not working properly, the body is vulnerable to infection by all sorts of different, dangerous pathogens. The most serious of all disorders that affect the immune system is the virus that causes AIDS, called HIV. But other, more common diseases, such as arthritis, are also disorders of the immune system. Even hay fever and other allergies are merely the result of an immune system that recognizes normally "harmless" foreign cells (such as pollens and dust) as pathogenic. These conditions are all immune system disorders, and all involve a malfunction in the way the immune system recognizes and destroys pathogens.

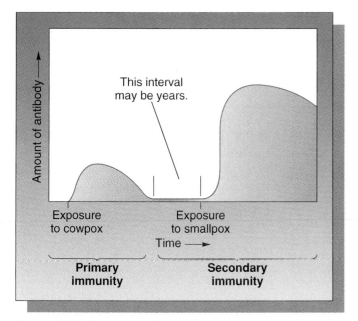

Figure 41.18 Primary and secondary immunity.
Primary immunity is conferred during the initial infection with a pathogen. Secondary immunity is conferred as a result of primary immunity. Exposure to cowpox confers primary immunity to cowpox and secondary immunity to both cowpox and smallpox.

Defeat of the Immune System: HIV

Several microbes have developed strategies, some of them quite successful, for defeating vertebrate immune defenses. As might be expected, these strategies are responsible for very serious diseases.

One of these strategies consists of a direct attack on the immune mechanism itself through helper T-cells. Helper T-cells are the key to the entire immune response, responsible for initiating proliferation of both T-cells and B-cells. Without helper T-cells, the immune system is unable to mount a response to *any* foreign antigen (figure 41.19).

AIDS (acquired immune deficiency syndrome) is a deadly disease for just this reason. The virus that causes AIDS, called **HIV (human immunodeficiency virus),** mounts a direct attack on all cells that have a specific protein called **CD4** on their surfaces. CD4 is found on a class of lymphocyte cells called T4-cells. This class includes helper T-cells, which is why HIV destroys the body's population of helper T-cells. CD4 receptors are also found on macrophage surfaces, and as a result, macrophages also become infected with HIV. Much of the HIV transmission from one individual to another is thought to occur within macrophages passed as part of body fluids.

HIV-infected T4-cells die, but only after releasing progeny viruses that infect other T4-cells, until the entire population of T4-cells is destroyed (figure 41.20). In a normal individual, T4-cells make up 60 to 80 percent of circulating T-cells; in AIDS patients, T4-cells often become too rare to detect.

T4-cell destruction by HIV infection wipes out the human immune defense. An effective immune response—either a T-cell–mediated immune response or a B-cell–mediated antibody immune response—is impossible without helper T-cells to

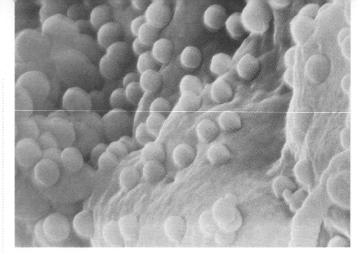

Figure 41.20 HIV, the cause of AIDS.
Large numbers of virus particles bud out of an infected lymphocyte. All of theses particles will be fully able to infect neighboring cells bearing CD4 receptors.

initiate the response. With no defense against infection, any of a variety of otherwise commonplace infections proves fatal. This is why AIDS is a particularly devastating disease.

> HIV destroys the immune system's ability to mount a defense against any infection. HIV attacks and destroys T4-cells, without which no immune response can be initiated.

Although HIV became a prominent cause of human disease only recently, possibly transmitted to humans from chimpanzees in Central Africa, AIDS is clearly one of the most serious diseases in human history. The fatality rate of AIDS is about 100 percent; few patients exhibiting AIDS symptoms survive more than a few years. The disease is *not* highly infectious; it is transmitted from one individual to another during the direct transfer of internal body fluids, typically in semen or vaginal fluid during sex and in blood during transfusions or by contaminated hypodermic needles. Not all individuals exposed to HIV (as judged by antibodies in their blood directed against HIV) have yet come down with the disease. There were about 1.5 million such individuals in the United States by 1993, and an estimated 15 million worldwide. Almost all of them will eventually contract AIDS. Most will die within two years of onset of the symptoms unless additional strategies for the treatment of AIDS are discovered first. Over 208,000 Americans have already died (figure 41.21). Efforts to develop a vaccine against HIV continue and are discussed in Sidelight 41.1.

Allergy

Although the human immune system provides very effective protection against viruses, bacteria, parasites, and other microorganisms, sometimes it does its job too well, mounting a major defense against a harmless antigen. Such immune responses are called **allergic reactions.** Hay

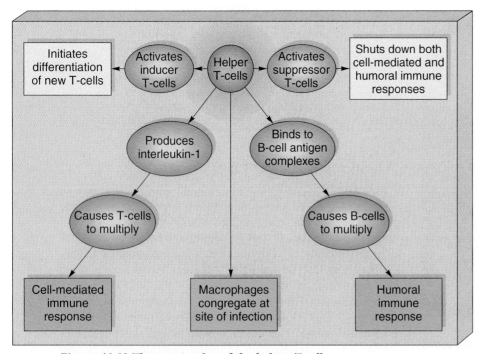

Figure 41.19 The many roles of the helper T-cell.
Targeting the helper T-cell causes the immune system to collapse. This is exactly what happens in HIV infection.

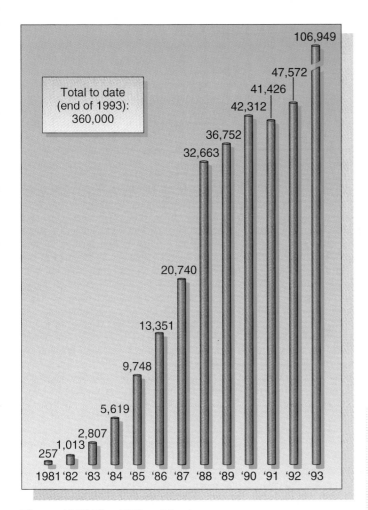

Figure 41.21 The AIDS epidemic.
These numbers show only the new *cases and deaths for each year represented on the histogram. The* total *number of U.S. AIDS cases from 1981 to the present is about 360,000; the total number of deaths is about 210,000.*

Source: Data from the U. S. Centers for Disease Control and Prevention.

Figure 41.22 Invisible housemate.
The house-dust mite Dermatophagoides *lives in dust particles. Many humans are allergic to the feces of this dust mite.*

fever—the sensitivity that many people exhibit to proteins released from plant pollen—is a familiar example of an allergy. In response to as few as twenty pollen grains per cubic meter, a sensitive person's immune system will swiftly mount a defense. Many other people are sensitive to proteins released from the feces of a minute house-dust mite called *Dermatophagoides*, which lives in the house dust present on mattresses and pillows and consumes the dead skin cells that everyone sheds in large quantities daily. Many people sensitive to feather pillows are in reality allergic to the mites who reside in the feathers (figure 41.22).

What makes an allergic reaction uncomfortable, and sometimes dangerous, is the involvement of antibodies with a kind of heavy chain called "E." Antibodies with class E heavy chains are typically attached to mast cells. The binding of antigen to these antibodies initiates an inflammatory response; histamines and prostaglandins are released from the mast cells, causing dilation of blood vessels and a host of other physiological changes: sneezing, runny noses, fever—all the symptoms of

hay fever. In some instances, when the body possesses substantial amounts of class E antibodies directed against an antigen, allergic reactions can be far more dangerous than hay fever, resulting in anaphylactic shock, in which swelling makes breathing difficult.

Not all antigens are **allergens** (initiators of strong immune responses). Nettle pollen, for example, is as abundant in the air as ragweed pollen, but few people are allergic to it. Also, not all people develop allergies; the sensitivity seems to run in families. Allergies appear to require both a particular kind of antigen and a high level of class E antibodies. The antigen must be able to bind simultaneously to two adjacent E antibodies on the surface of the mast cell to trigger the mast cell's inflammatory response, and only certain antigens are able to do this. The class E antibodies must be produced in large enough amounts that many mast cells will have antibody molecules spaced close to one another, rather than the few such cells typical of a normal immune response. Only certain people churn out these high levels of class E antibodies. This combination of appropriate antigen on the one hand and inappropriately high levels of particular class E antibodies on the other is what produces the allergic response.

Hay fever and other allergies are often treated by injecting sufferers with extracts of the antigen, a process called **desensitization.** Allergy shots work best for pollen allergies and for allergies to the venoms of bee and wasp stings; they are not effective against food or drug allergies. The strategy of desensitization is to produce high levels of normal (class G) antibodies in the bloodstream, so that when a particular antigen is encountered, the G antibodies mop it up before the antigen encounters E antibodies on mast cells. In reality, levels of

Sidelight 41.1
THE HUNT FOR AN AIDS VACCINE

The search for an AIDS vaccine is one of the most active areas of scientific research today. However, despite continued efforts and much funding, experts believe that an AIDS vaccine will not be available before the end of the 1990s, if then. What is it about HIV that makes this search for a vaccine so difficult?

A vaccine is a substance that is deliberately introduced into the body to provide an immune response. Jenner's smallpox vaccine, for example, consisted of the fluid extracted from the pustules of persons afflicted with cowpox. To vaccinate a person against smallpox, Jenner made a scratch in his patient's arm and dribbled the cowpox fluid into the scratch. Once inside the patient, the cowpox virus roused the immune response and caused antibodies to be made that attacked the cowpox virus. After the infection subsided, memory B-cells remained in the body to attack the same virus if the body again encountered it. These same memory B-cells would also attack the smallpox virus, which was similar in form to the less virulent cowpox virus.

The cowpox virus, called vaccinia, was an ideal vaccine to use because it was relatively safe and (usually) easy to administer. Infection with cowpox causes a short-term infection with some mild symptoms, none of which are life-threatening. Other vaccines, such as the vaccines used to protect against measles and polio, contain inactivated virus that cannot cause infection in the body. These vaccines elicit an immune response because of the antigens on the surface of the virus—the body is tricked into thinking that these viruses are alive, when in fact, they are not. Some vaccines, such as tetanus, require that "booster" shots be given to bolster the amount of memory B-cells in the blood. Thus, to develop a vaccine, scientists must understand the way that a particular virus or bacterium causes infection and be able to estimate the number of memory B-cells needed to prevent subsequent infection.

These two criteria have not, as yet, been met for AIDS. Researchers still do not understand why certain HIV-infected persons do not develop AIDS symptoms until as many as fifteen years later. They also do not understand why an AIDS patient who has survived for five years or more with relatively few symptoms can suddenly worsen and die. Even more puzzling are those individuals who are continually exposed to HIV but do not develop an infection—their blood shows no trace of HIV antibodies! Scientists are anxious to study the immune responses of these individuals for clues to an effective vaccine.

Perhaps the most disturbing aspect of HIV is its ability to rapidly mutate in the body. Like flu, HIV indulges in a form of antigen shifting. A single HIV-infected person may harbor many different strains of the virus, each strain resistant to different immune attacks. Researchers do not know how to effectively combat HIV's mutational ability, which is essential before a vaccine can even be tested.

Research efforts, however groping, are nonetheless utilizing some very creative ideas. Several groups of researchers have implanted the glycoprotein *gp160*, present on the viral envelope of HIV, into vaccinia and other viral carriers. The hope is that the protein will act as an effective antigen, eliciting an immune response by unleashing antibodies against HIV. Since only the viral envelope proteins are introduced into the body, there is no chance that the vaccinated individual will become infected with HIV. Although this approach has been repeatedly tested and retested, it has not yet yielded promising results (Sidelight figure 41.1).

Another approach to an AIDS vaccine tries to circumvent the body's antibody immune response in favor of the body's cell-mediated immune response. A major problem in developing an effective vaccine is sensitivity: The antibody response is roused in HIV infection only when large amounts of HIV are introduced. By then, the antibodies that are generated are too few to combat the large number of HIV particles that have replicated within the body's T-cells. Some researchers are convinced that low doses of live HIV will initiate the body's cell-mediated immune response, complete with memory cell formation. As this is inherently dangerous—it can result in complete HIV infection—investigators are trying inactivated HIV (key genes have been removed), which also elicits a cell-mediated immune response but cannot lead to AIDS.

Other vaccine attempts include: (1) genetic engineering techniques that attempt to introduce a gene into the body's T-cells that destroys HIV; (2) the use of HIV MHC markers as the vaccine antigen; and (3) the introduction of DNA complementary to HIV into T-cells, in an attempt to confuse HIV replication.

Even if a vaccine is eventually tested and approved, experts contend that the extremely virulent nature of HIV warrants the continued avoidance of high-risk behaviors. The vaccine will be no "magic bullet," but one prong in the arsenal that individuals can use to protect themselves against the deadly disease of AIDS.

---

circulating G antibodies do not seem to be highly correlated with successful desensitization, and why the procedure works as well as it does is not clear. A more ideal therapy would be to lower the amounts of class E antibodies produced during the immune response, an approach that is being actively investigated.

Autoimmune Diseases

Autoimmune diseases are conditions in which the immune system cannot tell the difference between "self" and "nonself" cells. As a result of this confusion, the immune system attacks "self" cells as if they were invading "nonself" cells. While experts are not sure about the exact mechanism that causes the immune system to become confused, autoimmune diseases seem to be related to suppressor T-cells that malfunction and can no longer distinguish a "self" cell as harmless.

Rheumatoid arthritis is a common autoimmune disease. In this condition, the immune system attacks the cells of tissues that reside in the body's joints, such as cartilage and the synovial membranes. This attack causes the tissues to become inflamed, resulting in swelling, pain, and loss of function. If the

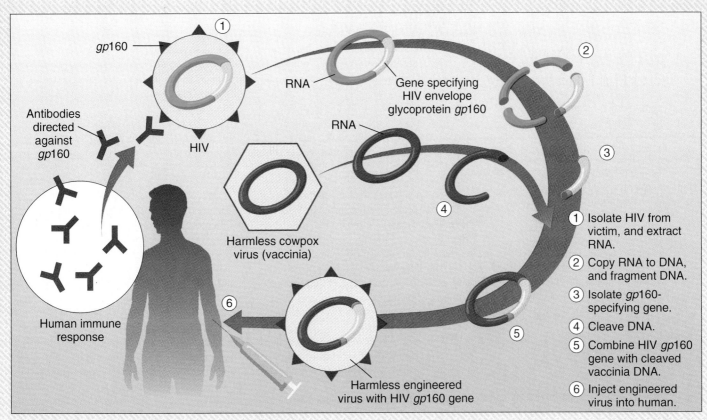

gp160

RNA

Gene specifying
HIV envelope
glycoprotein *gp*160

RNA

Antibodies
directed
against
*gp*160

HIV

Harmless cowpox
virus (vaccinia)

Human immune
response

Harmless engineered
virus with HIV *gp*160 gene

1. Isolate HIV from
   victim, and extract
   RNA.
2. Copy RNA to DNA,
   and fragment DNA.
3. Isolate *gp*160-
   specifying gene.
4. Cleave DNA.
5. Combine HIV *gp*160
   gene with cleaved
   vaccinia DNA.
6. Inject engineered
   virus into human.

Sidelight figure 41.1 How scientists are attempting to construct a vaccine for AIDS.

Of the several genes of the human immunodeficiency virus (HIV), one (gp160) is selected that encodes a surface feature of the virus. All of the other HIV genes are discarded. This one gene is not in itself harmful to humans; it is simply the shape of one of the HIV surface proteins. This one gene, or fragment of it, is inserted into the DNA of a harmless vaccinia cowpox virus, resulting in a harmless virus whose surface imitates the AIDS virus. Persons injected with the vaccinia virus do not become ill (the vaccinia virus is harmless), but they do develop antibodies directed against the infecting virus surface. Because the surface contains gp160 HIV proteins, the new antibodies serve to protect the infected person against any subsequent exposure to the HIV virus.

immune system completely destroys the cartilage, tough, fibrous tissue is sometimes deposited in the joint, making the joint immovable and malformed. This malformation can be seen in the finger distortion commonly associated with sufferers of rheumatoid arthritis. Currently, this condition is not curable.

Recently, scientists have proposed that diabetes mellitus might also be an autoimmune disease. In this case, the immune system might attack the pancreatic cells that make insulin. Although scientists are not sure about the exact mechanism, research in this area is ongoing.

> *Disorders of the immune system compromise the body's ability to fight off pathogens and can be devastating (AIDS). An autoimmune disease is one in which the cells of the immune system cannot tell the difference between certain "self" and "nonself" cells and so attack normal tissue.*

EVOLUTIONARY VIEWPOINT

In the ongoing coevolutionary relationship between the vertebrate immune system and the many pathogens and chemicals that assault it from the exterior world (some as microbes; others as chemicals that induce cancers), all parties have found ways to adapt, accelerating the warfare or accommodating the change. But human society has been changing the rules of engagement—in ways that favor the opposition. Increasing use of air travel and the emergence of megacities all over the globe facilitate the spread of disease as never before. This will severely test the abilities of the human immune system to adapt quickly enough to new challenges.

SUMMARY

1. The body has two lines of defense against pathogens: the nonspecific defenses and the specific defenses. Nonspecific defenses are directed against *any* invading pathogens. Specific defenses are tailored to specific pathogens. Nonspecific defenses include barriers, chemicals, killer cells, complement system proteins, the inflammatory response, and the temperature response.

2. Barrier nonspecific defenses include the skin, mucous membranes, and ciliated membranes. The skin provides a nearly impenetrable line of defense that is reinforced with chemical weapons on the surface. Mucous and ciliated membranes entrap invading microbes so that the invaders can be removed from the body.

3. Chemical nonspecific defenses include lysozyme, the hydrochloric acid found in the stomach, enzymes in the small intestine, and histamine.

4. The killer cells of the nonspecific defenses are macrophages, which ingest foreign cells; neutrophils, which ingest bacteria but also release chemicals that kill any other bacteria in the area and themselves as well; and natural killer cells, which punch holes in foreign cells, causing them to rupture.

5. Complement system proteins form a membrane attack complex (MAC) that ruptures the cell membranes of pathogens, causing the invading cells to swell and burst.

6. The inflammatory response, in which surface blood vessels expand their diameter, is the nonspecific defense response that produces the redness and swelling often associated with infection.

7. The temperature response results in a fever that stimulates phagocytosis, inhibits microbial growth, and reduces blood iron levels.

8. Edward Jenner initially discovered the specific defenses when he vaccinated patients against smallpox with the harmless cowpox virus.

9. The specific defenses are of two types: the cell-mediated immune response and the humoral (antibody) immune response. The cell-mediated response destroys foreign cells directly. The humoral response marks foreign cells with antibodies for destruction by the nonspecific defenses.

10. The immune system employs several types of white blood cells, including phagocytes, T-cells, B-cells, and natural killer cells.

11. The "alarm" response to pathogen invasion is initiated by macrophages. Helper T-cells activated by the macrophage alarm initiate the two specific defenses.

12. In the cell-mediated immune response, cytotoxic T-cells attack infected body cells.

13. The humoral immune response is a longer-range defense in which B-cells secrete antibodies that bind to circulating antigens, thereby marking these foreign cells for destruction.

14. The exquisite specificity of antibodies for particular antigens reflects the three-dimensional shape of a cleft in antibody molecules. Slight changes in the amino-acid sequence alter the cleft's shape and thus the identity of molecules able to fit into it.

15. Vertebrates can recognize many different foreign antigens because vertebrate bone marrow contains many different stem cells. During maturation, a stem cell assembles the two genes encoding its particular antibody by splicing together component parts, randomly selecting each part from a large library of possibilities.

16. The primary immune response is the initial response to pathogen invasion. The secondary immune response is the response to a second invasion by the same pathogen. In this response, memory B-cells from the first invasion attack and destroy the pathogen.

17. Active immunity depends on responses by a person's own immune system and is often lifelong. Passive immunity does not involve the person's own immune system and is temporary.

18. HIV, the virus that causes AIDS, attacks the helper T-cells of the immune system and prevents the immune system from mounting an attack against invading pathogens.

19. Allergies are the result of an immune system that recognizes harmless foreign cells (such as pollen and dust) as pathogenic.

20. Autoimmune diseases are the result of an immune system that fails to distinguish "self" from "nonself" cells. Consequently, the immune system attacks the body's own cells.

REVIEWING THE CHAPTER

1. What are the body's defenses against infection by microbes and viruses?

2. What are the body's nonspecific defenses against infection?

3. What are the three layers of the skin, and how do they function?

4. Why is the skin surface like a battlefield?

5. How do mucous membranes function?

6. Why are ciliated membranes important?

7. What are the body's chemical defenses?

8. Which three types of cells kill invading microbes?

9. How are proteins that kill invading microbes a complement system?

10. What is the inflammatory response, and how does it work?

11. What is the body's temperature response to invading microbes?

12. What are the body's specific defenses against infection?

13. How did Jenner and Pasteur become aware of the immune response?

14. What are the four types of white blood cells in the immune system?

15. How is the alarm sounded to initiate a response to invading pathogens?

16. What is the cell-mediated immune response?

17. What is the humoral immune response?

18. How does the antibody defense work?

19. How do antibodies recognize antigens?

20. How can the immune system respond to so many different foreign antigens?

21. What is the nature of immunity?

22. How do the primary and secondary immune responses differ?

23. What are active and passive immunity?

24. How does HIV defeat the immune system?

25. Why is an AIDS vaccine so elusive?

26. What are allergic reactions?

27. What is an autoimmune disease?

COMPLETING YOUR UNDERSTANDING

1. Which microbial infection is currently inflicting more than one million deaths in the human population annually?

2. What percent of the total weight of the adult human is skin?

3. What is psoriasis, and what causes it?

4. Why do the oil and sweat glands lower the pH to 3 to 5 at the skin's surface?

5. How can sneezing or coughing benefit the body?

6. How would cigarette smoking interfere with the human defense system?

7. What are the similarities and differences in function between macrophages and neutrophils?

8. Which of the following cells is one of the body's most potent defenses against cancer?
   a. Macrophages   c. Phagocytes
   b. Neutrophils   d. Natural killer cells

9. What is an antigen, and how is it recognized?

10. How do interferons play a key role in body defenses?

11. What exactly is the pus associated with some infections?

12. How can a fever be good for the body's defense, but too high of a fever be dangerous?

13. Why do we only rarely catch some childhood diseases, such as measles, chickenpox, and mumps, a second time?

14. What is a vaccination, and how does it assist with the body's defenses?

15. What is an antibody, and how does it differ from an antigen?

16. Which cells of the immune system are lymphocytes? [*two answers required*]

   a. Phagocytes
   b. Macrophages
   c. T-cells
   d. Natural killer cells
   e. B-cells
   f. Erythrocytes

17. What are the functions of gamma-interferon and interleukin-1?

18. What are memory cells, where are they found, what do they do, and how long do they survive?

19. How do gamma-globulins and immunoglobulins differ?

20. Why is bone marrow so critical to our defense system?

21. Why is it necessary to get flu shots periodically and to get vaccinated against measles, polio, and smallpox only once?

22. Which cell is the key to the entire immune system?
   a. Macrophages
   b. Helper T-cells
   c. Erythrocytes
   d. B-cells
   e. Inducer T-cells
   f. Suppressor T-cells

23. AIDS transmission from one individual to another may occur through [*four answers required*]
   a. casual skin contact.
   b. semen.
   c. blood transfusions.
   d. vaginal secretions.
   e. contaminated syringes.
   f. saliva.

24. What would scientists gain from studying individuals who are continually exposed to HIV but who do not develop an infection?

25. Are some humans really allergic to feather pillows, or is this a "cover-up" for something else?

26. Why is rheumatoid arthritis an autoimmune disease?

27. When Pasteur injected a two-week-old culture of fowl cholera bacteria into chickens, the vaccinated birds did not die. What would have happened if he had heat-killed the two-week-old culture before injecting it?
   a. All the vaccinated chickens would have died.
   b. The heat killing would not have changed the result.
   c. Some of the vaccinated chickens would have died.
   d. None of the above.

28. Why do you suppose there have been outbreaks of measles on college campuses in recent years?

*29. Why do you suppose that the human immune system encodes only about a hundred basic antibody-encoding genes, when much more diversity could be generated by encoding thousands of copies of each?

30. Suppose you are an avid swimmer but do most of your swimming in an indoor pool. Over the past month, you have noticed that, after leaving the pool, your sinuses become congested, and you can feel a swelling in your nasal passages. Upon awakening the next morning, you have difficulty breathing through your nose and have to resort to breathing mostly through your mouth. By late afternoon, however, your breathing returns to normal, and the overall reaction subsides. What might be responsible for this sudden change to your sinuses, and how can this sequence of events be explained by your immune system?

*For discussion

FOR FURTHER READING

Boon, T. 1993. Teaching the immune system to fight cancer. *Scientific American,* March, 82–89. A report on antigens that have been discovered on some cancer cells that can be used to provoke attack by the immune system. The genes that specify these antigens are being isolated. The hope is that the immune system can be prodded into responding to antigens they normally ignore.

Gallo, R. C., and L. Montagnier. 1987. AIDS in 1988. *Scientific American,* Oct., 40. A collaboration by the two investigators who established the cause of AIDS that describes how HIV was isolated and linked to AIDS.

Greene, W. 1993. AIDS and the immune system. *Scientific American,* Sept., 98–105. A review of where we stand in our fight against the HIV pathogen. Vaccines and effective treatments are still beyond reach, although new findings offer some encouragement.

Jaret, P. 1986. Our immune system: The wars within. *National Geographic,* June, 702–36. A very readable account of how the human immune system works, with striking photographs by Lennart Nilsson.

Johnson, H. et al. 1992. Superantigens in human disease. *Scientific American,* April, 92–101. An account of some proteins that arouse the immune system to a destructive frenzy, millions of times stronger a response than a normal antigen elicits. Such "superantigens" have been implicated in toxic shock syndrome and may help explain the lethality of AIDS.

Nossal, G. et al. 1993. The immune system—A special issue. *Scientific American,* Sept. Ten articles summarizing what is now known about the immune system, written by experts in a very rapidly advancing area of biology.

Smith, K. A. 1990. Interleukin-2. *Scientific American,* March, 50–57. The first hormone of the immune system to be recognized. Interleukin-2 helps the body to mount a defense against microorganisms by triggering the multiplication of only those cells that attack an invader.

Tonegawa, S. 1985. The molecules of the immune system. *Scientific American,* Oct., 122–31. A lucid account of what is currently known of the structure of the receptors on B-cells and T-cells.

EXPLORATIONS

Interactive Software:
Immune Response

This interactive exercise allows students to explore the way the human body defends itself from cancer and against invasion by viruses and microbes by varying the effectiveness of the body's defenses. The exercise presents a cross section of the human body, with a variety of white blood cells circulating, whose numbers change in response to an infection. The "infection" may be a virus (HIV), a microbe (*Pneumocystis carinii*), or a cancer (Kaposi's sarcoma), each of which elicits a different immune system response.

Students can investigate how the defense works by altering the numbers of particular cell types; raising or lowering the number of macrophages, helper T-cells, memory T-cells; etc. When changes mimic those produced in an HIV infection, there is drastic damage to the immune system.

Questions to Explore:

1. Why are there *two* immune responses?

2. What happens if helper T-cells are removed?

3. What is the effect of reducing the macrophage population?

4. Is long-term immunity to past infections, like mumps, destroyed by HIV infection?

WATER BALANCE AND EXCRETION

An African elephant and its offspring lounge peacefully in the water. Water cycles through their bodies in enormous quantities each day.

FOR REVIEW

Here are some important terms and concepts that have been discussed in previous chapters and that you will encounter again in this chapter. Review them before proceeding if necessary.

Sodium chloride (*chapter 2*)
Diffusion (*chapter 5*)
Active transport (*chapter 5*)
Hypotonic and hypertonic solutions (*chapter 5*)
Hormone regulation of water retention (*chapters 33 and 37*)
Countercurrent exchange (*chapter 39*)

The first vertebrates evolved in water, and the physiology of all vertebrates still reflects this origin (figure 42.1). Approximately two-thirds of every vertebrate's body is water, and if water amounts fall much lower than this, the animal dies. Suspended within the water are wastes that the body is in the process of eliminating. This chapter discusses the various strategies vertebrates employ to rid their bodies of wastes and to regulate water intake and loss.

Excretion

The average person who lives seventy years will, in the course of his or her life, eat some 45,000 pounds of food (over 22 tons!) and drink over 7,250 gallons of fluid (enough to fill a tanker truck). What happens to all of this food and fluid? It is not just "burned up." Every atom in those tons of food and water still exists—if not in the body, then eliminated from it.

Figure 42.1 Water affects us all.
This African elephant is knee-deep in water and enjoys its tromp through the mud every bit as much as you might enjoy going to the beach. For him and us, water is both a necessity and a pleasure.

The body eliminates excess materials in two ways. One is respiration. Almost all the energy in food is found in carbon-hydrogen bonds, and when this energy has been extracted, left-over carbon and hydrogen atoms are combined with oxygen to form carbon dioxide and water. Most of the carbon dioxide is eliminated from the body through the lungs during exhalation.

All of the other kinds of atoms in food—for example, the nitrogen atoms in each amino acid of proteins and the sodium atoms in salt—eventually leave the body through **excretion.** Among the many kinds of substances that the body excretes are nitrogenous waste products, enzymes, acids, detergents, water, salts, and ions.

Nitrogenous Waste Products

Removal of amino groups from the amino acids of proteins produces the extremely toxic by-product ammonia (NH_3) (figure 42.2). Because even very low concentrations of ammonia can kill cells, it must be transported and excreted in very dilute solution. In freshwater fishes, which have, if anything, too much water, dilution presents no problem, and toxic ammonia is flushed out in highly diluted form. In saltwater fishes and terrestrial animals, however, water is precious, and excreting highly diluted ammonia is not practical.

Animals have developed three general solutions to the problem of ammonia excretion:

1. *Flushing.* Both freshwater and saltwater fishes carry out protein breakdown in the gills so that little ammonia actually enters the body. It is simply carried away in the water passing across the gills.

2. *Detoxification.* Most mammals and many other land animals solve the problem a different way. In the liver, pairs of amino groups are linked in a complicated series of chemical reactions to carbon dioxide to form the largely harmless molecule **urea,** which is far less toxic and so can be transported and excreted at far higher concentrations (see figure 42.2). Urea is carried by the bloodstream to the **kidneys,** a pair of bean-shaped, reddish-brown organs, each the size of a small fist, located in the lower back region (figure 42.3). The kidneys excrete urea as a principal component of **urine.** In this process, the kidneys receive a flow of about 2,000

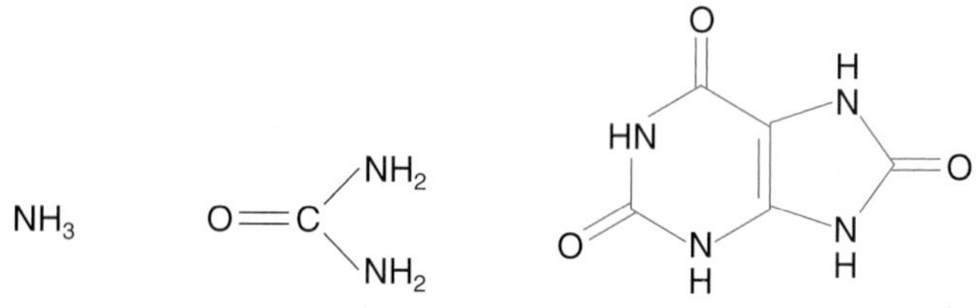

NH_3 **Ammonia**   **Urea**   **Uric acid**

Figure 42.2 Wastes.
When amino acids are metabolized, the immediate by-product is ammonia, which is quite toxic. Mammals convert ammonia to urea, which is less toxic. Animals that lay eggs convert it instead to uric acid, which is insoluble.

liters of blood each day—more than the volume of a car! Because the human body actually holds only 5.6 liters of blood, the blood obviously must go through the kidneys for cleaning many times during the day (about 350 times, or approximately once every four minutes).

3. *Insolubilization.* Birds and terrestrial reptiles face a special problem: Their eggs are encased within shells; thus, metabolic wastes build up as the embryo grows within the egg. The solution is to convert ammonia to **uric acid,** which is insoluble (see figure 42.2). The process, although lengthy and requiring considerable energy, produces a compound that crystallizes and precipitates as it becomes more concentrated. Adult birds and terrestrial reptiles carry out the same process, excreting the final result as a semisolid paste called guano.

The metabolic breakdown of protein produces ammonia as a by-product. Because ammonia is very toxic, animals have devised a variety of strategies for removing it. These include flushing, detoxification, and insolubilization.

Enzymes, Acids, and Detergents: Aids to Digestion

As discussed in chapter 38, the human pancreas secretes many enzymes into the stomach and intestines that aid in food digestion. Large quantities of hydrochloric acid (HCl) are also secreted into the stomach, where the acid rips proteins apart. All of this acid is neutralized by bicarbonate ions (HCO_3^-) secreted into the small intestine by the pancreas. The liver produces detergents called bile salts that it stores in the gallbladder and empties into the small intestine to break up fats. All of these enzymes, acids, and detergents are eventually eliminated from the body in the digestive stream.

Water

The average person drinks over a liter of fluid every day and would swell up like a balloon except that bodies lose an equal amount of water in sweat and urine. Sweat plays an important role in controlling body temperature, while urine is a vehicle for excreting urea and excess salts. Blood pressure is determined by the difference between the amount of fluid taken in and the amount excreted; the body constantly varies the amount of water it contributes to urine so as to keep blood pressure constant.

Salts and Ions

Proper operations of the many body systems requires that the blood's osmotic concentration—the concentration of solutes (salts and ions) dissolved within it—be kept within narrow bounds. After water consumption, the kidneys remove any excess water from the blood by increasing urine production. The body also monitors blood salt levels. When sodium ion levels

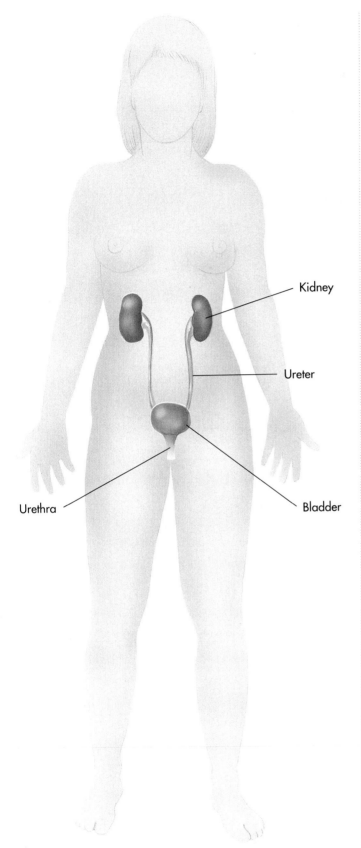

Figure 42.3 Urinary system of the human female.
The kidneys eliminate from the body a variety of potentially harmful substances that animals eat, drink, or inhale.

fall below normal, the brain directs the adrenal gland to send out hormones that cause the kidneys to extract more sodium ions from the water passing through them. On the other hand, if blood salt levels rise too high, hormone levels are decreased so that more salt is excreted via the urine. In the total absence of hormones, humans may excrete up to 25 grams of salt a day in their urine. The regulation of how much salt and water the blood contains is called **osmoregulation** and is discussed in more detail in the next section.

Osmoregulation

Plasma membranes are freely permeable to water but have very low permeability to salts and ions. This property of differential permeability forms the basis for many life processes, including the nerve conduction discussed in chapter 34.

If the concentration of salts and ions dissolved in the water surrounding a vertebrate's body were the same as that within the body, the differential permeability of the cells would present no problem; there would be no tendency for water to leave or enter the body. The osmotic pressure of the body fluids would be essentially the same as that of the surroundings. This is true of most marine invertebrates, but only of sharks and their relatives among the vertebrates. Such animals, called **osmoconformers,** vary the osmotic concentration of their body fluids as necessary to maintain an osmotic concentration that is close to that of the medium in which they are living.

Problems Faced by Osmoregulators

Among the aquatic vertebrates other than sharks, mechanisms have evolved for controlling the osmotic concentration of body fluids more precisely than other aquatic animals. These vertebrate **osmoregulators** maintain a constant internal solute concentration, regardless of their environment. While this has permitted osmoregulators to evolve complex patterns of internal metabolism, it requires constant internal water regulation.

Freshwater vertebrates must maintain much higher salt concentrations in their bodies than are present in the water surrounding them. In other words, their body fluids are **hypertonic** relative to their environment, and water tends to enter their bodies (figure 42.4*a*). They must therefore exclude water to prevent internal fluids from being diluted.

Marine vertebrates have only about one-third the osmotic concentration of the surrounding seawater in their body fluids. Their body fluids are therefore said to be **hypotonic** relative to the environment, and water tends to leave their bodies (figure 42.4*b*). For this reason, marine animals must retain water to prevent dehydration.

On land, surrounded by air, the bodies of vertebrates have a higher concentration of water than does the air surrounding them. They therefore tend to lose water to the air by evaporation. Amphibians, which live on land only part of the time, experience this to some degree, while all terrestrial reptiles, birds, and mammals must constantly conserve water to prevent dehydration. Figure 42.5 shows the ion concentrations in the body fluids of some representative vertebrates.

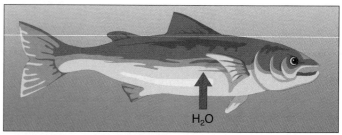

(a) Hypertonic fish in freshwater

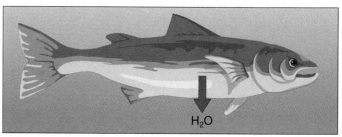

(b) Hypotonic fish in saltwater

Figure 42.4 Fish are osmoregulators.

In freshwater (a) *the fish body contains more salt than the surrounding water, and water diffuses into the body; in saltwater* (b) *the fish body contains less salt than surrounding water, and water diffuses out of the body. Freshwater and marine fish have developed a variety of measures to counter these tendencies.*

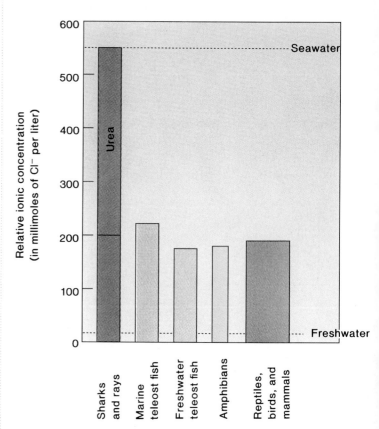

Figure 42.5 Ion concentration roughly similar in the bodies of different classes of vertebrates.

Sharks hold the concentration of solutes in their blood at about the level in seawater, or at a slightly higher level, by adding urea to their bloodstream. Terrestrial vertebrates have ion concentrations not unlike those of the fishes from which they evolved.

How Osmoregulation Is Achieved

Animals have evolved a variety of mechanisms to cope with osmoregulation problems, all of them based in one way or another on the animal's excretory system. In many animals, the removal of water or salts is coupled with the removal of metabolic wastes from the body. Simple organisms, such as many protists and sponges, employ contractile vacuoles for this purpose. Many freshwater invertebrates have **nephrid organs,** in which water and waste pass from the body across the membrane into a collecting organ, from which they are ultimately expelled to the outside through a pore. The membrane acts as a filter, retaining proteins and sugars within the body, while permitting water and dissolved waste products to leave.

Insects use a similar filtration system, with a significant improvement that helps them to guard against water loss. Insects' excretory organs, called **Malpighian tubules,** are tubular extensions of the digestive tract and branch off before the hindgut. Potassium (K⁺) ions are secreted into the tubules, causing body water and organic wastes to flow into them from the body's circulatory system because of the osmotic gradient. Blood cells and protein molecules are too large to pass across the membrane into the Malpighian tubules. Because the system of tubules empties into the hindgut, the water and potassium can be reabsorbed by the hindgut, and the insect excretes only small molecules and waste products. Malpighian tubules provide a very efficient means of water conservation (figure 42.6a, b).

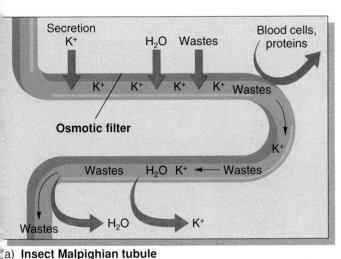

(a) **Insect Malpighian tubule**

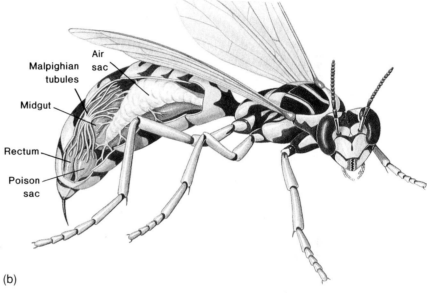

(b)

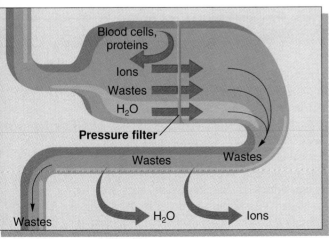

(c) **Vertebrate renal filtration**

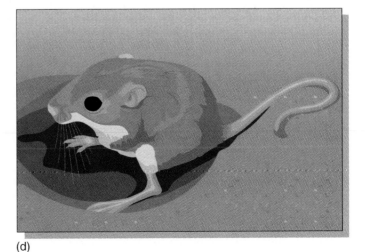

(d)

Figure 42.6 Two effective ways to conserve water.
(a,b) *Insects like this wasp add salt to Malpighian tubules to draw body fluid in, and then retrieve the water and useful ions;* (c,d) *Vertebrates like this kangaroo rat push the body fluid through a filter before retrieving water and other desirable molecules.*

Like insects, vertebrates use a strategy that couples water balance and salt concentration with waste excretion. Instead of relying on the secretion of salts into the excretory organ to establish an osmotic gradient, however, vertebrates employ pressure-driven filtration. Whereas insects use an osmotic gradient to *pull* the blood through the filter, vertebrates *push* blood through the filter, using the higher blood pressure that a closed circulatory system makes possible. Fluids are forced through a membrane that retains both proteins and large molecules within the blood vessel but allows the small molecules to exit. Water is then reabsorbed from the filtrate as it passes through a long tube (figure 42.6*c, d*).

> *Terrestrial animals require efficient means of water conservation. In eliminating metabolic wastes, both insects and vertebrates filter the blood to retain blood cells and protein. Then water is reabsorbed from the filtrate as the filtrate passes through a long tube.*

Like the osmotically collected waste fluid of insects, the filtered waste fluid, or urine, of vertebrates contains many small molecules that are of value to the organism, such as glucose, amino acids, and various salts or ions. Vertebrates have evolved a means of selectively reabsorbing these valuable small molecules without also reabsorbing the waste molecules. Selective reabsorption gives the vertebrates great flexibility, since the membranes of different groups of animals can and have evolved different active-transport channels and thus the ability to reabsorb different molecules. This flexibility is key to the ability of different vertebrates to function in many diverse environments: They can reabsorb small molecules that are especially valuable in their particular habitat and not reabsorb wastes. Among the vertebrates, the apparatus that performs filtration, reabsorption, and secretion is the kidney. It can function, with modifications, in freshwater, in seawater, and on land.

Structure of the Kidney

The kidney is a complex structure of roughly one million repeating elements called **nephrons,** each of which is composed of three elements (figure 42.7):

1. *A filter.* The filtration device at the top of each nephron tube is called **Bowman's capsule.** Within each capsule, an arteriole enters and splits into a fine network of vessels called a **glomerulus** (figure 42.8). The walls of these capillaries act as a filtration device. Blood pressure forces fluid through the capillary walls. These walls withhold proteins and other large molecules in the blood, while passing water and small molecules, such as glucose, sodium and other ions, and urea, the primary waste product of metabolism.

2. *A tube.* The Bowman's capsule is connected to a long, narrow tube, called a **renal tubule** in mammals. Between its proximal and distal arms, each mammalian renal tubule is folded into a hairpin loop called the **loop of Henle.** The loop of Henle is a reabsorption device.

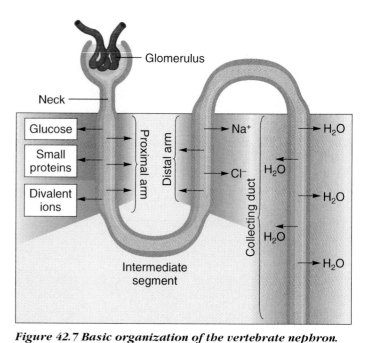

Figure 42.7 Basic organization of the vertebrate nephron.
The nephron tube of the freshwater fish is a basic design that has been retained in the kidneys of marine fishes and terrestrial vertebrates, which evolved later. Sugars, small proteins, and divalent ions such as Ca^{++} and PO_4 are recovered in the proximal arm; ions such as Na^+ and Cl^- are recovered in the distal arm; and water is recovered in the collecting duct. Mammals and birds achieve much greater water conservation by bending the nephron tube, producing what is known as the loop of Henle (not labeled here).

Figure 42.8 Bowman's capsules.
The spherical structures in this micrograph are Bowman's capsules. In each, a fine network of capillaries, the glomerulus, is connected to a nephron tube by an arteriole.

From R. G. Kessel and R. H. Kardon, *Tissues Organs: A Text-Atlas of Scanning Electron Microscopy.*

Like the small intestine, it extracts from the filtrate passing through the tube a variety of molecules useful to the body, such as glucose and many ions. Otherwise, these molecules would be lost in the urine.

3. *A duct.* The renal tubule empties into a large **collecting duct.** The collecting duct operates as a water conservation device, reclaiming water from the urine so that it is not lost to the body. Human urine is four times as concentrated as blood plasma—that is, the collecting ducts remove much of the water from the filtrate passing through the kidneys. Human kidneys achieve this remarkable degree of water conservation by a simple but superbly designed mechanism: The duct is bent back alongside the renal tubule and made permeable to urea. This greatly increases the local salt (urea) concentration in the tissue surrounding the tube, causing water in urine

to pass out of the tube by osmosis. The salty tissue sucks up water from the urine like blotting paper, passing it on to blood vessels that carry it out of the kidneys and back to the bloodstream.

The kidney is composed of many similar filtering units called nephrons. Blood pressure forces wastes in the blood through the filter into the tubular nephron, where they are then concentrated. Water is resorbed, leaving a very concentrated fluid waste, urine.

Evolution of the Vertebrate Kidney

All vertebrate kidneys have the same basic design, although there have been some modifications. The sections that follow discuss the structure and function of the kidney in the different major groups of vertebrates.

Freshwater Fishes

Kidneys are thought to have evolved first among the freshwater fishes. A freshwater fish drinks little and produces large amounts of urine. Because its body fluid is hypertonic, compared with the water in which it lives, water is not reabsorbed in the nephrons. The excess water that enters the fish's body passes instead through the nephron tubes to the bladder, from which it is eliminated as urine. The urine contains not only the excess water but also all the small molecules that were not reabsorbed while passing through the nephron tubes. Notable among these molecules is ammonia, the principal metabolic waste product of nitrogen metabolism (present in solution as ammonium ion NH_4^+). As discussed earlier, high concentrations of ammonia are toxic, but because the urine of freshwater fish contains so much water, the ammonia concentration is low enough not to harm the fish.

Marine Fishes

Although most groups of animals clearly seem to have evolved first in the sea, marine bony fishes probably evolved from freshwater ancestors, as was mentioned in chapter 18. In making the transition to the sea, they faced a significant new problem of water balance because their body fluids are hypotonic with respect to the water that surrounds them. That is, their body fluids are less concentrated osmotically than is seawater, and for this reason, water tends to leave their bodies. To compensate, marine fishes drink a lot of water, excrete salts instead of reabsorbing them, and reabsorb water. This places radically different demands on their kidneys than those faced by freshwater fishes, thus turning the tables on an organ that had originally evolved to eliminate water and reabsorb salts.

As a result, the kidneys of marine fishes have evolved important differences from their freshwater relatives. For example, they possess active ion transport channels that conduct ions with two or three charges—molecules that are particularly abundant in seawater—*out* of the body and *into* the nephron tube. In the sea, the water that the fish drinks is rich in ions, such as Ca^{++}, Mg^{++}, $SO_4^=$, and $PO_4^=$, all of which must be excreted. Because of the functioning of these ion transport channels, the direction of movement is reversed, compared with that found in the kidneys of freshwater ancestors.

Sharks

Except for one species of shark found in Lake Nicaragua, all sharks, rays, and their relatives live in the sea. Some of these members of the class Chondrichthyes have solved the osmotic problem posed by their environment in a different way than have the bony fishes. Instead of actively pumping ions out of their bodies through their kidneys, these fishes use their kidneys to reabsorb the metabolic waste product urea, creating and maintaining urea concentrations in their blood that are one hundred times as high as those that occur among the mammals. As a result, sharks and their relatives become isotonic with the surrounding sea. They have evolved enzymes and tissues that tolerate these high urea concentrations. Because they are isotonic with the water in which they swim, they avoid the problem of water loss that other marine fishes face (figure 42.9). Since sharks do not need to drink large amounts of seawater, their kidneys do not have to remove large amounts of divalent ions from their bodies.

Amphibians and Reptiles

The first terrestrial vertebrates were the amphibians. The amphibian kidney is identical to that of the freshwater fishes, the amphibians' ancestors. This is not surprising, since amphibians spend a significant portion of their time in freshwater, and when on land, they generally stay in wet places.

Figure 42.9 A great white shark.
Although the seawater in which it is swimming contains a far higher concentration of ions than does the shark's body, the shark avoids losing water osmotically by maintaining such a high concentration of urea that its body fluids are osmotically similar to the sea around it.

Reptiles, on the other hand, live in diverse habitats, many of them very dry. Reptiles that live mainly in freshwater, like some kinds of crocodiles and alligators, occupy a habitat similar to that of the freshwater fishes and amphibians and have similar kidneys. Marine reptiles, which consist of other crocodiles, some turtles, and a few lizards and snakes, possess kidneys similar to those of their freshwater relatives. They eliminate excess salts not by kidney excretion but rather by means of salt glands located near the nose or eye.

Terrestrial reptiles, which must conserve water to survive, reabsorb much of the water in kidney filtrate before the filtrate leaves the kidneys and so excrete a concentrated urine. This urine, however, cannot become any more concentrated than the blood plasma. Otherwise, the body water of reptiles would simply flow into the urine while it was in the kidneys.

In the relatively concentrated urine of most reptiles, nitrogenous waste is not excreted in the form of ammonia, but either as solid urea or as uric acid. These metabolic conversions take place in the liver.

Mammals and Birds

As mentioned earlier, the human body possesses two kidneys, each about the size of a small fist, located in the lower back region (see figure 42.3). These kidneys are much more efficient than those of reptiles. Mammalian, and to a lesser extent avian, kidneys can remove far more water from the glomerular filtrate than can the kidneys of reptiles and amphibians. Human urine is four times as concentrated as blood plasma. Some desert mammals achieve even greater efficiency: A camel's urine is eight times as concentrated as its plasma, a gerbil's is fourteen times as concentrated, and some desert rats and mice have urine that is more than twenty times as concentrated as their blood plasma.

Mammals and birds achieve this remarkable degree of water conservation by using a simple but superbly designed mechanism: They greatly increase the local salt concentration in the tissue through which the nephron tube passes and then use this osmotic gradient to draw the water out of the tube.

> *The vertebrate kidney has evolved to cope with its osmotic environment. Kidneys that function in hypertonic (freshwater) environments generate copious amounts of dilute urine. Those that function in hypotonic or marine environments either excrete salts and concentrate water or concentrate wastes to make the body isotonic to the surroundings.*

How the Mammalian Kidney Works

Remarkably, mammals and birds have brought about this major improvement in efficiency by a very simple change—a bend in the nephron tube. To review: A single mammalian kidney contains about a million nephrons. Each of these is composed of a glomerulus that is connected to a nephron tube, which is called a renal tubule in mammals. Between its proximal and distal segments, each renal tubule is folded into a hairpin loop called the loop of Henle (figure 42.10).

The kidney uses the hairpin loop of Henle to set up a countercurrent flow. Just as in the gills of a fish, as discussed in chapter 39, but with water being absorbed instead of oxygen, countercurrent flow enables water to be reabsorbed with high efficiency. In general, the longer the hairpin loop, the more

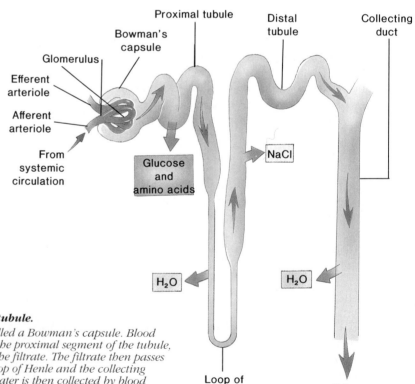

Figure 42.10 Organization of a mammalian renal tubule.
The glomerulus is enclosed within a filtration device called a Bowman's capsule. Blood pressure forces liquid through the glomerulus and into the proximal segment of the tubule, where glucose and small proteins are reabsorbed from the filtrate. The filtrate then passes through a double-loop arrangement consisting of the loop of Henle and the collecting duct, which act to remove water from the filtrate. The water is then collected by blood vessels and transported out of the kidney to the systemic (body) circulation.

water can be reabsorbed. Animals such as desert rodents that have highly concentrated urine have exceptionally long loops of Henle.

The countercurrent process involves the passage of *two* solutes across the membrane of the loop: salt (NaCl) and urea, the waste product of nitrogen metabolism. It has long been known that animals fed high-protein diets, yielding large amounts of urea as waste products, can concentrate their urine better than animals excreting lower amounts of urea, a clue that urea plays a pivotal role in kidney function. Figure 42.11 illustrates the five-stage flow of materials:

1. Driven by the blood pressure, which within the glomerulus is about 60 millimeters of mercury, small molecules are pushed across the thin walls of the glomerulus to the inside of the Bowman's capsule. This is a pressure filtration process because blood cells and large molecules like proteins cannot pass through. As a result, the bloodstream that enters the glomerulus is divided into two paths: nonfilterable blood components that are retained and leave the glomerulus in the bloodstream and filterable components that pass across and leave the glomerulus in the urine stream. This filterable stream, called the **glomerular filtrate,** contains water, nitrogenous wastes (principally urea), nutrients (principally glucose and amino acids), and a variety of ions. The glomerular filtrate passes down the descending arm of the loop of Henle. The walls of this portion of the tubule are impermeable to either salts or urea but are freely permeable to water. Because (for reasons described soon) the surrounding tissue has a high osmotic concentration of urea, water passes out of the descending arm by osmosis, leaving behind a more concentrated filtrate.

2. At the turn in the loop of Henle, the walls of the tubule become permeable to salts but much less permeable to water. As the concentrated filtrate passes up the ascending arm, salt diffuses into the surrounding tissue. (The surrounding tissue, although it has much urea, does not contain as much salt as the concentrated filtrate.) This makes salt more concentrated at the bottom of the loop and in the surrounding tissues.

3. Higher in the ascending arm are active-transport channels that pump out even more salt. This active removal of salt from the ascending arm encourages even more water to diffuse outward from the filtrate. Left behind in the filtrate is the urea that initially passed through the glomerulus as nitrogenous waste. Eventually, the urea concentration becomes very high in the tubule.

4. The tubule then empties into a collecting duct that passes back through the tissue. Unlike the tubule, the lower portions of the collecting duct are permeable to urea. During this final passage, the concentrated urea in the filtrate diffuses out into the surrounding tissue, which has a lower urea concentration.

A high urea concentration in the tissue results, which is what caused water to move out of the filtrate by osmosis when it first passed down the descending arm.

5. As the filtrate passes down the collecting duct, even more water passes outward by osmosis because the osmotic concentration of the surrounding tissue reflects both the urea diffusing out from the duct *and* the salt diffusing out from the ascending arm. The sum of these two is greater than the osmotic concentration of urea in the filtrate (the salt has already been removed).

In effect, the interior of the mammalian kidney is divided into two functional zones. The outer portion of the kidney, called the **cortex,** contains the upper portion of the loop, including the upper ascending arm where reabsorption of salt from the filtrate by active transport occurs. The inner portion of the kidney, or **medulla,** contains both the lower portion of the loop and the bottom of the collecting duct, which is permeable to urea (figure 42.12).

The active reabsorption of salt in the cortex of the kidney drives the process. This salt reabsorption from the filtrate in one arm of the loop establishes a gradient of salt concentration, with the salt concentration higher in the medulla at the bottom of the loop. It is this high salt concentration that raises the total tissue osmotic concentration so high that water passes by osmosis out of the collecting duct. Just as important is the active reabsorption of salt in the cortex that concentrates the renal filtrate with respect to urea. This high urea concentration in the filtrate causes urea to diffuse outward into surrounding tissue in the only zone where it is able to do so—the lower collecting duct—creating a high urea concentration in the medulla. This high urea concentration causes water to diffuse out from the filtrate in the initial descending arm. The water is then collected by blood vessels in the kidney, which carry it into the systemic circulation (figure 42.13).

The mammalian kidneys achieve a high degree of water reabsorption by using the salts and urea in the glomerular filtrate to increase the osmotic concentration of the kidney tissue. This facilitates the movement of water from the filtrate out into the surrounding tissue, where the water is collected by blood vessels impermeable to the high urea concentration but permeable to water.

The Kidneys As Regulatory Organs

Like most organs concerned with homeostasis (the maintenance of constant physiological conditions within the body), the kidneys are regulated by the central nervous system, which uses the voluntary, autonomic, and hormonal controls discussed in previous chapters.

Small molecules are pushed across the thin walls of the glomerulus to the inside of the Bowman's capsule; bloodstream that enters the glomerulus is divided into two paths: nonfilterable blood components that are retained and leave the glomerulus in the bloodstream and filterable components that pass across and leave the glomerulus in the urine stream.

Figure 42.11 Flow of materials in the human kidney.
The text provides a detailed discussion of the five stages in this process.

3. Active transport channels pump out salt (NaCl) in the upper region of the ascending arm; urea that initially passed through the glomerulus as nitrogenous waste is left behind in the filtrate; in the ascending loop, other nitrogenous wastes such as uric acid and ammonia, as well as excess hydrogen ions, are excreted into the urine.

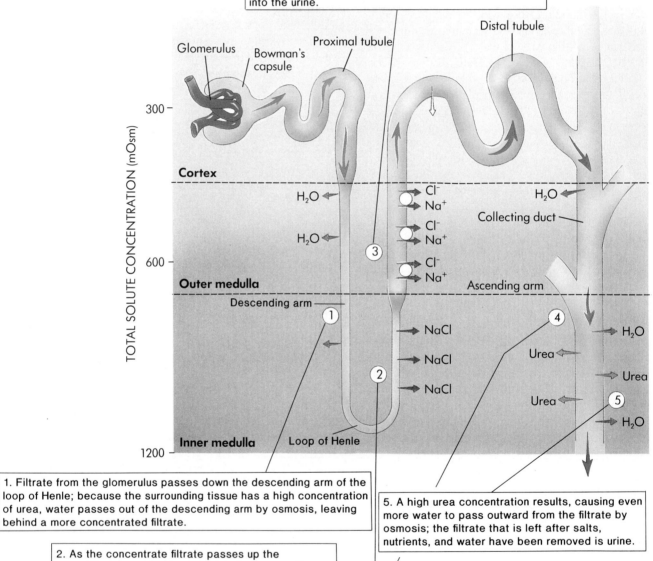

1. Filtrate from the glomerulus passes down the descending arm of the loop of Henle; because the surrounding tissue has a high concentration of urea, water passes out of the descending arm by osmosis, leaving behind a more concentrated filtrate.

5. A high urea concentration results, causing even more water to pass outward from the filtrate by osmosis; the filtrate that is left after salts, nutrients, and water have been removed is urine.

2. As the concentrate filtrate passes up the ascending arm, nutrients pass into the surrounding tissue, where they are carried away by blood vessels.

4. The tubule empties into a collecting duct that passes back through the tissue; the concentrated urea in the filtrate diffuses into the surrounding tissue.

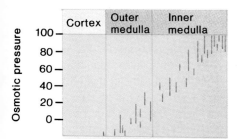

Figure 42.12 Structure of the human kidney.
*The cortex has an osmotic concentration like that of the rest of the body.
The outer medulla has a somewhat higher osmotic concentration,
primarily salt. The osmotic concentration within the inner medulla
becomes progressively higher at greater depths; this part of the kidney
maintains substantial concentrations of urea, which is a major
component of its high osmotic concentration.*

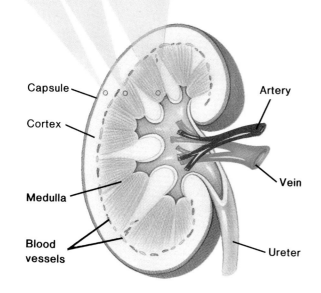

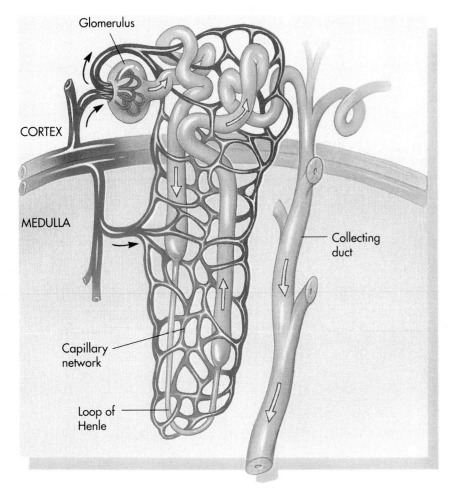

Why is regulating the operation of the kidneys necessary? As
described in chapter 33, proper operation of the body's many
organ systems requires that the osmotic concentration of the
blood be maintained within narrow bounds. For this reason, it
is not always desirable for the body to retain the same amount
of water. After consumption of an unusually large amount of
water, for example, the balance of ions in the blood can only
be preserved if the body retains less of the water than it would
otherwise.

Control of water and salt balance is centered in the hypothalamus. The hypothalamus produces the hormone vasopressin (antidiuretic hormone—ADH), which renders the
collecting ducts of the kidneys freely permeable to urea, which
passes out into the interstitial fluid of the medulla and thus
maximizes water retention by causing water to pass out of the
descending loop into the surrounding tissue. When the uptake
of water is excessive, blood pressure rises, which is detected
by pressure-sensitive neurons in the atria. In addition, osmore-
ceptors in the hypothalamus detect a decrease in osmolarity
and decrease the hypothalamus output of vasopressin. A decrease in vasopressin levels
renders the collecting ducts of the renal
tubules less permeable to water, lowering the
osmotic concentration within the medulla, and
in that way inhibits the reabsorption of water
and increases urine volume. As mentioned in
chapter 37, alcohol consumption is one factor
that suppresses vasopressin secretion.

Another example of the central nervous
system regulating the kidneys concerns the salt
balance in the body. The amount of salt in a
diet can vary considerably, and yet it is important to many physiological processes that salt
levels in the blood not vary widely. When
sodium ion levels in the blood fall, the adrenal
gland increases its production of the hormone
aldosterone, described in chapter 37. Aldosterone stimulates active sodium ion reabsorption across the walls of the ascending arms of
the kidneys' renal tubules, thereby decreasing
the amount of sodium lost in the urine. In the
total absence of aldosterone, human beings
may excrete up to 25 grams of salt a day.

**Figure 42.13 How the blood supply collects
water from the kidney.**
*The loop of Henle is surrounded by a capillary
network that is impermeable to urea. Water
drawn out of the loop by the high exterior urea
concentration passes into the capillaries and is
carried out of the kidney.*

The kidneys maintain osmotic balance in the body. ADH from the hypothalamus and aldosterone from the adrenal gland cooperate to conserve, respectively, water and salt; to keep electrolyte levels balanced; and to guard against drastic changes in blood pressure.

Kidney Failure

Because of the kidneys' great importance in maintaining body fluid homeostasis, kidney failure is a life-threatening event. The most common diseases affecting the kidneys are acute or chronic infection, long-standing diabetes, untreated long-standing high blood pressure, and damage by misdirected attack by the body's own immune system (autoimmune kidney disease). When kidneys stop working, toxic waste materials, such as urea and hydrogen (H^+) ions accumulate in the blood plasma. In addition, blood plasma ion levels (sometimes called "electrolytes") rapidly depart from their normal physiological values. If the kidneys fail, only two treatment options remain: dialysis or kidney transplant.

Kidney Dialysis

Dialysis is a method of removing toxic wastes from the blood, effectively substituting a machine for the kidneys' filtering and homeostatic mechanisms. Dialysis is performed in two ways: hemodialysis and continuous ambulatory peritoneal dialysis.

Hemodialysis

In **hemodialysis,** tubes called catheters are surgically inserted into an artery and a vein, usually on the lower arm. The catheters are equipped with valves. Every few days, the affected individual goes to a clinic, where the catheters are connected to a dialysis machine. Blood passes from the patient's artery into the dialysis machine and then back into the vein. Inside the dialysis machine, the blood passes through a disposable unit consisting of many hollow fibers surrounded by a thin, cellulose, acetate membrane. This dialysis membrane allows waste materials and ions that have accumulated in the plasma to pass by diffusion into the dialyzing fluid (which has the same composition as normal plasma, but no wastes). Dialysis patients must carefully manage their salt and water intake because the dialysis machine, unlike the kidney, does not regulate blood volume and total body sodium levels.

Continuous Ambulatory Peritoneal Dialysis

In **continuous ambulatory peritoneal dialysis**—a recent modification—the peritoneal membrane lining the patient's own abdominal cavity is used as the dialysis membrane. The patient is implanted with an abdominal catheter that allows the abdominal cavity to be filled with dialysis fluid. The dialysis fluid is changed several times daily. This procedure has the advantage of reducing the frequency of visits to an outpatient clinic, making work and travel easier. Diet and fluid restrictions are fewer, but great care is needed to prevent infections.

Kidney Transplants

Dialysis is not a permanent solution to kidney failure. Not only is dialysis expensive, but it also does not fully substitute for a functional kidney, since it has no regulatory homeostatic capability. A single healthy kidney can meet the entire homeostatic needs of the body, but no dialysis machine can. A more permanent solution to kidney failure is transplantation of a kidney from a healthy donor. Because the kidney can regulate body fluid homeostasis without any neural innervation, a kidney transplant solves the problem of homeostatic regulation permanently and cures the patient of kidney disease.

The only problem with kidney transplant is common to all organ transplants: rejection of the transplanted organ by the recipient's immune system. As discussed in chapter 41, all body cells are marked with "self-markers" on their surfaces to protect them from attack by the body's immune system. These markers are called **histocompatibility antigens.** The combination of these antigens displayed on body cells is as unique as fingerprints. Only identical twins have the same self-markers. The more closely related two individuals are to one another, the more likely they are to possess some common self-antigens. This is why tissue transplants are more likely to succeed if the donor and recipient are closely related—they are matched with respect to these antigens. To reduce chances of rejection, the recipient is treated with drugs, like cyclosporin, that suppress the immune system. However, such drugs increase the risk of infections and may be toxic to the liver and to bone marrow.

Since the kidney is critical for maintaining homeostasis, kidney failure is lethal. Kidney failure can be caused by infection, diabetes, high blood pressure, or autoimmune disease. If kidney function is severely impaired, life can be preserved by dialysis or a kidney transplant.

EVOLUTIONARY VIEWPOINT

The vertebrate kidney is an excellent example of evolutionary flexibility. It evolved in freshwater, with ion channels pointed inward for recovery of sodium and other ions. When fish reinvaded saltwater, the kidneys worked equally well—the ion channels simply turned around for discharge, rather than recovery, of sodium ions. This same transition is made by salmon, which are born in freshwater, grow to adulthood in saltwater, and then return to freshwater to spawn.

SUMMARY

1. The body eliminates excess materials in two ways: respiration and excretion. Excretion rids the body of nitrogenous waste products, enzymes, acids, detergents, water, salts, and ions.

2. Ammonia is the by-product of the metabolic breakdown of protein. Because of its high toxicity, ammonia in the body must be eliminated. Animals have evolved three general solutions to this problem: (1) flushing, (2) detoxification, and (3) insolubilization.

3. Osmoregulation is the maintenance of constant internal salt and water concentrations in an organism. Animals that live in freshwater tend to gain water from their surroundings, whereas those that live in saltwater or on land tend to lose water.

4. Insects solve the problem of dehydration by conserving water. In particular, they avoid excreting water along with body wastes. They pump ions into Malpighian tubules so that body fluids that contain the wastes created by metabolism are drawn in by osmosis. The membranes of Malpighian tubules act as filters, passing wastes but preventing proteins and blood cells from crossing into the tubules. Insects reabsorb the water and useful metabolites in the hindgut, leaving the wastes behind for excretion.

5. Instead of pulling body liquid through a filter, as insects do, vertebrates push it through. They can do this because they have a closed circulation system that operates under considerable pressure.

6. Among the vertebrates, the kidney is the organ that filters, reabsorbs, and secretes in the processes of excretion and water conservation. The vertebrate kidney is composed of many individual units called nephrons, each made up of three segments: a filter (Bowman's capsule), a renal tubule with a reabsorption device (loop of Henle), and a large collecting duct.

7. Freshwater fishes do not reabsorb water in the nephrons. Their bodies already gain too much by direct diffusion from the water.

Marine fishes excrete the salts in the water they drink and reabsorb water. Some marine vertebrates, notably sharks, maintain a high body level of urea so that they are isotonic with the sea and do not tend to gain or lose water. Amphibians and reptiles have kidneys much like those of freshwater fishes.

8. Birds and mammals achieve much greater water conservation by bending the nephron tube, producing what is known as the loop of Henle. This creates a countercurrent flow, which greatly increases the efficiency of water reabsorption. In general, the longer the loop of Henle, the greater the osmotic concentration that can be achieved and the more water that can be reclaimed from the urine.

9. The mammalian kidneys achieve a high degree of water reabsorption by using the salts and urea in the glomerular filtrate to increase the osmotic concentration of the kidney tissue. This facilitates the movement of water from the filtrate out into the surrounding tissue, where the water is collected by blood vessels impermeable to the high urea concentration but permeable to the water.

10. The kidneys are regulated by vasopressin (antidiuretic hormone), which affects osmolarity, and aldosterone, which regulates salt balance.

11. When kidneys stop working, toxic waste materials accumulate in the blood plasma. Treatment options are dialysis, in which a machine removes toxic wastes from the blood, and kidney transplant.

REVIEWING THE CHAPTER

1. How does the body excrete substances?

2. How does the body get rid of nitrogenous waste products?

3. How are aids to digestion—enzymes, acids, and detergents—excreted by the body?

4. How is water balance maintained in the body?

5. How are salts and ions regulated?

6. What is osmoregulation?

7. What are the problems faced by osmoregulators?

8. How is osmoregulation achieved?

9. What is the structure of the kidney?

10. How did the vertebrate kidney evolve?

11. How did kidneys first evolve among freshwater fishes?

12. How did kidneys evolve in marine fishes?

13. How does kidney evolution differ in sharks?

14. How are amphibian and reptile kidneys both similar and different?

15. How are the kidneys of birds and mammals more efficient?

16. How does the mammalian kidney work?

17. How do the kidneys function as regulatory organs?

18. Why do kidneys fail?

19. What is kidney dialysis?

20. What is hemodialysis?

21. What is continuous ambulatory peritoneal dialysis?

22. Why are kidney transplants sometimes rejected?

COMPLETING YOUR UNDERSTANDING

1. A toxic nitrogenous waste product that your body gets rid of is

    a. amino acids. d. urea.

    b. proteins.   e. nucleotide bases.

    c. ammonia.   f. nitrate.

2. Why does your blood pass through your kidneys about 350 times per day?

3. What is an osmoconformer, and to which animal groups does this term apply?

4. Animals with body fluids that are hypotonic relative to their environment are

    a. freshwater vertebrates.

    b. sharks.

    c. marine invertebrates.

    d. marine vertebrates other than sharks.

5. Excretory organs called Malpighian tubules are characteristic of

    a. amphibians. d. sharks.

    b. insects.   e. protists.

    c. freshwater   f. sponges.
       invertebrates.

6. Why is the movement of fluids through the kidneys of marine fishes the reverse of their freshwater fish ancestors?

7. Why do sharks not have to remove large amounts of divalent ions from their bodies like other marine fishes?

8. What role does the liver play in excretion in terrestrial reptiles?

9. Why are human kidneys so efficient in the conservation of water?

10. In the human kidney, active-transport channels that remove salts occur in the

    a. cortex.   d. loop of Henle.

    b. outer medulla. e. Bowman's
                     capsule.

    c. inner medulla. f. collecting duct.

11. If you drink excess amounts of fluid, how do your kidneys compensate?

12. What is vasopressin, where does it originate, and how does it function?

13. Under what conditions would your body produce the hormone aldosterone?

14. Why is kidney failure life-threatening? What are the treatment options to kidney failure?

15. Can your body survive with one healthy kidney?

16. How are kidney transplant rejections reduced? What is the problem with using this treatment?

*__17.__ Salt substitutes sold under such names as "light salt" usually consist in part of potassium chloride (KCl), which has a salty taste that is not as pleasant as that of sodium chloride (NaCl). Why would consumption of KCl as a substitute for NaCl be less likely to increase blood pressure?

*__18.__ Atherosclerosis of the renal artery, restricting blood flow to one kidney, results in hypertension. Explain why. **Hint:** The hypertension is not caused by the flow-impaired kidney's failure to form urine.

*For discussion

FOR FURTHER READING

Beauchamp, G. 1987. The human preference for excess salt. *American Scientist* 75: 27. Discusses salt appetite and the relationship of excess salt consumption to hypertension.

Beeuwkes, R. 1982. Renal countercurrent mechanisms, or how to get something for (almost) nothing. In *A companion to animal physiology,* edited by C. R. Taylor et al. New York: Cambridge University Press. A clearly presented summary of current ideas about how the human kidney works.

Heatwole, H. 1978. Adaptations of marine snakes. *American Scientist* 66: 594–604. Several groups of snakes that are able to live in the sea by clever adaptations that modify salt and water balance.

Marshall, E. 1988. Testing urine for drugs. *Science,* July, 150–52. How urinalysis now provides fast and accurate tests for the presence of many drugs.

Smith, H. W. 1961. *From fish to philosopher.* 2d ed. Boston: Little, Brown. A classic, well-written account of the evolution of the vertebrate kidney.

SEX AND REPRODUCTION

A scanning electron micrograph of human sperm cells migrating up the in-side of the uterus, traveling across a background of uterine mucosa cells. Only one of millions will reach the egg—if there's an egg there!

FOR REVIEW

Here are some important terms and concepts that have been discussed in previous chapters and that you will encounter again in this chapter. Review them before proceeding if necessary.

Meiosis (*chapter 9*)
Vertebrate evolution (*chapter 18*)
Amniotic egg (*chapter 18*)
Mammals (*chapter 18*)
Hormones (*chapter 37*)

Animals can reproduce in many ways. **Sexual reproduction,** in which two different individuals contribute to making an offspring, is a complex process that evolved late in the history of life.

This chapter deals with sex and reproduction among the vertebrates, including human beings. In this age of AIDS, sexually transmitted diseases, and innumerable unplanned pregnancies, all of us must make important, well-informed decisions about sex. Given that the average age at which Americans have sex for the first time is between fourteen and fifteen years of age, the subject is of far more than academic interest.

Sex Evolved in the Sea

Sexual reproduction first evolved among marine organisms. Most female marine fishes produce eggs in batches. When the eggs are ripe, they are simply released into the water. The male releases sperm into the water containing the eggs, thereby achieving fertilization. Seawater itself is not a hostile environment for gametes or for young organisms.

The effective union of free gametes in the sea, an example of **external fertilization,** poses a significant problem for all marine organisms. Eggs and sperm rapidly become diluted in seawater, so their release by females and males must be almost simultaneous for successful fertilization. For this reason, most fishes restrict egg and sperm release to a few brief, well-defined periods.

The ocean has few seasonal cues that organisms can use as signals, but one that is all-pervasive is the cycle of the moon. Approximately every twenty-eight days, the moon revolves around the earth. Variations in its gravitational attraction cause the differences in ocean level called tides. Many different kinds of marine organisms sense the changes in water pressure that accompany the tides, and much of the reproduction that takes place in the sea is timed by this lunar cycle.

Invasion of the land by organisms from the sea meant facing for the first time in the history of life on the earth the danger of drying out. This problem was all the more severe because of marine organisms' small and vulnerable reproductive cells. Obviously, gametes released near one another on land would soon dry up and perish. Terrestrial organisms developed many possible solutions to this problem. The seeds of plants, the spores of fungi, and the eggs of arthropods are all successful adaptations to a dry land existence. Vertebrates solved the problem in yet other ways.

Sex evolved in the sea. Animals' invasion of land raised significant challenges, the most severe of which was the drying out of gametes and fertilized zygotes.

Vertebrate Sex and Reproduction: Four Strategies

The five major classes of living vertebrates have evolved quite different reproductive strategies, ranging from external fertilization to bearing live young. These differences mostly reflect alternate approaches to protecting gametes from drying out, since four of these five major vertebrate groups are terrestrial.

Fishes: External Fertilization That Works Well in Water

Some vertebrates, of which the bony fishes are the most abundant, have remained aquatic. Fertilization in most fish species is external, with the eggs containing only enough yolk material to sustain the developing zygote for a short time. After the initial dowry of yolk has been exhausted, the growing individual must seek its food from the waters around it. Many thousands of eggs may be fertilized in an individual mating, but few of the resulting zygotes survive and grow to maturity. Some succumb to bacterial infection, many others to predation. Fertilized eggs develop quickly, and the young that survive mature rapidly.

Amphibians: Fertilization Still Tied to Water

A second group of vertebrates—the amphibians (frogs, toads, salamanders)—invaded the land without fully adapting to the terrestrial environment. The amphibian life cycle is still inextricably tied to the presence of free water. Among most amphibians, fertilization is still external, just as it is among the fishes and other aquatic animals. Many female amphibians lay their eggs in a puddle or a pond of water. Among the frogs and toads, the male grasps the female and discharges fluid containing sperm onto the eggs as she releases them (figure 43.1).

Amphibian development takes much longer than fish development, but amphibian eggs do not contain a significantly greater amount of yolk. Instead, the developmental process consists of two distinct life stages—a larval stage and an adult stage—like some of the life cycles found among the insects.

Amphibian larvae develop rapidly, using yolk supplied from the egg. They then function, often for a considerable period of time, as independent food-gathering machines. They scavenge nutrients from their environments and often grow rapidly. Tadpoles, which are the larvae of frogs, can grow in a matter of days from creatures the size of a pencil point into individuals as big as goldfish. When an individual larva has grown to a sufficient size, it undergoes a developmental transition, or **metamorphosis,** into the terrestrial adult form.

Figure 43.1 External fertilization.
When frogs mate, as these two are doing, the clasp of the male induces the female to release a large mass of mature eggs, over which the male discharges his sperm.

Figure 43.2 Internal fertilization.
The male injects sperm-containing semen into the female's body during copulation. Reptiles such as these turtles were the first terrestrial vertebrates to develop this form of reproduction, which is particularly suited to terrestrial existence.

> *The vertebrate invasion of land was tentative at first, with amphibians retaining the external fertilization and reproduction by means of eggs that is characteristic of most of the fishes, the group from which they evolved.*

Reptiles and Birds: Evolution of the Watertight Egg

Reptiles were the first group of vertebrates to abandon aquatic habitats completely. Reptile eggs are fertilized internally within the mother before they are laid, with the male introducing his **semen,** a fluid containing sperm and fluid secretions, directly into her body. This **internal fertilization** protects gametes from drying out, even though the adult animals are fully terrestrial (figure 43.2). Most vertebrates that fertilize internally utilize a tube—the **penis**—to inject semen into the female. Composed largely of erectile tissue, the penis can become quite rigid and penetrate far into the female reproductive tract.

Many reptiles are **oviparous,** the eggs being deposited outside the mother's body, while others are **viviparous,** forming eggs that hatch within the mother's body. The young of viviparous vertebrates, therefore, are born alive.

Most birds lack a penis (swans are an exception) and achieve internal fertilization by the male slapping semen against the female's reproductive opening (in birds, joined with the excretory opening in a common portal called the **cloaca**) before eggs form their hard shells. This kind of mating occurs more quickly than that of most reptiles.

All birds are oviparous, the young emerging from their eggs outside of the mother's body. Birds encase their eggs in a harder shell than do their reptilian ancestors. They also hasten embryonic development within each egg by warming the eggs with their bodies. The young that hatch from the eggs of most bird species are not able to survive unaided, since their development is still incomplete. Their parents feed and nurture them as they gradually mature.

The shelled eggs of reptiles and birds constitute one of the most important adaptations to life on land, since these eggs can be laid in dry places. Each egg is provided with a large amount of yolk and is encased within a membranous cover. The zygote develops within the egg, eventually achieving the form of a miniature adult before it completely uses up its supply of yolk and hatches from the egg.

> *The first major evolutionary change in reproductive biology among land vertebrates was that of the reptiles and their descendants, the birds. Although, like fishes, both groups are still oviparous, they practice internal fertilization, with the zygote encased within a watertight egg.*

Mammals: Mothers Who Nourish Their Young

The most primitive mammals—the monotremes—are oviparous like the reptiles from which they evolved. The living monotremes consist solely of the duck-billed platypus and the echidna. All other mammals are viviparous.

(a)

(b)

Figure 43.3 Mammalian reproductive strategies.
(a) *Marsupials. The kangaroo gives birth to a very immature offspring, which is nurtured and fed in a special pouch until it can exist independently.* (b) *Placental mammals. Humans nurture their embryos inside their bodies. The placenta is a special organ across which nutrients from the mother's body are transmitted to the developing embryo.*

The young of viviparous mammals are nourished and protected by their mother, an outstanding characteristic of the members of this class. Viviparous mammals nourish their young in two ways:

1. Marsupials give birth to live embryos at a very early stage of development. The tiny animals crawl to and enter pouches on the mother's body, where they continue to develop until they are able to function on their own (figure 43.3*a*).

2. Placental mammals retain their young for a much longer period within the mother's body. To nourish their young, placental mammals have evolved a specialized, massive network of blood vessels called a **placenta,** through which nutrients are channeled to the embryo from the mother's blood (figure 43.3*b*).

The second major evolutionary change in reproductive biology among land vertebrates was that of the marsupials and placental mammals. These groups nourish their young within a pouch or inside the body until the young reach a fairly advanced stage of development.

The Human Reproductive System

The human reproductive system, like those of all other vertebrates, produces gametes (figure 43.4). The male gametes are called **sperm cells,** and the female gametes are called

Figure 43.4 Human reproductive system.
Gamete production by the male and female reproductive systems is mediated by hormones from the pituitary and hypothalamus. Although these organs are not shown on the diagram, they are important parts of the human reproductive system.

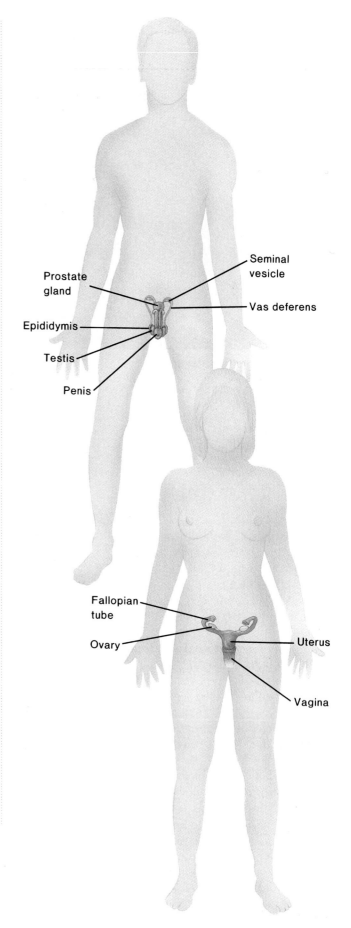

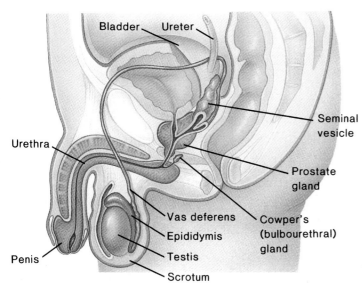

Figure 43.5 Male reproductive system.
The male reproductive organs are specialized for the production and delivery of sperm.

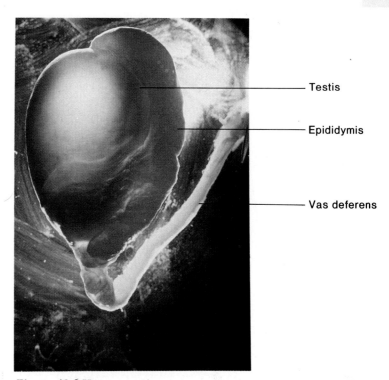

Figure 43.6 Human testis.
The testis is the darker sphere in the center of the photograph; within it, sperm are formed. Cupped above the testis is the epididymis, a highly coiled passageway within which sperm complete their maturation. Extending away from the epididymis is a long tube, the vas deferens.

secondary oocytes, or just oocytes. In biological terms, the human female gamete is not an "egg." The term ***ovum,*** which is Latin for "egg," only refers to a fertilized female gamete. The oocyte is fertilized within the female, and the zygote develops into a mature fetus there.

The Male Reproductive System

The human male gamete, or sperm, is highly specialized for its role as a carrier of genetic information. Produced after meiosis, the sperm cells have twenty-three chromosomes instead of the forty-six found in most human body cells. Unlike these other cells, sperm complete their development only at a temperature lower than the normal 37 degrees Celsius (98.6 degrees Fahrenheit) human body temperature. The sperm-producing organs, or **testes,** move during the course of fetal development out of the body proper and into a sac called the **scrotum** (figures 43.5 and 43.6). The scrotum, which hangs between the legs of the male, maintains the testes at a temperature about 3 degrees Celsius cooler than that of the rest of the body.

Male Gametes Formed in the Testes

The testes are composed of several hundred compartments, each of which is packed with a large number of tightly coiled **seminiferous tubules** (figure 43.7). The seminiferous tubules are the sites of sperm-cell production, or **spermatogenesis.** The full process of sperm development takes about two months. The number of sperm produced is truly incredible: A typical adult male produces several hundred million sperm each day of his life. Those that are not ejaculated from the body are broken down and their constituents reabsorbed, in a continual cycle of renewal.

The testes also contain interstitial cells that secrete the male sex hormone testosterone. All the cells of the testes—gametes and supporting cells—require a combination of **follicle-stimulating hormone (FSH)** and **luteinizing hormone (LH)** from the pituitary (see chapter 37) for normal function.

After sperm cells complete their differentiation within the testes, they are delivered to a long, coiled tube called the **epididymis,** where they are stored and mature further. The sperm cells are not motile when they arrive in the epididymis, and they must remain there for at least eighteen hours before their motility develops. Sperm remain in the epididymis and the very beginning of another long tube—the **vas deferens**—until they are ejaculated from the body. When delivered during intercourse, the sperm travel through the vas deferens to the urethra, where the reproductive and urinary tracts join, emptying through the penis.

Male Gametes Delivered by the Penis

The penis is an external tube composed of three cylinders of spongy tissue. In cross section, the arteries and veins can be seen along the dorsal surface, beneath which two of the cylinders sit side by side. Below the pair of cylinders is a third cylinder, which contains in its center the **urethra,** through which both semen (during ejaculation) and urine (during urination) pass (figure 43.8).

The spongy tissue that makes up the three cylinders is riddled with small spaces between cells. When nerve impulses from the central nervous system cause the arterioles leading

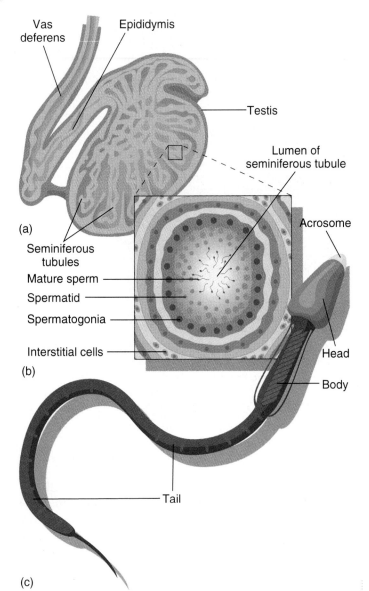

(a)

(b)

(c)

Figure 43.7 Structure of the testis interior.
(a) *The seminiferous tubules in the testis interior are the site of spermatogenesis.* (b) *Within the seminiferous tubules, spermatogonia develop into spermatids, which eventually develop and mature into sperm.* (c) *Each sperm cell is composed of a head region and a tail region. Mitochondria at the top of the tail region power the sperm's long journey into the uterus.*

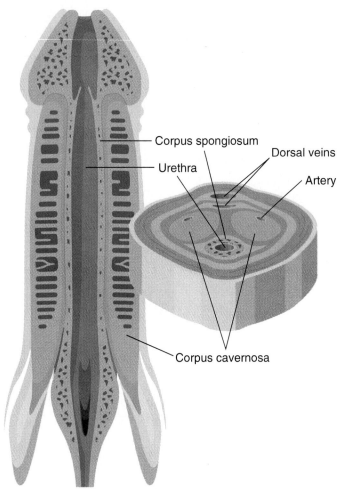

Figure 43.8 A penis in longitudinal and cross section.
The penis is filled with three cylinders of spongy tissue. Arteries carry blood to the penis, and when these cylinders fill with blood, erection is achieved.

into this tissue to dilate, blood collects within the spaces. This causes the tissue to become distended and the penis to become erect and rigid. Continued stimulation by the central nervous system is required for this erection to be maintained.

Erection can be achieved without any physical stimulation of the penis. Mental imagery is a common initiating factor. However, physical stimulation of the penis usually is required for semen delivery. Stimulation of the penis, as by repeated thrusts into the vagina of a female, leads first to sperm mobilization. In this process, muscles encircling the vas deferens contract, moving the sperm along the vas deferens into the

urethra. Eventually, stimulation leads to contraction of the muscles at the base of the penis. The result is **ejaculation,** the ejection of about 5 milliliters of semen out of the penis.

Semen is a collection of secretions from the **prostate gland, seminal vesicles,** and **Cowper's (bulbourethral) glands** (see figure 43.5). It provides metabolic energy sources for the sperm. Within the small volume of ejaculated semen are several hundred million sperm. The odds against any one individual sperm cell successfully completing the long journey to the female oocyte and fertilizing it are extraordinarily high. Successful fertilization requires a high sperm count: Males with fewer than twenty million sperm per milliliter are generally considered sterile.

An adult male produces sperm continuously, several hundred million each day of his life. The sperm are stored and then delivered during sexual intercourse.

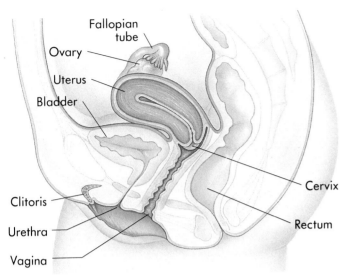

Figure 43.9 Female reproductive system.
The organs of the female reproductive system are specialized to produce gametes and to provide the site for embryo development if the gamete is fertilized.

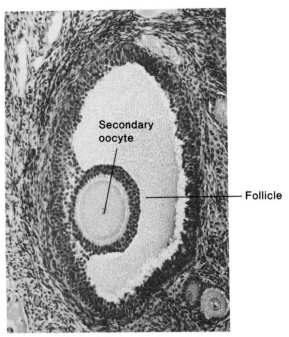

Figure 43.10 A mature secondary oocyte in an ovarian follicle of a cat.
This secondary oocyte awaits ovulation.

The Female Reproductive System

Fertilization requires more than insemination. There must also be a mature oocyte to fertilize.

Female Gametes Formed in the Ovaries

Secondary oocytes are produced within the ovaries of females (figure 43.9). **Ovaries** are compact masses of cells, 2 to 3 centimeters long, located within the abdominal cavity. Secondary oocytes develop from cells called **primary oocytes,** which are located in protective structures called **follicles** in the ovary's outer layer (figure 43.10). Unlike males, whose gamete-producing cells (spermatogonia) are constantly dividing, females have at birth all the primary oocytes that they will ever produce. At each cycle of ovulation, one or a few of these primary oocytes initiate development; the others remain in a developmental "holding pattern." This long maintenance period is one reason why developmental abnormalities crop up with increasing frequency in pregnancies of women who are over age thirty-five. The primary oocytes are continually exposed to mutation throughout life, and after thirty-five years, the odds of a harmful mutation having occurred become high enough to increase significantly the incidence of fetal abnormalities.

Maturation of Female Gametes: One Per Month

At birth, a female's ovaries contain some two million primary oocytes, all of which have begun the first meiotic division. Meiosis is arrested, however, in prophase of the first meiotic division. Very few primary oocytes ever develop further. With the onset of puberty, the female matures sexually. At this time, the release of FSH initiates resumption of the first meiotic division in a few primary oocytes, but a single primary oocyte soon becomes dominant, and the others regress. Approximately every twenty-eight days after that, another primary oocyte matures, although the exact timing may vary from month to month. Only about four hundred out of the approximately two million primary oocytes with which a female is born mature during her lifetime to become secondary oocytes.

Unlike spermatogenesis, the process of **oogenesis** does not result in the production of four haploid gametes. Instead, a single haploid secondary oocyte is produced, with the other meiotic products being discarded as **polar bodies.** The process of meiosis is stop-and-go rather than continuous (figure 43.11).

Fertilization on the Journey to the Uterus

When the secondary oocyte is released at **ovulation,** it is swept by beating cilia into one of the **fallopian tubes,** which lead away from the ovary and into the uterus (figure 43.12). Smooth muscles lining the fallopian tube contract rhythmically, which moves the oocyte down the tube to the uterus. The journey is a slow one, taking about three days to complete. If the oocyte is unfertilized, it soon loses its capacity to develop. It can only live approximately twenty-four hours unless it has been fertilized. For this reason, the sperm cannot simply lie in wait within the uterus. Any sperm cell that is to fertilize an oocyte must make its way up the fallopian tube, a long passage that few sperm survive.

Sperm are deposited within the **vagina,** a muscular tube about 7 centimeters long, that leads to the mouth of the uterus. This opening is bounded by a muscular sphincter called the **cervix.** The **uterus** is a hollow, pear-shaped organ about the size of a small fist. Its inner wall, the **endometrium,** has

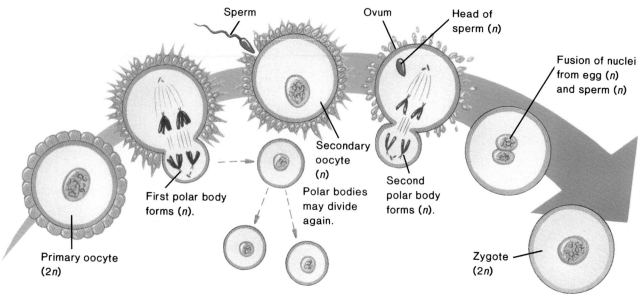

Figure 43.11 Oogenesis.
A primary oocyte is diploid. In its maturation, the first meiotic division is completed, and one division produced is eliminated as a polar body. The other product, the secondary oocyte, is released during ovulation. The second meiotic division does not occur until after fertilization and results in the production of a second polar body and a single haploid ovum. Fusion of the haploid ovum nucleus with a haploid sperm produces a diploid zygote, from which an embryo subsequently forms.

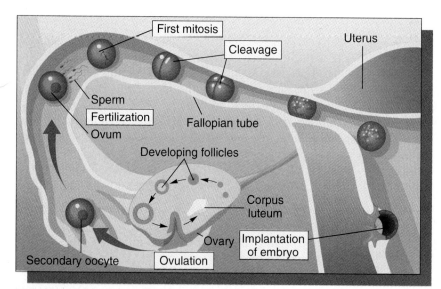

Figure 43.12 Journey of the secondary oocyte.
Produced within a follicle and released at ovulation, the secondary oocyte is swept up into a fallopian tube by the waves of contraction of the tube walls. Fertilization occurs within the tube by sperm journeying upward. The secondary oocyte is now called an ovum. Several mitotic divisions occur while the fertilized ovum continues its journey down the fallopian tube, so that by the time it enters the uterus, it is a hollow sphere of cells. The sphere, a new embryo, implants itself within the uterine wall, where it continues its development.

two layers. The outer of these layers is shed during **menstruation** while the one beneath it generates the next layer. Sperm entering the uterus swim and are carried upward by waves of motion in uterine walls, enter the fallopian tube, and then swim upward against the current generated by the peristaltic contractions that are carrying the ovum downward toward the uterus.

All the primary oocytes that a woman will produce during her life develop from cells that are already present at her birth. Their development is halted early in meiosis, and one or a few of these cells resume meiosis approximately every twenty-eight days to produce a mature oocyte. On maturation, secondary oocytes travel to the uterus. Fertilization by a sperm cell, if it occurs, happens en route.

If a secondary oocyte is fertilized, it completes the second meiotic division of meiosis and is called an ovum. When a successfully fertilized ovum reaches the uterus, the new embryo attaches itself to the endometrial lining and thus starts the long developmental journey that eventually leads to the birth of a child.

The Estrous Cycle

For efficient reproduction, mating or the release of sperm must occur as soon as the mature oocyte is available. Otherwise, few of the sperm will survive long enough to achieve successful fertilization. Among those vertebrates that practice internal fertilization, the females typically signal the successful development and release of a secondary oocyte by the release of chemical signals called **pheromones.** Female dogs signal their reproductive readiness in this manner, and the male is able to detect their pheromones, even at very low concentrations, in the air. In many mammals, such as the domestic cat, the female does not actually release a mature oocyte until after mating. For them, the physical stimulus of copulation causes the pituitary gland to release a signal that triggers ovulation.

When a female vertebrate does not possess a mature oocyte, she will often reject a male's sexual advances. Most female mammals are sexually receptive, or "in heat," for only a few short periods each year. The period in which the animal is in heat is called **estrus.** The periods of estrus correspond to ovulation events during a periodic cycle called the **estrous cycle.** In general, small mammals have many estrous cycles in rapid succession, whereas larger ones have fewer cycles spaced farther apart (figure 43.13). During estrus, the female is willing to mate and either has ovulated or will soon ovulate. Meanwhile, the endometrium of the uterus has developed into an ideal environment in which an embryo can implant and develop.

Human beings and some apes provide the exception to the rule of cycles of receptivity. Human females are sexually receptive throughout the reproductive cycle. The human reproductive cycle of oocyte production and release takes, on average, about twenty-eight days. One cycle follows another continuously, and a human female may mate at any time during a cycle. Successful fertilization, however, is possible only during a three- to four-day period starting the day before ovulation.

Sex Hormones

A lot of signaling goes on during sex. Not only is it necessary for most viviparous females to signal a male when a mature oocyte has been released at the beginning of an estrous cycle, but many different processes within the female and male must also be coordinated during the process of gametogenesis and reproduction. For example, the delayed sexual development that is common in mammals entails a nonsexual juvenile period, after which changes produce sexual maturity. These changes occur in many parts of the body, a process that requires the simultaneous coordination of further development in many different kinds of tissues. Gamete production is another carefully orchestrated process, involving a series of carefully timed developmental events. Successful fertilization begins yet another developmental "program," in which the female body prepares itself for the many changes of pregnancy.

All of this signaling is carried out by a portion of the brain—the hypothalamus. The signals by which the hypothalamus regulates reproduction are hormones produced by the brain, which are carried by the bloodstream to the various body organs. Most reproductive hormones, such as **estrogen** and **progesterone,** are steroids (complex carbon-ring lipids); others, such as FSH and LH, are peptides (table 43.1).

The body uses hormones as signals to control various body functions in the reproductive cycle. Most reproductive hormones are steroids.

(a)

(b)

(c)

Figure 43.13 The estrous cycle.
Smaller animals are receptive to mating more often during the year than larger animals. (a) Mice go into estrus several times a year and are capable of producing numerous offspring. (b) Cats go into estrus twice a year. One cat can produce up to twenty kittens per year. (c) A female elephant is only receptive to mating for two days about once every four years! Their long gestation and nursing periods prevent them from having more frequent estrus periods.

Table 43.1   Reproductive Hormones

Male

Follicle-stimulating hormone (FSH)	Stimulates spermatogenesis
Luteinizing hormone (LH)	Stimulates testosterone secretion
Testosterone	Stimulates development and maintenance of male secondary sexual characteristics

Female

Follicle-stimulating hormone (FSH)	Stimulates growth of ovarian follicle
Luteinizing hormone (LH)	Stimulates conversion of ovarian follicles into corpus luteum; stimulates estrogen secretion
Estrogen	Stimulates development and maintenance of female secondary sexual characteristics; prompts monthly preparation of uterus for pregnancy
Progesterone	Completes preparation of uterus for pregnancy; helps to maintain female secondary sexual characteristics
Oxytocin	Stimulates uterine contraction; initiates milk release
Prolactin	Stimulates milk production

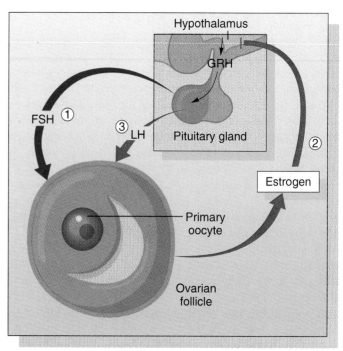

Figure 43.14 The maturation of secondary oocytes is under hormonal control.
(1) *Gonadotropic-releasing hormone (GRH) from the hypothalamus causes the pituitary gland to release follicle-stimulating hormone (FSH). FSH acts on the follicle and initiates the maturation of the primary oocyte. (2) FSH also causes the ovaries to produce estrogen. Rising estrogen levels in the blood cause the pituitary to shut down its FSH production and instead produce luteinizing hormone (LH). (3) LH inhibits estrogen production and initiates ovulation.*

The Human Reproductive Cycle

The reproductive cycle of female mammals, including that of human females, is composed of two distinct phases: the follicular phase and the luteal phase.

The Follicular Phase: Triggering the Maturation of a Primary Oocyte

The first or **follicular phase** of the reproductive cycle is marked by the hormonally controlled development of primary oocytes within the ovary. The anterior pituitary, after receiving a chemical signal in the form of **gonadotropic-releasing hormone** (GRH) from the hypothalamus, starts the cycle by secreting follicle-stimulating hormone (FSH), which binds to receptors on the surface of the follicles, initiating the final development and maturation of the primary oocyte (figure 43.14). Normally, only a few primary oocytes at any one time have developed far enough to respond immediately to the FSH. FSH levels are reduced before other oocytes reach maturity, so that in every cycle, only a few primary oocytes ripen.

In addition to starting the final development of a primary oocyte, FSH also triggers production of the female sex hormone estrogen by the ovary. Rising estrogen levels in the bloodstream feed back to the hypothalamus and cut off further FSH production. In this way, only the few oocytes that are already developed far enough for their maturation to be initiated by the FSH are taken into the final stage of development. The

rise in estrogen level and the maturation of one or more primary oocytes completes the follicular phase of the human reproductive cycle.

The Luteal Phase: Preparing the Body for Fertilization

The second or **luteal phase** of the reproductive cycle follows smoothly from the first. The hypothalamus responds to estrogen by causing the pituitary to secrete a second hormone, called luteinizing hormone (LH), which is carried in the bloodstream to the developing follicle (see figure 43.14). LH inhibits estrogen production and causes the wall of the mature follicle to burst. The secondary oocyte within the follicle is released into one of the fallopian tubes, which extend from the ovary to the uterus. This is called ovulation.

Meanwhile, the ruptured follicle repairs itself, filling in and becoming yellowish. In this condition, it is called the **corpus luteum,** which is simply the Latin phrase for "yellow body." The corpus luteum soon begins to secrete the hormone progesterone, which inhibits FSH, preventing further ovulations. The body is preparing itself for fertilization. The corpus luteum continues its progesterone production for two weeks after ovulation. If fertilization does not occur soon after ovulation, then progesterone production slows and eventually ceases, marking the end of the luteal phase.

The mammalian reproductive cycle is composed of two alternating phases. During the follicular phase, some of the primary oocytes within the ovary complete their development. During the following luteal phase, the mature secondary oocyte is released into the fallopian tubes, a process called ovulation. If fertilization does not occur, ovulation is followed by a new follicular phase, the start of another cycle.

When estrogen and progesterone levels are low, the pituitary can again initiate FSH production, thus starting another reproductive cycle. In human beings, the next cycle follows immediately after the end of the preceding one. A cycle usually occurs every twenty-eight days, or a little more frequently than once a month, although this varies in individual cases. The Latin word for month is *mensis,* which is why the reproductive cycle in humans is called the **menstrual cycle,** or monthly cycle.

In human beings and some other primates, the hormone progesterone has among its many effects a thickening of the uterine lining (endometrium) in preparation for the implantation of the developing embryo. When fertilization does not occur, the decreasing levels of progesterone cause this thickened layer of blood-rich tissue to be sloughed off, a process that results in the bleeding associated with **menstruation.** Menstruation, or "having a period," usually occurs about midway between successive ovulations, or roughly once every twenty-eight days, although its timing varies widely even for individual females (figure 43.15).

Two other hormones are important in the female reproductive system. Prolactin is needed for milk production and is secreted by the anterior pituitary. Oxytocin, secreted by the posterior pituitary, causes milk release. In combination with uterine prostaglandins, oxytocin also initiates labor and delivery.

The Physiology of Human Intercourse

Few physical activities are more pleasurable to humans than sexual intercourse. It is one of the strongest drives directing human behavior, and as such, is circumscribed by many rules and customs. Few subjects are at the same time more private and of more general interest.

Until relatively recently, the physiology of human sexual activity was largely unknown. Perhaps because of the prevalence of strong social taboos against the open discussion of sexual matters, research on the subject was not being carried

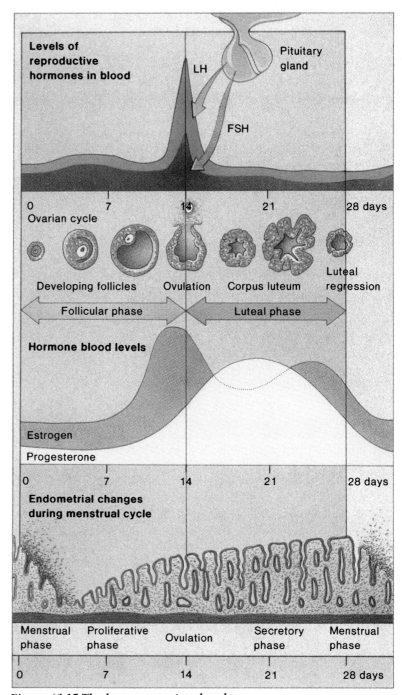

Figure 43.15 The human menstrual cycle.
The growth and thickening of the uterine lining (the endometrium) is governed by rising levels of progesterone. Menstruation (the sloughing off of blood-rich tissue) is initiated by falling progesterone levels.

out, and detailed information was lacking. Everyone learned from anecdote, from what parents or friends said, and eventually, from experience. Largely through the pioneering efforts of William Masters and Virginia Johnson in the last twenty-five years, and an army of workers who have followed them, this gap in the generally available information about the biological nature of human sexual lives has now largely been filled.

Sexual intercourse is referred to by a variety of names, including copulation and coitus, as well as a host of more informal ones. The physiological events that accompany intercourse are commonly partitioned into four periods, although the divisions are somewhat arbitrary. The four periods are **excitement, plateau, orgasm,** and **resolution.**

Excitement

The sexual response is initiated by commands from the brain that increase the heartbeat, blood pressure, and rate of breathing. These changes are very similar to the ones that the brain induces in response to alarm. Other changes increase the diameter of blood vessels, leading to increased peripheral circulation. The nipples commonly become erect and more sensitive. In the male genital area, this increased circulation leads to the vasocongestion in the penis that produces erection. Similar swelling occurs in the **clitoris** of the female, a small knob of tissue composed of a shaft and glans much like the male penis but without the urethra running through it. The female experiences additional changes that prepare the vagina for sexual intercourse: The increased circulation leads to swelling and parting of the lips of tissue, or **labia,** that cover the opening to the vagina; the vaginal walls become moist; and the muscles encasing the vagina relax.

Plateau

The continued stimulation of nerve endings in the clitoris and in the tip of the penis gradually leads to increased excitement in the female and in the male. The clitoris becomes swollen and very sensitive, withdrawing up into a sheath or "hood." Once it has withdrawn, stimulation of the clitoris is indirect, with the thrusting movements of the penis rubbing the clitoral hood against the clitoris. The nerve stimulation produced by the repeated movements of the penis within the vagina elicits a continuous sympathetic nervous system response, greatly intensifying the physiological changes that were initiated in the excitement phase. In this plateau phase, pelvic thrusts may begin in the female, whereas in the male, the penis maintains its rigidity.

Orgasm

The climax of intercourse is reached when stimulation is sufficient to initiate a series of reflexive muscular contractions. The nerve signals producing these contractions are associated with other nervous activity within the central nervous system, activity that is experienced as intense pleasure. In females, the contractions are initiated by impulses in the hypothalamus, which causes the pituitary to release large amounts of the hormone oxytocin. This hormone, in turn, causes the muscles in the uterus and around the vaginal opening to contract. Orgasmic contractions occur about one second apart. There may be one intense peak of contractions (an "orgasm") or several, or the peaks may be more numerous but less intense.

Analogous contractions occur in the male, initiated by nerve signals from the brain. These signals first cause **emission,** in which the rhythmic, peristaltic contractions of the vas deferens and of the prostate gland cause the sperm and seminal fluid to move to a collecting zone of the urethra. This collecting zone, which is located at the base of the penis, is called the bulbourethral area. Shortly after the sperm move into the bulbourethral zone, nerve signals from the brain induce violent contractions of the muscles at the base of the penis, resulting in the ejaculation of the collected semen out through the penis. As in the female orgasm, the contractions are spaced about a second apart, although in the male, they continue for a few seconds only. Unlike those of the female, orgasmic contractions in the male do not vary in their pattern; they are restricted to the single intense wave of contractions that is associated with ejaculation.

Resolution

After ejaculation, males lose their penile erection and enter a **refractory period,** often lasting twenty minutes or longer, in which sexual arousal is difficult to achieve and ejaculation is almost impossible. After orgasm, the bodies of both men and women return slowly, over a period of several minutes, to their original physiological state.

> *Human sexual intercourse can be said to involve four phases: excitement, plateau, orgasm, and resolution.*

Contraception and Birth Control

In most vertebrates, sexual intercourse is associated solely with reproduction, and deeply ingrained reflexive behavior in the female limits sexual receptivity to those periods of the sexual cycle when she is fertile. In human beings, however, sexual behavior serves a second important function: the reinforcement of pair bonding, the emotional relationship between two individuals. The evolution of strong pair bonding is not unique to human beings but probably was a necessary precondition for the evolution of humans' increased mental capacity. The associative activities that make up human "thinking" are largely based on learning, and learning takes time. Human children are very vulnerable during the extended period of learning that follows birth, and they require parental nurturing. This may explain why human pair bonding is a continuous process, not restricted to short periods coinciding with ovulation. Among all the vertebrates, human females and a few species of apes are the only ones in which the characteristic of sexual receptivity throughout the reproductive cycle has evolved and has come to play a role in pair bonding.

Not all human couples want to initiate a pregnancy every time they have sexual intercourse, yet sexual intercourse may be a necessary and important part of their emotional lives together. Among some religious groups, this problem does not arise, or is not recognized, because group members believe that sexual intercourse has only a reproductive function and thus should be limited to situations in which pregnancy is acceptable—that is, among married couples wishing to have children.

Most couples, however, do not limit sexual relations to procreation, and among them, unwanted pregnancy presents a real problem. The solution to this dilemma is to find a way to avoid reproduction without avoiding sexual intercourse, an approach that is commonly called birth control or **contraception.**

Several different contraceptive approaches are common (figure 43.16). These methods differ from one another in their effectiveness and in their acceptability to different couples (table 43.2).

Abstinence

The simplest and most reliable way to avoid pregnancy is not to have sex at all. Of all birth-control methods, this is the most certain—and the most limiting. For many individuals, abstinence is not a realistic birth-control option.

A variant of this approach is to avoid sexual relations only on the two days preceding and following ovulation because this is the only period during which successful fertilization is likely. The rest of the sexual cycle is relatively "safe" for intercourse. This approach, called the **rhythm method,** is satisfactory in principle but difficult in application because ovulation is not easy to predict and may occur unexpectedly. The failure rate of the rhythm method is estimated to be 12 to 40 percent (twelve to forty pregnancies per one hundred women practicing the rhythm method per year), as compared to a failure rate of 90 percent for unprotected sex with sexual intercourse at random.

Another variant of this approach is to have only incomplete sex—the penis is withdrawn before ejaculation, a procedure known as **coitus interruptus.** This requires considerable willpower, often destroys the emotional bonding of intercourse, and is not as reliable as it might seem. Prematurely released sperm can be secreted by the penis within its lubricating fluid, and a second sexual act may transfer sperm ejaculated earlier. The failure rate of this approach is estimated at 9 to 25 percent, which is not much better than that of the rhythm method.

(a)

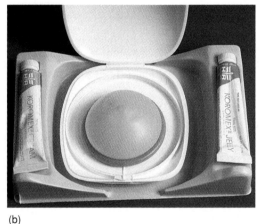

(b)

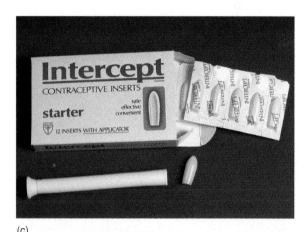

(c)

(d)

(e)

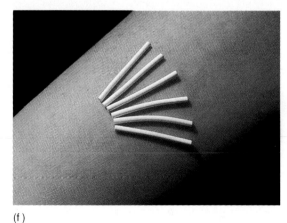

(f)

Figure 43.16 Methods of contraception.
Six of the common ways to achieve nonsurgical birth control are: (a) Condom. (b) Diaphragm and spermicidal jelly. (c) Foams. (d) Oral contraceptives. (e) Vaginal sponge. (f) Norplant implant.

T a b l e 4 3 . 2   Nonsurgical Methods of Birth Control*

Device	Action	Failure Rates	Advantages	Disadvantages
Intrauterine device	Small plastic or metal device placed in the uterus that somehow prevents fertilization or implantation; some contain copper, others release hormones	6	Convenient and highly effective; needs to be replaced infrequently	Can cause excess menstrual bleeding and pain; danger of perforation, infection, and expulsion; not recommended for those who are childless or not monogamous, risk of pelvic inflammatory disease or infertility; dangerous during pregnancy
Oral contraceptives	Hormones, either in combination or progestin only, that primarily prevent release of oocyte	3	Convenient and highly effective; provide significant noncontraceptive health benefits, such as protection against ovarian and endometrial cancers	Pills must be taken regularly; possible minor side effects, which new formulations have reduced; not for women with cardiovascular risks, smokers over thirty-five
Condom	Thin, rubber sheath for penis that collects semen	12	Easy to use, effective, and inexpensive; protects against some sexually transmitted diseases	Requires male cooperation; may diminish spontaneity; may deteriorate on the shelf
Diaphragm with spermicide	Soft rubber cup that covers entrance to uterus, prevents sperm from reaching oocyte, and holds spermicide	18	No dangerous side effects; reliable if used properly; provides some protection against sexually transmitted diseases and cervical cancer	Requires careful fitting; some inconvenience associated with insertion and removal; may be dislodged during sex
Cervical cap	Miniature diaphragm that covers cervix closely, prevents sperm from reaching oocyte, and holds spermicide	comparable to diaphragm	No dangerous side effects; fairly effective; can remain in place longer than diaphragm	Problems with fitting and insertion; comes in limited number of sizes
Foams, creams, jellies, vaginal suppositories	Chemical spermicides inserted in vagina before intercourse that also prevent sperm from entering uterus	21	Can be used by anyone who is not allergic; protect against some sexually transmitted diseases; no known side effects	Relatively unreliable, sometimes messy; must be used 5 to 10 minutes before each act of intercourse
Sponge	Acts as sperm barrier and releases spermicide	18	Safe; easy to insert; provides some protection against sexually transmitted diseases; can be left in place for 24 hours	Relatively unreliable; one size only; some sensitivity and removal problems; cannot be used during menstruation
Implant	Capsules surgically implanted under skin that slowly release a hormone that blocks release of oocytes	<1	Very safe, convenient, and effective; very long-lasting (5 years); may have nonreproductive health benefits like those of oral contraceptives	Irregular or absent periods; necessity of minor surgical procedure to insert and remove
Injectable contraceptive	Injection every 3 months of a hormone slowly released from the muscle that prevents ovulation	<1	Convenient and highly effective; no serious side effects other than occasional heavy menstrual bleeding	Animal studies suggest that it may cause cancer, though new studies of women are mostly encouraging

Source: American College of Obstetricians and Gynecologists: Benefits, Risks, and Effectiveness of Contraception 1990, Washington, D.C., ACOG.
*Approximate effectiveness of these reversible methods of birth control is measured in pregnancies per one hundred actual users per year.

Sperm Blockage

If sperm are not delivered to the uterus, fertilization cannot occur. One way to prevent sperm delivery is to encase the penis within a thin rubber bag, or **condom** (figure 43.16*a*). This method is easy and inexpensive, but condoms must be used correctly, by following the instructions included on the package, to be effective. Condoms have a failure rate of 12 percent (that is, 12 percent of women using this method become pregnant each year), mostly due to incorrect use (such as attempting to put it on with the wrong side out, then reversing the condom and using it—with sperm now adhering to the outside). Condoms are the most commonly employed form of birth control in the United States, with more than a billion sold in 1989. In addition, condoms are the only form of birth control that protects individuals against HIV infection.

A second way to prevent sperm from entering the uterus is to place a cover over the cervix. The cover may be a relatively tight-fitting **cervical cap,** which is worn for days at a time, or a rubber dome called a **diaphragm,** which is inserted immediately before intercourse (figure 43.16*b*). Because individual cervix dimensions vary, a cervical cap or diaphragm must be fitted by a physician. Failure rates average 18 percent for diaphragms, perhaps because of the propensity to insert them carelessly when in a hurry. Failure rates for cervical caps are about the same.

A method similar to the diaphragm is the contraceptive sponge (figure 43.16*e*). This sponge, which is inserted before intercourse and covers the cervix, contains contraceptive chemicals that kill sperm, thereby providing both a chemical and physical barrier. Sponges, since they can be bought over-the-counter, are convenient, but their failure rate is 18 percent.

Sperm Destruction

A third general approach to birth control is to remove or destroy the sperm after ejaculation. This can, in principle, be achieved by washing out the vagina immediately after intercourse, before the sperm have a chance to travel up into the uterus. Such a procedure is called by the French name for "wash," **douche.** This method is inconvenient in that it involves a rapid dash to the bathroom immediately after ejaculation and a very thorough washing. The failure rate has been estimated to be as high as 40 percent.

Sperm delivered to the vagina can be destroyed there with spermicidal jellies, sponges, or foams (figure 43.16*c*). These require application immediately before intercourse. The failure rate varies widely, averaging 21 percent.

Prevention of Primary Oocyte Maturation

Since about 1960, a widespread form of birth control in the United States has been the daily ingestion of hormones, or **birth-control pills** (figure 43.16*d*). These pills contain estrogen and progesterone, either taken together in the same pill or in separate pills taken sequentially. In the normal sexual cycle of a female, these hormones act to shut down the production of the pituitary hormones FSH and LH. The artificial maintenance of high levels of estrogen and progesterone in a woman's bloodstream fools the body into acting as if ovulation has already occurred, when in fact, it has not: The ovarian follicles do not ripen in the absence of FSH, and ovulation does not occur in the absence of LH. For these reasons, birth-control pills provide a very effective means of birth control, with a failure rate of 3 percent. A small number of women using birth-control pills experience undesirable side effects, such as blood clotting and nausea. The long-term consequences of prolonged use are not yet known, since birth-control pills have been in widespread use for only thirty-five years. To date, however, extensive studies have revealed no conclusive evidence of any serious side effects for the great majority of women.

Birth-Control Implants

The new form of birth control on the market is an **implant** called Norplant (figure 43.16*f*). Norplant consists of six small, plastic cylinders that contain progestin, the active ingredient in birth-control pills. Norplant prevents ovulation and also thickens the cervical mucus, making sperm entry into the uterus more difficult. The six cylinders are inserted directly under the skin of the upper arm in a relatively simple surgical procedure. Once implanted, Norplant is effective for five years. Studies have shown that the effectiveness of Norplant is about 99.7 percent. However, a majority of women using Norplant (75 percent) experience irregular menstrual bleeding, an irregularity that lessens the longer the device is in place.

The Injectable Contraceptive

A form of birth control that has been used in foreign countries for years was approved in 1993 for use in the United States. This contraceptive, called Depo-Provera, is an injectable form of the "birth-control pill hormones" that prevent ovulation. One injection lasts for about three months. Depo-Provera is convenient and relatively safe, although some studies on animals have linked this contraceptive with cancer.

Surgical Intervention

A completely effective, although usually permanent, means of birth control is the surgical removal of a portion of the tube through which gametes are delivered to the reproductive organs. The failure rate of such surgical approaches is 0 percent. Their great disadvantage is that they generally render the person permanently sterile.

In males, such an operation involves the removal of a portion of the vas deferens, the tube through which sperm travel to the penis. This simple procedure, called a **vasectomy,** can be performed in a physician's office (figure 43.17*a*).

In females, the comparable operation involves the removal of a section of each of the two fallopian tubes through which the secondary oocyte travels to the uterus. Since these tubes are located within the abdomen, the operation, called a **tubal ligation,** is more difficult than a vasectomy and is even more difficult to reverse (figure 43.17*b*).

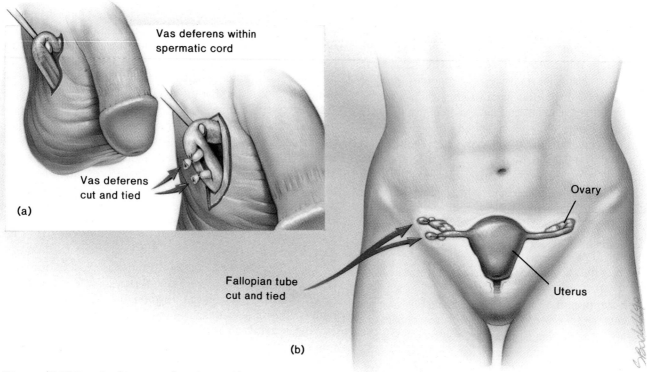

Figure 43.17 Surgical means of contraception.
(a) *Vasectomy.* (b) *Tubal ligation.*

Surgical removal of the entire uterus, an operation that is not uncommon but is usually performed for medical reasons other than birth control, is called a **hysterectomy.**

Prevention of Embryo Implantation
Insertion of a coil or other irregularly shaped object into the uterus is an effective means of birth control, since uterine irritation prevents implantation of the descending embryo within the uterine wall. Such **intrauterine devices (IUDs)** have a failure rate of only 0.5 to 4 percent. Their high degree of effectiveness probably reflects their convenience: Once inserted, they can be forgotten. The great disadvantage of this method is that almost a third of the women attempting to use IUDs cannot; the devices cause them cramps, pain, and sometimes bleeding. Currently, only one brand of IUD is available, and no American company is researching new IUDs. Because of recent lawsuits against IUD manufacturers, IUDs have fallen out of favor with most American women.

Another effective way to prevent embryo implantation is the use of the "morning-after" pill, which contains fifty times the dose of estrogen present in birth-control pills. The failure rate is only 4 to 5 percent, but many women are uneasy about taking such high hormone doses.

Abortion
Reproduction can be prevented after fertilization if the embryo is **aborted** (removed) before its development and birth. During the first trimester, this can be accomplished by **vacuum suction** or by **dilation and curettage,** in which the cervix is dilated and the uterine wall is scraped with a spoon-shaped surgical knife called a curette. Chemical methods are also being developed that cause abortion early in the first trimester, apparently with complete safety. One such drug, **RU 486,** is already in use in France and China, although it has not yet been approved for use in the United States. Administration of RU 486 followed by prostaglandins that induce uterine contractions is almost 100 percent effective when taken within forty-nine days of the patient's last menstrual period.

In the second trimester, the embryo can be removed by injecting a 20 percent saline solution into the uterus, which induces labor and delivery of the fetus. In general, the more advanced the pregnancy, the more difficult and dangerous the abortion is to the woman.

As a method of birth control, abortion takes a great emotional toll, both on the woman undergoing the abortion and often on others who know and care for her. Abortion also presents serious moral problems for some. Many people believe that the fetus is a living person from the time of conception and that abortion is simply murder. Many countries define abortion as a crime. In the United States, a fetus is not legally considered a person until birth, and abortions are permitted by law during the first two trimesters. They are illegal in the third trimester, however, except when the mother's life is endangered. The Supreme Court ruling on legalized abortion in the first two trimesters is a relatively recent one and is still the subject of intense controversy. The abortion rate in the United States (and most other developed countries) exceeds the birth rate.

A variety of safe and effective birth-control procedures are now widely available. In general, the most effective are those that do not require anything be done by the user at the time of sexual intercourse.

Sexually Transmitted Diseases

In addition to AIDS (discussed in detail in chapter 41 and Sidelight 43.1) are a number of **sexually transmitted diseases (STDs)** that every sexually active person must be aware of and take precautions against. Some of these STDs are caused by bacteria; others are caused by viruses. The viral STDs have no cures, and in some bacterial STDs, the bacteria exhibit the ability to mutate into drug-resistant strains. With all the media coverage given to AIDS, people sometimes forget about the other STDs that also are hazardous to health. Thus, all sexually active persons must learn about the nature and symptoms of STDs and take the appropriate measures to prevent their transmission.

Gonorrhea

Gonorrhea is one of the most prevalent communicable diseases in North America. Caused by a bacterium, gonorrhea can be transmitted through sexual intercourse or any other sexual contacts in which body fluids are exchanged, such as oral sex or anal intercourse. Gonorrhea can infect the throat and rectum and can then spread to the eyes and internal organs, causing conjunctivitis (a severe infection of the eyes) and arthritic meningitis (an infection of the joints). Left untreated in women, gonorrhea can cause **pelvic inflammatory disease (PID),** a condition in which the fallopian tubes become scarred, blocking the path of the secondary oocyte into the uterus. PID can eventually lead to sterility.

Gonorrhea symptoms in women are often so mild that they go unnoticed. Fifty to eighty percent of infected women experience no symptoms at all. Infected women who do experience symptoms notice a vaginal discharge, vaginal pain, and pain upon urination. If the infection has spread to the uterus, there may be irregular menstruation and lower abdominal pain.

Gonorrhea symptoms in men may include a thick, milky discharge from the penis and pain and irritation upon urination. Like women, men sometimes experience no symptoms at all.

The treatment for gonorrhea is a combination of antibiotics, usually amoxicillin and tetracycline. Both sexual partners must be treated, or reinfection could occur. Sometimes, however, the bacteria that cause gonorrhea are resistant to these antibiotics, and sufferers must try a battery of antibiotics to find one that is effective. More exotic strains of the gonorrhea bacteria that have been found in Asia are completely resistant to all known antibiotics. Drug companies are currently working on new antibiotics that are effective against these strains.

Syphilis

Syphilis, a very destructive STD, was once prevalent but is now less common due to the advent of blood-screening procedures and the development of antibiotics. Syphilis is caused by a spirochete bacteria, *Treponema pallidum,* that is transmitted during sexual intercourse or through direct contact with an open syphilis sore. The bacteria can also be transmitted from a mother to her fetus.

Once inside the body, the disease progresses in four distinct stages. The first, or primary stage, is characterized by the appearance of a small, painless, often unnoticed sore called a **chancre.** The chancre resembles a blister and occurs at the location where the bacterium entered the body. This stage of the disease is highly infectious, and an infected person may unwittingly transmit the disease to others.

The second stage of syphilis is marked by a rash that may cover the entire body, a sore throat, and sores in the mouth. The bacteria can be transmitted at this stage through kissing or through contact with a open sore.

The third stage of syphilis is characterized by no symptoms at all. This stage may last for several years, and at this point, the person is no longer infectious. The final stage of syphilis is the most debilitating, in which the damage done by the bacteria in the third stage becomes evident. Sufferers at this stage of syphilis experience heart disease, mental deficiency, and nerve damage, which may include a loss of motor functions or blindness.

Fortunately, few persons infected with syphilis experience the third and fourth stages of the disease. Two weeks after exposure to syphilis, the spirochetes appear in the bloodstream. Treatment for syphilis is a course of antibiotics. If a newborn has been infected with the bacteria inside the mother's uterus, antibiotics can halt the course of the disease but cannot undo the damage already done to the baby's internal organs.

Genital Herpes

Several years ago, the media publicized a "new" STD with no cure: **genital herpes.** Many people panicked at the thought of contracting this new disease. In reality, however, genital herpes had already been around for many years at the time it received its overload of media attention. Now, the media and the public have turned their interest to AIDS, an incurable STD that leads to death, and not much is heard anymore about genital herpes. But the disease is still prevalent, and there still is no cure.

Genital herpes is caused by two types of closely related viruses: **herpes simplex virus Type I** and **Type II.** The Type II virus directly causes the genital lesions of genital herpes, while the Type I virus causes cold sores and fever blisters that can be transmitted to the genital regions if the infected person engages in oral sex. The virus is transmitted through sexual contact with an infected person during a herpes outbreak. The fluid in the herpes sores can transmit the virus.

Once inside the body, the herpes virus travels along nerve endings until it reaches the base of the spine. There, the virus can remain dormant and cause no symptoms at all, or the virus can become active and produce painful symptoms. The initial outbreak of genital herpes is usually the most severe. The first signs are a burning, tingling sensation or a minor rash in the genital area. After a few days, small, measlelike lesions are seen, which eventually develop into moist, blisterlike sores with red edges. In women, the sores may appear on the labia, clitoris, vaginal opening, and occasionally on the buttocks,

Sidelight 43.1

AIDS ON THE COLLEGE CAMPUS

As most college students now know, AIDS is a serious disease that is rapidly becoming common in the United States and around the world. The disease, first reported in 1981, is transmitted by *human immunodeficiency virus* (*HIV*) (Sidelight figure 43.1) and is always fatal. Over 400,000 Americans had contracted AIDS by mid-1994. Over 210,000 Americans have died of AIDS since 1981. Some one hundred Americans will die each day of AIDS in 1994. Despite active research, AIDS has no known cure.

The acronym *AIDS* is shorthand for *acquired* (transmitted from another infected individual) *immunodeficiency* (a breakdown of the body's ability to defend itself against disease) *syndrome* (a spectrum of symptoms). The disease is fatal because no one can survive for long without an immune system to defend against viral and bacterial infections and to ward off cancer.

AIDS is not the only fatal disease to threaten humans, and it is not the most contagious. What makes AIDS an unusually serious threat is that recently infected people usually show no symptoms of the disease. Only much later, typically five years, does the virus begin to multiply and attack the immune system. During these years, however, the infected person is an unknowing carrier, able to transmit the virus to others. The large reservoir of undiagnosed, infected individuals casts a shadow over the future. Current estimates of the number of individuals in their twenties and thirties who are infected with the virus in the United States vary, but most estimates exceed one per 250, suggesting both a staggering load of future suffering and a great danger that the infection will spread further. College campuses will not escape infection, and every student should face that fact squarely.

WHAT CAUSES AIDS?

A fragile virus that does not survive outside body cells, HIV is present in infected individuals'

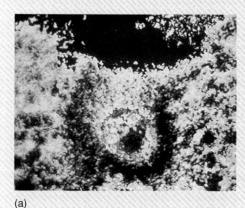

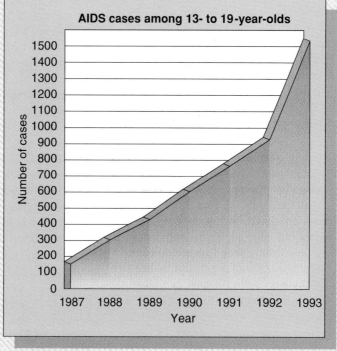

AIDS cases among 13- to 19-year-olds

Number of cases (y-axis: 0 to 1500)
Year (x-axis: 1987, 1988, 1989, 1990, 1991, 1992, 1993)

(a)   (b)

Sidelight figure 43.1 AIDS in action.
(a) *This white blood cell is being penetrated by HIV viruses. A protein present on the virus surface recognizes a receptor on the lymphocyte's plasma membrane, triggering endocytosis—the entire virus is engulfed by the membrane, introducing it into the cell's cytoplasm.* (b) *The number of AIDS cases among 13- to 19-year olds in the United States.*

anus, and thighs. In men, sores appear most often on the head and shaft of the penis, but also on the scrotum, anus, and thighs. In addition to these blisters, sufferers may experience flulike symptoms of general fatigue, nausea, and fever.

Once the initial outbreak of herpes heals, the virus may become dormant, only to reappear at intervals. Some genital herpes sufferers, like those who suffer from the cold-sore type of herpes, claim that their outbreaks occur during times of stress or after exposure to the sun.

Although antiviral drugs can lessen the frequency and severity of the attacks, genital herpes cannot be cured. Once the virus invades the body, there is no way to get rid of it. During outbreaks, the infected person can transmit the virus to his or her sexual partner. A mother may also infect her fetus if she is experiencing a herpes outbreak at the time of birth. In this case, a cesarean section is usually performed to protect the baby against infection.

Chlamydia

Sometimes called the "silent STD," **chlamydia** is caused by an unusual bacterium, *Chlamydia trachomatis*, that has both bacterial and viral characteristics: Like a bacterium, it is susceptible to antibiotics, and like a virus, it depends on its host to replicate its genetic material. The bacterium is transmitted through vaginal, anal, or oral sex with an infected person.

Chlamydia is called the "silent STD" because women usually experience no symptoms until after the infection has become established. The effects of an established chlamydia infection on the female body are extremely serious. Chlamydia can cause pelvic inflammatory disease, which as mentioned earlier, can lead to sterility. Symptoms of PID include painful urination, painful intercourse, vaginal discharge, and irregular vaginal bleeding (such as between menstrual periods or after intercourse). In men, the symptoms are a watery discharge from the penis and burning or itching around the urethra.

body fluids (such as blood, semen, and vaginal fluid) and is *only* transmitted from one individual to another via body fluid. It is *not* transmitted in the air or by casual contact. You cannot catch AIDS from a bathroom seat, from a hot tub shared with an AIDS victim, from kissing an AIDS carrier, or by being bitten by a mosquito that bit an AIDS victim. In studies of several thousand households of AIDS victims, only two family members had contracted the AIDS virus by the end of 1993.

College students must deal with two important routes of HIV infection:

1. *Sexual intercourse.* Both semen and vaginal fluid of infected individuals have high levels of HIV. This means that vaginal, anal, and oral sex with an infected individual can all transmit the virus successfully—and to either sex. HIV is absorbed across the vaginal wall or into the penis. The tiny tears produced during anal sex may facilitate entry even better. The microscopic abrasions that everybody has in their mouth from eating and chewing are a third easy means of entry, making oral sex also dangerous.

   Condoms offer the best available protection from HIV infection, but they must be used properly. Some 10 percent of couples using condoms for birth control end up with a pregnancy, almost always as a result of careless use. This suggests that mixing alcohol or drugs with sexual encounters can cloud your judgment and lead you to not using a condom or using it carelessly. Such a mistake could cost you your life.

2. *Drug use.* A needle used more than once by an infected individual typically harbors large quantities of HIV, both in the fluid that remains behind in the needle and in the body of the hypodermic syringe. Anyone who reuses the needle will become infected. The use of intravenous drugs is itself dangerous—both illegal and life-threatening—but employing a used needle or syringe greatly magnifies the danger.

WHO IS AT RISK?

You are. AIDS is commonly perceived as a disease of homosexual men because the disease first appeared in the United States among the gay community. Because homosexuals tend to confine their sexual interactions to one another, the disease initially spread among homosexuals without entering the larger heterosexual community. That initial segregation appears to be ending: Although only 9 percent of the AIDS cases diagnosed in 1992 were heterosexual nondrug users, the incidence of HIV among heterosexuals is now expanding. Estimates vary widely. AIDS statistics from the Centers for Disease Control show that, in certain categories, heterosexual transmission of HIV is higher than in other categories. For example, in 1992, 39 percent of all reported AIDS cases among women, 16 percent of all reported AIDS cases among individuals thirteen to nineteen years old, and 12 percent of all reported AIDS cases among individuals twenty to twenty-four years old were transmitted heterosexually. These statistics show that heterosexual contact is real and is gaining ground fast. In 1993, AIDS was the leading cause of death of American men ages twenty-five to forty-four.

HOW IS HIV INFECTION DETERMINED?

A simple test identifies infected individuals by detecting antibodies in their blood directed against HIV. Such antibodies, detectable only in the bodies of infected individuals, are the remnants of the body's attempt to ward off HIV infection.

The standard test for detecting the presence of antibodies directed against HIV is called the ELISA antibody test. In the test, about 5 milliliters of blood (roughly the amount that would fit in a paper drinking straw) is drawn and checked for antibodies that will interact with bits of HIV attached to the surface of a plastic dish. If such antibodies are present, the test is said to be positive.

The Red Cross offers an inexpensive walk-in AIDS antibody test at a variety of locations in most metropolitan areas. The Centers for Disease Control also maintains a national AIDS hotline that you can call for information and advice (1-800-342-7154).

At the Red Cross, the testing is completely confidential. The test is coded with a number that the individual selects—the test is not associated with anyone's name—and the individual telephones a few days later for the test results identified with that number. Maintaining the confidentiality of a positive result is very difficult, however. An individual with a positive AIDS test result must see a doctor frequently to monitor possibilities of disease progression. All of this has to be documented in the physician's office, along with billing information. Because these files may be accessed when the individual signs a medical release form or pays bills with insurance, maintaining confidentiality over a long period is impossible.

The only way to survive AIDS is not to contract it. The only way (short of contaminated needles) that any student will contract AIDS is by having unprotected sex with someone who has the virus. But the five-year lag prevents people from knowing who is infected and who is not (unless they are tested), and the disease continues to spread. While some have proposed a widespread antibody testing on college campuses to identify infected individuals and so dampen the spread of the virus while it is still confined to a relatively few individuals, the proposal is controversial, and the associated dangers of invasion of privacy a real concern. In the absence of such wide-scale, on-campus testing, any student you have sex with might be a carrier and not know it. To avoid becoming part of the epidemic, *you* have to accept the responsibility of protecting your health.

Within the last few years, two types of tests for chlamydia have been developed that look for the presence of the bacteria in the discharge from men and women. The treatment for chlamydia is antibiotics, usually tetracycline (penicillin is not effective against chlamydia). Any woman who experiences the symptoms associated with this STD should be tested for the presence of the chlamydia bacteria; otherwise, her fertility might be at risk.

Genital Warts

Genital warts are caused by the human papillomavirus (HPV) and are transmitted through intimate contact with an infected person (figure 43.18). Of the more than thirty known strains of HPV, at least three strains are known to cause cervical cancer in women. Early detection and removal of genital warts is thus essential in cancer prevention.

Genital warts in women appear on the outside of the vagina and may also occur inside the vagina and on the cervix. In

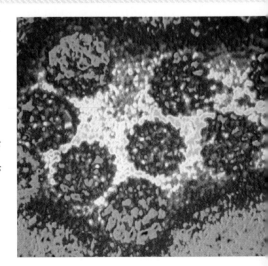

Figure 43.18 The papillomavirus.

This virus, a papovavirus, is responsible for warts. Cervical cancer, which kills about 7,000 American women annually, is associated with this virus, which can be spread sexually. Current estimates show that at least 10 percent of American adults are infected with papillomavirus of the genital tract.

men, the warts occur on the head and shaft of the penis. The warts themselves are small, painless bumps that, as they develop, take on a cauliflower-type appearance. They can appear three weeks to three months after exposure.

Treatment for genital warts involves their removal with any number of agents, including chemicals that freeze the warts, cauterization (burning), and laser treatments. Both partners must be treated, and all warts must be removed to prevent reinfection.

The preceding discussion of STDs may give the impression that sexual activity is fraught with danger, and in a way, it is. It is folly not to take precautions to avoid STDs. The best way to do this is to know sexual partners well enough so that the possible presence of an STD can be discussed. Condom use can also prevent the transmission of most of these diseases. Responsibility for protection lies with the individual.

Sexually transmitted diseases (STDs) are becoming increasingly widespread among American teens and college students, as sexual activity increases. Most can be prevented with condoms.

Overview of Vertebrate Reproduction

Vertebrates carry out reproduction in many different ways, reflecting the evolutionary course of their invasion of the land. These differences are reflected in whether fertilization is external or internal, in whether zygotes begin life as eggs or develop internally, and in the size of broods and the frequency with which they are produced. The story does not end here. The zygote's course of development is influenced in many ways by the organism's reproductive strategy. A zygote nourished by the yolk of an egg develops differently from one nourished by its mother's blood supply. The next chapter considers these differences in developmental processes in detail.

Much of the diversity seen among the vertebrates in reproduction and development, however, reflects the transition to the land. As discussed in previous chapters, the biology of all vertebrates is similar in most other respects. Humans employ a different mechanism than fishes do to obtain the oxygen necessary for metabolism because extracting oxygen from air presents different problems than extracting it from water. However, having obtained oxygen, humans circulate it to body tissues in much the same manner that fishes do. The similarities among vertebrates are more striking than their differences.

▼

EVOLUTIONARY VIEWPOINT

The evolution of reproduction among terrestrial vertebrates is a story of decreasing reliance on the environment for moisture, temperature, and safety. Because mammals represent the current end point of that journey, as the most reproductively independent of vertebrates, many might leap to the conclusion that mammals are the most successful vertebrates. In fact, fish (20,000 species), birds (8,900 species), and reptiles (7,400 species) far outnumber mammals (4,500 species). There are almost as many amphibians (4,200 species). Judged by numbers of species, mammals are not a highly successful group. But they dominate the land by virtue of their size: Virtually all large terrestrial vertebrates living today are mammals.

SUMMARY

1. Sexual reproduction evolved in the sea. Among most fishes and the amphibians, fertilization is external, with gametes being released into water. Amphibians and the great majority of fishes are oviparous, the young being nurtured by the egg rather than by the mother.

2. Successful invasion of the land by vertebrates involved major changes in reproductive strategy. The first change was the watertight egg of the reptiles, which could survive in dry places. Birds, as well as a primitive group of mammals called monotremes, have the same kinds of eggs as reptiles.

3. The second important change in reproductive adaptation to life on land was that of the marsupials and placental mammals. In them, fertilization is internal. The young are nourished within a pouch or inside the mother's body until they reach a fairly advanced stage of development.

4. The male gametes, or sperm, of mammals are produced within the testes. In human males, hundreds of millions of sperm are produced each day. Sperm mature and become motile in the epididymis, and they are stored in the vas deferens. Stimulation of the penis causes it to become distended and erect, and the sperm are then delivered from the vas deferens to the urethra at the base of the penis. Further stimulation causes violent muscle contractions, which ejaculate the sperm from the penis.

5. At birth, female mammals contain all the gametes, or primary oocytes, that they will ever have—approximately two million. All but a very few of these are arrested in meiotic prophase. At each ovulation, the first meiotic division of one or a few oocytes is completed. The second meiotic division does not occur until after fertilization.

6. Fertilization occurs within the fallopian tubes. The journey of an oocyte to the uterus takes three days, and the oocyte is viable for only twenty-four hours unless it is fertilized. Consequently, only those secondary oocytes that have been reached within twenty-four hours by sperm swimming up the fallopian tubes from the uterus can be fertilized successfully. Fertilized ovum continue their journey down the fallopian tubes and attach to the lining of the uterus, where their development proceeds.

7. Estrus is the period of maximum female receptivity to sexual advances. Periods of estrus correspond to ovulation events. Human females, however, are sexually receptive throughout the reproductive cycle.

8. Reproduction in mammals is regulated by hormones—typically produced by the pituitary on commands from the hypothalamus—that circulate in the bloodstream. When a hormone molecule reaches its target tissue, it binds to specific receptors of the target cell.

9. The mammalian reproductive cycle is composed of two phases: (1) a follicular phase, in which one or more oocytes in the ovary are hormonally signaled to complete their development; and (2) a luteal phase, in which one or more mature oocytes are released into the fallopian tube, a process called ovulation. A complete menstrual cycle in a human female takes about twenty-eight days.

10. Human intercourse is marked by four physiological periods: excitement, plateau, orgasm, and resolution. Orgasm in women is highly variable and may be prolonged. Orgasm in men is uniformly abrupt and coincides with sperm ejaculation.

11. Humans practice a variety of birth-control procedures, of which condoms (for men) and birth-control pills (for women) are perhaps the most common. Surgical procedures, such as vasectomies and tubal ligations, that block the delivery of gametes are increasingly chosen birth-control

methods. Birth control is not always effective, and some women terminate unwanted pregnancies. In the United States, as in other developed countries, the abortion rate exceeds the birth rate.

12. All sexually active persons should be aware of and take precautions against sexually transmitted diseases, such as AIDS, gonorrhea, syphilis, genital herpes, chlamydia, and genital warts. Some of these diseases can lead to sterility, while AIDS is always fatal.

REVIEWING THE CHAPTER

1. Where did sex first evolve?
2. What are the four strategies of vertebrate sex and reproduction?
3. Why do fishes use external fertilization?
4. How are amphibians still tied to the water?
5. What are the advantages of the watertight egg found in reptiles and birds?
6. How do mammalian mothers nourish their young?
7. How is the human reproductive system characterized?
8. What are the features of the male reproductive system?
9. How are male gametes formed in the testes?
10. How are male gametes conveyed through the penis?
11. What are the features of the female reproductive system?

12. How are female gametes formed in the ovaries?
13. How many female gametes mature each month?
14. Where does fertilization occur?
15. What is the estrous cycle?
16. How do sex hormones function?
17. How is the human reproductive cycle characterized?
18. What happens during the follicular phase of the human reproductive cycle?
19. What happens during the luteal phase of the human reproductive cycle?
20. What characterizes the physiology of human intercourse?
21. What characterizes the excitement phase?
22. What occurs during the plateau phase?
23. How is orgasm achieved?
24. What is the resolution period?
25. What is the rationale for contraception and birth control?
26. What is abstinence?
27. What birth-control measures block sperm?
28. How can sperm be destroyed?

29. How can primary oocyte maturation be prevented?
30. What are birth-control implants?
31. How does the injectable contraceptive work?
32. How is surgical intervention a completely effective means of birth control?
33. How can embryo implantation be prevented?
34. How is abortion achieved?
35. What are the causes of sexually transmitted diseases (STDs)?
36. What is gonorrhea, how is it transmitted, and how is it treated?
37. Why is syphilis a very destructive sexually transmitted disease?
38. Why is there no cure for genital herpes?
39. Why is chlamydia the "silent STD"?
40. What causes genital warts, and how are they cured?
41. How serious is AIDS on the college campus?
42. How can vertebrate reproduction be best summarized?

COMPLETING YOUR UNDERSTANDING

1. The average age at which Americans have sex for the first time is between —— years of age.
    a. 12–13   d. 18–19
    b. 14–15   e. 20–22
    c. 16–17   f. 23–24
2. Of the thousands of eggs fertilized in an individual fish mating, why do only a few survive and grow to maturity?
3. What is metamorphosis? Give an example of where it occurs.

4. Internal fertilization is characteristic of [*three answers required*]
    a. reptiles.   d. mammals.
    b. birds.   e. protists.
    c. fishes.   f. amphibians.
5. Which group of animals achieves internal fertilization without using a penis? How is fertilization accomplished in this group?
6. What is the difference between oviparous and viviparous? To which animal groups do these terms apply?
7. Which group of mammals lays eggs?

8. In the human male, why is the temperature in the testes 3 degrees Celsius lower than that of the rest of the body?
9. What hormones are required for normal functioning of the testes, and where do these substances come from?
10. What happens in seminiferous tubules?
11. How is an erection achieved in the male penis during arousal?
12. How many sperm per milliliter are necessary for a human male to be considered fertile?
13. How do men and women differ in gamete production?

14. Why is it a concern to women to bear children before age thirty-five?

15. Approximately what percent of a women's primary oocytes mature during her lifetime?

16. Where are oocytes fertilized in the female human reproductive system?

17. What are pheromones, and where are they found?

18. Why do smaller mammals produce more offspring than larger mammals?

19. What does the hypothalamus in the brain have to do with sex hormones?

20. What is the corpus luteum, where is it found, and what does it do?

21. During which phase of the human menstrual cycle is the progesterone level in the blood highest?

22. What is the correlation between the menstrual phase and ovulation in the human female?

23. Ejaculation occurs during which phase of sexual intercourse?

    a. Excitement   c. Orgasm

    b. Plateau   d. Resolution

24. The highest failure rate in birth control is associated with

    a. condoms.   d. the rhythm method.

    b. abstinence.   e. IUDs.

    c. foams.   f. injectable contraceptives.

25. What are the disadvantages of the different birth-control devices?

26. What is coitus interruptus and why is it a risky birth-control method?

27. What is the main reason for why condoms sometimes fail as a method of birth control?

28. In a vasectomy, what part of the male reproductive system is removed surgically?

29. What is a tubal ligation, and what does it involve?

30. What is RU 486, how does it work, and where is it available?

31. What is the only way that you can get infected with HIV?

32. What is the current evidence in the United States that AIDS should be of concern to the heterosexual population?

33. Which of the following STDs cannot be treated with antibiotics and why? [*three answers required*]

    a. Gonorrhea   d. Chlamydia

    b. Syphilis   e. Genital warts

    c. Genital herpes   f. AIDS

34. Why do you think that far more women than men attend most AIDS information sessions given on college campuses?

35. Probably the most controversial issue to arise in this consideration of sex and reproduction is abortion. List arguments for and against abortion. Under what conditions would you permit abortions? Forbid them? Do you think the disadvantages of abortion are in any way counterbalanced by the advantages it provides to underdeveloped countries with very high birth rates and swelling populations? Should the United States promote or oppose dissemination of information about abortion in underdeveloped countries?

36. Some fishes and many reptiles are viviparous, retaining fertilized eggs within a mother's body to protect them. Birds, however, although they evolved from reptiles, never employ this means of egg protection. Can you think of a reason why?

37. Relatively few kinds of animals have both male and female sex organs on the same individual, whereas most plants do. Propose an explanation for this.

FOR FURTHER READING

Aral, S., and K. Holmes. 1991. Sexually transmitted diseases in the AIDS era. *Scientific American,* Feb., 52–59. Gonorrhea, syphilis, and other infections that still exact a terrible toll.

Frisch, R. 1988. Fatness and fertility. *Scientific American,* March, 88–95. The argument that dieting and exercise can lead to infertility. The author believes that fat tissue exerts a regulatory effect on female reproductive ability.

Hrdy, S. 1988. Daughters or sons: Can parents influence the sex of their offspring? *Natural History,* April, 64–83. New research that says that it is distinctly possible to influence the odds.

Lagercrantz, H., and A. Slotkin. 1986. The "stress" of being born. *Scientific American,* April, 100–107. How passage through the narrow birth canal triggers the release of hormones important to the newborn's future survival.

Lein, A. 1979. *The cycling female.* San Francisco: W. H. Freeman. A short, informal description of the human menstrual cycle and its physical and emotional effects on women.

Leishman, K. 1987. Heterosexuals and AIDS. *The Atlantic,* Feb., 39–58. A chilling account of the difficulty of modifying sexual behavior, despite the knowledge of the dangers associated with AIDS.

Development

This fetus started out as a single cell 150 days ago, and still has four months of growth ahead before making its debut outside the womb.

FOR REVIEW

Here are some important terms and concepts that have been discussed in previous chapters and that you will encounter again in this chapter. Review them before proceeding if necessary.

The amniotic egg (*chapter 18*)
Deuterostomes (*chapter 29*)
Ectoderm, endoderm, mesoderm (*chapter 29*)
Coelom (*chapter 29*)
Radial and spiral cleavage (*chapter 29*)
Chordates (*chapter 29*)
Terrestrial reproductive strategies (*chapter 43*)

Vertebrate development is a complex, dynamic process, a symphony of cell movement and change that starts with the formation of a single diploid cell from sperm and egg. From this single cell will arise one hundred trillion others as the individual that cell is destined to become develops and grows to adulthood. All the complexity of the vertebrate body, the beauty and cleverness of design seen repeatedly in the preceding eleven chapters, are established during an elaborate developmental process described in this chapter. **Development** is not simply the beginning of life—-it is the process that determines life's form and function.

In almost all vertebrates, development begins with an encounter between two haploid gametes that unite to form a single diploid cell called a zygote. This zygote grows by a process of cell division and differentiation into a complex multicellular animal composed of many different tissues and organs. Although some details differ from group to group, development is fundamentally the same in all vertebrates.

In vertebrates, development occurs in six stages, outlined in table 44.1. This chapter discusses each of these stages in turn, then considers the mechanisms governing developmental changes, and concludes with a detailed description of human development.

Fertilization: Beginning the Action

In vertebrates, as in all sexual animals, the first step in reproduction is the union of male and female gametes, a process called **fertilization.** Fertilization consists of three stages: (1) penetration, (2) activation, and (3) pronuclear fusion. The male gametes of vertebrates, like those of other animals, are small, motile **sperm.** Each sperm is shaped like a tadpole, with a head containing a haploid nucleus and a long tail. Sperm are among the smallest cells in the body. The female gametes, called **secondary oocytes,** are large cells (figure 44.1). In many vertebrates, secondary oocytes contain significant amounts of yolk.

Table 44.1   Stages of Development

Stage	Development
1. Fertilization	The male and female gametes form a zygote.
2. Cleavage	The zygote rapidly divides into many cells, with no overall increase in size. These divisions set the stage for development, since different cells receive different portions of the ovum cytoplasm and hence different regulatory signals.
3. Gastrulation	The cells of the zygote move, forming three cell layers. These layers are the primary cell types: ectoderm, mesoderm, and endoderm.
4. Neurulation	In all chordates, the first organ to form is the notochord, followed by formation of the dorsal nerve cord.
5. Neural crest formation	The first uniquely vertebrate event is the formation of the neural crest. From it develop many of the uniquely vertebrate structures.
6. Organogenesis	The three primary cell types then proceed to combine in various ways to produce the organs of the body.

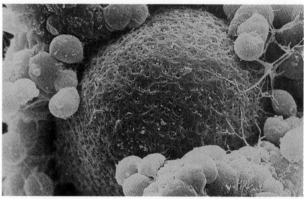

(a)

(b)

Figure 44.1 Male and female gametes.
(a) *Secondary oocyte surrounded by nutritional cells.* (b) *One sperm entering an egg.*

Penetration: How Sperm Work to Acquire Entry to Oocyte

In fishes and amphibians, fertilization is typically external, whereas in all other vertebrates, it occurs internally. Internal fertilization is achieved by the release of a mature secondary oocyte into a body cavity where the oocyte can be fertilized by one of the many sperm introduced into the female reproductive tract during mating. The actively swimming sperm migrate up the fallopian tube until they encounter the mature secondary oocyte.

Like a traveling princess, the mammalian secondary oocyte is surrounded by a great deal of baggage. The secondary oocyte itself is encased within an outer membrane called the **zona pellucida,** which is, in turn, surrounded by a protective layer of follicle cells. The first sperm to make its way through the zona pellucida adheres to the secondary oocyte membrane by the tip of the sperm cell head, the **acrosome.** From its acrosome, the sperm releases enzymes that cause the plasma membranes of the sperm and secondary oocyte to fuse. The oocyte cytoplasm bulges out at this point, engulfing the head of the sperm and permitting the sperm nucleus to enter the oocyte cytoplasm (figure 44.2).

Activation: The Secondary Oocyte's Quick Response to Sperm Entry

The series of events initiated by sperm penetration is collectively called **activation.** Three effects of sperm penetration are:

1. In general, the penetration of the first sperm initiates changes in the secondary oocyte membrane that prevents the entry of other sperm.

2. In mammals, secondary oocytes are produced by meiotic divisions that occur early in a female's life, but in most cases, meiosis stops at an intermediate stage. Sperm penetration triggers the resumption of meiosis: The chromosomes in the secondary oocyte nucleus complete meiosis, producing two egg nuclei. One of these two newly formed nuclei is extruded from the secondary oocyte as a **polar body,** leaving a single haploid nucleus. After the second meiotic division, the secondary oocyte is called an **ovum** (plural, *ova*) (from the Latin word meaning "egg").

3. In vertebrates, a third effect of sperm penetration is the rearrangement of ovum cytoplasm and its metabolic activation. A series of cytoplasmic movements is initiated within the ovum around the point of sperm entry. These movements ultimately establish the bilateral symmetry of the developing organism. In frogs, for example, sperm penetration causes an outer pigmented cap of ovum cytoplasm to rotate toward the point of entry, uncovering a **gray crescent** of interior cytoplasm opposite the point of penetration. The position of the gray crescent determines the orientation of initial cell division. A line drawn between the point of sperm entry and the gray crescent would bisect the right and left halves of the future adult (figure 44.3).

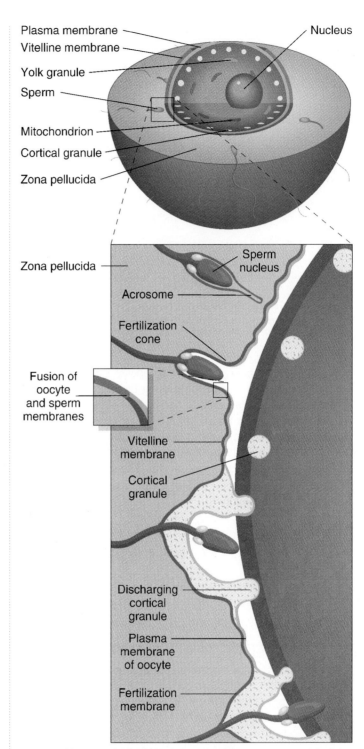

Figure 44.2 Fertilization in a sea urchin.
A sea urchin egg is surrounded by a membrane that encloses the jellylike zona pellucida. The actual oocyte sits in the middle of the zona pellucida. The oocyte is bounded by two membranes: the vitelline membrane and the plasma membrane. (1) The acrosome of the sperm cell contacts the vitelline membrane. (2) The sperm and oocyte membranes fuse, forming a fertilization cone. (3) Meanwhile, structures called cortical granules release their contents into the fertilization cone and prevent other sperm from fertilizing the oocyte. (4) The head of the sperm cell is completely enclosed in a fertilization membrane.

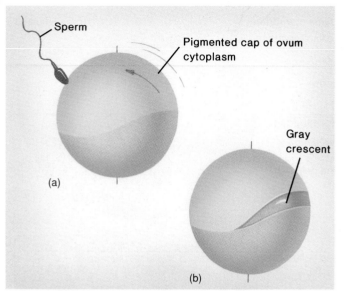

Figure 44.3 Gray crescent formation in frogs.
(a) *Sperm penetrates frog secondary oocyte.* (b) *The gray crescent usually appears at the point opposite the point of sperm penetration.*

In some vertebrates, a secondary oocyte can be activated without sperm entry by simply pricking the egg membrane. In these situations, the secondary oocyte may go on to develop parthenogenetically. In parthenogenetic reproduction, or **parthenogenesis,** an embryo develops from an unfertilized oocyte. A few kinds of amphibians, fishes, and reptiles rely entirely on parthenogenetic reproduction. The secondary oocytes of some domestic animals, such as turkeys, can be made to reproduce by parthenogenesis.

Pronuclear Fusion: Joining of Ovum and Sperm to Create a Diploid Nucleus

In the third stage of fertilization, called **pronuclear fusion,** the sperm nucleus unites with the haploid nucleus of the ovum to form a diploid zygotic nucleus. This pronuclear fusion is triggered by the activation of the secondary oocyte. If a sperm nucleus is introduced by microinjection (a procedure in which a sperm cell is artificially injected into a secondary oocyte with a very fine needle), without activation of the secondary oocyte, pronuclear fusion of the two nuclei will not take place. The nature of the signals exchanged between the two nuclei is not known. Sidelight 44.1 explores in vitro fertilization, in which a human secondary oocyte is fertilized on a glass plate with sperm, and the resulting zygote is implanted in the mother's uterus.

The three stages of fertilization are penetration, activation, and pronuclear fusion. Penetration initiates a complex series of developmental events, including major movements of cytoplasm, that eventually lead to the fusion of the ovum and sperm nuclei.

Cleavage: Setting the Stage for Development

The second major event in vertebrate development is the rapid division of the zygote into a larger and larger number of smaller and smaller cells—an **embryo.** This period of division, called **cleavage,** is not accompanied by any increase in the embryo's overall size. The resulting tightly packed mass of about thirty-two cells is called a **morula,** and each individual morula cell is referred to as a **blastomere.**

Blastomeres are by no means equivalent to one another. Different blastomeres may contain different components of the ovum cytoplasm, particularly at later stages of division, that dictate different developmental fates for the cells in which they are present. The cells of the morula continue to divide without an overall increase in size, each cell secreting a fluid into the center of the cell mass. Eventually, a hollow mass of five hundred to two thousand cells, called a **blastula,** surrounds a fluid-filled cavity called the **blastocoel.**

Cleavage Pattern Determined by Presence of Yolk

The pattern of cleavage division is greatly influenced by the amount of yolk present in the ovum. As discussed in chapter 43, vertebrates have embraced a variety of reproductive strategies that involve different patterns of yolk utilization (figure 44.4).

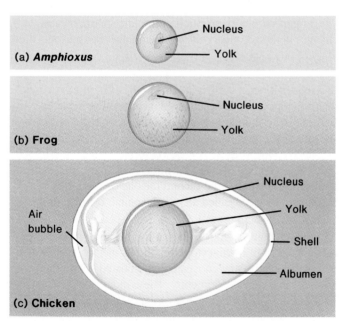

Figure 44.4 Three kinds of ova.
(a) *In primitive chordates like the lancelet, ovum organization is simple, with a central nucleus surrounded by yolk.* (b) *A frog ovum has much more yolk, and the nucleus is displaced toward one pole.* (c) *A bird ovum has a complex organization, with a nucleus astride the surface of a large, central yolk, like a spot painted on a balloon.*

Primitive Aquatic Vertebrates
When an ovum contains little or no yolk, cleavage occurs throughout the entire ovum. This pattern, called **holoblastic cleavage,** was characteristic of vertebrate ancestors and is still seen in such groups as the lancelets and agnathans. It results in the formation of a symmetrical blastula composed of cells of approximately equal size.

Amphibians and Advanced Fishes
The ova of bony fishes and frogs contain much more yolk in one hemisphere than in the other. Because yolk-rich cells divide much more slowly than do yolk-poor cells, unequal holoblastic cleavage results in a very asymmetrical blastula, with large cells containing a lot of yolk at one pole and a concentrated mass of small cells containing very little yolk at the other (figure 44.5).

Reptiles and Birds
The ova of reptiles, birds, and some fishes are composed almost entirely of yolk, with a small amount of cytoplasm concentrated at one pole. In such ova, cleavage occurs only in the tiny disk of polar cytoplasm, called the **blastodisc,** that lies astride the large ball of yolk material. Such a pattern of cleavage is called **meroblastic cleavage.** The resulting stage of the embryo, called a **blastoderm,** is not spherical, but rather has the form of a hollow cap perched on the yolk (figure 44.6).

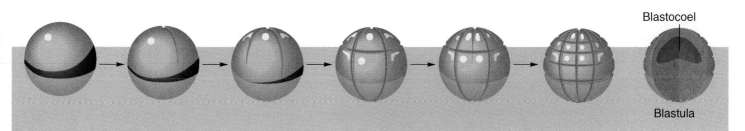

Figure 44.5 Holoblastic cleavage.
In the holoblastic cleavage shown here, cleavage occurs throughout the ovum, but in an asymmetrical pattern.

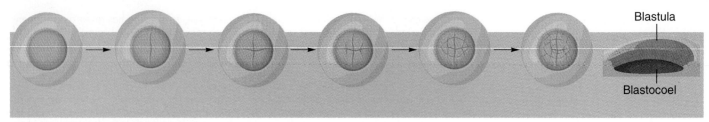

Figure 44.6 Meroblastic cleavage.
Some ova are composed almost entirely of yolk, with a small amount of cytoplasm concentrated at one pole. In meroblastic

cleavage, cleavage occurs only in the tiny disk of polar cytoplasm, called the blastodisc, that lies astride the yolk.

Mammals

Mammalian ova are in many ways similar to the reptilian ova from which they evolved, except that they contain very little yolk. Because no mass of yolk impedes cleavage in mammalian ova, the cleavage of the developing zygote is holoblastic. Such cleavage forms a ball of cells surrounding a blastocoel. In mammalian ova, an inner cell mass is concentrated at one pole. This interior plate of cells is analogous to the blastodisc of reptiles and goes on to form the developing embryo (figure 44.7). The outer sphere of cells, called a **trophoblast,** is analogous to the cells that form the membrane that functions as a water-tight covering for the reptilian ovum. During the course of mammalian evolution, these cells have changed to carry out a very different function. The trophoblast develops into the **chorion** and a complex series of membranes known as the **placenta,** which connects the developing embryo to the mother's blood supply.

> *Development is initiated in the zygote by a series of rapid cell divisions called cleavage, producing a ball of cells called a blastula. The evolution of the amniotic ovum in reptiles caused an alteration in the pattern of cleavage. The alteration was related to the presence and amount of yolk. This kind of cleavage pattern is carried on by birds, a reflection of their ancestry.*

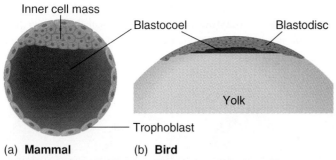

(a) Mammal   **(b) Bird**

Figure 44.7 Similarity of ova of mammals and birds.
(a) A mammalian blastula is composed of a sphere of cells (the trophoblast) surrounding a cavity (the blastocoel) and an inner cell mass. (b) A bird blastula is a disk rather than a sphere, and the blastodisc rests astride a large yolk mass. In these ova, the blastocoel is a cavity between the blastodisc and the yolk.

Blastula Cells Prepatterned for Development

Viewed from the outside, a blastula looks like a simple ball of cells that all resemble one another. But the apparent close similarity of these cells is misleading. In fact, the cells differ from one another in three essential respects:

1. Each cell contains a different portion of cytoplasm derived from the ovum.

2. Some cells are larger than others, containing more yolk and dividing more slowly.

3. Each cell is in contact with a different set of neighboring cells.

Ova contain many substances that act as genetic signals during the early stages of zygote development. These signal substances are not distributed uniformly within the ovum's cytoplasm, but instead are clustered at specific sites within the ovum. The location of each site is genetically determined by information encoded on the mother's chromosomes.

When the secondary oocyte is activated during fertilization, its cytoplasm reorients itself with respect to the site of sperm entry. During the cleavage divisions that follow, the signal substances within this cytoplasm are partitioned into different daughter cells. The signals endow the different daughter cells with distinct developmental instructions. The blastula is therefore **prepatterned** because the pattern of its cytoplasm determines the future orientation of the different embryonic cells.

Gastrulation: The Onset of Developmental Change

The first visible results of prepatterning and of the cell orientation within the blastula can be seen immediately after completion of the cleavage divisions. Certain groups of cells *move* inward from the sphere surface in a carefully orchestrated migration called **gastrulation.**

How can cells move within a cell mass? As explained in chapter 4, cell shape can readily be changed by microfilament contraction. Apparently, the migrating cells creep over the stationary ones by means of a series of microfilament contractions. The migrating cells move as a single mass because they adhere to one another. How do cells "know" to which other cell to adhere? Within an adhering cell, genes are expressed that cause the synthesis of specific adhesive molecules on the cell surface that attach to similar adhesive molecules on the surfaces of other adhering cells.

During gastrulation, about half of the blastula's cells move into the ball's hollow interior. By doing so, they form a structure that looks something like an indented tennis ball. Just as the pattern of cleavage divisions in different groups of vertebrates depends heavily on the amount and distribution of yolk in the ovum, so the pattern of gastrulation varies among vertebrates, depending on the blastula shapes produced by earlier cleavage divisions.

Aquatic Vertebrates

In fishes and other aquatic vertebrates with asymmetrical yolk distribution in their ova, the blastula produced by the cleavage divisions has two distinct poles, one more rich in yolk than the other. The blastula hemisphere comprising yolk-rich cells is called the **vegetal pole;** the opposite hemisphere, comprising relatively yolk-poor cells, is called the **animal pole.**

In primitive chordates such as lancelets, the animal hemisphere bulges inward, invaginating into the blastocoel cavity. Eventually, the inward-moving wall of cells pushes up against the opposite side of the blastula, and then it ceases to move. The resulting two-layered, cup-shaped embryo is the **gastrula.** The hollow crater resulting from the invagination is called the **archenteron,** and it becomes the progenitor of the gut. The opening of the archenteron, the future anus of the lancelet, is the **blastopore.**

Gastrulation in the lancelets produces an embryo with two cell layers: an outer **ectoderm** and an inner **endoderm.** A third cell layer—the **mesoderm**—forms soon afterward between these two layers from pouches pinched off of the endoderm. The formation of these three primary cell types sets the stage for all subsequent tissue and organ differentiation because the descendants of each cell type are destined to have very different developmental fates (table 44.2).

In amphibian blastulas, the yolk-laden cells of the vegetal pole are fewer and far larger than the yolk-poor cells of the animal pole. Because of this cell distribution, it is mechanically not feasible to invaginate the blastula at the vegetal pole. Instead, a layer of cells from the animal pole folds down over the yolk-rich cells and then invaginates inward. The place where the invagination begins is called the **dorsal lip.** As in the lancelets, the invaginating cell layer eventually eliminates the blastocoel cavity, its cells pressing against the inner surface of the opposite side of the embryo.

Table 44.2	Developmental Fates of the Primary Tissues
Tissue Type	**Develops into . . .**
Ectoderm	Skin, central nervous system, sense organs, neural crest
Mesoderm	Skeleton, muscles, blood vessels, heart, gonads
Endoderm	Digestive tract, lungs, many glands

In both fishes and amphibians, the opening of the cavity produced by the invagination is called the blastopore. In this case, the blastopore is filled with yolk-rich cells, the **yolk plug.** The outer layer of cells in the gastrula, which is formed as a result of these cell movements, is the ectoderm, and the inner layer is the endoderm. Some ectodermal cells migrate between the ectoderm and endoderm to form the mesoderm layer (figure 44.8).

Reptiles, Birds, and Mammals

In the blastodisc of a chick, the developing embryo is not shaped like a sphere. Instead, it is a hollow cap of cells situated over the animal pole of the large yolk mass. Despite this seemingly great difference between the developing embryos of birds, reptiles, and mammals on the one hand and amphibians on the other hand, the pattern of establishing the three primary cell layers is basically similar in all of these groups.

No yolk separates the two sides of the blastodisc in reptiles, birds, and mammals. Consequently and without cell movement, the lower cell layer is able to differentiate into endoderm, the upper layer into ectoderm. Just after this differentiation, much of the mesoderm and endoderm arises by the invagination of cells from the upper layer inward, along the edges of a furrow that appears at the embryo's longitudinal midline. The site of this invagination, which is analogous to an elongated blastopore, appears as a slit on the surface of the gastrula. Because of its appearance, it is called the **primitive streak** (figure 44.9). Gastrulation occurs at the site of formation of a primitive streak in the reptiles and in their descendants—the birds and mammals (figure 44.10).

The many cells of the blastula gain unequal portions of the ovum's cytoplasm during cleavage. This asymmetry results in the activation of different genes and a repositioning of cells with respect to one another, which establishes the three primary cell types: ectoderm, mesoderm, and endoderm.

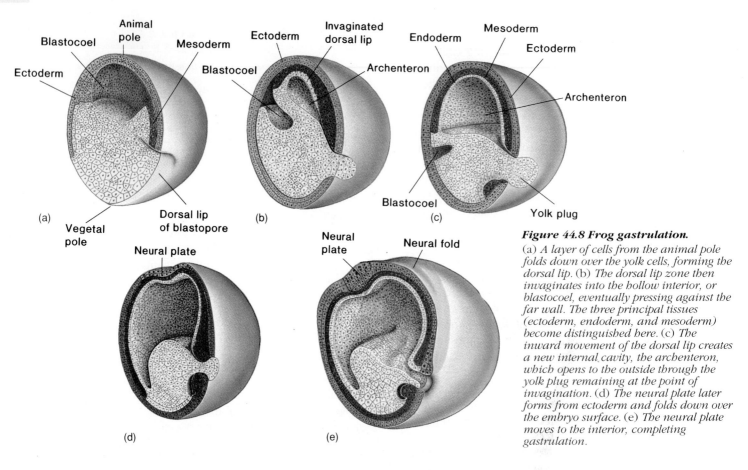

Figure 44.8 Frog gastrulation.

(a) *A layer of cells from the animal pole folds down over the yolk cells, forming the dorsal lip.* (b) *The dorsal lip zone then invaginates into the hollow interior, or blastocoel, eventually pressing against the far wall. The three principal tissues (ectoderm, endoderm, and mesoderm) become distinguished here.* (c) *The inward movement of the dorsal lip creates a new internal cavity, the archenteron, which opens to the outside through the yolk plug remaining at the point of invagination.* (d) *The neural plate later forms from ectoderm and folds down over the embryo surface.* (e) *The neural plate moves to the interior, completing gastrulation.*

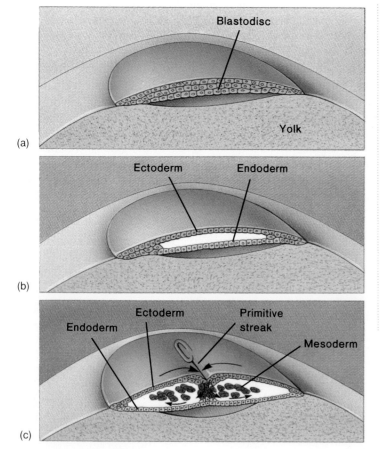

Figure 44.9 Gastrulation of the chick blastodisc.
(a) *The two sides of the chick blastodisc are not separated by yolk.* (b) *The upper layer of the blastodisc differentiates into ectoderm, the lower layer into endoderm.* (c) *Among the cells that migrate into the interior through the dorsal primitive streak are future mesodermal cells.*

The events of gastrulation determine the basic developmental pattern of the vertebrate embryo. By the end of gastrulation, cells have been distributed into the three primary cell types. Although the position of the yolk mass dictates changes in the details of gastrulation, the end result of the process is fundamentally the same in all deuterostomes (as defined in chapter 29, animals in which the anus forms from or near the blastopore and the mouth forms subsequently on another part of the blastula): The ectoderm is destined to form the epidermis and neural tissue; the mesoderm to form the connective tissue, muscle, and vascular elements; and the endoderm to form the lining of the gut and its derivatives.

Neurulation: The Determination of Body Architecture

In the next step of vertebrate development, the three primary cell types begin their development into the tissues and organs of the body. In all chordates, tissue differentiation begins with the formation of two characteristic morphological features: the

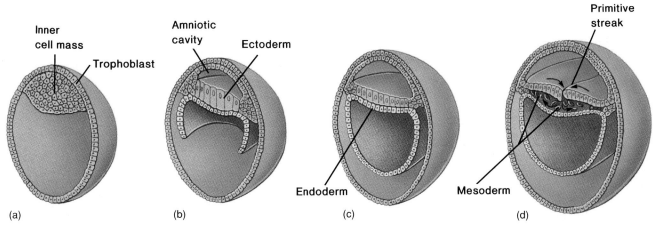

Inner
cell mass

Trophoblast

Amniotic
cavity

Ectoderm

Endoderm

Primitive
streak

Mesoderm

(a) (b) (c) (d)

Figure 44.10 Mammalian gastrulation.
(a) *The amniotic cavity forms within the inner cell mass.* (b) *and* (c) *In the base of the inner cell mass, layers of ectoderm and endoderm differentiate, as in the chick blastodisc.* (d) *A primitive* streak *develops, through which cells destined to become mesoderm migrate into the interior, again reminiscent of gastrulation in the chick.*

notochord and the hollow **dorsal nerve cord.** This stage in development, called **neurulation,** occurs only in the chordates.

The first structure to form—the notochord—is first visible soon after gastrulation is complete and is formed from mesoderm tissue along the embryo's midline, below its dorsal surface. After the notochord has been laid down, the dorsal nerve cord forms from the ectoderm. The region of the ectoderm that is located above the notochord later differentiates into the spinal cord and brain. The process is illustrated in figure 44.11: First, a layer of ectodermal cells situated above the notochord invaginates inward, forming the **neural groove** along the embryo's long axis. Then the edges of this groove move toward each other and fuse, creating the long, hollow **neural tube,** which runs beneath the surface of the embryo's back.

> *The key developmental event that marks the evolution of chordates is neurulation, the elaboration of a notochord and a dorsal nerve cord.*

While the neural tube is forming from ectoderm, the rest of the basic body architecture is being rapidly determined by changes in the mesoderm. On either side of the developing notochord, segmented blocks of tissue form. Ultimately, these blocks, or **somites,** give rise to the muscles, vertebrae, and connective tissue. As the process of development continues, more somites are formed progressively. Many of the significant glands of the body, including the kidneys, adrenal glands, and gonads, develop within another strip of mesoderm that runs alongside the somites. The remainder of the mesoderm layer moves out and around the inner endoderm layer of cells and eventually surrounds it entirely. As a result of this movement, the mesoderm forms a hollow tube within the ectoderm. The space within this tube is the coelom; it contains the endoderm layers that ultimately form the lining of the stomach and gut.

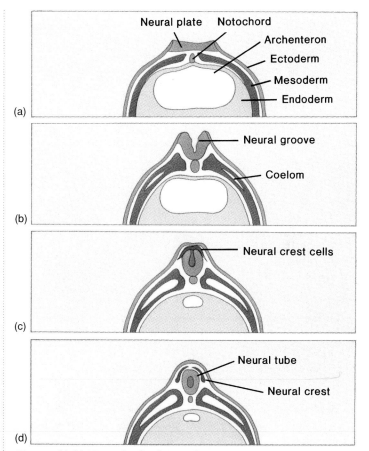

(a) Neural plate Notochord
Archenteron
Ectoderm
Mesoderm
Endoderm

(b) Neural groove
Coelom

(c) Neural crest cells

(d) Neural tube
Neural crest

Figure 44.11 Neural tube formation.
The neural tube forms above the notochord (a) *when ectodermal cells of the neural plate fold together to form the neural groove* (b), *which eventually closes* (c) *to form a hollow tube* (d). *As this is happening, some of the cells from the dorsal margin of the neural tube differentiate into the neural crest, which is characteristic of vertebrates.*

Neural Crest Formation: Evolutionary Origins of the Vertebrates

Neurulation occurs in all chordates; the process is much the same in a lancelet as it is in a human being. The next stage in the development of a vertebrate, however, is unique to that group and is largely responsible for the characteristic body architecture of its members. Just before the neural groove closes over to form the neural tube, its edges develop a special strip of cells, the **neural crest,** which becomes incorporated into the roof of the neural tube (see figure 44.11*c, d*). Subsequently, the cells of the neural crest shift laterally to the sides of the developing embryo. The appearance of the neural crest was a key event in the evolution of the vertebrates because neural crest cells, migrating to different parts of the embryo, ultimately develop into the structures characteristic of the vertebrate body (figure 44.12).

> *The appearance of the neural crest in the developing embryo marked the beginning of the first truly vertebrate phase of development, since many of the structures that are characteristic of vertebrates are derived directly or indirectly from neural crest cells.*

Organogenesis: Development of Organs and Organ Systems

The next process in vertebrate development is **organogenesis,** in which the embryo's organs and organ systems are formed. As already noted, the different organ systems are formed from the different tissue layers that have already been established in the embryo. The endoderm, for instance, gives rise to the inner endothelial linings of the respiratory and reproductive tracts as well as to the major endocrine glands. The mesoderm gives rise to the circulatory system, including the blood and blood vessels; the coverings of various organs; the sex organs; the inner layer of the skin and bones; and the kidney. The ectoderm gives rise to the outer layers of the skin as well as to the nervous system.

The development of these different organs and organ systems involves two intricately related processes: **morphogenesis** and **differentiation.** Morphogenesis involves changes in the *shape* and *size* of different cells that lead to the cells' specialization in the different organs. Differentiation refers to the change in *function* of the cells destined to become specialized in different organs. Examination of the development of a particular organ, such as the pancreas, demonstrates how these two processes work together: As discussed in chapter 37, the pancreas is both an endocrine and exocrine gland. It secretes

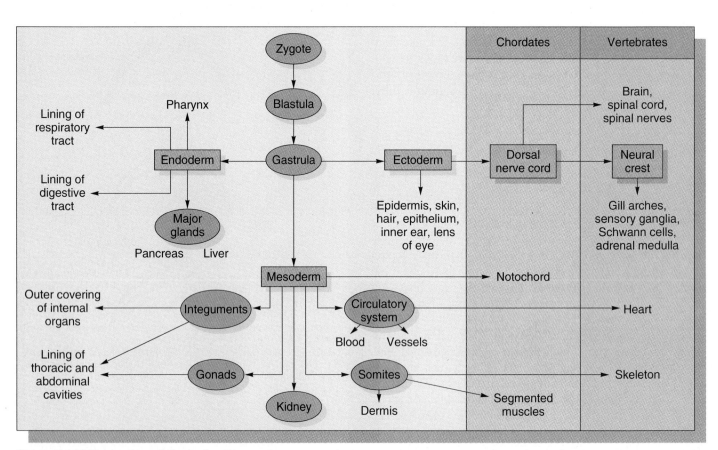

Figure 44.12 Derivation of the major tissue types.
The key role of the neural crest is evident from the many characteristically vertebrate features that derive from it.

digestive enzymes into the small intestine and also possesses cells—the alpha- and beta-cells—that release hormones. The endoderm cells from which the pancreas arises must therefore change their shape and size and differentiate into the different exocrine and endocrine cells that comprise this organ. Organogenesis, then, is a process of specialization in which undifferentiated cells of the three primary tissues change and differentiate into cells that have characteristic shapes and functions.

Morphogenesis begins with **cell migration,** when cells move from the primary tissue layers to the areas where the particular organs of the vertebrate body will be located. Cells move through the action of pseudopods, the cytoplasmic extensions that allow the cell to move forward.

The development of some organs requires that some of the cells die in order to fine-tune the structure. For example, the fingers of humans begin as paddlelike structures, with tissue present between the fingers. Separation of the fingers involves the **cell death** of the tissue between the fingers (figure 44.13).

> *During organogenesis, the organs and organ systems are formed. Organogenesis involves morphogenesis and differentiation.*

The Genetic Aspects of Development

So far, this chapter has examined only the larger aspects of development—how an embryo divides, how the primary tissue layers are organized, and how the organs and organ systems are formed. All of these processes depend on cells changing, migrating, and sometimes dying to form vertebrate body structures. A number of questions, however, remain unanswered: How do these cells know where to migrate? What controls cell death? How are cells specialized to perform the functions in a particular organ?

The general answer to all of these questions is that, like all other body processes, development is controlled by genes. These developmental genes, however, are special, and are turned "on" and "off " only in special circumstances. This section looks at how genes influence vertebrate development.

Induction: How Cells Communicate during Development

In the process of vertebrate development, the relative position of particular cell layers determines, to a large extent, the organs that develop from them (figure 44.14). But how do these cell layers know where they are? For example, when cells of the ectoderm situated above the developing notochord give rise to the neural groove, how do these cells know they are above the notochord?

The solution to this puzzle is one of the outstanding accomplishments of experimental embryology, the study of how embryos form. It was worked out by German biologist Hans Spemann and his student Hilde Mangold early in the twentieth century. Spemann and Mangold removed cells from the dorsal lip of an amphibian blastula and transplanted them to a different location on another blastula. The dorsal lip region of amphibian blastulas develops from the gray crescent zone and is the site of origin of those mesoderm cells that later produce the notochord. The new location corresponded to that of the future

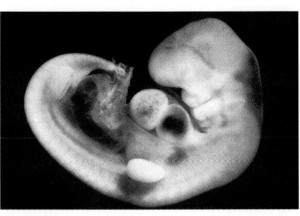
(a)

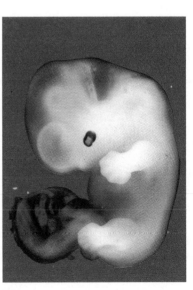
(b)

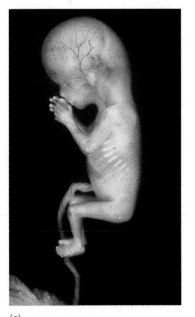
(c)

Figure 44.13 Cell death results in organ formation.
(a) *Human hands first appear at thirty days as blunt paddles.*
(b) *Ten days later the first indications of separation can be seen as faint indentations along the edges.* (c) *Sixty days later the fully-formed fingers are evident.*

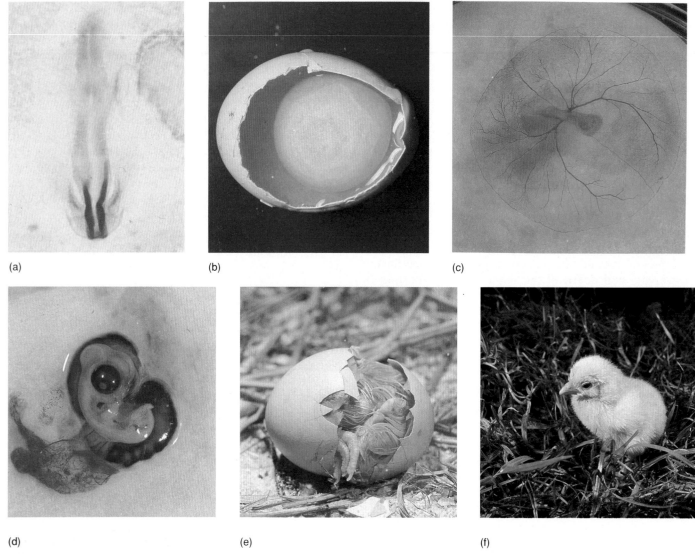

(a)   (b)   (c)

(d)   (e)   (f)

Figure 44.14 Development in a chick.

(a) *After twenty-four hours, development of the neural groove is well advanced.* (b) *After thirty-six hours, the embryo has grown much larger.* (c) *After seventy-two hours, the eyes have already formed. In this photograph, the shell has been removed to show the extensive blood circulation around the developing embryo.*

(d) *After seven days, most of the body's internal organs are present.* (e) *After twenty-one days, the chick pecks its way out of the shell. Note that it does not have much room to maneuver.* (f) *This pensive chick is one day old. Although not yet an adult, it is fully able to live on its own.*

belly of the animal. What happened? The embryo developed *two* notochords, one normal dorsal one and a second one along its belly (figure 44.15).

By using genetically different donor and host blastulas, Spemann and Mangold were able to show that the notochord produced by transplanting dorsal lip cells contained host cells as well as transplanted ones. The transplanted dorsal lip cells acted as primary organizers of notochord development. As such, these cells stimulated a developmental program in the belly cells of embryos to which they were transplanted: the development of a notochord. The belly cells clearly contained this developmental program but would not have expressed it in the normal course of their development. The transplantation of the dorsal lip cells *induced* the ectodermal cells of the belly to form a notochord. This phenomenon as a whole is known as **induction.**

Induction is one tissue's determination of the course of development of another tissue.

Primary and Secondary Induction

The process of induction that Spemann and Mangold discovered appears to be the basic mode of vertebrate development. Inductions between the three primary tissue types—ectoderm, mesoderm, and endoderm—are referred to as **primary inductions.** Inductions between tissues that have already differentiated are called **secondary inductions.** The differentiation of the central nervous system during neurulation by the interaction of dorsal ectoderm and dorsal mesoderm to form the neural tube is an example of primary induction. In contrast, the differentiation of the lens of the vertebrate eye from ectoderm

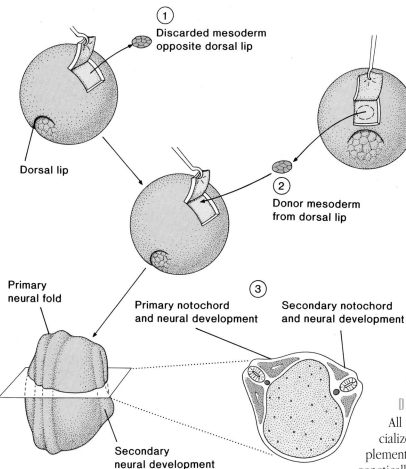

(1)
Discarded mesoderm
opposite dorsal lip

Dorsal lip

Donor mesoderm
from dorsal lip

(2)

Primary
neural fold

(3)

Primary notochord
and neural development

Secondary notochord
and neural development

Secondary
neural development

Figure 44.15 Spemann and Mangold's dorsal lip experiment.

(1) In the first step of the experiment, Spemann and Mangold removed cells from the mesoderm tissue opposite the dorsal lip of an amphibian blastula. The dorsal lip area is the site of notochord development. (2) They replaced the "hole" made by this removal with donor mesoderm from the dorsal lip of another blastula. However, this donor mesoderm was taken from a different region on the blastula. (3) The blastula that received the transplant developed two notochords. This result proved that certain cells induce other cells to form specific structures.

The chemical nature of the induction process is not known in detail. If a nonporous barrier, such as a layer of cellophane, is interposed between the inducer and the target tissue, induction does not occur. A porous filter, in contrast, does permit induction. Inducer cells are believed to produce a protein factor that binds to the cells of the target tissue, stimulating mitosis in them and initiating changes in gene expression.

The Nature of Developmental Decisions: Homeotic Genes

All the cells of the body, with the exception of a few specialized ones that have lost their nuclei, have an entire complement of genetic information. Even though all of its cells are genetically identical, an adult vertebrate contains hundreds of different cell types, each expressing different aspects of the total genetic information for that individual. What factors determine which genes are to be expressed in a particular cell and which are not? In a liver cell, what mechanism keeps the genetic information that specifies nerve cell characteristics turned off? Does the differentiation of that particular cell into a liver cell entail the physical loss of the information specifying other cell types? No, it does not, but cells progressively lose the capacity to *express* ever-larger portions of their genomes. *Development is a process of progressive restriction of gene expression.*

Scientists studying the genetic aspects of development have called the genes that control development **homeotic genes.** In fruit flies, mutations in some of these homeotic genes result in the displacement of certain body parts: For

by interaction with tissue from the central nervous system is an example of secondary induction.

The eye develops as an extension of the forebrain, a stalk that grows outward until it comes into contact with the ectoderm. At a point directly above the growing stalk, a layer of the ectoderm pinches off, forming a transparent lens (figure 44.16). When the optic stalks of the two eyes have just started to project from the brain and the lenses have not yet formed, one of the budding stalks can be transplanted to a region underneath a different epidermis, such as that of the belly. When Spemann performed this critical experiment, a lens still formed, this time from belly epidermis cells in the region above where the budding bulge had been transplanted.

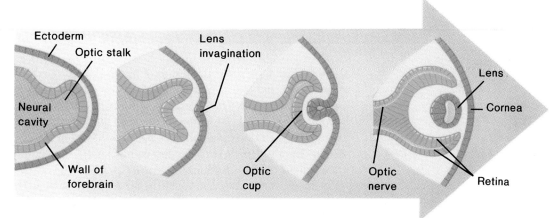

Ectoderm
Optic stalk

Lens
invagination

Lens

Cornea

Neural
cavity

Wall of
forebrain

Optic
cup

Optic
nerve

Retina

Figure 44.16 Vertebrate eye development proceeds by induction.

The eye develops as an extension of the forebrain called the optic stalk. The optic stalk grows outward until it contacts the ectoderm. This contact induces the formation of a lens from ectodermal tissue.

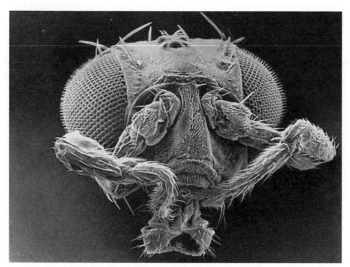

Figure 44.17 A developmental mutation.
Mutation can alter the positioning of body parts. In this
Drosophila, *a so-called homeotic mutation has resulted in a*
fruit fly with legs growing out of the front of its head!

example, a particular mutation in a homeotic gene causes the legs of a fruit fly to grow from the antennae sockets (figure 44.17).

In their attempts to pinpoint the nucleotide sequences in these homeotic genes, scientists identified a sequence of 180 nucleotides that seemed to be present in every homeotic gene they studied. This sequence, named the **homeobox,** is virtually the same in the homeotic genes of different animal species. This remarkable characteristic suggests that the homeobox has been essential not only in the development of individual vertebrate species, but in vertebrate evolution as well.

The homeobox works in a fashion similar to the control regions of the *lac* operon discussed in chapter 13. In short, the homeobox controls development by controlling the activity of homeotic genes. This hierarchical type of organization (homeobox, homeotic genes, and the genes they control) allows the activity of groups of genes to be switched "on" and "off." For example, a particular homeotic gene in amphibians produces a protein that controls the expression of another gene, one that directs the formation of the spinal cord. Researchers have found that, if they inject a developing amphibian with an antibody that works against the homeobox protein, the spinal cord will not develop. Clearly, the homeobox sequence at the beginning of each homeotic gene plays a "master" role in determining cells' developmental fates.

The homeobox sequence of homeotic genes controls cells' developmental fate. Homeobox proteins control the activity of other genes that direct the formation of different structures and organs.

Determination and Commitment

Some cells become **determined** very early. For example, all of the egg cells of a human female are set aside very early in the embryo's life, yet some of these cells will not achieve differentiation into functional oocytes for more than forty years. To a large degree, the fate of a particular cell is determined by its location in the developing embryo. By changing a cell's location, an experimenter can alter the cell's developmental destiny. However, this is only true up to a certain point in a cell's development. At some stage, the ultimate fate of every cell becomes fixed and irreversible, a process referred to as **commitment.**

When a cell is "determined," it is possible to predict its developmental fate. When a cell is "committed," that developmental fate cannot be altered. Determination often occurs very early in development, commitment somewhat later.

The Course of Human Development

Vertebrates seem to have evolved largely by the addition of new instructions to the developmental program. The development of the human embryo shows its evolutionary origins. The embryo proceeds through a series of stages, the earlier stages unchanged from those that occur in the development of more primitive vertebrates.

Without an evolutionary perspective, the fact that human development proceeds in much the same way as development in a chick would be difficult to explain. In both embryonic chickens and embryonic human beings, the blastodisc is flattened. In a chick egg, the blastodisc is pressed against a yolk mass; in a human embryo, the blastodisc is similarly flat despite the absence of a yolk mass. In human blastodiscs, a primitive streak forms and gives rise to the three primary cell types, just as it does in the chick blastodisc.

Human development takes much longer than chicken development, an average of 266 days from fertilization to birth, the familiar nine months of pregnancy. But what may not be so readily apparent is how very early the critical stages of development outlined in this chapter occur during the course of human pregnancy.

First Trimester: When Development Is Essentially Completed

The First Month

In the first week after fertilization, the fertilized ovum undergoes cleavage divisions. The first of these divisions occurs about thirty hours after the fusion of the secondary oocyte and the sperm, and the second occurs about thirty hours later. Cell divisions continue until a blastodisc forms within a ball of cells. During this period, the embryo continues the journey that the ovum initiated down the mother's fallopian tube. On about the sixth day, the embryo reaches the uterus, attaches to the uterine lining, or endometrium, and penetrates into the tissue of the lining. The trophoblast begins to grow rapidly, initiating the formation of membranes. One of these membranes—the **amnion**—will enclose the developing embryo, whereas another—the chorion—will interact with uterine tissue to form the placenta that will nourish the growing embryo.

Ten to eleven days after fertilization, gastrulation takes place. The primitive streak can be seen on the surface of the embryo, and the three primary tissue types are differentiated. Around the developing embryo, the placenta starts to form from the chorion.

In the third week, neurulation occurs. This stage is marked by the formation of the neural tube along the embryo's axis, as well as by the appearance of the first somites, from which the muscles, vertebrae, and connective tissue develop. By the end of the week, over a dozen somites are evident, the blood vessels and gut have begun to develop, and the neural crest has formed. At this point, the embryo is about 2 millimeters long.

In the fourth week, organogenesis (the formation of body organs) occurs (figure 44.18*a*). The eyes form. The tubular

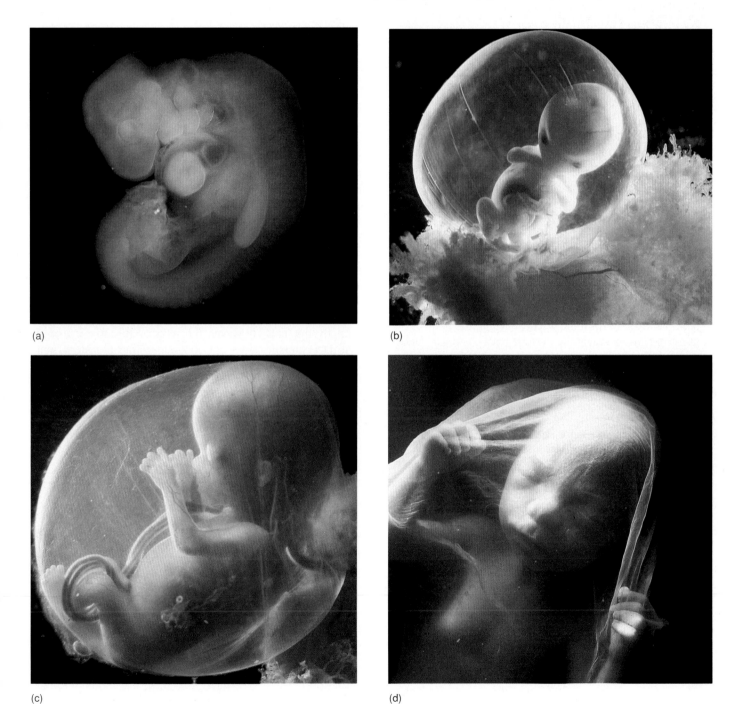

(a)

(b)

(c)

(d)

Figure 44.18 The developing human.
(a) *Four weeks.* (b) *Seven weeks.* (c) *Three months.* (d) *Four months.*

heart develops four chambers and begins to pulsate, its rhythmical beating stopping only with death. At seventy beats per minute, the little heart is destined to beat more than 2.5 billion times during a lifetime of about seventy years. Over thirty pairs of somites are visible by the end of the fourth week, and the arm and leg buds have begun to form. The embryo more than doubles in length during this week, to about 5 millimeters.

All of the major organs of the body have begun their formation by the end of the fourth week of development. Although the developmental scenario is now far advanced, many women are not even aware that they are pregnant at this stage.

Early pregnancy is a very critical time in development because the proper course of events can be interrupted easily. In the 1960s, for example, many pregnant women took the tranquilizer **thalidomide** to minimize discomforts associated with early pregnancy. Unfortunately, this drug interferes with fetus limb bud development, and its widespread use outside the United States resulted in many deformed babies. Also during the first and second months of pregnancy, a mother's contracting of rubella (German measles) can upset organogenesis in the developing embryo. Most spontaneous abortions occur in this period.

The Second Month

Morphogenesis (the formation of shape) takes place during the second month (figure 44.18*b*). The miniature limbs of the embryo assume their adult shapes. The arms, legs, knees, elbows, fingers, and toes can all be seen—as well as a short, bony tail. The bones of the embryonic tail, an evolutionary reminder of the past, later fuse to form the **coccyx.** Within the body cavity, the major organs, including the liver, pancreas, and gallbladder, become evident. By the end of the second month, the embryo has grown to about 25 millimeters in length, weighs perhaps a gram, and begins to look distinctly human.

The Third Month

The nervous system and sense organs develop during the third month (figure 44.18*c*). By the end of the month, the arms and legs begin to move. The embryo begins to show facial expressions and demonstrates primitive reflexes, such as the startle reflex and sucking. By the end of the third month, all of the major body organs have been established. Development of the embryo is essentially complete at eight weeks. From this point on, the developing human being is referred to as a **fetus,** rather than an embryo. What remains is essentially growth.

Second Trimester: When Fetal Growth Begins in Earnest

In the fourth and fifth months of pregnancy, the fetus grows to about 175 millimeters in length, with a body weight of about 225 grams (figure 44.18*d*). Bones enlarge actively during the fourth month. During the fifth month, the head and body become covered with fine hair. This downy body hair, called **lanugo,** is another evolutionary relict and is lost later in development. By the end of the fourth month, the mother can feel the baby kicking. By the end of the fifth month, she can hear its rapid heartbeat with a stethoscope. In the sixth month, growth begins in earnest. By the end of that month, the baby weighs 0.6 kilogram (about 1½ pounds) and is over 0.3 meter (1 foot) long, but most of its prebirth growth is still to come. The fetus cannot yet survive outside the uterus without special medical intervention.

Third Trimester: When the Pace of Fetal Growth Accelerates

The third trimester is predominantly a period of growth, rather than one of development. In the seventh, eighth, and ninth months of pregnancy, the weight of the fetus doubles several times. This increase in bulk is not the only kind of growth that occurs. Most of the major nerve tracks in the brain, as well as many new brain cells, are formed during this period. All of this growth is fueled by nutrients provided by the mother's bloodstream. Within the placenta, these nutrients pass into the fetal blood supply. Undernourishment of the fetus by a malnourished mother can adversely affect this growth and result in severe retardation of the infant. Retardation resulting from fetal malnourishment is a severe problem in many underdeveloped countries where poverty is common.

By the end of the third trimester, the fetus's neurological growth is far from complete and, in fact, continues long after birth. By this time, however, the fetus is able to exist on its own. Development does not continue within the uterus until neurological development is complete because physical growth would continue as well, and the fetus is probably as large as it can be for safe delivery through the pelvis without damage to mother or child (figure 44.19). Birth takes place as soon as the probability of survival is high. For better or worse, the infant is then on its own, a person.

> *The critical stages of human development take place quite early. All the major body organs have been established by the end of the third month. The six months that follow are essentially a period of growth.*

Postnatal Development

Growth continues rapidly after birth. Babies typically double their birth weight within two months. Different organs grow at different rates, however. Adult body proportions are different from those of infants because different body parts grow at different rates or stop growing at different times. The head, for

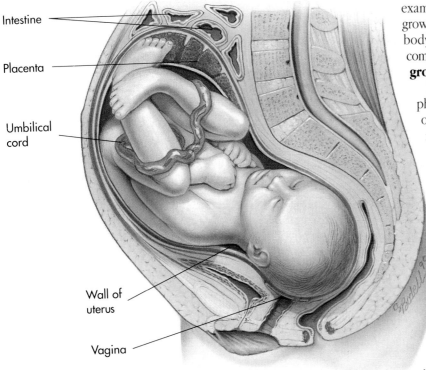

Intestine

Placenta

Umbilical
cord

Wall of
uterus

Vagina

Figure 44.19 Position of the fetus just before birth.
A developing fetus is a major addition to a woman's anatomy. The mother's stomach and intestines are pushed far up, and there is often considerable discomfort from pressure on the lower back. In a natural delivery, the fetus exits through the vagina, which must dilate considerably to permit passage.

example, which is disproportionately large in infants, grows more slowly in infants than does the rest of the body. Such a pattern of growth, in which different components grow at different rates, is called **allometric growth** (figure 44.20).

In most mammals, brain growth is entirely a fetal phenomenon. In chimpanzees, for example, growth of the brain and the cerebral portion of the skull rapidly decelerates after birth, while the bones of the jaw continue to grow. The skull of an adult chimpanzee, therefore, looks very different from that of a fetal chimpanzee.

In human beings, on the other hand, the brain and cerebral skull continue to grow at the same rate after birth as before. During gestation and after birth, the developing human brain generates neurons (nerve cells) at an average rate estimated at more than 250,000 per minute. This astonishing production of new neurons does not permanently cease until about six months after birth. Because both brain and jaw continue to grow, the jaw-skull proportions do not change after birth, and the skull of an adult human being looks very similar to that of a human fetus. It is primarily for this reason that a young human fetus seems so incredibly adultlike.

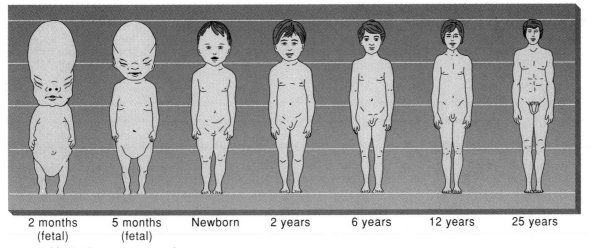

| 2 months (fetal) | 5 months (fetal) | Newborn | 2 years | 6 years | 12 years | 25 years |

Figure 44.20 Allometric growth.
This diagram shows the proportions of the different regions of the body at different ages. As you can see, the head of a newborn is disproportionally large in comparison to the body.

EVOLUTIONARY VIEWPOINT

The genes that control pattern formation in humans— the basic segmented architecture of the body—are essentially the same as those of the fruit fly, which says a great deal about the complex nature of development. The key developmental event that sets vertebrates apart from other animals is the evolution of neural crest tissue. Most other aspects of development seem to have been remarkably conserved by evolution. Among the vertebrates, humans still betray their evolutionary origins, exhibiting a tail in the second month of development and fur in the fifth month.

SUMMARY

1. Fertilization is the union of a secondary oocyte and a sperm to form a zygote. Fertilization is external in fishes and amphibians, internal in all more advanced vertebrates. The three stages of fertilization are: (1) penetration, in which the sperm cell moves past the cells surrounding the oocyte and penetrates the oocyte membrane; (2) activation, in which a series of cytoplasmic movements is initiated by penetration; and (3) pronuclear fusion, in which the sperm and oocyte nuclei fuse.

2. Cleavage is the rapid division of the newly formed zygote into a mass of five hundred to two thousand cells, without any increase in overall size. Because the ovum is structured with respect to the location of developmentally important regulating signals, the future embryo becomes structured by the cleavage divisions. In effect, these divisions partition the ovum cytoplasm into small portions with different regulatory elements.

3. Gastrulation is the mechanical repositioning of blastula cells to form the three basic cell types: ectoderm, endoderm, and mesoderm. In ovum that lack a yolk, the cell movement is one of simple invagination. When a yolk is present, cell movement is affected by it. In amphibians, the cell layers move down and around the yolk. In reptiles, birds, and mammals, cells establish the three primary cell types as an upper layer (ectoderm), a lower layer (endoderm), and a layer that invaginates inward from the upper layer (mesoderm).

4. Neurulation in chordates is the formation of the first tissues—particularly the notochord and the dorsal nerve cord—from the primary cell types.

5. The formation of the neural crest is the first developmental event unique to vertebrates. Most of the distinctive structures associated with vertebrates are derived from cells of the neural crest.

6. The next process in vertebrate development is organogenesis: the formation of the embryo's organs and organ systems. This involves two intricately related processes: morphogenesis and differentiation.

7. Cells influence one another during development by a process of induction. Induction is the determination of the course of development of one tissue by another tissue. In this process, substances that exist on the surface of one cell induce other cells to divide and to differentiate in a manner they otherwise would not have.

8. The homeobox, a sequence of 180 nucleotides in homeotic genes, controls development by switching other genes "on" and "off."

9. At some point during animal development, cells' ultimate developmental fate becomes fixed and unalterable. The cells are then said to be committed, even though they may not exhibit any of the characteristics they will eventually assume.

10. Most of the critical events in human development occur in the first month. Cleavage occurs during the first week, gastrulation during the second week, neurulation and neural crest formation during the third week, and organogenesis during the fourth week.

11. The second and third months of the first trimester are devoted to morphogenesis and to the elaboration of the nervous system and sense organs. By the end of this period, the embryo's development is essentially complete.

12. The last six months of human pregnancy are essentially a period of growth, devoted to increase in size and to the formation of nerve tracks within the brain. Most of the weight of a fetus is added in the final three months of pregnancy.

13. After birth, human babies display allometric growth in that different body parts grow at different rates or stop growing at different times. Humans are unique in that the brain and cerebral skull continue to grow at the same rate after birth as before.

REVIEWING THE CHAPTER

1. How does development begin with fertilization?

2. How does a sperm penetrate an egg?

3. How is the secondary oocyte activated by sperm entry?

4. What is pronuclear fusion?

5. Why would someone have in vitro fertilization performed?

6. How does cleavage set the stage for development?

7. How does the amount of yolk determine the pattern of cleavage?

8. What is the cleavage pattern in primitive aquatic vertebrates?

9. How does cleavage occur in amphibians and advanced fishes?

10. What is the cleavage pattern in reptiles and birds?

11. How is cleavage different in mammals?

12. How are blastula cells different from one another?

13. How does gastrulation indicate developmental change?

14. What is the pattern of gastrulation in aquatic vertebrates?

15. What is the gastrulation development in reptiles, birds, and mammals?

16. How does neurulation determine body architecture?

17. What is organogenesis, and how does it originate?

18. How does the formation of the neural crest mark the evolutionary origin of vertebrates?

19. What are the genetic aspects of development?

20. What is induction, and how does it affect cell communication during development?

21. How do primary and secondary induction differ?

22. What are homeotic genes, and how do they contribute to development?

23. What is the correlation between cell determination and commitment?

24. What is the course of human development?

25. What are the stages of human development during the first trimester?

26. What occurs during the human embryo's first month of development?

27. What are the events during the human embryo's second month of development?

28. What happens during the human embryo's third month of development?

29. What happens to the human fetus during the second trimester?

30. How does the pace of human fetal growth accelerate during the third trimester?

31. What happens during postnatal development of a human infant?

COMPLETING YOUR UNDERSTANDING

1. Why are male sperm some of the smallest cells, while female secondary oocytes are large cells?

2. What is an acrosome, and how does it function?

3. What is parthenogenesis?

4. Why cannot half a million American women have children?

5. Who is Louise Brown, and why did her birth create so many ethical questions?

6. What is meant by haploid gametes and a diploid zygote? In human reproduction, how are these terms employed?

7. What is an embryo, and what is its origin?

8. In vertebrate development, each individual cell in the morula is called a

   a. blastomere.   d. blastodisc.

   b. blastula.   e. blastocoel.

   c. blastopore.   f. archenteron.

9. How are meroblastic and holoblastic cleavage differentiated, and to which group of animals do these terms apply?

10. How does the placenta function, and in which animals is it found?

11. What is the significance of a prepatterned egg?

12. How do the terms *vegetal pole* and *animal pole* apply to animal development?

13. Which layer in the embryo gives rise to the central nervous system?

   a. Ectoderm   c. Endoderm

   b. Mesoderm   d. All the above

14. Which of the following originate from the endoderm? [*three answers required*]

   a. Heart   d. Blood vessels

   b. Lungs   e. Digestive tract

   c. Skin   f. Most glands

15. What is the primitive streak, and what is its function?

16. The notochord is derived from

   a. ectoderm.   d. mesoderm.

   b. ectoplasm.   e. a combination of

   c. endoderm.   *c* and *d*.

17. How does the coelom originate, and what does it give rise to?

18. How did Hans Spemann and Hilde Mangold contribute to the understanding of induction?

19. How do primary and secondary induction differ?

20. What does the homeobox tell about the evolution of vertebrate animals?

21. Are humans dramatically different from other animals in the course of their development? What evidence can you give to support your answer?

22. How do the amnion and the chorion differ from one another?

23. What is thalidomide, and what were the consequences of using it during the 1960s?

24. Why does the coccyx have evolutionary significance?

25. What distinguishes a fetus from an embryo?

26. What is the evolutionary significance of lanugo?

27. In what stages are the growth rates of chimpanzee and human brains and skulls similar and different?

28. In reptiles and birds, the fetus is basically masculine, and fetal estrogen hormones are necessary to induce the development of female characteristics. In mammals, the reverse is true, the fetus being basically female, with fetal hormones acting to induce the development of male characteristics. Can you suggest a reason why the pattern that occurs in reptiles and birds would not work in mammals?

29. Female armadillos always give birth to four offspring of the same sex. Can you suggest a mechanism that would account for this?

FOR FURTHER READING

Beardsley, T. 1991. Smart genes. *Scientific American*, Aug., 86–95. A very readable account of recent work on the molecular control of development.

Bogin, B. 1990. The evolution of human childhood. *BioScience,* Jan., 16–25. The argument that, among mammals, only the human species has childhood as a step in the life cycle.

Dale, B. 1983. *Fertilization in animals.* London: Edward Arnold, Publishers. A brief text devoted entirely to the process of animal fertilization, with good illustrations and up-to-date discussions of physical mechanisms.

De Robertis, E. et al. 1990. Homeobox genes and the vertebrate body plan. *Scientific American*, July, 46–52. Genes that determine the shape of the body by subdividing the embryo along the tail-to-head axis into groups of cells that eventually become limbs and other structures.

Dryden, R. 1978. *Before birth*. London: Heinemann Educational Books. A detailed description of human development.

Nilsson, L., and J. Lindberg. 1974. *Behold man*. Boston, Mass.: Brown. A wonderful collection of color photographs of the developing human fetus.

Trinkaus, J. 1984. *Cells into organs: The forces that shape the embryo.* 2d ed. Englewood Cliffs, N. J.: Prentice-Hall. An advanced but easily understood discussion of the physical mechanisms underlying developmental change.

Ulmann, A. et al. 1993. RU 486. *Scientific American,* June, 42–48. A controversial new drug now widely used in France to terminate unwanted pregnancies.

Wassarman, P. M. 1988. Fertilization in mammals. *Scientific American*. Dec., 78. Describes the cellular mechanisms of sperm penetration and fertilization.

Wessells, N. K. 1977. *Tissue interactions and development*. Menlo Park, Calif.: Benjamin-Cummings Publishing. A brief and lucid account of the experiments that led to the current understanding of the mechanisms of development.

Animal Behavior

On the lookout. These African meerkats are acting as sentries, perched on top of high termite mounds where they can see any approaching danger. It is an interesting biological question why some individuals of a group will expose themselves to danger like this to benefit the others. Does evolution favor such behavior?

FOR REVIEW

Here are some important terms and concepts that have been discussed in previous chapters and that you will encounter again in this chapter. Review them before proceeding if necessary.

Adaptation (*chapter 15*)
Reproductive isolation (*chapter 16*)
Memory and learning (*chapter 35*)
Neurons and interneurons (*chapter 35*)
Sensing the environment (*chapter 35*)
Hormonal control of physiological processes
  (*chapters 37 and 41*)

Birds sing. Honeybees extract nectar from flowers. Pets are housebroken. Humans interact socially with others. All of these are examples of animal behavior. **Behavior** is the way an organism responds to a stimulus in its environment. The stimulus might be as simple as the odor of food. In this sense, a bacterial cell "behaves" by swimming toward higher concentrations of sugar. This very simple behavioral response is just one of a variety of simple responses that are suited to the life of bacteria and allow these organisms to live and reproduce.

During the course of the evolution of multicellular animals, the nervous system became more complex, with more elaborate forms of behavior evolving to meet environmental demands. Peripheral and central nervous systems perceive and process the information of environmental stimuli and trigger an adaptive motor response, which is seen as a pattern of behavior.

Animal behavior can be explained in two ways: The first involves *how* the animal's senses, nerve networks, or internal state provide a physiological basis for the behavior. Analysis of the **proximate cause,** or mechanism, of behavior, involves measuring hormone levels or recording the firing patterns of neurons. The second explanation for animal behavior involves asking *why* the behavior evolved; that is, what is its adaptive value? Study of this **ultimate,** or evolutionary, **cause** of a behavior involves measuring how the behavior influences the animal's survival or reproductive success. Thus, a male dog's frenetic activity when he wants to mate can be explained by hormones, internal messengers released in the spring that cause him to seek out females; this is the proximate cause for his behavior. Evolutionarily speaking, however, the dog shows this behavior to pass on his genes. In effect, genes guide the dog to make more genes; this is the ultimate cause.

This chapter considers the mechanisms by which animals respond to their environment, as well as the adaptiveness of their behavior. Scientists have taken different approaches to the study of behavior, some focusing on instinct, some emphasizing learning, still others combining the elements of both approaches. The picture that emerges is one of behavioral biology as a diverse science that draws strongly from allied disciplines, such as neurobiology, physiology, psychology, and ecology. Today, the overarching theme of the study of behavior is evolution. The

chapter discussion of human behavior will demonstrate how such an evolutionary perspective can be controversial when applied to the social behavior of humans.

Approaches to the Study of Behavior

The study of behavior has had a long history of controversy. One theme of the controversy concerns the importance of genetics and learning in the shaping of behavior. Do genes determine behavior, or do animals learn from experience how to behave? Is behavior the result of **nature** (instinct) or **nurture** (learning)? Although in the past, the nature/nurture controversy has been an "either/or" argument, many studies have shown that both instinct and learning play significant roles, often interacting to produce the final behavioral product. The scientific study of instinct and learning, as well as their relationship, has led to the growth of several disciplines, such as ethology, behavioral genetics, behavioral neuroscience, and psychology.

Ethology

Ethology is the study of the natural history of behavior. Because of their training as zoologists and evolutionary biologists and their emphasis on the study of animal behavior under natural conditions (that is, in the field), ethologists believe that behavior is largely instinctive, or innate, and results from the programming of behavior by natural selection (figure 45.1). Ethologists emphasize that, because behavior is **stereotyped** (appearing in the same form in different individuals of a species), it is based on programmed neural circuits. These circuits are structured from genetic blueprints and allow an animal to show a relatively complete behavior the first time the behavior is produced.

For example, geese incubate their eggs in nests scooped out of the ground. If a goose notices that an egg has been knocked out of the nest accidentally, it will extend its neck toward the egg, get up, and roll the egg back into the nest with its bill. This behavior seems so reasonable that it is tempting to believe that the goose saw the problem and figured out what to do. In fact, however, the behavior is entirely stereotyped and largely instinctive. According to ethologists, egg retrieval behavior is triggered by detecting a **sign stimulus,** the egg out of the nest. A component of the goose's nervous system, the **innate releasing mechanism,** provides the neural instructions for the motor neuron response, or **fixed-action pattern** (figure 45.2). More generally, the sign stimulus is the environmental "signal" that triggers the behavior. The innate releasing mechanism refers to the sensory mechanism detecting the "signal," and the fixed-action pattern is the stereotyped act. Similarly, a frog unfolds its long, sticky tongue at the sight of an insect, and a male stickleback fish attacks another male showing a bright red underside. Such responses certainly appear to be programmed, but what evidence supports the underlying ethological view that behavior has a genetic and neural basis?

(a)

(b)

Figure 45.1 The founding fathers of ethology.
(a) *Karl von Frisch led the study of honeybee communication and sensory biology.* (b) *Konrad Lorenz focused on social development (imprinting) and the natural history of aggression.* (c) *Niko Tinbergen was the first behavioral ecologist. In 1973, these pioneers of ethology received the Nobel Prize for their pathbreaking contributions to behavioral science.*

(c)

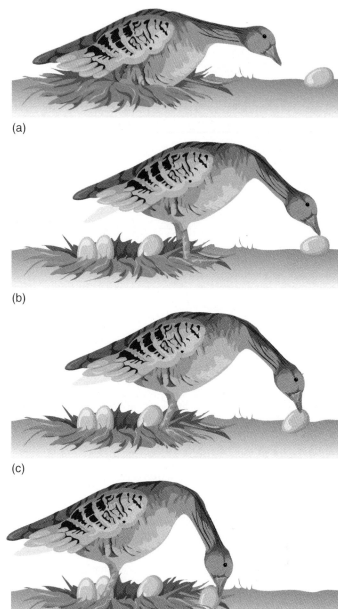

(a)

(b)

(c)

(d)

Figure 45.2 Fixed-action pattern in a graylag goose.
The goose's actions in retrieving an egg that has rolled from the nest are completely programmed behavior. (a) The goose notices that an egg has been knocked out of the nest. (b) The goose extends its neck toward the egg and stands up. (c and d) The goose uses its bill to roll the egg back into the nest.

The Genetic Basis of Behavior

In a famous 1940s experiment, Robert Tryon studied rats' ability to find their way through a maze with many blind alleys and only one exit, where a reward of food awaited. It took a while, as false avenues were tried and rejected, but eventually, some individuals learned to zip right through the maze to the food, making few incorrect turns. Other rats never seemed to learn the correct path. Tryon bred the "maze-bright" rats with one another, establishing a colony from the fast learners, and similarly established a second "maze-dull" colony by breeding the slowest learning rats with each other. He then tested the offspring in each colony to see how quickly they learned the maze. The offspring of maze-bright rats learned even more quickly than their parents had, whereas the offspring of maze-dull parents were even poorer at maze learning. After repeating this procedure over several generations, Tryon was able to produce two behaviorally distinct types of rat with very different maze-learning ability (figure 45.3). Clearly, the ability to learn the maze was to some degree hereditary, governed by genes passed from parent to offspring. The genes also were specific for this behavior, rather than being general ones that influence many behaviors. The abilities of the two groups of rats to per-

form other behavioral tasks, such as running a completely different kind of maze, did not differ. That is, selection was for ability to learn about a particular set of cues (left versus right turns, for example), and did not affect ability to learn about totally different cues (for example, a color-coded maze).

Tryon's research is an example of how a study can illustrate the genetic component of a behavior. Under natural conditions, rats with an ability to learn quickly may have some advantage that leads to increased survival and reproductive success.

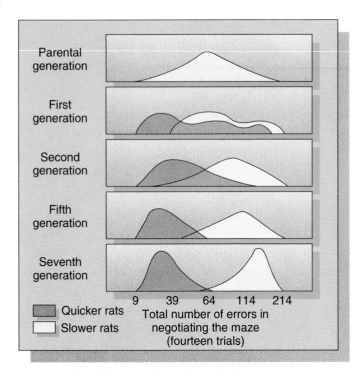

Figure 45.3 Tryon's rat maze experiments.
Tryon was able to select among rats for ability to negotiate a maze, demonstrating that this ability is directly influenced by genes. Tryon tested a large group of rats, selected the few that ran the maze in the shortest time, and let them breed with one another. He then tested their progenies and again selected those with the quickest maze-running times for breeding. By the seventh generation, he had succeeded in halving the average time an inexperienced rat required to negotiate the maze. Parallel selection for slow running time was also successful—it more than doubled the average running time.

Studies of species hybrids have shown that cricket and tree frog courtship song and lovebird nest-building behavior have a genetic basis. At a greater level of detail, research on the mating behavior of mutant fruit flies has shown that even single allele differences produce behaviorally different individuals.

The genetic basis of behavior can be shown by artificial selection and hybridization studies. Some studies have identified the alleles that control behavior.

The Neural Basis of Behavior

How we humans perceive the world depends on the structure of our sensory systems. We hear sounds of certain frequencies and see light of certain wavelengths. Our sense of smell decodes chemical information in the air. These sensory channels of hearing, vision, and smell govern our responses to environmental stimuli. Research on the neurobiology of stimulus detection has revealed how sensory and other aspects of the nervous system are organized to govern perception and behavior, thus providing support for the ethologist's concept of the innate releasing mechanism. The study of the neural basis of behavior is called **neuroethology.**

Neurons are often specialized for the detection of certain stimuli. Frogs and toads, which prey on insects by flipping out a sticky tongue, must be able to "track" prey in the environment. Light-sensitive neurons in the retina of each eye enable them to identify prey, such as insects or worms. For example, as an insect moves in front of a toad, the insect's image enters the toad's eye lens and is focused on the retina. The image thus moves over groups of retinal receptor cells that fire several nerve impulses and relay information via ganglion cells (which process information from several retinal cells) to the brain. The toad attempts to capture what it sees, depending on the size of the object on which is has focused and whether the image is moving horizontally. Thus, the innate releasing mechanism that responds to the sign stimuli provided by an insect and that turns on the fixed-action pattern of flipping out the tongue at the "target" may be thought of as a group of specialized "insect detector" neurons.

Some animals have been extremely useful as models for research on the neural basis of behavior because they have relatively simple nervous systems and display easy-to-record behaviors. A sea hare has a brain that features a small number of very large, individual nerve cells. A simple, yet important, sea hare behavior is its **escape response,** which consists of a series of muscle contractions and relaxations that causes its body to flex back and forth and thus move. The response occurs when the sea hare detects the presence of one of its predators, a sea star. The motor neuron responses that comprise the escape behavior result from impulses fired from three groups of neurons in the brain.

The neurobiology of escape behavior has also been examined in the cockroach. In this case, **mechanoreceptor hairs** (sensory hairs located on appendages at the end of the abdomen) detect microcurrents of air and fire nerve impulses that are transmitted rapidly along a giant interneuron to a thoracic ganglion, where they cause motor neurons to fire to initiate locomotion. The response time is extremely fast: It takes only about 60 milliseconds from the time the air movement is perceived to initiate the escape behavior.

Studies of the nerve networks that are the basis of response to stimuli support the concept of the innate nature of behavior. Components of the nervous system evolved in response to ecological pressures like predation and the need to capture food.

Psychology and the Study of Animal Behavior

In contrast to ethologists' instinct (nature) theory of behavior, other students of animal behavior proposed learning (nurture) as the primary behavior-shaping element. These **comparative psychologists** were not interested in naturalistic studies or evolutionary theory, and worked primarily in laboratory settings on rats. The main contribution of comparative psychology has been to identify how animals learn.

Learning is defined as a modification of behavior that arises as a result of experience, rather than as a result of maturation. Two broad categories of learning have been identified: The simplest type of learning is **nonassociative;** that is, it does not require the animal to form an association between two stimuli, or between a stimulus and a response. Learned behaviors that require such "pairing" within the central nervous system are termed **associative.**

Nonassociative Learning: Habituation and Sensitization

The two major forms of nonassociative learning are **habituation** and **sensitization.** Habituation is learning *not* to respond to a stimulus. Learning to ignore unimportant stimuli is a critical ability to an animal confronting a barrage of stimuli in a complex environment. In many cases, the stimulus evokes a strong response when it is first encountered, but then the magnitude of response gradually declines after repeated exposure. For example, young birds see many types of objects overhead. At first sight, they may crouch down and remain still in response to these stimuli. Some of these objects, like falling leaves or members of their own species flying by, are seen very frequently and have no positive or negative consequence to the nestling. Over time, the young bird may stop responding; it habituates to the stimuli.

Sensitization, by contrast, is learning to be hypersensitive to a stimulus. After encountering an intense stimulus, such as an electric shock, an animal may often react vigorously to a mild stimulus that it would previously have ignored.

Associative learning is more complex than habituation or sensitization. Thinking is associative, as are other **cognitive** behaviors (figure 45.4). Study of nonassociative behaviors in invertebrates such as the sea hare *Aplysia* suggest that many, if not all, associative behaviors may be built up from simpler nonassociative elements.

> *Associative learning is the alteration of behavior by experience, leading to the formation of an association between two stimuli, or between a stimulus and a response. Nonassociative learning involves no such associations.*

Associative Learning: Conditioning

Learning mechanisms can be grouped according to the way stimuli become associated when behavior is modified, or **conditioned.** The repeated presentation of a stimulus in association with a response can cause the brain to form an association between them, even if the stimulus and the response have never been associated before. If meat powder (stimulus) is presented to a dog, the dog will salivate (response). In his famous study of **classical conditioning,** Russian psychologist Ivan Pavlov also presented to a dog a second, unrelated stimulus (a light) at the same time that meat powder was blown into the dog's mouth. As expected, the dog salivated. After repeated trials, the dog would salivate in response to the light alone, having learned to associate the unrelated light stimulus with the

(a)

(b)

Figure 45.4 Associative behaviors.
(a) *This chimpanzee is fashioning a tool. It is stripping the leaves from a twig, which it will then use to dig into a termite nest. Advance preparations such as this strongly suggest that the chimpanzee is consciously planning ahead, with full knowledge of what it intends to do.* (b) *This sea otter is having dinner while swimming on its back. It is using the rock as a tool to break open a clam, bashing the clam on the rock "anvil." Often, a sea otter will keep a favorite rock for a long time, suggesting that the otter has a clear idea of how it is going to use the rock. The sea otter may learn this pattern of eating behavior from others while young, but the capacity to use tools and to consciously foresee their future use depends on inherited abilities.*

meat stimulus. Early experimenters believed that *any* stimulus could be linked in this way to any response. However, as is discussed shortly, researchers now know that this is not true.

In classical conditioning, learning does not influence whether or not the reinforcing stimulus is received. In **operant conditioning,** by contrast, the reward follows only after the animal shows the correct behavioral response. Thus, the animal

Figure 45.5 Skinner box.
The rat rapidly learns that pressing the lever results in the appearance of a food pellet. This kind of learning—trial and error with a reward for success—can also work for far more complex tasks.

must make the proper association before it receives the reinforcing stimulus, or reward. American psychologist B. F. Skinner studied such conditioning in rats by placing the rats in a box of a type that came to be called a "Skinner box." Once inside, the rat would explore the box. Occasionally, it would accidentally press a lever, and a pellet of food would appear. At first, the rat would ignore the lever and continue to move about, but soon it learned to press the lever to obtain food. When it was hungry, it would spend all of its time pushing the lever (figure 45.5). This sort of trial-and-error learning is of major importance to most vertebrates.

Comparative psychologists used to believe that animals could be conditioned to perform any learnable behavior in response to any stimulus by operant conditioning, but as in the case of classical conditioning, this is not so. Today, instinct is credited with the major role in guiding what type of information can be learned—that is, what types of stimuli can be paired through association.

Genetic Aspects of Learning

Roughly twenty years ago, most behavioral biologists came to believe that behavior has both genetic and learned components, and the polarization of the schools of ethology and psychology drew to an end. It has become clear, for example, that certain types of learning are not always possible. Some animals have innate predispositions toward forming certain associations. Such learning preparedness means that what an animal can learn is guided genetically. For example, rats easily learn to associate smells with foods that make them ill but are unable to associate sounds or colors with these foods, no matter how many trials they undergo. Similarly, pigeons associate colors with food but cannot make associations between sounds and food; they *can* associate sounds and danger—but cannot associate colors with danger. The sorts of associations possible are

genetically determined. That is, classical conditioning is possible only within boundaries set by instinct.

Neither is trial-and-error learning free of inherent limits imposed by an animal's genetic makeup. Rats in a Skinner box that learn to press a lever for food cannot learn to press the same lever to avoid an electric shock. They can learn to jump to avoid the shock, but not to obtain food. Animals are innately programmed to learn some things more readily than others. Instinct determines the boundaries of learning.

It now seems clear that animals are innately programmed to respond to specific clues in particular behavioral situations. These innate programs have evolved because they represent adaptive responses. Rats, which forage at night and have a highly developed sense of smell, are better able to identify dangerous food by odor than by color. The seed a pigeon eats may have a distinctive color that the pigeon can see, but the seed makes no sound that a pigeon can hear. Evolution has biased behavior with instincts that make adaptive response more likely.

The current view that learning is guided genetically was developed partially from a study by Peter Marler on how white-crowned sparrows acquire their courtship song. The song is sung by mature males and is characteristic only of the white-crowned sparrow species. By rearing male birds in soundproof incubators provided with a speaker and a microphone, Marler could completely control what a bird heard as it matured and could record on tape the song the bird produced as an adult. He found that birds that heard no song at all during growth sang a poorly developed song as adults. When played the song of the song sparrow, a related bird species, males also sang a poorly developed song as adults. But when both the white-crowned and song-sparrow songs were played to developed, mature males, they sang a fully developed, white-crowned sparrow song.

Marler's research suggests that birds have a **genetic template,** or instinctive program, to guide learning the appropriate song. Song acquisition is based on learning, but only the song of the *correct* species can be learned. The genetic template for learning is *selective* (figure 45.6).

> *Single genes are not responsible for complex behaviors, and complex behavior is not entirely governed by learning. Genes influence what can be learned. The way that genes and experience interact is adaptive. In different species, instinct and learning vary in importance, but in many species, genes seem to set limits on the extent to which behavior can be modified.*

The Physiology of Behavior

Psychologists criticized ethology's emphasis on instinct because it ignored the study of internal factors that control behavior. If asked why does a male bird defend a territory and sing only during the breeding season, ethologists would answer that birds sang when they were in the right *motivational state,* or

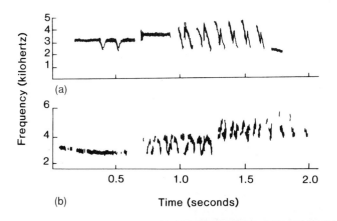

Figure 45.7 *Dewlap display in a male spiny lizard.*
Under hormonal stimulation, males extend the fleshy, colorful dewlap (loose skin hanging under the neck) to court females. This behavior also stimulates hormone release and egg-laying in the female.

Figure 45.6 *Song development in birds.*
(a) *Sonogram of song produced by white-crowned sparrow males exposed to their own species' song during development.*
(b) *Sonogram of song produced by white-crowned sparrow males*
(c) *that heard no song during rearing. The differences between the two sonograms illustrate that the genetic template itself is not enough to produce a normal song.*

mood, and had the appropriate *drive.* But what do these terms mean? They are simply "black-box" concepts that give a name to some internal control mechanism that remains unknown. Today, the internal control of behavior is understood from the study of physiology.

Reproductive Behaviors

One focus has been on the physiological control of reproductive behavior. Studies of lizards, birds, rats, and other animals have shown that hormones play an important role in the control of behavior and provide a chemical basis for motivation. Animals show reproductive behaviors such as courtship only during the breeding season. They monitor changes in day length to trigger a series of events that involve the release of hormones from the endocrine glands, hypothalamus, pituitary, and ovaries and testes. Ultimately, the steroid hormones estrogen and testosterone released from the ovaries and testes travel to the brain and cause animals to show behaviors associated with reproduction (figure 45.7). Bird song and territorial behavior depend on the level of testosterone in the male, and the receptivity of females to male courtship depends on estrogen.

Hormones are therefore a proximate cause of behavior. To control reproductive behavior, they are released at the time of the year during which conditions for the growth of young are favorable. Environmental stimuli that trigger hormone release and thus behavior include male courtship activities as well as changes in the physical environment, such as in temperature and day length.

Behavioral Rhythms

Many animals exhibit behaviors that vary in a regular fashion. Geese migrate south in the fall, birds sing in the early morning, bats fly at night rather than in daylight hours, and we humans get sleepy at night and are active in the daytime. Why do regularly repeating patterns of behavior occur, and what determines when they occur? Study of questions like these has revealed that rhythmic animal behaviors are based on both **exogenous** (external) timers and **endogenous** (internal) rhythms.

Much of the study of endogenous rhythms has focused on behaviors that seem keyed to a daily cycle, such as sleeping. Many of these behaviors have a strong endogenous component, as if they were driven by a **biological clock.** In the absence of any clues from the environment, the behaviors continue on a regular cycle. Such rhythms are termed **free-running.** Endogenous rhythms of about twenty-four hours that occur even in the absence of external cues are called **circadian** ("about a day") **rhythms.** Almost all fruit fly pupae hatch in the early morning, for example, even if kept in total darkness throughout their weeklong development. They keep track of time with an internal clock whose pattern is determined by a single gene.

The rhythms of most biological clocks do not exactly match that of the environment, so an exogenous cue is required to keep the behavior properly in time with the changing real-world environment. Because the duration of each individual cycle of the behavior deviates slightly from twenty-four hours, an individual kept under constant conditions gradually drifts out of phase with the outside world. Exposure to an environmental cue resets the clock. Light is the most common cue for resetting circadian rhythms.

The most obvious circadian rhythm in humans is the sleep/activity cycle. In controlled experiments, humans have lived for months underground in apartments where all light is artificial and there are no external cues of any kind. Left to set their own schedules, most people adopt daily activity patterns (one period of activity plus one period of sleep) of about twenty-five hours, although there is considerable variation. Some individuals exhibited fifty-hour clocks and were active for as long as thirty-six hours each period! In the real world, the day/night cycle resets the free-running clock every day to a cycle of twenty-four hours.

> *Circadian rhythms are endogenous cycles of about twenty-four hours that occur even in the absence of external clues.*

Many important biological rhythms occur on cycles longer than twenty-four hours. Annual cycles of breeding, hibernation, and migration are examples of behaviors that occur on a yearly cycle, so-called **circannual behaviors.** These behaviors seem to be largely timed by hormonal and other physiological changes that are keyed to exogenous factors, such as day length. The degree to which endogenous biological clocks underlie circannual rhythms is not known; constant-environment experiments of several years' duration are very difficult to perform.

The physiological mechanism of endogenous biological clocks is unknown, although a great deal of study and speculation has centered on regularly occurring molecular interactions. The mechanism of the biological clock remains one of biology's most tantalizing puzzles.

Animal Communication

The sign stimuli discussed earlier in the chapter are used by predators to locate prey or are used by prey to avoid predators. However, other stimuli are **social releasers,** or signals produced by one individual to communicate with another individual, usually of the same species. In this case, it is adaptive for both the signal sender and receiver to exchange information, perhaps concerning the readiness to mate or the location of a food source. Communication may occur through a number of sensory channels: Signals may be visual, acoustical, chemical, tactile, or electrical. Much of the research in animal behavior concerns analyzing the nature of the signal and determining how it is perceived.

Courtship

Depending on their reproductive condition, animals produce signals to communicate with potential mates. Courtship behaviors usually consist of a series of fixed-action patterns. A **stimulus/response chain** often occurs, in which a behavior released by some action of a partner, in turn releases another behavior (figure 45.8).

Courtship signals are often **species-specific:** They limit communication to members of the same species and thus play a key role in reproductive isolation. For example, the flash patterns of fireflies (which are actually beetles) are coded for species identity: Females recognize males of their own species by the number of flashes in a pattern; males recognize the female of their species by her flash response. The chemical sex attractant, or **pheromone,** of the female silk moth conveys information only to males of the species. The antennae of the male silk moth are covered with **chemoreceptors,** which are neurons modified to detect extraordinarily small quantities of the sex pheromone. Many insects, amphibians, and birds produce species-specific sound signals to attract mates (figure 45.9). Some fish use electrical currents for the same purpose.

Communication in Social Groups

Many insects, fishes, birds, and mammals live in social groups in which information is communicated among group members. For example, in mammalian societies, some individuals serve as "guards" and keep watch for predators. When danger occurs, an **alarm call** is given. Group members respond by absconding and seeking shelter (figure 45.10). Social insects, such as ants and honeybees, produce chemicals called **alarm pheromones** that trigger attack behavior. Ants deposit **trail pheromones** between the nest and a food source to induce cooperation during foraging. Honeybees have an extremely complex **dance language** behavior that directs nestmates to rich nectar sources (see Sidelight 45.1).

Human Language

Some primates have a "vocabulary" that allows individuals to communicate the identity of specific predators, such as eagles, leopards, and snakes. Chimps and gorillas can learn to recognize and use a large number of symbols to communicate abstract concepts. However, they cannot assemble symbols into sentences. This requires very complex "wiring" of the brain, a complexity that only humans have achieved.

Language develops at an early age in humans. Human infants are programmed to recognize the consonant sounds characteristic of human speech (including those not present in the particular language they learn), while ignoring a world full of other sounds. They learn by trial and error (the "babbling" phase) how to make these sounds.

Although human languages appear on the surface to be very different, in fact, they share many basic structural similarities. Researchers believe that these similarities reflect how our brains handle abstract information, a genetically determined characteristic all humans share. The differences among English, French, Japanese, and Swahili are learned—but any human can learn them. All human languages draw from the same set of forty consonant sounds (English uses two dozen of them), and every normal human baby can distinguish among all forty of them.

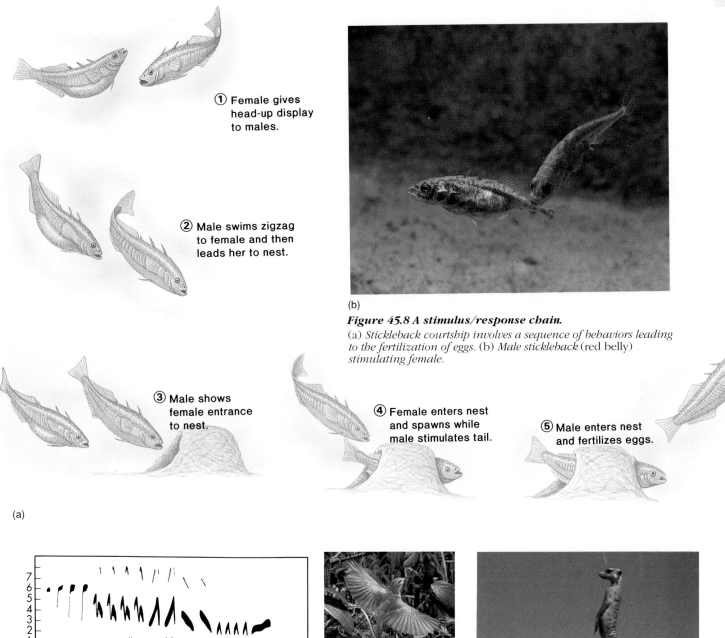

(b)

Figure 45.8 A stimulus/response chain.
(a) *Stickleback courtship involves a sequence of behaviors leading to the fertilization of eggs.* (b) *Male stickleback (red belly) stimulating female.*

① Female gives head-up display to males.

② Male swims zigzag to female and then leads her to nest.

③ Male shows female entrance to nest.

④ Female enters nest and spawns while male stimulates tail.

⑤ Male enters nest and fertilizes eggs.

(a)

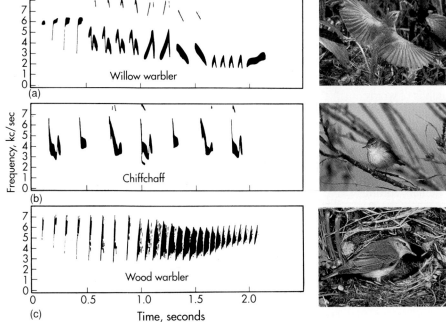

Figure 45.9 Singing a different tune.
Sonograms and photos of (a) *willow warbler,* (b) *chiffchaff, and* (c) *wood warbler. These sonograms illustrate how birds announce their species' names. At one time, these three species were considered to be one.*

Figure 45.10 Sentry duty.
This meerkat is acting as a sentinel. Meerkats are a species of highly social mongoose living in the semiarid sands of the Kalahari Desert. A field study by Oxford University zoologist Dr. David MacDonald revealed that theirs is an astonishingly complex and cooperative society. Under the security of this meerkat's vigilance, the other group members can focus their attention on foraging.

Sidelight 45.1

THE DANCE LANGUAGE OF THE HONEYBEE

A beehive is actually a society of thirty to forty thousand individuals whose behaviors are integrated into a complex unit. Worker bees may forage for miles from the hive, collecting nectar and pollen from a variety of plants, switching between species and populations according to how energetically rewarding their food is. A colony adjusts its food-harvesting effort through the reconnaissance behavior of scout bees that sample available food sources.

The nectar and pollen sources used by bees are clumped and potentially rich in food quality and quantity. Each clump offers much more food than can be transported to the hive by a single bee. A honeybee colony is able to exploit such rich resources through the behavior of the scout bees that locate a resource patch and communicate its location to nestmates. The life work of Nobel laureate Karl von Frisch was to unravel the details of this communication system that conveys information about the location of new food sources through a dance language.

To direct bees in the hive to a food source, scout bees communicate both the direction and distance of the patch by performing a remarkable behavior pattern called a *waggle dance.* The movement resembles a figure eight. It is called a waggle dance because, during one portion of the dance, called the *straight run,* the dancing bee vibrates, or waggles, her abdomen while producing bursts of sound. She may stop periodically to give her hivemates a sample of the nectar she has carried with her in her crop, a storage organ. Bees that closely follow the dancing bee's performance soon appear as foragers at the new food source.

How is the location of the food communicated through the waggle dance? Von Frisch and his colleagues found that the scout bee indicates the direction of the food source by transposing the visual angle between the food source and the nest in reference to the sun to the same angle given by the straight run of the dance relative to gravity on the vertical comb in the hive (Sidelight figure 45.1). Distance is given by the tempo, or degree of vigor, of the dance.

In spite of the many ingenious experiments carried out by von Frisch and his students, another scientist—Adrian Wenner from the University of California—did not believe that the dance language communicated anything, and he challenged von Frisch's results. Wenner believed that flower odor was the key cue in allowing recruited bees to arrive at a new food source. A heated controversy ensued as each group of researchers published articles supporting their position.

The "dance language controversy" was partially resolved in the mid-1970s by the creative research of James L. Gould, who devised an experiment in which a scout bee gave incorrect information to hivemates. Gould could thus cause a scout bee to tell nestmates of a "false" location—not where the scout had actually been but rather another place where automated traps could capture the "fooled" bees.

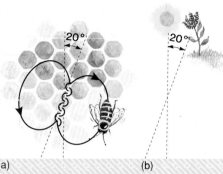

Sidelight figure 45.1 The waggle dance.
(a) *The straight run of the waggle dance is represented in the diagram by the wavy line in the center of the two spheres. The straight run is performed at the same angle to the vertical combs of the hive as the food source is to the sun. In this diagram, the straight run is performed at a 20-degree angle to the vertical combs. (b) This is the same angle that the sun is in relation to the food source.*

Gould's experiment conclusively showed that the nature of the dance itself, rather than some other clue, determined where bees went in search of food.

Recently, researchers were successful in building a computer-controlled robot bee that can perfectly reproduce the typical dance—the robot even stops to give food samples! Use of the robot bee has allowed researchers to determine precisely which cues are used to direct hivemates to food sources.

Most humans learn a particular language while young. Children who have not heard certain consonant sounds as infants can only rarely distinguish or produce them as adults. That is why Americans never master the throaty French /r/, whereas speakers of French typically replace the English /th/ with /z/, and why native Japanese substitute /l/ for the unfamiliar English /r/. Children quickly and effortlessly learn a vocabulary of thousands of words. This rapid-learning ability seems to be genetically programmed and disappears in most of us as we grow older.

Although language is the primary channel of human communication, much evidence suggests that odor and other nonverbal signals ("body language") may also be important. In an animal as socially complex and intelligent as a human, sorting out the relative contribution of the composite (that is, multichannel) signals humans produce is difficult.

> *Study of animal communication involves research on the specificity of a signal, its information content, and the methods used to produce and receive it. Communication plays an important ecological role in maintaining genetic isolation between species.*

Ecology and Behavior

This chapter often refers to behavior as being adaptive. Nobel laureate Niko Tinbergen was the first ethologist to study the **survival value** and **adaptive significance** of behavior. Survival value refers to the way a behavior may allow an animal to avoid a predator, while a behavior's adaptive significance is how the behavior contributes to reproductive success. Tinbergen noted that an animal's environment, or ecology, presents certain "problems," such as locating a nest or food, avoiding predators, or finding mates, and that behavior is a trait that evolves to "solve" such problems. Currently, **behavioral ecologists** study the ways in which behavior serves as adaptation and allows an animal to increase or even maximize its reproductive success. This section examines several categories of behavior that have been studied with respect to animal ecology.

Orientation and Migration

Animals may travel to and from a nest to feed, or may move regularly from one place to another. To do this, they must orient themselves by tracking environmental stimuli.

Movement toward or away from some stimulus is called a **taxis.** The crowd of flying insects around outdoor lights is a

Figure 45.11 Taxis.
This rainbow trout is exhibiting a taxis behavior. Trout always orient toward the current.

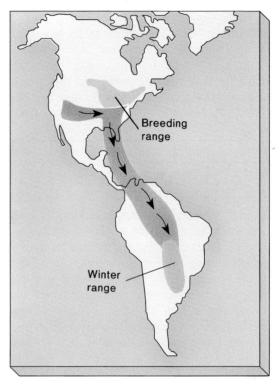

Figure 45.12 Birds on the move: The migratory path of California bobolinks.
These birds recently came to the Far West from their more established range in the Midwest. When they migrate to South America in the winter, they do not fly directly, but rather fly to the Midwest first and then use the ancestral flyway.

familiar example. Insects that are attracted to the light are said to be **positively phototactic.** Other insects, such as the common cockroach, avoid light and thus are **negatively phototactic.** Nor is light the only stimulus. Trout orient in a stream so as to face upstream, against the current (figure 45.11). Not all responses involve such a specific orientation. In some cases, an individual simply becomes more active under certain conditions. If an animal moves randomly but is active under poor conditions and quiet under favorable ones, then it will tend to stay in favorable areas. These changes in activity level are dependent on stimulus intensity and are called **kineses.**

Long-range, two-way movements are called **migrations.** In animals, many migrations are tied to a circannual clock, occurring once a year. Ducks and geese migrate down flyways from Canada across the United States each fall and return each spring. Monarch butterflies migrate from the eastern United States to Mexico, a journey of over 3,000 kilometers that takes from two to five generations of butterflies to complete. Perhaps the longest migration is that of the golden plover, which flies from Arctic breeding grounds to wintering areas in southeastern South America, a distance of some 13,000 kilometers.

Migration patterns may be genetically determined. When colonies of bobolinks became established in the western United States, far from their normal range in the Midwest and East, these birds did not migrate directly to their winter range in South America; rather, they migrated east to their ancestral range and then south along the old flyway. The old pattern was not changed, but rather, an additional pattern was added (figure 45.12).

Migrations are long-range, two-way movements by animals, often occurring once a year with the change of seasons.

Biologists now have a good idea of how these feats of orientation and navigation are achieved. Birds and other animals orient themselves by looking at the sun and the stars. The indigo bunting, for example, which flies during the day and uses the sun as a guide, compensates for the sun's movement in the sky as the day progresses by reference to the North Star, which does not move in the sky. Other birds, such as starlings, compensate for the sun's apparent movement by use of an internal clock. If captive birds are shown an experimental sun in a fixed position, they will change their orientation to it at a constant rate of about 15 degrees per hour.

The distinction between **orientation** (the ability to follow a bearing) and **navigation** (the ability to set or adjust a bearing, and then follow it) —"compass" versus "map and compass"—is important to note. Orientation mechanisms, such as those of the indigo bunting and the starling, are well understood; navigation map senses, such as that of the bobolink, are only poorly understood.

Many migrating bird species appear to use a compass. They have the ability to detect the earth's magnetic field and to orient themselves with respect to it. If such birds are studied in a closed indoor cage, they will attempt to move in the correct geographical direction, even though no external cues are visible. However, if a powerful electromagnet is placed near the cage to change the magnetic field, the direction in which the birds attempt to move can be altered at will. Little is known about the sensory receptors that birds employ to detect magnetic fields. Magnetite, a magnetized iron ore, has been found in the heads of some birds, but the exact nature of the receptor is unknown.

Although bird migration has been well studied, relatively little is known about how other migrating animals navigate. The green sea turtles that introduce chapter 1 of this text migrate from Brazil halfway across the Atlantic Ocean to Ascension Island—how do they find this tiny island, half an ocean away? How do the young that hatch on Ascension Island know how to find Brazil, thousands of miles away over the open sea? As adults, perhaps thirty years later, how do they find their way back to breed? Current studies suggest that wave action is an important cue.

Foraging Behavior

Feeding is one of the most basic of all animal behaviors. Because all animals are heterotrophs, all of them must eat to survive, and many complex **foraging behaviors** have evolved that influence what an animal eats and how it obtains its food.

Some foraging behaviors involve actively hunting prey. A chameleon will snap up an insect that comes close, a fixed-action pattern released by sight of the insect. Lions will hunt in groups, some driving wildebeests or other game past others waiting in ambush. Some animals, such as spiders, set traps.

What animals eat divides foraging behaviors into two broad categories, with considerable variation in each: (1) Some animals are specialists and feed on only one kind of organism, whereas (2) others are generalists, and feed on many different kinds of organisms. Oystercatchers, for example, are shorebirds that feed only on mussels. They have highly specialized foraging behavior, stabbing repeatedly into the mollusk to open its shell. This unusual behavior is of little use in catching fish or other prey, but it is a very efficient way of opening mussels. Generalists rarely exhibit such complex foraging behaviors and so are probably not as efficient at catching any one food item, but they can take advantage of more than one kind of prey.

Detailed studies of foraging behavior show that, in reality, generalists are not all that general. Usually, an animal will concentrate on one kind of organism, ignoring other potential food sources, and then switch to a second favored kind when the first becomes rare. An animal foraging in this way may have a **search image,** or "mental idea," of the food it is selecting. It may visualize in its brain the appearance of its prey. Search images are advantageous in that they permit individuals to focus on particular food items and to modify their foraging accordingly.

Foraging behavior is of obvious importance to growth and reproduction. Thus, strong evolutionary pressures would be expected to target this behavior. For example, does natural selection favor animals that are more efficient foragers, animals that maximize energy intake over expenditure? Evolution would certainly be expected to favor efficiency—but does it? To investigate this, biologists have studied situations in which foraging behavior involves a variety of trade-offs. For fish, food items come in a variety of sizes, larger ones containing more food energy but being harder to catch and less abundant. If the food value of different prey items, as well as the energy costs of pursuing and handling prey of different sizes, could be measured, it would, in principle, be possible to calculate the optimal foraging behavior. But in practice, measuring all of the variables is difficult, so precise predictions are usually impossible. However, a variety of studies indicate that the foraging behaviors of most animals are far from random and that individuals select prey in a highly efficient fashion. Animals seem to maximize their energy gain per unit of time spent foraging.

Territoriality

Territoriality is a form of behavior in which individual species members exclusively use an area holding some limited resource, such as forage or a group of females. The critical aspect of territorial behavior is defense against intrusion by other individuals. Animals defend territories by advertising through displays that the territories are occupied and by overt aggression. A bird sings from its perch within a territory to prevent a takeover by a neighboring bird. If an intruder persists, it is attacked. But singing is energetically expensive, and attacks can lead to injury. Moreover, advertisement through song or visual display can reveal the animal's position to a predator. Why bear these costs and take such risks?

Over the past two decades, it has become increasingly clear that an *economic* approach is insightful in studying the evolution of territoriality. While there are energy costs to defending a territory, there are also energy benefits. Studies of nectar-feeding birds, such as hummingbirds and sunbirds, make this point clear: A bird benefits from having exclusive use of a patch of flowers because it can efficiently harvest the nectar produced. But the benefits of exclusive use of such a resource outweigh the costs of defense only under certain conditions. For example, if flowers are very scarce, their nectar does not have enough calories to balance the number of calories used up in defense. In this case, being territorial is not advantageous. Similarly, if flowers are very abundant, a bird can efficiently meet its daily energy requirements *without* behaving territorially. Defending abundant resources is not energy efficient. But for immediate levels of flower availability, the benefits of territorial behavior outweigh the costs, and territorial behavior is favored.

> The "economic" approach has been useful in examining the evolution and ecology of territoriality and other behaviors, such as foraging. Animals that gain more energy from a behavior than they expend are assumed to have an advantage in survival and reproduction over animals that behave in less efficient ways. Thus, estimating a behavior's energy benefits and costs is one way of determining the adaptiveness of that behavior.

Reproductive Strategies

During the breeding season, animals make several important "decisions" concerning whom to mate with, how many mates to have, and how much time and energy should be devoted to rearing offspring. Males and females usually differ in their **reproductive strategies,** or sets of behaviors that have evolved to maximize reproductive success or fitness. Male and female reproductive behaviors have evolved in response to ecology, which, in this case, refers to how food resources, nest sites, or members of the opposite sex are spatially distributed in the environment.

Figure 45.13 Sexual dimorphism in deer.
This western mule deer buck (Odocoileus heimionus) *has a rack of antlers used in defending a harem of females from other males. Antlers are absent in females.*

Figure 45.14 Flamingos.
These birds are monogamous and show almost no size or plumage differences.

Darwin was the first to observe that females do not simply mate with the first male encountered, but seem first to somehow "evaluate" a male's quality. **Mate choice** occurs because individuals that select superior-quality mates leave behind more offspring. The benefits of choosing a mate may lie in acquiring "good genes" for offspring or in obtaining resources, such as favorable nest sites and energetically valuable food.

If members of one sex are selective in mating, then members of the opposite sex compete with one another to "demonstrate" that they are high-quality mates. This may result in aggressive competition among males (or females) to hold territories, if territory quality is considered in mate evaluation. Darwin referred to this competition for mates as a process of **sexual selection.** Because of sexual selection, males and females may look very different and show **sexual dimorphism.** Differences between males and females may be in size, or a trait may occur in one sex but not in the other. For example, stag deer have elaborate antlers; these are absent in female deer (figure 45.13). A peacock has a large, colorful tail that he struts in front of a female, but a peahen is drab in color and has no elaborate feathers.

Individuals may mate with only one individual during the breeding season and form a long-lasting pair bond or may have more than one mate. **Mating systems,** such as **monogamy** (a male mates with one female), **polygyny** (a male mates with more than one female), and **polyandry** (a female mates with more than one male), are other aspects of male and female reproductive strategy (figure 45.14). These mating systems have evolved to maximize male and female fitness and have a strong ecological component. For example, a male may defend a territory that holds nest sites or food sources necessary for a female to reproduce, and the territory might have resources sufficient for more than one female. If males differ in the quality of territories they hold, a female will do her best to mate with a male in a high-quality territory. Such a male may already have a mate, but it is still more advantageous for the female to breed with a mated male in a high-quality territory than with an unmated male in a low-quality territory. Thus, polygyny may evolve.

Mating decisions are also constrained by the needs of offspring. If both parents are needed to successfully rear young, then monogamy may be favored. This is generally the case in birds; over 90 percent of all bird species form monogamous pair bonds.

Natural selection has favored the evolution of behaviors that maximize the reproductive success of males and females. By evaluating and selecting a mate with superior qualities, an animal can increase the growth and survival chance of offspring.

The Evolution of Animal Societies

We humans tend to think of the word *society* in terms of the groups in which we live and our cultural environment. But species of organisms as diverse as the slime molds, cnidarians, insects, fishes, birds, prairie dogs, lions, whales, and chimpanzees exist in social groups. To encompass many types of social phenomena, a **society** is broadly defined here as a group of organisms of the same species that are organized in a cooperative manner. Why have individuals in some species given up a solitary existence to become members of a group? Recent research has focused on the advantages and disadvantages of group living, and much attention has been given to the evolution of the trait that defines social life: cooperation.

Sociobiology: The Biological Basis of Social Behavior

In the early 1970s, E. O. Wilson of Harvard University, Richard Alexander of the University of Michigan, and Robert Trivers, now at the University of California, initiated what has become a major movement within biology: the attempt to study and understand animal social behavior as a *biological* process, a process with a partly genetic basis that is shaped by evolution.

A biological view of social behavior being the result of evolution would predict that the behaviors characteristic of particular animals are, by and large, suited to their mode of living—that is, in the sense of Darwin, are adaptive. Natural selection would be expected to favor those gene combinations that allow animals to adapt more completely to particular habitats. The study of the biological basis of social behavior of animal societies is called **sociobiology.** Sociobiological insights into the origin of social behavior are described in the next sections. The chapter concludes with a discussion of one of the most contentious topics in the history of biology: Can biology relate anything about human nature?

Group Living

Living as a member of a group can actually be viewed as selfish behavior. A bird joins a flock because it may have greater protection from predators. In fact, as flock size increases, the risk of predation decreases because more individuals can vigilantly scan the environment. A flock member may also increase its feeding rate if it can acquire information from other flock members about the location of new, rich food sources. But group living also involves costs. For example, parasites and disease are more easily spread within groups, and the advantages of increased group size may be balanced by the disadvantages of loss of young to blood-sucking insect parasites.

Altruism

Altruism is often thought of as a heroic human behavior, such as jumping into a river to save a drowning person. But altruism, or self-sacrificing behavior, is not limited to humans and occurs in extreme forms in other animals (see figure 45.10). In many species, it is an important aspect of cooperation. For example, a honeybee worker has a barbed sting that remains in the skin of a vertebrate. As the bee flies off, the anchored sting causes the worker to eviscerate itself. Thus, any bee that tries to drive off an intruder dies as a consequence of its protective efforts. Another example of altruistic behavior is found in vampire bats, which may share blood meals at the roost. One of the most important aspects of altruism concerns assisting another individual in reproducing. In some species, nonreproductive or even sterile individuals exist to help other group members rear offspring. How can such behavior evolve?

One of the great misconceptions about social behavior is that altruism has evolved because a certain act benefits the group as a whole, or even the entire species. This **group selection** argument has been used (incorrectly!) to explain how animals regulate population size. For example, in some bird species, males display and compete with each other on the mating grounds to try to secure centrally located territories. Some males are able to hold territories; others are not. Females choose males with territories as mates, and males unable to hold territories may never mate. V. C. Wynne-Edwards, an animal ecologist, proposed that nonterritorial males were sacrificing their own reproduction to limit population growth. In other words, Wynne-Edwards believed that some males did not reproduce because it was "good for the species" in that too large a population might exhaust limited resources, with the result that the whole group or species might become extinct.

This explanation of the evolution of altruism has a very important flaw: How could the trait of altruism be passed from generation to generation if males that have the trait never leave any offspring? This argument has no Darwinian logic. Other possible evolutionary explanations of altruistic behavior include reciprocity and kindness.

Reciprocity

Robert Trivers proposed that individuals may form "partnerships" in which altruistic acts are continuously exchanged, or mutually reciprocated; hence, Trivers's theory is called **reciprocal altruism.** This theory requires that the individuals of an altruistic pair be unrelated—that is, that they share no genes in common. Studies of alliances among male baboons show that a male usually solicits the help of a particular unrelated male and that "favorite partners" occur. In the evolution of altruism through reciprocity, "cheaters" (nonreciprocators) are discriminated against and are cut off from receiving future aid. According to Trivers, cheating should not occur if the cost of not reciprocating exceeds the benefits of receiving future aid.

Kin Selection

The most influential theory of the origin of altruism was presented by sociobiologist William D. Hamilton in 1964. This theory is perhaps best introduced by quoting a passing remark made in a pub in 1932 by the great population geneticist J. B. S. Haldane. Haldane said that he would willingly lay down his life for *two brothers* or *eight first cousins*. What was Haldane getting at? Each brother and Haldane shared half of their genes in common—Haldane's brothers each had a 50 percent chance of receiving any given allele that Haldane had obtained. Consequently, it is statistically true that two of his brothers would carry as many of Haldane's particular combinations of alleles to the next generation as would Haldane himself. Similarly, Haldane and a first cousin would share an eighth of their alleles: Their sibling parents would each share half of their alleles, and each of their children would receive half of these, of which half, on the average, would be in common: $0.5 \times 0.5 \times 0.5 = 0.125$, or one-eighth. Eight first cousins would therefore pass on as many of those genes to the next generation as would Haldane himself. Hamilton saw Haldane's point clearly: *Evolution will favor any strategy that increases the net flow of a combination of genes to the next generation.*

Altruism has costs and benefits. Hamilton showed that, by directing aid toward kin, or close genetic relatives, the reduction in an altruist's (personal) fitness may be outweighed by the increased reproductive success of relatives. From an evolutionary perspective, selection will favor the behavior that maximizes the propagation of alleles. If giving up one's own reproduction to help relatives reproduce accomplishes this, then even sterility can be favored by natural selection. Selection acting to favor the propagation of genes by directing altruism toward relatives is called **kin selection.**

In Hamilton's theory of the evolution of altruism by kin selection, he coined the term ***inclusive fitness*** to describe the sum of genes propagated by personal reproduction and by the effect of help with relatives' reproduction. Inclusive fitness does not simply result from adding the number of genes passed on directly via an individual's own offspring and the number of genes passed on via relatives other than offspring. Rather, inclusive fitness is the sum of the number of genes directly passed on in an individual's offspring and those genes passed on indirectly by kin (other than offspring), whose existence results from the benefit of the individual's altruism.

Figure 45.15 Reproductive division of labor.
In a honeybee colony, the queen (with a red spot on her thorax) *is the sole egg-layer. Her daughters (workers) are sterile.*

> *The theory of kin selection proposed by W. D. Hamilton predicts that altruism is likely to be directed toward close relatives. The closer the degree of relatedness, the greater the potential genetic payoff in inclusive fitness.*

Insect Societies

As already mentioned, altruism can take striking forms in the social insects. The evolution of the honeybee's suicidal sting and of the sterility of worker bees was an enigma to Darwin, who believed that the insect societies were fatally damaging to his theory. But how does natural selection actually work? Evolution acts on *individuals,* not on populations, and selection favors the genes borne by those individuals that leave the most offspring. But what constitutes an individual in an insect society?

In truly social insects (some bees and wasps, all ants, and all termites), natural selection has acted on the *colony.* The society itself is the individual and the unit acted on by evolution. Each colony is made up of reproductive and sterile members. A honeybee hive, for example, has a single queen, who is the sole egg-layer, and tens of thousands of her offspring, who are female workers having generally nonfunctional ovaries. The sterility of workers is altruistic: During the course of evolution, these offspring gave up their personal reproduction to help their mother rear more of their sisters (figure 45.15).

Hamilton explained the origin of this trait with the theory of kin selection. He noted that, because of an unusual system of sex determination present in bees, wasps, and ants, workers share a very high proportion of genes—theoretically, as many as 75 percent. Because of this close genetic relatedness, *workers propagate more genes by giving up their own reproduction to assist their mother in rearing more of their sisters, some of which will start new colonies and thus reproduce.* This is how workers maximize their inclusive fitness.

> *The colony is the evolving entity in insect societies. Because workers share large fractions of genes, they propagate more genes by not directly reproducing but by helping their mother to reproduce.*

Social insect colonies are composed of highly integrated groups of individuals, called **castes,** each of which performs a set of tasks. The specialization is so extreme and the organization so rigid that these insect societies as a whole exhibit many of the properties of an individual organism and are sometimes considered a *superorganism.*

A human body relies on millions of individual cells that are specialized to perform many different tasks. In a similar way, a beehive or ant colony is a cohesively organized group of individuals in which certain individuals perform specialized tasks on which the survival and reproduction of the entire colony depend. Only one component of a human body—the gonads—is responsible for reproduction. In a similar way, only one component of the beehive or ant's nest—the queen—is involved in the reproduction of that colony. All the cells of a human body are related to one another by descent from one fertilized zygote. Similarly, all the members of a nest or hive are descended from an individual queen.

A honeybee colony may have up to fifty thousand sterile females and a single female queen who lays all the eggs. The queen maintains her dominance by secreting a pheromone, called "queen substance," that suppresses ovary development in other females, turning them into sterile workers. Drones (male bees) are produced in a hive only for purposes of mating.

When the colony becomes too large, some members do not receive a sufficient quantity of queen substance, and the colony begins preparations for swarming. Workers make several new

Figure 45.16 Leafcutter ants.
These leafcutter ants are cutting out sections of a leaf to carry back to the nest. The ants do not live on the leaf material. They feed it to fungi that they raise in underground gardens.

queen cells, in which new queens begin to develop. As these mature, the old queen acknowledges their presence by making a pulsating sound known as "quacking"; the most mature of the developing queens responds with a "tooting" sound. A scout worker returns with directions to a new hive site, and the old queen and a swarm of female workers leave to find the new hive. Left behind, the new queen emerges, kills the other candidate queens, flies out to mate, and returns to assume rule of the hive.

The lifestyles of social insects can border on the bizarre, none more so than the leafcutter ants. Leafcutters are farmers. They live in colonies of up to several million individuals, growing crops of fungi beneath the ground. Their moundlike nests are underground "cities" covering more than 100 square yards, with hundreds of entrances and chambers as deep as 16 feet underground. Long lines of leafcutters march daily from the mound to a tree or bush, cut its leaves into small pieces, and carry the pieces back in another long line to the mound. Small worker ants chew the leaf fragments into a mulch, which they spread like a carpet in underground chambers. Soon, a luxuriant garden of fungi is growing. Nurse ants carry the larvae of the nest around to browse on choice spots. Other workers weed out undesirable kinds of fungi. The **division of labor** among workers is related to worker size (figure 45.16).

Vertebrate Societies

In contrast to the highly structured and integrated insect societies and their remarkable forms of altruism, vertebrates form far less rigidly organized social groups. (Sidelight 45.2, however, examines an interesting exception to this general rule.) This seems paradoxical because vertebrates have larger brains and are capable of more complex behavior. Vertebrates show generally lower degrees of altruism apparently because of the

---

Sidelight 45.2
NAKED MOLE RATS—A RIGIDLY ORGANIZED VERTEBRATE SOCIETY

One exception to the general rule that vertebrate societies are not rigidly organized is the naked mole rat, a tiny, hairless rodent that lives in East Africa. Adult naked mole rats are about 8 to 13 centimeters long and weigh up to 60 grams—about the size of a sausage (Sidelight figure 45.2). Unlike other kinds of mole rats, which live alone or in small family groups, naked mole rats congregate in large underground colonies with a far-ranging system of tunnels and a central nesting area. Colonies often contain eighty or more animals.

Naked mole rats live by eating large underground roots and tubers, which they locate by tunneling in teams. Each mole rat has protruding front teeth, which make it look something like a pocket-sized walrus. It uses these teeth to chisel away the earth from the blind face at the end of a tunnel. When the leading mole rat has loosened a pile of earth, it pushes the pile between its feet and then scuttles backward through the tunnel, moving the pile of earth with its legs. When this animal finally reaches the opening, it gives the pile to another, who kicks the dirt out of the tunnel. Then, free from its pile of dirt, the tunneler returns to the end of the tunnel to dig again, crawling on tiptoe over the backs of a long train of other tunnelers who are moving backward with their own dirt piles.

In addition to being large and well-organized, naked mole rat colonies are unusual because they have a breeding structure that might normally be associated with truly social insects, such as ants and termites. All of the breeding is done by a single "queen," who has one or two male consorts. The worker group, composed of both sexes, keeps the tunnels clear and forages for food. This is very unusual in that few mammals surrender breeding rights without contest or challenge. But if the queen is removed from the colony, havoc breaks out among the workers: Individuals attack each other, and they all compete to become part of the new power structure. When another female becomes dominant and starts breeding, all of the mole rats settle down once again, and discord disappears.

Recent DNA fingerprinting studies have shown that colony members may share 80 percent of their genes. Kin selection may have been important in the evolution of this insectlike mammal society.

**Sidelight figure 45.2
A naked mole rat.**
These small rodents live in unusually large and well-organized colonies.

lower amount of gene sharing among group members (maximally 50 percent in nearly all vertebrates). Nevertheless, in many complex vertebrate social systems, individuals show both reciprocity and kin-selected altruism.

Vertebrate societies are also characterized by a greater degree of conflict and aggression among group members (although conflict does, at times, occur in insect societies). In vertebrates, conflict generally centers around access to food resources and mates.

Features of vertebrate societies that have been studied in detail in several species are cooperative breeding, alarm calling, and mating systems ecology.

Cooperative Breeding

Some species of birds, such as the African pied kingfisher and the Florida scrubjay, have evolved cooperative breeding systems. For example, a pair of scrubjays may have several other birds present in their territory that serve as helpers at the nest, assisting in feeding the pair's offspring, keeping watch for predators, and defending the territory. Helpers are fully capable of breeding on their own, but they remain as nonreproductive altruists for a time. Nests with helpers have more offspring than those that do not. Helpers are most often the fledged offspring of the pair they assist, so the situation resembles that of a family (figure 45.17). The evolution of this type of cooperative breeding in birds has been explained by the inclusive fitness concept discussed earlier.

Alarm Calling

Some aspects of vertebrate behavior, such as the cooperative breeding just discussed, are self-sacrificing and present a puzzle

Figure 45.17 Helpers at the nest.
In a scrubjay family, helpers at the nest cooperate with parents to rear younger siblings.

to evolutionists. Particularly perplexing is the fact that vertebrates are often organized into social groups in such a way that the activities of certain individuals benefit the group at the potential expense of those individuals themselves. The meerkat in figure 45.10, for example, is maintaining a lookout on a termite mound deep in the Kalahari Desert of southern Africa. Under the full glare of the sun, the lookout is exposed to predators and thus is in greater danger than if it were not a sentry. Such behaviors seem contrary to an individual's self-interest.

Individuals in some vertebrate species, such as the meerkat, not only keep watch but may also sound an alarm call when a predator is sighted. Sounding the alarm places the sentry in danger because the alarm alerts the predator to the sentry's location. Who benefits from hearing an alarm call?

Paul Sherman of Cornell University has answered this question through years of field observations of alarm calling in Belding's ground squirrel. The squirrels give an alarm call when a predator, such as a coyote or badger, is spotted. Such predators may attack the calling squirrel, so sounding the alarm signal places the caller at risk. The social unit of a ground squirrel colony is female-based; the group tends to be comprised of a female and her daughters, sisters, aunts, and nieces. Males in the colony are not genetically related to these females. By marking all squirrels in a colony with an individual dye pattern on their fur (using Lady Clairol hair color!) and recording which individuals gave calls and the social circumstances of calling, Sherman found that females with relatives living nearby were more likely to give alarm calls than were females without kin nearby. Males tended to call much less frequently. Alarm calling therefore seems to represent **nepotism;** that is, it favors relatives.

Mating Systems Ecology

Vertebrate societies, like insect societies, have a particular type of organization. A social group of vertebrates has a certain size and stability of members, and it may vary in the number of breeding males and females and in the type of mating system. Sociobiologists have learned that the way in which a group is organized is related to the species' ecology. Environmental features that often influence the evolution of a particular type of social organization are food type and predation.

African weaver birds provide an excellent example. The roughly ninety species of these finchlike birds construct nests of woven vegetation and can be divided according to the type of social group they form. One group builds camouflaged solitary nests in the forest. Males and females in these species have dull plumage, look alike, and are monogamous. They forage for insects to feed their young. The second group nests in colonies in trees on the African savanna. These species feed in flocks on seeds and are polygynous.

The feeding and nesting ecology is correlated with the type of breeding relationships. In the forest, insects are hard to find, and both parents must cooperate and make repeated trips to collect food for the young. Their drab feather coloration is an advantage in their many outings because it does not call them to the attention of predators. The cryptically camouflaged nests further reduce predation. On the open savanna, building a hidden nest is not an option. Rather, the young are protected in nests in spiny trees, which are not very abundant. This shortage of safe nest sites means that birds must nest together in a communal fashion. Because seeds occur abundantly, a female can acquire all the food needed to rear young without a male's help. The male, free from the duties of parenting, spends his time competing with other males for the best nest sites in the tree and courting many females. Bright male plumage has evolved because it is attractive to females, and a polygynous mating system is favored.

> *Social behavior in vertebrates is often characterized by kin-selected altruism. Altruistic behavior is involved in cooperative breeding in birds and alarm calling in mammals. The organization of a vertebrate society represents an adaptive response to ecological conditions.*

Human Sociobiology

Sociobiology is a comparative science. The same theory is applied to study the origin of social behavior in very different species. As was evident in the discussions of insect and vertebrate societies, altruism in bees, birds, and mammals can be explained with the same concept of kin selection. Almost all biologists agree that social behavior has a biological basis and the one unifying theory—evolution—explains it. But are humans just another social animal whose behavior can be explained and understood with Darwinian concepts?

As a social species, humans have an unparalleled complexity. Indeed, we are the only species with the intelligence to contemplate the social behavior of other animals. Intelligence and the ability to learn complex matters are just two human traits (figure 45.18). If an ethologist took an inventory of human behavior, he or she would record kin-selected altruism and reciprocity, other elaborate social contracts, extensive parental care, conflicts between parents and offspring, violence, and warfare. A variety of mating systems, such as monogamy, polygyny, and polyandry, would be described, along with a number of sexual behaviors, such as adultery and homosexuality. Behaviors such as adoption that appear to defy evolutionary explanation would also be

Figure 45.18 The young of primates and birds go through long periods of learning before they can survive on their own.
This young lady, not yet two years old, has much to learn before she can wear Daddy's hat.

a part of the ethologist's catalog. And this incredible variation in behavior all occurs *in one species,* and any of these traits can change within any *individual.* Are these behaviors rooted in human biology?

During the course of human evolution and the emergence of civilization, two processes led to adaptive change. One was **biological evolution.** Humans have a primate heritage, reflected in the extensive sharing of genetic material between humans and our closest relatives, chimpanzees. Our upright posture, bipedal locomotion, power, and precision hand grips are adaptations whose origins can be traced through our primate ancestors. Kin-selected and reciprocal altruism are also found in nonhuman primates, as are other shared traits, such as aggression, and different types of mating systems. Careful studies of nonhuman primates demonstrate that these social traits are adaptive. Researchers *speculate,* based on various lines of evidence, that similar traits evolved in early humans. If individuals showing certain behaviors had a reproductive advantage over other individuals showing alternative behaviors, and these social traits had a genetic basis, then the alleles for the expression of these traits are now part of the human genome and may influence human behavior.

The second process that underscored the emergence of civilization and led to adaptive change was **cultural evolution.** Cultural evolution refers to the transfer of information across generations that is necessary to survival. It is a nongenetic mode of adaptation. Many human adaptations—the use of tools, the formation of cooperative hunting groups, the construction of shelters, and marriage practices—do not follow Mendelian rules of inheritance and are passed from generation to generation by tradition. To anthropologists interested in the origin of human behavior, a culture is as real a way of conveying adaptations across generations as is the gene. Human cultures are also extraordinarily diverse. The ways in which children are socialized among Trobriand Islanders, Pygmies, and Yanomamo Indians are very different. Again, this fantastic variation occurs within *one* species, and individual behavior is very flexible.

Given this great flexibility, how can biological components of human behavior be identified? One way is to study behaviors that are cross-cultural. In spite of cultural variation, some traits characterize all human societies. For example, all cultures have an **incest taboo** forbidding marriages between close relatives. Incestuous matings lead to a greater chance of such disorders as mental retardation and hemophilia. Natural selection may have acted to create a cultural norm to avoid a serious biological problem. Genes responsible for guiding this behavior might have been fixed in human populations because of their adaptive effects.

Although human mating systems vary, polygyny is the most common when all cultures are surveyed. Most mammalian species are polygynous; the human pattern seems to reflect our mammalian evolutionary heritage. Nonverbal communication patterns, such as smiling and raising the hand as a greeting, also occur in many cultures. Perhaps these behaviors represent a common heritage of communication.

A significant number of biologists and social scientists vigorously resist any attempts to explain human behavior in evolutionary terms. In E. O. Wilson's 1975 landmark treatise *Sociobiology: The New Synthesis,* critics vehemently denounced the one chapter devoted to human behavior as an attempt to encourage thinking about human behavior in evolutionary terms. Biology has not been free from politics, and critics of human sociobiology believe that potential for abuse still exists. Moreover, if human behavior is considered to be the product of evolution—influenced by genes and at least in part "hard-wired"—does this not suggest that unpleasant aspects of human behavior, such as aggression and violence, cannot easily be modified? Such a view could affect how we perceive the prospects for positive social change.

Darwinian theory can provide an overarching evolutionary *perspective* on human nature, one that stresses the unity of the human species and the origin of some important, adaptive behaviors. But, because human behavior is affected by *both* innate and learned components, and because many human activities, such as art, music, and religion, are strongly influenced by culture and are not easy to study as adaptation, Darwinian theory is unlikely to offer any *resolution* of the fine details of human nature. The study of the biology of human behavior will always be provocative.

EVOLUTIONARY VIEWPOINT

The evidence now seems overwhelming that genes dictate or influence many if not most aspects of human behavior and personality. Heritability studies comparing identical and nonidentical twins raised apart and raised together clearly indicate that aggressiveness, intelligence, and many other key attributes are more influenced by what genes we inherit than by how we are raised. However, this does not mean that genes determine everything and that how children are raised and educated does not matter. Mendelian segregation only ensures that every child will be different, a unique challenge and opportunity.

SUMMARY

1. Behavior is an adaptive response to stimuli in the environment. An animal's sensory system monitors the environment and has specialized neural elements to detect and process environmental information.

2. Behavior is both instinctive and controlled by genes and is learned through experience. Genes are thought to limit the extent to which behavior can be modified and the types of associations that can be made.

3. The simplest forms of learning involve sensitization and habituation. More complex associative learning may also be built up in this way.

4. An animal's internal state influences when and how a response will occur. Hormones cause an animal's behavior and perception of stimuli to change in an adaptive way.

5. Animals communicate by producing visual, acoustical, chemical, and electrical signals. These signals are involved in mating, food finding, predator defense, and other social situations.

6. Many behaviors are important ecologically and serve as adaptations.

Animals use the position of the sun and stars and other stimuli to orient during daily activities and to navigate during long-range migrations. Foraging and territorial behaviors have evolved because they allow animals to use resources efficiently.

7. Male and female animals show different reproductive behaviors that maximize fitness. Usually, males are competitive and females show mate choice. Mating systems are related to a species' ecology.

8. Insects, vertebrates, and other animals show altruistic behavior. Altruism may evolve through reciprocity or may be directed toward genetic relatives. Cooperative behavior often increases an individual's inclusive fitness.

9. Individuals live in social groups because it is advantageous to do so. Animal societies are characterized by both cooperation and conflict. The organization of a society is related to the species' ecology.

10. Human behavior is extremely rich and varied and results from both biology and culture. Evolutionary theory offers limited but important insight into human nature.

REVIEWING THE CHAPTER

1. How does animal behavior work?

2. What are the approaches to the study of behavior?

3. What is ecology?

4. How does genetics relate to behavior?

5. What is the neural basis of behavior?

6. How do psychologists study animal behavior?

7. What is nonassociative learning, and why are associative behaviors more complex?

8. How does conditioning lead to associative learning?

9. What are the genetic aspects of learning?

10. What is the physiology of behavior?

11. How are reproductive behaviors influenced?

12. How do behavioral rhythms affect animals?

13. How do animals communicate?

14. What are the patterns of courtship behaviors?

15. How is communication achieved in social groups?

16. What is the significance of the dance language of honeybees?

17. What has allowed human language?

18. Is there a correlation between ecology and behavior?

19. What is the association between orientation and migration?

20. How do foraging behaviors differ?

21. What is territoriality?

22. How are ecology and reproduction associated?

23. How have animal societies evolved?

24. How is sociobiology the biological basis of social behavior?

25. What are the benefits and costs of group living?

26. What is altruism, and why did it evolve?

27. What are the consequences of reciprocity?

28. What is the rationale behind kin selection?

29. How are insect societies unique and sometimes bizarre?

30. How are vertebrate societies structured?

31. How do naked mole rats demonstrate a rigidly organized vertebrate society?

32. What are cooperative breeding systems?

33. How is alarm calling an altruistic behavior?

34. How do mating systems reflect species' ecology?

35. How complex is human sociobiology?

COMPLETING YOUR UNDERSTANDING

1. What contributions did von Frisch, Lorenz, and Tinbergen make to ethology?

2. What is the interrelationship between an innate releasing mechanism and a fixed-action pattern?

3. What is the evidence that single allele differences produce behaviorally different individuals?

4. The giant nerve cells that control escape behavior in the sea hare are an example of which concept in ethology?

    a. Sign stimuli

    b. Innate releasing mechanism

    c. Fixed-action pattern

    d. Motivation

    e. Associative learning

5. In the study of behavior, what are the different approaches used by neurologists and comparative psychologists?

6. Who was Ivan Pavlov, and how did he contribute to the understanding of associative learning?

7. What is the "Skinner box," and what did it tell about behavior?

8. How does instinct play a role in learning?

9. Marler's study of song acquisition in the white-crowned sparrow shows that bird song

    a. is entirely innate.

    b. is entirely learned.

    c. develops normally when a bird hears no song.

    d. develops normally when a bird hears another's species song.

    e. is learned, but guided by a genetic program.

10. What is a "black-box" concept?

11. What is the most common cue for resetting circadian rhythms? In the absence of external clues, what is the average time duration of a circadian rhythm endogenous cycle?

12. What are some examples of circannual behavior?

13. What types of "language" is found in chimpanzees and gorillas?

14. All human languages draw from the same set of how many consonant sounds?

    a. 5   d. 100

    b. 10   e. 500

    c. 40   f. 1,000

15. Behavioral ecologists study

    a. neural circuits.

    b. circadian rhythms.

    c. the adaptiveness of behavior.

    d. motivation.

    e. habituation.

16. Is there any evidence that birds have sensory receptors that detect magnetic fields?

17. What evidence suggests that natural selection favors animals that are more efficient foragers?

18. Under what circumstances is territoriality not advantageous?

19. Territorial behavior should evolve when

    a. resources are very scarce.

    b. resources are very abundant.

    c. species are monogamous.

    d. the cost of defense is less than the benefits of the energy gained from defense.

    e. the benefits of being territorial are less than the risks of injury during defense.

20. How has natural selection favored the evolution of behaviors that maximize reproductive success?

21. What did Robert Trivers contribute to the study of sociobiology?

22. In Hamilton's theory on the evolution of altruism by kin selection, what is inclusive fitness?

23. In truly social insects, how is the society itself an individual?

24. What are some analogies that can be made between human and insect societies?

25. Why does a greater degree of conflict arise in vertebrate societies compared to insect societies?

26. How is a naked mole rat society similar to that of an insect society?

27. How did Paul Sherman's research show the benefits of nepotism?

28. Why will the study of the biology of human behavior always be provocative?

29. A recent study of Yanomamo Indians has shown that high-ranking men have more wives, more children, and commit murder more often than other Yanomamo men. Does this mean that there is a gene for violence in humans?

30. There are many speculations about the evolutionary origin of human behavior, but it is usually not possible to rigorously test such hypotheses and critically evaluate alternative, nonevolutionary explanations. Does this mean that the evolutionary study of human behavior is not scientific?

31. Swallows often hunt in groups, whereas hawks and other predatory birds usually are solitary hunters. Can you suggest an explanation for this difference?

32. Can you suggest an evolutionary reason why many vertebrate reproductive groups are composed of one male and numerous females, rather than the reverse?

FOR FURTHER READING

Borgia, G. 1986. Sexual selection in bowerbirds. *Scientific American*, June, 92–100. Describes the often bizarre behavior of the Australasian bowerbirds, in which the females choose their mates depending on how well the males adorn their bowers or ritualized nests.

Cavalli-Sforza, L. 1991. Genes, peoples, and languages. *Scientific American*, Nov., 104–11. A study of the roots of language, using the methods of population genetics, lends support to the hypothesis of the African genesis of humanity.

Dawkins, R. 1989. *The selfish gene*. 2d ed. New York: Oxford University Press. An entertaining account of the sociobiologist's view of behavior.

Fitzgerald, G. 1993. The reproductive behavior of the stickleback. *Scientific American*, April, 80–85. An examination of the adaptive significance of the stickleback's reproductive behavior. This tiny fish was the subject of Niko Tinbergen, who shared the Nobel Prize in 1977 for his study of courtship practices.

Gould, J., and P. Marler. 1987. Learning by instinct. *Scientific American*, Jan., 74–85. A clear and interesting account of the relative roles of instinct and learning in behavior. The authors argue that learning is often limited or controlled by instinct.

Griffin, D. 1984. Animal thinking. *American Scientist* 72: 456–63. An exciting discussion of the possibility that animals have consciousness. This article describes many examples of what appears to be "awareness" by animals.

Heinrich, B. 1989. *Ravens in winter: A zoological detective story*. New York: Summit Books. An account of what ravens do in winter and how their social systems are organized. Vermont zoologist Bernd Heinrich discusses his outstanding field investigations of these intelligent animals.

Huber, F., and J. Thorson. 1985. Cricket auditory communication. *Scientific American*, Dec., 60–68. An unusually clear example of how nervous system activity underlies animal behavior.

Jolly, A. 1985. The evolution of primate behavior. *American Scientist* 73: 230–39. A fascinating survey of behavior among primates that indicates a progressive development of intelligence, rather than a sudden, full-blown appearance when humans evolved.

Lohmann, K. 1992. How sea turtles navigate. *Scientific American*, Jan., 100–107. The fascinating answer to the puzzle posed at the beginning of chapter 1.

Verrell, P. 1990. When males are choosy. *New Scientist*, January 20, 46–50. Article that discusses how, under certain circumstances, males can be choosy, too.

Williams, A. O. D. 1981. Giant brain cells in mollusks. *Scientific American* 244: 68–75. A good introduction to neural aspects of behavior and the use of the sea hare as a model system.

APPENDIX A

C LASSIFICATION OF ORGANISMS

The classification used in this book is explained in chapter 25. Responding to a wealth of recent molecular data, it recognizes two separate kingdoms for the bacteria (prokaryotes); archaebacteria and eubacteria. It divides the eukaryotes into four kingdoms: the diverse and predominantly unicellular Protista, and three large, characteristic multicellular groups derived from them: Fungi, Plantae, and Animalia. Viruses, which are considered nonliving, are not included in this appendix, but are treated in chapter 26.

Kingdom Archaebacteria

Prokaryotic bacteria; single-celled, cell walls lack muramic acid. Like all bacteria, they lack a membrane-bound nucleus, sexual recombination, and internal cell compartments. Archaebacterial cells have distinctive membranes, and unique rRNA and metabolic cofactors. Many are capable of living in an anaerobic environment rich in CO_2 and H_2.

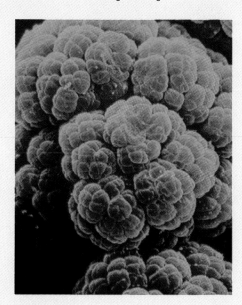

Methanogen

Kingdom Eubacteria

Prokaryotic bacteria; cell walls contain muramic acid. Like archaebacteria, they lack a true nucleus, sexual reproduction, and true internal cell compartments. Eubacteria often form filaments or other forms of colonies. A very diverse group metabolically, over 2,500 species have been named and far more undoubtedly exist.

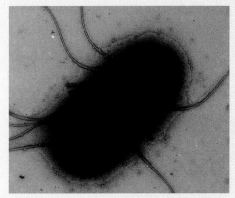

E. coli

Kingdom Protista

Eukaryotic organisms, including many evolutionary lines of primarily single-celled organisms. Eukaryotes have a membrane-bound nucleus and chromosomes, sexual recombination, and extensive internal compartmentalization of the cells; their flagella are complex, with 9 + 2 internal organization. They are diverse metabolically, but much less so than are bacteria; protists are heterotrophic or autotrophic and may capture prey, absorb their food, or photosynthesize. Reproduction in protists is either sexual, involving meiosis and syngamy, or asexual.

Phylum Caryoblastea

One species of primitive amoebalike organism, *Pelomyxa palustris,* which lacks mitosis, mitochondria, and chloroplasts.

Phylum Dinoflagellata

Dinoflagellates; unicellular, photosynthetic organisms, most of which are clad in stiff, cellulose plates and have two unequal flagella that beat in grooves encircling the body at right angles. About 1,000 species.

Phylum Rhizopoda

Amoebas; heterotrophic, unicellular organisms that move from place to place by cellular extensions called pseudopods and reproduce only asexually, by fission. Hundreds of species.

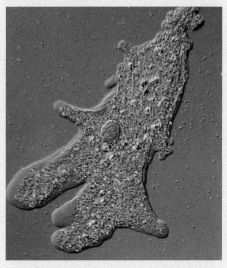

Amoeba

Phylum Sporozoa

Sporozoans; unicellular, heterotrophic, nonmotile, spore-forming parasites of animals. About 3,900 species.

Phylum Acrasiomycota

Cellular slime molds; unicellular, amoeba-like, heterotrophic organisms that aggregate in masses at certain stages of their cycle and form compound sporangia. About 65 species.

Phylum Myxomycota

Plasmodial slime molds; heterotrophic organisms that move from place to place as a multicellular, gelatinous mass, forming sporangia at times. About 450 species.

Phylum Zoomastigina

Zoomastigotes and euglenoids; a highly diverse phylum of mostly unicellular, heterotrophic or autotrophic, flagellated free-living or parasitic protists (flagella one to thousands). Thousands of species.

Phylum Phaeophyta

Brown algae; multicellular, photosynthetic, mostly marine protists with chlorophylls *a* and *c* and an abundant carotenoid (fucoxanthin) that colors the organisms brownish. About 1,500 species.

Phylum Chrysophyta

Diatoms and related groups; mostly unicellular, photosynthetic organisms with chlorophylls *a* and *c* and fucoxanthin. About 11,500 living species.

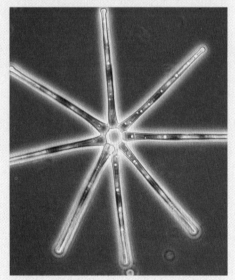

Diatom

Phylum Chlorophyta

Green algae; a large and diverse phylum of unicellular or multicellular, mostly aquatic organisms with chlorophylls *a* and *b*, carotenoids, and starch, accumulated within the plastids (as it also is in plants) as the food storage product. About 7,000 species.

Phylum Ciliophora

Ciliates; diverse, mostly unicellular, heterotrophic protists, characteristically with large numbers of cilia. About 8,000 species.

Phylum Oomycota

Oomycetes; water molds, white rusts, and downy mildews. Aquatic or terrestrial unicellular or multicellular parasites or saprobes that feed on dead organic matter. About 475 species.

Phylum Rhodophyta

Red algae; mostly marine, mostly multicellular protists with chloroplasts containing chlorophyll *a* and phycobilins. About 4,000 species.

Kingdom Fungi

Filamentous, multinucleate, heterotrophic eukaryotes with cell walls rich in chitin; no flagellated cells present. Mitosis in fungi takes place within the nuclei, the nuclear envelope never breaking down. The filaments of fungi grow through the substrate, secreting enzymes and digesting the products of their activity. Septa between the nuclei in the hyphae normally complete only when sexual or asexual reproductive structures are being cut off. Asexual reproduction is frequent in some groups. The nuclei of fungi are haploid, with the zygote the only diploid stage in life cycle. About 77,000 named species.

Division Zygomycota

Zygomycetes; bread molds and other microscopic fungi that occur on decaying organic matter. Hyphae aseptate except when forming sporangia or gametangia. About 665 species.

Division Ascomycota

Ascomycetes; yeasts, molds, many important plant pathogens, morels, cup fungi, and truffles. Hyphae divided by incomplete septa except when asci, the structures characteristic of sexual reproduction, are formed. Meiosis takes place within asci. About 30,000 named species.

Morel mushroom

Division Basidiomycota

Basidiomycetes; mushrooms, toadstools, bracket and shelf fungi, rusts, and smuts. Meiosis takes place within basidia. About 16,000 named species.

Amantia muscaria, a poisonous basidiomycete

Fungi Imperfecti

An artificial group of about 17,000 named species; the reproductive structures are not known.

Lichens

Lichens are symbiotic associations between an ascomycete (a few basidiomycetes are also involved) and either a green alga or a cyanobacterium. At least 13,500 species.

Lichens

Kingdom Plantae

Multicellular, photosynthetic, primarily terrestrial eukaryotes derived from the green algae (phylum Chlorophyta) and, like them, containing chlorophylls *a* and *b*, together with carotenoids, in chloroplasts and storing starch in chloroplasts. The cell walls of plants have a cellulose matrix and sometimes become lignified; cell division is by means of a cell plate that forms across the mitotic spindle. The vascular plants have an elaborate system of conducting cells consisting of xylem (in which water and minerals are transported) and phloem (in which carbohydrates are transported); the mosses have a reduced vascular system, which the liverworts and hornworts (which may not be directly related to the mosses) lack. Plants have a waxy cuticle that helps them to retain water. Most have stomata, flanked by specialized guard cells, which allow water to escape and carbon dioxide to reach the chloroplast-containing cells within their leaves and stems. All plants have an alternation of generations with reduced gametophytes and multicellular gametangia. About 270,000 species.

Division Bryophyta

Mosses, hornworts, and liverworts. Bryophytes have green photosynthetic gametophytes and usually brownish or yellowish sporophytes with little or no chlorophyll. About 16,600 species.

Liverwort

Division Psilophyta

Whisk ferns; a group of vascular plants. Two genera and several species.

Division Lycophyta

Lycopods (including clubmosses and quillworts); vascular plants. Five genera and about 1,000 species.

Division Sphenophyta

Horsetails; vascular plants. One genus (*Equisetum*), 15 species.

Division Pterophyta

Ferns; vascular plants, often with characteristically divided, feathery leaves (fronds). About 12,000 species.

Division Coniferophyta

Conifers; seed-forming vascular plants; mainly trees and shrubs. About 550 species.

Division Cycadophyta

Cycads; tropical and subtropical palmlike gymnosperms. Ten genera, about 100 species.

Division Ginkgophyta

One species, the ginkgo or maidenhair tree.

Division Anthophyta

Flowering plants, or angiosperms, the dominant group of plants; characterized by a specialized reproductive system involving flowers and fruits. About 235,000 species.

Kingdom Animalia

Animals are multicellular eukaryotes that characteristically ingest their food. Their cells are usually flexible. In all of the approximately 35 phyla except sponges, these cells are organized into structural and functional units called tissue, which in turn makes up organs in most animals. In animals, the cells move extensively during the development of the embryos; the blastula, a hollow ball of cells, forms early in this process and is characteristic of the group. Most animals reproduce sexually; their nonmotile eggs are much larger than their small, flagellated sperm. The gametes fuse directly to produce a zygote and do not divide by mitosis as in plants. More than a million species of animals have been described, and at least several times that many await discovery.

Phylum Porifera

Sponges; animals that mostly lack definite symmetry and possess neither tissues nor organs. About 10,000 species, mostly marine.

Barrel sponge

Phylum Cnidaria

Corals, jellyfish, hydras; mostly marine, radially symmetrical animals that usually have distinct tissues; two forms, polyps and medusae. About 10,100 species.

Jellyfish

Phylum Platyhelminthes

Flatworms; bilaterally symmetrical acoelomates; the simplest animals that have organs. About 13,000 species.

Phylum Nematoda

Nematodes, eelworms, and roundworms; ubiquitous, bilaterally symmetrical, cylindrical, unsegmented, pseudocoelomate worms, including many important parasites of plants and animals. More than 12,000 described species, but the actual number is probably 500,000 or more species.

Phylum Mollusca

Mollusks; bilaterally symmetrical, protostome coelomate animals that occur in marine, freshwater, and terrestrial habitats. Many mollusks possess a shell. At least 110,000 species.

Snail

Phylum Annelida

Annelids; segmented, bilaterally symmetrical, protostome coelomates; the segments are divided internally by septa. About 12,000 species.

Phylum Arthropoda

Arthropods; bilaterally symmetrical protostome coelomates with a segmented body, chitinous exoskeleton, complete digestive tract, dorsal brain and paired nerve cord, and jointed appendages. Arthropods are the largest phylum of animals, with nearly a million species described and many more to be found.

Desert tarantula

Phylum Echinodermata

Echinoderms; sea stars, brittle stars, sand dollars, sea cucumbers, and sea urchins. Complex deuterostome, coelomate, marine animals that are more or less radially symmetrical as adults. About 6,000 living species.

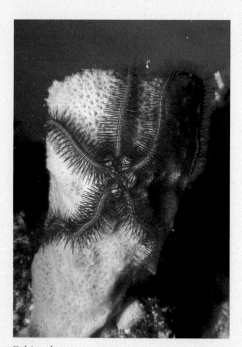

Echinoderm

Phylum Chordata

Chordates; bilaterally symmetrical, deuterostome, coelomate animals that have at some stage of their development a notochord, pharyngeal slits, a hollow nerve cord on their dorsal side, and a tail. The best-known group of animals; about 45,000 species.

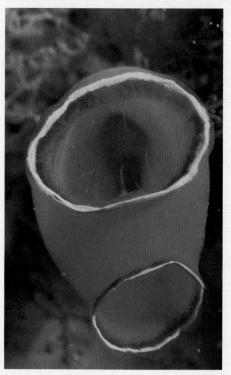

Tunicate

G LOSSARY

A

abortion (L. *abortum,* miscarried) The termination of pregnancy before the fetus reaches the stage of viability, which is approximately twenty to twenty-eight weeks of gestation.

abscission (L. *ab,* away, off + *scisso,* dividing) In vascular plants, the dropping of leaves, flowers, fruits, or stems at the end of a growing season, as the result of formation of a corky layer of young cells at the base.

absorption (L. *absorbere,* to swallow down) The movement of water and of substances dissolved in water into a cell, tissue, or organism.

absorption spectrum The range of photons that a given atom or molecule is capable of absorbing, depending on the electron energy levels available in the atom or molecule.

abyssal zone (Gr. *abyssos,* bottomless) The marine environment of the deep-water areas of the ocean.

accessory pigment A pigment, such as a carotenoid or chlorophyll *b,* that increases the percentage of the photons of sunlight that are harvested.

acetylcholine The most important of the numerous chemical neurotransmitters responsible for the passing of nerve impulses across synaptic junctions. The neurotransmitter in neuromuscular (nerve-muscle) junctions.

acetylcholinesterase An enzyme that removes the leftover acetylcholine from the synaptic cleft at the neuromuscular junction after the last impulse. One of the fastest-acting enzymes in the vertebrate body.

acid Any substance that dissociates to form H^+ ions when dissolved in water.

acid rain A blanket term for the process whereby industrial pollutants such as nitric and sulfuric acids—introduced into the upper atmosphere by factory smokestacks—are spread over wide areas by the prevailing winds and then fall to the earth with the precipitation, lowering the pH of groundwater and killing life.

acoelomate (Gr. *a,* not + *koiloma,* cavity) a bilaterally symmetrical animal not possessing a body cavity, such as a flatworm.

acquired immune deficiency syndrome (AIDS) An infectious and usually fatal human disease caused by a retrovirus, HIV, which attacks T-cells. The virus multiplies within and kills individual T-cells, releasing thousands of progeny that infect and kill other T-cells, until no T-cells remain, leaving the affected individual helpless in the face of microbial infections because his or her immune system is now incapable of marshaling a defense against them. *See* T-cell.

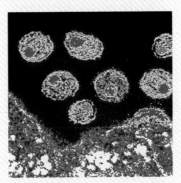

actin (Gr. *actis,* ray) One of the two major proteins that make up myofilaments (the other is myosin). It provides the cell with mechanical support and plays major roles in determining cell shape and cell movement.

actinin (Gr. *actis,* ray) A protein anchored to actin that enables myofilaments to convert the sliding of muscle fibers into cell movement.

action potential A single nerve impulse. A transient all-or-none reversal of the electric potential across a neuron membrane. Because it can activate nearby voltage-sensitive channels, an action potential propagates along a nerve cell.

activating enzyme Any of a battery of twenty enzymes, one or more of which recognizes a particular three-base anticodon sequence in a tRNA molecule.

activation The process by which the regulatory protein binds to DNA and turns on the transcription of specific genes.

activation energy The energy a molecule must possess to undergo a specific chemical reaction.

active transport The transport of a solute across a membrane by protein carrier molecules to a region of higher concentration by the expenditure of chemical energy. One of the most important functions of any cell.

actual rate of population increase The difference between the birth rate and the death rate per given number of individuals per unit of time.

adaptation (L. *adaptare,* to fit) Any peculiarity of structure, physiology, or behavior that promotes the likelihood of an organism's survival and reproduction in a particular environment.

adenine An organic molecule composed of two carbon-nitrogen rings.

adenosine diphosphate (ADP) The molecule resulting from the breaking off of a phosphate group from adenosine triphosphate.

adenosine triphosphate (ATP) A molecule composed of ribose, adenine, and a triphosphate group. ATP is the chief energy currency of all cells. Cells focus all of their energy resources on the manufacture of ATP from ADP and phosphate, which requires the cell to supply 7 kilocalories of energy obtained from photosynthesis or from electrons stripped from foodstuffs to form 1 mole of ATP. Cells then use this ATP to drive endergonic reactions.

adhesion (L. *adhaerere,* to stick to) The molecular attraction exerted between the surfaces of unlike bodies in contact, as water molecules to the walls of the narrow tubes that occur in plants.

ADP *See* adenosine diphosphate.

aerobic (Gr. *aer,* air + *bios,* life) Oxygen-requiring.

aerobic pathway A metabolic pathway, at least one step of which is an oxidation/reduction reaction that depends on oxygen gas as an electron acceptor. Includes the citric acid cycle and pyruvate oxidation.

afferent pathway *See* sensory pathway.

AIDS *See* acquired immunodeficiency syndrome.

albumen (L. *albus,* white) The white of an egg, which provides additional nutrients and water for the embryo.

aldosterone (from ald[ehyde] + ster[ol] + [horm]one) A steroid hormone of the adrenal cortex that controls the salt and water balance in the body.

alga, *pl.* **algae** (L.) A photosynthetic protist cell. Algae contain chloroplasts.

alkaptonuria (alkali + Gr. *haptein,* to possess + *ouron,* urine) A genetic disorder caused by the lack of an enzyme necessary to break down homogentisic acid (alkapton) in urine.

allantois (Gr. *allas*, sausage + *eidos*, form) A membrane of the amniotic egg that functions in respiration and excretion in birds and reptiles and plays an important role in placental development in most mammals.

allele (Gr. *allelon*, of one another) One of two or more alternative forms of a gene.

allele frequency The relative proportion of a particular allele among individuals of a population. Not equivalent to gene frequency, although the two terms are sometimes confused.

allergen Any substance that initiates a strong immune response in a particular individual.

allergy (Gr. *allos*, other + *ergon*, action) An unusual sensitivity to a certain substance that does not invoke a response in most people. Hay fever is a common manifestation of allergy.

allometric growth (Gr. *allos*, other + *meros*, part) A pattern of growth in which different components grow at different rates.

allosteric interaction (Gr. *allos*, other + *stereos*, shape) The change in shape that occurs when an activator or inhibitor binds to an enzyme. These changes result when specific, small molecules bind to the enzyme, molecules that are not substrates of that enzyme.

alpha-cell A cell in the islets of Langerhans that produces the hormone glucagon, which causes liver cells to release stored glucose and fat cells to break down triglycerides.

alternation of generations A reproductive life cycle in which the diploid phase produces spores that give rise to the haploid phase, and the haploid phase produces gametes that

fuse to give rise to the zygote. The zygote is the first cell of the multicellular diploid phase.

altruism (L. *alter*, other) Self-sacrificing behavior.

alveolus, *pl.* **alveoli** (L. *alveus*, a small cavity) One of the many small, thin-walled air sacs within the lungs in which the bronchioles terminate.

Alzheimer's disease A type of dementia seen in the elderly, probably resulting from a deficiency of acetylcholine, which leads to degeneration and death of neurons.

amino acids (Gr. *Ammon*, referring to the Egyptian sun god, near whose temple ammonium salts were first prepared from camel dung) Molecules containing an amino group ($-NH_2$), a carboxyl group ($-COOH$), a hydrogen atom, and a functional group designated R, all bonded to a central carbon atom. The twenty different amino acids that occur in proteins are grouped into five chemical classes.

amnion The innermost of the extraembryonic membranes. The amnion forms a fluid-filled sac around the embryo in amniotic eggs.

amniotic egg An egg that is isolated and protected from the environment by a more or less impervious shell. The shell protects the embryo from drying out, nourishes it, and enables it to develop outside of water.

ampulla, *pl.* **ampullae** (L. *ampulla*, pear-shaped bottle) A flask-shaped organ or bladder in an aquatic animal.

amylase (from *amyl* + -ase, enzyme suffix) An enzyme that breaks up starches and other carbohydrates into sugars.

amyloplast (Gr. *amylon*, starch + *plastos*, formed) A plant organelle called a plastid that specializes in storing starch.

anabolism (Gr. *ana*, up + *bolein*, to throw) A process in which more complex molecules are synthesized.

anaerobic (Gr. *an*, without + *aer*, air + *bios*, life) Any process that can occur without oxygen. Includes glycolysis and fermentation. Anaerobic organisms can live without free oxygen.

anaphase (Gr. *ana*, up + *phasis*, form) The third and shortest stage of mitosis, during which the daughter chromosomes move rapidly to the opposite poles of the cell.

animal pole In fishes and other aquatic vertebrates with asymmetrical yolk distribution in their eggs, the hemisphere of the blastula, comprising cells relatively poor in yolk.

anion (Gr. *anion*, to go up) A negatively charged ion.

annelid (L. *annulus*, ring) Any of a phylum (Annelida) of worms or wormlike animals characterized by a soft, elongated body composed of a series of similar ringlike segments. Earthworms and leeches are annelids.

annual (L. *annus*, year) A plant that germinates, flowers, produces seed, and dies within the same year or growing season.

annual ring Any of the concentric rings of wood seen when the stem of a tree or shrub is cut across. Each ring shows one year's growth.

anterior (L. *ante*, before) Located before or toward the front. In animals, the head end of an organism.

anther (Gr. *anthos*, flower) The part of the stamen of a flower that bears the pollen.

antheridium, *pl.* **antheridia** (Gr. *anthos*, flower) The male or sperm-producing organ in such plants as ferns and mosses.

antibody (Gr. *anti*, against) A protein substance produced in the blood by a B-cell lymphocyte in response to a foreign substance (antigen) and released into the bloodstream. Binding to the antigen, antibodies mark them for destruction by other elements of the immune system.

anticodon The three-nucleotide sequence at the end of a tRNA molecule that is complementary to, and base pairs with, an amino-acid-specifying codon in mRNA.

antidiuretic hormone (ADH) *See* vasopressin.

antigen (Gr. *anti*, against + *genos*, origin) A foreign substance, usually a protein, that stimulates lymphocytes to proliferate and secrete specific antibodies that bind to the foreign substance, labeling it as foreign and destined for destruction.

aorta (Gr. *aeirein*, to lift) The main artery of vertebrate systemic blood circulation. It carries the blood from the left side of the heart to all regions of the body except the lungs.

aortic valve A one-way valve that permits the flow of blood out of the ventricle, and then closes, thus preventing the reentry of blood from the aorta into the heart.

apical meristem (L. *apex*, top + Gr. *meristos*, divided) A region of active cell division that occurs at or near the tips of the roots and shoots of plants.

aposematic *See* warning coloration.

appendicular skeleton (L. *appendicula*, a small appendage) The skeleton of the limbs of the human body containing 126 bones.

archegonium, *pl.* **archegonia** (Gr. *archegonos*, first of a race)

The multicellular female reproductive organ in such plants as ferns and mosses.

archenteron (Gr. *arch*, early + *enteron*, intestine) The primitive intestinal or alimentary cavity of a gastrula.

arteriole A smaller artery, leading from the arteries to the capillaries.

artery, *pl.* **arteries** (Gr. *arteria*, artery) One of the tubular branching vessels that carries the blood from the heart through the body.

arthropod (Gr. *arthron*, joint + *pous*, foot) A phylum of hard-shelled animals with jointed legs and a segmented body.

artificial selection The differential reproduction of genotypes in response to demands imposed by human intervention.

asexual An organism that reproduces without forming gametes. Asexual reproduction does not involve sex. Its outstanding characteristic is that an individual offspring is genetically identical to its parent.

associative Learning behaviors that require associative activity within the central nervous system.

associative cortex The major portion of the cerebral cortex that appears to be the site of sensory information integration and of higher mental activities, such as planning and contemplation.

aster The array of microtubules that radiate from the centrioles when the latter reach the poles of the cells.

atom (Gr. *atomos*, indivisible) A core (nucleus) of protons and neutrons surrounded by an orbiting cloud of electrons. The chemical behavior of an atom is largely determined by the distribution of its electrons, particularly the number of electrons in its outermost level.

atomic mass The atomic mass of an atom consists of the combined weight of all of its protons and neutrons.

atomic number The number of protons in the nucleus of an atom. In an atom that does not bear an electric charge (that is, one that is not an ion), the atomic number is also equal to the number of electrons.

ATP *See* adenosine triphosphate.

atrioventricular (AV) node A slender connection of cardiac muscle cells that receives the heartbeat impulses from the sinoatrial node and conducts them by way of the bundle of His.

atrium Either of the chambers of the heart that receives blood from the veins and forces it into a ventricle. The heart of mammals, birds, and reptiles has two atria; that of fishes and amphibians has one.

auditory cortex A major sensory region on the cerebral cortex lying within the temporal lobe. Different surface regions of this cortex correspond to different sound frequencies.

autoimmune disease A disease in which antibodies are formed against the body's own cells because the immune system fails to distinguish between foreign and host tissue.

autonomic nervous system (Gr. *autos*, self + *nomos*, law) The motor pathways that carry commands from the central nervous system to regulate the glands and nonskeletal muscles of the body. Also called the involuntary nervous system.

autosome (Gr. *autos*, self + *soma*, body) Any of the twenty-two pairs of human chromosomes that are similar in size and morphology in both males and females.

autotroph (Gr. *autos*, self + *trophos*, feeder) Self-feeder. An organism that can harvest light energy from the sun or from the oxidation of inorganic compounds to make organic molecules.

auxin (Gr. *auxein*, to increase) A plant hormone that controls cell elongation, among other effects.

axial skeleton The skeleton of the head and trunk of the human body containing eighty bones.

axil (L. *axilla*, armpit) The angle between a branch or leaf and the stem from which it arises.

axon (Gr. *axon*, axis) A single, long process extending out from a neuron that conducts impulses away from the neuron cell body.

B

bacillus, *pl.* **bacilli** (L. *baculus*, rod) A straight or rod-shaped bacterium.

bacteriophage (Gr. *bakterion*, little rod + *phagein*, to eat) A virus that infects bacterial cells. Also called a phage.

bacterium, *pl.* **bacteria** (Gr. *bakterion*, dim. of *baktron*, a staff) The simplest cellular organism. Its cells are smaller and prokaryotic in structure, and they lack internal organization.

bark A term used to refer to all of the tissues of a mature stem or root outside of the vascular cambium.

barometric pressure The weight of the earth's atmosphere over a unit area of the earth's surface. Measured with a mercury barometer at sea level, this corresponds to the pressure required to lift a column of mercury 760 millimeters.

baroreceptor (Gr. *baros*, weight) A nerve cell or group of cells sensitive to changes in pressure, such as blood pressure.

Barr body After Murray L. Barr, Canadian anatomist. The inactivated X chromosome in female mammals that can be seen as a deeply staining body, which remains attached to the nuclear membrane.

basal body In cells that contain flagella or cilia, a form of centriole that anchors each flagellum.

base Any substance that combines with H+ ions. Having a pH value above 7.

basidium, *pl.* **basidia** (Gr. *basis*, base) A small, club-shaped structure on basidiomycetes that produces spores.

basilar membrane A membranous part of the cochlea that forms the fibrous base supporting the organ of Corti.

basophil A leukocyte containing granules that rupture and release chemicals that enhance the inflammatory response. Important in causing allergic responses.

Batesian mimicry After Henry W. Bates, English naturalist. A situation in which a palatable or nontoxic organism resembles another kind of organism that is distasteful or toxic. Both species exhibit warning coloration.

B-cell A lymphocyte that recognizes invading pathogens much as T-cells do, but instead of attacking the pathogens directly, it marks them for destruction by the nonspecific body defenses.

behavior A coordinated neuromotor response to changes in external or internal conditions. A product of the integration of sensory, neural, and hormonal factors.

beta-cell A cell in the islets of Langerhans that secretes insulin when a person eats, storing glucose to be used later.

biennial (L. *biennium,* a two-year period) A plant that normally requires two growing seasons to complete its life cycle. Biennials flower in the second year of their lives.

bilateral symmetry (L. *bi,* two + *lateris,* side; Gr. *symmetria,* symmetry) A body form in which the right and left halves of an organism are approximate mirror images of each other.

bile salt A molecule produced by the liver that acts as a superdetergent combining with fats in emulsification, a process that renders fats soluble.

binary fission (L. *binarius,* consisting of two things or parts + *fissus,* split) Asexual reproduction of a cell by division into two equal, or nearly equal, parts. Bacteria divide by binary fission.

binocular (L. *bi,* two + *ocularis,* the eye) Refers to vision in which both eyes are located at the front of the head and function together to provide a three-dimensional view of the object being examined.

binomial system (L. *bi,* twice, two + Gr. *nomos,* usage, law) A system of nomenclature that uses two words. The first names the genus, and the second designates the species.

biogeochemical cycle (Gr. *bios,* life + *geo,* earth + *chem,* chemical) A geological cycle that involves the controlled cycling of chemicals. Includes substances derived from weathering of rocks and those that occur in organisms.

biological evolution The extensive behavior patterns of humans rooted in evolutionary biology that have led to adaptive change.

biomass (Gr. *bios,* life + *maza,* lump or mass) The total weight of all of the organisms living in an ecosystem.

biome (Gr. *bios,* life + -*oma,* mass, group) A major terrestrial assemblage of plants, animals, and microorganisms that occur over wide geographical areas and have distinct characteristics. The largest ecological unit.

blade The wide, flat part of a leaf.

blastocoel (Gr. *blastos,* sprout + *koilia,* belly) The central cavity of a blastula.

blastodisc (Gr. *blastos,* sprout + *discos,* a round plate) A disklike aggregation of formative protoplasm at one pole of the yolk of a fertilized egg. Contains the nucleus.

blastomere (Gr. *blastos,* sprout + *meros,* part) An individual cell of the morula.

blastopore (Gr. *blastos,* sprout + *poros,* a path or passage) In vertebrate development, the opening that connects the archenteron cavity of a gastrula-stage embryo with the outside. Represents the future mouth in some animals (protostomes), the future anus in others (deuterostomes).

blastula (Gr. *blastos,* a little sprout) In vertebrates, an early embryonic stage consisting of a hollow, fluid-filled ball of cells one layer thick. A vertebrate embryo after cleavage and before gastrulation.

blood type In humans, the type of cell-surface antigens present on the red blood cells of an individual. Genetically determined, alternative alleles yield different surface antigens. When two different blood types are mixed, the cell surfaces often interact with surface antigen antibodies, leading to agglutination. One genetic locus encodes the ABO blood group, another encodes the Rh blood group, and still others encode other surface antigens.

bronchus, pl. bronchi (Gr. *bronchos,* windpipe) One of a pair of respiratory tubes branching from the lower end of the trachea (windpipe) into either lung.

C

calcitonin (L. *calcem,* lime) A thyroid hormone that stimulates calcium uptake.

calorie (L. *calor,* heat) The amount of energy in the form of heat required to raise the temperature of 1 gram of water 1 degree Celsius.

Calvin cycle After Melvin Calvin, American chemist. The series of dark reactions in which ATP and NADPH produced by the light reactions are used to fix carbon. In this process, ribulose 1,5-bisphosphate (RuBP) is carboxylated and the products run backward through a series of reactions also found in the glycolytic sequence to form fructose 6-phosphate molecules, some of which are used to reconstitute RuBP. The remainder enter the cell's metabolism as newly fixed carbon in glucose.

calyx (Gr. *kalyx,* a husk, cup) The sepals collectively. The outermost flower whorl.

cambium, pl. cambia (L. *combiare,* to exchange) In vascular plants, embryonic tissue zones (meristems) that run parallel to the sides of roots and stems. Consists of the cork cambium and the vascular cambium.

cancer Unrestrained invasive cell growth. A tumor or cell mass resulting from uncontrollable cell division.

capillary (L. *capillaris,* hairlike) A blood vessel with a very slender, hairlike opening. Blood exchanges gases and metabolites within capillaries. Capillaries join the end of an artery to the beginning of a vein.

capillary action The movement of a liquid along a surface as a result of the combined effects of cohesion and adhesion.

carbohydrate (L. *carbo,* charcoal + *hydro,* water) An organic compound consisting of a chain or ring of carbon atoms to which hydrogen and oxygen atoms are attached in a ratio of approximately 1:2:1. A compound of carbon, hydrogen, and oxygen having the generalized formula $(CH_2O)_n$, where n is the number of carbon atoms.

carbon cycle The worldwide circulation and reutilization of carbon atoms.

carbon fixation A process in which atmospheric carbon dioxide is incorporated into carbon-containing molecules.

carcinogen (Gr. *karkinos,* cancer + -*gen*) Any cancer-causing agent.

cardiovascular system (Gr. *kardia,* heart + L. *vasculum,* vessel) The blood circulatory system and the heart that pumps it. Collectively, the blood, heart, and blood vessels.

carpel (Gr. *karpos,* fruit) A leaflike organ in angiosperms that encloses one or more ovules. One of the members of the gynoecium.

carrying capacity The maximum population size that a habitat can support.

catabolism (Gr. *katabole*, throwing down) A process in which complex molecules are broken down into simpler ones.

catalysis (Gr. *katalysis*, dissolution + *lyein* to loosen) The enzyme-mediated process in which the subunits of polymers are held together and their bonds are stressed.

catalyst (Gr. *kata*, down + *lysis*, a loosening) A general term for a substance that speeds up a specific chemical reaction by lowering the energy required to activate or start the reaction. An enzyme is a biological catalyst.

cell (L. *cella*, a chamber or small room) The smallest unit of life. The basic organizational unit of all organisms. Composed of a nuclear region containing the hereditary apparatus within a larger volume called the cytoplasm bounded by a lipid membrane.

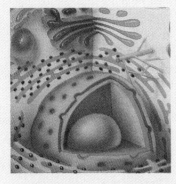

cell biology The study of how cells are constructed and how they grow, divide, and communicate at the molecular, subcellular, and cellular levels of organization.

cell-surface antigen A sugar molecule that has been encoded by an enzyme to a lipid on the surface of a blood cell and that acts as a recognition marker in the immune system.

cell-surface marker A specific set of proteins unique to a particular cell that enables each cell to signal to the environment what type of cell it is. Other cell-surface markers are glycolipids.

cell-surface receptor An information-transmitting protein that extends across a plasma membrane. The information it transmits includes the presence of hormones and the signals that pass from one nerve to another.

cellular respiration The process in which the energy stored in a glucose molecule is released by oxidation. Hydrogen atoms are lost by glucose and gained by oxygen.

central nervous system The portion of the nervous system in vertebrates composed of the brain and spinal cord that is the site of information processing and control within the nervous system.

centriole (Gr. *kentron*, center of a circle + L. *olus*, little one) An organelle associated with the assembly and organization of microtubules. An endosymbiont that occurs in animals and most protists, but not in plants or fungi.

centromere (Gr. *kentron*, center + *meros*, a part) A constricted region of the chromosome joining two sister chromatids, to which the kinetochore is attached. About 220 nucleotides in length, it is composed of highly repeated DNA sequences (satellite DNA).

cerebellum "Little brain." The hindbrain region of the vertebrate brain. It integrates information about body position and motion, coordinates muscular activities, and maintains equilibrium.

cerebral cortex The layer of gray matter covering the cerebrum. The seat of conscious sensations and voluntary muscular activity.

cerebrum (L. brain) The portion of the vertebrate brain that occupies the upper part of the skull, consisting of two cerebral hemispheres united by the corpus callosum. It is the primary association center of the brain, coordinating and processing sensory input and coordinating motor responses.

chaparral (Sp. *chaparro*, evergreen oak) Extensive communities of evergreen, often spiny shrubs and low trees in dry-summer areas of California and adjacent regions.

Chargaff's rule After Erwin Chargaff, American biochemist. The observation that, in all natural DNA molecules, the amount of adenine is always equal to the amount of thymine, and the amount of guanine is always equal to the amount of cytosine.

chemical bond The force holding two atoms together. The force can result from the attraction of opposite charges (ionic bond) or from the sharing of one or more pairs of electrons (a covalent bond).

chemically gated ion channel A transmembrane pathway for a particular ion that is opened or closed by a chemical, such as a neurotransmitter.

chemiosmosis The cellular process responsible for almost all of the adenosine triphosphate (ATP) harvested from eaten food and for all the ATP produced by photosynthesis.

chemoautotroph An autotrophic bacterium that uses chemical energy released by specific inorganic reactions to power its life processes, including the synthesis of organic molecules.

chemoreceptor In a male insect, a nerve cell modified to detect tiny quantities of the sex pheromone.

chiasma, *pl.* **chiasmata** (Gr. a cross) In meiosis, the points of crossing-over where portions of chromosomes have been exchanged during synapsis. A chiasma appears as an **X**-shaped structure under a light microscope.

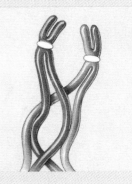

chief cell Within the mucosa of the stomach, an exocrine gland that secretes pepsinogen.

chloroplast (Gr. *chloros*, green + *plastos*, molded) An energy-producing organelle containing bacteria-like elements with vesicles containing chlorophyll. In plant cells, it is the site of photosynthesis.

chromatid (Gr. *chroma*, color + L. *-id*, daughters of) One of two daughter strands of a duplicated chromosome that is joined by a single centromere.

chromatin (Gr. *chroma*, color) The complex of DNA and proteins of which eukaryotic chromosomes are composed.

chromosomal rearrangement In eukaryotes, when large segments of chromosomes change their relative location or undergo duplication, often with drastic effects on the expression of the genetic message.

chromosomal theory of inheritance The theory that reproduction involves the initial union of only two cells—egg and sperm; that the two homologous chromosomes of each pair segregate during meiosis; that gametes have a copy of one member of each pair of homologous chromosomes while diploid individuals have a copy of both members of each pair; and that during meiosis, each pair of homologous chromosomes orients on the metaphase plate independent of any other pair.

chromosome (Gr. *chroma*, color + *soma*, body) The vehicle by which hereditary information is physically transmitted from one generation to the next. In a eukaryotic cell, long threads of DNA that are associated with protein and that contain hereditary information.

cilium, *pl.* **cilia** (L. eyelash) Refers to flagella, which are numerous and organized in dense rows. Cilia propel cells through water. In human tissue, they move water over the tissue surface.

citric acid cycle The cyclic series of reactions in which pyruvate, the product of glycolysis, enters the cycle to form citric acid and is then oxidized to carbon dioxide. Also called the Krebs cycle (after its discoverer) and the tricarboxylic acid (TCA) cycle (citric acid possesses three carboxyl groups).

class A taxonomic category ranking below a phylum (division) and above an order.

classical conditioning The repeated presentation of a stimulus in association with a response that causes the brain to form an association between the stimulus and the response, even if they have never been associated before.

cleavage The second stage of the ten reactions of glycolysis in which the six-carbon product of the first stage is split into two three-carbon molecules. One is G3P, and the other is converted

to G3P by another reaction. Also, the progressive division of cells during embryonic growth.

cleavage furrow During cytokinesis, the area where the cytoplasm is progressively pinched inward by the decreasing diameter of the microfilament belt.

climax community A self-perpetuating community in which populations remain stable and exist in balance with each other and the environment. The final stage of a succession.

clone (Gr. *klon*, twig) A line of cells, all of which have arisen from the same single cell by mitotic division. One of a population of individuals derived by asexual reproduction from a single ancestor. One of a population of genetically identical individuals.

clonal selection theory The mechanism that determines immune specificity and accounts for the immune system's memory of antigens; an antigen activates a few lymphocytes, which proliferate and form cells with specific activities toward the antigen.

cloning Producing a cell line or culture, all of whose members contain identical copies of a particular nucleotide sequence. An essential element in genetic engineering, cloning is usually carried out by inserting the desired gene into a virus or plasmid, infecting a cell culture or plasmid, infecting a cell culture with the hybrid virus, and selecting for culture a cell that has taken up the gene.

closed circulatory system A system in most animals in which the circulatory system fluid is separated from the rest of the body's fluids and does not mix with them.

co-dominance In genetics, a situation in which the effects of

both alleles at a particular locus are apparent in the phenotype of the heterozygote.

codon (L. code) The basic unit of the genetic code. A sequence of three adjacent nucleotides in DNA or mRNA that code for one amino acid or for polypeptide termination.

coelom (Gr. *koilos*, a hollow) A body cavity formed between layers of mesoderm and in which the digestive tract and other internal organs are suspended.

coenzyme A cofactor that is a nonprotein organic molecule.

coevolution (L. *co-*, together + *e-*, out + *volvere*, to fill) A term that describes the long-term evolutionary adjustment of one group of organisms to another.

cognitive (L. *cognoscere*, to know) Thinking. Using the mind.

collenchyma cell (Gr. *killa*, glue + *en-*, in + *chyma*, what is poured) Living plant tissue with cells whose walls are thickened and usually elongated.

commensalism (L. *cum*, together with + *mensa*, table) A symbiotic relationship in which one species benefits while the other neither benefits nor is harmed.

community (L. *communitas*, community, fellowship) The population of different species that live together and interact in a particular place.

companion cell A specialized parenchyma cell with a large nuclei adjacent to the sieve tube in the phloem of vascular plants responsible for loading and unloading glucose from the sieve tube.

competition Interaction between individuals of two or more species for the same scarce resources. Intraspecific competition is interaction for the same scarce resources between individuals of a single species.

competitive exclusion The hypothesis that, if two species are competing with one another for the same limited resource in the same place, one will be able to use that resource more efficiently than the other and eventually will drive that second species to extinction locally.

complement system The chemical defense of a vertebrate body that consists of a battery of proteins that become activated by the walls of bacteria and fungi. Complements the cellular defenses.

concentration gradient The concentration difference of a substance as a function of distance. In a cell, a greater concentration of its molecules in one region than in another.

condensation The coiling of the chromosomes into more and more tightly compacted bodies begun during the G2 phase of the cell cycle.

cone cell A specialized sensory cell of the retina of the eye that contains iodopsin and functions in color vision.

conjugation (L. *conjugare*, to yoke together) An unusual mode of reproduction that characterizes the ciliates, in which nuclei are exchanged between individuals through tubes connecting them during conjugation.

consumer In ecology, a heterotroph that derives its energy from living or freshly killed organisms or parts thereof. Primary consumers are herbivores; secondary consumers are carnivores or parasites.

conus arteriosus The anterior-most chamber of the embryonic heart in vertebrate animals. A conical structure in the upper left portion of the right ventrical in humans, supplying the pulmonary artery.

cooperative breeding A feature of some vertebrate bird societies, in which several birds of the same species will not breed but rather assist as helpers for a breeding pair.

cork cell A cell split off by the cork cambium, which contains a fatty substance and is nearly impermeable to water. Cork cells are dead at maturity.

corolla (L. *cornea*, crown) The petals, collectively. Usually, the conspicuously colored flower whorl.

corpus callosum (N.L. callous body) The band of nerve fibers that connect the two hemispheres of the cerebrum in humans and other primates.

corpus luteum (N.L. yellow body) A structure that develops from a ruptured follicle in the ovary after ovulation. It secretes the hormone progesterone, which maintains the uterus during pregnancy.

cortex (L. bark) In vascular plants, the primary ground tissue of a stem or root, bounded externally by the epidermis and internally by the central cylinder of vascular tissue. In animals, the outer, as opposed to the inner, part of an organ, as in the adrenal, kidney, and cerebral cortexes.

cotyledon (Gr. *kotyledon*, a cup-shaped hollow) Seed leaf. Monocot embryos have one cotyledon, and dicots have two.

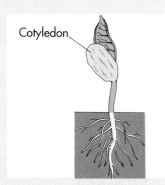

Cotyledon

countercurrent exchange In organisms, the passage of heat or of molecules (such as oxygen, water, or sodium ions) from one circulation path to another moving in the opposite direction. Because the flow of the two paths is in opposite directions, a concentration difference always exists between the two channels, facilitating transfer.

covalent bond (L. *co-*, together + *valare*, to be strong) A chemical bond formed by the sharing of one or more pairs of electrons.

crista, pl. cristae (L. crest) In mitochondria, the enfoldings of the inner mitochondrial membrane, which form a series of "shelves" containing the electron-transport chains involved in adenosine triphosphate formation.

cross A mating between two different strains of a plant.

crosscurrent flow In bird lungs, the latticework of capillaries arranged across the air flow, at a 90-degree angle.

crossing-over An essential element of meiosis occurring during prophase when nonsister chromatids exchange portions of DNA strands.

cryptic coloration An organism colored to blend in with its surroundings and thus be hidden from predators. Camouflage.

cuticle (L. *cutis*, skin) A very thin film covering the outer skin of many plants.

cutin (L. *cutis*, skin) A waxy, waterproof substance that is the chief ingredient of the cuticle of a plant.

cyanobacteria (Gr. *kyanos*, dark-blue + *bakterion*, dim. of *baktron*, a staff) Sometimes called "blue-green algae." A very important group of

photosynthetic bacteria in the history of life on earth. Producing oxygen, they played the decisive role in increasing the concentration of free oxygen in the earth's atmosphere from below 1 percent to the current level of 21 percent.

cyclic AMP (cAMP) A cyclic form of adenosine phosphate that acts as a chemical messenger and amplifies a hormonal signal.

cytokinesis (Gr. *kytos*, hollow vessel + *kinesis*, movement) The C phase of cell division in which the cell itself divides, creating two daughter cells.

cytoplasm (Gr. *kytos*, hollow vessel + *plasma*, anything molded) A semifluid matrix that occupies the volume between the nuclear region and the cell membrane. It contains the sugars, amino acids, proteins, and the organelles (in eukaryotes) with which the cell carries out its everyday activities of growth and reproduction.

cytoskeleton (Gr. *kytos*, hollow vessel + *skeleton*, a dried body) In the cytoplasm of all eukaryotic cells, a network of protein fibers that support the shape of the cell and anchor organelles, such as the nucleus, to fixed locations.

cytotoxic T-cell (Gr. *kytos*, hollow vessel + toxin) A special T-cell activated during cell-mediated immune response that recognizes and destroys infected body cells.

D

dark reactions Chemical reactions during the second phase of photosynthesis. So-called because, as long as adenosine triphosphate

generated during the first phase of photosynthesis is available, dark reactions occur as readily in the absence of light as in its presence.

deciduous (L. *decidere*, to fall off) In vascular plants, shedding all the leaves at a certain season.

decomposers Organisms that break down the organic matter accumulated in the bodies of other organisms.

dehydration reaction Water-losing. The process in which a hydroxyl (OH) group is removed from one subunit of a polymer and a hydrogen (H) group is removed from the other subunit.

demography (Gr. *demos*, people + *graphein*, to draw) The statistical study of population. The measurement of people, or, by extension, of the characteristics of people.

dendrite (Gr. *dendron*, tree) A process, typically branched, that extends from a neuron and conducts impulses inward toward the cell body. Neurons have many dendrites.

density The number of individuals in a population in a given area.

density-dependent effect An effect controlled by a factor that comes into play particularly when the population size is larger.

density-independent effect An effect controlled by a factor that operates regardless of population size.

dental plaque The film on teeth, composed largely of bacterial cells surrounded by a polysaccharide layer, that gives rise to dental caries, or cavities. Clear area in a sheet of bacterial cells growing in culture, resulting from the killing (lysis) of contiguous cells by viruses.

deoxyribonucleic acid (DNA) The basic storage vehicle or master plan of heredity information. It is stored as a sequence of nucleotides in a linear nucleotide polymer. Two of the polymers wind around each other like the outside and inside rails of a circular staircase.

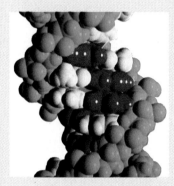

depolarization The movement of ions across a cell membrane that wipes out locally an electrical potential difference.

desert A hot, dry, barren region, usually sandy or rocky and without trees. Less than 25 centimeters of annual precipitation is the norm in such a region.

desmosome In adhering junctions, buttonlike welds that hold cells tightly together.

determinate Having flowers that arise from terminal buds and thus terminate a stem or branch.

detritivores (L. *detritus*, worn down + *vorare*, to devour) Organisms that live on dead organic matter. Included are large scavengers like vultures, smaller animals like crabs, and decomposers such as bacteria and fungi.

deuterostome (Gr. *deuteros*, second + *stoma*, mouth) An animal in whose embryonic development the anus forms from or near the blastopore, and the mouth forms later on another part of the blastula. Also characterized by radial cleavage.

development In a flowering plant, the entire series of events that occur between fertilization and maturity.

dicot Short for dicotyledon; a class of flowering plants generally characterized by having two cotyledons, netlike veins, and flower parts in fours or fives.

diffusion (L. *diffundere*, to pour out) The net movement of molecules to regions of lower concentration as a result of random, spontaneous molecular motions. The process tends to distribute molecules uniformly.

digestion (L. *digestio*, separating out, dividing) The process by which food is changed chemically into materials that the cells can assimilate, store, oxidize, or use as nourishment.

dihybrid (Gr. *dis*, twice + L. *hibrida*, mixed offspring) An individual heterozygous for two genes.

dioecious (Gr. *di*, two + *eikos*, house) Having male and female flowers on separate plants of the same species.

diploid (Gr. *diploos*, double + *eidos*, form) A cell, tissue, or individual with a double set of chromosomes.

directional selection A form of selection in which selection acts to eliminate one extreme from an array of phenotypes. Thus, the genes promoting this extreme become less frequent in the population.

disaccharide (Gr. *dis*, twice + *sakcharon*, sugar) A sugar formed by linking two monosaccharide molecules together. Sucrose (table sugar) is a disaccharide formed by linking a molecule of glucose to a molecule of fructose.

disruptive selection A form of selection in which selection acts to eliminate rather than favor the intermediate type.

diurnal (L. *diurnalis*, day) Active during the day.

divergence Increasing separation. Species that become progressively more different from one another as the result of each accumulating a different set of DNA mutations are said to diverge.

division A major taxonomic group of the plant kingdom, comparable to a phylum of the animal kingdom. Divisions are divided into classes.

DNA *See* deoxyribonucleic acid.

DNA polymerase An enzyme that catalyzes DNA replication only at the OH ends of DNA strands.

dominant allele An allele that dictates the appearance of heterozygotes. One allele is said to be dominant over another if an individual heterozygous for that allele has the same appearance as an individual homozygous for it.

dormancy (L. *dormire*, to sleep) A period during which growth ceases and is resumed only if certain requirements, involving temperature or day length, have been fulfilled.

dorsal (L. *dorsum*, the back) Toward the back, or upper surface. Opposite of ventral.

double bond A covalent bond sharing two pairs of electrons.

double fertilization A process unique to the angiosperms, in which one sperm nucleus fertilizes the egg and the second one fuses with the polar nuclei. These two events result in the formation of the zygote and the primary endosperm nucleus, respectively.

duodenum (L. *duodeni*, twelve each; with reference to its length, about twelve finger breadths) The initial, short segment of the small intestine, about 25 centimeters long, in which most digestion occurs.

duplex The DNA of a chromosome that exists as one very long, double-stranded fiber, which extends unbroken through the entire length of the chromosome.

E

ecdysis (Gr. *ekdysis*, stripping off) The shedding of the outer covering or skin of certain animals. Especially the shedding of the exoskeleton by arthropods.

ecological pyramid The relationships of the trophic structure of an ecosystem shown diagrammatically.

ecology (Gr. *oikos*, house + *logos*, word) The study of the relationships of organisms with one another and with their environment.

ecosystem (Gr. *oikos*, house + *systema*, that which is put together) A community, together with the nonliving factors with which it interacts.

ecotype (Gr. *oikos*, house + L. *typus*, image) A locally adapted variant of an organism, differing genetically from other ecotypes.

ectoderm (Gr. *ectos*, outside + *derma*, skin) The outer layer of cells formed during the development of animal embryos. Skin, hair, nails, tooth enamel, and the essential parts of the nervous system grow from the ectoderm.

efferent pathway *See* motor pathways.

egg activation A collective term for the series of events initiated by sperm penetration.

egg cell A female reproductive cell, ovum.

electron A subatomic particle with a negative electric charge. The negative charge of one electron exactly balances the positive charge of one proton. Electrons orbit the atom's positively charged nucleus and determine its chemical properties.

electron transport system A collective term describing the series of membrane-associated electron carriers generated by the citric acid cycle. It puts the electrons harvested from the oxidation of glucose to work driving proton-pumping channels.

element A substance that cannot be separated into different substances by ordinary chemical methods.

embryo (Gr. *en*, in + *bryein*, to swell) The early developmental stage of an organism produced from a fertilized egg. In plants, a young sporophyte; in animals, a young organism before it emerges from the egg or from the body of its mother; in humans, the first two months of intrauterine life.

endergonic (Gr. *endon*, within + *ergon*, work) Describing reactions in which the products contain more energy than the reactants and require an input of usable energy from an outside source before they can proceed. These reactions are not spontaneous.

endocrine gland (Gr. *endon*, within + *krinein*, to separate) A ductless gland producing hormonal secretions that pass directly into the bloodstream or lymph.

endocrine system The dozen or so major endocrine glands of a vertebrate.

endocytosis (Gr. *endon*, within + *kytos*, cell) The process by which the edges of plasma membranes fuse together and form an enclosed chamber called a vesicle. It involves the incorporation of a portion of an exterior medium into the cytoplasm of the cell by capturing it within the vesicle.

endoderm (Gr. *endon*, within + *derma*, skin) The inner layer of cells formed during development of early vertebrate embryos that is destined to give rise to the epithelium that lines certain internal structures, such as most of the digestive tract and its outgrowths, most of the respiratory tract, and the urinary bladder, liver, pancreas, and some endocrine glands.

endometrium (Gr. *endon*, within + *metrios*, of the womb) The two-layered lining of the uterus. The outer layer is shed during menstruation, while the one beneath it generates the next layer.

endoplasmic reticulum (Gr. *endon*, within + *plasma*, from cytoplasm; L. *reticulum*, network) A series of membranes that subdivides the interior of eukaryotic cells into separate compartments. It is the most distinctive feature of eukaryotic cells and one of the most important because it enables the cells to carry out different metabolic functions in separate compartments. Those portions containing a dense array of ribosomes are called rough ER, and other portions with fewer ribosomes are called smooth ER.

endoskeleton (Gr. *endon*, within + *skeletos*, hard) In vertebrates, an internal scaffold of bone to which muscles are attached.

endosperm (Gr. *endon*, within + *sperma*, seed) a nutritive tissue characteristic of the seeds of angiosperms that develops from the union of a male nucleus and the polar nuclei of the embryo sac. The endosperm is either digested by the growing embryo or retained in the mature seed to nourish the germinating seedling.

endosymbiont (Gr. *endon*, within + *bios*, life) An organism that is symbiotic within another. The major endosymbionts that occur in eukaryotic cells are mitochondria, chloroplasts, and centrioles.

endothelial (Gr. *endo*, within + *thele*, nipple) Describes the innermost layer of tissue that lines the arteries.

endothermic Referring to the ability of animals to maintain a constant body temperature.

energy The capacity to bring about change, to do work.

energy levels The placement of electrons in two orbitals that are different distances from the nucleus into separate concentric rings.

energy shells *See* energy levels.

entropy (Gr. *en*, in + *tropos*, change in manner) A measure of the disorder of a system. A measure of energy that has become so randomized and uniform in a system that the energy is no longer available to do work.

environmental science An applied science dedicated to finding solutions to environmental problems.

enzyme (Gr. *enzymos*, leavened; from *en*, in + *zyme*, leaven) A protein capable of speeding up specific chemical reactions by lowering the energy required to activate or start the reaction but that remains unaltered in the process.

epidermis (Gr. *epi*, on or over + *derma*, skin) The outermost layer of cells. In vertebrates, the nonvascular external layer of skin of ectodermal origin; in invertebrates, a single layer of ectodermal epithelium; in plants, the flattened, skinlike outer layer of cells.

epistasis (Gr. *epistasis*, a standing still) An interaction between the products of two genes in which one modifies the phenotypic expression produced by the other.

epithelium (Gr. *epi*, on + *thele*, nipple) A thin layer of cells forming a tissue that covers the internal and external surfaces of the body. Simple epithelium consists of the membranes that line the lungs and major body cavities and that are a single cell layer thick. Stratified epithelium (the skin or epidermis) is composed of more complex epithelial cells that are several cell layers thick.

erythrocyte (Gr. *erythros*, red + *kytos*, hollow vessel) A red blood cell, the carrier of hemoglobin. Erythrocytes act as the transporters of oxygen in the vertebrate body. During the process of their maturation in mammals, they lose their nucleus and mitochondria, and their endoplasmic reticulum is reabsorbed.

estrogen (Gr. *oestros*, frenzy + *genos*, origin) Any of various hormones that induce a series of physiological changes in females, especially in the reproductive or sexual organs.

estrous cycle The periodic cycle in which periods of estrus correspond to ovulation events.

estrus (L. *oestrus*, frenzy) The period of maximum female sexual receptivity. Associated with ovulation of the egg. Being "in heat."

estuary (L. *aestus*, tide) A partly enclosed body of water, such as those that often form at river mouths and in coastal bays, where the salinity is intermediate between that of saltwater and freshwater.

ethology (Gr. *ethos*, habit or custom + *logos*, discourse) The study of patterns of animal behavior in nature.

euchromatin (Gr. *eu*, good + *chroma*, color) Chromatin that is extended except during cell division, from which RNA is transcribed.

eukaryote (Gr. *eu*, good + *karyon*, kernel) A small, membrane-bound structure that possesses an internal chamber called the cell nucleus. The appearance of eukaryotes marks a major event in the evolution of life, since all organisms on earth other than bacteria are eukaryotes.

eumetazoan (Gr. *eu*, good + *meta*, with + *zoion*, animal) A "true animal." An animal with a definite shape and symmetry and nearly always distinct tissues.

eutrophic (Gr. *eutrophos*, thriving) Refers to a lake in which an abundant supply of minerals and organic matter exists.

evaporation The escape of water molecules from the liquid to the gas phase at the surface of a body of water.

evolution (L. *evolvere*, to unfold) Genetic change in a population of organisms over time (generations). Darwin proposed that natural selection was the mechanism of evolution.

exergonic (L. *ex*, out + Gr. *ergon*, work) Describes any reaction that produces products that contain less free energy than that possessed by the original reactants and that tends to proceed spontaneously.

exocytosis (Gr. *ex*, out of + *kytos*, cell) The extrusion of material from a cell by discharging it from vesicles at the cell surface. The reverse of endocytosis.

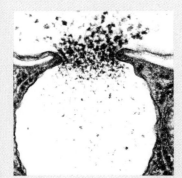

exoskeleton (Gr. *exo*, outside + *skeletos*, hard) An external hard shell that encases a body. In arthropods, comprised mainly of chitin; in vertebrates, comprised of bone.

experiment The test of a hypothesis. A successful experiment is one in which one or more alternative hypotheses are demonstrated to be inconsistent with experimental observation and are thus rejected.

external fertilization The fertilization of an egg outside the body of the female, as in many aquatic vertebrates, such as fish and amphibians.

exteroception (L. *exter*, outside + Eng. *(re)ceptive*) The sensing of information that relates to the body's external environment.

F

F₁ (first filial generation) The offspring resulting from a cross.

F₂ (second filial generation) The offspring resulting from a cross between members of the F₁ generation.

facilitated diffusion The transport of molecules across a membrane by a carrier protein in the direction of lowest concentration.

factor The term Mendel used to describe the bits of encoded information that parents transmit to their offspring and that act in the offspring to produce the trait. Now called a gene.

fall overturn A process in which the warm upper layer of water in a lake drops its temperature until it is the same as the cooler layer underneath and then the upper and lower layers mix, bringing up free supplies of dissolved nutrients.

fallopian tube After Gabriel Fallopius, Italian anatomist. Either of a pair of slender tubes through which ova from the ovaries pass to the uterus.

family A taxonomic group ranking below an order and above a genus.

fat A molecule containing many more C—H bonds than carbohydrates contain, thus providing more efficient energy storage.

fatty acid A long hydrocarbon chain ending with a —COOH group. Fatty acids are components of fats, oils, phospholipids, and waxes. A saturated fatty acid has two hydrogen side groups attached to all of its internal carbon atoms. An unsaturated fatty acid has a double bond. A polyunsaturated fatty acid has more than one double bond.

feedback inhibition A regulatory mechanism in which a biochemical pathway is regulated by the amount of the product that the pathway produces.

fermentation (L. *fermentum*, ferment) A catabolic process in which the final electron acceptor is an organic molecule.

fertilization (L. *ferre*, to bear) The union of male and female gametes to form a zygote.

fetus (L. pregnant) An animal embryo during the later stages of its development in the womb. In humans, a developing individual is referred to as a fetus from the end of the second month of gestation until birth.

fever (A.S. *fefer*) A higher-than-normal elevation of body temperature.

fiber (L. *fibra*) One of the narrow, elongated cells in the sclerenchyma of plants.

fibroblast (L. *fibra*, fiber + Gr. *blastos*, sprout) A flat, irregularly branching cell of connective tissue that secretes structurally strong proteins into the matrix between the cells.

fitness The genetic contribution of an individual to succeeding generations, relative to the contributions of other individuals in the population.

fixed-action pattern A stereotyped animal behavior response, thought by ethologists to be based on programmed neural circuits.

flagellum, pl. flagella (L. *flagellum*, whip) A fine, long, threadlike organelle protruding from the surface of a cell. In bacteria, a single protein fiber capable of rotary motion that propels the cell through the water. In eukaryotes, an array of microtubules with a characteristic internal 9 + 2 microtubule structure that is capable of vibratory but not rotary motion. Used in locomotion and feeding. Common in protists and motile gametes. A cilium is a small flagellum.

flavin adenine dinucleotide (FADH₂) A coenzyme used to carry less energetic electrons during the oxidation of glucose.

follicle (L. *folliculus*, small ball) In a mammalian ovary, one of the spherical chambers containing an oocyte.

follicle-stimulating hormone (FSH) A hormone secreted by the anterior lobe of the pituitary gland that stimulates the growth of the follicles of the ovaries.

food chain A series of organisms from each trophic level that feed on one another.

food web The food relationships within a community. A diagram of who eats whom.

foraging behaviors A collective term for the many complex, evolved behaviors that influence what an animal eats and how the food is obtained.

forebrain The third major division of a fish brain. Devoted to processing olfactory (smell) information.

founder principle The effect by which rare alleles and combinations of alleles may be enhanced in new populations.

frequency In statistics, defined as the proportion of individuals in a certain category, relative to the total number of individuals being considered.

frontal lobe The anterior part of either of the two lobes of the cerebrum.

fruit In angiosperms, a mature, ripened ovary (or group of ovaries) containing the seeds. Also applied informally to the reproductive structures of some other kinds of organisms.

functional group The special group of atoms attached to an organic molecule. Most chemical reactions that occur within organisms involve the transfer of a functional group from one molecule to another or the breaking of a carbon-carbon bond that attaches the functional group to the molecule.

Fungi Imperfecti A large group of fungi in which sexual reproduction is not known.

G

GABA *See* gamma-aminobutyric acid.

gallbladder A sac, attached to the liver, in which human beings concentrate and store bile manufactured in the liver.

gamete (Gr. wife) A haploid reproductive cell. Upon fertilization, its nucleus fuses with that of another gamete of the opposite sex. The resulting diploid cell (zygote) may develop into a new diploid individual, or in some protists and fungi, may undergo meiosis to form haploid somatic cells.

gametophyte (Gr. *gamete,* wife + *phyton,* plant) In plants, the haploid (N), gamete-producing generation, which alternates with the diploid (2N) sporophyte.

gamma-aminobutyric acid (GABA) One of many neurotransmitters that vertebrate nervous systems use, each with specific receptors on postsynaptic membranes. GABA opens the channel, which leads to the exit of positively charged potassium ions and a more negative interior.

ganglion, pl. ganglia (Gr. a swelling) A group of nerve cells forming a nerve center in the peripheral nervous system.

gap junction In communicating junctions, channels or pores through the two cell membranes and across the intercellular space that provide for electrical communication between cells and for flow of ions and small molecules.

gastric fluid (Gr. *gastros,* stomach) The digestive juice of the stomach.

gastric pit A deep depression in the upper epithelial surface of the stomach where exocrine glands are formed within the mucosa.

gastrin A polypeptide hormone that regulates the synthesis of HCl by the parietal cells of the gastric pits.

gastrula (Gr. little stomach) In vertebrates, the embryonic stage in which the blastula with its single layer of cells turns into a three-layered embryo made up of ectoderm, mesoderm, and endoderm, surrounding a cavity (archenteron) with one opening (blastopore).

gastrulation The inward movement of certain cell groups from the surface of the blastula.

gene (Gr. *genos,* birth, race) The basic unit of heredity. A sequence of DNA nucleotides on a chromosome that encodes a polypeptide or RNA molecule and so determines the nature of an individual's inherited traits.

gene expression The process in which an RNA copy of each active gene is made, and the RNA copy directs the sequential assembly of a chain of amino acids at a ribosome.

gene frequency The frequency with which individuals in a population possess a particular gene. Often confused with allele frequency.

genetic code The "language" of the genes. The mRNA codons specific for the twenty common amino acids constitute the genetic code.

genetic counseling The process of identifying parents at risk for producing children with genetic defects and of assessing the genetic state of early embryos.

genetic disorder The harmful effect produced when a detrimental allele occurs at a significant frequency in human populations.

genetic drift Random fluctuations in allele frequencies in a small population over time.

genetic engineering A collective term for the techniques of transferring genes from one kind of organism to another and multiplying them.

genetic map A diagram showing the relative positions of genes.

genetics (Gr. *genos,* birth, race) The study of the way in which an individual's traits are transmitted from one generation to the next.

genome (Gr. *genos,* offspring + L. *oma,* abstract group) The genetic information of an organism.

genotype (Gr. *genos,* offspring + *typos,* form) The total set of genes present in the cells of an organism. Also used to refer to the set of alleles at a single gene locus.

genus, pl. genera (L. race) A taxonomic group that ranks below a family and above a species.

germination (L. *germinare,* to sprout) The resumption of growth and development by a spore or seed.

gibberellin (*Gibberella,* a genus of fungi) A common and important class of plant hormones produced in the apical regions of shoots and roots. Play the major role in controlling stem elongation for most plants.

gill Part of the body of a fish or crab—for example, by which it breathes in water. Oxygen passes in and carbon dioxide passes out through the thin, membranous walls of the gills.

gland (L. *glandis,* acorn) Any of several organs in the body, such as exocrine or endocrine, that secrete substances for use in the body. Glands are composed of epithelial tissue.

glomerular filtrate The fluid that passes out of the capillaries of each glomerulus.

glomerulus (L. a little ball) A network of capillaries in a vertebrate kidney, whose walls act as a filtration device.

glucagon A hormone produced by the alpha-cells in the islets of Langerhans that raises blood sugar level by breaking down glycogen to glucose.

glucose A common six-carbon sugar. The most common monosaccharide in most organisms.

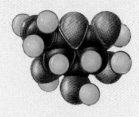

glycogen (Gr. *glykys*, sweet + *gen*, of a kind) Animal starch. A storage polymer occurring frequently in animals and characterized by complex branching.

glycolysis (Gr. *glykys*, sweet + *lyein*, to loosen) The harvesting of chemical energy by rearranging the chemical bonds of glucose to form two molecules of pyruvate and two molecules of ATP.

goiter (L. *guttur*, throat) An enlargement of the thyroid gland resulting from a deficiency of iodine in the diet.

Golgi body After Camillo Golgi, Italian physician. Flattened stacks of membranes in the cytoplasm that function in the collection, packaging, and distribution of molecules synthesized in the eukaryotic cell.

Golgi complex After Camillo Golgi, Italian physician. A collective term for Golgi bodies.

gradualism The gradualism model of evolution, which assumes that evolution proceeds gradually with progressive change in a given evolutionary line.

granum, *pl.* **grana** (L. grain or seed) In chloroplasts, stacks of membrane-bound disks (thylakoids). The thylakoids contain the chlorophylls and carotenoids and are the sites of the light reactions of photosynthesis.

gravitropism (L. *gravis*, heavy + *tropes*, turning) The response of a plant to gravity, which generally causes shoots to grow up and roots to grow down.

greenhouse effect The process in which carbon dioxide and certain other gases, such as methane, that occur in the earth's atmosphere transmit radiant energy from the sun but trap the longer wavelengths of infrared light, or heat, and prevent them from radiating into space.

ground tissue A type of tissue in which the vascular tissue of a plant is embedded.

group selection The argument about social behavior that altruism has evolved because a certain act benefits the group as a whole or even the entire species.

growth hormone (GH) A hormone secreted by the anterior pituitary, which regulates body growth.

guanine (Sp. from *Quechua*, huanu, dung) A purine base found in DNA and RNA. Its name derives from the fact that it occurs in high concentration as a white crystalline base ($C_5H_5H_5O$) in guano and other animal excrements.

guard cells Pairs of specialized epidermal cells that surround a stoma. When the guard cells are turgid, the stoma is open; when they are flaccid, it is closed.

gymnosperm (Gr. *gymnos*, naked + *sperma*, seed) A seed plant with seeds not enclosed in an ovary. The conifers are the most familiar group.

H

habitat (L. *habitare*, to inhabit) The place where individuals of a species live.

habituation (L. *habitus*, condition) A major form of nonassociate learning in which an individual learns not to respond to a stimulus.

half-life The length of time it takes for half of the carbon-14 present in a sample to be converted to carbon-12.

haploid (Gr. *haploos*, single + *eidos*, form) The gametes of a cell, tissue, or individual with only one set of chromosomes.

Hardy-Weinberg equilibrium After G. H. Hardy, English mathematician, and G. Weinberg, German physician. A mathematical description of the fact that the relative frequencies of two or more alleles in a population do not change because of Mendelian segregation. Allele and genotype frequencies remain constant in a random-mating population in the absence of inbreeding, selection, or other evolutionary forces. Usually stated as: If the frequency of allele A is p and the frequency of allele a is q, then the genotype frequencies after one generation of random mating will always be $(p + q)^2 = p^2 + 2pq + q^2$.

Haversian canal After Clopton Havers, English anatomist. Narrow channels that run parallel to the length of a bone and contain blood vessels and nerve cells.

heart The muscular organ that contracts and relaxes to pump blood throughout the vertebrate body.

helper T-cell A class of white blood cells that initiates both the cell-mediated immune response and the humoral immune response; helper T-cells are the targets of the AIDS virus (HIV).

hematopoietic stem cell (Gr. *haimatos*, blood + *poiesis*, a making) The cells in bone marrow where blood cells are formed.

hemoglobin (Gr. *haima*, blood + L. *globus*, a ball) A globular protein in vertebrate red blood cells and in the plasma of many invertebrates that carries oxygen and carbon dioxide. An essential part of each molecule is an iron-containing heme group, which both binds oxygen and carbon dioxide and gives blood its red color.

herbaceous plant (L. *herba*, herb) A plant in which secondary growth has been limited. Herbaceous plants produce new shoots each year.

herbivore (L. *herba*, grass + *vorare*, to devour) Any organism that eats plants.

heredity (L. *heredis*, heir) The transmission of characteristics from parent to offspring.

heterochromatin (Gr. *heteros*, different + *chroma*, color) That portion of a eukaryotic chromosome that remains permanently condensed and therefore is not transcribed into RNA. Most centromere regions are heterochromatic.

heterotroph (Gr. *heteros*, other + *trophos*, feeder) An organism that does not have the ability to produce its own food. *See also* autotroph.

heterozygote (Gr. *heteros*, other + *zygotos*, a pair) A diploid individual carrying two different alleles of a gene on its two homologous chromosomes.

hierarchical (Gr. *hieros*, sacred + *archos*, leader) Refers to a system of classification in which successively smaller units of classification are included within one another.

high-energy bond A chemical bond that has a low activation energy and is broken easily, which releases its energy.

histone (Gr. *histos*, tissue) A complex of small, very basic polypeptides rich in the amino acids arginine and lysine. Histones form the core of nucleosomes, around which DNA is wrapped.

holoblastic cleavage (Gr. *holos*, whole + *blastos*, germ) The pattern of cleavage that occurs throughout the whole egg when eggs contain little or no yolk.

homeostasis (Gr. *homeos*, similar + *stasis*, standing) The maintaining of a relatively stable internal physiological environment in an organism, or steady-state equilibrium in a population or ecosystem. Usually involves some form of feedback self-regulation.

homeotherm (Gr. *homeo*, similar + *therme*, heat) An organism, such as a bird or mammal, capable of maintaining a stable body temperature independent of the environmental temperature. "Warm-blooded."

hominid (L. *homo*, man) Human beings and their direct ancestors. A member of the family Hominidae. *Homo sapiens* is the only living member.

homologous chromosome (Gr. *homologia*, agreement) One of the two nearly identical versions of each chromosome. Chromosomes that associate in pairs in the first stage of meiosis. In diploid cells, one chromosome of a pair that carry equivalent genes.

homology (Gr. *homologia*, agreement) A condition in which the similarity between two structures or functions is indicative of a common evolutionary origin.

homozygote (Gr. *homos*, same or similar + *zygotos*, a pair) A diploid individual whose two copies of a gene are the same. An individual carrying identical alleles on both homologous chromosomes is said to be homozygous for that gene.

hormone (Gr. *hormaein*, to excite) A chemical messenger, often a steroid or peptide, produced in a small quantity in one part of an organism and then transported to another part of the organism, where it brings about a physiological response.

human immunodeficiency virus (HIV) The virus responsible for acquired immunodeficiency syndrome (AIDS), a deadly disease that destroys the human immune system. HIV is a retrovirus (its genetic material is RNA) that is thought to have been introduced to humans from African green monkeys.

hybrid (L. *hybrida*, the offspring of a tame sow and a wild boar) A plant that results from the crossing of dissimilar parents.

hybridization The mating of unlike parents of different taxa.

hydrogen bond A molecular force formed by the attraction of the partial positive charge of one hydrogen atom of a water molecule with the partial negative charge of the oxygen atom of another.

hydrolysis reaction (Gr. *hydro*, water, + *lyse*, break) The process of tearing down a polymer by adding a molecule of water. A hydrogen is attached to one subunit and a hydroxyl to the other, which breaks the covalent bond. Essentially the reverse of a dehydration reaction.

hydrophobic (Gr. *hydro*, water + *phobos*, hating) Refers to nonpolar molecules, which do not form hydrogen bonds with water and therefore are not soluble in water.

hydroskeleton (Gr. *hydro*, water + *skeletos*, hard) The skeleton of most soft-bodied invertebrates that have neither an internal nor an external skeleton. They use the relative incompressibility of the water within their bodies as a kind of skeleton.

hypertonic (Gr. *hyper*, above + *tonos*, tension) Refers to a cell that contains a higher concentration of solutes than its surrounding solution.

hypha, pl. hyphae (Gr. *hyphe*, web) A filament of a fungus. A mass of hyphae comprises a mycelium.

hypothalamus (Gr. *hypo*, under + *thalamos*, inner room) The region of the brain under the thalamus that controls temperature, hunger, and thirst, and that produces hormones that influence the pituitary gland.

hypothesis (Gr. *hypo*, under + *tithenai*, to put) A proposal that might be true. No hypothesis is ever proven correct. All hypotheses are provisional—proposals that are retained for the time being as useful but that may be rejected in the future if found to be inconsistent with new information. A hypothesis that stands the test of time—often tested and never rejected—is called a theory.

hypotonic (Gr. *hypo*, under + *tonos*, tension) Refers to the solution surrounding a cell that has a lower concentration of solutes than does the cell.

I

immune response The production of antibodies and T-cells directed against a specific antigen.

immune system A vertebrate's dual defense against non-self substances composed of white blood cells. One attacks and kills cells identified as foreign, while the other type marks the foreign invaders for elimination by the roaming patrols (nonspecific defenses).

inbreeding The breeding of genetically related plants or animals. In plants, inbreeding results from self-pollination. In animals, inbreeding results from matings between relatives. Inbreeding tends to increase homozygosity.

inclusive fitness Describes the sum of the number of genes directly passed on in an individual's offspring and those genes passed on indirectly by kin (other than offspring) whose existence results from the benefit of the individual's altruism.

incomplete dominance The ability of two alleles to produce a heterozygous phenotype that is different from either homozygous phenotype.

independent assortment Mendel's second law: the principle that segregation of alternative alleles at one locus into gametes is independent of the segregation of alleles at other loci. Only true for gene loci located on different chromosomes or those so far apart on one chromosome that crossing-over is very frequent between the loci. *See* Mendel's Second Law.

induced Describes how the *lac* operon is transcribed when lactose binding to the repressor protein changes its shape so that it can no longer sit on the operator site and block polymerase binding.

induction The determination of the course of development of one tissue by another tissue.

industrial melanism (Gr. *melas*, black) Phrase used to describe the evolutionary process in which initially light-colored organisms become dark as a result of natural selection.

inflammatory response (L. *inflammare*, to flame) A generalized nonspecific response to infection that acts to clear an infected area of infecting microbes and dead tissue cells so that tissue repair can begin.

inhalation (L. *in,* in + *halare,* to breathe) The act of breathing or drawing air into the lungs.

inhibitor A chemical whose binding alters the shape of a protein and shuts off enzyme activity.

initiation complex A complex consisting of a ribosome, mRNA, and a tRNA molecule, the formation of which begins polypeptide synthesis.

innate (L. *innatus,* born) Describing a characteristic based partly or wholly on inherited gene differences.

inner ear The innermost part of the ear, behind the middle ear, containing the essential organs of hearing and equilibrium.

inner medulla The inner portion of the kidney, which contains the lower portion of the loop of Henle and the bottom of the collecting duct, which is permeable to urea.

inositol phosphate A mediator molecule produced during insulin receptor-induced enzyme activity.

insertion The end of a muscle that is attached to a bone that moves if the muscle contracts.

instinct (L. *instinctus,* impelled) Stereotyped, predictable, genetically programmed behavior.

instructional theory A statement attempting to answer the question of how the human body is able to make such a great diversity of antibodies by proposing that the antigen elicited the appropriate antibody, like a shopper ordering a custom-made suit.

insulin (L. *insula,* island) A peptide hormone secreted by the islets of Langerhans that acts as a storage hormone. It enables the body to use sugar and other carbohydrates by regulating the body's sugar metabolism.

integration, neural The summation of the depolarizing and repolarizing effects contributed by all excitatory and inhibitory synapses acting on a neuron.

integument (L. *integumentum,* covering) The natural outer covering layers of an animal. Develops from the ectoderm.

interferon In vertebrates, a protein produced in virus-infected cells that inhibits viral multiplication.

interneuron A nerve cell found only in the middle of the spinal cord that acts as a functional link between sensory neurons and motor neurons.

internode The region of a plant stem between nodes where stems and leaves attach.

interoception (L. *interus,* inner + Eng. *(re)ceptive*) The sensing of information that relates to the body itself, its internal condition, and its position.

interphase That portion of the cell cycle preceding mitosis. It includes the G1 phase when cells grow, the S phase when a replica of the genome is synthesized, and a G2 phase when preparations are made for genomic separation.

intron (L. *intra,* within) A segment of DNA transcribed into mRNA but removed before translation. These untranslated regions make up the bulk of most eukaryotic genes.

involuntary nervous system *See* autonomic nervous system.

ion An atom in which the number of electrons does not equal the number of protons. An ion does carry an electrical charge.

ionic bond A chemical bond formed between ions as a result of the attraction of opposite electrical charges.

ionization The process of spontaneous ion formation. As when the covalent bonds of water sometimes break spontaneously, one of the protons dissociates from the molecule. Because the dissociate proton lacks the negatively charged electron that it shared in the covalent bond with oxygen, its own positive charge is not counterbalanced. It is a positively charged hydrogen ion, H^+. The remaining bit of the water molecule retains the shared electron from the covalent bond and has one less proton to counterbalance it. It is a negatively charged hydroxyl ion (OH^-).

ionizing radiation High-energy radiation, such as X rays and gamma rays.

iris (L. rainbow) A contractile disk, or shutter, between the cornea and the lens that controls the amount of light entering the eye.

islets of Langerhans After Paul Langerhans, German anatomist. The small, scattered endocrine glands in the pancreas that secrete insulin.

isolating mechanisms Mechanisms that prevent genetic exchange between individuals of different populations or species. May be behavioral, morphological, or physiological.

isotonic (Gr. *isos,* equal + *tonos,* tension) Refers to a cell with the same concentration of solutes as its environment.

isotope (Gr. *isos,* equal + *topos,* place) An atom that has the same number of protons but different numbers of neutrons.

J

joint The part of a vertebrate where one bone meets and moves on another.

H

karyotype (Gr. *karyon,* kernel + *typos,* stamp or print) The particular array of chromosomes that an individual possesses.

kidney In vertebrates, one of the pair of organs that carries out the processes of filtration, reabsorption, and secretion.

kin selection Selection that acts to favor the propagation of genes by directing altruism toward relatives.

kinetic energy The energy of motion.

kinetochore (Gr. *kinetikos,* putting in motion + *choros,* chorus) A disk of protein bound to the centromere to which microtubules attach during mitosis, linking each chromatid to the spindle.

kingdom The chief taxonomic category. This book recognizes six kingdoms: Archaebacteria, Eubacteria, Protista, Fungi, Animalia, and Plantae.

Krebs cycle After Hans A. Krebs, German-born English biochemist. The citric acid cycle. Also called the tricarboxylic acid (TCA) cycle. *See* citric acid cycle.

L

lac operon A cluster of genes encoding three proteins that bacteria use to obtain energy from the sugar lactose.

lamella, *pl.* **lamellae** (L. a little plate) A thin, platelike structure. In chloroplasts, a layer of chlorophyll-containing membranes. In bivalve mollusks, one of the two plates forming a gill. In vertebrates, one of the thin layers of bone laid concentrically around the Haversian canals.

larva, *pl.* **larvae** (L. a ghost) Immature form of an animal that is quite different from the adult and undergoes metamorphosis in reaching the adult form. Examples include caterpillars and tadpoles.

larynx (Gr.) The voice box. The upper end of the human windpipe, lying between the pharynx and trachea, that contains the vocal cords and acts as an organ of voice.

lateral meristems (L. *latus*, side + Gr. *meristos*, divided) In vascular plants, the meristems that give rise to secondary tissue. The vascular cambium and cork cambium.

Law of Independent Assortment See Mendel's Second Law.

Law of Segregation See Mendel's First Law.

leaf One of the thin, usually flat, green parts of a tree or other plant that grows on the stem or up from the roots. An expanded area of photosynthetically active tissue on a plant.

leaf primordium (L. *primordium*, beginning) A lateral outgrowth from the apical meristem that eventually becomes a leaf.

learning The creation of changes in behavior that arise as a result of experience, rather than as a result of maturation.

lens A transparent oval body in the eye, directly behind the iris, that focuses light rays upon the retina.

lenticels (L. *lenticella*, a small window) Spongy areas in the cork surfaces of stem, roots, and other plant parts that allow interchange of gases between internal tissues and the atmosphere through the periderm.

life cycle The sequence of phases in the growth and development of an organism, from zygote formation to gamete formation.

ligament (L. *ligare*, to bind) A band or sheet of connective tissue that links bone to bone.

light chains The two identical short strands of the four polypeptide chains of an antibody molecule.

light reactions The resultant synthesis of adenosine triphosphate, which takes place in the presence of light during the first process of photosynthesis.

limbic system The hypothalamus, together with the network of neurons that link the hypothalamus to some areas of the cerebral cortex. Responsible for many of the most deep-seated drives and emotions of vertebrates, including pain, anger, sex, hunger, thirst, and pleasure.

linkage The patterns of assortment of genes that are located on the same chromosome. Important because, if the genes are located relatively far apart, crossing-over is more likely to occur between them than if they are located close together.

lipase (Gr. *lipos*, fat + -ase, suffix used for enzymes) An enzyme that breaks up lipids and fats into small segments.

lipid (Gr. *lipos*, fat) A loosely defined group of molecules that are insoluble in water but soluble in oil. Oils such as olive, corn, and coconut are lipids, as well as waxes, such as beeswax and earwax.

lipid bilayer The basic foundation of all biological membranes. In such a layer, the nonpolar tails of phospholipid molecules point inward, forming a nonpolar zone in the interior of the bilayers. Lipid bilayers are selectively permeable and do not permit the diffusion of water-soluble molecules into the cell.

littoral (L. *litus*, shore) Referring to the shoreline zone of a lake or pond or the ocean that is exposed to the air whenever water recedes.

liver The largest internal organ of the human body and the body's principal metabolic factory, turning foodstuffs arriving from the digestive tract in the bloodstream into substances that are used by different body cells.

locus, pl. loci (L. place) The position on a chromosome where a gene is located.

long-term memory A functional type of memory that appears to involve changes in the way information is processed at neural connections within the brain.

loop of Henle After F. G. J. Henle, German anatomist. A hairpin loop formed by a urine-conveying tubule when it enters the inner layer of the kidney and then turns around to pass up again into the outer layer of the kidney.

luteal phase The second phase of the reproductive cycle, during which the mature eggs are released into the fallopian tubes, a process called ovulation.

luteinizing hormone (LH) A hormone produced by the anterior lobe of the pituitary gland, which in the female stimulates the development of the corpus luteum.

lymph (L. *lympha*, clear water) In animals, a colorless fluid derived from blood by filtration through capillary walls in the tissues.

lymph node Located throughout the lymph system, lymph nodes remove dead cells, debris, and foreign particles from the circulation.

lymphatic system An open circulatory system composed of a network of vessels that

function to collect the water within blood plasma forced out during passage through the capillaries and to return it to the bloodstream. The lymphatic system also returns proteins to the circulation, transports fats absorbed from the intestine, and carries bacteria and dead blood cells to the lymph nodes and spleen for destruction.

lymphocyte (Gr. *lympha*, water + Gr. *kytos*, hollow vessel) A white blood cell. A cell of the immune system that either synthesizes antibodies (B-cells) or attacks virus-infected cells (T-cells).

Lyonization After Mary F. Lyon, English geneticist. The inactivation of one X chromosome in female mammals.

lyse (Gr. *lysis*, loosening) To disintegrate a cell by rupturing its cell membrane.

lysosome (Gr. *lysis*, a loosening + *soma*, body) A membrane-bound organelle, formed by the Golgi complex, that contains digestive enzymes. Important for digesting worn-out cellular components, making way for newly formed ones while recycling the materials locked up in the old ones. A primary lysosome is one that is not actively functioning. A secondary lysosome is one that fuses with a food vacuole or other organelle so that its pH falls and hydrolytic enzymes are activated.

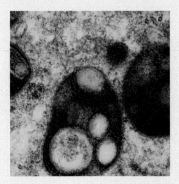

M

macroevolution (Gr. *makros*, large + L. *evolvere*, to unfold) The creation of new species and the extinction of old ones.

macromolecule (Gr. *makros*, large + L. *moliculus*, a little mass) An extremely large molecule. Refers specifically to carbohydrates, lipids, proteins, and nucleic acids.

macrophage (Gr. *makros*, large + -*phage*, eat) A phagocytic cell of the immune system able to engulf and digest invading bacteria, fungi, and other microorganisms, as well as cellular debris.

major histocompatibility complex (MHC) A set of protein cell-surface markers anchored in plasma membrane, which the immune system uses to identify "self." All the cells of a given individual have the same "self" marker, called a MHC protein.

Malpighian corpuscle After Marcello Malpighi, Italian anatomist. A renal corpuscle. A filtration apparatus at the front end of each nephron tube of a vertebrate kidney.

Malpighian tubule After Marcello Malpighi, Italian anatomist. A tubular extension of the digestive tract that opens into the hindgut of insects and functions as an excretory organ.

mantle In the body of a mollusk, a heavy fold of tissue that is wrapped around the visceral mass like a cape.

marginal meristem The method by which leaves grow. Marginal meristems grow outward and ultimately form the blade of the leaf, while the central portion becomes the midrib.

marrow The soft tissue that fills the cavities of most bones and is the source of red blood cells.

marsupial (L. *marsupium*, pouch) A mammal in which the young are born early in their development, sometimes as soon as eight days after fertilization, and are retained in a pouch. Opossums and kangaroos are marsupials.

marsupium (L. pouch) A pouch on the abdomen of a female marsupial for carrying its young.

mass In chemistry, the total number of protons and neutrons in the nucleus of an atom. Approximately equal to the atomic weight.

mass flow The overall process by which materials move in the phloem of plants.

mast cell A cell of the immune system that synthesizes the molecules involved in the body's response to trauma, including histamine and heparin.

medulla (L. marrow) The inner portion of an organ, in contrast to the cortex or outer portion, as in the kidney or adrenal gland. The part of the brain that controls breathing and other involuntary functions, located at the top end of the spinal cord. Also called medulla oblongata.

megagametophyte (Gr. *megas*, large + *gamos*, marriage + *phyton*, plant) In heterosporous plants, the female gametophyte, located within the ovule of seed plants.

megaspore (Gr. *megas*, large + *sporos*, seed) A spore of comparatively large size from which a female gametophyte develops.

meiosis (Gr. *meioun*, to make smaller) A special form of nuclear division that precedes gamete formation in sexually reproducing eukaryotes.

Meissner's corpuscles Receptors below the skin surface that fire in response to rapid changes in pressure.

melatonin (from mela[nin] + [sero] tonin) A hormone secreted by the pineal gland, whose function in humans is not well understood.

membrane attack complex (MAC) The battery of complement proteins that interacts like natural killer cells by inserting itself into the pathogen's cell membrane, forming a hole, and killing the invading cell.

Mendelian ratio After Gregor Mendel, Austrian monk. Refers to the characteristic 3:1 segregation ratio that Mendel observed, in which pairs of alternative traits are expressed in the F_2 generation in the ratio of three-fourths dominant to one-fourth recessive.

Mendel's First Law After Gregor Mendel, Austrian monk. The Law of Segregation: Central premises are that: (1) alleles do not blend in heterozygotes; (2) alleles segregate in heterozygous individuals; and (3) alleles have an equal probability of being included in either gamete.

Mendel's Second Law After Gregor Mendel, Austrian monk. The Law of Independent Assortment: Genes located on different chromosomes assort independently of one another. *See* independent assortment.

menstrual cycle (L. *mens*, month) Monthly cycle. The term used to describe the reproductive cycle, usually occurring every twenty-eight days, in women. The menstrual cycle in primates is the cycle of hormone-regulated changes in the condition of the uterine lining, which is marked by the periodic discharge of blood and disintegrated uterine lining through the vagina (menstruation).

menstruation (L. *mens*, month) Periodic sloughing off of the blood-enriched lining of the uterus when pregnancy does not occur.

meristem (Gr. *merizein*, to divide) In plants, a zone of unspecialized cells whose only function is to divide.

meroblastic cleavage (Gr. *meros*, part + *blastos*, sprout) A type of cleavage in the eggs of reptiles, birds, and some fishes. Occurs only in the blastodisc.

mesentery (Gr. *mesos*, middle + *enteron*, intestine) A double layer of mesoderm within the coelom.

mesoderm (Gr. *mesos*, middle + *derma*, skin) One of the three embryonic germ layers that form in the gastrula. Gives rise to muscle, bone, and other connective tissue; the peritoneum; the circulatory system; and most of the excretory and reproductive systems.

mesophyll (Gr. *mesos*, middle + *phyllon*, leaf) The photosynthetic parenchyma of a leaf, located within the epidermis. The vascular strands (veins) run through the mesophyll.

messenger RNA (mRNA) A class of RNA in which each molecule is a long, single strand of RNA that passes from the nucleus to the cytoplasm. During polypeptide synthesis, mRNA molecules bring information from the chromosomes to the ribosomes to direct which polypeptide is assembled.

metabolism (Gr. *metabole*, change) The process by which all living things assimilate energy and use it to grow.

metamorphosis (Gr. *meta*, after + *morphe*, form + *osis*, state of) Process in which form changes markedly during postembryonic development— for example, tadpole to frog or larval insect to adult.

metaphase (Gr. *meta*, middle + *phasis*, form) The stage of mitosis characterized by the alignment of the chromosomes on a plane in the center of the cell.

metaphase plate In metaphase, an imaginary plane passing through the circle around the spindle midpoint where the chromosomes array themselves.

metastasis, *pl.* ***metastases*** (Gr. to place in another way) The spread of cancerous cells to other parts of the body, forming new tumors at distant sites.

microbody A cellular organelle bounded by a single membrane and containing a variety of enzymes. Generally derived from endoplasmic reticulum.

microevolution (Gr. *mikros,* small + L. *evolvere,* to unfold) Refers to the evolutionary process itself. Evolution within a species. Also called adaptation.

microfilament (Gr. *mikros,* small + L. *filum,* a thread) In cells, a protein thread composed of parallel fibers of actin cross-connected by myosin. Their movement results from an ATP-driven shape change in myosin. The contraction of vertebrate muscles and many other kinds of cell movement in eukaryotes result from the movements of microfilaments within cells.

microgametophyte (Gr. *mikros,* small + *gamos,* marriage + *phyton,* plant) In heterosporous plants, the male gametophyte.

microsporangium, *pl.* ***microsporangia*** (Gr. *mikros,* small + *sporus,* seed + *angeion,* a vessel) A sporangium containing microspores. Homologous with the sac containing the pollen in flowering plants.

microspore (Gr. *mikros,* small + *sporus,* seed) In plants, a spore that develops into a male gametophyte. In seed plants, it develops into a pollen grain.

microtubule (Gr. *mikros,* small + L. *tubulus,* little pipe) In eukaryotic cells, a long, hollow cylinder about 25 nanometers in diameter and composed of the protein tubulin. Microtubules influence cell shape, move the chromosomes in cell division, and provide the functional internal structure of cilia and flagella.

microvillus, *pl.* ***microvilli*** (Gr. *mikros,* small + L. *villus,* tuft of hair) A microscopic hairlike projection growing on the surface of the epithelial cells that cover the villi.

middle ear The hollow space between the eardrum and the inner ear. In humans, it contains three small bones that transmit sound waves from the eardrum to the inner ear.

middle lamella The space, impregnated with pectins, between two new plant cells.

migration Long-range, two-way movements by animals that often occur once a year with the change of seasons.

mimicry (Gr. *mimos,* mime) The resemblance in form, color, or behavior of certain organisms (mimics) to other more powerful or more protected ones (models), which results in the mimics being protected in some way.

mitochondrion, *pl.* ***mitochondria*** (Gr. *mitos,* thread + *chondrion,* small grain) A tubular or sausage-shaped organelle 1 to 3 micrometers long. Bounded by two membranes, mitochondria closely resemble the aerobic bacteria from which they were originally derived. As chemical furnaces of the cell, they carry out its oxidative metabolism.

mitosis (Gr. *mitos,* thread) The M phase of cell division in which the microtubular apparatus is assembled, binds to the chromosomes, and moves them apart. This phase is the essential step in the separation of the two daughter cell genomes.

model Butterflies and moths provide many of the best-known examples of Batesian mimicry models; organisms that are mimicked by others to gain a selective advantage.

modified ratio A modified Mendelian ratio of 9:7 instead of the usual 9:3:3:1, which illustrates the effects of epistasis.

mole (L. *moles,* mass) The atomic weight of a substance, expressed in grams. One mole is defined as the mass of 6.0222×10^{23} atoms.

molecule (L. *moliculus,* a small mass) The smallest unit of a compound that displays the properties of that compound.

mollusk Any of a large phylum (Mollusca) of invertebrate animals usually having a soft, unsegmented body, a hard shell secreted by a covering mantle, and a muscular foot. Snails, clams, scallops, and oysters belong to the phylum.

monoamine An amine containing one amino group, especially one that functions as a neurotransmitter. Acetylcholine and norepinephrine are monoamines.

monoclonal antibody An antibody produced in the laboratory by fusing genetically distinct cells and cloning the resulting hybrids so that each hybrid cell produces the same antibody.

monocot Short for monocotyledon: A flowering plant in which the embryos have only one cotyledon, the flower parts are often in threes, and the leaves typically are parallel-veined.

monoculture (Gr. *monos,* one + L. *cultivare,* to cultivate) The exclusive cultivation of a single crop over a wide area.

monocyte (Gr. *monos,* single + *kytos,* hollow vessel) A type of circulating leukocyte that becomes a phagocytic cell (macrophage) after moving into tissues.

monosaccharide (Gr. *monos,* one + *sakcharon,* sugar) A simple sugar.

monosynaptic reflex arc A simple reflex arc in which the afferent nerve cell makes synaptic contacts directly with a motor neuron in the spinal cord, whose axon travels directly back to the muscle.

monotreme (Gr. *mono,* single + *treme,* hole) An egg-laying mammal. The only two are the duck-billed platypus and the echidna, or spiny anteater.

morphogenesis (Gr. *morphe,* form + *genesis,* origin) The formation of shape. The growth and differentiation of cells and tissues during development.

morphology (Gr. *morphe,* form + *logos,* discourse) The study of form and its development. Includes cytology (the study of cell structure), histology (the study of tissue structure), and anatomy (the study of gross structure).

morula (L. *morum,* mulberry) The mass of blastomeres forming the embryo of many animals, just after the segmentation of the ovum and before the formation of a blastula.

motor cortex A sensory region on the cerebral cortex containing neurons that control the movement of different body muscles.

motor endplate The point where a neuron attaches to a muscle. A neuromuscular synapse.

motor pathways (L. *mover*) The nerve pathways that transmit commands to the body from the central nervous system.

mRNA *See* messenger RNA.

mucosa, *pl.* **mucosae** A deep layer of connective tissue overlaid by epithelium and containing glands that secrete mucus.

Muellerian mimicry After Fritz Mueller, German biologist. A phenomenon in which two or more unrelated but protected species resemble one another, thus achieving a kind of group defense.

multicellularity A condition in which the activities of the individual cells are coordinated and the cells themselves are in contact. A property of eukaryotes alone and one of their major characteristics.

muscle (L. *musculus,* mouse) The tissue in the body of humans and animals that can be contracted and relaxed to make the body move.

muscle cell *See* muscle fiber.

muscle fiber Muscle cell. A long, cylindrical, multinucleated cell that contains numerous myofibrils and is capable of contraction when stimulated.

muscle spindle A sensory end organ that is attached to a muscle and sensitive to stretching.

mutagen (L. *mutare,* to change) A chemical capable of damaging DNA.

mutant (L. *mutare,* to change) A mutated gene. An organism carrying a gene that has undergone a mutation.

mutation (L. *mutare,* to change) A change in a cell's genetic message.

mutational repair Repairs undertaken by the cells, such as excising altered nucleotides or reforming single ruptured bonds. Not always accurate, and some of the mistakes are incorporated into the genetic message.

mutualism (L. *mutuus,* lent, borrowed) A symbiotic relationship in which both participating species benefit.

mycelium, *pl.* **mycelia** (Gr. *mykes,* fungus) In fungi, a mass of hyphae.

mycology (Gr. *mykes,* fungus) The study of fungi. A person who studies fungi is called a mycologist.

mycorrhiza, *pl.* **mycorrhizae** (Gr. *mykes,* fungus + *rhiza,* root) A symbiotic association between fungi and plant roots.

myelin sheath (Gr. *myelinos,* full of marrow) A flattened sheath of fatty material found in many, but not all, vertebrate neurons. Made up of the membranes of Schwann cells.

myelinated fiber The structure formed by an axon and its associated Schwann cells, or cells with similar properties.

myofibril (Gr. *myos,* muscle + L. *fibrilla,* little fiber) A contractile microfilament, composed of myosin and actin, within muscle.

myofilament (Gr. *myos,* muscle + L. *filare,* to spin) A contractile microfilament, composed largely of actin and myosin, within muscle. Sometimes called myofibril.

myosin (Gr. *myos,* muscle + *in,* belonging to) One of two protein components of myofilaments. (The other is actin.)

N

NADP *See* nicotinamide adenine dinucleotide phosphate.

NAD⁺ *See* nicotinamide adenine dinucleotide.

natural killer cell A cell that does not kill invading microbes, but rather, the cells infected by them.

natural selection The differential reproduction of genotypes caused by factors in the environment. Leads to evolutionary change.

nature or nurture Instinct or behavior: The question of which factor plays what role in shaping learning and behavior.

navigation (L. *naviage,* to move, direct) The ability to set or adjust a bearing and then follow it.

nectar (Gr. *nektar*) A sweet liquid rich in sugar and amino acids and found in many flowers. Attracts insects and birds that carry out pollination.

negative control The process of shutting off transcription by use of a regulatory site.

nematocyst (Gr. *nema,* thread + *kystos,* bladder) A coiled, threadlike stinging process of cnidarians that is discharged to capture prey and for defense.

nephrid organ A filtration system of many freshwater invertebrates in which water and waste pass from the body across the membrane into a collecting organ, from which they are expelled to the outside through a pore.

nephron (Gr. *nephros,* kidney) The functional unit of the vertebrate kidney. A human kidney has more than one million nephrons that filter waste matter from the blood. Each nephron consists of a Bowman's capsule, glomerulus, and tubule.

neritic zone (L. *nerita,* a sea mussel) The marine environment of shallow waters along the coasts of the continents.

nerve A bundle of axons with accompanying supportive cells, held together by connective tissue.

nerve cord The main trunk along the back to which the nerves that reach the different parts of the body are connected. A major characteristic of chordates.

nerve fiber An axon.

nerve impulse A rapid, transient, self-propagating reversal in electric potential that travels along the membrane of a neuron.

neural groove The long groove formed along the long axis of the embryo by a layer of ectodermal cells.

neural tube The dorsal tube, formed from the neural plate, that differentiates into the brain and spinal cord.

neuralation (Gr. *neuron,* nerve) The elaboration of a notochord and a dorsal nerve cord that marks the evolution of the chordates.

neuroendocrine system A network of endocrine glands whose hormone secretion is controlled by commands from the central nervous system.

neuroethology (Gr. *neuron,* nerve + *ethos,* habit or custom + *logos,* discourse) The study of the neural basis of behavior.

neuroglia (Gr. *neuron,* nerve + *glia,* glue) The delicate connective tissue forming a supporting network for the conducting elements of nervous tissue in the brain and spinal cord.

neurohormone A hormone, such as an enkephalin or endorphin, secreted by nerve cells.

neuromodulator A chemical transmitter that mediates effects that are slow and longer lasting and that typically involve second messengers within the cell.

neuromuscular junction The structure formed when the tips of axons contact (innervate) a muscle fiber.

neuron (Gr. nerve) A nerve cell specialized for signal transmission.

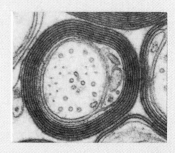

neurotransmitter (Gr. *neuron*, nerve + L. *trans*, across + *mitere*, to send) A chemical released at an axon tip that travels across the synapse and binds a specific receptor protein in the membrane on the far side.

neutron (L. *neuter*, neither) A subatomic particle located within the nucleus of an atom. Similar to a proton in mass, but as its name implies, a neutron is neutral and possesses no charge.

neutrophil An abundant type of granulocyte capable of engulfing microorganisms and other foreign particles; neutrophils comprise about 50% to 70% of the total number of white blood cells.

niche (L. *nidus*, nest) The role an organism plays in the environment; actual niche is the niche that an organism occupies under natural circumstances; theoretical niche is the niche an organism would occupy if competitors were not present.

nicotinamide adenine dinucleotide (NAD⁺) A composite molecule consisting of two nucleotides bound together; an important coenzyme that functions as an electron acceptor in many oxidative reactions.

nicotinamide adenine dinucleotide phosphate (NADP) A coenzyme that functions as an electron donor in many of the reduction reactions of biosynthesis. $NADP^+$ is the oxidized form of NADP, and $NADPH_2$ is the reduced form.

nitrogen fixation The incorporation of atmospheric nitrogen into nitrogen compounds, a process that can be carried out only by certain microorganisms.

nociceptor A naked dendrite that acts as a receptor in response to a pain stimulus.

nocturnal (L. *nocturnus*, night) Active primarily at night.

node (L. *nodus*, knot) The place on the stem where a leaf is formed.

node of Ranvier After L. A. Ranvier, French histologist. A gap formed at the point where two Schwann cells meet and where the axon is in direct contact with the surrounding intercellular fluid.

nonassociative A learned behavior that does not require an animal to form an association between two stimuli, or between a stimulus and a response.

nonrandom mating A phenomenon in which individuals with certain genotypes sometimes mate with one another more commonly than would be expected on a random basis.

nonsense codon A chain-terminating codon. A codon for which there is no tRNA with a complementary anticodon. There are three: UAA, UAG, and UGA.

notochord (Gr. *noto*, back + L. *chorda*, cord) In chordates, a dorsal rod of cartilage that forms between the nerve cord and the developing gut in the early embryo.

nuclear envelope The double membrane (outer and inner) surrounding the surface of the nucleus of eukaryotes.

nuclear pore Shallow depressions, like the craters of the moon, that are scattered over the surface of the nuclear envelope. Such pores contain many embedded proteins that act as molecular channels, permitting certain molecules to pass into and out of the nucleus.

nucleic acid A nucleotide polymer. A long chain of nucleotides. Chief types are deoxyribonucleic acid (DNA), which is double-stranded, and ribonucleic acid (RNA), which is typically single-stranded.

nucleolus, *pl.* **nucleoli** (L. a small nucleus) Aggregations of rRNA and some ribosomal proteins that are transported into the nucleus from the rough ER and accumulate at those regions on the chromosomes where active synthesis of rRNA is taking place.

nucleosome (L. *nucleus*, kernel + *soma*, body) The basic packaging unit of eukaryotic chromosomes, in which the DNA molecule is wound around a ball of histone proteins. Chromatin is composed of long strings of nucleosomes, like beads on a string.

nucleotide A single unit of nucleic acid, composed of a phosphate, a five-carbon sugar (either ribose or deoxyribose), and a purine or a pyrimidine.

nucleus (L. a kernel, dim. fr. *nux*, nut) A spherical organelle (structure) characteristic of eukaryotic cells. The repository of the genetic information that directs all activities of a living cell. In atoms, the central core, containing positively charged protons and (in all but hydrogen) electrically neutral neutrons.

O

occipital lobe The posterior lobe of each cerebral hemisphere.

olfaction (L. *olfactum*, smelled) The process or function of smelling.

oncogene (Gr. *oncos*, tumor) A cancer-causing gene.

oncogene theory (Gr. *oncos*, cancer) The hypothesis that cancer results from the action of a specific tumor-inducing *onc* gene.

one-gene/one-enzyme hypothesis The hypothesis that genes produce their effects by specifying the structure of enzymes and that each gene encodes the structure of a single enzyme.

oocyte (Gr. *oion*, egg + *kytos*, vessel) A cell in the outer layer of the ovary that gives rise to an ovum. A primary oocyte is any of the two million oocytes a female is born with, all of which have begun the first meiotic division.

open circulatory system A system within an organism that permits materials to pass from one cell to another without leaving the organism and in which generally no distinction exists between circulating fluid and body fluid.

operant conditioning A learning mechanism in which the reward follows only after the correct behavioral response.

operator A site of negative gene regulation. A sequence of nucleotides that may overlap the promoter, which is recognized by a repressor protein. Binding of the repressor protein to the operator prevents binding of the polymerase to the promoter and so blocks transcription of the structural genes of an operon.

operon (L. *operis*, work) A cluster of functionally related genes transcribed onto a single mRNA molecule. A common mode of gene regulation in prokaryotes, it is rare in eukaryotes other than fungi.

optic nerve The nerve of sight, which goes from the brain to the eyeball and terminates in the retina.

optimal yield Exploitation at the early, most productive part of the rising portion of the sigmoid growth curve to maximize harvesting of the population. Used by humans in agriculture and fisheries.

order A taxonomic category ranking below a class and above a family.

organ (L. *organon*, tool) A complex body structure composed of several different kinds of tissue grouped together in a structural and functional unit.

organ system A group of organs that function together to carry out the principal activities of the body.

organelle (Gr. *organella*, little tool) A specialized compartment of a cell. Mitochondria are organelles.

organism Any individual living creature, either unicellular or multicellular.

origin (L. *oriri*, to rise) The end of a muscle attached to a bone that remains stationary during contraction.

osmoconformer An animal that maintains the osmotic concentration of its body fluids at about the same level as that of the medium in which it is living.

osmoregulation The maintenance of a constant internal solute concentration by an organism, regardless of the environment in which it lives.

osmosis (Gr. *osmos*, act of pushing, thrust) The diffusion of water across a membrane that permits the free passage of water but not that of one or more solutes.

osmotic pressure The increase of hydrostatic water pressure within a cell as a result of water molecules that continue to diffuse inward

toward the area of lower water concentration (the water concentration is lower inside than outside the cell because of the dissolved solutes in the cell).

osteoblast (Gr. *osteon*, bone + *blastos*, bud) A bone-forming cell.

osteocyte (Gr. *osteon*, bone + *kytos*, hollow vessel) A mature osteoblast.

otolith *See* statocyst.

outcross A term used to describe species that interbreed with individuals other than those like themselves.

outer medulla The outer portion of the kidney, which contains the upper portion of the loop of Henle, including the upper ascending arm, where salt from the filtrate is reabsorbed by active transport.

oval window A membrane in the ear of a mammal or other vertebrate that connects the middle ear with the inner ear.

ovary (L. *ovum*, egg) In animals, the organ that produces eggs. In flowering plants, the enlarged basal portion of a carpel, which contains the ovule(s). The ovary matures to become the fruit.

oviparous (L. *ovum*, egg + *parere*, to bring forth) Refers to reproduction in which the eggs are developed after leaving the body of the mother, as in reptiles.

ovulation The successful development and release of an egg by the ovary.

ovule (L. *ovulum*, a little egg) A structure in a seed plant that becomes a seed when mature.

ovum, pl. ova (L. egg) A mature egg cell. A female gamete.

oxidation (Fr. *oxider*, to oxidize) The loss of an electron during a chemical reaction from one atom to another. Occurs simultaneously with reduction. Is the second stage of the ten reactions of glycolysis.

oxidative metabolism A collective term for metabolic reactions requiring oxygen.

oxidative respiration Respiration in which the final electron acceptor is molecular oxygen.

oxytocin (Gr. *oxys*, sharp + *tokos*, birth) A hormone of the posterior pituitary gland that affects uterine contraction during childbirth and stimulates lactation.

P

pain A stimulus the body receives that causes or is about to cause tissue damage.

palisade parenchyma (L. *palus*, stake + Gr. *para*, beside + *en*, in + *chein*, to pour) Parenchyma cells that are columnar, closely packed together, and located between the upper epidermis and the spongy mesophyll of a leaf.

pancreas (Gr. *pan*, all + *kreas*, flesh) In vertebrates, the principal digestive gland. A large gland situated between the stomach and the small intestine that secretes a host of digestive enzymes and the hormones insulin and glucagon.

Pangaea Name given to the giant landmass that comprised the major continents during the early Jurassic period.

parasite (Gr. *para*, beside + *sitos*, food) A symbiotic relationship in which one organism benefits and the other is harmed.

parasympathetic nervous system (Gr. *para*, beside + *syn*, with + *pathos*, feeling) One of two subdivisions of the autonomic nervous system. The other is the sympathetic. The two subdivisions operate antagonistically: The parasympathetic system stimulates resting activities, such as digestion and restoration of

the body to normal after emergencies by inhibiting alarm functions initiated by the sympathetic system.

parathyroid hormone (PTH) (Gr. *para*, beside + thyroid + hormone) A hormone produced by the parathyroid glands that regulates the way the body uses calcium.

parenchyma (Gr. *para*, beside + *en*, in + *chein*, to pour) The least specialized and most common cell of all plant cells. Alive at maturity and capable of further division. Usually photosynthetic or storage tissue.

parthenogenesis (Gr. *parthenos*, virgin + Eng. *genesis*, beginning) The development of an adult from an unfertilized egg. A common form of reproduction in insects.

partial pressures (P) The components of each individual gas—nitrogen, oxygen, and carbon—that together comprise the total air pressure.

pathogen (Gr. *pathos*, suffering + Eng. *genesis*, beginning) A disease-causing organism.

pectin (Gr. *pektos*, curdled, congealed, fr. *pegnynai*, to make fast or stiff) A complex form of plant starch having short, linear amylose branches consisting of twenty to thirty glucose subunits.

pedigree (L. *pes*, foot + *grus*, crane) A family tree. The patterns of inheritance observed in family histories. Used to determine the mode of inheritance of a particular trait.

pelvic girdle The body arch that provides strong connections and support for the legs.

penis (L. tail) The male organ of copulation. In mammals, it is also the male urinary organ.

peptide (Gr. *peptein*, to soften, digest) Two or more amino acids linked by peptide bonds.

peptide bond A covalent bond linking two amino acids.

Formed when the positive (amino, or NH$_3$) group at one end and a negative (carboxyl, or COO) group at the other end undergo a chemical reaction and lose a molecule of water.

peptide hormone A hormone that interacts with a receptor on the cell surface and initiates a chain of events within the cell by increasing the levels of secondary messengers.

perennial (L. *per*, through + *annus*, a year) A plant that lives for more than a year and produces flowers on more than one occasion.

peripheral nervous system (Gr. *peripherein*, to carry around) All of the neurons and nerve fibers outside the central nervous system, including motor neurons, sensory neurons, and the autonomic nervous system.

peristalsis (Gr. *peri*, around + *stellein*, to wrap) The rhythmic sequences of waves of muscular contraction in the walls of a tube.

peroxisome A membrane-bound spherical body, apparently derived from smooth endoplasmic reticulum (ER), that carries one set of enzymes active in converting fats to carbohydrates and another set that detoxifies various potentially harmful molecules—strong oxidants—that form in cells.

petal (Gr. *petalon*, leaf) A part of a flower that is usually colored. One of the leaves of a corolla.

petiole (L. *petiolus*, a little foot) The stalk of a leaf.

pH Refers to the concentration of H$^+$ ions in a solution. The numerical value of the pH is the negative of the exponent of the molar concentration. Low pH values indicate high concentrations of H$^+$ ions (acids), and high pH values indicate low concentrations.

phage *See* bacteriophage.

phagocyte (Gr. *phagein*, to eat + *kytos*, hollow vessel) A cell that kills invading cells by engulfing them. Includes neutrophils and macrophages.

phagocytosis (Gr. *phagein*, to eat + *kytos*, hollow vessel) A form of endocytosis in which cells engulf organisms or fragments of organisms.

pharynx (Gr. gullet) In vertebrates, a muscular tube that connects the mouth cavity and the esophagus. Serves as the gateway to the digestive tract and to the windpipe, or trachea.

phenotype (Gr. *phainein*, to show + *typos*, stamp or print) The realized expression of the genotype. The observable expression of a trait (affecting an individual's structure, physiology, or behavior) that results from the biological activity of proteins or RNA molecules transcribed from the DNA.

pheromone (Gr. *pherein*, to carry + [hor]mone) A chemical sex signal secreted by certain female animals that signals their reproductive readiness.

phloem (Gr. *phloos*, bark) In vascular plants, a food-conducting tissue basically composed of sieve elements, various kinds of parenchyma cells, fibers, and sclereids.

phosphate group (PO$_4$) A chemical group commonly involved in high-energy bonds.

phosphodiester The bond that results from the formation of a nucleic acid chain in which individual sugars are linked together in a line by the phosphate groups. The phosphate group of one sugar binds to the hydroxyl group of another, forming an —O—P—O bond.

phospholipid (Gr. *phosphoros*, light-bearer, + *lipos*, fat) The lipid molecule that forms the foundation of a plasma membrane. Similar to a fat molecule, but has only two fatty acids attached to its glycerol backbone, the third position being occupied by a highly polar phosphate group, often with additional small polar molecules attached, that readily forms hydrogen bonds with water. One end of a phospholipid molecule is therefore strongly nonpolar (water-insoluble), whereas the other end is extremely polar (water-soluble). The two nonpolar fatty acids extend in one direction, roughly parallel to each other, and the polar alcohol group points in the other direction.

phosphorus cycle A critical mineral cycle important in worldwide plant nutrition in which phosphates are transferred from the soil to the plants and then recycled when the plants die.

photon (Gr. *photos*, light) The unit of light energy.

photoperiodism (Gr. *photos*, light + *periodos*, a period) A mechanism that organisms use to measure seasonal changes in relative day and night length.

photophosphorylation (Gr. *photos*, light + *phosphoros*, bringing light) A fundamental photosynthetic light reaction in which organisms use a network of chlorophyll molecules (a photocenter) to channel photon excitation energy to one pigment molecule, referred to as P700. P700 then donates an electron to an electron transport chain, which drives a proton pump and returns the electron to P700.

photorespiration A process in which carbon dioxide is released without the production of ATP or NADPH. Because it produces neither ATP nor NADPH, photorespiration acts

to undo the work of photosynthesis.

photosynthesis (Gr. *photos*, light -*syn*, together + *tithenai*, to place) The process by which plants, algae, and some bacteria use the energy of sunlight to create from carbon dioxide (CO$_2$) and water (H$_2$O) the more complicated molecules that make up living organisms.

photosynthetic membrane The location where light reactions take place.

photosystem I A term to describe the more ancient bacterial photosystem of algae and plants.

photosystem II A photocenter system in which molecules of chlorophyll *a* are arranged with a different geometry in the photocenter so that more of the shorter-wavelength photons of higher energy are absorbed than in the more ancient bacterial photosystem.

phototropism (Gr. *photos*, light + *trope*, turning to light) A plant's growth response to a unidirectional light source.

phylogeny (Gr. *phylon*, race, tribe) The evolutionary relationships among any group of organisms.

phylum, pl. phyla (Gr. *phylon*, race, tribe) A major taxonomic category, ranking above a class.

physiology (Gr. *physis*, nature + *logos*, a discourse) The study of the function of cells, tissues, and organs.

pigment (L. *pigmentum*, paint) A molecule that absorbs light.

pineal gland (L. *pinus*, pine tree) A small endocrine gland in the center of the brain that secretes the hormone melatonin. It appears to function in humans as a light-sensing organ and in a variety of other roles concerning sexual development.

pinocytosis (Gr. *pinein*, to drink + *kytos*, cell) A form of endocytosis in which the material brought into the cell is a liquid containing dissolved molecules.

pistil (L. *pistillum*, pestle) The part of a flower that produces seeds, typically consisting of an ovary, a style, and a stigma. Such organs taken collectively are known as the gynoecium. A flower that has only ovules and no pollen is called pistillate. Functionally, it is female.

pith The ground tissue occupying the center of the stem or root within the vascular cylinder. Usually consists of parenchyma.

pituitary (L. *pituita*, phlegm) The major hormone-producing gland of the brain. Under the control of the hypothalamus. Secretes hormones that promote growth, stimulate glands, and regulate many other bodily functions.

placenta, pl. placentae (L. a flat cake) A specialized organ, held within the womb in the mother, across which she supplies the offspring with food, water, and oxygen and through which she removes wastes.

plankton (Gr. *planktos*, wandering) The small organisms that float or drift in water, especially at or near the surface.

plasma (Gr. form) The fluid of vertebrate blood. Contains dissolved salts, metabolic wastes, hormones, and a variety of proteins, including antibodies and albumen. Blood minus the blood cells.

plasma membrane A lipid bilayer with embedded proteins that control the cell's permeability to water and dissolved substances.

plasmid (Gr. *plasma*, a form or something molded) A small fragment of DNA that replicates independently of the bacterial chromosome.

plasmodesmata (Gr. *plasma*, something molded + *desma*, band) The cytoplasmic connections that extend through pairs of holes in the cell walls in plants.

plate tectonics The theory that the earth's crust is divided into a series of vast, platelike parts. Explains continental drift.

platelet (Gr. dim. of *plattus*, flat) A fragment of a megakaryocyte that floats in the blood and plays an important role in controlling blood clotting.

point mutation A mutation that changes one or a few nucleotides. May result from physical or chemical damage to the DNA or from spontaneous errors during DNA replication.

polar bodies The products resulting from the process of meiosis in the oocytes, with only a single haploid ovum being produced and the other meiotic products (called polar bodies) being discarded.

polar molecule A molecule with positively and negatively charged ends. One portion of a polar molecule attracts electrons more strongly than another portion, with the result that the molecule has electron-rich (−) and electron-poor (+) regions, giving it magnetlike positive and negative poles. Water is one of the most polar molecules known.

polar nuclei In flowering plants, two nuclei (usually), one derived from each end (pole) of the embryo sac, that become centrally located. They fuse with a male nucleus to form the primary (3N) endosperm nucleus.

polarization The charge difference of a neuron so that the interior of the cell is negative with respect to the exterior.

pollen (L. fine dust) A fine, yellowish powder consisting of grains or microspores, each of which contains a mature or immature male gametophyte. In flowering plants, pollen is released from the anthers of flowers and fertilizes the pistils.

pollen tube A tube that grows from a pollen grain. Male reproductive cells move through the pollen tube into the ovule.

pollination The transfer of pollen from the anthers to the stigmas of flowers for fertilization, as by insects or the wind.

polyclonal An antibody response in which an antigen elicits many different antibodies, each fitting a different portion of the antigen surface.

polygyny (Gr. *poly*, many + *gyne*, woman, wife) A mating choice in which a male mates with more than one female.

polymer (Gr. *polus*, many + *meris*, part) A large molecule formed of long chains of similar molecules.

polymerization (Gr. *polus*, many + *meris*, part + *izein*, to combine with) A process in which identical subunits are attracted to one another chemically and assemble into long chains.

polymorphism (Gr. *polys*, many + *morphe*, form) The presence in a population of more than one allele of a gene at a frequency greater than that of newly arising mutations.

polynomial system (Gr. *polys*, many + (bi)nomial) Before Linnaeus, naming a genus by use of a cumbersome string of Latin words and phrases.

polyp A cylindrical, pipe-shaped cnidarian usually attached to a rock with the mouth facing away from the rock on which it is growing. Coral is made up of polyps.

polypeptide (Gr. *polys*, many + *peptein*, to digest) A general term for a long chain of amino acids linked end to end by peptide bonds. A protein is a long, complex polypeptide.

polysaccharide (Gr. *polys*, many + *sakcharon*, sugar) A sugar polymer. A carbohydrate composed of many monosaccharide sugar subunits linked together in a long chain.

population (L. *populus*, the people) Any group of individuals, usually of a single species, occupying a given area at the same time.

population genetics The branch of genetics that deals with the behavior of genes in populations.

population pyramid A bar graph using five-year age categories by which the characteristics of a population can be illustrated graphically.

positive control The process of turning on transcription by use of a regulatory site.

posterior (L. *post*, after) Situated behind or farther back.

postsynaptic membrane The membrane of the target cell of the synaptic cleft.

potential difference A difference in electrical charge on two sides of a membrane caused by an unequal distribution of ions.

potential energy Energy with the potential to do work. Stored energy.

prairie A temperate grassland of the United States and southern Canada.

precapillary sphincter A ring of muscle that guards each capillary loop and that, when closed, blocks flow through the capillary.

predation (L. *praeda*, prey) The eating of other organisms. The one doing the eating is called a predator, and the one being consumed is called the prey.

pressure receptor See baroreceptor.

presynaptic membrane The membrane on the axonal side of the synaptic cleft.

prey (L. *prehendere*, to grasp, seize) An organism eaten by another organism.

primary induction Inductions between the three primary tissue types—ectoderm, mesoderm, and endoderm.

primary mRNA transcript *See* RNA transcript.

primary nondisjunction The failure of homologous chromosomes to separate in meiosis I. The cause of Down syndrome.

primary plant body The part of a plant that includes the young, soft shoots and roots, and that arises from the apical meristems.

primary producers Photosynthetic organisms, including plants, algae, and photosynthetic bacteria.

primary structure of a protein The sequence of amino acids that makes up a particular polypeptide chain.

primary tissue Tissue that comprises the primary plant body.

primitive streak The site of invagination on a blastodisc, which appears as a slit on the surface of the gastrula.

primordium, *pl.* **primordia** (L. *primus*, first + *ordiri*, begin) The first cells in the earliest stages of the development of an organ or structure.

productivity The total amount of energy of an ecosystem fixed by photosynthesis per unit of time. Net productivity is productivity minus that which is expended by the metabolic activity of the organisms in the community.

progesterone A steroid hormone secreted by the corpus luteum that makes the lining of the uterus more receptive to a fertilized ovum.

proglottid (Gr. *proglottis*, tip of the tongue) One of the segments or joints of a tapeworm. Contains both male and female sexual organs.

prokaryote (Gr. *pro*, before + *karyon*, kernel) A simple bacterial organism that is small and single-celled, lacks external appendages, and has little evidence of internal structure.

promoter An RNA polymerase binding site. The nucleotide sequence at the end of a gene to which RNA polymerase attaches to initiate transcription of mRNA.

prophase (Gr. *pro*, before + *phasis*, form) The first stage of mitosis during which the chromosomes become more condensed, the nuclear envelope is reabsorbed, and a network of microtubules (called the spindle) forms between opposite poles of the cell.

prostaglandin (from prosta[te] gland + -in) A modified lipid produced from membrane phospholipids by virtually all cells. It stimulates contraction and expansion of smooth muscles and contraction of blood vessels.

prostate gland (Gr. *prostates*, one standing in front) A large gland surrounding the male urethra just below the bladder. Its secretions, which transport sperm cells, make up a large part of the semen.

protease (Gr. *proteios*, primary + -ase, enzyme ending) An enzyme that breaks up proteins into amino acids.

protein (Gr. *proteios*, primary) A long chain of amino acids linked end to end by peptide bonds. Because the twenty amino acids that occur in proteins have side groups with very different chemical properties, the function and shape of a protein is critically affected by its particular sequence of amino acids.

protist (Gr. *protos*, first) A member of the kingdom Protista, which includes unicellular eukaryotic organisms and some multicellular lines derived from them.

proton A subatomic particle in the nucleus of an atom that carries a positive charge. The number of protons determines the chemical character of the atom because it dictates the number of electrons orbiting the nucleus and available for chemical activity.

proton pump A channel in the cell that, along with the sodium-potassium pump, transports molecules against a concentration gradient by expending energy.

protozoa (Gr. *protos*, first + *zoon*, animal) The traditional name given to heterotrophic protists.

proximate causation The question of how an animal behaves in a certain way. The mechanism of behavior.

pseudocoel (Gr. pseudos, false + *koiloma*, cavity) A body cavity similar to the coelom except that it is unlined.

pseudopod, *pl.* **pseudopodia** (Gr. *pseudos*, false + *pous*, foot) "False foot." A temporary protrusion of a one-celled organism that serves as a means of locomotion.

pulmonary circulation The part of the human circulatory system that carries the blood from the heart to the lungs and back again.

punctuated equilibrium A hypothesis of the mechanism of evolutionary change that proposes that long periods of little or no change are punctuated by periods of rapid evolution.

Punnett square After Reginald C. Punnett, English geneticist. A diagram useful in analyzing a Mendelian model of the outcome of an F_2 generation derived from a mating of F_1 heterozygous individuals.

pupil (L. *pupilla*, little doll) The opening in the center of the iris of the eye where light enters the eye.

Purkinje fiber After Johannes E. Purkinje, Bohemian physiologist. Any of the modified muscle fibers of the heart that make up a network by which muscle impulses are conducted through the heart.

pyloric sphincter (Gr. *pyloros*, gatekeeper) A muscular constriction between the stomach and the small intestine that functions as a kind of traffic light of the digestive system.

pyramid of biomass The structure obtained from weighing all the individuals at each trophic level of the ecosystem.

pyramid of energy The structure obtained from measuring the flow of energy through an ecosystem directly at each point of transfer.

pyramid of numbers The structure obtained from counting all the individuals in an ecosystem and assigning each to a trophic level.

pyruvate The three-carbon compound that is the end product of glycolysis and the starting material of the citric acid cycle.

pyruvate dehydrogenase The complex of enzymes that removes carbon dioxide from pyruvate. One of the largest known enzymes, containing forty-eight polypeptide chains.

Q

quaternary structure of a protein A term to describe the way the protein subunits are assembled into a whole.

R

radial symmetry (L. *radius*, a spoke of a wheel + Gr. *summetros*, symmetry) The regular arrangement of parts around a central axis so that any plane passing through the central axis divides the organism into halves that are approximate mirror images.

radicle (L. *radicula*, root) The part of the plant embryo that develops into the root.

radioactive An isotope, such as carbon-14, in which the nucleus tends to break up into elements with lower atomic numbers in a process called radioactive decay.

radioactivity The emission of nuclear particles and rays by unstable atoms as they decay into more stable forms. Measured in curies, with one curie equal to thirty-seven billion disintegrations a second.

radula (L. scraper) A rasping, tonguelike organ characteristic of most mollusks.

rain shadow effect The phenomenon in which the eastern sides of mountains are much drier than their western sides, and vegetation is often very different.

receptor protein A highly specific cell-surface receptor embedded in a cell membrane that responds only to a specific messenger molecule.

recessive allele An allele whose phenotype effects are masked in heterozygotes by the presence of a dominant allele.

recombinant DNA A DNA molecule created in the laboratory by molecular geneticists who join together bits of several genomes into a novel combination.

recombination The formation of new gene combinations. In bacteria, it is accomplished by the transfer of genes into cells, often in association with viruses. In eukaryotes, it is accomplished by reassortment of chromosomes during meiosis and by crossing-over.

recombination map The construction of a genetic map by using the frequencies of crossing-over events in crosses.

rectum The short and final segment of the large intestine, extending from the colon to the anus.

reducing power The use of light energy to extract hydrogen atoms from water.

reduction (L. *reductio*, a bringing back: originally, "bringing back" a metal from its oxide) The gain of an electron during a chemical reaction from one atom to another. Occurs simultaneously with oxidation.

reflex (L. *reflectere*, to bend back) An automatic consequence of a nerve stimulation. The motion that results from a nerve impulse passing through the system of neurons, eventually reaching the body muscles and causing them to contract.

reflex arc The nerve path in the body that leads from stimulus to reflex action.

refractory period The recovery period after membrane depolarization during which the membrane is unable to respond to additional stimulation. The period after ejaculation, lasting twenty minutes or longer, during which males lose their erection, arousal is difficult, and ejaculation is almost impossible.

regulatory site A special nucleotide sequence of a gene that acts as a point of control when the transcription of individual genes begins.

release factor A special protein that releases a newly made polypeptide from a ribosome when a stop codon is encountered during protein synthesis.

releasing hormone A peptide hormone produced by the hypothalamus that stimulates the secretion of specific hormones by the anterior pituitary.

renal (L. *renes*, kidneys) Pertaining to the kidney.

replication fork The structure formed where the double-stranded DNA molecule separates during DNA replication.

replication origin The sequence of DNA necessary for replication.

replication unit An individual zone of a eukaryotic chromosome that replicates as a discrete unit.

repression (L. *reprimere*, to press back, keep back) The process of blocking transcription by the placement of the regulatory protein between the polymerase and the gene, thus blocking movement of the polymerase to the gene.

repressor (L. *reprimere*, to press back, keep back) A protein that regulates transcription of mRNA from DNA by binding to the operator and so preventing RNA polymerase from attaching to the promoter.

resolving power The ability of a microscope to distinguish two lines as separate.

respiration (L. *respirare*, to breathe) The utilization of oxygen. In terrestrial vertebrates, the inhalation of oxygen and the exhalation of carbon dioxide.

resting membrane potential The charge difference that exists across a neuron's membrane at rest (about 70 millivolts).

restriction endonuclease A special kind of enzyme that can recognize and cleave DNA molecules into fragments. One of the basic tools of genetic engineering.

restriction enzyme An enzyme that cuts a DNA strand at a particular place.

restriction fragment-length polymorphism (RFLP) An associated genetic mutation marker detected because the mutation alters the length of DNA segments.

retina (L. a small net) The structure in the eye that is sensitive to light and receives optical images. Contains a field of receptor cells, rods, and cones.

retrovirus (L. *retro*, turning back) A virus whose genetic material is RNA rather than DNA. When a retrovirus infects a cell, it makes a DNA copy of itself, which it can then insert into the cellular DNA as if it were a cellular gene.

reverse transcriptase An enzyme that synthesizes a double strand of DNA complementary to viral RNA. Found only in association with retroviruses, such as the AIDS virus.

Rh blood group A set of cell-surface markers on human red blood cells. Named for the Rhesus monkey in which they were first described.

rhizome (Gr. *rhizoma*, mass of roots) In vascular plants, a usually more or less horizontal underground stem. May be enlarged for storage or may function in vegetative reproduction.

ribonucleic acid (RNA) The other principal form of nucleic acid that is similar in structure to and made as a template copy of portions of DNA. This copy is transported from nucleus to cytoplasm of the cell, where it provides a blueprint specifying the amino acid sequence of proteins.

ribose A five-carbon sugar.

ribosomal RNA (rRNA) A class of RNA molecules found, together with characteristic proteins, in ribosomes. During polypeptide synthesis, they provide the site on the ribosome where the polypeptide is assembled.

ribosome An organelle composed of protein and RNA that translates RNA copies of genes into protein.

RNA *See* ribonucleic acid.

RNA polymerase The enzyme that transcribes RNA from DNA.

RNA transcript The newly assembled mRNA chain, which is called the primary mRNA transcript, of the nucleotide sequence of the gene.

rod A specialized sensory cell in the retina of the eye. Responsible for black-and-white vision.

root The part of a plant that grows downward, usually into the ground, to hold the plant in place, to absorb water and mineral foods from the soil, and often to store food material.

root cap The thimblelike mass of relatively unorganized cells that covers and protects the root's apical meristem as it grows through the soil.

root hairs Fine projections from the epidermis, or outermost cell layer, of the terminal portion of roots.

root pressure In vascular plants, the pressure that develops in roots, primarily at night. Caused by the continued, active accumulation of ions by the roots of a plant at times when transpiration from the leaves is very low or absent.

S

saltatory conduction A very fast form of nerve impulse conduction in which the impulses leap from node to node over insulation portions.

sarcoma (Gr. *sarx,* flesh) A cancerous tumor that involves connective or hard tissue, such as muscle.

sarcomere (Gr. *sarx,* flesh + *meris,* part of) The fundamental unit of contraction in skeletal muscle. The repeating bands of actin and myosin that appear between two Z lines.

sarcoplasm (Gr. *sarx,* flesh + *plassein,* to form, mold) A special name for the cytoplasm in striated muscle.

sarcoplasmic reticulum (Gr. *sarx,* flesh + *plassein,* to form, mold; L. *reticulum,* network) The endoplasmic reticulum of a muscle cell. A sleeve of membrane that wraps around each myofilament.

savanna A tropical or subtropical region of open grassland that is a transitional biome between tropical rain forest and desert.

Schwann cells After Theodor Schwann, German anatomist. The supporting cells associated with projecting axons, along with all the other nerve cells that make up the peripheral nervous system.

scientific creationism A view that the biblical account of the origin of the earth is literally true, that the earth is much younger than most scientists believe, and that all species of organisms were individually created just as they are today.

sclereid (Gr. *skleros,* hard) In vascular plants, a type of sclerenchyma cell that is thick-walled or lignified, usually pitted, and not elongated. Also called stone cell.

sclerenchyma cell (Gr. *skleros,* hard + *en,* in + *chymein,* to pour) A cell of variable form and size with more or less thick, often lignified, secondary walls. May or may not be living at maturity. Includes fibers and sclereids. Collectively, sclerenchyma cells may make up a kind of tissue called sclerenchyma.

scolex (Gr. worm) The organ in a tapeworm that attaches to the intestinal wall.

scrotum The sac containing the testicles that hangs between the legs of the male.

scurvy A disease characterized by weakness, spongy gums, and bleeding of the skin and mucous membranes, resulting from insufficient quantities of vitamin C in the diet.

second messenger An intermediary compound that couples extracellular signals to intracellular processes and also amplifies a hormonal signal.

secondary cell wall A parenchymal cell that is deposited between the cytoplasm and primary wall of a fully expanded cell.

secondary chemical compound A chemical that is not involved in primary metabolic processes and that plays the dominant role in protecting plants from being eaten by herbivores or predators.

secondary growth In vascular plants, growth that results from the division of a cylinder of cells around the plant's periphery. Secondary growth causes a plant to grow in diameter.

secondary immune response The swifter response of the body the second time it is invaded by the same pathogen because of the presence of

memory cells, which quickly become antibody-producing plasma cells.

secondary induction An induction between tissues that have already differentiated.

secondary plant body The part of a plant characterized by thick accumulations of conducting tissue and the other cell types associated with it caused by secondary growth.

secondary sex characteristics External differences between male and female animals. Not directly involved in reproduction.

secondary structure of a protein The twisting or folding of a polypeptide chain. Results from the formation of hydrogen bonds between different amino acid side groups of a chain. The most common structures that form are a single-stranded helix, an extended sheet, or a cable containing three (as in collagen) or more strands.

secondary tissues The tissues that comprise the secondary plant body.

seed A structure that develops from the mature ovule of a seed plant. Contains an embryo surrounded by a protective coat.

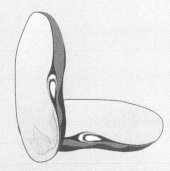

seed coat The outer layer of a seed. Developed from the integuments of the ovule.

seed plants A collective term for the groups of vascular plants that produce seeds.

segregation of alleles See Mendel's First Law.

segregation of alternative traits The finding that alternative traits segregate in crosses and may mask each other's appearance.

selection The process by which some organisms leave more offspring than competing ones, and their genetic traits tend to appear in greater proportions among members of succeeding generations than the traits of those individuals that leave fewer offspring.

selectively permeable membrane (L. *seligere,* to gather apart + *permeare,* to go through) Refers to the ability of a cell membrane to allow passage across the membrane of some solutes but not others. The result of specific protein channels extending across the membrane. Some molecules can pass through a specific kind of channel; others cannot.

self-fertilization The process in which a plant that has both male and female gametes can fertilize itself.

self-pollination The transfer of pollen from an anther to a stigma in the same flower or to another flower of the same plant, leading to self-fertilization.

semen (L. seed) The fluid produced in the male reproductive organs that contains sperm and fluid secretions.

semicircular canal Any of three fluid-filled canals in the inner ear that help to maintain balance.

semipermeable membrane See selectively permeable membrane.

sensory pathway The nerve pathway of the peripheral nervous system that transmits commands to the body from the central nervous system.

sepal (L. *sepalum,* a covering) A member of the outermost whorl of a flowering plant. Collectively, the sepals constitute the calyx.

septum, *pl.* **septa** (L. *saeptum,* a fence) A partition or cross-wall that divides fungal hyphae into cells.

seta, *pl.* **setae** (L. bristle) In an annelid, bristles of chitin that help to anchor the worm during locomotion or when it is in its burrow.

sex chromosomes The X and Y chromosomes that are different in the two sexes and that are involved in sex determination.

sex-linked characteristic A genetic characteristic that is determined by genes located on the sex chromosomes.

sexual reproduction Reproduction that involves the regular alternation between syngamy and meiosis. Its outstanding characteristic is that an individual offspring inherits genes from two parent individuals.

sexual selection Natural selection perpetuating certain characteristics that attract one sex to the other, such as bright feathers in birds.

shell The hard outer covering of a mollusk.

shifting agriculture A form of agriculture, common in tropical areas, in which people clear and cultivate a patch of forest, grow crops for a few years, and then move on.

shoot In vascular plants, the aboveground parts, such as the stem and leaves.

short-term memory A functional type of memory that is transient, lasting only a few moments.

sieve cell In the phloem (food-conducting tissue) of vascular plants, a long, slender sieve element with relatively unspecialized sieve areas and with tapering end walls that lack sieve plates. Found in all vascular plants except angiosperms, which have sieve-tube members.

sieve tube In the phloem of angiosperms, a series of sieve-tube members arranged end-to-end and interconnected by sieve plates.

sigmoid growth curve An S-shaped population growth curve that implies a relatively slow start in growth, a rapid increase, and then a leveling off when the carrying capacity of the species' environment is reached.

sign stimulus According to ethologists, the "signal" in the environment that triggers an animal's instinctive behavior.

single bond A covalent bond that shares only one electron pair.

sink An area in a vascular plant where sucrose is taken from a sieve tube.

sinus (L. curve) A reservoir or channel within the liver that contains venous blood.

skin The outer layer of tissue of the human or animal body.

smell A chemical sensory system in which the receptors are neurons whose cell bodies are embedded in the epithelium of the upper portion of the nasal passage.

smooth muscle Nonstriated muscle. Lines the walls of internal organs and arteries and is under involuntary control.

society A group of organisms of the same species that are organized in a cooperative manner.

sociobiology The study of the biological basis of social behavior of animal societies.

soluble Refers to polar molecules that dissolve in water and are surrounded by a hydration shell.

solute The molecules dissolved in a solution. *See also* solution, solvent.

solution A mixture of molecules, such as sugars, amino acids, and ions, dissolved in water.

solvent The most common of the molecules dissolved in a solution. Usually a liquid, commonly water.

somatic cells (Gr. *soma,* body) All the diploid body cells of an animal that are not involved in gamete formation.

somatic nervous system (Gr. *soma,* body) A motor pathway of the peripheral nervous system whose nerve fibers stimulate secretion from glands and excrete or inhibit the smooth muscles of the body. Known as the voluntary system, as contrasted with the involuntary, or autonomic, nervous system.

somatosensory cortex A sensory region on the leading edge of the parietal lobe that receives afferent input from sensory receptors of many different parts of the body.

somite A segmented block of tissue on either side of a developing notochord.

source An area in a vascular plant where sucrose is made.

species, *pl.* **species** (L. kind, sort) A level of taxonomic hierarchy; a species ranks next below a genus.

species specific Limited in reaction or effect to one species.

speech centers Regions in both hemispheres of the brain that are responsible for different associative speech activities.

sperm (Gr. *sperma,* sperm, seed) A sperm cell. The male gamete.

spermatogenesis (Gr. *sperma,* sperm, seed + *gignesthai,* to be born) The formation and development of spermatozoa.

sphincter (Gr. *sphinkter,* band, from *sphingein,* to bind tight) In vertebrate animals, a ringlike muscle that surrounds an opening or passage of the body and can contract to close it.

spinal cord The thick, whitish cord of nerve tissue extending from the medulla oblongata down through most of the spinal column and from which nerves to various parts of the body branch off.

spindle The mitotic assembly that carries out the separation of chromosomes during cell division. Composed of microtubules and assembled during prophase at the equator of the dividing cell.

spindle fibers An axis of microtubules formed by separating pairs of centrioles.

spine The flexible supporting column of bone along the middle of the back in vertebrate bodies.

spongin A tough protein that helps to strengthen the body of a sponge either by itself or together with spicules.

spongy bone tissue *See* marrow.

spongy parenchyma A leaf tissue composed of loosely arranged chloroplast-bearing cells. *See* palisade parenchyma.

sporangium, *pl.* **sporangia** (Gr. *spora,* seed + *angeion,* a vessel) A structure in which asexual spores are produced in plants and certain protists and fungi.

spore (Gr. *spora,* seed) A haploid reproductive cell,

usually unicellular, that is capable of developing into an adult without fusion with another cell. Spores result from meiosis, as do gametes, but gametes fuse immediately to produce a new diploid cell.

sporic meiosis The type of life cycle of a plant in which both diploid and haploid cells divide by mitosis.

sporophyte (Gr. *spora,* seed + *phyton,* plant) The spore-producing, diploid (2N) phase in the life cycle of a plant having alternation of generations.

stabilizing selection A form of selection in which selection acts to eliminate both extremes from a range of phenotypes. The result is the increase in the frequency of the intermediate type, which is already the most common.

stable population A population whose size remains the same through time.

stamen (L. thread) The part of the flower that contains the pollen. Consists of a slender filament that supports the anther. A flower that produces only pollen is called staminate and is functionally male.

stasis (Gr. a standing still) Refers to a lack of evolutionary change.

statocyst (Gr. *statos,* standing + *kystis,* bladder, sac) A gravity receptor located in a series of hollow chambers within the human inner ear that provides information the brain uses to perceive balance.

stem The aboveground axis of vascular plants. Stems are sometimes below ground (as in rhizomes and corms).

steppe (Russ. step) A temperate grassland of Eastern Europe and Central Asia.

stereotyped Animal behavior that appears in the same form in different individuals of a species.

steroid (Gr. *stereos,* solid + L. *ol,* from oleum, oil) A kind of lipid. Many of the molecules that function as messengers and pass across cell membranes are steroids, such as the male and female sex hormones and cholesterol.

steroid hormone A hormone derived from cholesterol. Those that promote the development of the secondary sexual characteristics are steroids.

stigma (Gr. mark) A specialized area of the carpel of a flowering plant that receives the pollen.

stoma, *pl.* **stomata** (Gr. mouth) A specialized opening in the leaves of some plants that allows carbon dioxide to pass into the plant body and allows water and oxygen to pass out of them.

stratum basal layer The cells of the innermost layer of the epidermis of the skin of the vertebrate body.

stratum corneum The outer layer of the epidermis of the skin of the vertebrate body.

stretch receptor A sense organ that is sensitive to any stretching of the tissue in which it is found. The muscle spindle is a stretch receptor.

striated muscle (L. *striare,* to groove) A type of muscle with fibers of cross-bands usually contracted by voluntary action.

stroma (Gr. anything spread out) The fluid matrix inside the chloroplast within which the thylakoids are embedded.

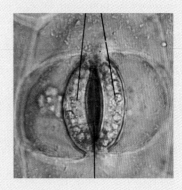

stromatolite (Gr. *stromatos,* a spread + Eng. -lite) A massive limestone deposit produced by cyanobacteria.

subcutaneous tissue The underlying layer of vertebrate skin.

subspecies A subdivision of a species. Often, a geographically distinct race.

substrate (L. *substratus,* strewn under) A molecule on which an enzyme acts.

substrate-level phosphorylation The generation of ATP by coupling its synthesis to a strongly exergonic (energy-yielding) reaction.

succession In ecology, the slow, orderly progression of changes in community composition that takes place through time. Primary succession occurs in nature over long periods of time. Secondary succession occurs when a climax community has been disturbed.

sucrose Table sugar. A common disaccharide found in many plants. A molecule of glucose linked to a molecule of fructose.

sugar Any monosaccharide or disaccharide.

supercoil Excessive helical turns in DNA that occur when the string of nucleosomes wraps up into higher-order coils.

supporting glial cell A nerve cell that supports and insulates the neurons.

surface layers The marine environment of the top layers of the open sea.

surface tension A tautness of the surface of a liquid, caused by the cohesion of the liquid molecules. Water has an extremely high surface tension.

surface-to-volume ratio Describes cell size increases. Cell volume grows much more rapidly than surface area.

survivorship The percentage of an original population that is living at a given age.

sympathetic nervous system A subdivision of the autonomic nervous system that functions as an alarm response. Increases heartbeat while slowing down everyday functions. Produces responses opposite to those of the parasympathetic nervous system.

synapse (Gr. *synapsis*, a union) A junction between a neuron and another neuron or muscle cell. The two cells do not touch. Instead, neurotransmitters cross the narrow space between them.

synapse, excitatory A synapse in which the receptor protein is a chemically gated sodium channel. Binding of a neurotransmitter opens the channel and initiates an excitatory electrical potential that increases the ease with which the membrane can be depolarized.

synapse, inhibitory A synapse in which the receptor protein is a chemically gated potassium or chloride channel. Binding of a neurotransmitter opens the channel and produces an inhibitory electrical potential that reduces the ability of the membrane to depolarize.

synapsis (Gr. *synapsis*, contact, union) The close pairing of homologous chromosomes that occurs early in prophase I of meiosis. With the genes of the chromosomes thus aligned, a DNA strand of one homologue can pair with the complimentary DNA strand of the other.

synaptic cleft The space between two adjacent neurons.

syngamy (Gr. *syn*, together with + *gamos*, marriage) Fertilization. The union of male and female gametes.

systemic circulation The circulation of blood between the heart and the rest of the body except the lungs.

systolic period (Gr. *systole*, contraction) The pushing period of heart contraction, which ends with the closing of the aortic valve, during which a pulse of blood is forced into the systemic arterial system, immediately raising the blood pressure within these vessels.

T

taiga (Russ.) A swampy, coniferous evergreen forestland of subarctic Siberia between the tundra and the steppes.

taste The specialized sensory receptors located in the mouth by which the flavor of a substance is perceived by the taste buds.

taste bud One of the groups of receptor cells in the mouth that are organs of taste.

taxonomy (Gr. *taxis*, arrangement + *nomos*, law) The science of the classification of organisms.

T-cell A type of lymphocyte involved in cell-mediated immune responses and interactions with B-cells. Also called a T-lymphocyte.

telophase (Gr. *telos*, end + *phasis*, form) The fourth and final stage of mitosis, during which the mitotic apparatus is disassembled, the nuclear envelope re-forms, and the chromosomes uncoil.

temperate grassland A region whose soil tends to be deep and fertile and whose vegetation consists mainly of grasses or grasslike plants.

temporal lobe (L. *temporalis*, the temples) The part of each cerebral hemisphere, in front of the occipital lobe, that contains the center of hearing in the brain.

tendon (Gr. *tenon*, stretch) A strap of cartilage that attaches muscle to bone.

territoriality A form of behavior in which an animal establishes an area as its own and defends it from encroachments by others, usually of its own species.

tertiary structure of a protein The three-dimensional shape of a protein. Primarily the result of hydrophobic interactions of amino acid side groups and, to a lesser extent, of hydrogen bonds between them. Forms spontaneously.

test cross A cross between a heterozygote and a recessive homozygote. A procedure Mendel used to further test his hypotheses.

testis, *pl.* testes (L. witness) In male mammals, the sperm-producing organ.

testosterone (Gr. *testis*, testicle + *steiras*, barren) A hormone secreted by the testes that is responsible for the secondary sex characteristics of males.

thalamus (Gr. *thalamos*, chamber) That part of the vertebrate forebrain just posterior to the cerebrum. Governs the flow of information from all other parts of the nervous system to the cerebral cortex.

theory (Gr. *theorein*, to look at) A well-tested hypothesis supported by a great deal of evidence.

thermal stratification A process characteristic of larger lakes in temperate regions in which water at a temperature of 4 degrees Celsius sinks beneath water that is either warmer or cooler.

thermodynamics (Gr. *therme*, heat + *dynamis*, power) The study of transformations of energy, using heat as the most convenient form of energy measurement. The first law of thermodynamics states that the total energy of the universe remains constant. The second law of thermodynamics states that the entropy, or degree of disorder, tends to increase.

thigmotropism (Gr. *thigma*, touch + *trope*, a turning) The growth response of a plant to touch.

thorax (Gr. a breastplate) The part of the body between the neck and the abdomen.

thoroughfare channel A capillary that connects arterioles and venules directly.

threshold value The minimal change in membrane potential necessary to produce an action potential.

thrombus (Gr. clot) A coagulation (clotting) that forms in a blood vessel and obstructs circulation.

thylakoid (Gr. *thylakos*, sac + *-oides*, like) A flattened, saclike membrane in the chloroplast of a eukaryote. Thylakoids are stacked on top of one another in arrangements called grana and are the sites of photosystem reactions.

thymine A pyrimidine occurring in DNA but not in RNA. *See also* uracil.

thyroxine (Gr. *thyros*, shield) A hormone secreted by the thyroid that increases metabolic rate and promotes growth.

tight junction A type of intercellular interaction characterized by belts of protein that isolate parts of plasma membrane and form a barrier separating surfaces of cells.

tissue (L. *texere*, to weave) A group of similar cells organized into a structural and functional unit.

trachea, *pl.* tracheae (L. windpipe) In vertebrates, the windpipe.

tracheid (Gr. *tracheia*, rough) An elongated cell with thick, perforated walls that carries water and dissolved minerals through a plant and provides support. Tracheids form an essential element of the xylem of vascular plants.

tracts The bundles of nerve fibers within the central nervous system.

transcription (L. *trans,* across + *scribere,* to write) The first stage of gene expression in which the RNA polymerase enzyme synthesizes an mRNA molecule whose sequence is complementary to the DNA.

transcription unit The portion of a gene segment that is transcribed into mRNA. It consists of the coding sequence and start and stop codons.

transfection (L. *trans,* across + [in]fect) A technique used in studying tumors that consists of isolating nuclear DNA from human tumor cells, cleaving the DNA into random fragments by using enzymes, and testing the fragments individually for the ability of any particular fragment to induce cancer in the cells that assimilate it.

transfer RNA (tRNA) (L. *trans,* across + *ferre,* to bear or carry) A second class of RNA that floats free in the cytoplasm. During polypeptide synthesis, tRNA molecules transport amino acids to the ribosome and position each amino acid at the correct place on the polypeptide chain.

transformation (L. *trans,* across + *formare,* to shape) Refers to the transfer of naked DNA from one organism to another. First observed as uptake of DNA fragments among pneumococcal bacteria.

translation (L. *trans,* across + *latus,* that which is carried) The second stage of gene expression in which a ribosome assembles a polypeptide, using the mRNA to specify the amino acids.

translocation (L. *trans,* across + *locare,* to put or place) In

plants, the process in which most of the carbohydrates manufactured in the leaves and other green parts of the plant are moved through the phloem to other parts of the plant.

transmission electron microscope A microscope that uses a beam of electrons rather than a light beam. A scanning electron microscope is a microscope that beams electrons on the surface of a specimen as a fine probe that rapidly passes back and forth.

transpiration (L. *trans,* across + *spirare,* to breathe) The loss of water vapor by plant parts, primarily through the stomata.

transport form Refers to sugars that are converted before they are moved within an organism. In many organisms, glucose is converted before it is moved and is less readily consumed (metabolized) while being moved.

transposon (L. *transponere,* to change the position of) A DNA sequence carrying one or more genes and flanked by insertion sequences that confer the ability to move from one DNA molecule to another. An element capable of transposition (the changing of chromosomal location).

triphosphate group Three phosphate groups linked in a chain.

trisomic (Gr. *tri,* three + *soma,* body) A chromosome abnormality in which one chromosome is present in triplicate.

trophic level (Gr. *trophos,* feeder) A step in the flow of energy through an ecosystem where all species in an ecosystem have the same main nutritional sources.

trophoblast (Gr. *trophe,* nourishment + *blastos,* germ, sprout) An outer sphere of cells,

external to the embryo in many mammals, that supplements the embryo with nourishment.

tropical rain forest A forest in a region near the equator. Characterized by year-round warmth and very heavy rainfall. The largest is in the Amazon Basin of South America. Rain forests are the richest of all biomes in terms of the number of species.

tropism (Gr. *trop,* turning) A plant's response to external stimuli. A positive tropism is one in which the movement or reaction is in the direction of the source of the stimulus. A negative tropism is one in which the movement or growth is in the opposite direction.

tropomyosin (Gr. *tropos,* turn + *myos,* muscle) A major protein element of muscle tissue. Tropomyosin binds to actin and helps to regulate the interaction of actin and myosin.

troponin (Gr. *tropos,* turn) A protein in muscle tissue regulated by calcium ions. Important in muscle contraction.

true-breeding Strains of plants that produce offspring resembling their parents in all respects.

tumor (L. swollen) A mass of cells, growing in an uncontrolled manner.

tundra (Russ.) A vast, level, treeless plain in the Arctic regions of Siberia and North America. The ground beneath is frozen even in summer. Tundra covers one-fifth of the earth's land surface.

turgor (L. *turgere,* to swell) The pressure exerted on the inside of a plant cell wall by the cell's fluid contents. The interior of the cell is hypertonic in relation to the fluids surrounding it and so gains water by osmosis.

turgor pressure (L. *turgor,* a swelling) The pressure within a

cell that results from the movement of water into the cell. A cell with high turgor pressure is said to be turgid.

twofold rotational symmetry Describes the nucleotides at one end of the recognition sequence that are complementary to those at the other end so that the two strands of the DNA duplex have the same nucleotide sequence running in opposite directions for the length of the recognition sequence.

tympanic membrane The large membrane separating the outer and middle ear; the eardrum.

U

ultimate causation The question of why a certain animal behavior evolved or what its adaptive value was.

ultrasound A noninvasive procedure that uses sound waves to produce an image of the fetus but that harms neither the mother nor the fetus.

undulipodia (L. *undulatus,* wavy + Gr. *pous,* foot) Describes the flagella of eukaryotic cells to underscore their complete distinctiveness from bacterial flagella, which they resemble only to a limited degree in external form and function.

unicellular Composed of a single cell.

universal donor A type O blood donor, whose blood is compatible with any individual's immune system.

universal recipient An individual with an AB blood type who may receive any type of blood.

uracil A pyrimidine found in RNA but not in DNA. *See also* thymine.

urea (Gr. *ouron,* urine) An organic molecule formed in the vertebrate liver. The principal form of disposal of nitrogenous wastes by mammals.

urethra (Gr. *ourein*, to urinate) A duct, which in males both semen (during ejaculation) and urine (during liquid elimination) pass, and in females, urine passes.

uric acid A waste product derived from ammonia found in the urine of reptiles and birds.

urine (Gr. *ouron*, urine) The liquid waste filtered from the blood by the kidneys.

useful energy The amount of energy remaining to do work as progressively more energy is degraded to heat.

uterus (L. womb) In mammals, a chamber in which the developing embryo is contained and nurtured during pregnancy.

V

vaccination The injection of a harmless microbe into a person or animal to confer resistance to a dangerous microbe.

vaccine (L. *vacca*, cow) A substance that prevents a disease when injected into the body of an individual who does not already have the disease.

vaccinia A cowpox-related virus that is the focus of modern researchers in their attempt to develop vaccines against malaria, herpes, and other diseases.

vacuole (L. *vacuus*, empty) A cavity in the cytoplasm of a cell that is bound by a single membrane and contains water and waste products of cell metabolism. Typically found in plant cells.

vacuum suction In humans, the removal of uterine contents by using a hollow curette or catheter to which a suction

apparatus is attached. Used prior to the twelfth week of pregnancy.

vagina (L. sheath) The membranous passage in female animals that leads to the mouth of the uterus.

variable Any factor that influences a process. In evaluating alternative hypotheses about one variable, all other variables are held constant so that the investigator is not misled or confused by other influences.

vas deferens (L. *vas*, a vessel + *deferre*, to carry down) In mammals, the tube that carries sperm from the testes to the urethra.

vascular bundle In vascular plants, a strand of tissue containing primary xylem and primary phloem. These bundles of elongated cells conduct water with dissolved minerals and carbohydrates throughout the plant body.

vascular cambium In vascular plants, the meristematic layer of cells that gives rise to secondary phloem and secondary xylem. The activity of the vascular cambium increases stem or root diameter.

vascular tissue In vascular plants, a major tissue type consisting of cells joined into tubes that conducts water and dissolved minerals up through the plant and the products of photosynthesis throughout.

vasopressin A posterior pituitary hormone that regulates the kidney's retention of water; also referred to as antidiuretic hormone (ADH).

vector The infected genome that harbors the foreign DNA and carries it into bacterial targets. An organism that transmits disease-causing microorganisms.

vegetal pole The hemisphere of the zygote, comprising cells rich in yolk.

vein (L. *vena*, vein) In plants, a vascular bundle forming a part of the framework of the conducting and supporting tissue of a stem or leaf. In animals, a blood vessel carrying blood from the tissues to the heart.

ventral (L. *venter*, belly) Refers to the bottom portion of an animal.

ventricle (L. *ventriculus*, belly) Either of the two lower chambers of the heart that receive blood from the atria and force it into the arteries.

venule (L. *vena*, vein) A small vein, especially one that begins at the capillaries and connects them with the larger veins.

vertebral column *See* spine.

vertebrate An animal having a backbone made of bony segments called vertebrae.

vesicle (L. *vesicula*, a little bladder) Membrane-enclosed sacs within eukaryotic organisms created by weaving sheets of endoplasmic reticulum through the cell's interior.

vessel (L. *vas*, a vessel) A tubelike element in the xylem of angiosperms. Composed of dead cells (vessel elements) arranged end to end. Conducts water and minerals from the soil.

vessel element In vascular plants, a typically elongated cell, dead at maturity, that conducts water and solutes in the xylem. Vessel elements make up vessels. Tracheids are less specialized conducting cells.

villus, pl. villi (L. a tuft of hair) In vertebrates, fine, microscopic, fingerlike projections lining the small

intestine that serve to increase the absorptive surface area of the intestine.

visible light The range of colors from violet (380 nanometers) to red (750 nanometers) that a human can see.

vision The perception of light, carried out by a specialized sensory apparatus called an eye.

visual cortex A major sensory region of the cerebral cortex lying on the occipital lobe, with different points in the visual field corresponding to different positions on the retina.

vitamin (L. *vita*, life + *amine*, of chemical origin) An organic substance required in minute quantities by an organism for growth and activity but that the organism cannot synthesize.

viviparous (L. *vivus*, alive + *parere*, to bring forth) Refers to reproduction in which eggs develop within the mother's body and young are born free-living.

voltage-gated channel A transmembrane pathway for an ion that is opened or closed by a change in the voltage, or charge difference, across the cell membrane.

voluntary nervous system *See* somatic nervous system.

W

warning coloration An ecological strategy of some organisms that "advertise" their poisonous nature by the use of bright colors.

water cycle The most familiar of the biogeochemical cycles, in which free water circulates between the atmosphere and the earth.

water vascular system The system of water-filled canals connecting the tube feet of echinoderms.

whorl A circle of leaves or of flower parts present at a single level along an axis.

wood Accumulated secondary xylem. Heartwood is the central, nonliving wood in the trunk of a tree. Hardwood is the wood of dicots, regardless of how hard or soft it actually is. Softwood is the wood of conifers.

woody plant A plant, such as a tree or shrub, in which secondary growth has been extensive.

X

xylem (Gr. *xylon,* wood) In vascular plants, a specialized tissue, composed primarily of elongate, thick-walled conducting cells, that transports water and solutes through the plant body.

Y

yolk (O.E. *geolu,* yellow) The stored substance in egg cells that provides the embryo's primary food supply.

Z

zona pellucida An outer membrane that encases a mammalian egg.

zygomycete (Gr. *zygon,* yoke + *mykes,* fungus) A type of fungus whose chief characteristic is the production of sexual structures called zygosporangia, which result from the fusion of two of its simple reproductive organs.

zygosporangium, *pl.* **zygosporangia** (Gr. *zygon,* yoke + *spora,* seed + *angeion,* a vessel) A sexual structure formed by the fusion of two simple reproductive organs, called gametangia, in a zygomycete.

zygote (Gr. *zygotos,* paired together) The diploid (2N) cell resulting from the fusion of male and female gametes (fertilization). A zygote may either develop into a diploid individual by mitotic divisions or undergo meiosis to form haploid (N) individuals that divide mitotically to form a population of cells.

Credits

Photographs

Part Openers

1: © Rod Planck/Photo Researchers, Inc.; 2: © Dr. Gopal Murti/SPL/Photo Researchers, Inc.; 3: © Horst Schafer/Peter Arnold, Inc.; 4: © Erika Stone/Peter Arnold, Inc.; 5: © Kevin Schafer/Peter Arnold, Inc.; 6: © Peter May/Peter Arnold, Inc.; 7: © Kenneth Fink/Photo Researchers, Inc.; 8: © Werner H. Muller/Peter Arnold, Inc.; 9: © Carl R. Sams II/Peter Arnold, Inc.

Chapter 1

Opener: © Michel Viard/Peter Arnold, Inc.; 1.1: © Jeanne A. Mortimer, Ph.D.; 1.2: © T. E. Adams/Visuals Unlimited; 1.3b: NASA; 1.3c: © Douglas Waugh/Peter Arnold, Inc.; 1.3d: © Jane Schreibman/Photo Researchers, Inc.; 1.3e: © Edward S. Ross; 1.3f: © Englebert/Photo Researchers, Inc.; 1.3h: © John D. Cunningham/Visuals Unlimited; 1.3i: © Edwin A. Reschke; 1.4: © Tom J. Ulrich/Visuals Unlimited; 1.5: © Michael Fogden; Sidelight 1.1 (left): © Frans Lanting/Minden Pictures; Sidelight 1.1 (top): Centers for Disease Control, Atlanta, GA; Sidelight 1.1 (bottom): © James A. Prince/Photo Researchers, Inc.; Sidelight 1.1 (right): NASA; 1.7: © George Bridgeman/Art Resource; 1.8a: © Christopher Ralling; 1.10a: © Christian Grzimek/Okapia/Photo Researchers, Inc.; 1.10b: © George Holton/Photo Researchers, Inc.; 1.11: The Bettmann Archive; 1.12: © Mary Evans Picture Library/Photo Researchers, Inc.; 1.13: © John D. Cunningham/Visuals Unlimited

Chapter 2

Opener: © James H. Karales/Peter Arnold, Inc.; 2.1: Courtesy of D. Marti, B. Drake, P. K. Hansma, University of California, Santa Barbara; 2.7b: © Cabisco/Visuals Unlimited; 2.8b: © Archive Photos; 2.10a: © Jurgen Schmitt/The Image Bank; 2.10b: © Frank Awbrey/Visuals Unlimited; 2.10c: © Nicholas Foster/The Image Bank; 2.10d: © John P. Kelly/The Image Bank; 2.12a: © Joseph Devenney/The Image Bank; 2.12b: © John Eastcott/Yva Momatiuk/The Image Works; 2.12c: © George I. Bernard/Animals Animals/Earth Scenes; 2.15: © D. Like/Visuals Unlimited; 2.17b: © Edward S. Ross; 2.18b: © Stanley Flegler/Visuals Unlimited; 2.18d: © Manfred Kage/Peter Arnold, Inc.; 2.19a: © J. D. Litvay/Visuals Unlimited; 2.20: © Scott Johnson/Animals Animals/Earth Scenes; 2.27a: © Manfred Kage/Peter Arnold, Inc.; 2.27b: © Michael Pasdzior/The Image Bank; 2.27c: © George Bernard/Animals Animals/Earth Scenes; 2.27d: © Oxford Scientific Films/Animals Animals/Earth Scenes; 2.27e: © Scott Blackman/Tom Stack and Associates; 2.29c: Courtesy of Laurence-Berkeley Laboratory; Sidelight 2.1: © Ray Pfortner/Peter Arnold, Inc.

Chapter 3

Opener: © Keith Kent/Peter Arnold, Inc.; 3.1: © John D. Cunningham/Visuals Unlimited; 3.2: NASA/Goodard Space Flight Center; Sidelight 3.1: Science VU/Whoi, D. Foster/Visuals Unlimited; Sidelight 3.2: Courtesy of Prof. W. D. Keller; 3.3: © Bob McKeever/Tom Stack and Associates; Page 52: © 1990 Keven Walsh, University of CA, San Diego; 3.5: © Sidney Fox/Visuals Unlimited; 3.6a: © John Cunningham/Visuals Unlimited; 3.6b: Elso S. Barghoorn, courtesy of the Botanical Museum, Harvard University; 3.6c: © Fred Bavendam/Peter Arnold, Inc.; 3.6d: © J. W. Schopf, *Journal of Paleontology*, 45:925–60, 1971; 3.7: © Edward S. Ross; 3.8a,b,c,d: Courtesy of Dr. J. William Schopf, Center for Study of Evolution and the Origin of Life, UCLA; 3.10: From J. W. Schopf, *Journal of Paleontology*, 45:925–60, 1971; 3.12: NASA

Chapter 4

Opener: © P. M. Motta and S. Correr/SPL/Photo Researchers, Inc.; 4.1a: © Ralph Slepecky/Visuals Unlimited; 4.1b: © Biophoto Associates/Science Source/Photo Researchers, Inc.; Sidelight 4.2a,b: Courtesy of Rothay House; Sidelight 4.3a: © John Walsh/SPL/Photo Researchers, Inc.; Sidelight 4.3b,c: © David M. Phillips/Visuals Unlimited; 4.2a: © Manfred Kage/Peter Arnold, Inc.; 4.2b: © M. Abbey/Visuals Unlimited; 4.2c: © Dr. David Scott/CNRI/Phototake; 4.2d: © Gerald Van Dyke/Visuals Unlimited; 4.2e: © David Scharf/Peter Arnold, Inc.; 4.3: © L. L. Sims/Visuals Unlimited; 4.5a,b: © David M. Phillips/Visuals Unlimited; 4.5c: © John D. Cunningham/Visuals Unlimited; 4.6a: © David M. Phillips/Visuals Unlimited; 4.7: Courtesy T. D. Pugh and E. H. Newcomb, University of Wisconsin Botany Department; 4.8 (background): © David Phillips/Visuals Unlimited; 4.9b (background): © Martha J. Powell/Visuals Unlimited; 4.11a: © J. David Robertson; 4.12a: © Dr. Jerry Burgess/SPL/Photo Researchers, Inc.; 4.13: © Edwin A. Reschke; 4.14b: © Alfred Pasiexa/SPL/Photo Researchers, Inc.; 4.15b: Courtesy of Dr. Charles Flickinger; 4.17b: © Don W. Fawcett/Visuals Unlimited; 4.18b: Courtesy of Kenneth Miller, Brown University; 4.19a: © Dr. Arnold Brody/SPL/Photo Researchers, Inc.; 4.21: © James A. Spudich; 4.22: © L. M. Pope/BPS/Tom Stack & Associates; 4.23a (both): Courtesy of Dr. Bessie Huang, Department of Cell Biology, The Scripps Research Institute; 4.23b: © Stanley Flegler/Visuals Unlimited; 4.24a: © Don W. Fawcett/Visuals Unlimited

Chapter 5

Opener: © Stephen Dalton/Photo Researchers, Inc.; Sidelight 5.1: Courtesy of Steve Walleg, Media Resources, University of CA-Riverside; 5.1 (left): © Doug Sokell/Visuals Unlimited; 5.1 (right): VU/Visuals Unlimited; 5.11b,c: © Gary Grimes/Martin M. Rotker; 5.12b: Courtesy of Birgit A. Satir, Ph.D.

Chapter 6

Opener: © Gunter Ziesler/Peter Arnold, Inc.; 6.1: © CO Rentmeester/Time Inc. 1970; 6.2: © Gunter Ziesler/Peter Arnold, Inc.; 6.3: © John Sohlden/Visuals Unlimited

Chapter 7

Opener: © Kim Heacox/Peter Arnold, Inc.; 7.1: © Link/Visuals Unlimited; Sidelight 7.1 (all): © Edward S. Ross

Chapter 8

Opener: © Jacques Jangoux/Peter Arnold, Inc.; 8.1: © Grant Heilman; 8.13a: © Michael Giannechini/Photo Researchers, Inc.; 8.16a: © Walt Anderson/Visuals Unlimited; 8.16b: © John Cancalosi/Peter Arnold, Inc.

Chapter 9

Opener: © CNRI/SPL/Photo Researchers, Inc.; 9.1: © Zig Leszczynski/Animals Animals/Earth Scenes; 9.3: Courtesy of Ulrich K.

Chapter 34

Opener: © Manfred Kage/Peter Arnold, Inc.; **34.2b:** © C. S. Raines/Visuals Unlimited; **34.3:** © Edward S. Ross; **34.4:** © E. R. Lewis/BPS/Tom Stack & Associates; **34.12:** Courtesy of Dr. Lennart Heimer; **Sidelight 34.1:** Reuters/Bettmann; **34.19:** © C. Bradley Simmons/Bruce Coleman, Inc.; **34.20:** © Seth Joel/SPL/Photo Researchers, Inc.; **Sidelight 34.2:** Nigel Unwin and Chikashi Toyoshima

Chapter 35

Opener: © Mehau Kulyk/SPL/Photo Researchers, Inc.; **35.3:** © W. Rosenberg/Tom Stack & Associates; **35.9:** Courtesy of *Washington University Magazine* 58(2), Summer, 1988.; **35.12a:** © Martha Swope Photography, Inc.; **35.12b:** © Michael and Patricia Fogden; **35.19b:** Courtesy of Beckman Vision Center at UCSF School of Medicine/Copenhagen, S. Mittman & M. Maglio; **35.25:** © Stephen Dalton/OSF/Animals Animals/Earth Scenes; **35.26a:** © Leonard Lee Rue III

Chapter 36

Opener: © David Scharf/Peter Arnold, Inc.; **36.1:** © Shane Anderson, UCSB; **36.3 (top left):** © Des & Jen Bartlett/Bruce Coleman Inc.; **36.3 (middle left):** © Anthony Bannister/Animals Animals/Earth Scenes; **36.3 (bottom left):** © Gunter Ziesler/Peter Arnold, Inc.; **36.3 (top right):** © Stephen Dalton/Natural History Photographic Agency; **36.3 (middle center):** © Charles Summers/Tom Stack & Associates; **36.3 (middle right and bottom center):** © Frans Lanting/Minden Pictures; **36.3 (bottom right):** © Dwight R. Kuhn; **36.6b:** © Lennart Nilsson, *Behold Man,*, Bonnier Fakta; **36.10b:** © D. W. Fawcett and J. Venable/Visuals Unlimited; **36.13a:** © D. W. Fawcett/Visuals Unlimited

Chapter 37

Opener: © Francois Gohicr/Photo Researchers, Inc.; **37.1:** © Willie Hill, Jr./The Image Works; **Sidelight 37.1:** © Mieke Maas/The Image Bank; **37.9:** The National Museum at Berlin, DDR, Egyptian Museum #14512; **37.10:** From L. H. Behrens and D. P. Barr, "Hyperpituitarism Beginning in Infancy," *Endocrinology* 16:125,

January 1932 © The Endocrine Society; **37.11:** © John Paul Kay/Peter Arnold, Inc.; **37.14:** © Edwin A. Reschke; **37.16:** © Lennart Nilsson

Chapter 38

Opener: © Tom & Pat Leeson/Photo Researchers, Inc.

Chapter 39

Opener: © Chris Bjornberg/Photo Researchers, Inc.; **Sidelight 39.1:** UPI/Bettmann; **39.9:** © Ellen Dirkson/Visuals Unlimited

Chapter 40

Opener: © David Scharf/Peter Arnold, Inc.; **40.3:** © Makio Murayama/BPS/Tom Stack & Associates; **40.4b:** © David M. Phillips/Visuals Unlimited; **40.6, 40.8, 40.12:** © Edwin A. Reschke; **40.18 a–c:** © Frank B. Sloop, Jr. M.D.; **40.19:** © Lennart Nilsson, *The Incredible Machine,*, Bonnier Fakta

Chapter 41

Opener: © CDC/RG/Peter Arnold, Inc.; **41.1:** Science VU-National Library of Medicine/Visuals Unlimited; **41.3 a,b:** George Johnson; **41.4:** © Manfred Kage/Peter Arnold, Inc.; **41.6:** © Lennart Nilsson, *The Body Victorious,* Bonnier Fakta; **41.9:** North Wind Picture Archives; **41.15 a,b:** Courtesy of Dale E. Bockman; **41.20:** © Lennart Nilsson, *The Body Victorious,* Bonnier Fakta; **41.22:** Courtesy of Dr. Larry G. Arlian

Chapter 42

Opener: © Kjell Sandved/Visuals Unlimited; **42.1:** Courtesy of C. P. Hickman; **42.8:** From Richard G. Kessel and Randy H. Kardon, *Tissues and Organs: A Text-Atlas of Scanning Electron Microscopy,* 1979; **42.9:** © Marty Snyderman

Chapter 43

Opener: © P. Motta/SPL/Photo Researchers, Inc.; **43.1:** © Hans Pfletschinger/Peter Arnold, Inc.; **43.2:** Courtesy of C. P. Hickman; **43.3a:** © Tom McHugh/Photo Researchers, Inc.; **43.3b:** © Dennis MacDonald/Unicorn Stock Photos; **43.6:** © Lennart Nilsson, *Behold Man,* Little Brown and

Company, Bonnier Fakta; **43.10:** © Edwin A. Reschke; **43.13a:** © C. Andrew Henley/Biofotos; **43.13b:** © Fritz Prenzel/Animals Animals/Earth Scenes; **43.13c:** © Hans Reinhard/Bruce Coleman Inc.; **43.16a:** © SIU Biomedical Communications/Photo Researchers, Inc.; **43.16 b–d:** © Ellis/Photo Researchers, Inc.; **43.16e:** © ISU Biomedical Communications/Photo Researchers, Inc.; **43.16f:** © Scott Camazine & Sue Trainor/Photo Researchers, Inc.; **43.18:** Science VU/ Boehringer Ingelheim GmbH/Visuals Unlimited; **Sidelight 43.1 (left):** © Charles Dauguet/Petit Format Institute Pasteur/Photo Researchers, Inc.

Chapter 44

Opener: © Alex Bartel/SPL/Photo Researchers, Inc.; **44.1 a,b:** © David M. Phillips/Photo Researchers, Inc.; **44.13a:** © Omikron/Photo Researchers, Inc.; **44.13 b,c:** © Cabisco/Visuals Unlimited; **44.14 a–f:** © Heather Angel; **44.17:** © F. R. Turner, Indiana University/Biological Photo Service; **44.18 a,b:** © Lennart Nilsson, *A Child is Born,* Dell Publishing Company, Bonnier Fakta; **44.18 c,d:** © Lennart Nilsson, *Behold Man,*, Little Brown and Company, Bonnier Fakta

Chapter 45

Opener: © J & B Photo/Animals Animals/Earth Scenes; **45.1 a–c:** The Bettmann Archive; **45.4a:** © Linda Koebner/Bruce Coleman Inc.; **45.4b:** © Jeff Foott/Bruce Coleman Inc.; **45.5:** © Will Rapport/Omikron/Photo Researchers, Inc.; **45.6c:** © Steve Maslowski/Photo Researchers, Inc.; **45.7:** © Dan Kline/Visuals Unlimited; **45.8b:** © Oxford Scientific Films/Animals Animals/Earth Scenes; **45.9 a-2:** © Stephen Dalton/Animals Animals/Earth Scenes; **45.9 b-2:** © Noah Satat/Animals Animals/Earth Scenes; **45.9 c-2:** © Mark Hamblin/OSF/Animals Animals/Earth Scenes; **45.10:** © D. W. Macdonald; **45.11:** © Clyde H. Smith/Peter Arnold, Inc.; **45.13:** © Ray Richardson/Animals Animals/Earth Scenes; **45.14:** © Zig Lesczcynski/Animals Animals/Earth Scenes; **45.15:** © Edward S. Ross; **45.16:** © Paulette Brunner/Tom Stack & Associates; **Sidelight 45.2:** © Christopher Springmann; **45.18:** George Johnson